BIG IDEAS MATH®

Algebra 2

A Common Core Curriculum

Ron Larson and Laurie Boswell

Erie, Pennsylvania
BigIdeasLearning.com

Big Ideas Learning, LLC
1762 Norcross Road
Erie, PA 16510-3838
USA

For product information and customer support, contact Big Ideas Learning at **1-877-552-7766** or visit us at ***BigIdeasLearning.com***.

Cover Image
CVADRAT/Shutterstock.com

Printed in the U.S.A.

ISBN 13: 978-1-60840-840-5
ISBN 10: 1-60840-840-X

8 9 10 WEB 18 17 16 15

Authors

Ron Larson, Ph.D., is well known as the lead author of a comprehensive program for mathematics that spans middle school, high school, and college courses. He holds the distinction of Professor Emeritus from Penn State Erie, The Behrend College, where he taught for nearly 40 years. He received his Ph.D. in mathematics from the University of Colorado. Dr. Larson's numerous professional activities keep him actively involved in the mathematics education community and allow him to fully understand the needs of students, teachers, supervisors, and administrators.

Ron Larson

Laurie Boswell, Ed.D., is the Head of School and a mathematics teacher at the Riverside School in Lyndonville, Vermont. Dr. Boswell is a recipient of the Presidential Award for Excellence in Mathematics Teaching and has taught mathematics to students at all levels, from elementary through college. Dr. Boswell was a Tandy Technology Scholar and served on the NCTM Board of Directors from 2002 to 2005. She currently serves on the board of NCSM and is a popular national speaker.

Laurie Boswell

Dr. Ron Larson and **Dr. Laurie Boswell** began writing together in 1992. Since that time, they have authored over two dozen textbooks. In their collaboration, Ron is primarily responsible for the student edition while Laurie is primarily responsible for the teaching edition.

For the Student

Welcome to *Big Ideas Math Algebra 2*. From start to finish, this program was designed with you, the learner, in mind.

As you work through the chapters in your Algebra 2 course, you will be encouraged to think and to make conjectures while you persevere through challenging problems and exercises. You will make errors—and that is ok! Learning and understanding occur when you make errors and push through mental roadblocks to comprehend and solve new and challenging problems.

In this program, you will also be required to explain your thinking and your analysis of diverse problems and exercises. Being actively involved in learning will help you develop mathematical reasoning and use it to solve math problems and work through other everyday challenges.

We wish you the best of luck as you explore Algebra 2. We are excited to be a part of your preparation for the challenges you will face in the remainder of your high school career and beyond.

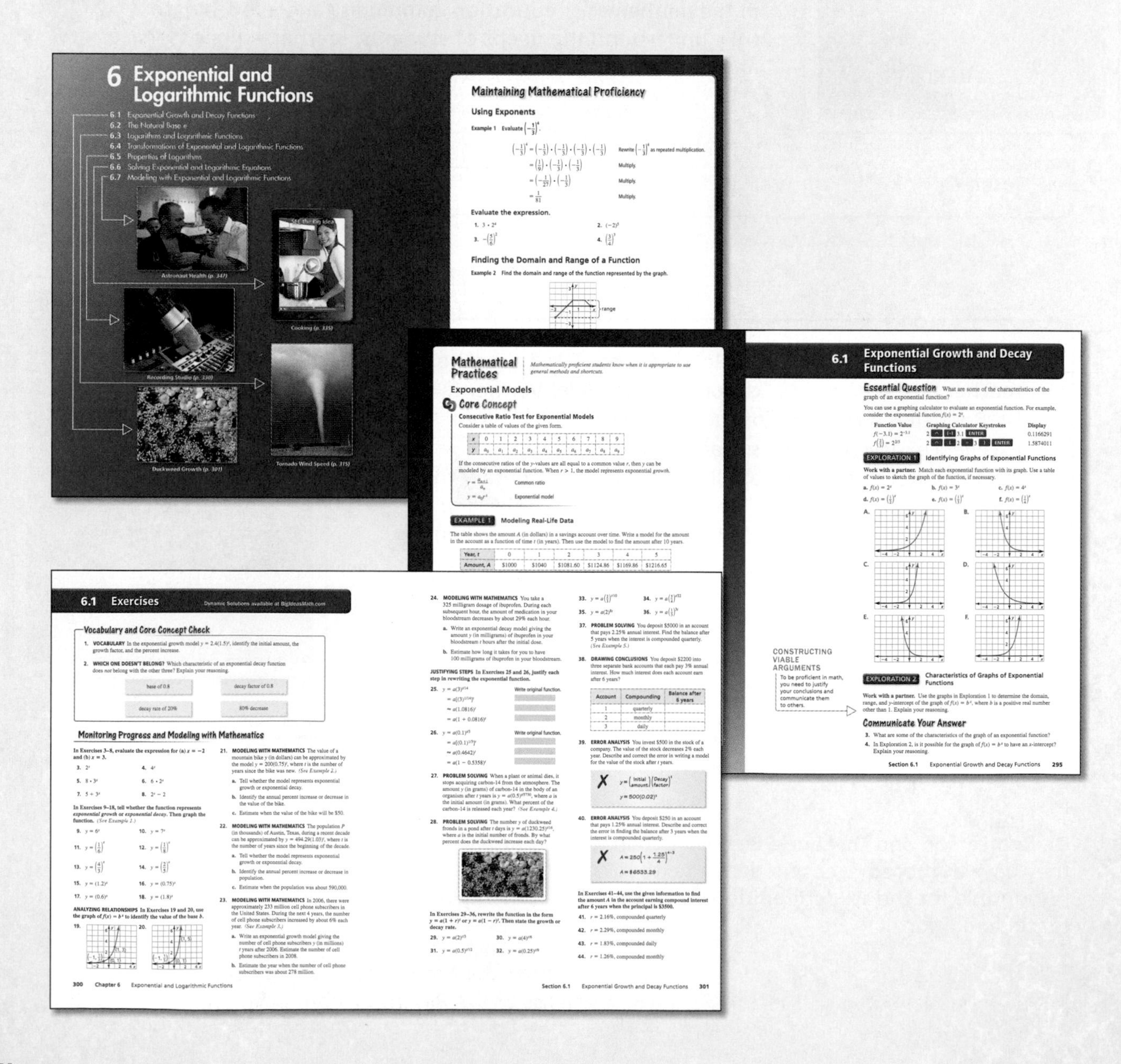

Big Ideas Math High School Research

Big Ideas Math Algebra 1, *Geometry*, and *Algebra 2* is a research-based program providing a rigorous, focused, and coherent curriculum for high school students. Ron Larson and Laurie Boswell utilized their expertise as well as the body of knowledge collected by additional expert mathematicians and researchers to develop each course.

The pedagogical approach to this program follows the best practices outlined in the most prominent and widely-accepted educational research and standards, including:

Achieve, ACT, and The College Board

Adding It Up: Helping Children Learn Mathematics
National Research Council ©2001

Common Core State Standards for Mathematics
National Governors Association Center for Best Practices and the Council of Chief State School Officers ©2010

Curriculum Focal Points and the *Principles and Standards for School Mathematics* ©2000
National Council of Teachers of Mathematics (NCTM)

Project Based Learning
The Buck Institute

Rigor/Relevance Framework™
International Center for Leadership in Education

Universal Design for Learning Guidelines
CAST ©2011

Big Ideas Math would like to express our gratitude to the mathematics education and instruction experts who served as consultants during the writing of *Big Ideas Math Algebra 1*, *Geometry*, and *Algebra 2*. Their input was an invaluable asset during the development of this program.

Kristen Karbon
Curriculum and Assessment Coordinator
Troy School District
Troy, Michigan

Jean Carwin
Math Specialist/TOSA
Snohomish School District
Snohomish, Washington

Carolyn Briles
Performance Tasks Consultant
Mathematics Teacher, Loudoun County Public Schools
Leesburg, Virginia

Bonnie Spence
Differentiated Instruction Consultant
Mathematics Lecturer, The University of Montana
Missoula, Montana

Connie Schrock, Ph.D.
Performance Tasks Consultant
Mathematics Professor, Emporia State University
Emporia, Kansas

We would also like to thank all of our reviewers who took the time to provide feedback during the final development phases. For a complete list of the *Big Ideas Math* program reviewers, please visit *www.BigIdeasLearning.com*.

Common Core State Standards for Mathematical Practice

Make sense of problems and persevere in solving them.

- *Essential Questions* help students focus on core concepts as they analyze and work through each *Exploration*.
- Section opening *Explorations* allow students to struggle with new mathematical concepts and explain their reasoning in the *Communicate Your Answer* questions.

Reason abstractly and quantitatively.

- *Reasoning*, *Critical Thinking*, *Abstract Reasoning*, and *Problem Solving* exercises challenge students to apply their acquired knowledge and reasoning skills to solve each problem.
- *Thought Provoking* exercises test the reasoning skills of students as they analyze and interpret perplexing scenarios.

Construct viable arguments and critique the reasoning of others.

- Students must justify their responses to each *Essential Question* in the *Communicate Your Answer* questions at the end of each *Exploration* set.
- Students are asked to construct arguments and critique the reasoning of others in specialized exercises, including *Making an Argument*, *How Do You See It?*, *Drawing Conclusions*, *Reasoning*, *Error Analysis*, *Problem Solving*, and *Writing*.

Model with mathematics.

- Real-life scenarios are utilized in *Explorations*, *Examples*, *Exercises*, and *Assessments* so students have opportunities to apply the mathematical concepts they have learned to realistic situations.
- *Modeling with Mathematics* exercises allow students to interpret a problem in the context of a real-life situation, often utilizing tables, graphs, visual representations, and formulas.

Use appropriate tools strategically.

- Students are provided opportunities for selecting and utilizing the appropriate mathematical tool in *Using Tools* exercises. Students work with graphing calculators, dynamic geometry software, models, and more.
- A variety of tool papers and manipulatives are available for students to use in problems as strategically appropriate.

Attend to precision.

- *Vocabulary and Core Concept Check* exercises require students to use clear, precise mathematical language in their solutions and explanations.
- The many opportunities for cooperative learning in this program, including working with partners for each *Exploration*, support precise, explicit mathematical communication.

Look for and make use of structure.

- *Using Structure* exercises provide students with the opportunity to explore patterns and structure in mathematics.
- Students analyze structure in problems through *Justifying Steps* and *Analyzing Equations* exercises.

Look for and express regularity in repeated reasoning.

- Students are continually encouraged to evaluate the reasonableness of their solutions and their steps in the problem-solving process.
- Stepped-out *Examples* encourage students to maintain oversight of their problem-solving process and pay attention to the relevant details in each step.

Go to *BigIdeasLearning.com* for more information on the Common Core State Standards for Mathematical Practice.

Common Core State Standards for Mathematical Content for Algebra 2

Chapter Coverage for Standards

1 2 3 4 5 6 7 8 9 10 11

Conceptual Category Number and Quantity

- The Real Number System
- The Complex Number System
- Quantities

1 2 3 4 5 6 7 8 9 10 11

Conceptual Category Algebra

- Seeing Structure in Expressions
- Creating Equations
- Arithmetic with Polynomials and Rational Expressions
- Reasoning with Equations and Inequalities

1 2 3 4 5 6 7 8 9 10 11

Conceptual Category Functions

- Interpreting Functions
- Linear, Quadratic, and Exponential Models
- Building Functions
- Trigonometric Functions

1 2 3 4 5 6 7 8 9 10 11

Conceptual Category Geometry

- Expressing Geometric Properties with Equations

1 2 3 4 5 6 7 8 9 10 11

Conceptual Category Statistics and Probability

- Interpreting Categorical and Quantitative Data
- Making Inferences and Justifying Conclusions
- Conditional Probability and the Rules of Probability
- Using Probability to Make Decisions

Go to *BigIdeasLearning.com* for more information on the Common Core State Standards for Mathematical Content.

1 Linear Functions

See the Big Idea
Analyze the trajectory of a dirt bike after it is launched off a ramp.

Quadratic Functions 2

See the Big Idea
Learn how to build your own parabolic mirror that uses sunlight to generate electricity.

3 Quadratic Equations and Complex Numbers

See the Big Idea
Explore imaginary numbers in the context of electrical circuits.

4 Polynomial Functions

See the Big Idea
Discover how Quonset Huts (and the related Nissen Huts) were utilized in World War II.

5 Rational Exponents and Radical Functions

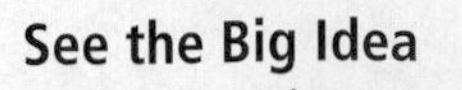

See the Big Idea
Explore heartbeat rates and life spans for different animals.

Exponential and Logarithmic Functions

See the Big Idea
Explore how the USDA uses Newton's Law of Cooling to develop safe cooking regulations using rules based on time and temperature.

7 Rational Functions

See the Big Idea
Analyze how 3-D printing compares economically with traditional manufacturing.

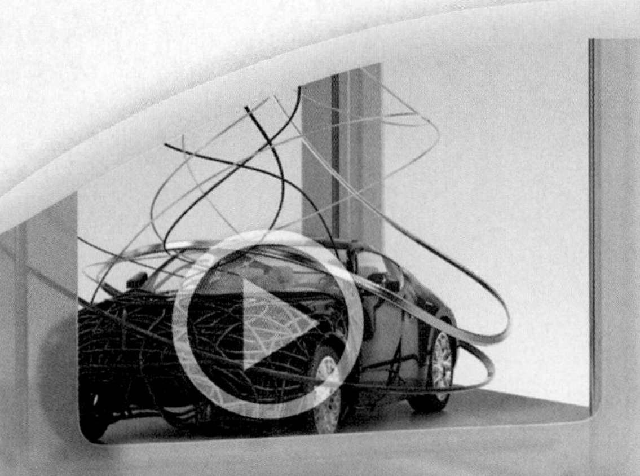

8 Sequences and Series

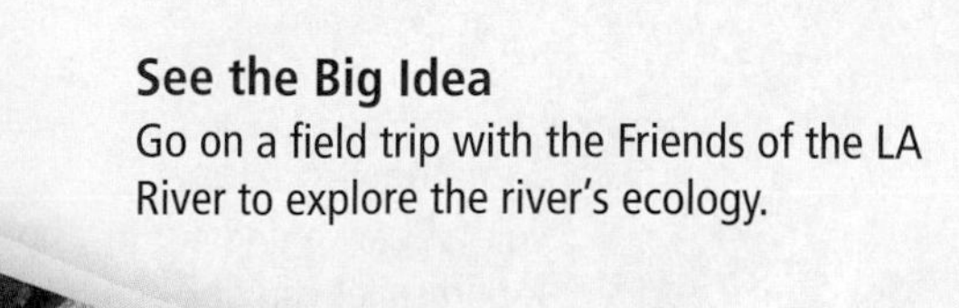

See the Big Idea
Go on a field trip with the Friends of the LA River to explore the river's ecology.

Trigonometric Ratios and Functions

See the Big Idea
Join us as we find out just how high a parasailer can go.

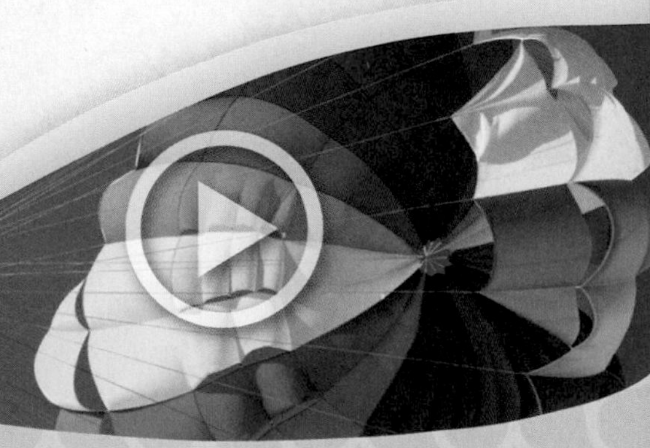

Probability 10

See the Big Idea
Learn about caring for trees at an arboretum.

11 Data Analysis and Statistics

See the Big Idea
Learn what kind of damage actually occurs after a volcanic eruption and how the information is collected.

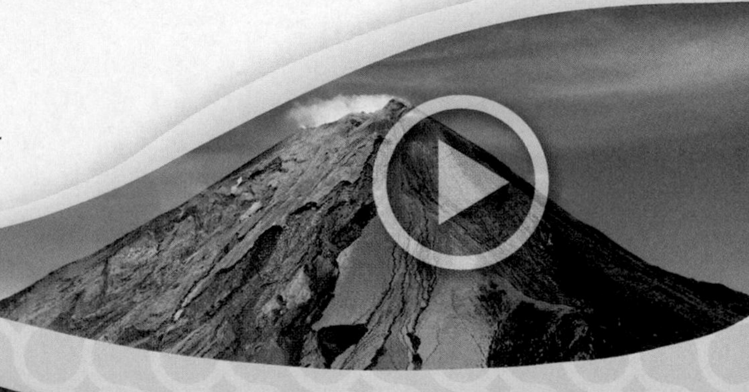

How to Use Your Math Book

Get ready for each chapter by **Maintaining Mathematical Proficiency** and reviewing the **Mathematical Practices**. Begin each section by working through the **EXPLORATIONS** to **Communicate Your Answer** to the **Essential Question**. Each **Lesson** will explain **What You Will Learn** through **EXAMPLES**, **Core Concepts**, and **Core Vocabulary**.

Answer the **Monitoring Progress** questions as you work through each lesson. Look for STUDY TIPS, COMMON ERRORS, and suggestions for looking at a problem ANOTHER WAY throughout the lessons. We will also provide you with guidance for accurate mathematical READING and concept details you should REMEMBER.

Sharpen your newly acquired skills with **Exercises** at the end of every section. Halfway through each chapter you will be asked **What Did You Learn?** and you can use the Mid-Chapter **Quiz** to check your progress. You can also use the **Chapter Review** and **Chapter Test** to review and assess yourself after you have completed a chapter.

Apply what you learned in each chapter to a **Performance Task** and build your confidence for taking standardized tests with each chapter's **Cumulative Assessment**.

For extra practice in any chapter, use your *Online Resources*, *Skills Review Handbook*, or your *Student Journal*.

1 Linear Functions

Pizza Shop *(p. 34)*

Prom *(p. 23)*

Café Expenses *(p. 16)*

Dirt Bike *(p. 7)*

Swimming *(p. 10)*

Maintaining Mathematical Proficiency

Evaluating Expressions

Example 1 **Evaluate the expression $36 \div (3^2 \times 2) - 3$.**

$36 \div (3^2 \times 2) - 3 = 36 \div (9 \times 2) - 3$ Evaluate the power within parentheses.

$= 36 \div 18 - 3$ Multiply within parentheses.

$= 2 - 3$ Divide.

$= -1$ Subtract.

Evaluate.

1. $5 \cdot 2^3 + 7$

2. $4 - 2(3 + 2)^2$

3. $48 \div 4^2 + \frac{3}{5}$

4. $50 \div 5^2 \cdot 2$

5. $\frac{1}{2}(2^2 + 22)$

6. $\frac{1}{6}(6 + 18) - 2^2$

Transformations of Figures

Example 2 **Reflect the black rectangle in the *x*-axis. Then translate the new rectangle 5 units to the left and 1 unit down.**

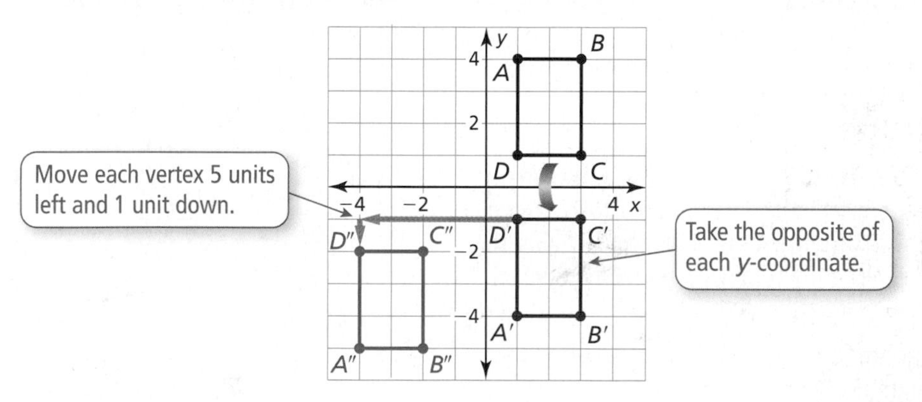

Graph the transformation of the figure.

7. Translate the rectangle 1 unit right and 4 units up.

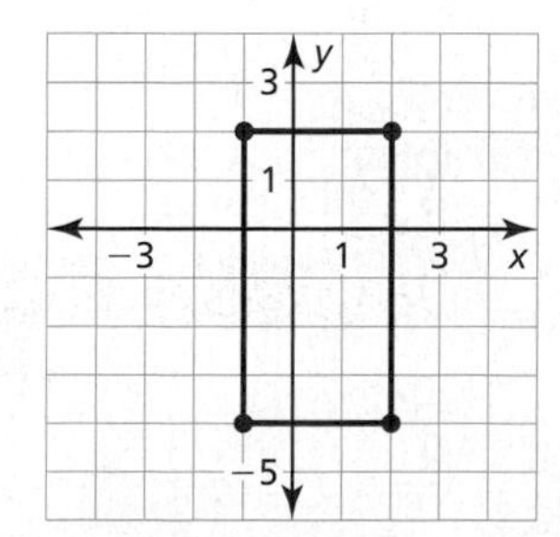

8. Reflect the triangle in the *y*-axis. Then translate 2 units left.

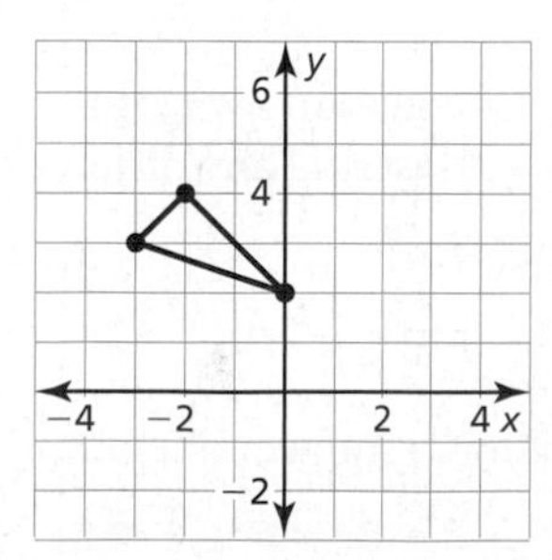

9. Translate the trapezoid 3 units down. Then reflect in the *x*-axis.

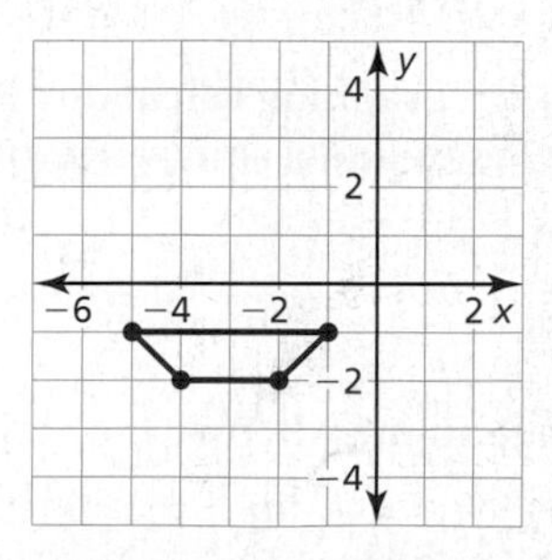

10. **ABSTRACT REASONING** Give an example to show why the order of operations is important when evaluating a numerical expression. Is the order of transformations of figures important? Justify your answer.

Mathematical Practices

Mathematically proficient students use technological tools to explore concepts.

Using a Graphing Calculator

Core Concept

Standard and Square Viewing Windows

A typical screen on a graphing calculator has a height-to-width ratio of 2 to 3. This means that when you view a graph using the *standard viewing window* of −10 to 10 (on each axis), the graph will not be shown in its true perspective.

To view a graph in its true perspective, you need to change to a *square viewing window*, where the tick marks on the x-axis are spaced the same as the tick marks on the y-axis.

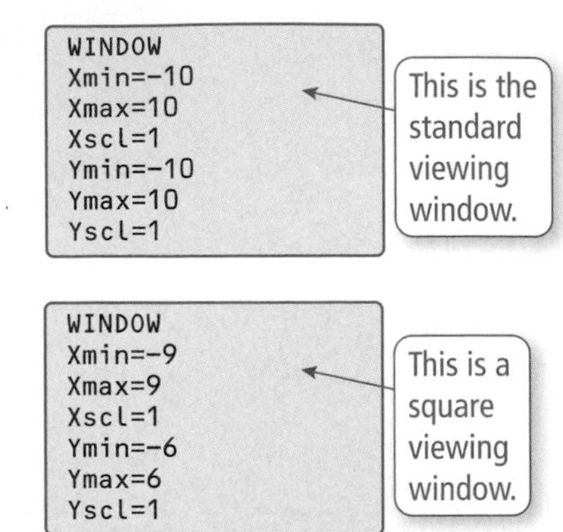

EXAMPLE 1 Using a Graphing Calculator

Use a graphing calculator to graph $y = |x| - 3$.

SOLUTION

In the standard viewing window, notice that the tick marks on the y-axis are closer together than those on the x-axis. This implies that the graph is not shown in its true perspective.

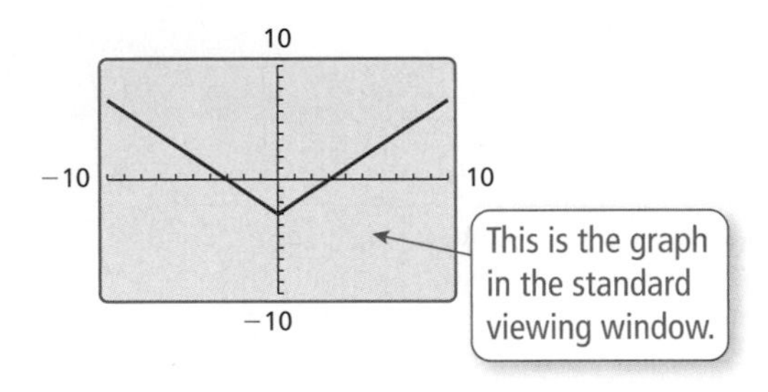

In a square viewing window, notice that the tick marks on both axes have the same spacing. This implies that the graph is shown in its true perspective.

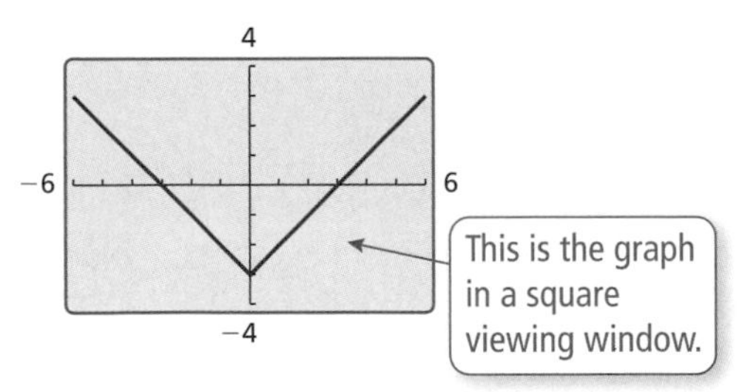

Monitoring Progress

Use a graphing calculator to graph the equation using the standard viewing window and a square viewing window. Describe any differences in the graphs.

1. $y = 2x - 3$ **2.** $y = |x + 2|$ **3.** $y = -x^2 + 1$

4. $y = \sqrt{x - 1}$ **5.** $y = x^3 - 2$ **6.** $y = 0.25x^3$

Determine whether the viewing window is square. Explain.

7. $-8 \le x \le 8,\ -2 \le y \le 8$ **8.** $-7 \le x \le 8,\ -2 \le y \le 8$

9. $-6 \le x \le 9,\ -2 \le y \le 8$ **10.** $-2 \le x \le 2,\ -3 \le y \le 3$

11. $-4 \le x \le 5,\ -3 \le y \le 3$ **12.** $-4 \le x \le 4,\ -3 \le y \le 3$

1.1 Parent Functions and Transformations

Essential Question What are the characteristics of some of the basic parent functions?

EXPLORATION 1 Identifying Basic Parent Functions

Work with a partner. Graphs of eight basic parent functions are shown below. Classify each function as *constant*, *linear*, *absolute value*, *quadratic*, *square root*, *cubic*, *reciprocal*, or *exponential*. Justify your reasoning.

> **JUSTIFYING CONCLUSIONS**
> To be proficient in math, you need to justify your conclusions and communicate them clearly to others.

a.

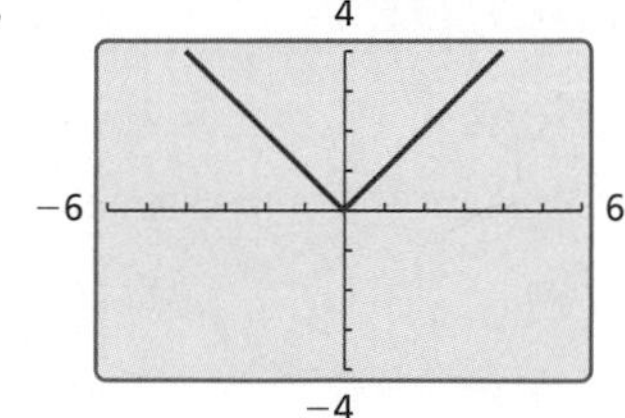

b.

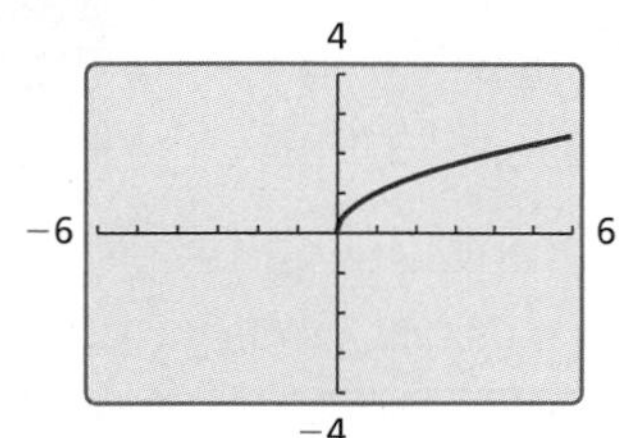

c.

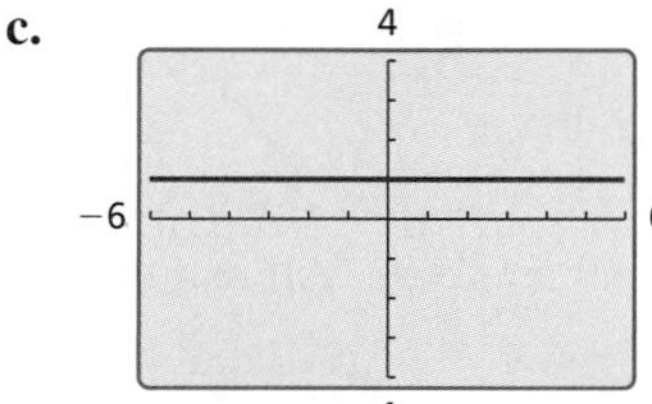

d.

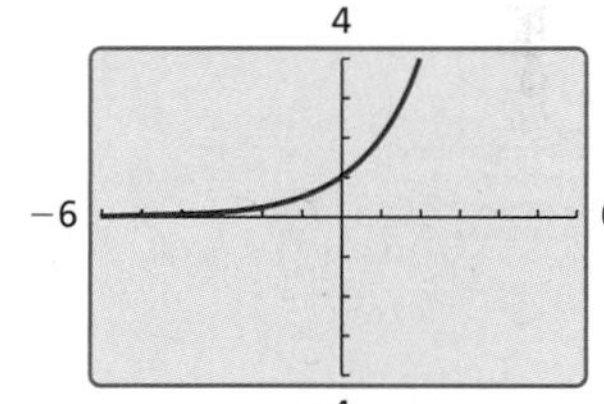

e.

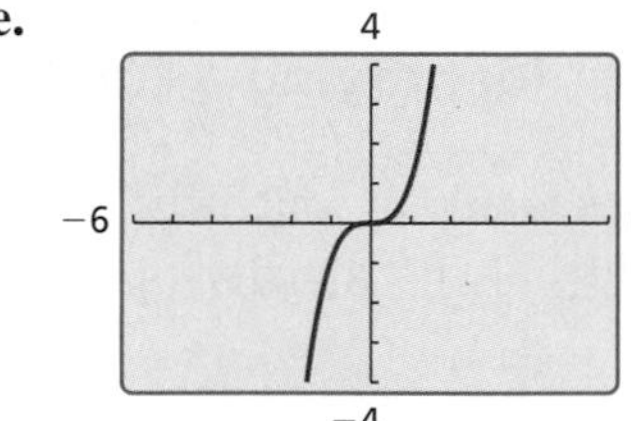

f.

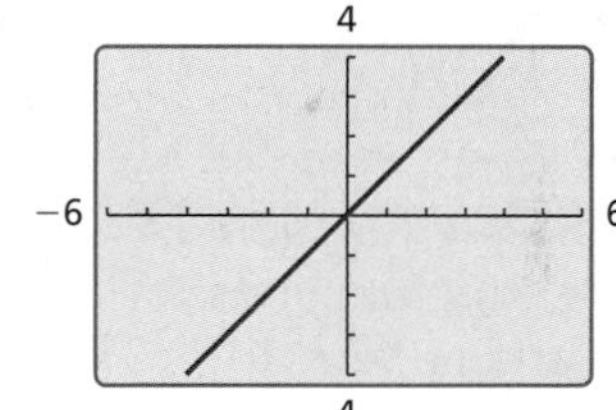

g.

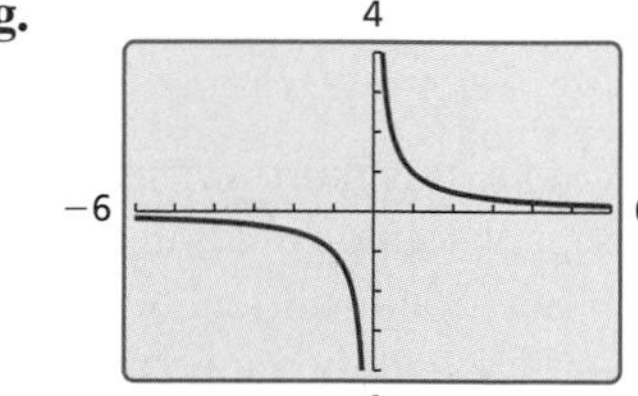

h.

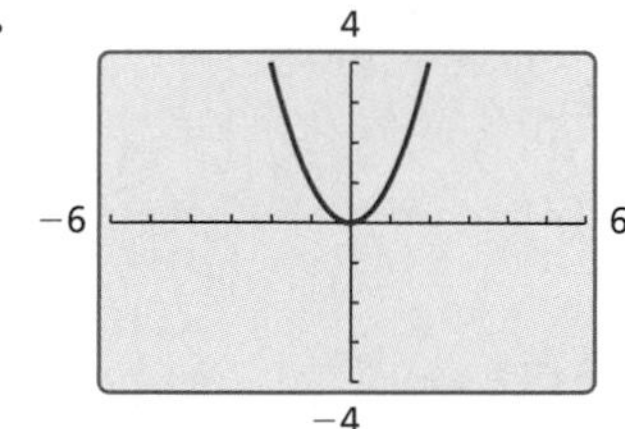

Communicate Your Answer

2. What are the characteristics of some of the basic parent functions?

3. Write an equation for each function whose graph is shown in Exploration 1. Then use a graphing calculator to verify that your equations are correct.

1.1 Lesson

What You Will Learn

- Identify families of functions.
- Describe transformations of parent functions.
- Describe combinations of transformations.

Core Vocabulary

parent function, p. 4
transformation, p. 5
translation, p. 5
reflection, p. 5
vertical stretch, p. 6
vertical shrink, p. 6

Previous
function
domain
range
slope
scatter plot

Identifying Function Families

Functions that belong to the same *family* share key characteristics. The **parent function** is the most basic function in a family. Functions in the same family are *transformations* of their parent function.

Core Concept

Parent Functions

Family	Constant	Linear	Absolute Value	Quadratic
Rule	$f(x) = 1$	$f(x) = x$	$f(x) = \|x\|$	$f(x) = x^2$
Graph	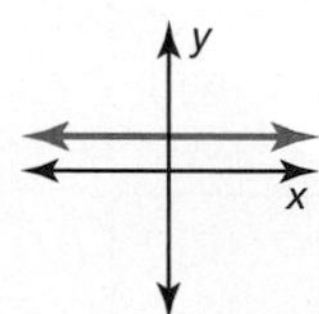	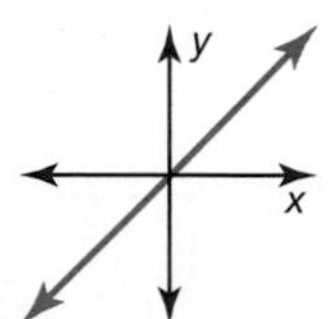	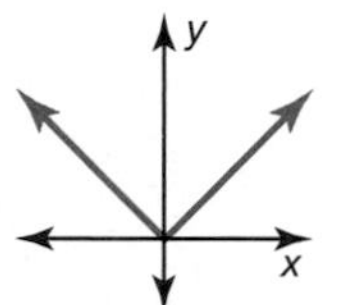	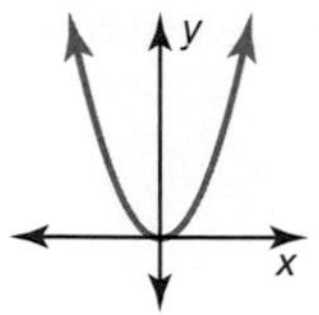
Domain	All real numbers	All real numbers	All real numbers	All real numbers
Range	$y = 1$	All real numbers	$y \geq 0$	$y \geq 0$

LOOKING FOR STRUCTURE

You can also use function rules to identify functions. The only variable term in f is an $|x|$-term, so it is an absolute value function.

EXAMPLE 1 Identifying a Function Family

Identify the function family to which f belongs. Compare the graph of f to the graph of its parent function.

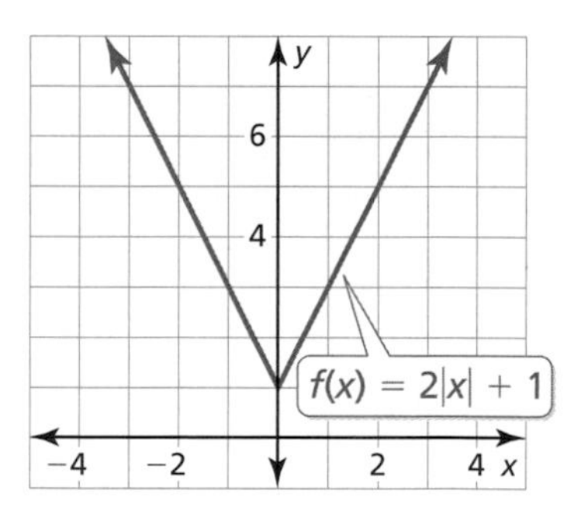

SOLUTION

The graph of f is V-shaped, so f is an absolute value function.

The graph is shifted up and is narrower than the graph of the parent absolute value function. The domain of each function is all real numbers, but the range of f is $y \geq 1$ and the range of the parent absolute value function is $y \geq 0$.

Monitoring Progress  Help in English and Spanish at *BigIdeasMath.com*

1. Identify the function family to which g belongs. Compare the graph of g to the graph of its parent function.

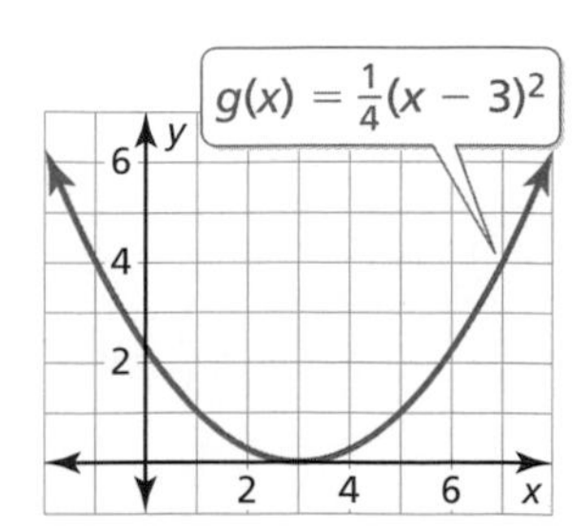

Describing Transformations

A **transformation** changes the size, shape, position, or orientation of a graph. A **translation** is a transformation that shifts a graph horizontally and/or vertically but does not change its size, shape, or orientation.

REMEMBER

The slope-intercept form of a linear equation is $y = mx + b$, where m is the slope and b is the y-intercept.

EXAMPLE 2 Graphing and Describing Translations

Graph $g(x) = x - 4$ and its parent function. Then describe the transformation.

SOLUTION

The function g is a linear function with a slope of 1 and a y-intercept of -4. So, draw a line through the point $(0, -4)$ with a slope of 1.

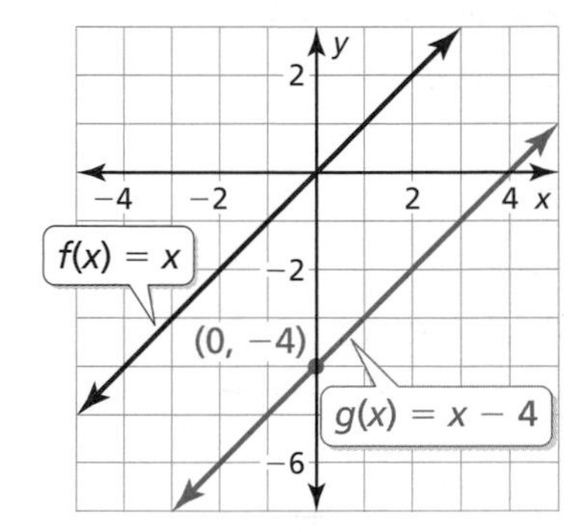

The graph of g is 4 units below the graph of the parent linear function f.

▶ So, the graph of $g(x) = x - 4$ is a vertical translation 4 units down of the graph of the parent linear function.

A **reflection** is a transformation that flips a graph over a line called the *line of reflection*. A reflected point is the same distance from the line of reflection as the original point but on the opposite side of the line.

REMEMBER

The function $p(x) = -x^2$ is written in *function notation*, where $p(x)$ is another name for y.

EXAMPLE 3 Graphing and Describing Reflections

Graph $p(x) = -x^2$ and its parent function. Then describe the transformation.

SOLUTION

The function p is a quadratic function. Use a table of values to graph each function.

x	$y = x^2$	$y = -x^2$
-2	4	-4
-1	1	-1
0	0	0
1	1	-1
2	4	-4

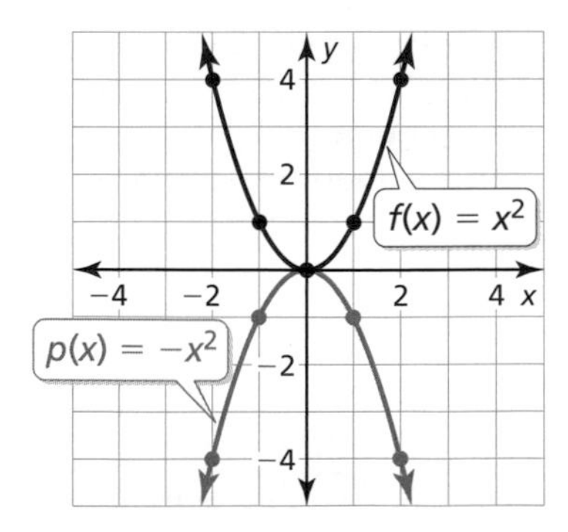

The graph of p is the graph of the parent function flipped over the x-axis.

▶ So, $p(x) = -x^2$ is a reflection in the x-axis of the parent quadratic function.

Monitoring Progress  Help in English and Spanish at *BigIdeasMath.com*

Graph the function and its parent function. Then describe the transformation.

2. $g(x) = x + 3$ **3.** $h(x) = (x - 2)^2$ **4.** $n(x) = -|x|$

Another way to transform the graph of a function is to multiply all of the y-coordinates by the same positive factor (other than 1). When the factor is greater than 1, the transformation is a **vertical stretch**. When the factor is greater than 0 and less than 1, it is a **vertical shrink**.

EXAMPLE 4 Graphing and Describing Stretches and Shrinks

Graph each function and its parent function. Then describe the transformation.

a. $g(x) = 2|x|$ **b.** $h(x) = \frac{1}{2}x^2$

SOLUTION

a. The function g is an absolute value function. Use a table of values to graph the functions.

x	$y = \|x\|$	$y = 2\|x\|$
−2	2	4
−1	1	2
0	0	0
1	1	2
2	2	4

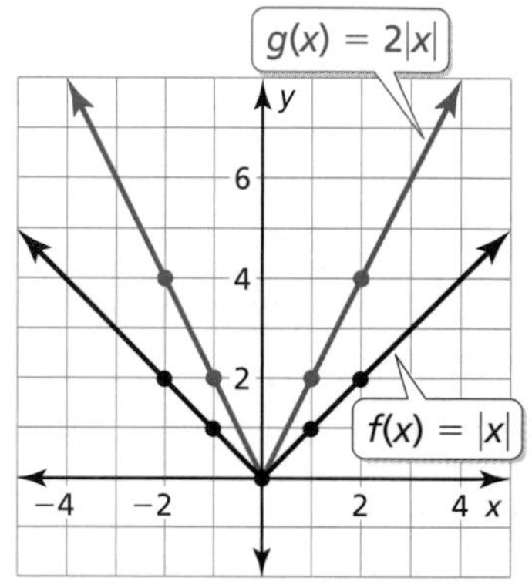

The y-coordinate of each point on g is two times the y-coordinate of the corresponding point on the parent function.

▶ So, the graph of $g(x) = 2|x|$ is a vertical stretch of the graph of the parent absolute value function.

b. The function h is a quadratic function. Use a table of values to graph the functions.

x	$y = x^2$	$y = \frac{1}{2}x^2$
−2	4	2
−1	1	$\frac{1}{2}$
0	0	0
1	1	$\frac{1}{2}$
2	4	2

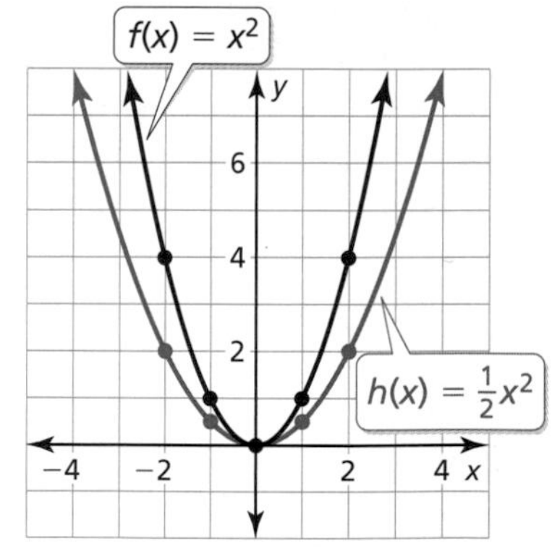

The y-coordinate of each point on h is one-half of the y-coordinate of the corresponding point on the parent function.

▶ So, the graph of $h(x) = \frac{1}{2}x^2$ is a vertical shrink of the graph of the parent quadratic function.

REASONING ABSTRACTLY

To visualize a vertical stretch, imagine *pulling* the points away from the x-axis.

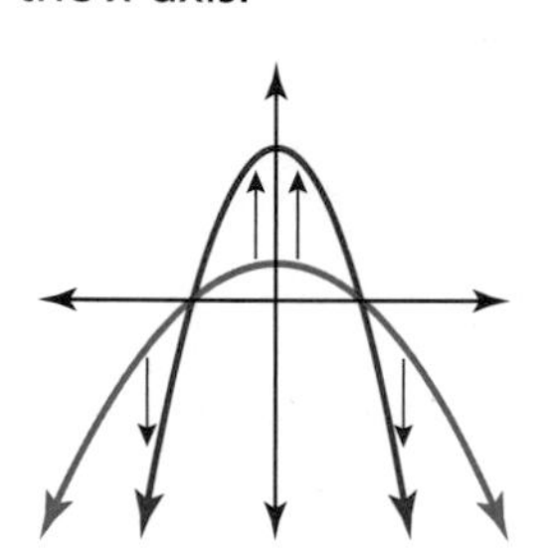

To visualize a vertical shrink, imagine *pushing* the points toward the x-axis.

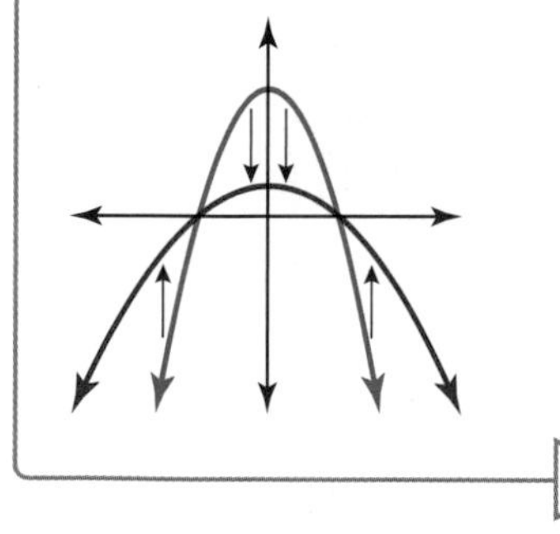

Monitoring Progress

Help in English and Spanish at *BigIdeasMath.com*

Graph the function and its parent function. Then describe the transformation.

5. $g(x) = 3x$ **6.** $h(x) = \frac{3}{2}x^2$ **7.** $c(x) = 0.2|x|$

Combinations of Transformations

You can use more than one transformation to change the graph of a function.

EXAMPLE 5 Describing Combinations of Transformations

Use a graphing calculator to graph $g(x) = -|x + 5| - 3$ and its parent function. Then describe the transformations.

SOLUTION

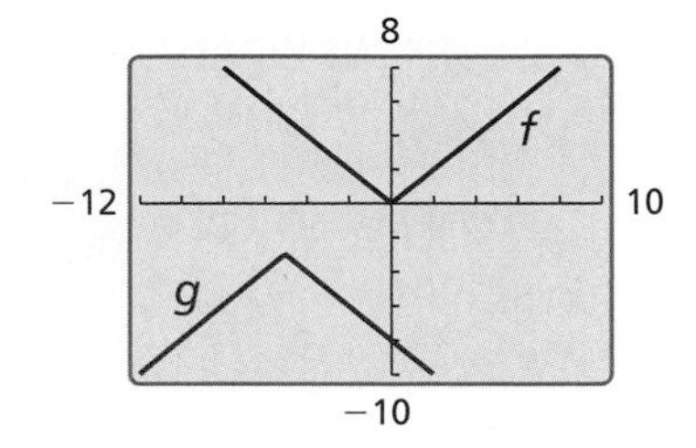

The function g is an absolute value function.

▶ The graph shows that $g(x) = -|x + 5| - 3$ is a reflection in the x-axis followed by a translation 5 units left and 3 units down of the graph of the parent absolute value function.

EXAMPLE 6 Modeling with Mathematics

Time (seconds), x	Height (feet), y
0	8
0.5	20
1	24
1.5	20
2	8

The table shows the height y of a dirt bike x seconds after jumping off a ramp. What type of function can you use to model the data? Estimate the height after 1.75 seconds.

SOLUTION

1. **Understand the Problem** You are asked to identify the type of function that can model the table of values and then to find the height at a specific time.
2. **Make a Plan** Create a scatter plot of the data. Then use the relationship shown in the scatter plot to estimate the height after 1.75 seconds.
3. **Solve the Problem** Create a scatter plot.

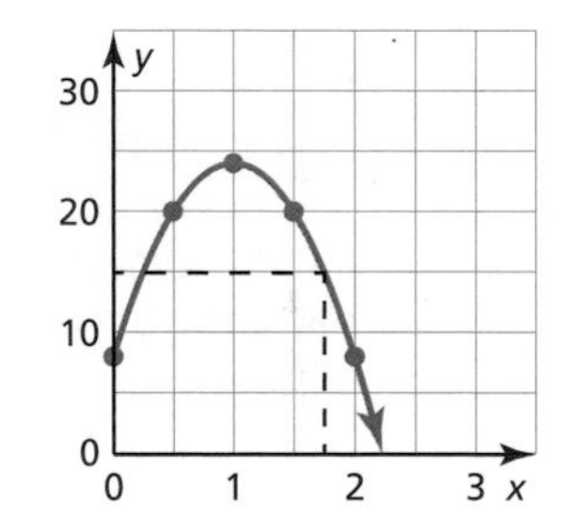

The data appear to lie on a curve that resembles a quadratic function. Sketch the curve.

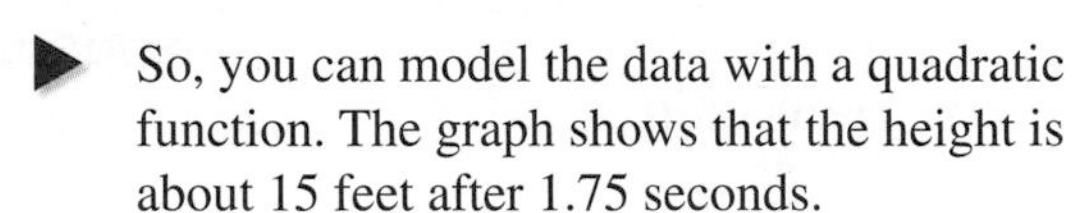

▶ So, you can model the data with a quadratic function. The graph shows that the height is about 15 feet after 1.75 seconds.

4. **Look Back** To check that your solution is reasonable, analyze the values in the table. Notice that the heights decrease after 1 second. Because 1.75 is between 1.5 and 2, the height must be between 20 feet and 8 feet.

$$8 < 15 < 20 \checkmark$$

Monitoring Progress  Help in English and Spanish at *BigIdeasMath.com*

Use a graphing calculator to graph the function and its parent function. Then describe the transformations.

8. $h(x) = -\frac{1}{4}x + 5$

9. $d(x) = 3(x - 5)^2 - 1$

10. The table shows the amount of fuel in a chainsaw over time. What type of function can you use to model the data? When will the tank be empty?

Time (minutes), x	0	10	20	30	40
Fuel remaining (fluid ounces), y	15	12	9	6	3

1.1 Exercises

Dynamic Solutions available at *BigIdeasMath.com*

Vocabulary and Core Concept Check

1. **COMPLETE THE SENTENCE** The function $f(x) = x^2$ is the ________ of $f(x) = 2x^2 - 3$.

2. **DIFFERENT WORDS, SAME QUESTION** Which is different? Find "both" answers.

What are the vertices of the figure after a reflection in the x-axis, followed by a translation 2 units right?

What are the vertices of the figure after a translation 6 units up and 2 units right?

What are the vertices of the figure after a translation 2 units right, followed by a reflection in the x-axis?

What are the vertices of the figure after a translation 6 units up, followed by a reflection in the x-axis?

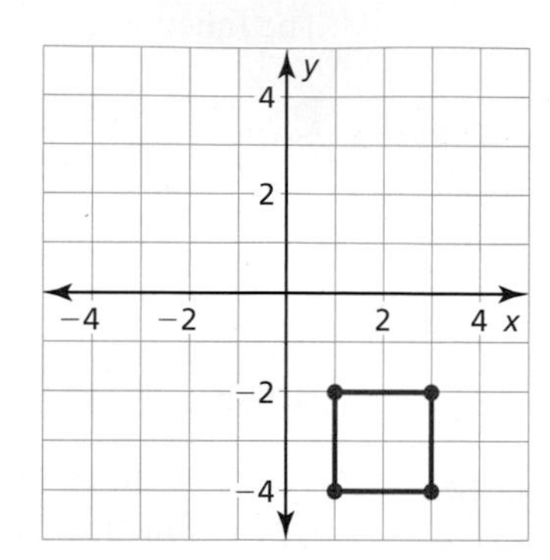

Monitoring Progress and Modeling with Mathematics

In Exercises 3–6, identify the function family to which f belongs. Compare the graph of f to the graph of its parent function. *(See Example 1.)*

3.

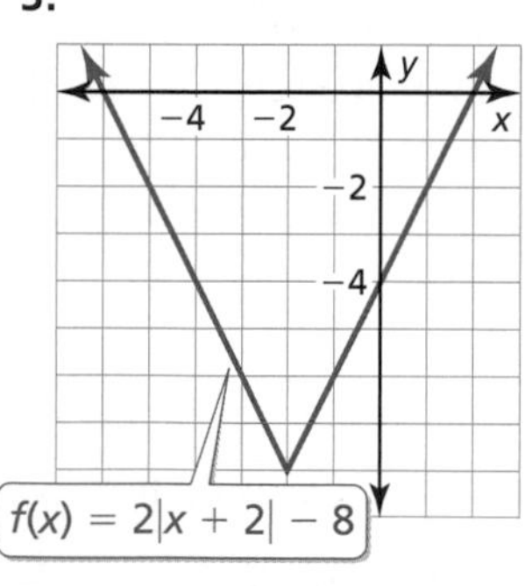

4.

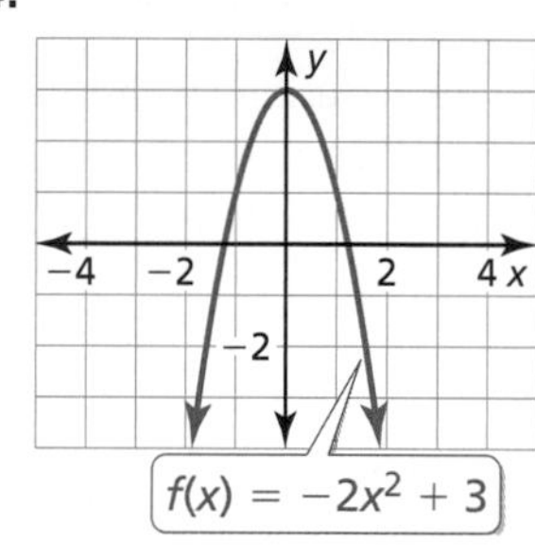

5.

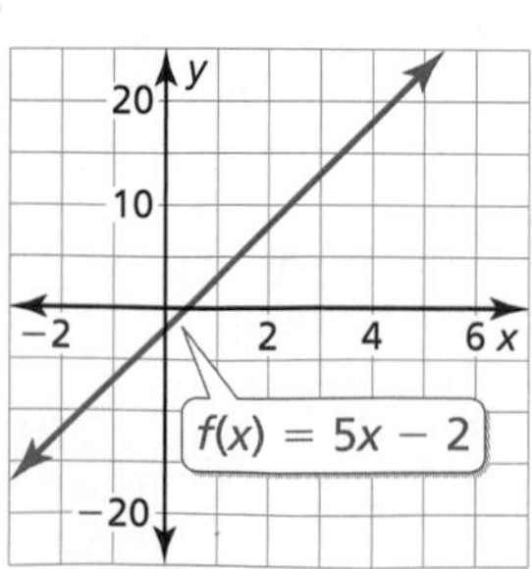

6.

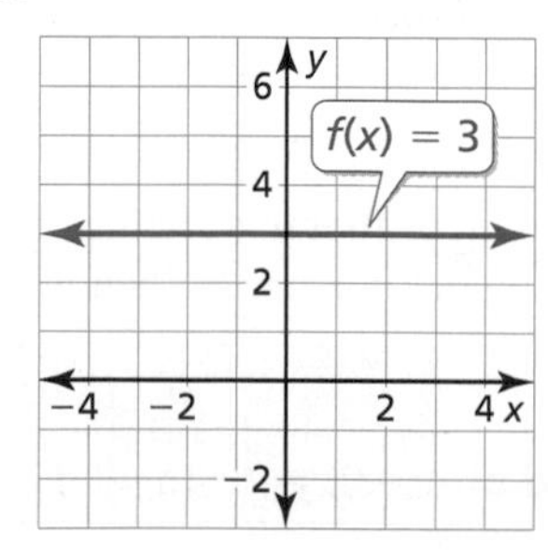

7. **MODELING WITH MATHEMATICS** At 8:00 A.M., the temperature is 43°F. The temperature increases 2°F each hour for the next 7 hours. Graph the temperatures over time t ($t = 0$ represents 8:00 A.M.). What type of function can you use to model the data? Explain.

8. **MODELING WITH MATHEMATICS** You purchase a car from a dealership for \$10,000. The trade-in value of the car each year after the purchase is given by the function $f(x) = 10{,}000 - 250x^2$. What type of function models the trade-in value?

In Exercises 9–18, graph the function and its parent function. Then describe the transformation. *(See Examples 2 and 3.)*

9. $g(x) = x + 4$
10. $f(x) = x - 6$
11. $f(x) = x^2 - 1$
12. $h(x) = (x + 4)^2$
13. $g(x) = |x - 5|$
14. $f(x) = 4 + |x|$
15. $h(x) = -x^2$
16. $g(x) = -x$
17. $f(x) = 3$
18. $f(x) = -2$

In Exercises 19–26, graph the function and its parent function. Then describe the transformation. *(See Example 4.)*

19. $f(x) = \frac{1}{3}x$

20. $g(x) = 4x$

21. $f(x) = 2x^2$

22. $h(x) = \frac{1}{3}x^2$

23. $h(x) = \frac{3}{4}x$

24. $g(x) = \frac{4}{3}x$

25. $h(x) = 3|x|$

26. $f(x) = \frac{1}{2}|x|$

In Exercises 27–34, use a graphing calculator to graph the function and its parent function. Then describe the transformations. *(See Example 5.)*

27. $f(x) = 3x + 2$

28. $h(x) = -x + 5$

29. $h(x) = -3|x| - 1$

30. $f(x) = \frac{3}{4}|x| + 1$

31. $g(x) = \frac{1}{2}x^2 - 6$

32. $f(x) = 4x^2 - 3$

33. $f(x) = -(x + 3)^2 + \frac{1}{4}$

34. $g(x) = -|x - 1| - \frac{1}{2}$

ERROR ANALYSIS In Exercises 35 and 36, identify and correct the error in describing the transformation of the parent function.

35.

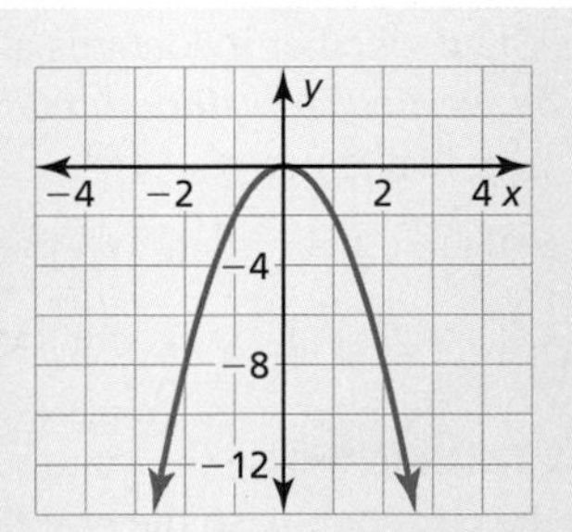

The graph is a reflection in the x-axis and a vertical shrink of the parent quadratic function.

36.

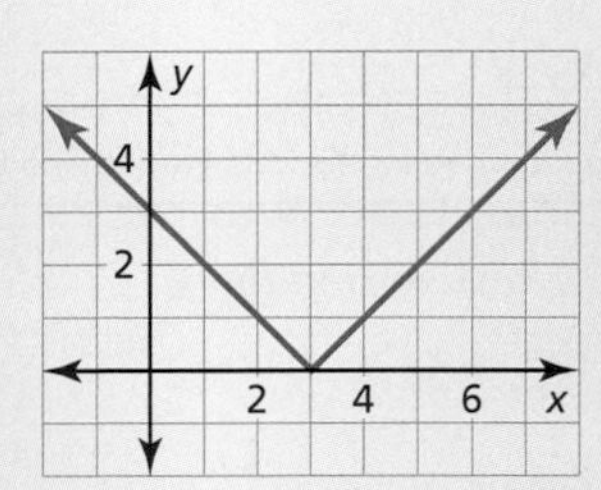

The graph is a translation 3 units right of the parent absolute value function, so the function is $f(x) = |x + 3|$.

MATHEMATICAL CONNECTIONS In Exercises 37 and 38, find the coordinates of the figure after the transformation.

37. Translate 2 units down.

38. Reflect in the x-axis.

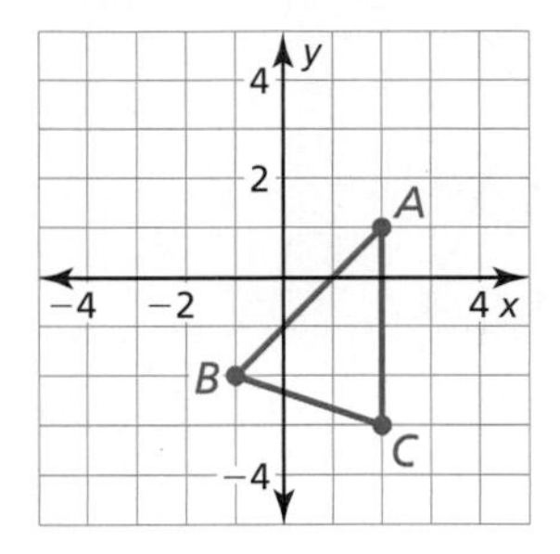

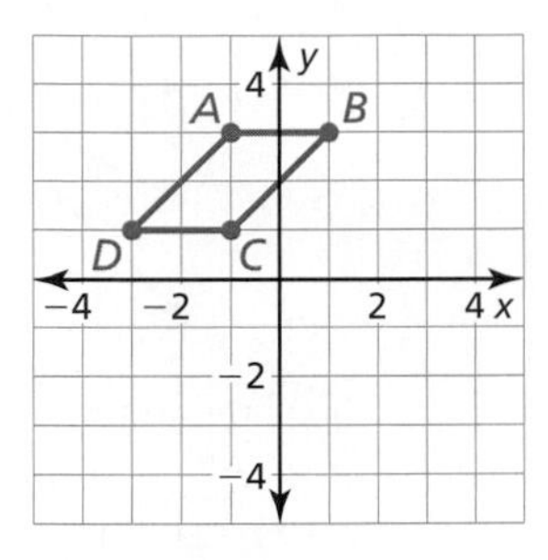

USING TOOLS In Exercises 39–44, identify the function family and describe the domain and range. Use a graphing calculator to verify your answer.

39. $g(x) = |x + 2| - 1$

40. $h(x) = |x - 3| + 2$

41. $g(x) = 3x + 4$

42. $f(x) = -4x + 11$

43. $f(x) = 5x^2 - 2$

44. $f(x) = -2x^2 + 6$

45. MODELING WITH MATHEMATICS The table shows the speeds of a car as it travels through an intersection with a stop sign. What type of function can you use to model the data? Estimate the speed of the car when it is 20 yards past the intersection. *(See Example 6.)*

Displacement from sign (yards), x	Speed (miles per hour), y
−100	40
−50	20
−10	4
0	0
10	4
50	20
100	40

46. THOUGHT PROVOKING In the same coordinate plane, sketch the graph of the parent quadratic function and the graph of a quadratic function that has no x-intercepts. Describe the transformation(s) of the parent function.

47. USING STRUCTURE Graph the functions $f(x) = |x - 4|$ and $g(x) = |x| - 4$. Are they equivalent? Explain.

48. **HOW DO YOU SEE IT?** Consider the graphs of f, g, and h.

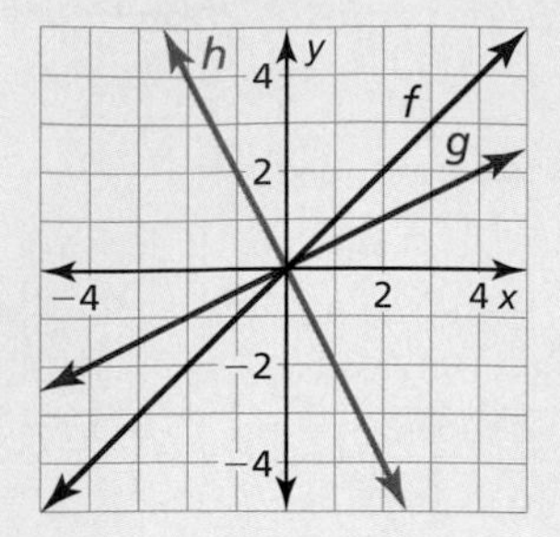

a. Does the graph of g represent a vertical stretch or a vertical shrink of the graph of f? Explain your reasoning.

b. Describe how to transform the graph of f to obtain the graph of h.

49. **MAKING AN ARGUMENT** Your friend says two different translations of the graph of the parent linear function can result in the graph of $f(x) = x - 2$. Is your friend correct? Explain.

50. **DRAWING CONCLUSIONS** A person swims at a constant speed of 1 meter per second. What type of function can be used to model the distance the swimmer travels? If the person has a 10-meter head start, what type of transformation does this represent? Explain.

51. **PROBLEM SOLVING** You are playing basketball with your friends. The height (in feet) of the ball above the ground t seconds after a shot is released from your hand is modeled by the function $f(t) = -16t^2 + 32t + 5.2$.

a. Without graphing, identify the type of function that models the height of the basketball.

b. What is the value of t when the ball is released from your hand? Explain your reasoning.

c. How many feet above the ground is the ball when it is released from your hand? Explain.

52. **MODELING WITH MATHEMATICS** The table shows the battery lives of a computer over time. What type of function can you use to model the data? Interpret the meaning of the x-intercept in this situation.

Time (hours), x	Battery life remaining, y
1	80%
3	40%
5	0%
6	20%
8	60%

53. **REASONING** Compare each function with its parent function. State whether it contains a *horizontal translation*, *vertical translation*, *both*, or *neither*. Explain your reasoning.

a. $f(x) = 2|x| - 3$

b. $f(x) = (x - 8)^2$

c. $f(x) = |x + 2| + 4$

d. $f(x) = 4x^2$

54. **CRITICAL THINKING** Use the values −1, 0, 1, and 2 in the correct box so the graph of each function intersects the x-axis. Explain your reasoning.

a. $f(x) = 3x \square + 1$

b. $f(x) = |2x - 6| - \square$

c. $f(x) = \square x^2 + 1$

d. $f(x) = \square$

Maintaining Mathematical Proficiency

Reviewing what you learned in previous grades and lessons

Determine whether the ordered pair is a solution of the equation. *(Skills Review Handbook)*

55. $f(x) = |x + 2|$; (1, −3)

56. $f(x) = |x| - 3$; (−2, −5)

57. $f(x) = x - 3$; (5, 2)

58. $f(x) = x - 4$; (12, 8)

Find the x-intercept and the y-intercept of the graph of the equation. *(Skills Review Handbook)*

59. $y = x$

60. $y = x + 2$

61. $3x + y = 1$

62. $x - 2y = 8$

1.2 Transformations of Linear and Absolute Value Functions

Essential Question

How do the graphs of $y = f(x) + k$, $y = f(x - h)$, and $y = -f(x)$ compare to the graph of the parent function f?

EXPLORATION 1 Transformations of the Parent Absolute Value Function

USING TOOLS STRATEGICALLY

To be proficient in math, you need to use technological tools to visualize results and explore consequences.

Work with a partner. Compare the graph of the function

$y = |x| + k$ Transformation

to the graph of the parent function

$f(x) = |x|$. Parent function

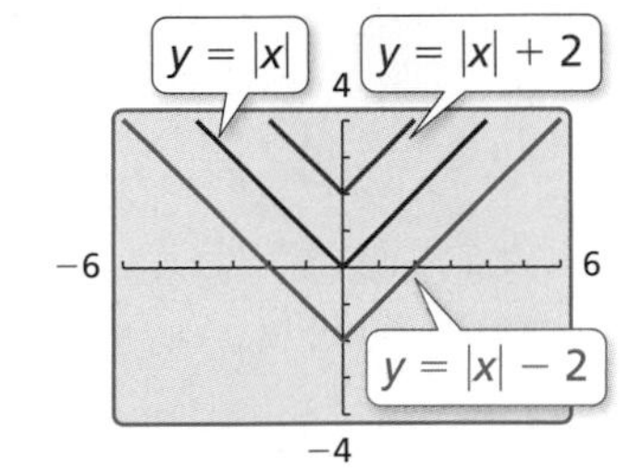

EXPLORATION 2 Transformations of the Parent Absolute Value Function

Work with a partner. Compare the graph of the function

$y = |x - h|$ Transformation

to the graph of the parent function

$f(x) = |x|$. Parent function

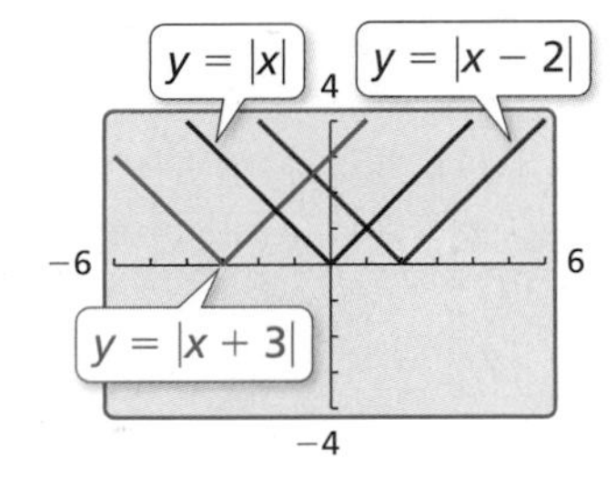

EXPLORATION 3 Transformation of the Parent Absolute Value Function

Work with a partner. Compare the graph of the function

$y = -|x|$ Transformation

to the graph of the parent function

$f(x) = |x|$. Parent function

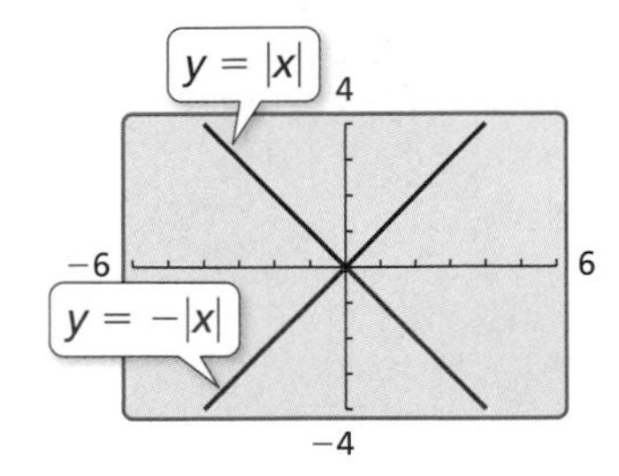

Communicate Your Answer

4. How do the graphs of $y = f(x) + k$, $y = f(x - h)$, and $y = -f(x)$ compare to the graph of the parent function f?

5. Compare the graph of each function to the graph of its parent function f. Use a graphing calculator to verify your answers are correct.

a. $y = \sqrt{x} - 4$ **b.** $y = \sqrt{x + 4}$ **c.** $y = -\sqrt{x}$

d. $y = x^2 + 1$ **e.** $y = (x - 1)^2$ **f.** $y = -x^2$

1.2 Lesson

What You Will Learn

- Write functions representing translations and reflections.
- Write functions representing stretches and shrinks.
- Write functions representing combinations of transformations.

Translations and Reflections

You can use function notation to represent transformations of graphs of functions.

Core Concept

Horizontal Translations

The graph of $y = f(x - h)$ is a horizontal translation of the graph of $y = f(x)$, where $h \neq 0$.

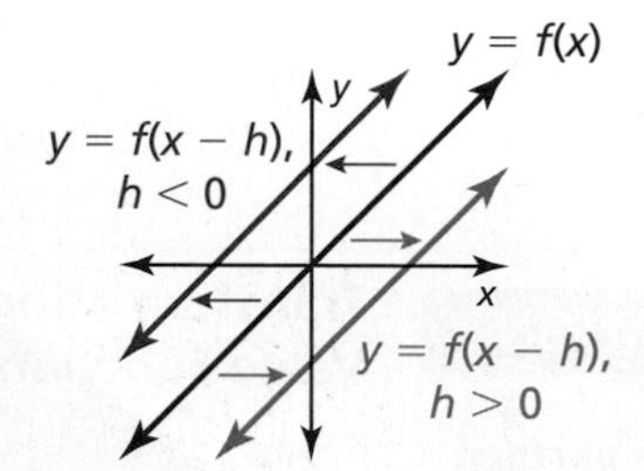

Subtracting h from the **inputs** before evaluating the function shifts the graph left when $h < 0$ and right when $h > 0$.

Vertical Translations

The graph of $y = f(x) + k$ is a vertical translation of the graph of $y = f(x)$, where $k \neq 0$.

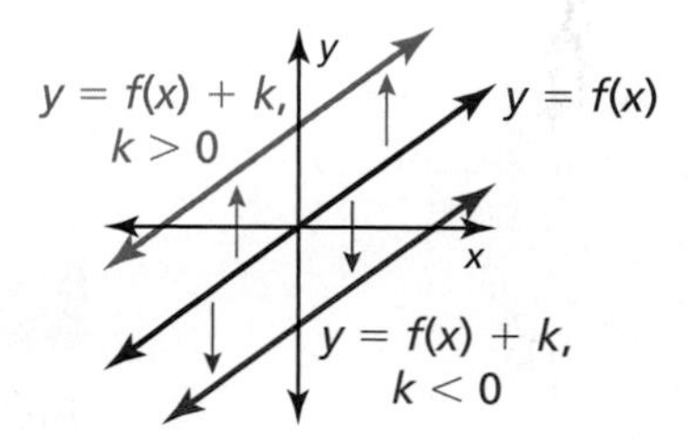

Adding k to the **outputs** shifts the graph down when $k < 0$ and up when $k > 0$.

EXAMPLE 1 **Writing Translations of Functions**

Let $f(x) = 2x + 1$.

a. Write a function g whose graph is a translation 3 units down of the graph of f.

b. Write a function h whose graph is a translation 2 units to the left of the graph of f.

SOLUTION

a. A translation 3 units down is a vertical translation that adds -3 to each output value.

$g(x) = f(x) + (-3)$	Add -3 to the output.
$= 2x + 1 + (-3)$	Substitute $2x + 1$ for $f(x)$.
$= 2x - 2$	Simplify.

▶ The translated function is $g(x) = 2x - 2$.

b. A translation 2 units to the left is a horizontal translation that subtracts -2 from each input value.

$h(x) = f(x - (-2))$	Subtract -2 from the input.
$= f(x + 2)$	Add the opposite.
$= 2(x + 2) + 1$	Replace x with $x + 2$ in $f(x)$.
$= 2x + 5$	Simplify.

▶ The translated function is $h(x) = 2x + 5$.

Check

Core Concept

STUDY TIP

When you reflect a function in a line, the graphs are symmetric about that line.

Reflections in the *x*-axis

The graph of $y = -f(x)$ is a reflection in the x-axis of the graph of $y = f(x)$.

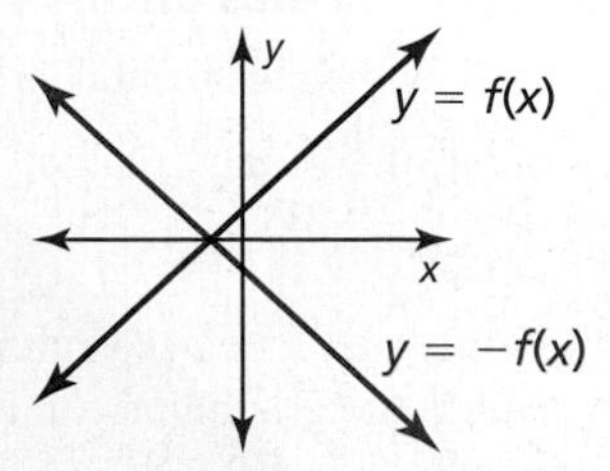

Multiplying the **outputs** by -1 changes their signs.

Reflections in the *y*-axis

The graph of $y = f(-x)$ is a reflection in the y-axis of the graph of $y = f(x)$.

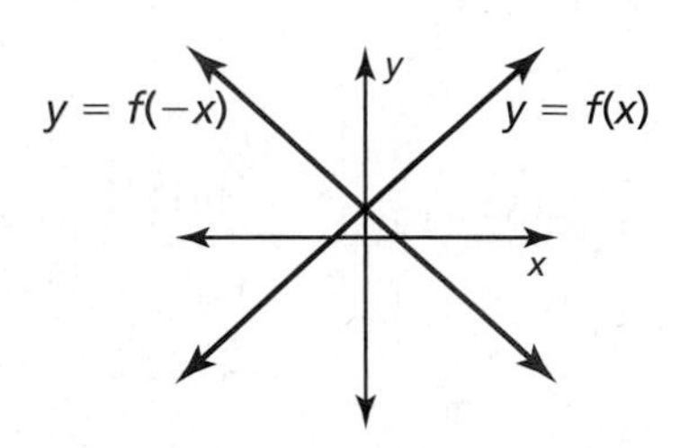

Multiplying the **inputs** by -1 changes their signs.

EXAMPLE 2 Writing Reflections of Functions

Let $f(x) = |x + 3| + 1$.

a. Write a function g whose graph is a reflection in the x-axis of the graph of f.

b. Write a function h whose graph is a reflection in the y-axis of the graph of f.

SOLUTION

a. A reflection in the x-axis changes the sign of each output value.

$g(x) = -f(x)$	Multiply the output by -1.
$= -(\lvert x + 3\rvert + 1)$	Substitute $\lvert x + 3\rvert + 1$ for $f(x)$.
$= -\lvert x + 3\rvert - 1$	Distributive Property

▶ The reflected function is $g(x) = -|x + 3| - 1$.

b. A reflection in the y-axis changes the sign of each input value.

$h(x) = f(-x)$	Multiply the input by -1.
$= \lvert -x + 3\rvert + 1$	Replace x with $-x$ in $f(x)$.
$= \lvert -(x - 3)\rvert + 1$	Factor out -1.
$= \lvert -1\rvert \cdot \lvert x - 3\rvert + 1$	Product Property of Absolute Value
$= \lvert x - 3\rvert + 1$	Simplify.

▶ The reflected function is $h(x) = |x - 3| + 1$.

Check

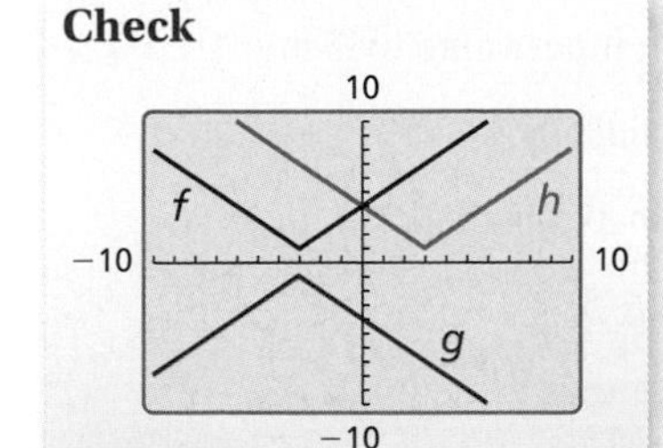

Monitoring Progress  Help in English and Spanish at *BigIdeasMath.com*

Write a function *g* whose graph represents the indicated transformation of the graph of *f*. Use a graphing calculator to check your answer.

1. $f(x) = 3x$; translation 5 units up
2. $f(x) = |x| - 3$; translation 4 units to the right
3. $f(x) = -|x + 2| - 1$; reflection in the x-axis
4. $f(x) = \frac{1}{2}x + 1$; reflection in the y-axis

Stretches and Shrinks

In the previous section, you learned that vertical stretches and shrinks transform graphs. You can also use *horizontal* stretches and shrinks to transform graphs.

Core Concept

Horizontal Stretches and Shrinks

The graph of $y = f(ax)$ is a horizontal stretch or shrink by a factor of $\frac{1}{a}$ of the graph of $y = f(x)$, where $a > 0$ and $a \neq 1$.

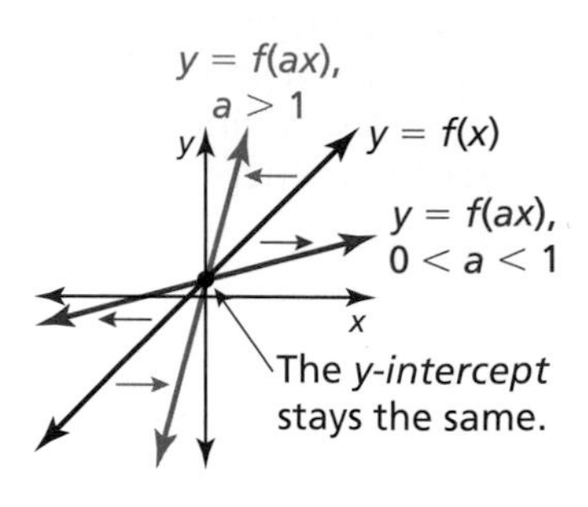

Multiplying the **inputs** by a before evaluating the function stretches the graph horizontally (away from the y-axis) when $0 < a < 1$, and shrinks the graph horizontally (toward the y-axis) when $a > 1$.

STUDY TIP

The graphs of $y = f(-ax)$ and $y = -a \cdot f(x)$ represent a stretch or shrink *and* a reflection in the x- or y-axis of the graph of $y = f(x)$.

Vertical Stretches and Shrinks

The graph of $y = a \cdot f(x)$ is a vertical stretch or shrink by a factor of a of the graph of $y = f(x)$, where $a > 0$ and $a \neq 1$.

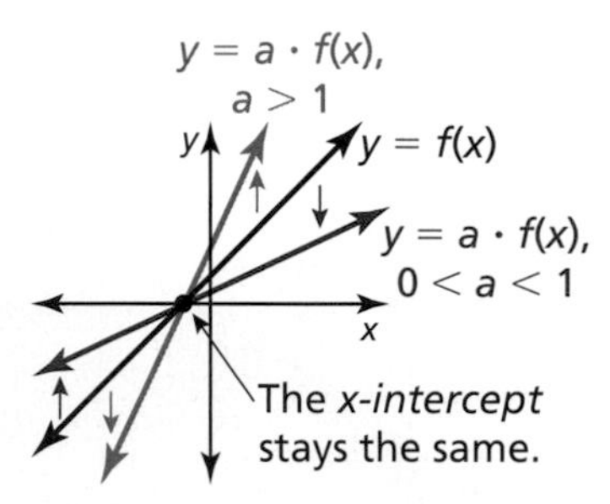

Multiplying the **outputs** by a stretches the graph vertically (away from the x-axis) when $a > 1$, and shrinks the graph vertically (toward the x-axis) when $0 < a < 1$.

EXAMPLE 3 Writing Stretches and Shrinks of Functions

Let $f(x) = |x - 3| - 5$. Write (a) a function g whose graph is a horizontal shrink of the graph of f by a factor of $\frac{1}{3}$, and (b) a function h whose graph is a vertical stretch of the graph of f by a factor of 2.

SOLUTION

a. A horizontal shrink by a factor of $\frac{1}{3}$ multiplies each input value by 3.

$g(x) = f(3x)$ — Multiply the input by 3.

$= |3x - 3| - 5$ — Replace x with $3x$ in $f(x)$.

▶ The transformed function is $g(x) = |3x - 3| - 5$.

Check

4, −10, 14, −12; g h f

b. A vertical stretch by a factor of 2 multiplies each output value by 2.

$h(x) = 2 \cdot f(x)$ — Multiply the output by 2.

$= 2 \cdot (|x - 3| - 5)$ — Substitute $|x - 3| - 5$ for $f(x)$.

$= 2|x - 3| - 10$ — Distributive Property

▶ The transformed function is $h(x) = 2|x - 3| - 10$.

Monitoring Progress  Help in English and Spanish at *BigIdeasMath.com*

Write a function g whose graph represents the indicated transformation of the graph of f. Use a graphing calculator to check your answer.

5. $f(x) = 4x + 2$; horizontal stretch by a factor of 2

6. $f(x) = |x| - 3$; vertical shrink by a factor of $\frac{1}{3}$

Combinations of Transformations

You can write a function that represents a series of transformations on the graph of another function by applying the transformations one at a time in the stated order.

EXAMPLE 4 Combining Transformations

Let the graph of g be a vertical shrink by a factor of 0.25 followed by a translation 3 units up of the graph of $f(x) = x$. Write a rule for g.

SOLUTION

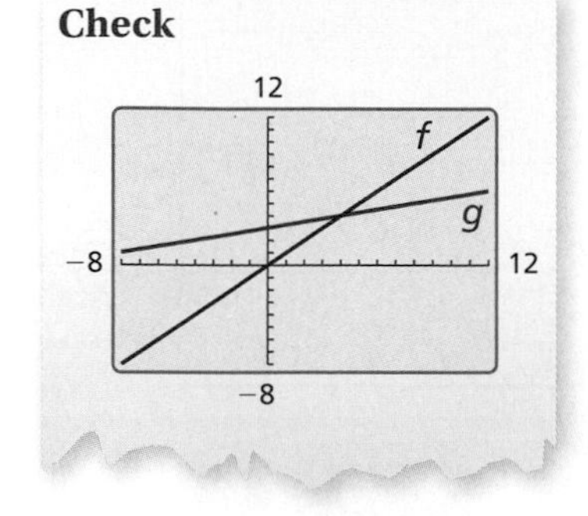

Step 1 First write a function h that represents the vertical shrink of f.

$h(x) = 0.25 \cdot f(x)$ Multiply the output by 0.25.

$= 0.25x$ Substitute x for $f(x)$.

Step 2 Then write a function g that represents the translation of h.

$g(x) = h(x) + 3$ Add 3 to the output.

$= 0.25x + 3$ Substitute $0.25x$ for $h(x)$.

▶ The transformed function is $g(x) = 0.25x + 3$.

EXAMPLE 5 Modeling with Mathematics

You design a computer game. Your revenue for x downloads is given by $f(x) = 2x$. Your profit is \$50 less than 90% of the revenue for x downloads. Describe how to transform the graph of f to model the profit. What is your profit for 100 downloads?

SOLUTION

1. **Understand the Problem** You are given a function that represents your revenue and a verbal statement that represents your profit. You are asked to find the profit for 100 downloads.
2. **Make a Plan** Write a function p that represents your profit. Then use this function to find the profit for 100 downloads.
3. **Solve the Problem** profit = 90% • revenue − 50

$p(x) = 0.9 \cdot f(x) - 50$

Vertical shrink by a factor of 0.9 (0.9); Translation 50 units down (− 50)

$= 0.9 \cdot 2x - 50$ Substitute $2x$ for $f(x)$.

$= 1.8x - 50$ Simplify.

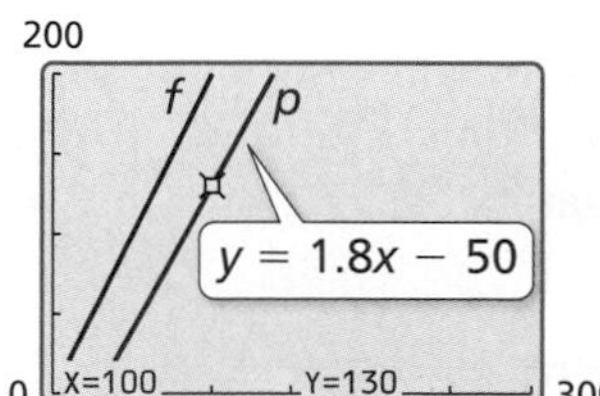

To find the profit for 100 downloads, evaluate p when $x = 100$.

$p(100) = 1.8(100) - 50 = 130$

▶ Your profit is \$130 for 100 downloads.

4. **Look Back** The vertical shrink decreases the slope, and the translation shifts the graph 50 units down. So, the graph of p is below and not as steep as the graph of f.

Monitoring Progress  Help in English and Spanish at *BigIdeasMath.com*

7. Let the graph of g be a translation 6 units down followed by a reflection in the x-axis of the graph of $f(x) = |x|$. Write a rule for g. Use a graphing calculator to check your answer.
8. **WHAT IF?** In Example 5, your revenue function is $f(x) = 3x$. How does this affect your profit for 100 downloads?

1.2 Exercises

Dynamic Solutions available at *BigIdeasMath.com*

Vocabulary and Core Concept Check

1. **COMPLETE THE SENTENCE** The function $g(x) = |5x| - 4$ is a horizontal ________ of the function $f(x) = |x| - 4$.

2. **WHICH ONE DOESN'T BELONG?** Which transformation does *not* belong with the other three? Explain your reasoning.

Translate the graph of $f(x) = 2x + 3$ up 2 units.	Shrink the graph of $f(x) = x + 5$ horizontally by a factor of $\frac{1}{2}$.
Stretch the graph of $f(x) = x + 3$ vertically by a factor of 2.	Translate the graph of $f(x) = 2x + 3$ left 1 unit.

Monitoring Progress and Modeling with Mathematics

In Exercises 3–8, write a function g whose graph represents the indicated transformation of the graph of f. Use a graphing calculator to check your answer. *(See Example 1.)*

3. $f(x) = x - 5$; translation 4 units to the left

4. $f(x) = x + 2$; translation 2 units to the right

5. $f(x) = |4x + 3| + 2$; translation 2 units down

6. $f(x) = 2x - 9$; translation 6 units up

7. $f(x) = 4 - |x + 1|$

8. $f(x) = |4x| + 5$

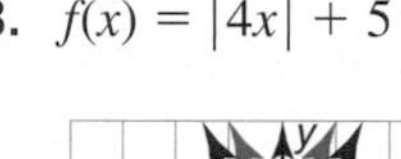

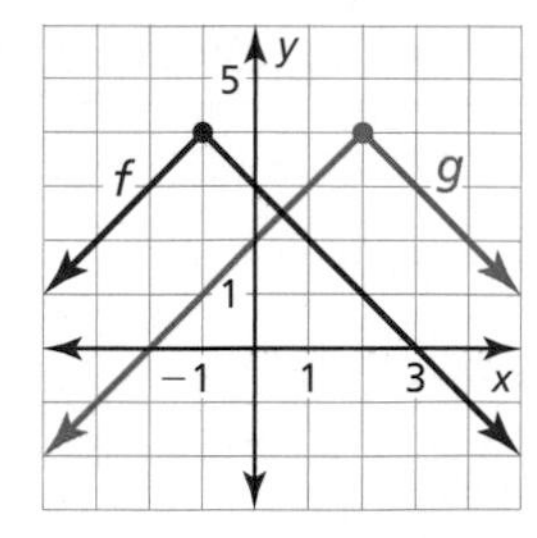

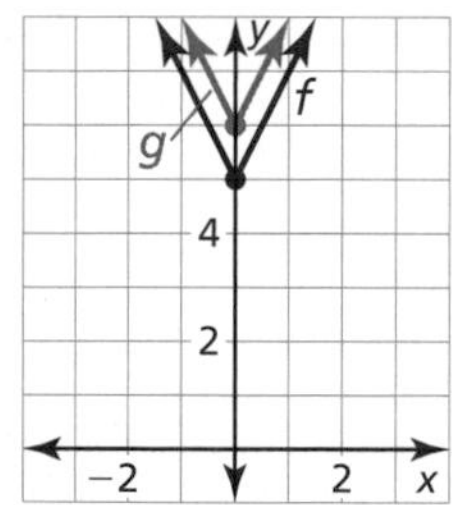

9. **WRITING** Describe two different translations of the graph of f that result in the graph of g.

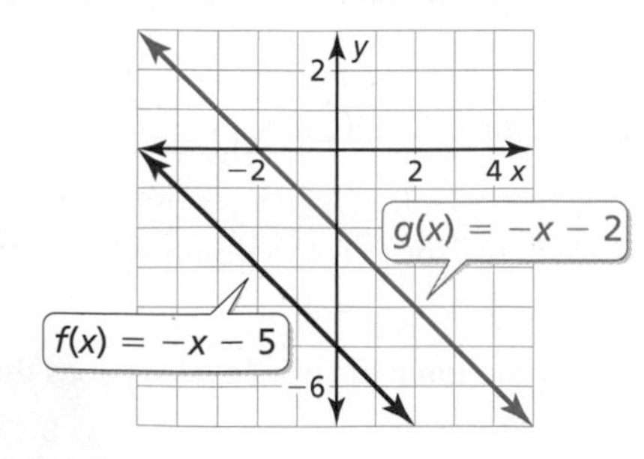

10. **PROBLEM SOLVING** You open a café. The function $f(x) = 4000x$ represents your expected net income (in dollars) after being open x weeks. Before you open, you incur an extra expense of \$12,000. What transformation of f is necessary to model this situation? How many weeks will it take to pay off the extra expense?

In Exercises 11–16, write a function g whose graph represents the indicated transformation of the graph of f. Use a graphing calculator to check your answer. *(See Example 2.)*

11. $f(x) = -5x + 2$; reflection in the x-axis

12. $f(x) = \frac{1}{2}x - 3$; reflection in the x-axis

13. $f(x) = |6x| - 2$; reflection in the y-axis

14. $f(x) = |2x - 1| + 3$; reflection in the y-axis

15. $f(x) = -3 + |x - 11|$; reflection in the y-axis

16. $f(x) = -x + 1$; reflection in the y-axis

In Exercises 17–22, write a function g whose graph represents the indicated transformation of the graph of f. Use a graphing calculator to check your answer. *(See Example 3.)*

17. $f(x) = x + 2$; vertical stretch by a factor of 5

18. $f(x) = 2x + 6$; vertical shrink by a factor of $\frac{1}{2}$

19. $f(x) = |2x| + 4$; horizontal shrink by a factor of $\frac{1}{2}$

20. $f(x) = |x + 3|$; horizontal stretch by a factor of 4

21. $f(x) = -2|x - 4| + 2$

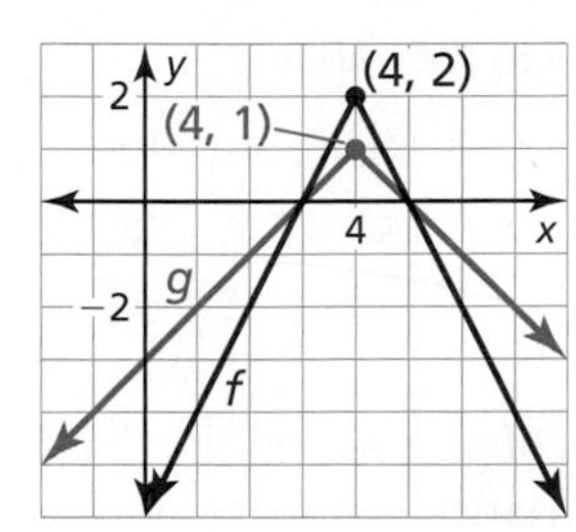

22. $f(x) = 6 - x$

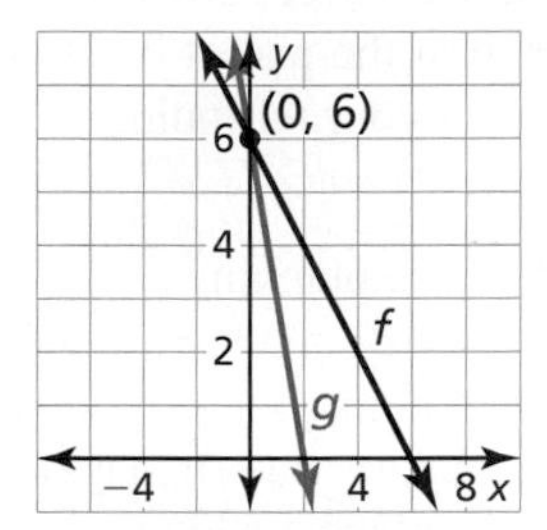

ANALYZING RELATIONSHIPS **In Exercises 23–26, match the graph of the transformation of f with the correct equation shown. Explain your reasoning.**

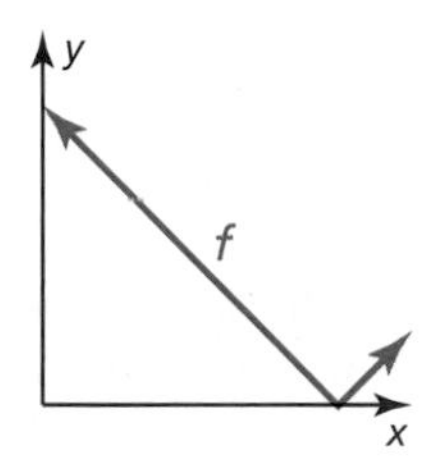

23.

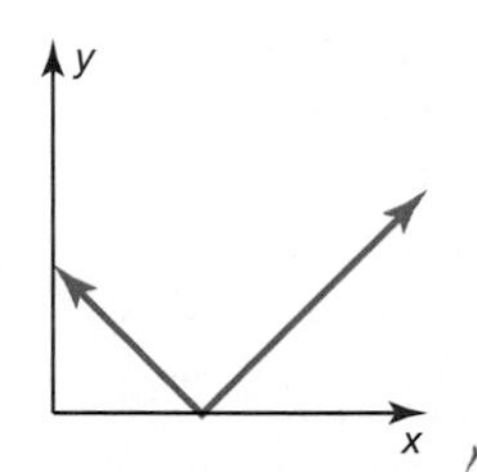

24.

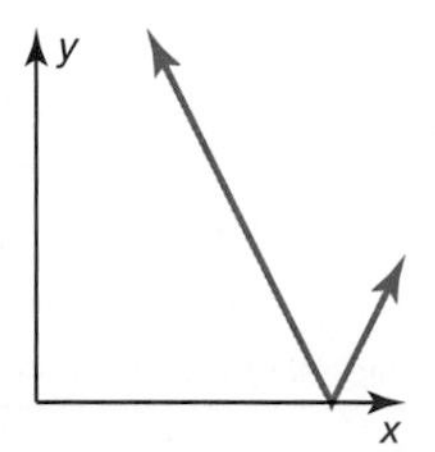

25.

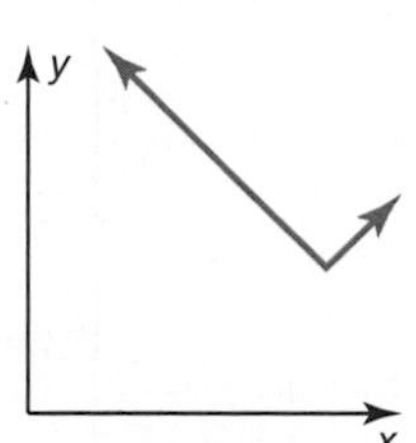

26.

A. $y = 2f(x)$ **B.** $y = f(2x)$

C. $y = f(x + 2)$ **D.** $y = f(x) + 2$

In Exercises 27–32, write a function g whose graph represents the indicated transformations of the graph of f. *(See Example 4.)*

27. $f(x) = x$; vertical stretch by a factor of 2 followed by a translation 1 unit up

28. $f(x) = x$; translation 3 units down followed by a vertical shrink by a factor of $\frac{1}{3}$

29. $f(x) = |x|$; translation 2 units to the right followed by a horizontal stretch by a factor of 2

30. $f(x) = |x|$; reflection in the y-axis followed by a translation 3 units to the right

31. $f(x) = |x|$

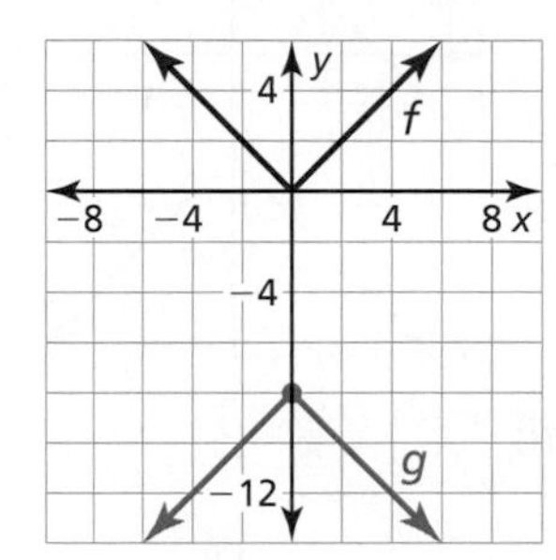

32. $f(x) = |x|$

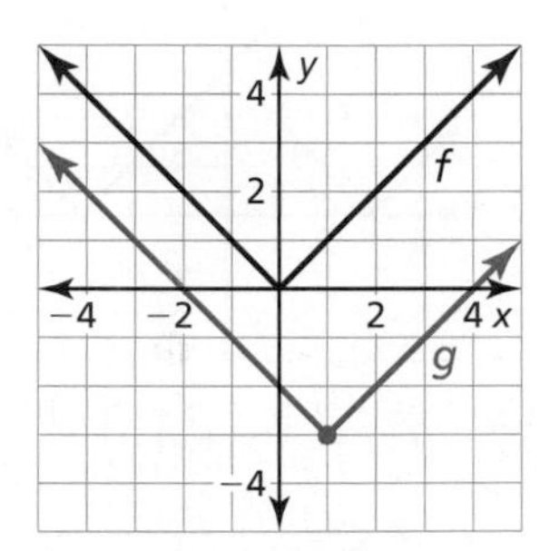

ERROR ANALYSIS **In Exercises 33 and 34, identify and correct the error in writing the function g whose graph represents the indicated transformations of the graph of f.**

33.

$f(x) = |x|$; translation 3 units to the right followed by a translation 2 units up

$g(x) = |x + 3| + 2$

34.

$f(x) = x$; translation 6 units down followed by a vertical stretch by a factor of 5

$g(x) = 5x - 6$

35. **MAKING AN ARGUMENT** Your friend claims that when writing a function whose graph represents a combination of transformations, the order is not important. Is your friend correct? Justify your answer.

36. **MODELING WITH MATHEMATICS** During a recent period of time, bookstore sales have been declining. The sales (in billions of dollars) can be modeled by the function $f(t) = -\frac{7}{5}t + 17.2$, where t is the number of years since 2006. Suppose sales decreased at twice the rate. How can you transform the graph of f to model the sales? Explain how the sales in 2010 are affected by this change. *(See Example 5.)*

MATHEMATICAL CONNECTIONS For Exercises 37–40, describe the transformation of the graph of f to the graph of g. Then find the area of the shaded triangle.

37. $f(x) = |x - 3|$

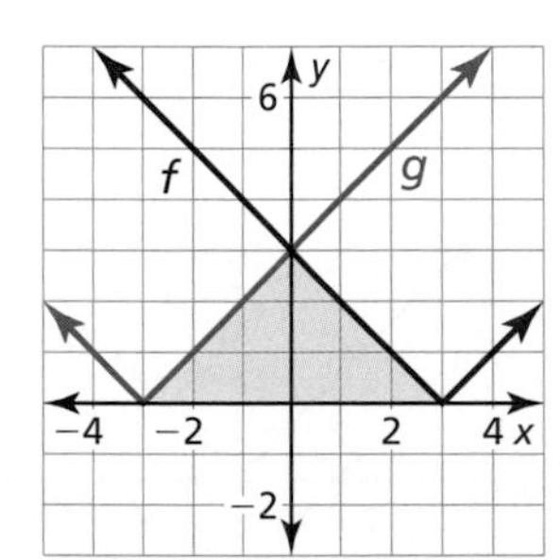

38. $f(x) = -|x| - 2$

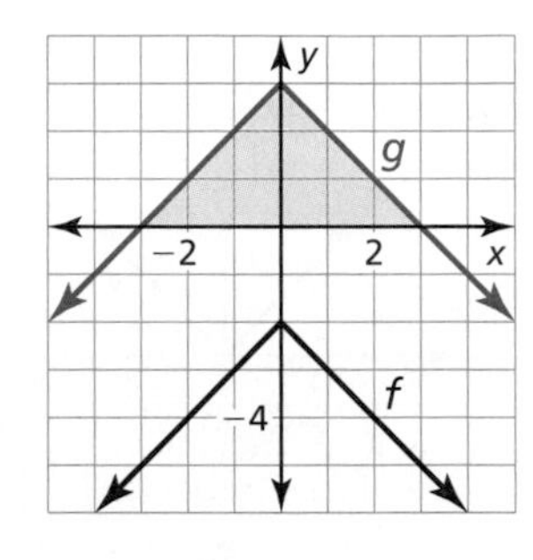

39. $f(x) = -x + 4$

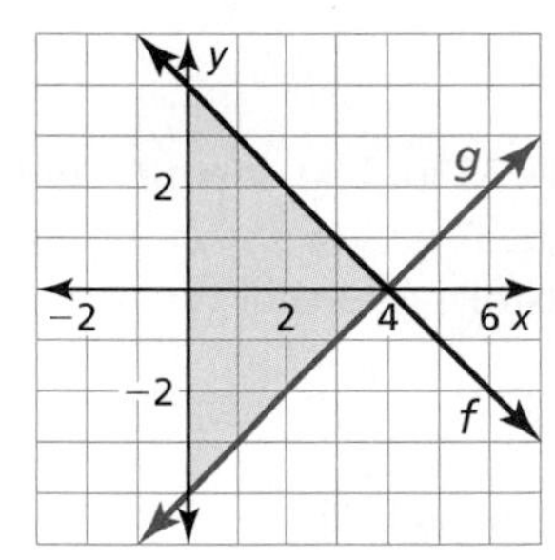

40. $f(x) = x - 5$

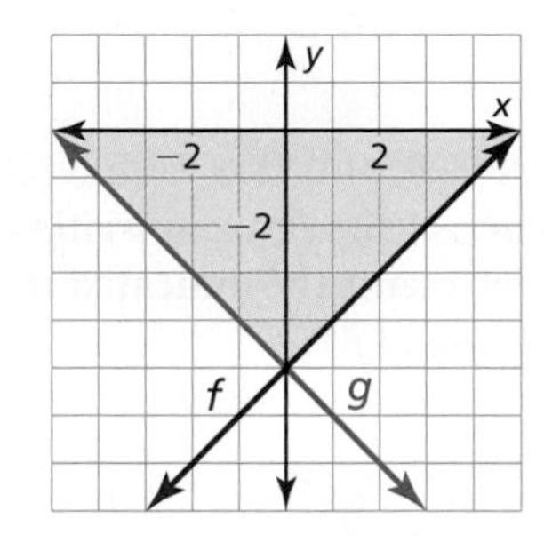

41. **ABSTRACT REASONING** The functions $f(x) = mx + b$ and $g(x) = mx + c$ represent two parallel lines.

    a. Write an expression for the vertical translation of the graph of f to the graph of g.

    b. Use the definition of slope to write an expression for the horizontal translation of the graph of f to the graph of g.

42. **HOW DO YOU SEE IT?** Consider the graph of $f(x) = mx + b$. Describe the effect each transformation has on the slope of the line and the intercepts of the graph.

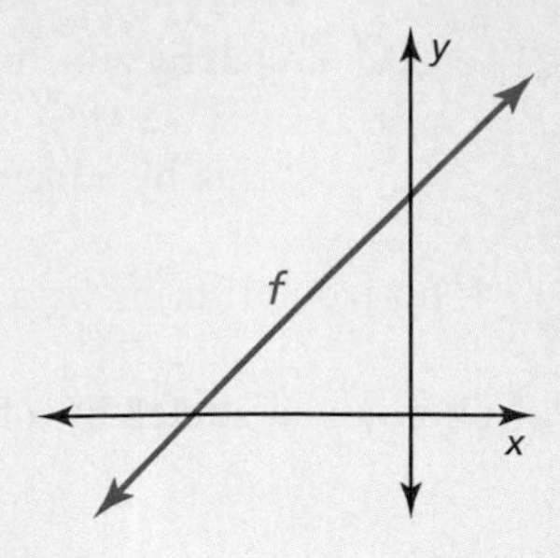

    a. Reflect the graph of f in the y-axis.

    b. Shrink the graph of f vertically by a factor of $\frac{1}{3}$.

    c. Stretch the graph of f horizontally by a factor of 2.

43. **REASONING** The graph of $g(x) = -4|x| + 2$ is a reflection in the x-axis, vertical stretch by a factor of 4, and a translation 2 units down of the graph of its parent function. Choose the correct order for the transformations of the graph of the parent function to obtain the graph of g. Explain your reasoning.

44. **THOUGHT PROVOKING** You are planning a cross-country bicycle trip of 4320 miles. Your distance d (in miles) from the halfway point can be modeled by $d = 72|x - 30|$, where x is the time (in days) and $x = 0$ represents June 1. Your plans are altered so that the model is now a right shift of the original model. Give an example of how this can happen. Sketch both the original model and the shifted model.

45. **CRITICAL THINKING** Use the correct value 0, −2, or 1 with a, b, and c so the graph of $g(x) = a|x - b| + c$ is a reflection in the x-axis followed by a translation one unit to the left and one unit up of the graph of $f(x) = 2|x - 2| + 1$. Explain your reasoning.

Maintaining Mathematical Proficiency

Reviewing what you learned in previous grades and lessons

Evaluate the function for the given value of x. *(Skills Review Handbook)*

46. $f(x) = x + 4; x = 3$

47. $f(x) = 4x - 1; x = -1$

48. $f(x) = -x + 3; x = 5$

49. $f(x) = -2x - 2; x = -1$

Create a scatter plot of the data. *(Skills Review Handbook)*

50.

x	8	10	11	12	15
$f(x)$	4	9	10	12	12

51.

x	2	5	6	10	13
$f(x)$	22	13	15	12	6

1.1–1.2 What Did You Learn?

Core Vocabulary

parent function, *p. 4*
transformation, *p. 5*
translation, *p. 5*
reflection, *p. 5*
vertical stretch, *p. 6*
vertical shrink, *p. 6*

Core Concepts

Section 1.1

Parent Functions, *p. 4*
Describing Transformations, *p. 5*

Section 1.2

Horizontal Translations, *p. 12*
Vertical Translations, *p. 12*
Reflections in the x-axis, *p. 13*
Reflections in the y-axis, *p. 13*
Horizontal Stretches and Shrinks, *p. 14*
Vertical Stretches and Shrinks, *p. 14*

Mathematical Practices

1. How can you analyze the values given in the table in Exercise 45 on page 9 to help you determine what type of function models the data?
2. Explain how you would round your answer in Exercise 10 on page 16 if the extra expense is $13,500.

Study Skills

Taking Control of Your Class Time

1. Sit where you can easily see and hear the teacher, and the teacher can see you.
2. Pay attention to what the teacher says about math, not just what is written on the board.
3. Ask a question if the teacher is moving through the material too fast.
4. Try to memorize new information while learning it.
5. Ask for clarification if you do not understand something.
6. Think as intensely as if you were going to take a quiz on the material at the end of class.
7. Volunteer when the teacher asks for someone to go up to the board.
8. At the end of class, identify concepts or problems for which you still need clarification.
9. Use the tutorials at *BigIdeasMath.com* for additional help.

1.1–1.2 Quiz

Identify the function family to which *g* belongs. Compare the graph of the function to the graph of its parent function. *(Section 1.1)*

1.

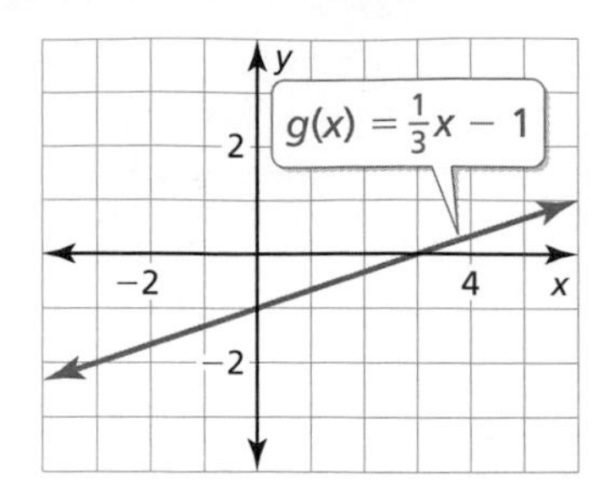

2.

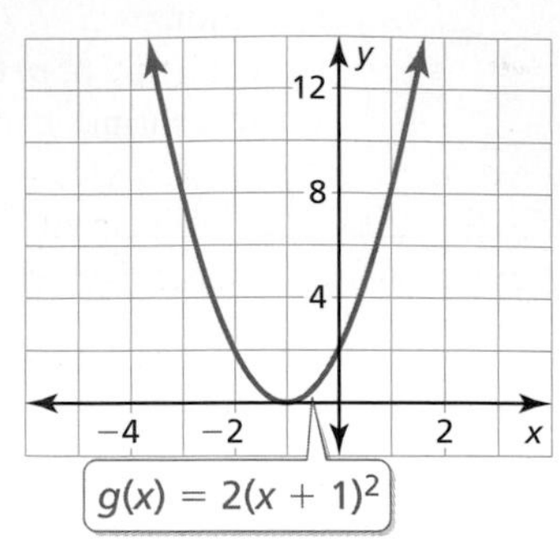

3.

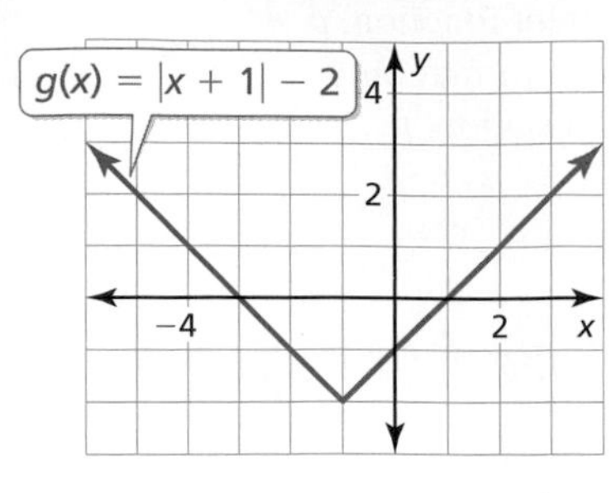

Graph the function and its parent function. Then describe the transformation. *(Section 1.1)*

4. $f(x) = \frac{3}{2}$

5. $f(x) = 3x$

6. $f(x) = 2(x - 1)^2$

7. $f(x) = -|x + 2| - 7$

8. $f(x) = \frac{1}{4}x^2 + 1$

9. $f(x) = -\frac{1}{2}x - 4$

Write a function *g* whose graph represents the indicated transformation of the graph of *f*. *(Section 1.2)*

10. $f(x) = 2x + 1$; translation 3 units up

11. $f(x) = -3|x - 4|$; vertical shrink by a factor of $\frac{1}{2}$

12. $f(x) = 3|x + 5|$; reflection in the x-axis

13. $f(x) = \frac{1}{3}x - \frac{2}{3}$; translation 4 units left

Write a function *g* whose graph represents the indicated transformations of the graph of *f*. *(Section 1.2)*

14. Let g be a translation 2 units down and a horizontal shrink by a factor of $\frac{2}{3}$ of the graph of $f(x) = x$.

15. Let g be a translation 9 units down followed by a reflection in the y-axis of the graph of $f(x) = x$.

16. Let g be a reflection in the x-axis and a vertical stretch by a factor of 4 followed by a translation 7 units down and 1 unit right of the graph of $f(x) = |x|$.

17. Let g be a translation 1 unit down and 2 units left followed by a vertical shrink by a factor of $\frac{1}{2}$ of the graph of $f(x) = |x|$.

18. The table shows the total distance a new car travels each month after it is purchased. What type of function can you use to model the data? Estimate the mileage after 1 year. *(Section 1.1)*

Time (months), x	0	2	5	6	9
Distance (miles), y	0	2300	5750	6900	10,350

19. The total cost of an annual pass plus camping for x days in a National Park can be modeled by the function $f(x) = 20x + 80$. Senior citizens pay half of this price and receive an additional \$30 discount. Describe how to transform the graph of f to model the total cost for a senior citizen. What is the total cost for a senior citizen to go camping for three days? *(Section 1.2)*

1.3 Modeling with Linear Functions

Essential Question How can you use a linear function to model and analyze a real-life situation?

EXPLORATION 1 Modeling with a Linear Function

Work with a partner. A company purchases a copier for $12,000. The spreadsheet shows how the copier depreciates over an 8-year period.

	A	B
1	Year, *t*	Value, *V*
2	0	$12,000
3	1	$10,750
4	2	$9,500
5	3	$8,250
6	4	$7,000
7	5	$5,750
8	6	$4,500
9	7	$3,250
10	8	$2,000
11		

a. Write a linear function to represent the value V of the copier as a function of the number t of years.

b. Sketch a graph of the function. Explain why this type of depreciation is called *straight line depreciation.*

c. Interpret the slope of the graph in the context of the problem.

MODELING WITH MATHEMATICS

To be proficient in math, you need to routinely interpret your results in the context of the situation.

EXPLORATION 2 Modeling with Linear Functions

Work with a partner. Match each description of the situation with its corresponding graph. Explain your reasoning.

a. A person gives $20 per week to a friend to repay a $200 loan.

b. An employee receives $12.50 per hour plus $2 for each unit produced per hour.

c. A sales representative receives $30 per day for food plus $0.565 for each mile driven.

d. A computer that was purchased for $750 depreciates $100 per year.

A.

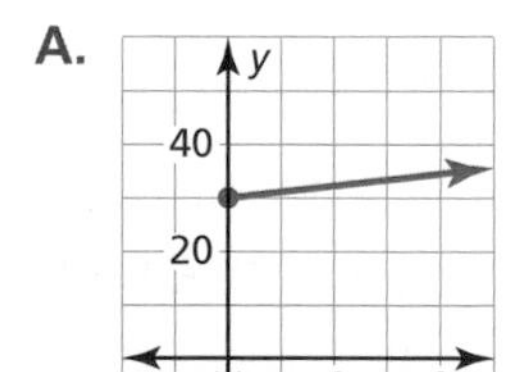

B.

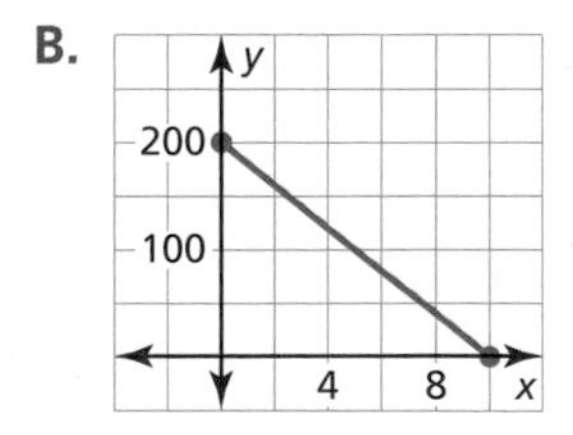

C.

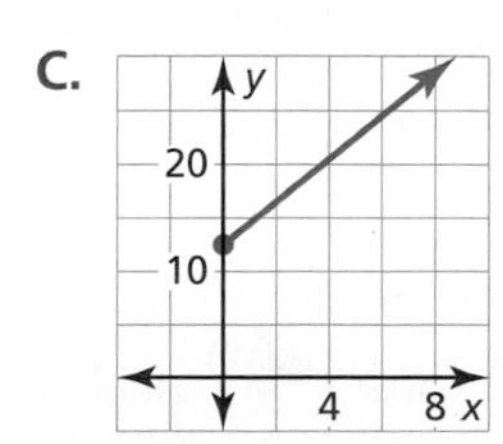

D.

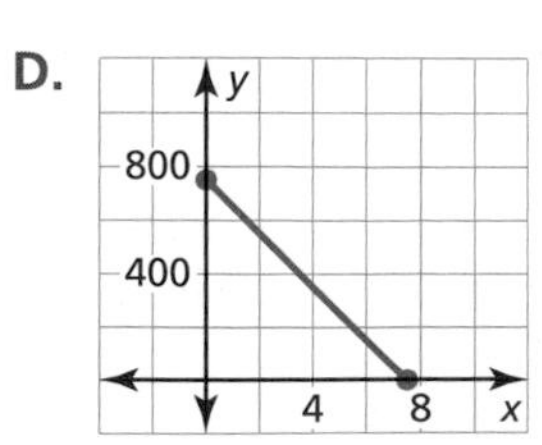

Communicate Your Answer

3. How can you use a linear function to model and analyze a real-life situation?

4. Use the Internet or some other reference to find a real-life example of straight line depreciation.

a. Use a spreadsheet to show the depreciation.

b. Write a function that models the depreciation.

c. Sketch a graph of the function.

1.3 Lesson

What You Will Learn

- Write equations of linear functions using points and slopes.
- Find lines of fit and lines of best fit.

Core Vocabulary

line of fit, *p. 24*
line of best fit, *p. 25*
correlation coefficient, *p. 25*

Previous
slope
slope-intercept form
point-slope form
scatter plot

Writing Linear Equations

Core Concept

Writing an Equation of a Line

Given slope m and y-intercept b — Use slope-intercept form:
$$y = mx + b$$

Given slope m and a point (x_1, y_1) — Use point-slope form:
$$y - y_1 = m(x - x_1)$$

Given points (x_1, y_1) and (x_2, y_2) — First use the slope formula to find m. Then use point-slope form with either given point.

EXAMPLE 1 Writing a Linear Equation from a Graph

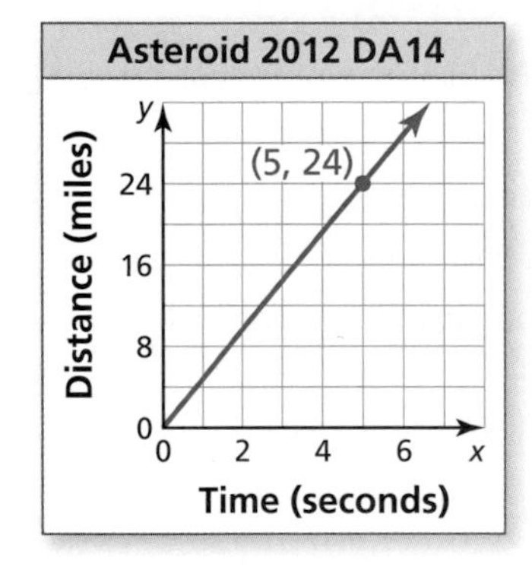

The graph shows the distance Asteroid 2012 DA14 travels in x seconds. Write an equation of the line and interpret the slope. The asteroid came within 17,200 miles of Earth in February, 2013. About how long does it take the asteroid to travel that distance?

SOLUTION

From the graph, you can see the slope is $m = \frac{24}{5} = 4.8$ and the y-intercept is $b = 0$. Use slope-intercept form to write an equation of the line.

$y = mx + b$ — Slope-intercept form

$= 4.8x + 0$ — Substitute 4.8 for m and 0 for b.

The equation is $y = 4.8x$. The slope indicates that the asteroid travels 4.8 miles per second. Use the equation to find how long it takes the asteroid to travel 17,200 miles.

$17{,}200 = 4.8x$ — Substitute 17,200 for y.

$3583 \approx x$ — Divide each side by 4.8.

- Because there are 3600 seconds in 1 hour, it takes the asteroid about 1 hour to travel 17,200 miles.

REMEMBER

An equation of the form $y = mx$ indicates that x and y are in a proportional relationship.

Monitoring Progress  Help in English and Spanish at *BigIdeasMath.com*

1. The graph shows the remaining balance y on a car loan after making x monthly payments. Write an equation of the line and interpret the slope and y-intercept. What is the remaining balance after 36 payments?

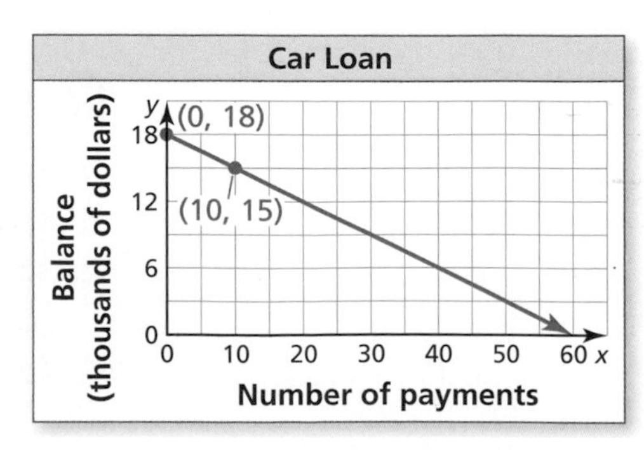

Lakeside Inn	
Number of students, x	Total cost, y
100	\$1500
125	\$1800
150	\$2100
175	\$2400
200	\$2700

EXAMPLE 2 Modeling with Mathematics

Two prom venues charge a rental fee plus a fee per student. The table shows the total costs for different numbers of students at Lakeside Inn. The total cost y (in dollars) for x students at Sunview Resort is represented by the equation

$$y = 10x + 600.$$

Which venue charges less per student? How many students must attend for the total costs to be the same?

SOLUTION

1. **Understand the Problem** You are given an equation that represents the total cost at one venue and a table of values showing total costs at another venue. You need to compare the costs.

2. **Make a Plan** Write an equation that models the total cost at Lakeside Inn. Then compare the slopes to determine which venue charges less per student. Finally, equate the cost expressions and solve to determine the number of students for which the total costs are equal.

3. **Solve the Problem** First find the slope using any two points from the table. Use $(x_1, y_1) = (100, 1500)$ and $(x_2, y_2) = (125, 1800)$.

$$m = \frac{y_2 - y_1}{x_2 - x_1} = \frac{1800 - 1500}{125 - 100} = \frac{300}{25} = 12$$

Write an equation that represents the total cost at Lakeside Inn using the slope of 12 and a point from the table. Use $(x_1, y_1) = (100, 1500)$.

$y - y_1 = m(x - x_1)$ Point-slope form

$y - 1500 = 12(x - 100)$ Substitute for m, x_1, and y_1.

$y - 1500 = 12x - 1200$ Distributive Property

$y = 12x + 300$ Add 1500 to each side.

Equate the cost expressions and solve.

$10x + 600 = 12x + 300$ Set cost expressions equal.

$300 = 2x$ Combine like terms.

$150 = x$ Divide each side by 2.

▶ Comparing the slopes of the equations, Sunview Resort charges \$10 per student, which is less than the \$12 per student that Lakeside Inn charges. The total costs are the same for 150 students.

4. **Look Back** Notice that the table shows the total cost for 150 students at Lakeside Inn is \$2100. To check that your solution is correct, verify that the total cost at Sunview Resort is also \$2100 for 150 students.

$y = 10(150) + 600$ Substitute 150 for x.

$= 2100$ ✓ Simplify.

Monitoring Progress

Help in English and Spanish at *BigIdeasMath.com*

2. **WHAT IF?** Maple Ridge charges a rental fee plus a \$10 fee per student. The total cost is \$1900 for 140 students. Describe the number of students that must attend for the total cost at Maple Ridge to be less than the total costs at the other two venues. Use a graph to justify your answer.

Finding Lines of Fit and Lines of Best Fit

Data do not always show an *exact* linear relationship. When the data in a scatter plot show an approximately linear relationship, you can model the data with a **line of fit**.

Core Concept

Finding a Line of Fit

Step 1 Create a scatter plot of the data.

Step 2 Sketch the line that most closely appears to follow the trend given by the data points. There should be about as many points above the line as below it.

Step 3 Choose two points on the line and estimate the coordinates of each point. These points do not have to be original data points.

Step 4 Write an equation of the line that passes through the two points from Step 3. This equation is a model for the data.

EXAMPLE 3 Finding a Line of Fit

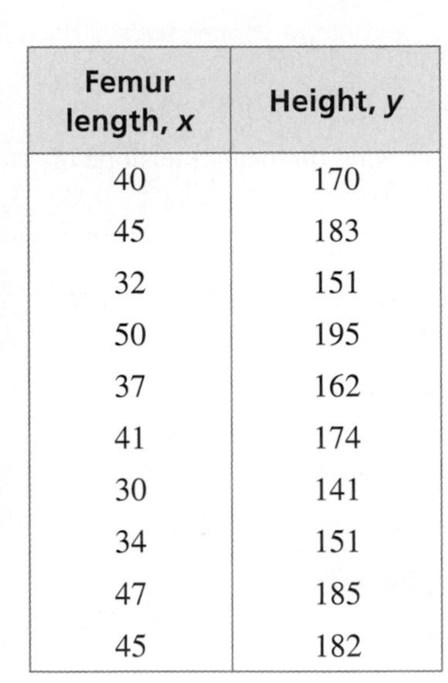

Femur length, x	Height, y
40	170
45	183
32	151
50	195
37	162
41	174
30	141
34	151
47	185
45	182

The table shows the femur lengths (in centimeters) and heights (in centimeters) of several people. Do the data show a linear relationship? If so, write an equation of a line of fit and use it to estimate the height of a person whose femur is 35 centimeters long.

SOLUTION

Step 1 Create a scatter plot of the data. The data show a linear relationship.

Step 2 Sketch the line that most closely appears to fit the data. One possibility is shown.

Step 3 Choose two points on the line. For the line shown, you might choose (40, 170) and (50, 195).

Step 4 Write an equation of the line.

First, find the slope.

$$m = \frac{y_2 - y_1}{x_2 - x_1} = \frac{195 - 170}{50 - 40} = \frac{25}{10} = 2.5$$

Use point-slope form to write an equation. Use $(x_1, y_1) = (40, 170)$.

$y - y_1 = m(x - x_1)$ Point-slope form

$y - 170 = 2.5(x - 40)$ Substitute for m, x_1, and y_1.

$y - 170 = 2.5x - 100$ Distributive Property

$y = 2.5x + 70$ Add 170 to each side.

Use the equation to estimate the height of the person.

$y = 2.5(35) + 70$ Substitute 35 for x.

$= 157.5$ Simplify.

▶ The approximate height of a person with a 35-centimeter femur is 157.5 centimeters.

The **line of best fit** is the line that lies as close as possible to all of the data points. Many technology tools have a *linear regression* feature that you can use to find the line of best fit for a set of data.

The **correlation coefficient**, denoted by r, is a number from -1 to 1 that measures how well a line fits a set of data pairs (x, y). When r is near 1, the points lie close to a line with a positive slope. When r is near -1, the points lie close to a line with a negative slope. When r is near 0, the points do not lie close to any line.

humerus

femur

EXAMPLE 4 Using a Graphing Calculator

Use the *linear regression* feature on a graphing calculator to find an equation of the line of best fit for the data in Example 3. Estimate the height of a person whose femur is 35 centimeters long. Compare this height to your estimate in Example 3.

SOLUTION

Step 1 Enter the data into two lists.

L1	L2	L3
40	170	------
45	183	
32	151	
50	195	
37	162	
41	174	
30	141	

L1(1)=40

Step 2 Use the *linear regression* feature. The line of best fit is $y = 2.6x + 65$.

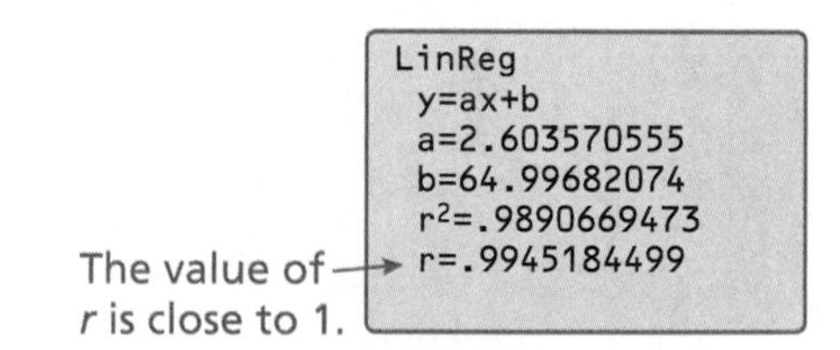

The value of r is close to 1.

Step 3 Graph the regression equation with the scatter plot.

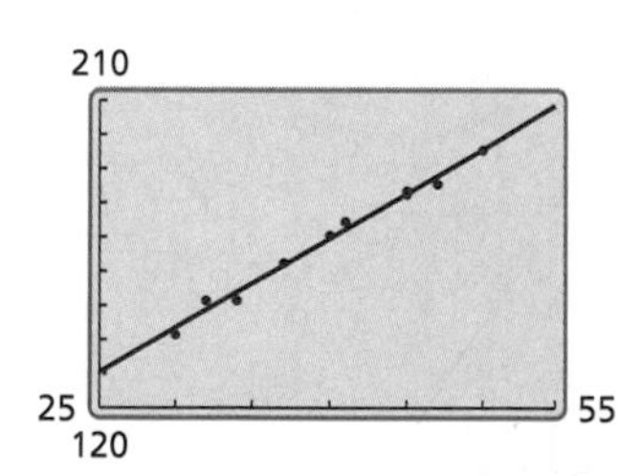

Step 4 Use the *trace* feature to find the value of y when $x = 35$.

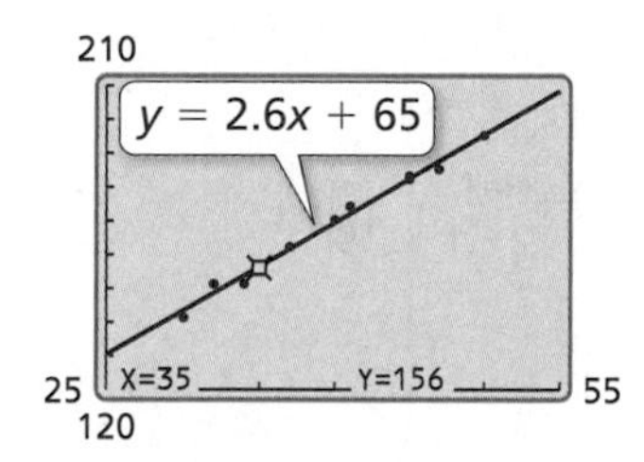

▶ The approximate height of a person with a 35-centimeter femur is 156 centimeters. This is less than the estimate found in Example 3.

ATTENDING TO PRECISION

Be sure to analyze the data values to help you select an appropriate viewing window for your graph.

Monitoring Progress  Help in English and Spanish at *BigIdeasMath.com*

3. The table shows the humerus lengths (in centimeters) and heights (in centimeters) of several females.

Humerus length, *x*	33	25	22	30	28	32	26	27
Height, *y*	166	142	130	154	152	159	141	145

a. Do the data show a linear relationship? If so, write an equation of a line of fit and use it to estimate the height of a female whose humerus is 40 centimeters long.

b. Use the *linear regression* feature on a graphing calculator to find an equation of the line of best fit for the data. Estimate the height of a female whose humerus is 40 centimeters long. Compare this height to your estimate in part (a).

1.3 Exercises

Dynamic Solutions available at *BigIdeasMath.com*

Vocabulary and Core Concept Check

1. **COMPLETE THE SENTENCE** The linear equation $y = \frac{1}{2}x + 3$ is written in ___________ form.

2. **VOCABULARY** A line of best fit has a correlation coefficient of -0.98. What can you conclude about the slope of the line?

Monitoring Progress and Modeling with Mathematics

In Exercises 3–8, use the graph to write an equation of the line and interpret the slope. *(See Example 1.)*

3. 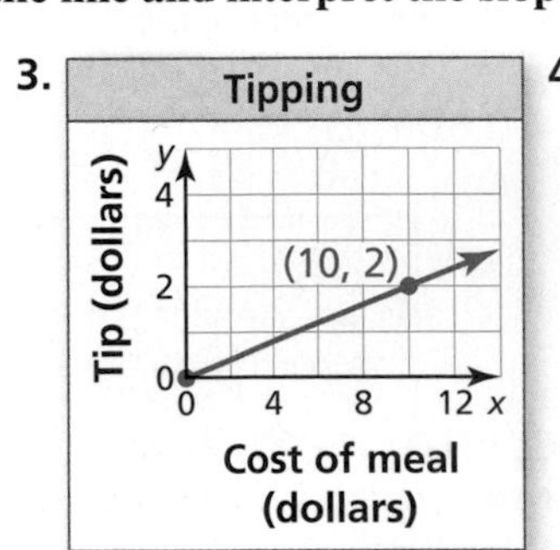

4. 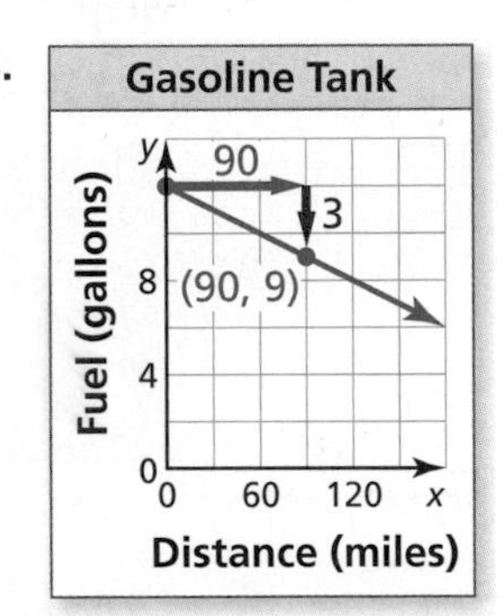

5. 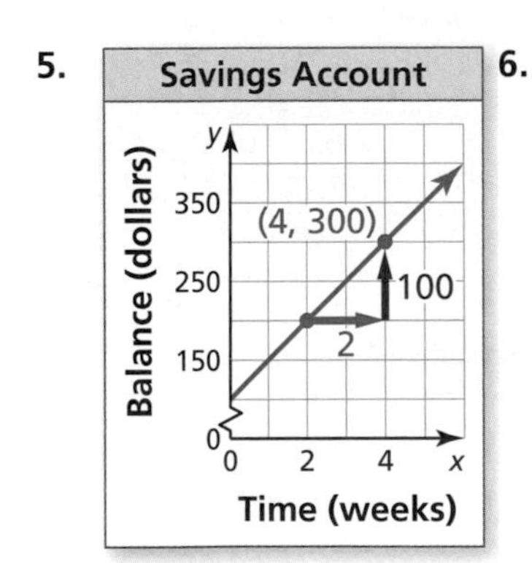

6. 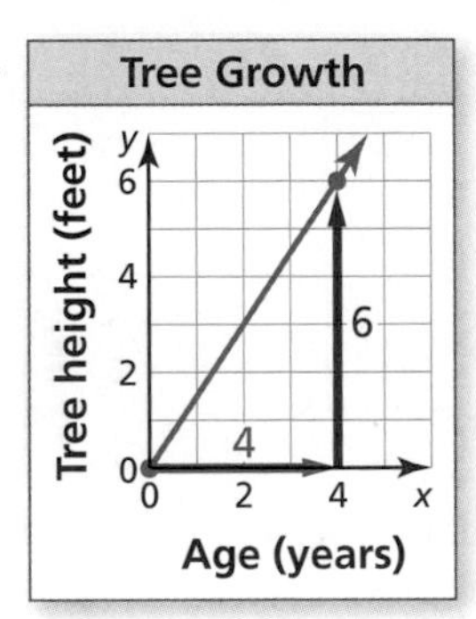

7. 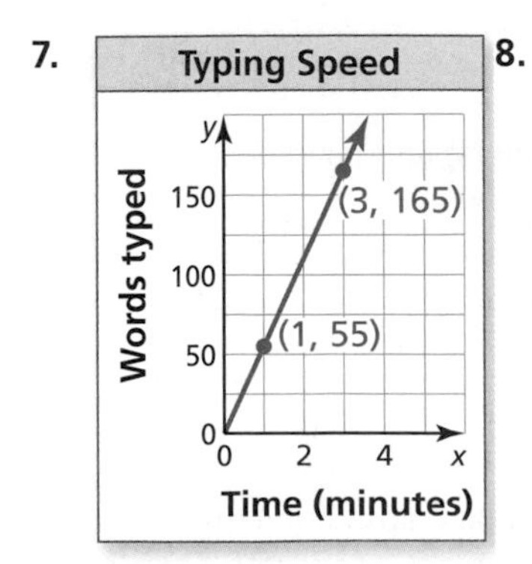

8. 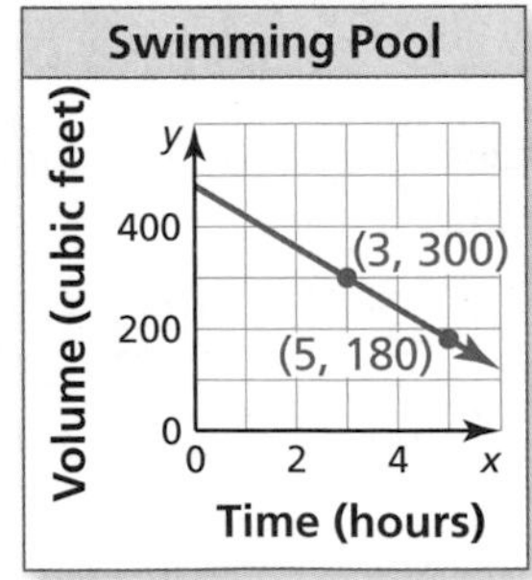

9. **MODELING WITH MATHEMATICS** Two newspapers charge a fee for placing an advertisement in their paper plus a fee based on the number of lines in the advertisement. The table shows the total costs for different length advertisements at the Daily Times. The total cost y (in dollars) for an advertisement that is x lines long at the Greenville Journal is represented by the equation $y = 2x + 20$. Which newspaper charges less per line? How many lines must be in an advertisement for the total costs to be the same? *(See Example 2.)*

Daily Times	
Number of lines, x	Total cost, y
4	27
5	30
6	33
7	36
8	39

10. **PROBLEM SOLVING** While on vacation in Canada, you notice that temperatures are reported in degrees Celsius. You know there is a linear relationship between Fahrenheit and Celsius, but you forget the formula. From science class, you remember the freezing point of water is 0°C or 32°F, and its boiling point is 100°C or 212°F.

   a. Write an equation that represents degrees Fahrenheit in terms of degrees Celsius.

   b. The temperature outside is 22°C. What is this temperature in degrees Fahrenheit?

   c. Rewrite your equation in part (a) to represent degrees Celsius in terms of degrees Fahrenheit.

   d. The temperature of the hotel pool water is 83°F. What is this temperature in degrees Celsius?

ERROR ANALYSIS **In Exercises 11 and 12, describe and correct the error in interpreting the slope in the context of the situation.**

11.

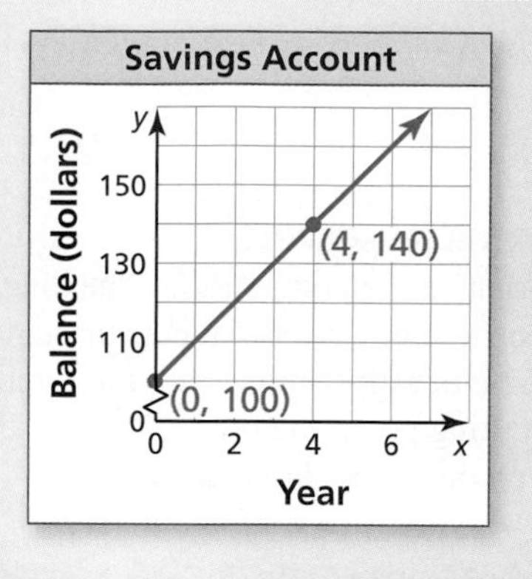

The slope of the line is 10, so after 7 years, the balance is \$70.

12.

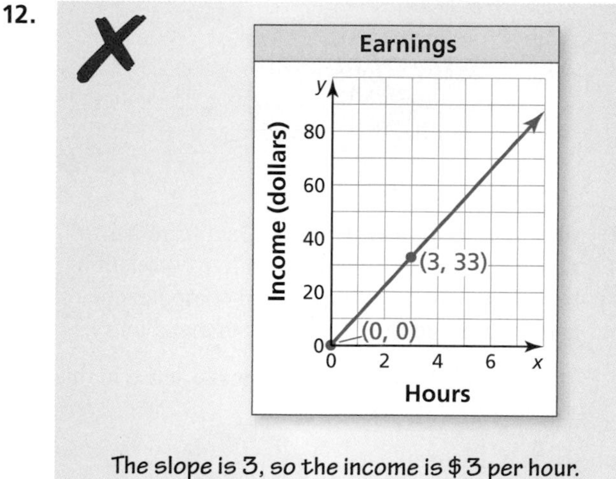

In Exercises 13–16, determine whether the data show a linear relationship. If so, write an equation of a line of fit. Estimate y when $x = 15$ and explain its meaning in the context of the situation. *(See Example 3.)*

13.

Minutes walking, x	1	6	11	13	16
Calories burned, y	6	27	50	56	70

14.

Months, x	9	13	18	22	23
Hair length (in.), y	3	5	7	10	11

15.

Hours, x	3	7	9	17	20
Battery life (%), y	86	61	50	26	0

16.

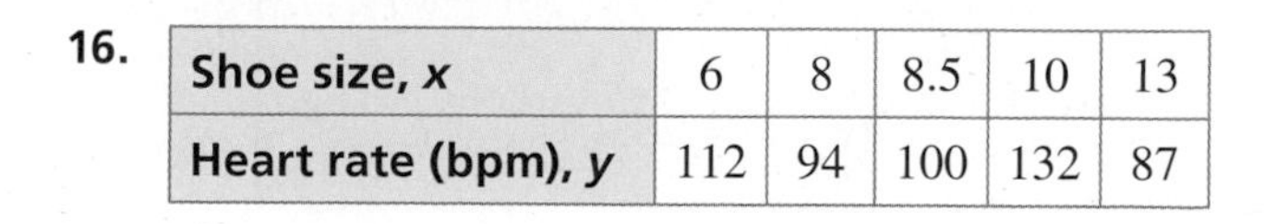

Shoe size, x	6	8	8.5	10	13
Heart rate (bpm), y	112	94	100	132	87

17. MODELING WITH MATHEMATICS The data pairs (x, y) represent the average annual tuition y (in dollars) for public colleges in the United States x years after 2005. Use the *linear regression* feature on a graphing calculator to find an equation of the line of best fit. Estimate the average annual tuition in 2020. Interpret the slope and y-intercept in this situation. *(See Example 4.)*

(0, 11,386), (1, 11,731), (2, 11,848)
(3, 12,375), (4, 12,804), (5, 13,297)

18. MODELING WITH MATHEMATICS The table shows the numbers of tickets sold for a concert when different prices are charged. Write an equation of a line of fit for the data. Does it seem reasonable to use your model to predict the number of tickets sold when the ticket price is \$85? Explain.

Ticket price (dollars), x	17	20	22	26
Tickets sold, y	450	423	400	395

USING TOOLS **In Exercises 19–24, use the *linear regression* feature on a graphing calculator to find an equation of the line of best fit for the data. Find and interpret the correlation coefficient.**

19.

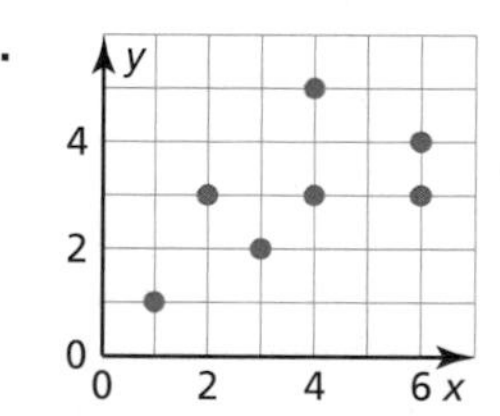

20.

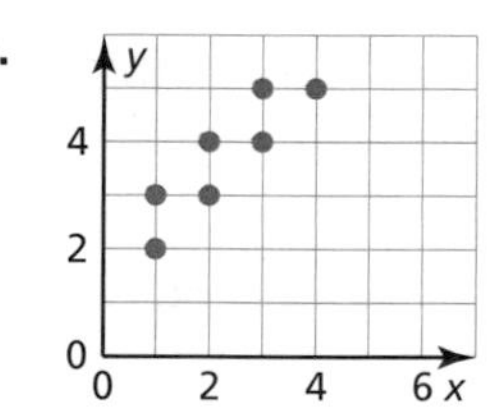

21.

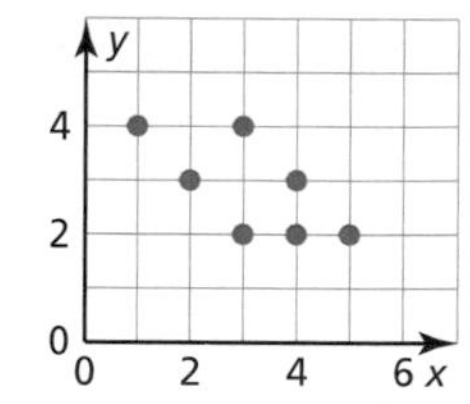

22.

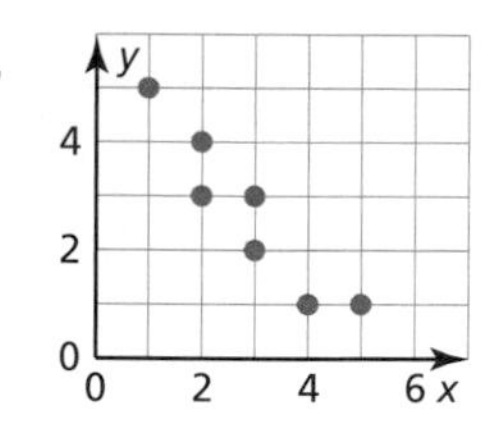

23.

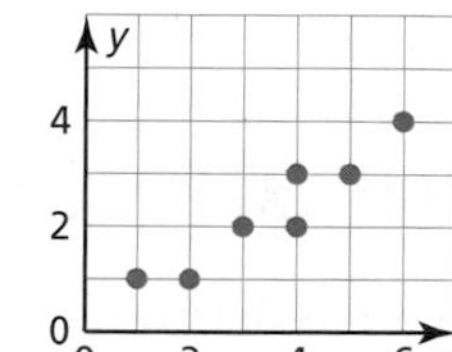

24.

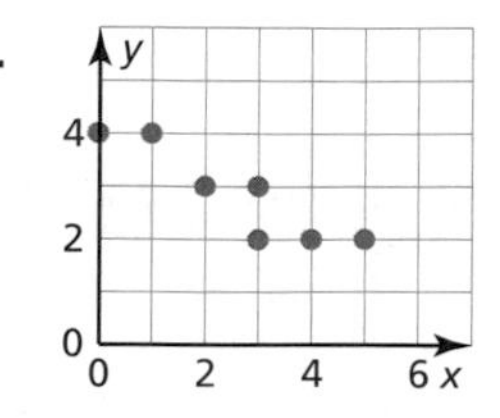

25. OPEN-ENDED Give two real-life quantities that have (a) a positive correlation, (b) a negative correlation, and (c) approximately no correlation. Explain.

26. HOW DO YOU SEE IT? You secure an interest-free loan to purchase a boat. You agree to make equal monthly payments for the next two years. The graph shows the amount of money you still owe.

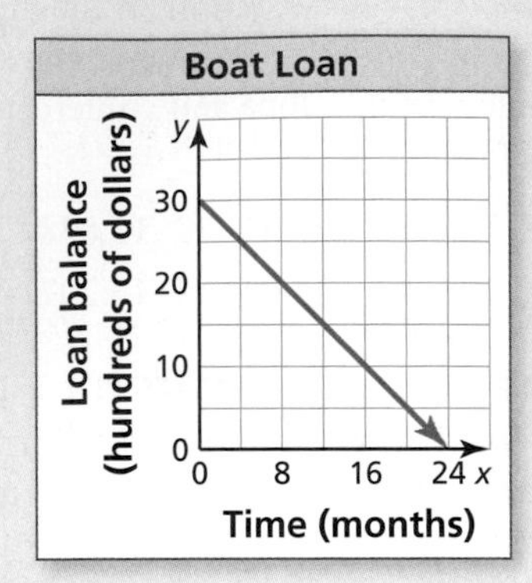

a. What is the slope of the line? What does the slope represent?

b. What is the domain and range of the function? What does each represent?

c. How much do you still owe after making payments for 12 months?

27. MAKING AN ARGUMENT A set of data pairs has a correlation coefficient $r = 0.3$. Your friend says that because the correlation coefficient is positive, it is logical to use the line of best fit to make predictions. Is your friend correct? Explain your reasoning.

28. THOUGHT PROVOKING Points A and B lie on the line $y = -x + 4$. Choose coordinates for points A, B, and C where point C is the same distance from point A as it is from point B. Write equations for the lines connecting points A and C and points B and C.

29. ABSTRACT REASONING If x and y have a positive correlation, and y and z have a negative correlation, then what can you conclude about the correlation between x and z? Explain.

30. MATHEMATICAL CONNECTIONS Which equation has a graph that is a line passing through the point $(8, -5)$ and is perpendicular to the graph of $y = -4x + 1$?

Ⓐ $y = \frac{1}{4}x - 5$ Ⓑ $y = -4x + 27$

Ⓒ $y = -\frac{1}{4}x - 7$ Ⓓ $y = \frac{1}{4}x - 7$

31. PROBLEM SOLVING You are participating in an orienteering competition. The diagram shows the position of a river that cuts through the woods. You are currently 2 miles east and 1 mile north of your starting point, the origin. What is the shortest distance you must travel to reach the river?

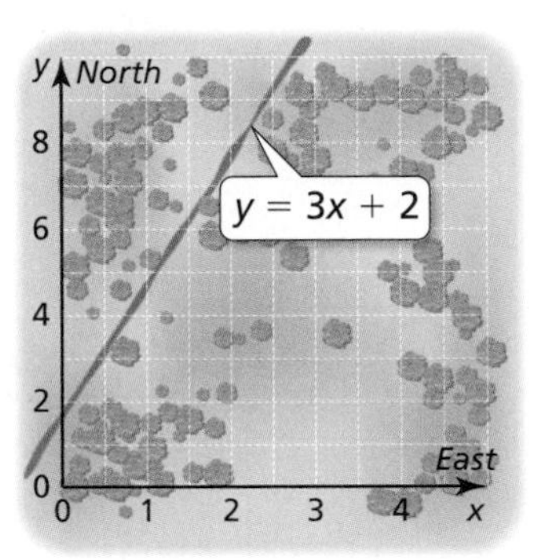

32. ANALYZING RELATIONSHIPS Data from North American countries show a positive correlation between the number of personal computers per capita and the average life expectancy in the country.

a. Does a positive correlation make sense in this situation? Explain.

b. Is it reasonable to conclude that giving residents of a country personal computers will lengthen their lives? Explain.

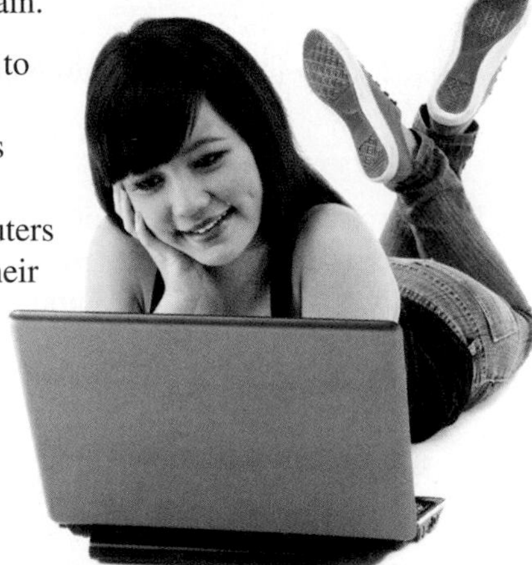

Maintaining Mathematical Proficiency

Reviewing what you learned in previous grades and lessons

Solve the system of linear equations in two variables by elimination or substitution. *(Skills Review Handbook)*

33. $3x + y = 7$
$-2x - y = 9$

34. $4x + 3y = 2$
$2x - 3y = 1$

35. $2x + 2y = 3$
$x = 4y - 1$

36. $y = 1 + x$
$2x + y = -2$

37. $\frac{1}{2}x + 4y = 4$
$2x - y = 1$

38. $y = x - 4$
$4x + y = 26$

1.4 Solving Linear Systems

Essential Question

How can you determine the number of solutions of a linear system?

A linear system is *consistent* when it has at least one solution. A linear system is *inconsistent* when it has no solution.

EXPLORATION 1 Recognizing Graphs of Linear Systems

Work with a partner. Match each linear system with its corresponding graph. Explain your reasoning. Then classify the system as *consistent* or *inconsistent*.

a. $2x - 3y = 3$
$-4x + 6y = 6$

b. $2x - 3y = 3$
$x + 2y = 5$

c. $2x - 3y = 3$
$-4x + 6y = -6$

A.

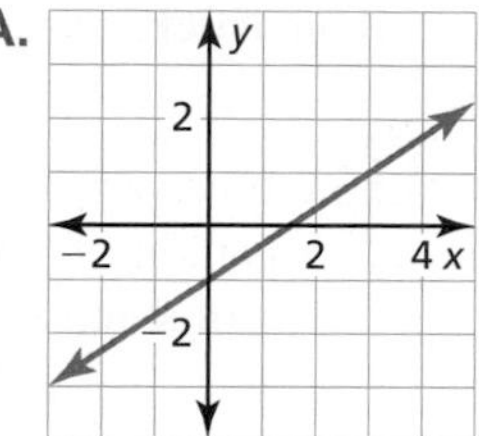

B.

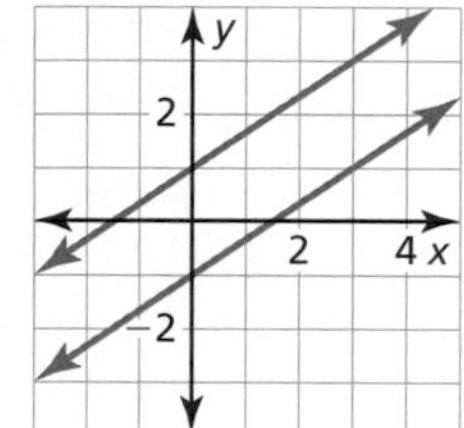

C.

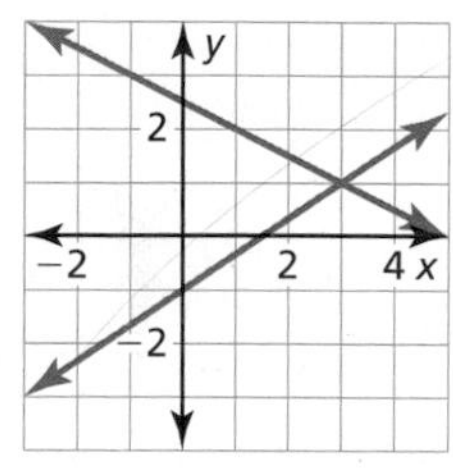

EXPLORATION 2 Solving Systems of Linear Equations

Work with a partner. Solve each linear system by substitution or elimination. Then use the graph of the system below to check your solution.

a. $2x + y = 5$
$x - y = 1$

b. $x + 3y = 1$
$-x + 2y = 4$

c. $x + y = 0$
$3x + 2y = 1$

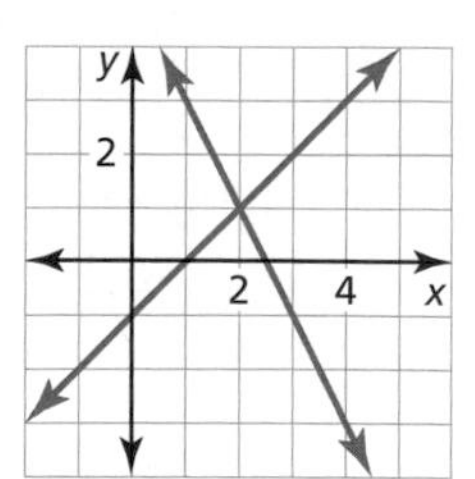

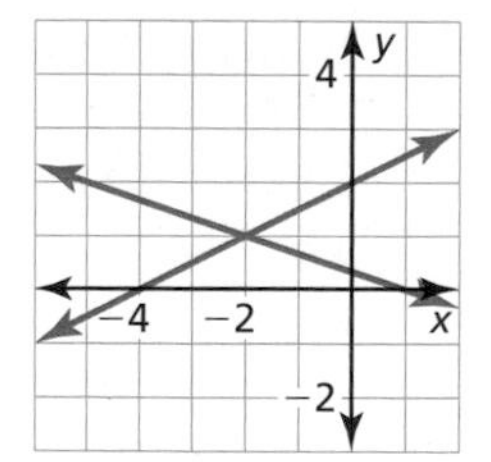

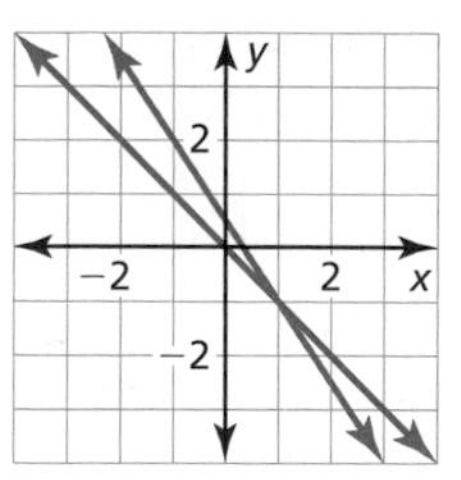

FINDING AN ENTRY POINT

To be proficient in math, you need to look for entry points to the solution of a problem.

Communicate Your Answer

3. How can you determine the number of solutions of a linear system?

4. Suppose you were given a system of *three* linear equations in *three* variables. Explain how you would approach solving such a system.

5. Apply your strategy in Question 4 to solve the linear system.

$x + y + z = 1$ Equation 1

$x - y - z = 3$ Equation 2

$-x - y + z = -1$ Equation 3

1.4 Lesson

Core Vocabulary

linear equation in three variables, *p. 30*
system of three linear equations, *p. 30*
solution of a system of three linear equations, *p. 30*
ordered triple, *p. 30*

Previous
system of two linear equations

What You Will Learn

- Visualize solutions of systems of linear equations in three variables.
- Solve systems of linear equations in three variables algebraically.
- Solve real-life problems.

Visualizing Solutions of Systems

A **linear equation in three variables** x, y, and z is an equation of the form $ax + by + cz = d$, where a, b, and c are not all zero.

The following is an example of a **system of three linear equations** in three variables.

$3x + 4y - 8z = -3$ Equation 1

$x + y + 5z = -12$ Equation 2

$4x - 2y + z = 10$ Equation 3

A **solution** of such a system is an **ordered triple** (x, y, z) whose coordinates make each equation true.

The graph of a linear equation in three variables is a plane in three-dimensional space. The graphs of three such equations that form a system are three planes whose intersection determines the number of solutions of the system, as shown in the diagrams below.

Exactly One Solution
The planes intersect in a single point, which is the solution of the system.

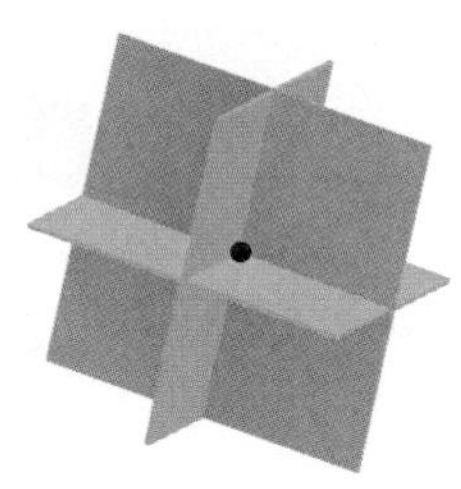

Infinitely Many Solutions
The planes intersect in a line. Every point on the line is a solution of the system.

The planes could also be the same plane. Every point in the plane is a solution of the system.

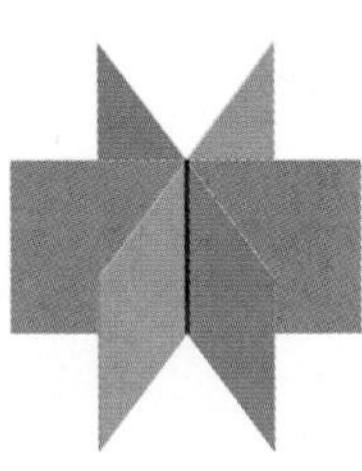

No Solution
There are no points in common with all three planes.

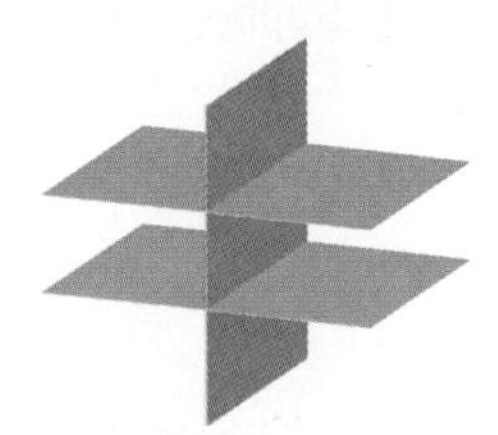

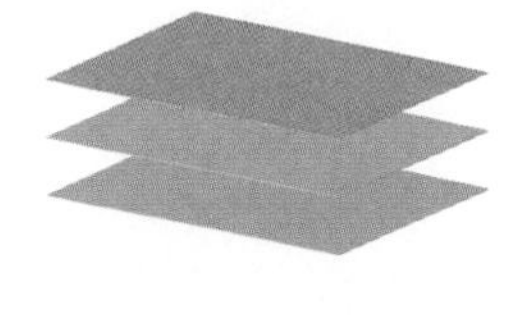

Solving Systems of Equations Algebraically

The algebraic methods you used to solve systems of linear equations in two variables can be extended to solve a system of linear equations in three variables.

Core Concept

Solving a Three-Variable System

Step 1 Rewrite the linear system in three variables as a linear system in two variables by using the substitution or elimination method.

Step 2 Solve the new linear system for both of its variables.

Step 3 Substitute the values found in Step 2 into one of the original equations and solve for the remaining variable.

When you obtain a false equation, such as $0 = 1$, in any of the steps, the system has no solution.

When you do not obtain a false equation, but obtain an identity such as $0 = 0$, the system has infinitely many solutions.

LOOKING FOR STRUCTURE

The coefficient of -1 in Equation 3 makes y a convenient variable to eliminate.

EXAMPLE 1 Solving a Three-Variable System (One Solution)

Solve the system.

$4x + 2y + 3z = 12$ Equation 1

$2x - 3y + 5z = -7$ Equation 2

$6x - y + 4z = -3$ Equation 3

SOLUTION

Step 1 Rewrite the system as a linear system in *two* variables.

$$\begin{array}{r} 4x + 2y + 3z = 12 \\ \underline{12x - 2y + 8z = -6} \\ 16x \qquad + 11z = 6 \end{array}$$

Add 2 times Equation 3 to Equation 1 (to eliminate y).

New Equation 1

$$\begin{array}{r} 2x - 3y + 5z = -7 \\ \underline{-18x + 3y - 12z = 9} \\ -16x \qquad - 7z = 2 \end{array}$$

Add -3 times Equation 3 to Equation 2 (to eliminate y).

New Equation 2

ANOTHER WAY

In Step 1, you could also eliminate x to get two equations in y and z, or you could eliminate z to get two equations in x and y.

Step 2 Solve the new linear system for both of its variables.

$$\begin{array}{r} 16x + 11z = 6 \\ \underline{-16x - 7z = 2} \\ 4z = 8 \end{array}$$

Add new Equation 1 and new Equation 2.

$z = 2$ Solve for z.

$x = -1$ Substitute into new Equation 1 or 2 to find x.

Step 3 Substitute $x = -1$ and $z = 2$ into an original equation and solve for y.

$6x - y + 4z = -3$ Write original Equation 3.

$6(-1) - y + 4(2) = -3$ Substitute -1 for x and 2 for z.

$y = 5$ Solve for y.

▶ The solution is $x = -1$, $y = 5$, and $z = 2$, or the ordered triple $(-1, 5, 2)$. Check this solution in each of the original equations.

EXAMPLE 2 Solving a Three-Variable System (No Solution)

Solve the system.

$x + y + z = 2$ Equation 1

$5x + 5y + 5z = 3$ Equation 2

$4x + y - 3z = -6$ Equation 3

SOLUTION

Step 1 Rewrite the system as a linear system in *two* variables.

$$\begin{array}{r} -5x - 5y - 5z = -10 \\ \underline{5x + 5y + 5z = 3} \\ 0 = -7 \end{array}$$

Add −5 times Equation 1 to Equation 2.

▶ Because you obtain a false equation, the original system has no solution.

EXAMPLE 3 Solving a Three-Variable System (Many Solutions)

ANOTHER WAY

Subtracting Equation 2 from Equation 1 gives $z = 0$. After substituting 0 for z in each equation, you can see that each is equivalent to $y = x + 3$.

Solve the system.

$x - y + z = -3$ Equation 1

$x - y - z = -3$ Equation 2

$5x - 5y + z = -15$ Equation 3

SOLUTION

Step 1 Rewrite the system as a linear system in *two* variables.

$$\begin{array}{r} x - y + z = -3 \\ \underline{x - y - z = -3} \\ 2x - 2y = -6 \end{array}$$

Add Equation 1 to Equation 2 (to eliminate z).

New Equation 2

$$\begin{array}{r} x - y - z = -3 \\ \underline{5x - 5y + z = -15} \\ 6x - 6y = -18 \end{array}$$

Add Equation 2 to Equation 3 (to eliminate z).

New Equation 3

Step 2 Solve the new linear system for both of its variables.

$$\begin{array}{r} -6x + 6y = 18 \\ \underline{6x - 6y = -18} \\ 0 = 0 \end{array}$$

Add −3 times new Equation 2 to new Equation 3.

Because you obtain the identity $0 = 0$, the system has infinitely many solutions.

Step 3 Describe the solutions of the system using an ordered triple. One way to do this is to solve new Equation 2 for y to obtain $y = x + 3$. Then substitute $x + 3$ for y in original Equation 1 to obtain $z = 0$.

▶ So, any ordered triple of the form $(x, x + 3, 0)$ is a solution of the system.

Monitoring Progress  Help in English and Spanish at *BigIdeasMath.com*

Solve the system. Check your solution, if possible.

1. $x - 2y + z = -11$
   $3x + 2y - z = 7$
   $-x + 2y + 4z = -9$

2. $x + y - z = -1$
   $4x + 4y - 4z = -2$
   $3x + 2y + z = 0$

3. $x + y + z = 8$
   $x - y + z = 8$
   $2x + y + 2z = 16$

4. In Example 3, describe the solutions of the system using an ordered triple in terms of y.

Solving Real-Life Problems

EXAMPLE 4 Solving a Multi-Step Problem

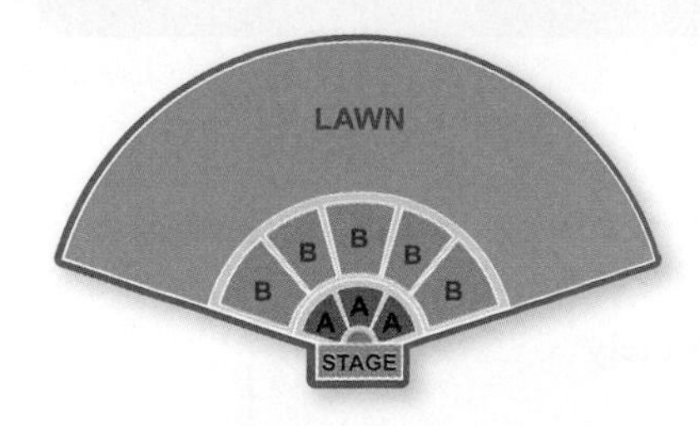

An amphitheater charges \$75 for each seat in Section A, \$55 for each seat in Section B, and \$30 for each lawn seat. There are three times as many seats in Section B as in Section A. The revenue from selling all 23,000 seats is \$870,000. How many seats are in each section of the amphitheater?

SOLUTION

Step 1 Write a verbal model for the situation.

[Number of seats in B, *y*] = 3 • [Number of seats in A, *x*]

[Number of seats in A, *x*] + [Number of seats in B, *y*] + [Number of lawn seats, *z*] = [Total number of seats]

75 • [Number of seats in A, *x*] + 55 • [Number of seats in B, *y*] + 30 • [Number of lawn seats, *z*] = [Total revenue]

Step 2 Write a system of equations.

$y = 3x$ Equation 1

$x + y + z = 23{,}000$ Equation 2

$75x + 55y + 30z = 870{,}000$ Equation 3

Step 3 Rewrite the system in Step 2 as a linear system in *two* variables by substituting $3x$ for y in Equations 2 and 3.

$x + y + z = 23{,}000$ Write Equation 2.

$x + 3x + z = 23{,}000$ Substitute $3x$ for y.

$4x + z = 23{,}000$ New Equation 2

$75x + 55y + 30z = 870{,}000$ Write Equation 3.

$75x + 55(3x) + 30z = 870{,}000$ Substitute $3x$ for y.

$240x + 30z = 870{,}000$ New Equation 3

Step 4 Solve the new linear system for both of its variables.

$$\begin{array}{rl} -120x - 30z &= -690{,}000 \\ 240x + 30z &= 870{,}000 \\ \hline 120x \quad &= 180{,}000 \end{array}$$ Add -30 times new Equation 2 to new Equation 3.

$x = 1500$ Solve for x.

$y = 4500$ Substitute into Equation 1 to find y.

$z = 17{,}000$ Substitute into Equation 2 to find z.

STUDY TIP

When substituting to find values of other variables, choose original or new equations that are easiest to use.

▶ The solution is $x = 1500$, $y = 4500$, and $z = 17{,}000$, or $(1500, 4500, 17{,}000)$. So, there are 1500 seats in Section A, 4500 seats in Section B, and 17,000 lawn seats.

Monitoring Progress  Help in English and Spanish at *BigIdeasMath.com*

5. **WHAT IF?** On the first day, 10,000 tickets sold, generating \$356,000 in revenue. The number of seats sold in Sections A and B are the same. How many lawn seats are still available?

 Dynamic Solutions available at *BigIdeasMath.com*

Vocabulary and Core Concept Check

1. **VOCABULARY** The solution of a system of three linear equations is expressed as a(n)________.
2. **WRITING** Explain how you know when a linear system in three variables has infinitely many solutions.

Monitoring Progress and Modeling with Mathematics

In Exercises 3–8, solve the system using the elimination method. *(See Example 1.)*

3. $x + y - 2z = 5$
   $-x + 2y + z = 2$
   $2x + 3y - z = 9$

4. $x + 4y - 6z = -1$
   $2x - y + 2z = -7$
   $-x + 2y - 4z = 5$

5. $2x + y - z = 9$
   $-x + 6y + 2z = -17$
   $5x + 7y + z = 4$

6. $3x + 2y - z = 8$
   $-3x + 4y + 5z = -14$
   $x - 3y + 4z = -14$

7. $2x + 2y + 5z = -1$
   $2x - y + z = 2$
   $2x + 4y - 3z = 14$

8. $3x + 2y - 3z = -2$
   $7x - 2y + 5z = -14$
   $2x + 4y + z = 6$

ERROR ANALYSIS In Exercises 9 and 10, describe and correct the error in the first step of solving the system of linear equations.

$$4x - y + 2z = -18$$
$$-x + 2y + z = 11$$
$$3x + 3y - 4z = 44$$

9. ✗
$$\begin{array}{r} 4x - y + 2z = -18 \\ -4x + 2y + z = 11 \\ \hline y + 3z = -7 \end{array}$$

10. ✗
$$\begin{array}{r} 12x - 3y + 6z = -18 \\ 3x + 3y - 4z = 44 \\ \hline 15x + 2z = 26 \end{array}$$

In Exercises 11–16, solve the system using the elimination method. *(See Examples 2 and 3.)*

11. $3x - y + 2z = 4$
    $6x - 2y + 4z = -8$
    $2x - y + 3z = 10$

12. $5x + y - z = 6$
    $x + y + z = 2$
    $12x + 4y = 10$

13. $x + 3y - z = 2$
    $x + y - z = 0$
    $3x + 2y - 3z = -1$

14. $x + 2y - z = 3$
    $-2x - y + z = -1$
    $6x - 3y - z = -7$

15. $x + 2y + 3z = 4$
    $-3x + 2y - z = 12$
    $-2x - 2y - 4z = -14$

16. $-2x - 3y + z = -6$
    $x + y - z = 5$
    $7x + 8y - 6z = 31$

17. **MODELING WITH MATHEMATICS** Three orders are placed at a pizza shop. Two small pizzas, a liter of soda, and a salad cost \$14; one small pizza, a liter of soda, and three salads cost \$15; and three small pizzas, a liter of soda, and two salads cost \$22. How much does each item cost?

18. **MODELING WITH MATHEMATICS** Sam's Furniture Store places the following advertisement in the local newspaper. Write a system of equations for the three combinations of furniture. What is the price of each piece of furniture? Explain.

SALE
SAM'S
Furniture Store
\$1300
Sofa and love seat
\$1400
Sofa and two chairs
\$1600
Sofa, love seat, and one chair

In Exercises 19–28, solve the system of linear equations using the substitution method. *(See Example 4.)*

19. $-2x + y + 6z = 1$
$3x + 2y + 5z = 16$
$7x + 3y - 4z = 11$

20. $x - 6y - 2z = -8$
$-x + 5y + 3z = 2$
$3x - 2y - 4z = 18$

21. $x + y + z = 4$
$5x + 5y + 5z = 12$
$x - 4y + z = 9$

22. $x + 2y = -1$
$-x + 3y + 2z = -4$
$-x + y - 4z = 10$

23. $2x - 3y + z = 10$
$y + 2z = 13$
$z = 5$

24. $x = 4$
$x + y = -6$
$4x - 3y + 2z = 26$

25. $x + y - z = 4$
$3x + 2y + 4z = 17$
$-x + 5y + z = 8$

26. $2x - y - z = 15$
$4x + 5y + 2z = 10$
$-x - 4y + 3z = -20$

27. $4x + y + 5z = 5$
$8x + 2y + 10z = 10$
$x - y - 2z = -2$

28. $x + 2y - z = 3$
$2x + 4y - 2z = 6$
$-x - 2y + z = -6$

29. PROBLEM SOLVING The number of left-handed people in the world is one-tenth the number of right-handed people. The percent of right-handed people is nine times the percent of left-handed people and ambidextrous people combined. What percent of people are ambidextrous?

30. MODELING WITH MATHEMATICS Use a system of linear equations to model the data in the following newspaper article. Solve the system to find how many athletes finished in each place.

> Lawrence High prevailed in Saturday's track meet with the help of 20 individual-event placers earning a combined 68 points. A first-place finish earns 5 points, a second-place finish earns 3 points, and a third-place finish earns 1 point. Lawrence had a strong second-place showing, with as many second place finishers as first- and third-place finishers combined.

31. WRITING Explain when it might be more convenient to use the elimination method than the substitution method to solve a linear system. Give an example to support your claim.

32. REPEATED REASONING Using what you know about solving linear systems in two and three variables, plan a strategy for how you would solve a system that has *four* linear equations in *four* variables.

MATHEMATICAL CONNECTIONS In Exercises 33 and 34, write and use a linear system to answer the question.

33. The triangle has a perimeter of 65 feet. What are the lengths of sides ℓ, m, and n?

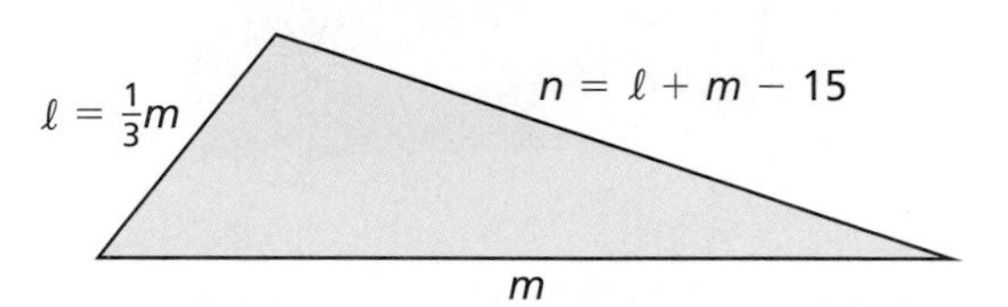

34. What are the measures of angles A, B, and C?

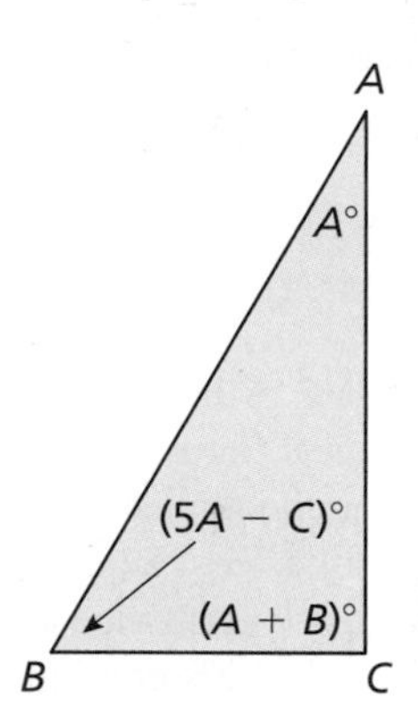

35. OPEN-ENDED Consider the system of linear equations below. Choose nonzero values for a, b, and c so the system satisfies the given condition. Explain your reasoning.

$$x + y + z = 2$$
$$ax + by + cz = 10$$
$$x - 2y + z = 4$$

a. The system has no solution.

b. The system has exactly one solution.

c. The system has infinitely many solutions.

36. MAKING AN ARGUMENT A linear system in three variables has no solution. Your friend concludes that it is not possible for two of the three equations to have any points in common. Is your friend correct? Explain your reasoning.

37. **PROBLEM SOLVING** A contractor is hired to build an apartment complex. Each 840-square-foot unit has a bedroom, kitchen, and bathroom. The bedroom will be the same size as the kitchen. The owner orders 980 square feet of tile to completely cover the floors of two kitchens and two bathrooms. Determine how many square feet of carpet is needed for each bedroom.

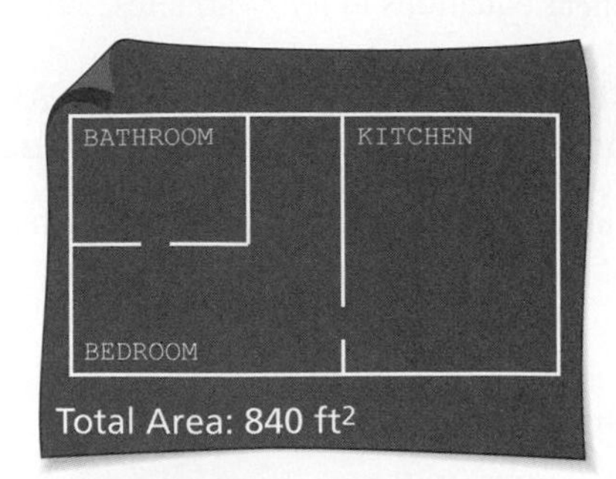

38. **THOUGHT PROVOKING** Does the system of linear equations have more than one solution? Justify your answer.

$$4x + y + z = 0$$
$$2x + \tfrac{1}{2}y - 3z = 0$$
$$-x - \tfrac{1}{4}y - z = 0$$

39. **PROBLEM SOLVING** A florist must make 5 identical bridesmaid bouquets for a wedding. The budget is \$160, and each bouquet must have 12 flowers. Roses cost \$2.50 each, lilies cost \$4 each, and irises cost \$2 each. The florist wants twice as many roses as the other two types of flowers combined.

 a. Write a system of equations to represent this situation, assuming the florist plans to use the maximum budget.

 b. Solve the system to find how many of each type of flower should be in each bouquet.

 c. Suppose there is no limitation on the total cost of the bouquets. Does the problem still have exactly one solution? If so, find the solution. If not, give three possible solutions.

40. **HOW DO YOU SEE IT?** Determine whether the system of equations that represents the circles has *no solution*, *one solution*, or *infinitely many solutions*. Explain your reasoning.

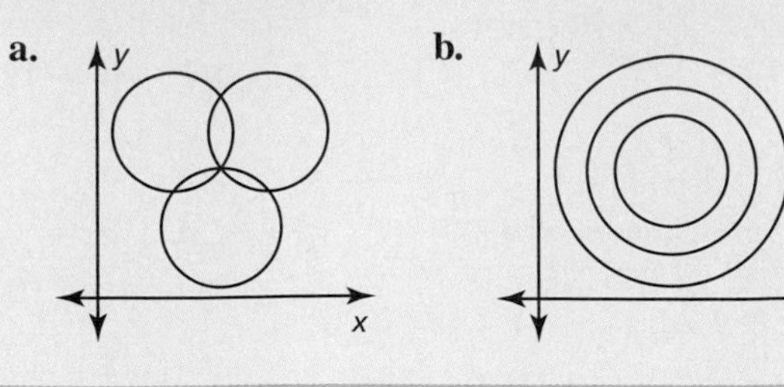

41. **CRITICAL THINKING** Find the values of a, b, and c so that the linear system shown has $(-1, 2, -3)$ as its only solution. Explain your reasoning.

$$x + 2y - 3z = a$$
$$-x - y + z = b$$
$$2x + 3y - 2z = c$$

42. **ANALYZING RELATIONSHIPS** Determine which arrangement(s) of the integers -5, 2, and 3 produce a solution of the linear system that consist of only integers. Justify your answer.

$$x - 3y + 6z = 21$$
$$\_\_x + \_\_y + \_\_z = -30$$
$$2x - 5y + 2z = -6$$

43. **ABSTRACT REASONING** Write a linear system to represent the first three pictures below. Use the system to determine how many tangerines are required to balance the apple in the fourth picture. *Note:* The first picture shows that one tangerine and one apple balance one grapefruit.

Maintaining Mathematical Proficiency

Reviewing what you learned in previous grades and lessons

Simplify. *(Skills Review Handbook)*

44. $(x - 2)^2$ 45. $(3m + 1)^2$ 46. $(2z - 5)^2$ 47. $(4 - y)^2$

Write a function g described by the given transformation of $f(x) = |x| - 5$. *(Section 1.2)*

48. translation 2 units to the left

49. reflection in the x-axis

50. translation 4 units up

51. vertical stretch by a factor of 3

1.3–1.4 What Did You Learn?

Core Vocabulary

line of fit, *p. 24*
line of best fit, *p. 25*
correlation coefficient, *p. 25*
linear equation in three variables, *p. 30*
system of three linear equations, *p. 30*
solution of a system of three linear equations, *p. 30*
ordered triple, *p. 30*

Core Concepts

Section 1.3

Writing an Equation of a Line, *p. 22*
Finding a Line of Fit, *p. 24*

Section 1.4

Solving a Three-Variable System, *p. 31*
Solving Real-Life Problems, *p. 33*

Mathematical Practices

1. Describe how you can write the equation of the line in Exercise 7 on page 26 using only one of the labeled points.
2. How did you use the information in the newspaper article in Exercise 30 on page 35 to write a system of three linear equations?
3. Explain the strategy you used to choose the values for a, b, and c in Exercise 35 part (a) on page 35.

Performance Task

Secret of the Hanging Baskets

A carnival game uses two baskets hanging from springs at different heights. Next to the higher basket is a pile of baseballs. Next to the lower basket is a pile of golf balls. The object of the game is to add the same number of balls to each basket so that the baskets have the same height. But there is a catch—you only get one chance. What is the secret to winning the game?

To explore the answers to this question and more, go to ***BigIdeasMath.com***.

1 Chapter Review

Dynamic Solutions available at *BigIdeasMath.com*

1.1 Parent Functions and Transformations *(pp. 3–10)*

Graph $g(x) = (x - 2)^2 + 1$ and its parent function. Then describe the transformation.

The function g is a quadratic function.

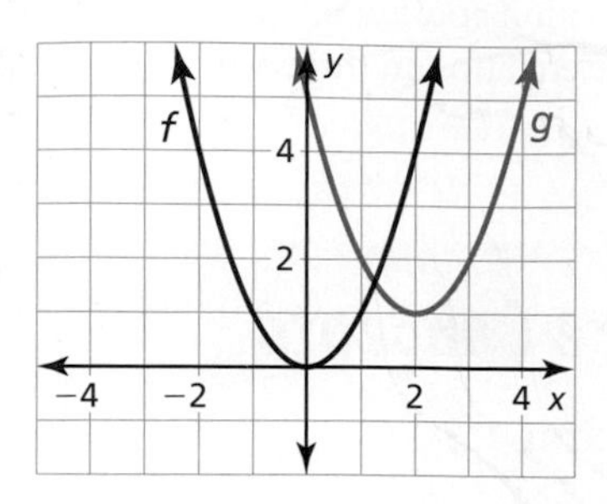

▶ The graph of g is a translation 2 units right and 1 unit up of the graph of the parent quadratic function.

Graph the function and its parent function. Then describe the transformation.

1. $f(x) = x + 3$

2. $g(x) = |x| - 1$

3. $h(x) = \frac{1}{2}x^2$

4. $h(x) = 4$

5. $f(x) = -|x| - 3$

6. $g(x) = -3(x + 3)^2$

1.2 Transformations of Linear and Absolute Value Functions *(pp. 11–18)*

Let the graph of g be a translation 2 units to the right followed by a reflection in the y-axis of the graph of $f(x) = |x|$. Write a rule for g.

Step 1 First write a function h that represents the translation of f.

$h(x) = f(x - 2)$ Subtract 2 from the input.

$= |x - 2|$ Replace x with $x - 2$ in $f(x)$.

Step 2 Then write a function g that represents the reflection of h.

$g(x) = h(-x)$ Multiply the input by -1.

$= |-x - 2|$ Replace x with $-x$ in $h(x)$.

$= |-(x + 2)|$ Factor out -1.

$= |-1| \cdot |x + 2|$ Product Property of Absolute Value

$= |x + 2|$ Simplify.

▶ The transformed function is $g(x) = |x + 2|$.

Write a function g whose graph represents the indicated transformations of the graph of f. Use a graphing calculator to check your answer.

7. $f(x) = |x|$; reflection in the x-axis followed by a translation 4 units to the left

8. $f(x) = |x|$; vertical shrink by a factor of $\frac{1}{2}$ followed by a translation 2 units up

9. $f(x) = x$; translation 3 units down followed by a reflection in the y-axis

1.3 Modeling with Linear Functions (pp. 21–28)

The table shows the numbers of ice cream cones sold for different outside temperatures (in degrees Fahrenheit). Do the data show a linear relationship? If so, write an equation of a line of fit and use it to estimate how many ice cream cones are sold when the temperature is 60°F.

Temperature, *x*	53	62	70	82	90
Number of cones, *y*	90	105	117	131	147

Step 1 Create a scatter plot of the data. The data show a linear relationship.

Step 2 Sketch the line that appears to most closely fit the data. One possibility is shown.

Step 3 Choose two points on the line. For the line shown, you might choose (70, 117) and (90, 147).

Step 4 Write an equation of the line. First, find the slope.

$$m = \frac{y_2 - y_1}{x_2 - x_1} = \frac{147 - 117}{90 - 70} = \frac{30}{20} = 1.5$$

Use point-slope form to write an equation.
Use $(x_1, y_1) = (70, 117)$.

$y - y_1 = m(x - x_1)$ — Point-slope form

$y - 117 = 1.5(x - 70)$ — Substitute for m, x_1, and y_1.

$y - 117 = 1.5x - 105$ — Distributive Property

$y = 1.5x + 12$ — Add 117 to each side.

Use the equation to estimate the number of ice cream cones sold.

$y = 1.5(60) + 12$ — Substitute 60 for x.

$= 102$ — Simplify.

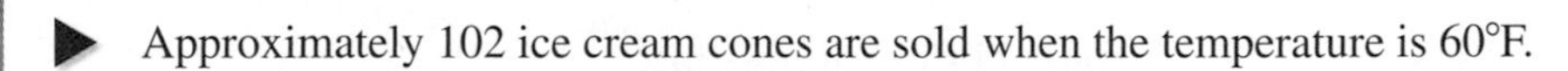

Approximately 102 ice cream cones are sold when the temperature is 60°F.

Write an equation of the line.

10. The table shows the total number y (in billions) of U.S. movie admissions each year for x years. Use a graphing calculator to find an equation of the line of best fit for the data.

Year, *x*	0	2	4	6	8	10
Admissions, *y*	1.24	1.26	1.39	1.47	1.49	1.57

11. You ride your bike and measure how far you travel. After 10 minutes, you travel 3.5 miles. After 30 minutes, you travel 10.5 miles. Write an equation to model your distance. How far can you ride your bike in 45 minutes?

1.4 Solving Linear Systems (pp. 29–36)

Solve the system.

$x - y + z = -3$	Equation 1
$2x - y + 5z = 4$	Equation 2
$4x + 2y - z = 2$	Equation 3

Step 1 Rewrite the system as a linear system in two variables.

$x - y + z = -3$ $\underline{4x + 2y - z = 2}$	Add Equation 1 to Equation 3 (to eliminate *z*).
$5x + y = -1$	New Equation 3
$-5x + 5y - 5z = 15$ $\underline{2x - y + 5z = 4}$	Add -5 times Equation 1 to Equation 2 (to eliminate *z*).
$-3x + 4y = 19$	New Equation 2

Step 2 Solve the new linear system for both of its variables.

$-20x - 4y = 4$ $\underline{-3x + 4y = 19}$	Add -4 times new Equation 3 to new Equation 2.
$-23x = 23$	
$x = -1$	Solve for *x*.
$y = 4$	Substitute into new Equation 2 or 3 to find *y*.

Step 3 Substitute $x = -1$ and $y = 4$ into an original equation and solve for *z*.

$x - y + z = -3$	Write original Equation 1.
$(-1) - 4 + z = -3$	Substitute -1 for *x* and 4 for *y*.
$z = 2$	Solve for *z*.

▶ The solution is $x = -1$, $y = 4$, and $z = 2$, or the ordered triple $(-1, 4, 2)$.

Solve the system. Check your solution, if possible.

12. $x + y + z = 3$
$-x + 3y + 2z = -8$
$x = 4z$

13. $2x - 5y - z = 17$
$x + y + 3z = 19$
$-4x + 6y + z = -20$

14. $x + y + z = 2$
$2x - 3y + z = 11$
$-3x + 2y - 2z = -13$

15. $x + 4y - 2z = 3$
$x + 3y + 7z = 1$
$2x + 9y - 13z = 2$

16. $x - y + 3z = 6$
$x - 2y = 5$
$2x - 2y + 5z = 9$

17. $x + 2z = 4$
$x + y + z = 6$
$3x + 3y + 4z = 28$

18. A school band performs a spring concert for a crowd of 600 people. The revenue for the concert is $3150. There are 150 more adults at the concert than students. How many of each type of ticket are sold?

1 Chapter Test

Write an equation of the line and interpret the slope and y-intercept.

1.

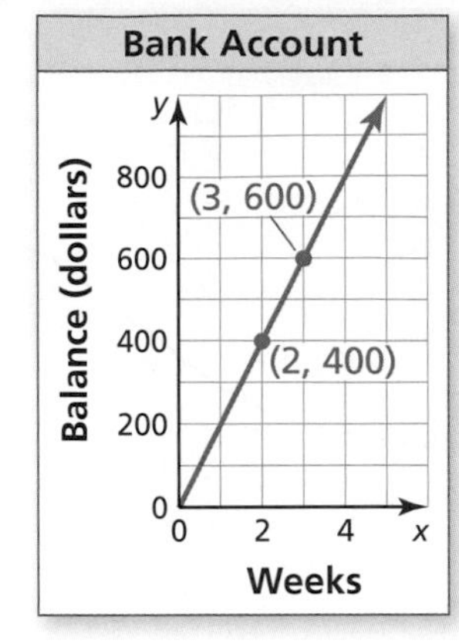

2.

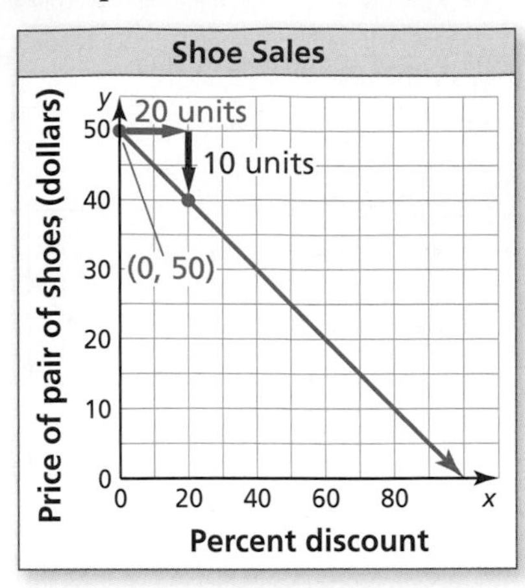

Solve the system. Check your solution, if possible.

3. $-2x + y + 4z = 5$

$x + 3y - z = 2$

$4x + y - 6z = 11$

4. $y = \frac{1}{2}z$

$x + 2y + 5z = 2$

$3x + 6y - 3z = 9$

5. $x - y + 5z = 3$

$2x + 3y - z = 2$

$-4x - y - 9z = -8$

Graph the function and its parent function. Then describe the transformation.

6. $f(x) = |x - 1|$

7. $f(x) = (3x)^2$

8. $f(x) = 4$

Match the transformation of $f(x) = x$ with its graph. Then write a rule for g.

9. $g(x) = 2f(x) + 3$

10. $g(x) = 3f(x) - 2$

11. $g(x) = -2f(x) - 3$

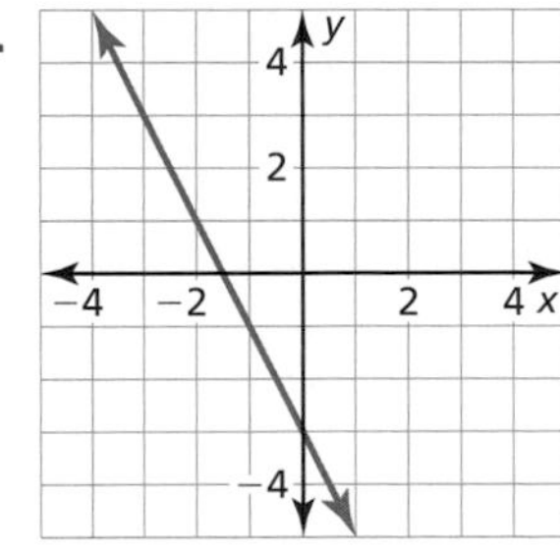

B.

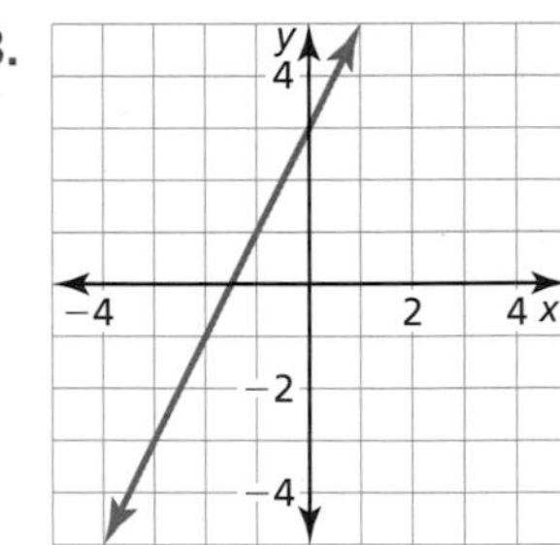

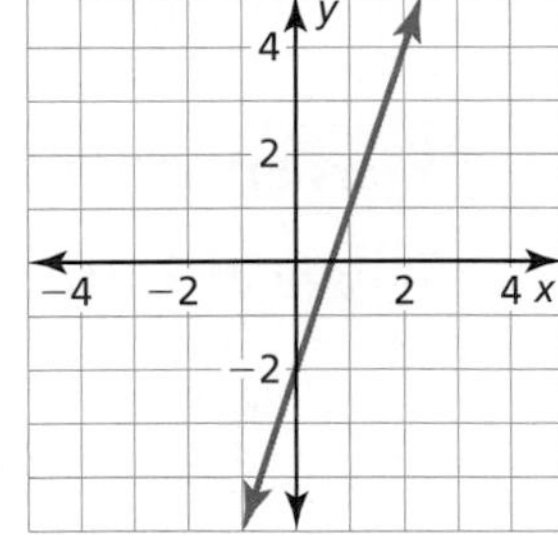

12. A bakery sells doughnuts, muffins, and bagels. The bakery makes three times as many doughnuts as bagels. The bakery earns a total of \$150 when all 130 baked items in stock are sold. How many of each item are in stock? Justify your answer.

Breakfast Specials

Doughnuts	\$1.00
Muffins	\$1.50
Bagels	\$1.20

13. A fountain with a depth of 5 feet is drained and then refilled. The water level (in feet) after t minutes can be modeled by $f(t) = \frac{1}{4}|t - 20|$. A second fountain with the same depth is drained and filled twice as quickly as the first fountain. Describe how to transform the graph of f to model the water level in the second fountain after t minutes. Find the depth of each fountain after 4 minutes. Justify your answers.

1 Cumulative Assessment

1. Describe the transformation of the graph of $f(x) = 2x - 4$ represented in each graph.

a.
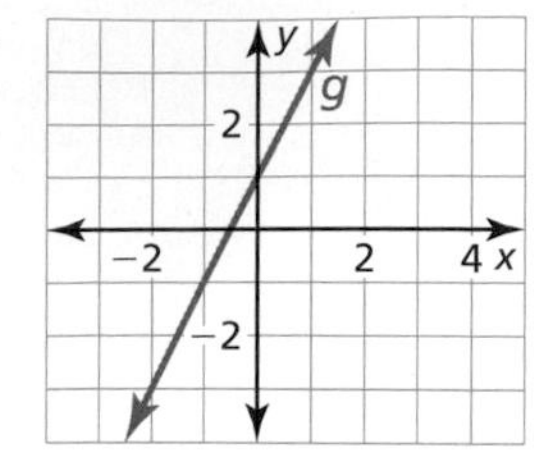

b.
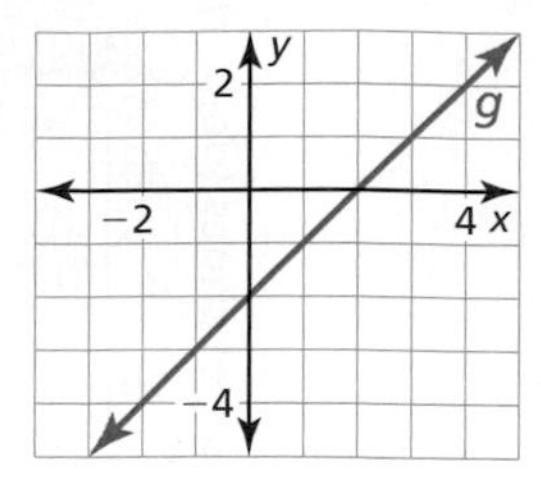

c.
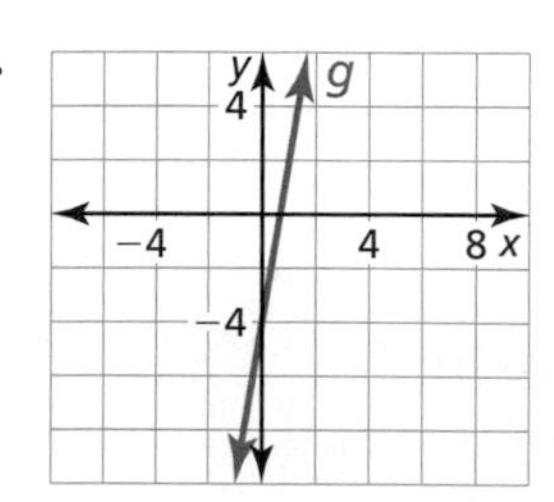

d.
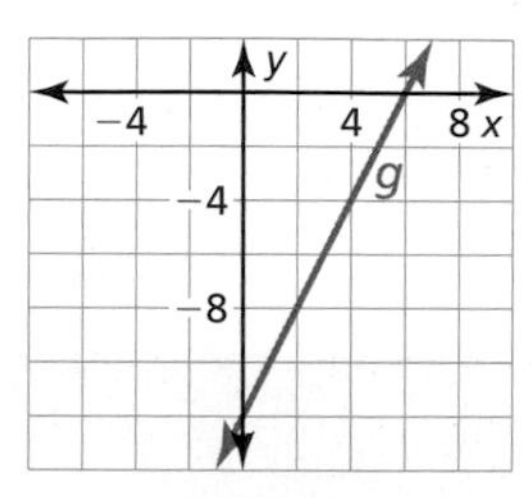

e.
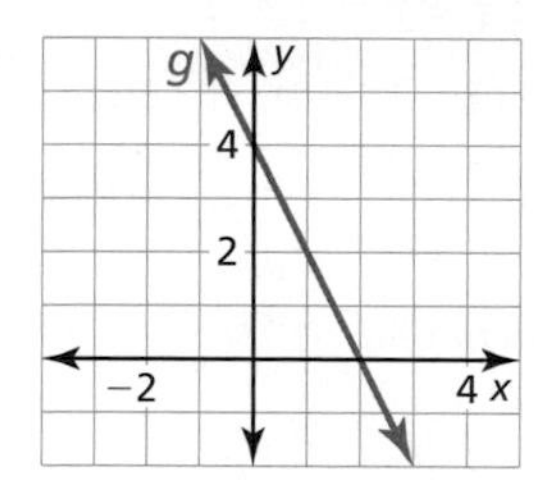

f.
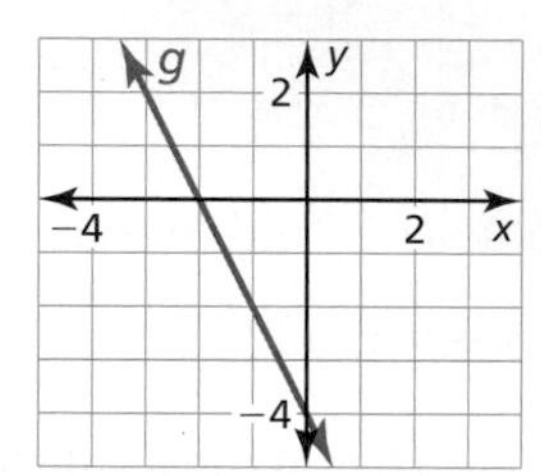

2. The table shows the tuition costs for a private school between the years 2010 and 2013.

Years after 2010, x	0	1	2	3
Tuition (dollars), y	36,208	37,620	39,088	40,594

a. Verify that the data show a linear relationship. Then write an equation of a line of fit.

b. Interpret the slope and y-intercept in this situation.

c. Predict the cost of tuition in 2015.

3. Your friend claims the line of best fit for the data shown in the scatter plot has a correlation coefficient close to 1. Is your friend correct? Explain your reasoning.

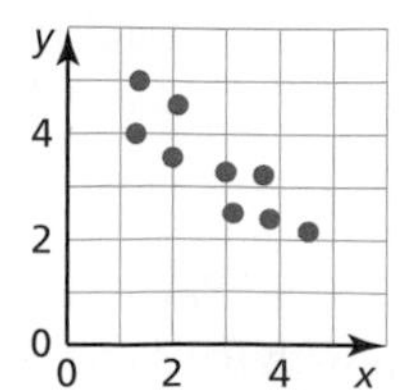

4. Order the following linear systems from least to greatest according to the number of solutions.

A. $2x + 4y - z = 7$
$14x + 28y - 7z = 49$
$-x + 6y + 12z = 13$

B. $3x - 3y + 3z = 5$
$-x + y - z = 8$
$14x - 3y + 12z = 108$

C. $4x - y + 2z = 18$
$-x + 2y + z = 11$
$3x + 3y - 4z = 44$

5. You make DVDs of three types of shows: comedy, drama, and reality-based. An episode of a comedy lasts 30 minutes, while a drama and a reality-based episode each last 60 minutes. The DVDs can hold 360 minutes of programming.

a. You completely fill a DVD with seven episodes and include twice as many episodes of a drama as a comedy. Create a system of equations that models the situation.

b. How many episodes of each type of show are on the DVD in part (a)?

c. You completely fill a second DVD with only six episodes. Do the two DVDs have a different number of comedies? dramas? reality-based episodes? Explain.

6. The graph shows the height of a hang glider over time. Which equation models the situation?

Ⓐ $y + 450 = 10x$

Ⓑ $10y = -x + 450$

Ⓒ $\frac{1}{10}y = -x + 450$

Ⓓ $10x + y = 450$

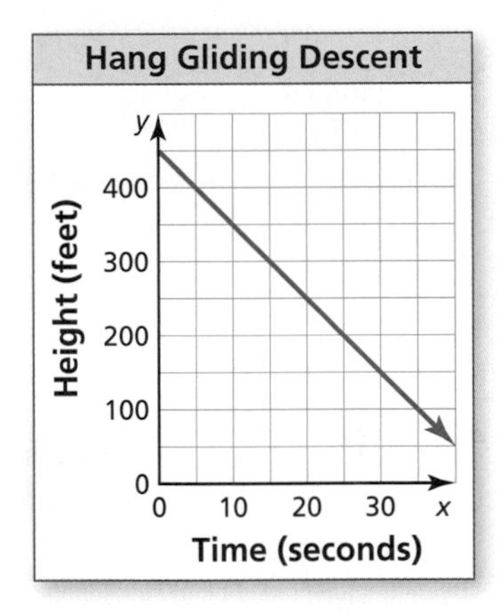

7. Let $f(x) = x$ and $g(x) = -3x - 4$. Select the possible transformations (in order) of the graph of f represented by the function g.

Ⓐ reflection in the x-axis

Ⓑ reflection in the y-axis

Ⓒ vertical translation 4 units down

Ⓓ horizontal translation 4 units right

Ⓔ horizontal shrink by a factor of $\frac{1}{3}$

Ⓕ vertical stretch by a factor of 3

8. Choose the correct equality or inequality symbol which completes the statement below about the linear functions f and g. Explain your reasoning.

x	f(x)
−5	−23
−4	−20
−3	−17
−2	−14

x	g(x)
−2	−18
−1	−14
0	−10
1	−6

$f(22)$ ▭ $g(22)$

2 Quadratic Functions

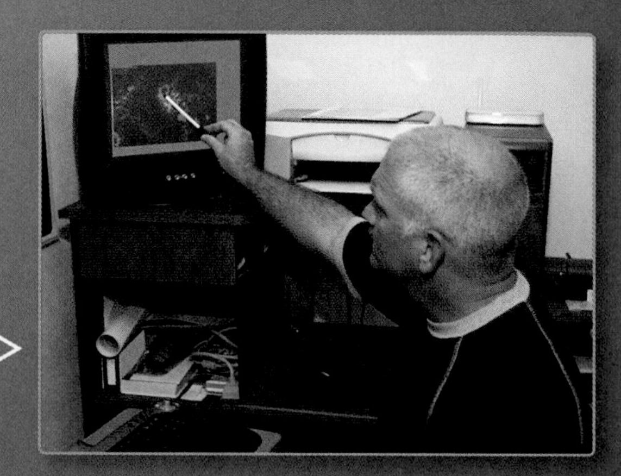

Meteorologist *(p. 77)*

Electricity-Generating Dish *(p. 71)*

Gateshead Millennium Bridge *(p. 64)*

Soccer *(p. 63)*

Kangaroo *(p. 53)*

Maintaining Mathematical Proficiency

Finding *x*-Intercepts

Example 1 **Find the *x*-intercept of the graph of the linear equation $y = 3x - 12$.**

$y = 3x - 12$	Write the equation.
$0 = 3x - 12$	Substitute 0 for *y*.
$12 = 3x$	Add 12 to each side.
$4 = x$	Divide each side by 3.

▶ The x-intercept is 4.

Find the *x*-intercept of the graph of the linear equation.

1. $y = 2x + 7$

2. $y = -6x + 8$

3. $y = -10x - 36$

4. $y = 3(x - 5)$

5. $y = -4(x + 10)$

6. $3x + 6y = 24$

The Distance Formula

The distance d between any two points (x_1, y_1) and (x_2, y_2) is given by the formula

$$d = \sqrt{(x_2 - x_1)^2 + (y_2 - y_1)^2}.$$

Example 2 **Find the distance between (1, 4) and (−3, 6).**

Let $(x_1, y_1) = (1, 4)$ and $(x_2, y_2) = (-3, 6)$.

$d = \sqrt{(x_2 - x_1)^2 + (y_2 - y_1)^2}$	Write the Distance Formula.
$= \sqrt{(-3 - 1)^2 + (6 - 4)^2}$	Substitute.
$= \sqrt{(-4)^2 + 2^2}$	Simplify.
$= \sqrt{16 + 4}$	Evaluate powers.
$= \sqrt{20}$	Add.
≈ 4.47	Use a calculator.

Find the distance between the two points.

7. $(2, 5), (-4, 7)$

8. $(-1, 0), (-8, 4)$

9. $(3, 10), (5, 9)$

10. $(7, -4), (-5, 0)$

11. $(4, -8), (4, 2)$

12. $(0, 9), (-3, -6)$

13. **ABSTRACT REASONING** Use the Distance Formula to write an expression for the distance between the two points (a, c) and (b, c). Is there an easier way to find the distance when the x-coordinates are equal? Explain your reasoning.

Mathematical Practices

Mathematically proficient students distinguish correct reasoning from flawed reasoning.

Using Correct Logic

Core Concept

Deductive Reasoning

In *deductive reasoning*, you start with two or more statements that you know or assume to be true. From these, you *deduce* or *infer* the truth of another statement. Here is an example.

1. Premise: If this traffic does not clear, then I will be late for work.

2. Premise: The traffic has not cleared.

3. Conclusion: I will be late for work.

This pattern for deductive reasoning is called a *syllogism*.

EXAMPLE 1 Recognizing Flawed Reasoning

The syllogisms below represent common types of *flawed reasoning*. Explain why each conclusion is not valid.

a. When it rains, the ground gets wet.
The ground is wet.
Therefore, it must have rained.

b. When it rains, the ground gets wet.
It is not raining.
Therefore, the ground is not wet.

c. Police, schools, and roads are necessary.
Taxes fund police, schools, and roads.
Therefore, taxes are necessary.

d. All students use cell phones.
My uncle uses a cell phone.
Therefore, my uncle is a student.

SOLUTION

a. The ground may be wet for another reason.

b. The ground may still be wet when the rain stops.

c. The services could be funded another way.

d. People other than students use cell phones.

Monitoring Progress

Decide whether the syllogism represents correct or flawed reasoning. If flawed, explain why the conclusion is not valid.

1. All mammals are warm-blooded.
All dogs are mammals.
Therefore, all dogs are warm-blooded.

2. All mammals are warm-blooded.
My pet is warm-blooded.
Therefore, my pet is a mammal.

3. If I am sick, then I will miss school.
I missed school.
Therefore, I am sick.

4. If I am sick, then I will miss school.
I did not miss school.
Therefore, I am not sick.

2.1 Transformations of Quadratic Functions

Essential Question How do the constants a, h, and k affect the graph of the quadratic function $g(x) = a(x - h)^2 + k$?

The parent function of the quadratic family is $f(x) = x^2$. A transformation of the graph of the parent function is represented by the function $g(x) = a(x - h)^2 + k$, where $a \neq 0$.

EXPLORATION 1 Identifying Graphs of Quadratic Functions

Work with a partner. Match each quadratic function with its graph. Explain your reasoning. Then use a graphing calculator to verify that your answer is correct.

a. $g(x) = -(x - 2)^2$ **b.** $g(x) = (x - 2)^2 + 2$ **c.** $g(x) = -(x + 2)^2 - 2$

d. $g(x) = 0.5(x - 2)^2 - 2$ **e.** $g(x) = 2(x - 2)^2$ **f.** $g(x) = -(x + 2)^2 + 2$

A.

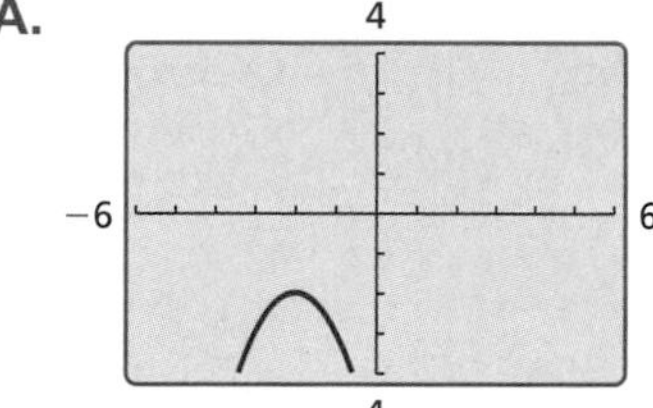

B.

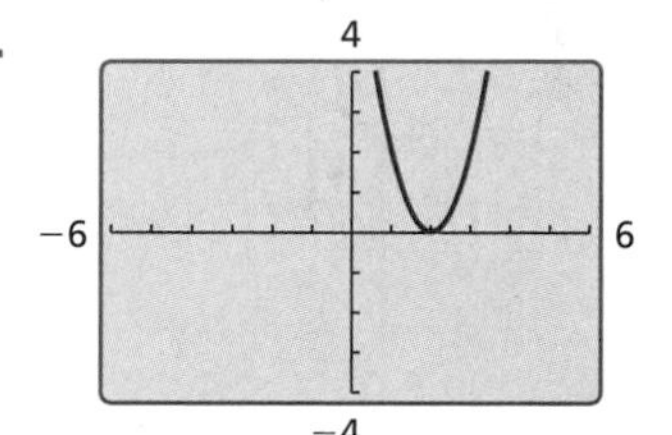

C.

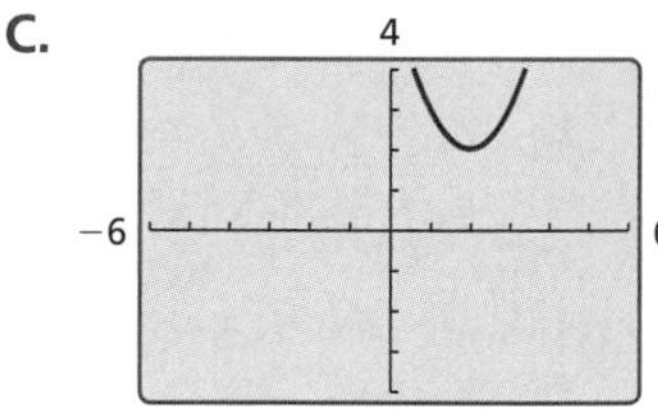

D.

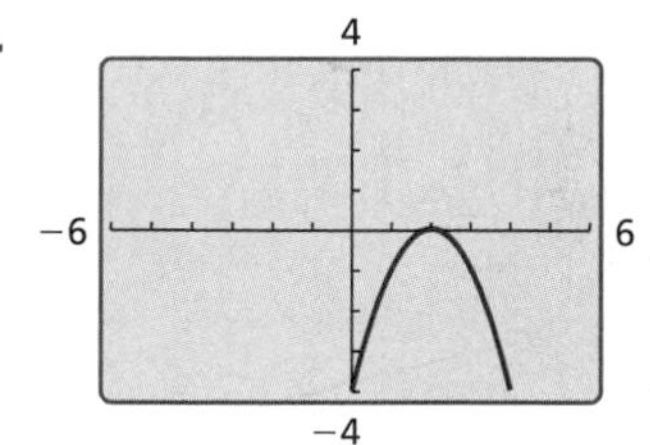

E.

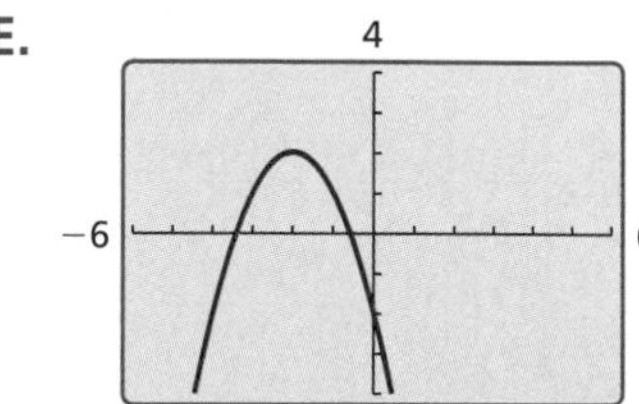

F.

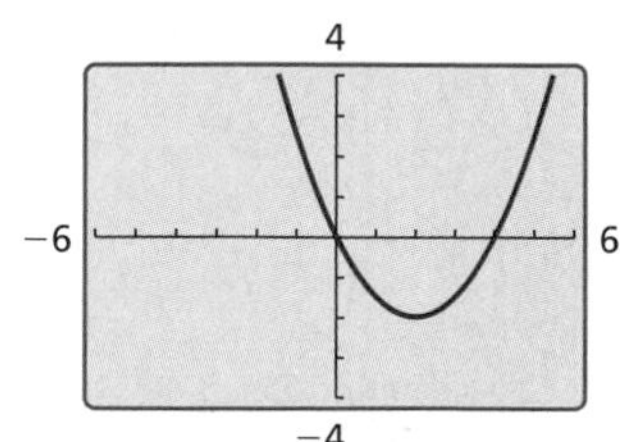

LOOKING FOR STRUCTURE

To be proficient in math, you need to look closely to discern a pattern or structure.

Communicate Your Answer

2. How do the constants a, h, and k affect the graph of the quadratic function $g(x) = a(x - h)^2 + k$?

3. Write the equation of the quadratic function whose graph is shown at the right. Explain your reasoning. Then use a graphing calculator to verify that your equation is correct.

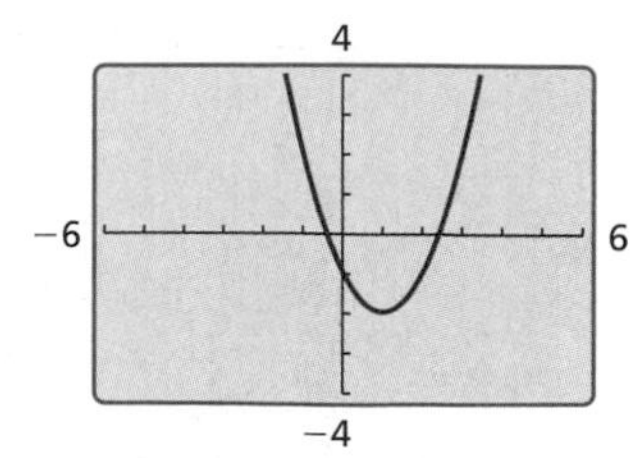

2.1 Lesson

What You Will Learn

- Describe transformations of quadratic functions.
- Write transformations of quadratic functions.

Core Vocabulary

quadratic function, *p. 48*
parabola, *p. 48*
vertex of a parabola, *p. 50*
vertex form, *p. 50*

Previous
transformations

Describing Transformations of Quadratic Functions

A **quadratic function** is a function that can be written in the form $f(x) = a(x - h)^2 + k$, where $a \neq 0$. The U-shaped graph of a quadratic function is called a **parabola**.

In Section 1.1, you graphed quadratic functions using tables of values. You can also graph quadratic functions by applying transformations to the graph of the parent function $f(x) = x^2$.

Core Concept

Horizontal Translations

$$f(x) = x^2$$

$$f(x - h) = (x - h)^2$$

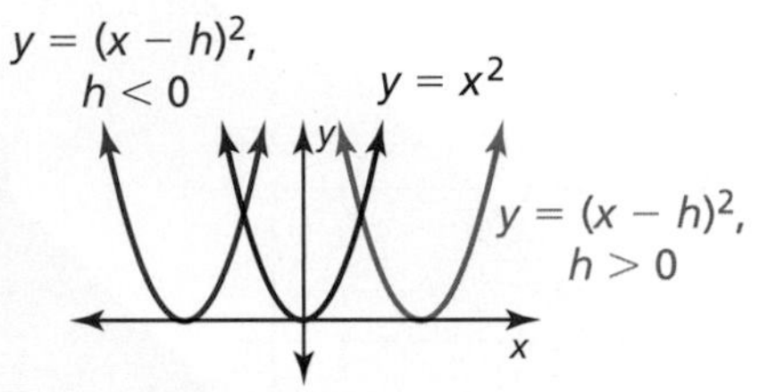

y = (x − h)², h < 0
y = x²
y = (x − h)², h > 0
y
x

- shifts left when $h < 0$
- shifts right when $h > 0$

Vertical Translations

$$f(x) = x^2$$

$$f(x) + k = x^2 + k$$

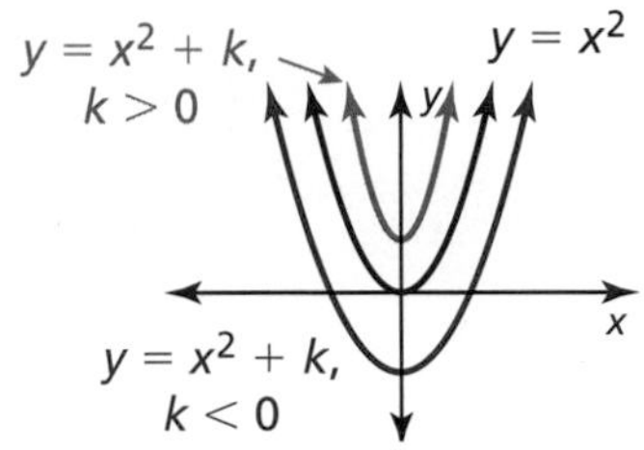

y = x² + k, k > 0
y = x²
y
x
y = x² + k, k < 0

- shifts down when $k < 0$
- shifts up when $k > 0$

EXAMPLE 1 Translations of a Quadratic Function

Describe the transformation of $f(x) = x^2$ represented by $g(x) = (x + 4)^2 - 1$. Then graph each function.

SOLUTION

Notice that the function is of the form $g(x) = (x - h)^2 + k$. Rewrite the function to identify h and k.

$$g(x) = (x - \underset{\substack{\uparrow \\ h}}{(-4)})^2 + \underset{\substack{\uparrow \\ k}}{(-1)}$$

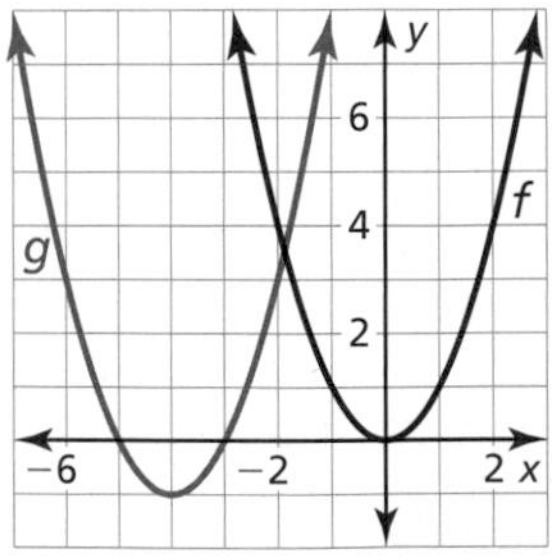

y
6
4
2
g
f
−6
−2
2 x

▶ Because $h = -4$ and $k = -1$, the graph of g is a translation 4 units left and 1 unit down of the graph of f.

Monitoring Progress

Help in English and Spanish at *BigIdeasMath.com*

Describe the transformation of $f(x) = x^2$ represented by g. Then graph each function.

1. $g(x) = (x - 3)^2$

2. $g(x) = (x - 2)^2 - 2$

3. $g(x) = (x + 5)^2 + 1$

Core Concept

Reflections in the x-Axis

$f(x) = x^2$

$-f(x) = -(x^2) = -x^2$

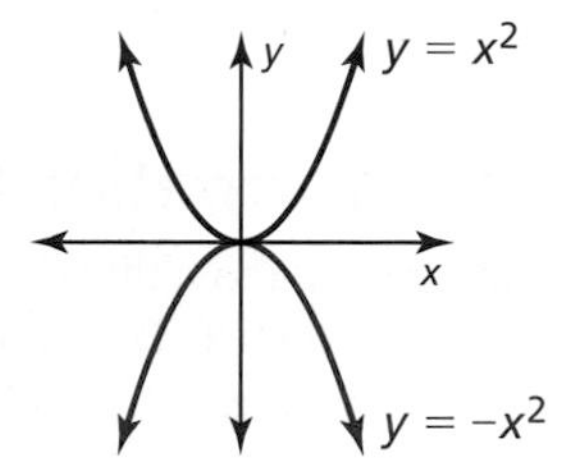

flips over the x-axis

Reflections in the y-Axis

$f(x) = x^2$

$f(-x) = (-x)^2 = x^2$

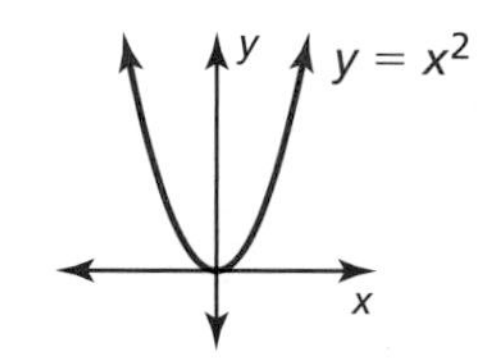

$y = x^2$ is its own reflection in the y-axis.

Horizontal Stretches and Shrinks

$f(x) = x^2$

$f(ax) = (ax)^2$

- horizontal stretch (away from y-axis) when $0 < a < 1$
- horizontal shrink (toward y-axis) when $a > 1$

Vertical Stretches and Shrinks

$f(x) = x^2$

$a \cdot f(x) = ax^2$

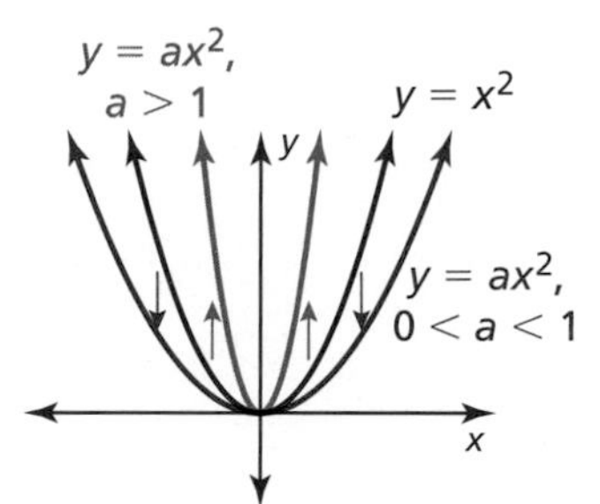

- vertical stretch (away from x-axis) when $a > 1$
- vertical shrink (toward x-axis) when $0 < a < 1$

EXAMPLE 2 **Transformations of Quadratic Functions**

Describe the transformation of $f(x) = x^2$ represented by g. Then graph each function.

a. $g(x) = -\frac{1}{2}x^2$ **b.** $g(x) = (2x)^2 + 1$

SOLUTION

a. Notice that the function is of the form $g(x) = -ax^2$, where $a = \frac{1}{2}$.

▶ So, the graph of g is a reflection in the x-axis and a vertical shrink by a factor of $\frac{1}{2}$ of the graph of f.

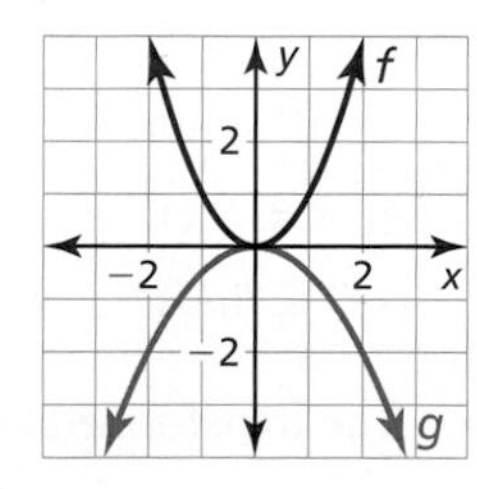

b. Notice that the function is of the form $g(x) = (ax)^2 + k$, where $a = 2$ and $k = 1$.

▶ So, the graph of g is a horizontal shrink by a factor of $\frac{1}{2}$ followed by a translation 1 unit up of the graph of f.

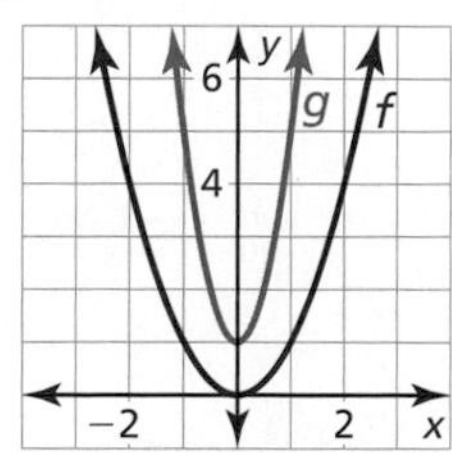

LOOKING FOR STRUCTURE

In Example 2b, notice that $g(x) = 4x^2 + 1$. So, you can also describe the graph of g as a vertical stretch by a factor of 4 followed by a translation 1 unit up of the graph of f.

Monitoring Progress  Help in English and Spanish at *BigIdeasMath.com*

Describe the transformation of $f(x) = x^2$ represented by g. Then graph each function.

4. $g(x) = \left(\frac{1}{3}x\right)^2$ **5.** $g(x) = 3(x - 1)^2$ **6.** $g(x) = -(x + 3)^2 + 2$

Writing Transformations of Quadratic Functions

The lowest point on a parabola that opens up or the highest point on a parabola that opens down is the **vertex**. The **vertex form** of a quadratic function is $f(x) = a(x - h)^2 + k$, where $a \neq 0$ and the vertex is (h, k).

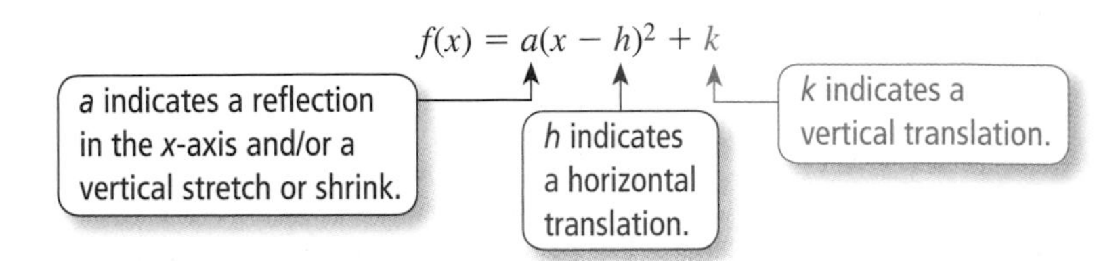

EXAMPLE 3 Writing a Transformed Quadratic Function

Let the graph of g be a vertical stretch by a factor of 2 and a reflection in the x-axis, followed by a translation 3 units down of the graph of $f(x) = x^2$. Write a rule for g and identify the vertex.

SOLUTION

Method 1 Identify how the transformations affect the constants in vertex form.

reflection in x-axis, vertical stretch by 2 } $a = -2$

translation 3 units down } $k = -3$

Write the transformed function.

$g(x) = a(x - h)^2 + k$ — Vertex form of a quadratic function

$= -2(x - 0)^2 + (-3)$ — Substitute −2 for *a*, 0 for *h*, and −3 for *k*.

$= -2x^2 - 3$ — Simplify.

▶ The transformed function is $g(x) = -2x^2 - 3$. The vertex is $(0, -3)$.

Method 2 Begin with the parent function and apply the transformations one at a time in the stated order.

First write a function h that represents the reflection and vertical stretch of f.

$h(x) = -2 \cdot f(x)$ — Multiply the output by −2.

$= -2x^2$ — Substitute x^2 for $f(x)$.

Then write a function g that represents the translation of h.

$g(x) = h(x) - 3$ — Subtract 3 from the output.

$= -2x^2 - 3$ — Substitute $-2x^2$ for $h(x)$.

▶ The transformed function is $g(x) = -2x^2 - 3$. The vertex is $(0, -3)$.

Check

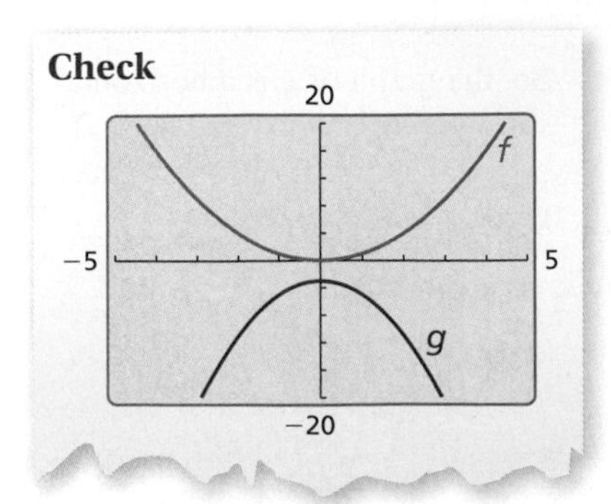

EXAMPLE 4 Writing a Transformed Quadratic Function

REMEMBER

To multiply two binomials, use the FOIL Method.

First Inner

$(x + 1)(x + 2) = x^2 + 2x + x + 2$

Outer Last

Let the graph of g be a translation 3 units right and 2 units up, followed by a reflection in the y-axis of the graph of $f(x) = x^2 - 5x$. Write a rule for g.

SOLUTION

Step 1 First write a function h that represents the translation of f.

$h(x) = f(x - 3) + 2$	Subtract 3 from the input. Add 2 to the output.
$= (x - 3)^2 - 5(x - 3) + 2$	Replace x with $x - 3$ in $f(x)$.
$= x^2 - 11x + 26$	Simplify.

Step 2 Then write a function g that represents the reflection of h.

$g(x) = h(-x)$	Multiply the input by -1.
$= (-x)^2 - 11(-x) + 26$	Replace x with $-x$ in $h(x)$.
$= x^2 + 11x + 26$	Simplify.

EXAMPLE 5 Modeling with Mathematics

The height h (in feet) of water spraying from a fire hose can be modeled by $h(x) = -0.03x^2 + x + 25$, where x is the horizontal distance (in feet) from the fire truck. The crew raises the ladder so that the water hits the ground 10 feet farther from the fire truck. Write a function that models the new path of the water.

SOLUTION

1. **Understand the Problem** You are given a function that represents the path of water spraying from a fire hose. You are asked to write a function that represents the path of the water after the crew raises the ladder.
2. **Make a Plan** Analyze the graph of the function to determine the translation of the ladder that causes water to travel 10 feet farther. Then write the function.
3. **Solve the Problem** Use a graphing calculator to graph the original function.

60

$y = -0.03x^2 + x + 25$

0 80

X=50 Y=0

−30

Because $h(50) = 0$, the water originally hits the ground 50 feet from the fire truck. The range of the function in this context does not include negative values. However, by observing that $h(60) = -23$, you can determine that a translation 23 units (feet) up causes the water to travel 10 feet farther from the fire truck.

$g(x) = h(x) + 23$	Add 23 to the output.
$= -0.03x^2 + x + 48$	Substitute for $h(x)$ and simplify.

▶ The new path of the water can be modeled by $g(x) = -0.03x^2 + x + 48$.

4. **Look Back** To check that your solution is correct, verify that $g(60) = 0$.

$$g(60) = -0.03(60)^2 + 60 + 48 = -108 + 60 + 48 = 0 \checkmark$$

Monitoring Progress

Help in English and Spanish at *BigIdeasMath.com*

7. Let the graph of g be a vertical shrink by a factor of $\frac{1}{2}$ followed by a translation 2 units up of the graph of $f(x) = x^2$. Write a rule for g and identify the vertex.
8. Let the graph of g be a translation 4 units left followed by a horizontal shrink by a factor of $\frac{1}{3}$ of the graph of $f(x) = x^2 + x$. Write a rule for g.
9. **WHAT IF?** In Example 5, the water hits the ground 10 feet closer to the fire truck after lowering the ladder. Write a function that models the new path of the water.

2.1 Exercises

Dynamic Solutions available at *BigIdeasMath.com*

Vocabulary and Core Concept Check

1. **COMPLETE THE SENTENCE** The graph of a quadratic function is called a(n) ________.

2. **VOCABULARY** Identify the vertex of the parabola given by $f(x) = (x + 2)^2 - 4$.

Monitoring Progress and Modeling with Mathematics

In Exercises 3–12, describe the transformation of $f(x) = x^2$ represented by g. Then graph each function. *(See Example 1.)*

3. $g(x) = x^2 - 3$
4. $g(x) = x^2 + 1$
5. $g(x) = (x + 2)^2$
6. $g(x) = (x - 4)^2$
7. $g(x) = (x - 1)^2$
8. $g(x) = (x + 3)^2$
9. $g(x) = (x + 6)^2 - 2$
10. $g(x) = (x - 9)^2 + 5$
11. $g(x) = (x - 7)^2 + 1$
12. $g(x) = (x + 10)^2 - 3$

ANALYZING RELATIONSHIPS **In Exercises 13–16, match the function with the correct transformation of the graph of f. Explain your reasoning.**

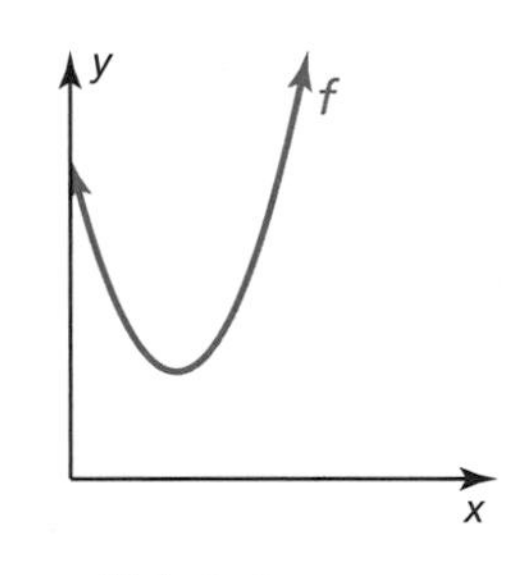

13. $y = f(x - 1)$
14. $y = f(x) + 1$
15. $y = f(x - 1) + 1$
16. $y = f(x + 1) - 1$

A. 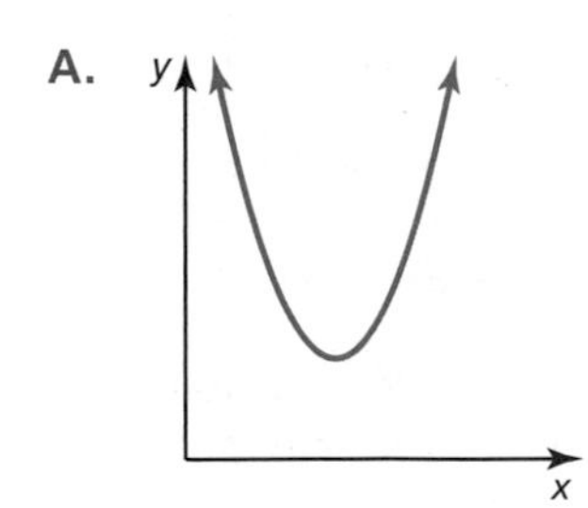

B. 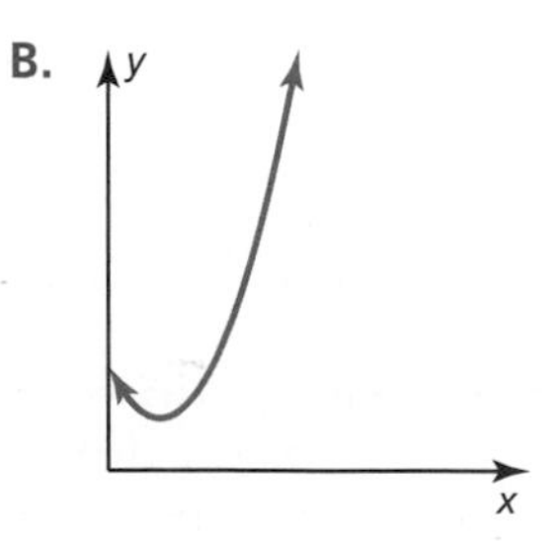

C. 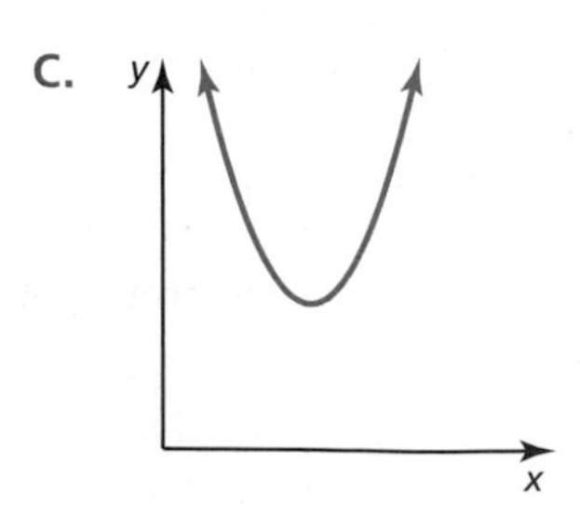

D. 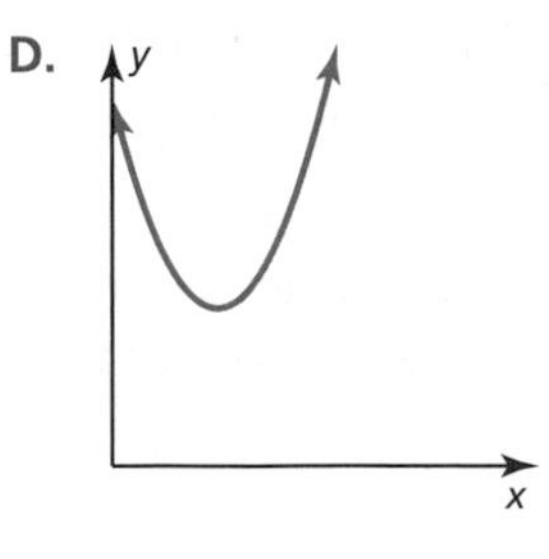

In Exercises 17–24, describe the transformation of $f(x) = x^2$ represented by g. Then graph each function. *(See Example 2.)*

17. $g(x) = -x^2$
18. $g(x) = (-x)^2$
19. $g(x) = 3x^2$
20. $g(x) = \frac{1}{3}x^2$
21. $g(x) = (2x)^2$
22. $g(x) = -(2x)^2$
23. $g(x) = \frac{1}{5}x^2 - 4$
24. $g(x) = \frac{1}{2}(x - 1)^2$

ERROR ANALYSIS **In Exercises 25 and 26, describe and correct the error in analyzing the graph of $f(x) = -6x^2 + 4$.**

25. 

The graph is a reflection in the y-axis and a vertical stretch by a factor of 6, followed by a translation 4 units up of the graph of the parent quadratic function.

26. 

The graph is a translation 4 units down, followed by a vertical stretch by a factor of 6 and a reflection in the x-axis of the graph of the parent quadratic function.

USING STRUCTURE **In Exercises 27–30, describe the transformation of the graph of the parent quadratic function. Then identify the vertex.**

27. $f(x) = 3(x + 2)^2 + 1$
28. $f(x) = -4(x + 1)^2 - 5$
29. $f(x) = -2x^2 + 5$
30. $f(x) = \frac{1}{2}(x - 1)^2$

In Exercises 31–34, write a rule for g described by the transformations of the graph of f. Then identify the vertex. *(See Examples 3 and 4.)*

31. $f(x) = x^2$; vertical stretch by a factor of 4 and a reflection in the x-axis, followed by a translation 2 units up

32. $f(x) = x^2$; vertical shrink by a factor of $\frac{1}{3}$ and a reflection in the y-axis, followed by a translation 3 units right

33. $f(x) = 8x^2 - 6$; horizontal stretch by a factor of 2 and a translation 2 units up, followed by a reflection in the y-axis

34. $f(x) = (x + 6)^2 + 3$; horizontal shrink by a factor of $\frac{1}{2}$ and a translation 1 unit down, followed by a reflection in the x-axis

USING TOOLS In Exercises 35–40, match the function with its graph. Explain your reasoning.

35. $g(x) = 2(x - 1)^2 - 2$

36. $g(x) = \frac{1}{2}(x + 1)^2 - 2$

37. $g(x) = -2(x - 1)^2 + 2$

38. $g(x) = 2(x + 1)^2 + 2$

39. $g(x) = -2(x + 1)^2 - 2$

40. $g(x) = 2(x - 1)^2 + 2$

A.
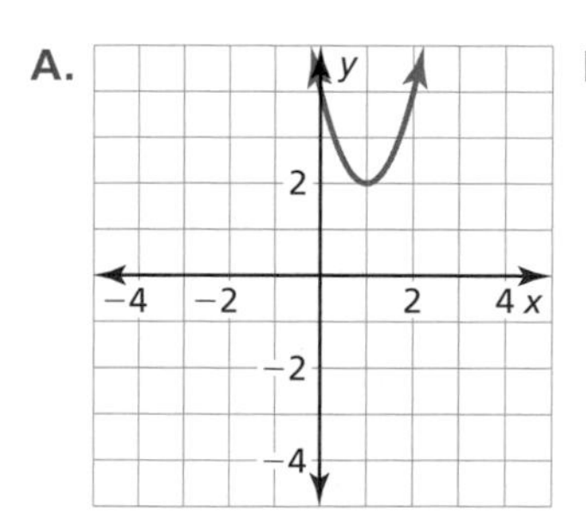

B.
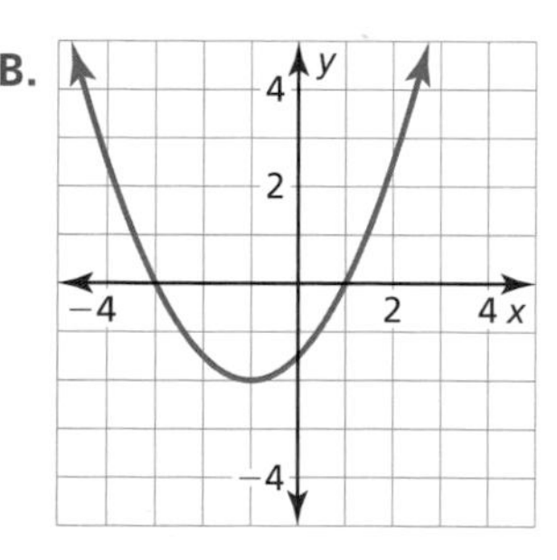

C.
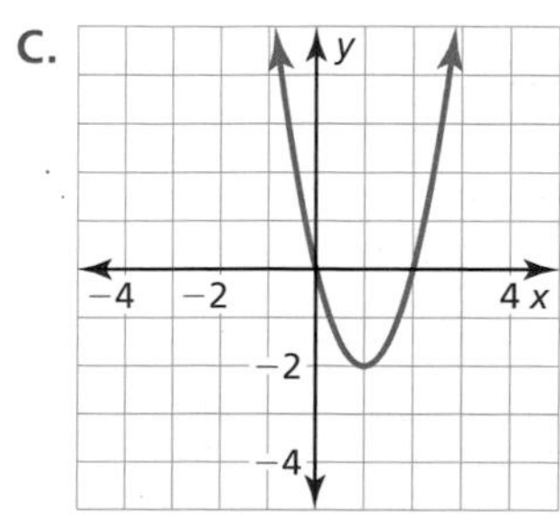

D.
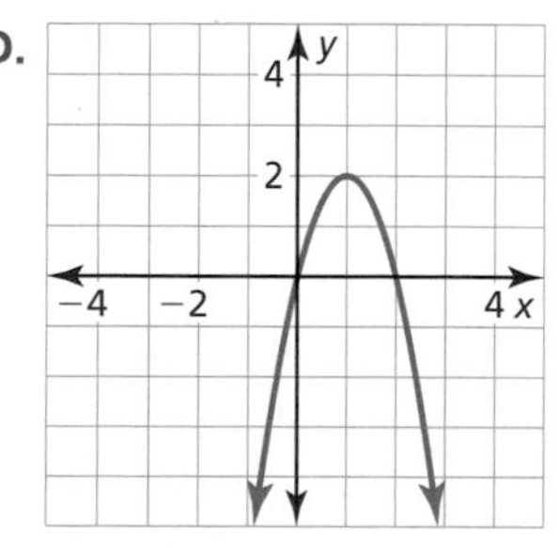

E.
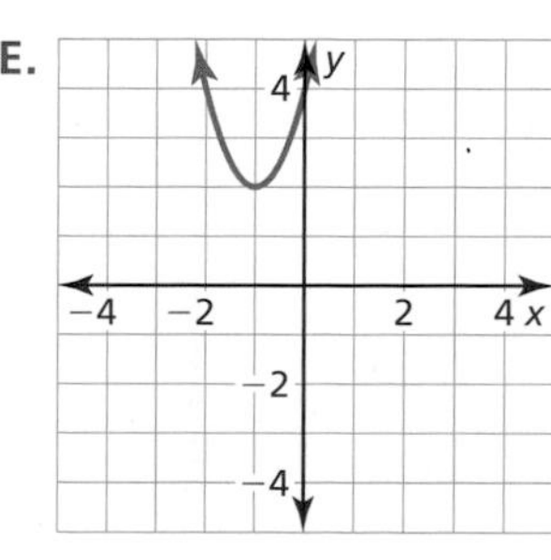

F.
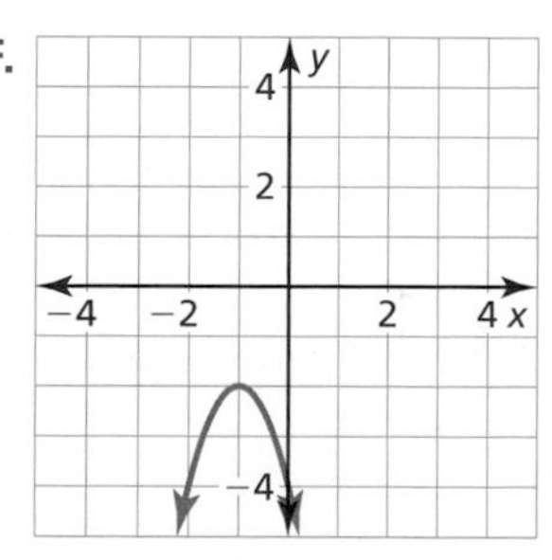

JUSTIFYING STEPS In Exercises 41 and 42, justify each step in writing a function g based on the transformations of $f(x) = 2x^2 + 6x$.

41. translation 6 units down followed by a reflection in the x-axis

$$h(x) = f(x) - 6$$
$$= 2x^2 + 6x - 6$$
$$g(x) = -h(x)$$
$$= -(2x^2 + 6x - 6)$$
$$= -2x^2 - 6x + 6$$

42. reflection in the y-axis followed by a translation 4 units right

$$h(x) = f(-x)$$
$$= 2(-x)^2 + 6(-x)$$
$$= 2x^2 - 6x$$
$$g(x) = h(x - 4)$$
$$= 2(x - 4)^2 - 6(x - 4)$$
$$= 2x^2 - 22x + 56$$

43. MODELING WITH MATHEMATICS The function $h(x) = -0.03(x - 14)^2 + 6$ models the jump of a red kangaroo, where x is the horizontal distance traveled (in feet) and $h(x)$ is the height (in feet). When the kangaroo jumps from a higher location, it lands 5 feet farther away. Write a function that models the second jump. *(See Example 5.)*

44. MODELING WITH MATHEMATICS The function $f(t) = -16t^2 + 10$ models the height (in feet) of an object t seconds after it is dropped from a height of 10 feet on Earth. The same object dropped from the same height on the moon is modeled by $g(t) = -\frac{8}{3}t^2 + 10$. Describe the transformation of the graph of f to obtain g. From what height must the object be dropped on the moon so it hits the ground at the same time as on Earth?

45. MODELING WITH MATHEMATICS Flying fish use their pectoral fins like airplane wings to glide through the air.

a. Write an equation of the form $y = a(x - h)^2 + k$ with vertex (33, 5) that models the flight path, assuming the fish leaves the water at (0, 0).

b. What are the domain and range of the function? What do they represent in this situation?

c. Does the value of a change when the flight path has vertex (30, 4)? Justify your answer.

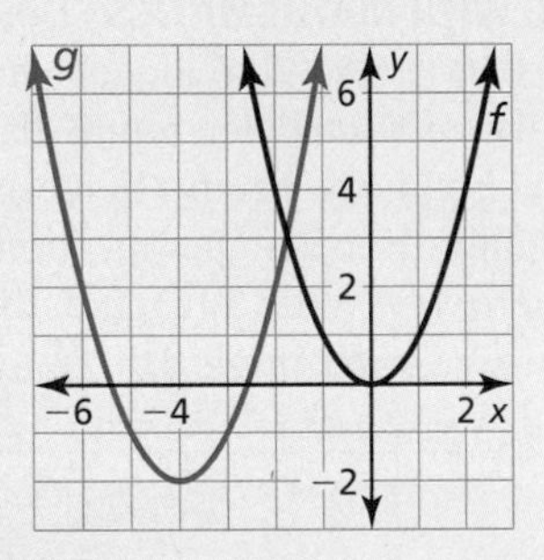

46. HOW DO YOU SEE IT? Describe the graph of g as a transformation of the graph of $f(x) = x^2$.

47. COMPARING METHODS Let the graph of g be a translation 3 units up and 1 unit right followed by a vertical stretch by a factor of 2 of the graph of $f(x) = x^2$.

a. Identify the values of a, h, and k and use vertex form to write the transformed function.

b. Use function notation to write the transformed function. Compare this function with your function in part (a).

c. Suppose the vertical stretch was performed first, followed by the translations. Repeat parts (a) and (b).

d. Which method do you prefer when writing a transformed function? Explain.

48. THOUGHT PROVOKING A jump on a pogo stick with a conventional spring can be modeled by $f(x) = -0.5(x - 6)^2 + 18$, where x is the horizontal distance (in inches) and $f(x)$ is the vertical distance (in inches). Write at least one transformation of the function and provide a possible reason for your transformation.

49. MATHEMATICAL CONNECTIONS The area of a circle depends on the radius, as shown in the graph. A circular earring with a radius of r millimeters has a circular hole with a radius of $\frac{3r}{4}$ millimeters. Describe a transformation of the graph below that models the area of the blue portion of the earring.

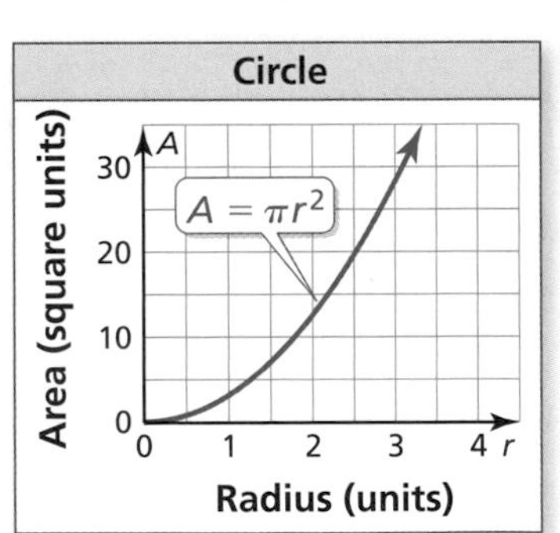

Maintaining Mathematical Proficiency

Reviewing what you learned in previous grades and lessons

A line of symmetry for the figure is shown in red. Find the coordinates of point A. *(Skills Review Handbook)*

50.

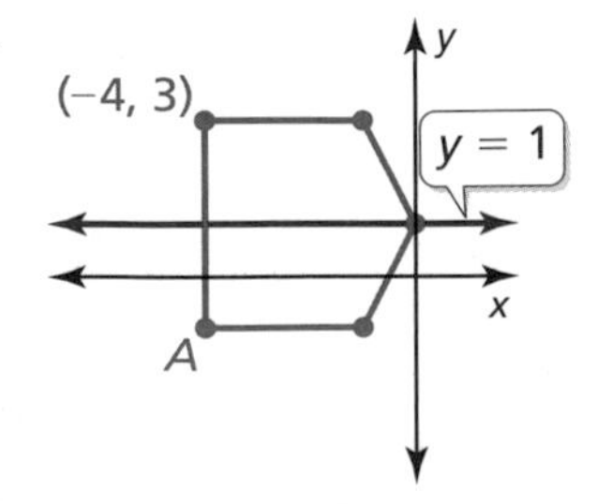

51.

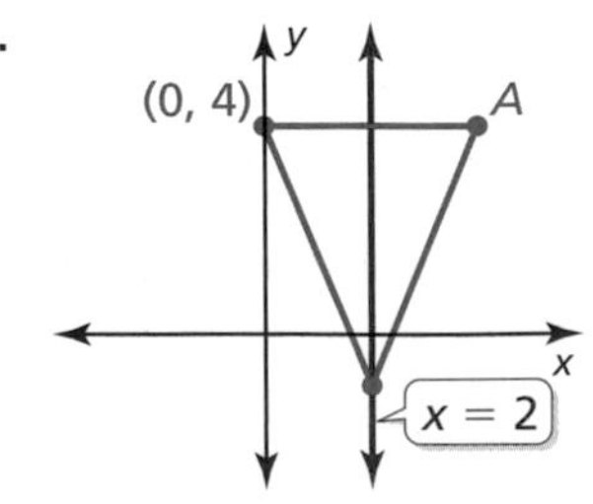

52.

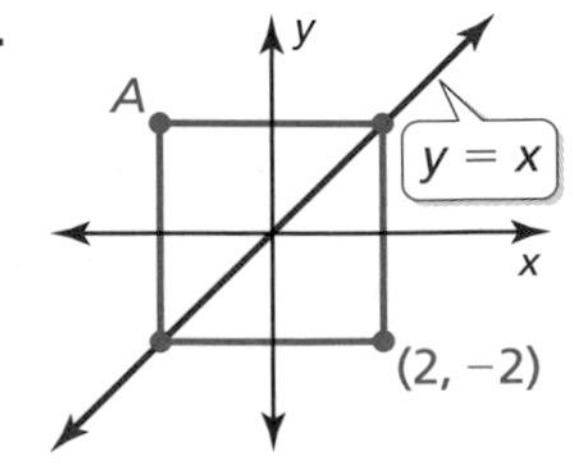

2.2 Characteristics of Quadratic Functions

Essential Question What type of symmetry does the graph of $f(x) = a(x - h)^2 + k$ have and how can you describe this symmetry?

EXPLORATION 1 Parabolas and Symmetry

Work with a partner.

a. Complete the table. Then use the values in the table to sketch the graph of the function

$$f(x) = \frac{1}{2}x^2 - 2x - 2$$

on graph paper.

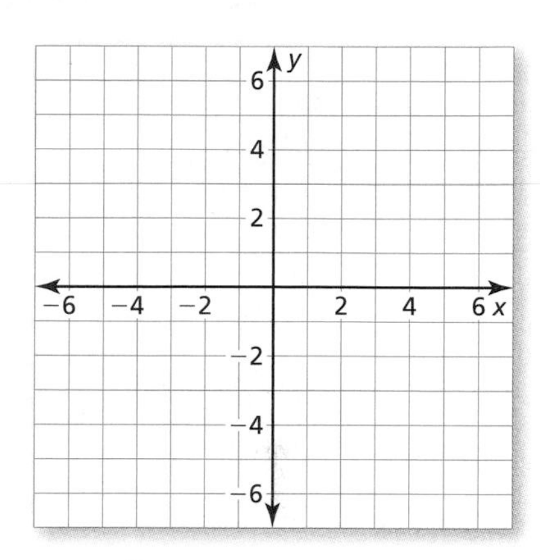

x	−2	−1	0	1	2
f(x)					

x	3	4	5	6
f(x)				

b. Use the results in part (a) to identify the vertex of the parabola.

c. Find a vertical line on your graph paper so that when you fold the paper, the left portion of the graph coincides with the right portion of the graph. What is the equation of this line? How does it relate to the vertex?

d. Show that the vertex form

$$f(x) = \frac{1}{2}(x - 2)^2 - 4$$

is equivalent to the function given in part (a).

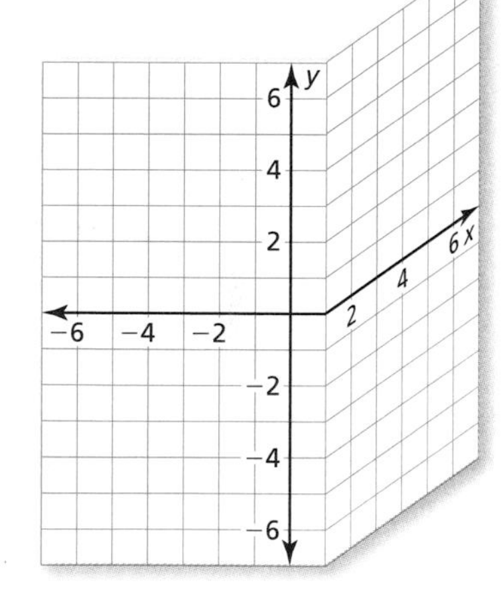

EXPLORATION 2 Parabolas and Symmetry

Work with a partner. Repeat Exploration 1 for the function given by

$$f(x) = -\frac{1}{3}x^2 + 2x + 3 = -\frac{1}{3}(x - 3)^2 + 6.$$

ATTENDING TO PRECISION

To be proficient in math, you need to use clear definitions in your reasoning and discussions with others.

Communicate Your Answer

3. What type of symmetry does the graph of $f(x) = a(x - h)^2 + k$ have and how can you describe this symmetry?

4. Describe the symmetry of each graph. Then use a graphing calculator to verify your answer.

a. $f(x) = -(x - 1)^2 + 4$ **b.** $f(x) = (x + 1)^2 - 2$ **c.** $f(x) = 2(x - 3)^2 + 1$

d. $f(x) = \frac{1}{2}(x + 2)^2$ **e.** $f(x) = -2x^2 + 3$ **f.** $f(x) = 3(x - 5)^2 + 2$

2.2 Lesson

Core Vocabulary

axis of symmetry, *p. 56*
standard form, *p. 56*
minimum value, *p. 58*
maximum value, *p. 58*
intercept form, *p. 59*

Previous
x-intercept

What You Will Learn

- Explore properties of parabolas.
- Find maximum and minimum values of quadratic functions.
- Graph quadratic functions using x-intercepts.
- Solve real-life problems.

Exploring Properties of Parabolas

An **axis of symmetry** is a line that divides a parabola into mirror images and passes through the vertex. Because the vertex of $f(x) = a(x - h)^2 + k$ is (h, k), the axis of symmetry is the vertical line $x = h$.

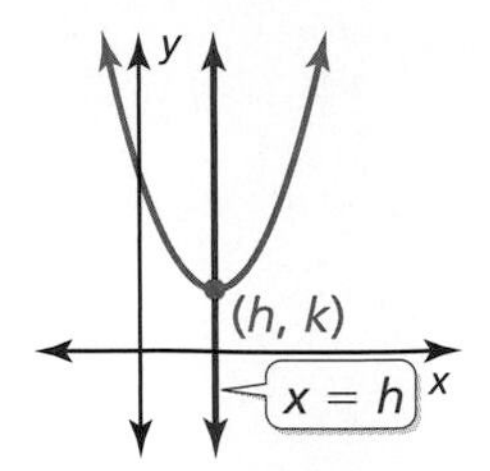

Previously, you used transformations to graph quadratic functions in vertex form. You can also use the axis of symmetry and the vertex to graph quadratic functions written in vertex form.

EXAMPLE 1 Using Symmetry to Graph Quadratic Functions

Graph $f(x) = -2(x + 3)^2 + 4$. Label the vertex and axis of symmetry.

SOLUTION

Step 1 Identify the constants $a = -2$, $h = -3$, and $k = 4$.

Step 2 Plot the vertex $(h, k) = (-3, 4)$ and draw the axis of symmetry $x = -3$.

Step 3 Evaluate the function for two values of x.

$$x = -2:\ f(-2) = -2(-2 + 3)^2 + 4 = 2$$
$$x = -1:\ f(-1) = -2(-1 + 3)^2 + 4 = -4$$

Plot the points $(-2, 2)$, $(-1, -4)$, and their reflections in the axis of symmetry.

Step 4 Draw a parabola through the plotted points.

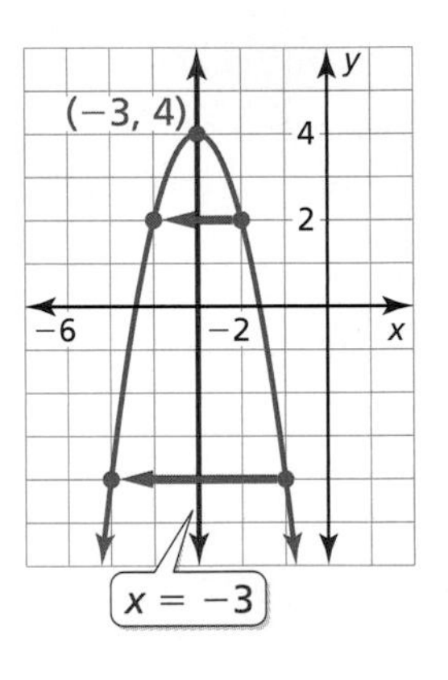

Quadratic functions can also be written in **standard form**, $f(x) = ax^2 + bx + c$, where $a \neq 0$. You can derive standard form by expanding vertex form.

$f(x) = a(x - h)^2 + k$	Vertex form
$f(x) = a(x^2 - 2hx + h^2) + k$	Expand $(x - h)^2$.
$f(x) = ax^2 - 2ahx + ah^2 + k$	Distributive Property
$f(x) = ax^2 + (-2ah)x + (ah^2 + k)$	Group like terms.
$f(x) = ax^2 + bx + c$	Let $b = -2ah$ and let $c = ah^2 + k$.

This allows you to make the following observations.

$a = a$: So, a has the same meaning in vertex form and standard form.

$b = -2ah$: Solve for h to obtain $h = -\dfrac{b}{2a}$. So, the axis of symmetry is $x = -\dfrac{b}{2a}$.

$c = ah^2 + k$: In vertex form $f(x) = a(x - h)^2 + k$, notice that $f(0) = ah^2 + k$. So, c is the y-intercept.

Core Concept

Properties of the Graph of $f(x) = ax^2 + bx + c$

$y = ax^2 + bx + c, a > 0$ $y = ax^2 + bx + c, a < 0$

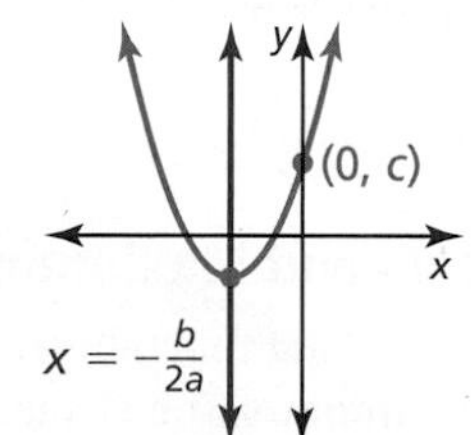

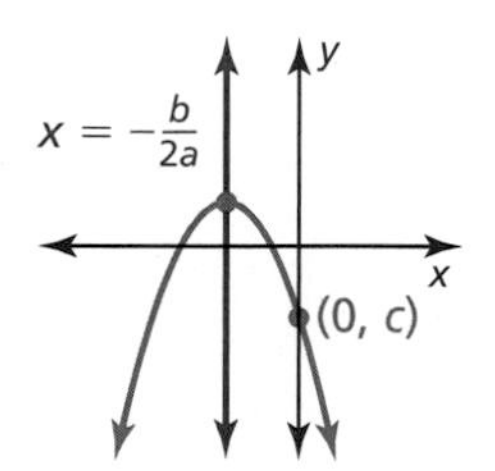

- The parabola opens up when $a > 0$ and opens down when $a < 0$.
- The graph is narrower than the graph of $f(x) = x^2$ when $|a| > 1$ and wider when $|a| < 1$.
- The axis of symmetry is $x = -\frac{b}{2a}$ and the vertex is $\left(-\frac{b}{2a}, f\left(-\frac{b}{2a}\right)\right)$.
- The y-intercept is c. So, the point $(0, c)$ is on the parabola.

EXAMPLE 2 Graphing a Quadratic Function in Standard Form

Graph $f(x) = 3x^2 - 6x + 1$. Label the vertex and axis of symmetry.

COMMON ERROR

Be sure to include the negative sign when writing the expression for the x-coordinate of the vertex.

SOLUTION

Step 1 Identify the coefficients $a = 3$, $b = -6$, and $c = 1$. Because $a > 0$, the parabola opens up.

Step 2 Find the vertex. First calculate the x-coordinate.

$$x = -\frac{b}{2a} = -\frac{-6}{2(3)} = 1$$

Then find the y-coordinate of the vertex.

$$f(1) = 3(1)^2 - 6(1) + 1 = -2$$

So, the vertex is $(1, -2)$. Plot this point.

Step 3 Draw the axis of symmetry $x = 1$.

Step 4 Identify the y-intercept c, which is 1. Plot the point $(0, 1)$ and its reflection in the axis of symmetry, $(2, 1)$.

Step 5 Evaluate the function for another value of x, such as $x = 3$.

$$f(3) = 3(3)^2 - 6(3) + 1 = 10$$

Plot the point $(3, 10)$ and its reflection in the axis of symmetry, $(-1, 10)$.

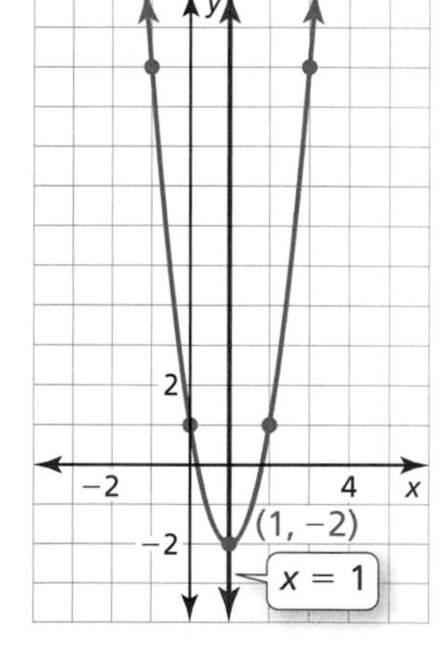

Step 6 Draw a parabola through the plotted points.

Monitoring Progress

Help in English and Spanish at *BigIdeasMath.com*

Graph the function. Label the vertex and axis of symmetry.

1. $f(x) = -3(x + 1)^2$

2. $g(x) = 2(x - 2)^2 + 5$

3. $h(x) = x^2 + 2x - 1$

4. $p(x) = -2x^2 - 8x + 1$

Finding Maximum and Minimum Values

Because the vertex is the highest or lowest point on a parabola, its y-coordinate is the *maximum value* or *minimum value* of the function. The vertex lies on the axis of symmetry, so the function is *increasing* on one side of the axis of symmetry and *decreasing* on the other side.

Core Concept

Minimum and Maximum Values

For the quadratic function $f(x) = ax^2 + bx + c$, the y-coordinate of the vertex is the **minimum value** of the function when $a > 0$ and the **maximum value** when $a < 0$.

$a > 0$

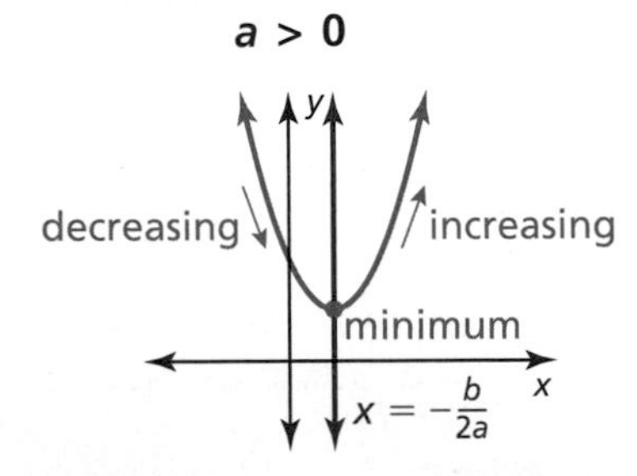

- Minimum value: $f\left(-\frac{b}{2a}\right)$
- Domain: All real numbers
- Range: $y \geq f\left(-\frac{b}{2a}\right)$
- Decreasing to the left of $x = -\frac{b}{2a}$
- Increasing to the right of $x = -\frac{b}{2a}$

$a < 0$

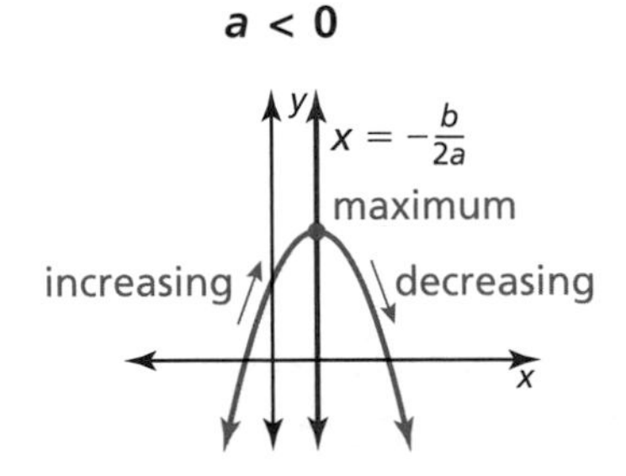

- Maximum value: $f\left(-\frac{b}{2a}\right)$
- Domain: All real numbers
- Range: $y \leq f\left(-\frac{b}{2a}\right)$
- Increasing to the left of $x = -\frac{b}{2a}$
- Decreasing to the right of $x = -\frac{b}{2a}$

STUDY TIP

When a function f is written in vertex form, you can use $h = -\frac{b}{2a}$ and $k = f\left(-\frac{b}{2a}\right)$ to state the properties shown.

EXAMPLE 3 **Finding a Minimum or a Maximum Value**

Find the minimum value or maximum value of $f(x) = \frac{1}{2}x^2 - 2x - 1$. Describe the domain and range of the function, and where the function is increasing and decreasing.

SOLUTION

Identify the coefficients $a = \frac{1}{2}$, $b = -2$, and $c = -1$. Because $a > 0$, the parabola opens up and the function has a minimum value. To find the minimum value, calculate the coordinates of the vertex.

$$x = -\frac{b}{2a} = -\frac{-2}{2\left(\frac{1}{2}\right)} = 2 \quad \Rightarrow \quad f(2) = \frac{1}{2}(2)^2 - 2(2) - 1 = -3$$

▶ The minimum value is -3. So, the domain is all real numbers and the range is $y \geq -3$. The function is decreasing to the left of $x = 2$ and increasing to the right of $x = 2$.

Check

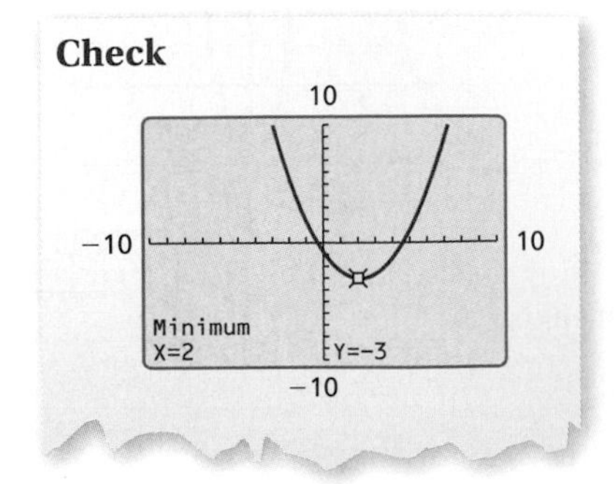

Monitoring Progress  Help in English and Spanish at *BigIdeasMath.com*

5. Find the minimum value or maximum value of (a) $f(x) = 4x^2 + 16x - 3$ and (b) $h(x) = -x^2 + 5x + 9$. Describe the domain and range of each function, and where each function is increasing and decreasing.

Graphing Quadratic Functions Using x-Intercepts

When the graph of a quadratic function has at least one x-intercept, the function can be written in **intercept form**, $f(x) = a(x - p)(x - q)$, where $a \neq 0$.

REMEMBER

An x-intercept of a graph is the x-coordinate of a point where the graph intersects the x-axis. It occurs where $f(x) = 0$.

Core Concept

Properties of the Graph of $f(x) = a(x - p)(x - q)$

- Because $f(p) = 0$ and $f(q) = 0$, p and q are the x-intercepts of the graph of the function.
- The axis of symmetry is halfway between $(p, 0)$ and $(q, 0)$. So, the axis of symmetry is $x = \frac{p + q}{2}$.
- The parabola opens up when $a > 0$ and opens down when $a < 0$.

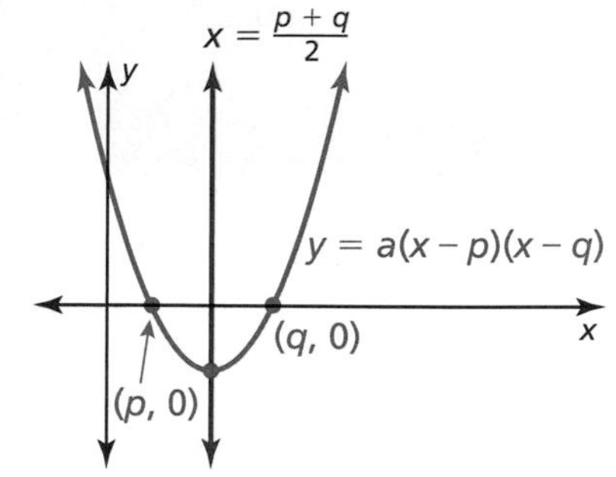

COMMON ERROR

Remember that the x-intercepts of the graph of $f(x) = a(x - p)(x - q)$ are p and q, not $-p$ and $-q$.

EXAMPLE 4 Graphing a Quadratic Function in Intercept Form

Graph $f(x) = -2(x + 3)(x - 1)$. Label the x-intercepts, vertex, and axis of symmetry.

SOLUTION

Step 1 Identify the x-intercepts. The x-intercepts are $p = -3$ and $q = 1$, so the parabola passes through the points $(-3, 0)$ and $(1, 0)$.

Step 2 Find the coordinates of the vertex.

$$x = \frac{p + q}{2} = \frac{-3 + 1}{2} = -1$$

$$f(-1) = -2(-1 + 3)(-1 - 1) = 8$$

So, the axis of symmetry is $x = -1$ and the vertex is $(-1, 8)$.

Step 3 Draw a parabola through the vertex and the points where the x-intercepts occur.

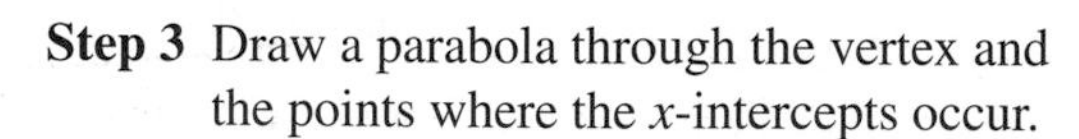

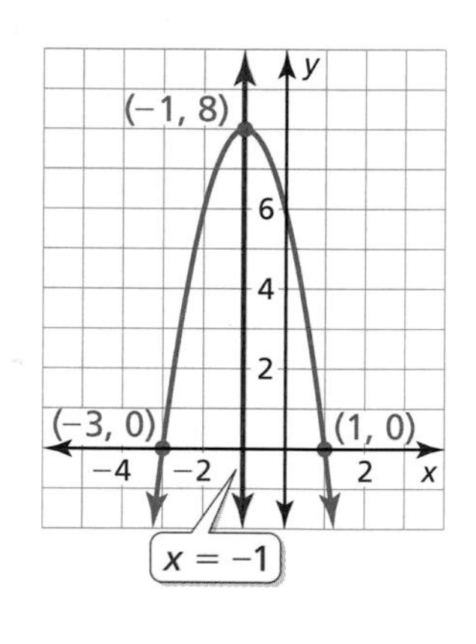

Check You can check your answer by generating a table of values for f on a graphing calculator.

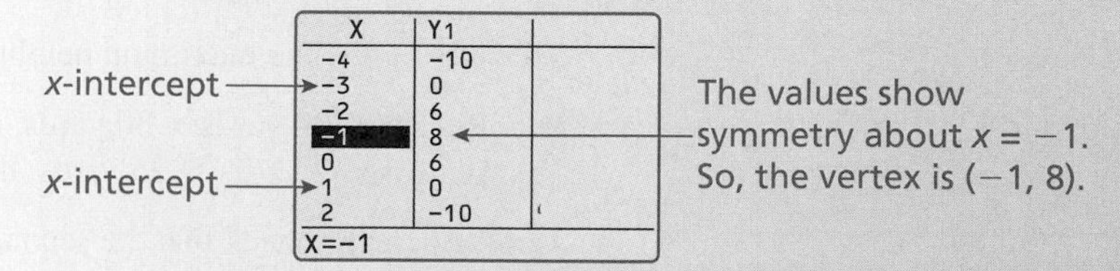

X	Y1
-4	-10
-3	0
-2	6
-1	8
0	6
1	0
2	-10

X=-1

Monitoring Progress

Help in English and Spanish at *BigIdeasMath.com*

Graph the function. Label the x-intercepts, vertex, and axis of symmetry.

6. $f(x) = -(x + 1)(x + 5)$

7. $g(x) = \frac{1}{4}(x - 6)(x - 2)$

Solving Real-Life Problems

EXAMPLE 5 Modeling with Mathematics

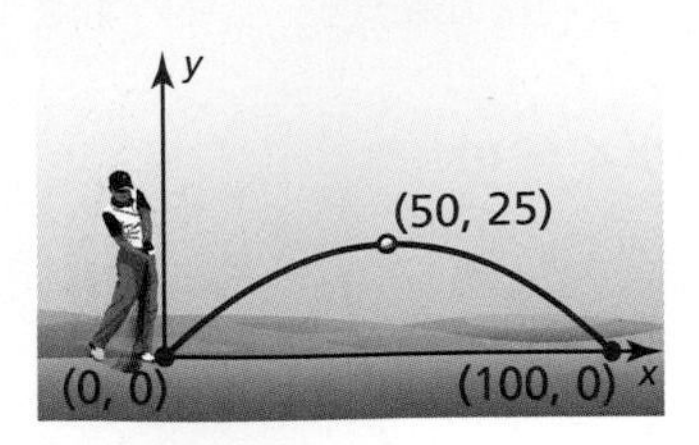

The parabola shows the path of your first golf shot, where x is the horizontal distance (in yards) and y is the corresponding height (in yards). The path of your second shot can be modeled by the function $f(x) = -0.02x(x - 80)$. Which shot travels farther before hitting the ground? Which travels higher?

SOLUTION

1. **Understand the Problem** You are given a graph and a function that represent the paths of two golf shots. You are asked to determine which shot travels farther before hitting the ground and which shot travels higher.
2. **Make a Plan** Determine how far each shot travels by interpreting the x-intercepts. Determine how high each shot travels by finding the maximum value of each function. Then compare the values.
3. **Solve the Problem**

First shot: The graph shows that the x-intercepts are 0 and 100. So, the ball travels 100 yards before hitting the ground.

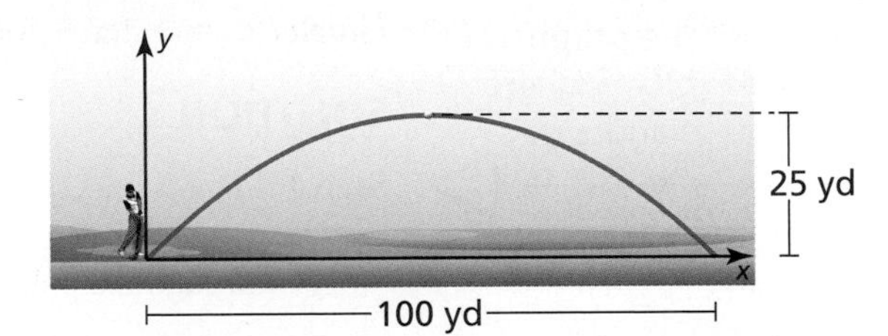

Because the axis of symmetry is halfway between (0, 0) and (100, 0), the axis of symmetry is $x = \frac{0 + 100}{2} = 50$. So, the vertex is (50, 25) and the maximum height is 25 yards.

Second shot: By rewriting the function in intercept form as $f(x) = -0.02(x - 0)(x - 80)$, you can see that $p = 0$ and $q = 80$. So, the ball travels 80 yards before hitting the ground.

To find the maximum height, find the coordinates of the vertex.

$$x = \frac{p + q}{2} = \frac{0 + 80}{2} = 40$$

$$f(40) = -0.02(40)(40 - 80) = 32$$

The maximum height of the second shot is 32 yards.

▶ Because 100 yards > 80 yards, the first shot travels farther.
Because 32 yards > 25 yards, the second shot travels higher.

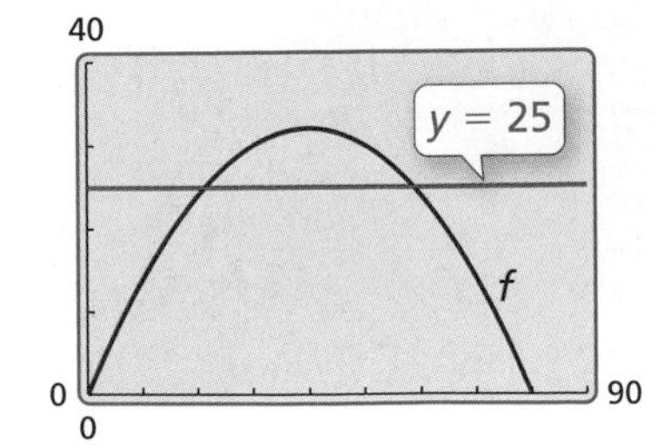

4. **Look Back** To check that the second shot travels higher, graph the function representing the path of the second shot and the line $y = 25$, which represents the maximum height of the first shot.

The graph rises above $y = 25$, so the second shot travels higher. ✓

Monitoring Progress

Help in English and Spanish at *BigIdeasMath.com*

8. **WHAT IF?** The graph of your third shot is a parabola through the origin that reaches a maximum height of 28 yards when $x = 45$. Compare the distance it travels before it hits the ground with the distances of the first two shots.

2.2 Exercises

Dynamic Solutions available at *BigIdeasMath.com*

Vocabulary and Core Concept Check

1. **WRITING** Explain how to determine whether a quadratic function will have a minimum value or a maximum value.

2. **WHICH ONE DOESN'T BELONG?** The graph of which function does *not* belong with the other three? Explain.

$f(x) = 3x^2 + 6x - 24$	$f(x) = 3x^2 + 24x - 6$
$f(x) = 3(x - 2)(x + 4)$	$f(x) = 3(x + 1)^2 - 27$

Monitoring Progress and Modeling with Mathematics

In Exercises 3–14, graph the function. Label the vertex and axis of symmetry. *(See Example 1.)*

3. $f(x) = (x - 3)^2$
4. $h(x) = (x + 4)^2$
5. $g(x) = (x + 3)^2 + 5$
6. $y = (x - 7)^2 - 1$
7. $y = -4(x - 2)^2 + 4$
8. $g(x) = 2(x + 1)^2 - 3$
9. $f(x) = -2(x - 1)^2 - 5$
10. $h(x) = 4(x + 4)^2 + 6$
11. $y = -\frac{1}{4}(x + 2)^2 + 1$
12. $y = \frac{1}{2}(x - 3)^2 + 2$
13. $f(x) = 0.4(x - 1)^2$
14. $g(x) = 0.75x^2 - 5$

ANALYZING RELATIONSHIPS In Exercises 15–18, use the axis of symmetry to match the equation with its graph.

15. $y = 2(x - 3)^2 + 1$
16. $y = (x + 4)^2 - 2$
17. $y = \frac{1}{2}(x + 1)^2 + 3$
18. $y = (x - 2)^2 - 1$

A.

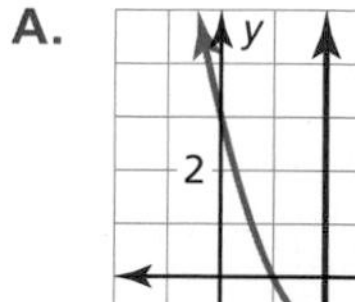

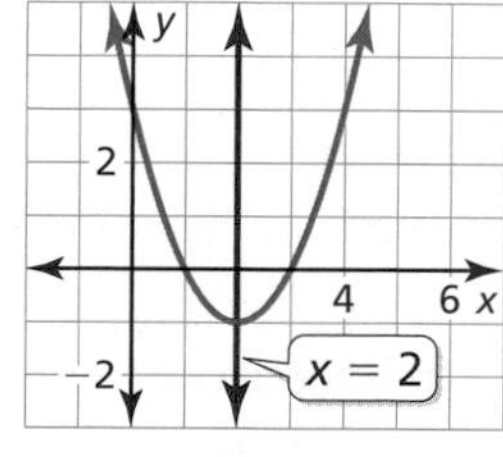

B.

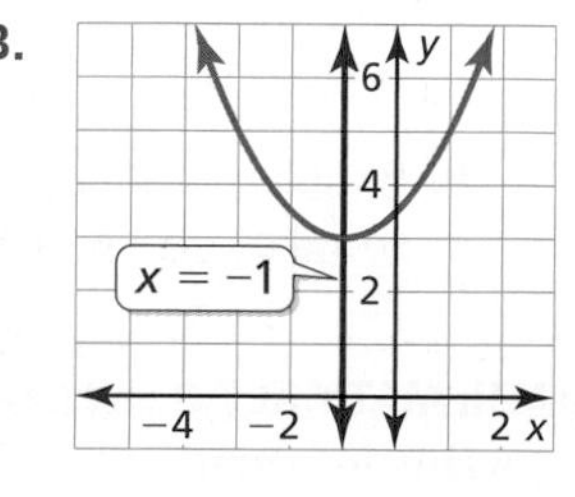

C.

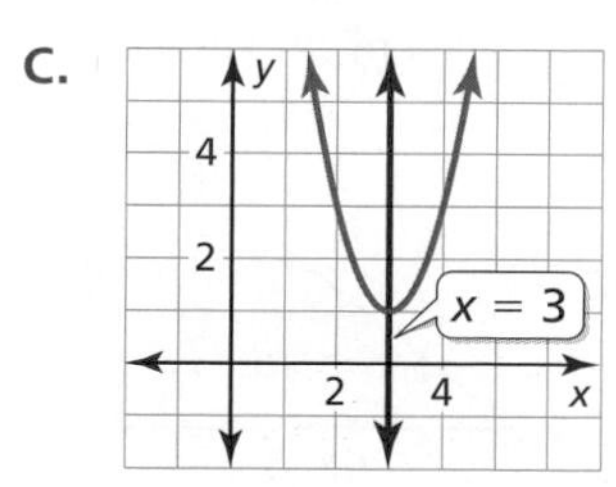

D.

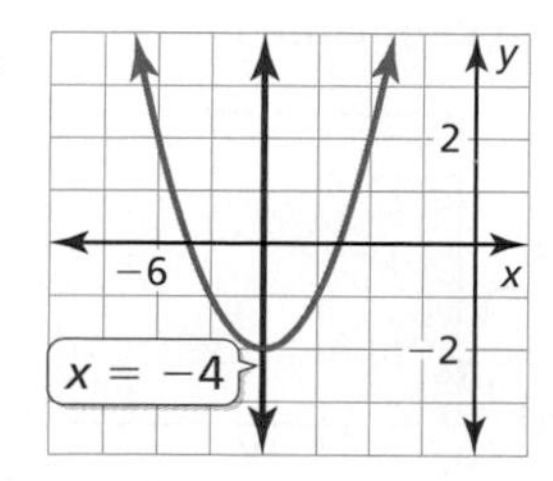

REASONING In Exercises 19 and 20, use the axis of symmetry to plot the reflection of each point and complete the parabola.

19.

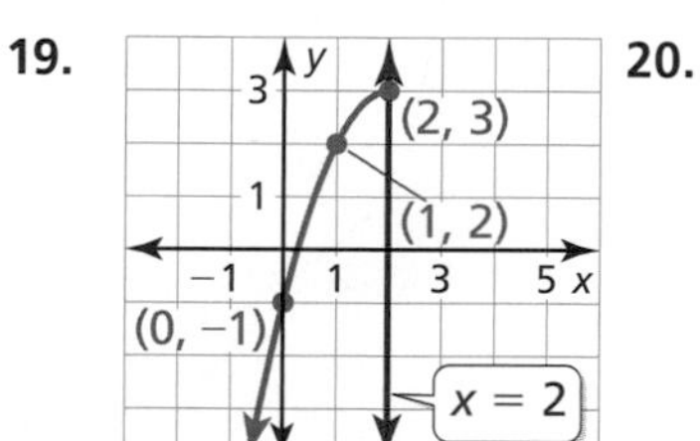

20.

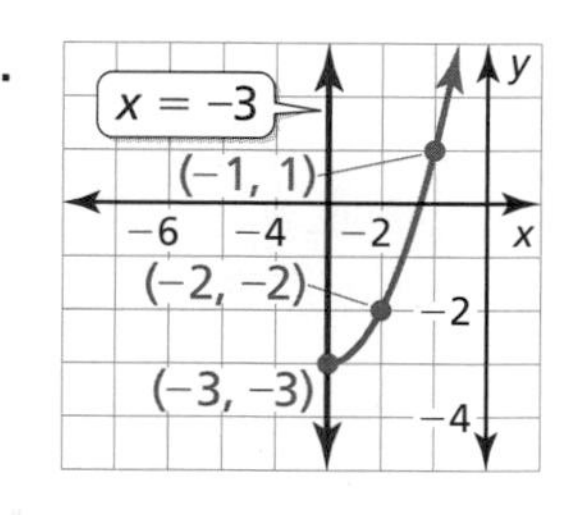

In Exercises 21–30, graph the function. Label the vertex and axis of symmetry. *(See Example 2.)*

21. $y = x^2 + 2x + 1$
22. $y = 3x^2 - 6x + 4$
23. $y = -4x^2 + 8x + 2$
24. $f(x) = -x^2 - 6x + 3$
25. $g(x) = -x^2 - 1$
26. $f(x) = 6x^2 - 5$
27. $g(x) = -1.5x^2 + 3x + 2$
28. $f(x) = 0.5x^2 + x - 3$
29. $y = \frac{3}{2}x^2 - 3x + 6$
30. $y = -\frac{5}{2}x^2 - 4x - 1$

31. **WRITING** Two quadratic functions have graphs with vertices $(2, 4)$ and $(2, -3)$. Explain why you can not use the axes of symmetry to distinguish between the two functions.

32. **WRITING** A quadratic function is increasing to the left of $x = 2$ and decreasing to the right of $x = 2$. Will the vertex be the highest or lowest point on the graph of the parabola? Explain.

ERROR ANALYSIS In Exercises 33 and 34, describe and correct the error in analyzing the graph of $y = 4x^2 + 24x - 7$.

33. 

The x-coordinate of the vertex is

$x = \frac{b}{2a} = \frac{24}{2(4)} = 3.$

34. 

The y-intercept of the graph is the value of c, which is 7.

MODELING WITH MATHEMATICS In Exercises 35 and 36, x is the horizontal distance (in feet) and y is the vertical distance (in feet). Find and interpret the coordinates of the vertex.

35. The path of a basketball thrown at an angle of 45° can be modeled by $y = -0.02x^2 + x + 6$.

36. The path of a shot put released at an angle of 35° can be modeled by $y = -0.01x^2 + 0.7x + 6$.

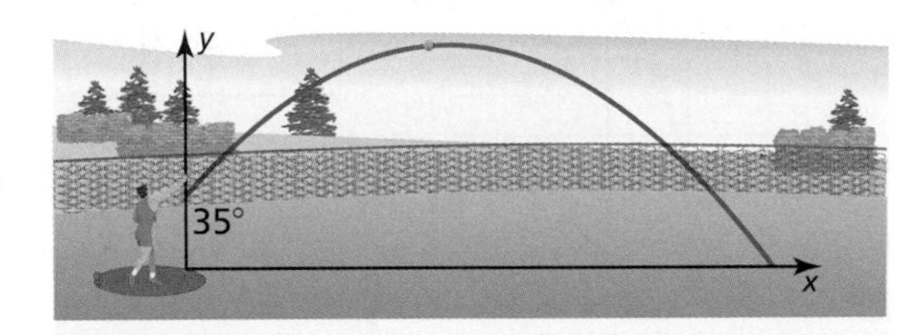

37. **ANALYZING EQUATIONS** The graph of which function has the same axis of symmetry as the graph of $y = x^2 + 2x + 2$?

Ⓐ $y = 2x^2 + 2x + 2$

Ⓑ $y = -3x^2 - 6x + 2$

Ⓒ $y = x^2 - 2x + 2$

Ⓓ $y = -5x^2 + 10x + 2$

38. **USING STRUCTURE** Which function represents the widest parabola? Explain your reasoning.

Ⓐ $y = 2(x + 3)^2$

Ⓑ $y = x^2 - 5$

Ⓒ $y = 0.5(x - 1)^2 + 1$

Ⓓ $y = -x^2 + 6$

In Exercises 39–48, find the minimum or maximum value of the function. Describe the domain and range of the function, and where the function is increasing and decreasing. *(See Example 3.)*

39. $y = 6x^2 - 1$

40. $y = 9x^2 + 7$

41. $y = -x^2 - 4x - 2$

42. $g(x) = -3x^2 - 6x + 5$

43. $f(x) = -2x^2 + 8x + 7$

44. $g(x) = 3x^2 + 18x - 5$

45. $h(x) = 2x^2 - 12x$

46. $h(x) = x^2 - 4x$

47. $y = \frac{1}{4}x^2 - 3x + 2$

48. $f(x) = \frac{3}{2}x^2 + 6x + 4$

49. **PROBLEM SOLVING** The path of a diver is modeled by the function $f(x) = -9x^2 + 9x + 1$, where $f(x)$ is the height of the diver (in meters) above the water and x is the horizontal distance (in meters) from the end of the diving board.

a. What is the height of the diving board?

b. What is the maximum height of the diver?

c. Describe where the diver is ascending and where the diver is descending.

50. **PROBLEM SOLVING** The engine torque y (in foot-pounds) of one model of car is given by $y = -3.75x^2 + 23.2x + 38.8$, where x is the speed (in thousands of revolutions per minute) of the engine.

a. Find the engine speed that maximizes torque. What is the maximum torque?

b. Explain what happens to the engine torque as the speed of the engine increases.

MATHEMATICAL CONNECTIONS In Exercises 51 and 52, write an equation for the area of the figure. Then determine the maximum possible area of the figure.

51. 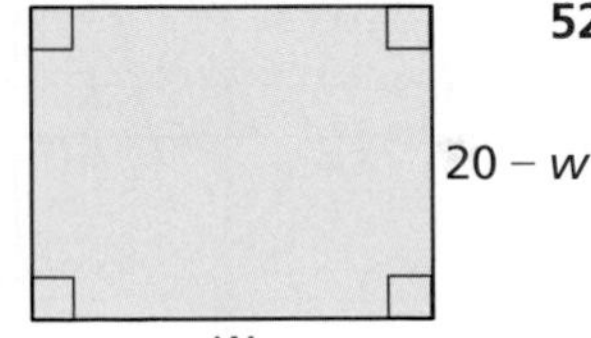

52. 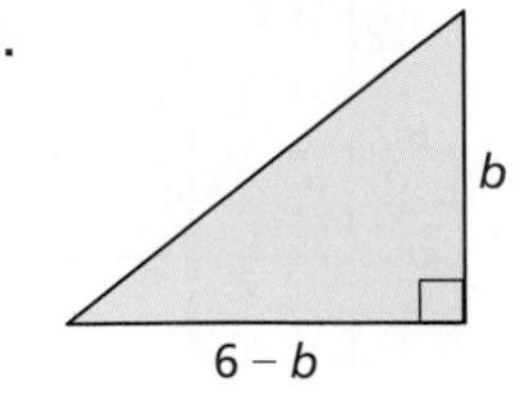

In Exercises 53–60, graph the function. Label the x-intercept(s), vertex, and axis of symmetry. *(See Example 4.)*

53. $y = (x + 3)(x - 3)$ **54.** $y = (x + 1)(x - 3)$

55. $y = 3(x + 2)(x + 6)$ **56.** $f(x) = 2(x - 5)(x - 1)$

57. $g(x) = -x(x + 6)$ **58.** $y = -4x(x + 7)$

59. $f(x) = -2(x - 3)^2$ **60.** $y = 4(x - 7)^2$

USING TOOLS In Exercises 61–64, identify the x-intercepts of the function and describe where the graph is increasing and decreasing. Use a graphing calculator to verify your answer.

61. $f(x) = \frac{1}{2}(x - 2)(x + 6)$

62. $y = \frac{3}{4}(x + 1)(x - 3)$

63. $g(x) = -4(x - 4)(x - 2)$

64. $h(x) = -5(x + 5)(x + 1)$

65. MODELING WITH MATHEMATICS A soccer player kicks a ball downfield. The height of the ball increases until it reaches a maximum height of 8 yards, 20 yards away from the player. A second kick is modeled by $y = x(0.4 - 0.008x)$. Which kick travels farther before hitting the ground? Which kick travels higher? *(See Example 5.)*

66. MODELING WITH MATHEMATICS Although a football field appears to be flat, some are actually shaped like a parabola so that rain runs off to both sides. The cross section of a field can be modeled by $y = -0.000234x(x - 160)$, where x and y are measured in feet. What is the width of the field? What is the maximum height of the surface of the field?

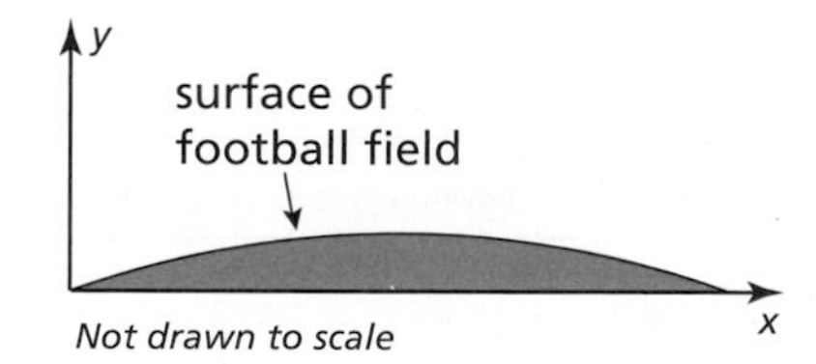

67. REASONING The points $(2, 3)$ and $(-4, 2)$ lie on the graph of a quadratic function. Determine whether you can use these points to find the axis of symmetry. If not, explain. If so, write the equation of the axis of symmetry.

68. OPEN-ENDED Write two different quadratic functions in intercept form whose graphs have the axis of symmetry $x = 3$.

69. PROBLEM SOLVING An online music store sells about 4000 songs each day when it charges \$1 per song. For each \$0.05 increase in price, about 80 fewer songs per day are sold. Use the verbal model and quadratic function to determine how much the store should charge per song to maximize daily revenue.

Revenue (dollars) = Price (dollars/song) • Sales (songs)

$R(x) = (1 + 0.05x) \cdot (4000 - 80x)$

70. PROBLEM SOLVING An electronics store sells 70 digital cameras per month at a price of \$320 each. For each \$20 decrease in price, about 5 more cameras per month are sold. Use the verbal model and quadratic function to determine how much the store should charge per camera to maximize monthly revenue.

Revenue (dollars) = Price (dollars/camera) • Sales (cameras)

$R(x) = (320 - 20x) \cdot (70 + 5x)$

71. DRAWING CONCLUSIONS Compare the graphs of the three quadratic functions. What do you notice? Rewrite the functions f and g in standard form to justify your answer.

$$f(x) = (x + 3)(x + 1)$$
$$g(x) = (x + 2)^2 - 1$$
$$h(x) = x^2 + 4x + 3$$

72. USING STRUCTURE Write the quadratic function $f(x) = x^2 + x - 12$ in intercept form. Graph the function. Label the x-intercepts, y-intercept, vertex, and axis of symmetry.

73. PROBLEM SOLVING A woodland jumping mouse hops along a parabolic path given by $y = -0.2x^2 + 1.3x$, where x is the mouse's horizontal distance traveled (in feet) and y is the corresponding height (in feet). Can the mouse jump over a fence that is 3 feet high? Justify your answer.

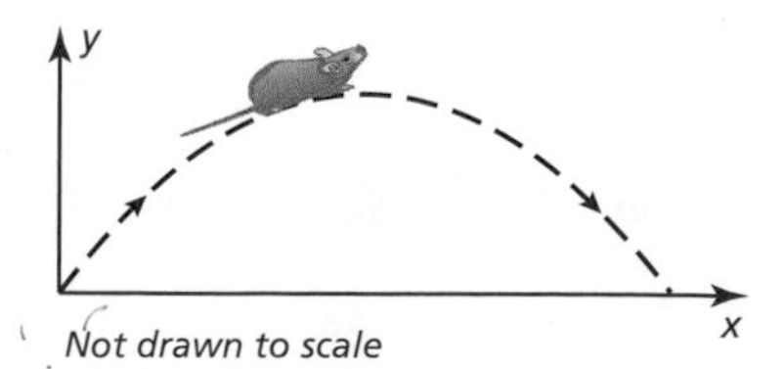

74. **HOW DO YOU SEE IT?** Consider the graph of the function $f(x) = a(x - p)(x - q)$.

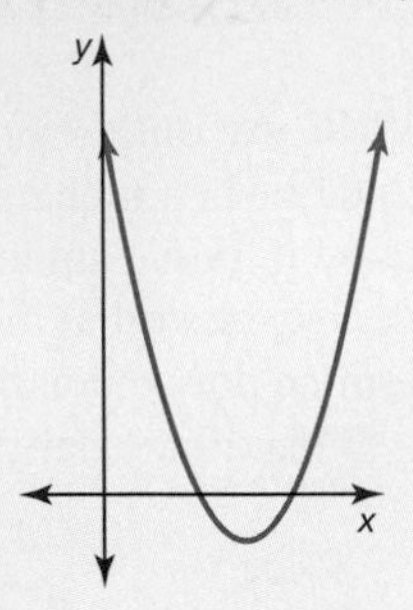

a. What does $f\left(\frac{p+q}{2}\right)$ represent in the graph?

b. If $a < 0$, how does your answer in part (a) change? Explain.

75. **MODELING WITH MATHEMATICS** The Gateshead Millennium Bridge spans the River Tyne. The arch of the bridge can be modeled by a parabola. The arch reaches a maximum height of 50 meters at a point roughly 63 meters across the river. Graph the curve of the arch. What are the domain and range? What do they represent in this situation?

76. **THOUGHT PROVOKING** You have 100 feet of fencing to enclose a rectangular garden. Draw three possible designs for the garden. Of these, which has the greatest area? Make a conjecture about the dimensions of the rectangular garden with the greatest possible area. Explain your reasoning.

77. **MAKING AN ARGUMENT** The point (1, 5) lies on the graph of a quadratic function with axis of symmetry $x = -1$. Your friend says the vertex could be the point (0, 5). Is your friend correct? Explain.

78. **CRITICAL THINKING** Find the y-intercept in terms of a, p, and q for the quadratic function $f(x) = a(x - p)(x - q)$.

79. **MODELING WITH MATHEMATICS** A kernel of popcorn contains water that expands when the kernel is heated, causing it to pop. The equations below represent the "popping volume" y (in cubic centimeters per gram) of popcorn with moisture content x (as a percent of the popcorn's weight).

Hot-air popping: $y = -0.761(x - 5.52)(x - 22.6)$

Hot-oil popping: $y = -0.652(x - 5.35)(x - 21.8)$

a. For hot-air popping, what moisture content maximizes popping volume? What is the maximum volume?

b. For hot-oil popping, what moisture content maximizes popping volume? What is the maximum volume?

c. Use a graphing calculator to graph both functions in the same coordinate plane. What are the domain and range of each function in this situation? Explain.

80. **ABSTRACT REASONING** A function is written in intercept form with $a > 0$. What happens to the vertex of the graph as a increases? as a approaches 0?

Maintaining Mathematical Proficiency

Reviewing what you learned in previous grades and lessons

Solve the equation. Check for extraneous solutions. *(Skills Review Handbook)*

81. $3\sqrt{x} - 6 = 0$

82. $2\sqrt{x - 4} - 2 = 2$

83. $\sqrt{5x} + 5 = 0$

84. $\sqrt{3x + 8} = \sqrt{x + 4}$

Solve the proportion. *(Skills Review Handbook)*

85. $\frac{1}{2} = \frac{x}{4}$

86. $\frac{2}{3} = \frac{x}{9}$

87. $\frac{-1}{4} = \frac{3}{x}$

88. $\frac{5}{2} = \frac{-20}{x}$

2.1–2.2 What Did You Learn?

Core Vocabulary

quadratic function, *p. 48*
parabola, *p. 48*
vertex of a parabola, *p. 50*
vertex form, *p. 50*
axis of symmetry, *p. 56*
standard form, *p. 56*
minimum value, *p. 58*
maximum value, *p. 58*
intercept form, *p. 59*

Core Concepts

Section 2.1

Horizontal Translations, *p. 48*
Vertical Translations, *p. 48*
Reflections in the x-Axis, *p. 49*
Reflections in the y-Axis, *p. 49*
Horizontal Stretches and Shrinks, *p. 49*
Vertical Stretches and Shrinks, *p. 49*

Section 2.2

Properties of the Graph of $f(x) = ax^2 + bx + c$, *p. 57*
Minimum and Maximum Values, *p. 58*
Properties of the Graph of $f(x) = a(x - p)(x - q)$, *p. 59*

Mathematical Practices

1. Why does the height you found in Exercise 44 on page 53 make sense in the context of the situation?
2. How can you effectively communicate your preference in methods to others in Exercise 47 on page 54?
3. How can you use technology to deepen your understanding of the concepts in Exercise 79 on page 64?

Study Skills

Using the Features of Your Textbook to Prepare for Quizzes and Tests

- Read and understand the core vocabulary and the contents of the Core Concept boxes.
- Review the Examples and the Monitoring Progress questions. Use the tutorials at *BigIdeasMath.com* for additional help.
- Review previously completed homework assignments.

2.1–2.2 Quiz

Describe the transformation of $f(x) = x^2$ represented by g. *(Section 2.1)*

1.

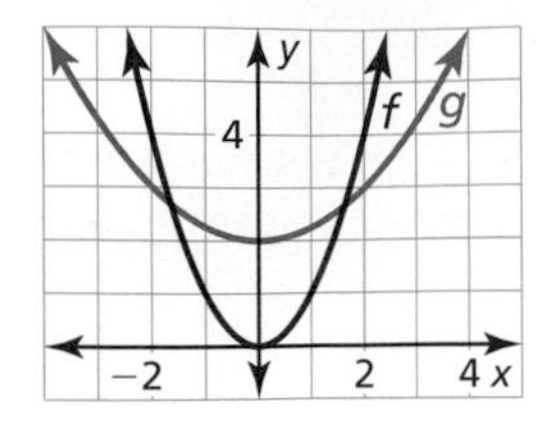

2.

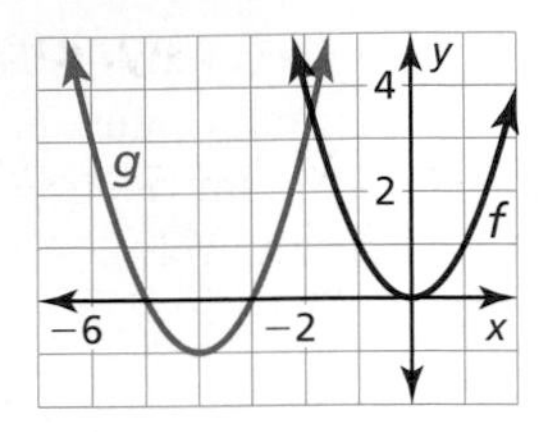

3.

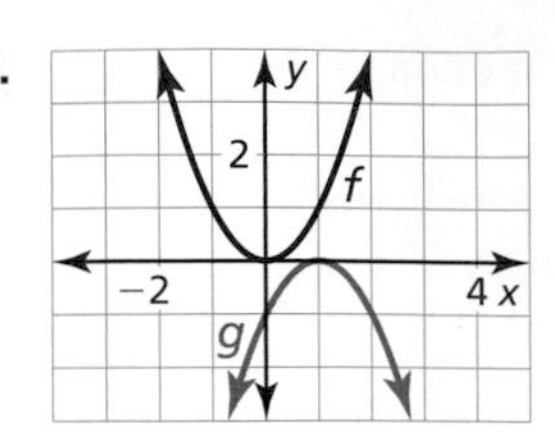

Write a rule for g and identify the vertex. *(Section 2.1)*

4. Let g be a translation 2 units up followed by a reflection in the x-axis and a vertical stretch by a factor of 6 of the graph of $f(x) = x^2$.

5. Let g be a translation 1 unit left and 6 units down, followed by a vertical shrink by a factor of $\frac{1}{2}$ of the graph of $f(x) = 3(x + 2)^2$.

6. Let g be a horizontal shrink by a factor of $\frac{1}{4}$, followed by a translation 1 unit up and 3 units right of the graph of $f(x) = (2x + 1)^2 - 11$.

Graph the function. Label the vertex and axis of symmetry. *(Section 2.2)*

7. $f(x) = 2(x - 1)^2 - 5$

8. $h(x) = 3x^2 + 6x - 2$

9. $f(x) = 7 - 8x - x^2$

Find the x-intercepts of the graph of the function. Then describe where the function is increasing and decreasing. *(Section 2.2)*

10. $g(x) = -3(x + 2)(x + 4)$

11. $g(x) = \frac{1}{2}(x - 5)(x + 1)$

12. $f(x) = 0.4x(x - 6)$

13. A grasshopper can jump incredible distances, up to 20 times its length. The height (in inches) of the jump above the ground of a 1-inch-long grasshopper is given by $h(x) = -\frac{1}{20}x^2 + x$, where x is the horizontal distance (in inches) of the jump. When the grasshopper jumps off a rock, it lands on the ground 2 inches farther. Write a function that models the new path of the jump. *(Section 2.1)*

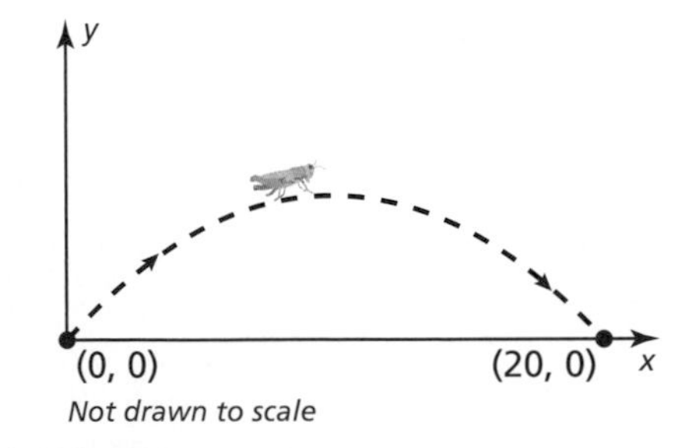

14. A passenger on a stranded lifeboat shoots a distress flare into the air. The height (in feet) of the flare above the water is given by $f(t) = -16t(t - 8)$, where t is time (in seconds) since the flare was shot. The passenger shoots a second flare, whose path is modeled in the graph. Which flare travels higher? Which remains in the air longer? Justify your answer. *(Section 2.2)*

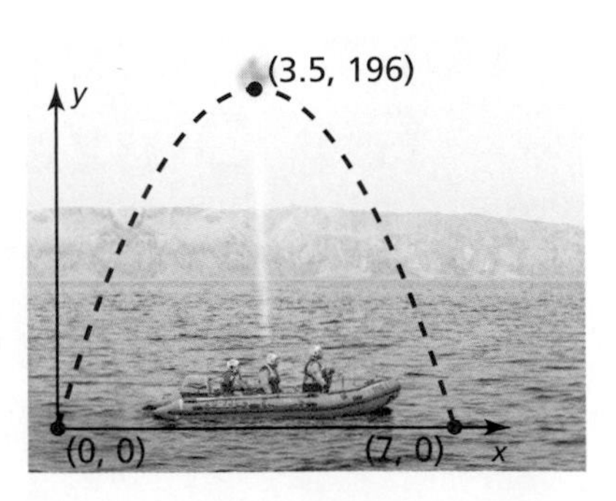

2.3 Focus of a Parabola

Essential Question What is the focus of a parabola?

EXPLORATION 1 Analyzing Satellite Dishes

Work with a partner. Vertical rays enter a satellite dish whose cross section is a parabola. When the rays hit the parabola, they reflect at the same angle at which they entered. (See Ray 1 in the figure.)

a. Draw the reflected rays so that they intersect the y-axis.

b. What do the reflected rays have in common?

c. The optimal location for the receiver of the satellite dish is at a point called the *focus* of the parabola. Determine the location of the focus. Explain why this makes sense in this situation.

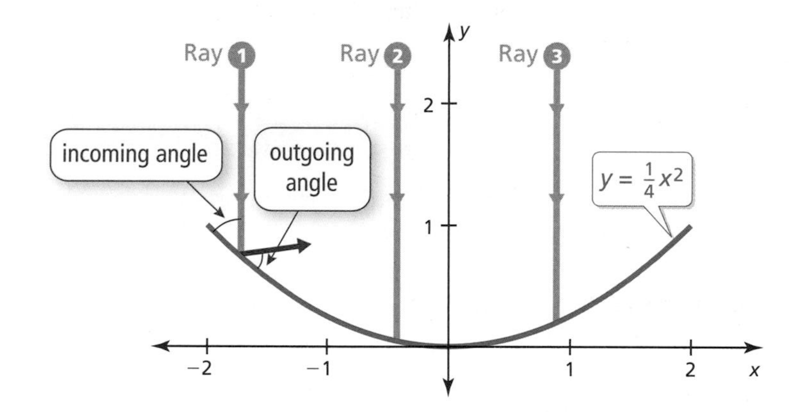

CONSTRUCTING VIABLE ARGUMENTS

To be proficient in math, you need to make conjectures and build logical progressions of statements to explore the truth of your conjectures.

EXPLORATION 2 Analyzing Spotlights

Work with a partner. Beams of light are coming from the bulb in a spotlight, located at the focus of the parabola. When the beams hit the parabola, they reflect at the same angle at which they hit. (See Beam 1 in the figure.) Draw the reflected beams. What do they have in common? Would you consider this to be the optimal result? Explain.

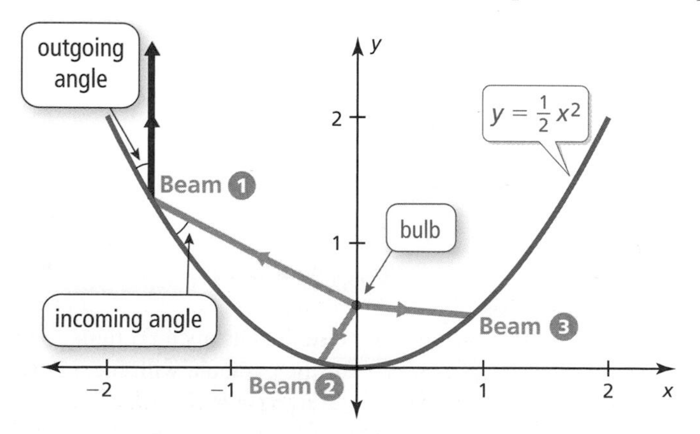

Communicate Your Answer

3. What is the focus of a parabola?

4. Describe some of the properties of the focus of a parabola.

2.3 Lesson

What You Will Learn

- Explore the focus and the directrix of a parabola.
- Write equations of parabolas.
- Solve real-life problems.

Core Vocabulary

focus, *p. 68*
directrix, *p. 68*

Previous
perpendicular
Distance Formula
congruent

Exploring the Focus and Directrix

Previously, you learned that the graph of a quadratic function is a parabola that opens up or down. A parabola can also be defined as the set of all points (x, y) in a plane that are equidistant from a fixed point called the **focus** and a fixed line called the **directrix**.

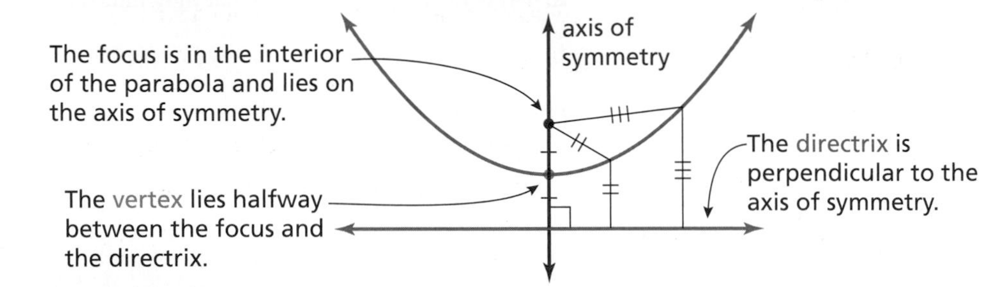
The focus is in the interior of the parabola and lies on the axis of symmetry.
axis of symmetry
The directrix is perpendicular to the axis of symmetry.
The vertex lies halfway between the focus and the directrix.

STUDY TIP

The distance from a point to a line is defined as the length of the perpendicular segment from the point to the line.

EXAMPLE 1 Using the Distance Formula to Write an Equation

Use the Distance Formula to write an equation of the parabola with focus $F(0, 2)$ and directrix $y = -2$.

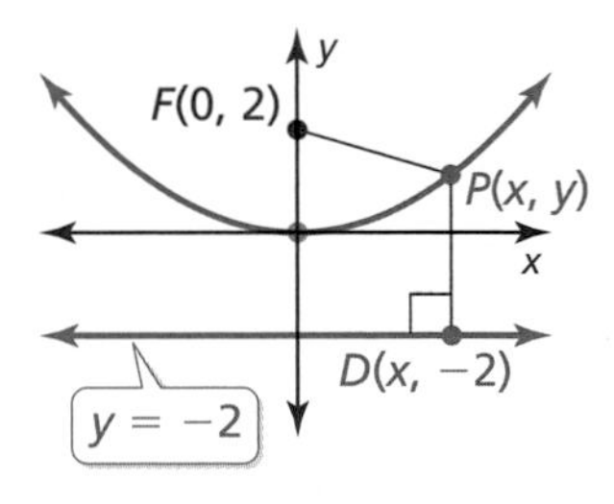
y
F(0, 2)
P(x, y)
x
D(x, −2)
y = −2

SOLUTION

Notice the line segments drawn from point F to point P and from point P to point D. By the definition of a parabola, these line segments must be congruent.

$PD = PF$	Definition of a parabola
$\sqrt{(x - x_1)^2 + (y - y_1)^2} = \sqrt{(x - x_2)^2 + (y - y_2)^2}$	Distance Formula
$\sqrt{(x - x)^2 + (y - (-2))^2} = \sqrt{(x - 0)^2 + (y - 2)^2}$	Substitute for x_1, y_1, x_2, and y_2.
$\sqrt{(y + 2)^2} = \sqrt{x^2 + (y - 2)^2}$	Simplify.
$(y + 2)^2 = x^2 + (y - 2)^2$	Square each side.
$y^2 + 4y + 4 = x^2 + y^2 - 4y + 4$	Expand.
$8y = x^2$	Combine like terms.
$y = \frac{1}{8}x^2$	Divide each side by 8.

Monitoring Progress

Help in English and Spanish at *BigIdeasMath.com*

1. Use the Distance Formula to write an equation of the parabola with focus $F(0, -3)$ and directrix $y = 3$.

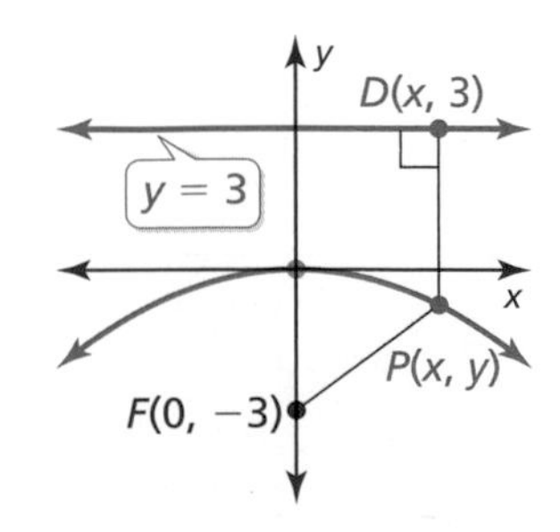
y
D(x, 3)
y = 3
x
P(x, y)
F(0, −3)

You can derive the equation of a parabola that opens up or down with vertex $(0, 0)$, focus $(0, p)$, and directrix $y = -p$ using the procedure in Example 1.

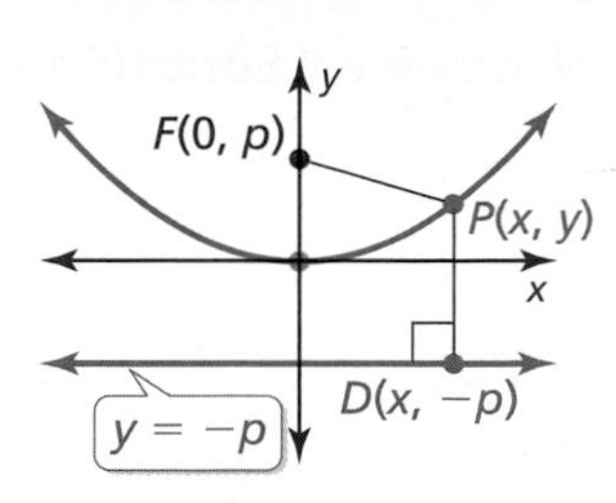

$$\sqrt{(x - x)^2 + (y - (-p))^2} = \sqrt{(x - 0)^2 + (y - p)^2}$$

$$(y + p)^2 = x^2 + (y - p)^2$$

$$y^2 + 2py + p^2 = x^2 + y^2 - 2py + p^2$$

$$4py = x^2$$

$$y = \frac{1}{4p}x^2$$

The focus and directrix each lie $|p|$ units from the vertex. Parabolas can also open left or right, in which case the equation has the form $x = \frac{1}{4p}y^2$ when the vertex is $(0, 0)$.

LOOKING FOR STRUCTURE

Notice that $y = \frac{1}{4p}x^2$ is of the form $y = ax^2$. So, changing the value of p vertically stretches or shrinks the parabola.

Core Concept

Standard Equations of a Parabola with Vertex at the Origin

Vertical axis of symmetry ($x = 0$)

Equation: $y = \frac{1}{4p}x^2$

Focus: $(0, p)$

Directrix: $y = -p$

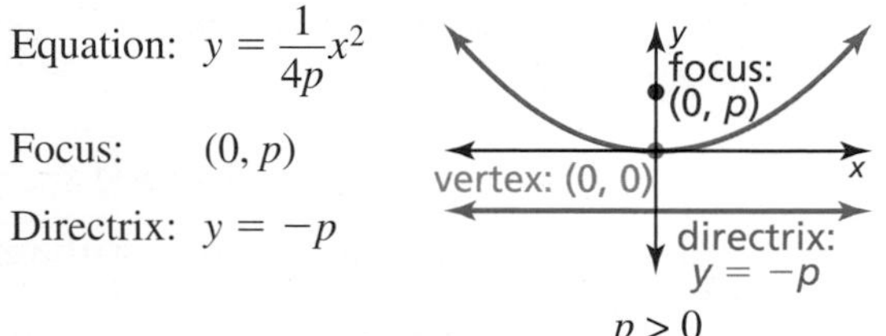

$p > 0$

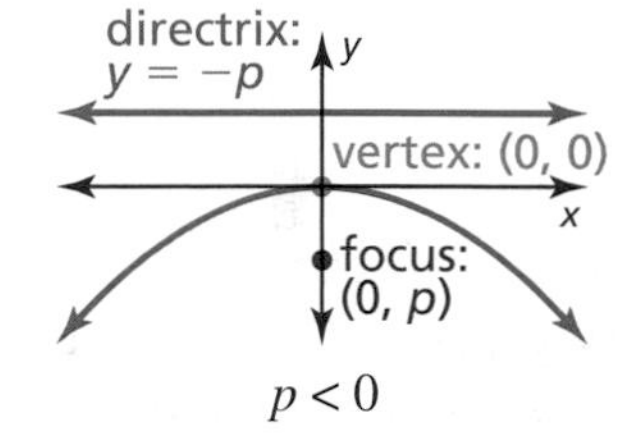

$p < 0$

Horizontal axis of symmetry ($y = 0$)

Equation: $x = \frac{1}{4p}y^2$

Focus: $(p, 0)$

Directrix: $x = -p$

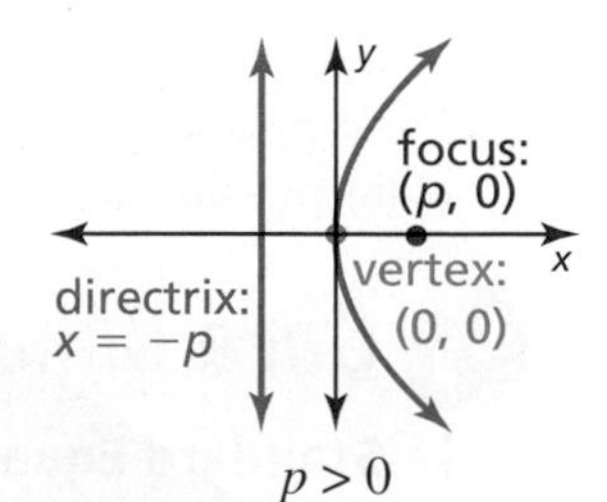

$p > 0$

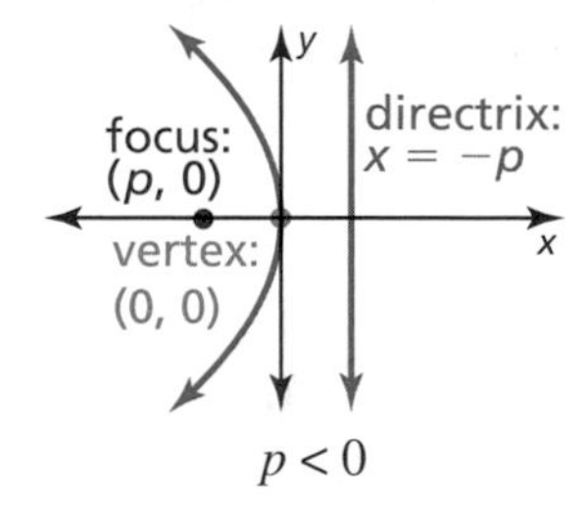

$p < 0$

STUDY TIP

Notice that parabolas opening left or right do *not* represent functions.

EXAMPLE 2 Graphing an Equation of a Parabola

Identify the focus, directrix, and axis of symmetry of $-4x = y^2$. Graph the equation.

SOLUTION

Step 1 Rewrite the equation in standard form.

$-4x = y^2$ Write the original equation.

$x = -\frac{1}{4}y^2$ Divide each side by −4.

Step 2 Identify the focus, directrix, and axis of symmetry. The equation has the form $x = \frac{1}{4p}y^2$, where $p = -1$. The focus is $(p, 0)$, or $(-1, 0)$. The directrix is $x = -p$, or $x = 1$. Because y is squared, the axis of symmetry is the x-axis.

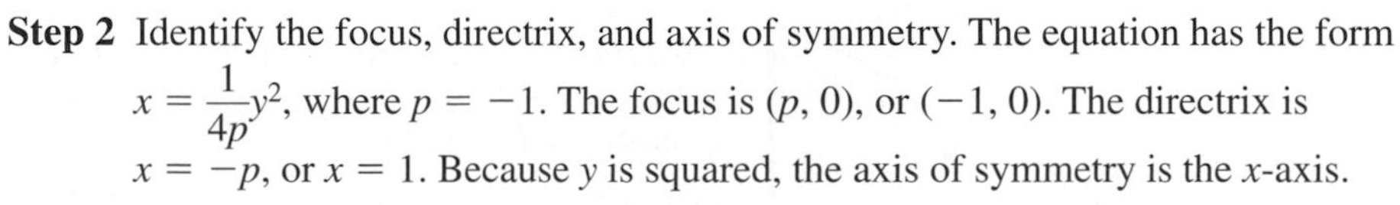

Step 3 Use a table of values to graph the equation. Notice that it is easier to substitute y-values and solve for x. Opposite y-values result in the same x-value.

y	0	±1	±2	±3	±4
x	0	−0.25	−1	−2.25	−4

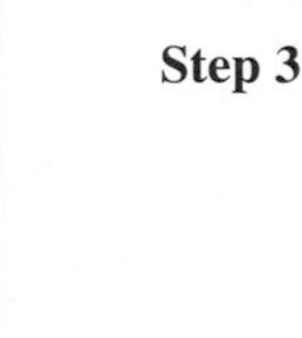

Writing Equations of Parabolas

EXAMPLE 3 Writing an Equation of a Parabola

Write an equation of the parabola shown.

SOLUTION

Because the vertex is at the origin and the axis of symmetry is vertical, the equation has the form $y = \frac{1}{4p}x^2$. The directrix is $y = -p = 3$, so $p = -3$. Substitute -3 for p to write an equation of the parabola.

$$y = \frac{1}{4(-3)}x^2 = -\frac{1}{12}x^2$$

▶ So, an equation of the parabola is $y = -\frac{1}{12}x^2$.

Monitoring Progress  Help in English and Spanish at *BigIdeasMath.com*

Identify the focus, directrix, and axis of symmetry of the parabola. Then graph the equation.

2. $y = 0.5x^2$ **3.** $-y = x^2$ **4.** $y^2 = 6x$

Write an equation of the parabola with vertex at (0, 0) and the given directrix or focus.

5. directrix: $x = -3$ **6.** focus: $(-2, 0)$ **7.** focus: $\left(0, \frac{3}{2}\right)$

The vertex of a parabola is not always at the origin. As in previous transformations, adding a value to the input or output of a function translates its graph.

Core Concept

Standard Equations of a Parabola with Vertex at (*h*, *k*)

Vertical axis of symmetry ($x = h$)

Equation: $y = \frac{1}{4p}(x - h)^2 + k$

Focus: $(h, k + p)$

Directrix: $y = k - p$

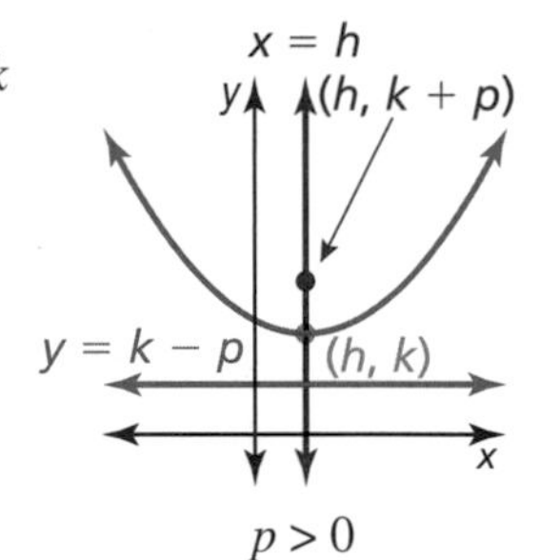

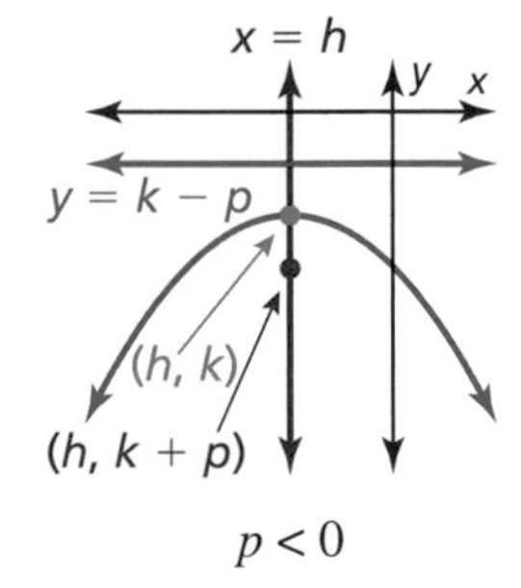

STUDY TIP

The standard form for a vertical axis of symmetry looks like vertex form. To remember the standard form for a horizontal axis of symmetry, switch *x* and *y*, and *h* and *k*.

Horizontal axis of symmetry ($y = k$)

Equation: $x = \frac{1}{4p}(y - k)^2 + h$

Focus: $(h + p, k)$

Directrix: $x = h - p$

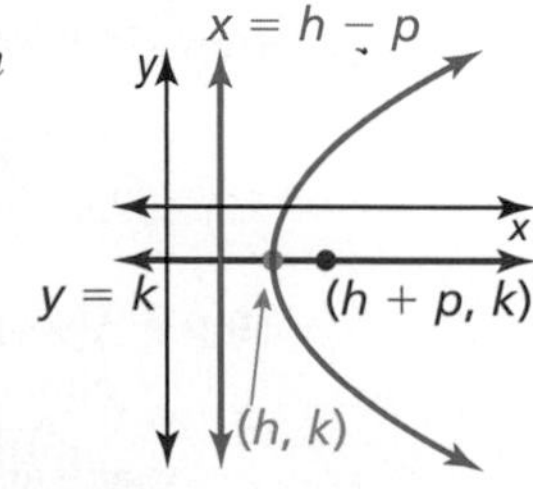

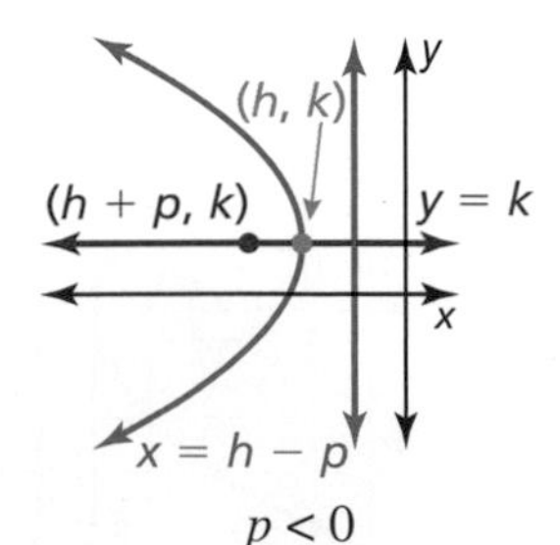

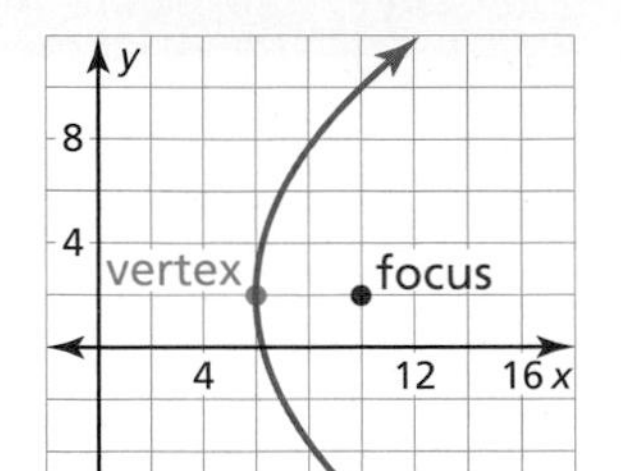

EXAMPLE 4 Writing an Equation of a Translated Parabola

Write an equation of the parabola shown.

SOLUTION

Because the vertex is not at the origin and the axis of symmetry is horizontal, the equation has the form $x = \frac{1}{4p}(y - k)^2 + h$. The vertex (h, k) is $(6, 2)$ and the focus $(h + p, k)$ is $(10, 2)$, so $h = 6$, $k = 2$, and $p = 4$. Substitute these values to write an equation of the parabola.

$$x = \frac{1}{4(4)}(y - 2)^2 + 6 = \frac{1}{16}(y - 2)^2 + 6$$

▶ So, an equation of the parabola is $x = \frac{1}{16}(y - 2)^2 + 6$.

Solving Real-Life Problems

Parabolic reflectors have cross sections that are parabolas. Incoming sound, light, or other energy that arrives at a parabolic reflector parallel to the axis of symmetry is directed to the focus (Diagram 1). Similarly, energy that is emitted from the focus of a parabolic reflector and then strikes the reflector is directed parallel to the axis of symmetry (Diagram 2).

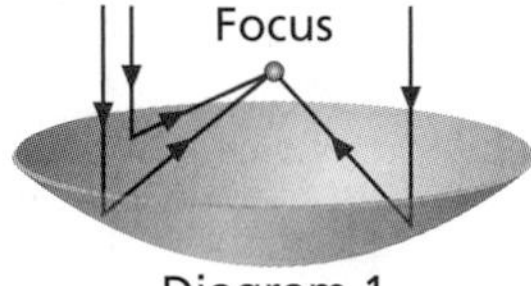

Diagram 1

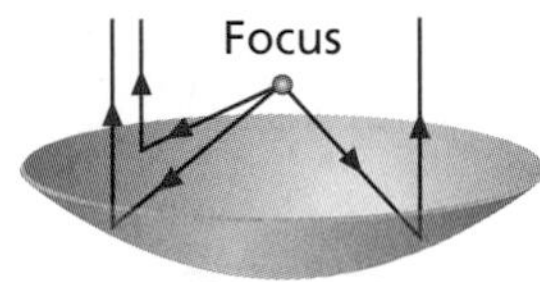

Diagram 2

EXAMPLE 5 Solving a Real-Life Problem

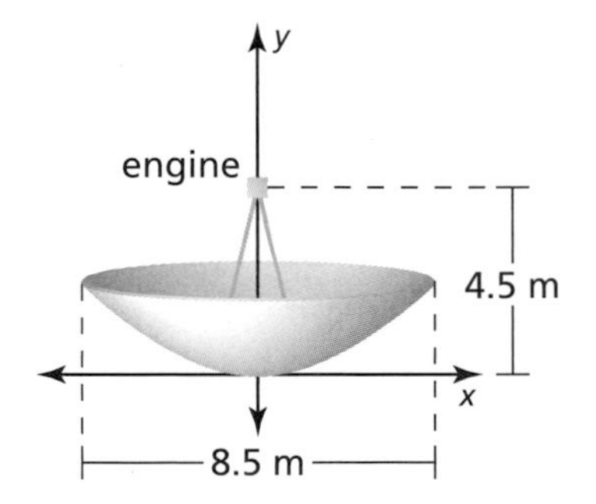

An electricity-generating dish uses a parabolic reflector to concentrate sunlight onto a high-frequency engine located at the focus of the reflector. The sunlight heats helium to 650°C to power the engine. Write an equation that represents the cross section of the dish shown with its vertex at (0, 0). What is the depth of the dish?

SOLUTION

Because the vertex is at the origin, and the axis of symmetry is vertical, the equation has the form $y = \frac{1}{4p}x^2$. The engine is at the focus, which is 4.5 meters above the vertex. So, $p = 4.5$. Substitute 4.5 for p to write the equation.

$$y = \frac{1}{4(4.5)}x^2 = \frac{1}{18}x^2$$

The depth of the dish is the y-value at the dish's outside edge. The dish extends $\frac{8.5}{2} = 4.25$ meters to either side of the vertex (0, 0), so find y when $x = 4.25$.

$$y = \frac{1}{18}(4.25)^2 \approx 1$$

▶ The depth of the dish is about 1 meter.

Monitoring Progress  Help in English and Spanish at *BigIdeasMath.com*

8. Write an equation of a parabola with vertex $(-1, 4)$ and focus $(-1, 2)$.

9. A parabolic microwave antenna is 16 feet in diameter. Write an equation that represents the cross section of the antenna with its vertex at (0, 0) and its focus 10 feet to the right of the vertex. What is the depth of the antenna?

2.3 Exercises

Dynamic Solutions available at *BigIdeasMath.com*

Vocabulary and Core Concept Check

1. **COMPLETE THE SENTENCE** A parabola is the set of all points in a plane equidistant from a fixed point called the _______ and a fixed line called the ___________ .

2. **WRITING** Explain how to find the coordinates of the focus of a parabola with vertex $(0, 0)$ and directrix $y = 5$.

Monitoring Progress and Modeling with Mathematics

In Exercises 3–10, use the Distance Formula to write an equation of the parabola. *(See Example 1.)*

3.
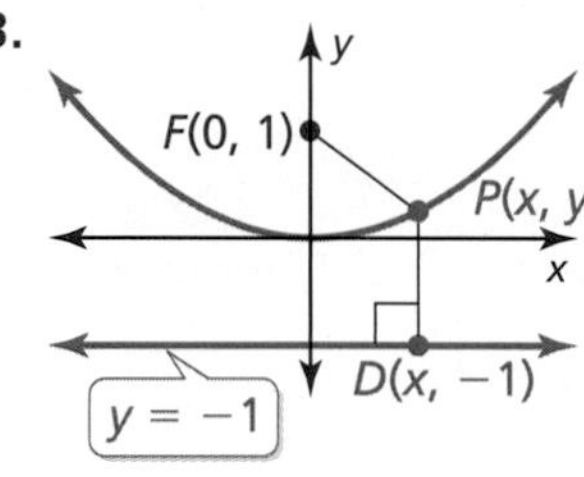

4.
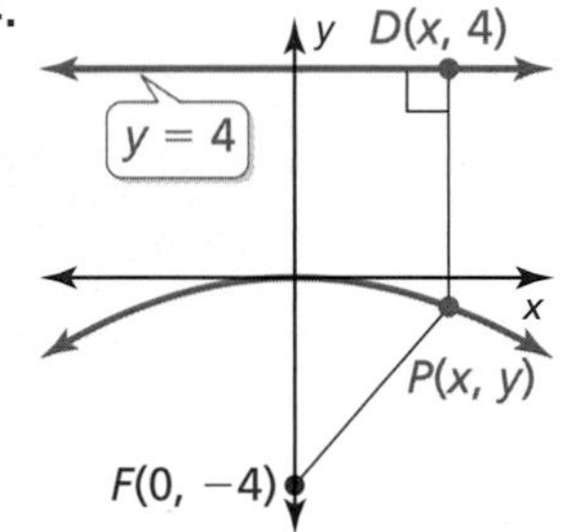

5. focus: $(0, -2)$
directrix: $y = 2$

6. directrix: $y = 7$
focus: $(0, -7)$

7. vertex: $(0, 0)$
directrix: $y = -6$

8. vertex: $(0, 0)$
focus: $(0, 5)$

9. vertex: $(0, 0)$
focus: $(0, -10)$

10. vertex: $(0, 0)$
directrix: $y = -9$

11. **ANALYZING RELATIONSHIPS** Which of the given characteristics describe parabolas that open down? Explain your reasoning.

(A) focus: $(0, -6)$
directrix: $y = 6$

(B) focus: $(0, -2)$
directrix: $y = 2$

(C) focus: $(0, 6)$
directrix: $y = -6$

(D) focus: $(0, -1)$
directrix: $y = 1$

12. **REASONING** Which of the following are possible coordinates of the point P in the graph shown? Explain.

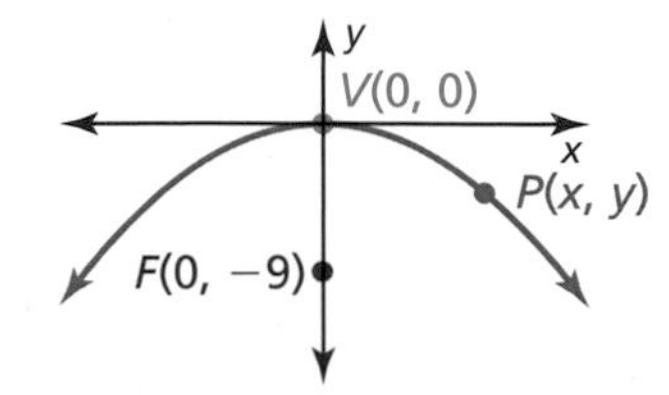

(A) $(-6, -1)$ (B) $\left(3, -\frac{1}{4}\right)$ (C) $\left(4, -\frac{4}{9}\right)$

(D) $\left(1, \frac{1}{36}\right)$ (E) $(6, -1)$ (F) $\left(2, -\frac{1}{18}\right)$

In Exercises 13–20, identify the focus, directrix, and axis of symmetry of the parabola. Graph the equation. *(See Example 2.)*

13. $y = \frac{1}{8}x^2$

14. $y = -\frac{1}{12}x^2$

15. $x = -\frac{1}{20}y^2$

16. $x = \frac{1}{24}y^2$

17. $y^2 = 16x$

18. $-x^2 = 48y$

19. $6x^2 + 3y = 0$

20. $8x^2 - y = 0$

ERROR ANALYSIS In Exercises 21 and 22, describe and correct the error in graphing the parabola.

21.
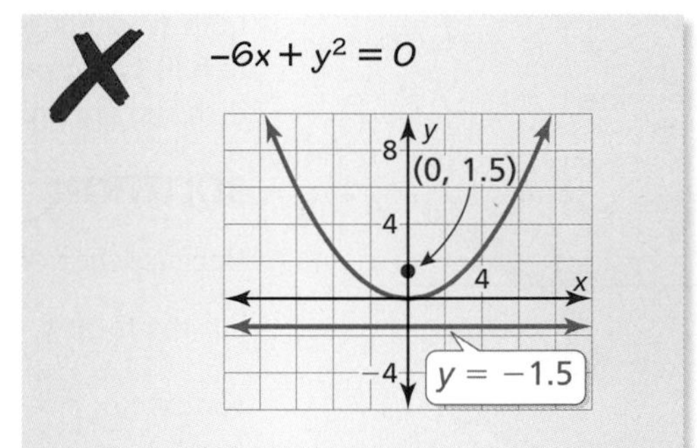

22.
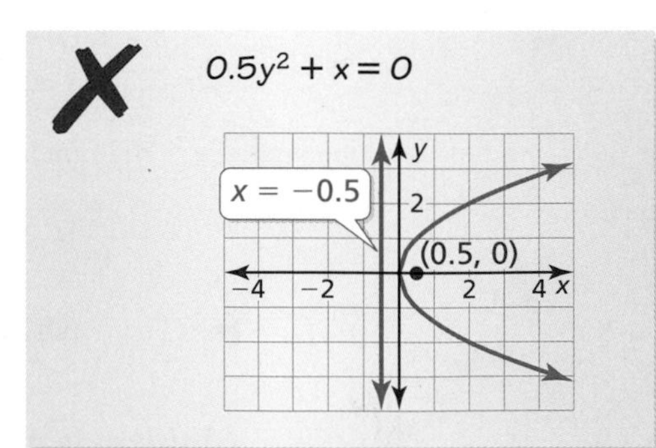

23. **ANALYZING EQUATIONS** The cross section (with units in inches) of a parabolic satellite dish can be modeled by the equation $y = \frac{1}{38}x^2$. How far is the receiver from the vertex of the cross section? Explain.

24. ANALYZING EQUATIONS The cross section (with units in inches) of a parabolic spotlight can be modeled by the equation $x = \frac{1}{20}y^2$. How far is the bulb from the vertex of the cross section? Explain.

In Exercises 25–28, write an equation of the parabola shown. *(See Example 3.)*

25.

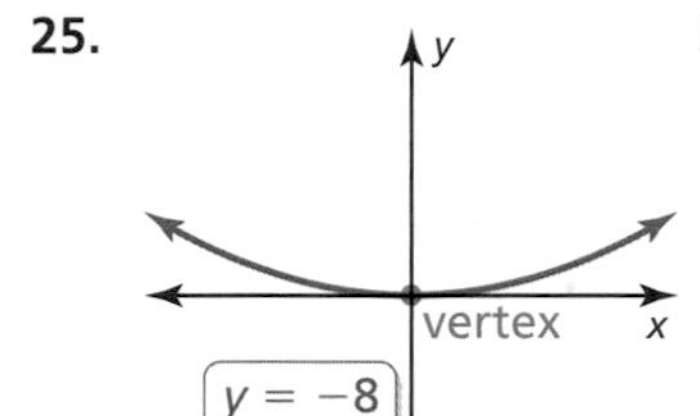

26.

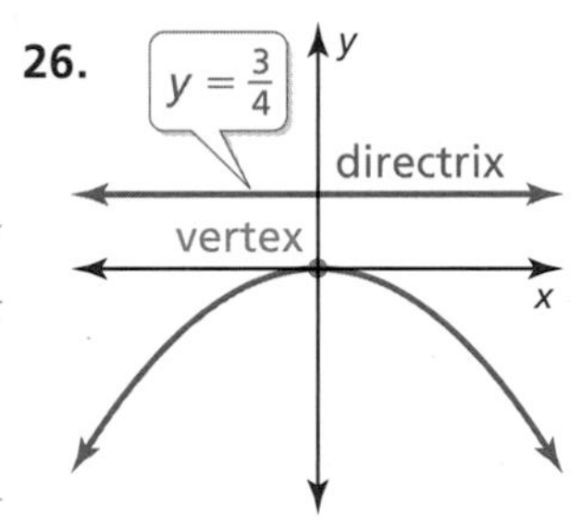

27.

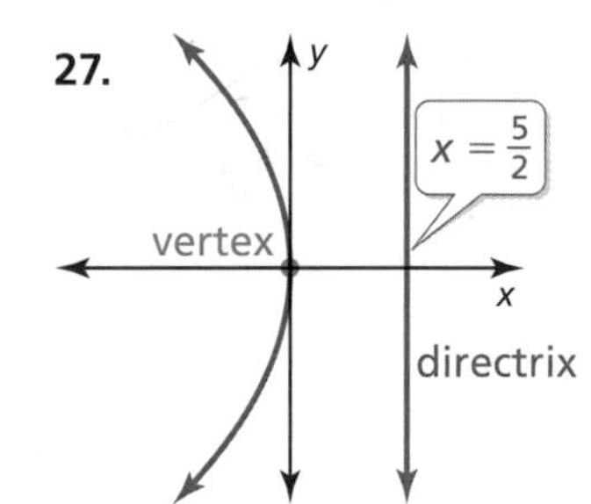

28.

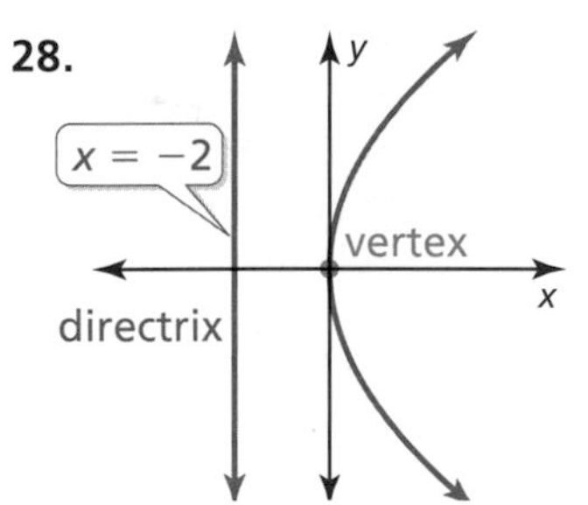

In Exercises 29–36, write an equation of the parabola with the given characteristics.

29. focus: $(3, 0)$
directrix: $x = -3$

30. focus: $\left(\frac{2}{3}, 0\right)$
directrix: $x = -\frac{2}{3}$

31. directrix: $x = -10$
vertex: $(0, 0)$

32. directrix: $y = \frac{8}{3}$
vertex: $(0, 0)$

33. focus: $\left(0, -\frac{5}{3}\right)$
directrix: $y = \frac{5}{3}$

34. focus: $\left(0, \frac{5}{4}\right)$
directrix: $y = -\frac{5}{4}$

35. focus: $\left(0, \frac{6}{7}\right)$
vertex: $(0, 0)$

36. focus: $\left(-\frac{4}{5}, 0\right)$
vertex: $(0, 0)$

In Exercises 37–40, write an equation of the parabola shown. *(See Example 4.)*

37.

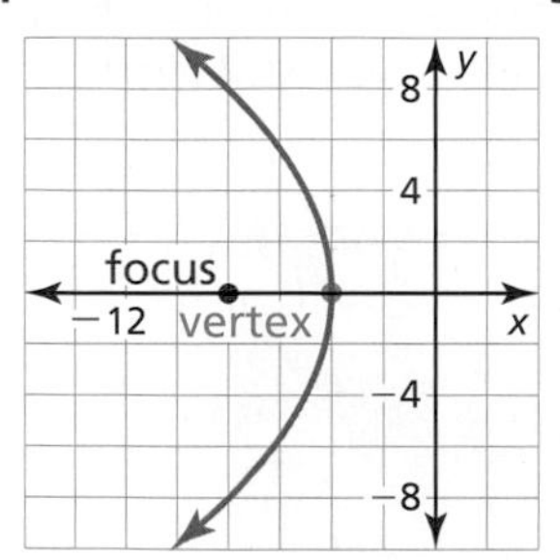

38.

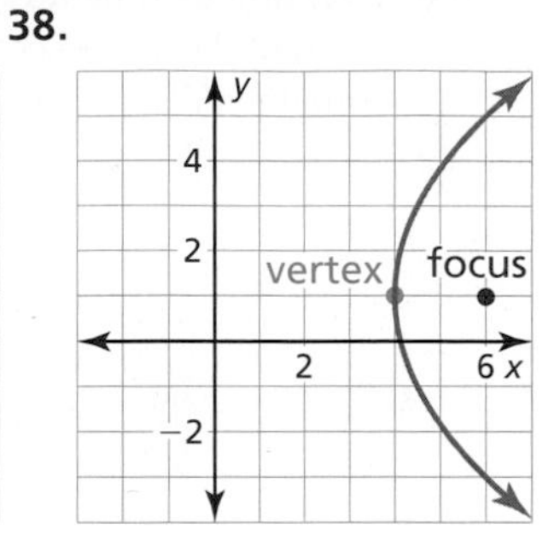

39.

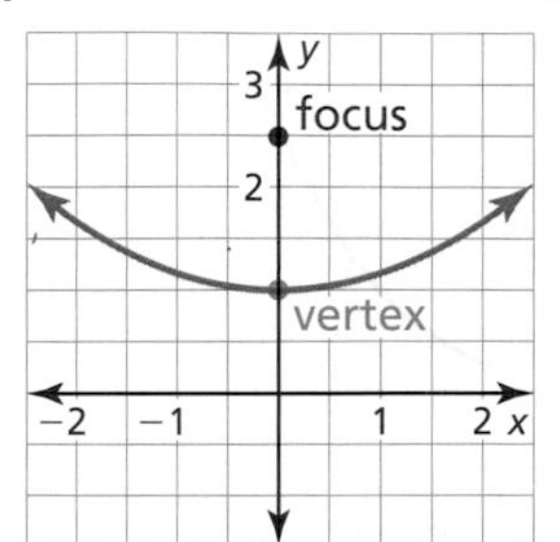

40.

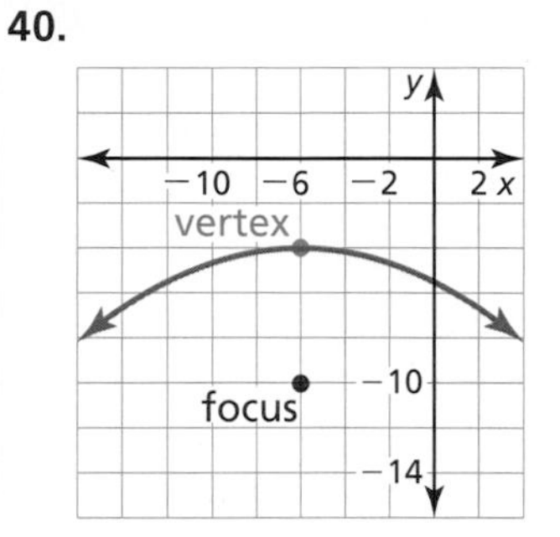

In Exercises 41–46, identify the vertex, focus, directrix, and axis of symmetry of the parabola. Describe the transformations of the graph of the standard equation with $p = 1$ and vertex $(0, 0)$.

41. $y = \frac{1}{8}(x - 3)^2 + 2$

42. $y = -\frac{1}{4}(x + 2)^2 + 1$

43. $x = \frac{1}{16}(y - 3)^2 + 1$

44. $y = (x + 3)^2 - 5$

45. $x = -3(y + 4)^2 + 2$

46. $x = 4(y + 5)^2 - 1$

47. MODELING WITH MATHEMATICS Scientists studying dolphin echolocation simulate the projection of a bottlenose dolphin's clicking sounds using computer models. The models originate the sounds at the focus of a parabolic reflector. The parabola in the graph shows the cross section of the reflector with focal length of 1.3 inches and aperture width of 8 inches. Write an equation to represent the cross section of the reflector. What is the depth of the reflector? *(See Example 5.)*

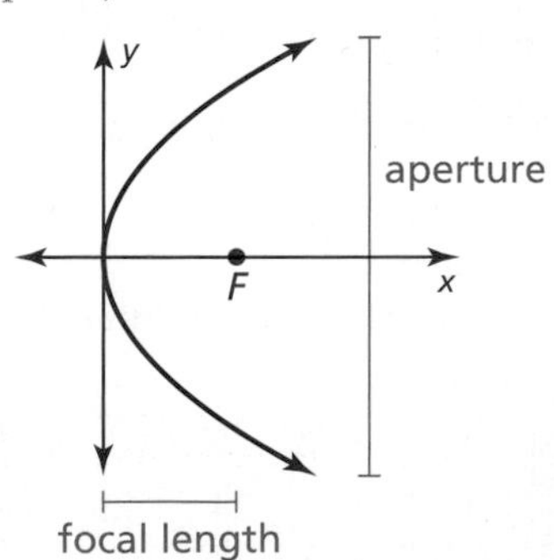

48. MODELING WITH MATHEMATICS Solar energy can be concentrated using long troughs that have a parabolic cross section as shown in the figure. Write an equation to represent the cross section of the trough. What are the domain and range in this situation? What do they represent?

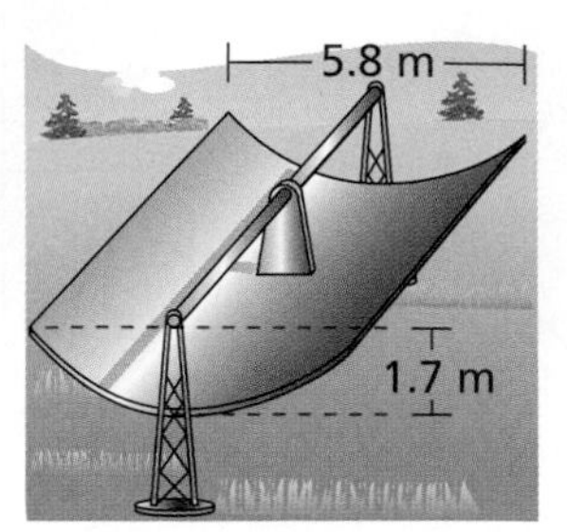

49. ABSTRACT REASONING As $|p|$ increases, how does the width of the graph of the equation $y = \frac{1}{4p}x^2$ change? Explain your reasoning.

50. HOW DO YOU SEE IT? The graph shows the path of a volleyball served from an initial height of 6 feet as it travels over a net.

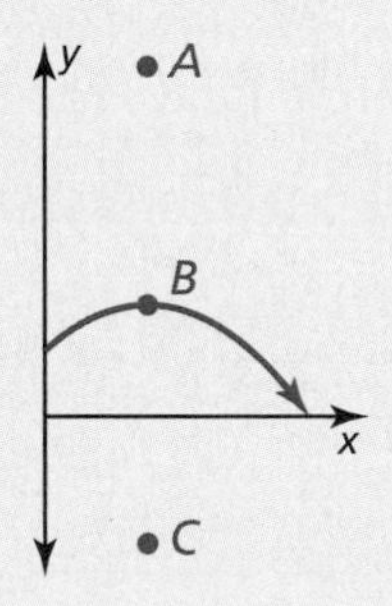

a. Label the vertex, focus, and a point on the directrix.

b. An underhand serve follows the same parabolic path but is hit from a height of 3 feet. How does this affect the focus? the directrix?

51. CRITICAL THINKING The distance from point P to the directrix is 2 units. Write an equation of the parabola.

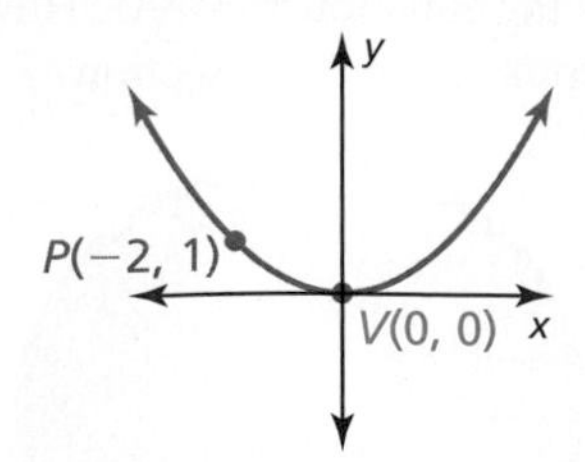

52. THOUGHT PROVOKING Two parabolas have the same focus (a, b) and focal length of 2 units. Write an equation of each parabola. Identify the directrix of each parabola.

53. REPEATED REASONING Use the Distance Formula to derive the equation of a parabola that opens to the right with vertex $(0, 0)$, focus $(p, 0)$, and directrix $x = -p$.

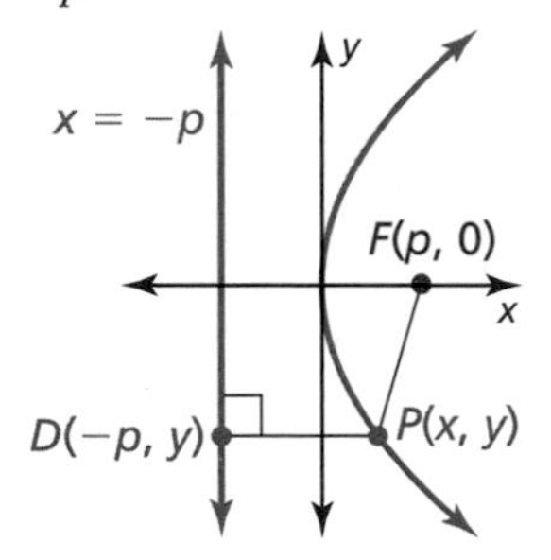

54. PROBLEM SOLVING The *latus rectum* of a parabola is the line segment that is parallel to the directrix, passes through the focus, and has endpoints that lie on the parabola. Find the length of the latus rectum of the parabola shown.

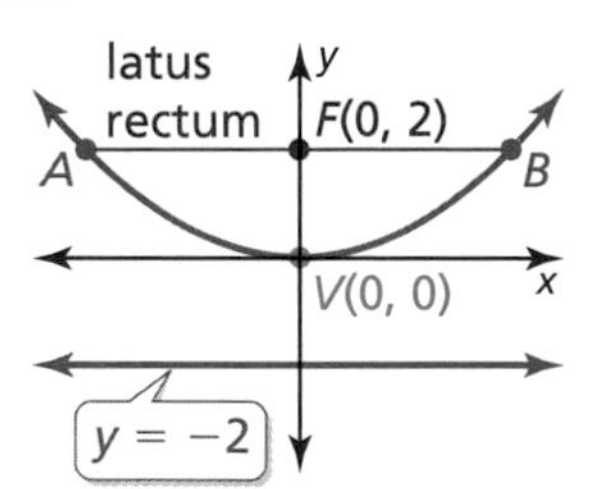

Maintaining Mathematical Proficiency

Reviewing what you learned in previous grades and lessons

Write an equation of the line that passes through the points. *(Section 1.3)*

55. $(1, -4), (2, -1)$ **56.** $(-3, 12), (0, 6)$ **57.** $(3, 1), (-5, 5)$ **58.** $(2, -1), (0, 1)$

Use a graphing calculator to find an equation for the line of best fit. *(Section 1.3)*

59.

x	0	3	6	7	11
y	4	9	24	29	46

60.

x	0	5	10	12	16
y	18	15	9	7	2

2.4 Modeling with Quadratic Functions

Essential Question How can you use a quadratic function to model a real-life situation?

EXPLORATION 1 Modeling with a Quadratic Function

Work with a partner. The graph shows a quadratic function of the form

$$P(t) = at^2 + bt + c$$

which approximates the yearly profits for a company, where $P(t)$ is the profit in year t.

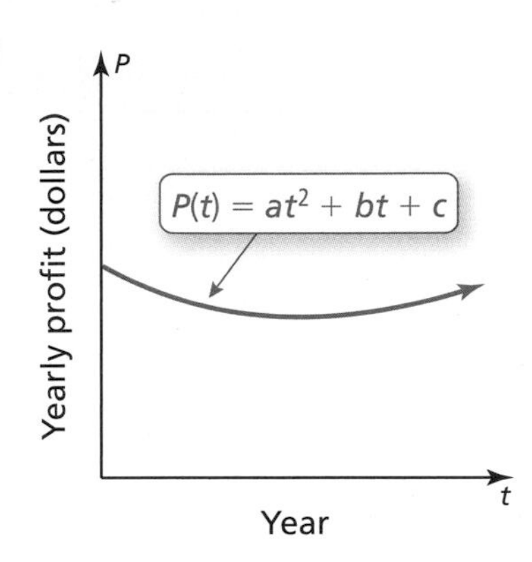

a. Is the value of a positive, negative, or zero? Explain.

b. Write an expression in terms of a and b that represents the year t when the company made the least profit.

c. The company made the same yearly profits in 2004 and 2012. Estimate the year in which the company made the least profit.

d. Assume that the model is still valid today. Are the yearly profits currently increasing, decreasing, or constant? Explain.

EXPLORATION 2 Modeling with a Graphing Calculator

Work with a partner. The table shows the heights h (in feet) of a wrench t seconds after it has been dropped from a building under construction.

Time, t	0	1	2	3	4
Height, h	400	384	336	256	144

a. Use a graphing calculator to create a scatter plot of the data, as shown at the right. Explain why the data appear to fit a quadratic model.

400

0 5

0

b. Use the *quadratic regression* feature to find a quadratic model for the data.

c. Graph the quadratic function on the same screen as the scatter plot to verify that it fits the data.

d. When does the wrench hit the ground? Explain.

MODELING WITH MATHEMATICS

To be proficient in math, you need to routinely interpret your results in the context of the situation.

Communicate Your Answer

3. How can you use a quadratic function to model a real-life situation?

4. Use the Internet or some other reference to find examples of real-life situations that can be modeled by quadratic functions.

2.4 Lesson

Core Vocabulary

Previous
average rate of change
system of three linear equations

What You Will Learn

- Write equations of quadratic functions using vertices, points, and x-intercepts.
- Write quadratic equations to model data sets.

Writing Quadratic Equations

Core Concept

Writing Quadratic Equations

Given a point and the vertex (h, k)	Use vertex form: $y = a(x - h)^2 + k$
Given a point and x-intercepts p and q	Use intercept form: $y = a(x - p)(x - q)$
Given three points	Write and solve a system of three equations in three variables.

EXAMPLE 1 Writing an Equation Using a Vertex and a Point

Human Cannonball

Height (feet): 0, 10, 20, 30, 40; Horizontal distance (feet): 0, 20, 40, 60, 80; points (0, 15), (50, 35)

The graph shows the parabolic path of a performer who is shot out of a cannon, where y is the height (in feet) and x is the horizontal distance traveled (in feet). Write an equation of the parabola. The performer lands in a net 90 feet from the cannon. What is the height of the net?

SOLUTION

From the graph, you can see that the vertex (h, k) is (50, 35) and the parabola passes through the point (0, 15). Use the vertex and the point to solve for a in vertex form.

$y = a(x - h)^2 + k$ Vertex form

$15 = a(0 - 50)^2 + 35$ Substitute for h, k, x, and y.

$-20 = 2500a$ Simplify.

$-0.008 = a$ Divide each side by 2500.

Because $a = -0.008$, $h = 50$, and $k = 35$, the path can be modeled by the equation $y = -0.008(x - 50)^2 + 35$, where $0 \le x \le 90$. Find the height when $x = 90$.

$y = -0.008(90 - 50)^2 + 35$ Substitute 90 for x.

$= -0.008(1600) + 35$ Simplify.

$= 22.2$ Simplify.

▶ So, the height of the net is about 22 feet.

Monitoring Progress

Help in English and Spanish at *BigIdeasMath.com*

1. **WHAT IF?** The vertex of the parabola is (50, 37.5). What is the height of the net?
2. Write an equation of the parabola that passes through the point $(-1, 2)$ and has vertex $(4, -9)$.

EXAMPLE 2 Writing an Equation Using a Point and x-Intercepts

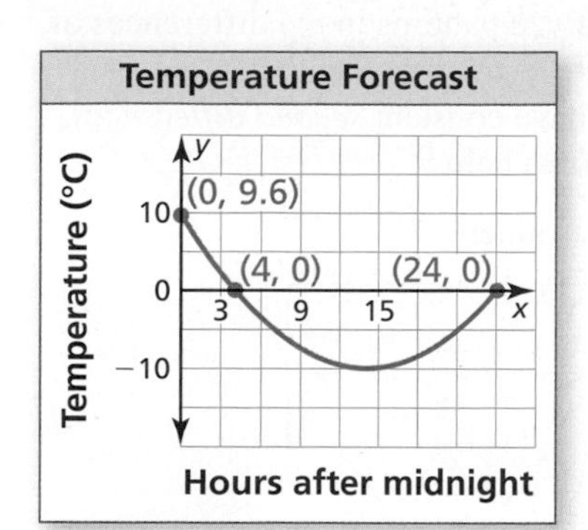

A meteorologist creates a parabola to predict the temperature tomorrow, where x is the number of hours after midnight and y is the temperature (in degrees Celsius).

a. Write a function f that models the temperature over time. What is the coldest temperature?

b. What is the average rate of change in temperature over the interval in which the temperature is decreasing? increasing? Compare the average rates of change.

SOLUTION

a. The x-intercepts are 4 and 24 and the parabola passes through (0, 9.6). Use the x-intercepts and the point to solve for a in intercept form.

$y = a(x - p)(x - q)$	Intercept form
$9.6 = a(0 - 4)(0 - 24)$	Substitute for p, q, x, and y.
$9.6 = 96a$	Simplify.
$0.1 = a$	Divide each side by 96.

Because $a = 0.1$, $p = 4$, and $q = 24$, the temperature over time can be modeled by $f(x) = 0.1(x - 4)(x - 24)$, where $0 \le x \le 24$. The coldest temperature is the minimum value. So, find $f(x)$ when $x = \frac{4 + 24}{2} = 14$.

$f(14) = 0.1(14 - 4)(14 - 24)$	Substitute 14 for x.
$= -10$	Simplify.

▶ So, the coldest temperature is −10°C at 14 hours after midnight, or 2 P.M.

b. The parabola opens up and the axis of symmetry is $x = 14$. So, the function is decreasing over the interval $0 < x < 14$ and increasing over the interval $14 < x < 24$.

REMEMBER

The average rate of change of a function f from x_1 to x_2 is the slope of the line connecting $(x_1, f(x_1))$ and $(x_2, f(x_2))$:

$$\frac{f(x_2) - f(x_1)}{x_2 - x_1}.$$

Average rate of change over $0 < x < 14$:

$$\frac{f(14) - f(0)}{14 - 0} = \frac{-10 - 9.6}{14} = -1.4$$

Average rate of change over $14 < x < 24$:

$$\frac{f(24) - f(14)}{24 - 14} = \frac{0 - (-10)}{10} = 1$$

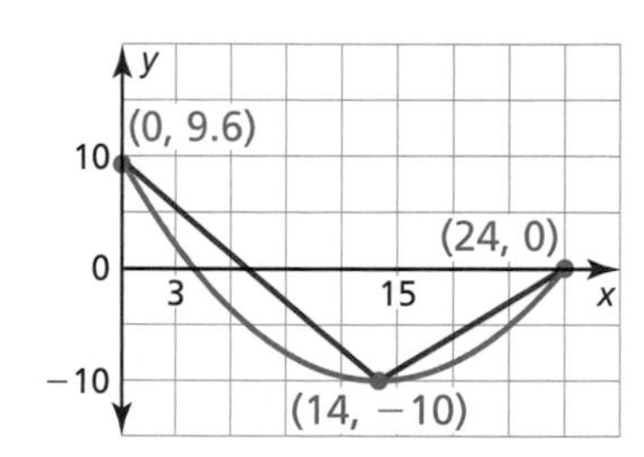

▶ Because $|-1.4| > |1|$, the average rate at which the temperature decreases from midnight to 2 P.M. is greater than the average rate at which it increases from 2 P.M. to midnight.

Monitoring Progress  Help in English and Spanish at *BigIdeasMath.com*

3. WHAT IF? The y-intercept is 4.8. How does this change your answers in parts (a) and (b)?

4. Write an equation of the parabola that passes through the point (2, 5) and has x-intercepts −2 and 4.

Writing Equations to Model Data

When data have equally-spaced inputs, you can analyze patterns in the differences of the outputs to determine what type of function can be used to model the data. Linear data have constant *first differences*. Quadratic data have constant *second differences*. The first and second differences of $f(x) = x^2$ are shown below.

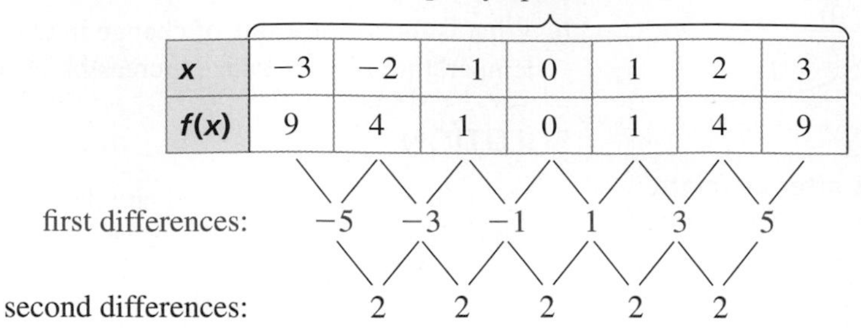

Equally-spaced x-values
x −3 −2 −1 0 1 2 3
f(x) 9 4 1 0 1 4 9
first differences: −5 −3 −1 1 3 5
second differences: 2 2 2 2 2

EXAMPLE 3 **Writing a Quadratic Equation Using Three Points**

Time, *t*	Height, *h*
10	26,900
15	29,025
20	30,600
25	31,625
30	32,100
35	32,025
40	31,400

NASA can create a weightless environment by flying a plane in parabolic paths. The table shows heights h (in feet) of a plane t seconds after starting the flight path. After about 20.8 seconds, passengers begin to experience a weightless environment. Write and evaluate a function to approximate the height at which this occurs.

SOLUTION

Step 1 The input values are equally spaced. So, analyze the differences in the outputs to determine what type of function you can use to model the data.

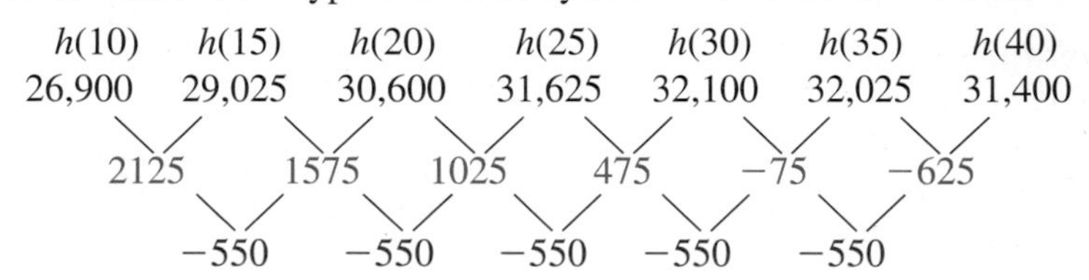

Because the second differences are constant, you can model the data with a quadratic function.

Step 2 Write a quadratic function of the form $h(t) = at^2 + bt + c$ that models the data. Use any three points (t, h) from the table to write a system of equations.

Use (10, 26,900): $100a + 10b + c = 26,900$ Equation 1
Use (20, 30,600): $400a + 20b + c = 30,600$ Equation 2
Use (30, 32,100): $900a + 30b + c = 32,100$ Equation 3

Use the elimination method to solve the system.

Subtract Equation 1 from Equation 2. → $300a + 10b = 3700$ New Equation 1

Subtract Equation 1 from Equation 3. → $800a + 20b = 5200$ New Equation 2

$200a = -2200$ Subtract 2 times new Equation 1 from new Equation 2.

$a = -11$ Solve for a.

$b = 700$ Substitute into new Equation 1 to find b.

$c = 21,000$ Substitute into Equation 1 to find c.

The data can be modeled by the function $h(t) = -11t^2 + 700t + 21,000$.

Step 3 Evaluate the function when $t = 20.8$.

$$h(20.8) = -11(20.8)^2 + 700(20.8) + 21,000 = 30,800.96$$

▶ Passengers begin to experience a weightless environment at about 30,800 feet.

Real-life data that show a quadratic relationship usually do not have constant second differences because the data are not *exactly* quadratic. Relationships that are *approximately* quadratic have second differences that are relatively "close" in value. Many technology tools have a *quadratic regression* feature that you can use to find a quadratic function that best models a set of data.

EXAMPLE 4 Using Quadratic Regression

Miles per hour, x	Miles per gallon, y
20	14.5
24	17.5
30	21.2
36	23.7
40	25.2
45	25.8
50	25.8
56	25.1
60	24.0
70	19.5

The table shows fuel efficiencies of a vehicle at different speeds. Write a function that models the data. Use the model to approximate the optimal driving speed.

SOLUTION

Because the x-values are not equally spaced, you cannot analyze the differences in the outputs. Use a graphing calculator to find a function that models the data.

Step 1 Enter the data in a graphing calculator using two lists and create a scatter plot. The data show a quadratic relationship.

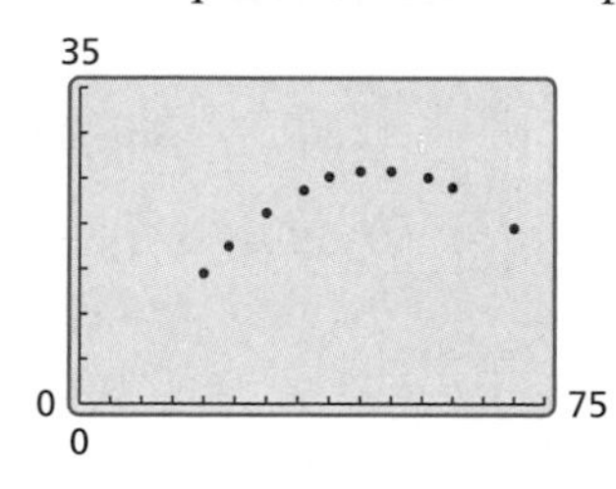

Step 2 Use the *quadratic regression* feature. A quadratic model that represents the data is $y = -0.014x^2 + 1.37x - 7.1$.

```
QuadReg
 y=ax²+bx+c
 a=-.014097349
 b=1.366218867
 c=-7.144052413
 R²=.9992475882
```

STUDY TIP

The *coefficient of determination* R^2 shows how well an equation fits a set of data. The closer R^2 is to 1, the better the fit.

Step 3 Graph the regression equation with the scatter plot.

In this context, the "optimal" driving speed is the speed at which the mileage per gallon is maximized. Using the *maximum* feature, you can see that the maximum mileage per gallon is about 26.4 miles per gallon when driving about 48.9 miles per hour.

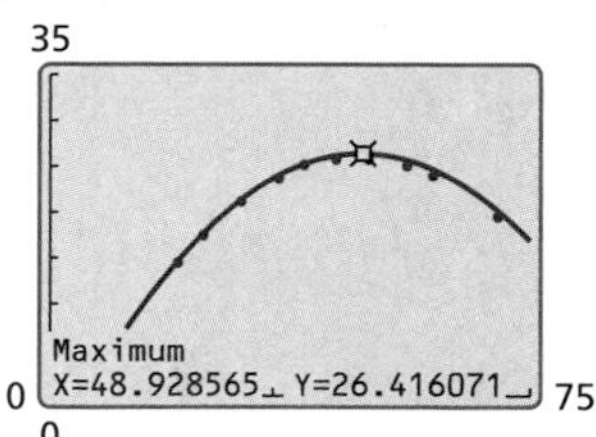

▶ So, the optimal driving speed is about 49 miles per hour.

Monitoring Progress  Help in English and Spanish at *BigIdeasMath.com*

5. Write an equation of the parabola that passes through the points $(-1, 4)$, $(0, 1)$, and $(2, 7)$.

6. The table shows the estimated profits y (in dollars) for a concert when the charge is x dollars per ticket. Write and evaluate a function to determine what the charge per ticket should be to maximize the profit.

Ticket price, x	2	5	8	11	14	17
Profit, y	2600	6500	8600	8900	7400	4100

7. The table shows the results of an experiment testing the maximum weights y (in tons) supported by ice x inches thick. Write a function that models the data. How much weight can be supported by ice that is 22 inches thick?

Ice thickness, x	12	14	15	18	20	24	27
Maximum weight, y	3.4	7.6	10.0	18.3	25.0	40.6	54.3

2.4 Exercises

Dynamic Solutions available at *BigIdeasMath.com*

Vocabulary and Core Concept Check

1. **WRITING** Explain when it is appropriate to use a quadratic model for a set of data.

2. **DIFFERENT WORDS, SAME QUESTION** Which is different? Find "both" answers.

What is the average rate of change over $0 \le x \le 2$?	What is the distance from $f(0)$ to $f(2)$?
What is the slope of the line segment?	What is $\frac{f(2) - f(0)}{2 - 0}$?

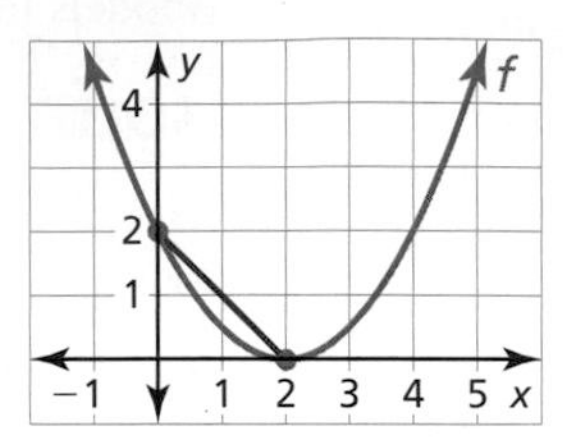

Monitoring Progress and Modeling with Mathematics

In Exercises 3–8, write an equation of the parabola in vertex form. *(See Example 1.)*

3.

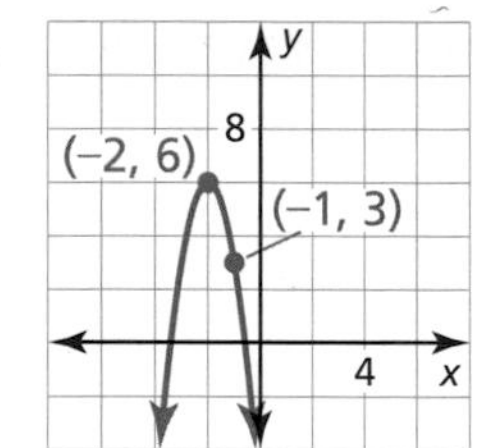

4.

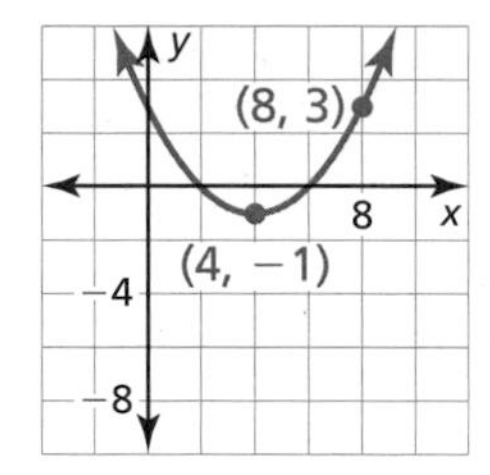

5. passes through (13, 8) and has vertex (3, 2)

6. passes through (−7, −15) and has vertex (−5, 9)

7. passes through (0, −24) and has vertex (−6, −12)

8. passes through (6, 35) and has vertex (−1, 14)

In Exercises 9–14, write an equation of the parabola in intercept form. *(See Example 2.)*

9.

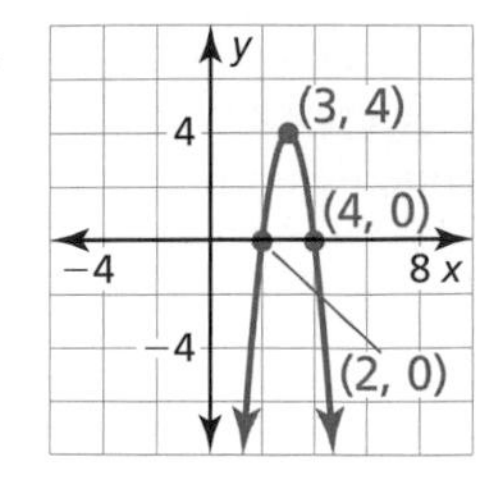

10.

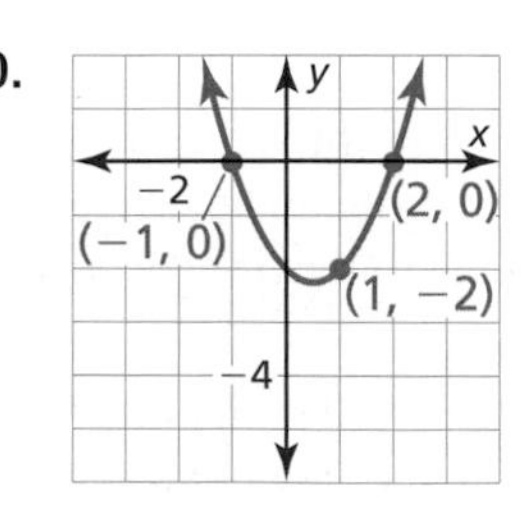

11. x-intercepts of 12 and −6; passes through (14, 4)

12. x-intercepts of 9 and 1; passes through (0, −18)

13. x-intercepts of −16 and −2; passes through (−18, 72)

14. x-intercepts of −7 and −3; passes through (−2, 0.05)

15. **WRITING** Explain when to use intercept form and when to use vertex form when writing an equation of a parabola.

16. **ANALYZING EQUATIONS** Which of the following equations represent the parabola?

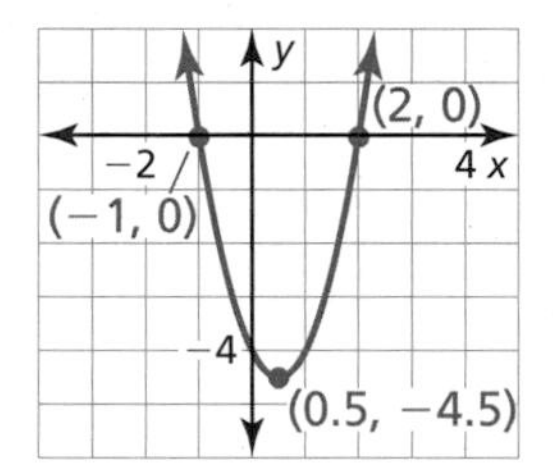

Ⓐ $y = 2(x - 2)(x + 1)$

Ⓑ $y = 2(x + 0.5)^2 - 4.5$

Ⓒ $y = 2(x - 0.5)^2 - 4.5$

Ⓓ $y = 2(x + 2)(x - 1)$

In Exercises 17–20, write an equation of the parabola in vertex form or intercept form.

17.

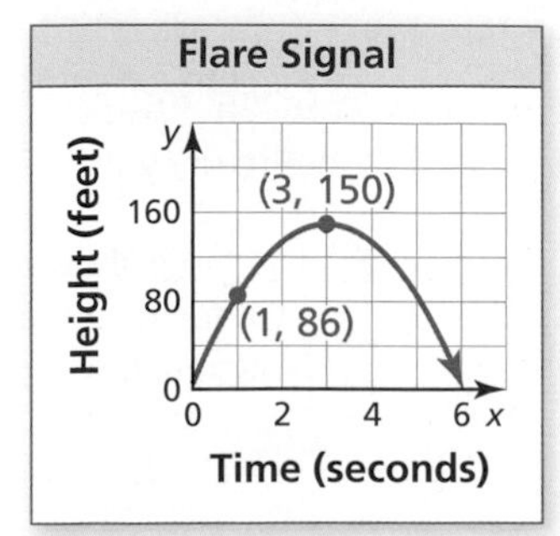

18.

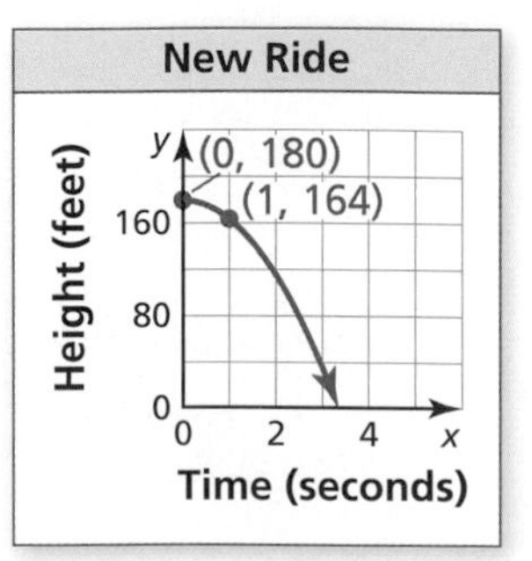

19. 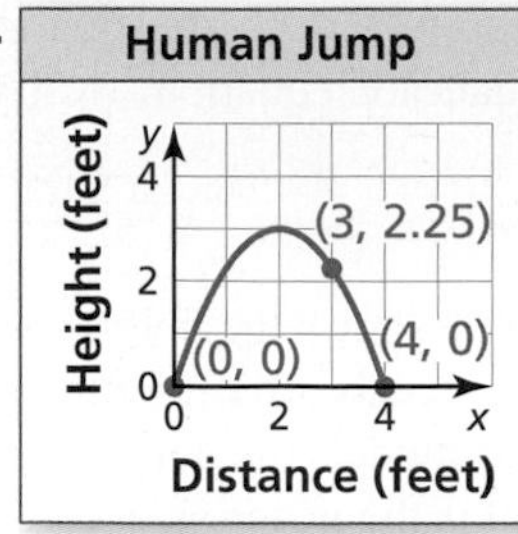

20. 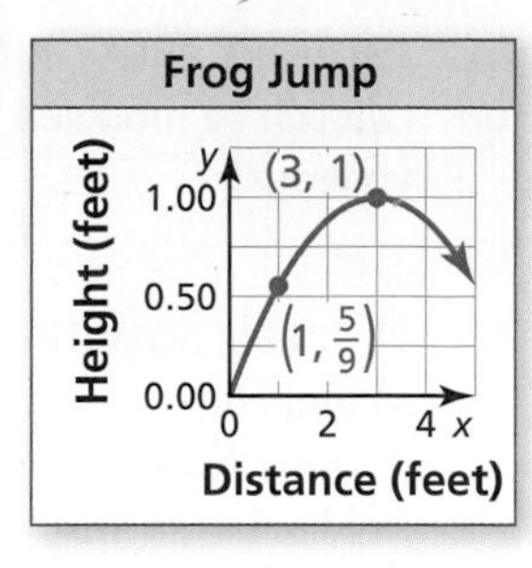

21. ERROR ANALYSIS Describe and correct the error in writing an equation of the parabola.

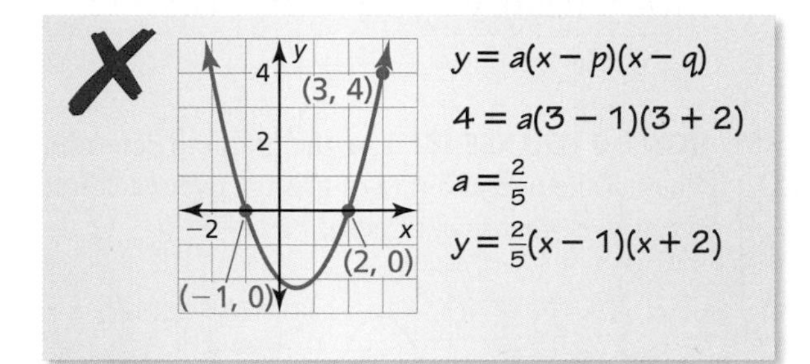

22. MATHEMATICAL CONNECTIONS The area of a rectangle is modeled by the graph where y is the area (in square meters) and x is the width (in meters). Write an equation of the parabola. Find the dimensions and corresponding area of one possible rectangle. What dimensions result in the maximum area?

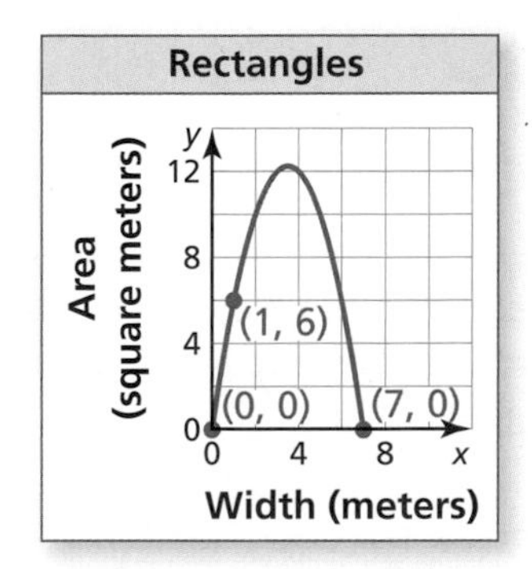

23. MODELING WITH MATHEMATICS Every rope has a safe working load. A rope should not be used to lift a weight greater than its safe working load. The table shows the safe working loads S (in pounds) for ropes with circumference C (in inches). Write an equation for the safe working load for a rope. Find the safe working load for a rope that has a circumference of 10 inches. *(See Example 3.)*

Circumference, C	0	1	2	3
Safe working load, S	0	180	720	1620

24. MODELING WITH MATHEMATICS A baseball is thrown up in the air. The table shows the heights y (in feet) of the baseball after x seconds. Write an equation for the path of the baseball. Find the height of the baseball after 5 seconds.

Time, x	0	2	4	6
Baseball height, y	6	22	22	6

25. COMPARING METHODS You use a system with three variables to find the equation of a parabola that passes through the points $(-8, 0)$, $(2, -20)$, and $(1, 0)$. Your friend uses intercept form to find the equation. Whose method is easier? Justify your answer.

26. MODELING WITH MATHEMATICS The table shows the distances y a motorcyclist is from home after x hours.

Time (hours), x	0	1	2	3
Distance (miles), y	0	45	90	135

a. Determine what type of function you can use to model the data. Explain your reasoning.

b. Write and evaluate a function to determine the distance the motorcyclist is from home after 6 hours.

27. USING TOOLS The table shows the heights h (in feet) of a sponge t seconds after it was dropped by a window cleaner on top of a skyscraper. *(See Example 4.)*

Time, t	0	1	1.5	2.5	3
Height, h	280	264	244	180	136

a. Use a graphing calculator to create a scatter plot. Which better represents the data, a line or a parabola? Explain.

b. Use the *regression* feature of your calculator to find the model that best fits the data.

c. Use the model in part (b) to predict when the sponge will hit the ground.

d. Identify and interpret the domain and range in this situation.

28. MAKING AN ARGUMENT Your friend states that quadratic functions with the same x-intercepts have the same equations, vertex, and axis of symmetry. Is your friend correct? Explain your reasoning.

In Exercises 29–32, analyze the differences in the outputs to determine whether the data are *linear*, *quadratic*, or *neither*. Explain. If linear or quadratic, write an equation that fits the data.

29.

Price decrease (dollars), x	0	5	10	15	20
Revenue ($1000s), y	470	630	690	650	510

30.

Time (hours), x	0	1	2	3	4
Height (feet), y	40	42	44	46	48

31.

Time (hours), x	1	2	3	4	5
Population (hundreds), y	2	4	8	16	32

32.

Time (days), x	0	1	2	3	4
Height (feet), y	320	303	254	173	60

33. PROBLEM SOLVING The graph shows the number y of students absent from school due to the flu each day x.

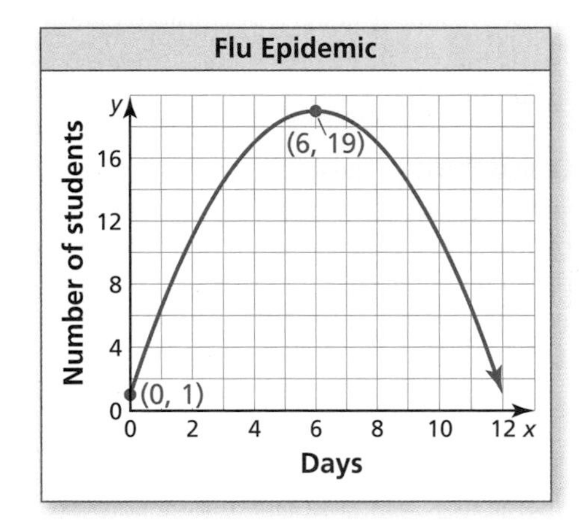

a. Interpret the meaning of the vertex in this situation.

b. Write an equation for the parabola to predict the number of students absent on day 10.

c. Compare the average rates of change in the students with the flu from 0 to 6 days and 6 to 11 days.

34. THOUGHT PROVOKING Describe a real-life situation that can be modeled by a quadratic equation. Justify your answer.

35. PROBLEM SOLVING The table shows the heights y of a competitive water-skier x seconds after jumping off a ramp. Write a function that models the height of the water-skier over time. When is the water-skier 5 feet above the water? How long is the skier in the air?

Time (seconds), x	0	0.25	0.75	1	1.1
Height (feet), y	22	22.5	17.5	12	9.24

36. HOW DO YOU SEE IT? Use the graph to determine whether the average rate of change over each interval is *positive*, *negative*, or *zero*.

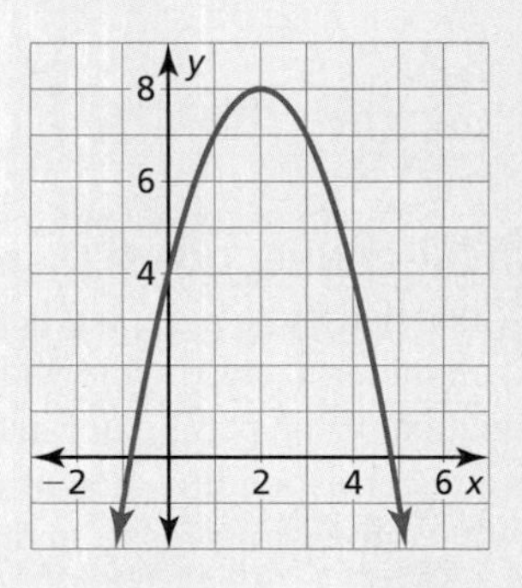

a. $0 \le x \le 2$

b. $2 \le x \le 5$

c. $2 \le x \le 4$

d. $0 \le x \le 4$

37. REPEATED REASONING The table shows the number of tiles in each figure. Verify that the data show a quadratic relationship. Predict the number of tiles in the 12th figure.

Figure 1

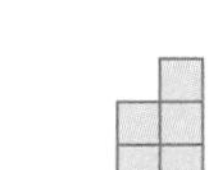

Figure 2

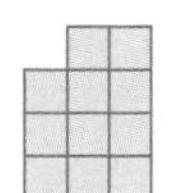

Figure 3

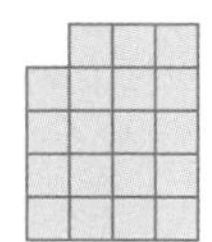

Figure 4

Figure	1	2	3	4
Number of Tiles	1	5	11	19

Maintaining Mathematical Proficiency

Reviewing what you learned in previous grades and lessons

Factor the trinomial. *(Skills Review Handbook)*

38. $x^2 + 4x + 3$

39. $x^2 - 3x + 2$

40. $3x^2 - 15x + 12$

41. $5x^2 + 5x - 30$

2.3–2.4 What Did You Learn?

Core Vocabulary

focus, *p. 68*
directrix, *p. 68*

Core Concepts

Section 2.3

Standard Equations of a Parabola with Vertex at the Origin, *p. 69*
Standard Equations of a Parabola with Vertex at (h, k), *p. 70*

Section 2.4

Writing Quadratic Equations, *p. 76*
Writing Quadratic Equations to Model Data, *p. 78*

Mathematical Practices

1. Explain the solution pathway you used to solve Exercise 47 on page 73.
2. Explain how you used definitions to derive the equation in Exercise 53 on page 74.
3. Explain the shortcut you found to write the equation in Exercise 25 on page 81.
4. Describe how you were able to construct a viable argument in Exercise 28 on page 81.

Performance Task

Accident Reconstruction

Was the driver of a car speeding when the brakes were applied? What do skid marks at the scene of an accident reveal about the moments before the collision?

To explore the answers to these questions and more, go to ***BigIdeasMath.com***.

2 Chapter Review

Dynamic Solutions available at *BigIdeasMath.com*

2.1 Transformations of Quadratic Functions (pp. 47–54)

Let the graph of g be a translation 1 unit left and 2 units up of the graph of $f(x) = x^2 + 1$. Write a rule for g.

$g(x) = f(x - (-1)) + 2$ — Subtract -1 from the input. Add 2 to the output.

$= (x + 1)^2 + 1 + 2$ — Replace x with $x + 1$ in $f(x)$.

$= x^2 + 2x + 4$ — Simplify.

▶ The transformed function is $g(x) = x^2 + 2x + 4$.

Describe the transformation of $f(x) = x^2$ represented by g. Then graph each function.

1. $g(x) = (x + 4)^2$
2. $g(x) = (x - 7)^2 + 2$
3. $g(x) = -3(x + 2)^2 - 1$

Write a rule for g.

4. Let the graph of g be a horizontal shrink by a factor of $\frac{2}{3}$, followed by a translation 5 units left and 2 units down of the graph of $f(x) = x^2$.
5. Let the graph of g be a translation 2 units left and 3 units up, followed by a reflection in the y-axis of the graph of $f(x) = x^2 - 2x$.

2.2 Characteristics of Quadratic Functions (pp. 55–64)

Graph $f(x) = 2x^2 - 8x + 1$. Label the vertex and axis of symmetry.

Step 1 Identify the coefficients $a = 2$, $b = -8$, and $c = 1$. Because $a > 0$, the parabola opens up.

Step 2 Find the vertex. First calculate the x-coordinate.

$$x = -\frac{b}{2a} = -\frac{-8}{2(2)} = 2$$

Then find the y-coordinate of the vertex.

$$f(2) = 2(2)^2 - 8(2) + 1 = -7$$

So, the vertex is $(2, -7)$. Plot this point.

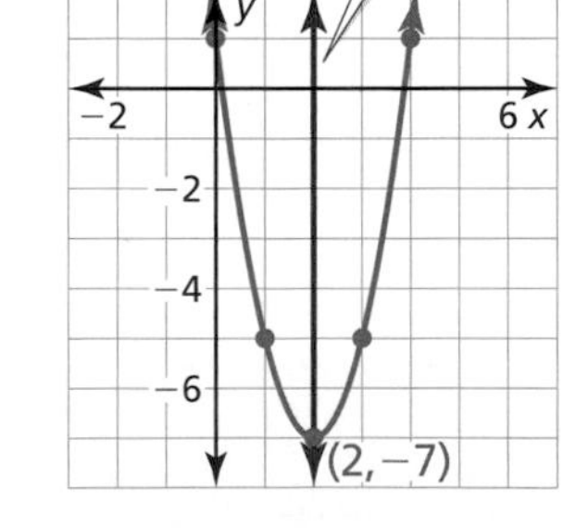

Step 3 Draw the axis of symmetry $x = 2$.

Step 4 Identify the y-intercept c, which is 1. Plot the point $(0, 1)$ and its reflection in the axis of symmetry, $(4, 1)$.

Step 5 Evaluate the function for another value of x, such as $x = 1$.

$$f(1) = 2(1)^2 - 8(1) + 1 = -5$$

Plot the point $(1, -5)$ and its reflection in the axis of symmetry, $(3, -5)$.

Step 6 Draw a parabola through the plotted points.

Graph the function. Label the vertex and axis of symmetry. Find the minimum or maximum value of f. Describe where the function is increasing and decreasing.

6. $f(x) = 3(x - 1)^2 - 4$
7. $g(x) = -2x^2 + 16x + 3$
8. $h(x) = (x - 3)(x + 7)$

2.3 Focus of a Parabola (pp. 67–74)

a. Identify the focus, directrix, and axis of symmetry of $8x = y^2$. Graph the equation.

Step 1 Rewrite the equation in standard form.

$8x = y^2$ Write the original equation.

$x = \frac{1}{8}y^2$ Divide each side by 8.

Step 2 Identify the focus, directrix, and axis of symmetry. The equation has the form $x = \frac{1}{4p}y^2$, where $p = 2$. The focus is $(p, 0)$, or $(2, 0)$. The directrix is $x = -p$, or $x = -2$. Because y is squared, the axis of symmetry is the x-axis.

Step 3 Use a table of values to graph the equation. Notice that it is easier to substitute y-values and solve for x.

y	0	±2	±4	±6
x	0	0.5	2	4.5

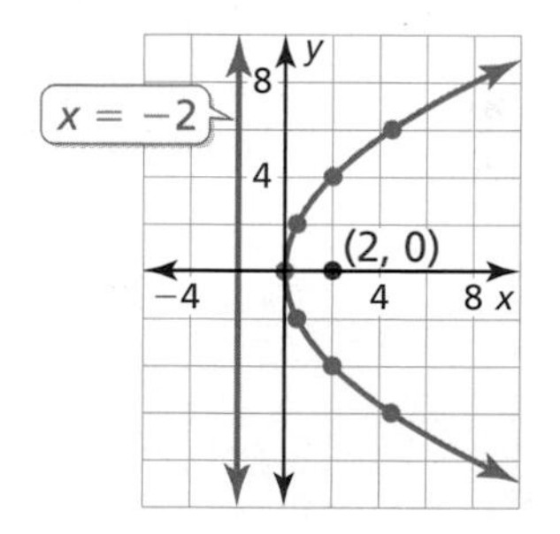

b. Write an equation of the parabola shown.

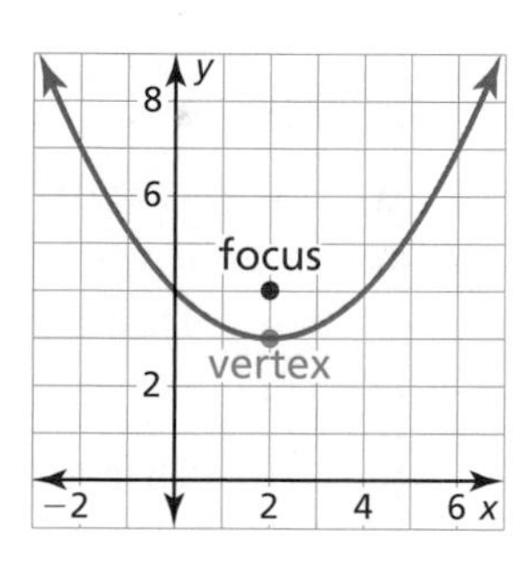

Because the vertex is not at the origin and the axis of symmetry is vertical, the equation has the form $y = \frac{1}{4p}(x - h)^2 + k$. The vertex (h, k) is $(2, 3)$ and the focus $(h, k + p)$ is $(2, 4)$, so $h = 2$, $k = 3$, and $p = 1$. Substitute these values to write an equation of the parabola.

$$y = \frac{1}{4(1)}(x - 2)^2 + 3 = \frac{1}{4}(x - 2)^2 + 3$$

▶ An equation of the parabola is $y = \frac{1}{4}(x - 2)^2 + 3$.

9. You can make a solar hot-dog cooker by shaping foil-lined cardboard into a parabolic trough and passing a wire through the focus of each end piece. For the trough shown, how far from the bottom should the wire be placed?

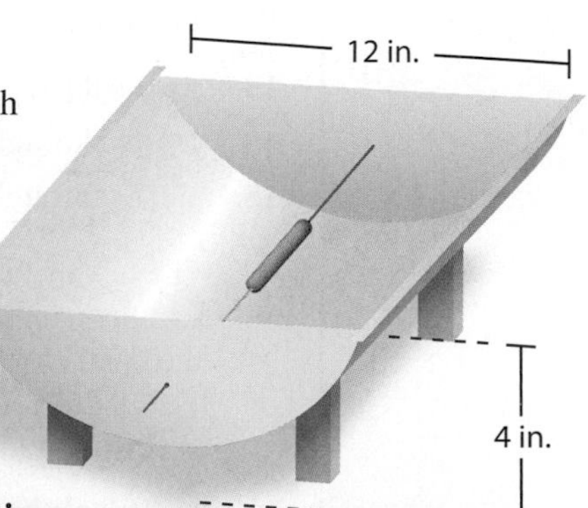

10. Graph the equation $36y = x^2$. Identify the focus, directrix, and axis of symmetry.

Write an equation of the parabola with the given characteristics.

11. vertex: $(0, 0)$
directrix: $x = 2$

12. focus: $(2, 2)$
vertex: $(2, 6)$

2.4 Modeling with Quadratic Functions *(pp. 75–82)*

The graph shows the parabolic path of a stunt motorcyclist jumping off a ramp, where y is the height (in feet) and x is the horizontal distance traveled (in feet). Write an equation of the parabola. The motorcyclist lands on another ramp 160 feet from the first ramp. What is the height of the second ramp?

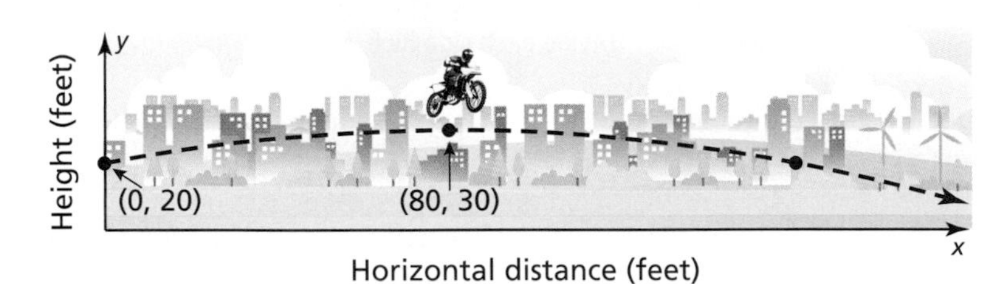

Step 1 First write an equation of the parabola.

From the graph, you can see that the vertex (h, k) is $(80, 30)$ and the parabola passes through the point $(0, 20)$. Use the vertex and the point to solve for a in vertex form.

$$y = a(x - h)^2 + k \qquad \text{Vertex form}$$

$$20 = a(0 - 80)^2 + 30 \qquad \text{Substitute for } h, k, x, \text{ and } y.$$

$$-10 = 6400a \qquad \text{Simplify.}$$

$$-\frac{1}{640} = a \qquad \text{Divide each side by 6400.}$$

Because $a = -\frac{1}{640}$, $h = 80$, and $k = 30$, the path can be modeled by

$$y = -\frac{1}{640}(x - 80)^2 + 30, \text{ where } 0 \le x \le 160.$$

Step 2 Then find the height of the second ramp.

$$y = -\frac{1}{640}(160 - 80)^2 + 30 \qquad \text{Substitute 160 for } x.$$

$$= 20 \qquad \text{Simplify.}$$

▶ So, the height of the second ramp is 20 feet.

Write an equation for the parabola with the given characteristics.

13. passes through $(1, 12)$ and has vertex $(10, -4)$

14. passes through $(4, 3)$ and has x-intercepts of -1 and 5

15. passes through $(-2, 7)$, $(1, 10)$, and $(2, 27)$

16. The table shows the heights y of a dropped object after x seconds. Verify that the data show a quadratic relationship. Write a function that models the data. How long is the object in the air?

Time (seconds), x	0	0.5	1	1.5	2	2.5
Height (feet), y	150	146	134	114	86	50

2 Chapter Test

1. A parabola has an axis of symmetry $y = 3$ and passes through the point $(2, 1)$. Find another point that lies on the graph of the parabola. Explain your reasoning.

2. Let the graph of g be a translation 2 units left and 1 unit down, followed by a reflection in the y-axis of the graph of $f(x) = (2x + 1)^2 - 4$. Write a rule for g.

3. Identify the focus, directrix, and axis of symmetry of $x = 2y^2$. Graph the equation.

4. Explain why a quadratic function models the data. Then use a linear system to find the model.

x	2	4	6	8	10
$f(x)$	0	−13	−34	−63	−100

Write an equation of the parabola. Justify your answer.

5. 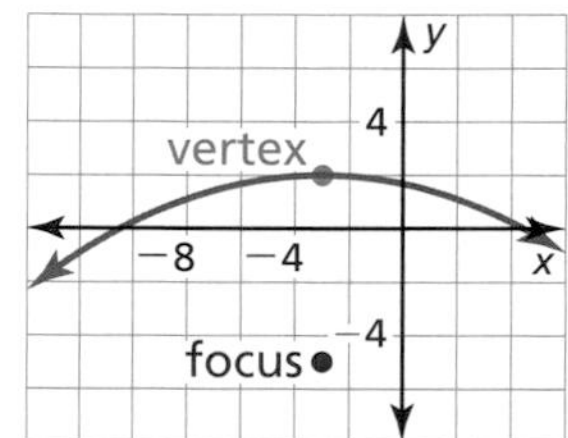

6. 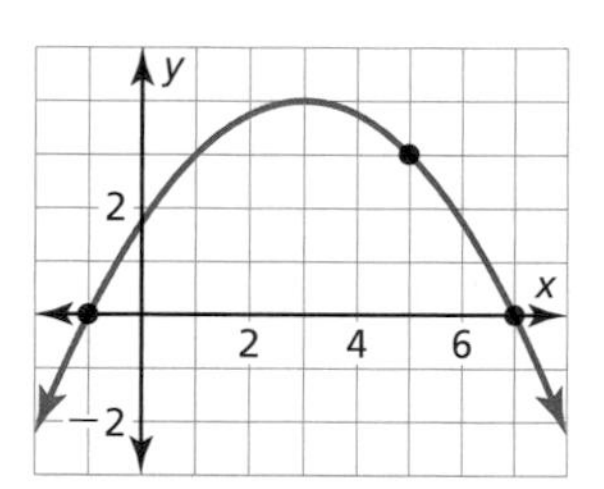

7. 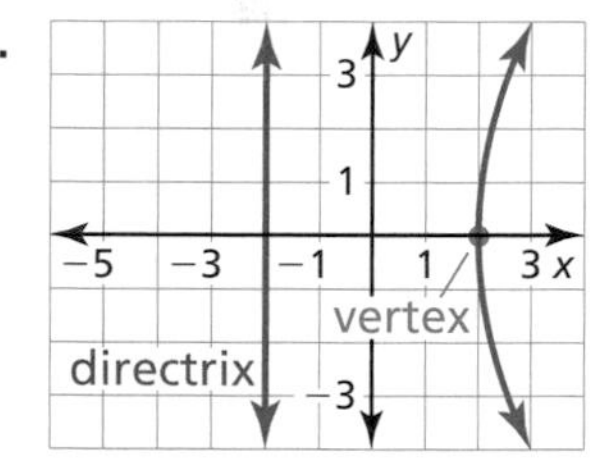

8. A surfboard shop sells 40 surfboards per month when it charges $500 per surfboard. Each time the shop decreases the price by $10, it sells 1 additional surfboard per month. How much should the shop charge per surfboard to maximize the amount of money earned? What is the maximum amount the shop can earn per month? Explain.

9. Graph $f(x) = 8x^2 - 4x + 3$. Label the vertex and axis of symmetry. Describe where the function is increasing and decreasing.

10. Sunfire is a machine with a parabolic cross section used to collect solar energy. The Sun's rays are reflected from the mirrors toward two boilers located at the focus of the parabola. The boilers produce steam that powers an alternator to produce electricity.

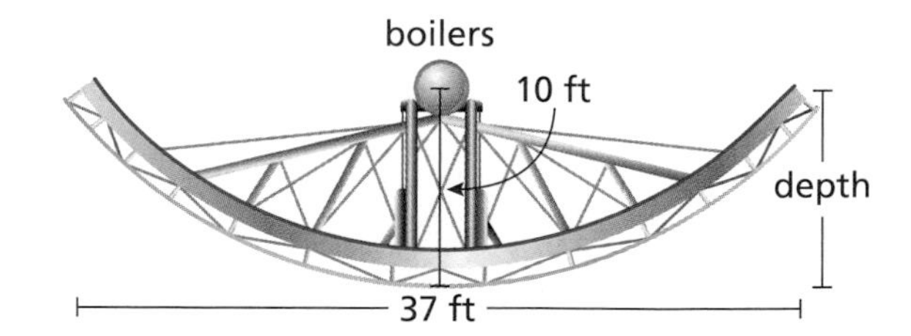

   a. Write an equation that represents the cross section of the dish shown with its vertex at $(0, 0)$.

   b. What is the depth of Sunfire? Justify your answer.

11. In 2011, the price of gold reached an all-time high. The table shows the prices (in dollars per troy ounce) of gold each year since 2006 ($t = 0$ represents 2006). Find a quadratic function that best models the data. Use the model to predict the price of gold in the year 2016.

Year, t	0	1	2	3	4	5
Price, p	$603.46	$695.39	$871.96	$972.35	$1224.53	$1571.52

2 Cumulative Assessment

1. You and your friend are throwing a football. The parabola shows the path of your friend's throw, where x is the horizontal distance (in feet) and y is the corresponding height (in feet). The path of your throw can be modeled by $h(x) = -16x^2 + 65x + 5$. Choose the correct inequality symbol to indicate whose throw travels higher. Explain your reasoning.

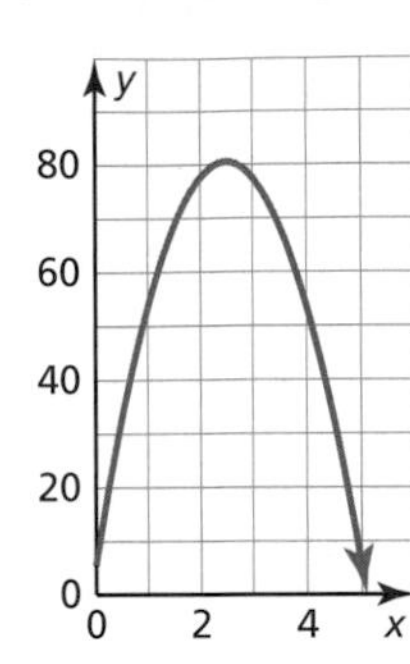

height of your throw ▢ height of your friend's throw

2. The function $g(x) = \frac{1}{2}|x - 4| + 4$ is a combination of transformations of $f(x) = |x|$. Which combinations describe the transformation from the graph of f to the graph of g?

   (A) translation 4 units right and vertical shrink by a factor of $\frac{1}{2}$, followed by a translation 4 units up

   (B) translation 4 units right and 4 units up, followed by a vertical shrink by a factor of $\frac{1}{2}$

   (C) vertical shrink by a factor of $\frac{1}{2}$, followed by a translation 4 units up and 4 units right

   (D) translation 4 units right and 8 units up, followed by a vertical shrink by a factor of $\frac{1}{2}$

3. Your school decides to sell tickets to a dance in the school cafeteria to raise money. There is no fee to use the cafeteria, but the DJ charges a fee of \$750. The table shows the profits (in dollars) when x students attend the dance.

Students, x	Profit, y
100	−250
200	250
300	750
400	1250
500	1750

   a. What is the cost of a ticket?

   b. Your school expects 400 students to attend and finds another DJ who only charges \$650. How much should your school charge per ticket to still make the same profit?

   c. Your school decides to charge the amount in part (a) and use the less expensive DJ. How much more money will the school raise?

4. Order the following parabolas from widest to narrowest.

   A. focus: $(0, -3)$; directrix: $y = 3$

   B. $y = \frac{1}{16}x^2 + 4$

   C. $x = \frac{1}{8}y^2$

   D. $y = \frac{1}{4}(x - 2)^2 + 3$

5. Your friend claims that for $g(x) = b$, where b is a real number, there is a transformation in the graph that is impossible to notice. Is your friend correct? Explain your reasoning.

6. Let the graph of g represent a vertical stretch and a reflection in the x-axis, followed by a translation left and down of the graph of $f(x) = x^2$. Use the tiles to write a rule for g.

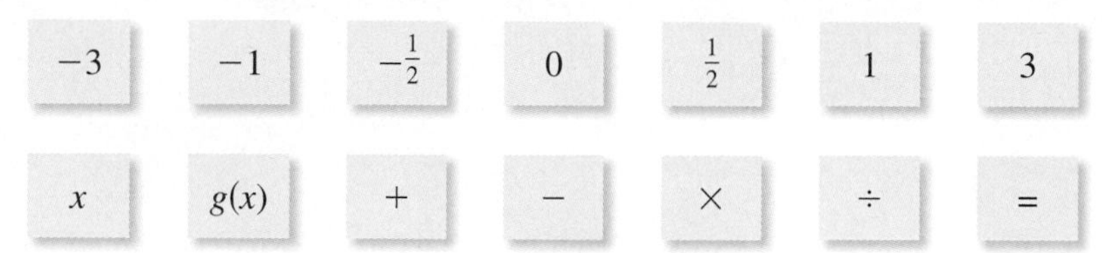

7. Two balls are thrown in the air. The path of the first ball is represented in the graph. The second ball is released 1.5 feet higher than the first ball and after 3 seconds reaches its maximum height 5 feet lower than the first ball.

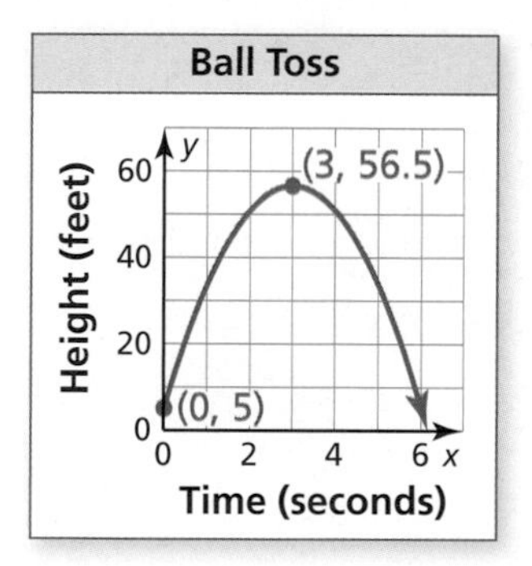

a. Write an equation for the path of the second ball.

b. Do the balls hit the ground at the same time? If so, how long are the balls in the air? If not, which ball hits the ground first? Explain your reasoning.

8. Let the graph of g be a translation 3 units right of the graph of f. The points $(-1, 6)$, $(3, 14)$, and $(6, 41)$ lie on the graph of f. Which points lie on the graph of g?

Ⓐ (2, 6) Ⓑ (2, 11) Ⓒ (6, 14)

Ⓓ (6, 19) Ⓔ (9, 41) Ⓕ (9, 46)

9. Gym A charges \$10 per month plus an initiation fee of \$100. Gym B charges \$30 per month, but due to a special promotion, is not currently charging an initiation fee.

a. Write an equation for each gym modeling the total cost y for a membership lasting x months.

b. When is it more economical for a person to choose Gym A over Gym B?

c. Gym A lowers its initiation fee to \$25. Describe the transformation this change represents and how it affects your decision in part (b).

3 Quadratic Equations and Complex Numbers

Robot-Building Competition *(p. 145)*

Broadcast Tower *(p. 137)*

Feeding Gannet *(p. 129)*

Baseball *(p. 115)*

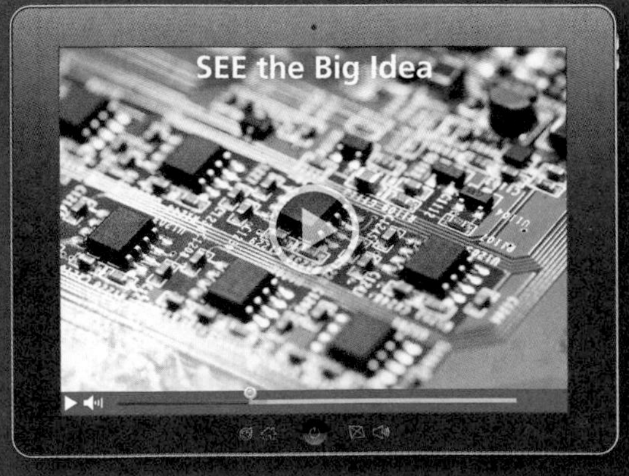

Electrical Circuits *(p. 106)*

Maintaining Mathematical Proficiency

Simplifying Square Roots

Example 1 **Simplify $\sqrt{8}$.**

$\sqrt{8} = \sqrt{4 \cdot 2}$ Factor using the greatest perfect square factor.

$= \sqrt{4} \cdot \sqrt{2}$ Product Property of Square Roots

$= 2\sqrt{2}$ Simplify.

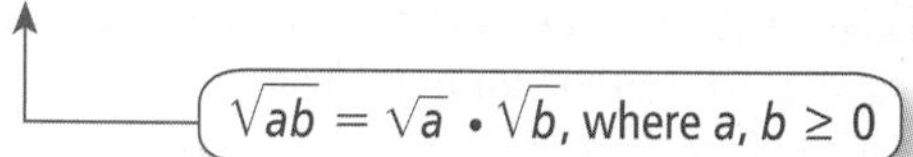
$\sqrt{ab} = \sqrt{a} \cdot \sqrt{b}$, where $a, b \geq 0$

Example 2 **Simplify $\sqrt{\frac{7}{36}}$.**

$\sqrt{\frac{7}{36}} = \frac{\sqrt{7}}{\sqrt{36}}$ Quotient Property of Square Roots

$= \frac{\sqrt{7}}{6}$ Simplify.

$\sqrt{\frac{a}{b}} = \frac{\sqrt{a}}{\sqrt{b}}$, where $a \geq 0$ and $b > 0$

Simplify the expression.

1. $\sqrt{27}$ **2.** $-\sqrt{112}$ **3.** $\sqrt{\frac{11}{64}}$ **4.** $\sqrt{\frac{147}{100}}$

5. $\sqrt{\frac{18}{49}}$ **6.** $-\sqrt{\frac{65}{121}}$ **7.** $-\sqrt{80}$ **8.** $\sqrt{32}$

Factoring Special Products

Example 3 **Factor (a) $x^2 - 4$ and (b) $x^2 - 14x + 49$.**

a. $x^2 - 4 = x^2 - 2^2$ Write as $a^2 - b^2$.

$= (x + 2)(x - 2)$ Difference of Two Squares Pattern

▶ So, $x^2 - 4 = (x + 2)(x - 2)$.

b. $x^2 - 14x + 49 = x^2 - 2(x)(7) + 7^2$ Write as $a^2 - 2ab + b^2$.

$= (x - 7)^2$ Perfect Square Trinomial Pattern

▶ So, $x^2 - 14x + 49 = (x - 7)^2$.

Factor the polynomial.

9. $x^2 - 36$ **10.** $x^2 - 9$ **11.** $4x^2 - 25$

12. $x^2 - 22x + 121$ **13.** $x^2 + 28x + 196$ **14.** $49x^2 + 210x + 225$

15. **ABSTRACT REASONING** Determine the possible integer values of a and c for which the trinomial $ax^2 + 8x + c$ is factorable using the Perfect Square Trinomial Pattern. Explain your reasoning.

Mathematical Practices

Mathematically proficient students recognize the limitations of technology.

Recognizing the Limitations of Technology

Core Concept

Graphing Calculator Limitations

Graphing calculators have a limited number of pixels to display the graph of a function. The result may be an inaccurate or misleading graph.

To correct this issue, use a viewing window setting based on the dimensions of the screen (in pixels).

EXAMPLE 1 Recognizing an Incorrect Graph

Use a graphing calculator to draw the circle given by the equation $x^2 + y^2 = 6.25$.

SOLUTION

Begin by solving the equation for y.

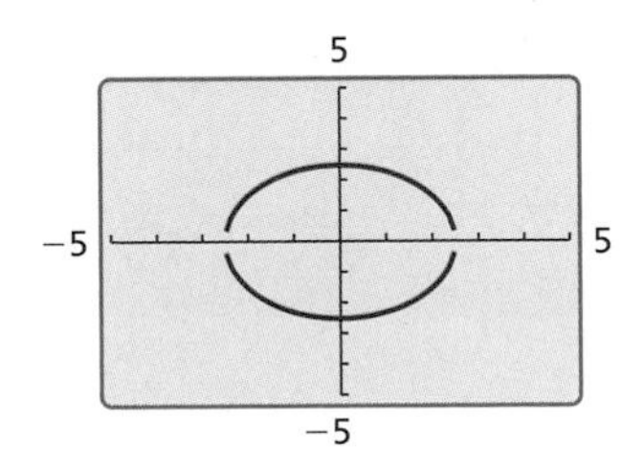

$y = \sqrt{6.25 - x^2}$ Equation of upper semicircle

$y = -\sqrt{6.25 - x^2}$ Equation of lower semicircle

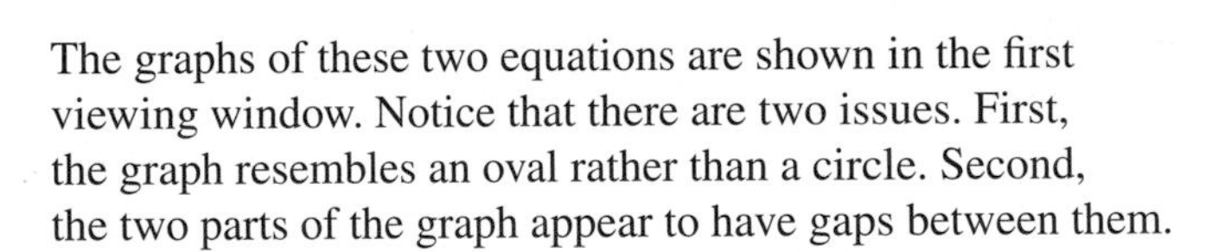

The graphs of these two equations are shown in the first viewing window. Notice that there are two issues. First, the graph resembles an oval rather than a circle. Second, the two parts of the graph appear to have gaps between them.

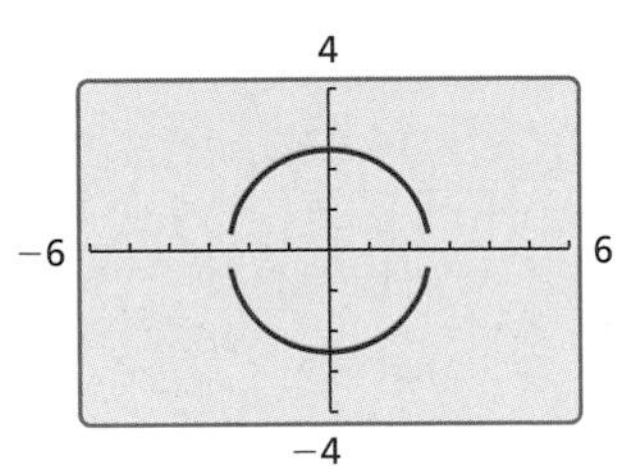

You can correct the first issue by using a *square viewing window*, as shown in the second viewing window.

To correct the second issue, you need to know the dimensions of the graphing calculator screen in terms of the number of pixels. For instance, for a screen that is 63 pixels high and 95 pixels wide, use a viewing window setting as shown at the right.

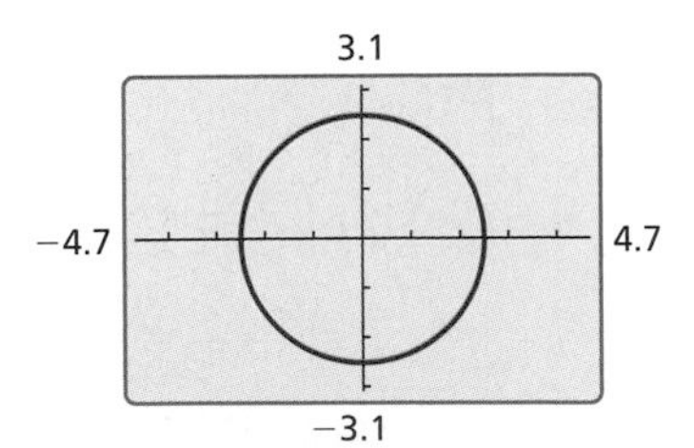

Monitoring Progress

1. Explain why the second viewing window in Example 1 shows gaps between the upper and lower semicircles, but the third viewing window does not show gaps.

Use a graphing calculator to draw an accurate graph of the equation. Explain your choice of viewing window.

2. $y = \sqrt{x^2 - 1.5}$
3. $y = \sqrt{x - 2.5}$
4. $x^2 + y^2 = 12.25$
5. $x^2 + y^2 = 20.25$
6. $x^2 + 4y^2 = 12.25$
7. $4x^2 + y^2 = 20.25$

3.1 Solving Quadratic Equations

Essential Question How can you use the graph of a quadratic equation to determine the number of real solutions of the equation?

EXPLORATION 1 **Matching a Quadratic Function with Its Graph**

Work with a partner. Match each quadratic function with its graph. Explain your reasoning. Determine the number of x-intercepts of the graph.

a. $f(x) = x^2 - 2x$ **b.** $f(x) = x^2 - 2x + 1$ **c.** $f(x) = x^2 - 2x + 2$

d. $f(x) = -x^2 + 2x$ **e.** $f(x) = -x^2 + 2x - 1$ **f.** $f(x) = -x^2 + 2x - 2$

A.

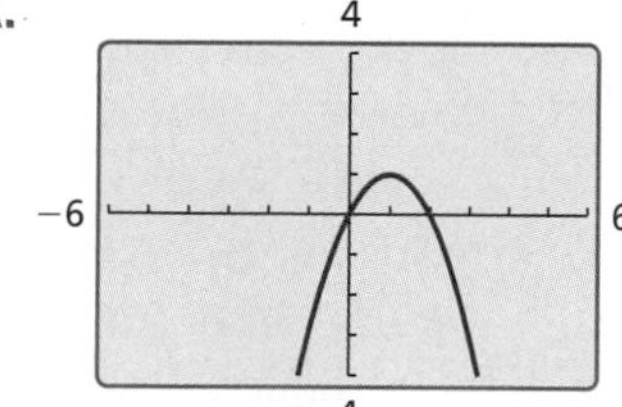

B.

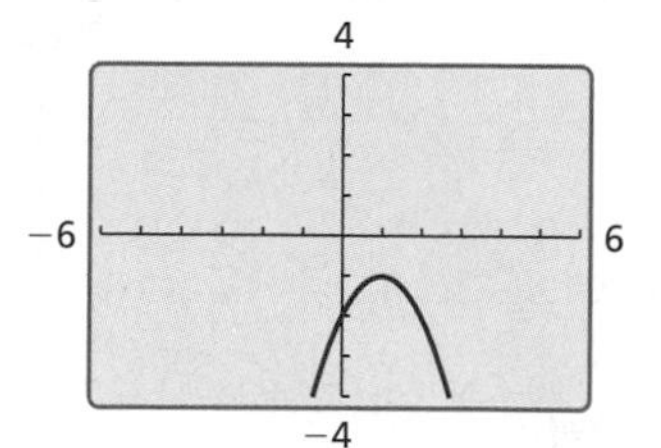

C.

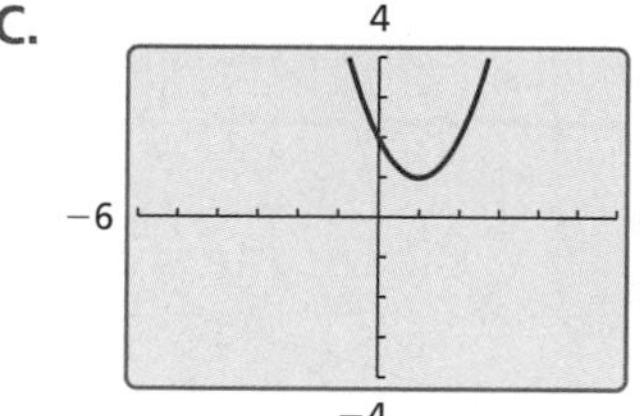

D.

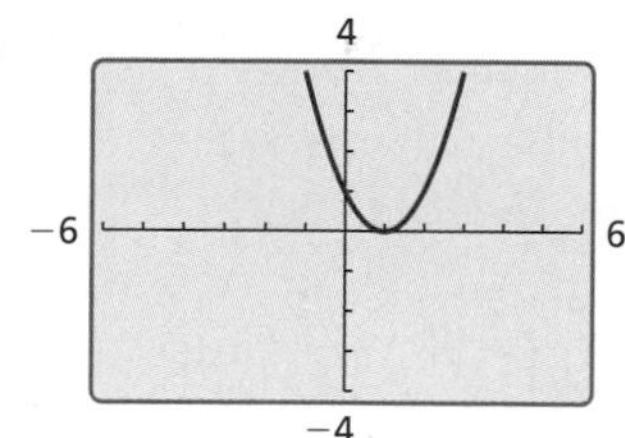

E.

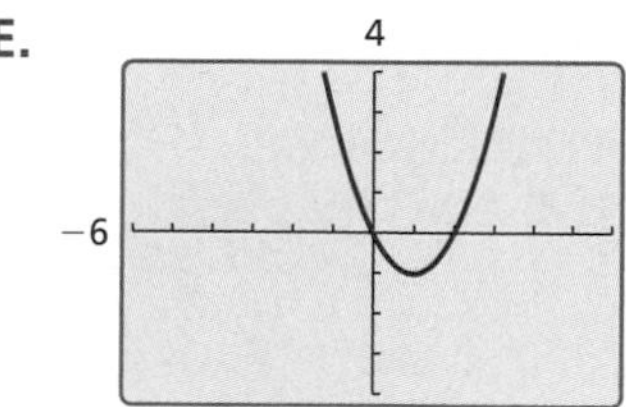

F.

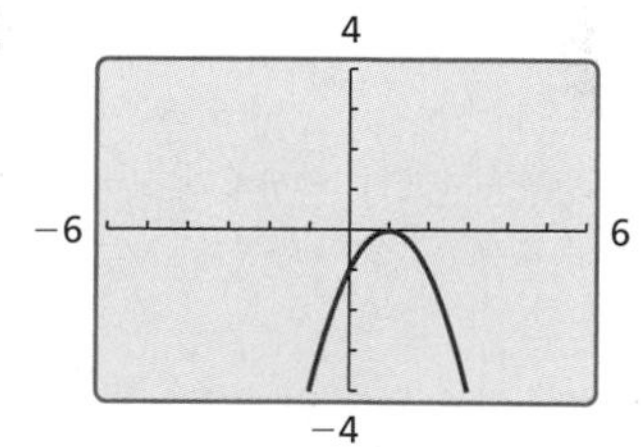

EXPLORATION 2 **Solving Quadratic Equations**

Work with a partner. Use the results of Exploration 1 to find the real solutions (if any) of each quadratic equation.

a. $x^2 - 2x = 0$ **b.** $x^2 - 2x + 1 = 0$ **c.** $x^2 - 2x + 2 = 0$

d. $-x^2 + 2x = 0$ **e.** $-x^2 + 2x - 1 = 0$ **f.** $-x^2 + 2x - 2 = 0$

MAKING SENSE OF PROBLEMS

To be proficient in math, you need to make conjectures about the form and meaning of solutions.

Communicate Your Answer

3. How can you use the graph of a quadratic equation to determine the number of real solutions of the equation?

4. How many real solutions does the quadratic equation $x^2 + 3x + 2 = 0$ have? How do you know? What are the solutions?

3.1 Lesson

Core Vocabulary

quadratic equation in one variable, *p. 94*
root of an equation, *p. 94*
zero of a function, *p. 96*

Previous
properties of square roots
factoring
rationalizing the denominator

What You Will Learn

- Solve quadratic equations by graphing.
- Solve quadratic equations algebraically.
- Solve real-life problems.

Solving Quadratic Equations by Graphing

A **quadratic equation in one variable** is an equation that can be written in the standard form $ax^2 + bx + c = 0$, where a, b, and c are real numbers and $a \neq 0$. A **root of an equation** is a solution of the equation. You can use various methods to solve quadratic equations.

STUDY TIP

Quadratic equations can have zero, one, or two real solutions.

Core Concept

Solving Quadratic Equations

By graphing Find the x-intercepts of the related function $y = ax^2 + bx + c$.

Using square roots Write the equation in the form $u^2 = d$, where u is an algebraic expression, and solve by taking the square root of each side.

By factoring Write the polynomial equation $ax^2 + bx + c = 0$ in factored form and solve using the Zero-Product Property.

EXAMPLE 1 Solving Quadratic Equations by Graphing

Solve each equation by graphing.

a. $x^2 - x - 6 = 0$

b. $-2x^2 - 2 = 4x$

SOLUTION

a. The equation is in standard form. Graph the related function $y = x^2 - x - 6$.

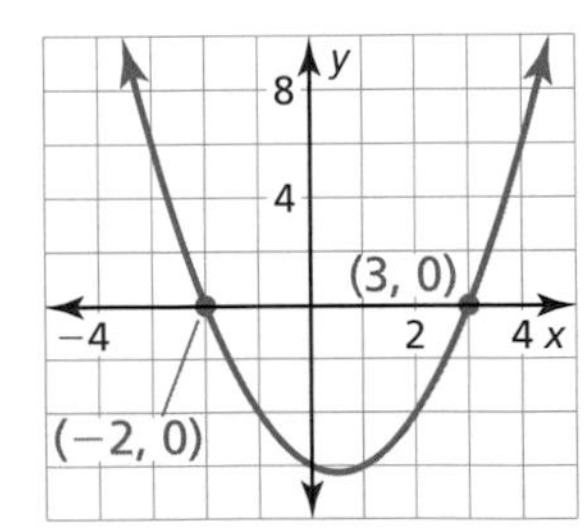

The x-intercepts are -2 and 3.

▶ The solutions, or roots, are $x = -2$ and $x = 3$.

Check

$$x^2 - x - 6 = 0$$
$$(-2)^2 - (-2) - 6 \stackrel{?}{=} 0$$
$$4 + 2 - 6 \stackrel{?}{=} 0$$
$$0 = 0 \checkmark$$

$$x^2 - x - 6 = 0$$
$$3^2 - 3 - 6 \stackrel{?}{=} 0$$
$$9 - 3 - 6 \stackrel{?}{=} 0$$
$$0 = 0 \checkmark$$

b. Add $-4x$ to each side to obtain $-2x^2 - 4x - 2 = 0$. Graph the related function $y = -2x^2 - 4x - 2$.

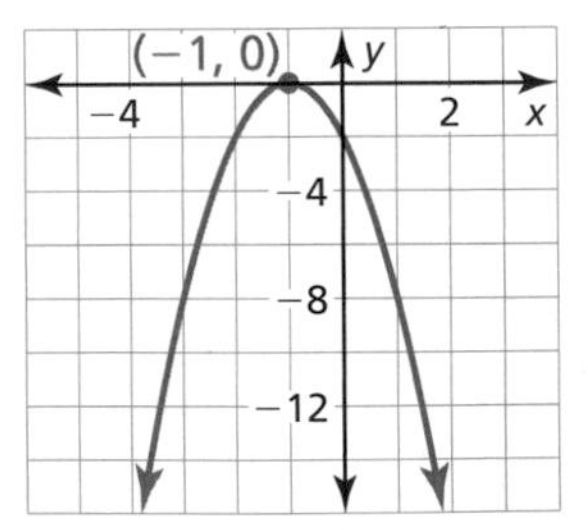

The x-intercept is -1.

▶ The solution, or root, is $x = -1$.

Monitoring Progress

Help in English and Spanish at *BigIdeasMath.com*

Solve the equation by graphing.

1. $x^2 - 8x + 12 = 0$

2. $4x^2 - 12x + 9 = 0$

3. $\frac{1}{2}x^2 = 6x - 20$

Solving Quadratic Equations Algebraically

When solving quadratic equations using square roots, you can use properties of square roots to write your solutions in different forms.

When a radicand in the denominator of a fraction is not a perfect square, you can multiply the fraction by an appropriate form of 1 to eliminate the radical from the denominator. This process is called *rationalizing the denominator*.

EXAMPLE 2 Solving Quadratic Equations Using Square Roots

Solve each equation using square roots.

a. $4x^2 - 31 = 49$ **b.** $3x^2 + 9 = 0$ **c.** $\frac{2}{5}(x + 3)^2 = 5$

SOLUTION

a. $4x^2 - 31 = 49$ Write the equation.

$4x^2 = 80$ Add 31 to each side.

$x^2 = 20$ Divide each side by 4.

$x = \pm\sqrt{20}$ Take square root of each side.

$x = \pm\sqrt{4} \cdot \sqrt{5}$ Product Property of Square Roots

$x = \pm 2\sqrt{5}$ Simplify.

▶ The solutions are $x = 2\sqrt{5}$ and $x = -2\sqrt{5}$.

b. $3x^2 + 9 = 0$ Write the equation.

$3x^2 = -9$ Subtract 9 from each side.

$x^2 = -3$ Divide each side by 3.

▶ The square of a real number cannot be negative. So, the equation has no real solution.

LOOKING FOR STRUCTURE

Notice that $(x + 3)^2 = \frac{25}{2}$ is of the form $u^2 = d$, where $u = x + 3$.

c. $\frac{2}{5}(x + 3)^2 = 5$ Write the equation.

$(x + 3)^2 = \frac{25}{2}$ Multiply each side by $\frac{5}{2}$.

$x + 3 = \pm\sqrt{\frac{25}{2}}$ Take square root of each side.

$x = -3 \pm \sqrt{\frac{25}{2}}$ Subtract 3 from each side.

$x = -3 \pm \frac{\sqrt{25}}{\sqrt{2}}$ Quotient Property of Square Roots

STUDY TIP

Because $\frac{\sqrt{2}}{\sqrt{2}} = 1$, the value of $\frac{\sqrt{25}}{\sqrt{2}}$ does not change when you multiply by $\frac{\sqrt{2}}{\sqrt{2}}$.

$x = -3 \pm \frac{\sqrt{25}}{\sqrt{2}} \cdot \frac{\sqrt{2}}{\sqrt{2}}$ Multiply by $\frac{\sqrt{2}}{\sqrt{2}}$.

$x = -3 \pm \frac{5\sqrt{2}}{2}$ Simplify.

▶ The solutions are $x = -3 + \frac{5\sqrt{2}}{2}$ and $x = -3 - \frac{5\sqrt{2}}{2}$.

Monitoring Progress  Help in English and Spanish at *BigIdeasMath.com*

Solve the equation using square roots.

4. $\frac{2}{3}x^2 + 14 = 20$ **5.** $-2x^2 + 1 = -6$ **6.** $2(x - 4)^2 = -5$

When the left side of $ax^2 + bx + c = 0$ is factorable, you can solve the equation using the *Zero-Product Property*.

Core Concept

Zero-Product Property

Words If the product of two expressions is zero, then one or both of the expressions equal zero.

Algebra If A and B are expressions and $AB = 0$, then $A = 0$ or $B = 0$.

EXAMPLE 3 Solving a Quadratic Equation by Factoring

Solve $x^2 - 4x = 45$ by factoring.

SOLUTION

UNDERSTANDING MATHEMATICAL TERMS

If a real number k is a zero of the function $f(x) = ax^2 + bx + c$, then k is an x-intercept of the graph of the function, and k is also a root of the equation $ax^2 + bx + c = 0$.

$x^2 - 4x = 45$		Write the equation.
$x^2 - 4x - 45 = 0$		Write in standard form.
$(x - 9)(x + 5) = 0$		Factor the polynomial.
$x - 9 = 0$ or	$x + 5 = 0$	Zero-Product Property
$x = 9$ or	$x = -5$	Solve for x.

▶ The solutions are $x = -5$ and $x = 9$.

You know the x-intercepts of the graph of $f(x) = a(x - p)(x - q)$ are p and q. Because the value of the function is zero when $x = p$ and when $x = q$, the numbers p and q are also called *zeros* of the function. A **zero of a function** f is an x-value for which $f(x) = 0$.

EXAMPLE 4 Finding the Zeros of a Quadratic Function

Find the zeros of $f(x) = 2x^2 - 11x + 12$.

SOLUTION

To find the zeros of the function, find the x-values for which $f(x) = 0$.

$2x^2 - 11x + 12 = 0$		Set $f(x)$ equal to 0.
$(2x - 3)(x - 4) = 0$		Factor the polynomial.
$2x - 3 = 0$ or	$x - 4 = 0$	Zero-Product Property
$x = 1.5$ or	$x = 4$	Solve for x.

▶ The zeros of the function are $x = 1.5$ and $x = 4$. You can check this by graphing the function. The x-intercepts are 1.5 and 4.

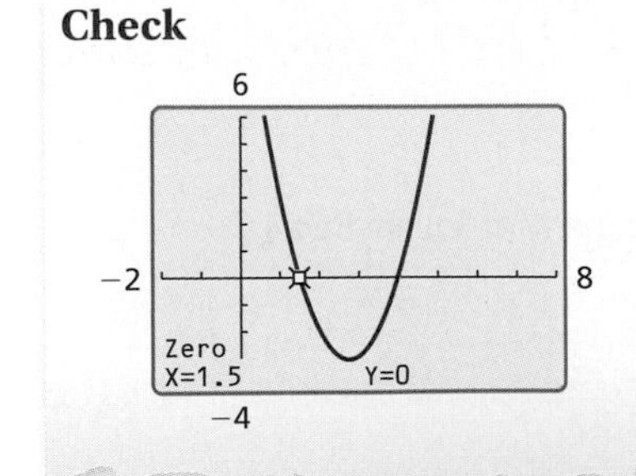

Monitoring Progress  Help in English and Spanish at *BigIdeasMath.com*

Solve the equation by factoring.

7. $x^2 + 12x + 35 = 0$

8. $3x^2 - 5x = 2$

Find the zero(s) of the function.

9. $f(x) = x^2 - 8x$

10. $f(x) = 4x^2 + 28x + 49$

Solving Real-Life Problems

To find the maximum value or minimum value of a quadratic function, you can first use factoring to write the function in intercept form $f(x) = a(x - p)(x - q)$. Because the vertex of the function lies on the axis of symmetry, $x = \frac{p + q}{2}$, the maximum value or minimum value occurs at the average of the zeros p and q.

EXAMPLE 5 **Solving a Multi-Step Problem**

A monthly teen magazine has 48,000 subscribers when it charges \$20 per annual subscription. For each \$1 increase in price, the magazine loses about 2000 subscribers. How much should the magazine charge to maximize annual revenue? What is the maximum annual revenue?

SOLUTION

Step 1 Define the variables. Let x represent the price increase and $R(x)$ represent the annual revenue.

Step 2 Write a verbal model. Then write and simplify a quadratic function.

Annual revenue (dollars)	$\overset{?}{=}$	Number of subscribers (people)	•	Subscription price (dollars/person)
↓		↓		↓
$R(x)$	=	$(48{,}000 - 2000x)$	•	$(20 + x)$

$$R(x) = (48{,}000 - 2000x) \cdot (20 + x)$$
$$R(x) = (-2000x + 48{,}000)(x + 20)$$
$$R(x) = -2000(x - 24)(x + 20)$$

Step 3 Identify the zeros and find their average. Then find how much each subscription should cost to maximize annual revenue.

The zeros of the revenue function are 24 and -20. The average of the zeros is $\frac{24 + (-20)}{2} = 2$.

To maximize revenue, each subscription should cost \$20 + \$2 = \$22.

Step 4 Find the maximum annual revenue.

$$R(2) = -2000(2 - 24)(2 + 20) = \$968{,}000$$

▶ So, the magazine should charge \$22 per subscription to maximize annual revenue. The maximum annual revenue is \$968,000.

Monitoring Progress  Help in English and Spanish at *BigIdeasMath.com*

11. WHAT IF? The magazine initially charges \$21 per annual subscription. How much should the magazine charge to maximize annual revenue? What is the maximum annual revenue?

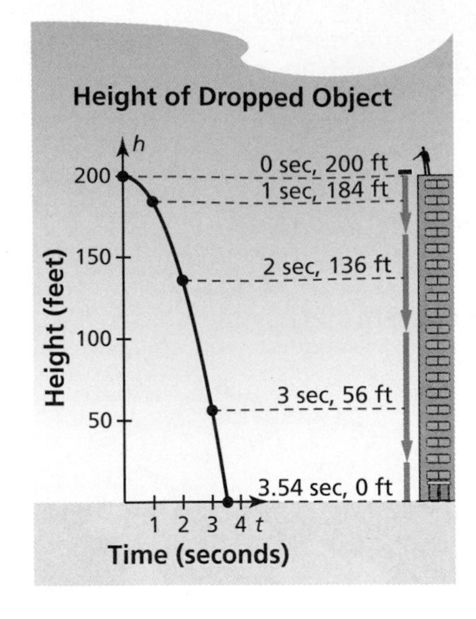

When an object is dropped, its height h (in feet) above the ground after t seconds can be modeled by the function $h = -16t^2 + h_0$, where h_0 is the initial height (in feet) of the object. The graph of $h = -16t^2 + 200$, representing the height of an object dropped from an initial height of 200 feet, is shown at the left.

The model $h = -16t^2 + h_0$ assumes that the force of air resistance on the object is negligible. Also, this model applies only to objects dropped on Earth. For planets with stronger or weaker gravitational forces, different models are used.

EXAMPLE 6 Modeling a Dropped Object

For a science competition, students must design a container that prevents an egg from breaking when dropped from a height of 50 feet.

a. Write a function that gives the height h (in feet) of the container after t seconds. How long does the container take to hit the ground?

b. Find and interpret $h(1) - h(1.5)$.

SOLUTION

a. The initial height is 50, so the model is $h = -16t^2 + 50$. Find the zeros of the function.

$h = -16t^2 + 50$	Write the function.
$0 = -16t^2 + 50$	Substitute 0 for h.
$-50 = -16t^2$	Subtract 50 from each side.
$\frac{-50}{-16} = t^2$	Divide each side by -16.
$\pm\sqrt{\frac{50}{16}} = t$	Take square root of each side.
$\pm 1.8 \approx t$	Use a calculator.

▶ Reject the negative solution, -1.8, because time must be positive. The container will fall for about 1.8 seconds before it hits the ground.

INTERPRETING EXPRESSIONS

In the model for the height of a dropped object, the term $-16t^2$ indicates that an object has fallen $16t^2$ feet after t seconds.

b. Find $h(1)$ and $h(1.5)$. These represent the heights after 1 and 1.5 seconds.

$$h(1) = -16(1)^2 + 50 = -16 + 50 = 34$$

$$h(1.5) = -16(1.5)^2 + 50 = -16(2.25) + 50 = -36 + 50 = 14$$

$$h(1) - h(1.5) = 34 - 14 = 20$$

▶ So, the container fell 20 feet between 1 and 1.5 seconds. You can check this by graphing the function. The points appear to be about 20 feet apart. So, the answer is reasonable.

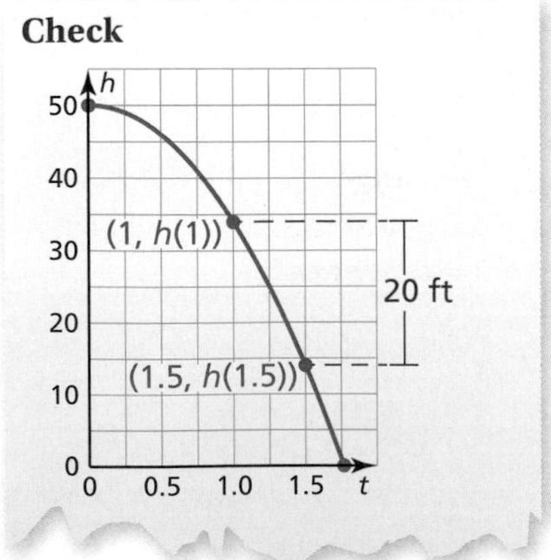

Monitoring Progress

Help in English and Spanish at BigIdeasMath.com

12. WHAT IF? The egg container is dropped from a height of 80 feet. How does this change your answers in parts (a) and (b)?

3.1 Exercises

Dynamic Solutions available at *BigIdeasMath.com*

Vocabulary and Core Concept Check

1. **WRITING** Explain how to use graphing to find the roots of the equation $ax^2 + bx + c = 0$.

2. **DIFFERENT WORDS, SAME QUESTION** Which is different? Find "both" answers.

What are the zeros of $f(x) = x^2 + 3x - 10$?	What are the solutions of $x^2 + 3x - 10 = 0$?
What are the roots of $10 - x^2 = 3x$?	What is the y-intercept of the graph of $y = (x + 5)(x - 2)$?

Monitoring Progress and Modeling with Mathematics

In Exercises 3–12, solve the equation by graphing. *(See Example 1.)*

3. $x^2 + 3x + 2 = 0$
4. $-x^2 + 2x + 3 = 0$
5. $0 = x^2 - 9$
6. $-8 = -x^2 - 4$
7. $8x = -4 - 4x^2$
8. $3x^2 = 6x - 3$
9. $7 = -x^2 - 4x$
10. $2x = x^2 + 2$
11. $\frac{1}{5}x^2 + 6 = 2x$
12. $3x = \frac{1}{4}x^2 + 5$

In Exercises 13–20, solve the equation using square roots. *(See Example 2.)*

13. $s^2 = 144$
14. $a^2 = 81$
15. $(z - 6)^2 = 25$
16. $(p - 4)^2 = 49$
17. $4(x - 1)^2 + 2 = 10$
18. $2(x + 2)^2 - 5 = 8$
19. $\frac{1}{2}r^2 - 10 = \frac{3}{2}r^2$
20. $\frac{1}{5}x^2 + 2 = \frac{3}{5}x^2$

21. **ANALYZING RELATIONSHIPS** Which equations have roots that are equivalent to the x-intercepts of the graph shown?

    Ⓐ $-x^2 - 6x - 8 = 0$

    Ⓑ $0 = (x + 2)(x + 4)$

    Ⓒ $0 = -(x + 2)^2 + 4$

    Ⓓ $2x^2 - 4x - 6 = 0$

    Ⓔ $4(x + 3)^2 - 4 = 0$

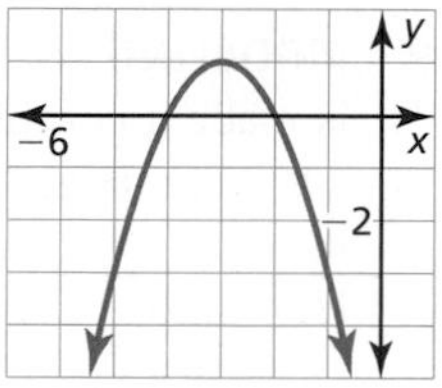

22. **ANALYZING RELATIONSHIPS** Which graph has x-intercepts that are equivalent to the roots of the equation $\left(x - \frac{3}{2}\right)^2 = \frac{25}{4}$? Explain your reasoning.

Ⓐ 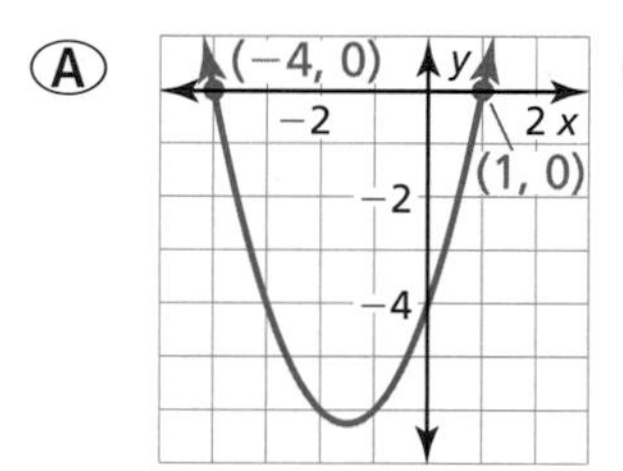

Ⓑ 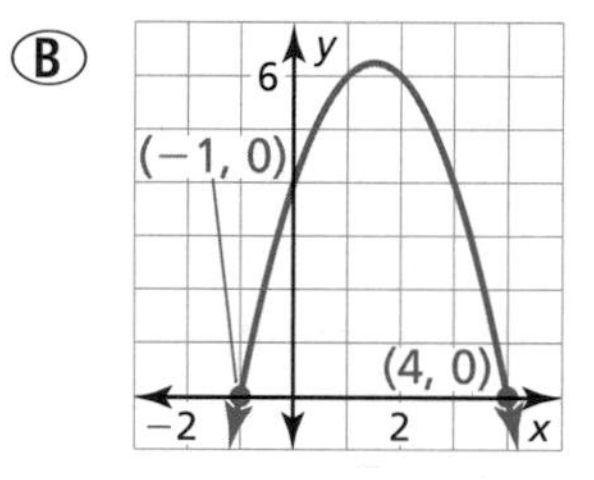

Ⓒ 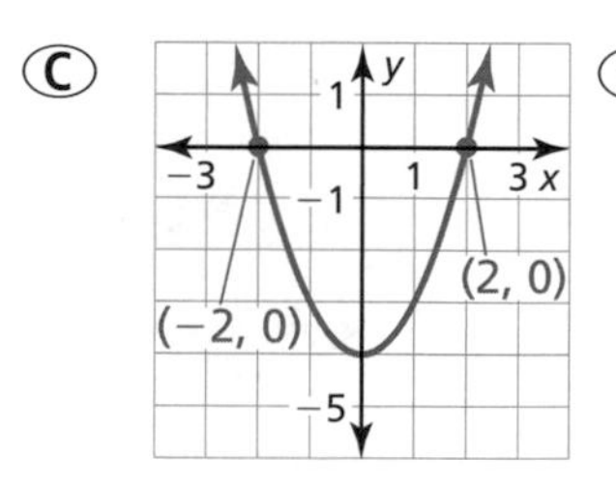

Ⓓ 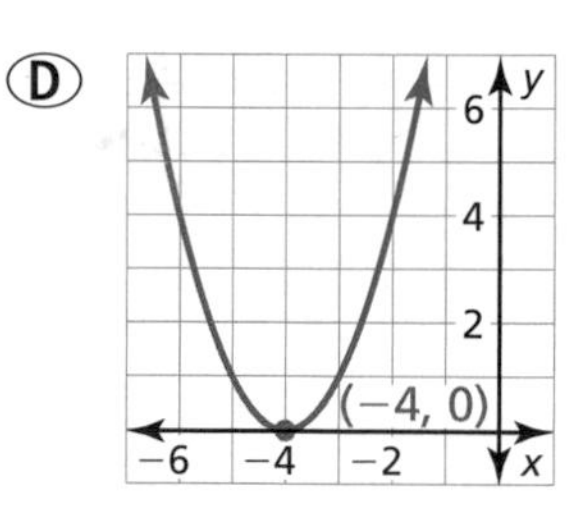

ERROR ANALYSIS In Exercises 23 and 24, describe and correct the error in solving the equation.

23. 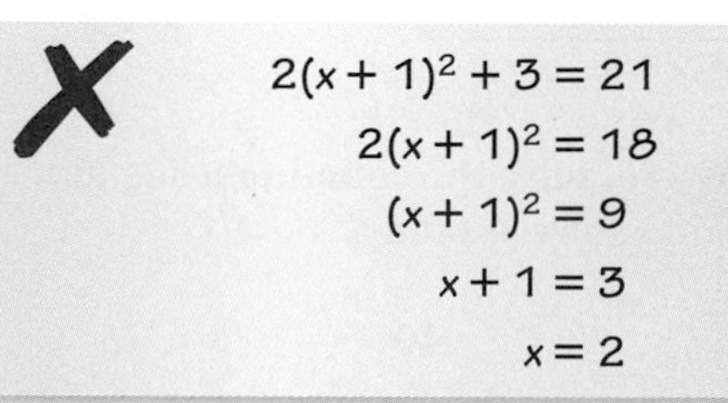

24. 

$-2x^2 - 8 = 0$

$-2x^2 = 8$

$x^2 = -4$

$x = \pm 2$

25. **OPEN-ENDED** Write an equation of the form $x^2 = d$ that has (a) two real solutions, (b) one real solution, and (c) no real solution.

26. **ANALYZING EQUATIONS** Which equation has one real solution? Explain.

Ⓐ $3x^2 + 4 = -2(x^2 + 8)$

Ⓑ $5x^2 - 4 = x^2 - 4$

Ⓒ $2(x + 3)^2 = 18$

Ⓓ $\frac{3}{2}x^2 - 5 = 19$

In Exercises 27–34, solve the equation by factoring. *(See Example 3.)*

27. $0 = x^2 + 6x + 9$ **28.** $0 = z^2 - 10z + 25$

29. $x^2 - 8x = -12$ **30.** $x^2 - 11x = -30$

31. $n^2 - 6n = 0$ **32.** $a^2 - 49 = 0$

33. $2w^2 - 16w = 12w - 48$

34. $-y + 28 + y^2 = 2y + 2y^2$

MATHEMATICAL CONNECTIONS In Exercises 35–38, find the value of *x*.

35. Area of rectangle = 36 **36.** Area of circle = 25π

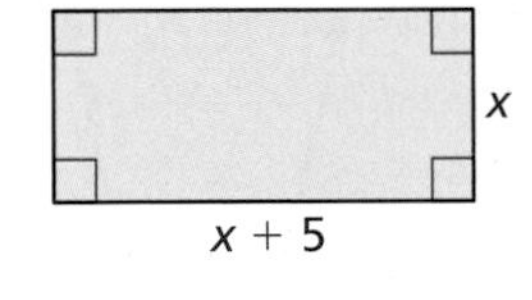

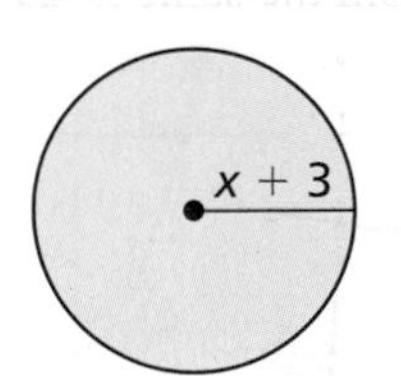

37. Area of triangle = 42 **38.** Area of trapezoid = 32

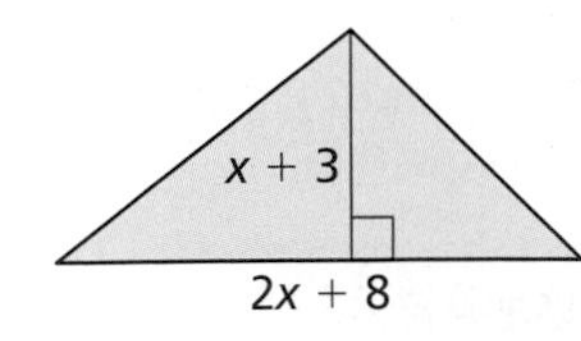

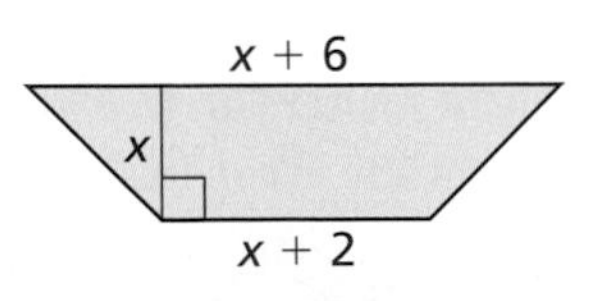

In Exercises 39–46, solve the equation using any method. Explain your reasoning.

39. $u^2 = -9u$ **40.** $\frac{t^2}{20} + 8 = 15$

41. $-(x + 9)^2 = 64$ **42.** $-2(x + 2)^2 = 5$

43. $7(x - 4)^2 - 18 = 10$ **44.** $t^2 + 8t + 16 = 0$

45. $x^2 + 3x + \frac{5}{4} = 0$ **46.** $x^2 - 1.75 = 0.5$

In Exercises 47–54, find the zero(s) of the function. *(See Example 4.)*

47. $g(x) = x^2 + 6x + 8$ **48.** $f(x) = x^2 - 8x + 16$

49. $h(x) = x^2 + 7x - 30$ **50.** $g(x) = x^2 + 11x$

51. $f(x) = 2x^2 - 2x - 12$ **52.** $f(x) = 4x^2 - 12x + 9$

53. $g(x) = x^2 + 22x + 121$

54. $h(x) = x^2 + 19x + 84$

55. **REASONING** Write a quadratic function in the form $f(x) = x^2 + bx + c$ that has zeros 8 and 11.

56. **NUMBER SENSE** Write a quadratic equation in standard form that has roots equidistant from 10 on the number line.

57. **PROBLEM SOLVING** A restaurant sells 330 sandwiches each day. For each \$0.25 decrease in price, the restaurant sells about 15 more sandwiches. How much should the restaurant charge to maximize daily revenue? What is the maximum daily revenue? *(See Example 5.)*

58. **PROBLEM SOLVING** An athletic store sells about 200 pairs of basketball shoes per month when it charges \$120 per pair. For each \$2 increase in price, the store sells two fewer pairs of shoes. How much should the store charge to maximize monthly revenue? What is the maximum monthly revenue?

59. **MODELING WITH MATHEMATICS** Niagara Falls is made up of three waterfalls. The height of the Canadian Horseshoe Falls is about 188 feet above the lower Niagara River. A log falls from the top of Horseshoe Falls. *(See Example 6.)*

a. Write a function that gives the height h (in feet) of the log after t seconds. How long does the log take to reach the river?

b. Find and interpret $h(2) - h(3)$.

60. MODELING WITH MATHEMATICS According to legend, in 1589, the Italian scientist Galileo Galilei dropped rocks of different weights from the top of the Leaning Tower of Pisa to prove his conjecture that the rocks would hit the ground at the same time. The height h (in feet) of a rock after t seconds can be modeled by $h(t) = 196 - 16t^2$.

a. Find and interpret the zeros of the function. Then use the zeros to sketch the graph.

b. What do the domain and range of the function represent in this situation?

61. PROBLEM SOLVING You make a rectangular quilt that is 5 feet by 4 feet. You use the remaining 10 square feet of fabric to add a border of uniform width to the quilt. What is the width of the border?

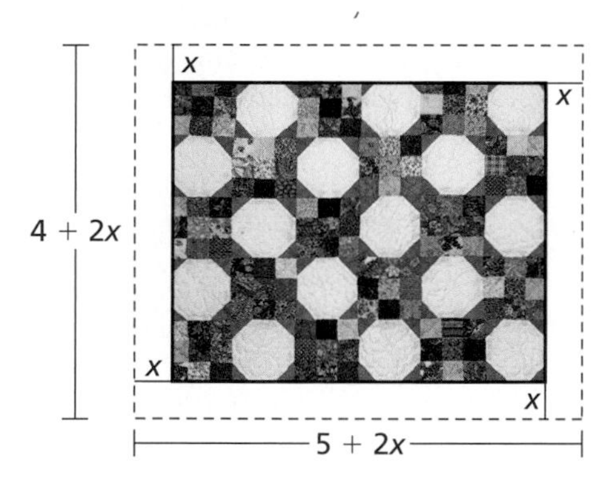

62. MODELING WITH MATHEMATICS You drop a seashell into the ocean from a height of 40 feet. Write an equation that models the height h (in feet) of the seashell above the water after t seconds. How long is the seashell in the air?

63. WRITING The equation $h = 0.019s^2$ models the height h (in feet) of the largest ocean waves when the wind speed is s knots. Compare the wind speeds required to generate 5-foot waves and 20-foot waves.

64. CRITICAL THINKING Write and solve an equation to find two consecutive odd integers whose product is 143.

65. MATHEMATICAL CONNECTIONS A quadrilateral is divided into two right triangles as shown in the figure. What is the length of each side of the quadrilateral?

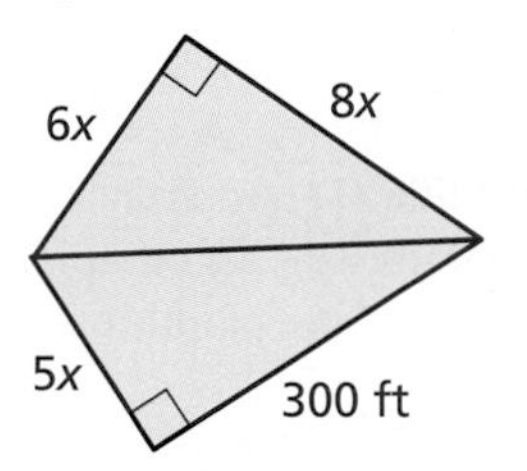

66. ABSTRACT REASONING Suppose the equation $ax^2 + bx + c = 0$ has no real solution and a graph of the related function has a vertex that lies in the second quadrant.

a. Is the value of a positive or negative? Explain your reasoning.

b. Suppose the graph is translated so the vertex is in the fourth quadrant. Does the graph have any x-intercepts? Explain.

67. REASONING When an object is dropped on *any* planet, its height h (in feet) after t seconds can be modeled by the function $h = -\frac{g}{2}t^2 + h_0$, where h_0 is the object's initial height and g is the planet's acceleration due to gravity. Suppose a rock is dropped from the same initial height on the three planets shown. Make a conjecture about which rock will hit the ground first. Justify your answer.

Earth: $g = 32$ ft/sec²   Mars: $g = 12$ ft/sec²   Jupiter: $g = 76$ ft/sec²

68. PROBLEM SOLVING A café has an outdoor, rectangular patio. The owner wants to add 329 square feet to the area of the patio by expanding the existing patio as shown. Write and solve an equation to find the value of x. By what distance should the patio be extended?

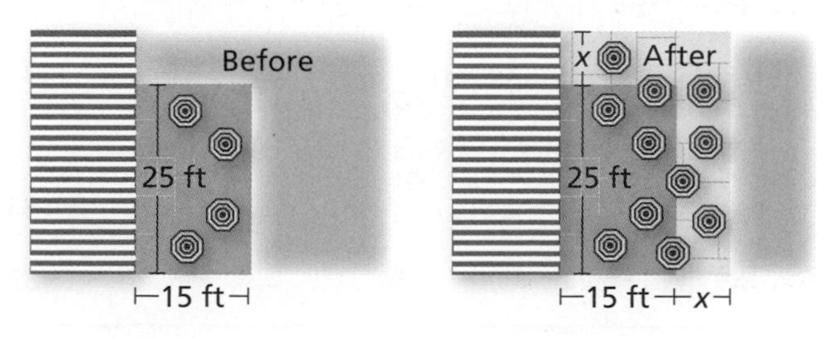

69. PROBLEM SOLVING A flea can jump very long distances. The path of the jump of a flea can be modeled by the graph of the function $y = -0.189x^2 + 2.462x$, where x is the horizontal distance (in inches) and y is the vertical distance (in inches). Graph the function. Identify the vertex and zeros and interpret their meanings in this situation.

70. HOW DO YOU SEE IT? An artist is painting a mural and drops a paintbrush. The graph represents the height h (in feet) of the paintbrush after t seconds.

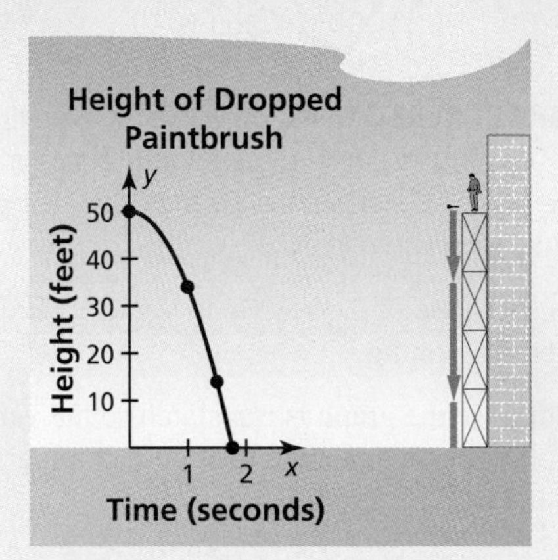

a. What is the initial height of the paintbrush?

b. How long does it take the paintbrush to reach the ground? Explain.

71. MAKING AN ARGUMENT Your friend claims the equation $x^2 + 7x = -49$ can be solved by factoring and has a solution of $x = 7$. You solve the equation by graphing the related function and claim there is no solution. Who is correct? Explain.

72. ABSTRACT REASONING Factor the expressions $x^2 - 4$ and $x^2 - 9$. Recall that an expression in this form is called a difference of two squares. Use your answers to factor the expression $x^2 - a^2$. Graph the related function $y = x^2 - a^2$. Label the vertex, x-intercepts, and axis of symmetry.

73. DRAWING CONCLUSIONS Consider the expression $x^2 + a^2$, where $a > 0$.

a. You want to rewrite the expression as $(x + m)(x + n)$. Write two equations that m and n must satisfy.

b. Use the equations you wrote in part (a) to solve for m and n. What can you conclude?

74. THOUGHT PROVOKING You are redesigning a rectangular raft. The raft is 6 feet long and 4 feet wide. You want to double the area of the raft by adding to the existing design. Draw a diagram of the new raft. Write and solve an equation you can use to find the dimensions of the new raft.

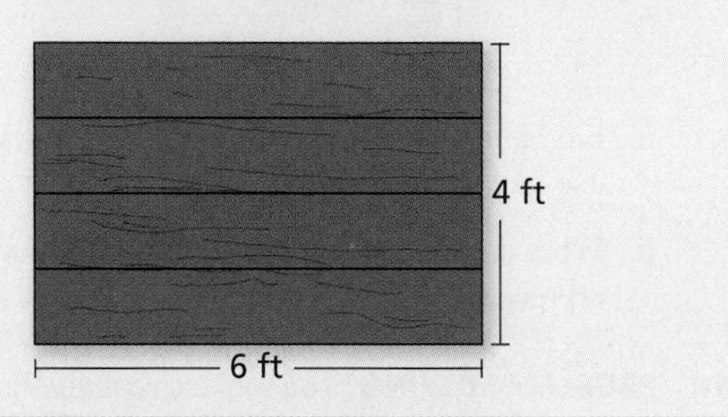

75. MODELING WITH MATHEMATICS A high school wants to double the size of its parking lot by expanding the existing lot as shown. By what distance x should the lot be expanded?

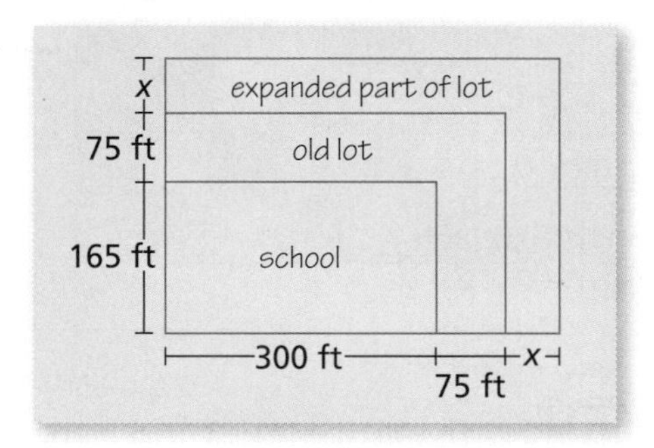

Maintaining Mathematical Proficiency

Reviewing what you learned in previous grades and lessons

Find the sum or difference. *(Skills Review Handbook)*

76. $(x^2 + 2) + (2x^2 - x)$

77. $(x^3 + x^2 - 4) + (3x^2 + 10)$

78. $(-2x + 1) - (-3x^2 + x)$

79. $(-3x^3 + x^2 - 12x) - (-6x^2 + 3x - 9)$

Find the product. *(Skills Review Handbook)*

80. $(x + 2)(x - 2)$

81. $2x(3 - x + 5x^2)$

82. $(7 - x)(x - 1)$

83. $11x(-4x^2 + 3x + 8)$

3.2 Complex Numbers

Essential Question

What are the subsets of the set of complex numbers?

In your study of mathematics, you have probably worked with only *real numbers*, which can be represented graphically on the real number line. In this lesson, the system of numbers is expanded to include *imaginary numbers*. The real numbers and imaginary numbers compose the set of *complex numbers*.

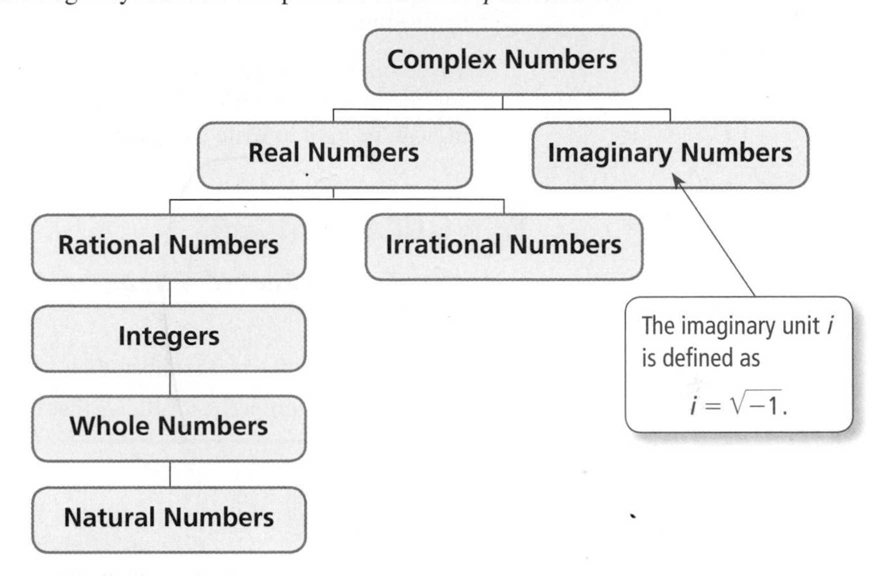

EXPLORATION 1 Classifying Numbers

Work with a partner. Determine which subsets of the set of complex numbers contain each number.

a. $\sqrt{9}$ **b.** $\sqrt{0}$ **c.** $-\sqrt{4}$

d. $\sqrt{\frac{4}{9}}$ **e.** $\sqrt{2}$ **f.** $\sqrt{-1}$

ATTENDING TO PRECISION

To be proficient in math, you need to use clear definitions in your reasoning and discussions with others.

EXPLORATION 2 Complex Solutions of Quadratic Equations

Work with a partner. Use the definition of the imaginary unit i to match each quadratic equation with its complex solution. Justify your answers.

a. $x^2 - 4 = 0$ **b.** $x^2 + 1 = 0$ **c.** $x^2 - 1 = 0$

d. $x^2 + 4 = 0$ **e.** $x^2 - 9 = 0$ **f.** $x^2 + 9 = 0$

A. i **B.** $3i$ **C.** 3

D. $2i$ **E.** 1 **F.** 2

Communicate Your Answer

3. What are the subsets of the set of complex numbers? Give an example of a number in each subset.

4. Is it possible for a number to be both whole and natural? natural and rational? rational and irrational? real and imaginary? Explain your reasoning.

3.2 Lesson

Core Vocabulary

imaginary unit *i*, *p. 104*
complex number, *p. 104*
imaginary number, *p. 104*
pure imaginary number, *p. 104*

What You Will Learn

- Define and use the imaginary unit *i*.
- Add, subtract, and multiply complex numbers.
- Find complex solutions and zeros.

The Imaginary Unit *i*

Not all quadratic equations have real-number solutions. For example, $x^2 = -3$ has no real-number solutions because the square of any real number is never a negative number.

To overcome this problem, mathematicians created an expanded system of numbers using the **imaginary unit *i***, defined as $i = \sqrt{-1}$. Note that $i^2 = -1$. The imaginary unit i can be used to write the square root of *any* negative number.

Core Concept

The Square Root of a Negative Number

Property	Example
1. If r is a positive real number, then $\sqrt{-r} = i\sqrt{r}$.	$\sqrt{-3} = i\sqrt{3}$
2. By the first property, it follows that $(i\sqrt{r})^2 = -r$.	$(i\sqrt{3})^2 = i^2 \cdot 3 = -3$

EXAMPLE 1 Finding Square Roots of Negative Numbers

Find the square root of each number.

a. $\sqrt{-25}$ **b.** $\sqrt{-72}$ **c.** $-5\sqrt{-9}$

SOLUTION

a. $\sqrt{-25} = \sqrt{25} \cdot \sqrt{-1} = 5i$

b. $\sqrt{-72} = \sqrt{72} \cdot \sqrt{-1} = \sqrt{36} \cdot \sqrt{2} \cdot i = 6\sqrt{2}\, i = 6i\sqrt{2}$

c. $-5\sqrt{-9} = -5\sqrt{9} \cdot \sqrt{-1} = -5 \cdot 3 \cdot i = -15i$

Monitoring Progress  Help in English and Spanish at *BigIdeasMath.com*

Find the square root of the number.

1. $\sqrt{-4}$ **2.** $\sqrt{-12}$ **3.** $-\sqrt{-36}$ **4.** $2\sqrt{-54}$

A **complex number** written in *standard form* is a number $a + bi$ where a and b are real numbers. The number a is the *real part*, and the number bi is the *imaginary part*.

$$a + bi$$

If $b \neq 0$, then $a + bi$ is an **imaginary number**. If $a = 0$ and $b \neq 0$, then $a + bi$ is a **pure imaginary number**. The diagram shows how different types of complex numbers are related.

Complex Numbers ($a + bi$)

Real Numbers ($a + 0i$)	Imaginary Numbers ($a + bi, b \neq 0$)
-1 $\frac{5}{3}$	$2 + 3i$ $9 - 5i$
π $\sqrt{2}$	Pure Imaginary Numbers ($0 + bi, b \neq 0$): $-4i$ $6i$

Two complex numbers $a + bi$ and $c + di$ are equal if and only if $a = c$ and $b = d$.

EXAMPLE 2 Equality of Two Complex Numbers

Find the values of x and y that satisfy the equation $2x - 7i = 10 + yi$.

SOLUTION

Set the real parts equal to each other and the imaginary parts equal to each other.

$2x = 10$	Equate the real parts.	$-7i = yi$	Equate the imaginary parts.
$x = 5$	Solve for x.	$-7 = y$	Solve for y.

▶ So, $x = 5$ and $y = -7$.

Monitoring Progress  Help in English and Spanish at *BigIdeasMath.com*

Find the values of x and y that satisfy the equation.

5. $x + 3i = 9 - yi$

6. $9 + 4yi = -2x + 3i$

Operations with Complex Numbers

Core Concept

Sums and Differences of Complex Numbers

To add (or subtract) two complex numbers, add (or subtract) their real parts and their imaginary parts separately.

Sum of complex numbers: $(a + bi) + (c + di) = (a + c) + (b + d)i$

Difference of complex numbers: $(a + bi) - (c + di) = (a - c) + (b - d)i$

EXAMPLE 3 Adding and Subtracting Complex Numbers

Add or subtract. Write the answer in standard form.

a. $(8 - i) + (5 + 4i)$

b. $(7 - 6i) - (3 - 6i)$

c. $13 - (2 + 7i) + 5i$

SOLUTION

a. $(8 - i) + (5 + 4i) = (8 + 5) + (-1 + 4)i$ — Definition of complex addition

$= 13 + 3i$ — Write in standard form.

b. $(7 - 6i) - (3 - 6i) = (7 - 3) + (-6 + 6)i$ — Definition of complex subtraction

$= 4 + 0i$ — Simplify.

$= 4$ — Write in standard form.

c. $13 - (2 + 7i) + 5i = [(13 - 2) - 7i] + 5i$ — Definition of complex subtraction

$= (11 - 7i) + 5i$ — Simplify.

$= 11 + (-7 + 5)i$ — Definition of complex addition

$= 11 - 2i$ — Write in standard form.

EXAMPLE 4 Solving a Real-Life Problem

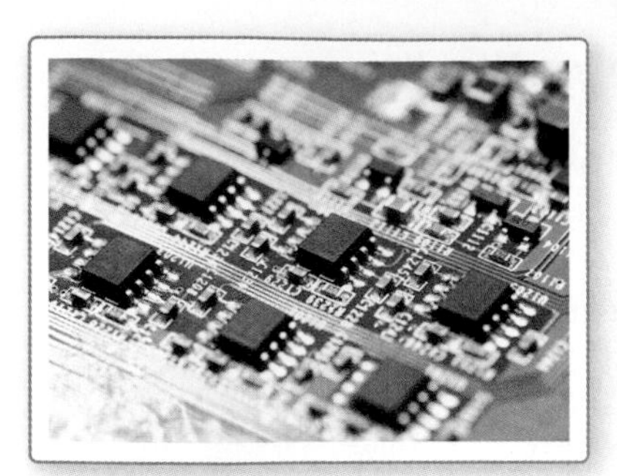

Electrical circuit components, such as resistors, inductors, and capacitors, all oppose the flow of current. This opposition is called *resistance* for resistors and *reactance* for inductors and capacitors. Each of these quantities is measured in ohms. The symbol used for ohms is Ω, the uppercase Greek letter omega.

Component and symbol	Resistor	Inductor	Capacitor
Resistance or reactance (in ohms)	R	L	C
Impedance (in ohms)	R	Li	$-Ci$

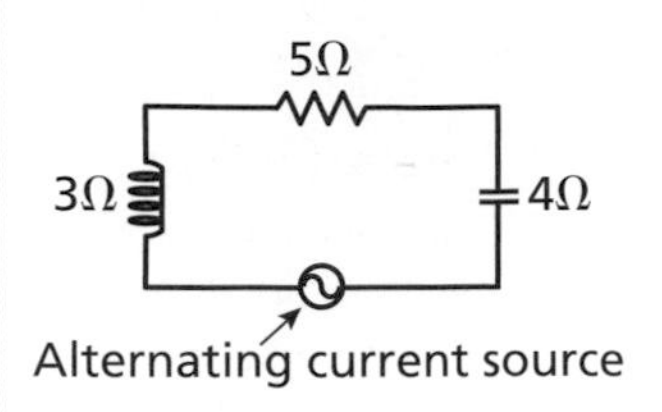

The table shows the relationship between a component's resistance or reactance and its contribution to impedance. A *series circuit* is also shown with the resistance or reactance of each component labeled. The impedance for a series circuit is the sum of the impedances for the individual components. Find the impedance of the circuit.

SOLUTION

The resistor has a resistance of 5 ohms, so its impedance is 5 ohms. The inductor has a reactance of 3 ohms, so its impedance is $3i$ ohms. The capacitor has a reactance of 4 ohms, so its impedance is $-4i$ ohms.

$$\text{Impedance of circuit} = 5 + 3i + (-4i) = 5 - i$$

▶ The impedance of the circuit is $(5 - i)$ ohms.

To multiply two complex numbers, use the Distributive Property, or the FOIL method, just as you do when multiplying real numbers or algebraic expressions.

EXAMPLE 5 Multiplying Complex Numbers

Multiply. Write the answer in standard form.

a. $4i(-6 + i)$ **b.** $(9 - 2i)(-4 + 7i)$

STUDY TIP

When simplifying an expression that involves complex numbers, be sure to simplify i^2 as -1.

SOLUTION

a. $4i(-6 + i) = -24i + 4i^2$ Distributive Property

$= -24i + 4(-1)$ Use $i^2 = -1$.

$= -4 - 24i$ Write in standard form.

b. $(9 - 2i)(-4 + 7i) = -36 + 63i + 8i - 14i^2$ Multiply using FOIL.

$= -36 + 71i - 14(-1)$ Simplify and use $i^2 = -1$.

$= -36 + 71i + 14$ Simplify.

$= -22 + 71i$ Write in standard form.

Monitoring Progress  Help in English and Spanish at *BigIdeasMath.com*

7. WHAT IF? In Example 4, what is the impedance of the circuit when the capacitor is replaced with one having a reactance of 7 ohms?

Perform the operation. Write the answer in standard form.

8. $(9 - i) + (-6 + 7i)$ **9.** $(3 + 7i) - (8 - 2i)$ **10.** $-4 - (1 + i) - (5 + 9i)$

11. $(-3i)(10i)$ **12.** $i(8 - i)$ **13.** $(3 + i)(5 - i)$

Complex Solutions and Zeros

EXAMPLE 6 Solving Quadratic Equations

Solve (a) $x^2 + 4 = 0$ and (b) $2x^2 - 11 = -47$.

SOLUTION

a. $x^2 + 4 = 0$ Write original equation.

$x^2 = -4$ Subtract 4 from each side.

$x = \pm\sqrt{-4}$ Take square root of each side.

$x = \pm 2i$ Write in terms of i.

▶ The solutions are $2i$ and $-2i$.

LOOKING FOR STRUCTURE

Notice that you can use the solutions in Example 6(a) to factor $x^2 + 4$ as $(x + 2i)(x - 2i)$.

b. $2x^2 - 11 = -47$ Write original equation.

$2x^2 = -36$ Add 11 to each side.

$x^2 = -18$ Divide each side by 2.

$x = \pm\sqrt{-18}$ Take square root of each side.

$x = \pm i\sqrt{18}$ Write in terms of i.

$x = \pm 3i\sqrt{2}$ Simplify radical.

▶ The solutions are $3i\sqrt{2}$ and $-3i\sqrt{2}$.

EXAMPLE 7 Finding Zeros of a Quadratic Function

Find the zeros of $f(x) = 4x^2 + 20$.

SOLUTION

$4x^2 + 20 = 0$ Set $f(x)$ equal to 0.

$4x^2 = -20$ Subtract 20 from each side.

$x^2 = -5$ Divide each side by 4.

$x = \pm\sqrt{-5}$ Take square root of each side.

$x = \pm i\sqrt{5}$ Write in terms of i.

▶ So, the zeros of f are $i\sqrt{5}$ and $-i\sqrt{5}$.

FINDING AN ENTRY POINT

The graph of f does not intersect the x-axis, which means f has no real zeros. So, f must have complex zeros, which you can find algebraically.

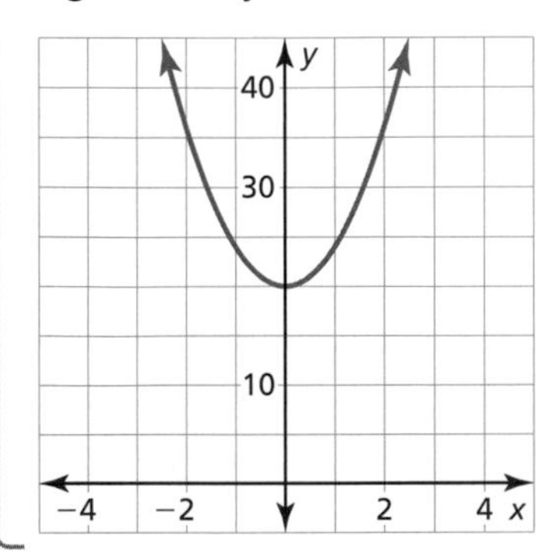

Check

$f(i\sqrt{5}) = 4(i\sqrt{5})^2 + 20 = 4 \cdot 5i^2 + 20 = 4(-5) + 20 = 0$ ✓

$f(-i\sqrt{5}) = 4(-i\sqrt{5})^2 + 20 = 4 \cdot 5i^2 + 20 = 4(-5) + 20 = 0$ ✓

Monitoring Progress  Help in English and Spanish at *BigIdeasMath.com*

Solve the equation.

14. $x^2 = -13$ **15.** $x^2 = -38$ **16.** $x^2 + 11 = 3$

17. $x^2 - 8 = -36$ **18.** $3x^2 - 7 = -31$ **19.** $5x^2 + 33 = 3$

Find the zeros of the function.

20. $f(x) = x^2 + 7$ **21.** $f(x) = -x^2 - 4$ **22.** $f(x) = 9x^2 + 1$

3.2 Exercises

Dynamic Solutions available at *BigIdeasMath.com*

Vocabulary and Core Concept Check

1. **VOCABULARY** What is the imaginary unit i defined as and how can you use i?

2. **COMPLETE THE SENTENCE** For the complex number $5 + 2i$, the imaginary part is ____ and the real part is ____.

3. **WRITING** Describe how to add complex numbers.

4. **WHICH ONE DOESN'T BELONG?** Which number does *not* belong with the other three? Explain your reasoning.

   $3 + 0i$ $\quad$ $2 + 5i$ $\quad$ $\sqrt{3} + 6i$ $\quad$ $0 - 7i$

Monitoring Progress and Modeling with Mathematics

In Exercises 5–12, find the square root of the number. *(See Example 1.)*

5. $\sqrt{-36}$
6. $\sqrt{-64}$
7. $\sqrt{-18}$
8. $\sqrt{-24}$
9. $2\sqrt{-16}$
10. $-3\sqrt{-49}$
11. $-4\sqrt{-32}$
12. $6\sqrt{-63}$

In Exercises 13–20, find the values of x and y that satisfy the equation. *(See Example 2.)*

13. $4x + 2i = 8 + yi$
14. $3x + 6i = 27 + yi$
15. $-10x + 12i = 20 + 3yi$
16. $9x - 18i = -36 + 6yi$
17. $2x - yi = 14 + 12i$
18. $-12x + yi = 60 - 13i$
19. $54 - \frac{1}{7}yi = 9x - 4i$
20. $15 - 3yi = \frac{1}{2}x + 2i$

In Exercises 21–30, add or subtract. Write the answer in standard form. *(See Example 3.)*

21. $(6 - i) + (7 + 3i)$
22. $(9 + 5i) + (11 + 2i)$
23. $(12 + 4i) - (3 - 7i)$
24. $(2 - 15i) - (4 + 5i)$
25. $(12 - 3i) + (7 + 3i)$
26. $(16 - 9i) - (2 - 9i)$
27. $7 - (3 + 4i) + 6i$
28. $16 - (2 - 3i) - i$
29. $-10 + (6 - 5i) - 9i$
30. $-3 + (8 + 2i) + 7i$

31. **USING STRUCTURE** Write each expression as a complex number in standard form.

    a. $\sqrt{-9} + \sqrt{-4} - \sqrt{16}$

    b. $\sqrt{-16} + \sqrt{8} + \sqrt{-36}$

32. **REASONING** The additive inverse of a complex number z is a complex number z_a such that $z + z_a = 0$. Find the additive inverse of each complex number.

    a. $z = 1 + i$ $\quad$ b. $z = 3 - i$ $\quad$ c. $z = -2 + 8i$

In Exercises 33–36, find the impedance of the series circuit. *(See Example 4.)*

33.

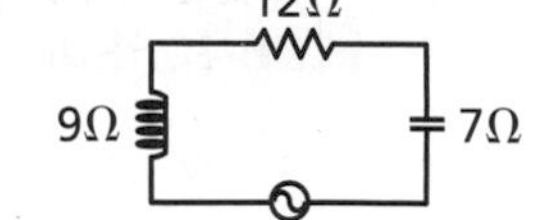

34.

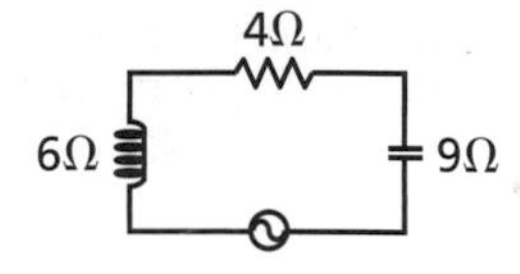

35.

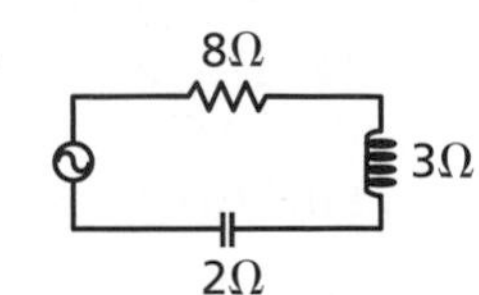

36.

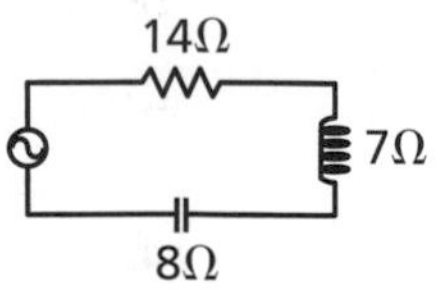

14Ω
7Ω
8Ω

In Exercises 37–44, multiply. Write the answer in standard form. *(See Example 5.)*

37. $3i(-5 + i)$ **38.** $2i(7 - i)$

39. $(3 - 2i)(4 + i)$ **40.** $(7 + 5i)(8 - 6i)$

41. $(4 - 2i)(4 + 2i)$ **42.** $(9 + 5i)(9 - 5i)$

43. $(3 - 6i)^2$ **44.** $(8 + 3i)^2$

JUSTIFYING STEPS In Exercises 45 and 46, justify each step in performing the operation.

45. $11 - (4 + 3i) + 5i$

$= [(11 - 4) - 3i] + 5i$

$= (7 - 3i) + 5i$

$= 7 + (-3 + 5)i$

$= 7 + 2i$

46. $(3 + 2i)(7 - 4i)$

$= 21 - 12i + 14i - 8i^2$

$= 21 + 2i - 8(-1)$

$= 21 + 2i + 8$

$= 29 + 2i$

REASONING In Exercises 47 and 48, place the tiles in the expression to make a true statement.

47. $(\_\_\_\_ - \_\_\_\_i) - (\_\_\_\_ - \_\_\_\_i) = 2 - 4i$

7 4 3 6

48. $\_\_\_\_i(\_\_\_\_ + \_\_\_\_i) = -18 - 10i$

−5 9 2

In Exercises 49–54, solve the equation. Check your solution(s). *(See Example 6.)*

49. $x^2 + 9 = 0$ **50.** $x^2 + 49 = 0$

51. $x^2 - 4 = -11$

52. $x^2 - 9 = -15$

53. $2x^2 + 6 = -34$

54. $x^2 + 7 = -47$

In Exercises 55–62, find the zeros of the function. *(See Example 7.)*

55. $f(x) = 3x^2 + 6$ **56.** $g(x) = 7x^2 + 21$

57. $h(x) = 2x^2 + 72$ **58.** $k(x) = -5x^2 - 125$

59. $m(x) = -x^2 - 27$ **60.** $p(x) = x^2 + 98$

61. $r(x) = -\frac{1}{2}x^2 - 24$ **62.** $f(x) = -\frac{1}{5}x^2 - 10$

ERROR ANALYSIS In Exercises 63 and 64, describe and correct the error in performing the operation and writing the answer in standard form.

63.

✗ $(3 + 2i)(5 - i) = 15 - 3i + 10i - 2i^2$

$= 15 + 7i - 2i^2$

$= -2i^2 + 7i + 15$

64.

✗ $(4 + 6i)^2 = (4)^2 + (6i)^2$

$= 16 + 36i^2$

$= 16 + (36)(-1)$

$= -20$

65. NUMBER SENSE Simplify each expression. Then classify your results in the table below.

a. $(-4 + 7i) + (-4 - 7i)$

b. $(2 - 6i) - (-10 + 4i)$

c. $(25 + 15i) - (25 - 6i)$

d. $(5 + i)(8 - i)$

e. $(17 - 3i) + (-17 - 6i)$

f. $(-1 + 2i)(11 - i)$

g. $(7 + 5i) + (7 - 5i)$

h. $(-3 + 6i) - (-3 - 8i)$

Real numbers	Imaginary numbers	Pure imaginary numbers

66. MAKING AN ARGUMENT The Product Property of Square Roots states $\sqrt{a} \cdot \sqrt{b} = \sqrt{ab}$. Your friend concludes $\sqrt{-4} \cdot \sqrt{-9} = \sqrt{36} = 6$. Is your friend correct? Explain.

67. FINDING A PATTERN Make a table that shows the powers of i from i^1 to i^8 in the first row and the simplified forms of these powers in the second row. Describe the pattern you observe in the table. Verify the pattern continues by evaluating the next four powers of i.

68. HOW DO YOU SEE IT? The graphs of three functions are shown. Which function(s) has real zeros? imaginary zeros? Explain your reasoning.

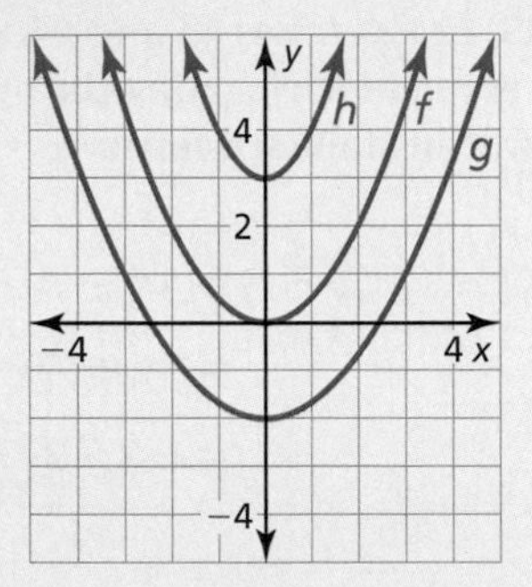

In Exercises 69–74, write the expression as a complex number in standard form.

69. $(3 + 4i) - (7 - 5i) + 2i(9 + 12i)$

70. $3i(2 + 5i) + (6 - 7i) - (9 + i)$

71. $(3 + 5i)(2 - 7i^4)$

72. $2i^3(5 - 12i)$

73. $(2 + 4i^5) + (1 - 9i^6) - (3 + i^7)$

74. $(8 - 2i^4) + (3 - 7i^8) - (4 + i^9)$

75. OPEN-ENDED Find two imaginary numbers whose sum and product are real numbers. How are the imaginary numbers related?

76. COMPARING METHODS Describe the two different methods shown for writing the complex expression in standard form. Which method do you prefer? Explain.

Method 1

$$4i(2 - 3i) + 4i(1 - 2i) = 8i - 12i^2 + 4i - 8i^2$$
$$= 8i - 12(-1) + 4i - 8(-1)$$
$$= 20 + 12i$$

Method 2

$$4i(2 - 3i) + 4i(1 - 2i) = 4i[(2 - 3i) + (1 - 2i)]$$
$$= 4i[3 - 5i]$$
$$= 12i - 20i^2$$
$$= 12i - 20(-1)$$
$$= 20 + 12i$$

77. CRITICAL THINKING Determine whether each statement is *true* or *false*. If it is true, give an example. If it is false, give a counterexample.

a. The sum of two imaginary numbers is an imaginary number.

b. The product of two pure imaginary numbers is a real number.

c. A pure imaginary number is an imaginary number.

d. A complex number is a real number.

78. THOUGHT PROVOKING Create a circuit that has an impedance of $14 - 3i$.

Maintaining Mathematical Proficiency

Reviewing what you learned in previous grades and lessons

Determine whether the given value of x is a solution to the equation. *(Skills Review Handbook)*

79. $3(x - 2) + 4x - 1 = x - 1;\ x = 1$

80. $x^3 - 6 = 2x^2 + 9 - 3x;\ x = -5$

81. $-x^2 + 4x = \frac{19}{3}x^2;\ x = -\frac{3}{4}$

Write a quadratic function in vertex form whose graph is shown. *(Section 2.4)*

82.

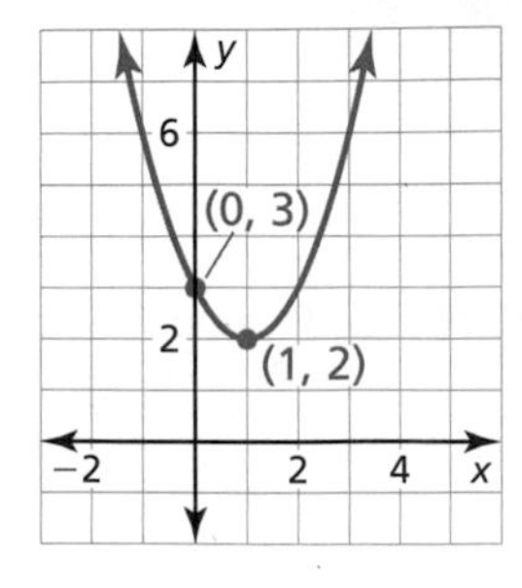

83.

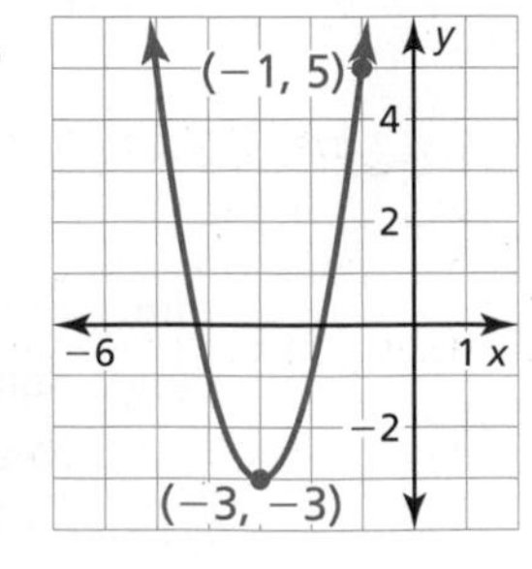

84.

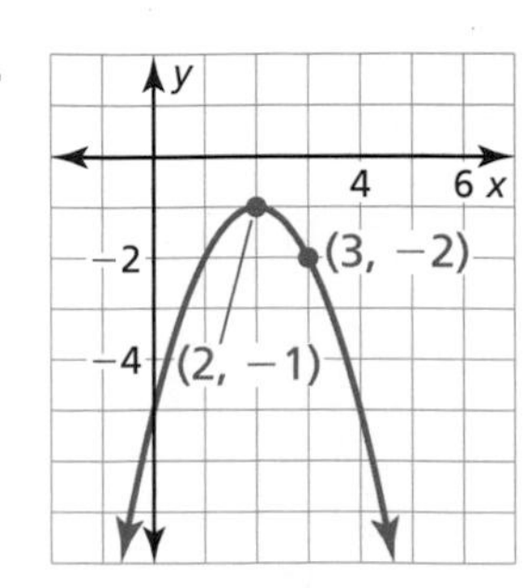

3.3 Completing the Square

Essential Question How can you complete the square for a quadratic expression?

EXPLORATION 1 Using Algebra Tiles to Complete the Square

Work with a partner. Use algebra tiles to complete the square for the expression

$x^2 + 6x$.

a. You can model $x^2 + 6x$ using one x^2-tile and six x-tiles. Arrange the tiles in a square. Your arrangement will be incomplete in one of the corners.

b. How many 1-tiles do you need to complete the square?

c. Find the value of c so that the expression

$x^2 + 6x + c$

is a perfect square trinomial.

d. Write the expression in part (c) as the square of a binomial.

EXPLORATION 2 Drawing Conclusions

Work with a partner.

a. Use the method outlined in Exploration 1 to complete the table.

Expression	Value of c needed to complete the square	Expression written as a binomial squared
$x^2 + 2x + c$		
$x^2 + 4x + c$		
$x^2 + 8x + c$		
$x^2 + 10x + c$		

LOOKING FOR STRUCTURE

To be proficient in math, you need to look closely to discern a pattern or structure.

b. Look for patterns in the last column of the table. Consider the general statement $x^2 + bx + c = (x + d)^2$. How are d and b related in each case? How are c and d related in each case?

c. How can you obtain the values in the second column directly from the coefficients of x in the first column?

Communicate Your Answer

3. How can you complete the square for a quadratic expression?

4. Describe how you can solve the quadratic equation $x^2 + 6x = 1$ by completing the square.

3.3 Lesson

Core Vocabulary

completing the square, *p. 112*

Previous
perfect square trinomial
vertex form

What You Will Learn

- Solve quadratic equations using square roots.
- Solve quadratic equations by completing the square.
- Write quadratic functions in vertex form.

Solving Quadratic Equations Using Square Roots

Previously, you have solved equations of the form $u^2 = d$ by taking the square root of each side. This method also works when one side of an equation is a perfect square trinomial and the other side is a constant.

EXAMPLE 1 Solving a Quadratic Equation Using Square Roots

Solve $x^2 - 16x + 64 = 100$ using square roots.

SOLUTION

$x^2 - 16x + 64 = 100$ — Write the equation.

$(x - 8)^2 = 100$ — Write the left side as a binomial squared.

$x - 8 = \pm 10$ — Take square root of each side.

$x = 8 \pm 10$ — Add 8 to each side.

▶ So, the solutions are $x = 8 + 10 = 18$ and $x = 8 - 10 = -2$.

ANOTHER WAY

You can also solve the equation by writing it in standard form as $x^2 - 16x - 36 = 0$ and factoring.

Monitoring Progress  Help in English and Spanish at *BigIdeasMath.com*

Solve the equation using square roots. Check your solution(s).

1. $x^2 + 4x + 4 = 36$ **2.** $x^2 - 6x + 9 = 1$ **3.** $x^2 - 22x + 121 = 81$

In Example 1, the expression $x^2 - 16x + 64$ is a perfect square trinomial because it equals $(x - 8)^2$. Sometimes you need to add a term to an expression $x^2 + bx$ to make it a perfect square trinomial. This process is called **completing the square**.

Core Concept

Completing the Square

Words To complete the square for the expression $x^2 + bx$, add $\left(\frac{b}{2}\right)^2$.

Diagrams In each diagram, the combined area of the shaded regions is $x^2 + bx$. Adding $\left(\frac{b}{2}\right)^2$ completes the square in the second diagram.

Algebra $x^2 + bx + \left(\frac{b}{2}\right)^2 = \left(x + \frac{b}{2}\right)\left(x + \frac{b}{2}\right) = \left(x + \frac{b}{2}\right)^2$

Solving Quadratic Equations by Completing the Square

EXAMPLE 2 Making a Perfect Square Trinomial

Find the value of c that makes $x^2 + 14x + c$ a perfect square trinomial. Then write the expression as the square of a binomial.

SOLUTION

Step 1 Find half the coefficient of x. $\frac{14}{2} = 7$

Step 2 Square the result of Step 1. $7^2 = 49$

Step 3 Replace c with the result of Step 2. $x^2 + 14x + 49$

	x	7
x	x^2	$7x$
7	$7x$	49

▶ The expression $x^2 + 14x + c$ is a perfect square trinomial when $c = 49$. Then $x^2 + 14x + 49 = (x + 7)(x + 7) = (x + 7)^2$.

Monitoring Progress  Help in English and Spanish at *BigIdeasMath.com*

Find the value of c that makes the expression a perfect square trinomial. Then write the expression as the square of a binomial.

4. $x^2 + 8x + c$ **5.** $x^2 - 2x + c$ **6.** $x^2 - 9x + c$

The method of completing the square can be used to solve *any* quadratic equation. When you complete the square as part of solving an equation, you must add the same number to *both* sides of the equation.

LOOKING FOR STRUCTURE

Notice you cannot solve the equation by factoring because $x^2 - 10x + 7$ is not factorable as a product of binomials.

EXAMPLE 3 Solving $ax^2 + bx + c = 0$ when $a = 1$

Solve $x^2 - 10x + 7 = 0$ by completing the square.

SOLUTION

$x^2 - 10x + 7 = 0$ Write the equation.

$x^2 - 10x = -7$ Write left side in the form $x^2 + bx$.

$x^2 - 10x + 25 = -7 + 25$ Add $\left(\frac{b}{2}\right)^2 = \left(\frac{-10}{2}\right)^2 = 25$ to each side.

$(x - 5)^2 = 18$ Write left side as a binomial squared.

$x - 5 = \pm\sqrt{18}$ Take square root of each side.

$x = 5 \pm \sqrt{18}$ Add 5 to each side.

$x = 5 \pm 3\sqrt{2}$ Simplify radical.

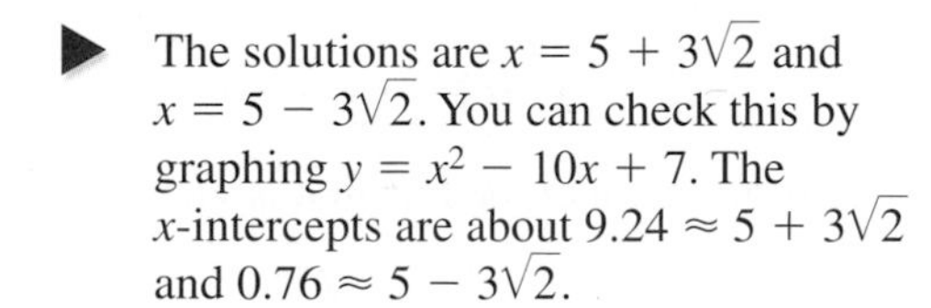

▶ The solutions are $x = 5 + 3\sqrt{2}$ and $x = 5 - 3\sqrt{2}$. You can check this by graphing $y = x^2 - 10x + 7$. The x-intercepts are about $9.24 \approx 5 + 3\sqrt{2}$ and $0.76 \approx 5 - 3\sqrt{2}$.

Check

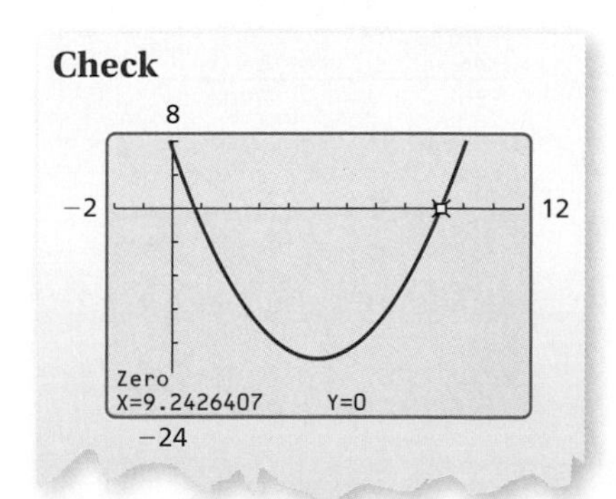

EXAMPLE 4 Solving $ax^2 + bx + c = 0$ when $a \neq 1$

Solve $3x^2 + 12x + 15 = 0$ by completing the square.

SOLUTION

The coefficient a is not 1, so you must first divide each side of the equation by a.

$3x^2 + 12x + 15 = 0$	Write the equation.
$x^2 + 4x + 5 = 0$	Divide each side by 3.
$x^2 + 4x = -5$	Write left side in the form $x^2 + bx$.
$x^2 + 4x + 4 = -5 + 4$	Add $\left(\frac{b}{2}\right)^2 = \left(\frac{4}{2}\right)^2 = 4$ to each side.
$(x + 2)^2 = -1$	Write left side as a binomial squared.
$x + 2 = \pm\sqrt{-1}$	Take square root of each side.
$x = -2 \pm \sqrt{-1}$	Subtract 2 from each side.
$x = -2 \pm i$	Write in terms of i.

▶ The solutions are $x = -2 + i$ and $x = -2 - i$.

Monitoring Progress

Help in English and Spanish at *BigIdeasMath.com*

Solve the equation by completing the square.

7. $x^2 - 4x + 8 = 0$
8. $x^2 + 8x - 5 = 0$
9. $-3x^2 - 18x - 6 = 0$
10. $4x^2 + 32x = -68$
11. $6x(x + 2) = -42$
12. $2x(x - 2) = 200$

Writing Quadratic Functions in Vertex Form

Recall that the vertex form of a quadratic function is $y = a(x - h)^2 + k$, where (h, k) is the vertex of the graph of the function. You can write a quadratic function in vertex form by completing the square.

EXAMPLE 5 Writing a Quadratic Function in Vertex Form

Write $y = x^2 - 12x + 18$ in vertex form. Then identify the vertex.

SOLUTION

$y = x^2 - 12x + 18$	Write the function.
$y + ? = (x^2 - 12x + ?) + 18$	Prepare to complete the square.
$y + 36 = (x^2 - 12x + 36) + 18$	Add $\left(\frac{b}{2}\right)^2 = \left(\frac{-12}{2}\right)^2 = 36$ to each side.
$y + 36 = (x - 6)^2 + 18$	Write $x^2 - 12x + 36$ as a binomial squared.
$y = (x - 6)^2 - 18$	Solve for y.

Check

4

−1 12

Minimum X=6 Y=−18

−26

▶ The vertex form of the function is $y = (x - 6)^2 - 18$. The vertex is $(6, -18)$.

Monitoring Progress

Help in English and Spanish at *BigIdeasMath.com*

Write the quadratic function in vertex form. Then identify the vertex.

13. $y = x^2 - 8x + 18$
14. $y = x^2 + 6x + 4$
15. $y = x^2 - 2x - 6$

EXAMPLE 6 Modeling with Mathematics

The height y (in feet) of a baseball t seconds after it is hit can be modeled by the function

$$y = -16t^2 + 96t + 3.$$

Find the maximum height of the baseball. How long does the ball take to hit the ground?

SOLUTION

1. **Understand the Problem** You are given a quadratic function that represents the height of a ball. You are asked to determine the maximum height of the ball and how long it is in the air.

2. **Make a Plan** Write the function in vertex form to identify the maximum height. Then find and interpret the zeros to determine how long the ball takes to hit the ground.

3. **Solve the Problem** Write the function in vertex form by completing the square.

$y = -16t^2 + 96t + 3$	Write the function.
$y = -16(t^2 - 6t) + 3$	Factor -16 from first two terms.
$y + ? = -16(t^2 - 6t + ?) + 3$	Prepare to complete the square.
$y + (-16)(9) = -16(t^2 - 6t + 9) + 3$	Add $(-16)(9)$ to each side.
$y - 144 = -16(t - 3)^2 + 3$	Write $t^2 - 6t + 9$ as a binomial squared.
$y = -16(t - 3)^2 + 147$	Solve for y.

The vertex is (3, 147). Find the zeros of the function.

$0 = -16(t - 3)^2 + 147$	Substitute 0 for y.
$-147 = -16(t - 3)^2$	Subtract 147 from each side.
$9.1875 = (t - 3)^2$	Divide each side by -16.
$\pm\sqrt{9.1875} = t - 3$	Take square root of each side.
$3 \pm \sqrt{9.1875} = t$	Add 3 to each side.

Reject the negative solution, $3 - \sqrt{9.1875} \approx -0.03$, because time must be positive.

▶ So, the maximum height of the ball is 147 feet, and it takes $3 + \sqrt{9.1875} \approx 6$ seconds for the ball to hit the ground.

4. **Look Back** The vertex indicates that the maximum height of 147 feet occurs when $t = 3$. This makes sense because the graph of the function is parabolic with zeros near $t = 0$ and $t = 6$. You can use a graph to check the maximum height.

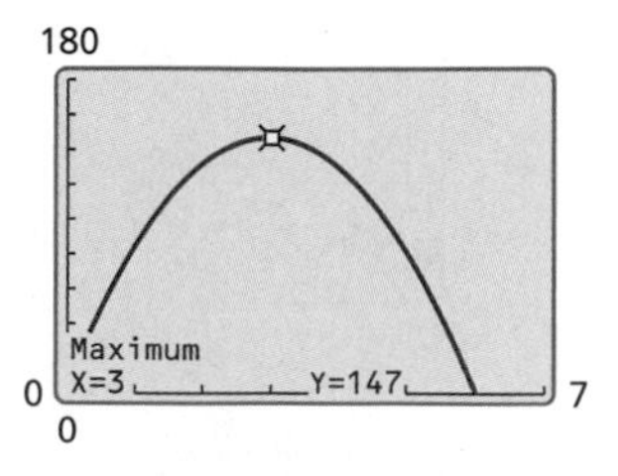

ANOTHER WAY

You can use the coefficients of the original function $y = f(x)$ to find the maximum height.

$$f\left(-\frac{b}{2a}\right) = f\left(-\frac{96}{2(-16)}\right)$$
$$= f(3)$$
$$= 147$$

LOOKING FOR STRUCTURE

You could write the zeros as $3 \pm \frac{7\sqrt{3}}{4}$, but it is easier to recognize that $3 - \sqrt{9.1875}$ is negative because $\sqrt{9.1875}$ is greater than 3.

Monitoring Progress

Help in English and Spanish at *BigIdeasMath.com*

16. **WHAT IF?** The height of the baseball can be modeled by $y = -16t^2 + 80t + 2$. Find the maximum height of the baseball. How long does the ball take to hit the ground?

3.3 Exercises

Dynamic Solutions available at *BigIdeasMath.com*

Vocabulary and Core Concept Check

1. **VOCABULARY** What must you add to the expression $x^2 + bx$ to complete the square?

2. **COMPLETE THE SENTENCE** The trinomial $x^2 - 6x + 9$ is a ____ because it equals ____.

Monitoring Progress and Modeling with Mathematics

In Exercises 3–10, solve the equation using square roots. Check your solution(s). *(See Example 1.)*

3. $x^2 - 8x + 16 = 25$
4. $r^2 - 10r + 25 = 1$
5. $x^2 - 18x + 81 = 5$
6. $m^2 + 8m + 16 = 45$
7. $y^2 - 24y + 144 = -100$
8. $x^2 - 26x + 169 = -13$
9. $4w^2 + 4w + 1 = 75$
10. $4x^2 - 8x + 4 = 1$

In Exercises 11–20, find the value of c that makes the expression a perfect square trinomial. Then write the expression as the square of a binomial. *(See Example 2.)*

11. $x^2 + 10x + c$
12. $x^2 + 20x + c$
13. $y^2 - 12y + c$
14. $t^2 - 22t + c$
15. $x^2 - 6x + c$
16. $x^2 + 24x + c$
17. $z^2 - 5z + c$
18. $x^2 + 9x + c$
19. $w^2 + 13w + c$
20. $s^2 - 26s + c$

In Exercises 21–24, find the value of c. Then write an expression represented by the diagram.

21.

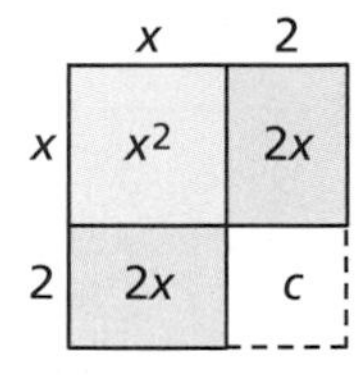

22.

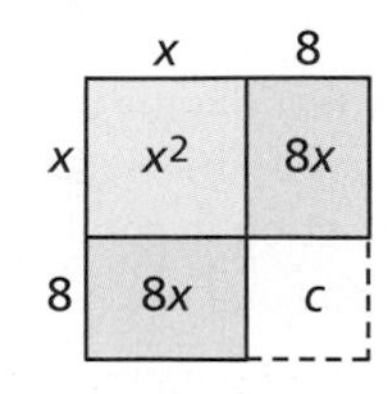

23.

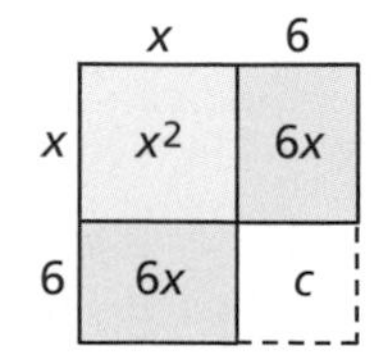

24.

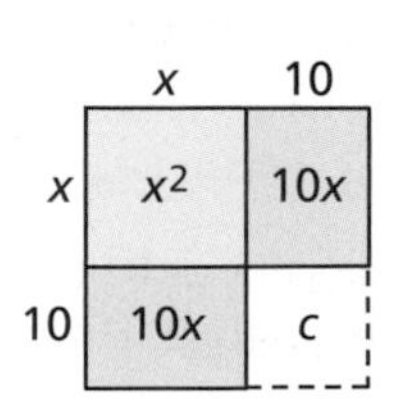

In Exercises 25–36, solve the equation by completing the square. *(See Examples 3 and 4.)*

25. $x^2 + 6x + 3 = 0$
26. $s^2 + 2s - 6 = 0$
27. $x^2 + 4x - 2 = 0$
28. $t^2 - 8t - 5 = 0$
29. $z(z + 9) = 1$
30. $x(x + 8) = -20$
31. $7t^2 + 28t + 56 = 0$
32. $6r^2 + 6r + 12 = 0$
33. $5x(x + 6) = -50$
34. $4w(w - 3) = 24$
35. $4x^2 - 30x = 12 + 10x$
36. $3s^2 + 8s = 2s - 9$

37. **ERROR ANALYSIS** Describe and correct the error in solving the equation.

$$4x^2 + 24x - 11 = 0$$
$$4(x^2 + 6x) = 11$$
$$4(x^2 + 6x + 9) = 11 + 9$$
$$4(x + 3)^2 = 20$$
$$(x + 3)^2 = 5$$
$$x + 3 = \pm\sqrt{5}$$
$$x = -3 \pm \sqrt{5}$$

38. **ERROR ANALYSIS** Describe and correct the error in finding the value of c that makes the expression a perfect square trinomial.

$$x^2 + 30x + c$$
$$x^2 + 30x + \frac{30}{2}$$
$$x^2 + 30x + 15$$

39. **WRITING** Can you solve an equation by completing the square when the equation has two imaginary solutions? Explain.

40. ABSTRACT REASONING Which of the following are solutions of the equation $x^2 - 2ax + a^2 = b^2$? Justify your answers.

Ⓐ ab Ⓑ $-a - b$

Ⓒ b Ⓓ a

Ⓔ $a - b$ Ⓕ $a + b$

USING STRUCTURE In Exercises 41–50, determine whether you would use factoring, square roots, or completing the square to solve the equation. Explain your reasoning. Then solve the equation.

41. $x^2 - 4x - 21 = 0$ **42.** $x^2 + 13x + 22 = 0$

43. $(x + 4)^2 = 16$ **44.** $(x - 7)^2 = 9$

45. $x^2 + 12x + 36 = 0$

46. $x^2 - 16x + 64 = 0$

47. $2x^2 + 4x - 3 = 0$

48. $3x^2 + 12x + 1 = 0$

49. $x^2 - 100 = 0$ **50.** $4x^2 - 20 = 0$

MATHEMATICAL CONNECTIONS In Exercises 51–54, find the value of *x*.

51. Area of rectangle = 50

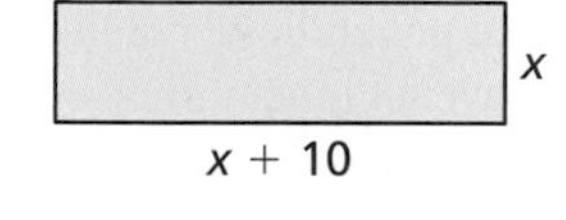

52. Area of parallelogram = 48

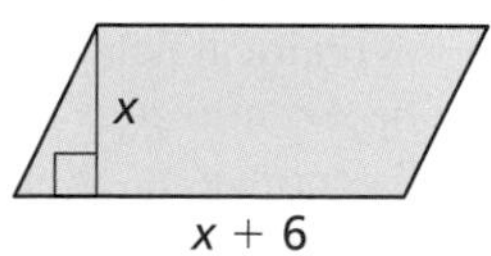

53. Area of triangle = 40

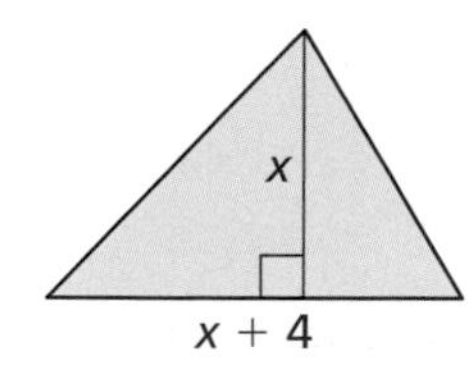

54. Area of trapezoid = 20

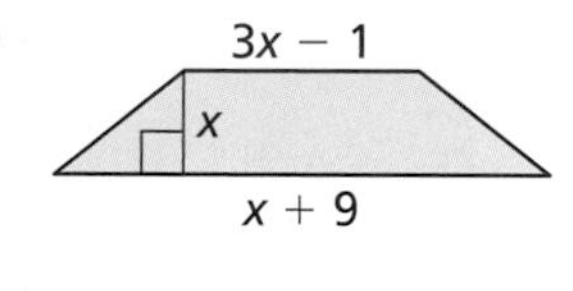

In Exercises 55–62, write the quadratic function in vertex form. Then identify the vertex. *(See Example 5.)*

55. $f(x) = x^2 - 8x + 19$

56. $g(x) = x^2 - 4x - 1$

57. $g(x) = x^2 + 12x + 37$

58. $h(x) = x^2 + 20x + 90$

59. $h(x) = x^2 + 2x - 48$

60. $f(x) = x^2 + 6x - 16$

61. $f(x) = x^2 - 3x + 4$

62. $g(x) = x^2 + 7x + 2$

63. MODELING WITH MATHEMATICS While marching, a drum major tosses a baton into the air and catches it. The height h (in feet) of the baton t seconds after it is thrown can be modeled by the function $h = -16t^2 + 32t + 6$. *(See Example 6.)*

a. Find the maximum height of the baton.

b. The drum major catches the baton when it is 4 feet above the ground. How long is the baton in the air?

64. MODELING WITH MATHEMATICS A firework explodes when it reaches its maximum height. The height h (in feet) of the firework t seconds after it is launched can be modeled by $h = -\frac{500}{9}t^2 + \frac{1000}{3}t + 10$. What is the maximum height of the firework? How long is the firework in the air before it explodes?

65. COMPARING METHODS A skateboard shop sells about 50 skateboards per week when the advertised price is charged. For each \$1 decrease in price, one additional skateboard per week is sold. The shop's revenue can be modeled by $y = (70 - x)(50 + x)$.

a. Use the intercept form of the function to find the maximum weekly revenue.

b. Write the function in vertex form to find the maximum weekly revenue.

c. Which way do you prefer? Explain your reasoning.

66. HOW DO YOU SEE IT? The graph of the function $f(x) = (x - h)^2$ is shown. What is the x-intercept? Explain your reasoning.

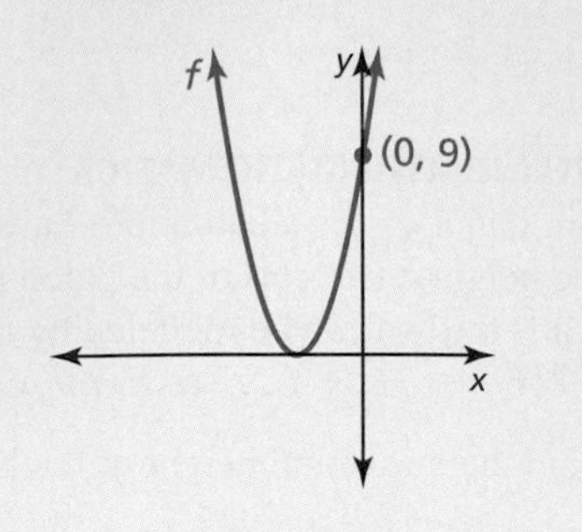

67. WRITING At Buckingham Fountain in Chicago, the height h (in feet) of the water above the main nozzle can be modeled by $h = -16t^2 + 89.6t$, where t is the time (in seconds) since the water has left the nozzle. Describe three different ways you could find the maximum height the water reaches. Then choose a method and find the maximum height of the water.

68. PROBLEM SOLVING A farmer is building a rectangular pen along the side of a barn for animals. The barn will serve as one side of the pen. The farmer has 120 feet of fence to enclose an area of 1512 square feet and wants each side of the pen to be at least 20 feet long.

a. Write an equation that represents the area of the pen.

b. Solve the equation in part (a) to find the dimensions of the pen.

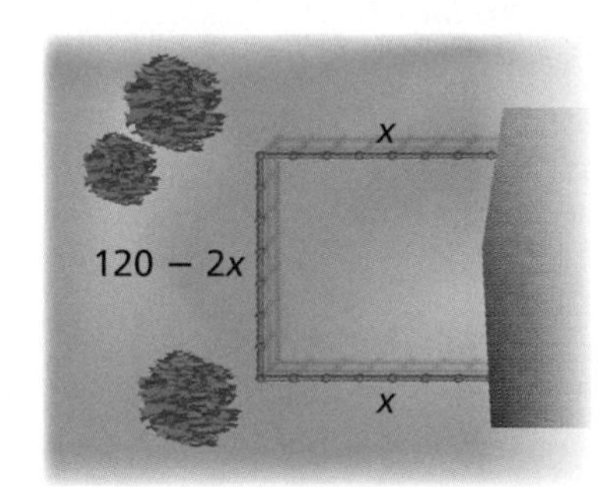

69. MAKING AN ARGUMENT Your friend says the equation $x^2 + 10x = -20$ can be solved by either completing the square or factoring. Is your friend correct? Explain.

70. THOUGHT PROVOKING Write a function g in standard form whose graph has the same x-intercepts as the graph of $f(x) = 2x^2 + 8x + 2$. Find the zeros of each function by completing the square. Graph each function.

71. CRITICAL THINKING Solve $x^2 + bx + c = 0$ by completing the square. Your answer will be an expression for x in terms of b and c.

72. DRAWING CONCLUSIONS In this exercise, you will investigate the graphical effect of completing the square.

a. Graph each pair of functions in the same coordinate plane.

$y = x^2 + 2x$ $\qquad$ $y = x^2 - 6x$

$y = (x + 1)^2$ $\qquad$ $y = (x - 3)^2$

b. Compare the graphs of $y = x^2 + bx$ and $y = \left(x + \frac{b}{2}\right)^2$. Describe what happens to the graph of $y = x^2 + bx$ when you complete the square.

73. MODELING WITH MATHEMATICS In your pottery class, you are given a lump of clay with a volume of 200 cubic centimeters and are asked to make a cylindrical pencil holder. The pencil holder should be 9 centimeters high and have an inner radius of 3 centimeters. What thickness x should your pencil holder have if you want to use all of the clay?

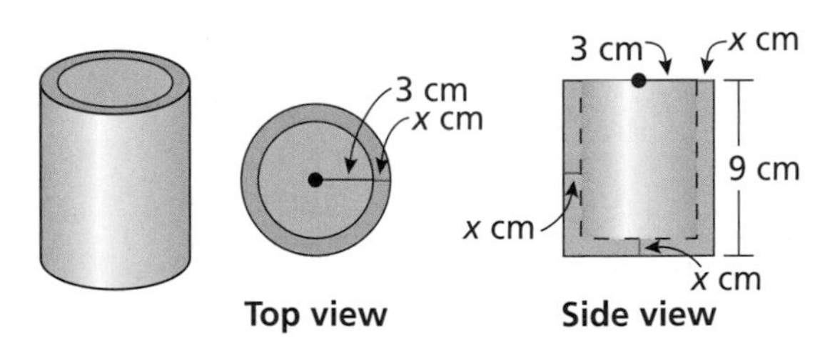

Maintaining Mathematical Proficiency

Reviewing what you learned in previous grades and lessons

Solve the inequality. Graph the solution. *(Skills Review Handbook)*

74. $2x - 3 < 5$ **75.** $4 - 8y \geq 12$ **76.** $\frac{n}{3} + 6 > 1$ **77.** $-\frac{2s}{5} \leq 8$

Graph the function. Label the vertex, axis of symmetry, and x-intercepts. *(Section 2.2)*

78. $g(x) = 6(x - 4)^2$ **79.** $h(x) = 2x(x - 3)$

80. $f(x) = x^2 + 2x + 5$ **81.** $f(x) = 2(x + 10)(x - 12)$

3.1–3.3 What Did You Learn?

Core Vocabulary

quadratic equation in one variable, *p. 94*
root of an equation, *p. 94*
zero of a function, *p. 96*
imaginary unit i, *p. 104*
complex number, *p. 104*
imaginary number, *p. 104*
pure imaginary number, *p. 104*
completing the square, *p. 112*

Core Concepts

Section 3.1

Solving Quadratic Equations Graphically, *p. 94*
Solving Quadratic Equations Algebraically, *p. 95*
Zero-Product Property, *p. 96*

Section 3.2

The Square Root of a Negative Number, *p. 104*
Operations with Complex Numbers, *p. 105*

Section 3.3

Solving Quadratic Equations by Completing the Square, *p. 113*
Writing Quadratic Functions in Vertex Form, *p. 114*

Mathematical Practices

1. Analyze the givens, constraints, relationships, and goals in Exercise 61 on page 101.
2. Determine whether it would be easier to find the zeros of the function in Exercise 63 on page 117 or Exercise 67 on page 118.

Study Skills

Creating a Positive Study Environment

- Set aside an appropriate amount of time for reviewing your notes and the textbook, reworking your notes, and completing homework.
- Set up a place for studying at home that is comfortable, but not too comfortable. The place needs to be away from all potential distractions.
- Form a study group. Choose students who study well together, help out when someone misses school, and encourage positive attitudes.

3.1–3.3 Quiz

Solve the equation by using the graph. Check your solution(s). *(Section 3.1)*

1. $x^2 - 10x + 25 = 0$

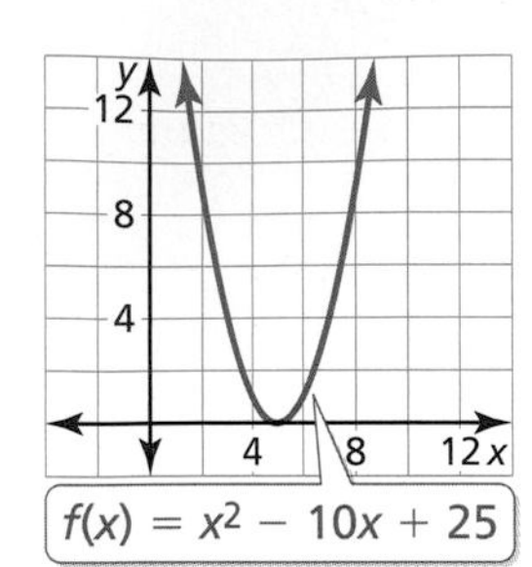

2. $2x^2 + 16 = 12x$

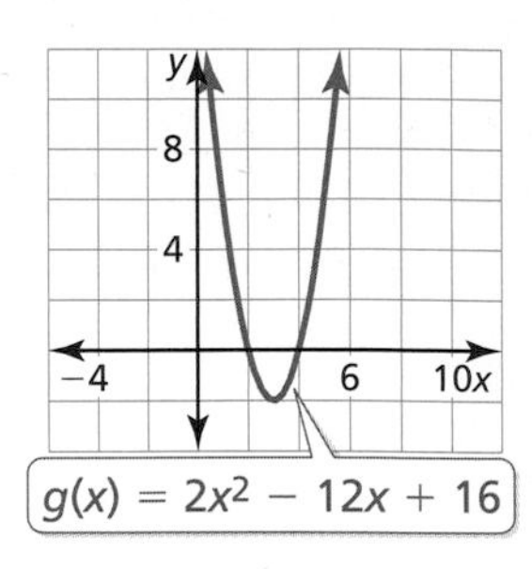

3. $x^2 = -2x + 8$

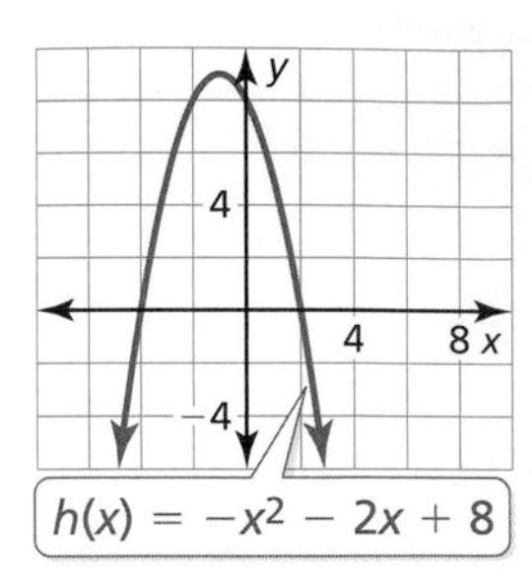

Solve the equation using square roots or by factoring. Explain the reason for your choice. *(Section 3.1)*

4. $2x^2 - 15 = 0$

5. $3x^2 - x - 2 = 0$

6. $(x + 3)^2 = 8$

7. Find the values of x and y that satisfy the equation $7x - 6i = 14 + yi$. *(Section 3.2)*

Perform the operation. Write your answer in standard form. *(Section 3.2)*

8. $(2 + 5i) + (-4 + 3i)$

9. $(3 + 9i) - (1 - 7i)$

10. $(2 + 4i)(-3 - 5i)$

11. Find the zeros of the function $f(x) = 9x^2 + 2$. Does the graph of the function intersect the x-axis? Explain your reasoning. *(Section 3.2)*

Solve the equation by completing the square. *(Section 3.3)*

12. $x^2 - 6x + 10 = 0$

13. $x^2 + 12x + 4 = 0$

14. $4x(x + 6) = -40$

15. Write $y = x^2 - 10x + 4$ in vertex form. Then identify the vertex. *(Section 3.3)*

16. A museum has a café with a rectangular patio. The museum wants to add 464 square feet to the area of the patio by expanding the existing patio as shown. *(Section 3.1)*

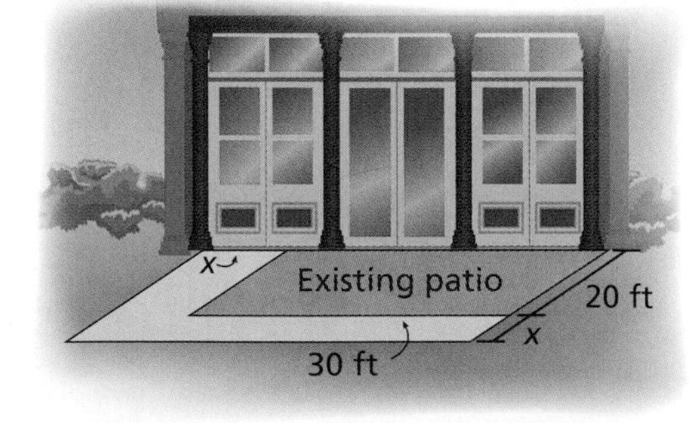

a. Find the area of the existing patio.

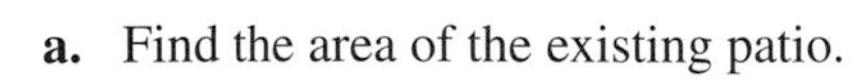

b. Write an equation to model the area of the new patio.

c. By what distance x should the length of the patio be expanded?

17. Find the impedance of the series circuit. *(Section 3.2)*

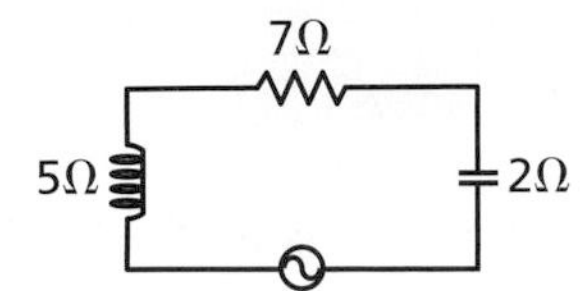

18. The height h (in feet) of a badminton birdie t seconds after it is hit can be modeled by the function $h = -16t^2 + 32t + 4$. *(Section 3.3)*

a. Find the maximum height of the birdie.

b. How long is the birdie in the air?

3.4 Using the Quadratic Formula

Essential Question How can you derive a general formula for solving a quadratic equation?

EXPLORATION 1 Deriving the Quadratic Formula

Work with a partner. Analyze and describe what is done in each step in the development of the Quadratic Formula.

Step	Justification
$ax^2 + bx + c = 0$	
$ax^2 + bx = -c$	
$x^2 + \frac{b}{a}x = -\frac{c}{a}$	
$x^2 + \frac{b}{a}x + \left(\frac{b}{2a}\right)^2 = -\frac{c}{a} + \left(\frac{b}{2a}\right)^2$	
$x^2 + \frac{b}{a}x + \left(\frac{b}{2a}\right)^2 = -\frac{4ac}{4a^2} + \frac{b^2}{4a^2}$	
$\left(x + \frac{b}{2a}\right)^2 = \frac{b^2 - 4ac}{4a^2}$	
$x + \frac{b}{2a} = \pm\sqrt{\frac{b^2 - 4ac}{4a^2}}$	
$x = -\frac{b}{2a} \pm \frac{\sqrt{b^2 - 4ac}}{2\lvert a \rvert}$	
$x = \frac{-b \pm \sqrt{b^2 - 4ac}}{2a}$	

The result is the Quadratic Formula.

REASONING ABSTRACTLY

To be proficient in math, you need to create a coherent representation of the problem at hand.

EXPLORATION 2 Using the Quadratic Formula

Work with a partner. Use the Quadratic Formula to solve each equation.

a. $x^2 - 4x + 3 = 0$

b. $x^2 - 2x + 2 = 0$

c. $x^2 + 2x - 3 = 0$

d. $x^2 + 4x + 4 = 0$

e. $x^2 - 6x + 10 = 0$

f. $x^2 + 4x + 6 = 0$

Communicate Your Answer

3. How can you derive a general formula for solving a quadratic equation?

4. Summarize the following methods you have learned for solving quadratic equations: graphing, using square roots, factoring, completing the square, and using the Quadratic Formula.

3.4 Lesson

Core Vocabulary

Quadratic Formula, *p. 122*
discriminant, *p. 124*

What You Will Learn

- Solve quadratic equations using the Quadratic Formula.
- Analyze the discriminant to determine the number and type of solutions.
- Solve real-life problems.

Solving Equations Using the Quadratic Formula

Previously, you solved quadratic equations by completing the square. In the Exploration, you developed a formula that gives the solutions of any quadratic equation by completing the square once for the general equation $ax^2 + bx + c = 0$. The formula for the solutions is called the **Quadratic Formula**.

Core Concept

The Quadratic Formula

Let a, b, and c be real numbers such that $a \neq 0$. The solutions of the quadratic equation $ax^2 + bx + c = 0$ are $x = \dfrac{-b \pm \sqrt{b^2 - 4ac}}{2a}$.

COMMON ERROR

Remember to write the quadratic equation in standard form before applying the Quadratic Formula.

EXAMPLE 1 Solving an Equation with Two Real Solutions

Solve $x^2 + 3x = 5$ using the Quadratic Formula.

SOLUTION

$x^2 + 3x = 5$ — Write original equation.

$x^2 + 3x - 5 = 0$ — Write in standard form.

$x = \dfrac{-b \pm \sqrt{b^2 - 4ac}}{2a}$ — Quadratic Formula

$x = \dfrac{-3 \pm \sqrt{3^2 - 4(1)(-5)}}{2(1)}$ — Substitute 1 for *a*, 3 for *b*, and −5 for *c*.

$x = \dfrac{-3 \pm \sqrt{29}}{2}$ — Simplify.

So, the solutions are $x = \dfrac{-3 + \sqrt{29}}{2} \approx 1.19$ and $x = \dfrac{-3 - \sqrt{29}}{2} \approx -4.19$.

Check Graph $y = x^2 + 3x - 5$. The x-intercepts are about -4.19 and about 1.19. ✓

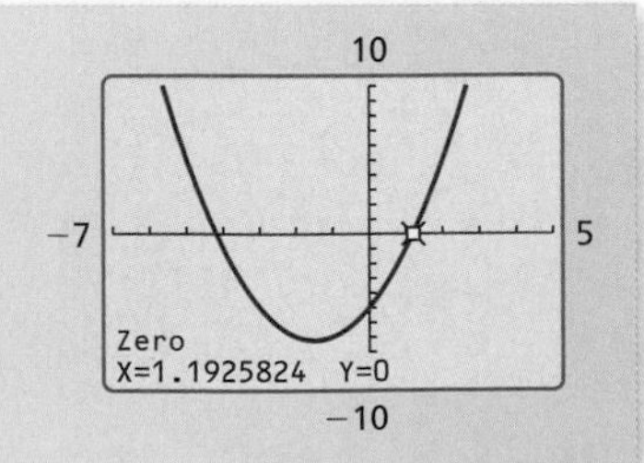

Monitoring Progress

Help in English and Spanish at *BigIdeasMath.com*

Solve the equation using the Quadratic Formula.

1. $x^2 - 6x + 4 = 0$ **2.** $2x^2 + 4 = -7x$ **3.** $5x^2 = x + 8$

ANOTHER WAY

You can also use factoring to solve $25x^2 - 20x + 4 = 0$ because the left side factors as $(5x - 2)^2$.

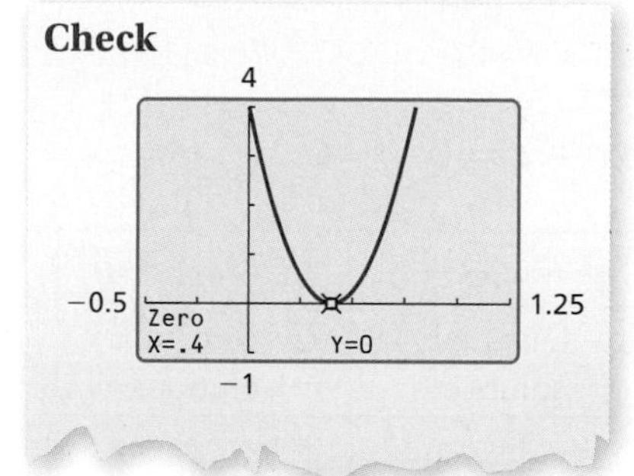

EXAMPLE 2 Solving an Equation with One Real Solution

Solve $25x^2 - 8x = 12x - 4$ using the Quadratic Formula.

SOLUTION

$$25x^2 - 8x = 12x - 4$$ Write original equation.

$$25x^2 - 20x + 4 = 0$$ Write in standard form.

$$x = \frac{-(-20) \pm \sqrt{(-20)^2 - 4(25)(4)}}{2(25)}$$ $a = 25, b = -20, c = 4$

$$x = \frac{20 \pm \sqrt{0}}{50}$$ Simplify.

$$x = \frac{2}{5}$$ Simplify.

▶ So, the solution is $x = \frac{2}{5}$. You can check this by graphing $y = 25x^2 - 20x + 4$. The only x-intercept is $\frac{2}{5}$.

EXAMPLE 3 Solving an Equation with Imaginary Solutions

Solve $-x^2 + 4x = 13$ using the Quadratic Formula.

SOLUTION

$$-x^2 + 4x = 13$$ Write original equation.

$$-x^2 + 4x - 13 = 0$$ Write in standard form.

$$x = \frac{-4 \pm \sqrt{4^2 - 4(-1)(-13)}}{2(-1)}$$ $a = -1, b = 4, c = -13$

$$x = \frac{-4 \pm \sqrt{-36}}{-2}$$ Simplify.

$$x = \frac{-4 \pm 6i}{-2}$$ Write in terms of i.

$$x = 2 \pm 3i$$ Simplify.

▶ The solutions are $x = 2 + 3i$ and $x = 2 - 3i$.

COMMON ERROR

Remember to divide the real part *and* the imaginary part by −2 when simplifying.

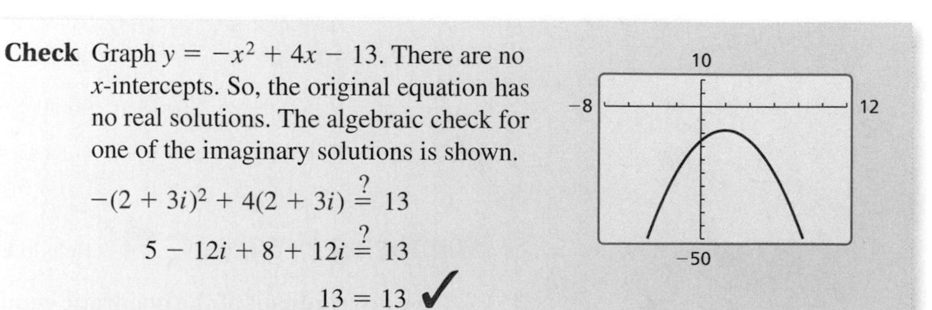

Check Graph $y = -x^2 + 4x - 13$. There are no x-intercepts. So, the original equation has no real solutions. The algebraic check for one of the imaginary solutions is shown.

$$-(2 + 3i)^2 + 4(2 + 3i) \stackrel{?}{=} 13$$

$$5 - 12i + 8 + 12i \stackrel{?}{=} 13$$

$$13 = 13 \checkmark$$

Monitoring Progress

Help in English and Spanish at *BigIdeasMath.com*

Solve the equation using the Quadratic Formula.

4. $x^2 + 41 = -8x$ **5.** $-9x^2 = 30x + 25$ **6.** $5x - 7x^2 = 3x + 4$

Analyzing the Discriminant

In the Quadratic Formula, the expression $b^2 - 4ac$ is called the **discriminant** of the associated equation $ax^2 + bx + c = 0$.

$$x = \frac{-b \pm \sqrt{b^2 - 4ac}}{2a}$$ ← discriminant

You can analyze the discriminant of a quadratic equation to determine the number and type of solutions of the equation.

Core Concept

Analyzing the Discriminant of $ax^2 + bx + c = 0$

Value of discriminant	$b^2 - 4ac > 0$	$b^2 - 4ac = 0$	$b^2 - 4ac < 0$
Number and type of solutions	Two real solutions	One real solution	Two imaginary solutions
Graph of $y = ax^2 + bx + c$	Two x-intercepts	One x-intercept	No x-intercept

EXAMPLE 4 Analyzing the Discriminant

Find the discriminant of the quadratic equation and describe the number and type of solutions of the equation.

a. $x^2 - 6x + 10 = 0$ **b.** $x^2 - 6x + 9 = 0$ **c.** $x^2 - 6x + 8 = 0$

SOLUTION

Equation	Discriminant	Solution(s)
$ax^2 + bx + c = 0$	$b^2 - 4ac$	$x = \frac{-b \pm \sqrt{b^2 - 4ac}}{2a}$
a. $x^2 - 6x + 10 = 0$	$(-6)^2 - 4(1)(10) = -4$	Two imaginary: $3 \pm i$
b. $x^2 - 6x + 9 = 0$	$(-6)^2 - 4(1)(9) = 0$	One real: 3
c. $x^2 - 6x + 8 = 0$	$(-6)^2 - 4(1)(8) = 4$	Two real: 2, 4

Monitoring Progress

Help in English and Spanish at *BigIdeasMath.com*

Find the discriminant of the quadratic equation and describe the number and type of solutions of the equation.

7. $4x^2 + 8x + 4 = 0$

8. $\frac{1}{2}x^2 + x - 1 = 0$

9. $5x^2 = 8x - 13$

10. $7x^2 - 3x = 6$

11. $4x^2 + 6x = -9$

12. $-5x^2 + 1 = 6 - 10x$

EXAMPLE 5 Writing an Equation

Find a possible pair of integer values for a and c so that the equation $ax^2 - 4x + c = 0$ has one real solution. Then write the equation.

SOLUTION

In order for the equation to have one real solution, the discriminant must equal 0.

$b^2 - 4ac = 0$ Write the discriminant.

$(-4)^2 - 4ac = 0$ Substitute -4 for b.

$16 - 4ac = 0$ Evaluate the power.

$-4ac = -16$ Subtract 16 from each side.

$ac = 4$ Divide each side by -4.

Because $ac = 4$, choose two integers whose product is 4, such as $a = 1$ and $c = 4$.

▶ So, one possible equation is $x^2 - 4x + 4 = 0$.

ANOTHER WAY

Another possible equation in Example 5 is $4x^2 - 4x + 1 = 0$. You can obtain this equation by letting $a = 4$ and $c = 1$.

Check Graph $y = x^2 - 4x + 4$. The only x-intercept is 2. You can also check by factoring.

$x^2 - 4x + 4 = 0$

$(x - 2)^2 = 0$

$x = 2$ 

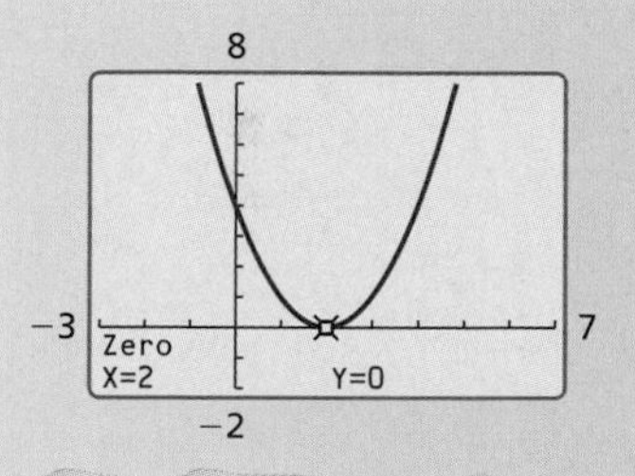

Monitoring Progress

Help in English and Spanish at *BigIdeasMath.com*

13. Find a possible pair of integer values for a and c so that the equation $ax^2 + 3x + c = 0$ has two real solutions. Then write the equation.

The table shows five methods for solving quadratic equations. For a given equation, it may be more efficient to use one method instead of another. Suggestions about when to use each method are shown below.

Concept Summary

Methods for Solving Quadratic Equations

Method	When to Use
Graphing	Use when approximate solutions are adequate.
Using square roots	Use when solving an equation that can be written in the form $u^2 = d$, where u is an algebraic expression.
Factoring	Use when a quadratic equation can be factored easily.
Completing the square	Can be used for *any* quadratic equation $ax^2 + bx + c = 0$ but is simplest to apply when $a = 1$ and b is an even number.
Quadratic Formula	Can be used for *any* quadratic equation.

Solving Real-Life Problems

The function $h = -16t^2 + h_0$ is used to model the height of a *dropped* object. For an object that is *launched* or *thrown*, an extra term v_0t must be added to the model to account for the object's initial vertical velocity v_0 (in feet per second). Recall that h is the height (in feet), t is the time in motion (in seconds), and h_0 is the initial height (in feet).

$h = -16t^2 + h_0$ Object is dropped.

$h = -16t^2 + v_0t + h_0$ Object is launched or thrown.

As shown below, the value of v_0 can be positive, negative, or zero depending on whether the object is launched upward, downward, or parallel to the ground.

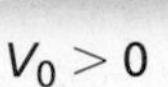

$V_0 > 0$ $V_0 < 0$ $V_0 = 0$

EXAMPLE 6 **Modeling a Launched Object**

A juggler tosses a ball into the air. The ball leaves the juggler's hand 4 feet above the ground and has an initial vertical velocity of 30 feet per second. The juggler catches the ball when it falls back to a height of 3 feet. How long is the ball in the air?

SOLUTION

Because the ball is *thrown*, use the model $h = -16t^2 + v_0t + h_0$. To find how long the ball is in the air, solve for t when $h = 3$.

$h = -16t^2 + v_0t + h_0$ Write the height model.

$3 = -16t^2 + 30t + 4$ Substitute 3 for h, 30 for v_0, and 4 for h_0.

$0 = -16t^2 + 30t + 1$ Write in standard form.

This equation is not factorable, and completing the square would result in fractions. So, use the Quadratic Formula to solve the equation.

$t = \dfrac{-30 \pm \sqrt{30^2 - 4(-16)(1)}}{2(-16)}$ $a = -16, b = 30, c = 1$

$t = \dfrac{-30 \pm \sqrt{964}}{-32}$ Simplify.

$t \approx -0.033$ or $t \approx 1.9$ Use a calculator.

▶ Reject the negative solution, −0.033, because the ball's time in the air cannot be negative. So, the ball is in the air for about 1.9 seconds.

Monitoring Progress  Help in English and Spanish at *BigIdeasMath.com*

14. WHAT IF? The ball leaves the juggler's hand with an initial vertical velocity of 40 feet per second. How long is the ball in the air?

3.4 Exercises

Dynamic Solutions available at *BigIdeasMath.com*

Vocabulary and Core Concept Check

1. **COMPLETE THE SENTENCE** When a, b, and c are real numbers such that $a \neq 0$, the solutions of the quadratic equation $ax^2 + bx + c = 0$ are $x =$ ____________.

2. **COMPLETE THE SENTENCE** You can use the ____________ of a quadratic equation to determine the number and type of solutions of the equation.

3. **WRITING** Describe the number and type of solutions when the value of the discriminant is negative.

4. **WRITING** Which two methods can you use to solve *any* quadratic equation? Explain when you might prefer to use one method over the other.

Monitoring Progress and Modeling with Mathematics

In Exercises 5–18, solve the equation using the Quadratic Formula. Use a graphing calculator to check your solution(s). *(See Examples 1, 2, and 3.)*

5. $x^2 - 4x + 3 = 0$
6. $3x^2 + 6x + 3 = 0$
7. $x^2 + 6x + 15 = 0$
8. $6x^2 - 2x + 1 = 0$
9. $x^2 - 14x = -49$
10. $2x^2 + 4x = 30$
11. $3x^2 + 5 = -2x$
12. $-3x = 2x^2 - 4$
13. $-10x = -25 - x^2$
14. $-5x^2 - 6 = -4x$
15. $-4x^2 + 3x = -5$
16. $x^2 + 121 = -22x$
17. $-z^2 = -12z + 6$
18. $-7w + 6 = -4w^2$

In Exercises 19–26, find the discriminant of the quadratic equation and describe the number and type of solutions of the equation. *(See Example 4.)*

19. $x^2 + 12x + 36 = 0$
20. $x^2 - x + 6 = 0$
21. $4n^2 - 4n - 24 = 0$
22. $-x^2 + 2x + 12 = 0$
23. $4x^2 = 5x - 10$
24. $-18p = p^2 + 81$
25. $24x = -48 - 3x^2$
26. $-2x^2 - 6 = x$

27. **USING EQUATIONS** What are the complex solutions of the equation $2x^2 - 16x + 50 = 0$?

   (A) $4 + 3i, 4 - 3i$ (B) $4 + 12i, 4 - 12i$

   (C) $16 + 3i, 16 - 3i$ (D) $16 + 12i, 16 - 12i$

28. **USING EQUATIONS** Determine the number and type of solutions to the equation $x^2 + 7x = -11$.

   (A) two real solutions

   (B) one real solution

   (C) two imaginary solutions

   (D) one imaginary solution

ANALYZING EQUATIONS In Exercises 29–32, use the discriminant to match each quadratic equation with the correct graph of the related function. Explain your reasoning.

29. $x^2 - 6x + 25 = 0$
30. $2x^2 - 20x + 50 = 0$
31. $3x^2 + 6x - 9 = 0$
32. $5x^2 - 10x - 35 = 0$

A.

B.

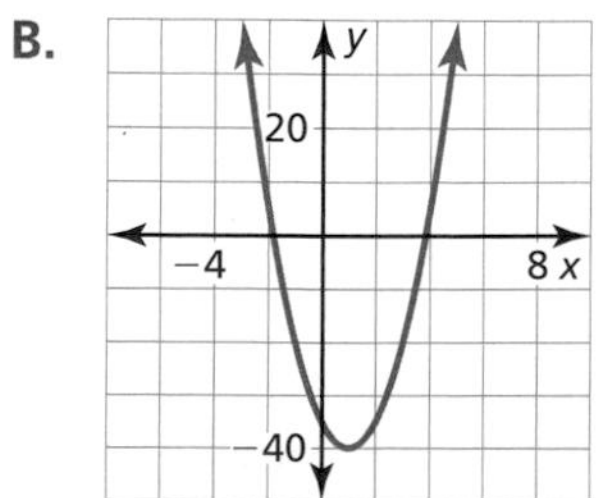

C.

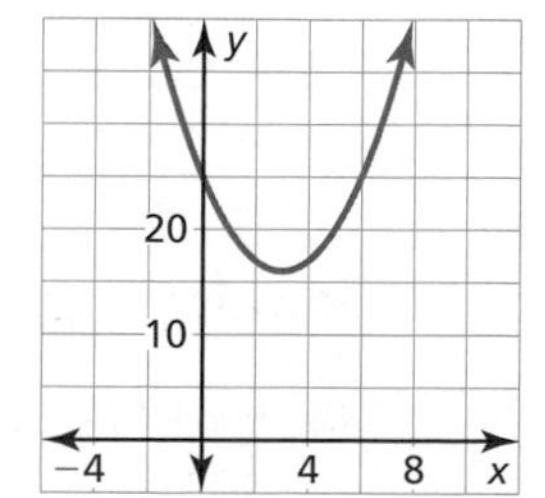

D.

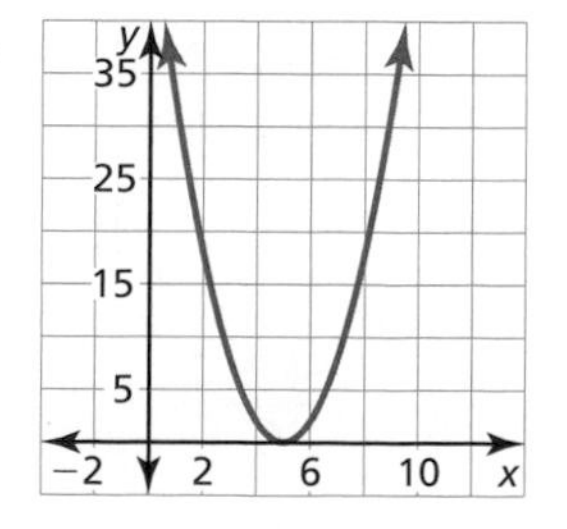

ERROR ANALYSIS **In Exercises 33 and 34, describe and correct the error in solving the equation.**

33.

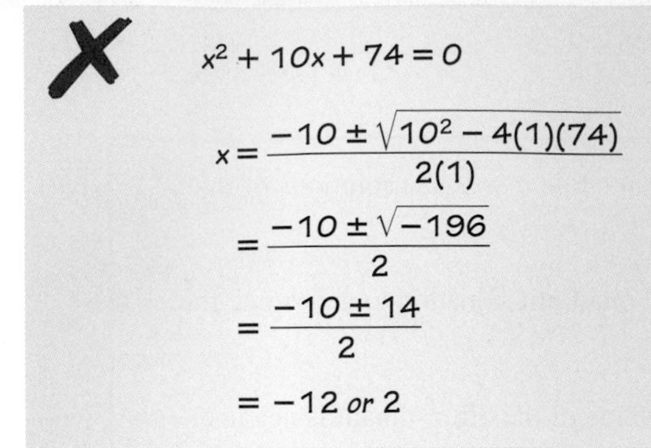

$x^2 + 10x + 74 = 0$
$x = \frac{-10 \pm \sqrt{10^2 - 4(1)(74)}}{2(1)}$
$= \frac{-10 \pm \sqrt{-196}}{2}$
$= \frac{-10 \pm 14}{2}$
$= -12$ or 2

34.

$x^2 + 6x + 8 = 2$

$x = \frac{-6 \pm \sqrt{6^2 - 4(1)(8)}}{2(1)}$

$= \frac{-6 \pm \sqrt{4}}{2}$

$= \frac{-6 \pm 2}{2}$

$= -2$ or -4

OPEN-ENDED **In Exercises 35–40, find a possible pair of integer values for a and c so that the quadratic equation has the given solution(s). Then write the equation.** *(See Example 5.)*

35. $ax^2 + 4x + c = 0$; two imaginary solutions

36. $ax^2 + 6x + c = 0$; two real solutions

37. $ax^2 - 8x + c = 0$; two real solutions

38. $ax^2 - 6x + c = 0$; one real solution

39. $ax^2 + 10x = c$; one real solution

40. $-4x + c = -ax^2$; two imaginary solutions

USING STRUCTURE **In Exercises 41–46, use the Quadratic Formula to write a quadratic equation that has the given solutions.**

41. $x = \frac{-8 \pm \sqrt{-176}}{-10}$

42. $x = \frac{15 \pm \sqrt{-215}}{22}$

43. $x = \frac{-4 \pm \sqrt{-124}}{-14}$

44. $x = \frac{-9 \pm \sqrt{137}}{4}$

45. $x = \frac{-4 \pm 2}{6}$

46. $x = \frac{2 \pm 4}{-2}$

COMPARING METHODS **In Exercises 47–58, solve the quadratic equation using the Quadratic Formula. Then solve the equation using another method. Which method do you prefer? Explain.**

47. $3x^2 - 21 = 3$

48. $5x^2 + 38 = 3$

49. $2x^2 - 54 = 12x$

50. $x^2 = 3x + 15$

51. $x^2 - 7x + 12 = 0$

52. $x^2 + 8x - 13 = 0$

53. $5x^2 - 50x = -135$

54. $8x^2 + 4x + 5 = 0$

55. $-3 = 4x^2 + 9x$

56. $-31x + 56 = -x^2$

57. $x^2 = 1 - x$

58. $9x^2 + 36x + 72 = 0$

MATHEMATICAL CONNECTIONS **In Exercises 59 and 60, find the value for x.**

59. Area of the rectangle = 24 m^2

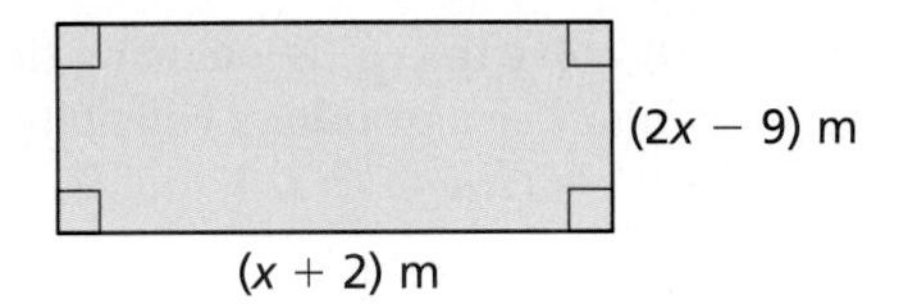

60. Area of the triangle = 8 ft^2

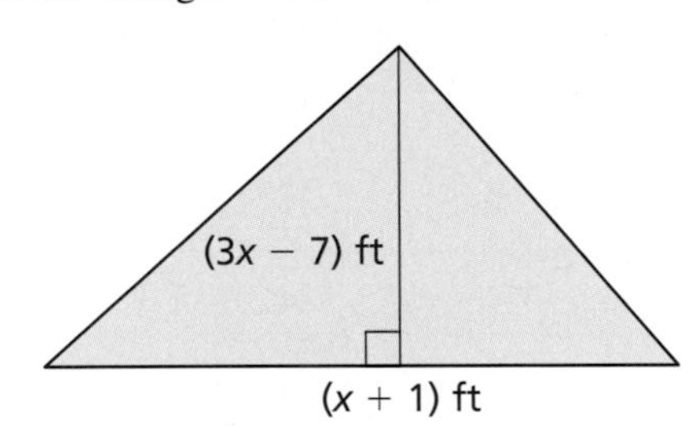

61. **MODELING WITH MATHEMATICS** A lacrosse player throws a ball in the air from an initial height of 7 feet. The ball has an initial vertical velocity of 90 feet per second. Another player catches the ball when it is 3 feet above the ground. How long is the ball in the air? *(See Example 6.)*

62. **NUMBER SENSE** Suppose the quadratic equation $ax^2 + 5x + c = 0$ has one real solution. Is it possible for a and c to be integers? rational numbers? Explain your reasoning. Then describe the possible values of a and c.

63. MODELING WITH MATHEMATICS In a volleyball game, a player on one team spikes the ball over the net when the ball is 10 feet above the court. The spike drives the ball downward with an initial vertical velocity of 55 feet per second. How much time does the opposing team have to return the ball before it touches the court?

64. MODELING WITH MATHEMATICS An archer is shooting at targets. The height of the arrow is 5 feet above the ground. Due to safety rules, the archer must aim the arrow parallel to the ground.

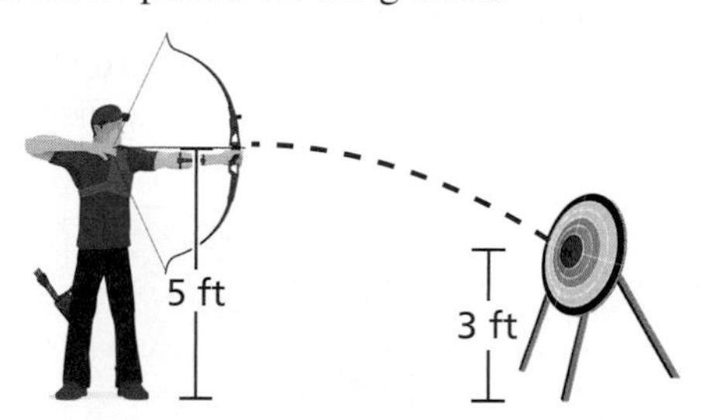

a. How long does it take for the arrow to hit a target that is 3 feet above the ground?

b. What method did you use to solve the quadratic equation? Explain.

65. PROBLEM SOLVING A rocketry club is launching model rockets. The launching pad is 30 feet above the ground. Your model rocket has an initial vertical velocity of 105 feet per second. Your friend's model rocket has an initial vertical velocity of 100 feet per second.

a. Use a graphing calculator to graph the equations of both model rockets. Compare the paths.

b. After how many seconds is your rocket 119 feet above the ground? Explain the reasonableness of your answer(s).

66. PROBLEM SOLVING The number A of tablet computers sold (in millions) can be modeled by the function $A = 4.5t^2 + 43.5t + 17$, where t represents the year after 2010.

a. In what year did the tablet computer sales reach 65 million?

b. Find the average rate of change from 2010 to 2012 and interpret the meaning in the context of the situation.

c. Do you think this model will be accurate after a new, innovative computer is developed? Explain.

67. MODELING WITH MATHEMATICS A gannet is a bird that feeds on fish by diving into the water. A gannet spots a fish on the surface of the water and dives 100 feet to catch it. The bird plunges toward the water with an initial vertical velocity of -88 feet per second.

a. How much time does the fish have to swim away?

b. Another gannet spots the same fish, and it is only 84 feet above the water and has an initial vertical velocity of -70 feet per second. Which bird will reach the fish first? Justify your answer.

68. USING TOOLS You are asked to find a possible pair of integer values for a and c so that the equation $ax^2 - 3x + c = 0$ has two real solutions. When you solve the inequality for the discriminant, you obtain $ac < 2.25$. So, you choose the values $a = 2$ and $c = 1$. Your graphing calculator displays the graph of your equation in a standard viewing window. Is your solution correct? Explain.

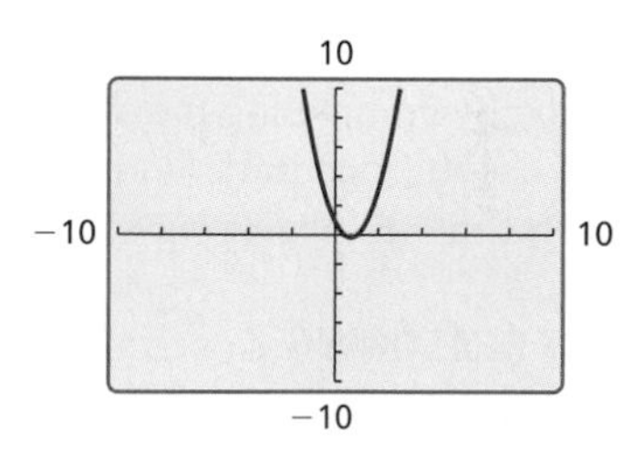

69. PROBLEM SOLVING Your family has a rectangular pool that measures 18 feet by 9 feet. Your family wants to put a deck around the pool but is not sure how wide to make the deck. Determine how wide the deck should be when the total area of the pool and deck is 400 square feet. What is the width of the deck?

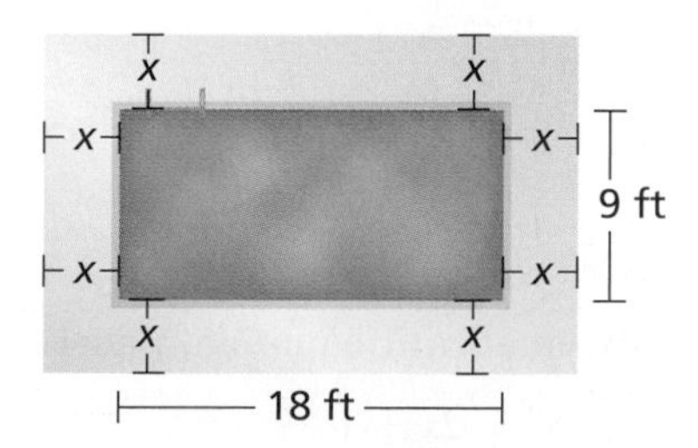

70. HOW DO YOU SEE IT? The graph of a quadratic function $y = ax^2 + bx + c$ is shown. Determine whether each discriminant of $ax^2 + bx + c = 0$ is *positive*, *negative*, or *zero*. Then state the number and type of solutions for each graph. Explain your reasoning.

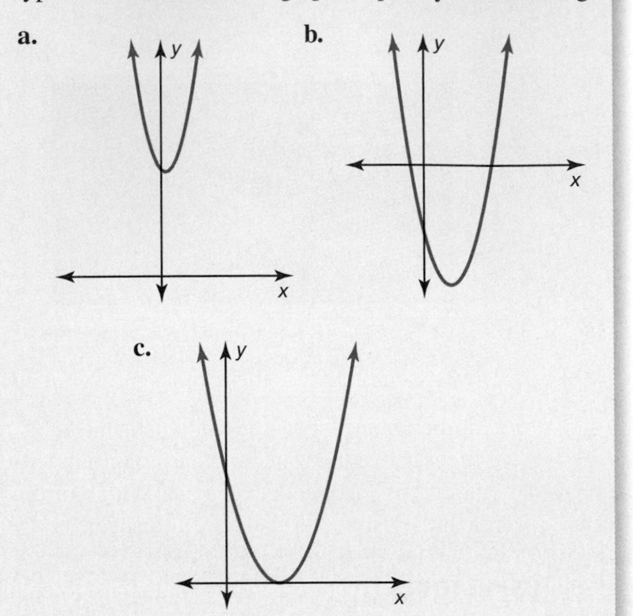

71. CRITICAL THINKING Solve each absolute value equation.

a. $|x^2 - 3x - 14| = 4$ **b.** $x^2 = |x| + 6$

72. MAKING AN ARGUMENT The class is asked to solve the equation $4x^2 + 14x + 11 = 0$. You decide to solve the equation by completing the square. Your friend decides to use the Quadratic Formula. Whose method is more efficient? Explain your reasoning.

73. ABSTRACT REASONING For a quadratic equation $ax^2 + bx + c = 0$ with two real solutions, show that the mean of the solutions is $-\frac{b}{2a}$. How is this fact related to the symmetry of the graph of $y = ax^2 + bx + c$?

74. THOUGHT PROVOKING Describe a real-life story that could be modeled by $h = -16t^2 + v_0t + h_0$. Write the height model for your story and determine how long your object is in the air.

75. REASONING Show there is no quadratic equation $ax^2 + bx + c = 0$ such that a, b, and c are real numbers and $3i$ and $-2i$ are solutions.

76. MODELING WITH MATHEMATICS The Stratosphere Tower in Las Vegas is 921 feet tall and has a "needle" at its top that extends even higher into the air. A thrill ride called Big Shot catapults riders 160 feet up the needle and then lets them fall back to the launching pad.

a. The height h (in feet) of a rider on the Big Shot can be modeled by $h = -16t^2 + v_0t + 921$, where t is the elapsed time (in seconds) after launch and v_0 is the initial vertical velocity (in feet per second). Find v_0 using the fact that the maximum value of h is $921 + 160 = 1081$ feet.

b. A brochure for the Big Shot states that the ride up the needle takes 2 seconds. Compare this time to the time given by the model $h = -16t^2 + v_0t + 921$, where v_0 is the value you found in part (a). Discuss the accuracy of the model.

Maintaining Mathematical Proficiency

Reviewing what you learned in previous grades and lessons

Solve the system of linear equations by graphing. *(Skills Review Handbook)*

77. $-x + 2y = 6$
$x + 4y = 24$

78. $y = 2x - 1$
$y = x + 1$

79. $3x + y = 4$
$6x + 2y = -4$

80. $y = -x + 2$
$-5x + 5y = 10$

Graph the quadratic equation. Label the vertex and axis of symmetry. *(Section 2.2)*

81. $y = -x^2 + 2x + 1$

82. $y = 2x^2 - x + 3$

83. $y = 0.5x^2 + 2x + 5$

84. $y = -3x^2 - 2$

3.5 Solving Nonlinear Systems

Essential Question

How can you solve a nonlinear system of equations?

EXPLORATION 1 Solving Nonlinear Systems of Equations

Work with a partner. Match each system with its graph. Explain your reasoning. Then solve each system using the graph.

a. $y = x^2$
$y = x + 2$

b. $y = x^2 + x - 2$
$y = x + 2$

c. $y = x^2 - 2x - 5$
$y = -x + 1$

d. $y = x^2 + x - 6$
$y = -x^2 - x + 6$

e. $y = x^2 - 2x + 1$
$y = -x^2 + 2x - 1$

f. $y = x^2 + 2x + 1$
$y = -x^2 + x + 2$

A.

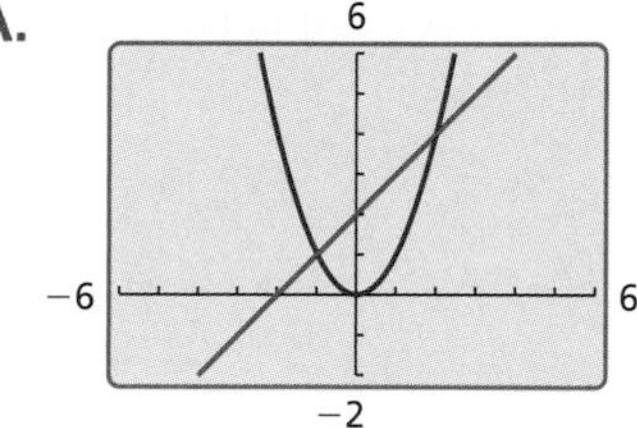

B.

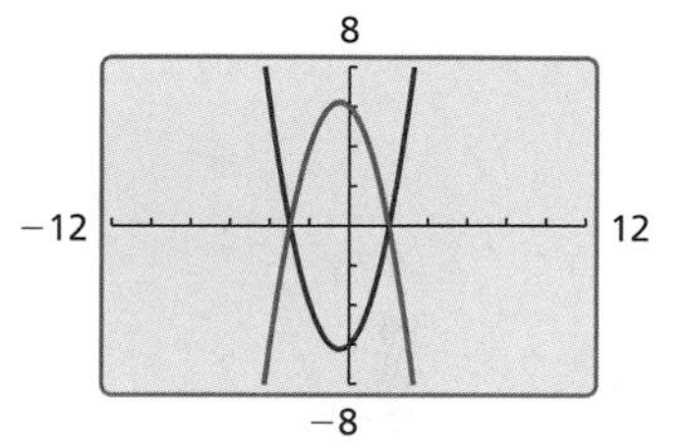

C.

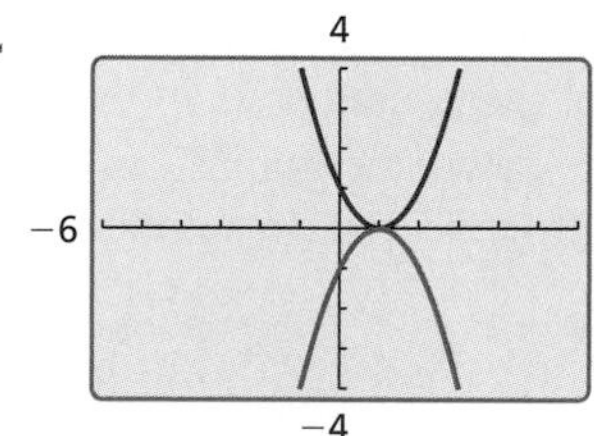

D.

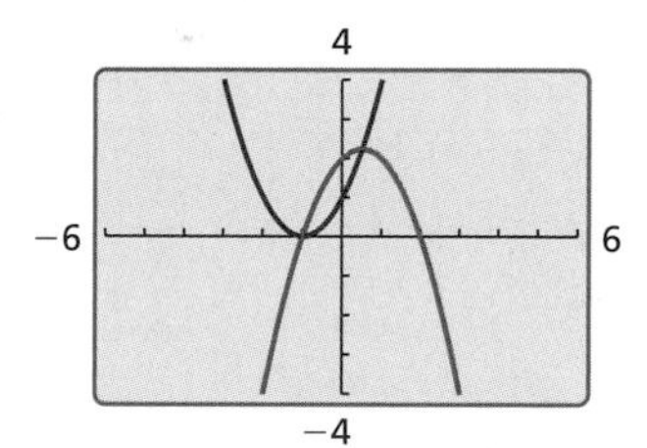

E.

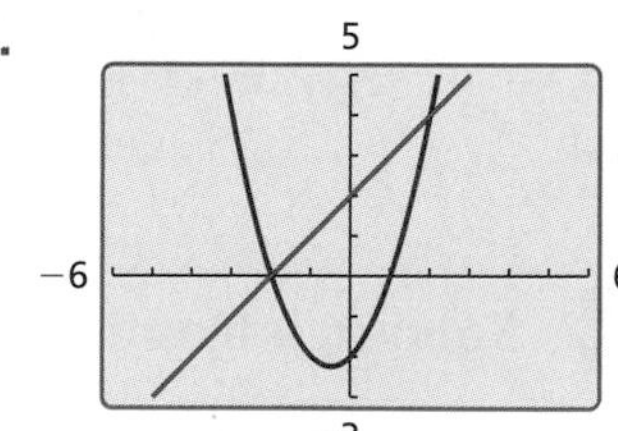

F.

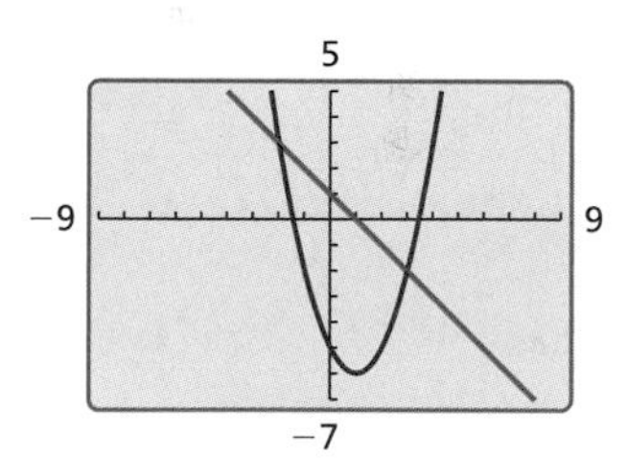

MAKING SENSE OF PROBLEMS

To be proficient in math, you need to plan a solution pathway rather than simply jumping into a solution attempt.

EXPLORATION 2 Solving Nonlinear Systems of Equations

Work with a partner. Look back at the nonlinear system in Exploration 1(f). Suppose you want a more accurate way to solve the system than using a graphical approach.

a. Show how you could use a *numerical approach* by creating a table. For instance, you might use a spreadsheet to solve the system.

b. Show how you could use an *analytical approach*. For instance, you might try solving the system by substitution or elimination.

Communicate Your Answer

3. How can you solve a nonlinear system of equations?

4. Would you prefer to use a graphical, numerical, or analytical approach to solve the given nonlinear system of equations? Explain your reasoning.

$y = x^2 + 2x - 3$
$y = -x^2 - 2x + 4$

3.5 Lesson

Core Vocabulary

system of nonlinear equations, p. 132

Previous
system of linear equations
circle

What You Will Learn

- Solve systems of nonlinear equations.
- Solve quadratic equations by graphing.

Systems of Nonlinear Equations

Previously, you solved systems of *linear* equations by graphing, substitution, and elimination. You can also use these methods to solve a system of *nonlinear* equations. In a **system of nonlinear equations**, at least one of the equations is nonlinear. For instance, the nonlinear system shown has a quadratic equation and a linear equation.

$y = x^2 + 2x - 4$ Equation 1 is nonlinear.

$y = 2x + 5$ Equation 2 is linear.

When the graphs of the equations in a system are a line and a parabola, the graphs can intersect in zero, one, or two points. So, the system can have zero, one, or two solutions, as shown.

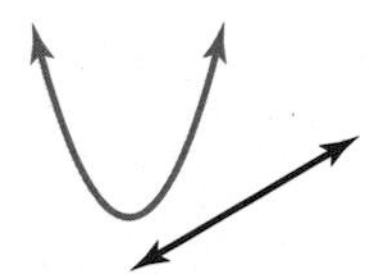

No solution

One solution

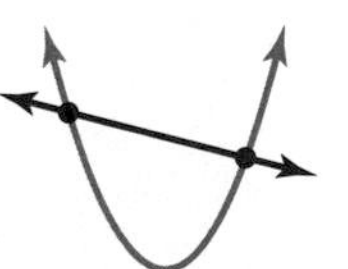

Two solutions

When the graphs of the equations in a system are a parabola that opens up and a parabola that opens down, the graphs can intersect in zero, one, or two points. So, the system can have zero, one, or two solutions, as shown.

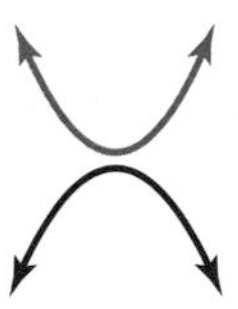

No solution

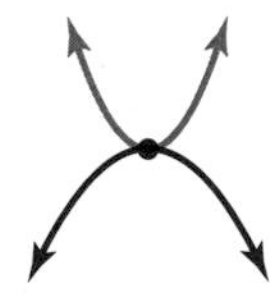

One solution

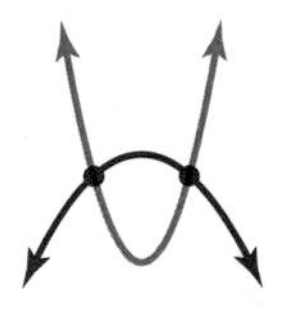

Two solutions

EXAMPLE 1 Solving a Nonlinear System by Graphing

Solve the system by graphing.

$y = x^2 - 2x - 1$ Equation 1

$y = -2x - 1$ Equation 2

SOLUTION

Graph each equation. Then estimate the point of intersection. The parabola and the line appear to intersect at the point $(0, -1)$. Check the point by substituting the coordinates into each of the original equations.

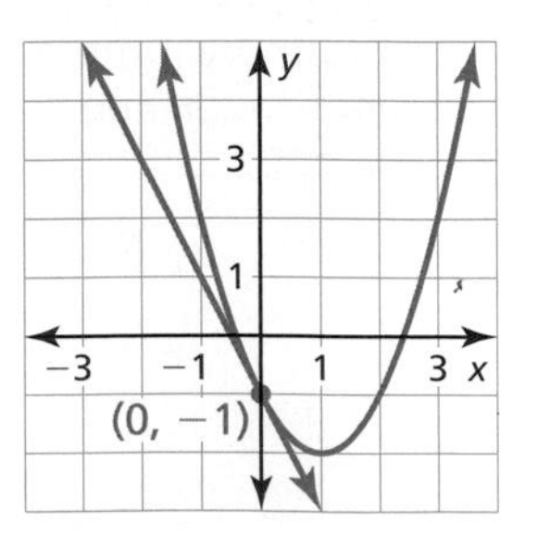

Equation 1

$y = x^2 - 2x - 1$

$-1 \stackrel{?}{=} (0)^2 - 2(0) - 1$

$-1 = -1$ ✓

Equation 2

$y = -2x - 1$

$-1 \stackrel{?}{=} -2(0) - 1$

$-1 = -1$ ✓

▶ The solution is $(0, -1)$.

EXAMPLE 2 Solving a Nonlinear System by Substitution

Solve the system by substitution.

$x^2 + x - y = -1$ Equation 1

$x + y = 4$ Equation 2

SOLUTION

Begin by solving for y in Equation 2.

$y = -x + 4$ Solve for y in Equation 2.

Next, substitute $-x + 4$ for y in Equation 1 and solve for x.

$x^2 + x - y = -1$	Write Equation 1.
$x^2 + x - (-x + 4) = -1$	Substitute $-x + 4$ for y.
$x^2 + 2x - 4 = -1$	Simplify.
$x^2 + 2x - 3 = 0$	Write in standard form.
$(x + 3)(x - 1) = 0$	Factor.
$x + 3 = 0$ or $x - 1 = 0$	Zero-Product Property
$x = -3$ or $x = 1$	Solve for x.

To solve for y, substitute $x = -3$ and $x = 1$ into the equation $y = -x + 4$.

$y = -x + 4 = -(-3) + 4 = 7$	Substitute -3 for x.
$y = -x + 4 = -1 + 4 = 3$	Substitute 1 for x.

▶ The solutions are $(-3, 7)$ and $(1, 3)$. Check the solutions by graphing the system.

Check

8

-6 6

Intersection X=-3 Y=7

-2

EXAMPLE 3 Solving a Nonlinear System by Elimination

Solve the system by elimination.

$2x^2 - 5x - y = -2$ Equation 1

$x^2 + 2x + y = 0$ Equation 2

SOLUTION

Add the equations to eliminate the y-term and obtain a quadratic equation in x.

$$\begin{aligned} 2x^2 - 5x - y &= -2 \\ x^2 + 2x + y &= 0 \\ \hline 3x^2 - 3x \quad &= -2 \end{aligned}$$ Add the equations.

$3x^2 - 3x + 2 = 0$ Write in standard form.

$x = \dfrac{3 \pm \sqrt{-15}}{6}$ Use the Quadratic Formula.

▶ Because the discriminant is negative, the equation $3x^2 - 3x + 2 = 0$ has no real solution. So, the original system has no real solution. You can check this by graphing the system and seeing that the graphs do not appear to intersect.

Check

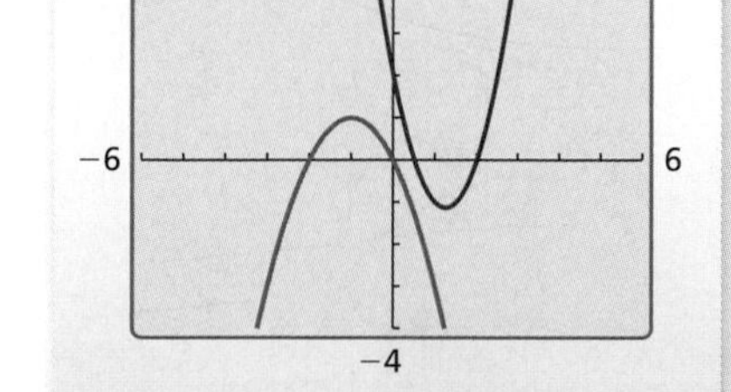

Monitoring Progress  Help in English and Spanish at *BigIdeasMath.com*

Solve the system using any method. Explain your choice of method.

1. $y = -x^2 + 4$
$y = -4x + 8$

2. $x^2 + 3x + y = 0$
$2x + y = 5$

3. $2x^2 + 4x - y = -2$
$x^2 + y = 2$

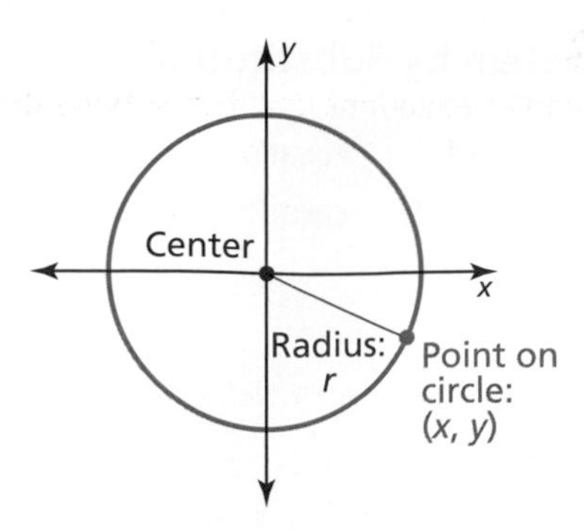

Some nonlinear systems have equations of the form

$$x^2 + y^2 = r^2.$$

This equation is the standard form of a circle with center (0, 0) and radius r.

When the graphs of the equations in a system are a line and a circle, the graphs can intersect in zero, one, or two points. So, the system can have zero, one, or two solutions, as shown.

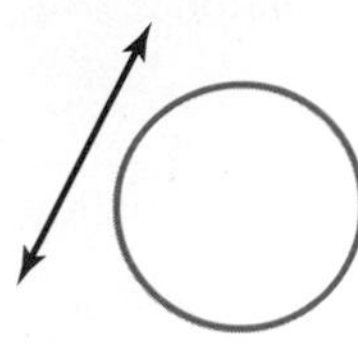

No solution

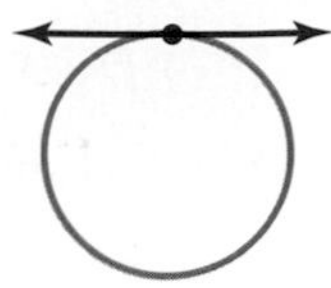

One solution

Two solutions

EXAMPLE 4 Solving a Nonlinear System by Substitution

Solve the system by substitution.

$x^2 + y^2 = 10$ Equation 1

$y = -3x + 10$ Equation 2

SOLUTION

Substitute $-3x + 10$ for y in Equation 1 and solve for x.

$x^2 + y^2 = 10$	Write Equation 1.
$x^2 + (-3x + 10)^2 = 10$	Substitute $-3x + 10$ for y.
$x^2 + 9x^2 - 60x + 100 = 10$	Expand the power.
$10x^2 - 60x + 90 = 0$	Write in standard form.
$x^2 - 6x + 9 = 0$	Divide each side by 10.
$(x - 3)^2 = 0$	Perfect Square Trinomial Pattern
$x = 3$	Zero-Product Property

COMMON ERROR

You can also substitute $x = 3$ in Equation 1 to find y. This yields two *apparent* solutions, (3, 1) and (3, −1). However, (3, −1) is *not* a solution because it does not satisfy Equation 2. You can also see (3, −1) is not a solution from the graph.

To find the y-coordinate of the solution, substitute $x = 3$ in Equation 2.

$$y = -3(3) + 10 = 1$$

▶ The solution is (3, 1). Check the solution by graphing the system. You can see that the line and the circle intersect only at the point (3, 1).

Check

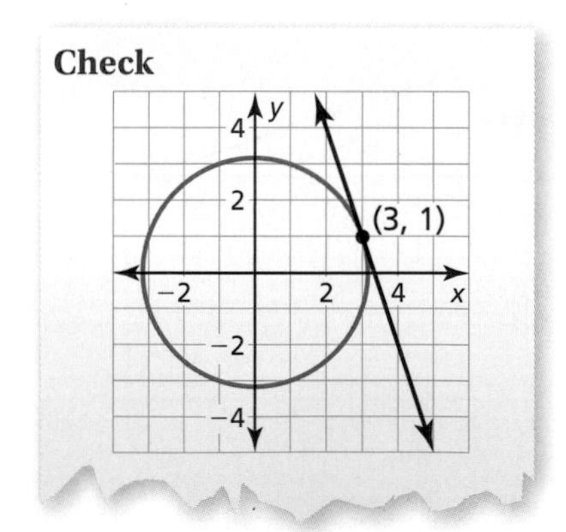

Monitoring Progress

Help in English and Spanish at *BigIdeasMath.com*

Solve the system.

4. $x^2 + y^2 = 16$
$y = -x + 4$

5. $x^2 + y^2 = 4$
$y = x + 4$

6. $x^2 + y^2 = 1$
$y = \frac{1}{2}x + \frac{1}{2}$

Solving Equations by Graphing

You can solve an equation by rewriting it as a system of equations and then solving the system by graphing.

Core Concept

Solving Equations by Graphing

Step 1 To solve the equation $f(x) = g(x)$, write a system of two equations, $y = f(x)$ and $y = g(x)$.

Step 2 Graph the system of equations $y = f(x)$ and $y = g(x)$. The x-value of each solution of the system is a solution of the equation $f(x) = g(x)$.

ANOTHER WAY

In Example 5(a), you can also find the solutions by writing the given equation as $4x^2 + 3x - 2 = 0$ and solving this equation using the Quadratic Formula.

EXAMPLE 5 Solving Quadratic Equations by Graphing

Solve (a) $3x^2 + 5x - 1 = -x^2 + 2x + 1$ and (b) $-(x - 1.5)^2 + 2.25 = 2x(x + 1.5)$ by graphing.

SOLUTION

a. Step 1 Write a system of equations using each side of the original equation.

Equation	*System*
$3x^2 + 5x - 1 = -x^2 + 2x + 1$	$y = 3x^2 + 5x - 1$ $y = -x^2 + 2x + 1$

Step 2 Use a graphing calculator to graph the system. Then use the *intersect* feature to find the x-value of each solution of the system.

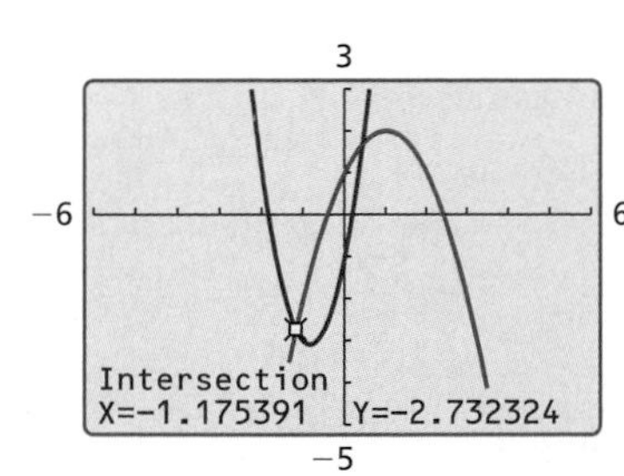

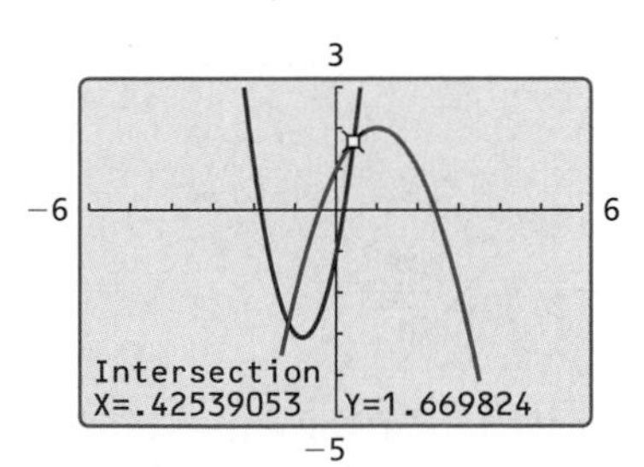

The graphs intersect when $x \approx -1.18$ and $x \approx 0.43$.

▶ The solutions of the equation are $x \approx -1.18$ and $x \approx 0.43$.

b. Step 1 Write a system of equations using each side of the original equation.

Equation	*System*
$-(x - 1.5)^2 + 2.25 = 2x(x + 1.5)$	$y = -(x - 1.5)^2 + 2.25$ $y = 2x(x + 1.5)$

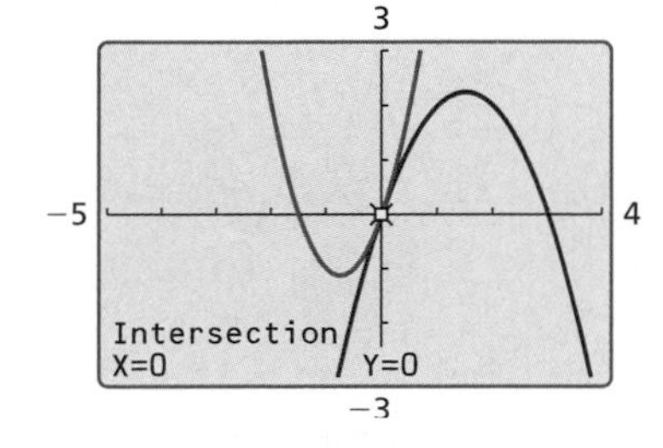

Step 2 Use a graphing calculator to graph the system, as shown at the left. Then use the *intersect* feature to find the x-value of each solution of the system. The graphs intersect when $x = 0$.

▶ The solution of the equation is $x = 0$.

Monitoring Progress

Help in English and Spanish at *BigIdeasMath.com*

Solve the equation by graphing.

7. $x^2 - 6x + 15 = -(x - 3)^2 + 6$

8. $(x + 4)(x - 1) = -x^2 + 3x + 4$

3.5 Exercises

Dynamic Solutions available at *BigIdeasMath.com*

Vocabulary and Core Concept Check

1. **WRITING** Describe the possible solutions of a system consisting of two quadratic equations.

2. **WHICH ONE DOESN'T BELONG?** Which system does *not* belong with the other three? Explain your reasoning.

$y = 3x + 4$ $y = x^2 + 1$	$y = 2x - 1$ $y = -3x + 6$	$y = 3x^2 + 4x + 1$ $y = -5x^2 - 3x + 1$	$x^2 + y^2 = 4$ $y = -x + 1$

Monitoring Progress and Modeling with Mathematics

In Exercises 3–10, solve the system by graphing. Check your solution(s). *(See Example 1.)*

3. $y = x + 2$
   $y = 0.5(x + 2)^2$

4. $y = (x - 3)^2 + 5$
   $y = 5$

5. $y = \frac{1}{3}x + 2$
   $y = -3x^2 - 5x - 4$

6. $y = -3x^2 - 30x - 71$
   $y = -3x - 17$

7. $y = x^2 + 8x + 18$
   $y = -2x^2 - 16x - 30$

8. $y = -2x^2 - 9$
   $y = -4x - 1$

9. $y = (x - 2)^2$
   $y = -x^2 + 4x - 2$

10. $y = \frac{1}{2}(x + 2)^2$
    $y = -\frac{1}{2}x^2 + 2$

In Exercises 11–14, solve the system of nonlinear equations using the graph.

11.
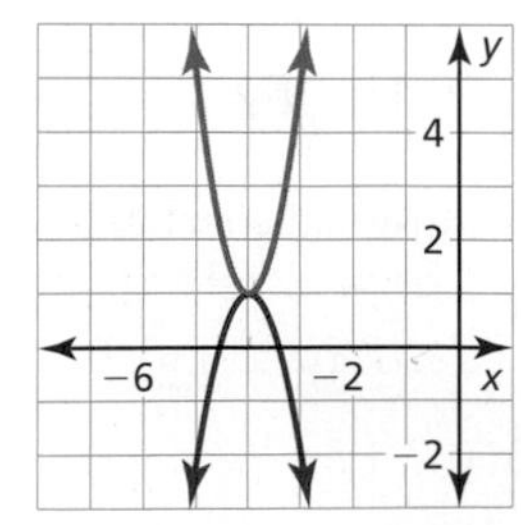

12.
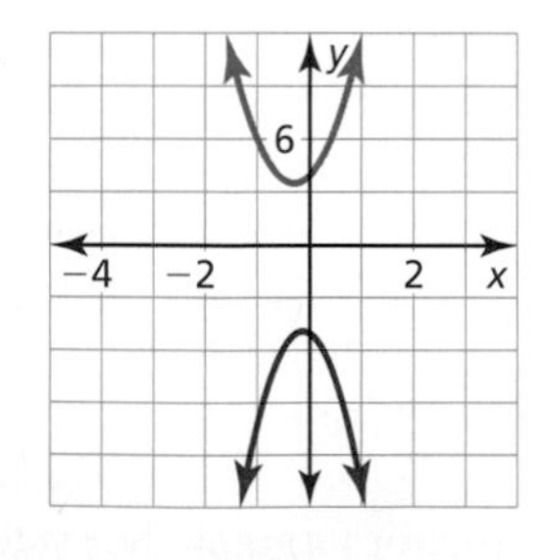

13.
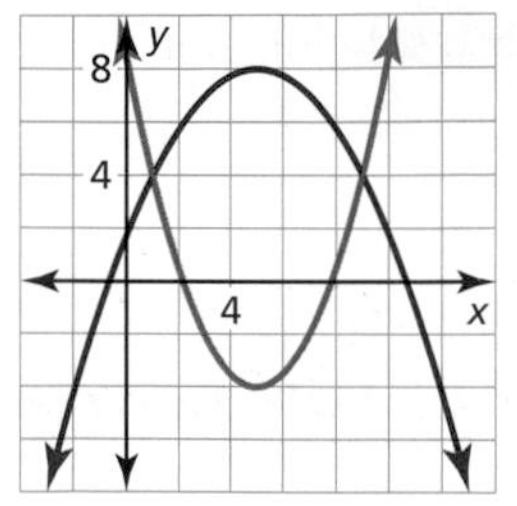

14.
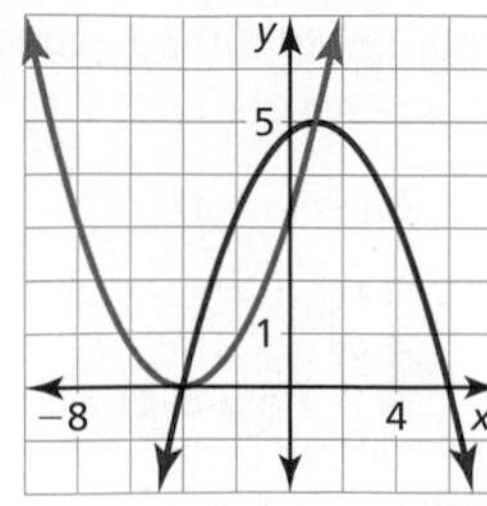

In Exercises 15–24, solve the system by substitution. *(See Examples 2 and 4.)*

15. $y = x + 5$
    $y = x^2 - x + 2$

16. $x^2 + y^2 = 49$
    $y = 7 - x$

17. $x^2 + y^2 = 64$
    $y = -8$

18. $x = 3$
    $-3x^2 + 4x - y = 8$

19. $2x^2 + 4x - y = -3$
    $-2x + y = -4$

20. $2x - 3 = y + 5x^2$
    $y = -3x - 3$

21. $y = x^2 - 1$
    $-7 = -x^2 - y$

22. $y + 16x - 22 = 4x^2$
    $4x^2 - 24x + 26 + y = 0$

23. $x^2 + y^2 = 7$
    $x + 3y = 21$

24. $x^2 + y^2 = 5$
    $-x + y = -1$

25. **USING EQUATIONS** Which ordered pairs are solutions of the nonlinear system?

$$y = \frac{1}{2}x^2 - 5x + \frac{21}{2}$$
$$y = -\frac{1}{2}x + \frac{13}{2}$$

(A) (1, 6) (B) (3, 0)
(C) (8, 2.5) (D) (7, 0)

26. **USING EQUATIONS** How many solutions does the system have? Explain your reasoning.

$$y = 7x^2 - 11x + 9$$
$$y = -7x^2 + 5x - 3$$

(A) 0 (B) 1
(C) 2 (D) 4

In Exercises 27–34, solve the system by elimination. *(See Example 3.)*

27. $2x^2 - 3x - y = -5$
$-x + y = 5$

28. $-3x^2 + 2x - 5 = y$
$-x + 2 = -y$

29. $-3x^2 + y = -18x + 29$
$-3x^2 - y = 18x - 25$

30. $y = -x^2 - 6x - 10$
$y = 3x^2 + 18x + 22$

31. $y + 2x = -14$
$-x^2 - y - 6x = 11$

32. $y = x^2 + 4x + 7$
$-y = 4x + 7$

33. $y = -3x^2 - 30x - 76$
$y = 2x^2 + 20x + 44$

34. $-10x^2 + y = -80x + 155$
$5x^2 + y = 40x - 85$

35. ERROR ANALYSIS Describe and correct the error in using elimination to solve a system.

> ✗
> $y = -2x^2 + 32x - 126$
> $-y = 2x - 14$
> $0 = 18x - 126$
> $126 = 18x$
> $x = 7$

36. NUMBER SENSE The table shows the inputs and outputs of two quadratic equations. Identify the solution(s) of the system. Explain your reasoning.

x	y_1	y_2
-3	29	-11
-1	9	9
1	-3	21
3	-7	25
7	9	9
11	57	-39

In Exercises 37–42, solve the system using any method. Explain your choice of method.

37. $y = x^2 - 1$
$-y = 2x^2 + 1$

38. $y = -4x^2 - 16x - 13$
$-3x^2 + y + 12x = 17$

39. $-2x + 10 + y = \frac{1}{3}x^2$
$y = 10$

40. $y = 0.5x^2 - 10$
$y = -x^2 + 14$

41. $y = -3(x - 4)^2 + 6$
$(x - 4)^2 + 2 - y = 0$

42. $-x^2 + y^2 = 100$
$y = -x + 14$

USING TOOLS In Exercises 43–48, solve the equation by graphing. *(See Example 5.)*

43. $x^2 + 2x = -\frac{1}{2}x^2 + 2x$

44. $2x^2 - 12x - 16 = -6x^2 + 60x - 144$

45. $(x + 2)(x - 2) = -x^2 + 6x - 7$

46. $-2x^2 - 16x - 25 = 6x^2 + 48x + 95$

47. $(x - 2)^2 - 3 = (x + 3)(-x + 9) - 38$

48. $(-x + 4)(x + 8) - 42 = (x + 3)(x + 1) - 1$

49. REASONING A nonlinear system contains the equations of a constant function and a quadratic function. The system has one solution. Describe the relationship between the graphs.

50. PROBLEM SOLVING The range (in miles) of a broadcast signal from a radio tower is bounded by a circle given by the equation

$$x^2 + y^2 = 1620.$$

A straight highway can be modeled by the equation

$$y = -\frac{1}{3}x + 30.$$

For what lengths of the highway are cars able to receive the broadcast signal?

51. PROBLEM SOLVING A car passes a parked police car and continues at a constant speed r. The police car begins accelerating at a constant rate when it is passed. The diagram indicates the distance d (in miles) the police car travels as a function of time t (in minutes) after being passed. Write and solve a system of equations to find how long it takes the police car to catch up to the other car.

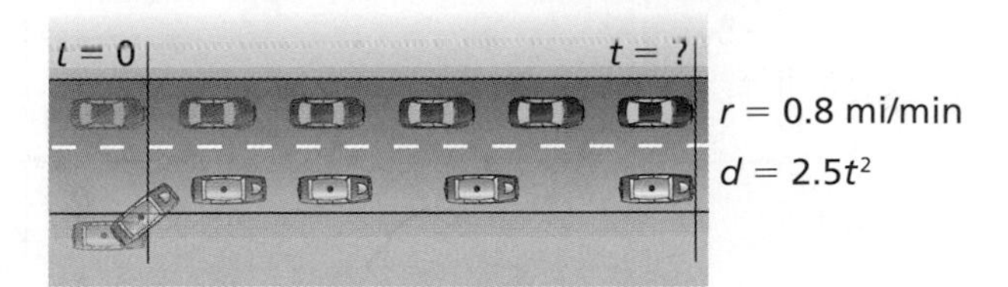

52. THOUGHT PROVOKING Write a nonlinear system that has two different solutions with the same y-coordinate. Sketch a graph of your system. Then solve the system.

53. OPEN-ENDED Find three values for m so the system has no solution, one solution, and two solutions. Justify your answer using a graph.

$$3y = -x^2 + 8x - 7$$
$$y = mx + 3$$

54. MAKING AN ARGUMENT You and a friend solve the system shown and determine that $x = 3$ and $x = -3$. You use Equation 1 to obtain the solutions $(3, 3)$, $(3, -3)$, $(-3, 3)$, and $(-3, -3)$. Your friend uses Equation 2 to obtain the solutions $(3, 3)$ and $(-3, -3)$. Who is correct? Explain your reasoning.

$x^2 + y^2 = 18$ Equation 1

$x - y = 0$ Equation 2

55. COMPARING METHODS Describe two different ways you could solve the quadratic equation. Which way do you prefer? Explain your reasoning.

$$-2x^2 + 12x - 17 = 2x^2 - 16x + 31$$

56. ANALYZING RELATIONSHIPS Suppose the graph of a line that passes through the origin intersects the graph of a circle with its center at the origin. When you know one of the points of intersection, explain how you can find the other point of intersection without performing any calculations.

57. WRITING Describe the possible solutions of a system that contains (a) one quadratic equation and one equation of a circle, and (b) two equations of circles. Sketch graphs to justify your answers.

58. HOW DO YOU SEE IT? The graph of a nonlinear system is shown. Estimate the solution(s). Then describe the transformation of the graph of the linear function that results in a system with no solution.

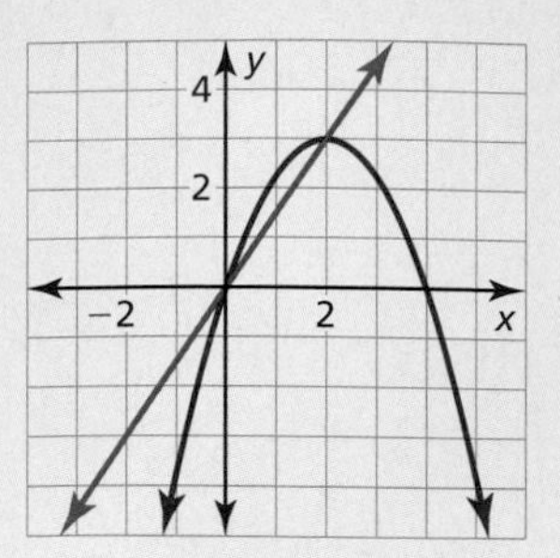

59. MODELING WITH MATHEMATICS To be eligible for a parking pass on a college campus, a student must live at least 1 mile from the campus center.

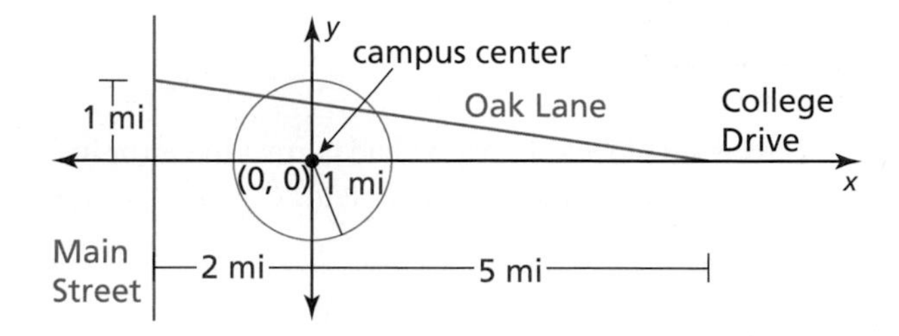

a. Write equations that represent the circle and Oak Lane.

b. Solve the system that consists of the equations in part (a).

c. For what length of Oak Lane are students *not* eligible for a parking pass?

60. CRITICAL THINKING Solve the system of three equations shown.

$$x^2 + y^2 = 4$$
$$2y = x^2 - 2x + 4$$
$$y = -x + 2$$

Maintaining Mathematical Proficiency

Reviewing what you learned in previous grades and lessons

Solve the inequality. Graph the solution on a number line. *(Skills Review Handbook)*

61. $4x - 4 > 8$

62. $-x + 7 \leq 4 - 2x$

63. $-3(x - 4) \geq 24$

Write an inequality that represents the graph. *(Skills Review Handbook)*

64.

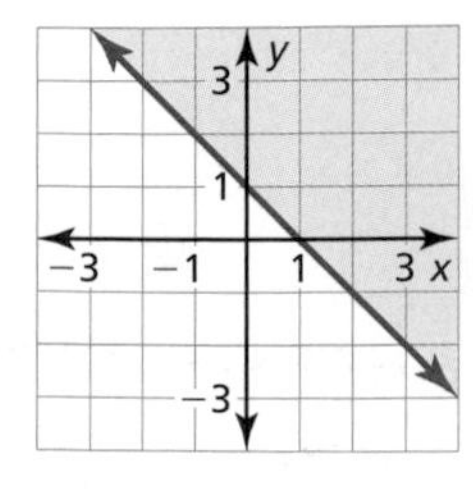

65.

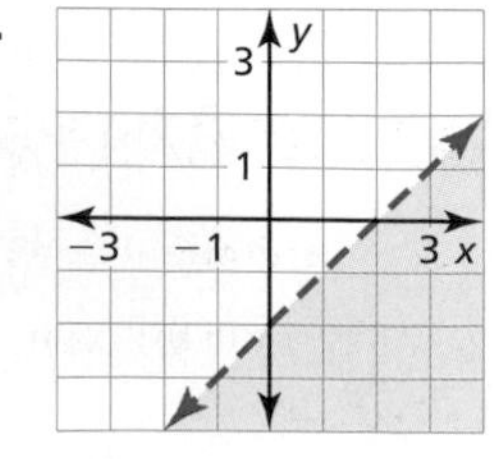

66.

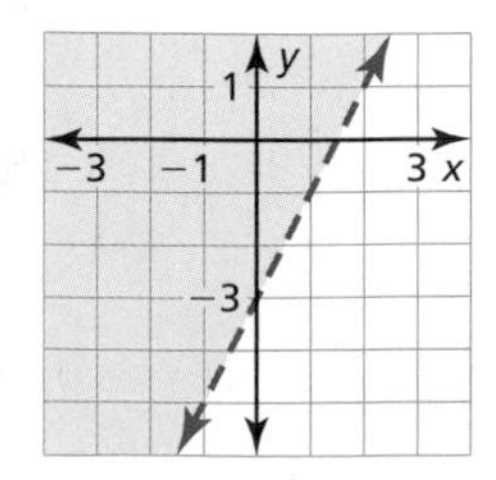

3.6 Quadratic Inequalities

Essential Question

How can you solve a quadratic inequality?

EXPLORATION 1 Solving a Quadratic Inequality

Work with a partner. The graphing calculator screen shows the graph of

$$f(x) = x^2 + 2x - 3.$$

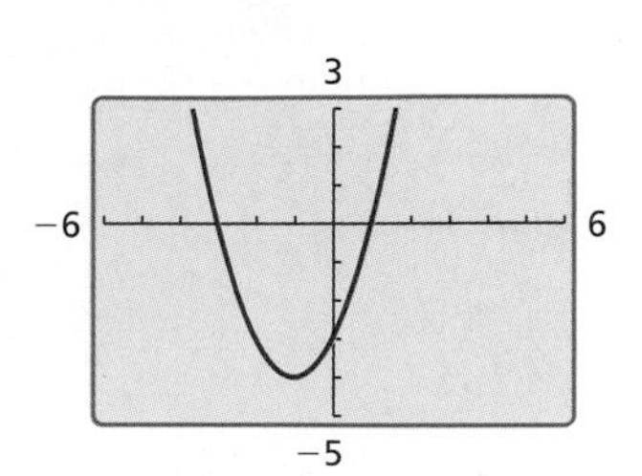

Explain how you can use the graph to solve the inequality

$$x^2 + 2x - 3 \le 0.$$

Then solve the inequality.

> **USING TOOLS STRATEGICALLY**
> To be proficient in math, you need to use technological tools to explore your understanding of concepts.

EXPLORATION 2 Solving Quadratic Inequalities

Work with a partner. Match each inequality with the graph of its related quadratic function. Then use the graph to solve the inequality.

a. $x^2 - 3x + 2 > 0$ **b.** $x^2 - 4x + 3 \le 0$ **c.** $x^2 - 2x - 3 < 0$

d. $x^2 + x - 2 \ge 0$ **e.** $x^2 - x - 2 < 0$ **f.** $x^2 - 4 > 0$

A.

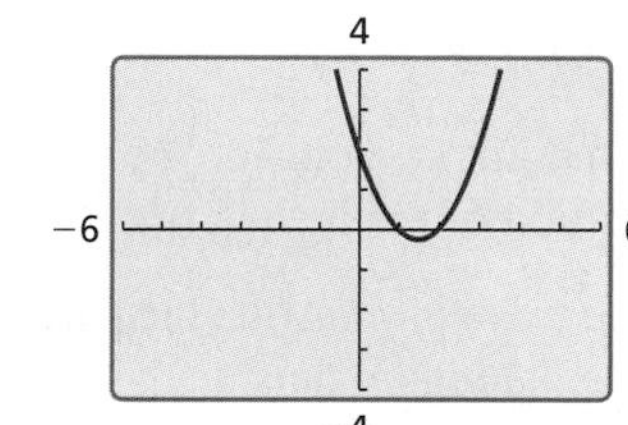

B.

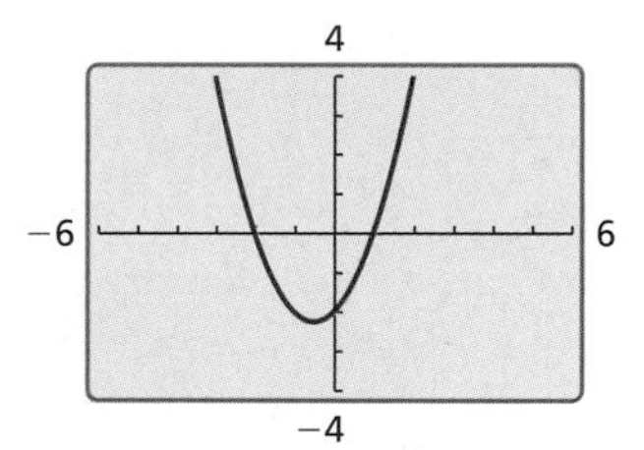

C.

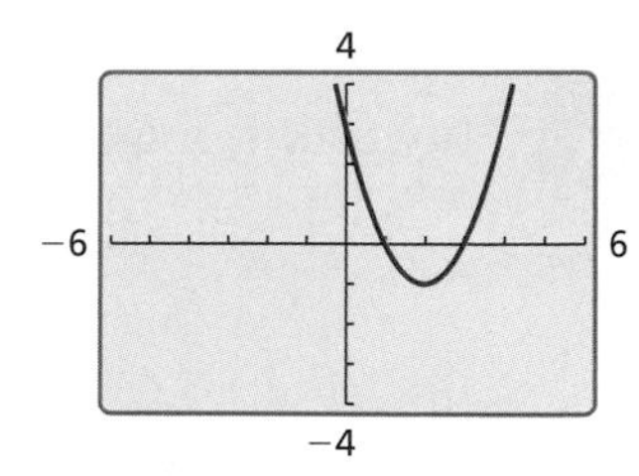

D.

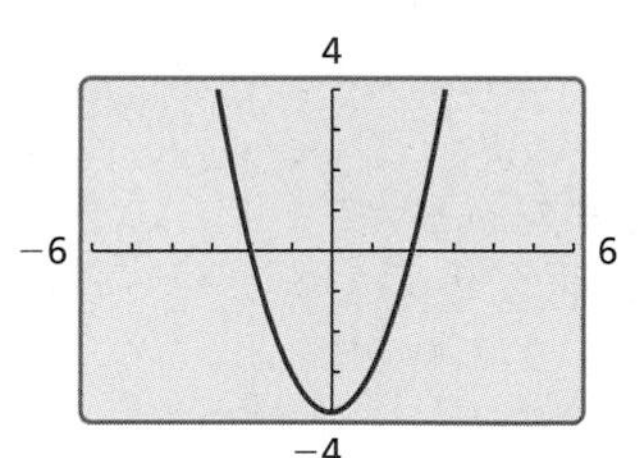

E.

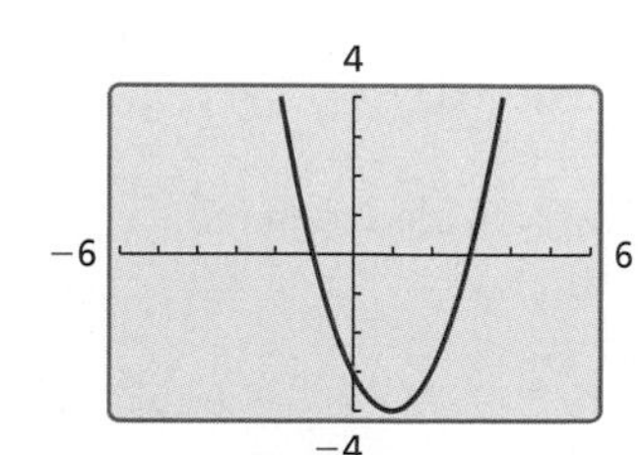

F.

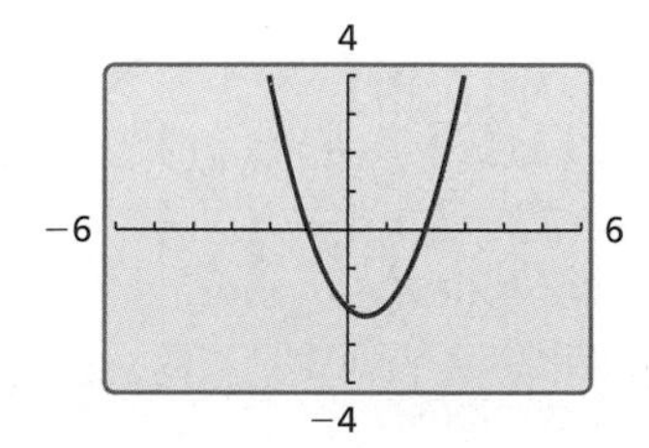

Communicate Your Answer

3. How can you solve a quadratic inequality?

4. Explain how you can use the graph in Exploration 1 to solve each inequality. Then solve each inequality.

a. $x^2 + 2x - 3 > 0$ **b.** $x^2 + 2x - 3 < 0$ **c.** $x^2 + 2x - 3 \ge 0$

3.6 Lesson

Core Vocabulary

quadratic inequality in two variables, *p. 140*
quadratic inequality in one variable, *p. 142*

Previous
linear inequality in two variables

What You Will Learn

- Graph quadratic inequalities in two variables.
- Solve quadratic inequalities in one variable.

Graphing Quadratic Inequalities in Two Variables

A **quadratic inequality in two variables** can be written in one of the following forms, where a, b, and c are real numbers and $a \neq 0$.

$$y < ax^2 + bx + c \qquad y > ax^2 + bx + c$$

$$y \leq ax^2 + bx + c \qquad y \geq ax^2 + bx + c$$

The graph of any such inequality consists of all solutions (x, y) of the inequality.

Previously, you graphed linear inequalities in two variables. You can use a similar procedure to graph quadratic inequalities in two variables.

Core Concept

Graphing a Quadratic Inequality in Two Variables

To graph a quadratic inequality in one of the forms above, follow these steps.

Step 1 Graph the parabola with the equation $y = ax^2 + bx + c$. Make the parabola *dashed* for inequalities with $<$ or $>$ and *solid* for inequalities with $\leq$ or $\geq$.

Step 2 Test a point (x, y) inside the parabola to determine whether the point is a solution of the inequality.

Step 3 Shade the region inside the parabola if the point from Step 2 is a solution. Shade the region outside the parabola if it is not a solution.

EXAMPLE 1 Graphing a Quadratic Inequality in Two Variables

Graph $y < -x^2 - 2x - 1$.

SOLUTION

Step 1 Graph $y = -x^2 - 2x - 1$. Because the inequality symbol is $<$, make the parabola dashed.

Step 2 Test a point inside the parabola, such as $(0, -3)$.

$$y < -x^2 - 2x - 1$$

$$-3 \overset{?}{<} -0^2 - 2(0) - 1$$

$$-3 < -1 \checkmark$$

So, $(0, -3)$ is a solution of the inequality.

Step 3 Shade the region inside the parabola.

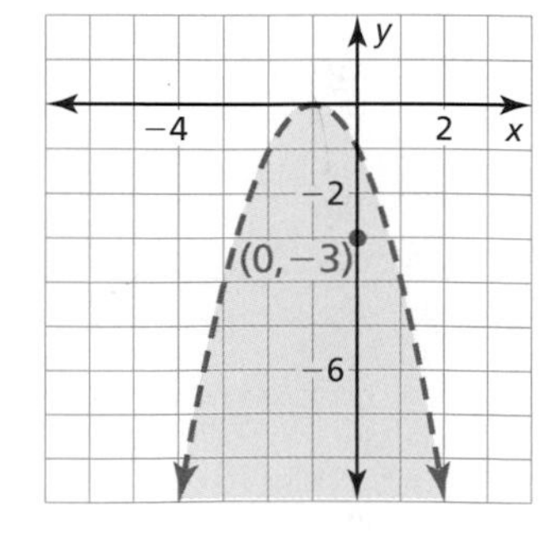

LOOKING FOR STRUCTURE

Notice that testing a point is less complicated when the x-value is 0 (the point is on the y-axis).

EXAMPLE 2 Using a Quadratic Inequality in Real Life

A manila rope used for rappelling down a cliff can safely support a weight W (in pounds) provided

$$W \leq 1480d^2$$

where d is the diameter (in inches) of the rope. Graph the inequality and interpret the solution.

SOLUTION

Graph $W = 1480d^2$ for nonnegative values of d. Because the inequality symbol is $\leq$, make the parabola solid. Test a point inside the parabola, such as (1, 3000).

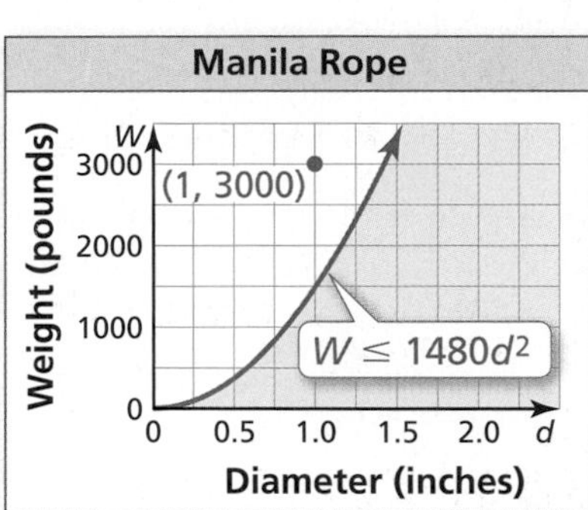

$$W \leq 1480d^2$$

$$3000 \overset{?}{\leq} 1480(1)^2$$

$$3000 \nleq 1480$$

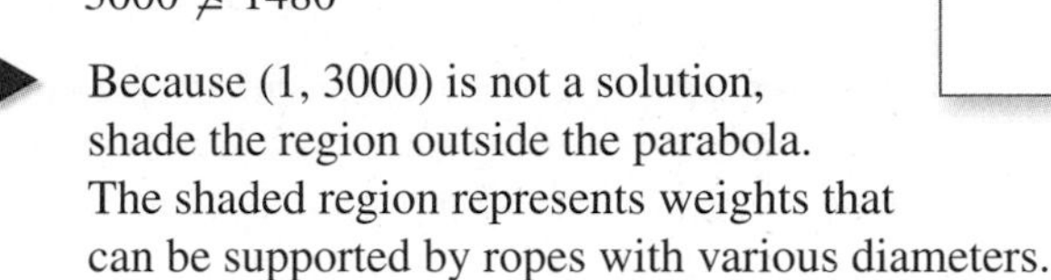

▶ Because (1, 3000) is not a solution, shade the region outside the parabola. The shaded region represents weights that can be supported by ropes with various diameters.

Graphing a *system* of quadratic inequalities is similar to graphing a system of linear inequalities. First graph each inequality in the system. Then identify the region in the coordinate plane common to all of the graphs. This region is called the *graph of the system*.

EXAMPLE 3 Graphing a System of Quadratic Inequalities

Graph the system of quadratic inequalities.

$y < -x^2 + 3$ Inequality 1

$y \geq x^2 + 2x - 3$ Inequality 2

SOLUTION

Step 1 Graph $y < -x^2 + 3$. The graph is the red region inside (but not including) the parabola $y = -x^2 + 3$.

Step 2 Graph $y \geq x^2 + 2x - 3$. The graph is the blue region inside and including the parabola $y = x^2 + 2x - 3$.

Step 3 Identify the purple region where the two graphs overlap. This region is the graph of the system.

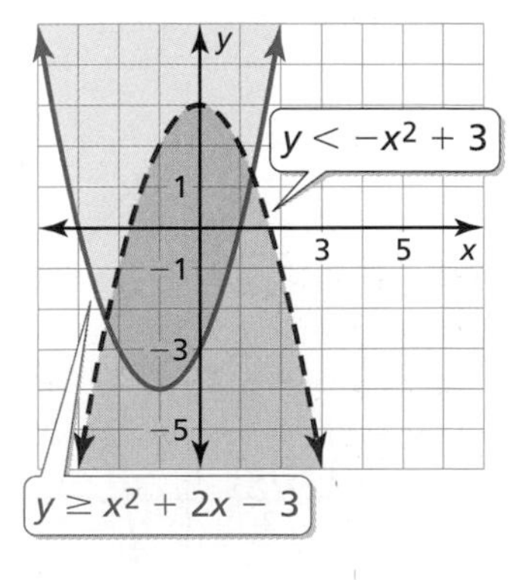

Check

Check that a point in the solution region, such as (0, 0), is a solution of the system.

$$y < -x^2 + 3$$

$$0 \overset{?}{<} -0^2 + 3$$

$$0 < 3 \checkmark$$

$$y \geq x^2 + 2x - 3$$

$$0 \overset{?}{\geq} 0^2 + 2(0) - 3$$

$$0 \geq -3 \checkmark$$

Monitoring Progress

Help in English and Spanish at *BigIdeasMath.com*

Graph the inequality.

1. $y \geq x^2 + 2x - 8$
2. $y \leq 2x^2 - x - 1$
3. $y > -x^2 + 2x + 4$
4. Graph the system of inequalities consisting of $y \leq -x^2$ and $y > x^2 - 3$.

Solving Quadratic Inequalities in One Variable

A **quadratic inequality in one variable** can be written in one of the following forms, where a, b, and c are real numbers and $a \neq 0$.

$ax^2 + bx + c < 0$ $\quad ax^2 + bx + c > 0$ $\quad ax^2 + bx + c \leq 0$ $\quad ax^2 + bx + c \geq 0$

You can solve quadratic inequalities using algebraic methods or graphs.

EXAMPLE 4 Solving a Quadratic Inequality Algebraically

Solve $x^2 - 3x - 4 < 0$ algebraically.

SOLUTION

First, write and solve the equation obtained by replacing $<$ with $=$.

$x^2 - 3x - 4 = 0$ Write the related equation.

$(x - 4)(x + 1) = 0$ Factor.

$x = 4$ or $x = -1$ Zero-Product Property

The numbers -1 and 4 are the *critical values* of the original inequality. Plot -1 and 4 on a number line, using open dots because the values do not satisfy the inequality. The critical x-values partition the number line into three intervals. Test an x-value in each interval to determine whether it satisfies the inequality.

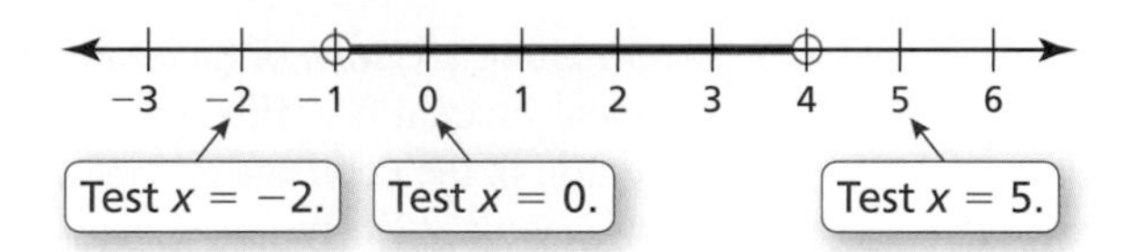

$(-2)^2 - 3(-2) - 4 = 6 \nless 0$ $\quad 0^2 - 3(0) - 4 = -4 < 0$ ✓ $\quad 5^2 - 3(5) - 4 = 6 \nless 0$

▶ So, the solution is $-1 < x < 4$.

Another way to solve $ax^2 + bx + c < 0$ is to first graph the related function $y = ax^2 + bx + c$. Then, because the inequality symbol is $<$, identify the x-values for which the graph lies *below* the x-axis. You can use a similar procedure to solve quadratic inequalities that involve $\leq$, $>$, or $\geq$.

EXAMPLE 5 Solving a Quadratic Inequality by Graphing

Solve $3x^2 - x - 5 \geq 0$ by graphing.

SOLUTION

The solution consists of the x-values for which the graph of $y = 3x^2 - x - 5$ lies on or above the x-axis. Find the x-intercepts of the graph by letting $y = 0$ and using the Quadratic Formula to solve $0 = 3x^2 - x - 5$ for x.

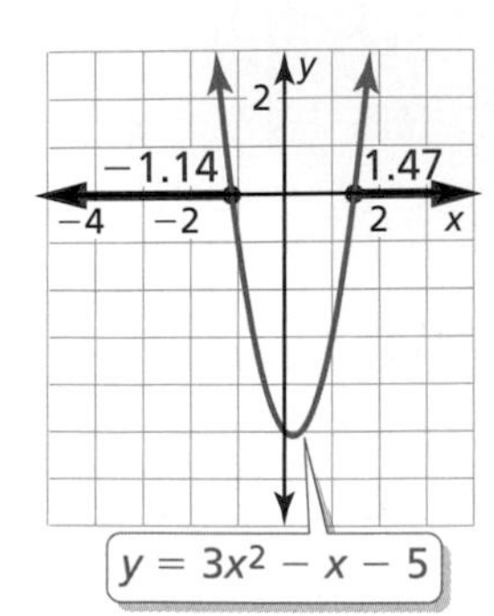

$$x = \frac{-(-1) \pm \sqrt{(-1)^2 - 4(3)(-5)}}{2(3)} \qquad a = 3, b = -1, c = -5$$

$$x = \frac{1 \pm \sqrt{61}}{6} \qquad \text{Simplify.}$$

The solutions are $x \approx -1.14$ and $x \approx 1.47$. Sketch a parabola that opens up and has -1.14 and 1.47 as x-intercepts. The graph lies on or above the x-axis to the left of (and including) $x = -1.14$ and to the right of (and including) $x = 1.47$.

▶ The solution of the inequality is approximately $x \leq -1.14$ or $x \geq 1.47$.

EXAMPLE 6 Modeling with Mathematics

A rectangular parking lot must have a perimeter of 440 feet and an area of at least 8000 square feet. Describe the possible lengths of the parking lot.

SOLUTION

1. **Understand the Problem** You are given the perimeter and the minimum area of a parking lot. You are asked to determine the possible lengths of the parking lot.
2. **Make a Plan** Use the perimeter and area formulas to write a quadratic inequality describing the possible lengths of the parking lot. Then solve the inequality.
3. **Solve the Problem** Let ℓ represent the length (in feet) and let w represent the width (in feet) of the parking lot.

Perimeter = 440	Area ≥ 8000
$2\ell + 2w = 440$	$\ell w \geq 8000$

> **ANOTHER WAY**
>
> You can graph each side of $220\ell - \ell^2 = 8000$ and use the intersection points to determine when $220\ell - \ell^2$ is greater than or equal to 8000.

Solve the perimeter equation for w to obtain $w = 220 - \ell$. Substitute this into the area inequality to obtain a quadratic inequality in one variable.

$\ell w \geq 8000$	Write the area inequality.
$\ell(220 - \ell) \geq 8000$	Substitute $220 - \ell$ for w.
$220\ell - \ell^2 \geq 8000$	Distributive Property
$-\ell^2 + 220\ell - 8000 \geq 0$	Write in standard form.

Use a graphing calculator to find the ℓ-intercepts of $y = -\ell^2 + 220\ell - 8000$.

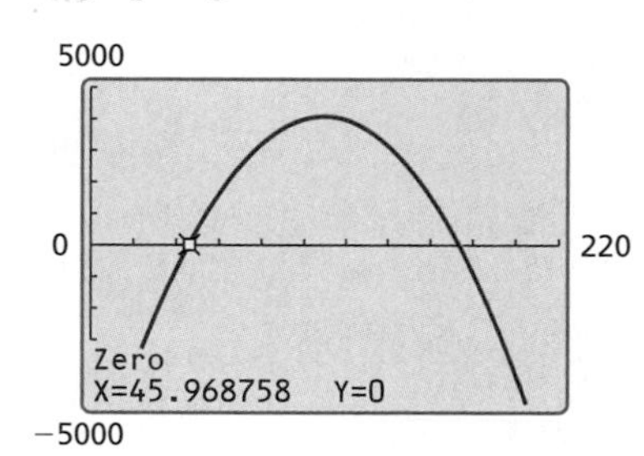

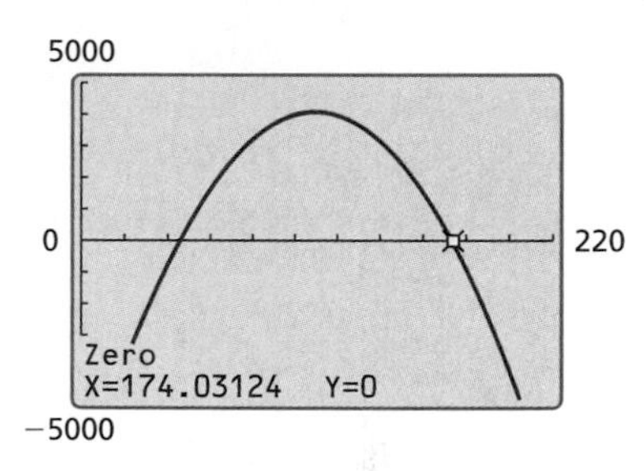

> **USING TECHNOLOGY**
>
> Variables displayed when using technology may not match the variables used in applications. In the graphs shown, the length ℓ corresponds to the independent variable x.

The ℓ-intercepts are $\ell \approx 45.97$ and $\ell \approx 174.03$. The solution consists of the ℓ-values for which the graph lies on or above the ℓ-axis. The graph lies on or above the ℓ-axis when $45.97 \leq \ell \leq 174.03$.

▶ So, the approximate length of the parking lot is at least 46 feet and at most 174 feet.

4. **Look Back** Choose a length in the solution region, such as $\ell = 100$, and find the width. Then check that the dimensions satisfy the original area inequality.

$2\ell + 2w = 440$	$\ell w \geq 8000$
$2(100) + 2w = 440$	$100(120) \overset{?}{\geq} 8000$
$w = 120$	$12{,}000 \geq 8000$ ✓

Monitoring Progress  Help in English and Spanish at *BigIdeasMath.com*

Solve the inequality.

5. $2x^2 + 3x \leq 2$
6. $-3x^2 - 4x + 1 < 0$
7. $2x^2 + 2 > -5x$
8. **WHAT IF?** In Example 6, the area must be at least 8500 square feet. Describe the possible lengths of the parking lot.

3.6 Exercises

Dynamic Solutions available at *BigIdeasMath.com*

Vocabulary and Core Concept Check

1. **WRITING** Compare the graph of a quadratic inequality in one variable to the graph of a quadratic inequality in two variables.
2. **WRITING** Explain how to solve $x^2 + 6x - 8 < 0$ using algebraic methods and using graphs.

Monitoring Progress and Modeling with Mathematics

In Exercises 3–6, match the inequality with its graph. Explain your reasoning.

3. $y \le x^2 + 4x + 3$
4. $y > -x^2 + 4x - 3$
5. $y < x^2 - 4x + 3$
6. $y \ge x^2 + 4x + 3$

A.
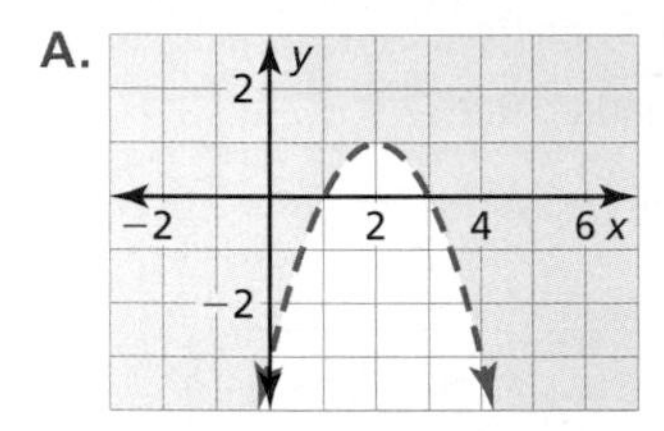

B.
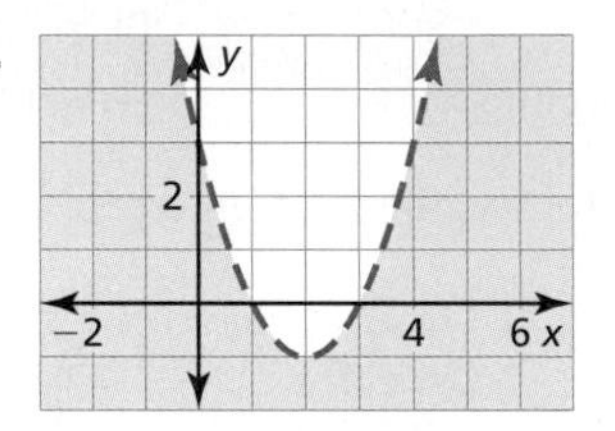

C.
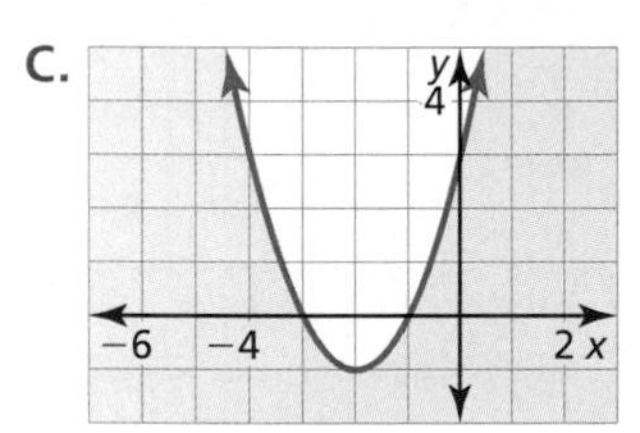

D.
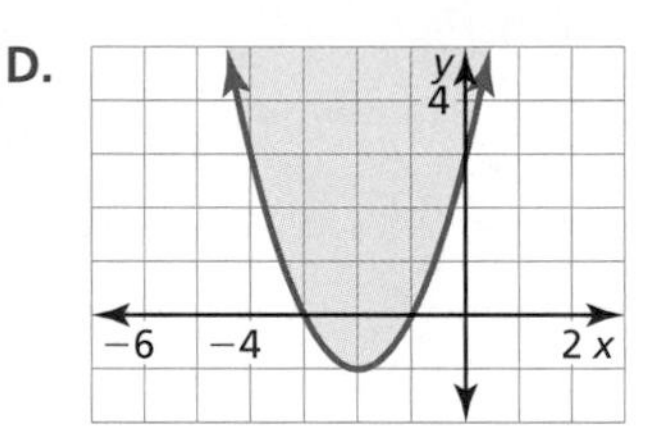

In Exercises 7–14, graph the inequality. *(See Example 1.)*

7. $y < -x^2$
8. $y \ge 4x^2$
9. $y > x^2 - 9$
10. $y < x^2 + 5$
11. $y \le x^2 + 5x$
12. $y \ge -2x^2 + 9x - 4$
13. $y > 2(x + 3)^2 - 1$
14. $y \le \left(x - \frac{1}{2}\right)^2 + \frac{5}{2}$

ANALYZING RELATIONSHIPS In Exercises 15 and 16, use the graph to write an inequality in terms of $f(x)$ so point P is a solution.

15.
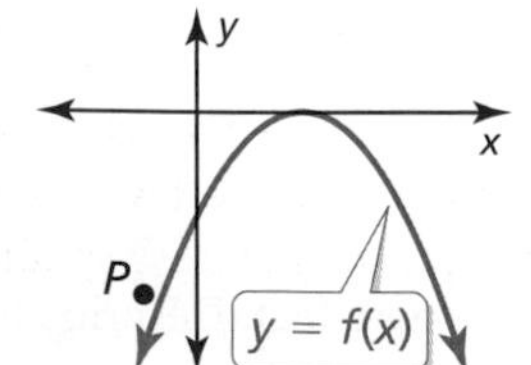

16.
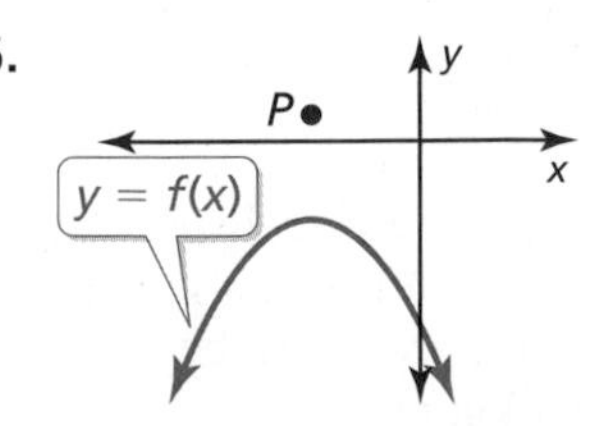

ERROR ANALYSIS In Exercises 17 and 18, describe and correct the error in graphing $y \ge x^2 + 2$.

17.
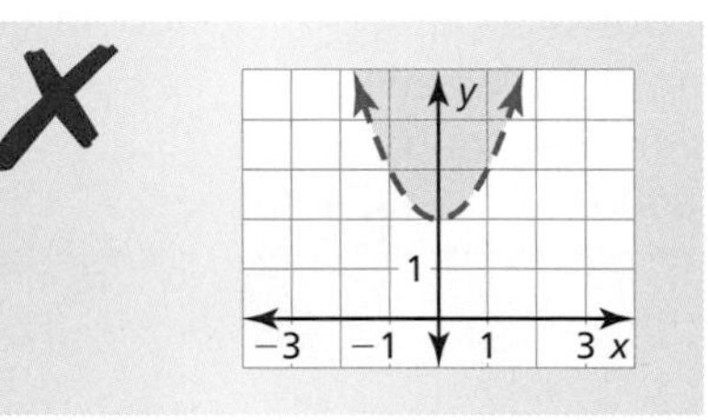

18.
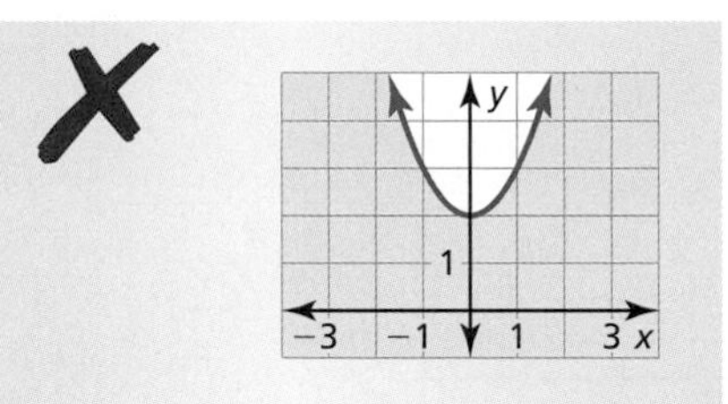

19. **MODELING WITH MATHEMATICS** A hardwood shelf in a wooden bookcase can safely support a weight W (in pounds) provided $W \le 115x^2$, where x is the thickness (in inches) of the shelf. Graph the inequality and interpret the solution. *(See Example 2.)*

20. **MODELING WITH MATHEMATICS** A wire rope can safely support a weight W (in pounds) provided $W \le 8000d^2$, where d is the diameter (in inches) of the rope. Graph the inequality and interpret the solution.

In Exercises 21–26, graph the system of quadratic inequalities. *(See Example 3.)*

21. $y \ge 2x^2$
    $y < -x^2 + 1$
22. $y > -5x^2$
    $y > 3x^2 - 2$
23. $y \le -x^2 + 4x - 4$
    $y < x^2 + 2x - 8$
24. $y \ge x^2 - 4$
    $y \le -2x^2 + 7x + 4$
25. $y \ge 2x^2 + x - 5$
    $y < -x^2 + 5x + 10$
26. $y \ge x^2 - 3x - 6$
    $y \ge x^2 + 7x + 6$

In Exercises 27–34, solve the inequality algebraically. *(See Example 4.)*

27. $4x^2 < 25$

28. $x^2 + 10x + 9 < 0$

29. $x^2 - 11x \geq -28$

30. $3x^2 - 13x > -10$

31. $2x^2 - 5x - 3 \leq 0$

32. $4x^2 + 8x - 21 \geq 0$

33. $\frac{1}{2}x^2 - x > 4$

34. $-\frac{1}{2}x^2 + 4x \leq 1$

In Exercises 35–42, solve the inequality by graphing. *(See Example 5.)*

35. $x^2 - 3x + 1 < 0$

36. $x^2 - 4x + 2 > 0$

37. $x^2 + 8x > -7$

38. $x^2 + 6x < -3$

39. $3x^2 - 8 \leq -2x$

40. $3x^2 + 5x - 3 < 1$

41. $\frac{1}{3}x^2 + 2x \geq 2$

42. $\frac{3}{4}x^2 + 4x \geq 3$

43. DRAWING CONCLUSIONS Consider the graph of the function $f(x) = ax^2 + bx + c$.

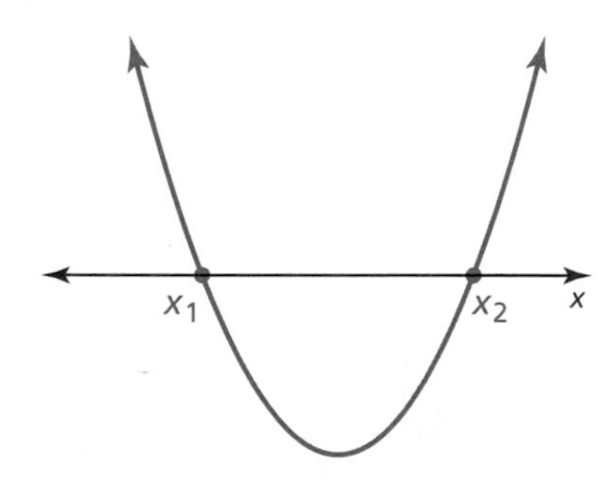

a. What are the solutions of $ax^2 + bx + c < 0$?

b. What are the solutions of $ax^2 + bx + c > 0$?

c. The graph of g represents a reflection in the x-axis of the graph of f. For which values of x is $g(x)$ positive?

44. MODELING WITH MATHEMATICS A rectangular fountain display has a perimeter of 400 feet and an area of at least 9100 feet. Describe the possible widths of the fountain. *(See Example 6.)*

45. MODELING WITH MATHEMATICS The arch of the Sydney Harbor Bridge in Sydney, Australia, can be modeled by $y = -0.00211x^2 + 1.06x$, where x is the distance (in meters) from the left pylons and y is the height (in meters) of the arch above the water. For what distances x is the arch above the road?

46. PROBLEM SOLVING The number T of teams that have participated in a robot-building competition for high-school students over a recent period of time x (in years) can be modeled by

$$T(x) = 17.155x^2 + 193.68x + 235.81, 0 \leq x \leq 6.$$

After how many years is the number of teams greater than 1000? Justify your answer.

47. PROBLEM SOLVING A study found that a driver's reaction time $A(x)$ to audio stimuli and his or her reaction time $V(x)$ to visual stimuli (both in milliseconds) can be modeled by

$$A(x) = 0.0051x^2 - 0.319x + 15, 16 \leq x \leq 70$$

$$V(x) = 0.005x^2 - 0.23x + 22, 16 \leq x \leq 70$$

where x is the age (in years) of the driver.

a. Write an inequality that you can use to find the x-values for which $A(x)$ is less than $V(x)$.

b. Use a graphing calculator to solve the inequality $A(x) < V(x)$. Describe how you used the domain $16 \leq x \leq 70$ to determine a reasonable solution.

c. Based on your results from parts (a) and (b), do you think a driver would react more quickly to a traffic light changing from green to yellow or to the siren of an approaching ambulance? Explain.

48. HOW DO YOU SEE IT? The graph shows a system of quadratic inequalities.

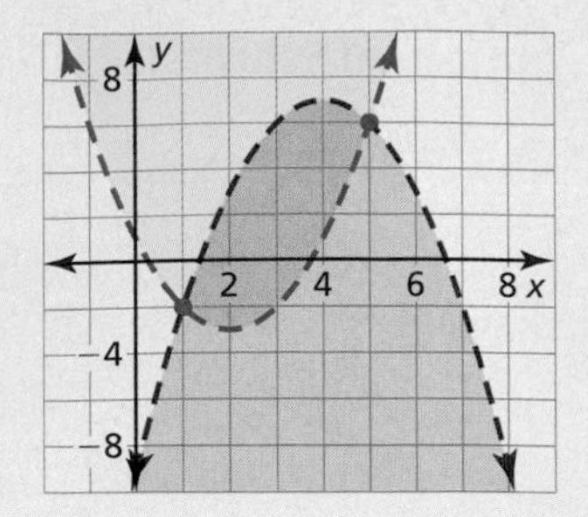

a. Identify two solutions of the system.

b. Are the points $(1, -2)$ and $(5, 6)$ solutions of the system? Explain.

c. Is it possible to change the inequality symbol(s) so that one, but not both of the points in part (b), is a solution of the system? Explain.

49. MODELING WITH MATHEMATICS The length L (in millimeters) of the larvae of the black porgy fish can be modeled by

$$L(x) = 0.00170x^2 + 0.145x + 2.35, 0 \leq x \leq 40$$

where x is the age (in days) of the larvae. Write and solve an inequality to find at what ages a larva's length tends to be greater than 10 millimeters. Explain how the given domain affects the solution.

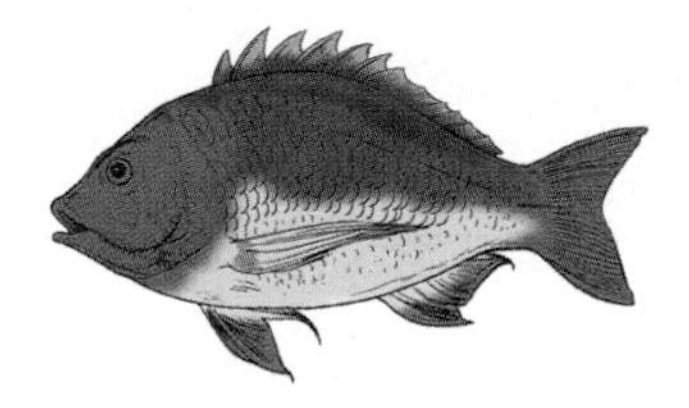

50. MAKING AN ARGUMENT You claim the system of inequalities below, where a and b are real numbers, has no solution. Your friend claims the system will always have at least one solution. Who is correct? Explain.

$$y < (x + a)^2$$
$$y < (x + b)^2$$

51. MATHEMATICAL CONNECTIONS The area A of the region bounded by a parabola and a horizontal line can be modeled by $A = \frac{2}{3}bh$, where b and h are as defined in the diagram. Find the area of the region determined by each pair of inequalities.

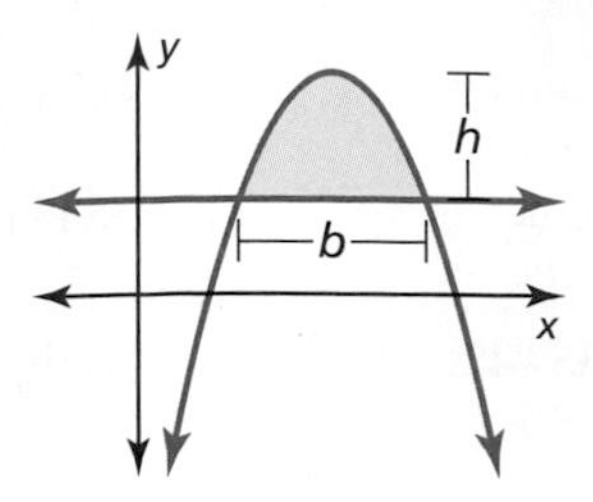

a. $y \leq -x^2 + 4x$
$y \geq 0$

b. $y \geq x^2 - 4x - 5$
$y \leq 7$

52. THOUGHT PROVOKING Draw a company logo that is created by the intersection of two quadratic inequalities. Justify your answer.

53. REASONING A truck that is 11 feet tall and 7 feet wide is traveling under an arch. The arch can be modeled by $y = -0.0625x^2 + 1.25x + 5.75$, where x and y are measured in feet.

a. Will the truck fit under the arch? Explain.

b. What is the maximum width that a truck 11 feet tall can have and still make it under the arch?

c. What is the maximum height that a truck 7 feet wide can have and still make it under the arch?

Maintaining Mathematical Proficiency

Reviewing what you learned in previous grades and lessons

Graph the function. Label the x-intercept(s) and the y-intercept. *(Section 2.2)*

54. $f(x) = (x + 7)(x - 9)$

55. $g(x) = (x - 2)^2 - 4$

56. $h(x) = -x^2 + 5x - 6$

Find the minimum value or maximum value of the function. Then describe where the function is increasing and decreasing. *(Section 2.2)*

57. $f(x) = -x^2 - 6x - 10$

58. $h(x) = \frac{1}{2}(x + 2)^2 - 1$

59. $f(x) = -(x - 3)(x + 7)$

60. $h(x) = x^2 + 3x - 18$

3.4–3.6 What Did You Learn?

Core Vocabulary

Quadratic Formula, *p. 122*
discriminant, *p. 124*
system of nonlinear equations, *p. 132*
quadratic inequality in two variables, *p. 140*
quadratic inequality in one variable, *p. 142*

Core Concepts

Section 3.4

Solving Equations Using the Quadratic Formula, *p. 122*
Analyzing the Discriminant of $ax^2 + bx + c = 0$, *p. 124*
Methods for Solving Quadratic Equations, *p. 125*
Modeling Launched Objects, *p. 126*

Section 3.5

Solving Systems of Nonlinear Equations, *p. 132*
Solving Equations by Graphing, *p. 135*

Section 3.6

Graphing a Quadratic Inequality in Two Variables, *p. 140*
Solving Quadratic Inequalities in One Variable, *p. 142*

Mathematical Practices

1. How can you use technology to determine whose rocket lands first in part (b) of Exercise 65 on page 129?
2. What question can you ask to help the person avoid making the error in Exercise 54 on page 138?
3. Explain your plan to find the possible widths of the fountain in Exercise 44 on page 145.

Performance Task

Algebra in Genetics: The Hardy-Weinberg Law

Some people have attached earlobes, the recessive trait. Some people have free earlobes, the dominant trait. What percent of people carry both traits?

To explore the answers to this question and more, go to ***BigIdeasMath.com***.

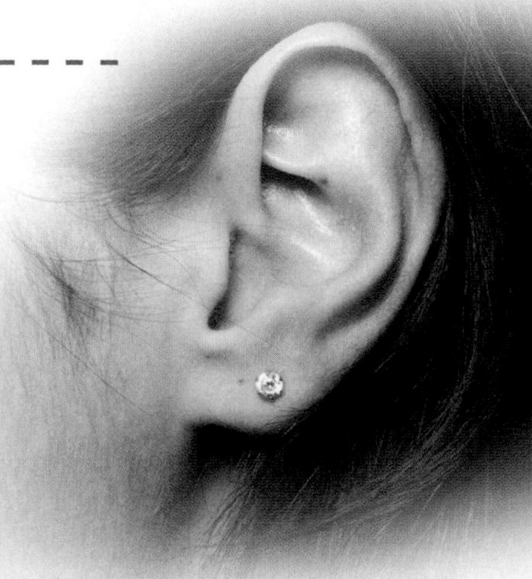

3 Chapter Review

Dynamic Solutions available at *BigIdeasMath.com*

3.1 Solving Quadratic Equations (pp. 93–102)

In a physics class, students must build a Rube Goldberg machine that drops a ball from a 3-foot table. Write a function h (in feet) of the ball after t seconds. How long is the ball in the air?

The initial height is 3, so the model is $h = -16t^2 + 3$. Find the zeros of the function.

$h = -16t^2 + 3$	Write the function.
$0 = -16t^2 + 3$	Substitute 0 for h.
$-3 = -16t^2$	Subtract 3 from each side.
$\dfrac{-3}{-16} = t^2$	Divide each side by -16.
$\pm\sqrt{\dfrac{3}{16}} = t$	Take square root of each side.
$\pm 0.3 \approx t$	Use a calculator.

▶ Reject the negative solution, -0.3, because time must be positive. The ball will fall for about 0.3 second before it hits the ground.

1. Solve $x^2 - 2x - 8 = 0$ by graphing.

Solve the equation using square roots or by factoring.

2. $3x^2 - 4 = 8$ **3.** $x^2 + 6x - 16 = 0$ **4.** $2x^2 - 17x = -30$

5. A rectangular enclosure at the zoo is 35 feet long by 18 feet wide. The zoo wants to double the area of the enclosure by adding the same distance x to the length and width. Write and solve an equation to find the value of x. What are the dimensions of the enclosure?

3.2 Complex Numbers (pp. 103–110)

Perform each operation. Write the answer in standard form.

a. $(3 - 6i) - (7 + 2i) = (3 - 7) + (-6 - 2)i$
$= -4 - 8i$

b. $5i(4 + 5i) = 20i + 25i^2$
$= 20i + 25(-1)$
$= -25 + 20i$

6. Find the values x and y that satisfy the equation $36 - yi = 4x + 3i$.

Perform the operation. Write the answer in standard form.

7. $(-2 + 3i) + (7 - 6i)$ **8.** $(9 + 3i) - (-2 - 7i)$ **9.** $(5 + 6i)(-4 + 7i)$

10. Solve $7x^2 + 21 = 0$.

11. Find the zeros of $f(x) = 2x^2 + 32$.

3.3 Completing the Square (pp. 111–118)

Solve $x^2 + 12x + 8 = 0$ by completing the square.

$x^2 + 12x + 8 = 0$	Write the equation.
$x^2 + 12x = -8$	Write left side in the form $x^2 + bx$.
$x^2 + 12x + 36 = -8 + 36$	Add $\left(\frac{b}{2}\right)^2 = \left(\frac{12}{2}\right)^2 = 36$ to each side.
$(x + 6)^2 = 28$	Write left side as a binomial squared.
$x + 6 = \pm\sqrt{28}$	Take square root of each side.
$x = -6 \pm \sqrt{28}$	Subtract 6 from each side.
$x = -6 \pm 2\sqrt{7}$	Simplify radical.

▶ The solutions are $x = -6 + 2\sqrt{7}$ and $x = -6 - 2\sqrt{7}$.

12. An employee at a local stadium is launching T-shirts from a T-shirt cannon into the crowd during an intermission of a football game. The height h (in feet) of the T-shirt after t seconds can be modeled by $h = -16t^2 + 96t + 4$. Find the maximum height of the T-shirt.

Solve the equation by completing the square.

13. $x^2 + 16x + 17 = 0$ **14.** $4x^2 + 16x + 25 = 0$ **15.** $9x(x - 6) = 81$

16. Write $y = x^2 - 2x + 20$ in vertex form. Then identify the vertex.

3.4 Using the Quadratic Formula (pp. 121–130)

Solve $-x^2 + 4x = 5$ using the Quadratic Formula.

$-x^2 + 4x = 5$	Write the equation.
$-x^2 + 4x - 5 = 0$	Write in standard form.
$x = \dfrac{-4 \pm \sqrt{4^2 - 4(-1)(-5)}}{2(-1)}$	$a = -1, b = 4, c = -5$
$x = \dfrac{-4 \pm \sqrt{-4}}{-2}$	Simplify.
$x = \dfrac{-4 \pm 2i}{-2}$	Write in terms of i.
$x = 2 \pm i$	Simplify.

▶ The solutions are $2 + i$ and $2 - i$.

Solve the equation using the Quadratic Formula.

17. $-x^2 + 5x = 2$ **18.** $2x^2 + 5x = 3$ **19.** $3x^2 - 12x + 13 = 0$

Find the discriminant of the quadratic equation and describe the number and type of solutions of the equation.

20. $-x^2 - 6x - 9 = 0$ **21.** $x^2 - 2x - 9 = 0$ **22.** $x^2 + 6x + 5 = 0$

3.5 Solving Nonlinear Systems (pp. 131–138)

Solve the system by elimination.

$2x^2 - 8x + y = -5$ Equation 1

$2x^2 - 16x - y = -31$ Equation 2

Add the equations to eliminate the y-term and obtain a quadratic equation in x.

$$\begin{array}{r} 2x^2 - 8x + y = -5 \\ 2x^2 - 16x - y = -31 \\ \hline 4x^2 - 24x \quad = -36 \end{array}$$ Add the equations.

$4x^2 - 24x + 36 = 0$ Write in standard form.

$x^2 - 6x + 9 = 0$ Divide each side by 4.

$(x - 3)^2 = 0$ Perfect Square Trinomial Pattern

$x = 3$ Zero-Product Property

To solve for y, substitute $x = 3$ in Equation 1 to obtain $y = 1$.

▶ So, the solution is $(3, 1)$.

Solve the system by any method. Explain your choice of method.

23. $2x^2 - 2 = y$
$-2x + 2 = y$

24. $x^2 - 6x + 13 = y$
$-y = -2x + 3$

25. $x^2 + y^2 = 4$
$-15x + 5 = 5y$

26. Solve $-3x^2 + 5x - 1 = 5x^2 - 8x - 3$ by graphing.

3.6 Quadratic Inequalities (pp. 139–146)

Graph the system of quadratic inequalities.

$y > x^2 - 2$ Inequality 1

$y \leq -x^2 - 3x + 4$ Inequality 2

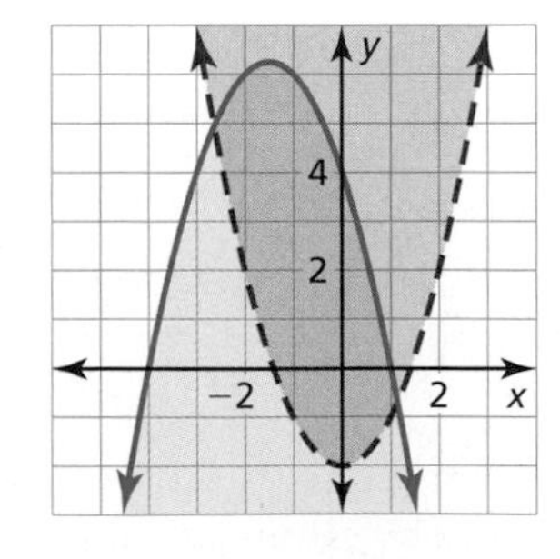

Step 1 Graph $y > x^2 - 2$. The graph is the red region inside (but not including) the parabola $y = x^2 - 2$.

Step 2 Graph $y \leq -x^2 - 3x + 4$. The graph is the blue region inside and including the parabola $y = -x^2 - 3x + 4$.

Step 3 Identify the purple region where the two graphs overlap. This region is the graph of the system.

Graph the inequality.

27. $y > x^2 + 8x + 16$

28. $y \geq x^2 + 6x + 8$

29. $x^2 + y \leq 7x - 12$

Graph the system of quadratic inequalities.

30. $x^2 - 4x + 8 > y$
$-x^2 + 4x + 2 \leq y$

31. $2x^2 - x \geq y - 5$
$0.5x^2 > y - 2x - 1$

32. $-3x^2 - 2x \leq y + 1$
$-2x^2 + x - 5 > -y$

Solve the inequality.

33. $3x^2 + 3x - 60 \geq 0$

34. $-x^2 - 10x < 21$

35. $3x^2 + 2 \leq 5x$

3 Chapter Test

Solve the equation using any method. Provide a reason for your choice.

1. $0 = x^2 + 2x + 3$

2. $6x = x^2 + 7$

3. $x^2 + 49 = 85$

4. $(x + 4)(x - 1) = -x^2 + 3x + 4$

Explain how to use the graph to find the number and type of solutions of the quadratic equation. Justify your answer by using the discriminant.

5. $\frac{1}{2}x^2 + 3x + \frac{9}{2} = 0$

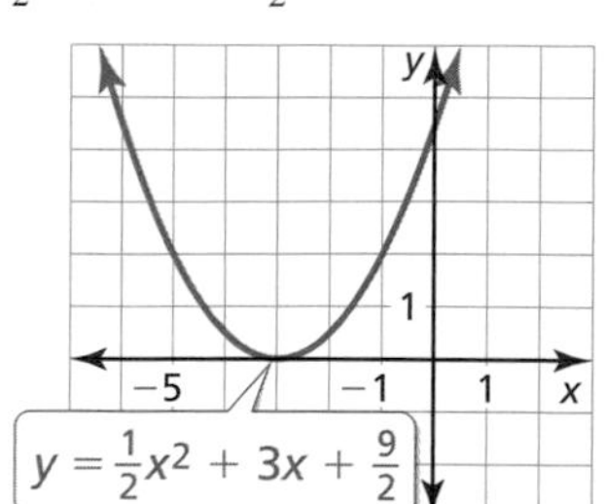

6. $4x^2 + 16x + 18 = 0$

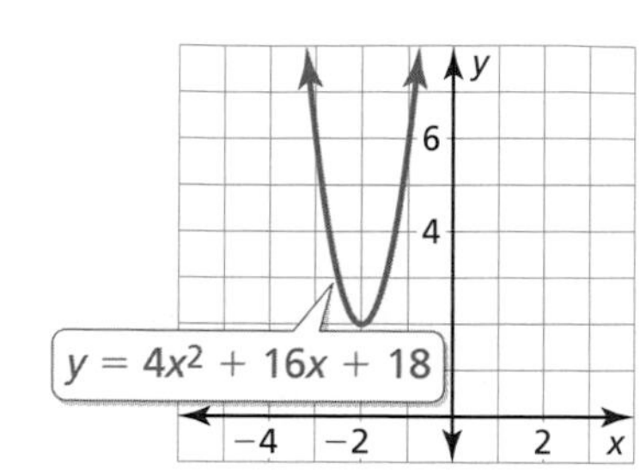

7. $-x^2 + \frac{1}{2}x + \frac{3}{2} = 0$

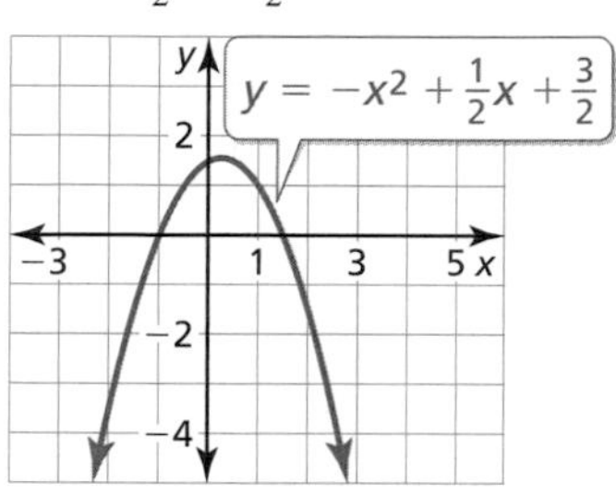

Solve the system of equations or inequalities.

8. $x^2 + 66 = 16x - y$
$2x - y = 18$

9. $y \geq \frac{1}{4}x^2 - 2$
$y < -(x + 3)^2 + 4$

10. $0 = x^2 + y^2 - 40$
$y = x + 4$

11. Write $(3 + 4i)(4 - 6i)$ as a complex number in standard form.

12. The *aspect ratio* of a widescreen TV is the ratio of the screen's width to its height, or 16 : 9. What are the width and the height of a 32-inch widescreen TV? Justify your answer. (*Hint:* Use the Pythagorean Theorem and the fact that TV sizes refer to the diagonal length of the screen.)

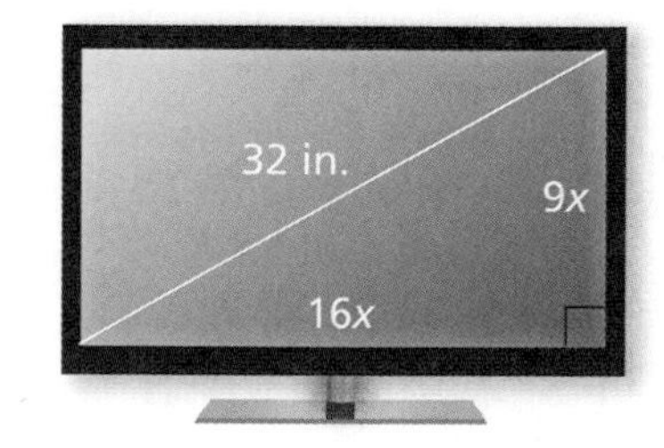

13. The shape of the Gateway Arch in St. Louis, Missouri, can be modeled by $y = -0.0063x^2 + 4x$, where x is the distance (in feet) from the left foot of the arch and y is the height (in feet) of the arch above the ground. For what distances x is the arch more than 200 feet above the ground? Justify your answer.

14. You are playing a game of horseshoes. One of your tosses is modeled in the diagram, where x is the horseshoe's horizontal position (in feet) and y is the corresponding height (in feet). Find the maximum height of the horseshoe. Then find the distance the horseshoe travels. Justify your answer.

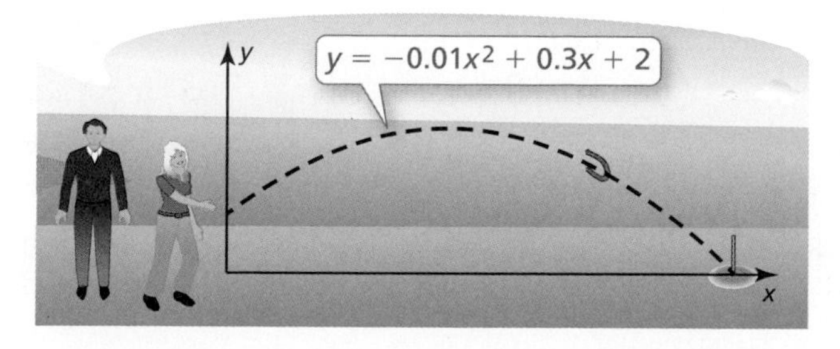

3 Cumulative Assessment

1. The graph of which inequality is shown?

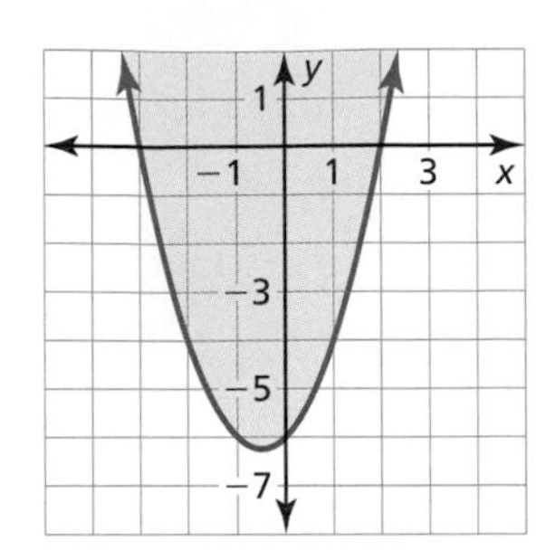

Ⓐ $y > x^2 + x - 6$

Ⓑ $y \geq x^2 + x - 6$

Ⓒ $y > x^2 - x - 6$

Ⓓ $y \geq x^2 - x - 6$

2. Classify each function by its function family. Then describe the transformation of the parent function.

a. $g(x) = x + 4$

b. $h(x) = 5$

c. $h(x) = x^2 - 7$

d. $g(x) = -|x + 3| - 9$

e. $g(x) = \frac{1}{4}(x - 2)^2 - 1$

f. $h(x) = 6x + 11$

3. Two baseball players hit back-to-back home runs. The path of each home run is modeled by the parabolas below, where x is the horizontal distance (in feet) from home plate and y is the height (in feet) above the ground. Choose the correct symbol for each inequality to model the possible locations of the top of the outfield wall.

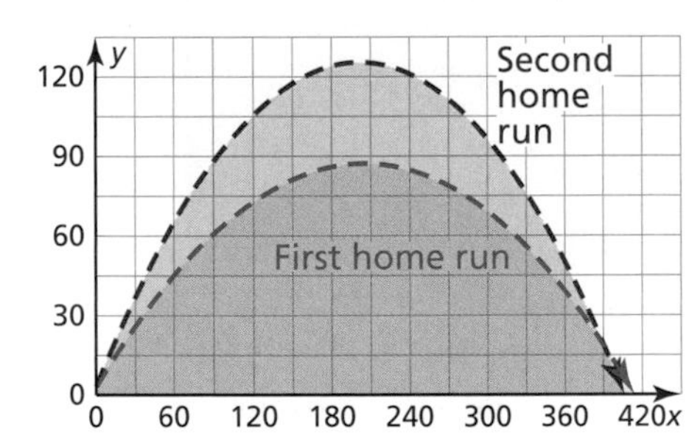

First home run: y ☐ $-0.002x^2 + 0.82x + 3.1$

Second home run: y ☐ $0.003x^2 + 1.21x + 3.3$

4. You claim it is possible to make a function from the given values that has an axis of symmetry at $x = 2$. Your friend claims it is possible to make a function that has an axis of symmetry at $x = -2$. What values can you use to support your claim? What values support your friend's claim?

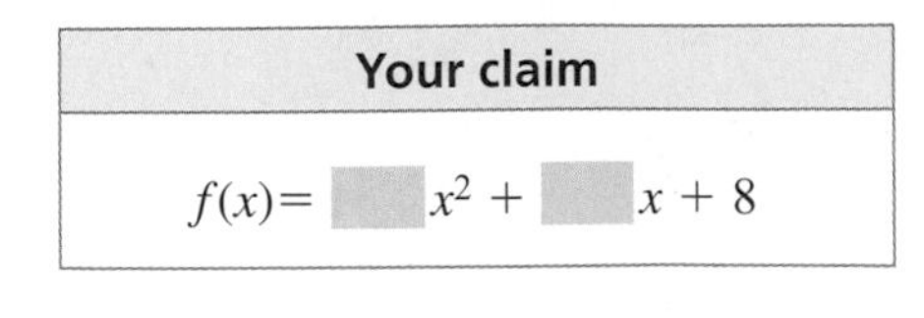

Your friend's claim

$f(x) =$ ☐ $x^2 +$ ☐ $x + 8$

8	−6	−2	0
1	3	5	12

5. Which of the following values are x-coordinates of the solutions of the system?

$$y = x^2 - 6x + 14$$
$$y = 2x + 7$$

−9	−7	−5	−3	−1
1	3	5	7	9

6. The table shows the altitudes of a hang glider that descends at a constant rate. How long will it take for the hang glider to descend to an altitude of 100 feet? Justify your answer.

Time (seconds), t	Altitude (feet), y
0	450
10	350
20	250
30	150

Ⓐ 25 seconds

Ⓑ 35 seconds

Ⓒ 45 seconds

Ⓓ 55 seconds

7. Use the numbers and symbols to write the expression $x^2 + 16$ as the product of two binomials. Some may be used more than once. Justify your answer.

+	−	16	x	)
4	i	2	8	(

8. Choose values for the constants h and k in the equation $x = \frac{1}{4}(y - k)^2 + h$ so that each statement is true.

a. The graph of $x = \frac{1}{4}\left(y - \square\right)^2 + \square$ is a parabola with its vertex in the second quadrant.

b. The graph of $x = \frac{1}{4}\left(y - \square\right)^2 + \square$ is a parabola with its focus in the first quadrant.

c. The graph of $x = \frac{1}{4}\left(y - \square\right)^2 + \square$ is a parabola with its focus in the third quadrant.

9. Which of the following graphs represent a perfect square trinomial? Write each function in vertex form by completing the square.

a.

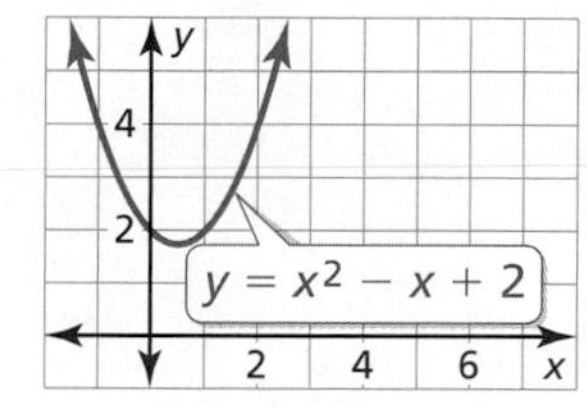

b.

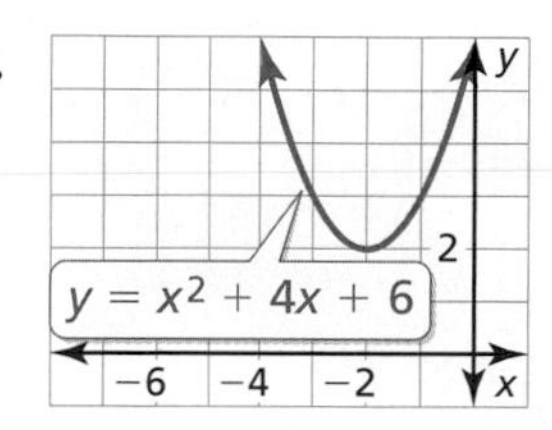

c.

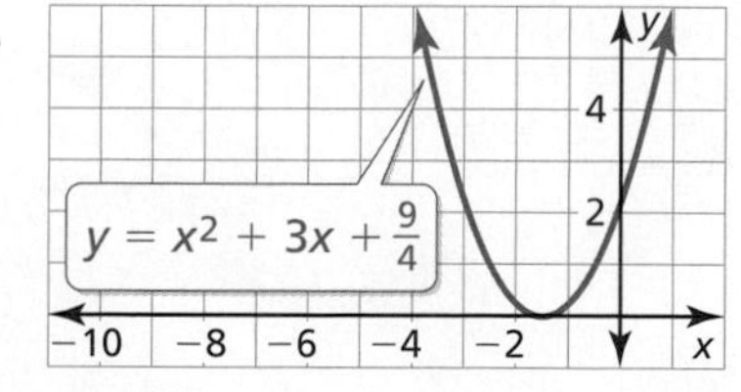

d.

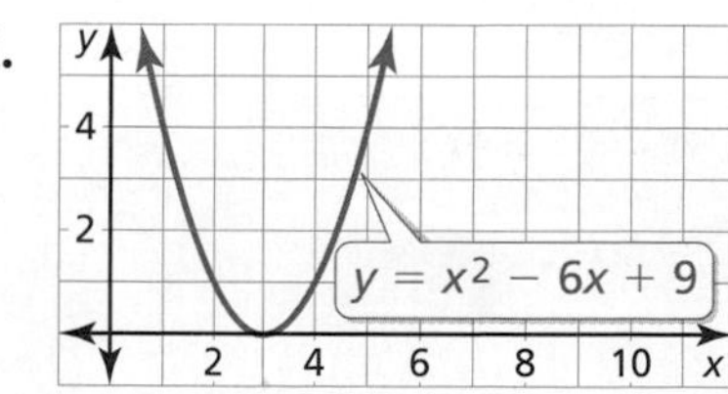

4 Polynomial Functions

Quonset Hut *(p. 218)*

Zebra Mussels *(p. 203)*

Ruins of Caesarea *(p. 195)*

Basketball *(p. 178)*

Electric Vehicles *(p. 161)*

Maintaining Mathematical Proficiency

Simplifying Algebraic Expressions

Example 1 **Simplify the expression $9x + 4x$.**

$9x + 4x = (9 + 4)x$ Distributive Property

$= 13x$ Add coefficients.

Example 2 **Simplify the expression $2(x + 4) + 3(6 - x)$.**

$2(x + 4) + 3(6 - x) = 2(x) + 2(4) + 3(6) + 3(-x)$ Distributive Property

$= 2x + 8 + 18 - 3x$ Multiply.

$= 2x - 3x + 8 + 18$ Group like terms.

$= -x + 26$ Combine like terms.

Simplify the expression.

1. $6x - 4x$

2. $12m - m - 7m + 3$

3. $3(y + 2) - 4y$

4. $9x - 4(2x - 1)$

5. $-(z + 2) - 2(1 - z)$

6. $-x^2 + 5x + x^2$

Finding Volume

Example 3 **Find the volume of a rectangular prism with length 10 centimeters, width 4 centimeters, and height 5 centimeters.**

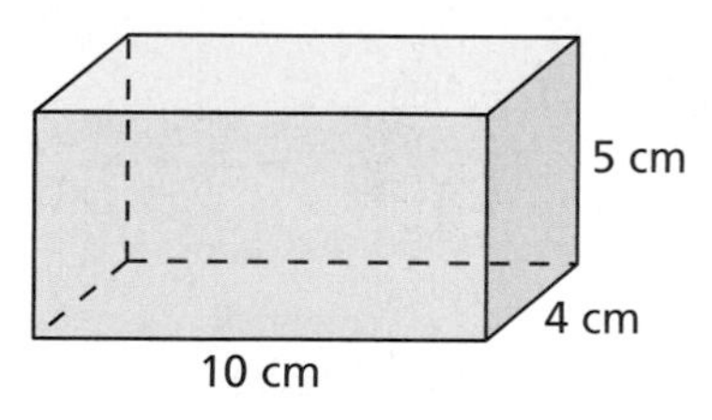

Volume $= \ell wh$ Write the volume formula.

$= (10)(4)(5)$ Substitute 10 for ℓ, 4 for w, and 5 for h.

$= 200$ Multiply.

▶ The volume is 200 cubic centimeters.

Find the volume of the solid.

7. cube with side length 4 inches

8. sphere with radius 2 feet

9. rectangular prism with length 4 feet, width 2 feet, and height 6 feet

10. right cylinder with radius 3 centimeters and height 5 centimeters

11. **ABSTRACT REASONING** Does doubling the volume of a cube have the same effect on the side length? Explain your reasoning.

Mathematical Practices

Mathematically proficient students use technological tools to explore concepts.

Using Technology to Explore Concepts

Core Concept

Continuous Functions

A function is *continuous* when its graph has no breaks, holes, or gaps.

Graph of a continuous function

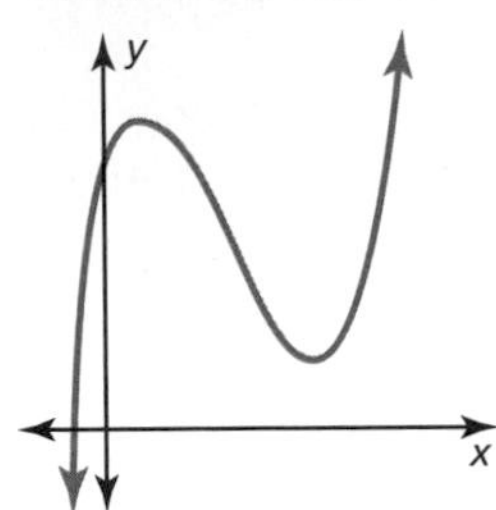

Graph of a function that is not continuous

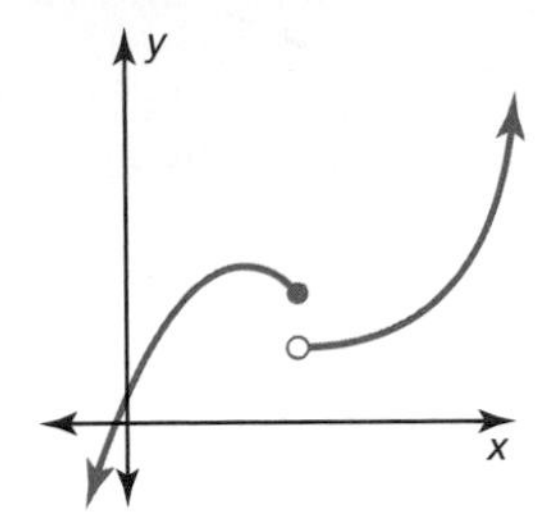

EXAMPLE 1 **Determining Whether Functions Are Continuous**

Use a graphing calculator to compare the two functions. What can you conclude? Which function is not continuous?

$f(x) = x^2$ $\qquad g(x) = \dfrac{x^3 - x^2}{x - 1}$

SOLUTION

The graphs appear to be identical, but g is not defined when $x = 1$. There is a *hole* in the graph of g at the point $(1, 1)$. Using the *table* feature of a graphing calculator, you obtain an error for $g(x)$ when $x = 1$. So, g is not continuous.

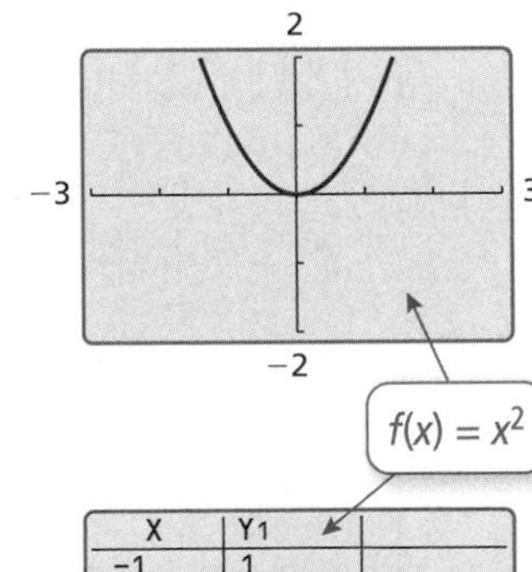

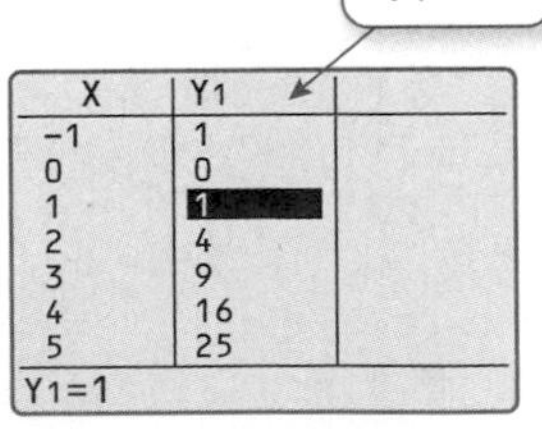

X	Y1
−1	1
0	0
1	1
2	4
3	9
4	16
5	25

Y1=1

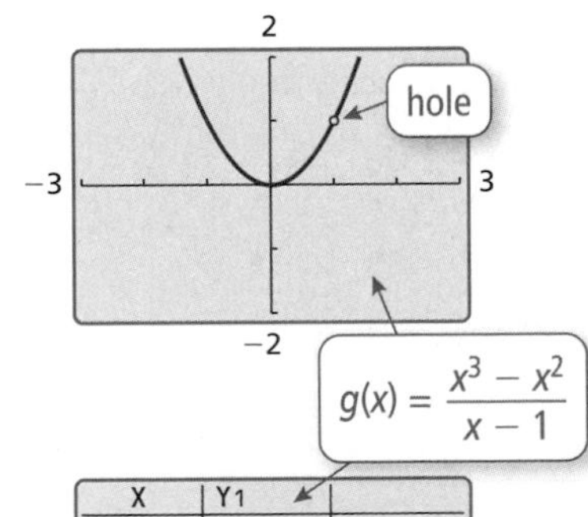

X	Y1
−1	1
0	0
1	ERROR
2	4
3	9
4	16
5	25

Y1=ERROR

Monitoring Progress

Use a graphing calculator to determine whether the function is continuous. Explain your reasoning.

1. $f(x) = \dfrac{x^2 - x}{x}$

2. $f(x) = x^3 - 3$

3. $f(x) = \sqrt{x^2 + 1}$

4. $f(x) = |x + 2|$

5. $f(x) = \dfrac{1}{x}$

6. $f(x) = \dfrac{1}{\sqrt{x^2 - 1}}$

7. $f(x) = x$

8. $f(x) = 2x - 3$

9. $f(x) = \dfrac{x}{x}$

4.1 Graphing Polynomial Functions

Essential Question What are some common characteristics of the graphs of cubic and quartic polynomial functions?

A *polynomial function* of the form

$$f(x) = a_n x^n + a_{n-1} x^{n-1} + \cdots + a_1 x + a_0$$

where $a_n \neq 0$, is *cubic* when $n = 3$ and *quartic* when $n = 4$.

EXPLORATION 1 Identifying Graphs of Polynomial Functions

Work with a partner. Match each polynomial function with its graph. Explain your reasoning. Use a graphing calculator to verify your answers.

a. $f(x) = x^3 - x$ **b.** $f(x) = -x^3 + x$ **c.** $f(x) = -x^4 + 1$

d. $f(x) = x^4$ **e.** $f(x) = x^3$ **f.** $f(x) = x^4 - x^2$

A.

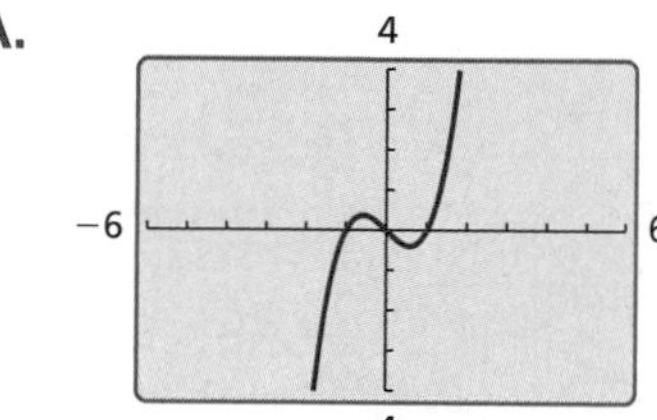

B.

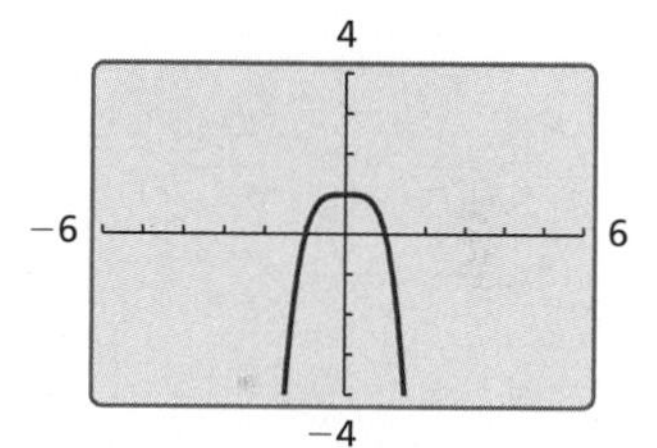

C.

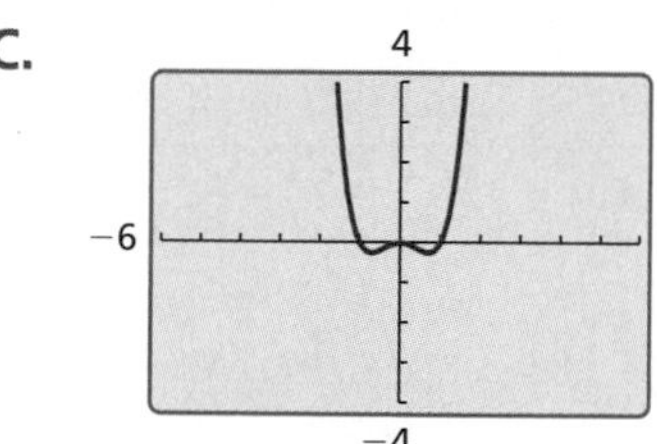

D.

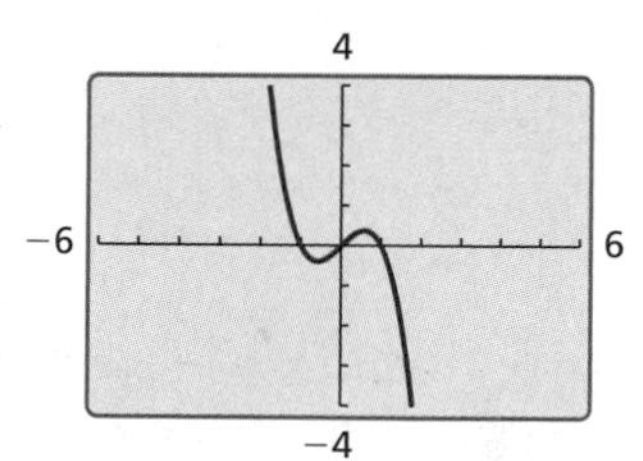

E.

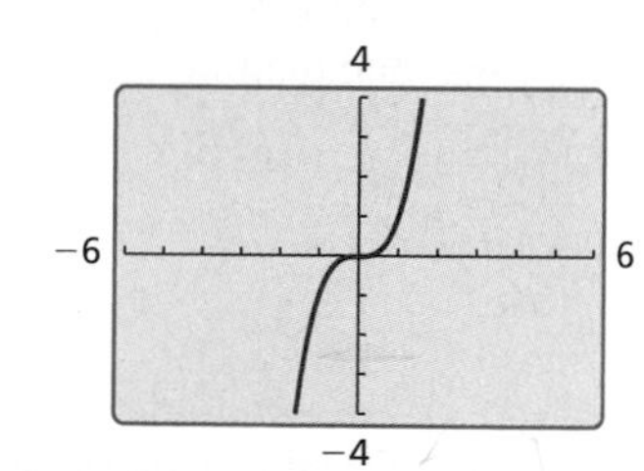

F.

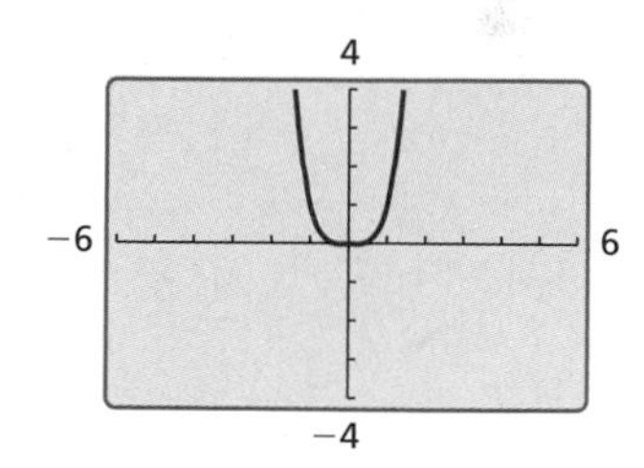

EXPLORATION 2 Identifying *x*-Intercepts of Polynomial Graphs

Work with a partner. Each of the polynomial graphs in Exploration 1 has x-intercept(s) of -1, 0, or 1. Identify the x-intercept(s) of each graph. Explain how you can verify your answers.

CONSTRUCTING VIABLE ARGUMENTS

To be proficient in math, you need to justify your conclusions and communicate them to others.

Communicate Your Answer

3. What are some common characteristics of the graphs of cubic and quartic polynomial functions?

4. Determine whether each statement is *true* or *false*. Justify your answer.

a. When the graph of a cubic polynomial function rises to the left, it falls to the right.

b. When the graph of a quartic polynomial function falls to the left, it rises to the right.

4.1 Lesson

Core Vocabulary

polynomial, *p. 158*
polynomial function, *p. 158*
end behavior, *p. 159*

Previous
monomial
linear function
quadratic function

What You Will Learn

- Identify polynomial functions.
- Graph polynomial functions using tables and end behavior.

Polynomial Functions

Recall that a monomial is a number, a variable, or the product of a number and one or more variables with whole number exponents. A **polynomial** is a monomial or a sum of monomials. A **polynomial function** is a function of the form

$$f(x) = a_nx^n + a_{n-1}x^{n-1} + \cdots + a_1x + a_0$$

where $a_n \neq 0$, the exponents are all whole numbers, and the coefficients are all real numbers. For this function, a_n is the leading coefficient, n is the degree, and a_0 is the constant term. A polynomial function is in *standard form* when its terms are written in descending order of exponents from left to right.

You are already familiar with some types of polynomial functions, such as linear and quadratic. Here is a summary of common types of polynomial functions.

Common Polynomial Functions

Degree	Type	Standard Form	Example
0	Constant	$f(x) = a_0$	$f(x) = -14$
1	Linear	$f(x) = a_1x + a_0$	$f(x) = 5x - 7$
2	Quadratic	$f(x) = a_2x^2 + a_1x + a_0$	$f(x) = 2x^2 + x - 9$
3	Cubic	$f(x) = a_3x^3 + a_2x^2 + a_1x + a_0$	$f(x) = x^3 - x^2 + 3x$
4	Quartic	$f(x) = a_4x^4 + a_3x^3 + a_2x^2 + a_1x + a_0$	$f(x) = x^4 + 2x - 1$

EXAMPLE 1 Identifying Polynomial Functions

Decide whether each function is a polynomial function. If so, write it in standard form and state its degree, type, and leading coefficient.

a. $f(x) = -2x^3 + 5x + 8$

b. $g(x) = -0.8x^3 + \sqrt{2}x^4 - 12$

c. $h(x) = -x^2 + 7x^{-1} + 4x$

d. $k(x) = x^2 + 3^x$

SOLUTION

a. The function is a polynomial function that is already written in standard form. It has degree 3 (cubic) and a leading coefficient of -2.

b. The function is a polynomial function written as $g(x) = \sqrt{2}x^4 - 0.8x^3 - 12$ in standard form. It has degree 4 (quartic) and a leading coefficient of $\sqrt{2}$.

c. The function is not a polynomial function because the term $7x^{-1}$ has an exponent that is not a whole number.

d. The function is not a polynomial function because the term 3^x does not have a variable base and an exponent that is a whole number.

Monitoring Progress  Help in English and Spanish at *BigIdeasMath.com*

Decide whether the function is a polynomial function. If so, write it in standard form and state its degree, type, and leading coefficient.

1. $f(x) = 7 - 1.6x^2 - 5x$

2. $p(x) = x + 2x^{-2} + 9.5$

3. $q(x) = x^3 - 6x + 3x^4$

EXAMPLE 2 Evaluating a Polynomial Function

Evaluate $f(x) = 2x^4 - 8x^2 + 5x - 7$ when $x = 3$.

SOLUTION

$f(x) = 2x^4 - 8x^2 + 5x - 7$	Write original equation.
$f(3) = 2(3)^4 - 8(3)^2 + 5(3) - 7$	Substitute 3 for x.
$= 162 - 72 + 15 - 7$	Evaluate powers and multiply.
$= 98$	Simplify.

The **end behavior** of a function's graph is the behavior of the graph as x approaches positive infinity ($+\infty$) or negative infinity ($-\infty$). For the graph of a polynomial function, the end behavior is determined by the function's degree and the sign of its leading coefficient.

READING

The expression "$x \to +\infty$" is read as "x approaches positive infinity."

Core Concept

End Behavior of Polynomial Functions

Degree: odd
Leading coefficient: positive

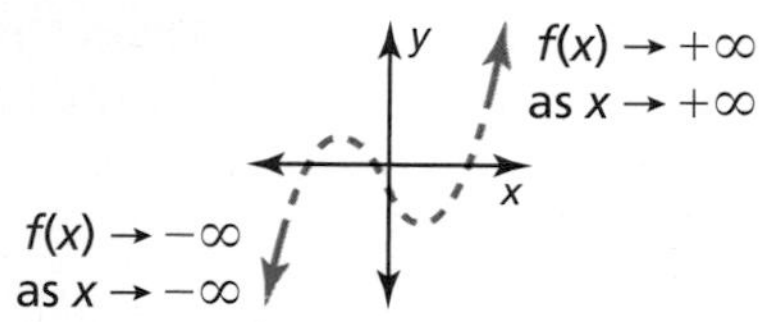

Degree: odd
Leading coefficient: negative

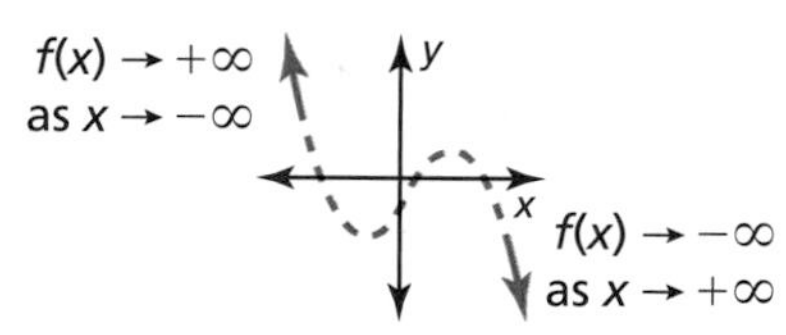

Degree: even
Leading coefficient: positive

f(x) → +∞ as x → −∞; f(x) → +∞ as x → +∞

Degree: even
Leading coefficient: negative

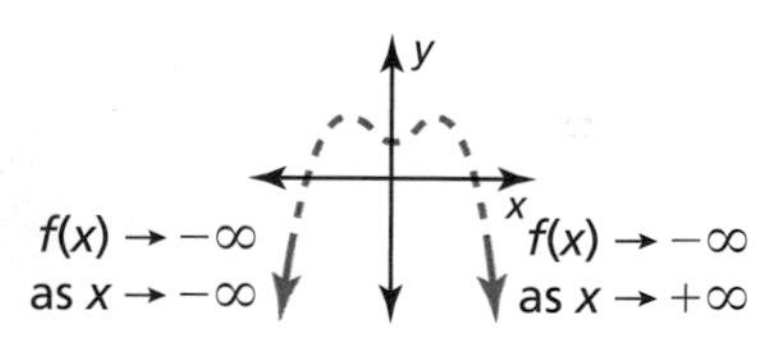

EXAMPLE 3 Describing End Behavior

Describe the end behavior of the graph of $f(x) = -0.5x^4 + 2.5x^2 + x - 1$.

SOLUTION

The function has degree 4 and leading coefficient -0.5. Because the degree is even and the leading coefficient is negative, $f(x) \to -\infty$ as $x \to -\infty$ and $f(x) \to -\infty$ as $x \to +\infty$. Check this by graphing the function on a graphing calculator, as shown.

Check

10, −10, 10, −10

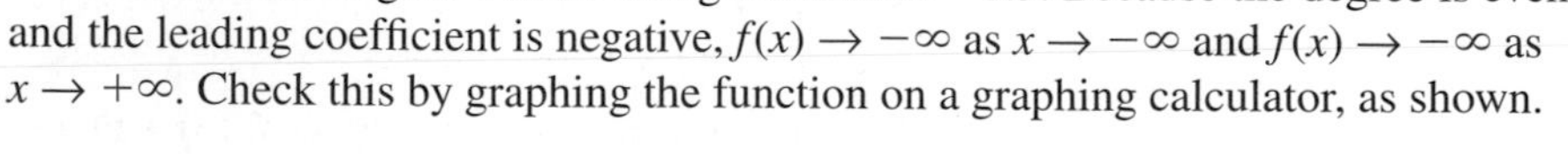

Evaluate the function for the given value of x.

4. $f(x) = -x^3 + 3x^2 + 9; x = 4$

5. $f(x) = 3x^5 - x^4 - 6x + 10; x = -2$

6. Describe the end behavior of the graph of $f(x) = 0.25x^3 - x^2 - 1$.

Graphing Polynomial Functions

To graph a polynomial function, first plot points to determine the shape of the graph's middle portion. Then connect the points with a smooth continuous curve and use what you know about end behavior to sketch the graph.

EXAMPLE 4 Graphing Polynomial Functions

Graph (a) $f(x) = -x^3 + x^2 + 3x - 3$ and (b) $f(x) = x^4 - x^3 - 4x^2 + 4$.

SOLUTION

a. To graph the function, make a table of values and plot the corresponding points. Connect the points with a smooth curve and check the end behavior.

x	-2	-1	0	1	2
$f(x)$	3	-4	-3	0	-1

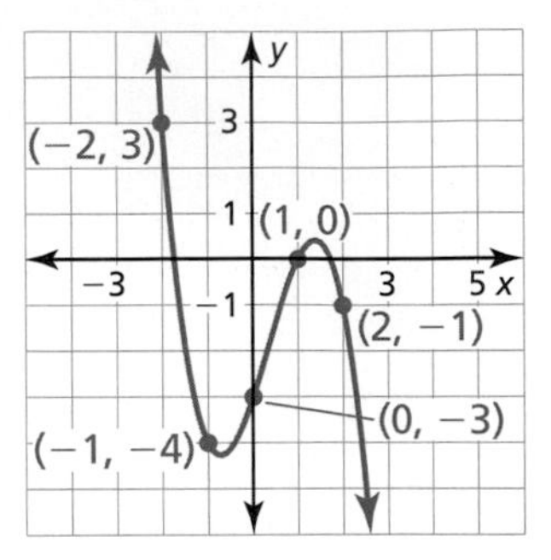

The degree is odd and the leading coefficient is negative. So, $f(x) \to +\infty$ as $x \to -\infty$ and $f(x) \to -\infty$ as $x \to +\infty$.

b. To graph the function, make a table of values and plot the corresponding points. Connect the points with a smooth curve and check the end behavior.

x	-2	-1	0	1	2
$f(x)$	12	2	4	0	-4

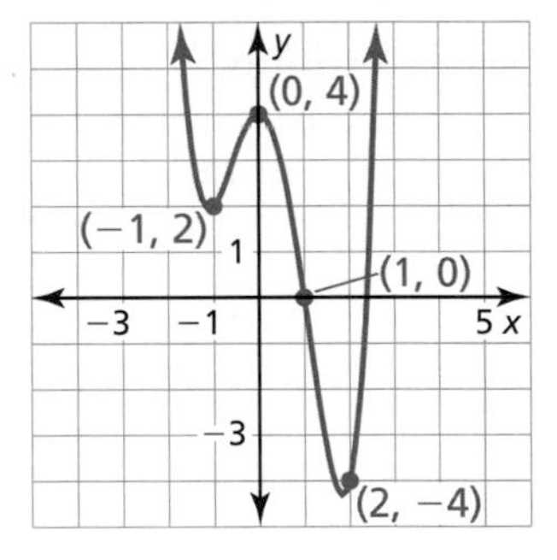

The degree is even and the leading coefficient is positive. So, $f(x) \to +\infty$ as $x \to -\infty$ and $f(x) \to +\infty$ as $x \to +\infty$.

EXAMPLE 5 Sketching a Graph

Sketch a graph of the polynomial function f having these characteristics.

- f is increasing when $x < 0$ and $x > 4$.
- f is decreasing when $0 < x < 4$.
- $f(x) > 0$ when $-2 < x < 3$ and $x > 5$.
- $f(x) < 0$ when $x < -2$ and $3 < x < 5$.

Use the graph to describe the degree and leading coefficient of f.

SOLUTION

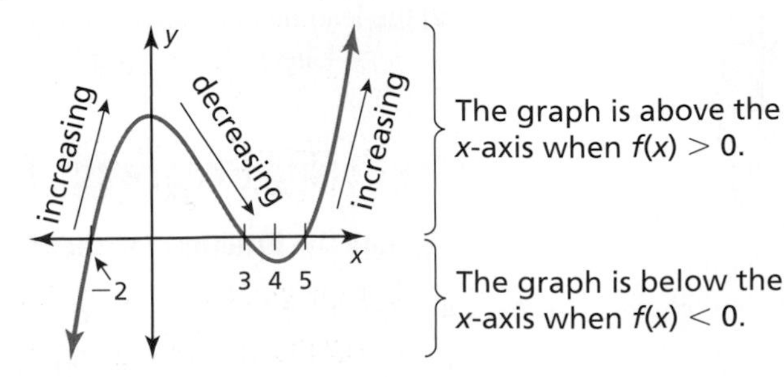

▶ From the graph, $f(x) \to -\infty$ as $x \to -\infty$ and $f(x) \to +\infty$ as $x \to +\infty$. So, the degree is odd and the leading coefficient is positive.

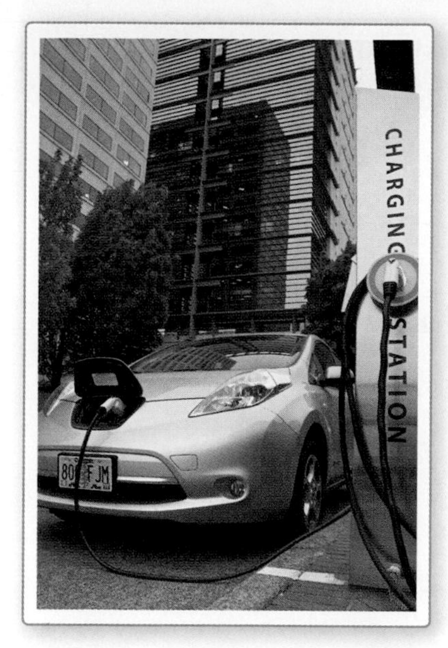

EXAMPLE 6 Solving a Real-Life Problem

The estimated number V (in thousands) of electric vehicles in use in the United States can be modeled by the polynomial function

$$V(t) = 0.151280t^3 - 3.28234t^2 + 23.7565t - 2.041$$

where t represents the year, with $t = 1$ corresponding to 2001.

a. Use a graphing calculator to graph the function for the interval $1 \leq t \leq 10$. Describe the behavior of the graph on this interval.

b. What was the average rate of change in the number of electric vehicles in use from 2001 to 2010?

c. Do you think this model can be used for years before 2001 or after 2010? Explain your reasoning.

SOLUTION

a. Using a graphing calculator and a viewing window of $1 \leq x \leq 10$ and $0 \leq y \leq 65$, you obtain the graph shown.

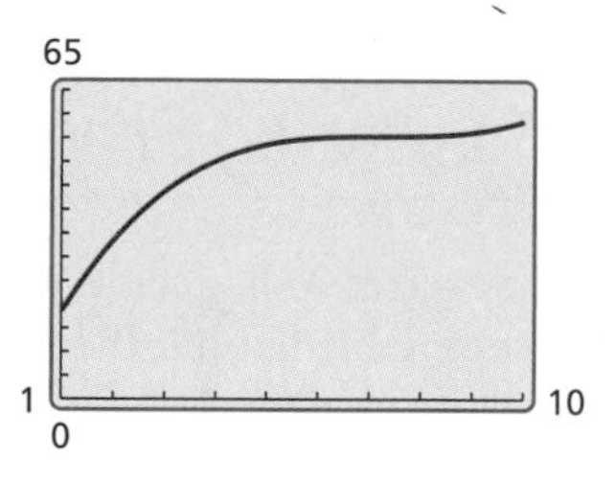

▶ From 2001 to 2004, the numbers of electric vehicles in use increased. Around 2005, the growth in the numbers in use slowed and started to level off. Then the numbers in use started to increase again in 2009 and 2010.

b. The years 2001 and 2010 correspond to $t = 1$ and $t = 10$.

Average rate of change over $1 \leq t \leq 10$:

$$\frac{V(10) - V(1)}{10 - 1} = \frac{58.57 - 18.58444}{9} \approx 4.443$$

▶ The average rate of change from 2001 to 2010 is about 4.4 thousand electric vehicles per year.

c. Because the degree is odd and the leading coefficient is positive, $V(t) \to -\infty$ as $t \to -\infty$ and $V(t) \to +\infty$ as $t \to +\infty$. The end behavior indicates that the model has unlimited growth as t increases. While the model may be valid for a few years after 2010, in the long run, unlimited growth is not reasonable. Notice in 2000 that $V(0) = -2.041$. Because negative values of $V(t)$ do not make sense given the context (electric vehicles in use), the model should not be used for years before 2001.

Monitoring Progress  Help in English and Spanish at *BigIdeasMath.com*

Graph the polynomial function.

7. $f(x) = x^4 + x^2 - 3$

8. $f(x) = 4 - x^3$

9. $f(x) = x^3 - x^2 + x - 1$

10. Sketch a graph of the polynomial function f having these characteristics.

- f is decreasing when $x < -1.5$ and $x > 2.5$; f is increasing when $-1.5 < x < 2.5$.
- $f(x) > 0$ when $x < -3$ and $1 < x < 4$; $f(x) < 0$ when $-3 < x < 1$ and $x > 4$.

Use the graph to describe the degree and leading coefficient of f.

11. WHAT IF? Repeat Example 6 using the alternative model for electric vehicles of

$$V(t) = -0.0290900t^4 + 0.791260t^3 - 7.96583t^2 + 36.5561t - 12.025.$$

4.1 Exercises

Dynamic Solutions available at *BigIdeasMath.com*

Vocabulary and Core Concept Check

1. **WRITING** Explain what is meant by the end behavior of a polynomial function.

2. **WHICH ONE DOESN'T BELONG?** Which function does *not* belong with the other three? Explain your reasoning.

$f(x) = 7x^5 + 3x^2 - 2x$	$g(x) = 3x^3 - 2x^8 + \frac{3}{4}$
$h(x) = -3x^4 + 5x^{-1} - 3x^2$	$k(x) = \sqrt{3}x + 8x^4 + 2x + 1$

Monitoring Progress and Modeling with Mathematics

In Exercises 3–8, decide whether the function is a polynomial function. If so, write it in standard form and state its degree, type, and leading coefficient. *(See Example 1.)*

3. $f(x) = -3x + 5x^3 - 6x^2 + 2$

4. $p(x) = \frac{1}{2}x^2 + 3x - 4x^3 + 6x^4 - 1$

5. $f(x) = 9x^4 + 8x^3 - 6x^{-2} + 2x$

6. $g(x) = \sqrt{3} - 12x + 13x^2$

7. $h(x) = \frac{5}{3}x^2 - \sqrt{7}x^4 + 8x^3 - \frac{1}{2} + x$

8. $h(x) = 3x^4 + 2x - \frac{5}{x} + 9x^3 - 7$

ERROR ANALYSIS In Exercises 9 and 10, describe and correct the error in analyzing the function.

9. $f(x) = 8x^3 - 7x^4 - 9x - 3x^2 + 11$

f is a polynomial function.
The degree is 3 and f is a cubic function.
The leading coefficient is 8.

10. $f(x) = 2x^4 + 4x - 9\sqrt{x} + 3x^2 - 8$

f is a polynomial function.
The degree is 4 and f is a quartic function.
The leading coefficient is 2.

In Exercises 11–16, evaluate the function for the given value of x. *(See Example 2.)*

11. $h(x) = -3x^4 + 2x^3 - 12x - 6;\ x = -2$

12. $f(x) = 7x^4 - 10x^2 + 14x - 26;\ x = -7$

13. $g(x) = x^6 - 64x^4 + x^2 - 7x - 51;\ x = 8$

14. $g(x) = -x^3 + 3x^2 + 5x + 1;\ x = -12$

15. $p(x) = 2x^3 + 4x^2 + 6x + 7;\ x = \frac{1}{2}$

16. $h(x) = 5x^3 - 3x^2 + 2x + 4;\ x = -\frac{1}{3}$

In Exercises 17–20, describe the end behavior of the graph of the function. *(See Example 3.)*

17. $h(x) = -5x^4 + 7x^3 - 6x^2 + 9x + 2$

18. $g(x) = 7x^7 + 12x^5 - 6x^3 - 2x - 18$

19. $f(x) = -2x^4 + 12x^8 + 17 + 15x^2$

20. $f(x) = 11 - 18x^2 - 5x^5 - 12x^4 - 2x$

In Exercises 21 and 22, describe the degree and leading coefficient of the polynomial function using the graph.

21.

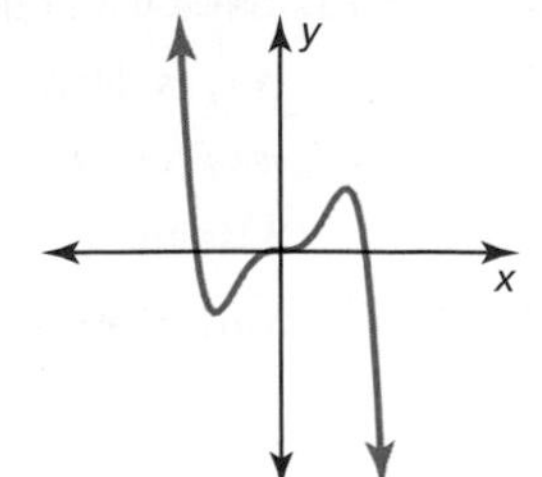

22.

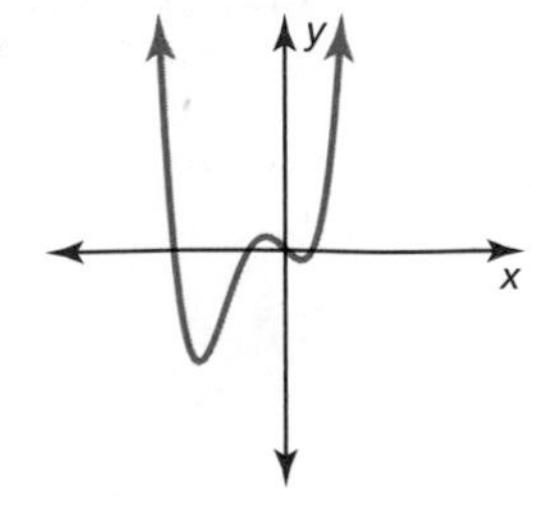

23. USING STRUCTURE Determine whether the function is a polynomial function. If so, write it in standard form and state its degree, type, and leading coefficient.

$$f(x) = 5x^3x + \frac{5}{2}x^3 - 9x^4 + \sqrt{2}x^2 + 4x - 1 - x^{-5}x^5 - 4$$

24. WRITING Let $f(x) = 13$. State the degree, type, and leading coefficient. Describe the end behavior of the function. Explain your reasoning.

In Exercises 25–32, graph the polynomial function. *(See Example 4.)*

25. $p(x) = 3 - x^4$

26. $g(x) = x^3 + x + 3$

27. $f(x) = 4x - 9 - x^3$

28. $p(x) = x^5 - 3x^3 + 2$

29. $h(x) = x^4 - 2x^3 + 3x$

30. $h(x) = 5 + 3x^2 - x^4$

31. $g(x) = x^5 - 3x^4 + 2x - 4$

32. $p(x) = x^6 - 2x^5 - 2x^3 + x + 5$

ANALYZING RELATIONSHIPS In Exercises 33–36, describe the x-values for which (a) f is increasing or decreasing, (b) $f(x) > 0$, and (c) $f(x) < 0$.

33.

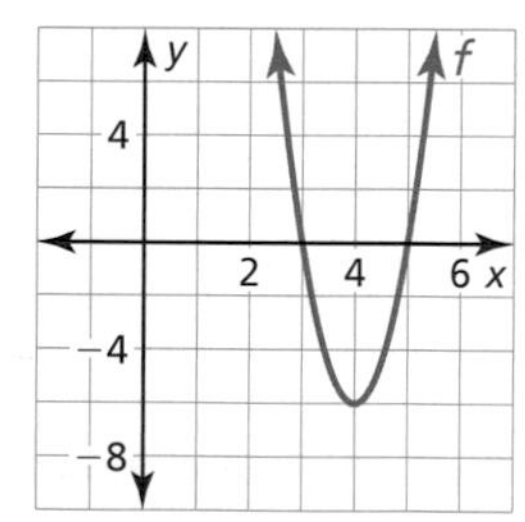

34.

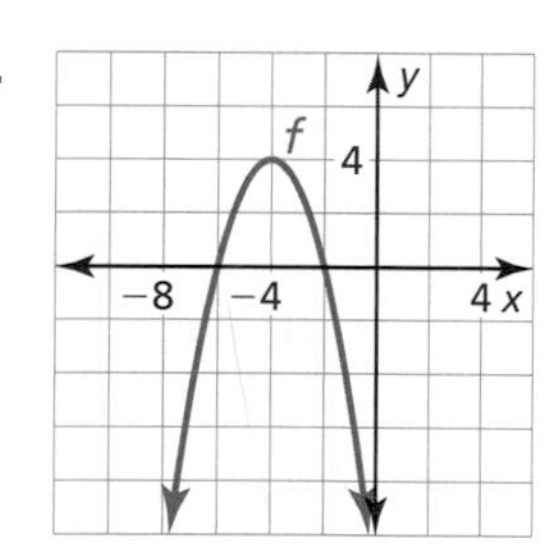

35.

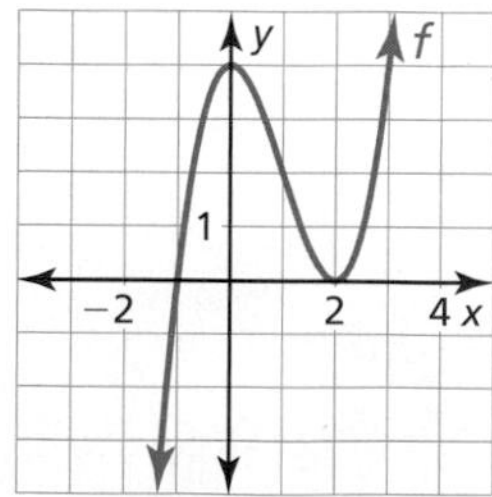

36.

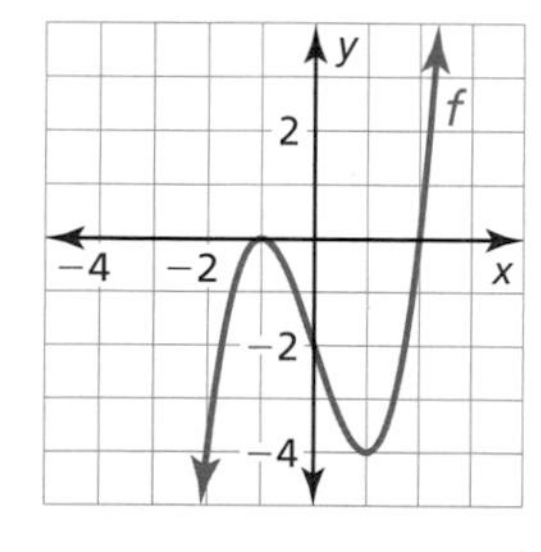

In Exercises 37–40, sketch a graph of the polynomial function f having the given characteristics. Use the graph to describe the degree and leading coefficient of the function f. *(See Example 5.)*

37.
- f is increasing when $x > 0.5$; f is decreasing when $x < 0.5$.
- $f(x) > 0$ when $x < -2$ and $x > 3$; $f(x) < 0$ when $-2 < x < 3$.

38.
- f is increasing when $-2 < x < 3$; f is decreasing when $x < -2$ and $x > 3$.
- $f(x) > 0$ when $x < -4$ and $1 < x < 5$; $f(x) < 0$ when $-4 < x < 1$ and $x > 5$.

39.
- f is increasing when $-2 < x < 0$ and $x > 2$; f is decreasing when $x < -2$ and $0 < x < 2$.
- $f(x) > 0$ when $x < -3$, $-1 < x < 1$, and $x > 3$; $f(x) < 0$ when $-3 < x < -1$ and $1 < x < 3$.

40.
- f is increasing when $x < -1$ and $x > 1$; f is decreasing when $-1 < x < 1$.
- $f(x) > 0$ when $-1.5 < x < 0$ and $x > 1.5$; $f(x) < 0$ when $x < -1.5$ and $0 < x < 1.5$.

41. MODELING WITH MATHEMATICS From 1980 to 2007 the number of drive-in theaters in the United States can be modeled by the function

$$d(t) = -0.141t^3 + 9.64t^2 - 232.5t + 2421$$

where $d(t)$ is the number of open theaters and t is the number of years after 1980. *(See Example 6.)*

a. Use a graphing calculator to graph the function for the interval $0 \leq t \leq 27$. Describe the behavior of the graph on this interval.

b. What is the average rate of change in the number of drive-in movie theaters from 1980 to 1995 and from 1995 to 2007? Interpret the average rates of change.

c. Do you think this model can be used for years before 1980 or after 2007? Explain.

42. PROBLEM SOLVING The weight of an ideal round-cut diamond can be modeled by

$$w = 0.00583d^3 - 0.0125d^2 + 0.022d - 0.01$$

where w is the weight of the diamond (in carats) and d is the diameter (in millimeters). According to the model, what is the weight of a diamond with a diameter of 12 millimeters?

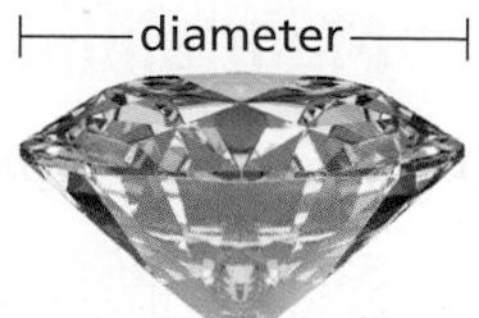

43. **ABSTRACT REASONING** Suppose $f(x) \to \infty$ as $x \to -\infty$ and $f(x) \to -\infty$ as $x \to \infty$. Describe the end behavior of $g(x) = -f(x)$. Justify your answer.

44. **THOUGHT PROVOKING** Write an even degree polynomial function such that the end behavior of f is given by $f(x) \to -\infty$ as $x \to -\infty$ and $f(x) \to -\infty$ as $x \to \infty$. Justify your answer by drawing the graph of your function.

45. **USING TOOLS** When using a graphing calculator to graph a polynomial function, explain how you know when the viewing window is appropriate.

46. **MAKING AN ARGUMENT** Your friend uses the table to speculate that the function f is an even degree polynomial and the function g is an odd degree polynomial. Is your friend correct? Explain your reasoning.

x	f(x)	g(x)
−8	4113	497
−2	21	5
0	1	1
2	13	−3
8	4081	−495

47. **DRAWING CONCLUSIONS** The graph of a function is symmetric with respect to the y-axis if for each point (a, b) on the graph, $(-a, b)$ is also a point on the graph. The graph of a function is symmetric with respect to the origin if for each point (a, b) on the graph, $(-a, -b)$ is also a point on the graph.

   a. Use a graphing calculator to graph the function $y = x^n$ when $n = 1, 2, 3, 4, 5$, and 6. In each case, identify the symmetry of the graph.

   b. Predict what symmetry the graphs of $y = x^{10}$ and $y = x^{11}$ each have. Explain your reasoning and then confirm your predictions by graphing.

48. **HOW DO YOU SEE IT?** The graph of a polynomial function is shown.

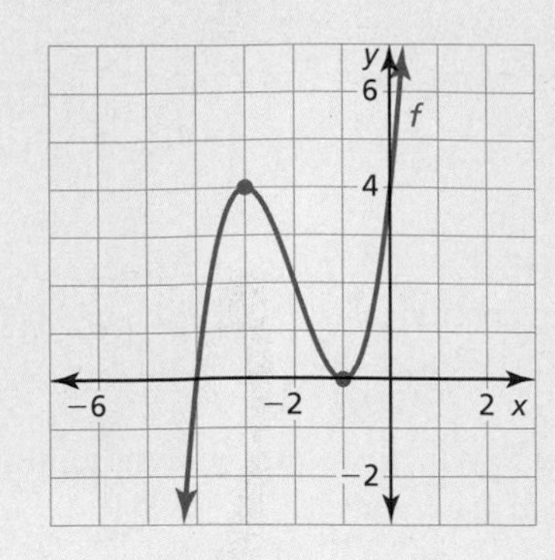

   a. Describe the degree and leading coefficient of f.

   b. Describe the intervals where the function is increasing and decreasing.

   c. What is the constant term of the polynomial function?

49. **REASONING** A cubic polynomial function f has a leading coefficient of 2 and a constant term of −5. When $f(1) = 0$ and $f(2) = 3$, what is $f(-5)$? Explain your reasoning.

50. **CRITICAL THINKING** The weight y (in pounds) of a rainbow trout can be modeled by $y = 0.000304x^3$, where x is the length (in inches) of the trout.

   a. Write a function that relates the weight y and length x of a rainbow trout when y is measured in kilograms and x is measured in centimeters. Use the fact that 1 kilogram ≈ 2.20 pounds and 1 centimeter ≈ 0.394 inch.

   b. Graph the original function and the function from part (a) in the same coordinate plane. What type of transformation can you apply to the graph of $y = 0.000304x^3$ to produce the graph from part (a)?

Maintaining Mathematical Proficiency

Reviewing what you learned in previous grades and lessons

Simplify the expression. *(Skills Review Handbook)*

51. $xy + x^2 + 2xy + y^2 - 3x^2$

52. $2h^3g + 3hg^3 + 7h^2g^2 + 5h^3g + 2hg^3$

53. $-wk + 3kz - 2kw + 9zk - kw$

54. $a^2(m - 7a^3) - m(a^2 - 10)$

55. $3x(xy - 4) + 3(4xy + 3) - xy(x^2y - 1)$

56. $cv(9 - 3c) + 2c(v - 4c) + 6c$

4.2 Adding, Subtracting, and Multiplying Polynomials

Essential Question How can you cube a binomial?

EXPLORATION 1 Cubing Binomials

Work with a partner. Find each product. Show your steps.

a. $(x + 1)^3 = (x + 1)(x + 1)^2$ Rewrite as a product of first and second powers.

$= (x + 1)$ ______ Multiply second power.

$=$ ______ Multiply binomial and trinomial.

$=$ ______ Write in standard form, $ax^3 + bx^2 + cx + d$.

b. $(a + b)^3 = (a + b)(a + b)^2$ Rewrite as a product of first and second powers.

$= (a + b)$ ______ Multiply second power.

$=$ ______ Multiply binomial and trinomial.

$=$ ______ Write in standard form.

c. $(x - 1)^3 = (x - 1)(x - 1)^2$ Rewrite as a product of first and second powers.

$= (x - 1)$ ______ Multiply second power.

$=$ ______ Multiply binomial and trinomial.

$=$ ______ Write in standard form.

d. $(a - b)^3 = (a - b)(a - b)^2$ Rewrite as a product of first and second powers.

$= (a - b)$ ______ Multiply second power.

$=$ ______ Multiply binomial and trinomial.

$=$ ______ Write in standard form.

LOOKING FOR STRUCTURE

To be proficient in math, you need to look closely to discern a pattern or structure.

EXPLORATION 2 Generalizing Patterns for Cubing a Binomial

Work with a partner.

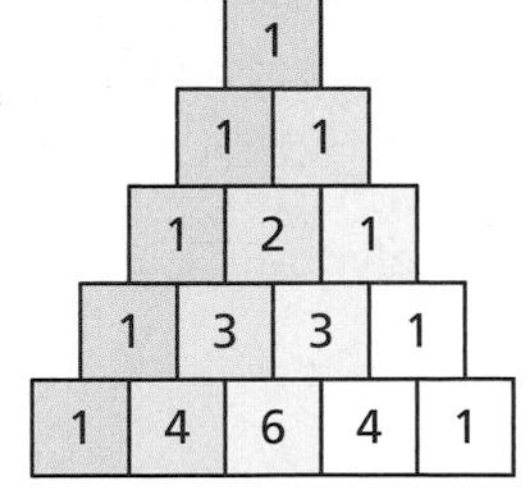

a. Use the results of Exploration 1 to describe a pattern for the coefficients of the terms when you expand the cube of a binomial. How is your pattern related to Pascal's Triangle, shown at the right?

b. Use the results of Exploration 1 to describe a pattern for the exponents of the terms in the expansion of a cube of a binomial.

c. Explain how you can use the patterns you described in parts (a) and (b) to find the product $(2x - 3)^3$. Then find this product.

Communicate Your Answer

3. How can you cube a binomial?

4. Find each product.

a. $(x + 2)^3$ **b.** $(x - 2)^3$ **c.** $(2x - 3)^3$

d. $(x - 3)^3$ **e.** $(-2x + 3)^3$ **f.** $(3x - 5)^3$

4.2 Lesson

Core Vocabulary

Pascal's Triangle, *p. 169*

Previous
like terms
identity

What You Will Learn

- Add and subtract polynomials.
- Multiply polynomials.
- Use Pascal's Triangle to expand binomials.

Adding and Subtracting Polynomials

Recall that the set of integers is *closed* under addition and subtraction because every sum or difference results in an integer. To add or subtract polynomials, you add or subtract the coefficients of like terms. Because adding or subtracting polynomials results in a polynomial, the set of polynomials is closed under addition and subtraction.

EXAMPLE 1 Adding Polynomials Vertically and Horizontally

a. Add $3x^3 + 2x^2 - x - 7$ and $x^3 - 10x^2 + 8$ in a vertical format.

b. Add $9y^3 + 3y^2 - 2y + 1$ and $-5y^2 + y - 4$ in a horizontal format.

SOLUTION

a.
$$\begin{array}{rrrrr} & 3x^3 + & 2x^2 & - x & - 7 \\ + & x^3 - & 10x^2 & & + 8 \\ \hline & 4x^3 - & 8x^2 & - x & + 1 \end{array}$$

b. $(9y^3 + 3y^2 - 2y + 1) + (-5y^2 + y - 4) = 9y^3 + 3y^2 - 5y^2 - 2y + y + 1 - 4$
$= 9y^3 - 2y^2 - y - 3$

To subtract one polynomial from another, add the opposite. To do this, change the sign of each term of the subtracted polynomial and then add the resulting like terms.

COMMON ERROR

A common mistake is to forget to change signs correctly when subtracting one polynomial from another. Be sure to add the opposite of *every* term of the subtracted polynomial.

EXAMPLE 2 Subtracting Polynomials Vertically and Horizontally

a. Subtract $2x^3 + 6x^2 - x + 1$ from $8x^3 - 3x^2 - 2x + 9$ in a vertical format.

b. Subtract $3z^2 + z - 4$ from $2z^2 + 3z$ in a horizontal format.

SOLUTION

a. Align like terms, then add the opposite of the subtracted polynomial.

$$\begin{array}{r} 8x^3 - 3x^2 - 2x + 9 \\ -\ (2x^3 + 6x^2 - \ \ x + 1) \\ \hline \end{array} \quad \Rightarrow \quad \begin{array}{r} 8x^3 - 3x^2 - 2x + 9 \\ +\ \ -2x^3 - 6x^2 + \ \ x - 1 \\ \hline 6x^3 - 9x^2 - \ \ x + 8 \end{array}$$

b. Write the opposite of the subtracted polynomial, then add like terms.

$(2z^2 + 3z) - (3z^2 + z - 4) = 2z^2 + 3z - 3z^2 - z + 4$
$= -z^2 + 2z + 4$

Monitoring Progress

Help in English and Spanish at *BigIdeasMath.com*

Find the sum or difference.

1. $(2x^2 - 6x + 5) + (7x^2 - x - 9)$
2. $(3t^3 + 8t^2 - t - 4) - (5t^3 - t^2 + 17)$

Multiplying Polynomials

To multiply two polynomials, you multiply each term of the first polynomial by each term of the second polynomial. As with addition and subtraction, the set of polynomials is closed under multiplication.

EXAMPLE 3 Multiplying Polynomials Vertically and Horizontally

a. Multiply $-x^2 + 2x + 4$ and $x - 3$ in a vertical format.

b. Multiply $y + 5$ and $3y^2 - 2y + 2$ in a horizontal format.

REMEMBER

Product of Powers Property

$a^m \cdot a^n = a^{m+n}$

a is a real number and m and n are integers.

SOLUTION

a.

$$\begin{array}{rl} -x^2 + 2x + 4 & \\ \times \qquad\quad x - 3 & \\ \hline 3x^2 - 6x - 12 & \text{Multiply } -x^2 + 2x + 4 \text{ by } -3. \\ -x^3 + 2x^2 + 4x \qquad & \text{Multiply } -x^2 + 2x + 4 \text{ by } x. \\ \hline -x^3 + 5x^2 - 2x - 12 & \text{Combine like terms.} \end{array}$$

b.

$$\begin{aligned} (y + 5)(3y^2 - 2y + 2) &= (y + 5)3y^2 - (y + 5)2y + (y + 5)2 \\ &= 3y^3 + 15y^2 - 2y^2 - 10y + 2y + 10 \\ &= 3y^3 + 13y^2 - 8y + 10 \end{aligned}$$

EXAMPLE 4 Multiplying Three Binomials

Multiply $x - 1$, $x + 4$, and $x + 5$ in a horizontal format.

SOLUTION

$$\begin{aligned} (x - 1)(x + 4)(x + 5) &= (x^2 + 3x - 4)(x + 5) \\ &= (x^2 + 3x - 4)x + (x^2 + 3x - 4)5 \\ &= x^3 + 3x^2 - 4x + 5x^2 + 15x - 20 \\ &= x^3 + 8x^2 + 11x - 20 \end{aligned}$$

Some binomial products occur so frequently that it is worth memorizing their patterns. You can verify these polynomial identities by multiplying.

COMMON ERROR

In general,

$(a \pm b)^2 \neq a^2 \pm b^2$

and

$(a \pm b)^3 \neq a^3 \pm b^3$.

Core Concept

Special Product Patterns

	Example
Sum and Difference	
$(a + b)(a - b) = a^2 - b^2$	$(x + 3)(x - 3) = x^2 - 9$
Square of a Binomial	
$(a + b)^2 = a^2 + 2ab + b^2$	$(y + 4)^2 = y^2 + 8y + 16$
$(a - b)^2 = a^2 - 2ab + b^2$	$(2t - 5)^2 = 4t^2 - 20t + 25$
Cube of a Binomial	
$(a + b)^3 = a^3 + 3a^2b + 3ab^2 + b^3$	$(z + 3)^3 = z^3 + 9z^2 + 27z + 27$
$(a - b)^3 = a^3 - 3a^2b + 3ab^2 - b^3$	$(m - 2)^3 = m^3 - 6m^2 + 12m - 8$

EXAMPLE 5 Proving a Polynomial Identity

a. Prove the polynomial identity for the cube of a binomial representing a sum:

$$(a + b)^3 = a^3 + 3a^2b + 3ab^2 + b^3.$$

b. Use the cube of a binomial in part (a) to calculate 11^3.

SOLUTION

a. Expand and simplify the expression on the left side of the equation.

$$\begin{aligned}(a + b)^3 &= (a + b)(a + b)(a + b)\\ &= (a^2 + 2ab + b^2)(a + b)\\ &= (a^2 + 2ab + b^2)a + (a^2 + 2ab + b^2)b\\ &= a^3 + a^2b + 2a^2b + 2ab^2 + ab^2 + b^3\\ &= a^3 + 3a^2b + 3ab^2 + b^3 \checkmark\end{aligned}$$

▶ The simplified left side equals the right side of the original identity. So, the identity $(a + b)^3 = a^3 + 3a^2b + 3ab^2 + b^3$ is true.

b. To calculate 11^3 using the cube of a binomial, note that $11 = 10 + 1$.

$11^3 = (10 + 1)^3$	Write 11 as 10 + 1.
$= 10^3 + 3(10)^2(1) + 3(10)(1)^2 + 1^3$	Cube of a binomial
$= 1000 + 300 + 30 + 1$	Expand.
$= 1331$	Simplify.

EXAMPLE 6 Using Special Product Patterns

REMEMBER

Power of a Product Property

$(ab)^m = a^mb^m$

a and b are real numbers and m is an integer.

Find each product.

a. $(4n + 5)(4n - 5)$ **b.** $(9y - 2)^2$ **c.** $(ab + 4)^3$

SOLUTION

a. $(4n + 5)(4n - 5) = (4n)^2 - 5^2$	Sum and difference
$= 16n^2 - 25$	Simplify.
b. $(9y - 2)^2 = (9y)^2 - 2(9y)(2) + 2^2$	Square of a binomial
$= 81y^2 - 36y + 4$	Simplify.
c. $(ab + 4)^3 = (ab)^3 + 3(ab)^2(4) + 3(ab)(4)^2 + 4^3$	Cube of a binomial
$= a^3b^3 + 12a^2b^2 + 48ab + 64$	Simplify.

Monitoring Progress  Help in English and Spanish at *BigIdeasMath.com*

Find the product.

3. $(4x^2 + x - 5)(2x + 1)$

4. $(y - 2)(5y^2 + 3y - 1)$

5. $(m - 2)(m - 1)(m + 3)$

6. $(3t - 2)(3t + 2)$

7. $(5a + 2)^2$

8. $(xy - 3)^3$

9. (a) Prove the polynomial identity for the cube of a binomial representing a difference: $(a - b)^3 = a^3 - 3a^2b + 3ab^2 - b^3$.

(b) Use the cube of a binomial in part (a) to calculate 9^3.

Pascal's Triangle

Consider the expansion of the binomial $(a + b)^n$ for whole number values of n. When you arrange the coefficients of the variables in the expansion of $(a + b)^n$, you will see a special pattern called **Pascal's Triangle**. Pascal's Triangle is named after French mathematician Blaise Pascal (1623–1662).

Core Concept

Pascal's Triangle

In Pascal's Triangle, the first and last numbers in each row are 1. Every number other than 1 is the sum of the closest two numbers in the row directly above it. The numbers in Pascal's Triangle are the same numbers that are the coefficients of binomial expansions, as shown in the first six rows.

	n	$(a + b)^n$	Binomial Expansion	Pascal's Triangle
0th row	0	$(a + b)^0 =$	1	1
1st row	1	$(a + b)^1 =$	$1a + 1b$	1 1
2nd row	2	$(a + b)^2 =$	$1a^2 + 2ab + 1b^2$	1 2 1
3rd row	3	$(a + b)^3 =$	$1a^3 + 3a^2b + 3ab^2 + 1b^3$	1 3 3 1
4th row	4	$(a + b)^4 =$	$1a^4 + 4a^3b + 6a^2b^2 + 4ab^3 + 1b^4$	1 4 6 4 1
5th row	5	$(a + b)^5 =$	$1a^5 + 5a^4b + 10a^3b^2 + 10a^2b^3 + 5ab^4 + 1b^5$	1 5 10 10 5 1

In general, the nth row in Pascal's Triangle gives the coefficients of $(a + b)^n$. Here are some other observations about the expansion of $(a + b)^n$.

1. An expansion has $n + 1$ terms.
2. The power of a begins with n, decreases by 1 in each successive term, and ends with 0.
3. The power of b begins with 0, increases by 1 in each successive term, and ends with n.
4. The sum of the powers of each term is n.

EXAMPLE 7 Using Pascal's Triangle to Expand Binomials

Use Pascal's Triangle to expand (a) $(x - 2)^5$ and (b) $(3y + 1)^3$.

SOLUTION

a. The coefficients from the fifth row of Pascal's Triangle are 1, 5, 10, 10, 5, and 1.

$$(x - 2)^5 = 1x^5 + 5x^4(-2) + 10x^3(-2)^2 + 10x^2(-2)^3 + 5x(-2)^4 + 1(-2)^5$$
$$= x^5 - 10x^4 + 40x^3 - 80x^2 + 80x - 32$$

b. The coefficients from the third row of Pascal's Triangle are 1, 3, 3, and 1.

$$(3y + 1)^3 = 1(3y)^3 + 3(3y)^2(1) + 3(3y)(1)^2 + 1(1)^3$$
$$= 27y^3 + 27y^2 + 9y + 1$$

Monitoring Progress

Help in English and Spanish at *BigIdeasMath.com*

10. Use Pascal's Triangle to expand (a) $(z + 3)^4$ and (b) $(2t - 1)^5$.

4.2 Exercises

Dynamic Solutions available at *BigIdeasMath.com*

Vocabulary and Core Concept Check

1. **WRITING** Describe three different methods to expand $(x + 3)^3$.

2. **WRITING** Is $(a + b)(a - b) = a^2 - b^2$ an identity? Explain your reasoning.

Monitoring Progress and Modeling with Mathematics

In Exercises 3–8, find the sum. *(See Example 1.)*

3. $(3x^2 + 4x - 1) + (-2x^2 - 3x + 2)$

4. $(-5x^2 + 4x - 2) + (-8x^2 + 2x + 1)$

5. $(12x^5 - 3x^4 + 2x - 5) + (8x^4 - 3x^3 + 4x + 1)$

6. $(8x^4 + 2x^2 - 1) + (3x^3 - 5x^2 + 7x + 1)$

7. $(7x^6 + 2x^5 - 3x^2 + 9x) + (5x^5 + 8x^3 - 6x^2 + 2x - 5)$

8. $(9x^4 - 3x^3 + 4x^2 + 5x + 7) + (11x^4 - 4x^2 - 11x - 9)$

In Exercises 9–14, find the difference. *(See Example 2.)*

9. $(3x^3 - 2x^2 + 4x - 8) - (5x^3 + 12x^2 - 3x - 4)$

10. $(7x^4 - 9x^3 - 4x^2 + 5x + 6) - (2x^4 + 3x^3 - x^2 + x - 4)$

11. $(5x^6 - 2x^4 + 9x^3 + 2x - 4) - (7x^5 - 8x^4 + 2x - 11)$

12. $(4x^5 - 7x^3 - 9x^2 + 18) - (14x^5 - 8x^4 + 11x^2 + x)$

13. $(8x^5 + 6x^3 - 2x^2 + 10x) - (9x^5 - x^3 - 13x^2 + 4)$

14. $(11x^4 - 9x^2 + 3x + 11) - (2x^4 + 6x^3 + 2x - 9)$

15. **MODELING WITH MATHEMATICS** During a recent period of time, the numbers (in thousands) of males M and females F that attend degree-granting institutions in the United States can be modeled by

$M = 19.7t^2 + 310.5t + 7539.6$

$F = 28t^2 + 368t + 10127.8$

where t is time in years. Write a polynomial to model the total number of people attending degree-granting institutions. Interpret its constant term.

16. **MODELING WITH MATHEMATICS** A farmer plants a garden that contains corn and pumpkins. The total area (in square feet) of the garden is modeled by the expression $2x^2 + 5x + 4$. The area of the corn is modeled by the expression $x^2 - 3x + 2$. Write an expression that models the area of the pumpkins.

In Exercises 17–24, find the product. *(See Example 3.)*

17. $7x^3(5x^2 + 3x + 1)$

18. $-4x^5(11x^3 + 2x^2 + 9x + 1)$

19. $(5x^2 - 4x + 6)(-2x + 3)$

20. $(-x - 3)(2x^2 + 5x + 8)$

21. $(x^2 - 2x - 4)(x^2 - 3x - 5)$

22. $(3x^2 + x - 2)(-4x^2 - 2x - 1)$

23. $(3x^3 - 9x + 7)(x^2 - 2x + 1)$

24. $(4x^2 - 8x - 2)(x^4 + 3x^2 + 4x)$

ERROR ANALYSIS In Exercises 25 and 26, describe and correct the error in performing the operation.

25.

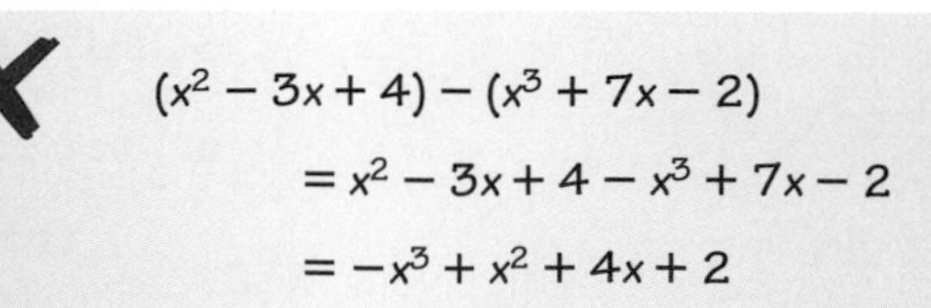

$(x^2 - 3x + 4) - (x^3 + 7x - 2)$

$= x^2 - 3x + 4 - x^3 + 7x - 2$

$= -x^3 + x^2 + 4x + 2$

26.

$(2x - 7)^3 = (2x)^3 - 7^3$

$= 8x^3 - 343$

In Exercises 27–32, find the product of the binomials. *(See Example 4.)*

27. $(x - 3)(x + 2)(x + 4)$

28. $(x - 5)(x + 2)(x - 6)$

29. $(x - 2)(3x + 1)(4x - 3)$

30. $(2x + 5)(x - 2)(3x + 4)$

31. $(3x - 4)(5 - 2x)(4x + 1)$

32. $(4 - 5x)(1 - 2x)(3x + 2)$

33. REASONING Prove the polynomial identity $(a + b)(a - b) = a^2 - b^2$. Then give an example of two whole numbers greater than 10 that can be multiplied using mental math and the given identity. Justify your answers. *(See Example 5.)*

34. NUMBER SENSE You have been asked to order textbooks for your class. You need to order 29 textbooks that cost \$31 each. Explain how you can use the polynomial identity $(a + b)(a - b) = a^2 - b^2$ and mental math to find the total cost of the textbooks.

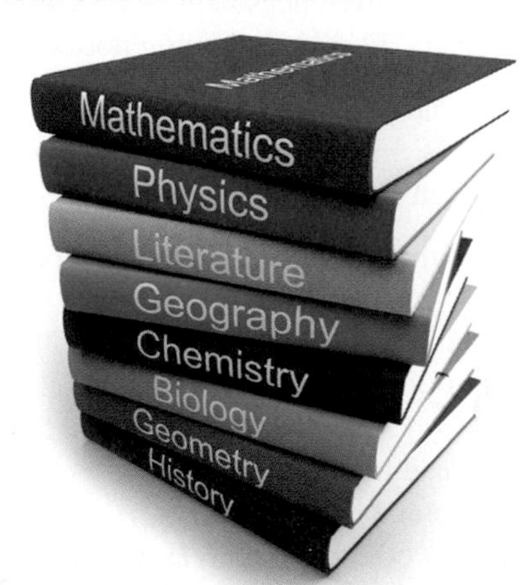

In Exercises 35–42, find the product. *(See Example 6.)*

35. $(x - 9)(x + 9)$

36. $(m + 6)^2$

37. $(3c - 5)^2$

38. $(2y - 5)(2y + 5)$

39. $(7h + 4)^2$

40. $(9g - 4)^2$

41. $(2k + 6)^3$

42. $(4n - 3)^3$

In Exercises 43–48, use Pascal's Triangle to expand the binomial. *(See Example 7.)*

43. $(2t + 4)^3$

44. $(6m + 2)^2$

45. $(2q - 3)^4$

46. $(g + 2)^5$

47. $(yz + 1)^5$

48. $(np - 1)^4$

49. COMPARING METHODS Find the product of the expression $(a^2 + 4b^2)^2(3a^2 - b^2)^2$ using two different methods. Which method do you prefer? Explain.

50. THOUGHT PROVOKING Adjoin one or more polygons to the rectangle to form a single new polygon whose perimeter is double that of the rectangle. Find the perimeter of the new polygon.

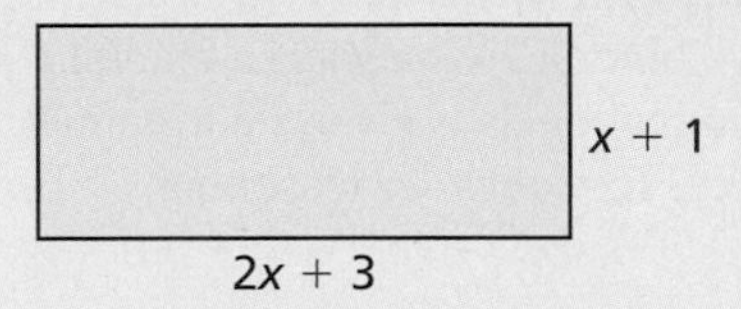

MATHEMATICAL CONNECTIONS In Exercises 51 and 52, write an expression for the volume of the figure as a polynomial in standard form.

51. $V = \ell wh$

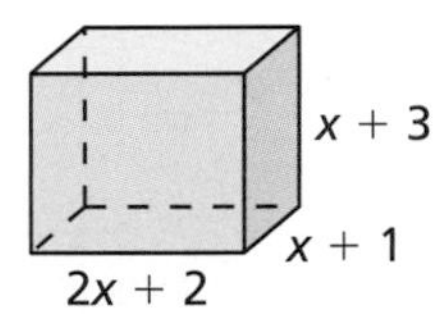

52. $V = \pi r^2 h$

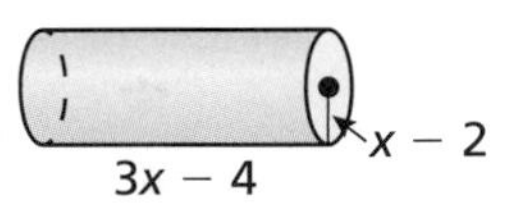

53. MODELING WITH MATHEMATICS Two people make three deposits into their bank accounts earning the same simple interest rate r.

Person A — Account No. 2-5384100608

Date	Transaction	Amount
01/01/2012	Deposit	\$2000.00
01/01/2013	Deposit	\$3000.00
01/01/2014	Deposit	\$1000.00

Person B — Account No. 1-5233032905

Date	Transaction	Amount
01/01/2012	Deposit	\$5000.00
01/01/2013	Deposit	\$1000.00
01/01/2014	Deposit	\$4000.00

Person A's account is worth

$$2000(1 + r)^3 + 3000(1 + r)^2 + 1000(1 + r)$$

on January 1, 2015.

a. Write a polynomial for the value of Person B's account on January 1, 2015.

b. Write the total value of the two accounts as a polynomial in standard form. Then interpret the coefficients of the polynomial.

c. Suppose their interest rate is 0.05. What is the total value of the two accounts on January 1, 2015?

54. PROBLEM SOLVING The sphere is centered in the cube. Find an expression for the volume of the cube outside the sphere.

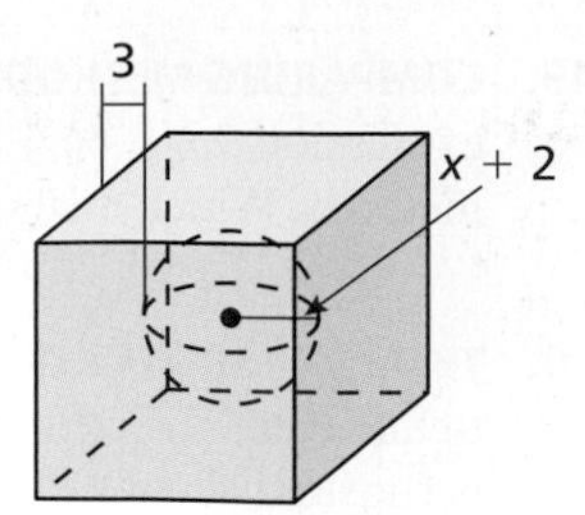

55. MAKING AN ARGUMENT Your friend claims the sum of two binomials is always a binomial and the product of two binomials is always a trinomial. Is your friend correct? Explain your reasoning.

56. HOW DO YOU SEE IT? You make a tin box by cutting x-inch-by-x-inch pieces of tin off the corners of a rectangle and folding up each side. The plan for your box is shown.

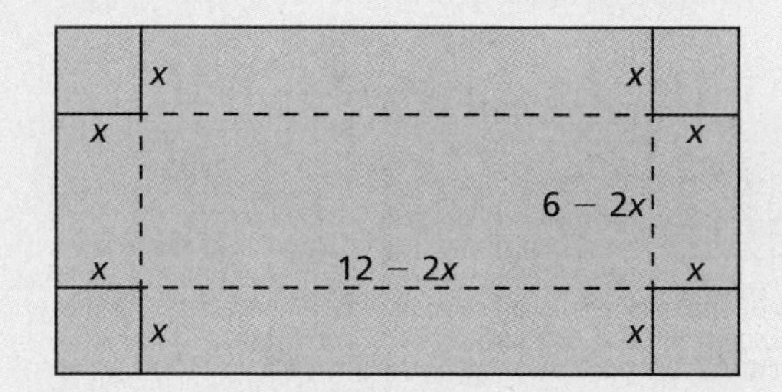

a. What are the dimensions of the original piece of tin?

b. Write a function that represents the volume of the box. Without multiplying, determine its degree.

USING TOOLS In Exercises 57–60, use a graphing calculator to make a conjecture about whether the two functions are equivalent. Explain your reasoning.

57. $f(x) = (2x - 3)^3$; $g(x) = 8x^3 - 36x^2 + 54x - 27$

58. $h(x) = (x + 2)^5$;
$k(x) = x^5 + 10x^4 + 40x^3 + 80x^2 + 64x$

59. $f(x) = (-x - 3)^4$;
$g(x) = x^4 + 12x^3 + 54x^2 + 108x + 80$

60. $f(x) = (-x + 5)^3$; $g(x) = -x^3 + 15x^2 - 75x + 125$

61. REASONING Copy Pascal's Triangle and add rows for $n = 6, 7, 8, 9$, and 10. Use the new rows to expand $(x + 3)^7$ and $(x - 5)^9$.

62. ABSTRACT REASONING You are given the function $f(x) = (x + a)(x + b)(x + c)(x + d)$. When $f(x)$ is written in standard form, show that the coefficient of x^3 is the sum of a, b, c, and d, and the constant term is the product of a, b, c, and d.

63. DRAWING CONCLUSIONS Let $g(x) = 12x^4 + 8x + 9$ and $h(x) = 3x^5 + 2x^3 - 7x + 4$.

a. What is the degree of the polynomial $g(x) + h(x)$?

b. What is the degree of the polynomial $g(x) - h(x)$?

c. What is the degree of the polynomial $g(x) \cdot h(x)$?

d. In general, if $g(x)$ and $h(x)$ are polynomials such that $g(x)$ has degree m and $h(x)$ has degree n, and $m > n$, what are the degrees of $g(x) + h(x)$, $g(x) - h(x)$, and $g(x) \cdot h(x)$?

64. FINDING A PATTERN In this exercise, you will explore the sequence of square numbers. The first four square numbers are represented below.

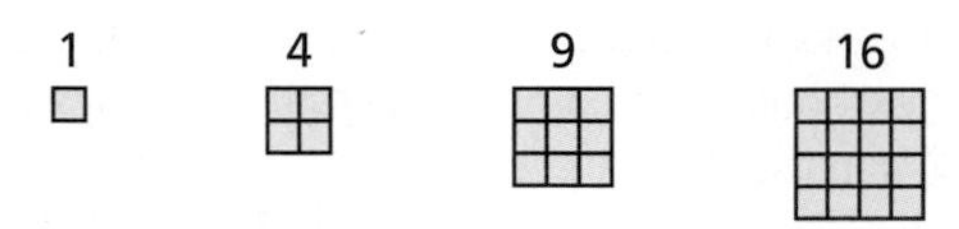

a. Find the differences between consecutive square numbers. Explain what you notice.

b. Show how the polynomial identity $(n + 1)^2 - n^2 = 2n + 1$ models the differences between square numbers.

c. Prove the polynomial identity in part (b).

65. CRITICAL THINKING Recall that a Pythagorean triple is a set of positive integers a, b, and c such that $a^2 + b^2 = c^2$. The numbers 3, 4, and 5 form a Pythagorean triple because $3^2 + 4^2 = 5^2$. You can use the polynomial identity $(x^2 - y^2)^2 + (2xy)^2 = (x^2 + y^2)^2$ to generate other Pythagorean triples.

a. Prove the polynomial identity is true by showing that the simplified expressions for the left and right sides are the same.

b. Use the identity to generate the Pythagorean triple when $x = 6$ and $y = 5$.

c. Verify that your answer in part (b) satisfies $a^2 + b^2 = c^2$.

Maintaining Mathematical Proficiency

Reviewing what you learned in previous grades and lessons

Perform the operation. Write the answer in standard form. *(Section 3.2)*

66. $(3 - 2i) + (5 + 9i)$

67. $(12 + 3i) - (7 - 8i)$

68. $(7i)(-3i)$

69. $(4 + i)(2 - i)$

4.3 Dividing Polynomials

Essential Question How can you use the factors of a cubic polynomial to solve a division problem involving the polynomial?

EXPLORATION 1 Dividing Polynomials

Work with a partner. Match each division statement with the graph of the related cubic polynomial $f(x)$. Explain your reasoning. Use a graphing calculator to verify your answers.

a. $\dfrac{f(x)}{x} = (x - 1)(x + 2)$

b. $\dfrac{f(x)}{x - 1} = (x - 1)(x + 2)$

c. $\dfrac{f(x)}{x + 1} = (x - 1)(x + 2)$

d. $\dfrac{f(x)}{x - 2} = (x - 1)(x + 2)$

e. $\dfrac{f(x)}{x + 2} = (x - 1)(x + 2)$

f. $\dfrac{f(x)}{x - 3} = (x - 1)(x + 2)$

A.

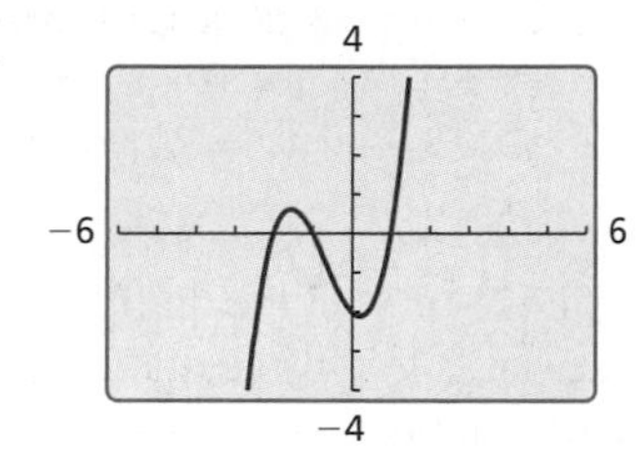

B.

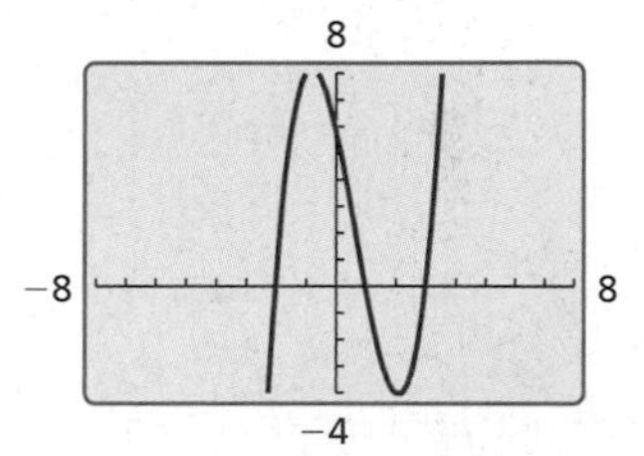

C.

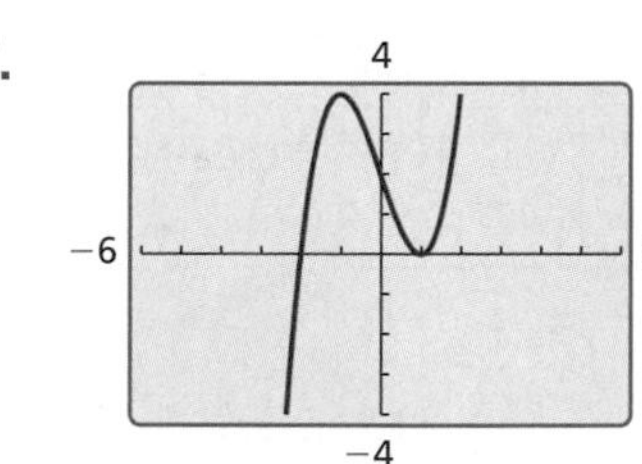

D.

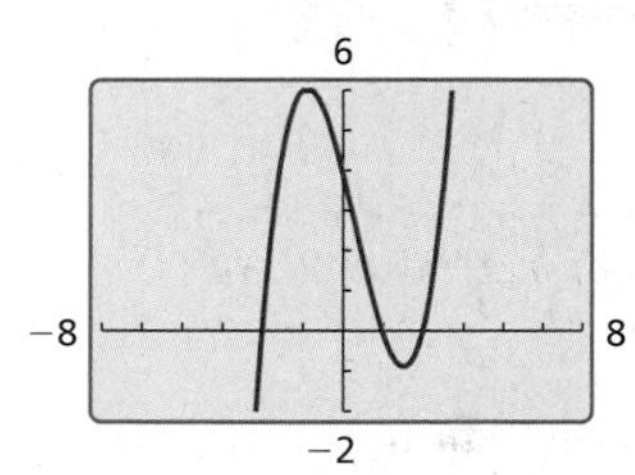

E.

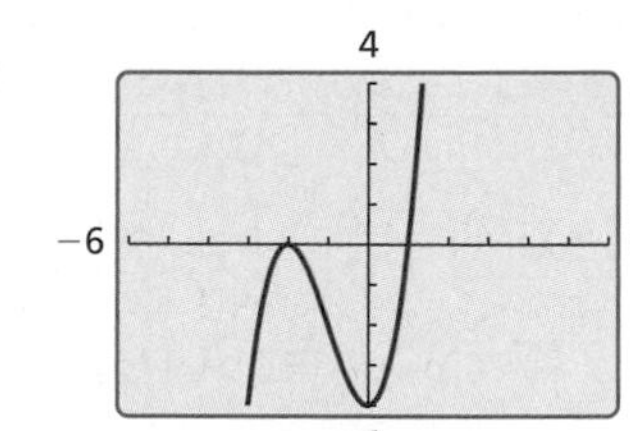

F.

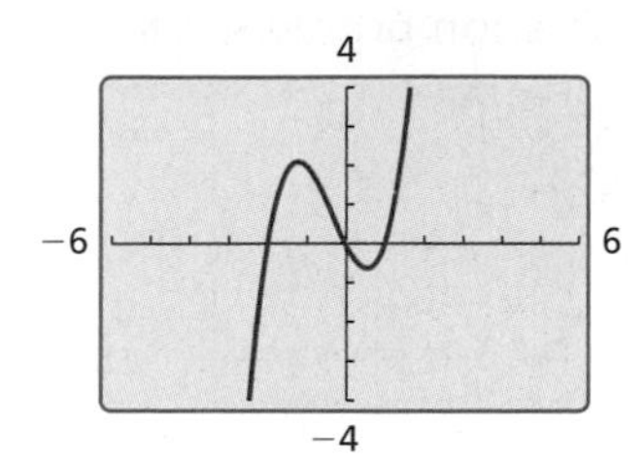

REASONING ABSTRACTLY

To be proficient in math, you need to understand a situation abstractly and represent it symbolically.

EXPLORATION 2 Dividing Polynomials

Work with a partner. Use the results of Exploration 1 to find each quotient. Write your answers in standard form. Check your answers by multiplying.

a. $(x^3 + x^2 - 2x) \div x$

b. $(x^3 - 3x + 2) \div (x - 1)$

c. $(x^3 + 2x^2 - x - 2) \div (x + 1)$

d. $(x^3 - x^2 - 4x + 4) \div (x - 2)$

e. $(x^3 + 3x^2 - 4) \div (x + 2)$

f. $(x^3 - 2x^2 - 5x + 6) \div (x - 3)$

Communicate Your Answer

3. How can you use the factors of a cubic polynomial to solve a division problem involving the polynomial?

4.3 Lesson

Core Vocabulary

polynomial long division, *p. 174*
synthetic division, *p. 175*

Previous
long division
divisor
quotient
remainder
dividend

What You Will Learn

- Use long division to divide polynomials by other polynomials.
- Use synthetic division to divide polynomials by binomials of the form $x - k$.
- Use the Remainder Theorem.

Long Division of Polynomials

When you divide a polynomial $f(x)$ by a nonzero polynomial divisor $d(x)$, you get a quotient polynomial $q(x)$ and a remainder polynomial $r(x)$.

$$\frac{f(x)}{d(x)} = q(x) + \frac{r(x)}{d(x)}$$

The degree of the remainder must be less than the degree of the divisor. When the remainder is 0, the divisor *divides evenly* into the dividend. Also, the degree of the divisor is less than or equal to the degree of the dividend $f(x)$. One way to divide polynomials is called **polynomial long division**.

EXAMPLE 1 Using Polynomial Long Division

Divide $2x^4 + 3x^3 + 5x - 1$ by $x^2 + 3x + 2$.

SOLUTION

Write polynomial division in the same format you use when dividing numbers. Include a "0" as the coefficient of x^2 in the dividend. At each stage, divide the term with the highest power in what is left of the dividend by the first term of the divisor. This gives the next term of the quotient.

$$\begin{array}{r l}
2x^2 - 3x + 5 & \leftarrow \text{quotient} \\
x^2 + 3x + 2 \,\big)\, 2x^4 + 3x^3 + 0x^2 + 5x - 1 & \\
2x^4 + 6x^3 + 4x^2 & \text{Multiply divisor by } \frac{2x^4}{x^2} = 2x^2. \\
-3x^3 - 4x^2 + 5x & \text{Subtract. Bring down next term.} \\
-3x^3 - 9x^2 - 6x & \text{Multiply divisor by } \frac{-3x^3}{x^2} = -3x. \\
5x^2 + 11x - 1 & \text{Subtract. Bring down next term.} \\
5x^2 + 15x + 10 & \text{Multiply divisor by } \frac{5x^2}{x^2} = 5. \\
-4x - 11 & \leftarrow \text{remainder}
\end{array}$$

COMMON ERROR

The expression added to the quotient in the result of a long division problem is $\frac{r(x)}{d(x)}$, not $r(x)$.

$$\frac{2x^4 + 3x^3 + 5x - 1}{x^2 + 3x + 2} = 2x^2 - 3x + 5 + \frac{-4x - 11}{x^2 + 3x + 2}$$

Check

You can check the result of a division problem by multiplying the quotient by the divisor and adding the remainder. The result should be the dividend.

$$(2x^2 - 3x + 5)(x^2 + 3x + 2) + (-4x - 11)$$
$$= (2x^2)(x^2 + 3x + 2) - (3x)(x^2 + 3x + 2) + (5)(x^2 + 3x + 2) - 4x - 11$$
$$= 2x^4 + 6x^3 + 4x^2 - 3x^3 - 9x^2 - 6x + 5x^2 + 15x + 10 - 4x - 11$$
$$= 2x^4 + 3x^3 + 5x - 1 \checkmark$$

Monitoring Progress

Help in English and Spanish at *BigIdeasMath.com*

Divide using polynomial long division.

1. $(x^3 - x^2 - 2x + 8) \div (x - 1)$
2. $(x^4 + 2x^2 - x + 5) \div (x^2 - x + 1)$

Synthetic Division

There is a shortcut for dividing polynomials by binomials of the form $x - k$. This shortcut is called **synthetic division**. This method is shown in the next example.

EXAMPLE 2 Using Synthetic Division

Divide $-x^3 + 4x^2 + 9$ by $x - 3$.

SOLUTION

Step 1 Write the coefficients of the dividend in order of descending exponents. Include a "0" for the missing x-term. Because the divisor is $x - 3$, use $k = 3$. Write the k-value to the left of the vertical bar.

$$k\text{-value} \rightarrow 3 \;\big|\; -1 \quad 4 \quad 0 \quad 9 \leftarrow \text{coefficients of } -x^3 + 4x^2 + 9$$

Step 2 Bring down the leading coefficient. Multiply the leading coefficient by the k-value. Write the product under the second coefficient. Add.

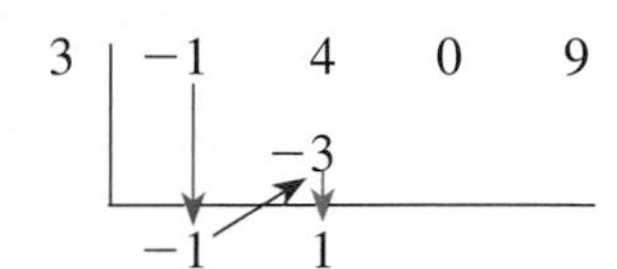

Step 3 Multiply the previous sum by the k-value. Write the product under the third coefficient. Add. Repeat this process for the remaining coefficient. The first three numbers in the bottom row are the coefficients of the quotient, and the last number is the remainder.

3	−1	4	0	9	
		−3	3	9	
coefficients of quotient →	−1	1	3	18	← remainder

▶ $\dfrac{-x^3 + 4x^2 + 9}{x - 3} = -x^2 + x + 3 + \dfrac{18}{x - 3}$

EXAMPLE 3 Using Synthetic Division

Divide $3x^3 - 2x^2 + 2x - 5$ by $x + 1$.

STUDY TIP

Note that dividing polynomials does not always result in a polynomial. This means that the set of polynomials is *not* closed under division.

SOLUTION

Use synthetic division. Because the divisor is $x + 1 = x - (-1)$, $k = -1$.

−1	3	−2	2	−5
		−3	5	−7
	3	−5	7	−12

▶ $\dfrac{3x^3 - 2x^2 + 2x - 5}{x + 1} = 3x^2 - 5x + 7 - \dfrac{12}{x + 1}$

Monitoring Progress  Help in English and Spanish at *BigIdeasMath.com*

Divide using synthetic division.

3. $(x^3 - 3x^2 - 7x + 6) \div (x - 2)$ **4.** $(2x^3 - x - 7) \div (x + 3)$

1-23 Odd

The Remainder Theorem

The remainder in the synthetic division process has an important interpretation. When you divide a polynomial $f(x)$ by $d(x) = x - k$, the result is

$$\frac{f(x)}{d(x)} = q(x) + \frac{r(x)}{d(x)} \qquad \text{Polynomial division}$$

$$\frac{f(x)}{x - k} = q(x) + \frac{r(x)}{x - k} \qquad \text{Substitute } x - k \text{ for } d(x).$$

$$f(x) = (x - k)q(x) + r(x). \qquad \text{Multiply both sides by } x - k.$$

Because either $r(x) = 0$ *or* the degree of $r(x)$ is less than the degree of $x - k$, you know that $r(x)$ is a constant function. So, let $r(x) = r$, where r is a real number, and evaluate $f(x)$ when $x = k$.

$$f(k) = (k - k)q(k) + r \qquad \text{Substitute } k \text{ for } x \text{ and } r \text{ for } r(x).$$

$$f(k) = r \qquad \text{Simplify.}$$

This result is stated in the *Remainder Theorem*.

Core Concept

The Remainder Theorem

If a polynomial $f(x)$ is divided by $x - k$, then the remainder is $r = f(k)$.

The Remainder Theorem tells you that synthetic division can be used to evaluate a polynomial function. So, to evaluate $f(x)$ when $x = k$, divide $f(x)$ by $x - k$. The remainder will be $f(k)$.

EXAMPLE 4 Evaluating a Polynomial

Use synthetic division to evaluate $f(x) = 5x^3 - x^2 + 13x + 29$ when $x = -4$.

SOLUTION

$$\begin{array}{r|rrrr} -4 & 5 & -1 & 13 & 29 \\ & & -20 & 84 & -388 \\ \hline & 5 & -21 & 97 & -359 \end{array}$$

▶ The remainder is -359. So, you can conclude from the Remainder Theorem that $f(-4) = -359$.

Check

Check this by substituting $x = -4$ in the original function.

$$f(-4) = 5(-4)^3 - (-4)^2 + 13(-4) + 29$$
$$= -320 - 16 - 52 + 29$$
$$= -359 \checkmark$$

Monitoring Progress

Help in English and Spanish at *BigIdeasMath.com*

Use synthetic division to evaluate the function for the indicated value of x.

5. $f(x) = 4x^2 - 10x - 21; x = 5$

6. $f(x) = 5x^4 + 2x^3 - 20x - 6; x = 2$

4.3 Exercises

Dynamic Solutions available at *BigIdeasMath.com*

Vocabulary and Core Concept Check

1. **WRITING** Explain the Remainder Theorem in your own words. Use an example in your explanation.

2. **VOCABULARY** What form must the divisor have to make synthetic division an appropriate method for dividing a polynomial? Provide examples to support your claim.

3. **VOCABULARY** Write the polynomial divisor, dividend, and quotient functions represented by the synthetic division shown at the right.

$$\begin{array}{r|rrrr} -3 & 1 & -2 & -9 & 18 \\ & & -3 & 15 & -18 \\ \hline & 1 & -5 & 6 & 0 \end{array}$$

4. **WRITING** Explain what the colored numbers represent in the synthetic division in Exercise 3.

Monitoring Progress and Modeling with Mathematics

In Exercises 5–10, divide using polynomial long division. *(See Example 1.)*

5. $(x^2 + x - 17) \div (x - 4)$

6. $(3x^2 - 14x - 5) \div (x - 5)$

7. $(x^3 + x^2 + x + 2) \div (x^2 - 1)$

8. $(7x^3 + x^2 + x) \div (x^2 + 1)$

9. $(5x^4 - 2x^3 - 7x^2 - 39) \div (x^2 + 2x - 4)$

10. $(4x^4 + 5x - 4) \div (x^2 - 3x - 2)$

In Exercises 11–18, divide using synthetic division. *(See Examples 2 and 3.)*

11. $(x^2 + 8x + 1) \div (x - 4)$

12. $(4x^2 - 13x - 5) \div (x - 2)$

13. $(2x^2 - x + 7) \div (x + 5)$

14. $(x^3 - 4x + 6) \div (x + 3)$

15. $(x^2 + 9) \div (x - 3)$

16. $(3x^3 - 5x^2 - 2) \div (x - 1)$

17. $(x^4 - 5x^3 - 8x^2 + 13x - 12) \div (x - 6)$

18. $(x^4 + 4x^3 + 16x - 35) \div (x + 5)$

ANALYZING RELATIONSHIPS In Exercises 19–22, match the equivalent expressions. Justify your answers.

19. $(x^2 + x - 3) \div (x - 2)$

20. $(x^2 - x - 3) \div (x - 2)$

21. $(x^2 - x + 3) \div (x - 2)$

22. $(x^2 + x + 3) \div (x - 2)$

A. $x + 1 - \dfrac{1}{x - 2}$ **B.** $x + 3 + \dfrac{9}{x - 2}$

C. $x + 1 + \dfrac{5}{x - 2}$ **D.** $x + 3 + \dfrac{3}{x - 2}$

ERROR ANALYSIS In Exercises 23 and 24, describe and correct the error in using synthetic division to divide $x^3 - 5x + 3$ by $x - 2$.

23.

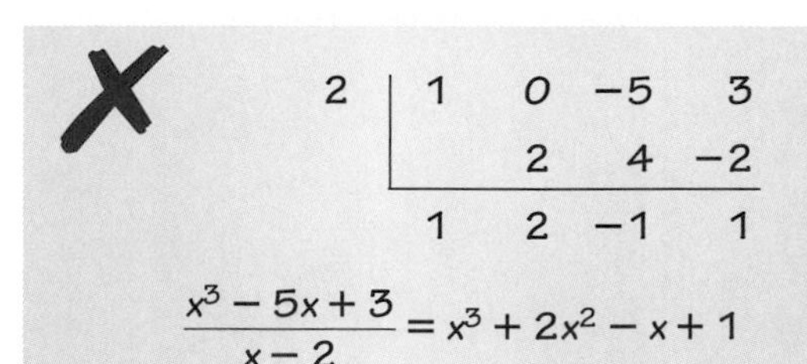

2 | 1 0 −5 3
2 4 −2
1 2 −1 1
$\dfrac{x^3 - 5x + 3}{x - 2} = x^3 + 2x^2 - x + 1$

24.

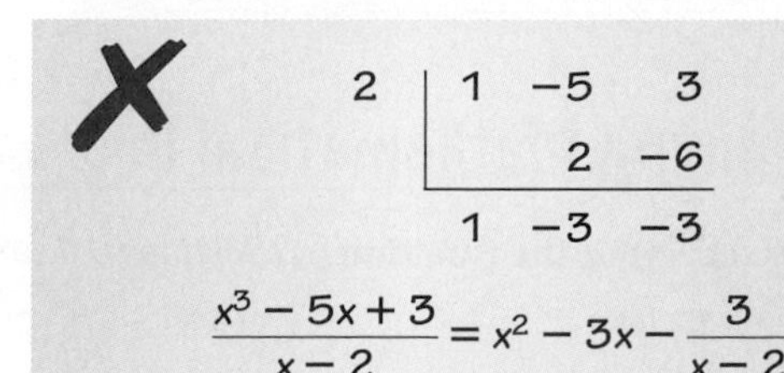

2 | 1 −5 3
2 −6
1 −3 −3
$\dfrac{x^3 - 5x + 3}{x - 2} = x^2 - 3x - \dfrac{3}{x - 2}$

In Exercises 25–32, use synthetic division to evaluate the function for the indicated value of x. *(See Example 4.)*

25. $f(x) = -x^2 - 8x + 30;\ x = -1$

26. $f(x) = 3x^2 + 2x - 20;\ x = 3$

27. $f(x) = x^3 - 2x^2 + 4x + 3;\ x = 2$

28. $f(x) = x^3 + x^2 - 3x + 9;\ x = -4$

29. $f(x) = x^3 - 6x + 1;\ x = 6$

30. $f(x) = x^3 - 9x - 7;\ x = 10$

31. $f(x) = x^4 + 6x^2 - 7x + 1;\ x = 3$

32. $f(x) = -x^4 - x^3 - 2;\ x = 5$

33. MAKING AN ARGUMENT You use synthetic division to divide $f(x)$ by $(x - a)$ and find that the remainder equals 15. Your friend concludes that $f(15) = a$. Is your friend correct? Explain your reasoning.

34. THOUGHT PROVOKING A polygon has an area represented by $A = 4x^2 + 8x + 4$. The figure has at least one dimension equal to $2x + 2$. Draw the figure and label its dimensions.

35. USING TOOLS The total attendance A (in thousands) at NCAA women's basketball games and the number T of NCAA women's basketball teams over a period of time can be modeled by

$$A = -1.95x^3 + 70.1x^2 - 188x + 2150$$
$$T = 14.8x + 725$$

where x is in years and $0 < x < 18$. Write a function for the average attendance per team over this period of time.

36. COMPARING METHODS The profit P (in millions of dollars) for a DVD manufacturer can be modeled by $P = -6x^3 + 72x$, where x is the number (in millions) of DVDs produced. Use synthetic division to show that the company yields a profit of \$96 million when 2 million DVDs are produced. Is there an easier method? Explain.

37. CRITICAL THINKING What is the value of k such that $(x^3 - x^2 + kx - 30) \div (x - 5)$ has a remainder of zero?

Ⓐ -14 Ⓑ -2

Ⓒ 26 Ⓓ 32

38. HOW DO YOU SEE IT? The graph represents the polynomial function $f(x) = x^3 + 3x^2 - x - 3$.

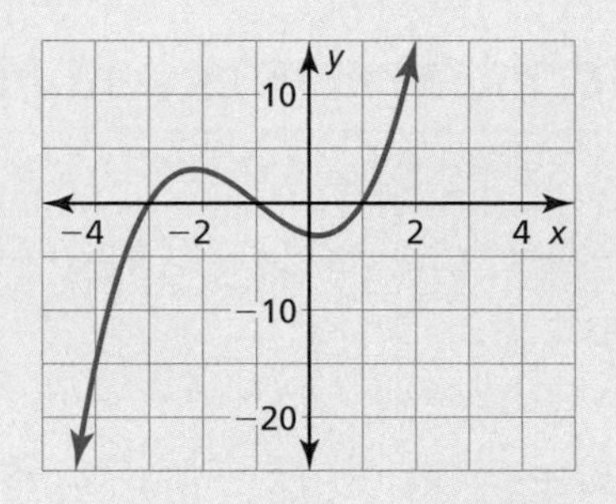

a. The expression $f(x) \div (x - k)$ has a remainder of -15. What is the value of k?

b. Use the graph to compare the remainders of $(x^3 + 3x^2 - x - 3) \div (x + 3)$ and $(x^3 + 3x^2 - x - 3) \div (x + 1)$.

39. MATHEMATICAL CONNECTIONS The volume V of the rectangular prism is given by $V = 2x^3 + 17x^2 + 46x + 40$. Find an expression for the missing dimension.

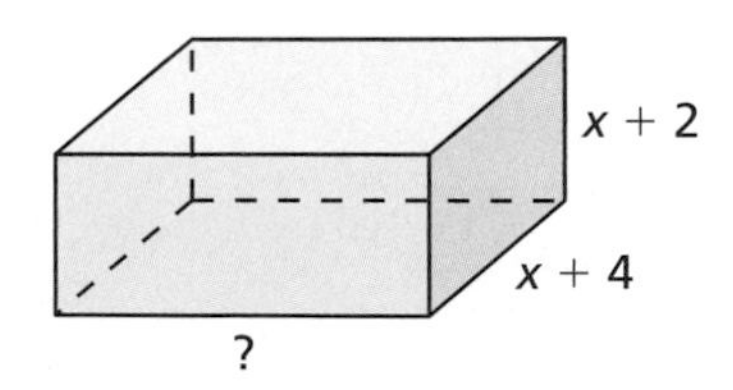

40. USING STRUCTURE You divide two polynomials and obtain the result $5x^2 - 13x + 47 - \frac{102}{x+2}$. What is the dividend? How did you find it?

Maintaining Mathematical Proficiency

Reviewing what you learned in previous grades and lessons

Find the zero(s) of the function. *(Sections 3.1 and 3.2)*

41. $f(x) = x^2 - 6x + 9$

42. $g(x) = 3(x + 6)(x - 2)$

43. $g(x) = x^2 + 14x + 49$

44. $h(x) = 4x^2 + 36$

4.4 Factoring Polynomials

Essential Question

How can you factor a polynomial?

EXPLORATION 1 Factoring Polynomials

Work with a partner. Match each polynomial equation with the graph of its related polynomial function. Use the x-intercepts of the graph to write each polynomial in factored form. Explain your reasoning.

a. $x^2 + 5x + 4 = 0$

b. $x^3 - 2x^2 - x + 2 = 0$

c. $x^3 + x^2 - 2x = 0$

d. $x^3 - x = 0$

e. $x^4 - 5x^2 + 4 = 0$

f. $x^4 - 2x^3 - x^2 + 2x = 0$

A.

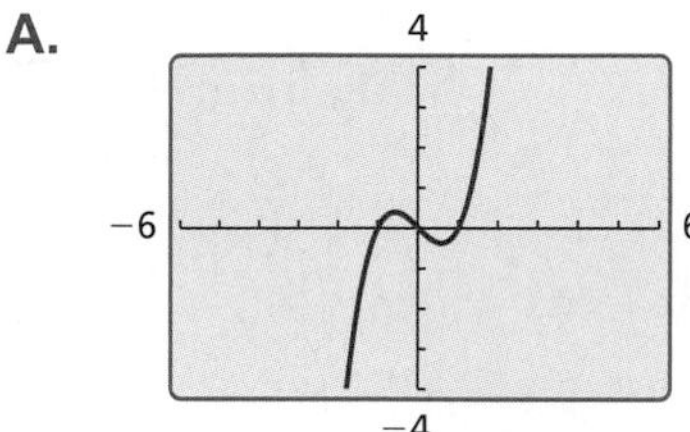

B.

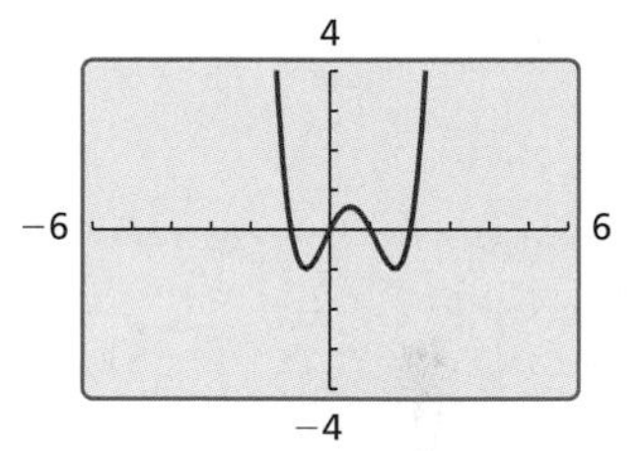

C.

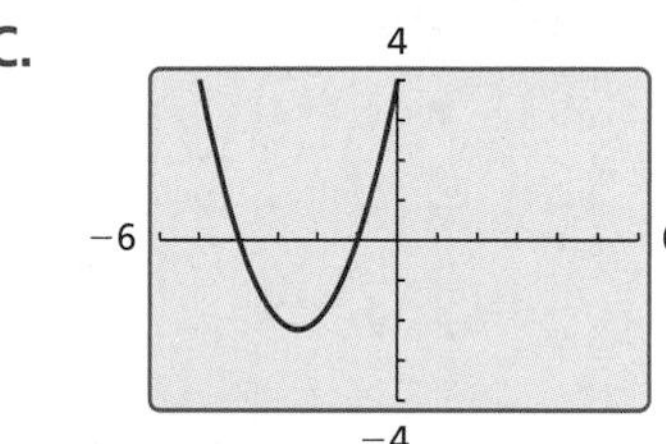

D.

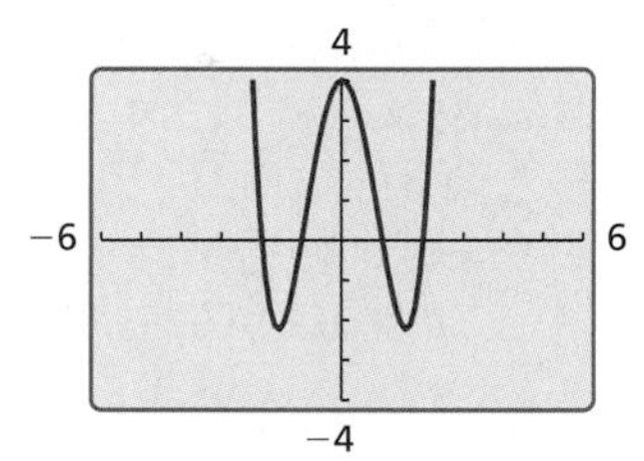

E.

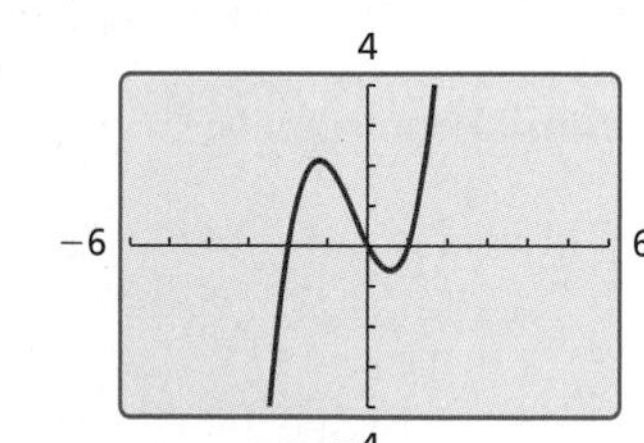

F.

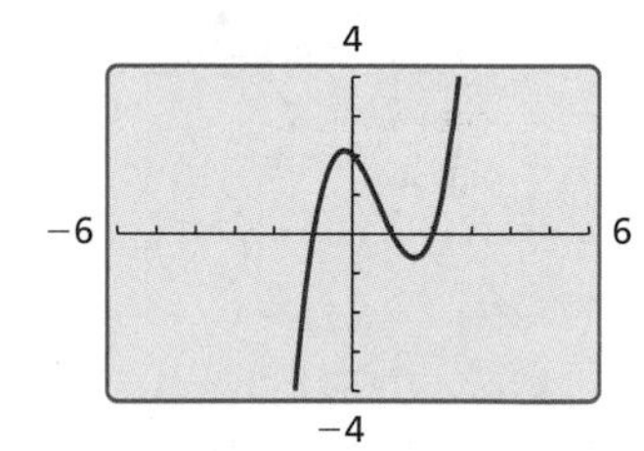

MAKING SENSE OF PROBLEMS

To be proficient in math, you need to check your answers to problems and continually ask yourself, "Does this make sense?"

EXPLORATION 2 Factoring Polynomials

Work with a partner. Use the x-intercepts of the graph of the polynomial function to write each polynomial in factored form. Explain your reasoning. Check your answers by multiplying.

a. $f(x) = x^2 - x - 2$

b. $f(x) = x^3 - x^2 - 2x$

c. $f(x) = x^3 - 2x^2 - 3x$

d. $f(x) = x^3 - 3x^2 - x + 3$

e. $f(x) = x^4 + 2x^3 - x^2 - 2x$

f. $f(x) = x^4 - 10x^2 + 9$

Communicate Your Answer

3. How can you factor a polynomial?

4. What information can you obtain about the graph of a polynomial function written in factored form?

4.4 Lesson

Core Vocabulary

factored completely, *p. 180*
factor by grouping, *p. 181*
quadratic form, *p. 181*

Previous
zero of a function
synthetic division

What You Will Learn

- Factor polynomials.
- Use the Factor Theorem.

Factoring Polynomials

Previously, you factored quadratic polynomials. You can also factor polynomials with degree greater than 2. Some of these polynomials can be *factored completely* using techniques you have previously learned. A factorable polynomial with integer coefficients is **factored completely** when it is written as a product of unfactorable polynomials with integer coefficients.

EXAMPLE 1 Finding a Common Monomial Factor

Factor each polynomial completely.

a. $x^3 - 4x^2 - 5x$ **b.** $3y^5 - 48y^3$ **c.** $5z^4 + 30z^3 + 45z^2$

SOLUTION

a. $x^3 - 4x^2 - 5x = x(x^2 - 4x - 5)$ Factor common monomial.

$= x(x - 5)(x + 1)$ Factor trinomial.

b. $3y^5 - 48y^3 = 3y^3(y^2 - 16)$ Factor common monomial.

$= 3y^3(y - 4)(y + 4)$ Difference of Two Squares Pattern

c. $5z^4 + 30z^3 + 45z^2 = 5z^2(z^2 + 6z + 9)$ Factor common monomial.

$= 5z^2(z + 3)^2$ Perfect Square Trinomial Pattern

Monitoring Progress  Help in English and Spanish at *BigIdeasMath.com*

Factor the polynomial completely.

1. $x^3 - 7x^2 + 10x$ **2.** $3n^7 - 75n^5$ **3.** $8m^5 - 16m^4 + 8m^3$

In part (b) of Example 1, the special factoring pattern for the difference of two squares was used to factor the expression completely. There are also factoring patterns that you can use to factor the sum or difference of two *cubes*.

Core Concept

Special Factoring Patterns

Sum of Two Cubes

$a^3 + b^3 = (a + b)(a^2 - ab + b^2)$

Example

$64x^3 + 1 = (4x)^3 + 1^3$

$= (4x + 1)(16x^2 - 4x + 1)$

Difference of Two Cubes

$a^3 - b^3 = (a - b)(a^2 + ab + b^2)$

Example

$27x^3 - 8 = (3x)^3 - 2^3$

$= (3x - 2)(9x^2 + 6x + 4)$

EXAMPLE 2 Factoring the Sum or Difference of Two Cubes

Factor (a) $x^3 - 125$ and (b) $16s^5 + 54s^2$ completely.

SOLUTION

a. $x^3 - 125 = x^3 - 5^3$ — Write as $a^3 - b^3$.

$= (x - 5)(x^2 + 5x + 25)$ — Difference of Two Cubes Pattern

b. $16s^5 + 54s^2 = 2s^2(8s^3 + 27)$ — Factor common monomial.

$= 2s^2[(2s)^3 + 3^3]$ — Write $8s^3 + 27$ as $a^3 + b^3$.

$= 2s^2(2s + 3)(4s^2 - 6s + 9)$ — Sum of Two Cubes Pattern

For some polynomials, you can **factor by grouping** pairs of terms that have a common monomial factor. The pattern for factoring by grouping is shown below.

$$ra + rb + sa + sb = r(a + b) + s(a + b)$$
$$= (r + s)(a + b)$$

EXAMPLE 3 Factoring by Grouping

Factor $z^3 + 5z^2 - 4z - 20$ completely.

SOLUTION

$z^3 + 5z^2 - 4z - 20 = z^2(z + 5) - 4(z + 5)$ — Factor by grouping.

$= (z^2 - 4)(z + 5)$ — Distributive Property

$= (z - 2)(z + 2)(z + 5)$ — Difference of Two Squares Pattern

An expression of the form $au^2 + bu + c$, where u is an algebraic expression, is said to be in **quadratic form**. The factoring techniques you have studied can sometimes be used to factor such expressions.

LOOKING FOR STRUCTURE

The expression $16x^4 - 81$ is in quadratic form because it can be written as $u^2 - 81$ where $u = 4x^2$.

EXAMPLE 4 Factoring Polynomials in Quadratic Form

Factor (a) $16x^4 - 81$ and (b) $3p^8 + 15p^5 + 18p^2$ completely.

SOLUTION

a. $16x^4 - 81 = (4x^2)^2 - 9^2$ — Write as $a^2 - b^2$.

$= (4x^2 + 9)(4x^2 - 9)$ — Difference of Two Squares Pattern

$= (4x^2 + 9)(2x - 3)(2x + 3)$ — Difference of Two Squares Pattern

b. $3p^8 + 15p^5 + 18p^2 = 3p^2(p^6 + 5p^3 + 6)$ — Factor common monomial.

$= 3p^2(p^3 + 3)(p^3 + 2)$ — Factor trinomial in quadratic form.

Factor the polynomial completely.

4. $a^3 + 27$
5. $6z^5 - 750z^2$
6. $x^3 + 4x^2 - x - 4$
7. $3y^3 + y^2 + 9y + 3$
8. $-16n^4 + 625$
9. $5w^6 - 25w^4 + 30w^2$

The Factor Theorem

When dividing polynomials in the previous section, the examples had nonzero remainders. Suppose the remainder is 0 when a polynomial $f(x)$ is divided by $x - k$. Then,

$$\frac{f(x)}{x - k} = q(x) + \frac{0}{x - k} = q(x)$$

where $q(x)$ is the quotient polynomial. Therefore, $f(x) = (x - k) \cdot q(x)$, so that $x - k$ is a factor of $f(x)$. This result is summarized by the *Factor Theorem*, which is a special case of the Remainder Theorem.

READING

In other words, $x - k$ is a factor of $f(x)$ if and only if k is a zero of f.

Core Concept

The Factor Theorem

A polynomial $f(x)$ has a factor $x - k$ if and only if $f(k) = 0$.

STUDY TIP

In part (b), notice that direct substitution would have resulted in more difficult computations than synthetic division.

EXAMPLE 5 Determining Whether a Linear Binomial Is a Factor

Determine whether (a) $x - 2$ is a factor of $f(x) = x^2 + 2x - 4$ and (b) $x + 5$ is a factor of $f(x) = 3x^4 + 15x^3 - x^2 + 25$.

SOLUTION

a. Find $f(2)$ by direct substitution.

$$\begin{aligned} f(2) &= 2^2 + 2(2) - 4 \\ &= 4 + 4 - 4 \\ &= 4 \end{aligned}$$

▶ Because $f(2) \neq 0$, the binomial $x - 2$ is not a factor of $f(x) = x^2 + 2x - 4$.

b. Find $f(-5)$ by synthetic division.

$$\begin{array}{r|rrrrr} -5 & 3 & 15 & -1 & 0 & 25 \\ & & -15 & 0 & 5 & -25 \\ \hline & 3 & 0 & -1 & 5 & 0 \end{array}$$

▶ Because $f(-5) = 0$, the binomial $x + 5$ is a factor of $f(x) = 3x^4 + 15x^3 - x^2 + 25$.

EXAMPLE 6 Factoring a Polynomial

Show that $x + 3$ is a factor of $f(x) = x^4 + 3x^3 - x - 3$. Then factor $f(x)$ completely.

SOLUTION

Show that $f(-3) = 0$ by synthetic division.

$$\begin{array}{r|rrrrr} -3 & 1 & 3 & 0 & -1 & -3 \\ & & -3 & 0 & 0 & 3 \\ \hline & 1 & 0 & 0 & -1 & 0 \end{array}$$

Because $f(-3) = 0$, you can conclude that $x + 3$ is a factor of $f(x)$ by the Factor Theorem. Use the result to write $f(x)$ as a product of two factors and then factor completely.

$$\begin{aligned} f(x) &= x^4 + 3x^3 - x - 3 && \text{Write original polynomial.} \\ &= (x + 3)(x^3 - 1) && \text{Write as a product of two factors.} \\ &= (x + 3)(x - 1)(x^2 + x + 1) && \text{Difference of Two Cubes Pattern} \end{aligned}$$

ANOTHER WAY

Notice that you can factor $f(x)$ by grouping.

$$\begin{aligned} f(x) &= x^3(x + 3) - 1(x + 3) \\ &= (x^3 - 1)(x + 3) \\ &= (x + 3)(x - 1) \cdot (x^2 + x + 1) \end{aligned}$$

Because the x-intercepts of the graph of a function are the zeros of the function, you can use the graph to approximate the zeros. You can check the approximations using the Factor Theorem.

Real-Life Application

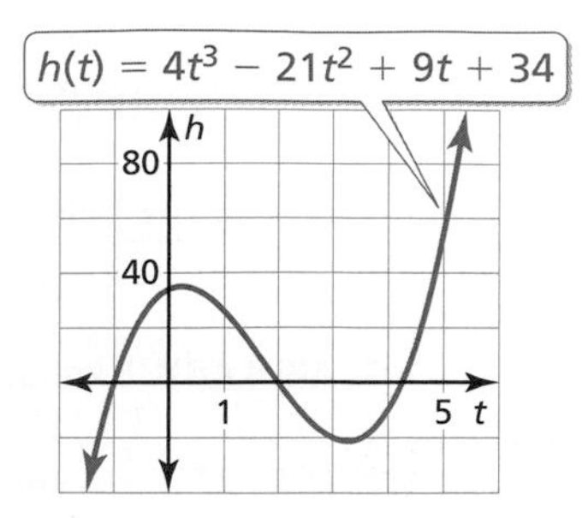

During the first 5 seconds of a roller coaster ride, the function $h(t) = 4t^3 - 21t^2 + 9t + 34$ represents the height h (in feet) of the roller coaster after t seconds. How long is the roller coaster at or below ground level in the first 5 seconds?

SOLUTION

1. **Understand the Problem** You are given a function rule that represents the height of a roller coaster. You are asked to determine how long the roller coaster is at or below ground during the first 5 seconds of the ride.
2. **Make a Plan** Use a graph to estimate the zeros of the function and check using the Factor Theorem. Then use the zeros to describe where the graph lies below the t-axis.
3. **Solve the Problem** From the graph, two of the zeros appear to be -1 and 2. The third zero is between 4 and 5.

Step 1 Determine whether -1 is a zero using synthetic division.

$$\begin{array}{r|rrrr} -1 & 4 & -21 & 9 & 34 \\ & & -4 & 25 & -34 \\ \hline & 4 & -25 & 34 & 0 \end{array}$$

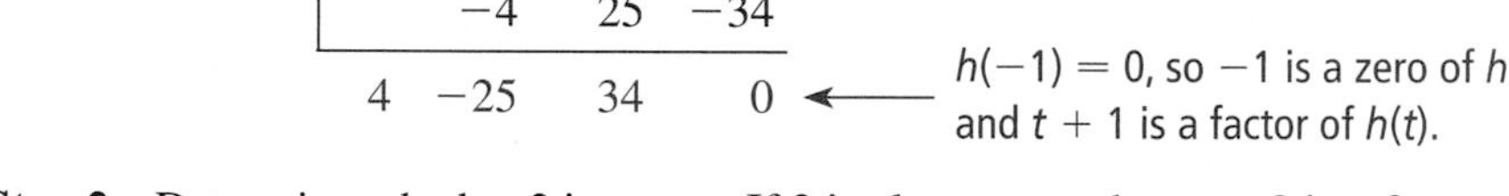

$h(-1) = 0$, so -1 is a zero of h and $t + 1$ is a factor of $h(t)$.

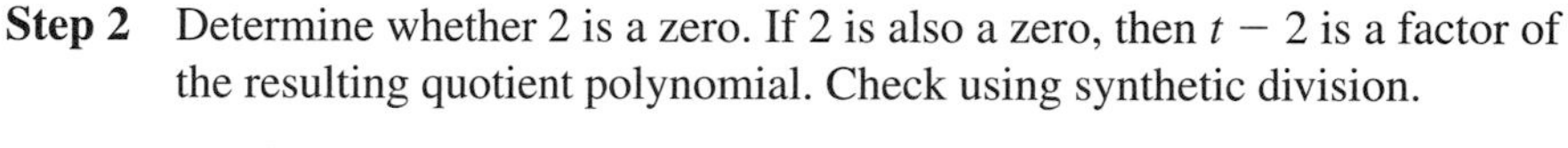

Step 2 Determine whether 2 is a zero. If 2 is also a zero, then $t - 2$ is a factor of the resulting quotient polynomial. Check using synthetic division.

$$\begin{array}{r|rrr} 2 & 4 & -25 & 34 \\ & & 8 & -34 \\ \hline & 4 & -17 & 0 \end{array}$$

The remainder is 0, so $t - 2$ is a factor of $h(t)$ and 2 is a zero of h.

STUDY TIP
You could also check that 2 is a zero using the original function, but using the quotient polynomial helps you find the remaining factor.

So, $h(t) = (t + 1)(t - 2)(4t - 17)$. The factor $4t - 17$ indicates that the zero between 4 and 5 is $\frac{17}{4}$, or 4.25.

▶ The zeros are -1, 2, and 4.25. Only $t = 2$ and $t = 4.25$ occur in the first 5 seconds. The graph shows that the roller coaster is at or below ground level for $4.25 - 2 = 2.25$ seconds.

4. **Look Back** Use a table of values to verify the positive zeros and heights between the zeros.

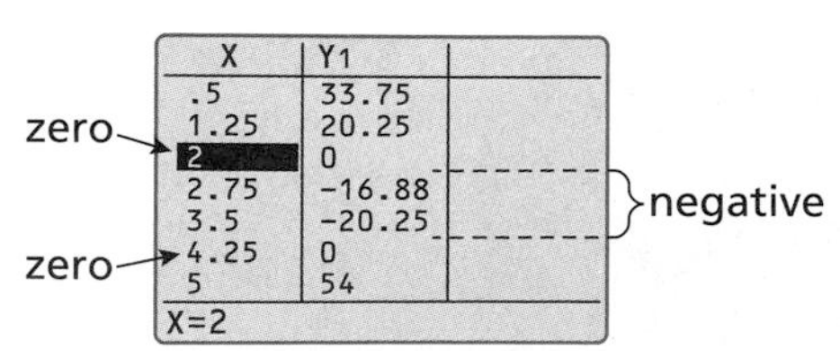

X	Y1
.5	33.75
1.25	20.25
2	0
2.75	-16.88
3.5	-20.25
4.25	0
5	54

Monitoring Progress

Help in English and Spanish at *BigIdeasMath.com*

10. Determine whether $x - 4$ is a factor of $f(x) = 2x^2 + 5x - 12$.
11. Show that $x - 6$ is a factor of $f(x) = x^3 - 5x^2 - 6x$. Then factor $f(x)$ completely.
12. In Example 7, does your answer change when you first determine whether 2 is a zero and then whether -1 is a zero? Justify your answer.

4.4 Exercises

Dynamic Solutions available at *BigIdeasMath.com*

Vocabulary and Core Concept Check

1. **COMPLETE THE SENTENCE** The expression $9x^4 - 49$ is in ________ form because it can be written as $u^2 - 49$ where $u =$ _____.

2. **VOCABULARY** Explain when you should try factoring a polynomial by grouping.

3. **WRITING** How do you know when a polynomial is factored completely?

4. **WRITING** Explain the Factor Theorem and why it is useful.

Monitoring Progress and Modeling with Mathematics

In Exercises 5–12, factor the polynomial completely. *(See Example 1.)*

5. $x^3 - 2x^2 - 24x$
6. $4k^5 - 100k^3$
7. $3p^5 - 192p^3$
8. $2m^6 - 24m^5 + 64m^4$
9. $2q^4 + 9q^3 - 18q^2$
10. $3r^6 - 11r^5 - 20r^4$
11. $10w^{10} - 19w^9 + 6w^8$
12. $18v^9 + 33v^8 + 14v^7$

In Exercises 13–20, factor the polynomial completely. *(See Example 2.)*

13. $x^3 + 64$
14. $y^3 + 512$
15. $g^3 - 343$
16. $c^3 - 27$
17. $3h^9 - 192h^6$
18. $9n^6 - 6561n^3$
19. $16t^7 + 250t^4$
20. $135z^{11} - 1080z^8$

ERROR ANALYSIS In Exercises 21 and 22, describe and correct the error in factoring the polynomial.

21. ✗
$$3x^3 + 27x = 3x(x^2 + 9)$$
$$= 3x(x + 3)(x - 3)$$

22. ✗
$$x^9 + 8x^3 = (x^3)^3 + (2x)^3$$
$$= (x^3 + 2x)[(x^3)^2 - (x^3)(2x) + (2x)^2]$$
$$= (x^3 + 2x)(x^6 - 2x^4 + 4x^2)$$

In Exercises 23–30, factor the polynomial completely. *(See Example 3.)*

23. $y^3 - 5y^2 + 6y - 30$
24. $m^3 - m^2 + 7m - 7$
25. $3a^3 + 18a^2 + 8a + 48$
26. $2k^3 - 20k^2 + 5k - 50$
27. $x^3 - 8x^2 - 4x + 32$
28. $z^3 - 5z^2 - 9z + 45$
29. $4q^3 - 16q^2 - 9q + 36$
30. $16n^3 + 32n^2 - n - 2$

In Exercises 31–38, factor the polynomial completely. *(See Example 4.)*

31. $49k^4 - 9$
32. $4m^4 - 25$
33. $c^4 + 9c^2 + 20$
34. $y^4 - 3y^2 - 28$
35. $16z^4 - 81$
36. $81a^4 - 256$
37. $3r^8 + 3r^5 - 60r^2$
38. $4n^{12} - 32n^7 + 48n^2$

In Exercises 39–44, determine whether the binomial is a factor of the polynomial. *(See Example 5.)*

39. $f(x) = 2x^3 + 5x^2 - 37x - 60; x - 4$
40. $g(x) = 3x^3 - 28x^2 + 29x + 140; x + 7$
41. $h(x) = 6x^5 - 15x^4 - 9x^3; x + 3$
42. $g(x) = 8x^5 - 58x^4 + 60x^3 + 140; x - 6$
43. $h(x) = 6x^4 - 6x^3 - 84x^2 + 144x; x + 4$
44. $t(x) = 48x^4 + 36x^3 - 138x^2 - 36x; x + 2$

In Exercises 45–50, show that the binomial is a factor of the polynomial. Then factor the polynomial completely. *(See Example 6.)*

45. $g(x) = x^3 - x^2 - 20x$; $x + 4$

46. $t(x) = x^3 - 5x^2 - 9x + 45$; $x - 5$

47. $f(x) = x^4 - 6x^3 - 8x + 48$; $x - 6$

48. $s(x) = x^4 + 4x^3 - 64x - 256$; $x + 4$

49. $r(x) = x^3 - 37x + 84$; $x + 7$

50. $h(x) = x^3 - x^2 - 24x - 36$; $x + 2$

ANALYZING RELATIONSHIPS **In Exercises 51–54, match the function with the correct graph. Explain your reasoning.**

51. $f(x) = (x - 2)(x - 3)(x + 1)$

52. $g(x) = x(x + 2)(x + 1)(x - 2)$

53. $h(x) = (x + 2)(x + 3)(x - 1)$

54. $k(x) = x(x - 2)(x - 1)(x + 2)$

A.

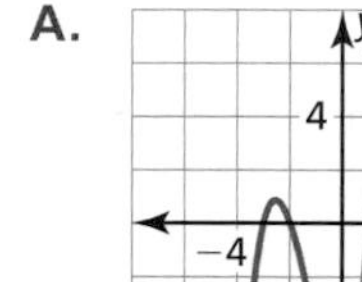

B.

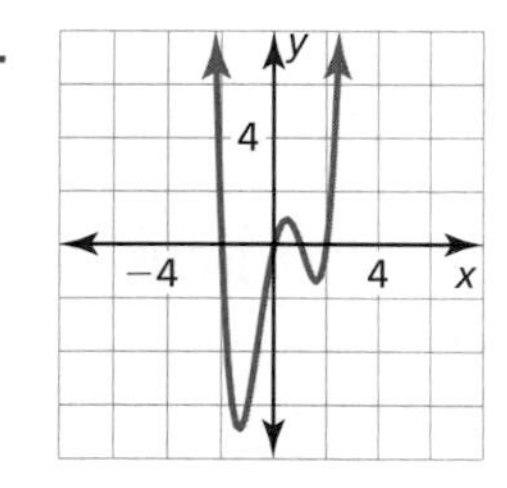

C.

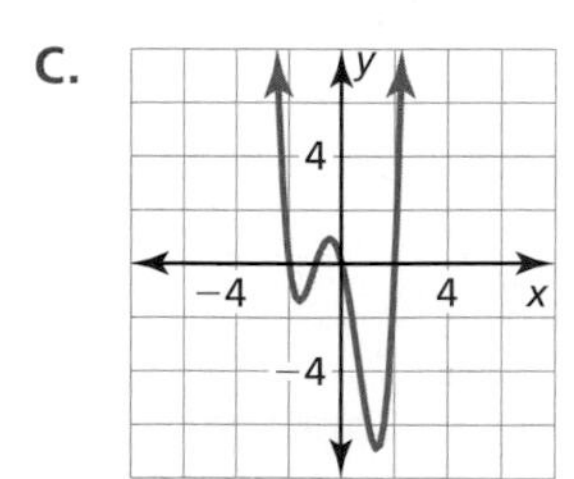

D.

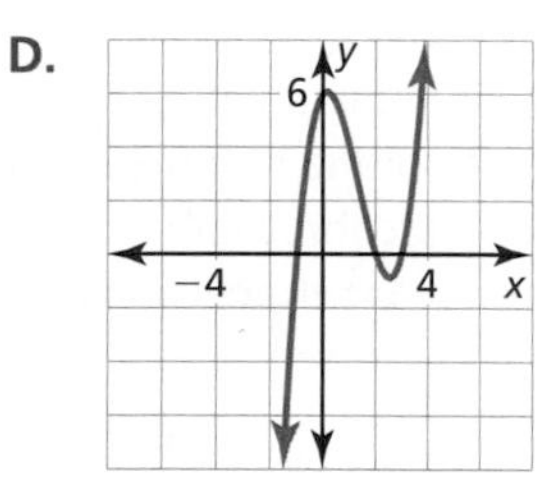

55. MODELING WITH MATHEMATICS The volume (in cubic inches) of a shipping box is modeled by $V = 2x^3 - 19x^2 + 39x$, where x is the length (in inches). Determine the values of x for which the model makes sense. Explain your reasoning. *(See Example 7.)*

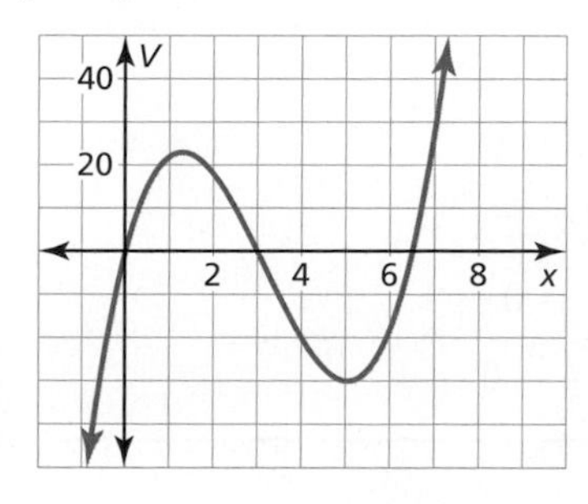

56. MODELING WITH MATHEMATICS The volume (in cubic inches) of a rectangular birdcage can be modeled by $V = 3x^3 - 17x^2 + 29x - 15$, where x is the length (in inches). Determine the values of x for which the model makes sense. Explain your reasoning.

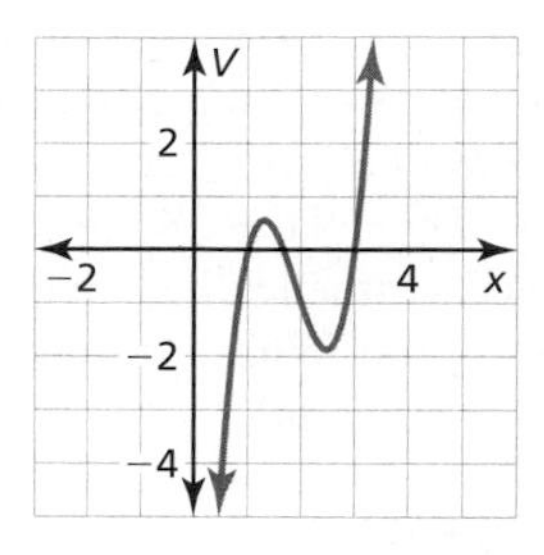

USING STRUCTURE **In Exercises 57–64, use the method of your choice to factor the polynomial completely. Explain your reasoning.**

57. $a^6 + a^5 - 30a^4$

58. $8m^3 - 343$

59. $z^3 - 7z^2 - 9z + 63$

60. $2p^8 - 12p^5 + 16p^2$

61. $64r^3 + 729$

62. $5x^5 - 10x^4 - 40x^3$

63. $16n^4 - 1$

64. $9k^3 - 24k^2 + 3k - 8$

65. REASONING Determine whether each polynomial is factored completely. If not, factor completely.

a. $7z^4(2z^2 - z - 6)$

b. $(2 - n)(n^2 + 6n)(3n - 11)$

c. $3(4y - 5)(9y^2 - 6y - 4)$

66. PROBLEM SOLVING The profit P (in millions of dollars) for a T-shirt manufacturer can be modeled by $P = -x^3 + 4x^2 + x$, where x is the number (in millions) of T-shirts produced. Currently the company produces 4 million T-shirts and makes a profit of \$4 million. What lesser number of T-shirts could the company produce and still make the same profit?

67. PROBLEM SOLVING The profit P (in millions of dollars) for a shoe manufacturer can be modeled by $P = -21x^3 + 46x$, where x is the number (in millions) of shoes produced. The company now produces 1 million shoes and makes a profit of \$25 million, but it would like to cut back production. What lesser number of shoes could the company produce and still make the same profit?

68. THOUGHT PROVOKING Find a value of k such that $\dfrac{f(x)}{x-k}$ has a remainder of 0. Justify your answer.

$$f(x) = x^3 - 3x^2 - 4x$$

69. COMPARING METHODS You are taking a test where calculators are not permitted. One question asks you to evaluate $g(7)$ for the function $g(x) = x^3 - 7x^2 - 4x + 28$. You use the Factor Theorem and synthetic division and your friend uses direct substitution. Whose method do you prefer? Explain your reasoning.

70. MAKING AN ARGUMENT You divide $f(x)$ by $(x - a)$ and find that the remainder does not equal 0. Your friend concludes that $f(x)$ cannot be factored. Is your friend correct? Explain your reasoning.

71. CRITICAL THINKING What is the value of k such that $x - 7$ is a factor of $h(x) = 2x^3 - 13x^2 - kx + 105$? Justify your answer.

72. HOW DO YOU SEE IT? Use the graph to write an equation of the cubic function in factored form. Explain your reasoning.

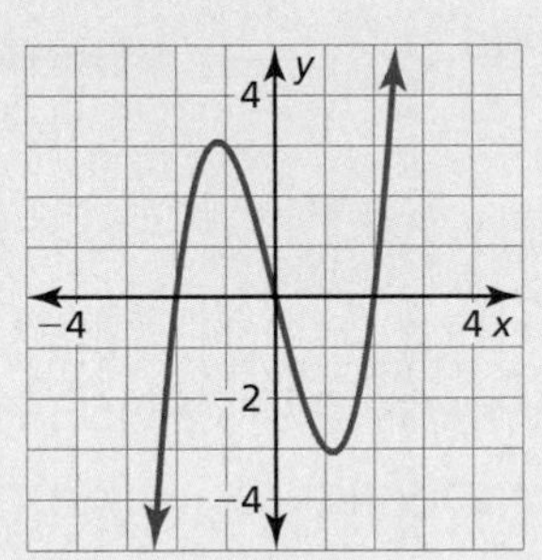

73. ABSTRACT REASONING Factor each polynomial completely.

a. $7ac^2 + bc^2 - 7ad^2 - bd^2$

b. $x^{2n} - 2x^n + 1$

c. $a^5b^2 - a^2b^4 + 2a^4b - 2ab^3 + a^3 - b^2$

74. REASONING The graph of the function $f(x) = x^4 + 3x^3 + 2x^2 + x + 3$ is shown. Can you use the Factor Theorem to factor $f(x)$? Explain.

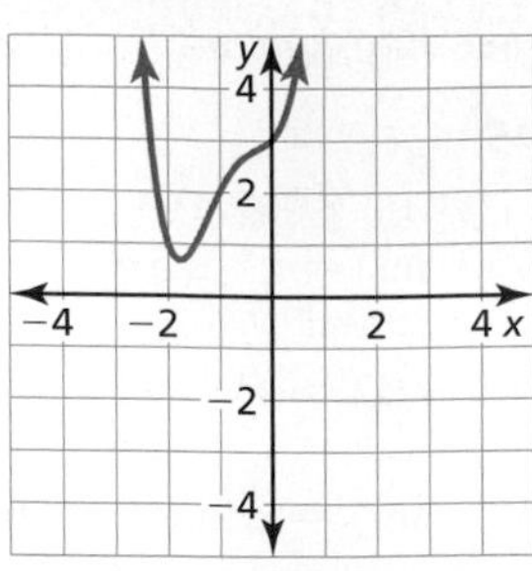

75. MATHEMATICAL CONNECTIONS The standard equation of a circle with radius r and center (h, k) is $(x - h)^2 + (y - k)^2 = r^2$. Rewrite each equation of a circle in standard form. Identify the center and radius of the circle. Then graph the circle.

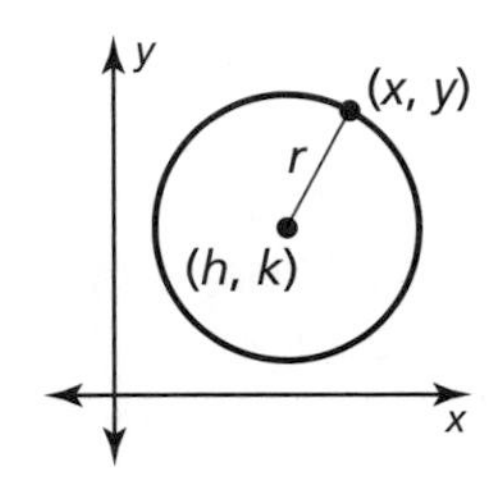

a. $x^2 + 6x + 9 + y^2 = 25$

b. $x^2 - 4x + 4 + y^2 = 9$

c. $x^2 - 8x + 16 + y^2 + 2y + 1 = 36$

76. CRITICAL THINKING Use the diagram to complete parts (a)–(c).

a. Explain why $a^3 - b^3$ is equal to the sum of the volumes of the solids I, II, and III.

b. Write an algebraic expression for the volume of each of the three solids. Leave your expressions in factored form.

c. Use the results from part (a) and part (b) to derive the factoring pattern $a^3 - b^3$.

Maintaining Mathematical Proficiency

Reviewing what you learned in previous grades and lessons

Solve the quadratic equation by factoring. *(Section 3.1)*

77. $x^2 - x - 30 = 0$

78. $2x^2 - 10x - 72 = 0$

79. $3x^2 - 11x + 10 = 0$

80. $9x^2 - 28x + 3 = 0$

Solve the quadratic equation by completing the square. *(Section 3.3)*

81. $x^2 - 12x + 36 = 144$

82. $x^2 - 8x - 11 = 0$

83. $3x^2 + 30x + 63 = 0$

84. $4x^2 + 36x - 4 = 0$

4.1–4.4 What Did You Learn?

Core Vocabulary

polynomial, *p. 158*
polynomial function, *p. 158*
end behavior, *p. 159*
Pascal's Triangle, *p. 169*
polynomial long division, *p. 174*
synthetic division, *p. 175*
factored completely, *p. 180*
factor by grouping, *p. 181*
quadratic form, *p. 181*

Core Concepts

Section 4.1

Common Polynomial Functions, *p. 158*
End Behavior of Polynomial Functions, *p. 159*
Graphing Polynomial Functions, *p. 160*

Section 4.2

Operations with Polynomials, *p. 166*
Special Product Patterns, *p. 167*
Pascal's Triangle, *p. 169*

Section 4.3

Polynomial Long Division, *p. 174*
Synthetic Division, *p. 175*
The Remainder Theorem, *p. 176*

Section 4.4

Factoring Polynomials, *p. 180*
Special Factoring Patterns, *p. 180*
The Factor Theorem, *p. 182*

Mathematical Practices

1. Describe the entry points you used to analyze the function in Exercise 43 on page 164.
2. Describe how you maintained oversight in the process of factoring the polynomial in Exercise 49 on page 185.

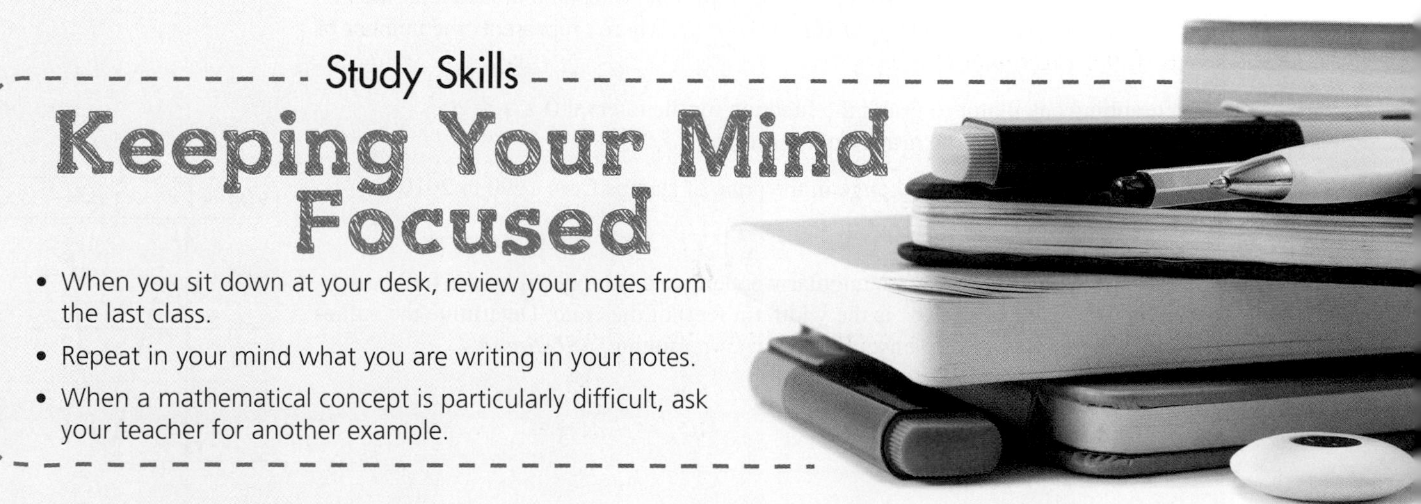

Study Skills

Keeping Your Mind Focused

- When you sit down at your desk, review your notes from the last class.
- Repeat in your mind what you are writing in your notes.
- When a mathematical concept is particularly difficult, ask your teacher for another example.

4.1–4.4 Quiz

Decide whether the function is a polynomial function. If so, write it in standard form and state its degree, type, and leading coefficient. *(Section 4.1)*

1. $f(x) = 5 + 2x^2 - 3x^4 - 2x - x^3$

2. $g(x) = \frac{1}{4}x^3 + 2x - 3x^2 + 1$

3. $h(x) = 3 - 6x^3 + 4x^{-2} + 6x$

4. Describe the x-values for which (a) f is increasing or decreasing, (b) $f(x) > 0$, and (c) $f(x) < 0$. *(Section 4.1)*

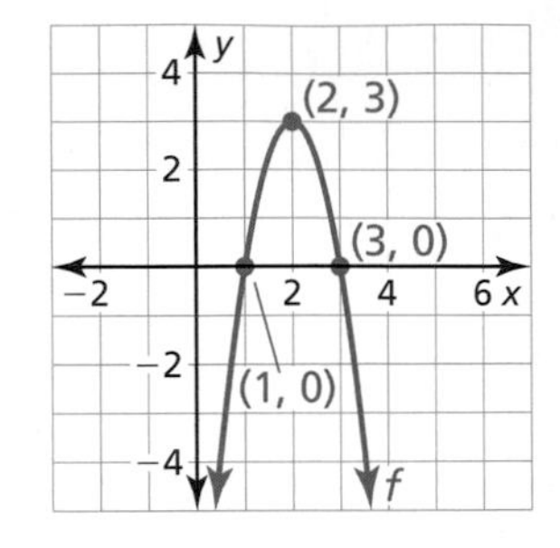

5. Write an expression for the area and perimeter for the figure shown. *(Section 4.2)*

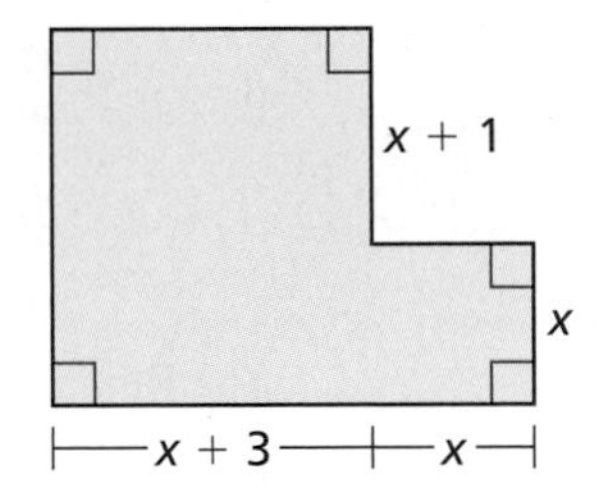

Perform the indicated operation. *(Section 4.2)*

6. $(7x^2 - 4) - (3x^2 - 5x + 1)$

7. $(x^2 - 3x + 2)(3x - 1)$

8. $(x - 1)(x + 3)(x - 4)$

9. Use Pascal's Triangle to expand $(x + 2)^5$. *(Section 4.2)*

10. Divide $4x^4 - 2x^3 + x^2 - 5x + 8$ by $x^2 - 2x - 1$. *(Section 4.3)*

Factor the polynomial completely. *(Section 4.4)*

11. $a^3 - 2a^2 - 8a$

12. $8m^3 + 27$

13. $z^3 + z^2 - 4z - 4$

14. $49b^4 - 64$

15. Show that $x + 5$ is a factor of $f(x) = x^3 - 2x^2 - 23x + 60$. Then factor $f(x)$ completely. *(Section 4.4)*

16. The estimated price P (in cents) of stamps in the United States can be modeled by the polynomial function $P(t) = 0.007t^3 - 0.16t^2 + 1t + 17$, where t represents the number of years since 1990. *(Section 4.1)*

a. Use a graphing calculator to graph the function for the interval $0 \leq t \leq 20$. Describe the behavior of the graph on this interval.

b. What was the average rate of change in the price of stamps from 1990 to 2010?

17. The volume V (in cubic feet) of a rectangular wooden crate is modeled by the function $V(x) = 2x^3 - 11x^2 + 12x$, where x is the width (in feet) of the crate. Determine the values of x for which the model makes sense. Explain your reasoning. *(Section 4.4)*

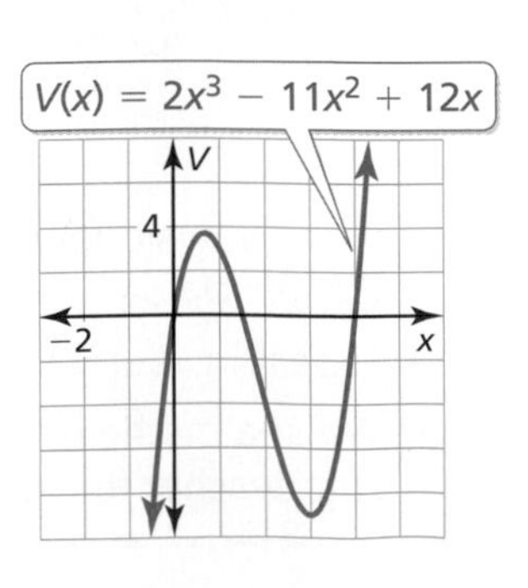

4.5 Solving Polynomial Equations

Essential Question

How can you determine whether a polynomial equation has a repeated solution?

EXPLORATION 1 Cubic Equations and Repeated Solutions

USING TOOLS STRATEGICALLY

To be proficient in math, you need to use technological tools to explore and deepen your understanding of concepts.

Work with a partner. Some cubic equations have three distinct solutions. Others have repeated solutions. Match each cubic polynomial equation with the graph of its related polynomial function. Then solve each equation. For those equations that have repeated solutions, describe the behavior of the related function near the repeated zero using the graph or a table of values.

a. $x^3 - 6x^2 + 12x - 8 = 0$

b. $x^3 + 3x^2 + 3x + 1 = 0$

c. $x^3 - 3x + 2 = 0$

d. $x^3 + x^2 - 2x = 0$

e. $x^3 - 3x - 2 = 0$

f. $x^3 - 3x^2 + 2x = 0$

A.

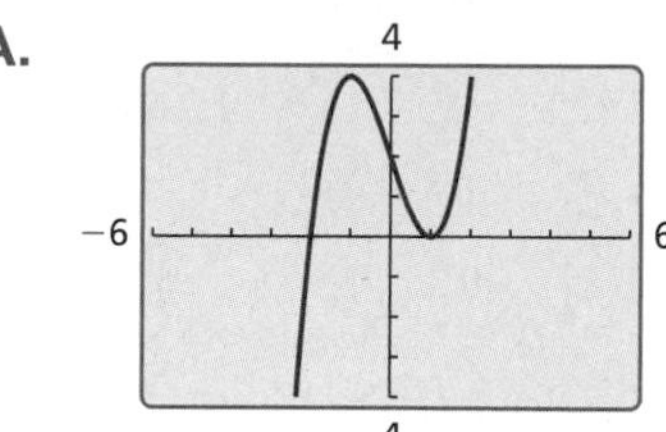

B.

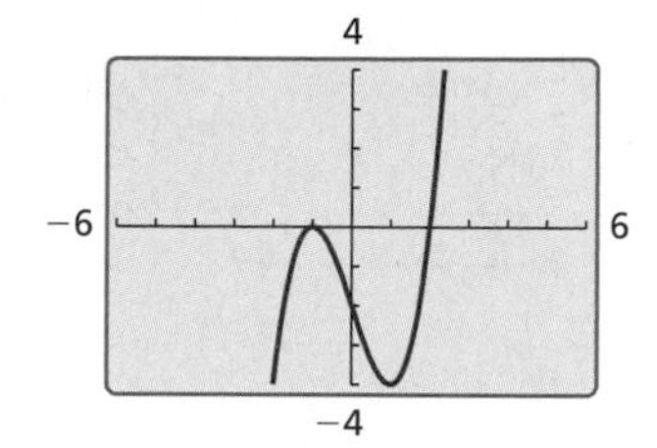

C.

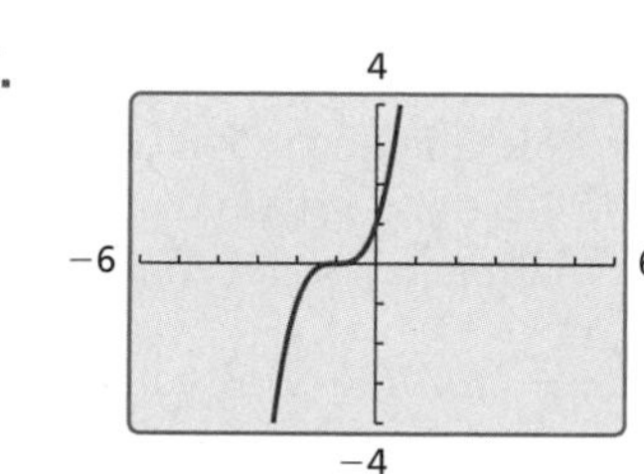

D.

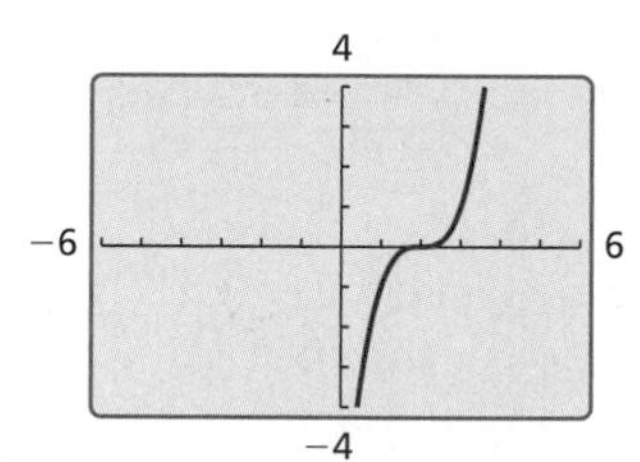

E.

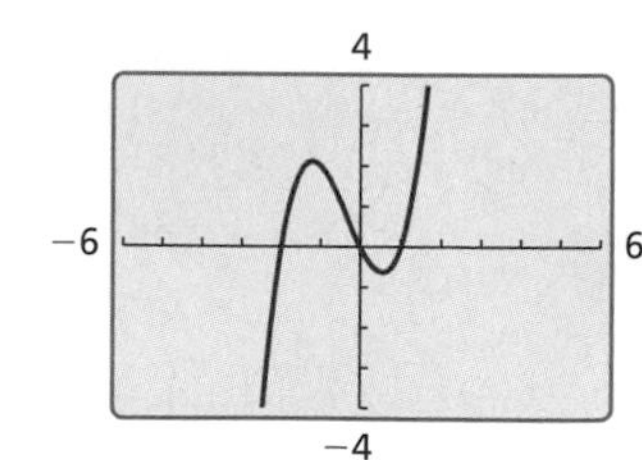

F.

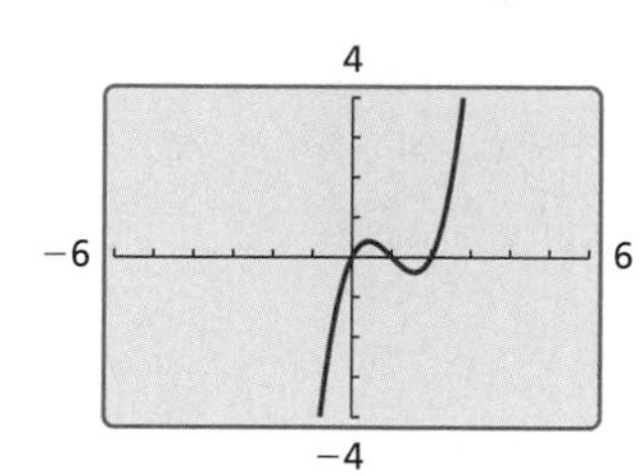

EXPLORATION 2 Quartic Equations and Repeated Solutions

Work with a partner. Determine whether each quartic equation has repeated solutions using the graph of the related quartic function or a table of values. Explain your reasoning. Then solve each equation.

a. $x^4 - 4x^3 + 5x^2 - 2x = 0$

b. $x^4 - 2x^3 - x^2 + 2x = 0$

c. $x^4 - 4x^3 + 4x^2 = 0$

d. $x^4 + 3x^3 = 0$

Communicate Your Answer

3. How can you determine whether a polynomial equation has a repeated solution?

4. Write a cubic or a quartic polynomial equation that is different from the equations in Explorations 1 and 2 and has a repeated solution.

4.5 Lesson

Core Vocabulary

repeated solution, *p. 190*

Previous
roots of an equation
real numbers
conjugates

What You Will Learn

- Find solutions of polynomial equations and zeros of polynomial functions.
- Use the Rational Root Theorem.
- Use the Irrational Conjugates Theorem.

Finding Solutions and Zeros

You have used the Zero-Product Property to solve factorable quadratic equations. You can extend this technique to solve some higher-degree polynomial equations.

EXAMPLE 1 Solving a Polynomial Equation by Factoring

Solve $2x^3 - 12x^2 + 18x = 0$.

SOLUTION

$2x^3 - 12x^2 + 18x = 0$	Write the equation.
$2x(x^2 - 6x + 9) = 0$	Factor common monomial.
$2x(x - 3)^2 = 0$	Perfect Square Trinomial Pattern
$2x = 0$ or $(x - 3)^2 = 0$	Zero-Product Property
$x = 0$ or $x = 3$	Solve for x.

▶ The solutions, or roots, are $x = 0$ and $x = 3$.

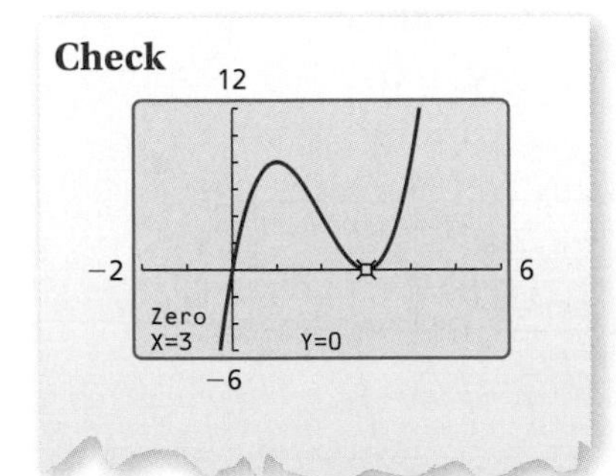

In Example 1, the factor $x - 3$ appears more than once. This creates a **repeated solution** of $x = 3$. Note that the graph of the related function touches the x-axis (but does not cross the x-axis) at the repeated zero $x = 3$, and crosses the x-axis at the zero $x = 0$. This concept can be generalized as follows.

- When a factor $x - k$ of $f(x)$ is raised to an odd power, the graph of f *crosses* the x-axis at $x = k$.
- When a factor $x - k$ of $f(x)$ is raised to an even power, the graph of f *touches* the x-axis (but does not cross the x-axis) at $x = k$.

STUDY TIP

Because the factor $x - 3$ appears twice, the root $x = 3$ has a *multiplicity* of 2.

EXAMPLE 2 Finding Zeros of a Polynomial Function

Find the zeros of $f(x) = -2x^4 + 16x^2 - 32$. Then sketch a graph of the function.

SOLUTION

$0 = -2x^4 + 16x^2 - 32$	Set $f(x)$ equal to 0.
$0 = -2(x^4 - 8x^2 + 16)$	Factor out −2.
$0 = -2(x^2 - 4)(x^2 - 4)$	Factor trinomial in quadratic form.
$0 = -2(x + 2)(x - 2)(x + 2)(x - 2)$	Difference of Two Squares Pattern
$0 = -2(x + 2)^2(x - 2)^2$	Rewrite using exponents.

Because both factors $x + 2$ and $x - 2$ are raised to an even power, the graph of f touches the x-axis at the zeros $x = -2$ and $x = 2$.

By analyzing the original function, you can determine that the y-intercept is −32. Because the degree is even and the leading coefficient is negative, $f(x) \to -\infty$ as $x \to -\infty$ and $f(x) \to -\infty$ as $x \to +\infty$. Use these characteristics to sketch a graph of the function.

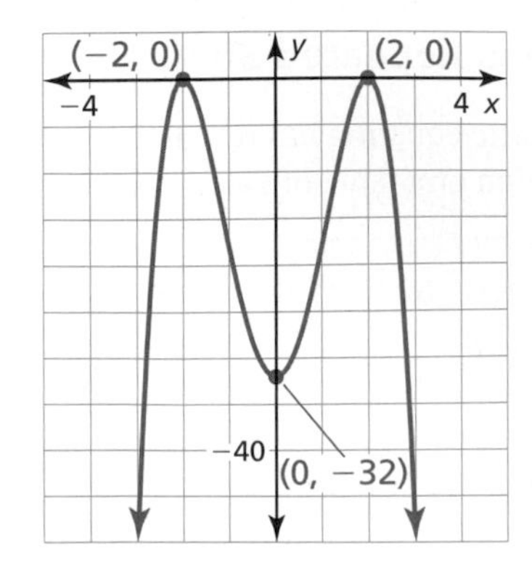

Monitoring Progress  Help in English and Spanish at *BigIdeasMath.com*

Solve the equation.

1. $4x^4 - 40x^2 + 36 = 0$

2. $2x^5 + 24x = 14x^3$

Find the zeros of the function. Then sketch a graph of the function.

3. $f(x) = 3x^4 - 6x^2 + 3$

4. $f(x) = x^3 + x^2 - 6x$

The Rational Root Theorem

The solutions of the equation $64x^3 + 152x^2 - 62x - 105 = 0$ are $-\frac{5}{2}$, $-\frac{3}{4}$, and $\frac{7}{8}$. Notice that the numerators (5, 3, and 7) of the zeros are factors of the constant term, -105. Also notice that the denominators (2, 4, and 8) are factors of the leading coefficient, 64. These observations are generalized by the *Rational Root Theorem*.

Core Concept

The Rational Root Theorem

If $f(x) = a_nx^n + \cdots + a_1x + a_0$ has *integer* coefficients, then every rational solution of $f(x) = 0$ has the following form:

$$\frac{p}{q} = \frac{\text{factor of constant term } a_0}{\text{factor of leading coefficient } a_n}$$

STUDY TIP

Notice that you can use the Rational Root Theorem to list possible zeros of polynomial functions.

The Rational Root Theorem can be a starting point for finding solutions of polynomial equations. However, the theorem lists only *possible* solutions. In order to find the *actual* solutions, you must test values from the list of possible solutions.

EXAMPLE 3 Using the Rational Root Theorem

Find all real solutions of $x^3 - 8x^2 + 11x + 20 = 0$.

SOLUTION

The polynomial $f(x) = x^3 - 8x^2 + 11x + 20$ is not easily factorable. Begin by using the Rational Root Theorem.

ANOTHER WAY

You can use direct substitution to test possible solutions, but synthetic division helps you identify other factors of the polynomial.

Step 1 List the possible rational solutions. The leading coefficient of $f(x)$ is 1 and the constant term is 20. So, the possible rational solutions of $f(x) = 0$ are

$$x = \pm\frac{1}{1}, \pm\frac{2}{1}, \pm\frac{4}{1}, \pm\frac{5}{1}, \pm\frac{10}{1}, \pm\frac{20}{1}.$$

Step 2 Test possible solutions using synthetic division until a solution is found.

Test $x = 1$:

$$\begin{array}{r|rrrr} 1 & 1 & -8 & 11 & 20 \\ & & 1 & -7 & 4 \\ \hline & 1 & -7 & 4 & 24 \end{array}$$

$f(1) \neq 0$, so $x - 1$ is not a factor of $f(x)$.

Test $x = -1$:

$$\begin{array}{r|rrrr} -1 & 1 & -8 & 11 & 20 \\ & & -1 & 9 & -20 \\ \hline & 1 & -9 & 20 & 0 \end{array}$$

$f(-1) = 0$, so $x + 1$ is a factor of $f(x)$.

Step 3 Factor completely using the result of the synthetic division.

$(x + 1)(x^2 - 9x + 20) = 0$ Write as a product of factors.

$(x + 1)(x - 4)(x - 5) = 0$ Factor the trinomial.

▶ So, the solutions are $x = -1$, $x = 4$, and $x = 5$.

In Example 3, the leading coefficient of the polynomial is 1. When the leading coefficient is not 1, the list of possible rational solutions or zeros can increase dramatically. In such cases, the search can be shortened by using a graph.

EXAMPLE 4 Finding Zeros of a Polynomial Function

Find all real zeros of $f(x) = 10x^4 - 11x^3 - 42x^2 + 7x + 12$.

SOLUTION

Step 1 List the possible rational zeros of f: $\pm\frac{1}{1}, \pm\frac{2}{1}, \pm\frac{3}{1}, \pm\frac{4}{1}, \pm\frac{6}{1}, \pm\frac{12}{1}$, $\pm\frac{1}{2}, \pm\frac{3}{2}, \pm\frac{1}{5}, \pm\frac{2}{5}, \pm\frac{3}{5}, \pm\frac{4}{5}, \pm\frac{6}{5}, \pm\frac{12}{5}, \pm\frac{1}{10}, \pm\frac{3}{10}$

Step 2 Choose reasonable values from the list above to test using the graph of the function. For f, the values

$$x = -\frac{3}{2}, x = -\frac{1}{2}, x = \frac{3}{5}, \text{ and } x = \frac{12}{5}$$

are reasonable based on the graph shown at the right.

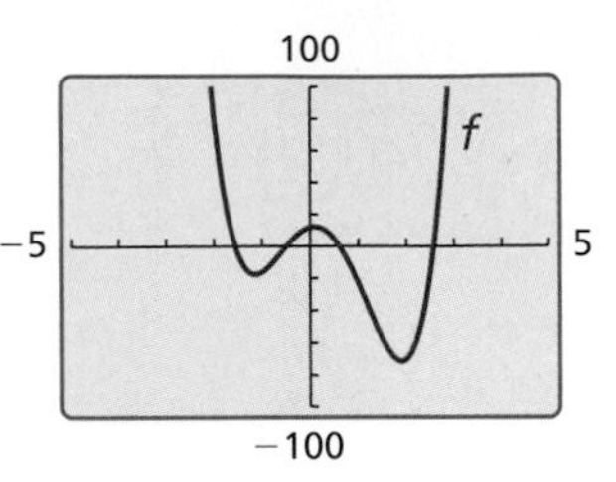

Step 3 Test the values using synthetic division until a zero is found.

$$\begin{array}{r|rrrrr} -\frac{3}{2} & 10 & -11 & -42 & 7 & 12 \\ & & -15 & 39 & \frac{9}{2} & -\frac{69}{4} \\ \hline & 10 & -26 & -3 & \frac{23}{2} & -\frac{21}{4} \end{array}$$

$$\begin{array}{r|rrrrr} -\frac{1}{2} & 10 & -11 & -42 & 7 & 12 \\ & & -5 & 8 & 17 & -12 \\ \hline & 10 & -16 & -34 & 24 & 0 \end{array}$$

$-\frac{1}{2}$ is a zero.

Step 4 Factor out a binomial using the result of the synthetic division.

$f(x) = \left(x + \frac{1}{2}\right)(10x^3 - 16x^2 - 34x + 24)$ Write as a product of factors.

$= \left(x + \frac{1}{2}\right)(2)(5x^3 - 8x^2 - 17x + 12)$ Factor 2 out of the second factor.

$= (2x + 1)(5x^3 - 8x^2 - 17x + 12)$ Multiply the first factor by 2.

Step 5 Repeat the steps above for $g(x) = 5x^3 - 8x^2 - 17x + 12$. Any zero of g will also be a zero of f. The possible rational zeros of g are:

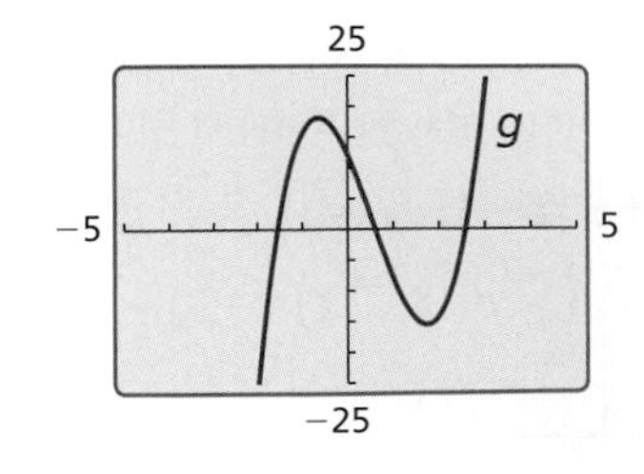

$$x = \pm 1, \pm 2, \pm 3, \pm 4, \pm 6, \pm 12, \pm\frac{1}{5}, \pm\frac{2}{5}, \pm\frac{3}{5}, \pm\frac{4}{5}, \pm\frac{6}{5}, \pm\frac{12}{5}$$

The graph of g shows that $\frac{3}{5}$ may be a zero. Synthetic division shows that $\frac{3}{5}$ *is* a zero and $g(x) = \left(x - \frac{3}{5}\right)(5x^2 - 5x - 20) = (5x - 3)(x^2 - x - 4)$. It follows that:

$$f(x) = (2x + 1) \cdot g(x) = (2x + 1)(5x - 3)(x^2 - x - 4)$$

Step 6 Find the remaining zeros of f by solving $x^2 - x - 4 = 0$.

$x = \frac{-(-1) \pm \sqrt{(-1)^2 - 4(1)(-4)}}{2(1)}$ Substitute 1 for a, -1 for b, and -4 for c in the Quadratic Formula.

$x = \frac{1 \pm \sqrt{17}}{2}$ Simplify.

▶ The real zeros of f are $-\frac{1}{2}, \frac{3}{5}, \frac{1 + \sqrt{17}}{2} \approx 2.56$, and $\frac{1 - \sqrt{17}}{2} \approx -1.56$.

Monitoring Progress  Help in English and Spanish at *BigIdeasMath.com*

5. Find all real solutions of $x^3 - 5x^2 - 2x + 24 = 0$.

6. Find all real zeros of $f(x) = 3x^4 - 2x^3 - 37x^2 + 24x + 12$.

The Irrational Conjugates Theorem

In Example 4, notice that the irrational zeros are *conjugates* of the form $a + \sqrt{b}$ and $a - \sqrt{b}$. This illustrates the theorem below.

Core Concept

The Irrational Conjugates Theorem

Let f be a polynomial function with rational coefficients, and let a and b be rational numbers such that $\sqrt{b}$ is irrational. If $a + \sqrt{b}$ is a zero of f, then $a - \sqrt{b}$ is also a zero of f.

EXAMPLE 5 Using Zeros to Write a Polynomial Function

Write a polynomial function f of least degree that has rational coefficients, a leading coefficient of 1, and the zeros 3 and $2 + \sqrt{5}$.

SOLUTION

Because the coefficients are rational and $2 + \sqrt{5}$ is a zero, $2 - \sqrt{5}$ must also be a zero by the Irrational Conjugates Theorem. Use the three zeros and the Factor Theorem to write $f(x)$ as a product of three factors.

$f(x) = (x - 3)[x - (2 + \sqrt{5})][x - (2 - \sqrt{5})]$	Write $f(x)$ in factored form.
$= (x - 3)[(x - 2) - \sqrt{5}][(x - 2) + \sqrt{5}]$	Regroup terms.
$= (x - 3)[(x - 2)^2 - 5]$	Multiply.
$= (x - 3)[(x^2 - 4x + 4) - 5]$	Expand binomial.
$= (x - 3)(x^2 - 4x - 1)$	Simplify.
$= x^3 - 4x^2 - x - 3x^2 + 12x + 3$	Multiply.
$= x^3 - 7x^2 + 11x + 3$	Combine like terms.

Check

You can check this result by evaluating f at each of its three zeros.

$f(3) = 3^3 - 7(3)^2 + 11(3) + 3 = 27 - 63 + 33 + 3 = 0$ ✓

$f(2 + \sqrt{5}) = (2 + \sqrt{5})^3 - 7(2 + \sqrt{5})^2 + 11(2 + \sqrt{5}) + 3$

$= 38 + 17\sqrt{5} - 63 - 28\sqrt{5} + 22 + 11\sqrt{5} + 3$

$= 0$ ✓

Because $f(2 + \sqrt{5}) = 0$, by the Irrational Conjugates Theorem $f(2 - \sqrt{5}) = 0$. ✓

Monitoring Progress  Help in English and Spanish at *BigIdeasMath.com*

7. Write a polynomial function f of least degree that has rational coefficients, a leading coefficient of 1, and the zeros 4 and $1 - \sqrt{5}$.

4.5 Exercises

Dynamic Solutions available at *BigIdeasMath.com*

Vocabulary and Core Concept Check

1. **COMPLETE THE SENTENCE** If a polynomial function f has integer coefficients, then every rational solution of $f(x) = 0$ has the form $\frac{p}{q}$, where p is a factor of the ____________ and q is a factor of the ____________.

2. **DIFFERENT WORDS, SAME QUESTION** Which is different? Find "both" answers.

Find the y-intercept of the graph of $y = x^3 - 2x^2 - x + 2$.	Find the x-intercepts of the graph of $y = x^3 - 2x^2 - x + 2$.
Find all the real solutions of $x^3 - 2x^2 - x + 2 = 0$.	Find the real zeros of $f(x) = x^3 - 2x^2 - x + 2$.

Monitoring Progress and Modeling with Mathematics

In Exercises 3–12, solve the equation. *(See Example 1.)*

3. $z^3 - z^2 - 12z = 0$
4. $a^3 - 4a^2 + 4a = 0$
5. $2x^4 - 4x^3 = -2x^2$
6. $v^3 - 2v^2 - 16v = -32$
7. $5w^3 = 50w$
8. $9m^5 = 27m^3$
9. $2c^4 - 6c^3 = 12c^2 - 36c$
10. $p^4 + 40 = 14p^2$
11. $12n^2 + 48n = -n^3 - 64$
12. $y^3 - 27 = 9y^2 - 27y$

In Exercises 13–20, find the zeros of the function. Then sketch a graph of the function. *(See Example 2.)*

13. $h(x) = x^4 + x^3 - 6x^2$
14. $f(x) = x^4 - 18x^2 + 81$
15. $p(x) = x^6 - 11x^5 + 30x^4$
16. $g(x) = -2x^5 + 2x^4 + 40x^3$
17. $g(x) = -4x^4 + 8x^3 + 60x^2$
18. $h(x) = -x^3 - 2x^2 + 15x$
19. $h(x) = -x^3 - x^2 + 9x + 9$
20. $p(x) = x^3 - 5x^2 - 4x + 20$

21. **USING EQUATIONS** According to the Rational Root Theorem, which is *not* a possible solution of the equation $2x^4 - 5x^3 + 10x^2 - 9 = 0$?

    Ⓐ -9 Ⓑ $-\frac{1}{2}$ Ⓒ $\frac{5}{2}$ Ⓓ 3

22. **USING EQUATIONS** According to the Rational Root Theorem, which is *not* a possible zero of the function $f(x) = 40x^5 - 42x^4 - 107x^3 + 107x^2 + 33x - 36$?

    Ⓐ $-\frac{2}{3}$ Ⓑ $-\frac{3}{8}$ Ⓒ $\frac{3}{4}$ Ⓓ $\frac{4}{5}$

ERROR ANALYSIS In Exercises 23 and 24, describe and correct the error in listing the possible rational zeros of the function.

23.

$f(x) = x^3 + 5x^2 - 9x - 45$

Possible rational zeros of f:

$1, 3, 5, 9, 15, 45$

24.

$f(x) = 3x^3 + 13x^2 - 41x + 8$

Possible rational zeros of f:

$\pm1, \pm3, \pm\frac{1}{2}, \pm\frac{1}{4}, \pm\frac{1}{8}, \pm\frac{3}{2}, \pm\frac{3}{4}, \pm\frac{3}{8}$

In Exercises 25–32, find all the real solutions of the equation. *(See Example 3.)*

25. $x^3 + x^2 - 17x + 15 = 0$
26. $x^3 - 2x^2 - 5x + 6 = 0$

27. $x^3 - 10x^2 + 19x + 30 = 0$

28. $x^3 + 4x^2 - 11x - 30 = 0$

29. $x^3 - 6x^2 - 7x + 60 = 0$

30. $x^3 - 16x^2 + 55x + 72 = 0$

31. $2x^3 - 3x^2 - 50x - 24 = 0$

32. $3x^3 + x^2 - 38x + 24 = 0$

In Exercises 33–38, find all the real zeros of the function. *(See Example 4.)*

33. $f(x) = x^3 - 2x^2 - 23x + 60$

34. $g(x) = x^3 - 28x - 48$

35. $h(x) = x^3 + 10x^2 + 31x + 30$

36. $f(x) = x^3 - 14x^2 + 55x - 42$

37. $p(x) = 2x^3 - x^2 - 27x + 36$

38. $g(x) = 3x^3 - 25x^2 + 58x - 40$

USING TOOLS In Exercises 39 and 40, use the graph to shorten the list of possible rational zeros of the function. Then find all real zeros of the function.

39. $f(x) = 4x^3 - 20x + 16$

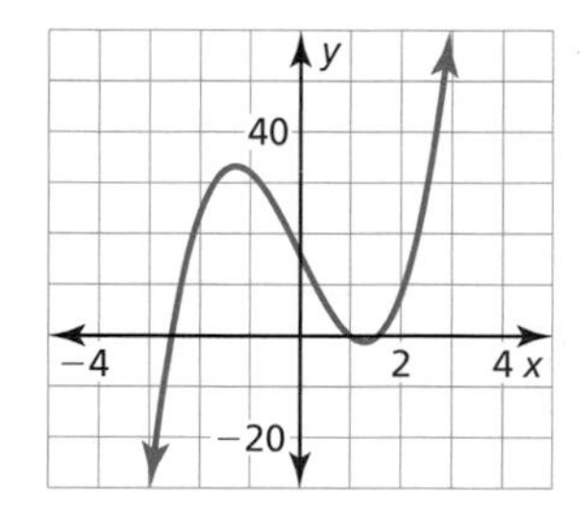

40. $f(x) = 4x^3 - 49x - 60$

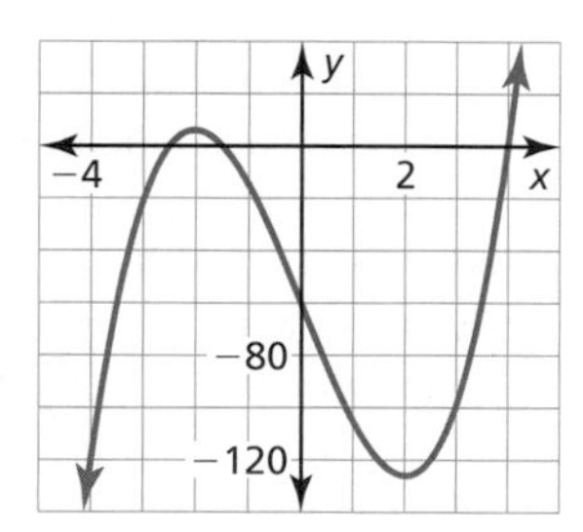

In Exercises 41–46, write a polynomial function *f* of least degree that has a leading coefficient of 1 and the given zeros. *(See Example 5.)*

41. $-2, 3, 6$

42. $-4, -2, 5$

43. $-2, 1 + \sqrt{7}$

44. $4, 6 - \sqrt{7}$

45. $-6, 0, 3 - \sqrt{5}$

46. $0, 5, -5 + \sqrt{8}$

47. **COMPARING METHODS** Solve the equation $x^3 - 4x^2 - 9x + 36 = 0$ using two different methods. Which method do you prefer? Explain your reasoning.

48. **REASONING** Is it possible for a cubic function to have more than three real zeros? Explain.

49. **PROBLEM SOLVING** At a factory, molten glass is poured into molds to make paperweights. Each mold is a rectangular prism with a height 3 centimeters greater than the length of each side of its square base. Each mold holds 112 cubic centimeters of glass. What are the dimensions of the mold?

50. **MATHEMATICAL CONNECTIONS** The volume of the cube shown is 8 cubic centimeters.

    a. Write a polynomial equation that you can use to find the value of x.

    b. Identify the possible rational solutions of the equation in part (a).

    c. Use synthetic division to find a rational solution of the equation. Show that no other real solutions exist.

    d. What are the dimensions of the cube?

51. **PROBLEM SOLVING** Archaeologists discovered a huge hydraulic concrete block at the ruins of Caesarea with a volume of 945 cubic meters. The block is x meters high by $12x - 15$ meters long by $12x - 21$ meters wide. What are the dimensions of the block?

52. **MAKING AN ARGUMENT** Your friend claims that when a polynomial function has a leading coefficient of 1 and the coefficients are all integers, every possible rational zero is an integer. Is your friend correct? Explain your reasoning.

53. **MODELING WITH MATHEMATICS** During a 10-year period, the amount (in millions of dollars) of athletic equipment E sold domestically can be modeled by $E(t) = -20t^3 + 252t^2 - 280t + 21{,}614$, where t is in years.

    a. Write a polynomial equation to find the year when about \$24,014,000,000 of athletic equipment is sold.

    b. List the possible whole-number solutions of the equation in part (a). Consider the domain when making your list of possible solutions.

    c. Use synthetic division to find when \$24,014,000,000 of athletic equipment is sold.

54. THOUGHT PROVOKING Write a third or fourth degree polynomial function that has zeros at $\pm\frac{3}{4}$. Justify your answer.

55. MODELING WITH MATHEMATICS You are designing a marble basin that will hold a fountain for a city park. The sides and bottom of the basin should be 1 foot thick. Its outer length should be twice its outer width and outer height. What should the outer dimensions of the basin be if it is to hold 36 cubic feet of water?

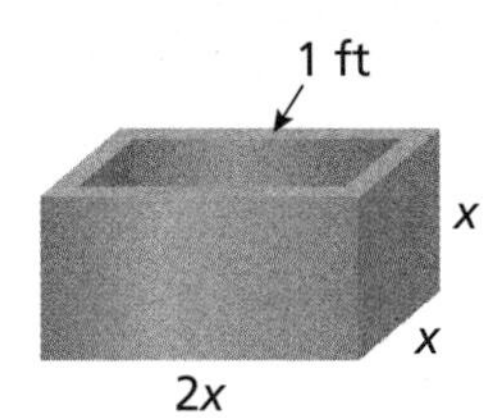

56. HOW DO YOU SEE IT? Use the information in the graph to answer the questions.

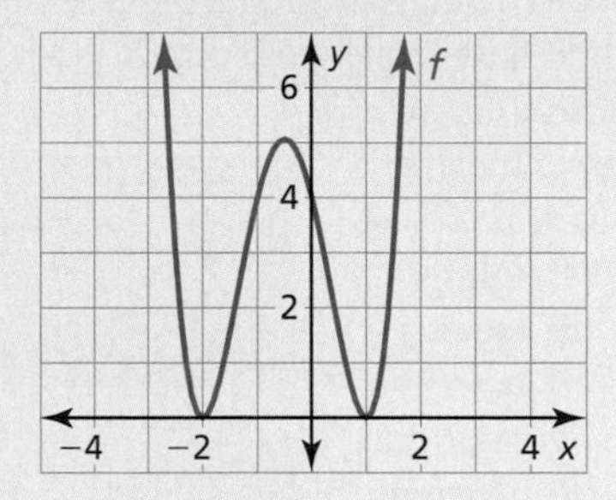

a. What are the real zeros of the function f?

b. Write an equation of the quartic function in factored form.

57. REASONING Determine the value of k for each equation so that the given x-value is a solution.

a. $x^3 - 6x^2 - 7x + k = 0;\ x = 4$

b. $2x^3 + 7x^2 - kx - 18 = 0;\ x = -6$

c. $kx^3 - 35x^2 + 19x + 30 = 0;\ x = 5$

58. WRITING EQUATIONS Write a polynomial function g of least degree that has rational coefficients, a leading coefficient of 1, and the zeros $-2 + \sqrt{7}$ and $3 + \sqrt{2}$.

In Exercises 59–62, solve $f(x) = g(x)$ by graphing and algebraic methods.

59. $f(x) = x^3 + x^2 - x - 1;\ g(x) = -x + 1$

60. $f(x) = x^4 - 5x^3 + 2x^2 + 8x;\ g(x) = -x^2 + 6x - 8$

61. $f(x) = x^3 - 4x^2 + 4x;\ g(x) = -2x + 4$

62. $f(x) = x^4 + 2x^3 - 11x^2 - 12x + 36;$
$g(x) = -x^2 - 6x - 9$

63. MODELING WITH MATHEMATICS You are building a pair of ramps for a loading platform. The left ramp is twice as long as the right ramp. If 150 cubic feet of concrete are used to build the ramps, what are the dimensions of each ramp?

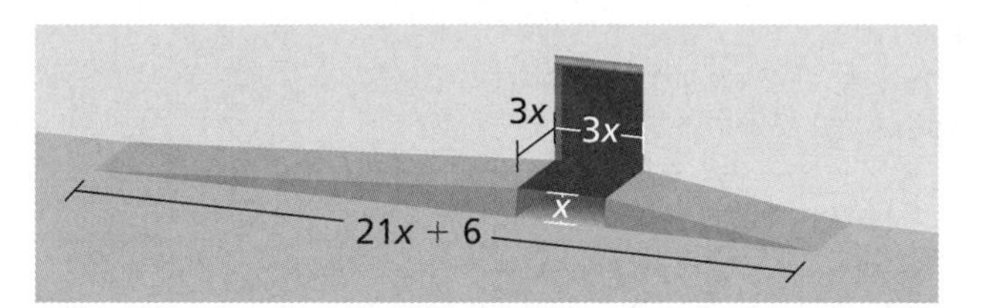

64. MODELING WITH MATHEMATICS Some ice sculptures are made by filling a mold and then freezing it. You are making an ice mold for a school dance. It is to be shaped like a pyramid with a height 1 foot greater than the length of each side of its square base. The volume of the ice sculpture is 4 cubic feet. What are the dimensions of the mold?

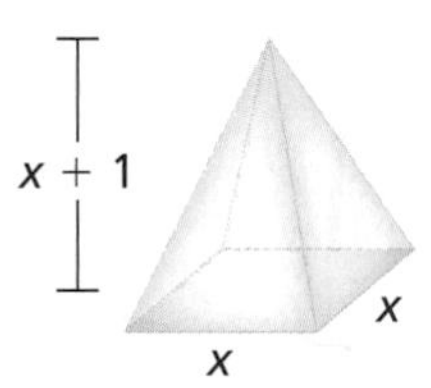

65. ABSTRACT REASONING Let a_n be the leading coefficient of a polynomial function f and a_0 be the constant term. If a_n has r factors and a_0 has s factors, what is the greatest number of possible rational zeros of f that can be generated by the Rational Zero Theorem? Explain your reasoning.

Maintaining Mathematical Proficiency

Reviewing what you learned in previous grades and lessons

Decide whether the function is a polynomial function. If so, write it in standard form and state its degree, type, and leading coefficient. *(Section 4.1)*

66. $h(x) = -3x^2 + 2x - 9 + \sqrt{4}x^3$

67. $g(x) = 2x^3 - 7x^2 - 3x^{-1} + x$

68. $f(x) = \frac{1}{3}x^2 + 2x^3 - 4x^4 - \sqrt{3}$

69. $p(x) = 2x - 5x^3 + 9x^2 + \sqrt[4]{x} + 1$

Find the zeros of the function. *(Section 3.2)*

70. $f(x) = 7x^2 + 42$

71. $g(x) = 9x^2 + 81$

72. $h(x) = 5x^2 + 40$

73. $f(x) = 8x^2 - 1$

4.6 The Fundamental Theorem of Algebra

Essential Question How can you determine whether a polynomial equation has imaginary solutions?

EXPLORATION 1 Cubic Equations and Imaginary Solutions

Work with a partner. Match each cubic polynomial equation with the graph of its related polynomial function. Then find *all* solutions. Make a conjecture about how you can use a graph or table of values to determine the number and types of solutions of a cubic polynomial equation.

a. $x^3 - 3x^2 + x + 5 = 0$

b. $x^3 - 2x^2 - x + 2 = 0$

c. $x^3 - x^2 - 4x + 4 = 0$

d. $x^3 + 5x^2 + 8x + 6 = 0$

e. $x^3 - 3x^2 + x - 3 = 0$

f. $x^3 - 3x^2 + 2x = 0$

A.

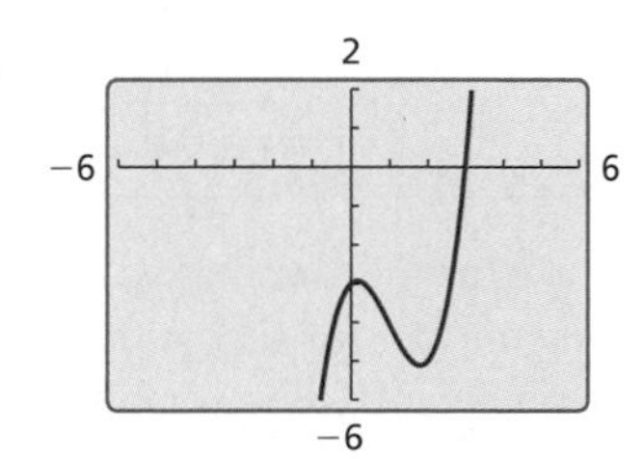

B.

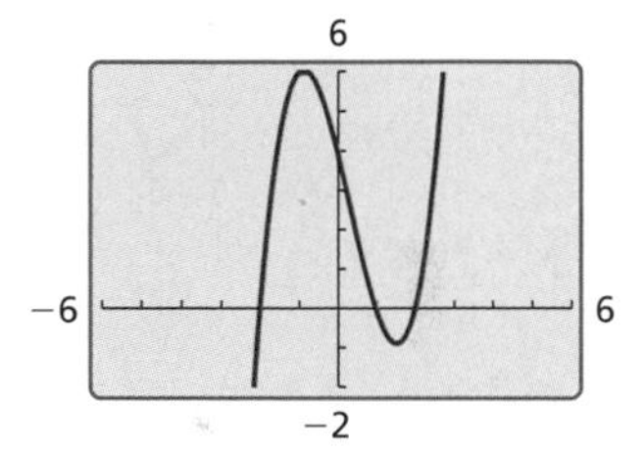

C.

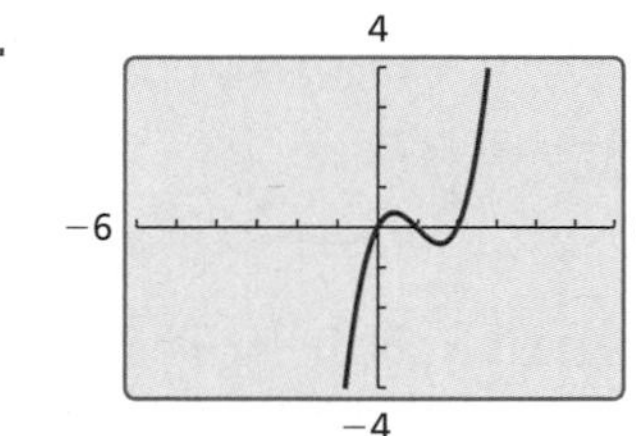

D.

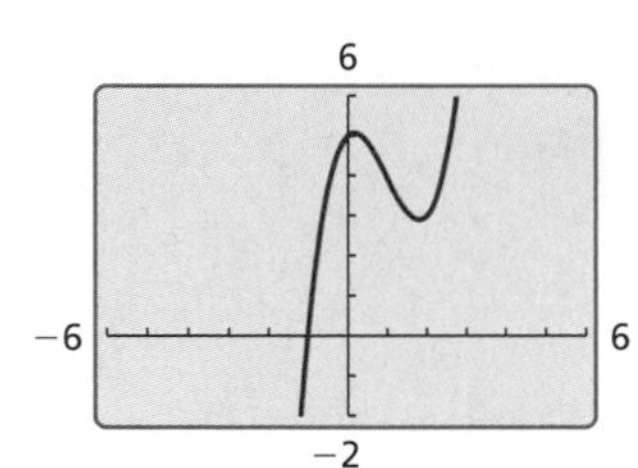

E.

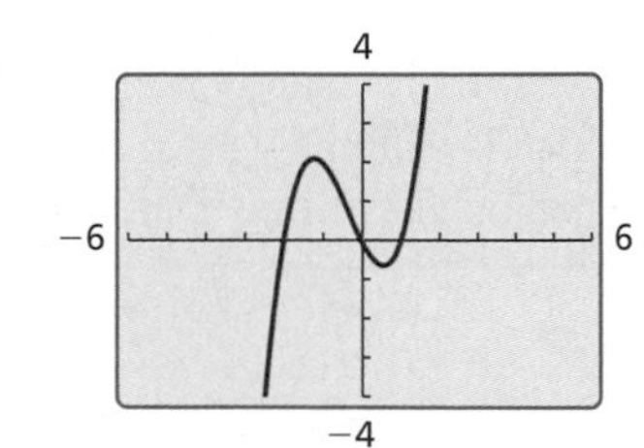

F.

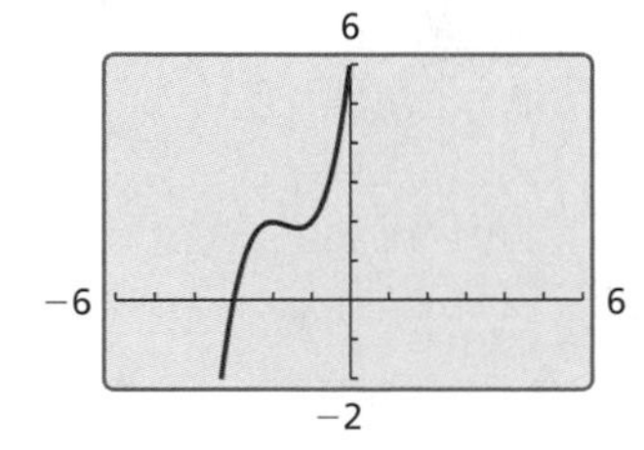

USING TOOLS STRATEGICALLY

To be proficient in math, you need to use technology to enable you to visualize results and explore consequences.

EXPLORATION 2 Quartic Equations and Imaginary Solutions

Work with a partner. Use the graph of the related quartic function, or a table of values, to determine whether each quartic equation has imaginary solutions. Explain your reasoning. Then find *all* solutions.

a. $x^4 - 2x^3 - x^2 + 2x = 0$

b. $x^4 - 1 = 0$

c. $x^4 + x^3 - x - 1 = 0$

d. $x^4 - 3x^3 + x^2 + 3x - 2 = 0$

Communicate Your Answer

3. How can you determine whether a polynomial equation has imaginary solutions?

4. Is it possible for a cubic equation to have three imaginary solutions? Explain your reasoning.

4.6 Lesson

Core Vocabulary

complex conjugates, *p. 199*

Previous
repeated solution
degree of a polynomial
solution of an equation
zero of a function
conjugates

What You Will Learn

- Use the Fundamental Theorem of Algebra.
- Find conjugate pairs of complex zeros of polynomial functions.
- Use Descartes's Rule of Signs.

The Fundamental Theorem of Algebra

The table shows several polynomial equations and their solutions, including repeated solutions. Notice that for the last equation, the repeated solution $x = -1$ is counted twice.

Equation	Degree	Solution(s)	Number of solutions
$2x - 1 = 0$	1	$\frac{1}{2}$	1
$x^2 - 2 = 0$	2	$\pm\sqrt{2}$	2
$x^3 - 8 = 0$	3	$2, -1 \pm i\sqrt{3}$	3
$x^3 + x^2 - x - 1 = 0$	3	$-1, -1, 1$	3

In the table, note the relationship between the degree of the polynomial $f(x)$ and the number of solutions of $f(x) = 0$. This relationship is generalized by the *Fundamental Theorem of Algebra*, first proven by German mathematician Carl Friedrich Gauss (1777–1855).

Core Concept

The Fundamental Theorem of Algebra

Theorem If $f(x)$ is a polynomial of degree n where $n > 0$, then the equation $f(x) = 0$ has at least one solution in the set of complex numbers.

Corollary If $f(x)$ is a polynomial of degree n where $n > 0$, then the equation $f(x) = 0$ has exactly n solutions provided each solution repeated twice is counted as two solutions, each solution repeated three times is counted as three solutions, and so on.

STUDY TIP

The statements "the polynomial equation $f(x) = 0$ has exactly n solutions" and "the polynomial function f has exactly n zeros" are equivalent.

The corollary to the Fundamental Theorem of Algebra also means that an nth-degree polynomial function f has exactly n zeros.

EXAMPLE 1 Finding the Number of Solutions or Zeros

a. How many solutions does the equation $x^3 + 3x^2 + 16x + 48 = 0$ have?

b. How many zeros does the function $f(x) = x^4 + 6x^3 + 12x^2 + 8x$ have?

SOLUTION

a. Because $x^3 + 3x^2 + 16x + 48 = 0$ is a polynomial equation of degree 3, it has three solutions. (The solutions are -3, $4i$, and $-4i$.)

b. Because $f(x) = x^4 + 6x^3 + 12x^2 + 8x$ is a polynomial function of degree 4, it has four zeros. (The zeros are -2, -2, -2, and 0.)

EXAMPLE 2 Finding the Zeros of a Polynomial Function

Find all zeros of $f(x) = x^5 + x^3 - 2x^2 - 12x - 8$.

SOLUTION

Step 1 Find the rational zeros of f. Because f is a polynomial function of degree 5, it has five zeros. The possible rational zeros are ± 1, ± 2, ± 4, and ± 8. Using synthetic division, you can determine that -1 is a zero repeated twice and 2 is also a zero.

Step 2 Write $f(x)$ in factored form. Dividing $f(x)$ by its known factors $x + 1$, $x + 1$, and $x - 2$ gives a quotient of $x^2 + 4$. So,

$$f(x) = (x + 1)^2(x - 2)(x^2 + 4).$$

Step 3 Find the complex zeros of f. Solving $x^2 + 4 = 0$, you get $x = \pm 2i$. This means $x^2 + 4 = (x + 2i)(x - 2i)$.

$$f(x) = (x + 1)^2(x - 2)(x + 2i)(x - 2i)$$

STUDY TIP

Notice that you can use imaginary numbers to write $(x^2 + 4)$ as $(x + 2i)(x - 2i)$. In general, $(a^2 + b^2) = (a + bi)(a - bi)$.

▶ From the factorization, there are five zeros. The zeros of f are

$-1, -1, 2, -2i$, and $2i$.

The graph of f and the real zeros are shown. Notice that only the *real* zeros appear as x-intercepts. Also, the graph of f touches the x-axis at the repeated zero $x = -1$ and crosses the x-axis at $x = 2$.

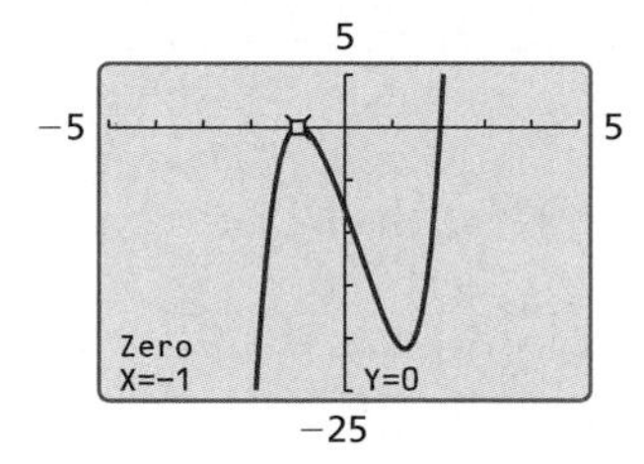

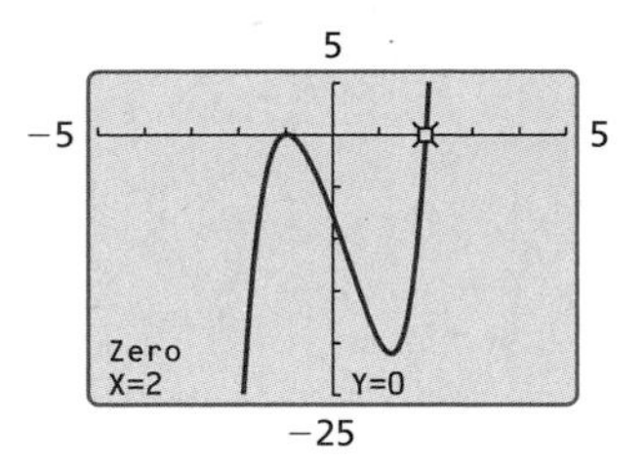

Monitoring Progress

Help in English and Spanish at *BigIdeasMath.com*

1. How many solutions does the equation $x^4 + 7x^2 - 144 = 0$ have?
2. How many zeros does the function $f(x) = x^3 - 5x^2 - 8x + 48$ have?

Find all zeros of the polynomial function.

3. $f(x) = x^3 + 7x^2 + 16x + 12$
4. $f(x) = x^5 - 3x^4 + 5x^3 - x^2 - 6x + 4$

Complex Conjugates

Pairs of complex numbers of the forms $a + bi$ and $a - bi$, where $b \neq 0$, are called **complex conjugates**. In Example 2, notice that the zeros $2i$ and $-2i$ are complex conjugates. This illustrates the next theorem.

Core Concept

The Complex Conjugates Theorem

If f is a polynomial function with real coefficients, and $a + bi$ is an imaginary zero of f, then $a - bi$ is also a zero of f.

EXAMPLE 3 Using Zeros to Write a Polynomial Function

Write a polynomial function f of least degree that has rational coefficients, a leading coefficient of 1, and the zeros 2 and $3 + i$.

SOLUTION

Because the coefficients are rational and $3 + i$ is a zero, $3 - i$ must also be a zero by the Complex Conjugates Theorem. Use the three zeros and the Factor Theorem to write $f(x)$ as a product of three factors.

$f(x) = (x - 2)[x - (3 + i)][x - (3 - i)]$	Write $f(x)$ in factored form.
$= (x - 2)[(x - 3) - i][(x - 3) + i]$	Regroup terms.
$= (x - 2)[(x - 3)^2 - i^2]$	Multiply.
$= (x - 2)[(x^2 - 6x + 9) - (-1)]$	Expand binomial and use $i^2 = -1$.
$= (x - 2)(x^2 - 6x + 10)$	Simplify.
$= x^3 - 6x^2 + 10x - 2x^2 + 12x - 20$	Multiply.
$= x^3 - 8x^2 + 22x - 20$	Combine like terms.

Check

You can check this result by evaluating f at each of its three zeros.

$f(2) = (2)^3 - 8(2)^2 + 22(2) - 20 = 8 - 32 + 44 - 20 = 0$ ✓

$f(3 + i) = (3 + i)^3 - 8(3 + i)^2 + 22(3 + i) - 20$

$= 18 + 26i - 64 - 48i + 66 + 22i - 20$

$= 0$ ✓

Because $f(3 + i) = 0$, by the Complex Conjugates Theorem $f(3 - i) = 0$. ✓

Monitoring Progress

Help in English and Spanish at *BigIdeasMath.com*

Write a polynomial function f of least degree that has rational coefficients, a leading coefficient of 1, and the given zeros.

5. $-1, 4i$ **6.** $3, 1 + i\sqrt{5}$ **7.** $\sqrt{2}, 1 - 3i$ **8.** $2, 2i, 4 - \sqrt{6}$

Descartes's Rule of Signs

French mathematician René Descartes (1596–1650) found the following relationship between the coefficients of a polynomial function and the number of positive and negative zeros of the function.

Core Concept

Descartes's Rule of Signs

Let $f(x) = a_nx^n + a_{n-1}x^{n-1} + \cdots + a_2x^2 + a_1x + a_0$ be a polynomial function with real coefficients.

- The number of *positive real zeros* of f is equal to the number of changes in sign of the coefficients of $f(x)$ or is less than this by an even number.
- The number of *negative real zeros* of f is equal to the number of changes in sign of the coefficients of $f(-x)$ or is less than this by an even number.

EXAMPLE 4 Using Descartes's Rule of Signs

Determine the possible numbers of positive real zeros, negative real zeros, and imaginary zeros for $f(x) = x^6 - 2x^5 + 3x^4 - 10x^3 - 6x^2 - 8x - 8$.

SOLUTION

$$f(x) = x^6 - 2x^5 + 3x^4 - 10x^3 - 6x^2 - 8x - 8.$$

The coefficients in $f(x)$ have 3 sign changes, so f has 3 or 1 positive real zero(s).

$$f(-x) = (-x)^6 - 2(-x)^5 + 3(-x)^4 - 10(-x)^3 - 6(-x)^2 - 8(-x) - 8$$
$$= x^6 + 2x^5 + 3x^4 + 10x^3 - 6x^2 + 8x - 8$$

The coefficients in $f(-x)$ have 3 sign changes, so f has 3 or 1 negative zero(s).

▶ The possible numbers of zeros for f are summarized in the table below.

Positive real zeros	Negative real zeros	Imaginary zeros	Total zeros
3	3	0	6
3	1	2	6
1	3	2	6
1	1	4	6

EXAMPLE 5 Real-Life Application

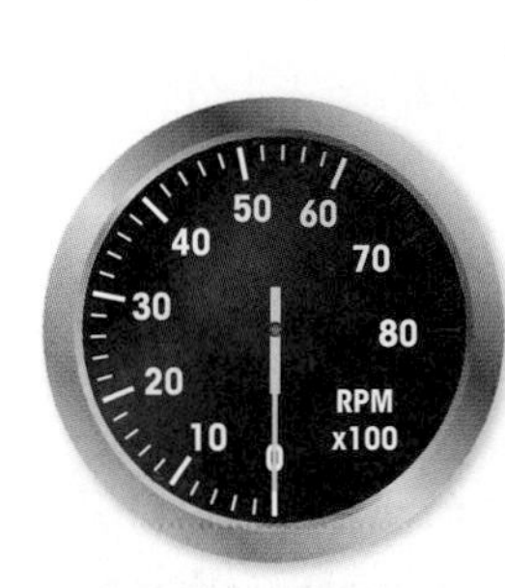

A tachometer measures the speed (in revolutions per minute, or RPMs) at which an engine shaft rotates. For a certain boat, the speed x (in hundreds of RPMs) of the engine shaft and the speed s (in miles per hour) of the boat are modeled by

$$s(x) = 0.00547x^3 - 0.225x^2 + 3.62x - 11.0.$$

What is the tachometer reading when the boat travels 15 miles per hour?

SOLUTION

Substitute 15 for $s(x)$ in the function. You can rewrite the resulting equation as

$$0 = 0.00547x^3 - 0.225x^2 + 3.62x - 26.0.$$

The related function to this equation is $f(x) = 0.00547x^3 - 0.225x^2 + 3.62x - 26.0$. By Descartes's Rule of Signs, you know f has 3 or 1 positive real zero(s). In the context of speed, negative real zeros and imaginary zeros do not make sense, so you do not need to check for them. To approximate the positive real zeros of f, use a graphing calculator. From the graph, there is 1 real zero, $x \approx 19.9$.

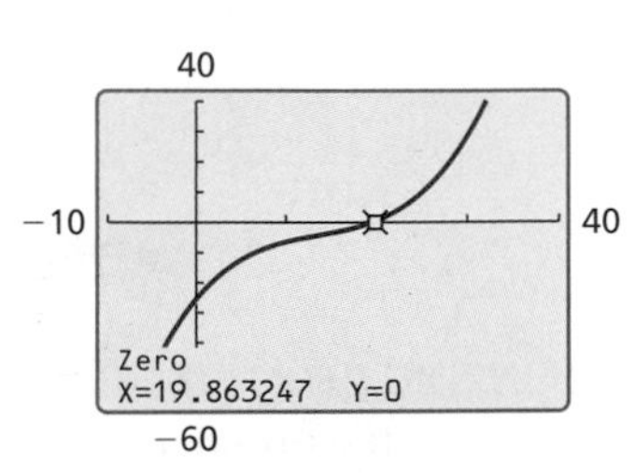

▶ The tachometer reading is about 1990 RPMs.

Monitoring Progress

Help in English and Spanish at *BigIdeasMath.com*

Determine the possible numbers of positive real zeros, negative real zeros, and imaginary zeros for the function.

9. $f(x) = x^3 + 9x - 25$

10. $f(x) = 3x^4 - 7x^3 + x^2 - 13x + 8$

11. WHAT IF? In Example 5, what is the tachometer reading when the boat travels 20 miles per hour?

4.6 Exercises

Dynamic Solutions available at *BigIdeasMath.com*

Vocabulary and Core Concept Check

1. **COMPLETE THE SENTENCE** The expressions $5 + i$ and $5 - i$ are ____________.

2. **WRITING** How many solutions does the polynomial equation $(x + 8)^3(x - 1) = 0$ have? Explain.

Monitoring Progress and Modeling with Mathematics

In Exercises 3–8, identify the number of solutions or zeros. *(See Example 1.)*

3. $x^4 + 2x^3 - 4x^2 + x = 0$

4. $5y^3 - 3y^2 + 8y = 0$

5. $9t^6 - 14t^3 + 4t - 1 = 0$

6. $f(z) = -7z^4 + z^2 - 25$

7. $g(s) = 4s^5 - s^3 + 2s^7 - 2$

8. $h(x) = 5x^4 + 7x^8 - x^{12}$

In Exercises 9–16, find all zeros of the polynomial function. *(See Example 2.)*

9. $f(x) = x^4 - 6x^3 + 7x^2 + 6x - 8$

10. $f(x) = x^4 + 5x^3 - 7x^2 - 29x + 30$

11. $g(x) = x^4 - 9x^2 - 4x + 12$

12. $h(x) = x^3 + 5x^2 - 4x - 20$

13. $g(x) = x^4 + 4x^3 + 7x^2 + 16x + 12$

14. $h(x) = x^4 - x^3 + 7x^2 - 9x - 18$

15. $g(x) = x^5 + 3x^4 - 4x^3 - 2x^2 - 12x - 16$

16. $f(x) = x^5 - 20x^3 + 20x^2 - 21x + 20$

ANALYZING RELATIONSHIPS In Exercises 17–20, determine the number of imaginary zeros for the function with the given degree and graph. Explain your reasoning.

17. Degree: 4

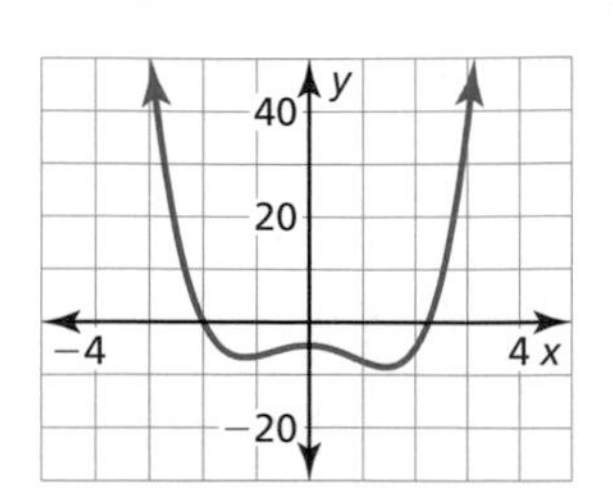

18. Degree: 5

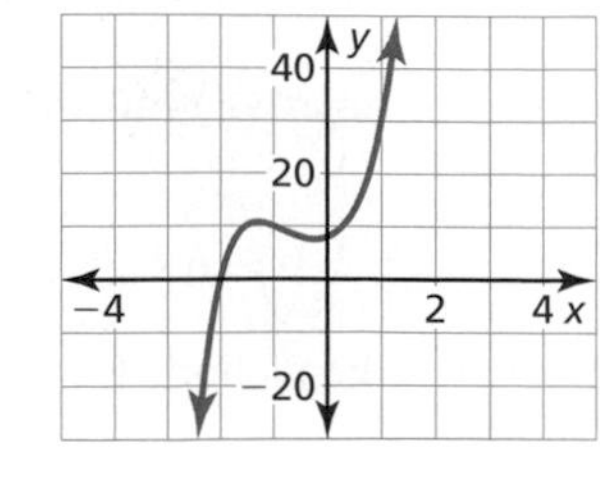

19. Degree: 2

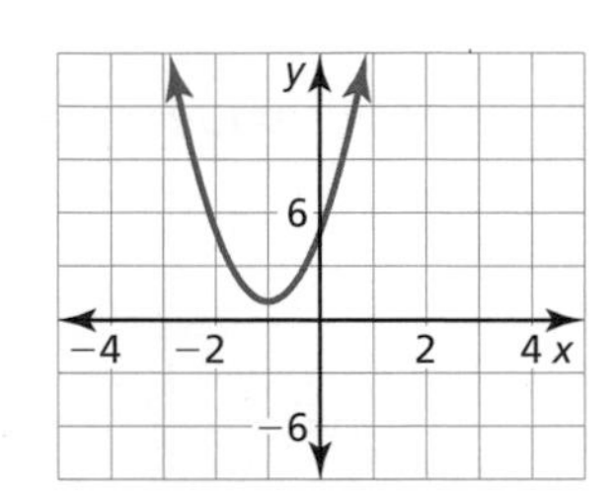

20. Degree: 3

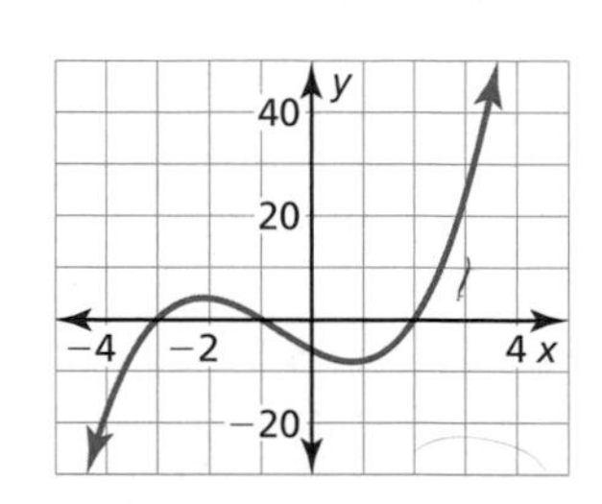

In Exercises 21–28, write a polynomial function f of least degree that has rational coefficients, a leading coefficient of 1, and the given zeros. *(See Example 3.)*

21. $-5, -1, 2$

22. $-2, 1, 3$

23. $3, 4 + i$

24. $2, 5 - i$

25. $4, -\sqrt{5}$

26. $3i, 2 - i$

27. $2, 1 + i, 2 - \sqrt{3}$

28. $3, 4 + 2i, 1 + \sqrt{7}$

ERROR ANALYSIS In Exercises 29 and 30, describe and correct the error in writing a polynomial function with rational coefficients and the given zero(s).

29. Zeros: $2, 1 + i$

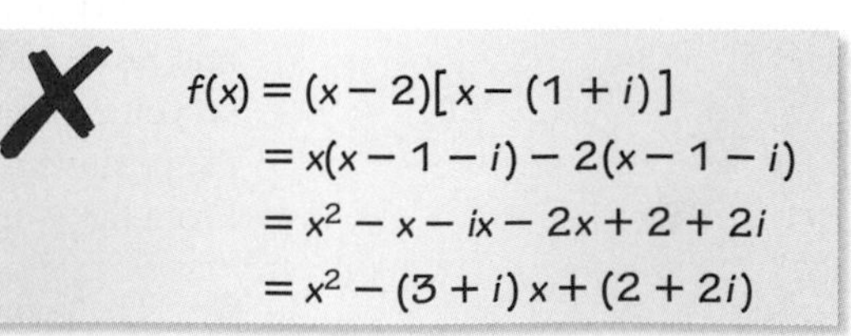

$f(x) = (x - 2)[x - (1 + i)]$
$= x(x - 1 - i) - 2(x - 1 - i)$
$= x^2 - x - ix - 2x + 2 + 2i$
$= x^2 - (3 + i)x + (2 + 2i)$

30. Zero: $2 + i$

$f(x) = [x - (2 + i)][x + (2 + i)]$
$= (x - 2 - i)(x + 2 + i)$
$= x^2 + 2x + ix - 2x - 4 - 2i - ix - 2i - i^2$
$= x^2 - 4i - 3$

31. **OPEN-ENDED** Write a polynomial function of degree 6 with zeros 1, 2, and $-i$. Justify your answer.

32. **REASONING** Two zeros of $f(x) = x^3 - 6x^2 - 16x + 96$ are 4 and -4. Explain why the third zero must also be a real number.

In Exercises 33–40, determine the possible numbers of positive real zeros, negative real zeros, and imaginary zeros for the function. *(See Example 4.)*

33. $g(x) = x^4 - x^2 - 6$

34. $g(x) = -x^3 + 5x^2 + 12$

35. $g(x) = x^3 - 4x^2 + 8x + 7$

36. $g(x) = x^5 - 2x^3 - x^2 + 6$

37. $g(x) = x^5 - 3x^3 + 8x - 10$

38. $g(x) = x^5 + 7x^4 - 4x^3 - 3x^2 + 9x - 15$

39. $g(x) = x^6 + x^5 - 3x^4 + x^3 + 5x^2 + 9x - 18$

40. $g(x) = x^7 + 4x^4 - 10x + 25$

41. **REASONING** Which is *not* a possible classification of zeros for $f(x) = x^5 - 4x^3 + 6x^2 + 2x - 6$? Explain.

Ⓐ three positive real zeros, two negative real zeros, and zero imaginary zeros

Ⓑ three positive real zeros, zero negative real zeros, and two imaginary zeros

Ⓒ one positive real zero, four negative real zeros, and zero imaginary zeros

Ⓓ one positive real zero, two negative real zeros, and two imaginary zeros

42. **USING STRUCTURE** Use Descartes's Rule of Signs to determine which function has at least 1 positive real zero.

Ⓐ $f(x) = x^4 + 2x^3 - 9x^2 - 2x - 8$

Ⓑ $f(x) = x^4 + 4x^3 + 8x^2 + 16x + 16$

Ⓒ $f(x) = -x^4 - 5x^2 - 4$

Ⓓ $f(x) = x^4 + 4x^3 + 7x^2 + 12x + 12$

43. **MODELING WITH MATHEMATICS** From 1890 to 2000, the American Indian, Eskimo, and Aleut population P (in thousands) can be modeled by the function $P = 0.004t^3 - 0.24t^2 + 4.9t + 243$, where t is the number of years since 1890. In which year did the population first reach 722,000? *(See Example 5.)*

44. **MODELING WITH MATHEMATICS** Over a period of 14 years, the number N of inland lakes infested with zebra mussels in a certain state can be modeled by

$$N = -0.0284t^4 + 0.5937t^3 - 2.464t^2 + 8.33t - 2.5$$

where t is time (in years). In which year did the number of infested inland lakes first reach 120?

45. **MODELING WITH MATHEMATICS** For the 12 years that a grocery store has been open, its annual revenue R (in millions of dollars) can be modeled by the function

$$R = 0.0001(-t^4 + 12t^3 - 77t^2 + 600t + 13{,}650)$$

where t is the number of years since the store opened. In which year(s) was the revenue \$1.5 million?

46. **MAKING AN ARGUMENT** Your friend claims that $2 - i$ is a complex zero of the polynomial function $f(x) = x^3 - 2x^2 + 2x + 5i$, but that its conjugate is *not* a zero. You claim that both $2 - i$ and its conjugate *must* be zeros by the Complex Conjugates Theorem. Who is correct? Justify your answer.

47. **MATHEMATICAL CONNECTIONS** A solid monument with the dimensions shown is to be built using 1000 cubic feet of marble. What is the value of x?

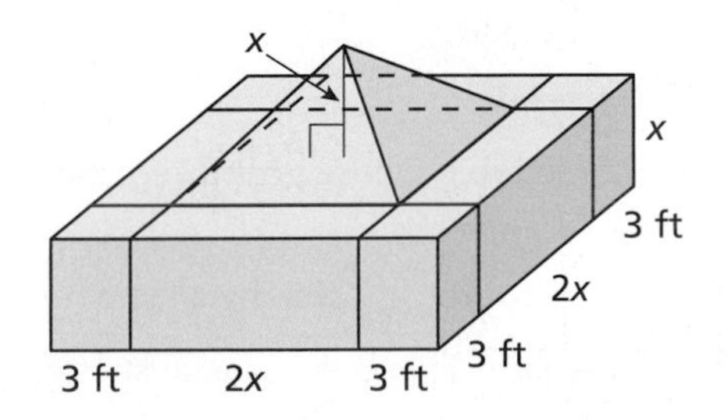

48. THOUGHT PROVOKING Write and graph a polynomial function of degree 5 that has all positive or negative real zeros. Label each x-intercept. Then write the function in standard form.

49. WRITING The graph of the constant polynomial function $f(x) = 2$ is a line that does not have any x-intercepts. Does the function contradict the Fundamental Theorem of Algebra? Explain.

50. HOW DO YOU SEE IT? The graph represents a polynomial function of degree 6.

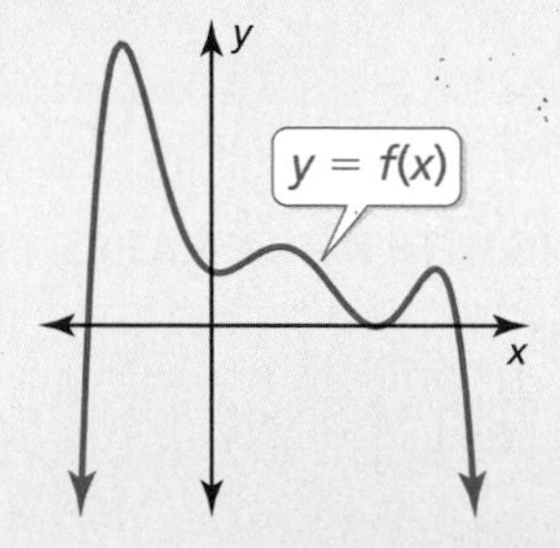

a. How many positive real zeros does the function have? negative real zeros? imaginary zeros?

b. Use Descartes's Rule of Signs and your answers in part (a) to describe the possible sign changes in the coefficients of $f(x)$.

51. FINDING A PATTERN Use a graphing calculator to graph the function $f(x) = (x + 3)^n$ for $n = 2, 3, 4, 5, 6$, and 7.

a. Compare the graphs when n is even and n is odd.

b. Describe the behavior of the graph near the zero $x = -3$ as n increases.

c. Use your results from parts (a) and (b) to describe the behavior of the graph of $g(x) = (x - 4)^{20}$ near $x = 4$.

52. DRAWING CONCLUSIONS Find the zeros of each function.

$$f(x) = x^2 - 5x + 6$$

$$g(x) = x^3 - 7x + 6$$

$$h(x) = x^4 + 2x^3 + x^2 + 8x - 12$$

$$k(x) = x^5 - 3x^4 - 9x^3 + 25x^2 - 6x$$

a. Describe the relationship between the sum of the zeros of a polynomial function and the coefficients of the polynomial function.

b. Describe the relationship between the product of the zeros of a polynomial function and the coefficients of the polynomial function.

53. PROBLEM SOLVING You want to save money so you can buy a used car in four years. At the end of each summer, you deposit \$1000 earned from summer jobs into your bank account. The table shows the value of your deposits over the four-year period. In the table, g is the growth factor $1 + r$, where r is the annual interest rate expressed as a decimal.

Deposit	Year 1	Year 2	Year 3	Year 4
1st Deposit	1000	$1000g$	$1000g^2$	$1000g^3$
2nd Deposit	–	1000		
3rd Deposit	–	–	1000	
4th Deposit	–	–	–	1000

a. Copy and complete the table.

b. Write a polynomial function that gives the value v of your account at the end of the fourth summer in terms of g.

c. You want to buy a car that costs about \$4300. What growth factor do you need to obtain this amount? What annual interest rate do you need?

Maintaining Mathematical Proficiency

Reviewing what you learned in previous grades and lessons

Describe the transformation of $f(x) = x^2$ represented by g. Then graph each function. *(Section 2.1)*

54. $g(x) = -3x^2$

55. $g(x) = (x - 4)^2 + 6$

56. $g(x) = -(x - 1)^2$

57. $g(x) = 5(x + 4)^2$

Write a function g whose graph represents the indicated transformation of the graph of f. *(Sections 1.2 and 2.1)*

58. $f(x) = x$; vertical shrink by a factor of $\frac{1}{3}$ and a reflection in the y-axis

59. $f(x) = |x + 1| - 3$; horizontal stretch by a factor of 9

60. $f(x) = x^2$; reflection in the x-axis, followed by a translation 2 units right and 7 units up

4.7 Transformations of Polynomial Functions

Essential Question

How can you transform the graph of a polynomial function?

EXPLORATION 1 Transforming the Graph of a Cubic Function

Work with a partner. The graph of the cubic function

$$f(x) = x^3$$

is shown. The graph of each cubic function g represents a transformation of the graph of f. Write a rule for g. Use a graphing calculator to verify your answers.

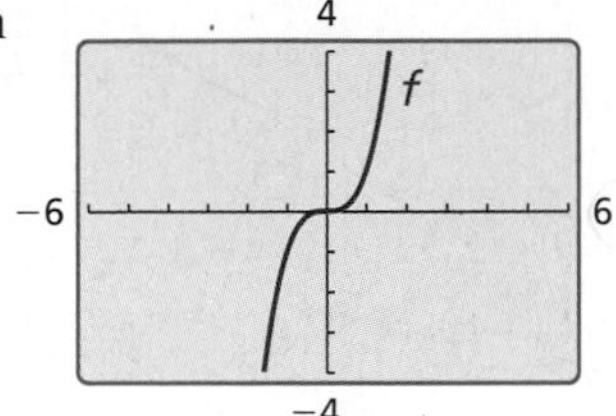

a.

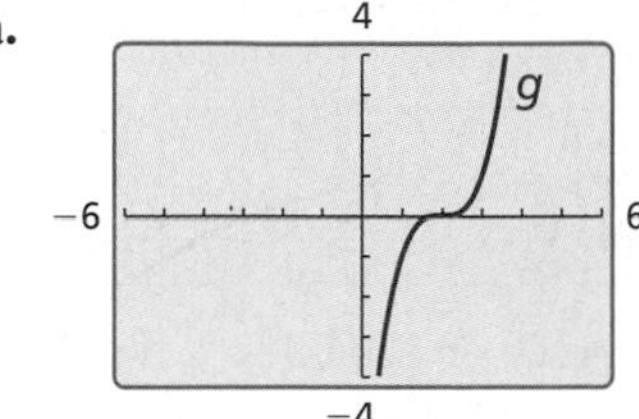

b.

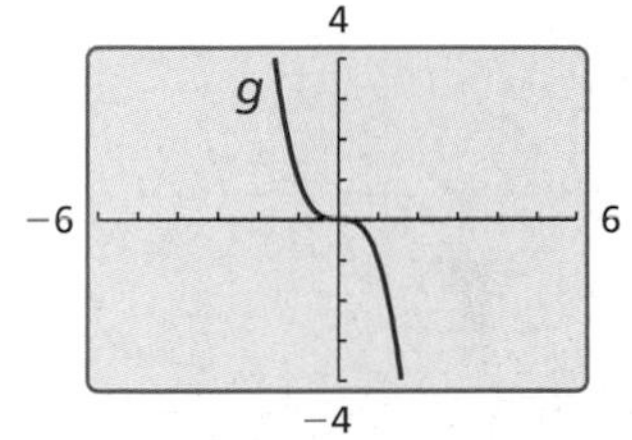

c.

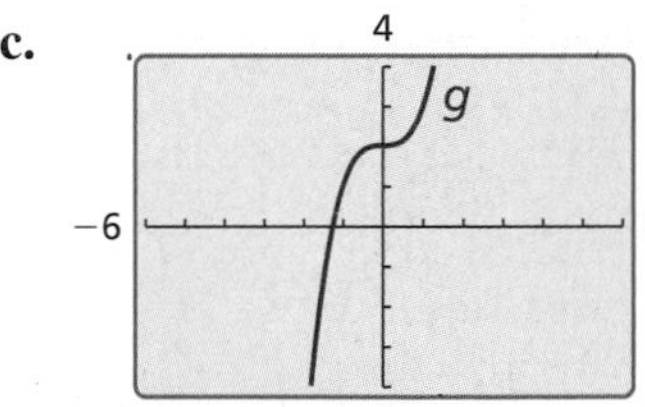

d.

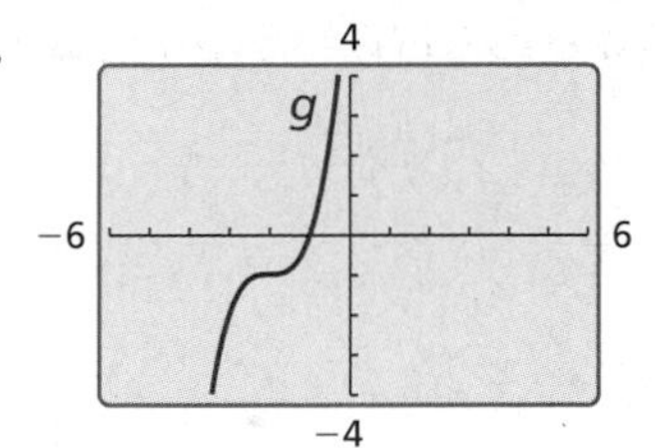

EXPLORATION 2 Transforming the Graph of a Quartic Function

Work with a partner. The graph of the quartic function

$$f(x) = x^4$$

is shown. The graph of each quartic function g represents a transformation of the graph of f. Write a rule for g. Use a graphing calculator to verify your answers.

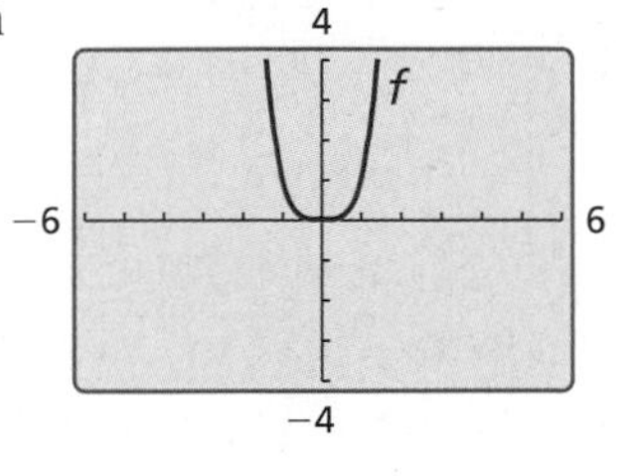

a.

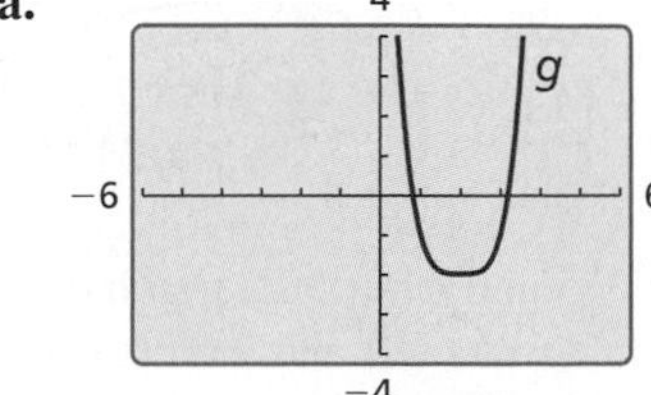

b.

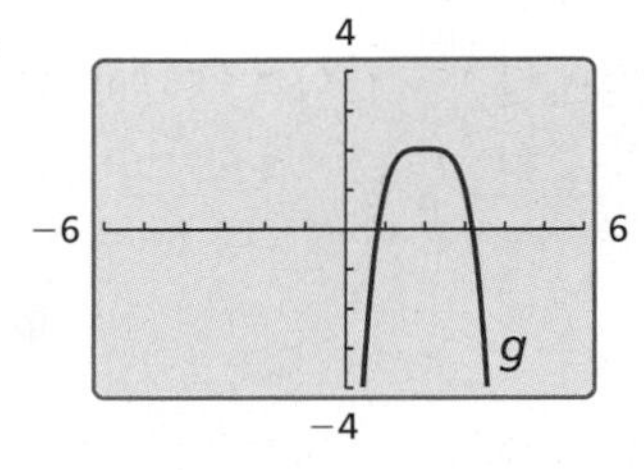

LOOKING FOR STRUCTURE

To be proficient in math, you need to see complicated things, such as some algebraic expressions, as being single objects or as being composed of several objects.

Communicate Your Answer

3. How can you transform the graph of a polynomial function?

4. Describe the transformation of $f(x) = x^4$ represented by $g(x) = (x + 1)^4 + 3$. Then graph g.

4.7 Lesson

Core Vocabulary

Previous
polynomial function
transformations

What You Will Learn

- Describe transformations of polynomial functions.
- Write transformations of polynomial functions.

Describing Transformations of Polynomial Functions

You can transform graphs of polynomial functions in the same way you transformed graphs of linear functions, absolute value functions, and quadratic functions. Examples of transformations of the graph of $f(x) = x^4$ are shown below.

Core Concept

Transformation	$f(x)$ Notation	Examples	
Horizontal Translation Graph shifts left or right.	$f(x - h)$	$g(x) = (x - 5)^4$ $g(x) = (x + 2)^4$	5 units right 2 units left
Vertical Translation Graph shifts up or down.	$f(x) + k$	$g(x) = x^4 + 1$ $g(x) = x^4 - 4$	1 unit up 4 units down
Reflection Graph flips over x- or y-axis.	$f(-x)$ $-f(x)$	$g(x) = (-x)^4 = x^4$ $g(x) = -x^4$	over y-axis over x-axis
Horizontal Stretch or Shrink Graph stretches away from or shrinks toward y-axis.	$f(ax)$	$g(x) = (2x)^4$ $g(x) = \left(\frac{1}{2}x\right)^4$	shrink by a factor of $\frac{1}{2}$ stretch by a factor of 2
Vertical Stretch or Shrink Graph stretches away from or shrinks toward x-axis.	$a \cdot f(x)$	$g(x) = 8x^4$ $g(x) = \frac{1}{4}x^4$	stretch by a factor of 8 shrink by a factor of $\frac{1}{4}$

EXAMPLE 1 Translating a Polynomial Function

Describe the transformation of $f(x) = x^3$ represented by $g(x) = (x + 5)^3 + 2$. Then graph each function.

SOLUTION

Notice that the function is of the form $g(x) = (x - h)^3 + k$. Rewrite the function to identify h and k.

$$g(x) = (x - (-5))^3 + 2$$

($h = -5$, $k = 2$)

- Because $h = -5$ and $k = 2$, the graph of g is a translation 5 units left and 2 units up of the graph of f.

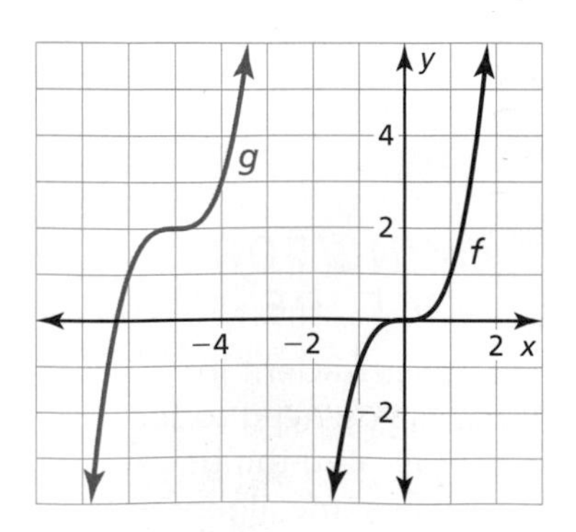

Monitoring Progress  Help in English and Spanish at BigIdeasMath.com

1. Describe the transformation of $f(x) = x^4$ represented by $g(x) = (x - 3)^4 - 1$. Then graph each function.

EXAMPLE 2 Transforming Polynomial Functions

Describe the transformation of f represented by g. Then graph each function.

a. $f(x) = x^4$, $g(x) = -\frac{1}{4}x^4$

b. $f(x) = x^5$, $g(x) = (2x)^5 - 3$

SOLUTION

a. Notice that the function is of the form $g(x) = -ax^4$, where $a = \frac{1}{4}$.

▶ So, the graph of g is a reflection in the x-axis and a vertical shrink by a factor of $\frac{1}{4}$ of the graph of f.

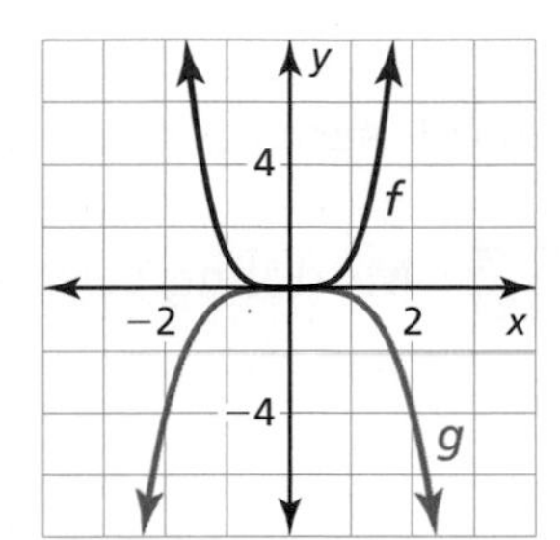

b. Notice that the function is of the form $g(x) = (ax)^5 + k$, where $a = 2$ and $k = -3$.

▶ So, the graph of g is a horizontal shrink by a factor of $\frac{1}{2}$ and a translation 3 units down of the graph of f.

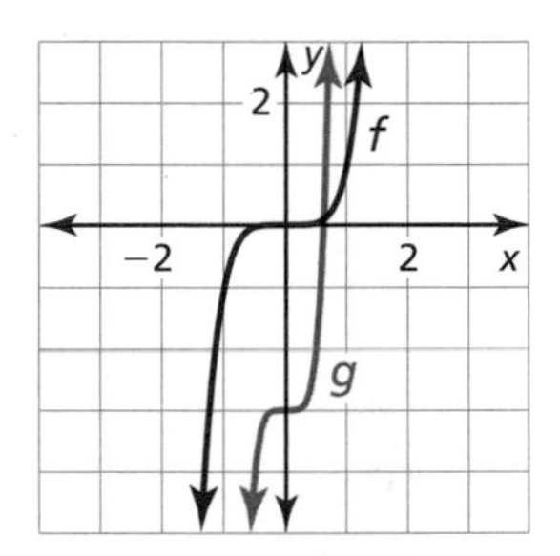

Monitoring Progress  Help in English and Spanish at *BigIdeasMath.com*

2. Describe the transformation of $f(x) = x^3$ represented by $g(x) = 4(x + 2)^3$. Then graph each function.

Writing Transformations of Polynomial Functions

EXAMPLE 3 Writing Transformed Polynomial Functions

Let $f(x) = x^3 + x^2 + 1$. Write a rule for g and then graph each function. Describe the graph of g as a transformation of the graph of f.

a. $g(x) = f(-x)$

b. $g(x) = 3f(x)$

SOLUTION

a. $g(x) = f(-x)$

$= (-x)^3 + (-x)^2 + 1$

$= -x^3 + x^2 + 1$

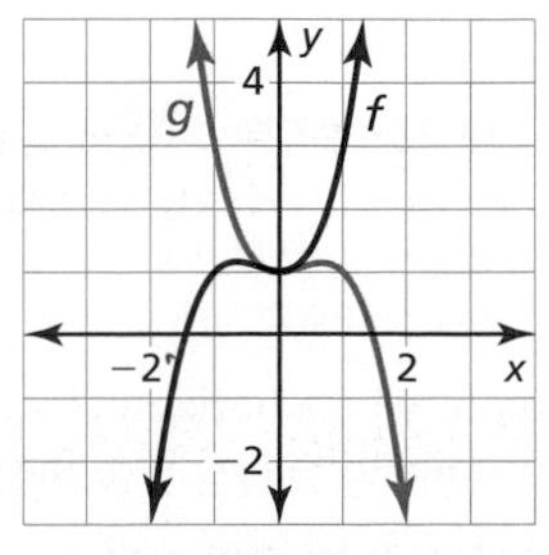

▶ The graph of g is a reflection in the y-axis of the graph of f.

b. $g(x) = 3f(x)$

$= 3(x^3 + x^2 + 1)$

$= 3x^3 + 3x^2 + 3$

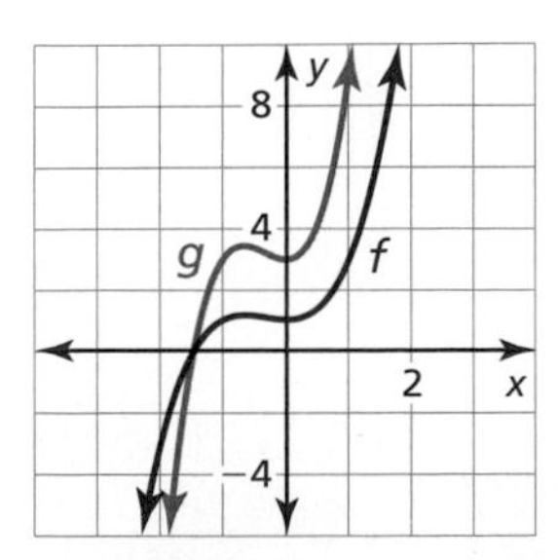

▶ The graph of g is a vertical stretch by a factor of 3 of the graph of f.

REMEMBER

Vertical stretches and shrinks do not change the x-intercept(s) of a graph. You can observe this using the graph in Example 3(b).

EXAMPLE 4 Writing a Transformed Polynomial Function

Let the graph of g be a vertical stretch by a factor of 2, followed by a translation 3 units up of the graph of $f(x) = x^4 - 2x^2$. Write a rule for g.

SOLUTION

Check

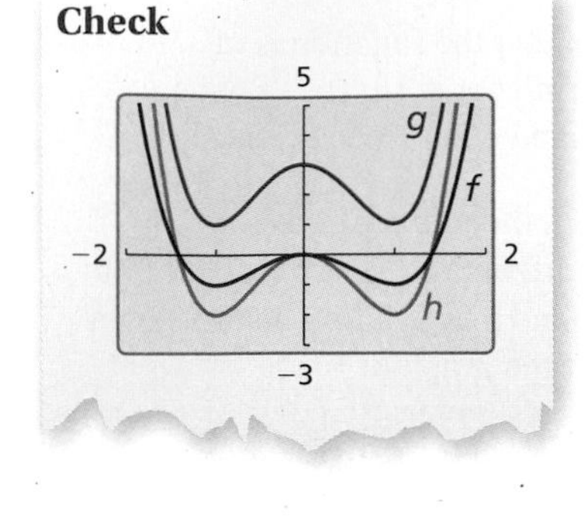

Step 1 First write a function h that represents the vertical stretch of f.

$h(x) = 2 \cdot f(x)$	Multiply the output by 2.
$= 2(x^4 - 2x^2)$	Substitute $x^4 - 2x^2$ for $f(x)$.
$= 2x^4 - 4x^2$	Distributive Property

Step 2 Then write a function g that represents the translation of h.

$g(x) = h(x) + 3$	Add 3 to the output.
$= 2x^4 - 4x^2 + 3$	Substitute $2x^4 - 4x^2$ for $h(x)$.

▶ The transformed function is $g(x) = 2x^4 - 4x^2 + 3$.

EXAMPLE 5 Modeling with Mathematics

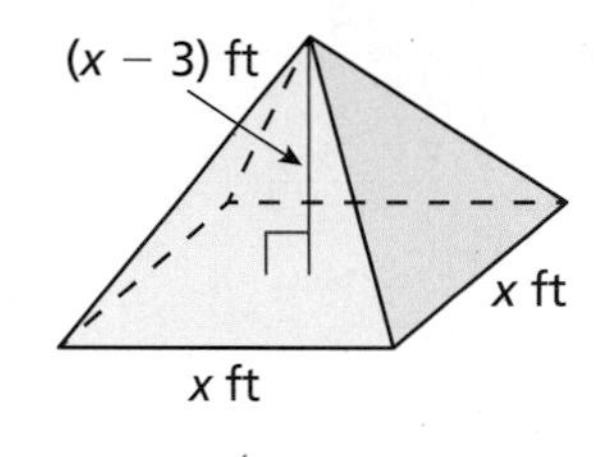

The function $V(x) = \frac{1}{3}x^3 - x^2$ represents the volume (in cubic feet) of the square pyramid shown. The function $W(x) = V(3x)$ represents the volume (in cubic feet) when x is measured in yards. Write a rule for W. Find and interpret $W(10)$.

SOLUTION

1. **Understand the Problem** You are given a function V whose inputs are in feet and whose outputs are in cubic feet. You are given another function W whose inputs are in yards and whose outputs are in cubic feet. The horizontal shrink shown by $W(x) = V(3x)$ makes sense because there are 3 feet in 1 yard. You are asked to write a rule for W and interpret the output for a given input.
2. **Make a Plan** Write the transformed function $W(x)$ and then find $W(10)$.
3. **Solve the Problem** $W(x) = V(3x)$

$= \frac{1}{3}(3x)^3 - (3x)^2$	Replace x with $3x$ in $V(x)$.
$= 9x^3 - 9x^2$	Simplify.

   Next, find $W(10)$.

   $$W(10) = 9(10)^3 - 9(10)^2 = 9000 - 900 = 8100$$

   ▶ When x is 10 yards, the volume of the pyramid is 8100 cubic feet.
4. **Look Back** Because $W(10) = V(30)$, you can check that your solution is correct by verifying that $V(30) = 8100$.

   $$V(30) = \frac{1}{3}(30)^3 - (30)^2 = 9000 - 900 = 8100 \checkmark$$

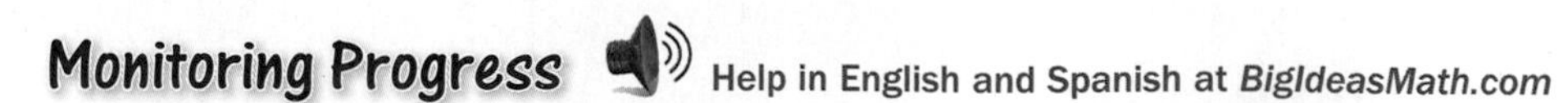

Monitoring Progress

Help in English and Spanish at BigIdeasMath.com

3. Let $f(x) = x^5 - 4x + 6$ and $g(x) = -f(x)$. Write a rule for g and then graph each function. Describe the graph of g as a transformation of the graph of f.
4. Let the graph of g be a horizontal stretch by a factor of 2, followed by a translation 3 units to the right of the graph of $f(x) = 8x^3 + 3$. Write a rule for g.
5. **WHAT IF?** In Example 5, the height of the pyramid is $6x$, and the volume (in cubic feet) is represented by $V(x) = 2x^3$. Write a rule for W. Find and interpret $W(7)$.

4.7 Exercises

Dynamic Solutions available at *BigIdeasMath.com*

Vocabulary and Core Concept Check

1. **COMPLETE THE SENTENCE** The graph of $f(x) = (x + 2)^3$ is a ____________ translation of the graph of $f(x) = x^3$.

2. **VOCABULARY** Describe how the vertex form of quadratic functions is similar to the form $f(x) = a(x - h)^3 + k$ for cubic functions.

Monitoring Progress and Modeling with Mathematics

In Exercises 3–6, describe the transformation of f represented by g. Then graph each function. *(See Example 1.)*

3. $f(x) = x^4,\ g(x) = x^4 + 3$

4. $f(x) = x^4,\ g(x) = (x - 5)^4$

5. $f(x) = x^5,\ g(x) = (x - 2)^5 - 1$

6. $f(x) = x^6,\ g(x) = (x + 1)^6 - 4$

ANALYZING RELATIONSHIPS In Exercises 7–10, match the function with the correct transformation of the graph of f. Explain your reasoning.

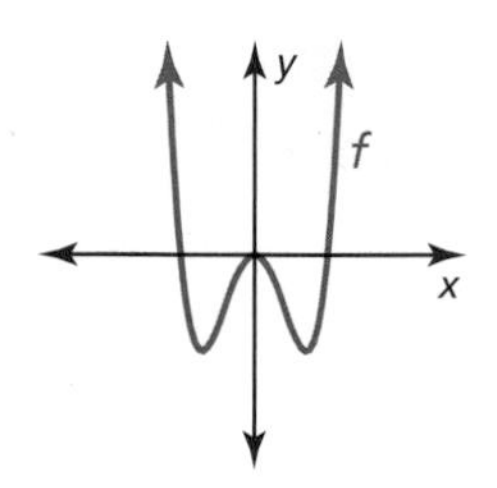

7. $y = f(x - 2)$

8. $y = f(x + 2) + 2$

9. $y = f(x - 2) + 2$

10. $y = f(x) - 2$

A.

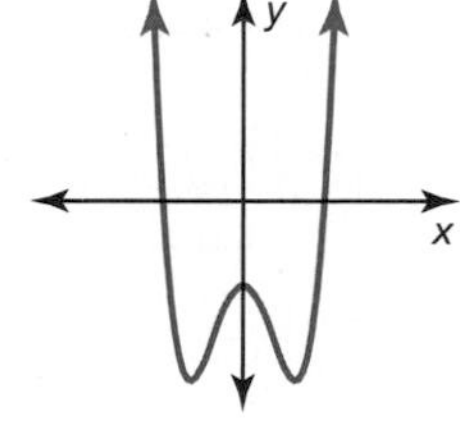

B.

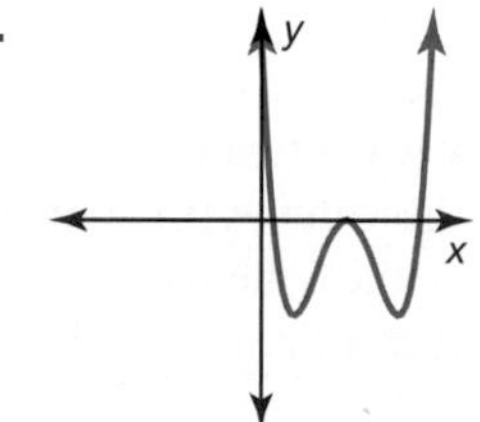

C.

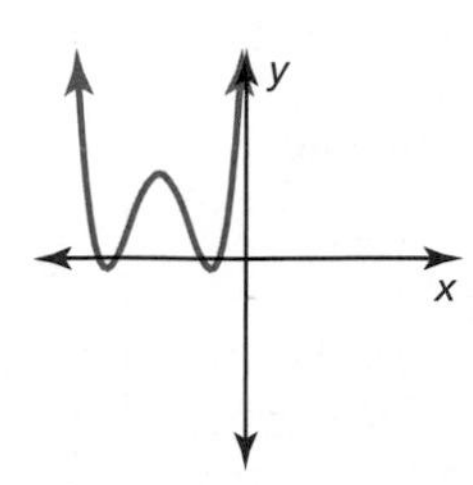

D.

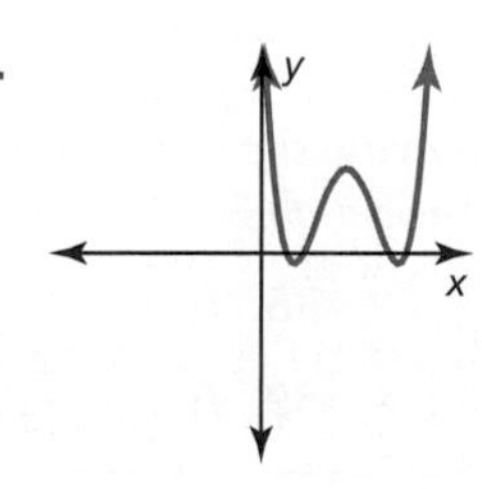

In Exercises 11–16, describe the transformation of f represented by g. Then graph each function. *(See Example 2.)*

11. $f(x) = x^4,\ g(x) = -2x^4$

12. $f(x) = x^6,\ g(x) = -3x^6$

13. $f(x) = x^3,\ g(x) = 5x^3 + 1$

14. $f(x) = x^4,\ g(x) = \frac{1}{2}x^4 + 1$

15. $f(x) = x^5,\ g(x) = \frac{3}{4}(x + 4)^5$

16. $f(x) = x^4,\ g(x) = (2x)^4 - 3$

In Exercises 17–20, write a rule for g and then graph each function. Describe the graph of g as a transformation of the graph of f. *(See Example 3.)*

17. $f(x) = x^4 + 1,\ g(x) = f(x + 2)$

18. $f(x) = x^5 - 2x + 3,\ g(x) = 3f(x)$

19. $f(x) = 2x^3 - 2x^2 + 6,\ g(x) = -\frac{1}{2}f(x)$

20. $f(x) = x^4 + x^3 - 1,\ g(x) = f(-x) - 5$

21. **ERROR ANALYSIS** Describe and correct the error in graphing the function $g(x) = (x + 2)^4 - 6$.

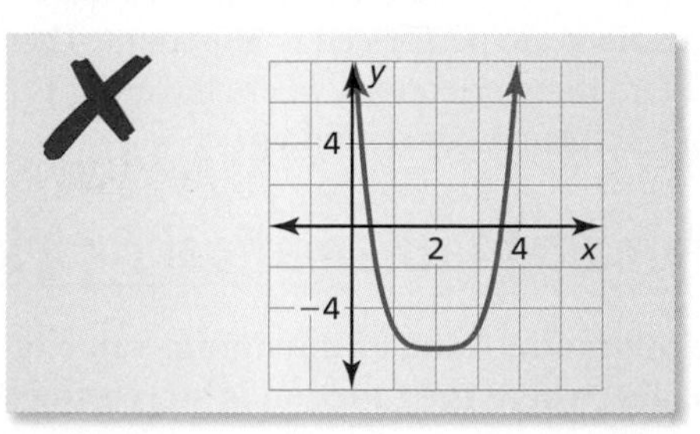

22. **ERROR ANALYSIS** Describe and correct the error in describing the transformation of the graph of $f(x) = x^5$ represented by the graph of $g(x) = (3x)^5 - 4$.

The graph of g is a horizontal shrink by a factor of 3, followed by a translation 4 units down of the graph of f.

In Exercises 23–26, write a rule for g that represents the indicated transformations of the graph of f. *(See Example 4.)*

23. $f(x) = x^3 - 6$; translation 3 units left, followed by a reflection in the y-axis

24. $f(x) = x^4 + 2x + 6$; vertical stretch by a factor of 2, followed by a translation 4 units right

25. $f(x) = x^3 + 2x^2 - 9$; horizontal shrink by a factor of $\frac{1}{3}$ and a translation 2 units up, followed by a reflection in the x-axis

26. $f(x) = 2x^5 - x^3 + x^2 + 4$; reflection in the y-axis and a vertical stretch by a factor of 3, followed by a translation 1 unit down

27. **MODELING WITH MATHEMATICS** The volume V (in cubic feet) of the pyramid is given by $V(x) = x^3 - 4x$. The function $W(x) = V(3x)$ gives the volume (in cubic feet) of the pyramid when x is measured in yards. Write a rule for W. Find and interpret $W(5)$. *(See Example 5.)*

x ft

$(2x - 4)$ ft

$(3x + 6)$ ft

28. **MAKING AN ARGUMENT** The volume of a cube with side length x is given by $V(x) = x^3$. Your friend claims that when you divide the volume in half, the volume decreases by a greater amount than when you divide each side length in half. Is your friend correct? Justify your answer.

29. **OPEN-ENDED** Describe two transformations of the graph of $f(x) = x^5$ where the order in which the transformations are performed is important. Then describe two transformations where the order is *not* important. Explain your reasoning.

30. **THOUGHT PROVOKING** Write and graph a transformation of the graph of $f(x) = x^5 - 3x^4 + 2x - 4$ that results in a graph with a y-intercept of -2.

31. **PROBLEM SOLVING** A portion of the path that a hummingbird flies while feeding can be modeled by the function

$$f(x) = -\frac{1}{5}x(x - 4)^2(x - 7),\ 0 \leq x \leq 7$$

where x is the horizontal distance (in meters) and $f(x)$ is the height (in meters). The hummingbird feeds each time it is at ground level.

a. At what distances does the hummingbird feed?

b. A second hummingbird feeds 2 meters farther away than the first hummingbird and flies twice as high. Write a function to model the path of the second hummingbird.

32. **HOW DO YOU SEE IT?** Determine the real zeros of each function. Then describe the transformation of the graph of f that results in the graph of g.

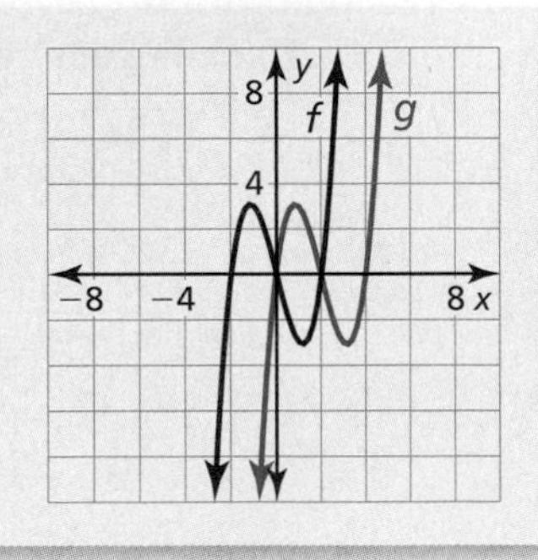

33. **MATHEMATICAL CONNECTIONS** Write a function V for the volume (in cubic yards) of the right circular cone shown. Then write a function W that gives the volume (in cubic yards) of the cone when x is measured in feet. Find and interpret $W(3)$.

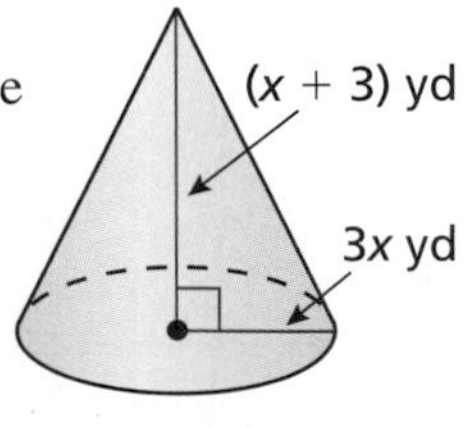

Maintaining Mathematical Proficiency

Reviewing what you learned in previous grades and lessons

Find the minimum value or maximum value of the function. Describe the domain and range of the function, and where the function is increasing and decreasing. *(Section 2.2)*

34. $h(x) = (x + 5)^2 - 7$

35. $f(x) = 4 - x^2$

36. $f(x) = 3(x - 10)(x + 4)$

37. $g(x) = -(x + 2)(x + 8)$

38. $h(x) = \frac{1}{2}(x - 1)^2 - 3$

39. $f(x) = -2x^2 + 4x - 1$

4.8 Analyzing Graphs of Polynomial Functions

Essential Question

How many turning points can the graph of a polynomial function have?

A *turning point* of the graph of a polynomial function is a point on the graph at which the function changes from

- increasing to decreasing, or
- decreasing to increasing.

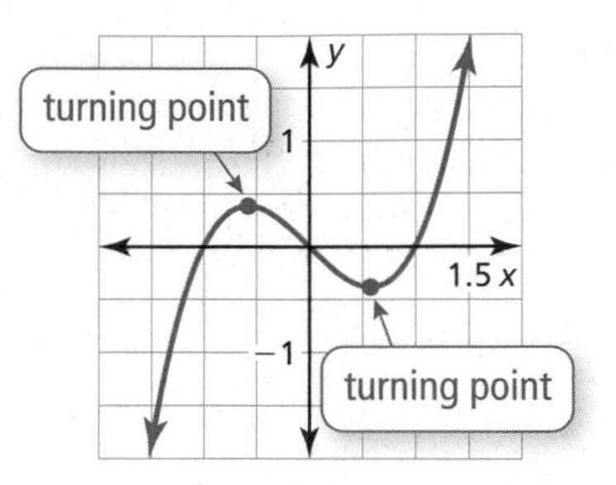

EXPLORATION 1 Approximating Turning Points

ATTENDING TO PRECISION

To be proficient in math, you need to express numerical answers with a degree of precision appropriate for the problem context.

Work with a partner. Match each polynomial function with its graph. Explain your reasoning. Then use a graphing calculator to approximate the coordinates of the turning points of the graph of the function. Round your answers to the nearest hundredth.

a. $f(x) = 2x^2 + 3x - 4$

b. $f(x) = x^2 + 3x + 2$

c. $f(x) = x^3 - 2x^2 - x + 1$

d. $f(x) = -x^3 + 5x - 2$

e. $f(x) = x^4 - 3x^2 + 2x - 1$

f. $f(x) = -2x^5 - x^2 + 5x + 3$

A.

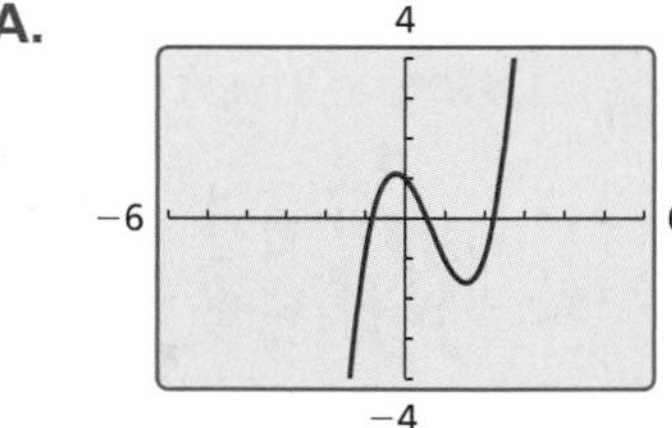

B.

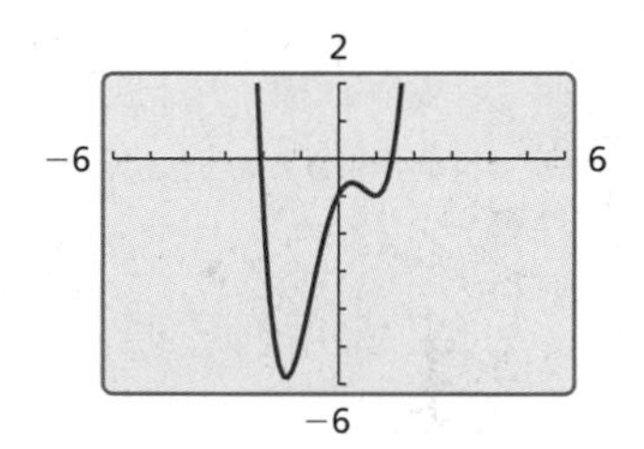

C.

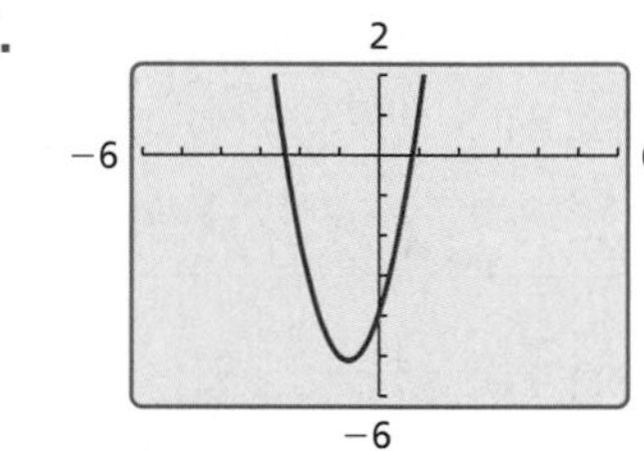

D.

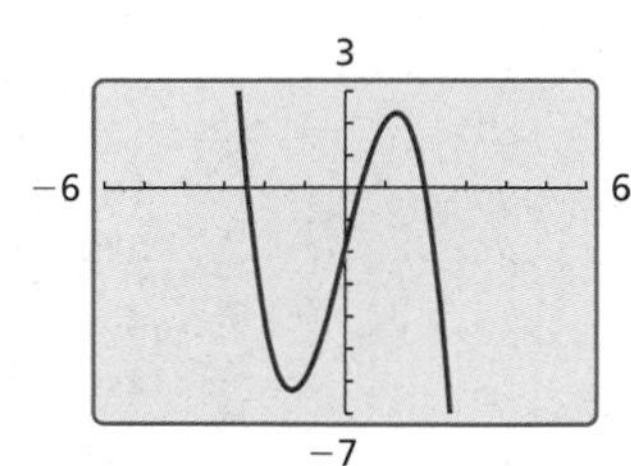

E.

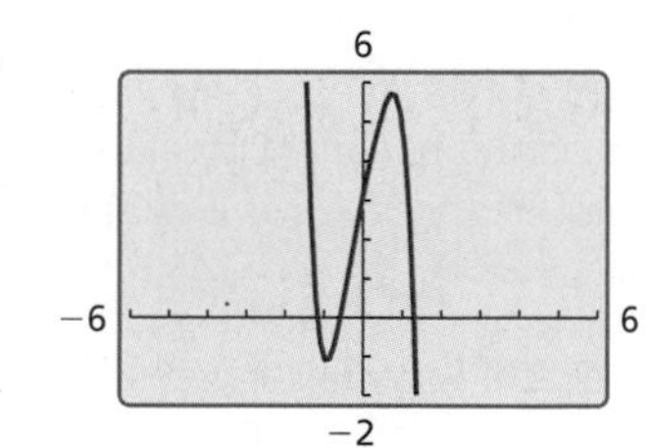

F.

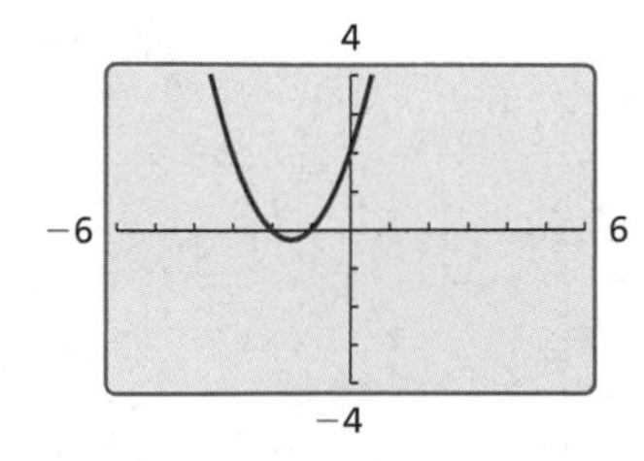

Communicate Your Answer

2. How many turning points can the graph of a polynomial function have?

3. Is it possible to sketch the graph of a cubic polynomial function that has *no* turning points? Justify your answer.

4.8 Lesson

Core Vocabulary

local maximum, *p. 214*
local minimum, *p. 214*
even function, *p. 215*
odd function, *p. 215*

Previous
end behavior
increasing
decreasing
symmetric about the *y*-axis

What You Will Learn

- Use *x*-intercepts to graph polynomial functions.
- Use the Location Principle to identify zeros of polynomial functions.
- Find turning points and identify local maximums and local minimums of graphs of polynomial functions.
- Identify even and odd functions.

Graphing Polynomial Functions

In this chapter, you have learned that zeros, factors, solutions, and *x*-intercepts are closely related concepts. Here is a summary of these relationships.

Concept Summary

Zeros, Factors, Solutions, and Intercepts

Let $f(x) = a_nx^n + a_{n-1}x^{n-1} + \cdots + a_1x + a_0$ be a polynomial function. The following statements are equivalent.

Zero: k is a zero of the polynomial function f.

Factor: $x - k$ is a factor of the polynomial $f(x)$.

Solution: k is a solution (or root) of the polynomial equation $f(x) = 0$.

x-Intercept: If k is a real number, then k is an x-intercept of the graph of the polynomial function f. The graph of f passes through $(k, 0)$.

EXAMPLE 1 Using *x*-Intercepts to Graph a Polynomial Function

Graph the function

$$f(x) = \frac{1}{6}(x + 3)(x - 2)^2.$$

SOLUTION

Step 1 Plot the x-intercepts. Because -3 and 2 are zeros of f, plot $(-3, 0)$ and $(2, 0)$.

Step 2 Plot points between and beyond the x-intercepts.

x	-2	-1	0	1	3
y	$\frac{8}{3}$	3	2	$\frac{2}{3}$	1

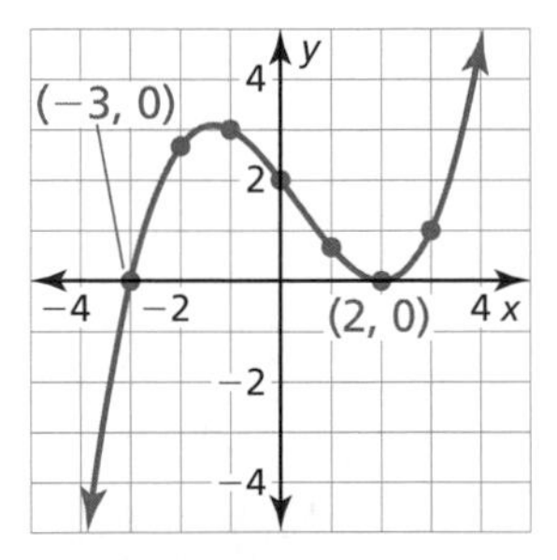

Step 3 Determine end behavior. Because $f(x)$ has three factors of the form $x - k$ and a constant factor of $\frac{1}{6}$, f is a cubic function with a positive leading coefficient. So, $f(x) \to -\infty$ as $x \to -\infty$ and $f(x) \to +\infty$ as $x \to +\infty$.

Step 4 Draw the graph so that it passes through the plotted points and has the appropriate end behavior.

Monitoring Progress

Help in English and Spanish at *BigIdeasMath.com*

Graph the function.

1. $f(x) = \frac{1}{2}(x + 1)(x - 4)^2$

2. $f(x) = \frac{1}{4}(x + 2)(x - 1)(x - 3)$

The Location Principle

You can use the *Location Principle* to help you find real zeros of polynomial functions.

Core Concept

The Location Principle

If f is a polynomial function, and a and b are two real numbers such that $f(a) < 0$ and $f(b) > 0$, then f has at least one real zero between a and b.

To use this principle to locate real zeros of a polynomial function, find a value a at which the polynomial function is negative and another value b at which the function is positive. You can conclude that the function has *at least* one real zero between a and b.

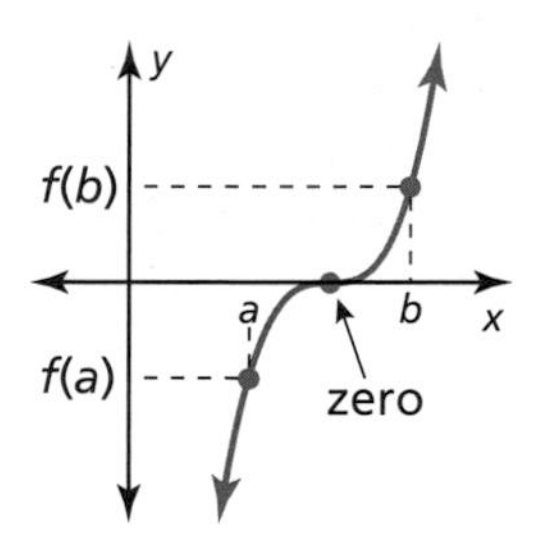

EXAMPLE 2 **Locating Real Zeros of a Polynomial Function**

Find all real zeros of

$$f(x) = 6x^3 + 5x^2 - 17x - 6.$$

SOLUTION

Step 1 Use a graphing calculator to make a table.

X	Y1
0	-6
1	-12
2	28
3	150
4	390
5	784
6	1368

X=1

Step 2 Use the Location Principle. From the table shown, you can see that $f(1) < 0$ and $f(2) > 0$. So, by the Location Principle, f has a zero between 1 and 2. Because f is a polynomial function of degree 3, it has three zeros. The only possible *rational* zero between 1 and 2 is $\frac{3}{2}$. Using synthetic division, you can confirm that $\frac{3}{2}$ is a zero.

Step 3 Write $f(x)$ in factored form. Dividing $f(x)$ by its known factor $x - \frac{3}{2}$ gives a quotient of $6x^2 + 14x + 4$. So, you can factor $f(x)$ as

$$f(x) = \left(x - \tfrac{3}{2}\right)(6x^2 + 14x + 4)$$

$$= 2\left(x - \tfrac{3}{2}\right)(3x^2 + 7x + 2)$$

$$= 2\left(x - \tfrac{3}{2}\right)(3x + 1)(x + 2).$$

Check

20
−5 5
Zero
X=1.5 Y=0
−20

▶ From the factorization, there are three zeros. The zeros of f are

$\frac{3}{2}$, $-\frac{1}{3}$, and -2.

Check this by graphing f.

Monitoring Progress

Help in English and Spanish at *BigIdeasMath.com*

3. Find all real zeros of $f(x) = 18x^3 + 21x^2 - 13x - 6$.

Turning Points

Another important characteristic of graphs of polynomial functions is that they have *turning points* corresponding to local maximum and minimum values.

- The y-coordinate of a turning point is a **local maximum** of the function when the point is higher than all nearby points.
- The y-coordinate of a turning point is a **local minimum** of the function when the point is lower than all nearby points.

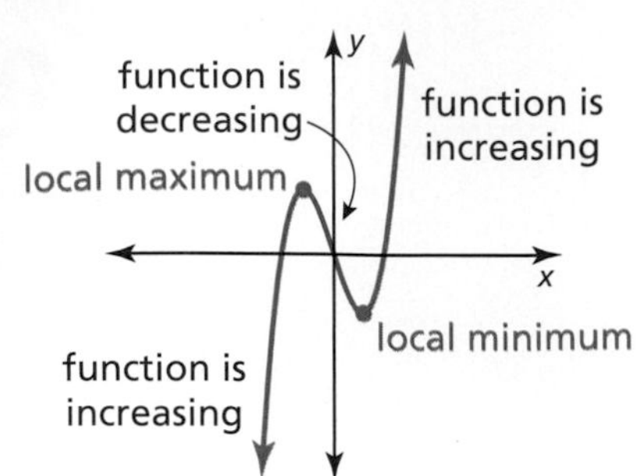

The turning points of a graph help determine the intervals for which a function is increasing or decreasing.

READING

Local maximum and local minimum are sometimes referred to as *relative maximum* and *relative minimum*.

Core Concept

Turning Points of Polynomial Functions

1. The graph of every polynomial function of degree n has *at most* $n - 1$ turning points.
2. If a polynomial function has n distinct real zeros, then its graph has *exactly* $n - 1$ turning points.

EXAMPLE 3 Finding Turning Points

Graph each function. Identify the x-intercepts and the points where the local maximums and local minimums occur. Determine the intervals for which each function is increasing or decreasing.

a. $f(x) = x^3 - 3x^2 + 6$ **b.** $g(x) = x^4 - 6x^3 + 3x^2 + 10x - 3$

SOLUTION

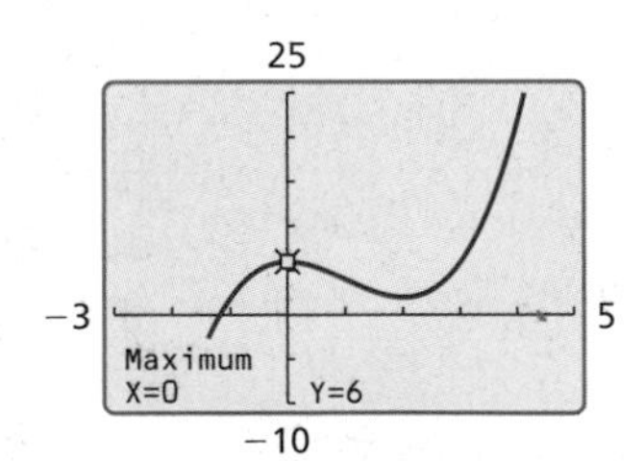

a. Use a graphing calculator to graph the function. The graph of f has one x-intercept and two turning points. Use the graphing calculator's *zero*, *maximum*, and *minimum* features to approximate the coordinates of the points.

▶ The x-intercept of the graph is $x \approx -1.20$. The function has a local maximum at $(0, 6)$ and a local minimum at $(2, 2)$. The function is increasing when $x < 0$ and $x > 2$ and decreasing when $0 < x < 2$.

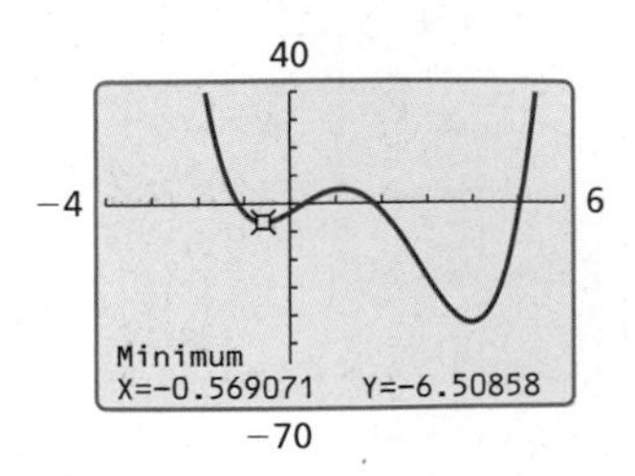

b. Use a graphing calculator to graph the function. The graph of g has four x-intercepts and three turning points. Use the graphing calculator's *zero*, *maximum*, and *minimum* features to approximate the coordinates of the points.

▶ The x-intercepts of the graph are $x \approx -1.14$, $x \approx 0.29$, $x \approx 1.82$, and $x \approx 5.03$. The function has a local maximum at $(1.11, 5.11)$ and local minimums at $(-0.57, -6.51)$ and $(3.96, -43.04)$. The function is increasing when $-0.57 < x < 1.11$ and $x > 3.96$ and decreasing when $x < -0.57$ and $1.11 < x < 3.96$.

Monitoring Progress  Help in English and Spanish at *BigIdeasMath.com*

4. Graph $f(x) = 0.5x^3 + x^2 - x + 2$. Identify the x-intercepts and the points where the local maximums and local minimums occur. Determine the intervals for which the function is increasing or decreasing.

Even and Odd Functions

Core Concept

Even and Odd Functions

A function f is an **even function** when $f(-x) = f(x)$ for all x in its domain. The graph of an even function is *symmetric about the y-axis*.

A function f is an **odd function** when $f(-x) = -f(x)$ for all x in its domain. The graph of an odd function is *symmetric about the origin*. One way to recognize a graph that is symmetric about the origin is that it looks the same after a 180° rotation about the origin.

Even Function

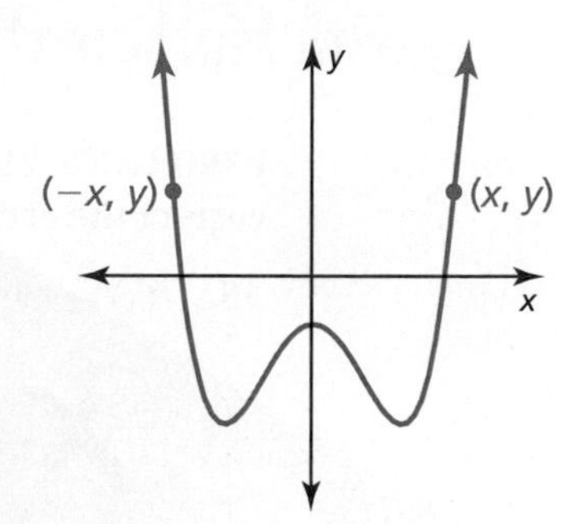

Odd Function

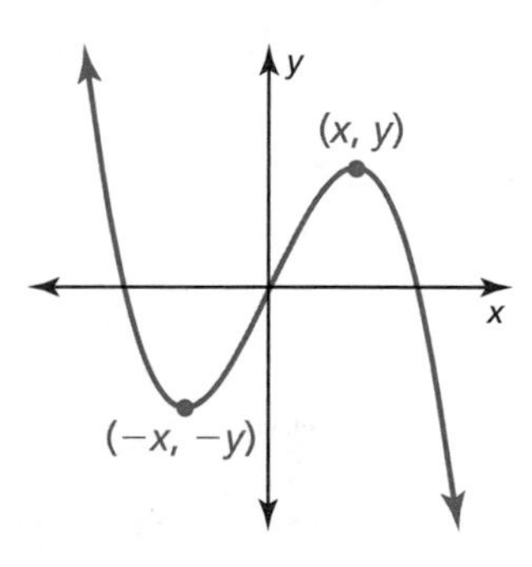

For an even function, if (x, y) is on the graph, then $(-x, y)$ is also on the graph.

For an odd function, if (x, y) is on the graph, then $(-x, -y)$ is also on the graph.

EXAMPLE 4 Identifying Even and Odd Functions

Determine whether each function is *even*, *odd*, or *neither*.

a. $f(x) = x^3 - 7x$ **b.** $g(x) = x^4 + x^2 - 1$ **c.** $h(x) = x^3 + 2$

SOLUTION

a. Replace x with $-x$ in the equation for f, and then simplify.

$$f(-x) = (-x)^3 - 7(-x) = -x^3 + 7x = -(x^3 - 7x) = -f(x)$$

▶ Because $f(-x) = -f(x)$, the function is odd.

b. Replace x with $-x$ in the equation for g, and then simplify.

$$g(-x) = (-x)^4 + (-x)^2 - 1 = x^4 + x^2 - 1 = g(x)$$

▶ Because $g(-x) = g(x)$, the function is even.

c. Replacing x with $-x$ in the equation for h produces

$$h(-x) = (-x)^3 + 2 = -x^3 + 2.$$

▶ Because $h(x) = x^3 + 2$ and $-h(x) = -x^3 - 2$, you can conclude that $h(-x) \neq h(x)$ and $h(-x) \neq -h(x)$. So, the function is neither even nor odd.

Monitoring Progress  Help in English and Spanish at *BigIdeasMath.com*

Determine whether the function is *even*, *odd*, or *neither*.

5. $f(x) = -x^2 + 5$ **6.** $f(x) = x^4 - 5x^3$ **7.** $f(x) = 2x^5$

4.8 Exercises

Dynamic Solutions available at *BigIdeasMath.com*

Vocabulary and Core Concept Check

1. **COMPLETE THE SENTENCE** A local maximum or local minimum of a polynomial function occurs at a ____________ point of the graph of the function.

2. **WRITING** Explain what a local maximum of a function is and how it may be different from the maximum value of the function.

Monitoring Progress and Modeling with Mathematics

ANALYZING RELATIONSHIPS In Exercises 3–6, match the function with its graph.

3. $f(x) = (x - 1)(x - 2)(x + 2)$

4. $h(x) = (x + 2)^2(x + 1)$

5. $g(x) = (x + 1)(x - 1)(x + 2)$

6. $f(x) = (x - 1)^2(x + 2)$

A.
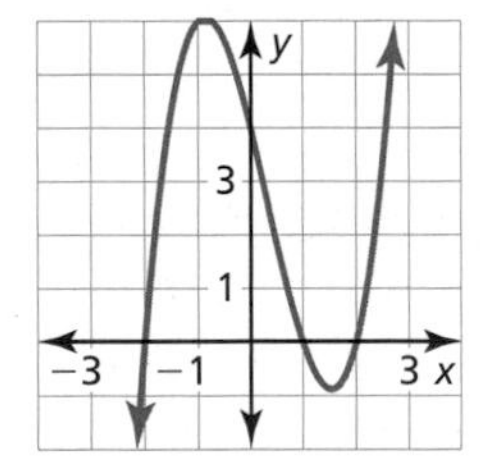

B.
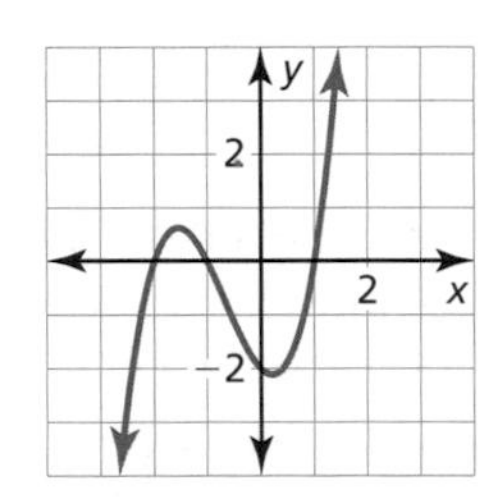

C.
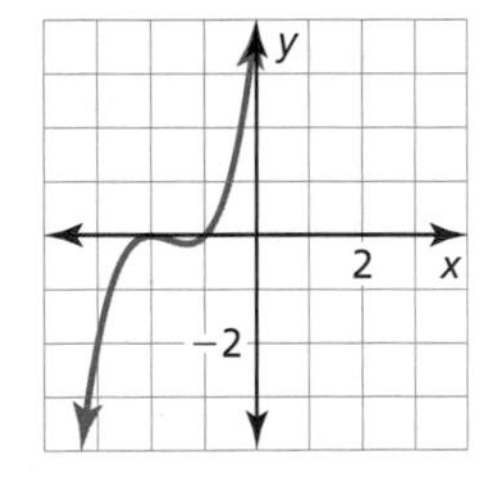

D.
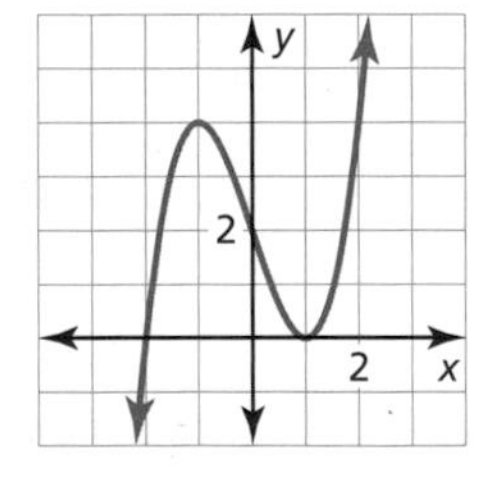

In Exercises 7–14, graph the function. *(See Example 1.)*

7. $f(x) = (x - 2)^2(x + 1)$

8. $f(x) = (x + 2)^2(x + 4)^2$

9. $h(x) = (x + 1)^2(x - 1)(x - 3)$

10. $g(x) = 4(x + 1)(x + 2)(x - 1)$

11. $h(x) = \frac{1}{3}(x - 5)(x + 2)(x - 3)$

12. $g(x) = \frac{1}{12}(x + 4)(x + 8)(x - 1)$

13. $h(x) = (x - 3)(x^2 + x + 1)$

14. $f(x) = (x - 4)(2x^2 - 2x + 1)$

ERROR ANALYSIS In Exercises 15 and 16, describe and correct the error in using factors to graph *f*.

15. $f(x) = (x + 2)(x - 1)^2$

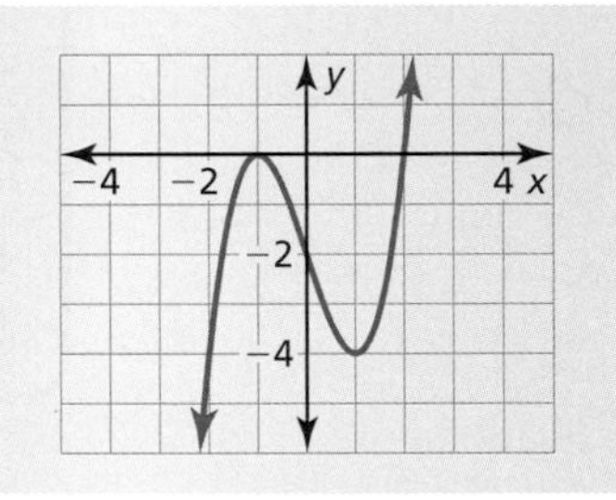

16. $f(x) = x^2(x - 3)^3$

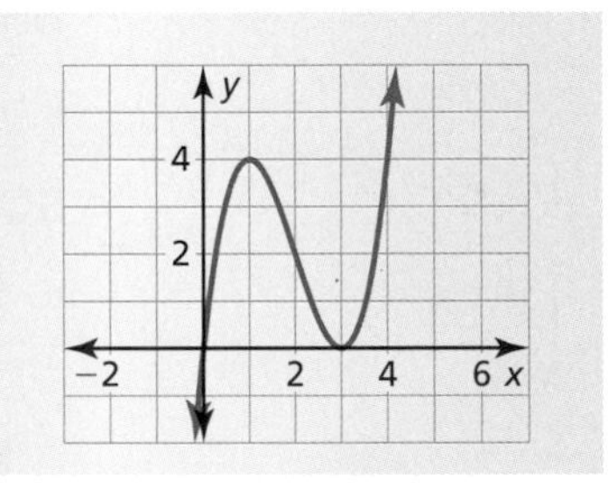

In Exercises 17–22, find all real zeros of the function. *(See Example 2.)*

17. $f(x) = x^3 - 4x^2 - x + 4$

18. $f(x) = x^3 - 3x^2 - 4x + 12$

19. $h(x) = 2x^3 + 7x^2 - 5x - 4$

20. $h(x) = 4x^3 - 2x^2 - 24x - 18$

21. $g(x) = 4x^3 + x^2 - 51x + 36$

22. $f(x) = 2x^3 - 3x^2 - 32x - 15$

In Exercises 23–30, graph the function. Identify the x-intercepts and the points where the local maximums and local minimums occur. Determine the intervals for which the function is increasing or decreasing. *(See Example 3.)*

23. $g(x) = 2x^3 + 8x^2 - 3$

24. $g(x) = -x^4 + 3x$

25. $h(x) = x^4 - 3x^2 + x$

26. $f(x) = x^5 - 4x^3 + x^2 + 2$

27. $f(x) = 0.5x^3 - 2x + 2.5$

28. $f(x) = 0.7x^4 - 3x^3 + 5x$

29. $h(x) = x^5 + 2x^2 - 17x - 4$

30. $g(x) = x^4 - 5x^3 + 2x^2 + x - 3$

In Exercises 31–36, estimate the coordinates of each turning point. State whether each corresponds to a local maximum or a local minimum. Then estimate the real zeros and find the least possible degree of the function.

31.

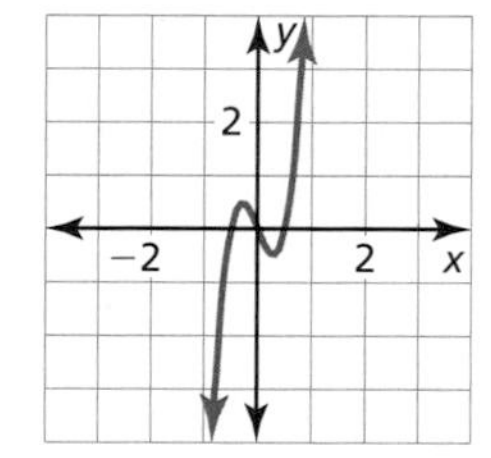

32.

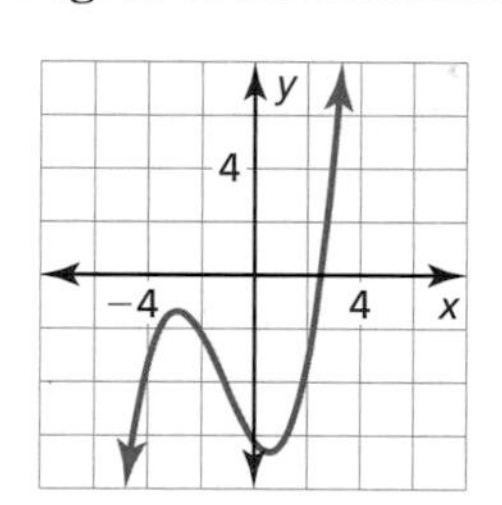

33.

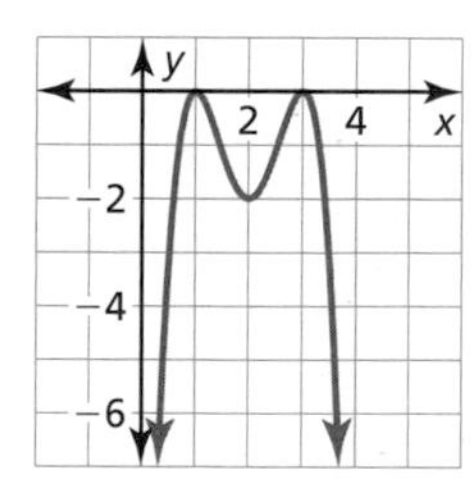

34.

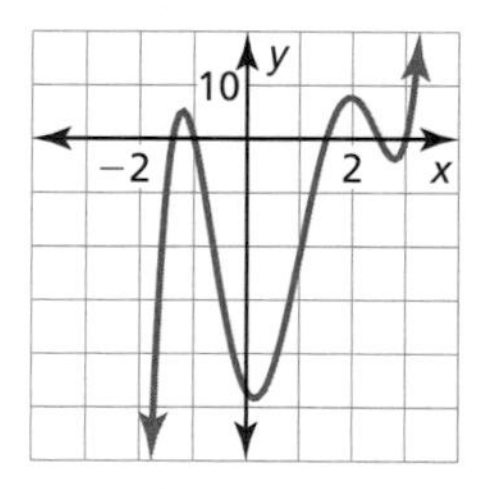

35.

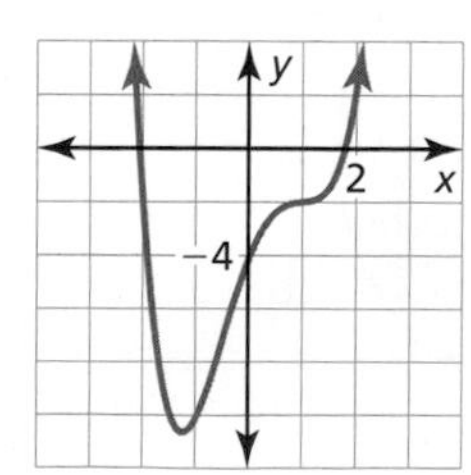

36.

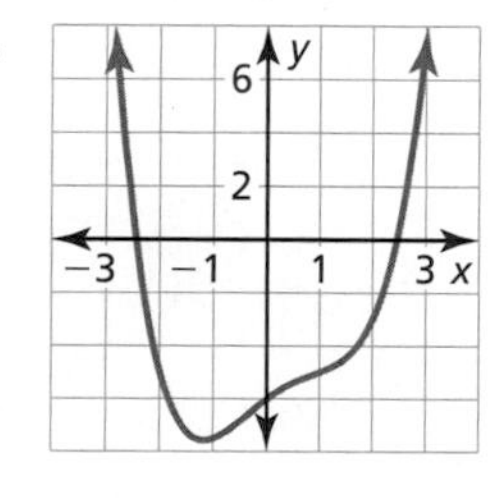

OPEN-ENDED In Exercises 37 and 38, sketch a graph of a polynomial function f having the given characteristics.

37.
- The graph of f has x-intercepts at $x = -4$, $x = 0$, and $x = 2$.
- f has a local maximum value when $x = 1$.
- f has a local minimum value when $x = -2$.

38.
- The graph of f has x-intercepts at $x = -3$, $x = 1$, and $x = 5$.
- f has a local maximum value when $x = 1$.
- f has a local minimum value when $x = -2$ and when $x = 4$.

In Exercises 39–46, determine whether the function is *even*, *odd*, or *neither*. *(See Example 4.)*

39. $h(x) = 4x^7$

40. $g(x) = -2x^6 + x^2$

41. $f(x) = x^4 + 3x^2 - 2$

42. $f(x) = x^5 + 3x^3 - x$

43. $g(x) = x^2 + 5x + 1$

44. $f(x) = -x^3 + 2x - 9$

45. $f(x) = x^4 - 12x^2$

46. $h(x) = x^5 + 3x^4$

47. USING TOOLS When a swimmer does the breaststroke, the function

$$S = -241t^7 + 1060t^6 - 1870t^5 + 1650t^4 - 737t^3 + 144t^2 - 2.43t$$

models the speed S (in meters per second) of the swimmer during one complete stroke, where t is the number of seconds since the start of the stroke and $0 \leq t \leq 1.22$. Use a graphing calculator to graph the function. At what time during the stroke is the swimmer traveling the fastest?

48. USING TOOLS During a recent period of time, the number S (in thousands) of students enrolled in public schools in a certain country can be modeled by $S = 1.64x^3 - 102x^2 + 1710x + 36{,}300$, where x is time (in years). Use a graphing calculator to graph the function for the interval $0 \leq x \leq 41$. Then describe how the public school enrollment changes over this period of time.

49. WRITING Why is the adjective *local*, used to describe the maximums and minimums of cubic functions, sometimes not required for quadratic functions?

50. HOW DO YOU SEE IT? The graph of a polynomial function is shown.

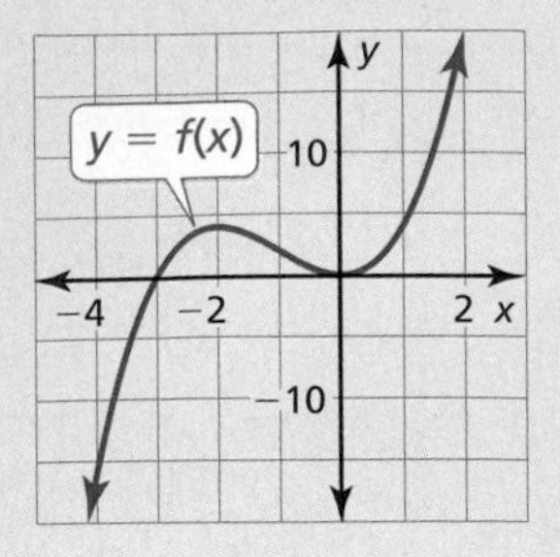

a. Find the zeros, local maximum, and local minimum values of the function.

b. Compare the x-intercepts of the graphs of $y = f(x)$ and $y = -f(x)$.

c. Compare the maximum and minimum values of the functions $y = f(x)$ and $y = -f(x)$.

51. MAKING AN ARGUMENT Your friend claims that the product of two odd functions is an odd function. Is your friend correct? Explain your reasoning.

52. MODELING WITH MATHEMATICS You are making a rectangular box out of a 16-inch-by-20-inch piece of cardboard. The box will be formed by making the cuts shown in the diagram and folding up the sides. You want the box to have the greatest volume possible.

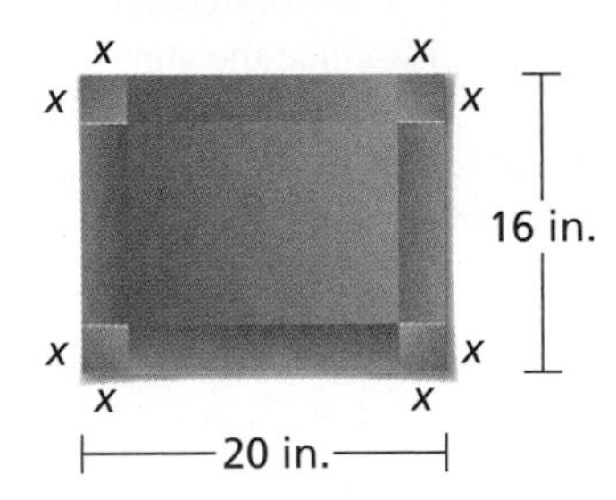

a. How long should you make the cuts?

b. What is the maximum volume?

c. What are the dimensions of the finished box?

53. PROBLEM SOLVING Quonset huts are temporary, all-purpose structures shaped like half-cylinders. You have 1100 square feet of material to build a quonset hut.

a. The surface area S of a quonset hut is given by $S = \pi r^2 + \pi r\ell$. Substitute 1100 for S and then write an expression for ℓ in terms of r.

b. The volume V of a quonset hut is given by $V = \frac{1}{2}\pi r^2\ell$. Write an equation that gives V as a function in terms of r only.

c. Find the value of r that maximizes the volume of the hut.

54. THOUGHT PROVOKING Write and graph a polynomial function that has one real zero in each of the intervals $-2 < x < -1$, $0 < x < 1$, and $4 < x < 5$. Is there a maximum degree that such a polynomial function can have? Justify your answer.

55. MATHEMATICAL CONNECTIONS A cylinder is inscribed in a sphere of radius 8 inches. Write an equation for the volume of the cylinder as a function of h. Find the value of h that maximizes the volume of the inscribed cylinder. What is the maximum volume of the cylinder?

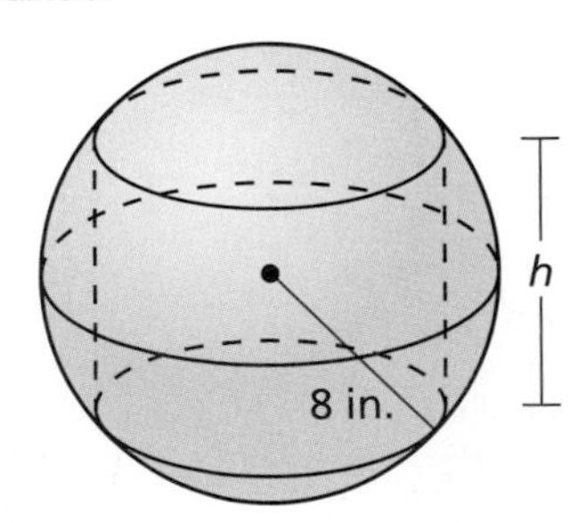

Maintaining Mathematical Proficiency

Reviewing what you learned in previous grades and lessons

State whether the table displays *linear data*, *quadratic data*, or *neither*. Explain. *(Section 2.4)*

56.

Months, x	0	1	2	3
Savings (dollars), y	100	150	200	250

57.

Time (seconds), x	0	1	2	3
Height (feet), y	300	284	236	156

4.9 Modeling with Polynomial Functions

Essential Question How can you find a polynomial model for real-life data?

EXPLORATION 1 **Modeling Real-Life Data**

Work with a partner. The distance a baseball travels after it is hit depends on the angle at which it was hit and the initial speed. The table shows the distances a baseball hit at an angle of 35° travels at various initial speeds.

Initial speed, x (miles per hour)	80	85	90	95	100	105	110	115
Distance, y (feet)	194	220	247	275	304	334	365	397

a. Recall that when data have equally-spaced x-values, you can analyze patterns in the differences of the y-values to determine what type of function can be used to model the data. If the first differences are constant, then the set of data fits a linear model. If the second differences are constant, then the set of data fits a quadratic model.

Find the first and second differences of the data. Are the data linear or quadratic? Explain your reasoning.

USING TOOLS STRATEGICALLY

To be proficient in math, you need to use technological tools to explore and deepen your understanding of concepts.

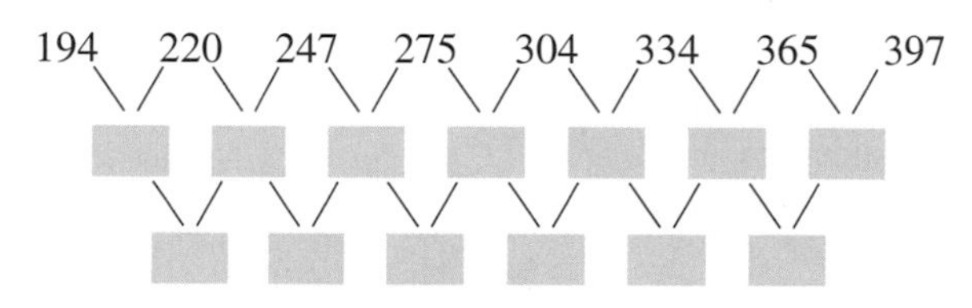

b. Use a graphing calculator to draw a scatter plot of the data. Do the data appear linear or quadratic? Use the *regression* feature of the graphing calculator to find a linear or quadratic model that best fits the data.

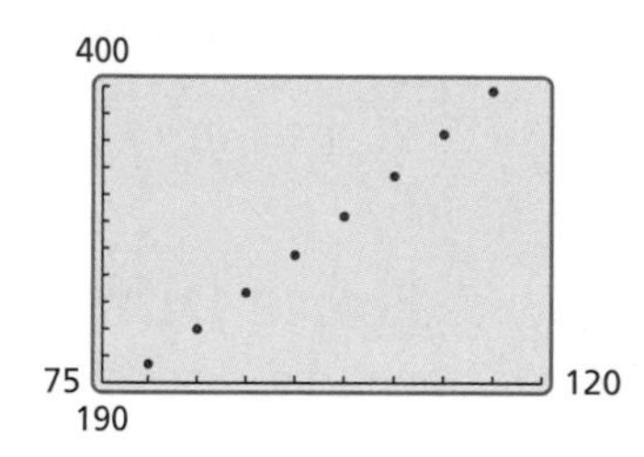

c. Use the model you found in part (b) to find the distance a baseball travels when it is hit at an angle of 35° and travels at an initial speed of 120 miles per hour.

d. According to the *Baseball Almanac*, "Any drive over 400 feet is noteworthy. A blow of 450 feet shows exceptional power, as the majority of major league players are unable to hit a ball that far. Anything in the 500-foot range is genuinely historic." Estimate the initial speed of a baseball that travels a distance of 500 feet.

Communicate Your Answer

2. How can you find a polynomial model for real-life data?

3. How well does the model you found in Exploration 1(b) fit the data? Do you think the model is valid for any initial speed? Explain your reasoning.

4.9 Lesson

Core Vocabulary

finite differences, *p. 220*

Previous
scatter plot

What You Will Learn

- Write polynomial functions for sets of points.
- Write polynomial functions using finite differences.
- Use technology to find models for data sets.

Writing Polynomial Functions for a Set of Points

You know that two points determine a line and three points not on a line determine a parabola. In Example 1, you will see that four points not on a line or a parabola determine the graph of a cubic function.

EXAMPLE 1 Writing a Cubic Function

Write the cubic function whose graph is shown.

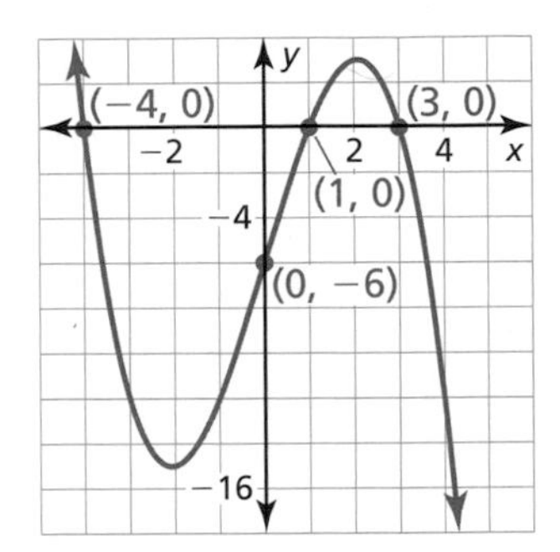

SOLUTION

Step 1 Use the three x-intercepts to write the function in factored form.

$$f(x) = a(x + 4)(x - 1)(x - 3)$$

Step 2 Find the value of a by substituting the coordinates of the point $(0, -6)$.

$$-6 = a(0 + 4)(0 - 1)(0 - 3)$$

$$-6 = 12a$$

$$-\frac{1}{2} = a$$

▶ The function is $f(x) = -\frac{1}{2}(x + 4)(x - 1)(x - 3)$.

Check
Check the end behavior of f. The degree of f is odd and $a < 0$. So, $f(x) \to +\infty$ as $x \to -\infty$ and $f(x) \to -\infty$ as $x \to +\infty$, which matches the graph. ✓

Monitoring Progress  Help in English and Spanish at *BigIdeasMath.com*

Write a cubic function whose graph passes through the given points.

1. $(-4, 0), (0, 10), (2, 0), (5, 0)$

2. $(-1, 0), (0, -12), (2, 0), (3, 0)$

Finite Differences

When the x-values in a data set are equally spaced, the differences of consecutive y-values are called **finite differences**. Recall from Section 2.4 that the first and second differences of $y = x^2$ are:

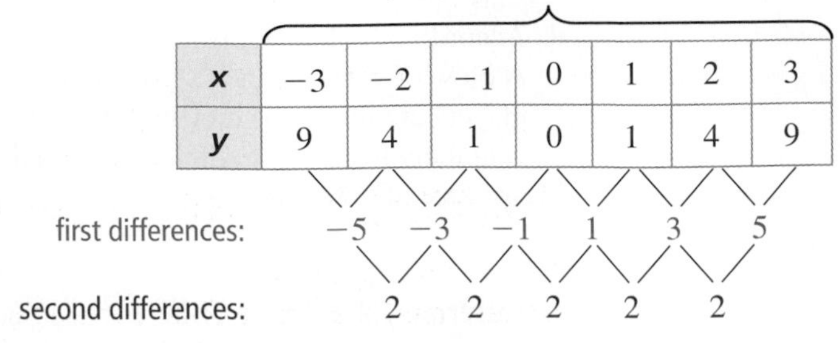

equally-spaced x-values
x | −3 | −2 | −1 | 0 | 1 | 2 | 3
y | 9 | 4 | 1 | 0 | 1 | 4 | 9
first differences: −5 −3 −1 1 3 5
second differences: 2 2 2 2 2

Notice that $y = x^2$ has degree *two* and that the *second* differences are constant and nonzero. This illustrates the first of the two properties of finite differences shown on the next page.

Core Concept

Properties of Finite Differences

1. If a polynomial function $y = f(x)$ has degree n, then the nth differences of function values for equally-spaced x-values are nonzero and constant.
2. Conversely, if the nth differences of equally-spaced data are nonzero and constant, then the data can be represented by a polynomial function of degree n.

The second property of finite differences allows you to write a polynomial function that models a set of equally-spaced data.

EXAMPLE 2 Writing a Function Using Finite Differences

Use finite differences to determine the degree of the polynomial function that fits the data. Then use technology to find the polynomial function.

x	1	2	3	4	5	6	7
$f(x)$	1	4	10	20	35	56	84

SOLUTION

Step 1 Write the function values. Find the first differences by subtracting consecutive values. Then find the second differences by subtracting consecutive first differences. Continue until you obtain differences that are nonzero and constant.

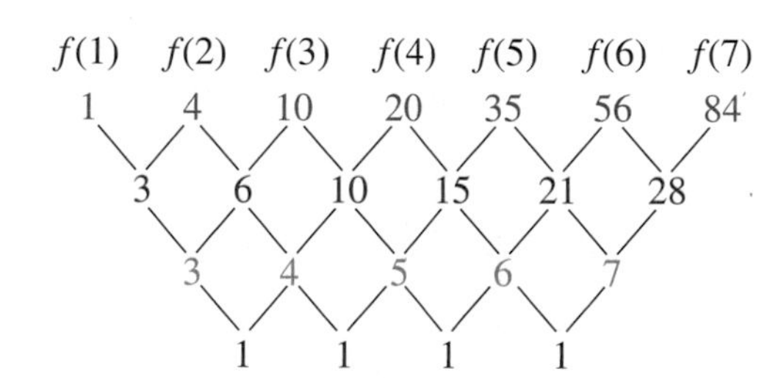

Write function values for equally-spaced x-values.

First differences

Second differences

Third differences

Because the third differences are nonzero and constant, you can model the data *exactly* with a cubic function.

Step 2 Enter the data into a graphing calculator and use cubic regression to obtain a polynomial function.

```
CubicReg
y=ax3+bx2+cx+d
a=.1666666667
b=.5
c=.3333333333
d=0
R2=1
```

▶ Because $\frac{1}{6} \approx 0.1666666667$, $\frac{1}{2} = 0.5$, and $\frac{1}{3} \approx 0.333333333$, a polynomial function that fits the data exactly is

$$f(x) = \frac{1}{6}x^3 + \frac{1}{2}x^2 + \frac{1}{3}x.$$

Monitoring Progress  Help in English and Spanish at *BigIdeasMath.com*

3. Use finite differences to determine the degree of the polynomial function that fits the data. Then use technology to find the polynomial function.

x	−3	−2	−1	0	1	2
$f(x)$	6	15	22	21	6	−29

Finding Models Using Technology

In Examples 1 and 2, you found a cubic model that *exactly* fits a set of data. In many real-life situations, you cannot find models to fit data exactly. Despite this limitation, you can still use technology to approximate the data with a polynomial model, as shown in the next example.

EXAMPLE 3 Real-Life Application

The table shows the total U.S. biomass energy consumptions y (in trillions of British thermal units, or Btus) in the year t, where $t = 1$ corresponds to 2001. Find a polynomial model for the data. Use the model to estimate the total U.S. biomass energy consumption in 2013.

t	1	2	3	4	5	6
y	2622	2701	2807	3010	3117	3267

t	7	8	9	10	11	12
y	3493	3866	3951	4286	4421	4316

According to the U.S. Department of Energy, *biomass* includes "agricultural and forestry residues, municipal solid wastes, industrial wastes, and terrestrial and aquatic crops grown solely for energy purposes." Among the uses for biomass is production of electricity and liquid fuels such as ethanol.

SOLUTION

Step 1 Enter the data into a graphing calculator and make a scatter plot. The data suggest a cubic model.

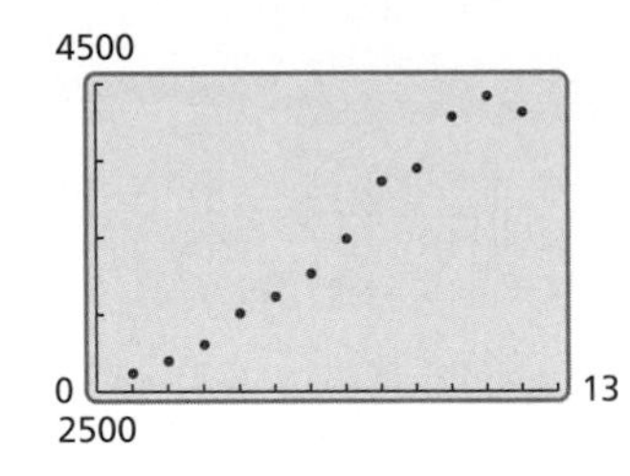

Step 2 Use the *cubic regression* feature. The polynomial model is

$$y = -2.545t^3 + 51.95t^2 - 118.1t + 2732.$$

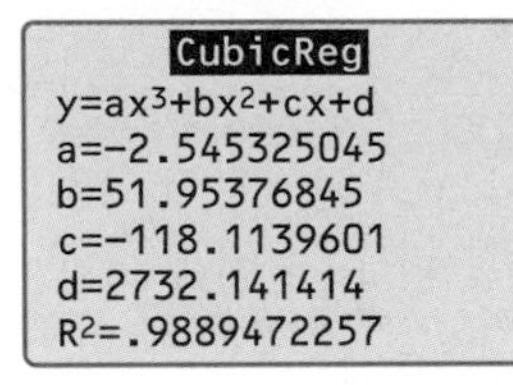

Step 3 Check the model by graphing it and the data in the same viewing window.

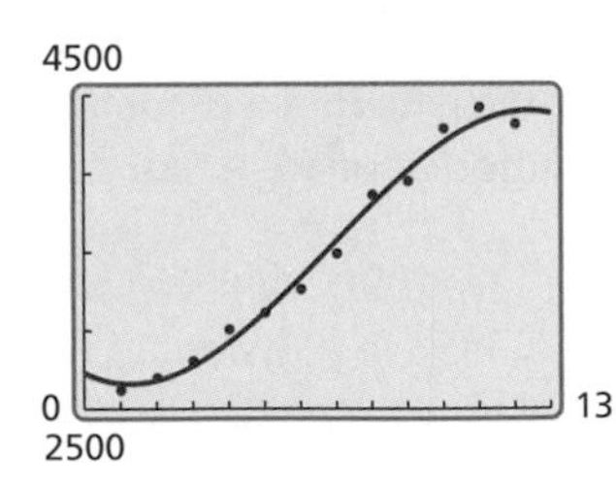

Step 4 Use the *trace* feature to estimate the value of the model when $t = 13$.

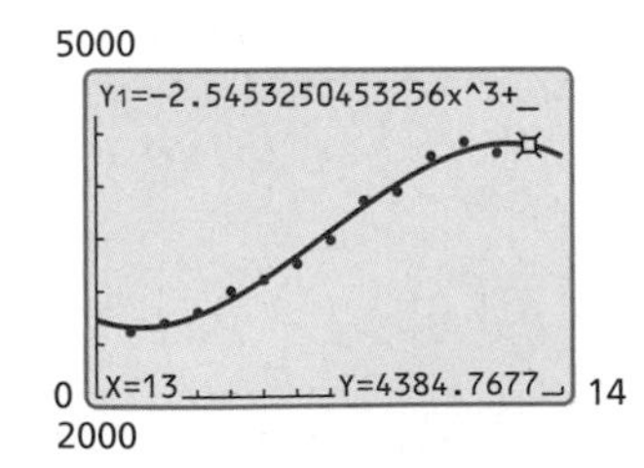

▶ The approximate total U.S. biomass energy consumption in 2013 was about 4385 trillion Btus.

Monitoring Progress  Help in English and Spanish at *BigIdeasMath.com*

Use a graphing calculator to find a polynomial function that fits the data.

4.

x	1	2	3	4	5	6
y	5	13	17	11	11	56

5.

x	0	2	4	6	8	10
y	8	0	15	69	98	87

4.9 Exercises

Dynamic Solutions available at *BigIdeasMath.com*

Vocabulary and Core Concept Check

1. **COMPLETE THE SENTENCE** When the x-values in a set of data are equally spaced, the differences of consecutive y-values are called ________________.

2. **WRITING** Explain how you know when a set of data could be modeled by a cubic function.

Monitoring Progress and Modeling with Mathematics

In Exercises 3–6, write a cubic function whose graph is shown. *(See Example 1.)*

3.

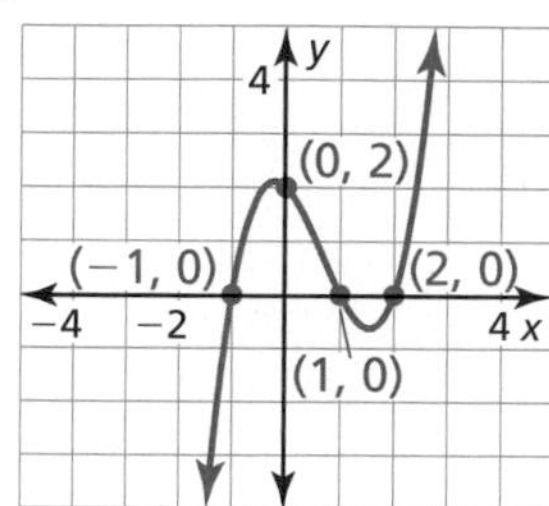

4.

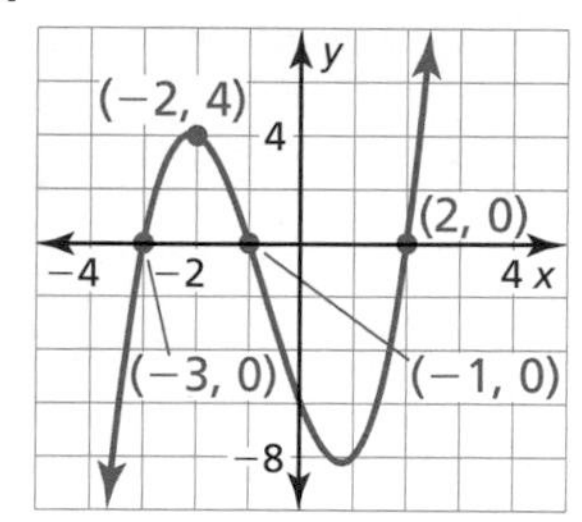

5.

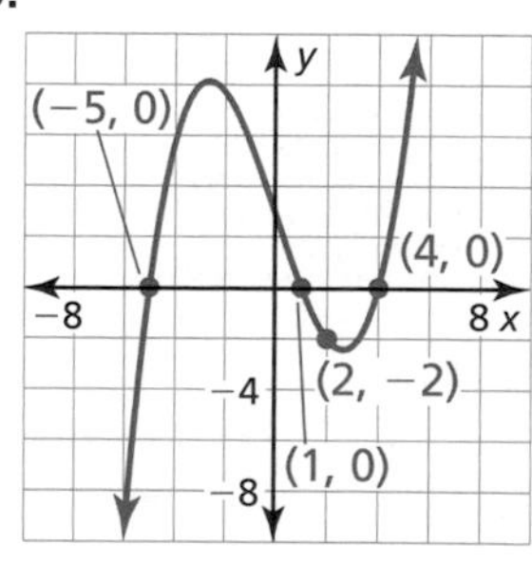

6.

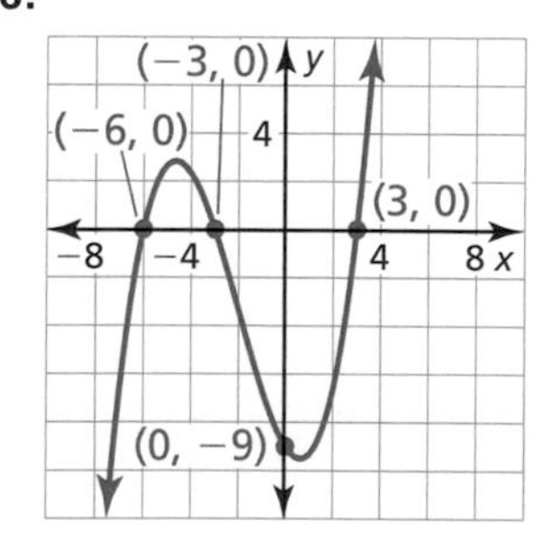

In Exercises 7–12, use finite differences to determine the degree of the polynomial function that fits the data. Then use technology to find the polynomial function. *(See Example 2.)*

7.

x	−6	−3	0	3	6	9
$f(x)$	−2	15	−4	49	282	803

8.

x	−1	0	1	2	3	4
$f(x)$	−14	−5	−2	7	34	91

9. $(-4, -317), (-3, -37), (-2, 21), (-1, 7), (0, -1), (1, 3), (2, -47), (3, -289), (4, -933)$

10. $(-6, 744), (-4, 154), (-2, 4), (0, -6), (2, 16), (4, 154), (6, 684), (8, 2074), (10, 4984)$

11. $(-2, 968), (-1, 422), (0, 142), (1, 26), (2, -4), (3, -2), (4, 2), (5, 2), (6, 16)$

12. $(1, 0), (2, 6), (3, 2), (4, 6), (5, 12), (6, -10), (7, -114), (8, -378), (9, -904)$

13. **ERROR ANALYSIS** Describe and correct the error in writing a cubic function whose graph passes through the given points.

✗ $(-6, 0), (1, 0), (3, 0), (0, 54)$

$$54 = a(0 - 6)(0 + 1)(0 + 3)$$
$$54 = -18a$$
$$a = -3$$
$$f(x) = -3(x - 6)(x + 1)(x + 3)$$

14. **MODELING WITH MATHEMATICS** The dot patterns show pentagonal numbers. The number of dots in the nth pentagonal number is given by $f(n) = \frac{1}{2}n(3n - 1)$. Show that this function has constant second-order differences.

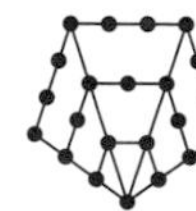
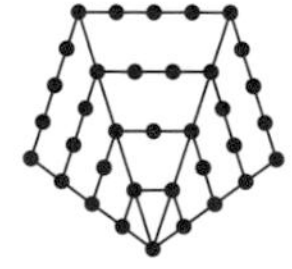

15. **OPEN-ENDED** Write three different cubic functions that pass through the points (3, 0), (4, 0), and (2, 6). Justify your answers.

16. **MODELING WITH MATHEMATICS** The table shows the ages of cats and their corresponding ages in human years. Find a polynomial model for the data for the first 8 years of a cat's life. Use the model to estimate the age (in human years) of a cat that is 3 years old. *(See Example 3.)*

Age of cat, x	1	2	4	6	7	8
Human years, y	15	24	32	40	44	48

17. **MODELING WITH MATHEMATICS** The data in the table show the average speeds y (in miles per hour) of a pontoon boat for several different engine speeds x (in hundreds of revolutions per minute, or RPMs). Find a polynomial model for the data. Estimate the average speed of the pontoon boat when the engine speed is 2800 RPMs.

x	10	20	25	30	45	55
y	4.5	8.9	13.8	18.9	29.9	37.7

18. **HOW DO YOU SEE IT?** The graph shows typical speeds y (in feet per second) of a space shuttle x seconds after it is launched.

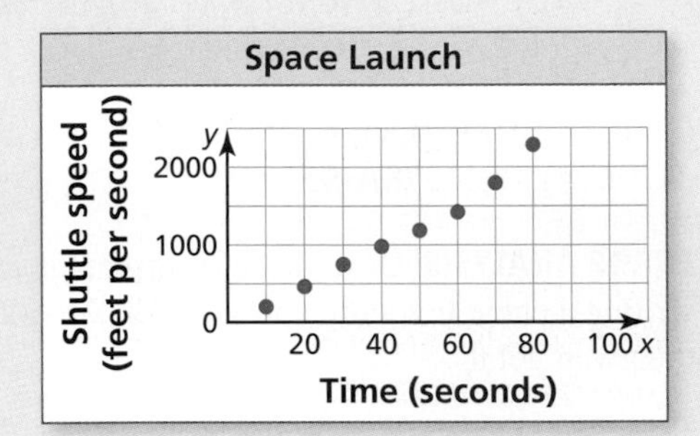

a. What type of polynomial function models the data? Explain.

b. Which nth-order finite difference should be constant for the function in part (a)? Explain.

19. **MATHEMATICAL CONNECTIONS** The table shows the number of diagonals for polygons with n sides. Find a polynomial function that fits the data. Determine the total number of diagonals in the decagon shown.

diagonal

Number of sides, n	3	4	5	6	7	8
Number of diagonals, d	0	2	5	9	14	20

20. **MAKING AN ARGUMENT** Your friend states that it is not possible to determine the degree of a function given the first-order differences. Is your friend correct? Explain your reasoning.

21. **WRITING** Explain why you cannot always use finite differences to find a model for real-life data sets.

22. **THOUGHT PROVOKING** A, B, and C are zeros of a cubic polynomial function. Choose values for A, B, and C such that the distance from A to B is less than or equal to the distance from A to C. Then write the function using the A, B, and C values you chose.

23. **MULTIPLE REPRESENTATIONS** Order the polynomial functions according to their degree, from least to greatest.

A. $f(x) = -3x + 2x^2 + 1$

B.

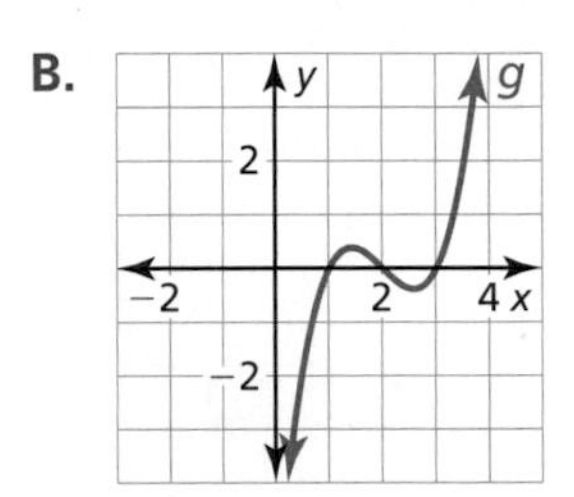

C.

x	−2	−1	0	1	2	3
$h(x)$	8	6	4	2	0	−2

D.

x	−2	−1	0	1	2	3
$k(x)$	25	6	7	4	−3	10

24. **ABSTRACT REASONING** Substitute the expressions $z, z + 1, z + 2, \ldots, z + 5$ for x in the function $f(x) = ax^3 + bx^2 + cx + d$ to generate six equally-spaced ordered pairs. Then show that the third-order differences are constant.

Maintaining Mathematical Proficiency

Reviewing what you learned in previous grades and lessons

Solve the equation using square roots. *(Section 3.1)*

25. $x^2 - 6 = 30$

26. $5x^2 - 38 = 187$

27. $2(x - 3)^2 = 24$

28. $\frac{4}{3}(x + 5)^2 = 4$

Solve the equation using the Quadratic Formula. *(Section 3.4)*

29. $2x^2 + 3x = 5$

30. $2x^2 + \frac{1}{2} = 2x$

31. $2x^2 + 3x = -3x^2 + 1$

32. $4x - 20 = x^2$

4.5–4.9 What Did You Learn?

Core Vocabulary

repeated solution, *p. 190*
complex conjugates, *p. 199*
local maximum, *p. 214*
local minimum, *p. 214*
even function, *p. 215*
odd function, *p. 215*
finite differences, *p. 220*

Core Concepts

Section 4.5

The Rational Root Theorem, *p. 191*
The Irrational Conjugates Theorem, *p. 193*

Section 4.6

The Fundamental Theorem of Algebra, *p. 198*
The Complex Conjugates Theorem, *p. 199*
Descartes's Rule of Signs, *p. 200*

Section 4.7

Transformations of Polynomial Functions, *p. 206*
Writing Transformed Polynomial Functions, *p. 207*

Section 4.8

Zeros, Factors, Solutions, and Intercepts, *p. 212*
The Location Principle, *p. 213*
Turning Points of Polynomial Functions, *p. 214*
Even and Odd Functions, *p. 215*

Section 4.9

Writing Polynomial Functions for Data Sets, *p. 220*
Properties of Finite Differences, *p. 221*

Mathematical Practices

1. Explain how understanding the Complex Conjugates Theorem allows you to construct your argument in Exercise 46 on page 203.
2. Describe how you use structure to accurately match each graph with its transformation in Exercises 7–10 on page 209.

Performance Task

For the Birds – Wildlife Management

How does the presence of humans affect the population of sparrows in a park? Do more humans mean fewer sparrows? Or does the presence of humans increase the number of sparrows up to a point? Are there a minimum number of sparrows that can be found in a park, regardless of how many humans there are? What can a mathematical model tell you?

To explore the answers to these questions and more, go to ***BigIdeasMath.com***.

4 Chapter Review

Dynamic Solutions available at *BigIdeasMath.com*

4.1 Graphing Polynomial Functions *(pp. 157–164)*

Graph $f(x) = x^3 + 3x^2 - 3x - 10$.

To graph the function, make a table of values and plot the corresponding points. Connect the points with a smooth curve and check the end behavior.

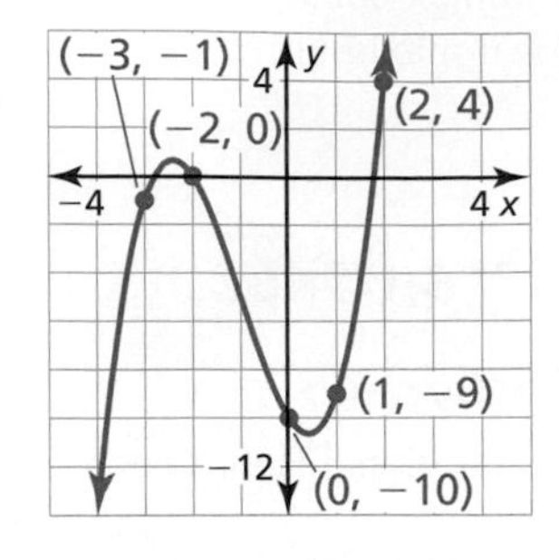

x	−3	−2	−1	0	1	2	3
$f(x)$	−1	0	−5	−10	−9	4	35

The degree is odd and the leading coefficient is positive. So, $f(x) \to -\infty$ as $x \to -\infty$ and $f(x) \to +\infty$ as $x \to +\infty$.

Decide whether the function is a polynomial function. If so, write it in standard form and state its degree, type, and leading coefficient.

1. $h(x) = -x^3 + 2x^2 - 15x^7$

2. $p(x) = x^3 - 5x^{0.5} + 13x^2 + 8$

Graph the polynomial function.

3. $h(x) = x^2 + 6x^5 - 5$

4. $f(x) = 3x^4 - 5x^2 + 1$

5. $g(x) = -x^4 + x + 2$

4.2 Adding, Subtracting, and Multiplying Polynomials *(pp. 165–172)*

a. Multiply $(x - 2)$, $(x - 1)$, and $(x + 3)$ in a horizontal format.

$$(x - 2)(x - 1)(x + 3) = (x^2 - 3x + 2)(x + 3)$$
$$= (x^2 - 3x + 2)x + (x^2 - 3x + 2)3$$
$$= x^3 - 3x^2 + 2x + 3x^2 - 9x + 6$$
$$= x^3 - 7x + 6$$

b. Use Pascal's Triangle to expand $(4x + 2)^4$.

The coefficients from the fourth row of Pascal's Triangle are 1, 4, 6, 4, and 1.

$$(4x + 2)^4 = 1(4x)^4 + 4(4x)^3(2) + 6(4x)^2(2)^2 + 4(4x)(2)^3 + 1(2)^4$$
$$= 256x^4 + 512x^3 + 384x^2 + 128x + 16$$

Find the sum or difference.

6. $(4x^3 - 12x^2 - 5) - (-8x^2 + 4x + 3)$

7. $(x^4 + 3x^3 - x^2 + 6) + (2x^4 - 3x + 9)$

8. $(3x^2 + 9x + 13) - (x^2 - 2x + 12)$

Find the product.

9. $(2y^2 + 4y - 7)(y + 3)$

10. $(2m + n)^3$

11. $(s + 2)(s + 4)(s - 3)$

Use Pascal's Triangle to expand the binomial.

12. $(m + 4)^4$

13. $(3s + 2)^5$

14. $(z + 1)^6$

4.3 Dividing Polynomials (pp. 173–178)

Use synthetic division to evaluate $f(x) = -2x^3 + 4x^2 + 8x + 10$ when $x = -3$.

$$\begin{array}{r|rrrr} -3 & -2 & 4 & 8 & 10 \\ & & 6 & -30 & 66 \\ \hline & -2 & 10 & -22 & 76 \end{array}$$

▶ The remainder is 76. So, you can conclude from the Remainder Theorem that $f(-3) = 76$. You can check this by substituting $x = -3$ in the original function.

Check

$$f(-3) = -2(-3)^3 + 4(-3)^2 + 8(-3) + 10$$
$$= 54 + 36 - 24 + 10$$
$$= 76 \ ✓$$

Divide using polynomial long division or synthetic division.

15. $(x^3 + x^2 + 3x - 4) \div (x^2 + 2x + 1)$

16. $(x^4 + 3x^3 - 4x^2 + 5x + 3) \div (x^2 + x + 4)$

17. $(x^4 - x^2 - 7) \div (x + 4)$

18. Use synthetic division to evaluate $g(x) = 4x^3 + 2x^2 - 4$ when $x = 5$.

4.4 Factoring Polynomials (pp. 179–186)

a. Factor $x^4 + 8x$ completely.

$x^4 + 8x = x(x^3 + 8)$ — Factor common monomial.

$= x(x^3 + 2^3)$ — Write $x^3 + 8$ as $a^3 + b^3$.

$= x(x + 2)(x^2 - 2x + 4)$ — Sum of Two Cubes Pattern

b. Determine whether $x + 4$ is a factor of $f(x) = x^5 + 4x^4 + 2x + 8$.

Find $f(-4)$ by synthetic division.

$$\begin{array}{r|rrrrrr} -4 & 1 & 4 & 0 & 0 & 2 & 8 \\ & & -4 & 0 & 0 & 0 & -8 \\ \hline & 1 & 0 & 0 & 0 & 2 & 0 \end{array}$$

▶ Because $f(-4) = 0$, the binomial $x + 4$ is a factor of $f(x) = x^5 + 4x^4 + 2x + 8$.

Factor the polynomial completely.

19. $64x^3 - 8$

20. $2z^5 - 12z^3 + 10z$

21. $2a^3 - 7a^2 - 8a + 28$

22. Show that $x + 2$ is a factor of $f(x) = x^4 + 2x^3 - 27x - 54$. Then factor $f(x)$ completely.

4.5 Solving Polynomial Equations (pp. 189–196)

a. Find all real solutions of $x^3 + x^2 - 8x - 12 = 0$.

Step 1 List the possible rational solutions. The leading coefficient of the polynomial $f(x) = x^3 + x^2 - 8x - 12$ is 1, and the constant term is -12. So, the possible rational solutions of $f(x) = 0$ are

$$x = \pm\frac{1}{1}, \pm\frac{2}{1}, \pm\frac{3}{1}, \pm\frac{4}{1}, \pm\frac{6}{1}, \pm\frac{12}{1}.$$

Step 2 Test possible solutions using synthetic division until a solution is found.

$$\begin{array}{r|rrrr} 2 & 1 & 1 & -8 & -12 \\ & & 2 & 6 & -4 \\ \hline & 1 & 3 & -2 & -16 \end{array} \qquad \begin{array}{r|rrrr} -2 & 1 & 1 & -8 & -12 \\ & & -2 & 2 & 12 \\ \hline & 1 & -1 & -6 & 0 \end{array}$$

$f(2) \neq 0$, so $x - 2$ is not a factor of $f(x)$. $f(-2) = 0$, so $x + 2$ is a factor of $f(x)$.

Step 3 Factor completely using the result of synthetic division.

$(x + 2)(x^2 - x - 6) = 0$ Write as a product of factors.

$(x + 2)(x + 2)(x - 3) = 0$ Factor the trinomial.

▶ So, the solutions are $x = -2$ and $x = 3$.

b. Write a polynomial function f of least degree that has rational coefficients, a leading coefficient of 1, and the zeros -4 and $1 + \sqrt{2}$.

By the Irrational Conjugates Theorem, $1 - \sqrt{2}$ must also be a zero of f.

$f(x) = (x + 4)\left[x - (1 + \sqrt{2})\right]\left[x - (1 - \sqrt{2})\right]$ Write $f(x)$ in factored form.

$= (x + 4)\left[(x - 1) - \sqrt{2}\right]\left[(x - 1) + \sqrt{2}\right]$ Regroup terms.

$= (x + 4)\left[(x - 1)^2 - 2\right]$ Multiply.

$= (x + 4)\left[(x^2 - 2x + 1) - 2\right]$ Expand binomial.

$= (x + 4)(x^2 - 2x - 1)$ Simplify.

$= x^3 - 2x^2 - x + 4x^2 - 8x - 4$ Multiply.

$= x^3 + 2x^2 - 9x - 4$ Combine like terms.

Find all real solutions of the equation.

23. $x^3 + 3x^2 - 10x - 24 = 0$

24. $x^3 + 5x^2 - 2x - 24 = 0$

Write a polynomial function f of least degree that has rational coefficients, a leading coefficient of 1, and the given zeros.

25. $1, 2 - \sqrt{3}$

26. $2, 3, \sqrt{5}$

27. $-2, 5, 3 + \sqrt{6}$

28. You use 240 cubic inches of clay to make a sculpture shaped as a rectangular prism. The width is 4 inches less than the length and the height is 2 inches more than three times the length. What are the dimensions of the sculpture? Justify your answer.

4.6 The Fundamental Theorem of Algebra (pp. 197–204)

Find all zeros of $f(x) = x^4 + 2x^3 + 6x^2 + 18x - 27$.

Step 1 Find the rational zeros of f. Because f is a polynomial function of degree 4, it has four zeros. The possible rational zeros are ±1, ±3, ±9, and ± 27. Using synthetic division, you can determine that 1 is a zero and -3 is also a zero.

Step 2 Write $f(x)$ in factored form. Dividing $f(x)$ by its known factors $x - 1$ and $x + 3$ gives a quotient of $x^2 + 9$. So,

$$f(x) = (x - 1)(x + 3)(x^2 + 9).$$

Step 3 Find the complex zeros of f. Solving $x^2 + 9 = 0$, you get $x = \pm 3i$. This means $x^2 + 9 = (x + 3i)(x - 3i)$.

$$f(x) = (x - 1)(x + 3)(x + 3i)(x - 3i)$$

▶ From the factorization, there are four zeros. The zeros of f are 1, -3, $-3i$, and $3i$.

Write a polynomial function f of least degree that has rational coefficients, a leading coefficient of 1, and the given zeros.

29. 3, $1 + 2i$

30. -1, 2, $4i$

31. -5, -4, $1 - i\sqrt{3}$

Determine the possible numbers of positive real zeros, negative real zeros, and imaginary zeros for the function.

32. $f(x) = x^4 - 10x + 8$

33. $f(x) = -6x^4 - x^3 + 3x^2 + 2x + 18$

4.7 Transformations of Polynomial Functions (pp. 205–210)

Describe the transformation of $f(x) = x^3$ represented by $g(x) = (x - 6)^3 - 2$. Then graph each function.

Notice that the function is of the form $g(x) = (x - h)^3 + k$. Rewrite the function to identify h and k.

$$g(x) = (x - 6)^3 + (-2)$$

↑ h ↑ k

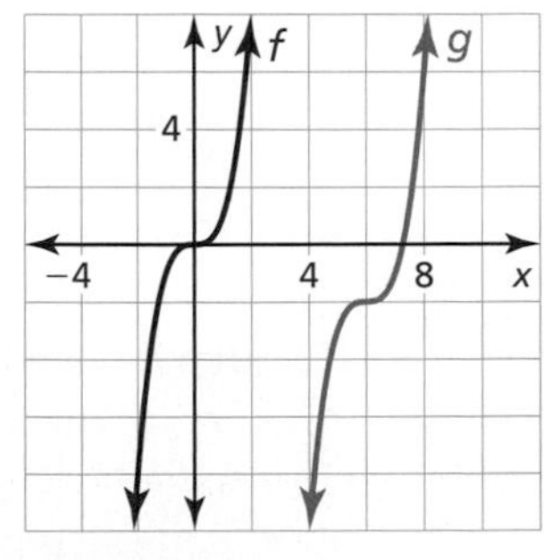

▶ Because $h = 6$ and $k = -2$, the graph of g is a translation 6 units right and 2 units down of the graph of f.

Describe the transformation of f represented by g. Then graph each function.

34. $f(x) = x^3$, $g(x) = (-x)^3 + 2$

35. $f(x) = x^4$, $g(x) = -(x + 9)^4$

Write a rule for g.

36. Let the graph of g be a horizontal stretch by a factor of 4, followed by a translation 3 units right and 5 units down of the graph of $f(x) = x^5 + 3x$.

37. Let the graph of g be a translation 5 units up, followed by a reflection in the y-axis of the graph of $f(x) = x^4 - 2x^3 - 12$.

4.8 Analyzing Graphs of Polynomial Functions (pp. 211–218)

Graph the function $f(x) = x(x + 2)(x - 2)$. Then estimate the points where the local maximums and local minimums occur.

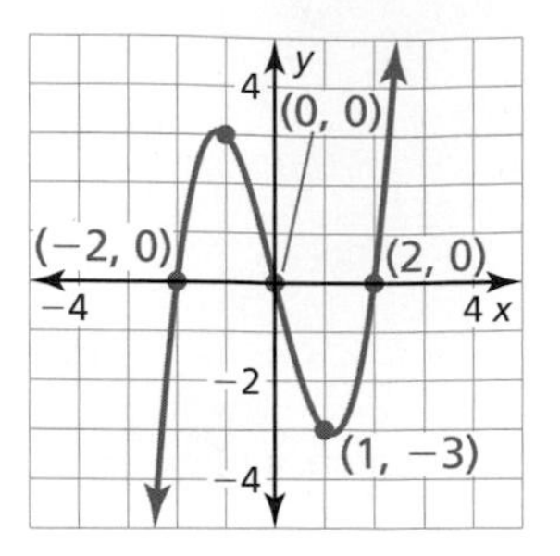

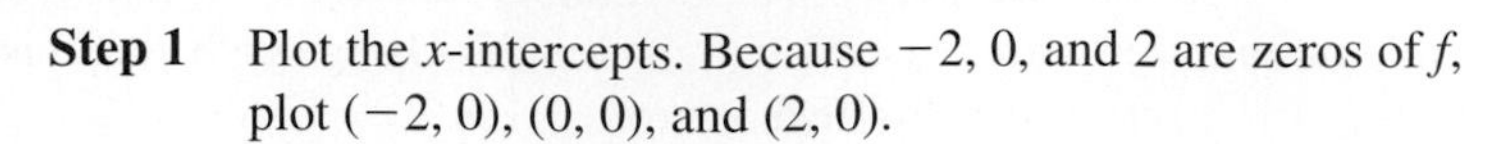

Step 1 Plot the x-intercepts. Because -2, 0, and 2 are zeros of f, plot $(-2, 0)$, $(0, 0)$, and $(2, 0)$.

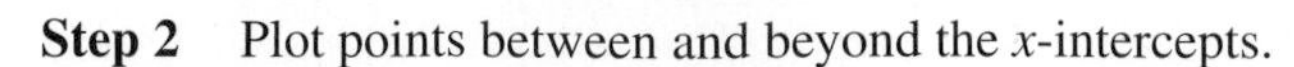

Step 2 Plot points between and beyond the x-intercepts.

x	−3	−2	−1	0	1	2	3
y	−15	0	3	0	−3	0	15

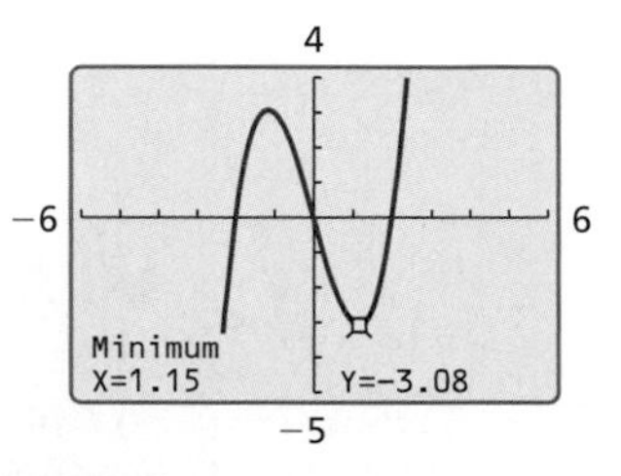

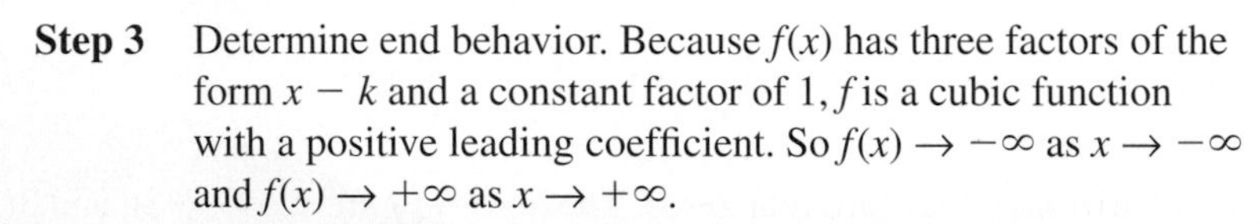

Step 3 Determine end behavior. Because $f(x)$ has three factors of the form $x - k$ and a constant factor of 1, f is a cubic function with a positive leading coefficient. So $f(x) \to -\infty$ as $x \to -\infty$ and $f(x) \to +\infty$ as $x \to +\infty$.

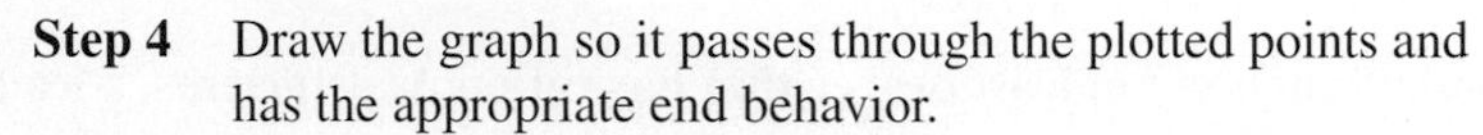

Step 4 Draw the graph so it passes through the plotted points and has the appropriate end behavior.

▶ The function has a local maximum at $(-1.15, 3.08)$ and a local minimum at $(1.15, -3.08)$.

Graph the function. Identify the x-intercepts and the points where the local maximums and local minimums occur. Determine the intervals for which the function is increasing or decreasing.

38. $f(x) = -2x^3 - 3x^2 - 1$

39. $f(x) = x^4 + 3x^3 - x^2 - 8x + 2$

Determine whether the function is *even*, *odd*, or *neither*.

40. $f(x) = 2x^3 + 3x$

41. $g(x) = 3x^2 - 7$

42. $h(x) = x^6 + 3x^5$

4.9 Modeling with Polynomial Functions (pp. 219–224)

Write the cubic function whose graph is shown.

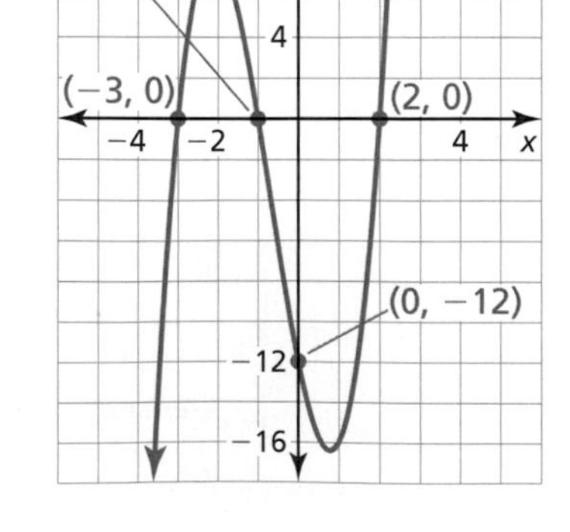

Step 1 Use the three x-intercepts to write the function in factored form.

$$f(x) = a(x + 3)(x + 1)(x - 2)$$

Step 2 Find the value of a by substituting the coordinates of the point $(0, -12)$.

$$-12 = a(0 + 3)(0 + 1)(0 - 2)$$

$$-12 = -6a$$

$$2 = a$$

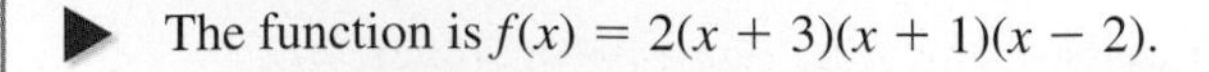

▶ The function is $f(x) = 2(x + 3)(x + 1)(x - 2)$.

43. Write a cubic function whose graph passes through the points $(-4, 0)$, $(4, 0)$, $(0, 6)$, and $(2, 0)$.

44. Use finite differences to determine the degree of the polynomial function that fits the data. Then use technology to find the polynomial function.

x	1	2	3	4	5	6	7
$f(x)$	−11	−24	−27	−8	45	144	301

4 Chapter Test

Write a polynomial function f of least degree that has rational coefficients, a leading coefficient of 1, and the given zeros.

1. $3, 1 - \sqrt{2}$

2. $-2, 4, 3i$

Find the product or quotient.

3. $(x^6 - 4)(x^2 - 7x + 5)$

4. $(3x^4 - 2x^3 - x - 1) \div (x^2 - 2x + 1)$

5. $(2x^3 - 3x^2 + 5x - 1) \div (x + 2)$

6. $(2x + 3)^3$

7. The graphs of $f(x) = x^4$ and $g(x) = (x - 3)^4$ are shown.

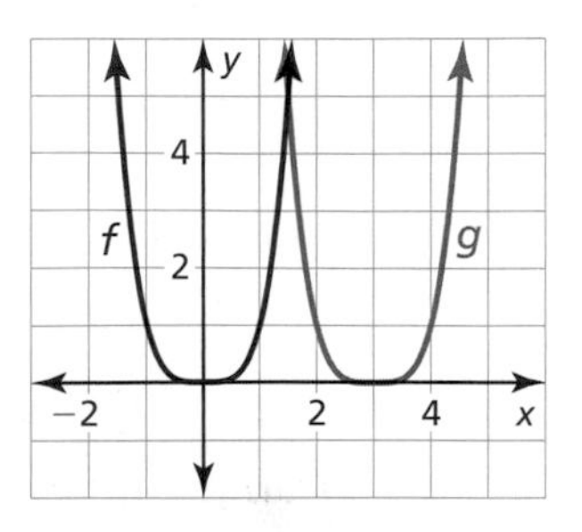

a. How many zeros does each function have? Explain.

b. Describe the transformation of f represented by g.

c. Determine the intervals for which the function g is increasing or decreasing.

8. The volume V (in cubic feet) of an aquarium is modeled by the polynomial function $V(x) = x^3 + 2x^2 - 13x + 10$, where x is the length of the tank.

a. Explain how you know $x = 4$ is *not* a possible rational zero.

b. Show that $x - 1$ is a factor of $V(x)$. Then factor $V(x)$ completely.

c. Find the dimensions of the aquarium shown.

Volume = 3 ft^3

9. One special product pattern is $(a - b)^2 = a^2 - 2ab + b^2$. Using Pascal's Triangle to expand $(a - b)^2$ gives $1a^2 + 2a(-b) + 1(-b)^2$. Are the two expressions equivalent? Explain.

10. Can you use the synthetic division procedure that you learned in this chapter to divide *any* two polynomials? Explain.

11. Let T be the number (in thousands) of new truck sales. Let C be the number (in thousands) of new car sales. During a 10-year period, T and C can be modeled by the following equations where t is time (in years).

$$T = 23t^4 - 330t^3 + 3500t^2 - 7500t + 9000$$

$$C = 14t^4 - 330t^3 + 2400t^2 - 5900t + 8900$$

a. Find a new model S for the total number of new vehicle sales.

b. Is the function S *even*, *odd*, or *neither*? Explain your reasoning.

12. Your friend has started a golf caddy business. The table shows the profits p (in dollars) of the business in the first 5 months. Use finite differences to find a polynomial model for the data. Then use the model to predict the profit after 7 months.

Month, t	1	2	3	4	5
Profit, p	4	2	6	22	56

4 Cumulative Assessment

1. The synthetic division below represents $f(x) \div (x - 3)$. Choose a value for m so that $x - 3$ is a factor of $f(x)$. Justify your answer.

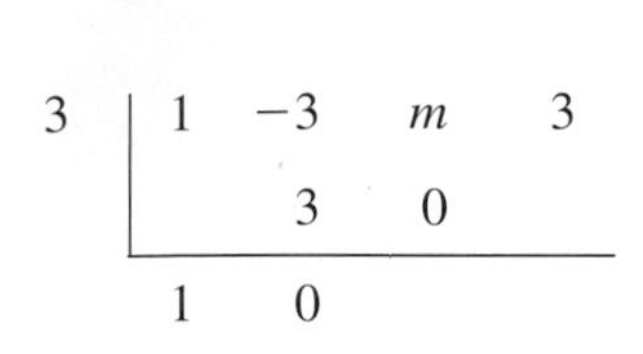

$$\begin{array}{r|rrrr} 3 & 1 & -3 & m & 3 \\ & & 3 & 0 & \\ \hline & 1 & 0 & & \end{array}$$

−3 3

−2 2

−1 1

2. Analyze the graph of the polynomial function to determine the sign of the leading coefficient, the degree of the function, and the number of real zeros. Explain.

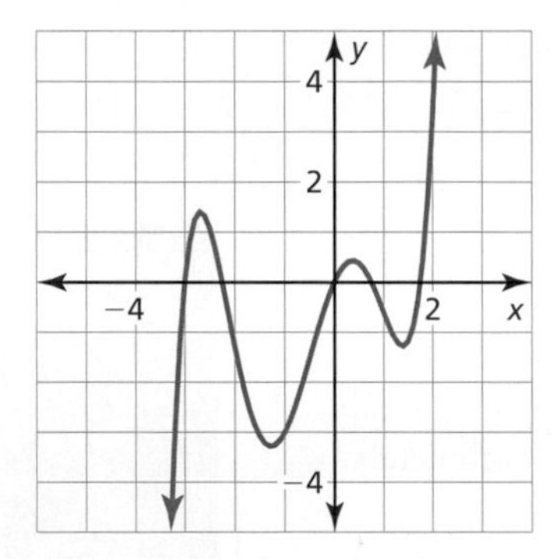

3. Which statement about the graph of the equation $12(x - 6) = -(y + 4)^2$ is *not* true?

Ⓐ The vertex is $(6, -4)$.

Ⓑ The axis of symmetry is $y = -4$.

Ⓒ The focus is $(3, -4)$.

Ⓓ The graph represents a function.

4. A parabola passes through the point shown in the graph. The equation of the axis of symmetry is $x = -a$. Which of the given points could lie on the parabola? If the axis of symmetry was $x = a$, then which points could lie on the parabola? Explain your reasoning.

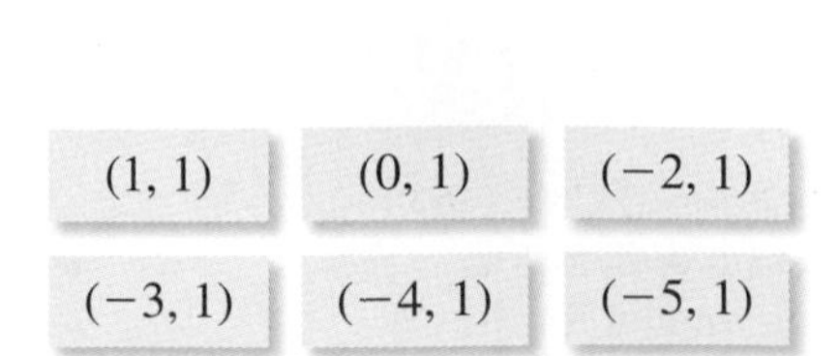

(1, 1) (0, 1) (−2, 1)

(−3, 1) (−4, 1) (−5, 1)

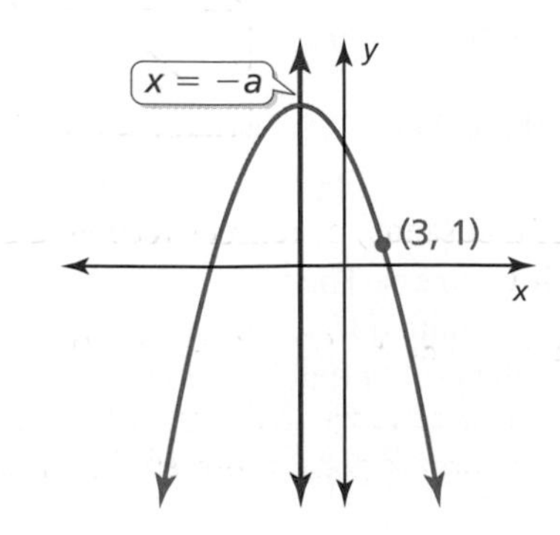

5. Select values for the function to model each transformation of the graph of $f(x) = x$.

$$g(x) = \square\left(x - \square\right) + \square$$

a. The graph is a translation 2 units up and 3 units left.

b. The graph is a translation 2 units right and 3 units down.

c. The graph is a vertical stretch by a factor of 2, followed by a translation 2 units up.

d. The graph is a translation 3 units right and a vertical shrink by a factor of $\frac{1}{2}$, followed by a translation 4 units down.

6. The diagram shows a circle inscribed in a square. The area of the shaded region is 21.5 square meters. To the nearest tenth of a meter, how long is each side of the square?

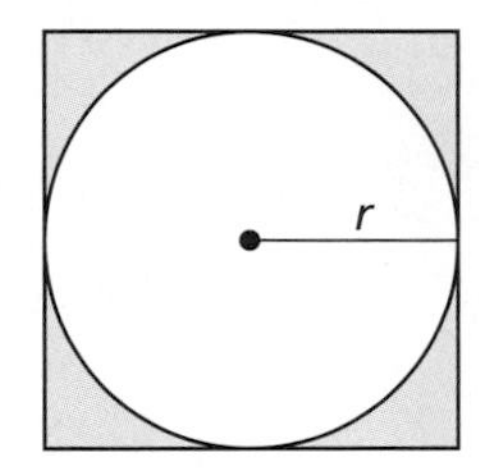

(A) 4.6 meters (B) 8.7 meters (C) 9.7 meters (D) 10.0 meters

7. Classify each function as *even*, *odd*, or *neither*. Justify your answer.

a. $f(x) = 3x^5$

b. $f(x) = 4x^3 + 8x$

c. $f(x) = 3x^3 + 12x^2 + 1$

d. $f(x) = 2x^4$

e. $f(x) = x^{11} - x^7$

f. $f(x) = 2x^8 + 4x^4 + x^2 - 5$

8. The volume of the rectangular prism shown is given by $V = 2x^3 + 7x^2 - 18x - 63$. Which polynomial represents the area of the base of the prism?

(A) $2x^2 + x - 21$

(B) $2x^2 + 21 - x$

(C) $13x + 21 + 2x^2$

(D) $2x^2 - 21 - 13x$

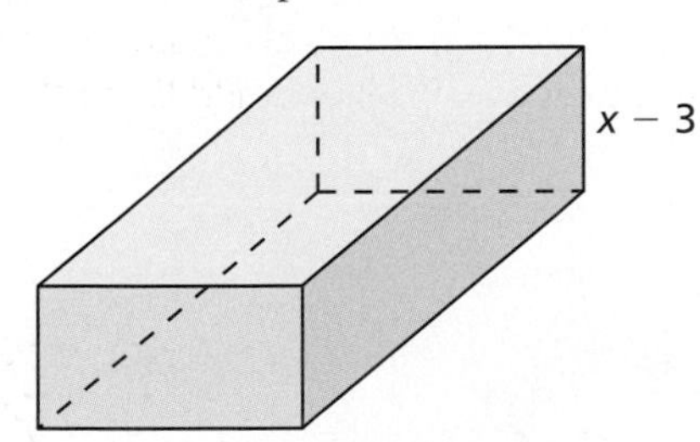

9. The number R (in tens of thousands) of retirees receiving Social Security benefits is represented by the function

$$R = 0.286t^3 - 4.68t^2 + 8.8t + 403, \quad 0 \leq t \leq 10$$

where t represents the number of years since 2000. Identify any turning points on the given interval. What does a turning point represent in this situation?

5 Rational Exponents and Radical Functions

Hull Speed *(p. 282)*

White Rhino *(p. 272)*

Concert *(p. 268)*

Mars Rover *(p. 254)*

Constellations *(p. 250)*

Maintaining Mathematical Proficiency

Properties of Integer Exponents

Example 1 **Simplify the expression $\dfrac{x^5 \cdot x^2}{x^3}$.**

$$\frac{x^5 \cdot x^2}{x^3} = \frac{x^{5+2}}{x^3} \qquad \text{Product of Powers Property}$$

$$= \frac{x^7}{x^3} \qquad \text{Add exponents.}$$

$$= x^{7-3} \qquad \text{Quotient of Powers Property}$$

$$= x^4 \qquad \text{Subtract exponents.}$$

Example 2 **Simplify the expression $\left(\dfrac{2s^3}{t}\right)^2$.**

$$\left(\frac{2s^3}{t}\right)^2 = \frac{(2s^3)^2}{t^2} \qquad \text{Power of a Quotient Property}$$

$$= \frac{2^2 \cdot (s^3)^2}{t^2} \qquad \text{Power of a Product Property}$$

$$= \frac{4s^6}{t^2} \qquad \text{Power of a Power Property}$$

Simplify the expression.

1. $y^6 \cdot y$

2. $\dfrac{n^4}{n^3}$

3. $\dfrac{x^5}{x^6 \cdot x^2}$

4. $\dfrac{x^6}{x^5} \cdot 3x^2$

5. $\left(\dfrac{4w^3}{2z^2}\right)^3$

6. $\left(\dfrac{m^7 \cdot m}{z^2 \cdot m^3}\right)^2$

Rewriting Literal Equations

Example 3 **Solve the literal equation $-5y - 2x = 10$ for y.**

$$-5y - 2x = 10 \qquad \text{Write the equation.}$$

$$-5y - 2x + 2x = 10 + 2x \qquad \text{Add } 2x \text{ to each side.}$$

$$-5y = 10 + 2x \qquad \text{Simplify.}$$

$$\frac{-5y}{-5} = \frac{10 + 2x}{-5} \qquad \text{Divide each side by } -5.$$

$$y = -2 - \frac{2}{5}x \qquad \text{Simplify.}$$

Solve the literal equation for *y*.

7. $4x + y = 2$

8. $x - \frac{1}{3}y = -1$

9. $2y - 9 = 13x$

10. $2xy + 6y = 10$

11. $8x - 4xy = 3$

12. $6x + 7xy = 15$

13. **ABSTRACT REASONING** Is the order in which you apply properties of exponents important? Explain your reasoning.

Mathematical Practices

Mathematically proficient students express numerical answers precisely.

Using Technology to Evaluate Roots

Core Concept

Evaluating Roots with a Calculator

	Example
Square root:	$\sqrt{64} = 8$
Cube root:	$\sqrt[3]{64} = 4$
Fourth root:	$\sqrt[4]{256} = 4$
Fifth root:	$\sqrt[5]{32} = 2$

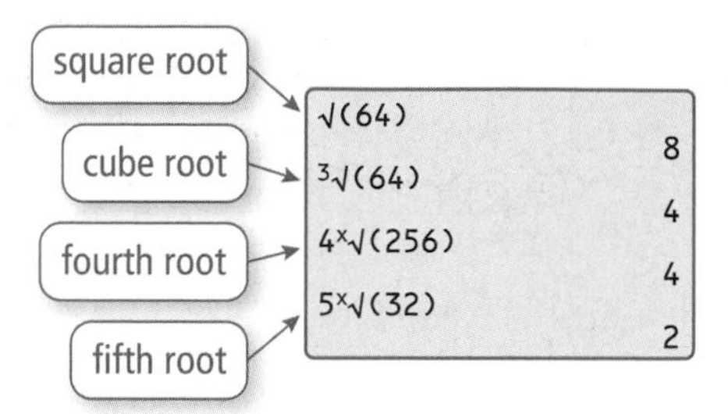

EXAMPLE 1 Approximating Roots

Evaluate each root using a calculator. Round your answer to two decimal places.

a. $\sqrt{50}$ **b.** $\sqrt[3]{50}$ **c.** $\sqrt[4]{50}$ **d.** $\sqrt[5]{50}$

SOLUTION

a. $\sqrt{50} \approx 7.07$ Round down.

b. $\sqrt[3]{50} \approx 3.68$ Round down.

c. $\sqrt[4]{50} \approx 2.66$ Round up.

d. $\sqrt[5]{50} \approx 2.19$ Round up.

√(50)	
	7.071067812
3√(50)	
	3.684031499
4x√(50)	
	2.659147948
5x√(50)	
	2.186724148

Monitoring Progress

1. Use the Pythagorean Theorem to find the exact lengths of a, b, c, and d in the figure.
2. Use a calculator to approximate each length to the nearest tenth of an inch.
3. Use a ruler to check the reasonableness of your answers.

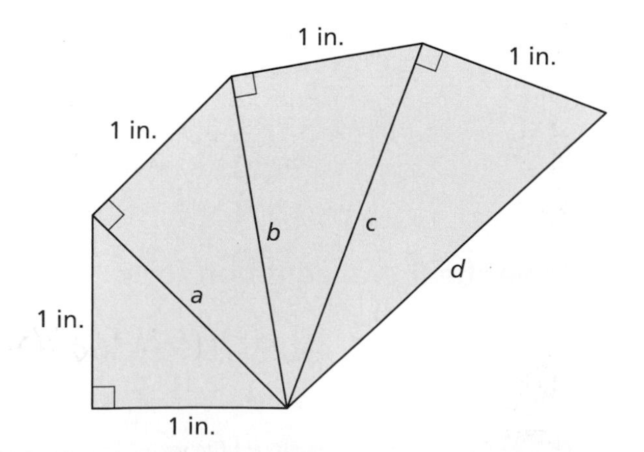

5.1 nth Roots and Rational Exponents

Essential Question How can you use a rational exponent to represent a power involving a radical?

Previously, you learned that the *n*th root of *a* can be represented as

$\sqrt[n]{a} = a^{1/n}$ Definition of rational exponent

for any real number *a* and integer *n* greater than 1.

EXPLORATION 1 Exploring the Definition of a Rational Exponent

Work with a partner. Use a calculator to show that each statement is true.

a. $\sqrt{9} = 9^{1/2}$ **b.** $\sqrt{2} = 2^{1/2}$ **c.** $\sqrt[3]{8} = 8^{1/3}$

d. $\sqrt[3]{3} = 3^{1/3}$ **e.** $\sqrt[4]{16} = 16^{1/4}$ **f.** $\sqrt[4]{12} = 12^{1/4}$

CONSTRUCTING VIABLE ARGUMENTS

To be proficient in math, you need to understand and use stated definitions and previously established results.

EXPLORATION 2 Writing Expressions in Rational Exponent Form

Work with a partner. Use the definition of a rational exponent and the properties of exponents to write each expression as a base with a single rational exponent. Then use a calculator to evaluate each expression. Round your answer to two decimal places.

Sample

$$\left(\sqrt[3]{4}\right)^2 = (4^{1/3})^2 = 4^{2/3} \approx 2.52$$

```
4^(2/3)
            2.5198421
```

a. $\left(\sqrt{5}\right)^3$ **b.** $\left(\sqrt[4]{4}\right)^2$ **c.** $\left(\sqrt[3]{9}\right)^2$

d. $\left(\sqrt[5]{10}\right)^4$ **e.** $\left(\sqrt{15}\right)^3$ **f.** $\left(\sqrt[3]{27}\right)^4$

EXPLORATION 3 Writing Expressions in Radical Form

Work with a partner. Use the properties of exponents and the definition of a rational exponent to write each expression as a radical raised to an exponent. Then use a calculator to evaluate each expression. Round your answer to two decimal places.

Sample $5^{2/3} = (5^{1/3})^2 = \left(\sqrt[3]{5}\right)^2 \approx 2.92$

a. $8^{2/3}$ **b.** $6^{5/2}$ **c.** $12^{3/4}$

d. $10^{3/2}$ **e.** $16^{3/2}$ **f.** $20^{6/5}$

Communicate Your Answer

4. How can you use a rational exponent to represent a power involving a radical?

5. Evaluate each expression *without* using a calculator. Explain your reasoning.

a. $4^{3/2}$ **b.** $32^{4/5}$ **c.** $625^{3/4}$

d. $49^{3/2}$ **e.** $125^{4/3}$ **f.** $100^{6/3}$

5.1 Lesson

Core Vocabulary

*n*th root of *a*, *p. 238*
index of a radical, *p. 238*

Previous
square root
cube root
exponent

What You Will Learn

- Find *n*th roots of numbers.
- Evaluate expressions with rational exponents.
- Solve equations using *n*th roots.

*n*th Roots

You can extend the concept of a square root to other types of roots. For example, 2 is a cube root of 8 because $2^3 = 8$. In general, for an integer n greater than 1, if $b^n = a$, then b is an ***n*th root of *a***. An nth root of a is written as $\sqrt[n]{a}$, where n is the **index** of the radical.

You can also write an nth root of a as a power of a. If you assume the Power of a Power Property applies to rational exponents, then the following is true.

$$(a^{1/2})^2 = a^{(1/2) \cdot 2} = a^1 = a$$

$$(a^{1/3})^3 = a^{(1/3) \cdot 3} = a^1 = a$$

$$(a^{1/4})^4 = a^{(1/4) \cdot 4} = a^1 = a$$

Because $a^{1/2}$ is a number whose square is a, you can write $\sqrt{a} = a^{1/2}$. Similarly, $\sqrt[3]{a} = a^{1/3}$ and $\sqrt[4]{a} = a^{1/4}$. In general, $\sqrt[n]{a} = a^{1/n}$ for any integer n greater than 1.

UNDERSTANDING MATHEMATICAL TERMS

When n is even and $a > 0$, there are two real roots. The positive root is called the *principal root*.

Core Concept

Real *n*th Roots of *a*

Let n be an integer ($n > 1$) and let a be a real number.

***n* is an even integer.**		***n* is an odd integer.**	
$a < 0$	No real nth roots	$a < 0$	One real nth root: $\sqrt[n]{a} = a^{1/n}$
$a = 0$	One real nth root: $\sqrt[n]{0} = 0$	$a = 0$	One real nth root: $\sqrt[n]{0} = 0$
$a > 0$	Two real nth roots: $\pm\sqrt[n]{a} = \pm a^{1/n}$	$a > 0$	One real nth root: $\sqrt[n]{a} = a^{1/n}$

EXAMPLE 1 Finding *n*th Roots

Find the indicated real nth root(s) of a.

a. $n = 3, a = -216$

b. $n = 4, a = 81$

SOLUTION

a. Because $n = 3$ is odd and $a = -216 < 0$, -216 has one real cube root. Because $(-6)^3 = -216$, you can write $\sqrt[3]{-216} = -6$ or $(-216)^{1/3} = -6$.

b. Because $n = 4$ is even and $a = 81 > 0$, 81 has two real fourth roots. Because $3^4 = 81$ and $(-3)^4 = 81$, you can write $\pm\sqrt[4]{81} = \pm 3$ or $\pm 81^{1/4} = \pm 3$.

Monitoring Progress

Help in English and Spanish at *BigIdeasMath.com*

Find the indicated real *n*th root(s) of *a*.

1. $n = 4, a = 16$

2. $n = 2, a = -49$

3. $n = 3, a = -125$

4. $n = 5, a = 243$

Rational Exponents

A rational exponent does not have to be of the form $\frac{1}{n}$. Other rational numbers, such as $\frac{3}{2}$ and $-\frac{1}{2}$, can also be used as exponents. Two properties of rational exponents are shown below.

Rational Exponents

Let $a^{1/n}$ be an nth root of a, and let m be a positive integer.

$$a^{m/n} = (a^{1/n})^m = (\sqrt[n]{a})^m$$

$$a^{-m/n} = \frac{1}{a^{m/n}} = \frac{1}{(a^{1/n})^m} = \frac{1}{(\sqrt[n]{a})^m},\ a \neq 0$$

EXAMPLE 2 Evaluating Expressions with Rational Exponents

Evaluate each expression.

a. $16^{3/2}$

b. $32^{-3/5}$

SOLUTION

	Rational Exponent Form	Radical Form
a.	$16^{3/2} = (16^{1/2})^3 = 4^3 = 64$	$16^{3/2} = (\sqrt{16})^3 = 4^3 = 64$
b.	$32^{-3/5} = \frac{1}{32^{3/5}} = \frac{1}{(32^{1/5})^3} = \frac{1}{2^3} = \frac{1}{8}$	$32^{-3/5} = \frac{1}{32^{3/5}} = \frac{1}{(\sqrt[5]{32})^3} = \frac{1}{2^3} = \frac{1}{8}$

When using a calculator to approximate an nth root, you may want to rewrite the nth root in rational exponent form.

COMMON ERROR

Be sure to use parentheses to enclose a rational exponent: 9^(1/5) ≈ 1.55. Without them, the calculator evaluates a power and then divides: 9^1/5 = 1.8.

EXAMPLE 3 Approximating Expressions with Rational Exponents

Evaluate each expression using a calculator. Round your answer to two decimal places.

a. $9^{1/5}$

b. $12^{3/8}$

c. $(\sqrt[4]{7})^3$

SOLUTION

a. $9^{1/5} \approx 1.55$

b. $12^{3/8} \approx 2.54$

c. Before evaluating $(\sqrt[4]{7})^3$, rewrite the expression in rational exponent form.

$$(\sqrt[4]{7})^3 = 7^{3/4} \approx 4.30$$

```
9^(1/5)
             1.551845574
12^(3/8)
             2.539176951
7^(3/4)
             4.303517071
```

Monitoring Progress  Help in English and Spanish at *BigIdeasMath.com*

Evaluate the expression without using a calculator.

5. $4^{5/2}$ **6.** $9^{-1/2}$ **7.** $81^{3/4}$ **8.** $1^{7/8}$

Evaluate the expression using a calculator. Round your answer to two decimal places when appropriate.

9. $6^{2/5}$ **10.** $64^{-2/3}$ **11.** $(\sqrt[4]{16})^5$ **12.** $(\sqrt[3]{-30})^2$

Solving Equations Using *n*th Roots

To solve an equation of the form $u^n = d$, where u is an algebraic expression, take the nth root of each side.

EXAMPLE 4 Solving Equations Using *n*th Roots

Find the real solution(s) of (a) $4x^5 = 128$ and (b) $(x - 3)^4 = 21$.

SOLUTION

a. $4x^5 = 128$ — Write original equation.

$x^5 = 32$ — Divide each side by 4.

$x = \sqrt[5]{32}$ — Take fifth root of each side.

$x = 2$ — Simplify.

▶ The solution is $x = 2$.

COMMON ERROR

When n is even and $a > 0$, be sure to consider both the positive and negative nth roots of a.

b. $(x - 3)^4 = 21$ — Write original equation.

$x - 3 = \pm\sqrt[4]{21}$ — Take fourth root of each side.

$x = 3 \pm \sqrt[4]{21}$ — Add 3 to each side.

$x = 3 + \sqrt[4]{21}$ or $x = 3 - \sqrt[4]{21}$ — Write solutions separately.

$x \approx 5.14$ or $x \approx 0.86$ — Use a calculator.

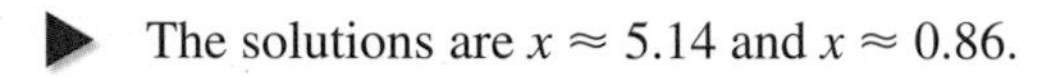

▶ The solutions are $x \approx 5.14$ and $x \approx 0.86$.

EXAMPLE 5 Real-Life Application

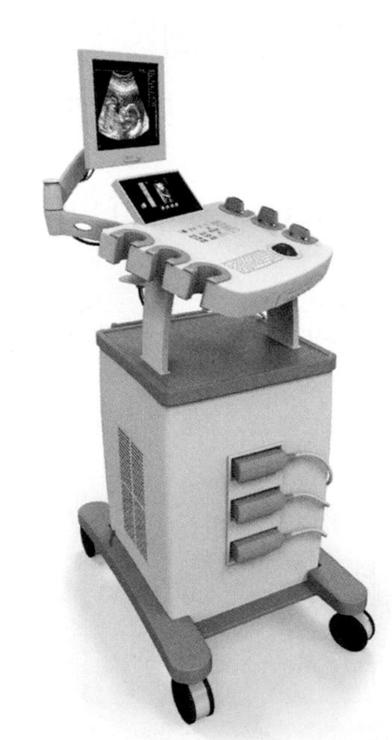

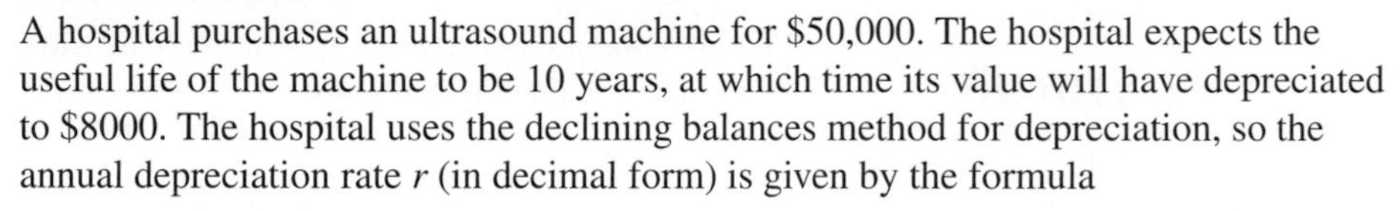

A hospital purchases an ultrasound machine for \$50,000. The hospital expects the useful life of the machine to be 10 years, at which time its value will have depreciated to \$8000. The hospital uses the declining balances method for depreciation, so the annual depreciation rate r (in decimal form) is given by the formula

$$r = 1 - \left(\frac{S}{C}\right)^{1/n}.$$

In the formula, n is the useful life of the item (in years), S is the salvage value (in dollars), and C is the original cost (in dollars). What annual depreciation rate did the hospital use?

SOLUTION

The useful life is 10 years, so $n = 10$. The machine depreciates to \$8000, so $S = 8000$. The original cost is \$50,000, so $C = 50{,}000$. So, the annual depreciation rate is

$$r = 1 - \left(\frac{S}{C}\right)^{1/n} = 1 - \left(\frac{8000}{50{,}000}\right)^{1/10} = 1 - \left(\frac{4}{25}\right)^{1/10} \approx 0.167.$$

▶ The annual depreciation rate is about 0.167, or 16.7%.

Monitoring Progress

Help in English and Spanish at *BigIdeasMath.com*

Find the real solution(s) of the equation. Round your answer to two decimal places when appropriate.

13. $8x^3 = 64$ **14.** $\frac{1}{2}x^5 = 512$ **15.** $(x + 5)^4 = 16$ **16.** $(x - 2)^3 = -14$

17. WHAT IF? In Example 5, what is the annual depreciation rate when the salvage value is \$6000?

5.1 Exercises

Dynamic Solutions available at *BigIdeasMath.com*

Vocabulary and Core Concept Check

1. **VOCABULARY** Rewrite the expression $a^{-s/t}$ in radical form. Then state the index of the radical.

2. **COMPLETE THE SENTENCE** For an integer n greater than 1, if $b^n = a$, then b is a(n) ___________ of a.

3. **WRITING** Explain how to use the sign of a to determine the number of real fourth roots of a and the number of real fifth roots of a.

4. **WHICH ONE DOESN'T BELONG?** Which expression does *not* belong with the other three? Explain your reasoning.

$(a^{1/n})^m$	$(\sqrt[n]{a})^m$	$(\sqrt[m]{a})^{-n}$	$a^{m/n}$

Monitoring Progress and Modeling with Mathematics

In Exercises 5–10, find the indicated real *n*th root(s) of *a*. *(See Example 1.)*

5. $n = 3, a = 8$
6. $n = 5, a = -1$
7. $n = 2, a = 0$
8. $n = 4, a = 256$
9. $n = 5, a = -32$
10. $n = 6, a = -729$

In Exercises 11–18, evaluate the expression without using a calculator. *(See Example 2.)*

11. $64^{1/6}$
12. $8^{1/3}$
13. $25^{3/2}$
14. $81^{3/4}$
15. $(-243)^{1/5}$
16. $(-64)^{4/3}$
17. $8^{-2/3}$
18. $16^{-7/4}$

ERROR ANALYSIS In Exercises 19 and 20, describe and correct the error in evaluating the expression.

19.

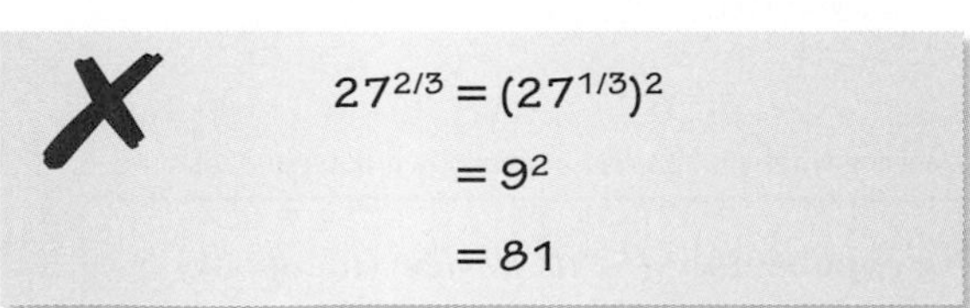

20.

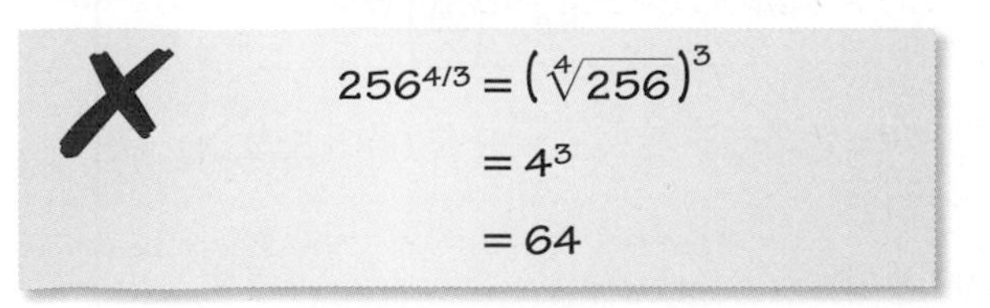

USING STRUCTURE In Exercises 21–24, match the equivalent expressions. Explain your reasoning.

21. $(\sqrt[3]{5})^4$
22. $(\sqrt[4]{5})^3$
23. $\dfrac{1}{\sqrt[4]{5}}$
24. $-\sqrt[4]{5}$

A. $5^{-1/4}$
B. $5^{4/3}$
C. $-5^{1/4}$
D. $5^{3/4}$

In Exercises 25–32, evaluate the expression using a calculator. Round your answer to two decimal places when appropriate. *(See Example 3.)*

25. $\sqrt[5]{32{,}768}$
26. $\sqrt[7]{1695}$
27. $25^{-1/3}$
28. $85^{1/6}$
29. $20{,}736^{4/5}$
30. $86^{-5/6}$
31. $(\sqrt[4]{187})^3$
32. $(\sqrt[5]{-8})^8$

MATHEMATICAL CONNECTIONS In Exercises 33 and 34, find the radius of the figure with the given volume.

33. $V = 216$ ft^3

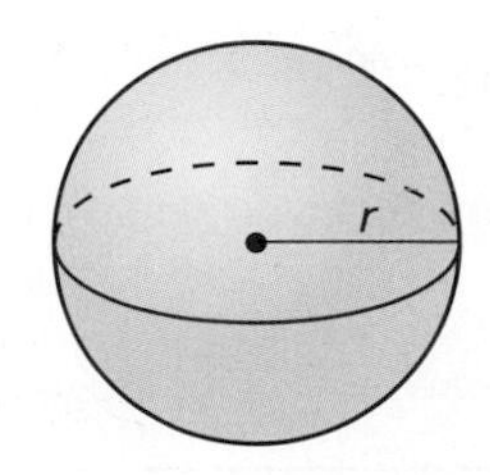

34. $V = 1332$ cm^3

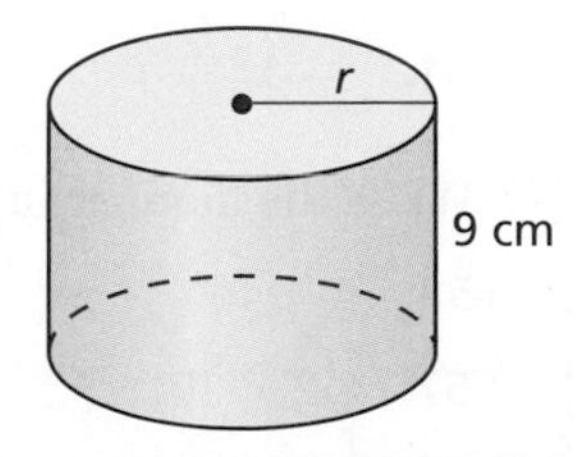

In Exercises 35–44, find the real solution(s) of the equation. Round your answer to two decimal places when appropriate. ***(See Example 4.)***

35. $x^3 = 125$

36. $5x^3 = 1080$

37. $(x + 10)^5 = 70$

38. $(x - 5)^4 = 256$

39. $x^5 = -48$

40. $7x^4 = 56$

41. $x^6 + 36 = 100$

42. $x^3 + 40 = 25$

43. $\frac{1}{3}x^4 = 27$

44. $\frac{1}{6}x^3 = -36$

45. MODELING WITH MATHEMATICS When the average price of an item increases from p_1 to p_2 over a period of n years, the annual rate of inflation r (in decimal form) is given by $r = \left(\frac{p_2}{p_1}\right)^{1/n} - 1$. Find the rate of inflation for each item in the table. ***(See Example 5.)***

Item	Price in 1913	Price in 2013
Potatoes (lb)	\$0.016	\$0.627
Ham (lb)	\$0.251	\$2.693
Eggs (dozen)	\$0.373	\$1.933

46. HOW DO YOU SEE IT? The graph of $y = x^n$ is shown in red. What can you conclude about the value of n? Determine the number of real nth roots of a. Explain your reasoning.

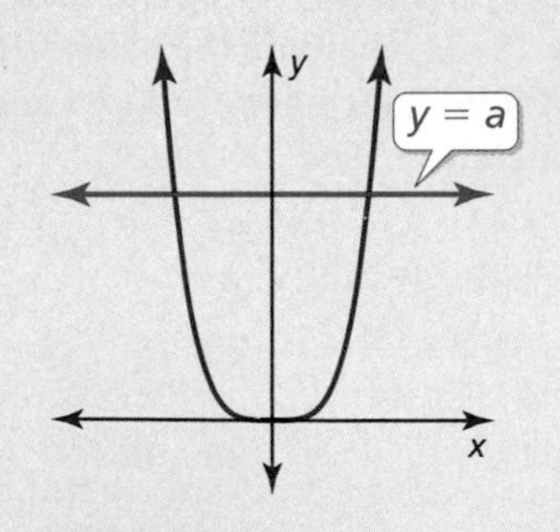

47. NUMBER SENSE Between which two consecutive integers does $\sqrt[4]{125}$ lie? Explain your reasoning.

48. THOUGHT PROVOKING In 1619, Johannes Kepler published his third law, which can be given by $d^3 = t^2$, where d is the mean distance (in astronomical units) of a planet from the Sun and t is the time (in years) it takes the planet to orbit the Sun. It takes Mars 1.88 years to orbit the Sun. Graph a possible location of Mars. Justify your answer. (The diagram shows the Sun at the origin of the xy-plane and a possible location of Earth.)

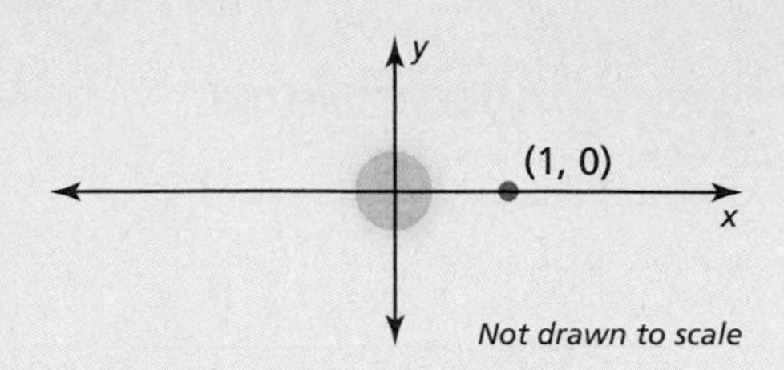

49. PROBLEM SOLVING A *weir* is a dam that is built across a river to regulate the flow of water. The flow rate Q (in cubic feet per second) can be calculated using the formula $Q = 3.367\ell h^{3/2}$, where ℓ is the length (in feet) of the bottom of the spillway and h is the depth (in feet) of the water on the spillway. Determine the flow rate of a weir with a spillway that is 20 feet long and has a water depth of 5 feet.

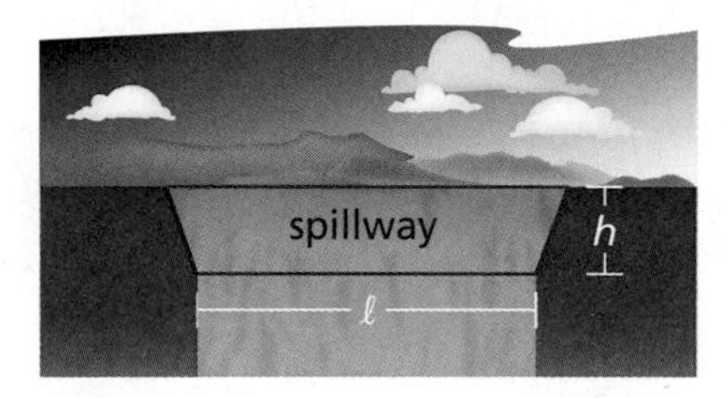

50. REPEATED REASONING The mass of the particles that a river can transport is proportional to the sixth power of the speed of the river. A certain river normally flows at a speed of 1 meter per second. What must its speed be in order to transport particles that are twice as massive as usual? 10 times as massive? 100 times as massive?

Maintaining Mathematical Proficiency

Reviewing what you learned in previous grades and lessons

Simplify the expression. Write your answer using only positive exponents. *(Skills Review Handbook)*

51. $5 \cdot 5^4$

52. $\frac{4^2}{4^7}$

53. $(z^2)^{-3}$

54. $\left(\frac{3x}{2}\right)^4$

Write the number in standard form. *(Skills Review Handbook)*

55. 5×10^3

56. 4×10^{-2}

57. 8.2×10^{-1}

58. 6.93×10^6

5.2 Properties of Rational Exponents and Radicals

Essential Question How can you use properties of exponents to simplify products and quotients of radicals?

EXPLORATION 1 Reviewing Properties of Exponents

Work with a partner. Let a and b be real numbers. Use the properties of exponents to complete each statement. Then match each completed statement with the property it illustrates.

Statement	Property
a. $a^{-2} = \_\_\_\_\_\_$, $a \neq 0$	**A.** Product of Powers
b. $(ab)^4 = \_\_\_\_\_\_$	**B.** Power of a Power
c. $(a^3)^4 = \_\_\_\_\_\_$	**C.** Power of a Product
d. $a^3 \cdot a^4 = \_\_\_\_\_\_$	**D.** Negative Exponent
e. $\left(\frac{a}{b}\right)^3 = \_\_\_\_\_\_$, $b \neq 0$	**E.** Zero Exponent
f. $\frac{a^6}{a^2} = \_\_\_\_\_\_$, $a \neq 0$	**F.** Quotient of Powers
g. $a^0 = \_\_\_\_\_\_$, $a \neq 0$	**G.** Power of a Quotient

USING TOOLS STRATEGICALLY

To be proficient in math, you need to consider the tools available to help you check your answers. For instance, the following calculator screen shows that $\sqrt[3]{4} \cdot \sqrt[3]{2}$ and $\sqrt[3]{8}$ are equivalent.

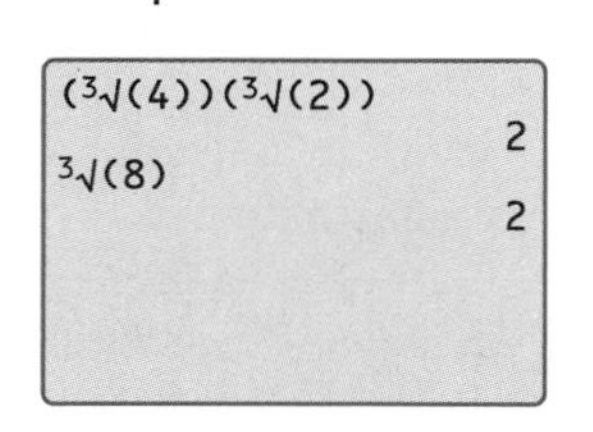

EXPLORATION 2 Simplifying Expressions with Rational Exponents

Work with a partner. Show that you can apply the properties of integer exponents to rational exponents by simplifying each expression. Use a calculator to check your answers.

a. $5^{2/3} \cdot 5^{4/3}$ **b.** $3^{1/5} \cdot 3^{4/5}$ **c.** $(4^{2/3})^3$

d. $(10^{1/2})^4$ **e.** $\frac{8^{5/2}}{8^{1/2}}$ **f.** $\frac{7^{2/3}}{7^{5/3}}$

EXPLORATION 3 Simplifying Products and Quotients of Radicals

Work with a partner. Use the properties of exponents to write each expression as a single radical. Then evaluate each expression. Use a calculator to check your answers.

a. $\sqrt{3} \cdot \sqrt{12}$ **b.** $\sqrt[3]{5} \cdot \sqrt[3]{25}$ **c.** $\sqrt[4]{27} \cdot \sqrt[4]{3}$

d. $\frac{\sqrt{98}}{\sqrt{2}}$ **e.** $\frac{\sqrt[4]{4}}{\sqrt[4]{1024}}$ **f.** $\frac{\sqrt[3]{625}}{\sqrt[3]{5}}$

Communicate Your Answer

4. How can you use properties of exponents to simplify products and quotients of radicals?

5. Simplify each expression.

a. $\sqrt{27} \cdot \sqrt{6}$ **b.** $\frac{\sqrt[3]{240}}{\sqrt[3]{15}}$ **c.** $(5^{1/2} \cdot 16^{1/4})^2$

5.2 Lesson

Core Vocabulary

simplest form of a radical, *p. 245*
conjugate, *p. 246*
like radicals, *p. 246*

Previous
properties of integer exponents
rationalizing the denominator
absolute value

What You Will Learn

- Use properties of rational exponents to simplify expressions with rational exponents.
- Use properties of radicals to simplify and write radical expressions in simplest form.

Properties of Rational Exponents

The properties of integer exponents that you have previously learned can also be applied to rational exponents.

COMMON ERROR

When you multiply powers, *do not* multiply the exponents. For example, $3^2 \cdot 3^5 \neq 3^{10}$.

Core Concept

Properties of Rational Exponents

Let a and b be real numbers and let m and n be rational numbers, such that the quantities in each property are real numbers.

Property Name	Definition	Example
Product of Powers	$a^m \cdot a^n = a^{m+n}$	$5^{1/2} \cdot 5^{3/2} = 5^{(1/2+3/2)} = 5^2 = 25$
Power of a Power	$(a^m)^n = a^{mn}$	$(3^{5/2})^2 = 3^{(5/2 \cdot 2)} = 3^5 = 243$
Power of a Product	$(ab)^m = a^m b^m$	$(16 \cdot 9)^{1/2} = 16^{1/2} \cdot 9^{1/2} = 4 \cdot 3 = 12$
Negative Exponent	$a^{-m} = \dfrac{1}{a^m}, a \neq 0$	$36^{-1/2} = \dfrac{1}{36^{1/2}} = \dfrac{1}{6}$
Zero Exponent	$a^0 = 1, a \neq 0$	$213^0 = 1$
Quotient of Powers	$\dfrac{a^m}{a^n} = a^{m-n}, a \neq 0$	$\dfrac{4^{5/2}}{4^{1/2}} = 4^{(5/2-1/2)} = 4^2 = 16$
Power of a Quotient	$\left(\dfrac{a}{b}\right)^m = \dfrac{a^m}{b^m}, b \neq 0$	$\left(\dfrac{27}{64}\right)^{1/3} = \dfrac{27^{1/3}}{64^{1/3}} = \dfrac{3}{4}$

EXAMPLE 1 Using Properties of Exponents

Use the properties of rational exponents to simplify each expression.

a. $7^{1/4} \cdot 7^{1/2} = 7^{(1/4+1/2)} = 7^{3/4}$

b. $(6^{1/2} \cdot 4^{1/3})^2 = (6^{1/2})^2 \cdot (4^{1/3})^2 = 6^{(1/2 \cdot 2)} \cdot 4^{(1/3 \cdot 2)} = 6^1 \cdot 4^{2/3} = 6 \cdot 4^{2/3}$

c. $(4^5 \cdot 3^5)^{-1/5} = [(4 \cdot 3)^5]^{-1/5} = (12^5)^{-1/5} = 12^{[5 \cdot (-1/5)]} = 12^{-1} = \dfrac{1}{12}$

d. $\dfrac{5}{5^{1/3}} = \dfrac{5^1}{5^{1/3}} = 5^{(1-1/3)} = 5^{2/3}$

e. $\left(\dfrac{42^{1/3}}{6^{1/3}}\right)^2 = \left[\left(\dfrac{42}{6}\right)^{1/3}\right]^2 = (7^{1/3})^2 = 7^{(1/3 \cdot 2)} = 7^{2/3}$

Monitoring Progress

Help in English and Spanish at *BigIdeasMath.com*

Simplify the expression.

1. $2^{3/4} \cdot 2^{1/2}$

2. $\dfrac{3}{3^{1/4}}$

3. $\left(\dfrac{20^{1/2}}{5^{1/2}}\right)^3$

4. $(5^{1/3} \cdot 7^{1/4})^3$

Simplifying Radical Expressions

The Power of a Product and Power of a Quotient properties can be expressed using radical notation when $m = \frac{1}{n}$ for some integer n greater than 1.

Core Concept

Properties of Radicals

Let a and b be real numbers and let n be an integer greater than 1.

Property Name	Definition	Example
Product Property	$\sqrt[n]{a \cdot b} = \sqrt[n]{a} \cdot \sqrt[n]{b}$	$\sqrt[3]{4} \cdot \sqrt[3]{2} = \sqrt[3]{8} = 2$
Quotient Property	$\sqrt[n]{\frac{a}{b}} = \frac{\sqrt[n]{a}}{\sqrt[n]{b}}, b \neq 0$	$\frac{\sqrt[4]{162}}{\sqrt[4]{2}} = \sqrt[4]{\frac{162}{2}} = \sqrt[4]{81} = 3$

EXAMPLE 2 Using Properties of Radicals

Use the properties of radicals to simplify each expression.

a. $\sqrt[3]{12} \cdot \sqrt[3]{18} = \sqrt[3]{12 \cdot 18} = \sqrt[3]{216} = 6$ Product Property of Radicals

b. $\frac{\sqrt[4]{80}}{\sqrt[4]{5}} = \sqrt[4]{\frac{80}{5}} = \sqrt[4]{16} = 2$ Quotient Property of Radicals

An expression involving a radical with index n is in **simplest form** when these three conditions are met.

- No radicands have perfect nth powers as factors other than 1.
- No radicands contain fractions.
- No radicals appear in the denominator of a fraction.

To meet the last two conditions, rationalize the denominator by multiplying the expression by an appropriate form of 1 that eliminates the radical from the denominator.

EXAMPLE 3 Writing Radicals in Simplest Form

Write each expression in simplest form.

a. $\sqrt[3]{135}$

b. $\frac{\sqrt[5]{7}}{\sqrt[5]{8}}$

SOLUTION

a. $\sqrt[3]{135} = \sqrt[3]{27 \cdot 5}$ Factor out perfect cube.

$= \sqrt[3]{27} \cdot \sqrt[3]{5}$ Product Property of Radicals

$= 3\sqrt[3]{5}$ Simplify.

b. $\frac{\sqrt[5]{7}}{\sqrt[5]{8}} = \frac{\sqrt[5]{7}}{\sqrt[5]{8}} \cdot \frac{\sqrt[5]{4}}{\sqrt[5]{4}}$ Make the radicand in the denominator a perfect fifth power.

$= \frac{\sqrt[5]{28}}{\sqrt[5]{32}}$ Product Property of Radicals

$= \frac{\sqrt[5]{28}}{2}$ Simplify.

For a denominator that is a sum or difference involving square roots, multiply both the numerator and denominator by the **conjugate** of the denominator. The expressions

$$a\sqrt{b} + c\sqrt{d} \qquad \text{and} \qquad a\sqrt{b} - c\sqrt{d}$$

are conjugates of each other, where a, b, c, and d are rational numbers.

EXAMPLE 4 Writing a Radical Expression in Simplest Form

Write $\frac{1}{5 + \sqrt{3}}$ in simplest form.

SOLUTION

$$\frac{1}{5+\sqrt{3}} = \frac{1}{5+\sqrt{3}} \cdot \frac{5-\sqrt{3}}{5-\sqrt{3}}$$ The conjugate of $5 + \sqrt{3}$ is $5 - \sqrt{3}$.

$$= \frac{1(5-\sqrt{3})}{5^2 - (\sqrt{3})^2}$$ Sum and Difference Pattern

$$= \frac{5-\sqrt{3}}{22}$$ Simplify.

Radical expressions with the same index and radicand are **like radicals**. To add or subtract like radicals, use the Distributive Property.

EXAMPLE 5 Adding and Subtracting Like Radicals and Roots

Simplify each expression.

a. $\sqrt[4]{10} + 7\sqrt[4]{10}$ **b.** $2(8^{1/5}) + 10(8^{1/5})$ **c.** $\sqrt[3]{54} - \sqrt[3]{2}$

SOLUTION

a. $\sqrt[4]{10} + 7\sqrt[4]{10} = (1 + 7)\sqrt[4]{10} = 8\sqrt[4]{10}$

b. $2(8^{1/5}) + 10(8^{1/5}) = (2 + 10)(8^{1/5}) = 12(8^{1/5})$

c. $\sqrt[3]{54} - \sqrt[3]{2} = \sqrt[3]{27} \cdot \sqrt[3]{2} - \sqrt[3]{2} = 3\sqrt[3]{2} - \sqrt[3]{2} = (3 - 1)\sqrt[3]{2} = 2\sqrt[3]{2}$

Monitoring Progress

Help in English and Spanish at *BigIdeasMath.com*

Simplify the expression.

5. $\sqrt[4]{27} \cdot \sqrt[4]{3}$ **6.** $\frac{\sqrt[3]{250}}{\sqrt[3]{2}}$ **7.** $\sqrt[3]{104}$ **8.** $\sqrt[5]{\frac{3}{4}}$

9. $\frac{3}{6 - \sqrt{2}}$ **10.** $7\sqrt[5]{12} - \sqrt[5]{12}$ **11.** $4(9^{2/3}) + 8(9^{2/3})$ **12.** $\sqrt[3]{5} + \sqrt[3]{40}$

The properties of rational exponents and radicals can also be applied to expressions involving variables. Because a variable can be positive, negative, or zero, sometimes absolute value is needed when simplifying a variable expression.

	Rule	Example
When *n* is odd	$\sqrt[n]{x^n} = x$	$\sqrt[7]{5^7} = 5$ and $\sqrt[7]{(-5)^7} = -5$
When *n* is even	$\sqrt[n]{x^n} = \lvert x \rvert$	$\sqrt[4]{3^4} = 3$ and $\sqrt[4]{(-3)^4} = 3$

Absolute value is not needed when all variables are assumed to be positive.

EXAMPLE 6 Simplifying Variable Expressions

Simplify each expression.

a. $\sqrt[3]{64y^6}$ **b.** $\sqrt[4]{\dfrac{x^4}{y^8}}$

STUDY TIP

You do not need to take the absolute value of *y* because *y* is being squared.

SOLUTION

a. $\sqrt[3]{64y^6} = \sqrt[3]{4^3(y^2)^3} = \sqrt[3]{4^3} \cdot \sqrt[3]{(y^2)^3} = 4y^2$

b. $\sqrt[4]{\dfrac{x^4}{y^8}} = \dfrac{\sqrt[4]{x^4}}{\sqrt[4]{y^8}} = \dfrac{\sqrt[4]{x^4}}{\sqrt[4]{(y^2)^4}} = \dfrac{|x|}{y^2}$

EXAMPLE 7 Writing Variable Expressions in Simplest Form

Write each expression in simplest form. Assume all variables are positive.

a. $\sqrt[5]{4a^8b^{14}c^5}$ **b.** $\dfrac{x}{\sqrt[3]{y^8}}$ **c.** $\dfrac{14xy^{1/3}}{2x^{3/4}z^{-6}}$

COMMON ERROR

You must multiply both the numerator *and* denominator of the fraction by $\sqrt[3]{y}$ so that the value of the fraction does not change.

SOLUTION

a. $\sqrt[5]{4a^8b^{14}c^5} = \sqrt[5]{4a^5a^3b^{10}b^4c^5}$ Factor out perfect fifth powers.

$= \sqrt[5]{a^5b^{10}c^5} \cdot \sqrt[5]{4a^3b^4}$ Product Property of Radicals

$= ab^2c\sqrt[5]{4a^3b^4}$ Simplify.

b. $\dfrac{x}{\sqrt[3]{y^8}} = \dfrac{x}{\sqrt[3]{y^8}} \cdot \dfrac{\sqrt[3]{y}}{\sqrt[3]{y}}$ Make denominator a perfect cube.

$= \dfrac{x\sqrt[3]{y}}{\sqrt[3]{y^9}}$ Product Property of Radicals

$= \dfrac{x\sqrt[3]{y}}{y^3}$ Simplify.

c. $\dfrac{14xy^{1/3}}{2x^{3/4}z^{-6}} = 7x^{(1-3/4)}y^{1/3}z^{-(-6)} = 7x^{1/4}y^{1/3}z^6$

EXAMPLE 8 Adding and Subtracting Variable Expressions

Perform each indicated operation. Assume all variables are positive.

a. $5\sqrt{y} + 6\sqrt{y}$ **b.** $12\sqrt[3]{2z^5} - z\sqrt[3]{54z^2}$

SOLUTION

a. $5\sqrt{y} + 6\sqrt{y} = (5 + 6)\sqrt{y} = 11\sqrt{y}$

b. $12\sqrt[3]{2z^5} - z\sqrt[3]{54z^2} = 12z\sqrt[3]{2z^2} - 3z\sqrt[3]{2z^2} = (12z - 3z)\sqrt[3]{2z^2} = 9z\sqrt[3]{2z^2}$

Monitoring Progress  Help in English and Spanish at *BigIdeasMath.com*

Simplify the expression. Assume all variables are positive.

13. $\sqrt[3]{27q^9}$ **14.** $\sqrt[5]{\dfrac{x^{10}}{y^5}}$ **15.** $\dfrac{6xy^{3/4}}{3x^{1/2}y^{1/2}}$ **16.** $\sqrt{9w^5} - w\sqrt{w^3}$

5.2 Exercises

Dynamic Solutions available at *BigIdeasMath.com*

Vocabulary and Core Concept Check

1. **WRITING** How do you know when a radical expression is in simplest form?

2. **WHICH ONE DOESN'T BELONG?** Which radical expression does *not* belong with the other three? Explain your reasoning.

$\sqrt[3]{\frac{4}{5}}$	$2\sqrt{x}$	$\sqrt[4]{11}$	$3\sqrt[5]{9x}$

Monitoring Progress and Modeling with Mathematics

In Exercises 3–12, use the properties of rational exponents to simplify the expression. *(See Example 1.)*

3. $(9^2)^{1/3}$
4. $(12^2)^{1/4}$
5. $\frac{6}{6^{1/4}}$
6. $\frac{7}{7^{1/3}}$
7. $\left(\frac{8^4}{10^4}\right)^{-1/4}$
8. $\left(\frac{9^3}{6^3}\right)^{-1/3}$
9. $(3^{-2/3} \cdot 3^{1/3})^{-1}$
10. $(5^{1/2} \cdot 5^{-3/2})^{-1/4}$
11. $\frac{2^{2/3} \cdot 16^{2/3}}{4^{2/3}}$
12. $\frac{49^{3/8} \cdot 49^{7/8}}{7^{5/4}}$

In Exercises 13–20, use the properties of radicals to simplify the expression. *(See Example 2.)*

13. $\sqrt{2} \cdot \sqrt{72}$
14. $\sqrt[3]{16} \cdot \sqrt[3]{32}$
15. $\sqrt[4]{6} \cdot \sqrt[4]{8}$
16. $\sqrt[4]{8} \cdot \sqrt[4]{8}$
17. $\frac{\sqrt[5]{486}}{\sqrt[5]{2}}$
18. $\frac{\sqrt{2}}{\sqrt{32}}$
19. $\frac{\sqrt[3]{6} \cdot \sqrt[3]{72}}{\sqrt[3]{2}}$
20. $\frac{\sqrt[3]{3} \cdot \sqrt[3]{18}}{\sqrt[6]{2} \cdot \sqrt[6]{2}}$

In Exercises 21–28, write the expression in simplest form. *(See Example 3.)*

21. $\sqrt[4]{567}$
22. $\sqrt[5]{288}$
23. $\frac{\sqrt[3]{5}}{\sqrt[3]{4}}$
24. $\frac{\sqrt[4]{4}}{\sqrt[4]{27}}$
25. $\sqrt{\frac{3}{8}}$
26. $\sqrt[3]{\frac{7}{4}}$
27. $\sqrt[3]{\frac{64}{49}}$
28. $\sqrt[4]{\frac{1296}{25}}$

In Exercises 29–36, write the expression in simplest form. *(See Example 4.)*

29. $\frac{1}{1 + \sqrt{3}}$
30. $\frac{1}{2 + \sqrt{5}}$
31. $\frac{5}{3 - \sqrt{2}}$
32. $\frac{11}{9 - \sqrt{6}}$
33. $\frac{9}{\sqrt{3} + \sqrt{7}}$
34. $\frac{2}{\sqrt{8} + \sqrt{7}}$
35. $\frac{\sqrt{6}}{\sqrt{3} - \sqrt{5}}$
36. $\frac{\sqrt{7}}{\sqrt{10} - \sqrt{2}}$

In Exercises 37–46, simplify the expression. *(See Example 5.)*

37. $9\sqrt[3]{11} + 3\sqrt[3]{11}$
38. $8\sqrt[6]{5} - 12\sqrt[6]{5}$
39. $3(11^{1/4}) + 9(11^{1/4})$
40. $13(8^{3/4}) - 4(8^{3/4})$
41. $5\sqrt{12} - 19\sqrt{3}$
42. $27\sqrt{6} + 7\sqrt{150}$
43. $\sqrt[5]{224} + 3\sqrt[5]{7}$
44. $7\sqrt[3]{2} - \sqrt[3]{128}$
45. $5(24^{1/3}) - 4(3^{1/3})$
46. $5^{1/4} + 6(405^{1/4})$

47. **ERROR ANALYSIS** Describe and correct the error in simplifying the expression.

$$\begin{aligned} 3\sqrt[3]{12} + 5\sqrt[3]{12} &= (3 + 5)\sqrt[3]{24} \\ &= 8\sqrt[3]{24} \\ &= 8\sqrt[3]{8 \cdot 3} \\ &= 8 \cdot 2\sqrt[3]{3} \\ &= 16\sqrt[3]{3} \end{aligned}$$

48. **MULTIPLE REPRESENTATIONS** Which radical expressions are like radicals?

Ⓐ $(5^{2/9})^{3/2}$ Ⓑ $\dfrac{5^3}{(\sqrt[3]{5})^8}$

Ⓒ $\sqrt[3]{625}$ Ⓓ $\sqrt[3]{5145} - \sqrt[3]{875}$

Ⓔ $\sqrt[3]{5} + 3\sqrt[3]{5}$ Ⓕ $7\sqrt[4]{80} - 2\sqrt[4]{405}$

In Exercises 49–54, simplify the expression. *(See Example 6.)*

49. $\sqrt[4]{81y^8}$

50. $\sqrt[3]{64r^3t^6}$

51. $\sqrt[5]{\dfrac{m^{10}}{n^5}}$

52. $\sqrt[4]{\dfrac{k^{16}}{16z^4}}$

53. $\sqrt[6]{\dfrac{g^6h}{h^7}}$

54. $\sqrt[8]{\dfrac{n^{18}p^7}{n^2p^{-1}}}$

55. **ERROR ANALYSIS** Describe and correct the error in simplifying the expression.

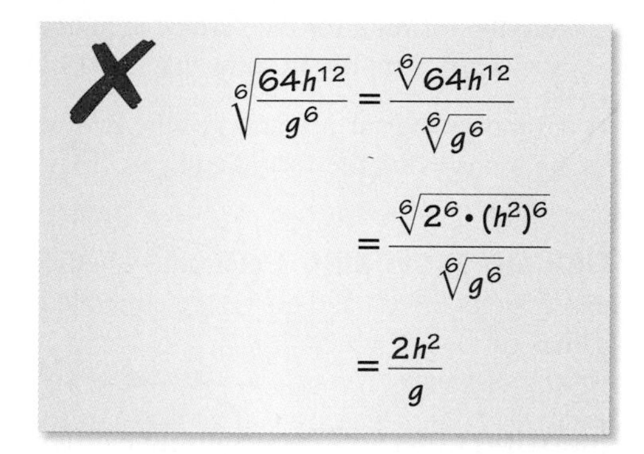

$\sqrt[6]{\dfrac{64h^{12}}{g^6}} = \dfrac{\sqrt[6]{64h^{12}}}{\sqrt[6]{g^6}}$
$= \dfrac{\sqrt[6]{2^6 \cdot (h^2)^6}}{\sqrt[6]{g^6}}$
$= \dfrac{2h^2}{g}$

56. **OPEN-ENDED** Write two variable expressions involving radicals, one that needs absolute value in simplifying and one that does not need absolute value. Justify your answers.

In Exercises 57–64, write the expression in simplest form. Assume all variables are positive. *(See Example 7.)*

57. $\sqrt{81a^7b^{12}c^9}$

58. $\sqrt[3]{125r^4s^9t^7}$

59. $\sqrt[5]{\dfrac{160m^6}{n^7}}$

60. $\sqrt[4]{\dfrac{405x^3y^3}{5x^{-1}y}}$

61. $\dfrac{\sqrt[3]{w} \cdot \sqrt{w^5}}{\sqrt{25w^{16}}}$

62. $\dfrac{\sqrt[4]{v^6}}{\sqrt[7]{v^5}}$

63. $\dfrac{18w^{1/3}v^{5/4}}{27w^{4/3}v^{1/2}}$

64. $\dfrac{7x^{-3/4}y^{5/2}z^{-2/3}}{56x^{-1/2}y^{1/4}}$

In Exercises 65–70, perform the indicated operation. Assume all variables are positive. *(See Example 8.)*

65. $12\sqrt[3]{y} + 9\sqrt[3]{y}$

66. $11\sqrt{2z} - 5\sqrt{2z}$

67. $3x^{7/2} - 5x^{7/2}$

68. $7\sqrt[3]{m^7} + 3m^{7/3}$

69. $\sqrt[4]{16w^{10}} + 2w\sqrt[4]{w^6}$

70. $(p^{1/2} \cdot p^{1/4}) - \sqrt[4]{16p^3}$

MATHEMATICAL CONNECTIONS In Exercises 71 and 72, find simplified expressions for the perimeter and area of the given figure.

71.

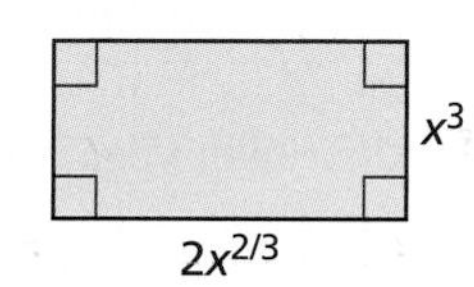

72.

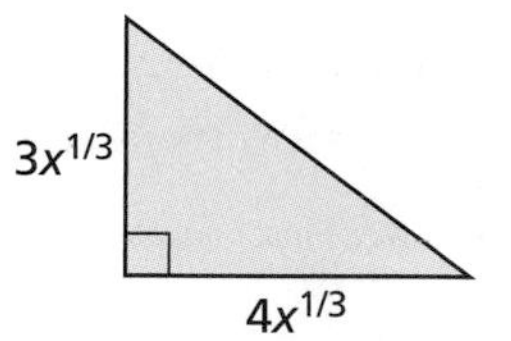

73. **MODELING WITH MATHEMATICS** The optimum diameter d (in millimeters) of the pinhole in a pinhole camera can be modeled by $d = 1.9[(5.5 \times 10^{-4})\ell]^{1/2}$, where ℓ is the length (in millimeters) of the camera box. Find the optimum pinhole diameter for a camera box with a length of 10 centimeters.

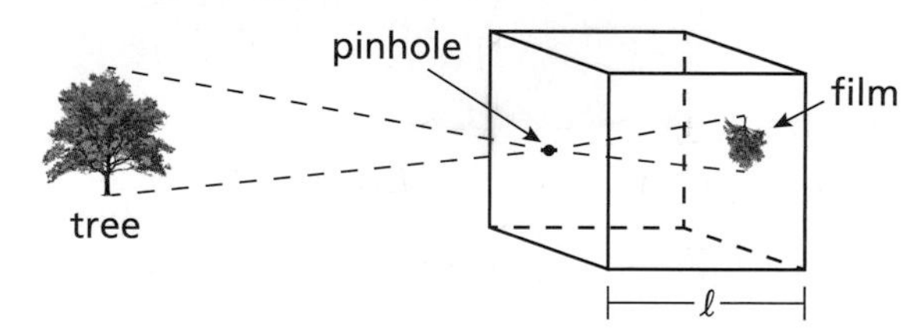

74. **MODELING WITH MATHEMATICS** The surface area S (in square centimeters) of a mammal can be modeled by $S = km^{2/3}$, where m is the mass (in grams) of the mammal and k is a constant. The table shows the values of k for different mammals.

Mammal	Rabbit	Human	Bat
Value of k	9.75	11.0	57.5

a. Find the surface area of a bat whose mass is 32 grams.

b. Find the surface area of a rabbit whose mass is 3.4 kilograms (3.4×10^3 grams).

c. Find the surface area of a human whose mass is 59 kilograms.

75. MAKING AN ARGUMENT Your friend claims it is not possible to simplify the expression $7\sqrt{11} - 9\sqrt{44}$ because it does not contain like radicals. Is your friend correct? Explain your reasoning.

76. PROBLEM SOLVING The apparent magnitude of a star is a number that indicates how faint the star is in relation to other stars. The expression $\frac{2.512^{m_1}}{2.512^{m_2}}$ tells how many times fainter a star with apparent magnitude m_1 is than a star with apparent magnitude m_2.

Star	Apparent magnitude	Constellation
Vega	0.03	Lyra
Altair	0.77	Aquila
Deneb	1.25	Cygnus

a. How many times fainter is Altair than Vega?

b. How many times fainter is Deneb than Altair?

c. How many times fainter is Deneb than Vega?

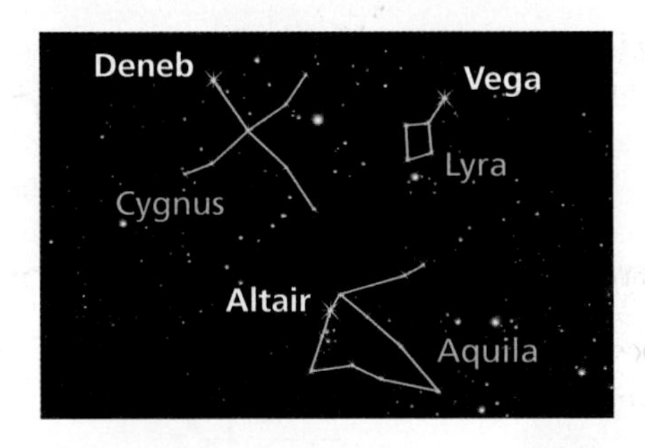

77. CRITICAL THINKING Find a radical expression for the perimeter of the triangle inscribed in the square shown. Simplify the expression.

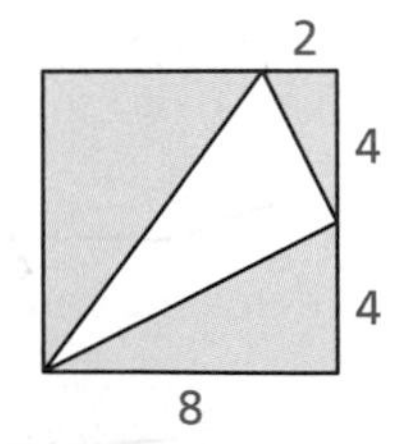

78. HOW DO YOU SEE IT? Without finding points, match the functions $f(x) = \sqrt{64x^2}$ and $g(x) = \sqrt[3]{64x^6}$ with their graphs. Explain your reasoning.

A.

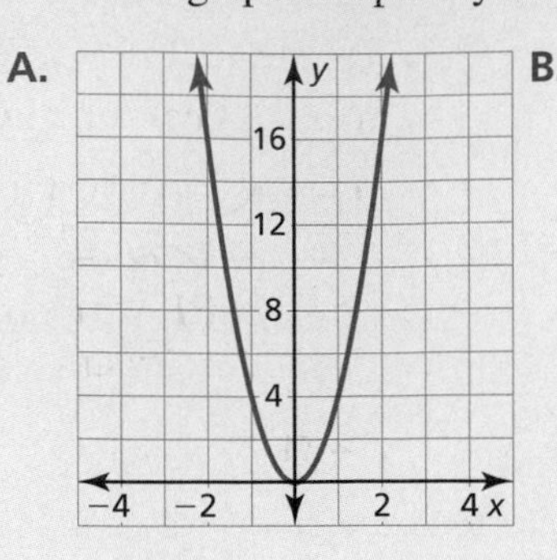

B.

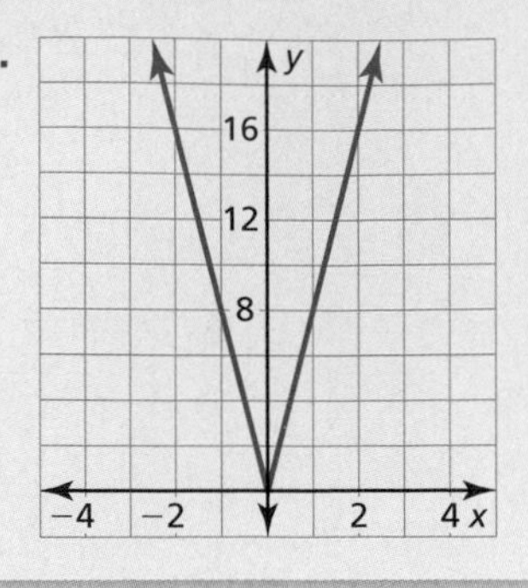

79. REWRITING A FORMULA You have filled two round balloons with water. One balloon contains twice as much water as the other balloon.

a. Solve the formula for the volume of a sphere, $V = \frac{4}{3}\pi r^3$, for r.

b. Substitute the expression for r from part (a) into the formula for the surface area of a sphere, $S = 4\pi r^2$. Simplify to show that $S = (4\pi)^{1/3}(3V)^{2/3}$.

c. Compare the surface areas of the two water balloons using the formula in part (b).

80. THOUGHT PROVOKING Determine whether the expressions $(x^2)^{1/6}$ and $(x^{1/6})^2$ are equivalent for all values of x.

81. DRAWING CONCLUSIONS Substitute different combinations of odd and even positive integers for m and n in the expression $\sqrt[n]{x^m}$. When you cannot assume x is positive, explain when absolute value is needed in simplifying the expression.

Maintaining Mathematical Proficiency

Reviewing what you learned in previous grades and lessons

Identify the focus, directrix, and axis of symmetry of the parabola. Then graph the equation. *(Section 2.3)*

82. $y = 2x^2$

83. $y^2 = -x$

84. $y^2 = 4x$

Write a rule for g. Describe the graph of g as a transformation of the graph of f. *(Section 4.7)*

85. $f(x) = x^4 - 3x^2 - 2x,\ g(x) = -f(x)$

86. $f(x) = x^3 - x,\ g(x) = f(x) - 3$

87. $f(x) = x^3 - 4,\ g(x) = f(x - 2)$

88. $f(x) = x^4 + 2x^3 - 4x^2,\ g(x) = f(2x)$

5.3 Graphing Radical Functions

Essential Question

How can you identify the domain and range of a radical function?

EXPLORATION 1 Identifying Graphs of Radical Functions

Work with a partner. Match each function with its graph. Explain your reasoning. Then identify the domain and range of each function.

a. $f(x) = \sqrt{x}$ **b.** $f(x) = \sqrt[3]{x}$ **c.** $f(x) = \sqrt[4]{x}$ **d.** $f(x) = \sqrt[5]{x}$

A.

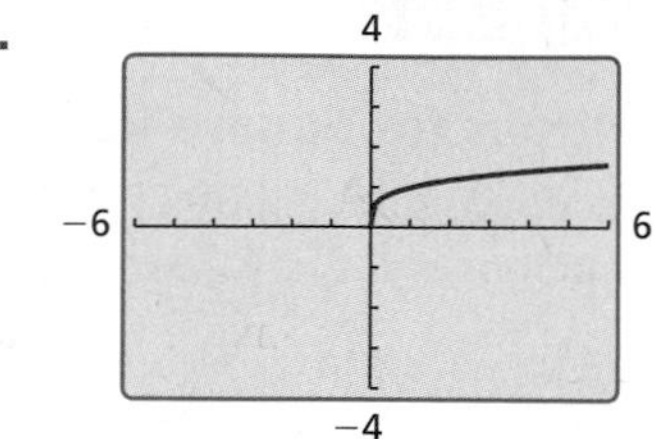

B.

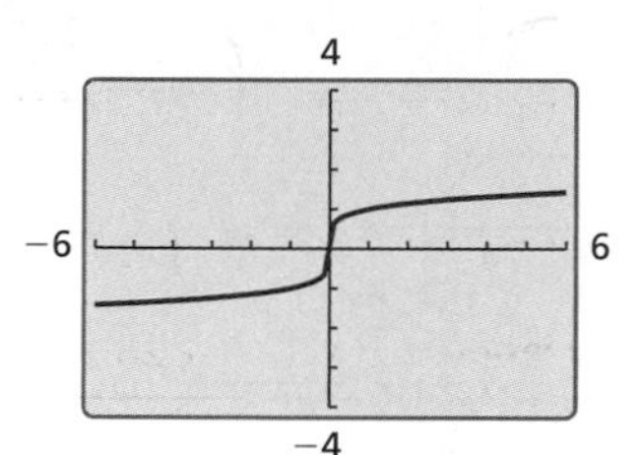

C.

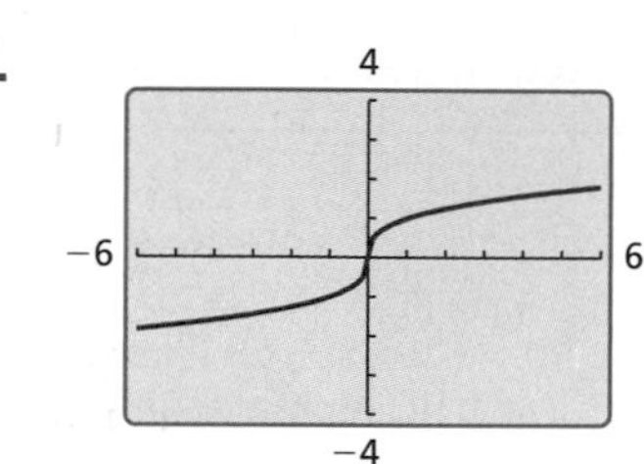

D.

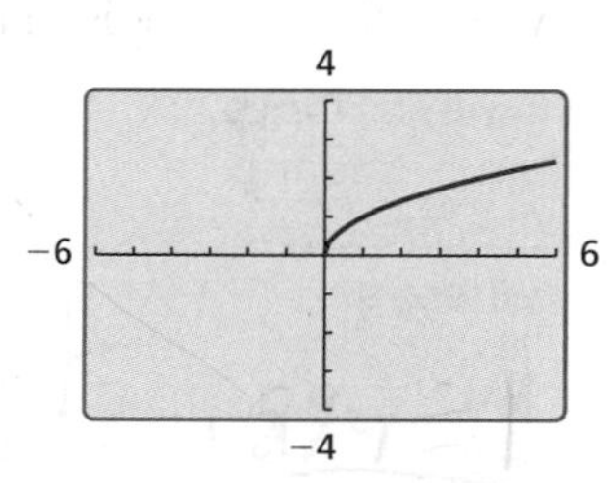

EXPLORATION 2 Identifying Graphs of Transformations

Work with a partner. Match each transformation of $f(x) = \sqrt{x}$ with its graph. Explain your reasoning. Then identify the domain and range of each function.

a. $g(x) = \sqrt{x+2}$ **b.** $g(x) = \sqrt{x-2}$ **c.** $g(x) = \sqrt{x+2} - 2$ **d.** $g(x) = -\sqrt{x+2}$

A.

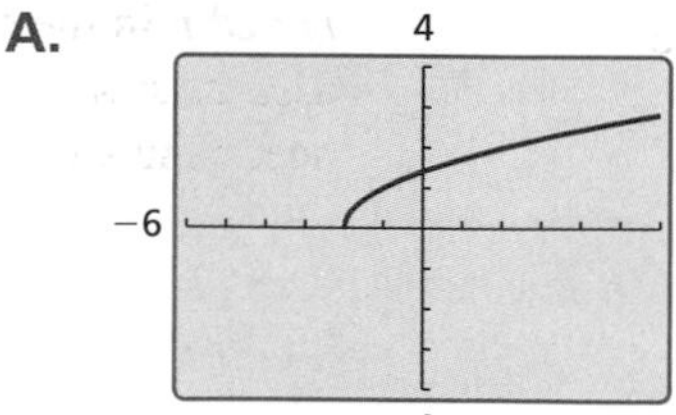

B.

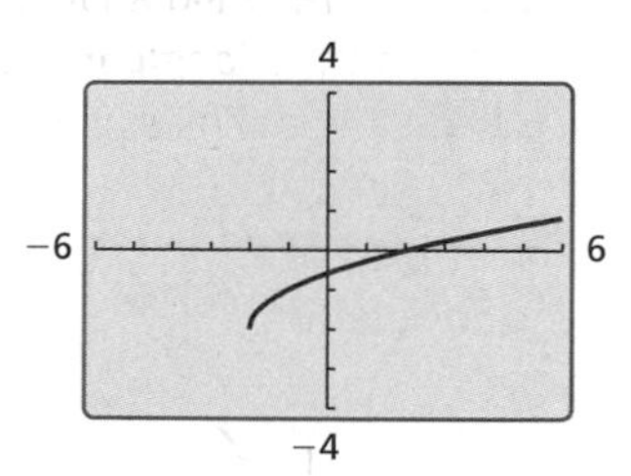

C.

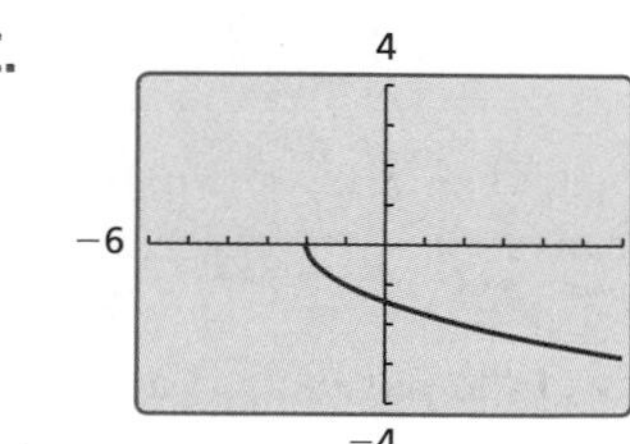

D.

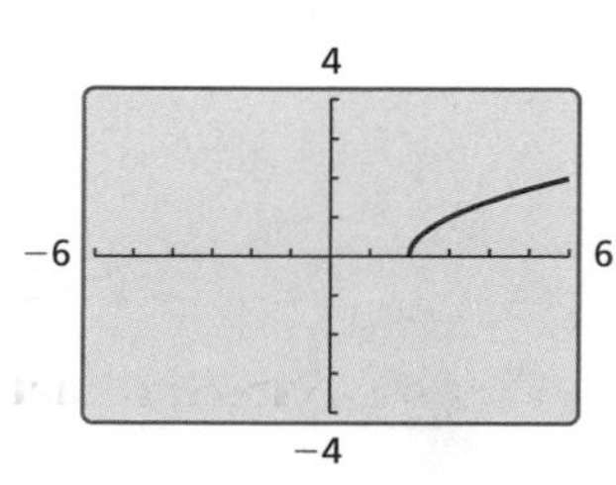

LOOKING FOR STRUCTURE

To be proficient in math, you need to look closely to discern a pattern or structure.

Communicate Your Answer

3. How can you identify the domain and range of a radical function?

4. Use the results of Exploration 1 to describe how the domain and range of a radical function are related to the index of the radical.

5.3 Lesson

Core Vocabulary

radical function, *p. 252*

Previous
transformations
parabola
circle

What You Will Learn

- Graph radical functions.
- Write transformations of radical functions.
- Graph parabolas and circles.

Graphing Radical Functions

A **radical function** contains a radical expression with the independent variable in the radicand. When the radical is a square root, the function is called a *square root function*. When the radical is a cube root, the function is called a *cube root function*.

STUDY TIP

A *power function* has the form $y = ax^b$, where a is a real number and b is a rational number. Notice that the parent square root function is a power function, where $a = 1$ and $b = \frac{1}{2}$.

Core Concept

Parent Functions for Square Root and Cube Root Functions

The parent function for the family of square root functions is $f(x) = \sqrt{x}$.

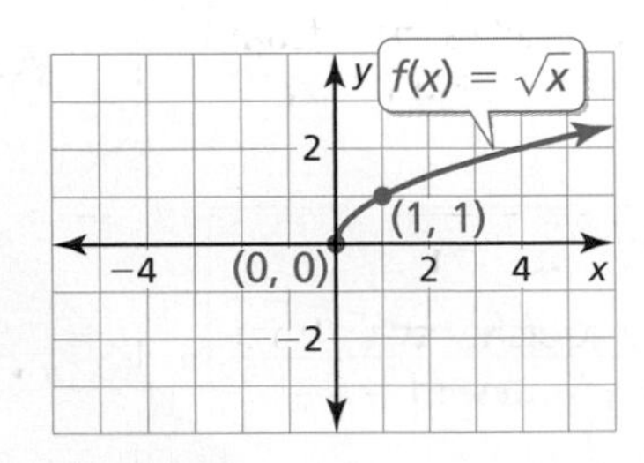

Domain: $x \geq 0$, Range: $y \geq 0$

The parent function for the family of cube root functions is $f(x) = \sqrt[3]{x}$.

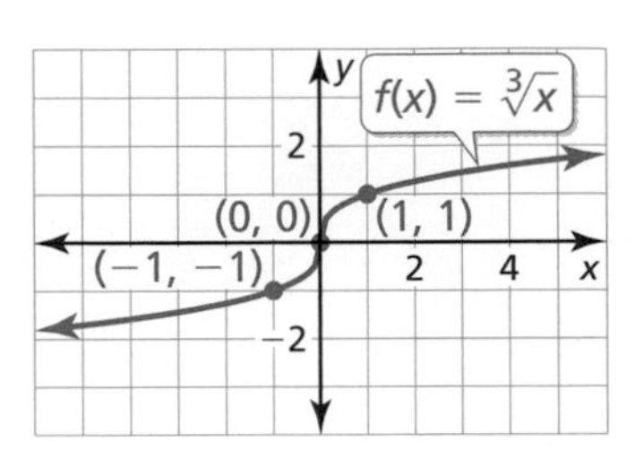

Domain and range: All real numbers

EXAMPLE 1 Graphing Radical Functions

Graph each function. Identify the domain and range of each function.

a. $f(x) = \sqrt{\frac{1}{4}x}$

b. $g(x) = -3\sqrt[3]{x}$

LOOKING FOR STRUCTURE

Example 1(a) uses x-values that are multiples of 4 so that the radicand is an integer.

SOLUTION

a. Make a table of values and sketch the graph.

x	0	4	8	12	16
y	0	1	1.41	1.73	2

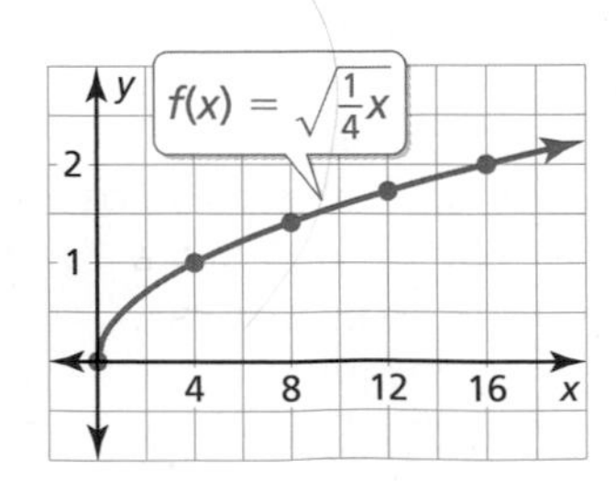

▶ The radicand of a square root must be nonnegative. So, the domain is $x \geq 0$. The range is $y \geq 0$.

b. Make a table of values and sketch the graph.

x	−2	−1	0	1	2
y	3.78	3	0	−3	−3.78

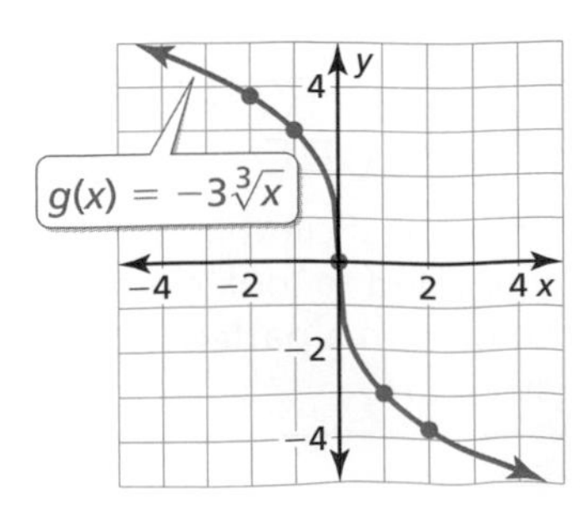

▶ The radicand of a cube root can be any real number. So, the domain and range are all real numbers.

In Example 1, notice that the graph of f is a horizontal stretch of the graph of the parent square root function. The graph of g is a vertical stretch and a reflection in the x-axis of the graph of the parent cube root function. You can transform graphs of radical functions in the same way you transformed graphs of functions previously.

Core Concept

Transformation	$f(x)$ Notation	Examples	
Horizontal Translation Graph shifts left or right.	$f(x - h)$	$g(x) = \sqrt{x - 2}$ $g(x) = \sqrt{x + 3}$	2 units right 3 units left
Vertical Translation Graph shifts up or down.	$f(x) + k$	$g(x) = \sqrt{x} + 7$ $g(x) = \sqrt{x} - 1$	7 units up 1 unit down
Reflection Graph flips over x- or y-axis.	$f(-x)$ $-f(x)$	$g(x) = \sqrt{-x}$ $g(x) = -\sqrt{x}$	in the y-axis in the x-axis
Horizontal Stretch or Shrink Graph stretches away from or shrinks toward y-axis.	$f(ax)$	$g(x) = \sqrt{3x}$ $g(x) = \sqrt{\frac{1}{2}x}$	shrink by a factor of $\frac{1}{3}$ stretch by a factor of 2
Vertical Stretch or Shrink Graph stretches away from or shrinks toward x-axis.	$a \cdot f(x)$	$g(x) = 4\sqrt{x}$ $g(x) = \frac{1}{5}\sqrt{x}$	stretch by a factor of 4 shrink by a factor of $\frac{1}{5}$

EXAMPLE 2 Transforming Radical Functions

Describe the transformation of f represented by g. Then graph each function.

a. $f(x) = \sqrt{x}$, $g(x) = \sqrt{x - 3} + 4$

b. $f(x) = \sqrt[3]{x}$, $g(x) = \sqrt[3]{-8x}$

LOOKING FOR STRUCTURE

In Example 2(b), you can use the Product Property of Radicals to write $g(x) = -2\sqrt[3]{x}$. So, you can also describe the graph of g as a vertical stretch by a factor of 2 and a reflection in the x-axis of the graph of f.

SOLUTION

a. Notice that the function is of the form $g(x) = \sqrt{x - h} + k$, where $h = 3$ and $k = 4$.

▶ So, the graph of g is a translation 3 units right and 4 units up of the graph of f.

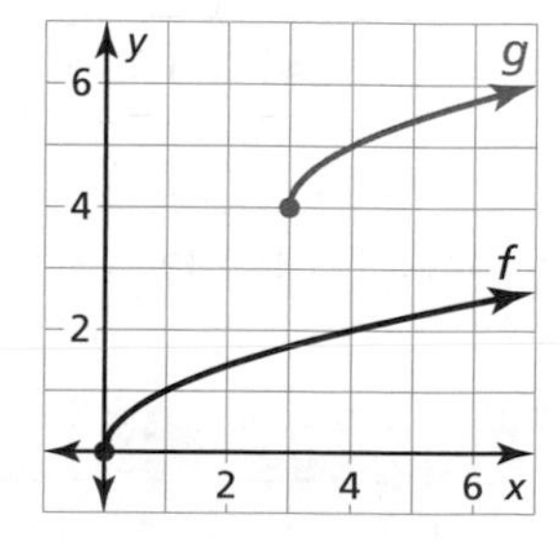

b. Notice that the function is of the form $g(x) = \sqrt[3]{ax}$, where $a = -8$.

▶ So, the graph of g is a horizontal shrink by a factor of $\frac{1}{8}$ and a reflection in the y-axis of the graph of f.

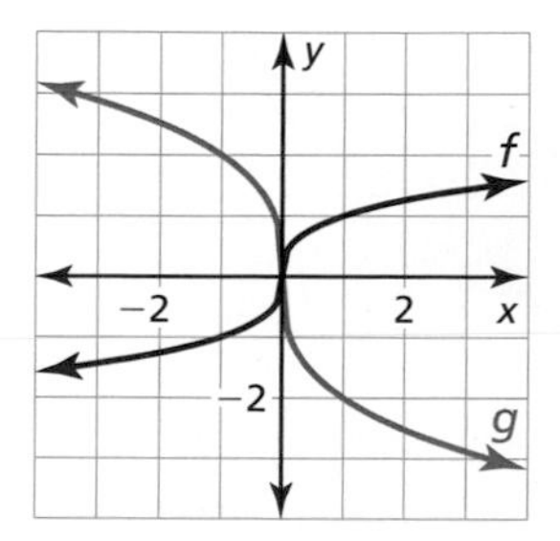

Monitoring Progress

Help in English and Spanish at *BigIdeasMath.com*

1. Graph $g(x) = \sqrt{x + 1}$. Identify the domain and range of the function.
2. Describe the transformation of $f(x) = \sqrt[3]{x}$ represented by $g(x) = -\sqrt[3]{x} - 2$. Then graph each function.

Writing Transformations of Radical Functions

EXAMPLE 3 Modeling with Mathematics

Self-Portrait of
NASA's Mars Rover Curiosity

The function $E(d) = 0.25\sqrt{d}$ approximates the number of seconds it takes a dropped object to fall d feet on Earth. The function $M(d) = 1.6 \cdot E(d)$ approximates the number of seconds it takes a dropped object to fall d feet on Mars. Write a rule for M. How long does it take a dropped object to fall 64 feet on Mars?

SOLUTION

1. **Understand the Problem** You are given a function that represents the number of seconds it takes a dropped object to fall d feet on Earth. You are asked to write a similar function for Mars and then evaluate the function for a given input.

2. **Make a Plan** Multiply $E(d)$ by 1.6 to write a rule for M. Then find $M(64)$.

3. **Solve the Problem**

$M(d) = 1.6 \cdot E(d)$

$= 1.6 \cdot 0.25\sqrt{d}$ Substitute $0.25\sqrt{d}$ for $E(d)$.

$= 0.4\sqrt{d}$ Simplify.

Next, find $M(64)$.

$$M(64) = 0.4\sqrt{64} = 0.4(8) = 3.2$$

▶ It takes a dropped object about 3.2 seconds to fall 64 feet on Mars.

4. **Look Back** Use the original functions to check your solution.

$E(64) = 0.25\sqrt{64} = 2$ $M(64) = 1.6 \cdot E(64) = 1.6 \cdot 2 = 3.2$ 

EXAMPLE 4 Writing a Transformed Radical Function

Let the graph of g be a horizontal shrink by a factor of $\frac{1}{6}$ followed by a translation 3 units to the left of the graph of $f(x) = \sqrt[3]{x}$. Write a rule for g.

SOLUTION

Step 1 First write a function h that represents the horizontal shrink of f.

$h(x) = f(6x)$ Multiply the input by $1 \div \frac{1}{6} = 6$.

$= \sqrt[3]{6x}$ Replace x with $6x$ in $f(x)$.

Step 2 Then write a function g that represents the translation of h.

$g(x) = h(x + 3)$ Subtract -3, or add 3, to the input.

$= \sqrt[3]{6(x + 3)}$ Replace x with $x + 3$ in $h(x)$.

$= \sqrt[3]{6x + 18}$ Distributive Property

▶ The transformed function is $g(x) = \sqrt[3]{6x + 18}$.

Check

3
g h f
−7 4
−3

  Help in English and Spanish at *BigIdeasMath.com*

3. **WHAT IF?** In Example 3, the function $N(d) = 2.4 \cdot E(d)$ approximates the number of seconds it takes a dropped object to fall d feet on the Moon. Write a rule for N. How long does it take a dropped object to fall 25 feet on the Moon?

4. In Example 4, is the transformed function the same when you perform the translation followed by the horizontal shrink? Explain your reasoning.

Graphing Parabolas and Circles

To graph parabolas and circles using a graphing calculator, first solve their equations for y to obtain radical functions. Then graph the functions.

EXAMPLE 5 Graphing a Parabola (Horizontal Axis of Symmetry)

Use a graphing calculator to graph $\frac{1}{2}y^2 = x$. Identify the vertex and the direction that the parabola opens.

SOLUTION

Step 1 Solve for y.

$\frac{1}{2}y^2 = x$ Write the original equation.

$y^2 = 2x$ Multiply each side by 2.

$y = \pm\sqrt{2x}$ Take square root of each side.

STUDY TIP

Notice y_1 is a function and y_2 is a function, but $\frac{1}{2}y^2 = x$ is not a function.

Step 2 Graph both radical functions.

$y_1 = \sqrt{2x}$

$y_2 = -\sqrt{2x}$

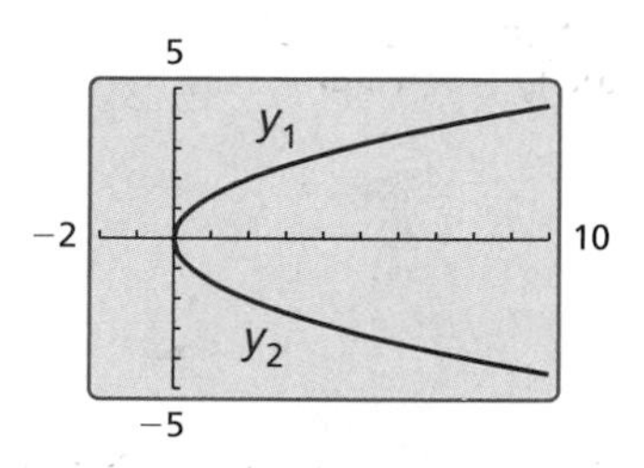

▶ The vertex is (0, 0) and the parabola opens right.

EXAMPLE 6 Graphing a Circle (Center at the Origin)

Use a graphing calculator to graph $x^2 + y^2 = 16$. Identify the radius and the intercepts.

SOLUTION

Step 1 Solve for y.

$x^2 + y^2 = 16$ Write the original equation.

$y^2 = 16 - x^2$ Subtract x^2 from each side.

$y = \pm\sqrt{16 - x^2}$ Take square root of each side.

Step 2 Graph both radical functions using a square viewing window.

$y_1 = \sqrt{16 - x^2}$

$y_2 = -\sqrt{16 - x^2}$

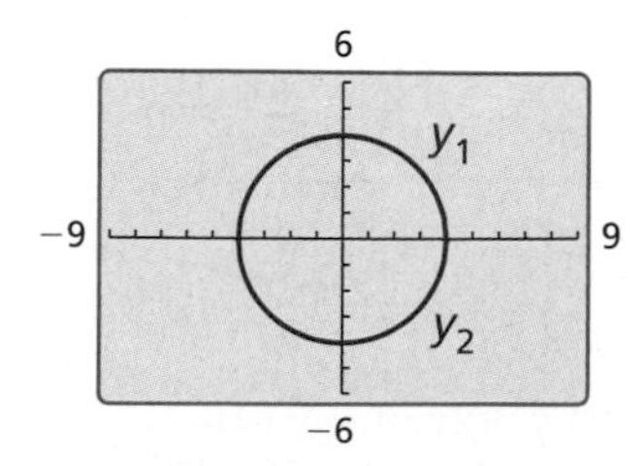

▶ The radius is 4 units. The x-intercepts are ± 4. The y-intercepts are also ± 4.

Monitoring Progress  Help in English and Spanish at *BigIdeasMath.com*

5. Use a graphing calculator to graph $-4y^2 = x + 1$. Identify the vertex and the direction that the parabola opens.

6. Use a graphing calculator to graph $x^2 + y^2 = 25$. Identify the radius and the intercepts.

5.3 Exercises

Dynamic Solutions available at *BigIdeasMath.com*

Vocabulary and Core Concept Check

1. **COMPLETE THE SENTENCE** Square root functions and cube root functions are examples of __________ functions.

2. **COMPLETE THE SENTENCE** When graphing $y = a\sqrt[3]{x - h} + k$, translate the graph of $y = a\sqrt[3]{x}$ h units __________ and k units __________.

Monitoring Progress and Modeling with Mathematics

In Exercises 3–8, match the function with its graph.

3. $f(x) = \sqrt{x + 3}$
4. $h(x) = \sqrt{x} + 3$
5. $f(x) = \sqrt{x - 3}$
6. $g(x) = \sqrt{x} - 3$
7. $h(x) = \sqrt{x + 3} - 3$
8. $f(x) = \sqrt{x - 3} + 3$

A. 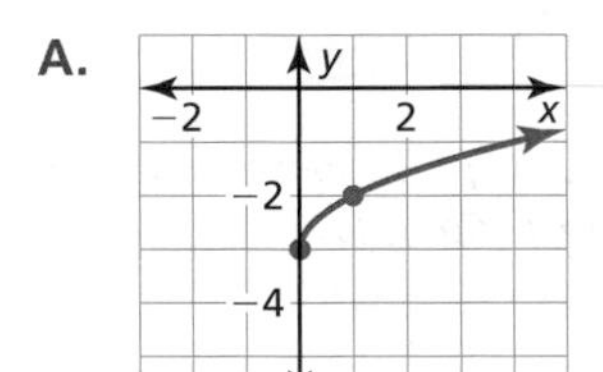

B. 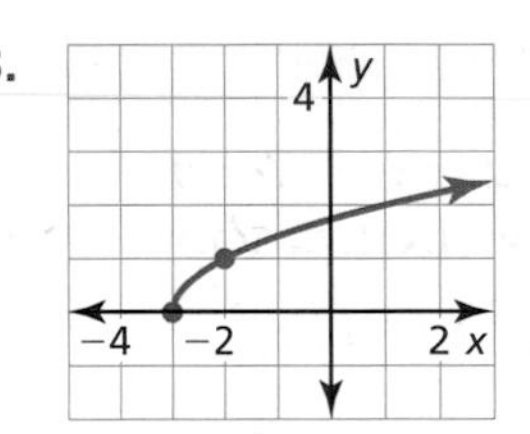

C. 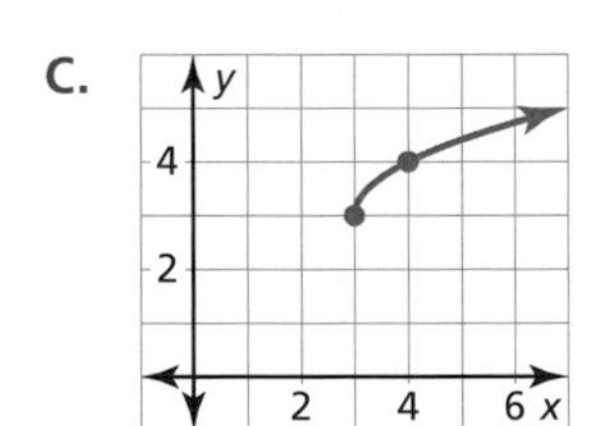

D. 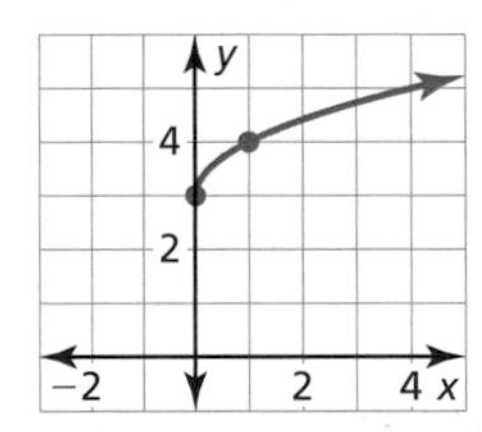

E. 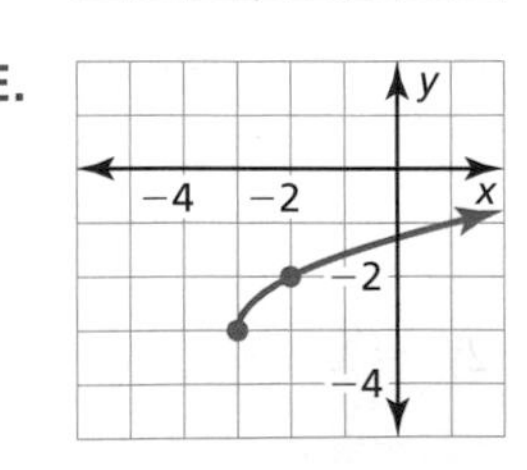

F. 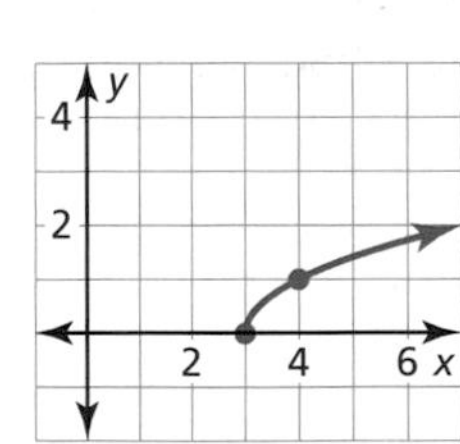

In Exercises 9–18, graph the function. Identify the domain and range of the function. *(See Example 1.)*

9. $h(x) = \sqrt{x} + 4$
10. $g(x) = \sqrt{x} - 5$
11. $g(x) = -\sqrt[3]{2x}$
12. $f(x) = \sqrt[3]{-5x}$
13. $g(x) = \frac{1}{5}\sqrt{x - 3}$
14. $f(x) = \frac{1}{2}\sqrt[3]{x + 6}$
15. $f(x) = (6x)^{1/2} + 3$
16. $g(x) = -3(x + 1)^{1/3}$
17. $h(x) = -\sqrt[4]{x}$
18. $h(x) = \sqrt[5]{2x}$

In Exercises 19–26, describe the transformation of f represented by g. Then graph each function. *(See Example 2.)*

19. $f(x) = \sqrt{x}$, $g(x) = \sqrt{x + 1} + 8$
20. $f(x) = \sqrt{x}$, $g(x) = 2\sqrt{x - 1}$
21. $f(x) = \sqrt[3]{x}$, $g(x) = -\sqrt[3]{x} - 1$
22. $f(x) = \sqrt[3]{x}$, $g(x) = \sqrt[3]{x + 4} - 5$
23. $f(x) = x^{1/2}$, $g(x) = \frac{1}{4}(-x)^{1/2}$
24. $f(x) = x^{1/3}$, $g(x) = \frac{1}{3}x^{1/3} + 6$
25. $f(x) = \sqrt[4]{x}$, $g(x) = 2\sqrt[4]{x + 5} - 4$
26. $f(x) = \sqrt[5]{x}$, $g(x) = \sqrt[5]{-32x} + 3$

27. **ERROR ANALYSIS** Describe and correct the error in graphing $f(x) = \sqrt{x - 2} - 2$.

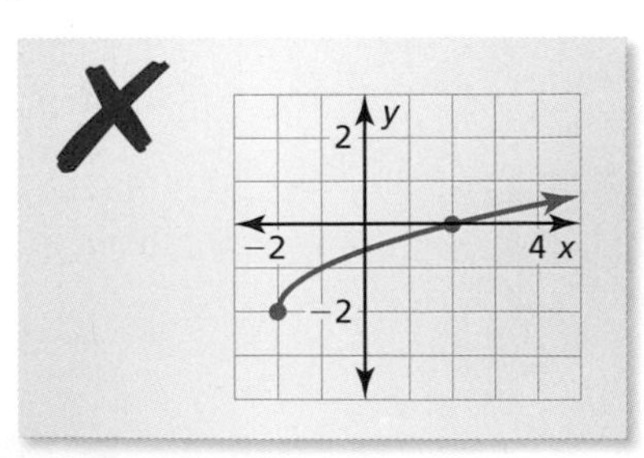

28. **ERROR ANALYSIS** Describe and correct the error in describing the transformation of the parent square root function represented by $g(x) = \sqrt{\frac{1}{2}x} + 3$.

The graph of g is a horizontal shrink by a factor of $\frac{1}{2}$ and a translation 3 units up of the parent square root function.

USING TOOLS In Exercises 29–34, use a graphing calculator to graph the function. Then identify the domain and range of the function.

29. $g(x) = \sqrt{x^2 + x}$

30. $h(x) = \sqrt{x^2 - 2x}$

31. $f(x) = \sqrt[3]{x^2 + x}$

32. $f(x) = \sqrt[3]{3x^2 - x}$

33. $f(x) = \sqrt{2x^2 + x + 1}$

34. $h(x) = \sqrt[3]{\frac{1}{2}x^2 - 3x + 4}$

ABSTRACT REASONING In Exercises 35–38, complete the statement with *sometimes*, *always*, or *never*.

35. The domain of the function $y = a\sqrt{x}$ is ______ $x \geq 0$.

36. The range of the function $y = a\sqrt{x}$ is ______ $y \geq 0$.

37. The domain and range of the function $y = \sqrt[3]{x - h} + k$ are ________ all real numbers.

38. The domain of the function $y = a\sqrt{-x} + k$ is ________ $x \geq 0$.

39. PROBLEM SOLVING The distance (in miles) a pilot can see to the horizon can be approximated by $E(n) = 1.22\sqrt{n}$, where n is the plane's altitude (in feet above sea level) on Earth. The function $M(n) = 0.75E(n)$ approximates the distance a pilot can see to the horizon n feet above the surface of Mars. Write a rule for M. What is the distance a pilot can see to the horizon from an altitude of 10,000 feet above Mars? *(See Example 3.)*

40. MODELING WITH MATHEMATICS The speed (in knots) of sound waves in air can be modeled by

$$v(K) = 643.855\sqrt{\frac{K}{273.15}}$$

where K is the air temperature (in kelvin). The speed (in meters per second) of sound waves in air can be modeled by

$$s(K) = \frac{v(K)}{1.944}.$$

Write a rule for s. What is the speed (in meters per second) of sound waves when the air temperature is 305 kelvin?

In Exercises 41–44, write a rule for g described by the transformations of the graph of f. *(See Example 4.)*

41. Let g be a vertical stretch by a factor of 2, followed by a translation 2 units up of the graph of $f(x) = \sqrt{x} + 3$.

42. Let g be a reflection in the y-axis, followed by a translation 1 unit right of the graph of $f(x) = 2\sqrt[3]{x - 1}$.

43. Let g be a horizontal shrink by a factor of $\frac{2}{3}$, followed by a translation 4 units left of the graph of $f(x) = \sqrt{6x}$.

44. Let g be a translation 1 unit down and 5 units right, followed by a reflection in the x-axis of the graph of $f(x) = -\frac{1}{2}\sqrt[4]{x} + \frac{3}{2}$.

In Exercises 45 and 46, write a rule for g.

45.

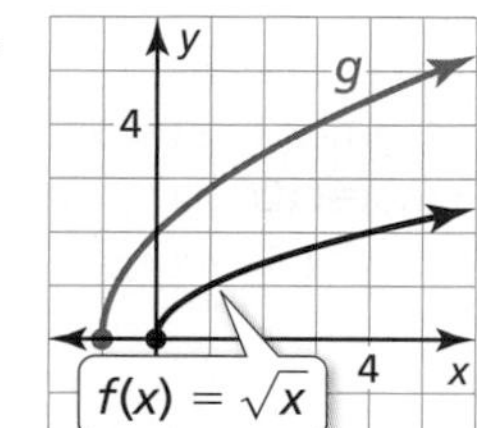

46.

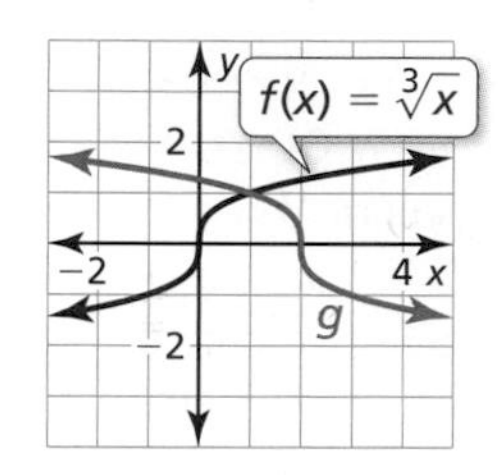

In Exercises 47–50, write a rule for g that represents the indicated transformation of the graph of f.

47. $f(x) = 2\sqrt{x},\ g(x) = f(x + 3)$

48. $f(x) = \frac{1}{3}\sqrt{x - 1},\ g(x) = -f(x) + 9$

49. $f(x) = -\sqrt{x^2 - 2},\ g(x) = -2f(x + 5)$

50. $f(x) = \sqrt[3]{x^2 + 10x},\ g(x) = \frac{1}{4}f(-x) + 6$

In Exercises 51–56, use a graphing calculator to graph the equation of the parabola. Identify the vertex and the direction that the parabola opens. *(See Example 5.)*

51. $\frac{1}{4}y^2 = x$

52. $3y^2 = x$

53. $-8y^2 + 2 = x$

54. $2y^2 = x - 4$

55. $x + 8 = \frac{1}{5}y^2$

56. $\frac{1}{2}x = y^2 - 4$

In Exercises 57–62, use a graphing calculator to graph the equation of the circle. Identify the radius and the intercepts. *(See Example 6.)*

57. $x^2 + y^2 = 9$

58. $x^2 + y^2 = 4$

59. $1 - y^2 = x^2$

60. $64 - x^2 = y^2$

61. $-y^2 = x^2 - 36$

62. $x^2 = 100 - y^2$

63. **MODELING WITH MATHEMATICS** The *period* of a pendulum is the time the pendulum takes to complete one back-and-forth swing. The period T (in seconds) can be modeled by the function $T = 1.11\sqrt{\ell}$, where ℓ is the length (in feet) of the pendulum. Graph the function. Estimate the length of a pendulum with a period of 2 seconds. Explain your reasoning.

64. **HOW DO YOU SEE IT?** Does the graph represent a square root function or a cube root function? Explain. What are the domain and range of the function?

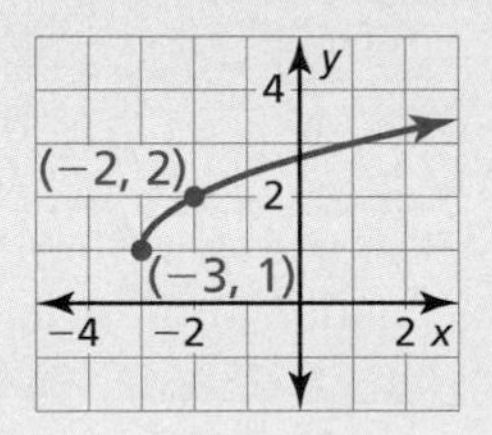

65. **PROBLEM SOLVING** For a drag race car with a total weight of 3500 pounds, the speed s (in miles per hour) at the end of a race can be modeled by $s = 14.8\sqrt[3]{p}$, where p is the power (in horsepower). Graph the function.

a. Determine the power of a 3500-pound car that reaches a speed of 200 miles per hour.

b. What is the average rate of change in speed as the power changes from 1000 horsepower to 1500 horsepower?

66. **THOUGHT PROVOKING** The graph of a radical function f passes through the points (3, 1) and (4, 0). Write two different functions that could represent $f(x + 2) + 1$. Explain.

67. **MULTIPLE REPRESENTATIONS** The terminal velocity v_t (in feet per second) of a skydiver who weighs 140 pounds is given by

$$v_t = 33.7\sqrt{\frac{140}{A}}$$

where A is the cross-sectional surface area (in square feet) of the skydiver. The table shows the terminal velocities (in feet per second) for various surface areas (in square feet) of a skydiver who weighs 165 pounds.

Cross-sectional surface area, A	Terminal velocity, v_t
1	432.9
3	249.9
5	193.6
7	163.6

a. Which skydiver has a greater terminal velocity for each value of A given in the table?

b. Describe how the different values of A given in the table relate to the possible positions of the falling skydiver.

68. **MATHEMATICAL CONNECTIONS** The surface area S of a right circular cone with a slant height of 1 unit is given by $S = \pi r + \pi r^2$, where r is the radius of the cone.

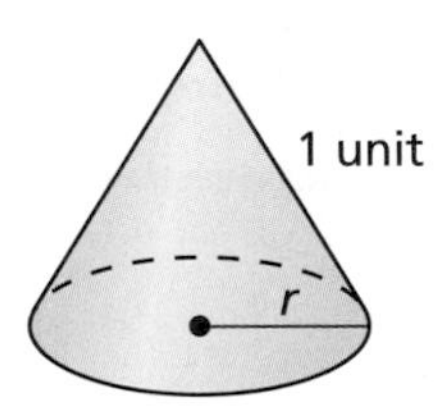

a. Use completing the square to show that

$$r = \frac{1}{\sqrt{\pi}}\sqrt{S + \frac{\pi}{4}} - \frac{1}{2}.$$

b. Graph the equation in part (a) using a graphing calculator. Then find the radius of a right circular cone with a slant height of 1 unit and a surface area of $\frac{3\pi}{4}$ square units.

Maintaining Mathematical Proficiency

Reviewing what you learned in previous grades and lessons

Solve the equation. Check your solutions. *(Skills Review Handbook)*

69. $|3x + 2| = 5$ **70.** $|4x + 9| = -7$ **71.** $|2x - 6| = |x|$ **72.** $|x + 8| = |2x + 2|$

Solve the inequality. *(Section 3.6)*

73. $x^2 + 7x + 12 < 0$ **74.** $x^2 - 10x + 25 \geq 4$ **75.** $2x^2 + 6 > 13x$ **76.** $\frac{1}{8}x^2 + x \leq -2$

5.1–5.3 What Did You Learn?

Core Vocabulary

*n*th root of *a*, *p. 238*
index of a radical, *p. 238*
simplest form of a radical, *p. 245*
conjugate, *p. 246*
like radicals, *p. 246*
radical function, *p. 252*

Core Concepts

Section 5.1

Real *n*th Roots of *a*, *p. 238*
Rational Exponents, *p. 239*

Section 5.2

Properties of Rational Exponents, *p. 244*
Properties of Radicals, *p. 245*

Section 5.3

Parent Functions for Square Root and Cube Root Functions, *p. 252*
Transformations of Radical Functions, *p. 253*

Mathematical Practices

1. How can you use definitions to explain your reasoning in Exercises 21–24 on page 241?
2. How did you use structure to solve Exercise 76 on page 250?
3. How can you check that your answer is reasonable in Exercise 39 on page 257?
4. How can you make sense of the terms of the surface area formula given in Exercise 68 on page 258?

Study Skills

Analyzing Your Errors

Application Errors

What Happens: You can do numerical problems, but you struggle with problems that have context.

How to Avoid This Error: Do not just mimic the steps of solving an application problem. Explain out loud what the question is asking and why you are doing each step. After solving the problem, ask yourself, "Does my solution make sense?"

5.1–5.3 Quiz

Find the indicated real nth root(s) of a. *(Section 5.1)*

1. $n = 4, a = 81$
2. $n = 5, a = -1024$

3. Evaluate (a) $16^{3/4}$ and (b) $125^{2/3}$ without using a calculator. Explain your reasoning. *(Section 5.1)*

Find the real solution(s) of the equation. Round your answer to two decimal places. *(Section 5.1)*

4. $2x^6 = 1458$
5. $(x + 6)^3 = 28$

Simplify the expression. *(Section 5.2)*

6. $\left(\dfrac{48^{1/4}}{6^{1/4}}\right)^6$
7. $\sqrt[4]{3} \cdot \sqrt[4]{432}$
8. $\dfrac{1}{3 + \sqrt{2}}$
9. $\sqrt[3]{16} - 5\sqrt[3]{2}$

10. Simplify $\sqrt[8]{x^9y^8z^{16}}$. *(Section 5.2)*

Write the expression in simplest form. Assume all variables are positive. *(Section 5.2)*

11. $\sqrt[3]{216p^9}$
12. $\dfrac{\sqrt[5]{32}}{\sqrt[5]{m^3}}$
13. $\sqrt[4]{n^4q} + 7n\sqrt[4]{q}$

14. Graph $f(x) = 2\sqrt[3]{x} + 1$. Identify the domain and range of the function. *(Section 5.3)*

Describe the transformation of the parent function represented by the graph of g. Then write a rule for g. *(Section 5.3)*

15.

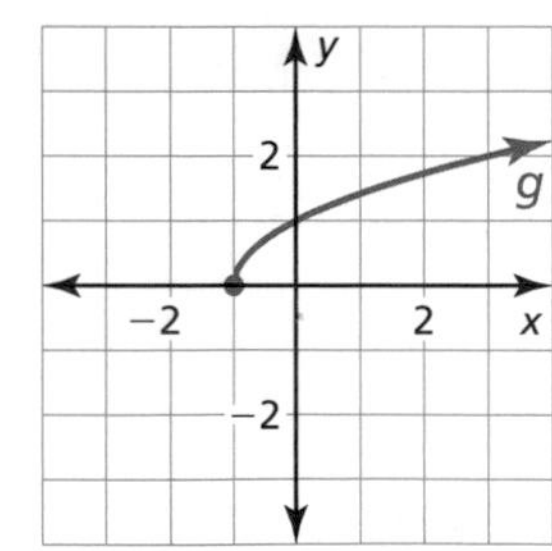

16.

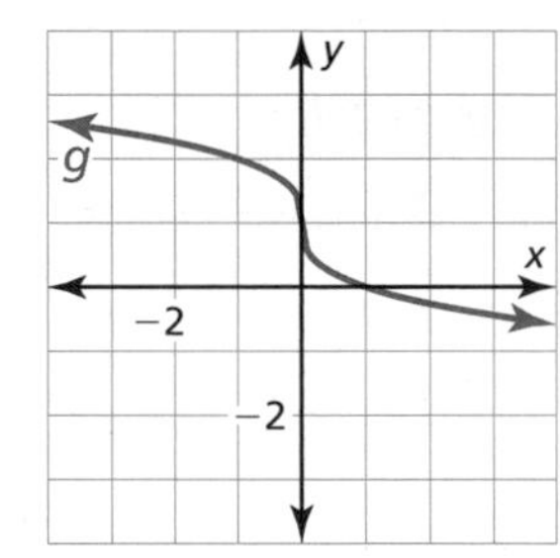

17.

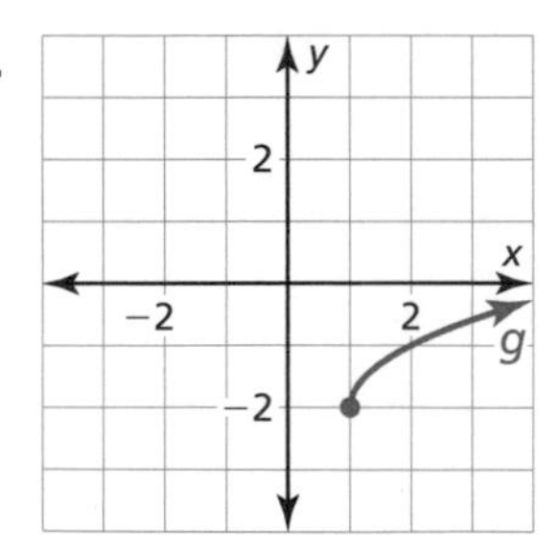

18. Use a graphing calculator to graph $x = 3y^2 - 6$. Identify the vertex and direction the parabola opens. *(Section 5.3)*

19. A jeweler is setting a stone cut in the shape of a regular octahedron. A regular octahedron is a solid with eight equilateral triangles as faces, as shown. The formula for the volume of the stone is $V = 0.47s^3$, where s is the side length (in millimeters) of an edge of the stone. The volume of the stone is 161 cubic millimeters. Find the length of an edge of the stone. *(Section 5.1)*

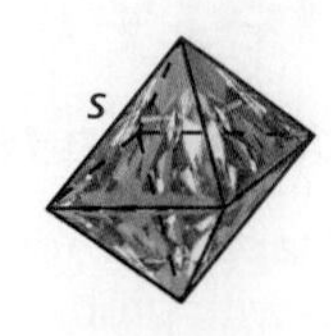

20. An investigator can determine how fast a car was traveling just prior to an accident using the model $s = 4\sqrt{d}$, where s is the speed (in miles per hour) of the car and d is the length (in feet) of the skid marks. Graph the model. The length of the skid marks of a car is 90 feet. Was the car traveling at the posted speed limit prior to the accident? Explain your reasoning. *(Section 5.3)*

5.4 Solving Radical Equations and Inequalities

Essential Question How can you solve a radical equation?

EXPLORATION 1 Solving Radical Equations

Work with a partner. Match each radical equation with the graph of its related radical function. Explain your reasoning. Then use the graph to solve the equation, if possible. Check your solutions.

a. $\sqrt{x-1} - 1 = 0$ **b.** $\sqrt{2x+2} - \sqrt{x+4} = 0$ **c.** $\sqrt{9-x^2} = 0$

d. $\sqrt{x+2} - x = 0$ **e.** $\sqrt{-x+2} - x = 0$ **f.** $\sqrt{3x^2+1} = 0$

A.

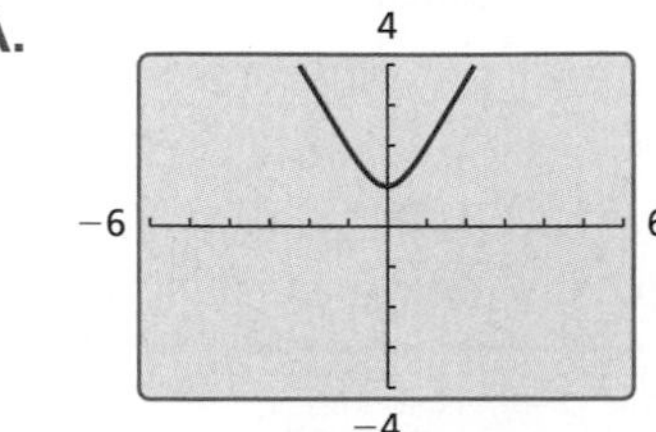

B.

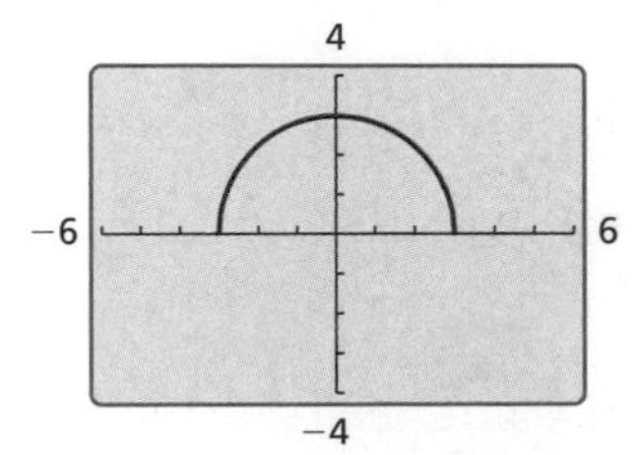

C.

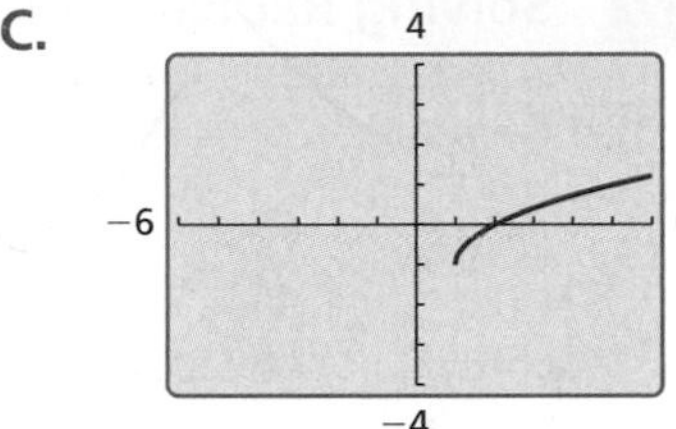

D.

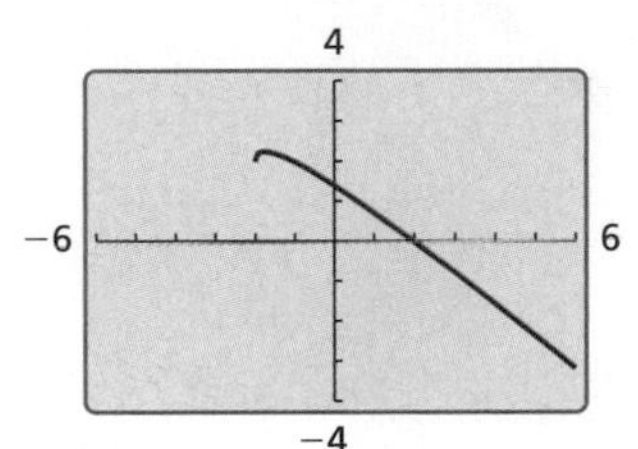

E.

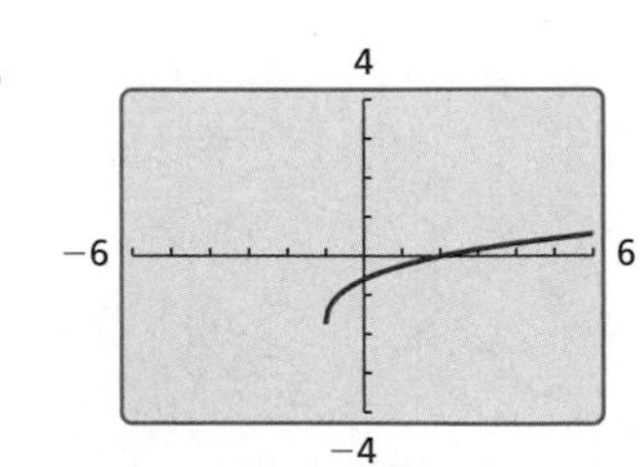

F.

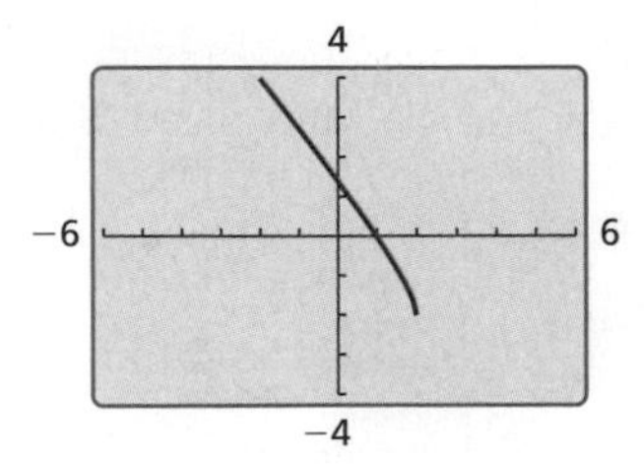

LOOKING FOR STRUCTURE

To be proficient in math, you need to look closely to discern a pattern or structure.

EXPLORATION 2 Solving Radical Equations

Work with a partner. Look back at the radical equations in Exploration 1. Suppose that you did not know how to solve the equations using a graphical approach.

a. Show how you could use a *numerical approach* to solve one of the equations. For instance, you might use a spreadsheet to create a table of values.

b. Show how you could use an *analytical approach* to solve one of the equations. For instance, look at the similarities between the equations in Exploration 1. What first step may be necessary so you could square each side to eliminate the radical(s)? How would you proceed to find the solution?

Communicate Your Answer

3. How can you solve a radical equation?

4. Would you prefer to use a graphical, numerical, or analytical approach to solve the given equation? Explain your reasoning. Then solve the equation.

$$\sqrt{x+3} - \sqrt{x-2} = 1$$

5.4 Lesson

Core Vocabulary

radical equation, *p. 262*
extraneous solutions, *p. 263*

Previous
rational exponents
radical expressions
solving quadratic equations

What You Will Learn

- Solve equations containing radicals and rational exponents.
- Solve radical inequalities.

Solving Equations

Equations with radicals that have variables in their radicands are called **radical equations**. An example of a radical equation is $2\sqrt{x+1} = 4$.

Core Concept

Solving Radical Equations

To solve a radical equation, follow these steps:

Step 1 Isolate the radical on one side of the equation, if necessary.

Step 2 Raise each side of the equation to the same exponent to eliminate the radical and obtain a linear, quadratic, or other polynomial equation.

Step 3 Solve the resulting equation using techniques you learned in previous chapters. Check your solution.

EXAMPLE 1 Solving Radical Equations

Solve (a) $2\sqrt{x+1} = 4$ and (b) $\sqrt[3]{2x-9} - 1 = 2$.

SOLUTION

a.

$2\sqrt{x+1} = 4$	Write the original equation.
$\sqrt{x+1} = 2$	Divide each side by 2.
$(\sqrt{x+1})^2 = 2^2$	Square each side to eliminate the radical.
$x + 1 = 4$	Simplify.
$x = 3$	Subtract 1 from each side.

▶ The solution is $x = 3$.

Check

$2\sqrt{3+1} \stackrel{?}{=} 4$

$2\sqrt{4} \stackrel{?}{=} 4$

$4 = 4$ ✓

b.

$\sqrt[3]{2x-9} - 1 = 2$	Write the original equation.
$\sqrt[3]{2x-9} = 3$	Add 1 to each side.
$(\sqrt[3]{2x-9})^3 = 3^3$	Cube each side to eliminate the radical.
$2x - 9 = 27$	Simplify.
$2x = 36$	Add 9 to each side.
$x = 18$	Divide each side by 2.

▶ The solution is $x = 18$.

Check

$\sqrt[3]{2(18) - 9} - 1 \stackrel{?}{=} 2$

$\sqrt[3]{27} - 1 \stackrel{?}{=} 2$

$2 = 2$ ✓

Monitoring Progress  Help in English and Spanish at *BigIdeasMath.com*

Solve the equation. Check your solution.

1. $\sqrt[3]{x} - 9 = -6$
2. $\sqrt{x+25} = 2$
3. $2\sqrt[3]{x-3} = 4$

EXAMPLE 2 Solving a Real-Life Problem

In a hurricane, the mean sustained wind velocity v (in meters per second) can be modeled by $v(p) = 6.3\sqrt{1013 - p}$, where p is the air pressure (in millibars) at the center of the hurricane. Estimate the air pressure at the center of the hurricane when the mean sustained wind velocity is 54.5 meters per second.

SOLUTION

$v(p) = 6.3\sqrt{1013 - p}$	Write the original function.
$54.5 = 6.3\sqrt{1013 - p}$	Substitute 54.5 for $v(p)$.
$8.65 \approx \sqrt{1013 - p}$	Divide each side by 6.3.
$8.65^2 \approx \left(\sqrt{1013 - p}\right)^2$	Square each side.
$74.8 \approx 1013 - p$	Simplify.
$-938.2 \approx -p$	Subtract 1013 from each side.
$938.2 \approx p$	Divide each side by -1.

▶ The air pressure at the center of the hurricane is about 938 millibars.

ATTEND TO PRECISION

To understand how extraneous solutions can be introduced, consider the equation $\sqrt{x} = -3$. This equation has no real solution; however, you obtain $x = 9$ after squaring each side.

Monitoring Progress  Help in English and Spanish at *BigIdeasMath.com*

4. **WHAT IF?** Estimate the air pressure at the center of the hurricane when the mean sustained wind velocity is 48.3 meters per second.

Raising each side of an equation to the same exponent may introduce solutions that are *not* solutions of the original equation. These solutions are called **extraneous solutions**. When you use this procedure, you should always check each apparent solution in the *original* equation.

EXAMPLE 3 Solving an Equation with an Extraneous Solution

Solve $x + 1 = \sqrt{7x + 15}$.

SOLUTION

$x + 1 = \sqrt{7x + 15}$	Write the original equation.
$(x + 1)^2 = \left(\sqrt{7x + 15}\right)^2$	Square each side.
$x^2 + 2x + 1 = 7x + 15$	Expand left side and simplify right side.
$x^2 - 5x - 14 = 0$	Write in standard form.
$(x - 7)(x + 2) = 0$	Factor.
$x - 7 = 0$ or $x + 2 = 0$	Zero-Product Property
$x = 7$ or $x = -2$	Solve for x.

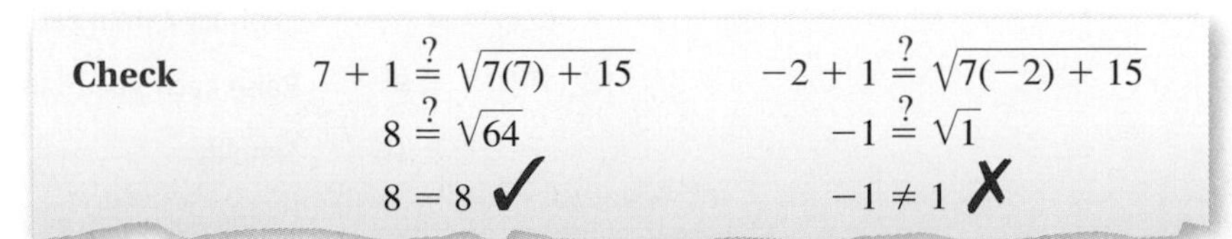

Check

$7 + 1 \stackrel{?}{=} \sqrt{7(7) + 15}$ $\qquad$ $-2 + 1 \stackrel{?}{=} \sqrt{7(-2) + 15}$

$8 \stackrel{?}{=} \sqrt{64}$ $\qquad$ $-1 \stackrel{?}{=} \sqrt{1}$

$8 = 8$ ✓ $\qquad$ $-1 \neq 1$ ✗

▶ The apparent solution $x = -2$ is extraneous. So, the only solution is $x = 7$.

EXAMPLE 4 Solving an Equation with Two Radicals

Solve $\sqrt{x+2} + 1 = \sqrt{3-x}$.

SOLUTION

$\sqrt{x+2} + 1 = \sqrt{3-x}$	Write the original equation.
$\left(\sqrt{x+2} + 1\right)^2 = \left(\sqrt{3-x}\right)^2$	Square each side.
$x + 2 + 2\sqrt{x+2} + 1 = 3 - x$	Expand left side and simplify right side.
$2\sqrt{x+2} = -2x$	Isolate radical expression.
$\sqrt{x+2} = -x$	Divide each side by 2.
$\left(\sqrt{x+2}\right)^2 = (-x)^2$	Square each side.
$x + 2 = x^2$	Simplify.
$0 = x^2 - x - 2$	Write in standard form.
$0 = (x-2)(x+1)$	Factor.
$x - 2 = 0$ or $x + 1 = 0$	Zero-Product Property
$x = 2$ or $x = -1$	Solve for x.

ANOTHER METHOD

You can also graph each side of the equation and find the x-value where the graphs intersect.

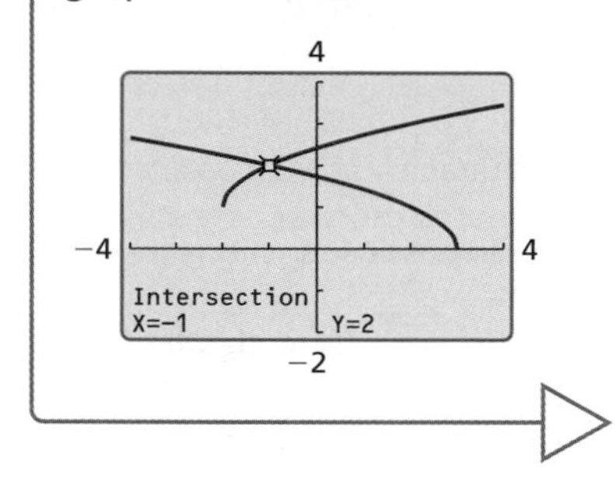

Check $\sqrt{2+2} + 1 \stackrel{?}{=} \sqrt{3-2}$ $\qquad$ $\sqrt{-1+2} + 1 \stackrel{?}{=} \sqrt{3-(-1)}$

$\sqrt{4} + 1 \stackrel{?}{=} \sqrt{1}$ $\qquad$ $\sqrt{1} + 1 \stackrel{?}{=} \sqrt{4}$

$3 \neq 1$ ✗ $\qquad$ $2 = 2$ ✓

▶ The apparent solution $x = 2$ is extraneous. So, the only solution is $x = -1$.

Monitoring Progress

Help in English and Spanish at *BigIdeasMath.com*

Solve the equation. Check your solution(s).

5. $\sqrt{10x+9} = x + 3$ $\qquad$ **6.** $\sqrt{2x+5} = \sqrt{x+7}$ $\qquad$ **7.** $\sqrt{x+6} - 2 = \sqrt{x-2}$

When an equation contains a power with a rational exponent, you can solve the equation using a procedure similar to the one for solving radical equations. In this case, you first isolate the power and then raise each side of the equation to the reciprocal of the rational exponent.

EXAMPLE 5 Solving an Equation with a Rational Exponent

Solve $(2x)^{3/4} + 2 = 10$.

SOLUTION

$(2x)^{3/4} + 2 = 10$	Write the original equation.
$(2x)^{3/4} = 8$	Subtract 2 from each side.
$[(2x)^{3/4}]^{4/3} = 8^{4/3}$	Raise each side to the four-thirds.
$2x = 16$	Simplify.
$x = 8$	Divide each side by 2.

▶ The solution is $x = 8$.

Check

$(2 \cdot 8)^{3/4} + 2 \stackrel{?}{=} 10$

$16^{3/4} + 2 \stackrel{?}{=} 10$

$10 = 10$ ✓

EXAMPLE 6 Solving an Equation with a Rational Exponent

Solve $(x + 30)^{1/2} = x$.

SOLUTION

$(x + 30)^{1/2} = x$	Write the original equation.
$[(x + 30)^{1/2}]^2 = x^2$	Square each side.
$x + 30 = x^2$	Simplify.
$0 = x^2 - x - 30$	Write in standard form.
$0 = (x - 6)(x + 5)$	Factor.
$x - 6 = 0$ or $x + 5 = 0$	Zero-Product Property
$x = 6$ or $x = -5$	Solve for x.

Check

$(6 + 30)^{1/2} \stackrel{?}{=} 6$

$36^{1/2} \stackrel{?}{=} 6$

$6 = 6$ ✓

$(-5 + 30)^{1/2} \stackrel{?}{=} -5$

$25^{1/2} \stackrel{?}{=} -5$

$5 \neq -5$ ✗

▶ The apparent solution $x = -5$ is extraneous. So, the only solution is $x = 6$.

Monitoring Progress

Help in English and Spanish at *BigIdeasMath.com*

Solve the equation. Check your solution(s).

8. $(3x)^{1/3} = -3$ **9.** $(x + 6)^{1/2} = x$ **10.** $(x + 2)^{3/4} = 8$

Solving Radical Inequalities

To solve a simple radical inequality of the form $\sqrt[n]{u} < d$, where u is an algebraic expression and d is a nonnegative number, raise each side to the exponent n. This procedure also works for $>$, $\leq$, and $\geq$. Be sure to consider the possible values of the radicand.

EXAMPLE 7 Solving a Radical Inequality

Solve $3\sqrt{x - 1} \leq 12$.

SOLUTION

Step 1 Solve for x.

$3\sqrt{x - 1} \leq 12$	Write the original inequality.
$\sqrt{x - 1} \leq 4$	Divide each side by 3.
$x - 1 \leq 16$	Square each side.
$x \leq 17$	Add 1 to each side.

Step 2 Consider the radicand.

$x - 1 \geq 0$	The radicand cannot be negative.
$x \geq 1$	Add 1 to each side.

Check

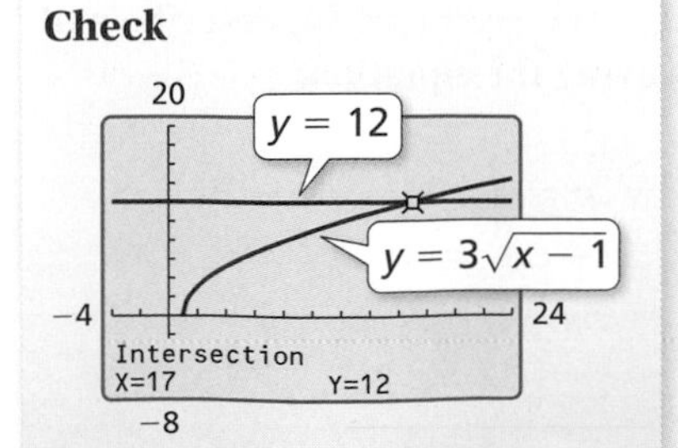

▶ So, the solution is $1 \leq x \leq 17$.

Monitoring Progress

Help in English and Spanish at *BigIdeasMath.com*

11. Solve (a) $2\sqrt{x} - 3 \geq 3$ and (b) $4\sqrt[3]{x + 1} < 8$.

5.4 Exercises

Dynamic Solutions available at *BigIdeasMath.com*

Vocabulary and Core Concept Check

1. **VOCABULARY** Is the equation $3x - \sqrt{2} = \sqrt{6}$ a radical equation? Explain your reasoning.

2. **WRITING** Explain the steps you should use to solve $\sqrt{x} + 10 < 15$.

Monitoring Progress and Modeling with Mathematics

In Exercises 3–12, solve the equation. Check your solution. *(See Example 1.)*

3. $\sqrt{5x + 1} = 6$

4. $\sqrt{3x + 10} = 8$

5. $\sqrt[3]{x - 16} = 2$

6. $\sqrt[3]{x} - 10 = -7$

7. $-2\sqrt{24x} + 13 = -11$

8. $8\sqrt[3]{10x} - 15 = 17$

9. $\frac{1}{5}\sqrt[3]{3x} + 10 = 8$

10. $\sqrt{2x} - \frac{2}{3} = 0$

11. $2\sqrt[5]{x} + 7 = 15$

12. $\sqrt[4]{4x} - 13 = -15$

13. **MODELING WITH MATHEMATICS** Biologists have discovered that the shoulder height h (in centimeters) of a male Asian elephant can be modeled by $h = 62.5\sqrt[3]{t} + 75.8$, where t is the age (in years) of the elephant. Determine the age of an elephant with a shoulder height of 250 centimeters. *(See Example 2.)*

h

14. **MODELING WITH MATHEMATICS** In an amusement park ride, a rider suspended by cables swings back and forth from a tower. The maximum speed v (in meters per second) of the rider can be approximated by $v = \sqrt{2gh}$, where h is the height (in meters) at the top of each swing and g is the acceleration due to gravity ($g \approx 9.8$ m/sec^2). Determine the height at the top of the swing of a rider whose maximum speed is 15 meters per second.

In Exercises 15–26, solve the equation. Check your solution(s). *(See Examples 3 and 4.)*

15. $x - 6 = \sqrt{3x}$

16. $x - 10 = \sqrt{9x}$

17. $\sqrt{44 - 2x} = x - 10$

18. $\sqrt{2x + 30} = x + 3$

19. $\sqrt[3]{8x^3 - 1} = 2x - 1$

20. $\sqrt[4]{3 - 8x^2} = 2x$

21. $\sqrt{4x + 1} = \sqrt{x + 10}$

22. $\sqrt{3x - 3} - \sqrt{x + 12} = 0$

23. $\sqrt[3]{2x - 5} - \sqrt[3]{8x + 1} = 0$

24. $\sqrt[3]{x + 5} = 2\sqrt[3]{2x + 6}$

25. $\sqrt{3x - 8} + 1 = \sqrt{x + 5}$

26. $\sqrt{x + 2} = 2 - \sqrt{x}$

In Exercises 27–34, solve the equation. Check your solution(s). *(See Examples 5 and 6.)*

27. $2x^{2/3} = 8$

28. $4x^{3/2} = 32$

29. $x^{1/4} + 3 = 0$

30. $2x^{3/4} - 14 = 40$

31. $(x + 6)^{1/2} = x$

32. $(5 - x)^{1/2} - 2x = 0$

33. $2(x + 11)^{1/2} = x + 3$

34. $(5x^2 - 4)^{1/4} = x$

ERROR ANALYSIS In Exercises 35 and 36, describe and correct the error in solving the equation.

35. ✗

$$\sqrt[3]{3x - 8} = 4$$
$$\left(\sqrt[3]{3x - 8}\right)^3 = 4$$
$$3x - 8 = 4$$
$$3x = 12$$
$$x = 4$$

36.

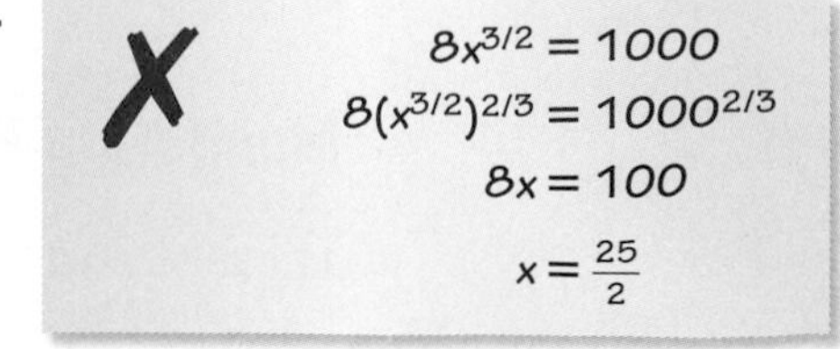

✗
$8x^{3/2} = 1000$
$8(x^{3/2})^{2/3} = 1000^{2/3}$
$8x = 100$
$x = \frac{25}{2}$

In Exercises 37–44, solve the inequality. *(See Example 7.)*

37. $2\sqrt[3]{x} - 5 \geq 3$

38. $\sqrt[3]{x - 4} \leq 5$

39. $4\sqrt{x - 2} > 20$

40. $7\sqrt{x} + 1 < 9$

41. $2\sqrt{x} + 3 \leq 8$

42. $\sqrt[3]{x + 7} \geq 3$

43. $-2\sqrt[3]{x + 4} < 12$

44. $-0.25\sqrt{x} - 6 \leq -3$

45. MODELING WITH MATHEMATICS The length ℓ (in inches) of a standard nail can be modeled by $\ell = 54d^{3/2}$, where d is the diameter (in inches) of the nail. What is the diameter of a standard nail that is 3 inches long?

46. DRAWING CONCLUSIONS "Hang time" is the time you are suspended in the air during a jump. Your hang time t (in seconds) is given by the function $t = 0.5\sqrt{h}$, where h is the height (in feet) of the jump. Suppose a kangaroo and a snowboarder jump with the hang times shown.

a. Find the heights that the snowboarder and the kangaroo jump.

b. Double the hang times of the snowboarder and the kangaroo and calculate the corresponding heights of each jump.

c. When the hang time doubles, does the height of the jump double? Explain.

USING TOOLS In Exercises 47–52, solve the nonlinear system. Justify your answer with a graph.

47. $y^2 = x - 3$
$y = x - 3$

48. $y^2 = 4x + 17$
$y = x + 5$

49. $x^2 + y^2 = 4$
$y = x - 2$

50. $x^2 + y^2 = 25$
$y = -\frac{3}{4}x + \frac{25}{4}$

51. $x^2 + y^2 = 1$
$y = \frac{1}{2}x^2 - 1$

52. $x^2 + y^2 = 4$
$y^2 = x + 2$

53. PROBLEM SOLVING The speed s (in miles per hour) of a car can be given by $s = \sqrt{30fd}$, where f is the coefficient of friction and d is the stopping distance (in feet). The table shows the coefficient of friction for different surfaces.

Surface	Coefficient of friction, f
dry asphalt	0.75
wet asphalt	0.30
snow	0.30
ice	0.15

a. Compare the stopping distances of a car traveling 45 miles per hour on the surfaces given in the table.

b. You are driving 35 miles per hour on an icy road when a deer jumps in front of your car. How far away must you begin to brake to avoid hitting the deer? Justify your answer.

54. MODELING WITH MATHEMATICS The Beaufort wind scale was devised to measure wind speed. The Beaufort numbers B, which range from 0 to 12, can be modeled by $B = 1.69\sqrt{s + 4.25} - 3.55$, where s is the wind speed (in miles per hour).

Beaufort number	Force of wind
0	calm
3	gentle breeze
6	strong breeze
9	strong gale
12	hurricane

a. What is the wind speed for $B = 0$? $B = 3$?

b. Write an inequality that describes the range of wind speeds represented by the Beaufort model.

55. USING TOOLS Solve the equation $x - 4 = \sqrt{2x}$. Then solve the equation $x - 4 = -\sqrt{2x}$.

a. How does changing $\sqrt{2x}$ to $-\sqrt{2x}$ change the solution(s) of the equation?

b. Justify your answer in part (a) using graphs.

56. MAKING AN ARGUMENT Your friend says it is impossible for a radical equation to have two extraneous solutions. Is your friend correct? Explain your reasoning.

57. USING STRUCTURE Explain how you know the radical equation $\sqrt{x+4} = -5$ has no real solution without solving it.

58. HOW DO YOU SEE IT? Use the graph to find the solution of the equation $2\sqrt{x-4} = -\sqrt{x-1} + 4$. Explain your reasoning.

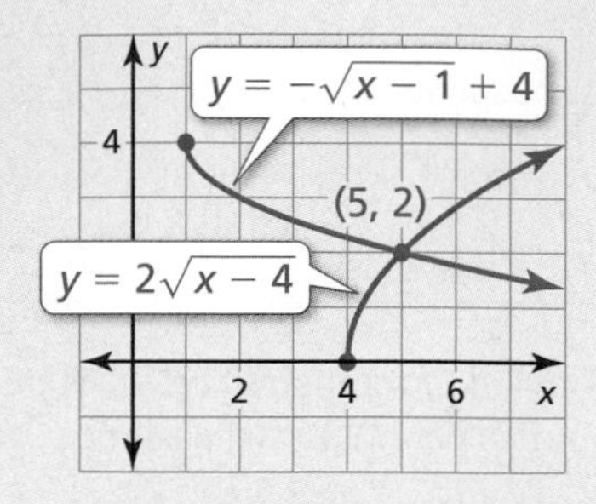

59. WRITING A company determines that the price p of a product can be modeled by $p = 70 - \sqrt{0.02x + 1}$, where x is the number of units of the product demanded per day. Describe the effect that raising the price has on the number of units demanded.

60. THOUGHT PROVOKING City officials rope off a circular area to prepare for a concert in the park. They estimate that each person occupies 6 square feet. Describe how you can use a radical inequality to determine the possible radius of the region when P people are expected to attend the concert.

61. MATHEMATICAL CONNECTIONS The Moeraki Boulders along the coast of New Zealand are stone spheres with radii of approximately 3 feet. A formula for the radius of a sphere is

$$r = \frac{1}{2}\sqrt{\frac{S}{\pi}}$$

where S is the surface area of the sphere. Find the surface area of a Moeraki Boulder.

62. PROBLEM SOLVING You are trying to determine the height of a truncated pyramid, which cannot be measured directly. The height h and slant height ℓ of the truncated pyramid are related by the formula below.

$$\ell = \sqrt{h^2 + \frac{1}{4}(b_2 - b_1)^2}$$

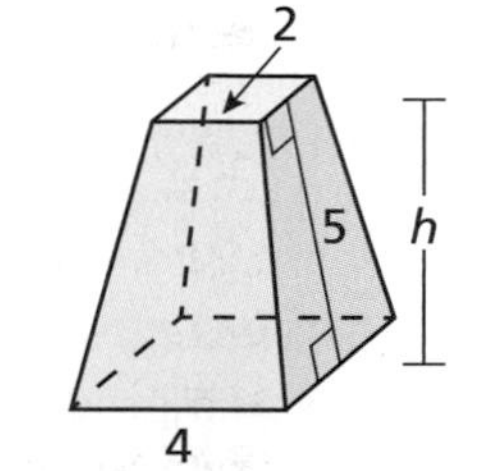

In the given formula, b_1 and b_2 are the side lengths of the upper and lower bases of the pyramid, respectively. When $\ell = 5$, $b_1 = 2$, and $b_2 = 4$, what is the height of the pyramid?

63. REWRITING A FORMULA A burning candle has a radius of r inches and was initially h_0 inches tall. After t minutes, the height of the candle has been reduced to h inches. These quantities are related by the formula

$$r = \sqrt{\frac{kt}{\pi(h_0 - h)}}$$

where k is a constant. Suppose the radius of a candle is 0.875 inch, its initial height is 6.5 inches, and $k = 0.04$.

a. Rewrite the formula, solving for h in terms of t.

b. Use your formula in part (a) to determine the height of the candle after burning 45 minutes.

Maintaining Mathematical Proficiency

Reviewing what you learned in previous grades and lessons

Perform the indicated operation. *(Section 4.2 and Section 4.3)*

64. $(x^3 - 2x^2 + 3x + 1) + (x^4 - 7x)$

65. $(2x^5 + x^4 - 4x^2) - (x^5 - 3)$

66. $(x^3 + 2x^2 + 1)(x^2 + 5)$

67. $(x^4 + 2x^3 + 11x^2 + 14x - 16) \div (x + 2)$

Let $f(x) = x^3 - 4x^2 + 6$. Write a rule for g. Describe the graph of g as a transformation of the graph of f. *(Section 4.7)*

68. $g(x) = f(-x) + 4$

69. $g(x) = \frac{1}{2}f(x) - 3$

70. $g(x) = -f(x - 1) + 6$

5.5 Performing Function Operations

Essential Question How can you use the graphs of two functions to sketch the graph of an arithmetic combination of the two functions?

Just as two real numbers can be combined by the operations of addition, subtraction, multiplication, and division to form other real numbers, two functions can be combined to form other functions. For example, the functions $f(x) = 2x - 3$ and $g(x) = x^2 - 1$ can be combined to form the sum, difference, product, or quotient of f and g.

$$f(x) + g(x) = (2x - 3) + (x^2 - 1) = x^2 + 2x - 4 \qquad \text{sum}$$

$$f(x) - g(x) = (2x - 3) - (x^2 - 1) = -x^2 + 2x - 2 \qquad \text{difference}$$

$$f(x) \cdot g(x) = (2x - 3)(x^2 - 1) = 2x^3 - 3x^2 - 2x + 3 \qquad \text{product}$$

$$\frac{f(x)}{g(x)} = \frac{2x - 3}{x^2 - 1} \qquad \text{quotient}$$

EXPLORATION 1 Graphing the Sum of Two Functions

Work with a partner. Use the graphs of f and g to sketch the graph of $f + g$. Explain your steps.

Sample Choose a point on the graph of g. Use a compass or a ruler to measure its distance above or below the x-axis. If above, add the distance to the y-coordinate of the point with the same x-coordinate on the graph of f. If below, subtract the distance. Plot the new point. Repeat this process for several points. Finally, draw a smooth curve through the new points to obtain the graph of $f + g$.

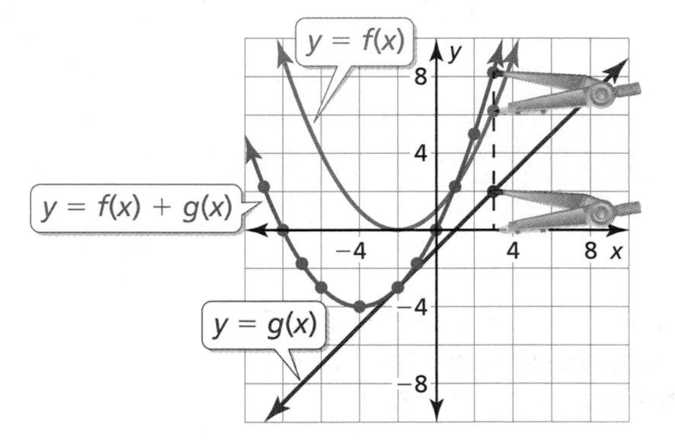

a.

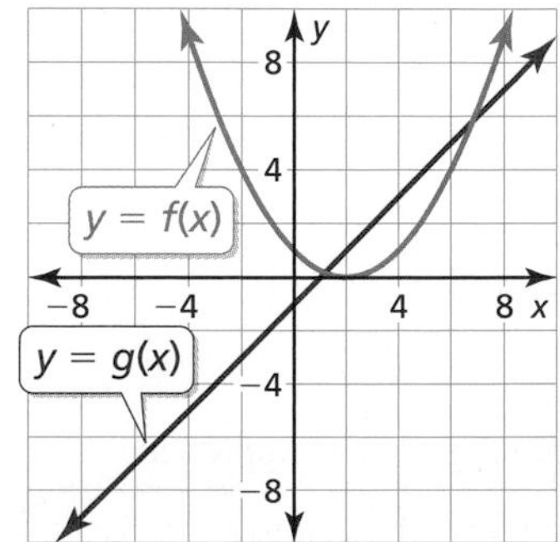

b.

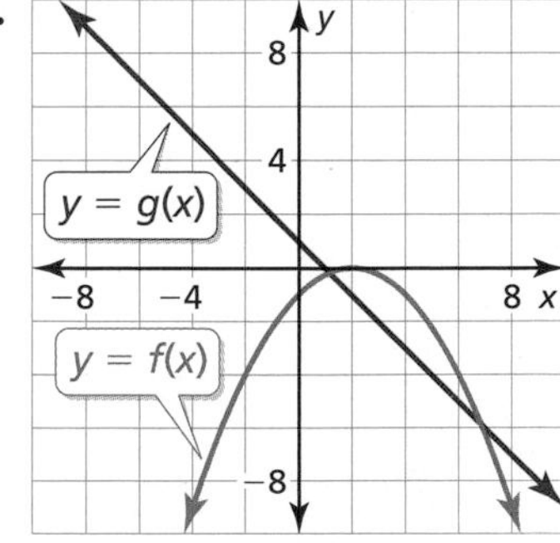

MAKING SENSE OF PROBLEMS

To be proficient in math, you need to check your answers to problems using a different method and continually ask yourself, "Does this make sense?"

Communicate Your Answer

2. How can you use the graphs of two functions to sketch the graph of an arithmetic combination of the two functions?
3. Check your answers in Exploration 1 by writing equations for f and g, adding the functions, and graphing the sum.

5.5 Lesson

What You Will Learn

- Add, subtract, multiply, and divide functions.

Core Vocabulary

Previous
domain
scientific notation

Operations on Functions

You have learned how to add, subtract, multiply, and divide polynomial expressions. These operations can also be defined for functions.

Core Concept

Operations on Functions

Let f and g be any two functions. A new function can be defined by performing any of the four basic operations on f and g.

Operation	Definition	Example: $f(x) = 5x$, $g(x) = x + 2$
Addition	$(f + g)(x) = f(x) + g(x)$	$(f + g)(x) = 5x + (x + 2) = 6x + 2$
Subtraction	$(f - g)(x) = f(x) - g(x)$	$(f - g)(x) = 5x - (x + 2) = 4x - 2$
Multiplication	$(fg)(x) = f(x) \cdot g(x)$	$(fg)(x) = 5x(x + 2) = 5x^2 + 10x$
Division	$\left(\frac{f}{g}\right)(x) = \frac{f(x)}{g(x)}$	$\left(\frac{f}{g}\right)(x) = \frac{5x}{x + 2}$

The domains of the sum, difference, product, and quotient functions consist of the x-values that are in the domains of both f and g. Additionally, the domain of the quotient does not include x-values for which $g(x) = 0$.

EXAMPLE 1 Adding Two Functions

Let $f(x) = 3\sqrt{x}$ and $g(x) = -10\sqrt{x}$. Find $(f + g)(x)$ and state the domain. Then evaluate the sum when $x = 4$.

SOLUTION

$$(f + g)(x) = f(x) + g(x) = 3\sqrt{x} + (-10\sqrt{x}) = (3 - 10)\sqrt{x} = -7\sqrt{x}$$

The functions f and g each have the same domain: all nonnegative real numbers. So, the domain of $f + g$ also consists of all nonnegative real numbers. To evaluate $f + g$ when $x = 4$, you can use several methods. Here are two:

Method 1 Use an algebraic approach.

When $x = 4$, the value of the sum is

$$(f + g)(4) = -7\sqrt{4} = -14.$$

Method 2 Use a graphical approach.

Enter the functions $y_1 = 3\sqrt{x}$, $y_2 = -10\sqrt{x}$, and $y_3 = y_1 + y_2$ in a graphing calculator. Then graph y_3, the sum of the two functions. Use the *trace* feature to find the value of $f + g$ when $x = 4$. From the graph, $(f + g)(4) = -14$.

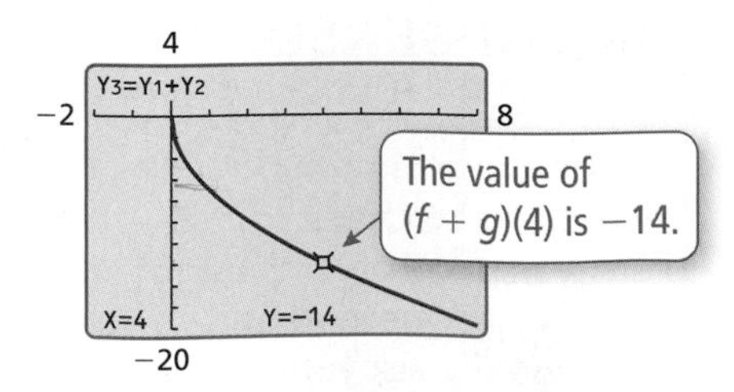

EXAMPLE 2 Subtracting Two Functions

Let $f(x) = 3x^3 - 2x^2 + 5$ and $g(x) = x^3 - 3x^2 + 4x - 2$. Find $(f - g)(x)$ and state the domain. Then evaluate the difference when $x = -2$.

SOLUTION

$$(f - g)(x) = f(x) - g(x) = 3x^3 - 2x^2 + 5 - (x^3 - 3x^2 + 4x - 2) = 2x^3 + x^2 - 4x + 7$$

The functions f and g each have the same domain: all real numbers. So, the domain of $f - g$ also consists of all real numbers. When $x = -2$, the value of the difference is

$$(f - g)(-2) = 2(-2)^3 + (-2)^2 - 4(-2) + 7 = 3.$$

EXAMPLE 3 Multiplying Two Functions

Let $f(x) = x^2$ and $g(x) = \sqrt{x}$. Find $(fg)(x)$ and state the domain. Then evaluate the product when $x = 9$.

SOLUTION

$$(fg)(x) = f(x) \cdot g(x) = x^2(\sqrt{x}) = x^2(x^{1/2}) = x^{(2+1/2)} = x^{5/2}$$

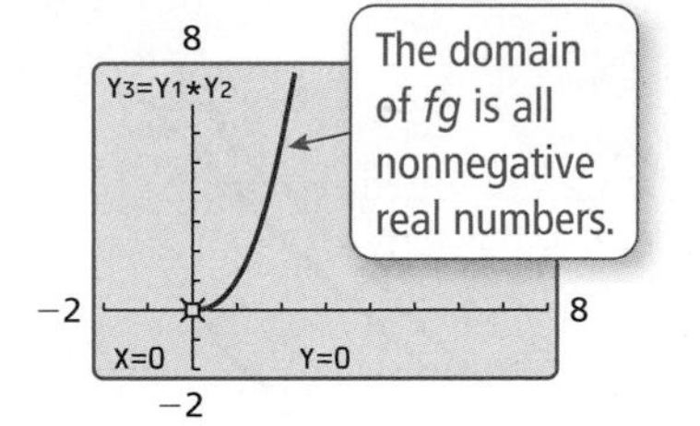

The domain of f consists of all real numbers, and the domain of g consists of all nonnegative real numbers. So, the domain of fg consists of all nonnegative real numbers. To confirm this, enter the functions $y_1 = x^2$, $y_2 = \sqrt{x}$, and $y_3 = y_1 \cdot y_2$ in a graphing calculator. Then graph y_3, the product of the two functions. It appears from the graph that the domain of fg consists of all nonnegative real numbers. When $x = 9$, the value of the product is

$$(fg)(9) = 9^{5/2} = (9^{1/2})^5 = 3^5 = 243.$$

EXAMPLE 4 Dividing Two Functions

Let $f(x) = 6x$ and $g(x) = x^{3/4}$. Find $\left(\frac{f}{g}\right)(x)$ and state the domain. Then evaluate the quotient when $x = 16$.

SOLUTION

$$\left(\frac{f}{g}\right)(x) = \frac{f(x)}{g(x)} = \frac{6x}{x^{3/4}} = 6x^{(1-3/4)} = 6x^{1/4}$$

The domain of f consists of all real numbers, and the domain of g consists of all nonnegative real numbers. Because $g(0) = 0$, the domain of $\frac{f}{g}$ is restricted to all *positive* real numbers. When $x = 16$, the value of the quotient is

$$\left(\frac{f}{g}\right)(16) = 6(16)^{1/4} = 6(2^4)^{1/4} = 12.$$

ANOTHER WAY

In Example 4, you can also evaluate $\left(\frac{f}{g}\right)(16)$ as

$$\left(\frac{f}{g}\right)(16) = \frac{f(16)}{g(16)} = \frac{6(16)}{(16)^{3/4}} = \frac{96}{8} = 12.$$

Monitoring Progress

Help in English and Spanish at *BigIdeasMath.com*

1. Let $f(x) = -2x^{2/3}$ and $g(x) = 7x^{2/3}$. Find $(f + g)(x)$ and $(f - g)(x)$ and state the domain of each. Then evaluate $(f + g)(8)$ and $(f - g)(8)$.

2. Let $f(x) = 3x$ and $g(x) = x^{1/5}$. Find $(fg)(x)$ and $\left(\frac{f}{g}\right)(x)$ and state the domain of each. Then evaluate $(fg)(32)$ and $\left(\frac{f}{g}\right)(32)$.

EXAMPLE 5 Performing Function Operations Using Technology

Let $f(x) = \sqrt{x}$ and $g(x) = \sqrt{9 - x^2}$. Use a graphing calculator to evaluate $(f + g)(x)$, $(f - g)(x)$, $(fg)(x)$, and $\left(\frac{f}{g}\right)(x)$ when $x = 2$. Round your answers to two decimal places.

SOLUTION

Enter the functions $y_1 = \sqrt{x}$ and $y_2 = \sqrt{9 - x^2}$ in a graphing calculator. On the home screen, enter $y_1(2) + y_2(2)$. The first entry on the screen shows that $y_1(2) + y_2(2) \approx 3.65$, so $(f + g)(2) \approx 3.65$. Enter the other function operations as shown. Here are the results of the other function operations rounded to two decimal places:

```
Y1(2)+Y2(2)
          3.65028154
Y1(2)-Y2(2)
         -.8218544151
Y1(2)*Y2(2)
          3.16227766
Y1(2)/Y2(2)
          .632455532
```

$(f - g)(2) \approx -0.82$ $\quad$ $(fg)(2) \approx 3.16$ $\quad$ $\left(\frac{f}{g}\right)(2) \approx 0.63$

EXAMPLE 6 Solving a Real-Life Problem

For a white rhino, heart rate r (in beats per minute) and life span s (in minutes) are related to body mass m (in kilograms) by the functions

$$r(m) = 241m^{-0.25}$$

and

$$s(m) = (6 \times 10^6)m^{0.2}.$$

a. Find $(rs)(m)$.

b. Explain what $(rs)(m)$ represents.

SOLUTION

a. $(rs)(m) = r(m) \cdot s(m)$ — Definition of multiplication

$= 241m^{-0.25}[(6 \times 10^6)m^{0.2}]$ — Write product of $r(m)$ and $s(m)$.

$= 241(6 \times 10^6)m^{-0.25+0.2}$ — Product of Powers Property

$= (1446 \times 10^6)m^{-0.05}$ — Simplify.

$= (1.446 \times 10^9)m^{-0.05}$ — Use scientific notation.

b. Multiplying heart rate by life span gives the total number of heartbeats over the lifetime of a white rhino with body mass m.

Monitoring Progress  Help in English and Spanish at BigIdeasMath.com

3. Let $f(x) = 8x$ and $g(x) = 2x^{5/6}$. Use a graphing calculator to evaluate $(f + g)(x)$, $(f - g)(x)$, $(fg)(x)$, and $\left(\frac{f}{g}\right)(x)$ when $x = 5$. Round your answers to two decimal places.

4. In Example 5, explain why you can evaluate $(f + g)(3)$, $(f - g)(3)$, and $(fg)(3)$ but not $\left(\frac{f}{g}\right)(3)$.

5. Use the answer in Example 6(a) to find the total number of heartbeats over the lifetime of a white rhino when its body mass is 1.7×10^5 kilograms.

5.5 Exercises

Dynamic Solutions available at *BigIdeasMath.com*

Vocabulary and Core Concept Check

1. **WRITING** Let f and g be any two functions. Describe how you can use f, g, and the four basic operations to create new functions.

2. **WRITING** What x-values are not included in the domain of the quotient of two functions?

Monitoring Progress and Modeling with Mathematics

In Exercises 3–6, find $(f+g)(x)$ and $(f-g)(x)$ and state the domain of each. Then evaluate $f+g$ and $f-g$ for the given value of x. *(See Examples 1 and 2.)*

3. $f(x) = -5\sqrt[4]{x}$, $g(x) = 19\sqrt[4]{x}$; $x = 16$

4. $f(x) = \sqrt[3]{2x}$, $g(x) = -11\sqrt[3]{2x}$; $x = -4$

5. $f(x) = 6x - 4x^2 - 7x^3$, $g(x) = 9x^2 - 5x$; $x = -1$

6. $f(x) = 11x + 2x^2$, $g(x) = -7x - 3x^2 + 4$; $x = 2$

In Exercises 7–12, find $(fg)(x)$ and $\left(\frac{f}{g}\right)(x)$ and state the domain of each. Then evaluate fg and $\frac{f}{g}$ for the given value of x. *(See Examples 3 and 4.)*

7. $f(x) = 2x^3$, $g(x) = \sqrt[3]{x}$; $x = -27$

8. $f(x) = x^4$, $g(x) = 3\sqrt{x}$; $x = 4$

9. $f(x) = 4x$, $g(x) = 9x^{1/2}$; $x = 9$

10. $f(x) = 11x^3$, $g(x) = 7x^{7/3}$; $x = -8$

11. $f(x) = 7x^{3/2}$, $g(x) = -14x^{1/3}$; $x = 64$

12. $f(x) = 4x^{5/4}$, $g(x) = 2x^{1/2}$; $x = 16$

USING TOOLS In Exercises 13–16, use a graphing calculator to evaluate $(f+g)(x)$, $(f-g)(x)$, $(fg)(x)$, and $\left(\frac{f}{g}\right)(x)$ when $x = 5$. Round your answers to two decimal places. *(See Example 5.)*

13. $f(x) = 4x^4$; $g(x) = 24x^{1/3}$

14. $f(x) = 7x^{5/3}$; $g(x) = 49x^{2/3}$

15. $f(x) = -2x^{1/3}$; $g(x) = 5x^{1/2}$

16. $f(x) = 4x^{1/2}$; $g(x) = 6x^{3/4}$

ERROR ANALYSIS In Exercises 17 and 18, describe and correct the error in stating the domain.

17. ✗ $f(x) = x^{1/2}$ and $g(x) = x^{3/2}$
The domain of fg is all real numbers.

18. ✗ $f(x) = x^3$ and $g(x) = x^2 - 4$
The domain of $\frac{f}{g}$ is all real numbers except $x = 2$.

19. **MODELING WITH MATHEMATICS** From 1990 to 2010, the numbers (in millions) of female F and male M employees from the ages of 16 to 19 in the United States can be modeled by $F(t) = -0.007t^2 + 0.10t + 3.7$ and $M(t) = 0.0001t^3 - 0.009t^2 + 0.11t + 3.7$, where t is the number of years since 1990. *(See Example 6.)*

   a. Find $(F + M)(t)$.

   b. Explain what $(F + M)(t)$ represents.

20. **MODELING WITH MATHEMATICS** From 2005 to 2009, the numbers of cruise ship departures (in thousands) from around the world W and Florida F can be modeled by the equations

$$W(t) = -5.8333t^3 + 17.43t^2 + 509.1t + 11496$$
$$F(t) = 12.5t^3 - 60.29t^2 + 136.6t + 4881$$

   where t is the number of years since 2005.

   a. Find $(W - F)(t)$.

   b. Explain what $(W - F)(t)$ represents.

21. **MAKING AN ARGUMENT** Your friend claims that the addition of functions and the multiplication of functions are commutative. Is your friend correct? Explain your reasoning.

22. **HOW DO YOU SEE IT?** The graphs of the functions $f(x) = 3x^2 - 2x - 1$ and $g(x) = 3x + 4$ are shown. Which graph represents the function $f + g$? the function $f - g$? Explain your reasoning.

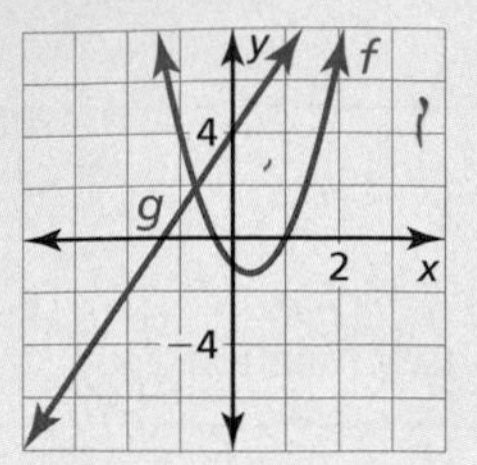

A.

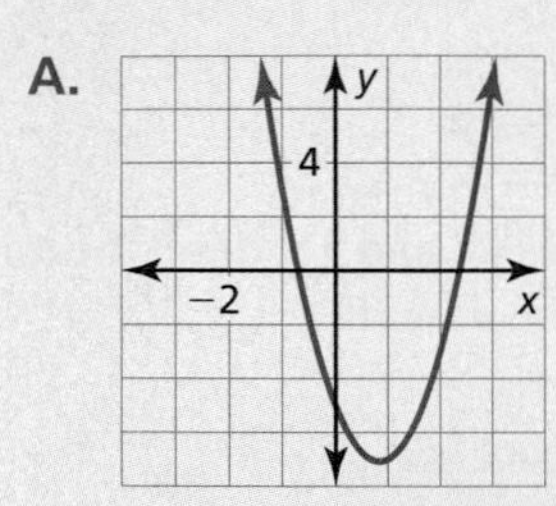

B.

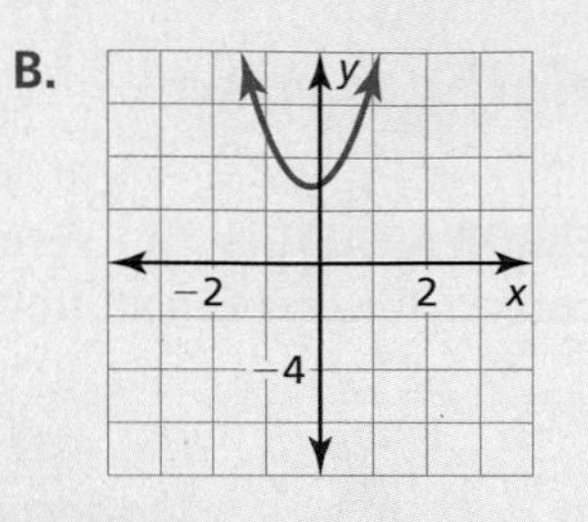

23. **REASONING** The table shows the outputs of the two functions f and g. Use the table to evaluate $(f + g)(3)$, $(f - g)(1)$, $(fg)(2)$, and $\left(\frac{f}{g}\right)(0)$.

x	0	1	2	3	4
f(x)	−2	−4	0	10	26
g(x)	−1	−3	−13	−31	−57

24. **THOUGHT PROVOKING** Is it possible to write two functions whose sum contains radicals, but whose product does not? Justify your answers.

25. **MATHEMATICAL CONNECTIONS** A triangle is inscribed in a square, as shown. Write and simplify a function r in terms of x that represents the area of the shaded region.

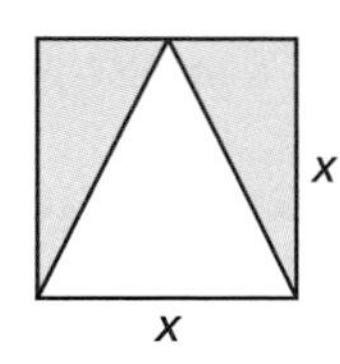

26. **REWRITING A FORMULA** For a mammal that weighs w grams, the volume b (in milliliters) of air breathed in and the volume d (in milliliters) of "dead space" (the portion of the lungs not filled with air) can be modeled by

$$b(w) = 0.007w \text{ and } d(w) = 0.002w.$$

The breathing rate r (in breaths per minute) of a mammal that weighs w grams can be modeled by

$$r(w) = \frac{1.1w^{0.734}}{b(w) - d(w)}.$$

Simplify $r(w)$ and calculate the breathing rate for body weights of 6.5 grams, 300 grams, and 70,000 grams.

27. **PROBLEM SOLVING** A mathematician at a lake throws a tennis ball from point A along the water's edge to point B in the water, as shown. His dog, Elvis, first runs along the beach from point A to point D and then swims to fetch the ball at point B.

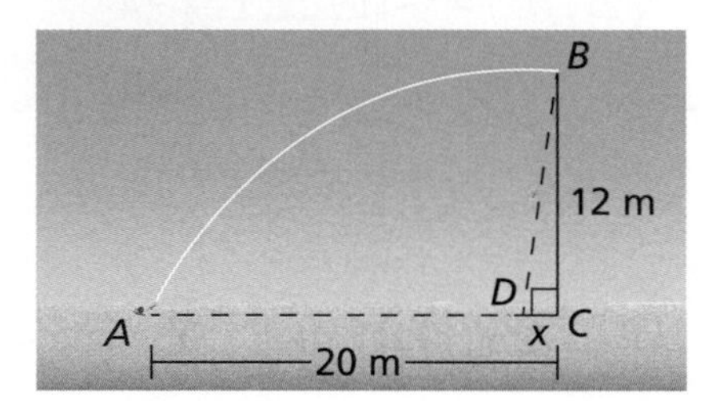

a. Elvis runs at a speed of about 6.4 meters per second. Write a function r in terms of x that represents the time he spends running from point A to point D. Elvis swims at a speed of about 0.9 meter per second. Write a function s in terms of x that represents the time he spends swimming from point D to point B.

b. Write a function t in terms of x that represents the total time Elvis spends traveling from point A to point D to point B.

c. Use a graphing calculator to graph t. Find the value of x that minimizes t. Explain the meaning of this value.

Maintaining Mathematical Proficiency

Reviewing what you learned in previous grades and lessons

Solve the literal equation for n. *(Skills Review Handbook)*

28. $3xn - 9 = 6y$

29. $5z = 7n + 8nz$

30. $3nb = 5n - 6z$

31. $\frac{3 + 4n}{n} = 7b$

Determine whether the relation is a function. Explain. *(Skills Review Handbook)*

32. (3, 4), (4, 6), (1, 4), (2, −1)

33. (−1, 2), (3, 7), (0, 2), (−1, −1)

34. (1, 6), (7, −3), (4, 0), (3, 0)

35. (3, 8), (2, 5), (9, 5), (2, −3)

5.6 Inverse of a Function

Essential Question How can you sketch the graph of the inverse of a function?

EXPLORATION 1 Graphing Functions and Their Inverses

CONSTRUCTING VIABLE ARGUMENTS

To be proficient in math, you need to reason inductively and make a plausible argument.

Work with a partner. Each pair of functions are *inverses* of each other. Use a graphing calculator to graph f and g in the same viewing window. What do you notice about the graphs?

a. $f(x) = 4x + 3$

$g(x) = \dfrac{x - 3}{4}$

b. $f(x) = x^3 + 1$

$g(x) = \sqrt[3]{x - 1}$

c. $f(x) = \sqrt{x - 3}$

$g(x) = x^2 + 3,\ x \geq 0$

d. $f(x) = \dfrac{4x + 4}{x + 5}$

$g(x) = \dfrac{4 - 5x}{x - 4}$

EXPLORATION 2 Sketching Graphs of Inverse Functions

Work with a partner. Use the graph of f to sketch the graph of g, the inverse function of f, on the same set of coordinate axes. Explain your reasoning.

a.

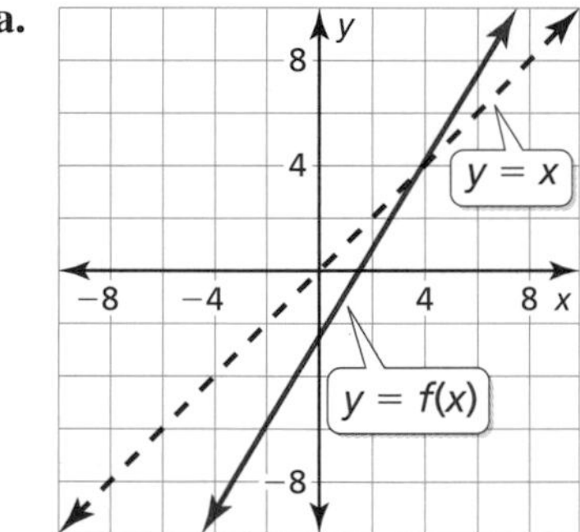

b.

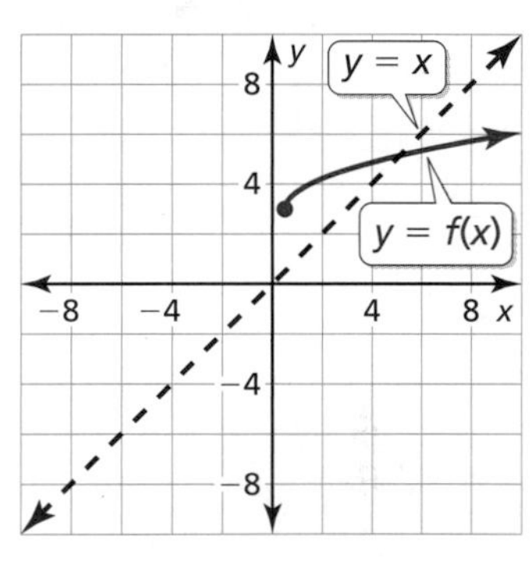

c.

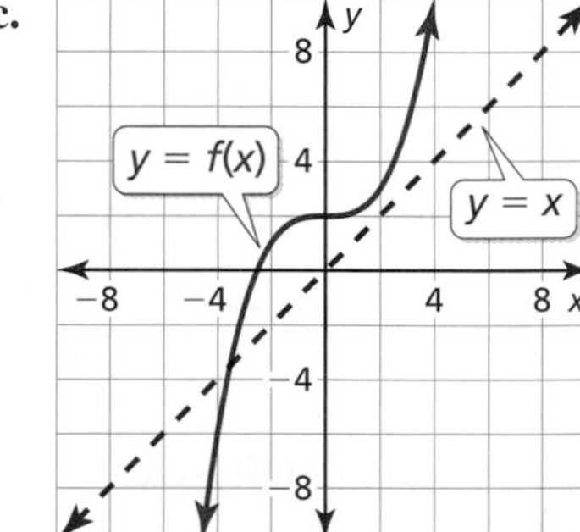

d.

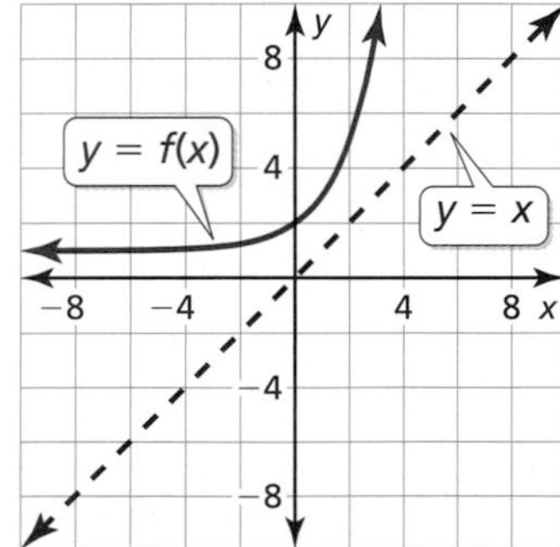

Communicate Your Answer

3. How can you sketch the graph of the inverse of a function?

4. In Exploration 1, what do you notice about the relationship between the equations of f and g? Use your answer to find g, the inverse function of

$f(x) = 2x - 3.$

Use a graph to check your answer.

5.6 Lesson

Core Vocabulary

inverse functions, *p. 277*

Previous
input
output
inverse operations
reflection
line of reflection

What You Will Learn

- Explore inverses of functions.
- Find and verify inverses of nonlinear functions.
- Solve real-life problems using inverse functions.

Exploring Inverses of Functions

You have used given inputs to find corresponding outputs of $y = f(x)$ for various types of functions. You have also used given outputs to find corresponding inputs. Now you will solve equations of the form $y = f(x)$ for x to obtain a general formula for finding the input given a specific output of a function f.

EXAMPLE 1 Writing a Formula for the Input of a Function

Let $f(x) = 2x + 3$.

a. Solve $y = f(x)$ for x.

b. Find the input when the output is -7.

SOLUTION

a. $y = 2x + 3$ — Set y equal to $f(x)$.

$y - 3 = 2x$ — Subtract 3 from each side.

$\frac{y - 3}{2} = x$ — Divide each side by 2.

b. Find the input when $y = -7$.

$x = \frac{-7 - 3}{2}$ — Substitute -7 for y.

$= \frac{-10}{2}$ — Subtract.

$= -5$ — Divide.

So, the input is -5 when the output is -7.

Check

$f(-5) = 2(-5) + 3$
$= -10 + 3$
$= -7$ ✓

Monitoring Progress

Help in English and Spanish at *BigIdeasMath.com*

Solve $y = f(x)$ for x. Then find the input(s) when the output is 2.

1. $f(x) = x - 2$ **2.** $f(x) = 2x^2$ **3.** $f(x) = -x^3 + 3$

In Example 1, notice the steps involved after substituting for x in $y = 2x + 3$ and after substituting for y in $x = \frac{y - 3}{2}$.

$y = 2x + 3$

Step 1 Multiply by 2.
Step 2 Add 3.

$x = \frac{y - 3}{2}$

Step 1 Subtract 3.
Step 2 Divide by 2.

inverse operations in the reverse order

UNDERSTANDING MATHEMATICAL TERMS

The term *inverse functions* does not refer to a new type of function. Rather, it describes any pair of functions that are inverses.

Notice that these steps *undo* each other. Functions that undo each other are called **inverse functions**. In Example 1, you can use the equation solved for x to write the inverse of f by switching the roles of x and y.

$f(x) = 2x + 3$ original function $\qquad g(x) = \dfrac{x-3}{2}$ inverse function

Because inverse functions interchange the input and output values of the original function, the domain and range are also interchanged.

Original function: $f(x) = 2x + 3$

x	−2	−1	0	1	2
y	−1	1	3	5	7

Inverse function: $g(x) = \dfrac{x-3}{2}$

x	−1	1	3	5	7
y	−2	−1	0	1	2

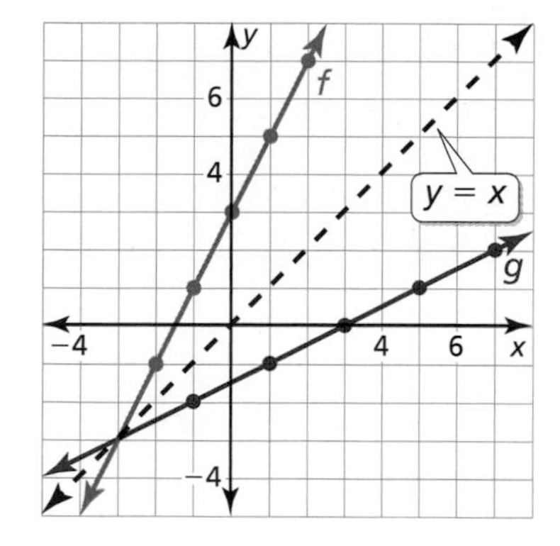

The graph of an inverse function is a *reflection* of the graph of the original function. The *line of reflection* is $y = x$. To find the inverse of a function algebraically, switch the roles of x and y, and then solve for y.

EXAMPLE 2 Finding the Inverse of a Linear Function

Find the inverse of $f(x) = 3x - 1$.

SOLUTION

Method 1 Use inverse operations in the reverse order.

$f(x) = 3x - 1$ Multiply the input x by 3 and then subtract 1.

To find the inverse, apply inverse operations in the reverse order.

$g(x) = \dfrac{x+1}{3}$ Add 1 to the input x and then divide by 3.

▶ The inverse of f is $g(x) = \dfrac{x+1}{3}$, or $g(x) = \dfrac{1}{3}x + \dfrac{1}{3}$.

Method 2 Set y equal to $f(x)$. Switch the roles of x and y and solve for y.

$y = 3x - 1$ Set y equal to $f(x)$.

$x = 3y - 1$ Switch x and y.

$x + 1 = 3y$ Add 1 to each side.

$\dfrac{x+1}{3} = y$ Divide each side by 3.

▶ The inverse of f is $g(x) = \dfrac{x+1}{3}$, or $g(x) = \dfrac{1}{3}x + \dfrac{1}{3}$.

Check

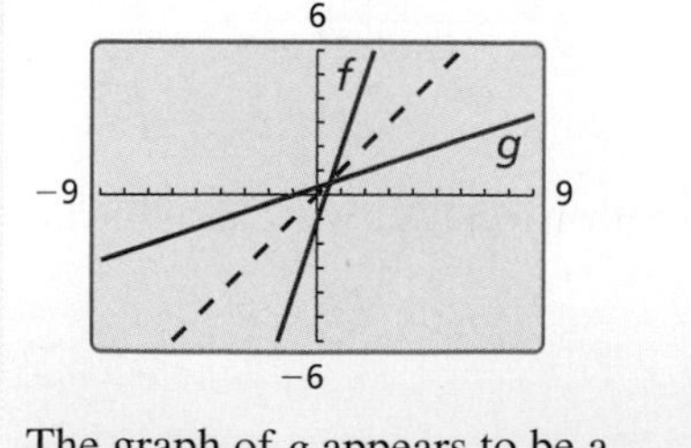

The graph of g appears to be a reflection of the graph of f in the line $y = x$. ✓

Monitoring Progress

Help in English and Spanish at *BigIdeasMath.com*

Find the inverse of the function. Then graph the function and its inverse.

4. $f(x) = 2x$ **5.** $f(x) = -x + 1$ **6.** $f(x) = \frac{1}{3}x - 2$

Inverses of Nonlinear Functions

In the previous examples, the inverses of the linear functions were also functions. However, inverses are not always functions. The graphs of $f(x) = x^2$ and $f(x) = x^3$ are shown along with their reflections in the line $y = x$. Notice that the inverse of $f(x) = x^3$ is a function, but the inverse of $f(x) = x^2$ is *not* a function.

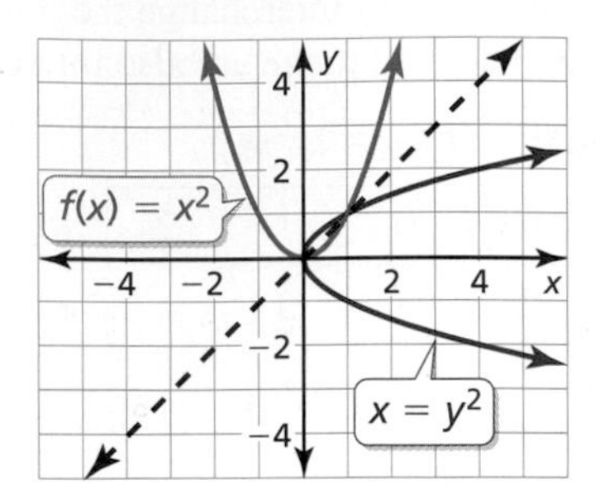

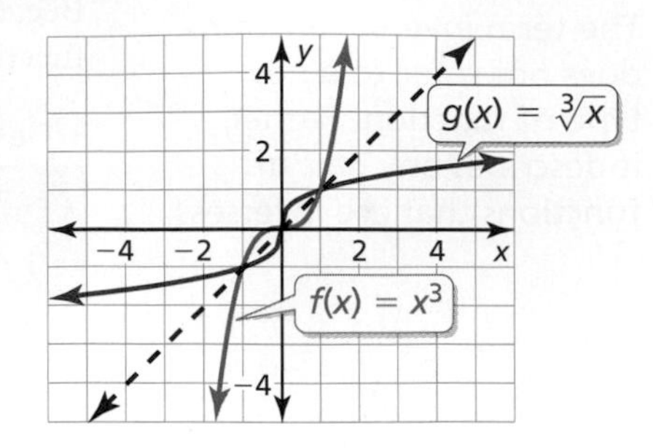

When the domain of $f(x) = x^2$ is *restricted* to only nonnegative real numbers, the inverse of f *is* a function.

EXAMPLE 3 Finding the Inverse of a Quadratic Function

Find the inverse of $f(x) = x^2$, $x \geq 0$. Then graph the function and its inverse.

SOLUTION

$f(x) = x^2$	Write the original function.
$y = x^2$	Set y equal to $f(x)$.
$x = y^2$	Switch x and y.
$\pm\sqrt{x} = y$	Take square root of each side.

The domain of f is restricted to nonnegative values of x. So, the range of the inverse must also be restricted to nonnegative values.

STUDY TIP

If the domain of f were restricted to $x \leq 0$, then the inverse would be $g(x) = -\sqrt{x}$.

So, the inverse of f is $g(x) = \sqrt{x}$.

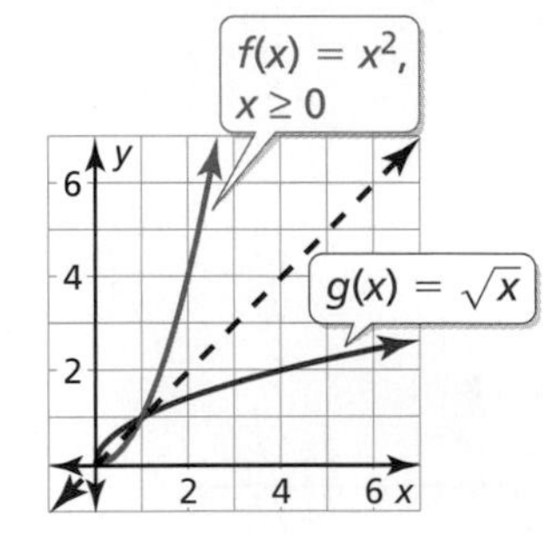

You can use the graph of a function f to determine whether the inverse of f is a function by applying the *horizontal line test*.

Core Concept

Horizontal Line Test

The inverse of a function f is also a function if and only if no horizontal line intersects the graph of f more than once.

Inverse is a function

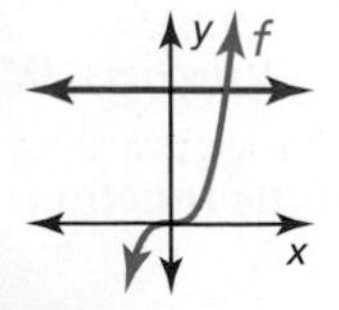

Inverse is not a function

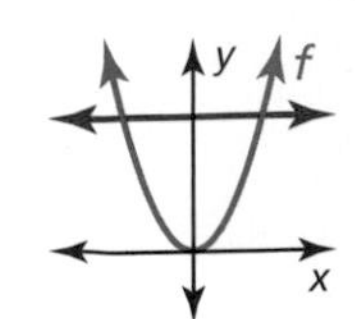

EXAMPLE 4 Finding the Inverse of a Cubic Function

Consider the function $f(x) = 2x^3 + 1$. Determine whether the inverse of f is a function. Then find the inverse.

SOLUTION

Graph the function f. Notice that no horizontal line intersects the graph more than once. So, the inverse of f is a function. Find the inverse.

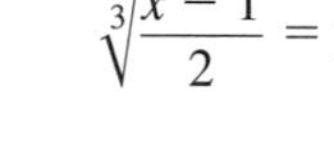

$y = 2x^3 + 1$	Set y equal to $f(x)$.
$x = 2y^3 + 1$	Switch x and y.
$x - 1 = 2y^3$	Subtract 1 from each side.
$\dfrac{x-1}{2} = y^3$	Divide each side by 2.
$\sqrt[3]{\dfrac{x-1}{2}} = y$	Take cube root of each side.

Check

▶ So, the inverse of f is $g(x) = \sqrt[3]{\dfrac{x-1}{2}}$.

EXAMPLE 5 Finding the Inverse of a Radical Function

Consider the function $f(x) = 2\sqrt{x-3}$. Determine whether the inverse of f is a function. Then find the inverse.

SOLUTION

Graph the function f. Notice that no horizontal line intersects the graph more than once. So, the inverse of f is a function. Find the inverse.

$y = 2\sqrt{x-3}$	Set y equal to $f(x)$.
$x = 2\sqrt{y-3}$	Switch x and y.
$x^2 = \left(2\sqrt{y-3}\right)^2$	Square each side.
$x^2 = 4(y-3)$	Simplify.
$x^2 = 4y - 12$	Distributive Property
$x^2 + 12 = 4y$	Add 12 to each side.
$\frac{1}{4}x^2 + 3 = y$	Divide each side by 4.

Check

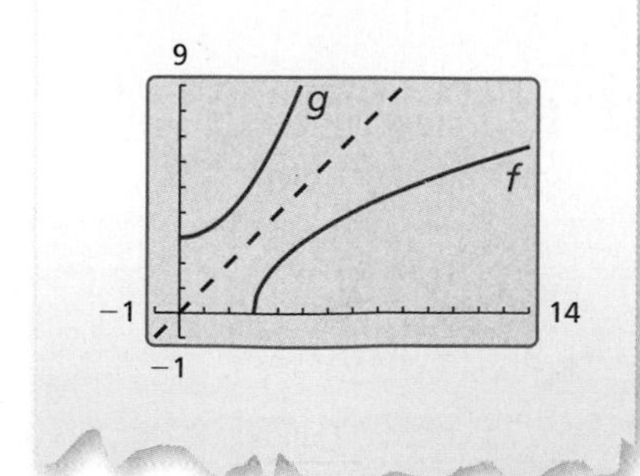

Because the range of f is $y \geq 0$, the domain of the inverse must be restricted to $x \geq 0$.

▶ So, the inverse of f is $g(x) = \frac{1}{4}x^2 + 3$, where $x \geq 0$.

Monitoring Progress

Help in English and Spanish at *BigIdeasMath.com*

Find the inverse of the function. Then graph the function and its inverse.

7. $f(x) = -x^2, x \leq 0$ **8.** $f(x) = -x^3 + 4$ **9.** $f(x) = \sqrt{x+2}$

Let f and g be inverse functions. If $f(a) = b$, then $g(b) = a$. So, in general,

$$f(g(x)) = x \quad \text{and} \quad g(f(x)) = x.$$

REASONING ABSTRACTLY

Inverse functions *undo* each other. So, when you evaluate a function for a specific input, and then evaluate its inverse using the output, you obtain the original input.

EXAMPLE 6 Verifying Functions Are Inverses

Verify that $f(x) = 3x - 1$ and $g(x) = \frac{x+1}{3}$ are inverse functions.

SOLUTION

Step 1 Show that $f(g(x)) = x$.

$$f(g(x)) = f\left(\frac{x+1}{3}\right)$$
$$= 3\left(\frac{x+1}{3}\right) - 1$$
$$= x + 1 - 1$$
$$= x \checkmark$$

Step 2 Show that $g(f(x)) = x$.

$$g(f(x)) = g(3x - 1)$$
$$= \frac{3x - 1 + 1}{3}$$
$$= \frac{3x}{3}$$
$$= x \checkmark$$

Monitoring Progress  Help in English and Spanish at *BigIdeasMath.com*

Determine whether the functions are inverse functions.

10. $f(x) = x + 5,\ g(x) = x - 5$

11. $f(x) = 8x^3,\ g(x) = \sqrt[3]{2x}$

Solving Real-Life Problems

In many real-life problems, formulas contain meaningful variables, such as the radius r in the formula for the surface area S of a sphere, $S = 4\pi r^2$. In this situation, switching the variables to find the inverse would create confusion by switching the meanings of S and r. So, when finding the inverse, solve for r without switching the variables.

EXAMPLE 7 Solving a Multi-Step Problem

Find the inverse of the function that represents the surface area of a sphere, $S = 4\pi r^2$. Then find the radius of a sphere that has a surface area of 100π square feet.

SOLUTION

Step 1 Find the inverse of the function.

$$S = 4\pi r^2$$
$$\frac{S}{4\pi} = r^2$$
$$\sqrt{\frac{S}{4\pi}} = r$$

The radius r must be positive, so disregard the negative square root.

Step 2 Evaluate the inverse when $S = 100\pi$.

$$r = \sqrt{\frac{100\pi}{4\pi}}$$
$$= \sqrt{25} = 5$$

▶ The radius of the sphere is 5 feet.

Monitoring Progress  Help in English and Spanish at *BigIdeasMath.com*

12. The distance d (in meters) that a dropped object falls in t seconds on Earth is represented by $d = 4.9t^2$. Find the inverse of the function. How long does it take an object to fall 50 meters?

5.6 Exercises

Dynamic Solutions available at *BigIdeasMath.com*

Vocabulary and Core Concept Check

1. **VOCABULARY** In your own words, state the definition of inverse functions.

2. **WRITING** Explain how to determine whether the inverse of a function is also a function.

3. **COMPLETE THE SENTENCE** Functions f and g are inverses of each other provided that $f(g(x)) =$ ____ and $g(f(x)) =$ ____.

4. **DIFFERENT WORDS, SAME QUESTION** Which is different? Find "both" answers.

Let $f(x) = 5x - 2$. Solve $y = f(x)$ for x and then switch the roles of x and y.

Write an equation that represents a reflection of the graph of $f(x) = 5x - 2$ in the x-axis.

Write an equation that represents a reflection of the graph of $f(x) = 5x - 2$ in the line $y = x$.

Find the inverse of $f(x) = 5x - 2$.

Monitoring Progress and Modeling with Mathematics

In Exercises 5–12, solve $y = f(x)$ for x. Then find the input(s) when the output is -3. *(See Example 1.)*

5. $f(x) = 3x + 5$

6. $f(x) = -7x - 2$

7. $f(x) = \frac{1}{2}x - 3$

8. $f(x) = -\frac{2}{3}x + 1$

9. $f(x) = 3x^3$

10. $f(x) = 2x^4 - 5$

11. $f(x) = (x - 2)^2 - 7$

12. $f(x) = (x - 5)^3 - 1$

In Exercises 13–20, find the inverse of the function. Then graph the function and its inverse. *(See Example 2.)*

13. $f(x) = 6x$

14. $f(x) = -3x$

15. $f(x) = -2x + 5$

16. $f(x) = 6x - 3$

17. $f(x) = -\frac{1}{2}x + 4$

18. $f(x) = \frac{1}{3}x - 1$

19. $f(x) = \frac{2}{3}x - \frac{1}{3}$

20. $f(x) = -\frac{4}{5}x + \frac{1}{5}$

21. **COMPARING METHODS** Find the inverse of the function $f(x) = -3x + 4$ by switching the roles of x and y and solving for y. Then find the inverse of the function f by using inverse operations in the reverse order. Which method do you prefer? Explain.

22. **REASONING** Determine whether each pair of functions f and g are inverses. Explain your reasoning.

a.

x	-2	-1	0	1	2
$f(x)$	-2	1	4	7	10

x	-2	1	4	7	10
$g(x)$	-2	-1	0	1	2

b.

x	2	3	4	5	6
$f(x)$	8	6	4	2	0

x	2	3	4	5	6
$g(x)$	-8	-6	-4	-2	0

c.

x	-4	-2	0	2	4
$f(x)$	2	10	18	26	34

x	-4	-2	0	2	4
$g(x)$	$\frac{1}{2}$	$\frac{1}{10}$	$\frac{1}{18}$	$\frac{1}{26}$	$\frac{1}{34}$

In Exercises 23–28, find the inverse of the function. Then graph the function and its inverse. *(See Example 3.)*

23. $f(x) = 4x^2, x \leq 0$

24. $f(x) = 9x^2, x \leq 0$

25. $f(x) = (x - 3)^3$

26. $f(x) = (x + 4)^3$

27. $f(x) = 2x^4, x \geq 0$

28. $f(x) = -x^6, x \geq 0$

ERROR ANALYSIS In Exercises 29 and 30, describe and correct the error in finding the inverse of the function.

29.

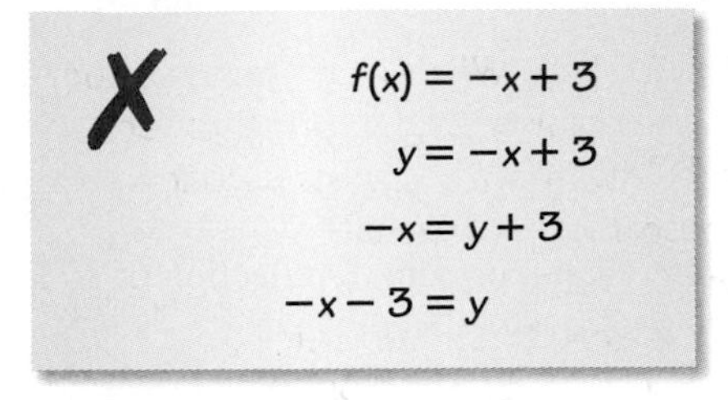

30.

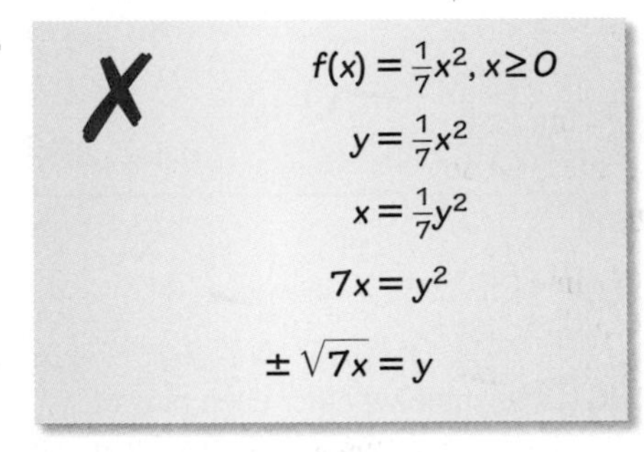

USING TOOLS In Exercises 31–34, use the graph to determine whether the inverse of *f* is a function. Explain your reasoning.

31.

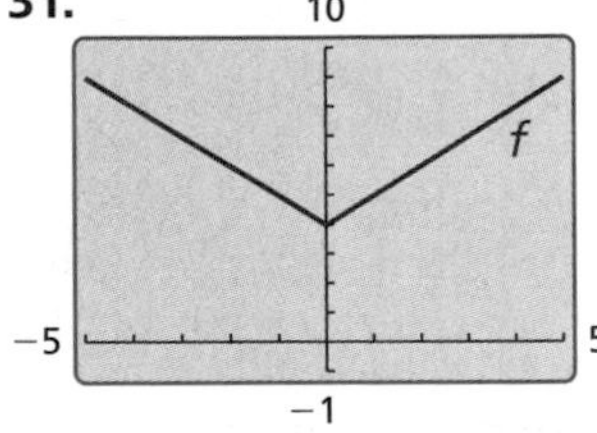

32.

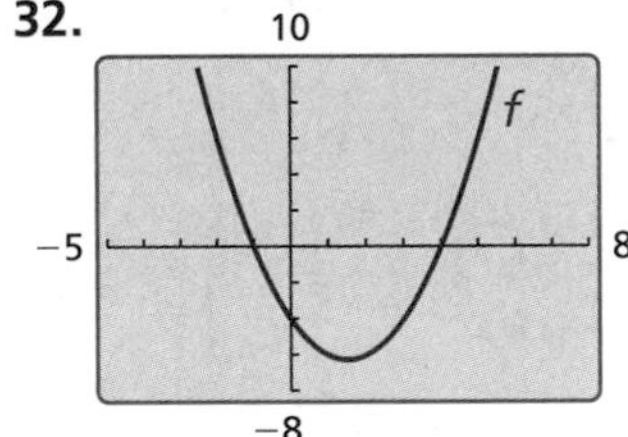

33.

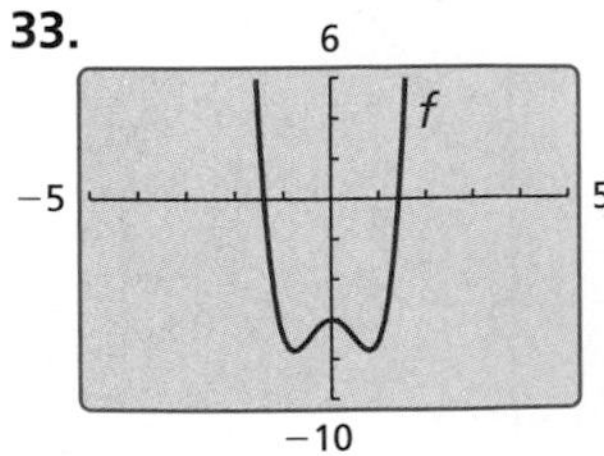

34.

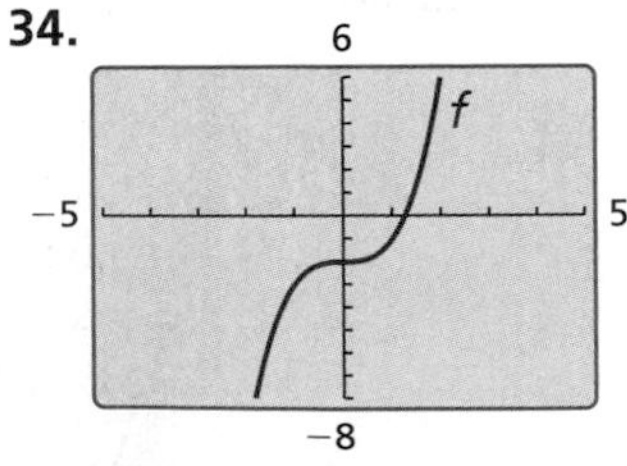

In Exercises 35–46, determine whether the inverse of *f* is a function. Then find the inverse. *(See Examples 4 and 5.)*

35. $f(x) = x^3 - 1$

36. $f(x) = -x^3 + 3$

37. $f(x) = \sqrt{x + 4}$

38. $f(x) = \sqrt{x - 6}$

39. $f(x) = 2\sqrt[3]{x - 5}$

40. $f(x) = 2x^2 - 5$

41. $f(x) = x^4 + 2$

42. $f(x) = 2x^3 - 5$

43. $f(x) = 3\sqrt[3]{x + 1}$

44. $f(x) = -\sqrt[3]{\frac{2x + 4}{3}}$

45. $f(x) = \frac{1}{2}x^5$

46. $f(x) = -3\sqrt{\frac{4x - 7}{3}}$

47. WRITING EQUATIONS What is the inverse of the function whose graph is shown?

Ⓐ $g(x) = \frac{3}{2}x - 6$

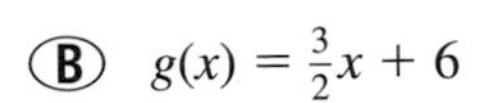
Ⓑ $g(x) = \frac{3}{2}x + 6$

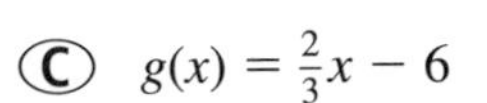
Ⓒ $g(x) = \frac{2}{3}x - 6$

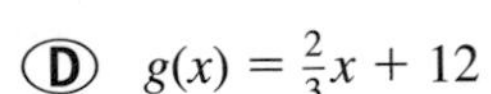
Ⓓ $g(x) = \frac{2}{3}x + 12$

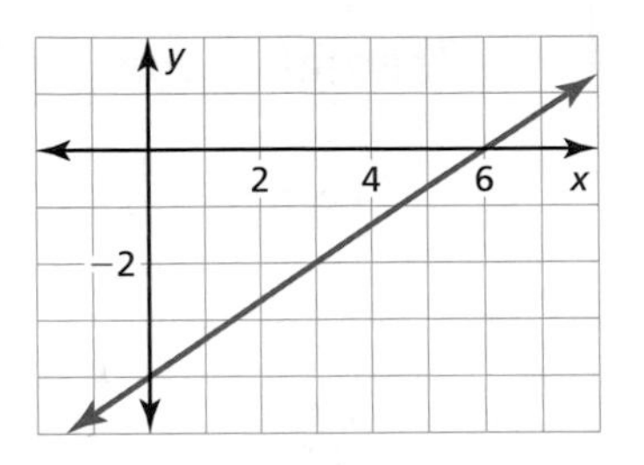

48. WRITING EQUATIONS What is the inverse of $f(x) = -\frac{1}{64}x^3$?

Ⓐ $g(x) = -4x^3$

Ⓑ $g(x) = 4\sqrt[3]{x}$

Ⓒ $g(x) = -4\sqrt[3]{x}$

Ⓓ $g(x) = \sqrt[3]{-4x}$

In Exercises 49–52, determine whether the functions are inverses. *(See Example 6.)*

49. $f(x) = 2x - 9, g(x) = \frac{x}{2} + 9$

50. $f(x) = \frac{x - 3}{4}, g(x) = 4x + 3$

51. $f(x) = \sqrt[5]{\frac{x + 9}{5}}, g(x) = 5x^5 - 9$

52. $f(x) = 7x^{3/2} - 4, g(x) = \left(\frac{x + 4}{7}\right)^{3/2}$

53. MODELING WITH MATHEMATICS The maximum hull speed v (in knots) of a boat with a displacement hull can be approximated by $v = 1.34\sqrt{\ell}$, where ℓ is the waterline length (in feet) of the boat. Find the inverse function. What waterline length is needed to achieve a maximum speed of 7.5 knots? *(See Example 7.)*

54. **MODELING WITH MATHEMATICS** Elastic bands can be used for exercising to provide a range of resistance. The resistance R (in pounds) of a band can be modeled by $R = \frac{3}{8}L - 5$, where L is the total length (in inches) of the stretched band. Find the inverse function. What length of the stretched band provides 19 pounds of resistance?

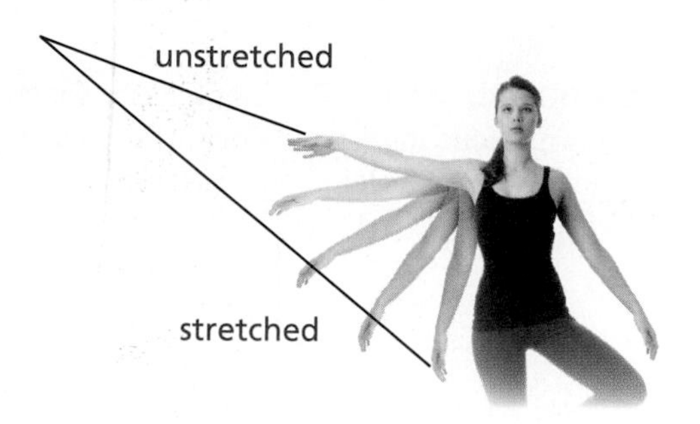

ANALYZING RELATIONSHIPS In Exercises 55–58, match the graph of the function with the graph of its inverse.

55.

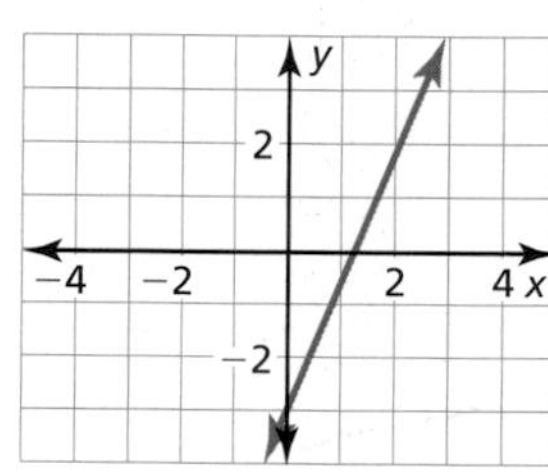

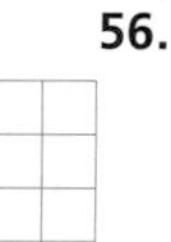

56.

57.

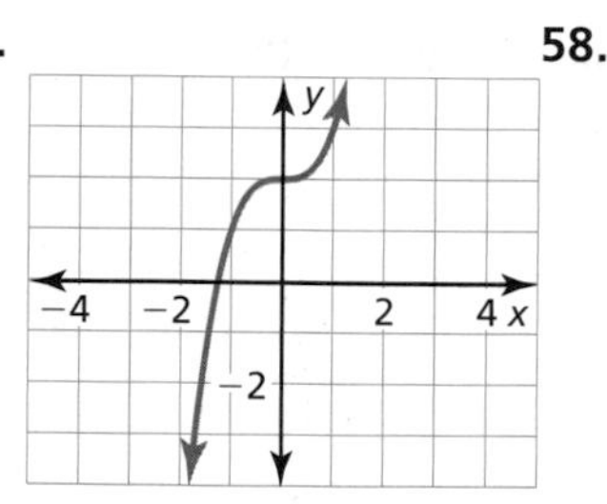

58.

A.

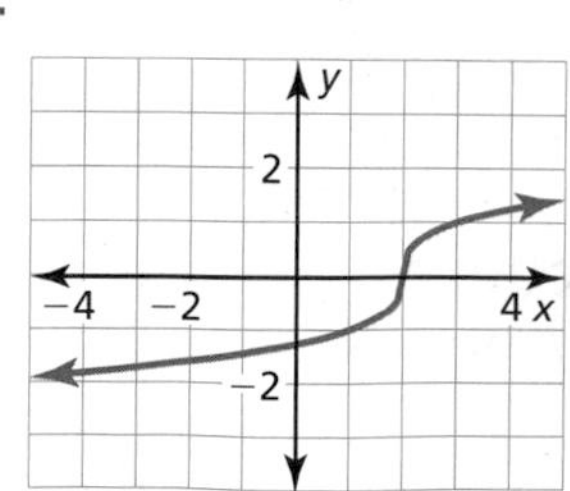

B.

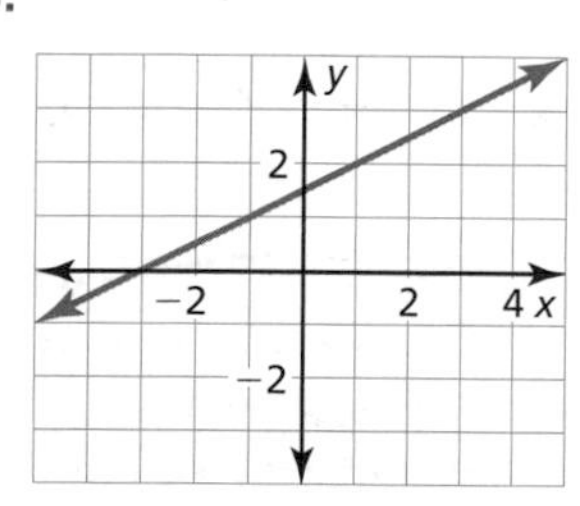

C.

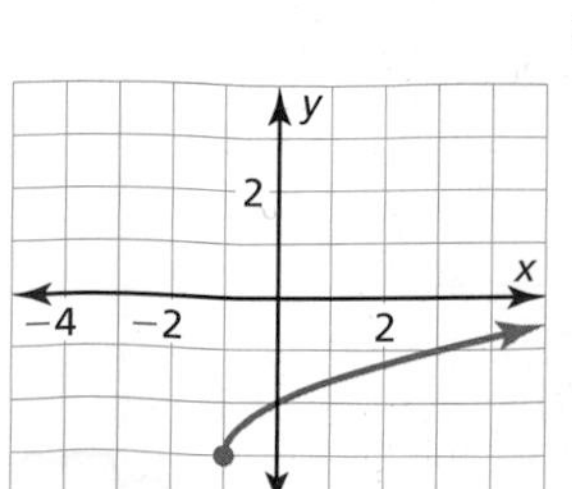

D.

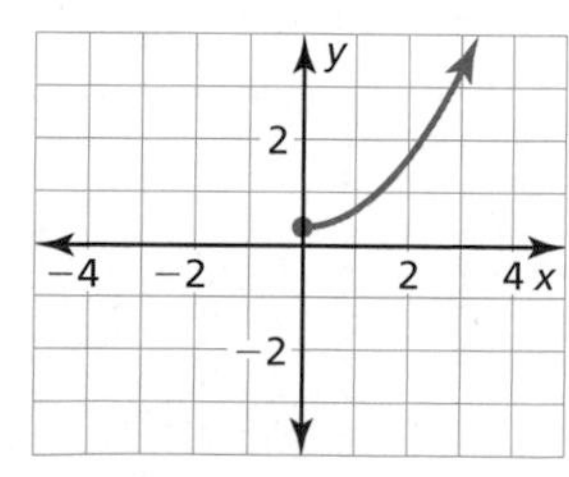

59. **REASONING** You and a friend are playing a number-guessing game. You ask your friend to think of a positive number, square the number, multiply the result by 2, and then add 3. Your friend's final answer is 53. What was the original number chosen? Justify your answer.

60. **MAKING AN ARGUMENT** Your friend claims that every quadratic function whose domain is restricted to nonnegative values has an inverse function. Is your friend correct? Explain your reasoning.

61. **PROBLEM SOLVING** When calibrating a spring scale, you need to know how far the spring stretches for various weights. Hooke's Law states that the length a spring stretches is proportional to the weight attached to it. A model for one scale is $\ell = 0.5w + 3$, where ℓ is the total length (in inches) of the stretched spring and w is the weight (in pounds) of the object.

unweighted spring | spring with weight attached | 3 | 0.5w | ℓ

Not drawn to scale

  a. Find the inverse function. Describe what it represents.

  b. You place a melon on the scale, and the spring stretches to a total length of 5.5 inches. Determine the weight of the melon.

  c. Verify that the function $\ell = 0.5w + 3$ and the inverse model in part (a) are inverse functions.

62. **THOUGHT PROVOKING** Do functions of the form $y = x^{m/n}$, where m and n are positive integers, have inverse functions? Justify your answer with examples.

63. **PROBLEM SOLVING** At the start of a dog sled race in Anchorage, Alaska, the temperature was 5°C. By the end of the race, the temperature was −10°C. The formula for converting temperatures from degrees Fahrenheit F to degrees Celsius C is $C = \frac{5}{9}(F - 32)$.

  a. Find the inverse function. Describe what it represents.

  b. Find the Fahrenheit temperatures at the start and end of the race.

  c. Use a graphing calculator to graph the original function and its inverse. Find the temperature that is the same on both temperature scales.

64. **PROBLEM SOLVING** The surface area A (in square meters) of a person with a mass of 60 kilograms can be approximated by $A = 0.2195h^{0.3964}$, where h is the height (in centimeters) of the person.

a. Find the inverse function. Then estimate the height of a 60-kilogram person who has a body surface area of 1.6 square meters.

b. Verify that function A and the inverse model in part (a) are inverse functions.

USING STRUCTURE In Exercises 65–68, match the function with the graph of its inverse.

65. $f(x) = \sqrt[3]{x - 4}$

66. $f(x) = \sqrt[3]{x + 4}$

67. $f(x) = \sqrt{x + 1} - 3$

68. $f(x) = \sqrt{x - 1} + 3$

A.
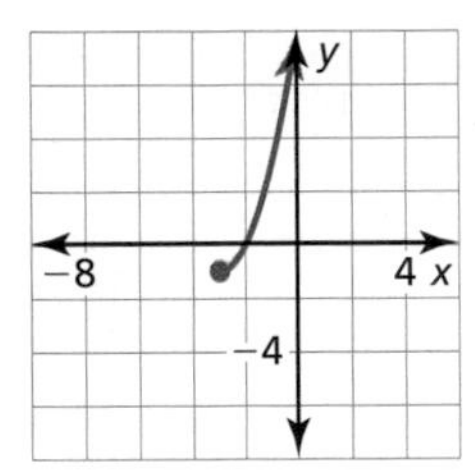

B.
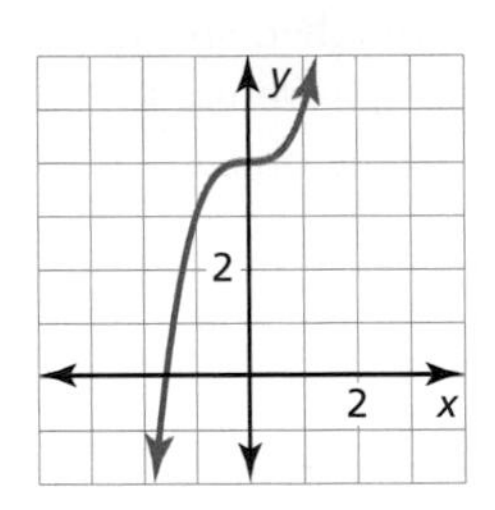

C.
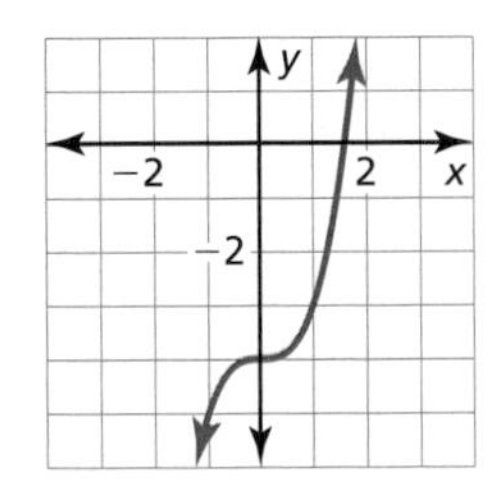

D.
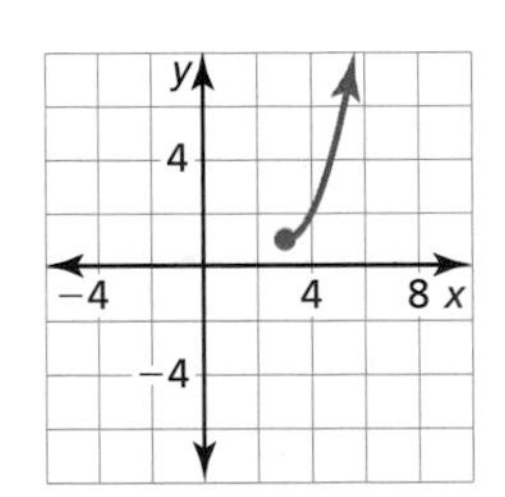

69. **DRAWING CONCLUSIONS** Determine whether the statement is *true* or *false*. Explain your reasoning.

a. If $f(x) = x^n$ and n is a positive even integer, then the inverse of f is a function.

b. If $f(x) = x^n$ and n is a positive odd integer, then the inverse of f is a function.

70. **HOW DO YOU SEE IT?** The graph of the function f is shown. Name three points that lie on the graph of the inverse of f. Explain your reasoning.

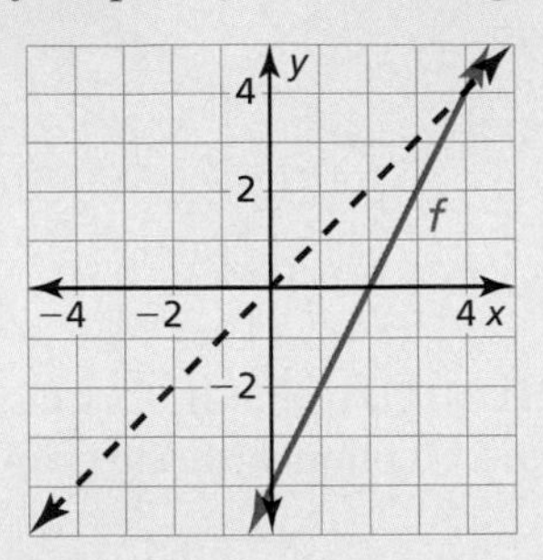

71. **ABSTRACT REASONING** Show that the inverse of any linear function $f(x) = mx + b$, where $m \neq 0$, is also a linear function. Identify the slope and y-intercept of the graph of the inverse function in terms of m and b.

72. **CRITICAL THINKING** Consider the function $f(x) = -x$.

a. Graph $f(x) = -x$ and explain why it is its own inverse. Also, verify that $f(x) = -x$ is its own inverse algebraically.

b. Graph other linear functions that are their own inverses. Write equations of the lines you graphed.

c. Use your results from part (b) to write a general equation describing the family of linear functions that are their own inverses.

Maintaining Mathematical Proficiency

Reviewing what you learned in previous grades and lessons

Simplify the expression. Write your answer using only positive exponents. *(Skills Review Handbook)*

73. $(-3)^{-3}$

74. $2^3 \cdot 2^2$

75. $\frac{4^5}{4^3}$

76. $\left(\frac{2}{3}\right)^4$

Describe the x-values for which the function is increasing, decreasing, positive, and negative. *(Section 4.1)*

77.
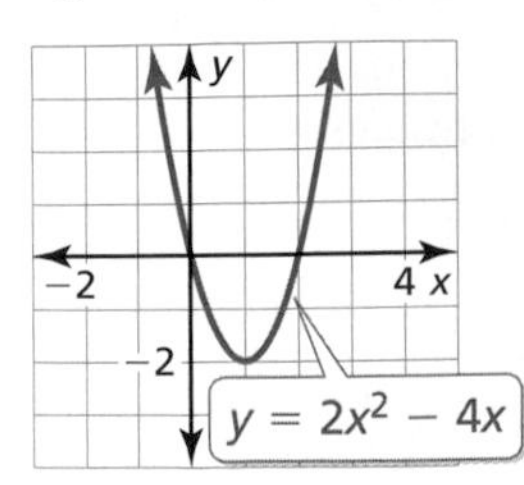

78.
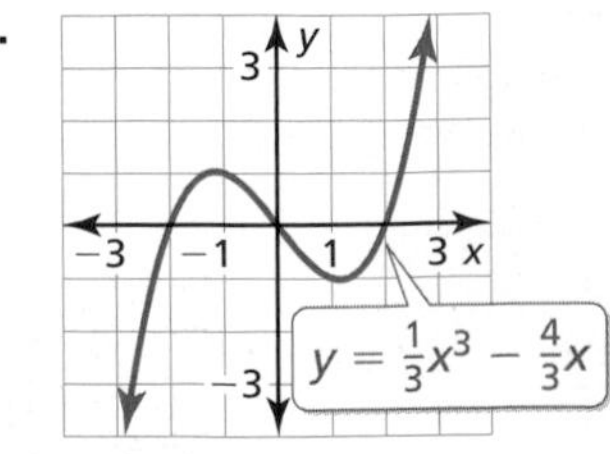

79.
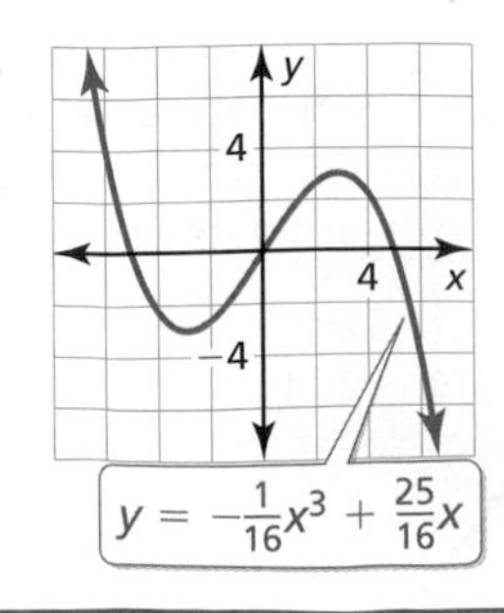

5.4–5.6 What Did You Learn?

Core Vocabulary

radical equation, *p. 262*
extraneous solutions, *p. 263*
inverse functions, *p. 277*

Core Concepts

Section 5.4

Solving Radical Equations, *p. 262*
Solving Radical Inequalities, *p. 265*

Section 5.5

Operations on Functions, *p. 270*

Section 5.6

Exploring Inverses of Functions, *p. 276*
Inverses of Nonlinear Functions, *p. 278*
Horizontal Line Test, *p. 278*

Mathematical Practices

1. How did you find the endpoints of the range in part (b) of Exercise 54 on page 267?
2. How did you use structure in Exercise 57 on page 268?
3. How can you evaluate the reasonableness of the results in Exercise 27 on page 274?
4. How can you use a graphing calculator to check your answers in Exercises 49–52 on page 282?

Performance Task

Turning the Tables

In this chapter, you have used properties of rational exponents and functions to find an answer to the problem. Using those same properties, can you find a problem to the answer? How many problems can you find?

To explore the answers to these questions and more, go to ***BigIdeasMath.com***.

5 Chapter Review

Dynamic Solutions available at *BigIdeasMath.com*

5.1 *n*th Roots and Rational Exponents (pp. 237–242)

a. Evaluate $8^{4/3}$ without using a calculator.

Rational Exponent Form	**Radical Form**
$8^{4/3} = (8^{1/3})^4 = 2^4 = 16$	$8^{4/3} = (\sqrt[3]{8})^4 = 2^4 = 16$

b. Find the real solution(s) of $x^4 - 45 = 580$.

$x^4 - 45 = 580$ Write original equation.

$x^4 = 625$ Add 45 to each side.

$x = \pm\sqrt[4]{625}$ Take fourth root of each side.

$x = 5$ or $x = -5$ Simplify.

▶ The solutions are $x = 5$ and $x = -5$.

Evaluate the expression without using a calculator.

1. $8^{7/3}$ **2.** $9^{5/2}$ **3.** $(-27)^{-2/3}$

Find the real solution(s) of the equation. Round your answer to two decimal places when appropriate.

4. $x^5 + 17 = 35$ **5.** $7x^3 = 189$ **6.** $(x + 8)^4 = 16$

5.2 Properties of Rational Exponents and Radicals (pp. 243–250)

a. Use the properties of rational exponents to simplify $\left(\dfrac{54^{1/3}}{2^{1/3}}\right)^4$.

$$\left(\frac{54^{1/3}}{2^{1/3}}\right)^4 = \left[\left(\frac{54}{2}\right)^{1/3}\right]^4 = (27^{1/3})^4 = 3^4 = 81$$

b. Write $\sqrt[4]{16x^{13}y^8z^7}$ in simplest form.

$\sqrt[4]{16x^{13}y^8z^7} = \sqrt[4]{16x^{12}xy^8z^4z^3}$ Factor out perfect fourth powers.

$= \sqrt[4]{16x^{12}y^8z^4} \cdot \sqrt[4]{xz^3}$ Product Property of Radicals

$= 2y^2|x^3z| \cdot \sqrt[4]{xz^3}$ Simplify.

Simplify the expression.

7. $\left(\dfrac{6^{1/5}}{6^{2/5}}\right)^3$ **8.** $\sqrt[4]{32} \cdot \sqrt[4]{8}$ **9.** $\dfrac{1}{2 - \sqrt[4]{9}}$

10. $4\sqrt[5]{8} + 3\sqrt[5]{8}$ **11.** $2\sqrt{48} - \sqrt{3}$ **12.** $(5^{2/3} \cdot 2^{3/2})^{1/2}$

Simplify the expression. Assume all variables are positive.

13. $\sqrt[3]{125z^9}$ **14.** $\dfrac{2^{1/4}z^{5/4}}{6z}$ **15.** $\sqrt{10z^5} - z^2\sqrt{40z}$

5.3 Graphing Radical Functions (pp. 251–258)

Describe the transformation of $f(x) = \sqrt{x}$ represented by $g(x) = 2\sqrt{x + 5}$. Then graph each function.

Notice that the function is of the form $g(x) = a\sqrt{x - h}$, where $a = 2$ and $h = -5$.

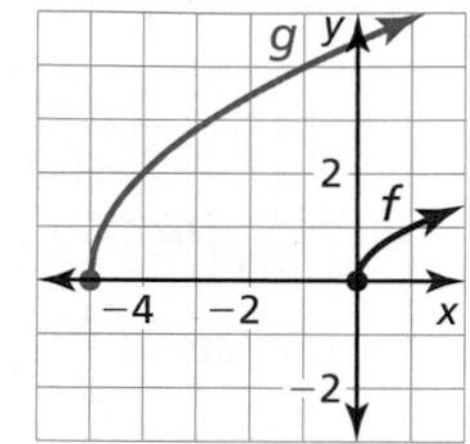

▶ So, the graph of g is a vertical stretch by a factor of 2 and a translation 5 units left of the graph of f.

Describe the transformation of f represented by g. Then graph each function.

16. $f(x) = \sqrt{x}$, $g(x) = -2\sqrt{x}$

17. $f(x) = \sqrt[3]{x}$, $g(x) = \sqrt[3]{-x} - 6$

18. Let the graph of g be a reflection in the y-axis, followed by a translation 7 units to the right of the graph of $f(x) = \sqrt[3]{x}$. Write a rule for g.

19. Use a graphing calculator to graph $2y^2 = x - 8$. Identify the vertex and the direction that the parabola opens.

20. Use a graphing calculator to graph $x^2 + y^2 = 81$. Identify the radius and the intercepts.

5.4 Solving Radical Equations and Inequalities (pp. 261–268)

Solve $6\sqrt{x + 2} < 18$.

Step 1 Solve for x.

$6\sqrt{x + 2} < 18$	Write the original inequality.
$\sqrt{x + 2} < 3$	Divide each side by 6.
$x + 2 < 9$	Square each side.
$x < 7$	Subtract 2 from each side.

Step 2 Consider the radicand.

$x + 2 \geq 0$	The radicand cannot be negative.
$x \geq -2$	Subtract 2 from each side.

Check

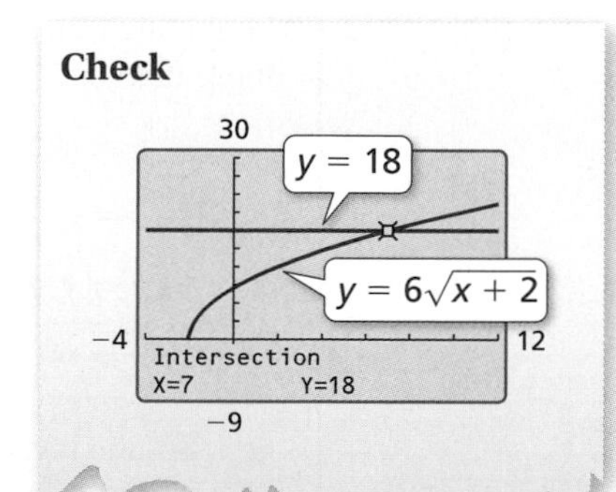

▶ So, the solution is $-2 \leq x < 7$.

Solve the equation. Check your solution.

21. $4\sqrt[3]{2x + 1} = 20$

22. $\sqrt{4x - 4} = \sqrt{5x - 1} - 1$

23. $(6x)^{2/3} = 36$

Solve the inequality.

24. $5\sqrt{x} + 2 > 17$

25. $2\sqrt{x - 8} < 24$

26. $7\sqrt[3]{x - 3} \geq 21$

27. In a tsunami, the wave speeds (in meters per second) can be modeled by $s(d) = \sqrt{9.8d}$, where d is the depth (in meters) of the water. Estimate the depth of the water when the wave speed is 200 meters per second.

5.5 Performing Function Operations (pp. 269–274)

Let $f(x) = 2x^{3/2}$ and $g(x) = x^{1/4}$. Find $\left(\frac{f}{g}\right)(x)$ and state the domain. Then evaluate the quotient when $x = 81$.

$$\left(\frac{f}{g}\right)(x) = \frac{f(x)}{g(x)} = \frac{2x^{3/2}}{x^{1/4}} = 2x^{(3/2-1/4)} = 2x^{5/4}$$

The functions f and g each have the same domain: all nonnegative real numbers. Because $g(0) = 0$, the domain of $\frac{f}{g}$ is restricted to all *positive* real numbers.

When $x = 81$, the value of the quotient is

$$\left(\frac{f}{g}\right)(81) = 2(81)^{5/4} = 2(81^{1/4})^5 = 2(3)^5 = 2(243) = 486.$$

28. Let $f(x) = 2\sqrt{3-x}$ and $g(x) = 4\sqrt[3]{3-x}$. Find $(fg)(x)$ and $\left(\frac{f}{g}\right)(x)$ and state the domain of each. Then evaluate $(fg)(2)$ and $\left(\frac{f}{g}\right)(2)$.

29. Let $f(x) = 3x^2 + 1$ and $g(x) = x + 4$. Find $(f + g)(x)$ and $(f - g)(x)$ and state the domain of each. Then evaluate $(f + g)(-5)$ and $(f - g)(-5)$.

5.6 Inverse of a Function (pp. 275–284)

Consider the function $f(x) = (x + 5)^3$. Determine whether the inverse of f is a function. Then find the inverse.

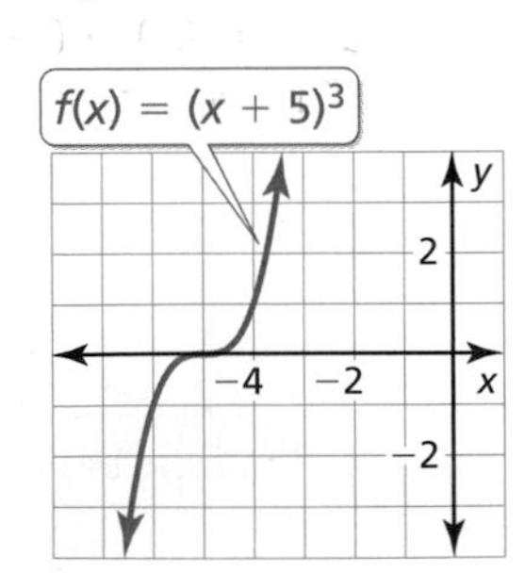

Graph the function f. Notice that no horizontal line intersects the graph more than once. So, the inverse of f is a function. Find the inverse.

$y = (x + 5)^3$	Set y equal to $f(x)$.
$x = (y + 5)^3$	Switch x and y.
$\sqrt[3]{x} = y + 5$	Take cube root of each side.
$\sqrt[3]{x} - 5 = y$	Subtract 5 from each side.

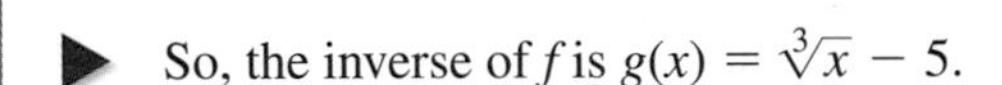

▶ So, the inverse of f is $g(x) = \sqrt[3]{x} - 5$.

Check

8, −12, 12, −8, f, g

Find the inverse of the function. Then graph the function and its inverse.

30. $f(x) = -\frac{1}{2}x + 10$

31. $f(x) = x^2 + 8, x \geq 0$

32. $f(x) = -x^3 - 9$

33. $f(x) = 3\sqrt{x} + 5$

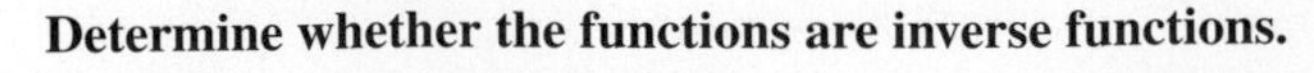

Determine whether the functions are inverse functions.

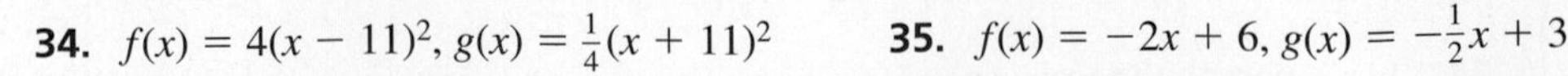

34. $f(x) = 4(x - 11)^2$, $g(x) = \frac{1}{4}(x + 11)^2$

35. $f(x) = -2x + 6$, $g(x) = -\frac{1}{2}x + 3$

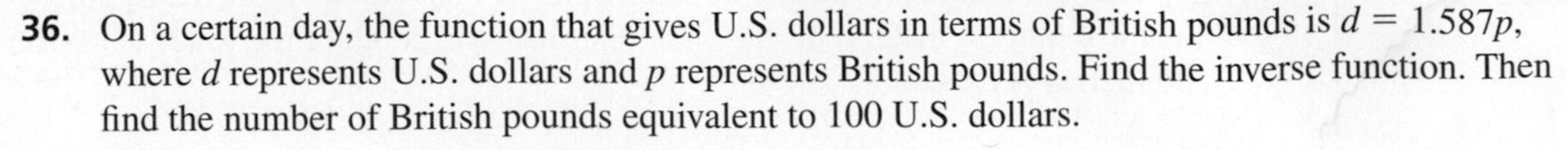

36. On a certain day, the function that gives U.S. dollars in terms of British pounds is $d = 1.587p$, where d represents U.S. dollars and p represents British pounds. Find the inverse function. Then find the number of British pounds equivalent to 100 U.S. dollars.

5 Chapter Test

1. Solve the inequality $5\sqrt{x-3} - 2 \leq 13$ and the equation $5\sqrt{x-3} - 2 = 13$. Describe the similarities and differences in solving radical equations and radical inequalities.

Describe the transformation of f represented by g. Then write a rule for g.

2. $f(x) = \sqrt{x}$

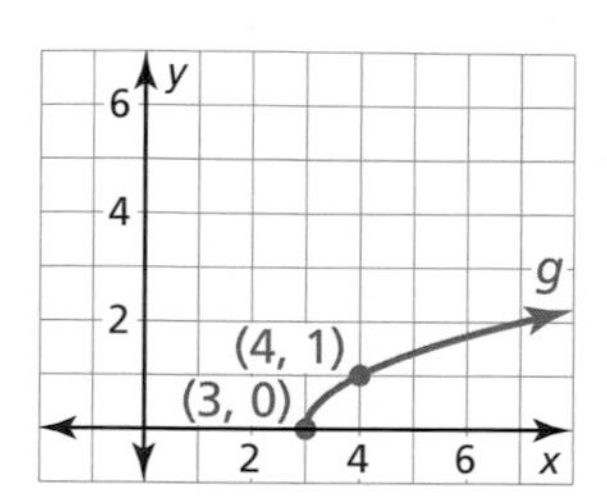

3. $f(x) = \sqrt[3]{x}$

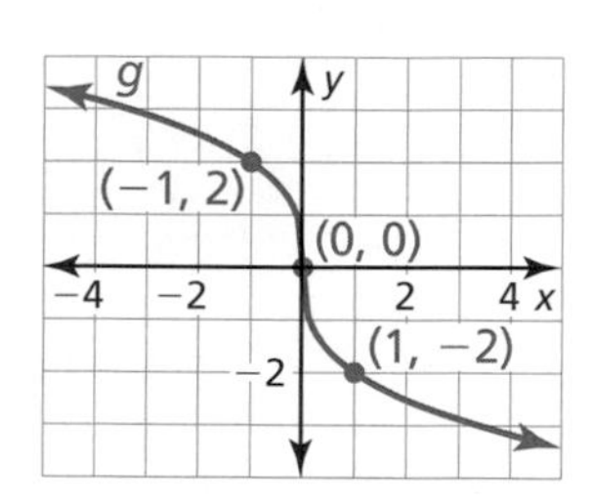

4. $f(x) = \sqrt[5]{x}$

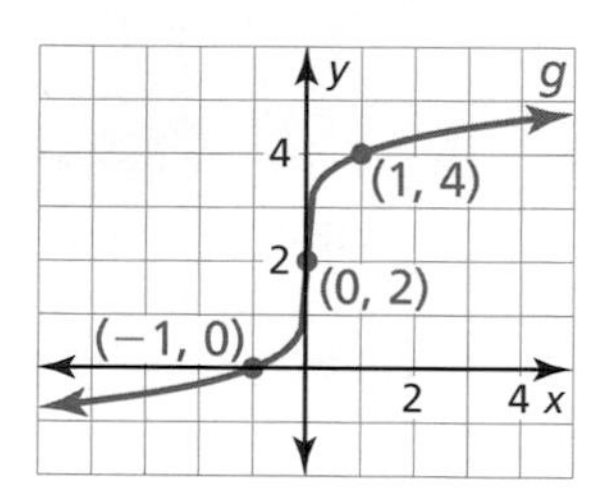

Simplify the expression. Explain your reasoning.

5. $64^{2/3}$

6. $(-27)^{5/3}$

7. $\sqrt[4]{48xy^{11}z^3}$

8. $\dfrac{\sqrt[3]{256}}{\sqrt[3]{32}}$

9. Write two functions whose graphs are translations of the graph of $y = \sqrt{x}$. The first function should have a domain of $x \geq 4$. The second function should have a range of $y \geq -2$.

10. In bowling, a handicap is a change in score to adjust for differences in the abilities of players. You belong to a bowling league in which your handicap h is determined using the formula $h = 0.9(200 - a)$, where a is your average score. Find the inverse of the model. Then find the average for a bowler whose handicap is 36.

11. The basal metabolic rate of an animal is a measure of the amount of calories burned at rest for basic functioning. Kleiber's law states that an animal's basal metabolic rate R (in kilocalories per day) can be modeled by $R = 73.3w^{3/4}$, where w is the mass (in kilograms) of the animal. Find the basal metabolic rates of each animal in the table.

Animal	Mass (kilograms)
rabbit	2.5
sheep	50
human	70
lion	210

12. Let $f(x) = 6x^{3/5}$ and $g(x) = -x^{3/5}$. Find $(f + g)(x)$ and $(f - g)(x)$ and state the domain of each. Then evaluate $(f + g)(32)$ and $(f - g)(32)$.

13. Let $f(x) = \frac{1}{2}x^{3/4}$ and $g(x) = 8x$. Find $(fg)(x)$ and $\left(\frac{f}{g}\right)(x)$ and state the domain of each. Then evaluate $(fg)(16)$ and $\left(\frac{f}{g}\right)(16)$.

14. A football player jumps to catch a pass. The maximum height h (in feet) of the player above the ground is given by the function $h = \frac{1}{64}s^2$, where s is the initial speed (in feet per second) of the player. Find the inverse of the function. Use the inverse to find the initial speed of the player shown. Verify that the functions are inverse functions.

5 Cumulative Assessment

1. Identify three pairs of equivalent expressions. Assume all variables are positive. Justify your answer.

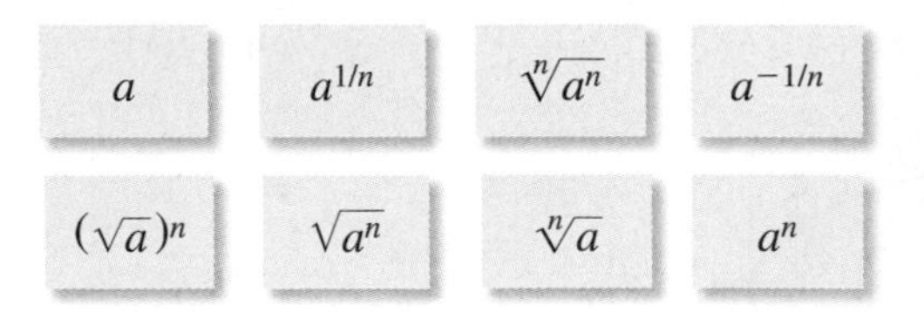

2. The graph represents the function $f(x) = (x - \square)^2 + \square$. Choose the correct values to complete the function.

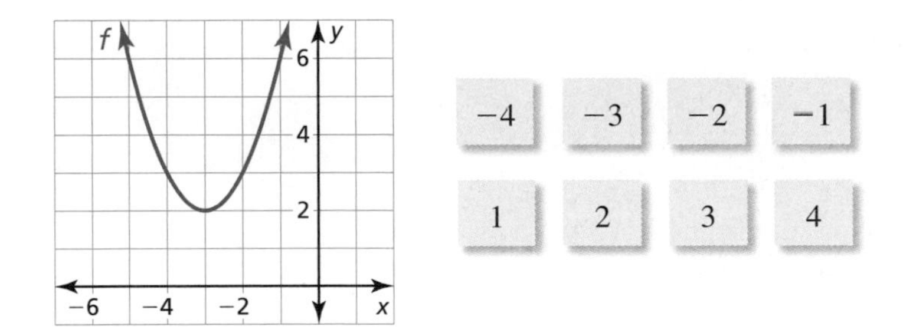

3. In rowing, the boat speed s (in meters per second) can be modeled by $s = 4.62\sqrt[9]{n}$, where n is the number of rowers.

   a. Find the boat speeds for crews of 2 people, 4 people, and 8 people.

   b. Does the boat speed double when the number of rowers doubles? Explain.

   c. Find the time (in minutes) it takes each crew in part (a) to complete a 2000-meter race.

4. A polynomial function fits the data in the table. Use finite differences to find the degree of the function and complete the table. Explain your reasoning.

x	−4	−3	−2	−1	0	1	2	3
f(x)	28	2	−6	−2	8	18		

5. The area of the triangle is 42 square inches. Find the value of x.

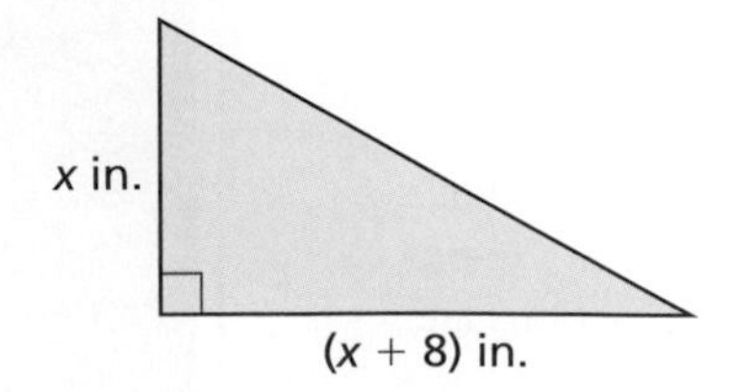

6. Which equations are represented by parabolas? Which equations are functions? Place check marks in the appropriate spaces. Explain your reasoning.

Equation	Parabola	Function
$y = (x + 3)^2$		
$x = 4y^2 - 2$		
$y = (x - 1)^{1/2} + 6$		
$y^2 = 10 - x^2$		

7. What is the solution of the inequality $2\sqrt{x + 3} - 1 < 3$?

(A) $x < 1$

(B) $-3 < x < 1$

(C) $-3 \leq x < 1$

(D) $x \geq -3$

8. Which function does the graph represent? Explain your reasoning.

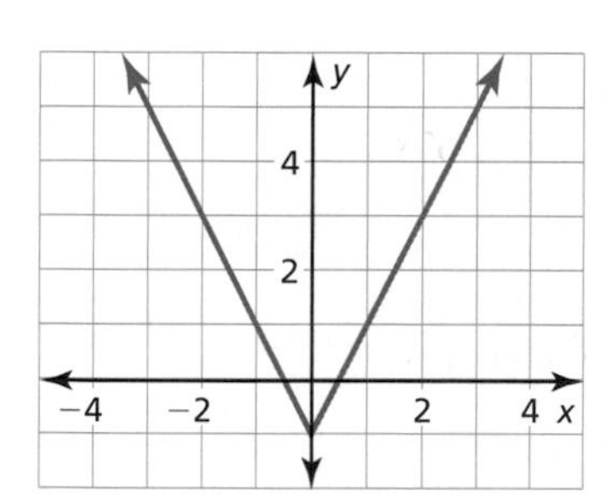

(A) $y = -|2x| - 1$

(B) $y = -2|x + 1|$

(C) $y = |2x| - 1$

(D) $y = 2|x + 1|$

9. Your friend releases a weather balloon 50 feet from you. The balloon rises vertically. When the balloon is at height h, the distance d between you and the balloon is given by $d = \sqrt{2500 + h^2}$, where h and d are measured in feet. Find the inverse of the function. What is the height of the balloon when the distance between you and the balloon is 100 feet?

10. The graphs of two functions f and g are shown. Are f and g inverse functions? Explain your reasoning.

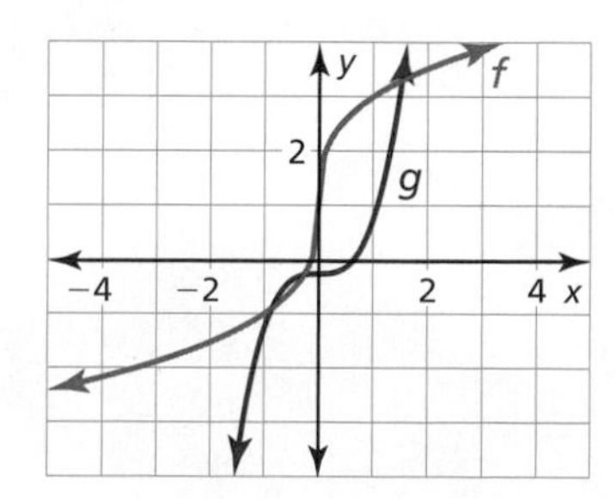

6 Exponential and Logarithmic Functions

Astronaut Health (p. 347)

Cooking (p. 335)

Recording Studio (p. 330)

Tornado Wind Speed (p. 315)

Duckweed Growth (p. 301)

Maintaining Mathematical Proficiency

Using Exponents

Example 1 **Evaluate** $\left(-\frac{1}{3}\right)^4$.

$$\left(-\frac{1}{3}\right)^4 = \left(-\frac{1}{3}\right) \cdot \left(-\frac{1}{3}\right) \cdot \left(-\frac{1}{3}\right) \cdot \left(-\frac{1}{3}\right)$$ Rewrite $\left(-\frac{1}{3}\right)^4$ as repeated multiplication.

$$= \left(\frac{1}{9}\right) \cdot \left(-\frac{1}{3}\right) \cdot \left(-\frac{1}{3}\right)$$ Multiply.

$$= \left(-\frac{1}{27}\right) \cdot \left(-\frac{1}{3}\right)$$ Multiply.

$$= \frac{1}{81}$$ Multiply.

Evaluate the expression.

1. $3 \cdot 2^4$

2. $(-2)^5$

3. $-\left(\frac{5}{6}\right)^2$

4. $\left(\frac{3}{4}\right)^3$

Finding the Domain and Range of a Function

Example 2 **Find the domain and range of the function represented by the graph.**

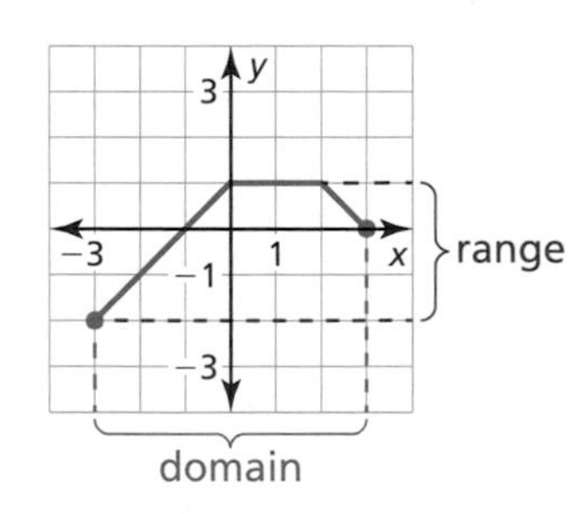

▶ The domain is $-3 \leq x \leq 3$.
The range is $-2 \leq y \leq 1$.

Find the domain and range of the function represented by the graph.

5.

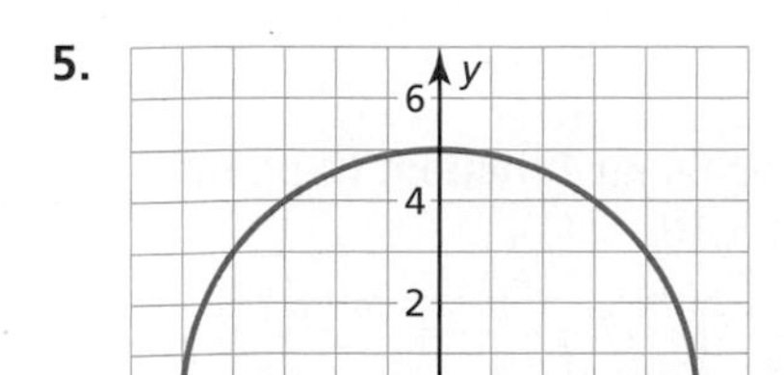

6.

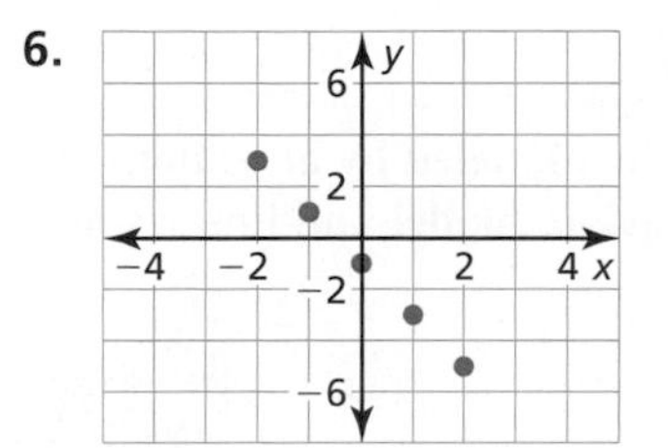

7.

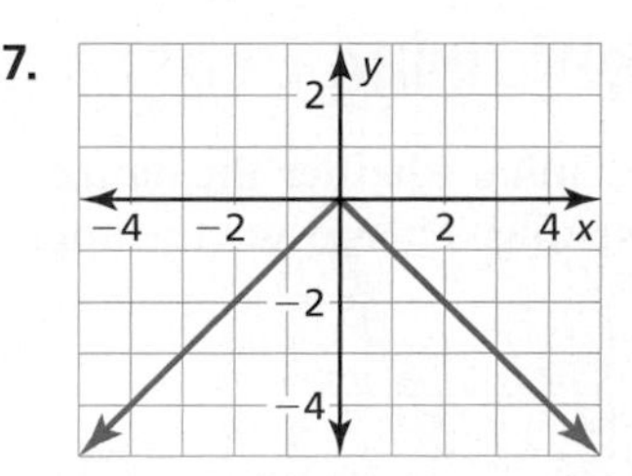

8. **ABSTRACT REASONING** Consider the expressions -4^n and $(-4)^n$, where n is an integer. For what values of n is each expression negative? positive? Explain your reasoning.

Mathematical Practices

Mathematically proficient students know when it is appropriate to use general methods and shortcuts.

Exponential Models

Core Concept

Consecutive Ratio Test for Exponential Models

Consider a table of values of the given form.

x	0	1	2	3	4	5	6	7	8	9
y	a_0	a_1	a_2	a_3	a_4	a_5	a_6	a_7	a_8	a_9

If the consecutive ratios of the y-values are all equal to a common value r, then y can be modeled by an exponential function. When $r > 1$, the model represents exponential *growth*.

$r = \frac{a_{n+1}}{a_n}$ Common ratio

$y = a_0 r^x$ Exponential model

EXAMPLE 1 Modeling Real-Life Data

The table shows the amount A (in dollars) in a savings account over time. Write a model for the amount in the account as a function of time t (in years). Then use the model to find the amount after 10 years.

Year, t	0	1	2	3	4	5
Amount, A	\$1000	\$1040	\$1081.60	\$1124.86	\$1169.86	\$1216.65

SOLUTION

Begin by determining whether the ratios of consecutive amounts are equal.

$$\frac{1040}{1000} = 1.04, \quad \frac{1081.60}{1040} = 1.04, \quad \frac{1124.86}{1081.60} \approx 1.04, \quad \frac{1169.86}{1124.86} \approx 1.04, \quad \frac{1216.65}{1169.86} \approx 1.04$$

The ratios of consecutive amounts are equal, so the amount A after t years can be modeled by

$$A = 1000(1.04)^t.$$

Using this model, the amount when $t = 10$ is $A = 1000(1.04)^{10} = \$1480.24$.

Monitoring Progress

Determine whether the data can be modeled by an exponential or linear function. Explain your reasoning. Then write the appropriate model and find y when $x = 10$.

1.

x	0	1	2	3	4
y	1	2	4	8	16

2.

x	0	1	2	3	4
y	0	4	8	12	16

3.

x	0	1	2	3	4
y	1	4	7	10	13

4.

x	0	1	2	3	4
y	1	3	9	27	81

6.1 Exponential Growth and Decay Functions

Essential Question

What are some of the characteristics of the graph of an exponential function?

You can use a graphing calculator to evaluate an exponential function. For example, consider the exponential function $f(x) = 2^x$.

Function Value	Graphing Calculator Keystrokes	Display
$f(-3.1) = 2^{-3.1}$	2 [^] [(−)] 3.1 [ENTER]	0.1166291
$f\left(\frac{2}{3}\right) = 2^{2/3}$	2 [^] [(] 2 [÷] 3 [)] [ENTER]	1.5874011

EXPLORATION 1 Identifying Graphs of Exponential Functions

Work with a partner. Match each exponential function with its graph. Use a table of values to sketch the graph of the function, if necessary.

a. $f(x) = 2^x$ **b.** $f(x) = 3^x$ **c.** $f(x) = 4^x$

d. $f(x) = \left(\frac{1}{2}\right)^x$ **e.** $f(x) = \left(\frac{1}{3}\right)^x$ **f.** $f(x) = \left(\frac{1}{4}\right)^x$

A.

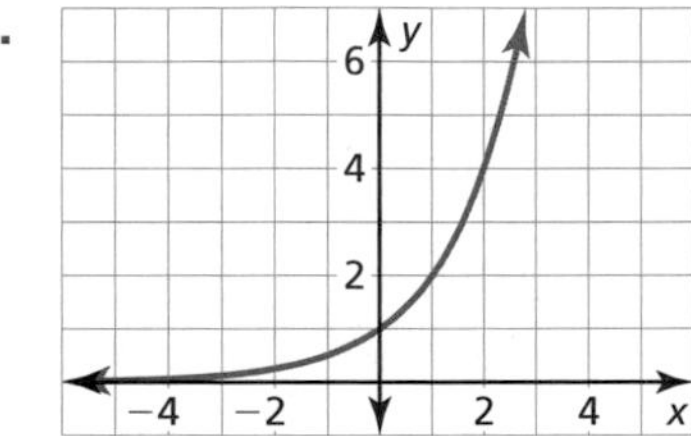

B.

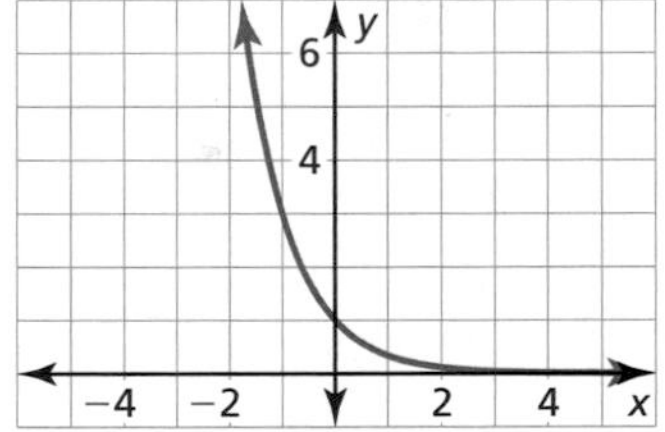

C.

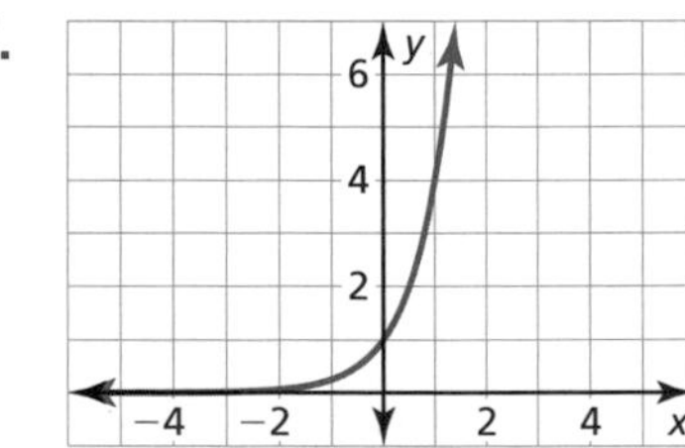

D.

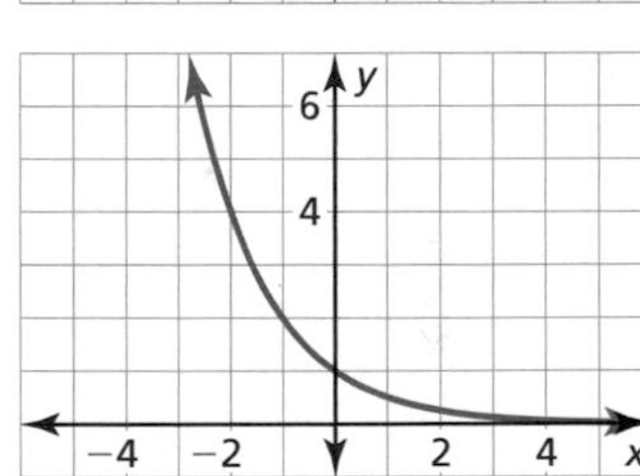

E.

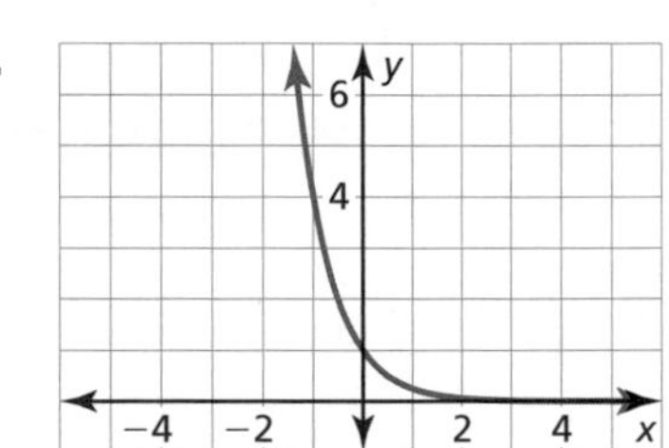

F.

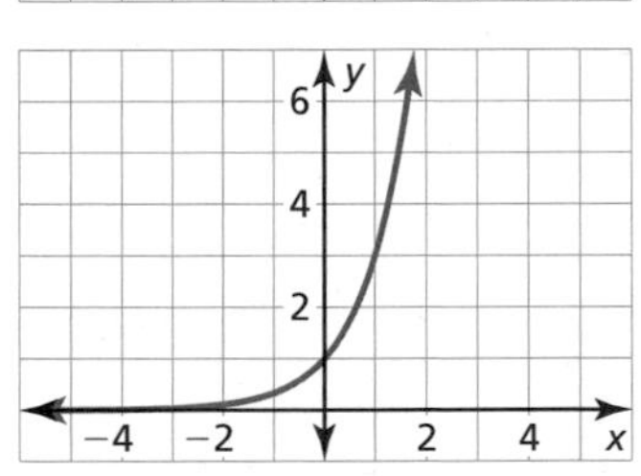

CONSTRUCTING VIABLE ARGUMENTS

To be proficient in math, you need to justify your conclusions and communicate them to others.

EXPLORATION 2 Characteristics of Graphs of Exponential Functions

Work with a partner. Use the graphs in Exploration 1 to determine the domain, range, and y-intercept of the graph of $f(x) = b^x$, where b is a positive real number other than 1. Explain your reasoning.

Communicate Your Answer

3. What are some of the characteristics of the graph of an exponential function?

4. In Exploration 2, is it possible for the graph of $f(x) = b^x$ to have an x-intercept? Explain your reasoning.

6.1 Lesson

Core Vocabulary

exponential function, *p. 296*
exponential growth function, *p. 296*
growth factor, *p. 296*
asymptote, *p. 296*
exponential decay function, *p. 296*
decay factor, *p. 296*

Previous
properties of exponents

What You Will Learn

- Graph exponential growth and decay functions.
- Use exponential models to solve real-life problems.

Exponential Growth and Decay Functions

An **exponential function** has the form $y = ab^x$, where $a \neq 0$ and the base b is a positive real number other than 1. If $a > 0$ and $b > 1$, then $y = ab^x$ is an **exponential growth function**, and b is called the **growth factor**. The simplest type of exponential growth function has the form $y = b^x$.

Core Concept

Parent Function for Exponential Growth Functions

The function $f(x) = b^x$, where $b > 1$, is the parent function for the family of exponential growth functions with base b. The graph shows the general shape of an exponential growth function.

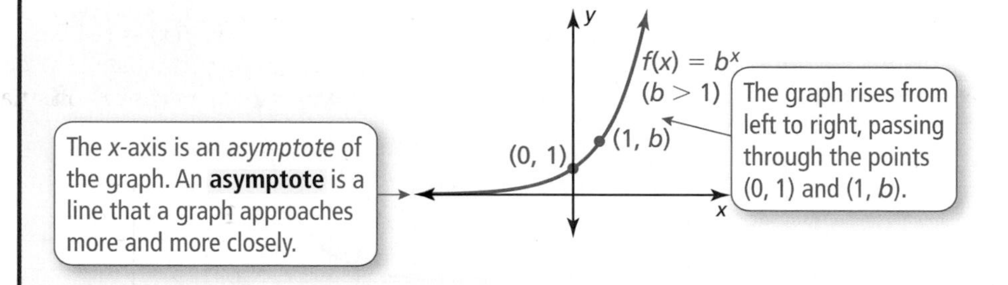

The x-axis is an asymptote of the graph. An asymptote is a line that a graph approaches more and more closely.
y
f(x) = b^x
(b > 1)
(0, 1)
(1, b)
x
The graph rises from left to right, passing through the points (0, 1) and (1, b).

The domain of $f(x) = b^x$ is all real numbers. The range is $y > 0$.

If $a > 0$ and $0 < b < 1$, then $y = ab^x$ is an **exponential decay function**, and b is called the **decay factor**.

Core Concept

Parent Function for Exponential Decay Functions

The function $f(x) = b^x$, where $0 < b < 1$, is the parent function for the family of exponential decay functions with base b. The graph shows the general shape of an exponential decay function.

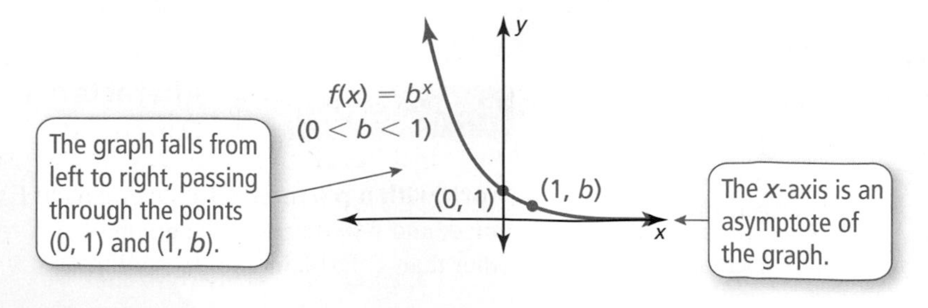

The domain of $f(x) = b^x$ is all real numbers. The range is $y > 0$.

EXAMPLE 1 Graphing Exponential Growth and Decay Functions

Tell whether each function represents *exponential growth* or *exponential decay*. Then graph the function.

a. $y = 2^x$

b. $y = \left(\frac{1}{2}\right)^x$

SOLUTION

a. Step 1 Identify the value of the base. The base, 2, is greater than 1, so the function represents exponential growth.

Step 2 Make a table of values.

x	-2	-1	0	1	2	3
y	$\frac{1}{4}$	$\frac{1}{2}$	1	2	4	8

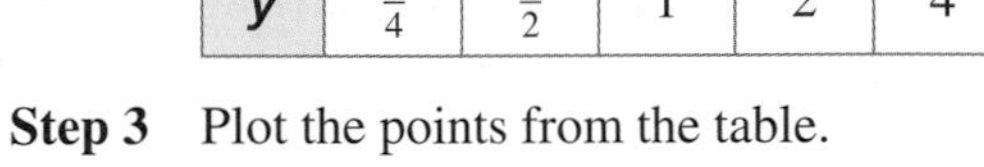

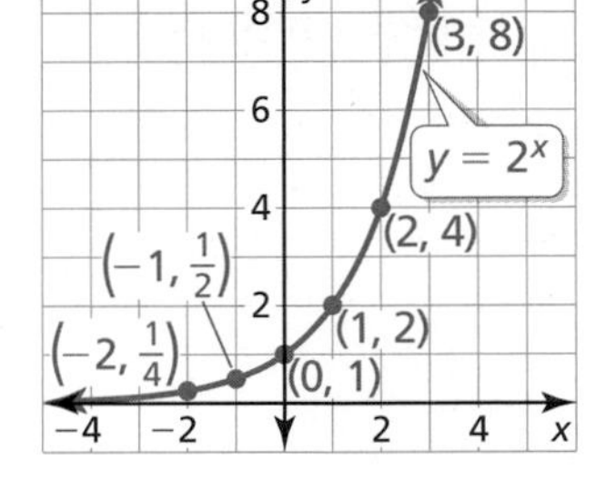

Step 3 Plot the points from the table.

Step 4 Draw, from *left to right*, a smooth curve that begins just above the x-axis, passes through the plotted points, and moves up to the right.

b. Step 1 Identify the value of the base. The base, $\frac{1}{2}$, is greater than 0 and less than 1, so the function represents exponential decay.

Step 2 Make a table of values.

x	-3	-2	-1	0	1	2
y	8	4	2	1	$\frac{1}{2}$	$\frac{1}{4}$

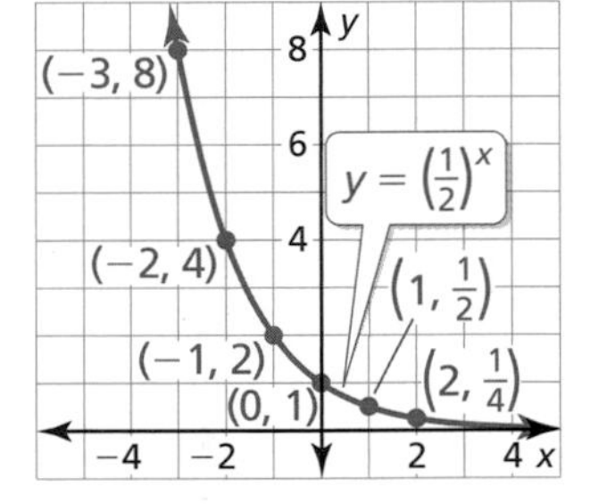

Step 3 Plot the points from the table.

Step 4 Draw, from *right to left*, a smooth curve that begins just above the x-axis, passes through the plotted points, and moves up to the left.

Monitoring Progress  Help in English and Spanish at *BigIdeasMath.com*

Tell whether the function represents *exponential growth* or *exponential decay*. Then graph the function.

1. $y = 4^x$

2. $y = \left(\frac{2}{3}\right)^x$

3. $f(x) = (0.25)^x$

4. $f(x) = (1.5)^x$

Exponential Models

Some real-life quantities increase or decrease by a fixed percent each year (or some other time period). The amount y of such a quantity after t years can be modeled by one of these equations.

Exponential Growth Model

$$y = a(1 + r)^t$$

Exponential Decay Model

$$y = a(1 - r)^t$$

Note that a is the initial amount and r is the percent increase or decrease written as a decimal. The quantity $1 + r$ is the growth factor, and $1 - r$ is the decay factor.

EXAMPLE 2 Solving a Real-Life Problem

The value of a car y (in thousands of dollars) can be approximated by the model $y = 25(0.85)^t$, where t is the number of years since the car was new.

a. Tell whether the model represents exponential growth or exponential decay.

b. Identify the annual percent increase or decrease in the value of the car.

c. Estimate when the value of the car will be $8000.

REASONING QUANTITATIVELY

The percent decrease, 15%, tells you how much value the car *loses* each year. The decay factor, 0.85, tells you what fraction of the car's value *remains* each year.

SOLUTION

a. The base, 0.85, is greater than 0 and less than 1, so the model represents exponential decay.

b. Because t is given in years and the decay factor $0.85 = 1 - 0.15$, the annual percent decrease is 0.15, or 15%.

c. Use the *trace* feature of a graphing calculator to determine that $y \approx 8$ when $t = 7$. After 7 years, the value of the car will be about $8000.

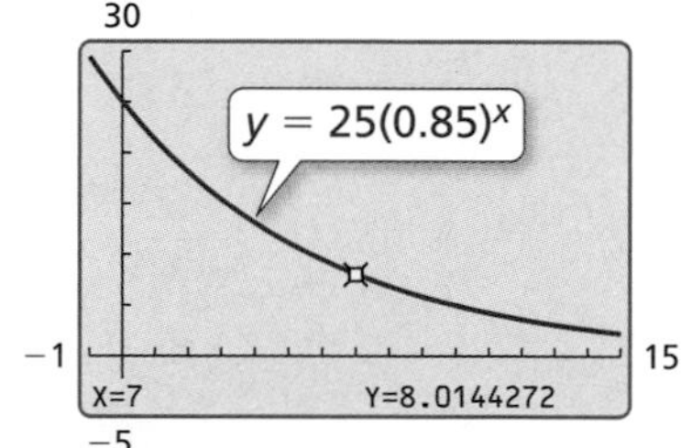

EXAMPLE 3 Writing an Exponential Model

In 2000, the world population was about 6.09 billion. During the next 13 years, the world population increased by about 1.18% each year.

a. Write an exponential growth model giving the population y (in billions) t years after 2000. Estimate the world population in 2005.

b. Estimate the year when the world population was 7 billion.

SOLUTION

a. The initial amount is $a = 6.09$, and the percent increase is $r = 0.0118$. So, the exponential growth model is

$y = a(1 + r)^t$ Write exponential growth model.

$= 6.09(1 + 0.0118)^t$ Substitute 6.09 for a and 0.0118 for r.

$= 6.09(1.0118)^t$. Simplify.

Using this model, you can estimate the world population in 2005 ($t = 5$) to be $y = 6.09(1.0118)^5 \approx 6.46$ billion.

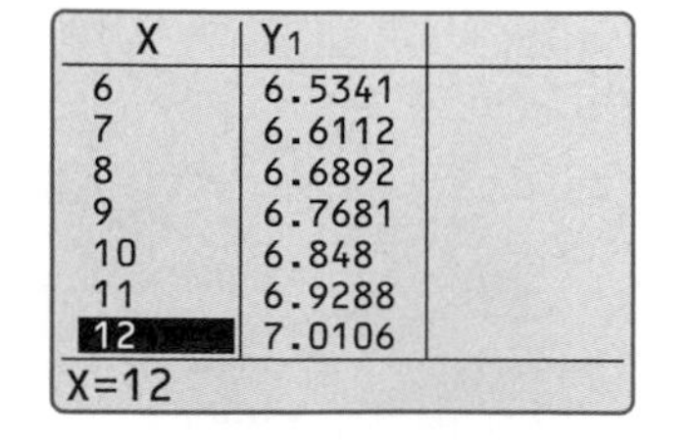

X	Y1
6	6.5341
7	6.6112
8	6.6892
9	6.7681
10	6.848
11	6.9288
12	7.0106

X=12

b. Use the *table* feature of a graphing calculator to determine that $y \approx 7$ when $t = 12$. So, the world population was about 7 billion in 2012.

Monitoring Progress  Help in English and Spanish at *BigIdeasMath.com*

5. WHAT IF? In Example 2, the value of the car can be approximated by the model $y = 25(0.9)^t$. Identify the annual percent decrease in the value of the car. Estimate when the value of the car will be $8000.

6. WHAT IF? In Example 3, assume the world population increased by 1.5% each year. Write an equation to model this situation. Estimate the year when the world population was 7 billion.

EXAMPLE 4 Rewriting an Exponential Function

The amount y (in grams) of the radioactive isotope chromium-51 remaining after t days is $y = a(0.5)^{t/28}$, where a is the initial amount (in grams). What percent of the chromium-51 decays each day?

SOLUTION

$y = a(0.5)^{t/28}$	Write original function.
$= a[(0.5)^{1/28}]^t$	Power of a Power Property
$\approx a(0.9755)^t$	Evaluate power.
$= a(1 - 0.0245)^t$	Rewrite in form $y = a(1 - r)^t$.

▶ The daily decay rate is about 0.0245, or 2.45%.

Compound interest is interest paid on an initial investment, called the *principal*, and on previously earned interest. Interest earned is often expressed as an *annual* percent, but the interest is usually compounded more than once per year. So, the exponential growth model $y = a(1 + r)^t$ must be modified for compound interest problems.

Core Concept

Compound Interest

Consider an initial principal P deposited in an account that pays interest at an annual rate r (expressed as a decimal), compounded n times per year. The amount A in the account after t years is given by

$$A = P\left(1 + \frac{r}{n}\right)^{nt}.$$

EXAMPLE 5 Finding the Balance in an Account

You deposit \$9000 in an account that pays 1.46% annual interest. Find the balance after 3 years when the interest is compounded quarterly.

SOLUTION

With interest compounded quarterly (4 times per year), the balance after 3 years is

$A = P\left(1 + \frac{r}{n}\right)^{nt}$	Write compound interest formula.
$= 9000\left(1 + \frac{0.0146}{4}\right)^{4 \cdot 3}$	$P = 9000, r = 0.0146, n = 4, t = 3$
$\approx 9402.21.$	Use a calculator.

▶ The balance at the end of 3 years is \$9402.21.

Monitoring Progress

Help in English and Spanish at *BigIdeasMath.com*

7. The amount y (in grams) of the radioactive isotope iodine-123 remaining after t hours is $y = a(0.5)^{t/13}$, where a is the initial amount (in grams). What percent of the iodine-123 decays each hour?
8. **WHAT IF?** In Example 5, find the balance after 3 years when the interest is compounded daily.

6.1 Exercises

Dynamic Solutions available at *BigIdeasMath.com*

Vocabulary and Core Concept Check

1. **VOCABULARY** In the exponential growth model $y = 2.4(1.5)^x$, identify the initial amount, the growth factor, and the percent increase.

2. **WHICH ONE DOESN'T BELONG?** Which characteristic of an exponential decay function does *not* belong with the other three? Explain your reasoning.

base of 0.8	decay factor of 0.8
decay rate of 20%	80% decrease

Monitoring Progress and Modeling with Mathematics

In Exercises 3–8, evaluate the expression for (a) $x = -2$ and (b) $x = 3$.

3. 2^x
4. 4^x
5. $8 \cdot 3^x$
6. $6 \cdot 2^x$
7. $5 + 3^x$
8. $2^x - 2$

In Exercises 9–18, tell whether the function represents *exponential growth* or *exponential decay*. Then graph the function. *(See Example 1.)*

9. $y = 6^x$
10. $y = 7^x$
11. $y = \left(\frac{1}{6}\right)^x$
12. $y = \left(\frac{1}{8}\right)^x$
13. $y = \left(\frac{4}{3}\right)^x$
14. $y = \left(\frac{2}{5}\right)^x$
15. $y = (1.2)^x$
16. $y = (0.75)^x$
17. $y = (0.6)^x$
18. $y = (1.8)^x$

ANALYZING RELATIONSHIPS In Exercises 19 and 20, use the graph of $f(x) = b^x$ to identify the value of the base b.

19.

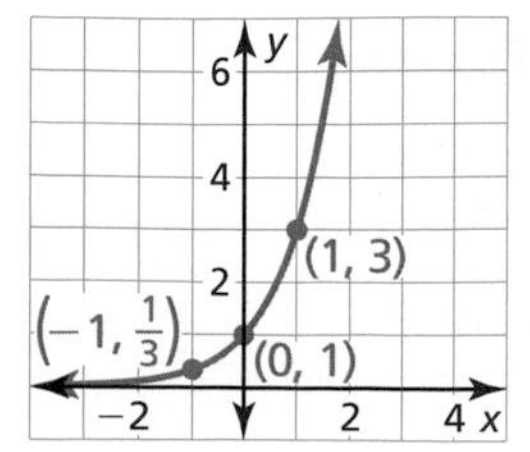

20.

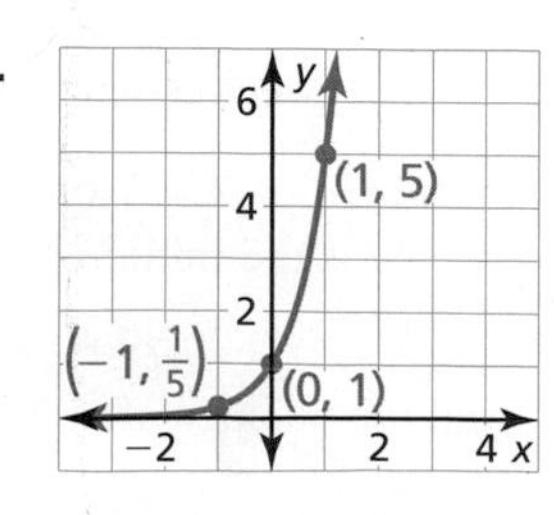

21. **MODELING WITH MATHEMATICS** The value of a mountain bike y (in dollars) can be approximated by the model $y = 200(0.75)^t$, where t is the number of years since the bike was new. *(See Example 2.)*

    a. Tell whether the model represents exponential growth or exponential decay.

    b. Identify the annual percent increase or decrease in the value of the bike.

    c. Estimate when the value of the bike will be $50.

22. **MODELING WITH MATHEMATICS** The population P (in thousands) of Austin, Texas, during a recent decade can be approximated by $y = 494.29(1.03)^t$, where t is the number of years since the beginning of the decade.

    a. Tell whether the model represents exponential growth or exponential decay.

    b. Identify the annual percent increase or decrease in population.

    c. Estimate when the population was about 590,000.

23. **MODELING WITH MATHEMATICS** In 2006, there were approximately 233 million cell phone subscribers in the United States. During the next 4 years, the number of cell phone subscribers increased by about 6% each year. *(See Example 3.)*

    a. Write an exponential growth model giving the number of cell phone subscribers y (in millions) t years after 2006. Estimate the number of cell phone subscribers in 2008.

    b. Estimate the year when the number of cell phone subscribers was about 278 million.

24. **MODELING WITH MATHEMATICS** You take a 325 milligram dosage of ibuprofen. During each subsequent hour, the amount of medication in your bloodstream decreases by about 29% each hour.

a. Write an exponential decay model giving the amount y (in milligrams) of ibuprofen in your bloodstream t hours after the initial dose.

b. Estimate how long it takes for you to have 100 milligrams of ibuprofen in your bloodstream.

JUSTIFYING STEPS In Exercises 25 and 26, justify each step in rewriting the exponential function.

25. $y = a(3)^{t/14}$ Write original function.

$= a[(3)^{1/14}]^t$

$\approx a(1.0816)^t$

$= a(1 + 0.0816)^t$

26. $y = a(0.1)^{t/3}$ Write original function.

$= a[(0.1)^{1/3}]^t$

$\approx a(0.4642)^t$

$= a(1 - 0.5358)^t$

27. **PROBLEM SOLVING** When a plant or animal dies, it stops acquiring carbon-14 from the atmosphere. The amount y (in grams) of carbon-14 in the body of an organism after t years is $y = a(0.5)^{t/5730}$, where a is the initial amount (in grams). What percent of the carbon-14 is released each year? *(See Example 4.)*

28. **PROBLEM SOLVING** The number y of duckweed fronds in a pond after t days is $y = a(1230.25)^{t/16}$, where a is the initial number of fronds. By what percent does the duckweed increase each day?

In Exercises 29–36, rewrite the function in the form $y = a(1 + r)^t$ or $y = a(1 - r)^t$. Then state the growth or decay rate.

29. $y = a(2)^{t/3}$

30. $y = a(4)^{t/6}$

31. $y = a(0.5)^{t/12}$

32. $y = a(0.25)^{t/9}$

33. $y = a\left(\frac{2}{3}\right)^{t/10}$

34. $y = a\left(\frac{5}{4}\right)^{t/22}$

35. $y = a(2)^{8t}$

36. $y = a\left(\frac{1}{3}\right)^{3t}$

37. **PROBLEM SOLVING** You deposit \$5000 in an account that pays 2.25% annual interest. Find the balance after 5 years when the interest is compounded quarterly. *(See Example 5.)*

38. **DRAWING CONCLUSIONS** You deposit \$2200 into three separate bank accounts that each pay 3% annual interest. How much interest does each account earn after 6 years?

Account	Compounding	Balance after 6 years
1	quarterly	
2	monthly	
3	daily	

39. **ERROR ANALYSIS** You invest \$500 in the stock of a company. The value of the stock decreases 2% each year. Describe and correct the error in writing a model for the value of the stock after t years.

✗ $y = \left(\text{Initial amount}\right)\left(\text{Decay factor}\right)^t$

$y = 500(0.02)^t$

40. **ERROR ANALYSIS** You deposit \$250 in an account that pays 1.25% annual interest. Describe and correct the error in finding the balance after 3 years when the interest is compounded quarterly.

✗ $A = 250\left(1 + \frac{1.25}{4}\right)^{4 \cdot 3}$

$A = \$6533.29$

In Exercises 41–44, use the given information to find the amount A in the account earning compound interest after 6 years when the principal is \$3500.

41. $r = 2.16\%$, compounded quarterly

42. $r = 2.29\%$, compounded monthly

43. $r = 1.83\%$, compounded daily

44. $r = 1.26\%$, compounded monthly

45. USING STRUCTURE A website recorded the number y of referrals it received from social media websites over a 10-year period. The results can be modeled by $y = 2500(1.50)^t$, where t is the year and $0 \leq t \leq 9$. Interpret the values of a and b in this situation. What is the annual percent increase? Explain.

46. HOW DO YOU SEE IT? Consider the graph of an exponential function of the form $f(x) = ab^x$.

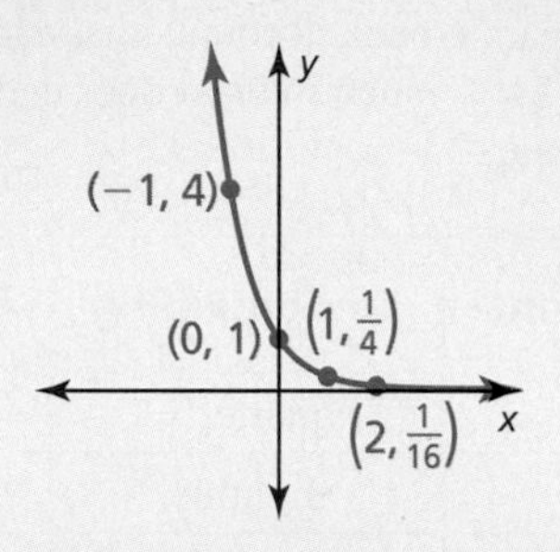

a. Determine whether the graph of f represents exponential growth or exponential decay.

b. What are the domain and range of the function? Explain.

47. MAKING AN ARGUMENT Your friend says the graph of $f(x) = 2^x$ increases at a faster rate than the graph of $g(x) = x^2$ when $x \geq 0$. Is your friend correct? Explain your reasoning.

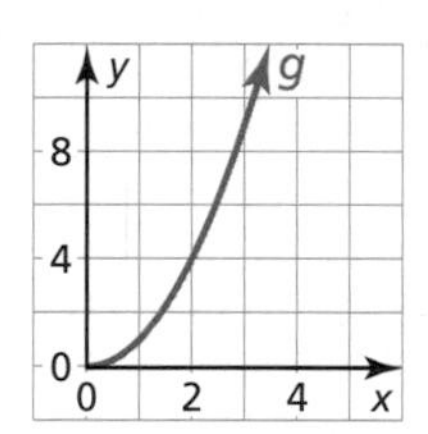

48. THOUGHT PROVOKING The function $f(x) = b^x$ represents an exponential decay function. Write a second exponential decay function in terms of b and x.

49. PROBLEM SOLVING The population p of a small town after x years can be modeled by the function $p = 6850(1.03)^x$. What is the average rate of change in the population over the first 6 years? Justify your answer.

50. REASONING Consider the exponential function $f(x) = ab^x$.

a. Show that $\dfrac{f(x + 1)}{f(x)} = b$.

b. Use the equation in part (a) to explain why there is no exponential function of the form $f(x) = ab^x$ whose graph passes through the points in the table below.

x	0	1	2	3	4
y	4	4	8	24	72

51. PROBLEM SOLVING The number E of eggs a Leghorn chicken produces per year can be modeled by the equation $E = 179.2(0.89)^{w/52}$, where w is the age (in weeks) of the chicken and $w \geq 22$.

a. Identify the decay factor and the percent decrease.

b. Graph the model.

c. Estimate the egg production of a chicken that is 2.5 years old.

d. Explain how you can rewrite the given equation so that time is measured in years rather than in weeks.

52. CRITICAL THINKING You buy a new stereo for \$1300 and are able to sell it 4 years later for \$275. Assume that the resale value of the stereo decays exponentially with time. Write an equation giving the resale value V (in dollars) of the stereo as a function of the time t (in years) since you bought it.

Maintaining Mathematical Proficiency

Reviewing what you learned in previous grades and lessons

Simplify the expression. *(Skills Review Handbook)*

53. $x^9 \cdot x^2$

54. $\dfrac{x^4}{x^3}$

55. $4x \cdot 6x$

56. $\left(\dfrac{4x^8}{2x^6}\right)^4$

57. $\dfrac{x + 3x}{2}$

58. $\dfrac{6x}{2} + 4x$

59. $\dfrac{12x}{4x} + 5x$

60. $(2x \cdot 3x^5)^3$

6.2 The Natural Base e

Essential Question

What is the natural base e?

So far in your study of mathematics, you have worked with special numbers such as π and i. Another special number is called the *natural base* and is denoted by e. The natural base e is irrational, so you cannot find its exact value.

EXPLORATION 1 Approximating the Natural Base e

Work with a partner. One way to approximate the natural base e is to approximate the sum

$$1 + \frac{1}{1} + \frac{1}{1 \cdot 2} + \frac{1}{1 \cdot 2 \cdot 3} + \frac{1}{1 \cdot 2 \cdot 3 \cdot 4} + \cdots.$$

Use a spreadsheet or a graphing calculator to approximate this sum. Explain the steps you used. How many decimal places did you use in your approximation?

USING TOOLS STRATEGICALLY

To be proficient in math, you need to use technological tools to explore and deepen your understanding of concepts.

EXPLORATION 2 Approximating the Natural Base e

Work with a partner. Another way to approximate the natural base e is to consider the expression

$$\left(1 + \frac{1}{x}\right)^x.$$

As x increases, the value of this expression approaches the value of e. Copy and complete the table. Then use the results in the table to approximate e. Compare this approximation to the one you obtained in Exploration 1.

x	10^1	10^2	10^3	10^4	10^5	10^6
$\left(1 + \frac{1}{x}\right)^x$						

EXPLORATION 3 Graphing a Natural Base Function

Work with a partner. Use your approximate value of e in Exploration 1 or 2 to complete the table. Then sketch the graph of the *natural base exponential function* $y = e^x$. You can use a graphing calculator and the e^x key to check your graph. What are the domain and range of $y = e^x$? Justify your answers.

x	-2	-1	0	1	2
$y = e^x$					

Communicate Your Answer

4. What is the natural base e?

5. Repeat Exploration 3 for the natural base exponential function $y = e^{-x}$. Then compare the graph of $y = e^x$ to the graph of $y = e^{-x}$.

6. The natural base e is used in a wide variety of real-life applications. Use the Internet or some other reference to research some of the real-life applications of e.

6.2 Lesson

Core Vocabulary

natural base e, *p. 304*

Previous
irrational number
properties of exponents
percent increase
percent decrease
compound interest

What You Will Learn

- Define and use the natural base e.
- Graph natural base functions.
- Solve real-life problems.

The Natural Base e

The history of mathematics is marked by the discovery of special numbers, such as π and i. Another special number is denoted by the letter e. The number is called the **natural base *e***. The expression $\left(1 + \frac{1}{x}\right)^x$ approaches e as x increases, as shown in the graph and table.

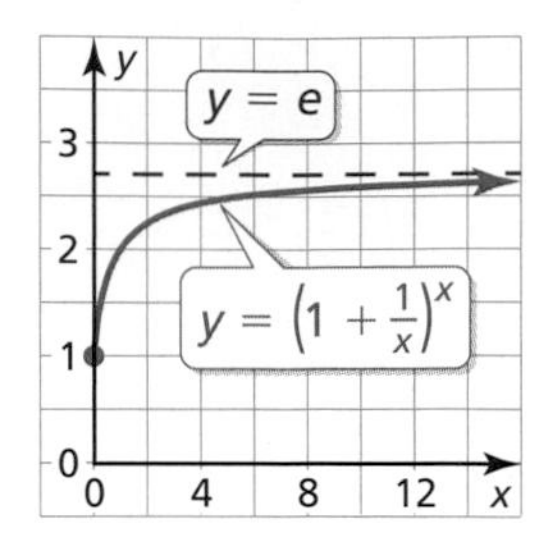

x	10^1	10^2	10^3	10^4	10^5	10^6
$\left(1 + \frac{1}{x}\right)^x$	2.59374	2.70481	2.71692	2.71815	2.71827	2.71828

Core Concept

The Natural Base e

The natural base e is irrational. It is defined as follows:

As x approaches $+\infty$, $\left(1 + \frac{1}{x}\right)^x$ approaches $e \approx 2.71828182846$.

EXAMPLE 1 Simplifying Natural Base Expressions

Simplify each expression.

a. $e^3 \cdot e^6$ **b.** $\frac{16e^5}{4e^4}$ **c.** $(3e^{-4x})^2$

SOLUTION

a. $e^3 \cdot e^6 = e^{3+6}$
$= e^9$

b. $\frac{16e^5}{4e^4} = 4e^{5-4}$
$= 4e$

c. $(3e^{-4x})^2 = 3^2(e^{-4x})^2$
$= 9e^{-8x}$
$= \frac{9}{e^{8x}}$

Check

You can use a calculator to check the equivalence of numerical expressions involving e.

```
e^(3)*e^(6)
      8103.083928
e^(9)
      8103.083928
```

Monitoring Progress

Help in English and Spanish at *BigIdeasMath.com*

Simplify the expression.

1. $e^7 \cdot e^4$ **2.** $\frac{24e^8}{8e^5}$ **3.** $(10e^{-3x})^3$

Graphing Natural Base Functions

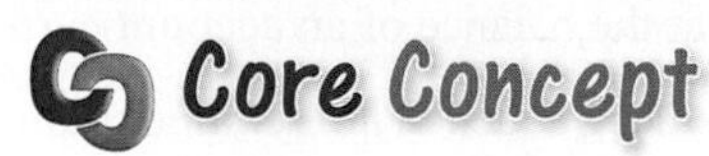

Natural Base Functions

A function of the form $y = ae^{rx}$ is called a *natural base exponential function*.

- When $a > 0$ and $r > 0$, the function is an exponential growth function.
- When $a > 0$ and $r < 0$, the function is an exponential decay function.

The graphs of the basic functions $y = e^x$ and $y = e^{-x}$ are shown.

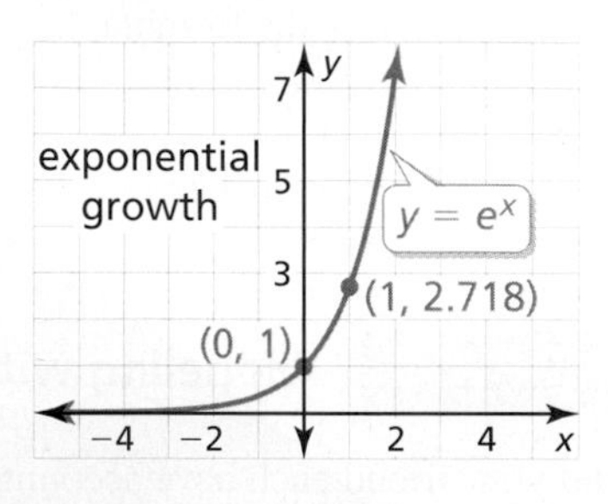

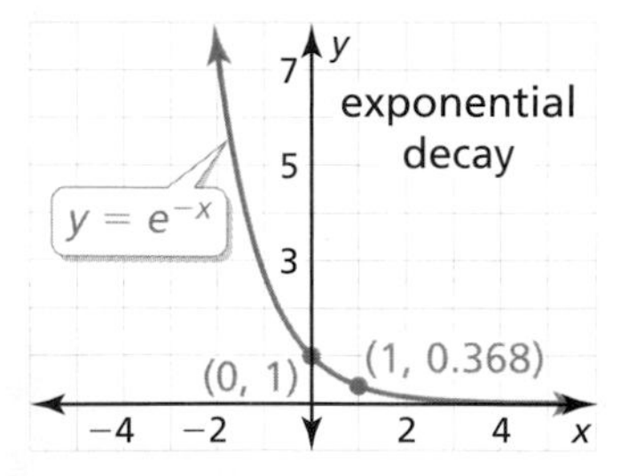

EXAMPLE 2 **Graphing Natural Base Functions**

Tell whether each function represents *exponential growth* or *exponential decay*. Then graph the function.

a. $y = 3e^x$

b. $f(x) = e^{-0.5x}$

SOLUTION

a. Because $a = 3$ is positive and $r = 1$ is positive, the function is an exponential growth function. Use a table to graph the function.

x	−2	−1	0	1
y	0.41	1.10	3	8.15

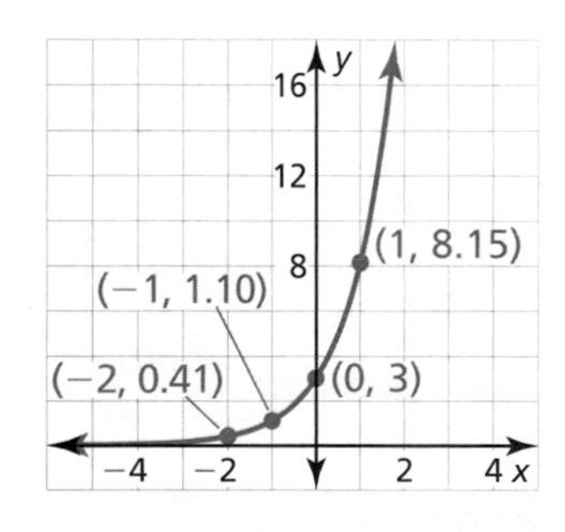

b. Because $a = 1$ is positive and $r = -0.5$ is negative, the function is an exponential decay function. Use a table to graph the function.

x	−4	−2	0	2
y	7.39	2.72	1	0.37

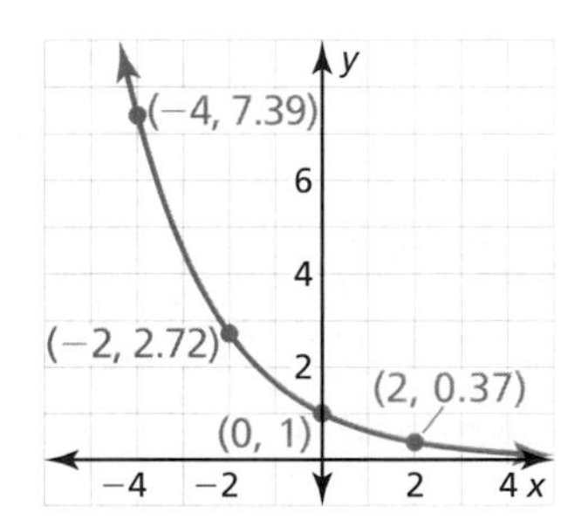

LOOKING FOR STRUCTURE

You can rewrite natural base exponential functions to find percent rates of change. In Example 2(b),

$$\begin{aligned} f(x) &= e^{-0.5x} \\ &= (e^{-0.5})^x \\ &\approx (0.6065)^x \\ &= (1 - 0.3935)^x. \end{aligned}$$

So, the percent decrease is about 39.35%.

Monitoring Progress  Help in English and Spanish at *BigIdeasMath.com*

Tell whether the function represents *exponential growth* or *exponential decay*. Then graph the function.

4. $y = \frac{1}{2}e^x$

5. $y = 4e^{-x}$

6. $f(x) = 2e^{2x}$

Solving Real-Life Problems

You have learned that the balance of an account earning compound interest is given by $A = P\left(1 + \frac{r}{n}\right)^{nt}$. As the frequency n of compounding approaches positive infinity, the compound interest formula approximates the following formula.

Core Concept

Continuously Compounded Interest

When interest is compounded *continuously*, the amount A in an account after t years is given by the formula

$$A = Pe^{rt}$$

where P is the principal and r is the annual interest rate expressed as a decimal.

EXAMPLE 3 Modeling with Mathematics

You and your friend each have accounts that earn annual interest compounded continuously. The balance A (in dollars) of your account after t years can be modeled by $A = 4500e^{0.04t}$. The graph shows the balance of your friend's account over time. Which account has a greater principal? Which has a greater balance after 10 years?

Your Friend's Account

(0, 4000)

Balance (dollars)

Year

SOLUTION

1. **Understand the Problem** You are given a graph and an equation that represent account balances. You are asked to identify the account with the greater principal and the account with the greater balance after 10 years.

2. **Make a Plan** Use the equation to find your principal and account balance after 10 years. Then compare these values to the graph of your friend's account.

3. **Solve the Problem** The equation $A = 4500e^{0.04t}$ is of the form $A = Pe^{rt}$, where $P = 4500$. So, your principal is \$4500. Your balance A when $t = 10$ is

$$A = 4500e^{0.04(10)} = \$6713.21.$$

Because the graph passes through (0, 4000), your friend's principal is \$4000. The graph also shows that the balance is about \$7250 when $t = 10$.

▶ So, your account has a greater principal, but your friend's account has a greater balance after 10 years.

4. **Look Back** Because your friend's account has a lesser principal but a greater balance after 10 years, the average rate of change from $t = 0$ to $t = 10$ should be greater for your friend's account than for your account.

MAKING CONJECTURES

You can also use this reasoning to conclude that your friend's account has a greater annual interest rate than your account.

Your account: $\frac{A(10) - A(0)}{10 - 0} = \frac{6713.21 - 4500}{10} = 221.321$

Your friend's account: $\frac{A(10) - A(0)}{10 - 0} \approx \frac{7250 - 4000}{10} = 325$ ✓

Monitoring Progress

Help in English and Spanish at *BigIdeasMath.com*

7. You deposit \$4250 in an account that earns 5% annual interest compounded continuously. Compare the balance after 10 years with the accounts in Example 3.

6.2 Exercises

Dynamic Solutions available at *BigIdeasMath.com*

Vocabulary and Core Concept Check

1. **VOCABULARY** What is the natural base e?

2. **WRITING** Tell whether the function $f(x) = \frac{1}{3}e^{4x}$ represents exponential growth or exponential decay. Explain.

Monitoring Progress and Modeling with Mathematics

In Exercises 3–12, simplify the expression. *(See Example 1.)*

3. $e^3 \cdot e^5$ **4.** $e^{-4} \cdot e^6$

5. $\dfrac{11e^9}{22e^{10}}$ **6.** $\dfrac{27e^7}{3e^4}$

7. $(5e^{7x})^4$ **8.** $(4e^{-2x})^3$

9. $\sqrt{9e^{6x}}$ **10.** $\sqrt[3]{8e^{12x}}$

11. $e^x \cdot e^{-6x} \cdot e^8$ **12.** $e^x \cdot e^4 \cdot e^{x+3}$

ERROR ANALYSIS In Exercises 13 and 14, describe and correct the error in simplifying the expression.

13.

$(4e^{3x})^2 = 4e^{(3x)(2)}$
$= 4e^{6x}$

14.

$\dfrac{e^{5x}}{e^{-2x}} = e^{5x-2x}$
$= e^{3x}$

In Exercises 15–22, tell whether the function represents *exponential growth* or *exponential decay*. Then graph the function. *(See Example 2.)*

15. $y = e^{3x}$ **16.** $y = e^{-2x}$

17. $y = 2e^{-x}$ **18.** $y = 3e^{2x}$

19. $y = 0.5e^x$ **20.** $y = 0.25e^{-3x}$

21. $y = 0.4e^{-0.25x}$ **22.** $y = 0.6e^{0.5x}$

ANALYZING EQUATIONS In Exercises 23–26, match the function with its graph. Explain your reasoning.

23. $y = e^{2x}$ **24.** $y = e^{-2x}$

25. $y = 4e^{-0.5x}$ **26.** $y = 0.75e^x$

A.

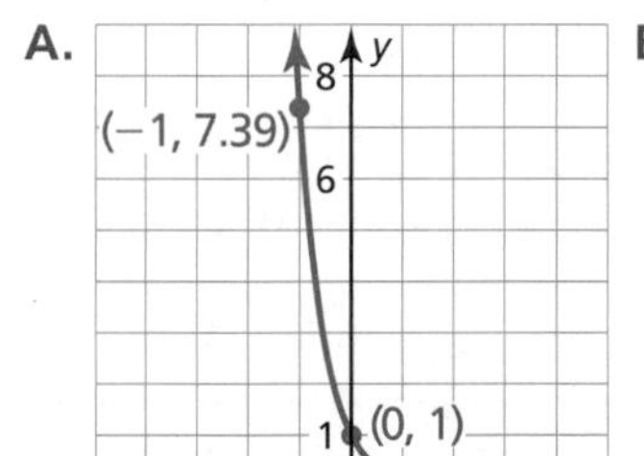

B.

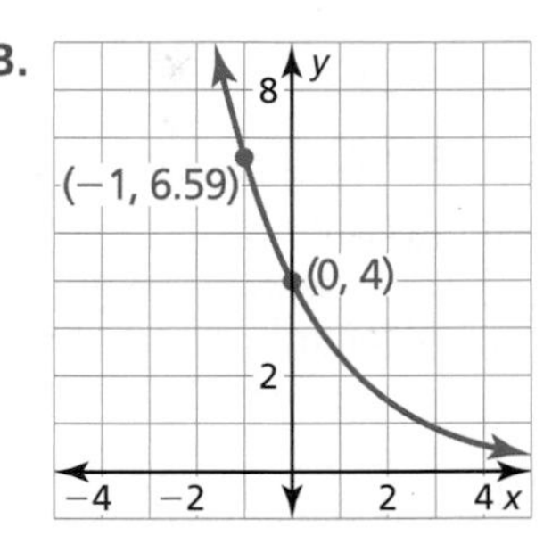

C.

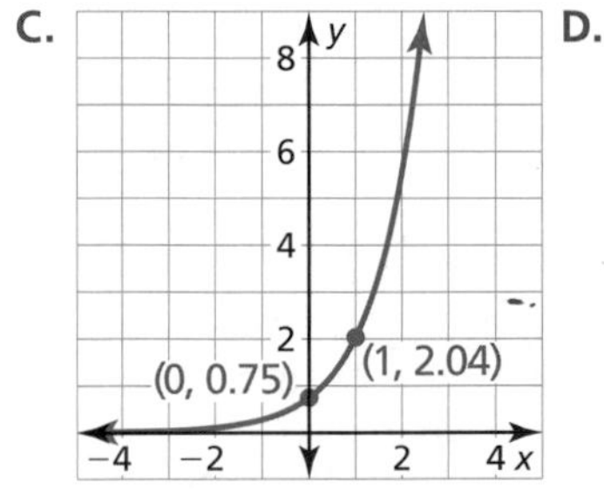

D.

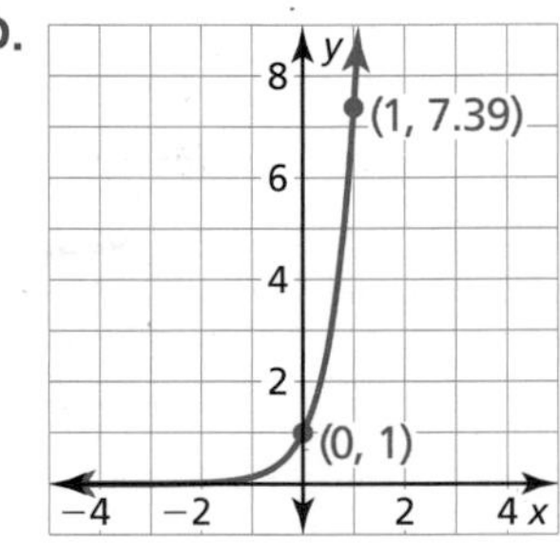

USING STRUCTURE In Exercises 27–30, use the properties of exponents to rewrite the function in the form $y = a(1 + r)^t$ or $y = a(1 - r)^t$. Then find the percent rate of change.

27. $y = e^{-0.25t}$ **28.** $y = e^{-0.75t}$

29. $y = 2e^{0.4t}$ **30.** $y = 0.5e^{0.8t}$

USING TOOLS In Exercises 31–34, use a table of values or a graphing calculator to graph the function. Then identify the domain and range.

31. $y = e^{x-2}$ **32.** $y = e^{x+1}$

33. $y = 2e^x + 1$ **34.** $y = 3e^x - 5$

35. MODELING WITH MATHEMATICS Investment accounts for a house and education earn annual interest compounded continuously. The balance H (in dollars) of the house fund after t years can be modeled by $H = 3224e^{0.05t}$. The graph shows the balance in the education fund over time. Which account has the greater principal? Which account has a greater balance after 10 years? *(See Example 3.)*

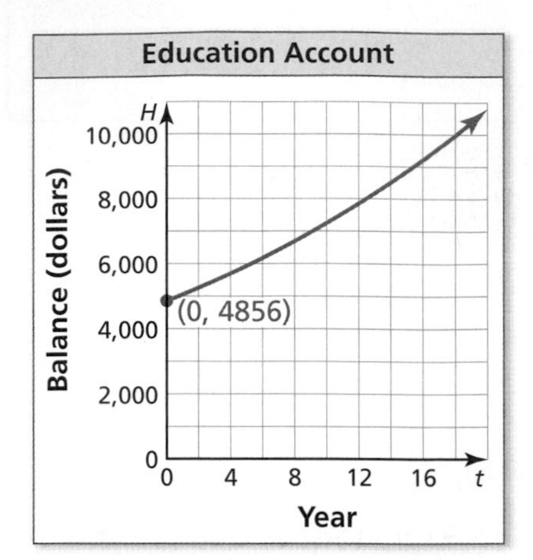

36. MODELING WITH MATHEMATICS Tritium and sodium-22 decay over time. In a sample of tritium, the amount y (in milligrams) remaining after t years is given by $y = 10e^{-0.0562t}$. The graph shows the amount of sodium-22 in a sample over time. Which sample started with a greater amount? Which has a greater amount after 10 years?

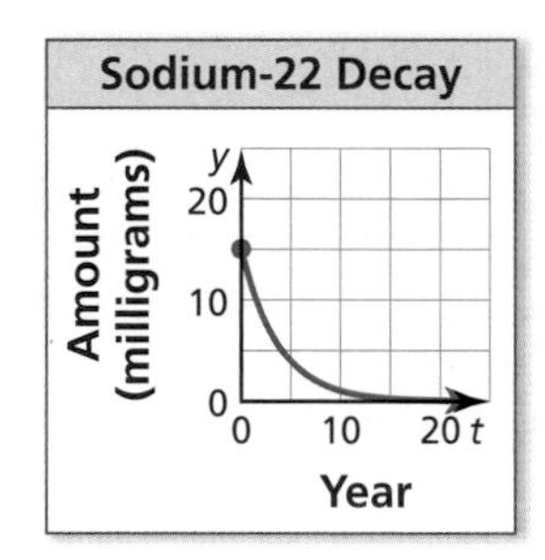

37. OPEN-ENDED Find values of a, b, r, and q such that $f(x) = ae^{rx}$ and $g(x) = be^{qx}$ are exponential decay functions, but $\frac{f(x)}{g(x)}$ represents exponential growth.

38. THOUGHT PROVOKING Explain why $A = P\left(1 + \frac{r}{n}\right)^{nt}$ approximates $A = Pe^{rt}$ as n approaches positive infinity.

39. WRITING Can the natural base e be written as a ratio of two integers? Explain.

40. MAKING AN ARGUMENT Your friend evaluates $f(x) = e^{-x}$ when $x = 1000$ and concludes that the graph of $y = f(x)$ has an x-intercept at $(1000, 0)$. Is your friend correct? Explain your reasoning.

41. DRAWING CONCLUSIONS You invest \$2500 in an account to save for college. Account 1 pays 6% annual interest compounded quarterly. Account 2 pays 4% annual interest compounded continuously. Which account should you choose to obtain the greater amount in 10 years? Justify your answer.

42. HOW DO YOU SEE IT? Use the graph to complete each statement.

a. $f(x)$ approaches _____ as x approaches $+\infty$.

b. $f(x)$ approaches _____ as x approaches $-\infty$.

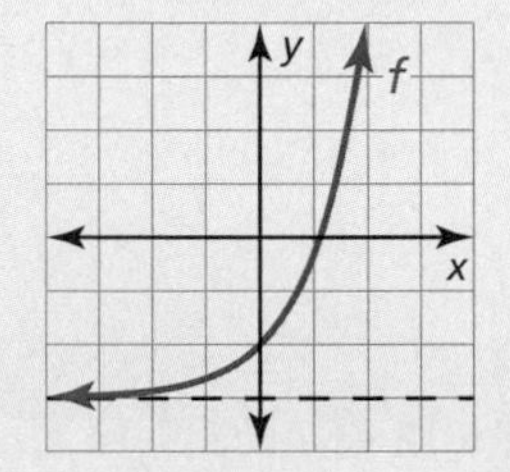

43. PROBLEM SOLVING The growth of *Mycobacterium tuberculosis* bacteria can be modeled by the function $N(t) = ae^{0.166t}$, where N is the number of cells after t hours and a is the number of cells when $t = 0$.

a. At 1:00 P.M., there are 30 *M. tuberculosis* bacteria in a sample. Write a function that gives the number of bacteria after 1:00 P.M.

b. Use a graphing calculator to graph the function in part (a).

c. Describe how to find the number of cells in the sample at 3:45 P.M.

Maintaining Mathematical Proficiency

Reviewing what you learned in previous grades and lessons

Write the number in scientific notation. *(Skills Review Handbook)*

44. 0.006 **45.** 5000 **46.** 26,000,000 **47.** 0.000000047

Find the inverse of the function. Then graph the function and its inverse. *(Section 5.6)*

48. $y = 3x + 5$ **49.** $y = x^2 - 1, x \leq 0$

50. $y = \sqrt{x + 6}$ **51.** $y = x^3 - 2$

6.3 Logarithms and Logarithmic Functions

Essential Question What are some of the characteristics of the graph of a logarithmic function?

Every exponential function of the form $f(x) = b^x$, where b is a positive real number other than 1, has an inverse function that you can denote by $g(x) = \log_b x$. This inverse function is called a *logarithmic function with base b*.

EXPLORATION 1 Rewriting Exponential Equations

Work with a partner. Find the value of x in each exponential equation. Explain your reasoning. Then use the value of x to rewrite the exponential equation in its equivalent logarithmic form, $x = \log_b y$.

a. $2^x = 8$ **b.** $3^x = 9$ **c.** $4^x = 2$

d. $5^x = 1$ **e.** $5^x = \frac{1}{5}$ **f.** $8^x = 4$

EXPLORATION 2 Graphing Exponential and Logarithmic Functions

Work with a partner. Complete each table for the given exponential function. Use the results to complete the table for the given logarithmic function. Explain your reasoning. Then sketch the graphs of f and g in the same coordinate plane.

a.

x	-2	-1	0	1	2
$f(x) = 2^x$					

x					
$g(x) = \log_2 x$	-2	-1	0	1	2

b.

x	-2	-1	0	1	2
$f(x) = 10^x$					

x					
$g(x) = \log_{10} x$	-2	-1	0	1	2

CONSTRUCTING VIABLE ARGUMENTS

To be proficient in math, you need to justify your conclusions and communicate them to others.

EXPLORATION 3 Characteristics of Graphs of Logarithmic Functions

Work with a partner. Use the graphs you sketched in Exploration 2 to determine the domain, range, x-intercept, and asymptote of the graph of $g(x) = \log_b x$, where b is a positive real number other than 1. Explain your reasoning.

Communicate Your Answer

4. What are some of the characteristics of the graph of a logarithmic function?

5. How can you use the graph of an exponential function to obtain the graph of a logarithmic function?

6.3 Lesson

Core Vocabulary

logarithm of y with base b, *p. 310*
common logarithm, *p. 311*
natural logarithm, *p. 311*

Previous
inverse functions

What You Will Learn

- Define and evaluate logarithms.
- Use inverse properties of logarithmic and exponential functions.
- Graph logarithmic functions.

Logarithms

You know that $2^2 = 4$ and $2^3 = 8$. However, for what value of x does $2^x = 6$? Mathematicians define this x-value using a *logarithm* and write $x = \log_2 6$. The definition of a logarithm can be generalized as follows.

Core Concept

Definition of Logarithm with Base *b*

Let b and y be positive real numbers with $b \neq 1$. The **logarithm of *y* with base *b*** is denoted by $\log_b y$ and is defined as

$$\log_b y = x \qquad \text{if and only if} \qquad b^x = y.$$

The expression $\log_b y$ is read as "log base b of y."

This definition tells you that the equations $\log_b y = x$ and $b^x = y$ are equivalent. The first is in *logarithmic form*, and the second is in *exponential form*.

EXAMPLE 1 Rewriting Logarithmic Equations

Rewrite each equation in exponential form.

a. $\log_2 16 = 4$ **b.** $\log_4 1 = 0$ **c.** $\log_{12} 12 = 1$ **d.** $\log_{1/4} 4 = -1$

SOLUTION

Logarithmic Form	Exponential Form
a. $\log_2 16 = 4$	$2^4 = 16$
b. $\log_4 1 = 0$	$4^0 = 1$
c. $\log_{12} 12 = 1$	$12^1 = 12$
d. $\log_{1/4} 4 = -1$	$\left(\frac{1}{4}\right)^{-1} = 4$

EXAMPLE 2 Rewriting Exponential Equations

Rewrite each equation in logarithmic form.

a. $5^2 = 25$ **b.** $10^{-1} = 0.1$ **c.** $8^{2/3} = 4$ **d.** $6^{-3} = \frac{1}{216}$

SOLUTION

Exponential Form	Logarithmic Form
a. $5^2 = 25$	$\log_5 25 = 2$
b. $10^{-1} = 0.1$	$\log_{10} 0.1 = -1$
c. $8^{2/3} = 4$	$\log_8 4 = \frac{2}{3}$
d. $6^{-3} = \frac{1}{216}$	$\log_6 \frac{1}{216} = -3$

Parts (b) and (c) of Example 1 illustrate two special logarithm values that you should learn to recognize. Let b be a positive real number such that $b \neq 1$.

Logarithm of 1	**Logarithm of b with Base b**
$\log_b 1 = 0$ because $b^0 = 1$.	$\log_b b = 1$ because $b^1 = b$.

EXAMPLE 3 Evaluating Logarithmic Expressions

Evaluate each logarithm.

a. $\log_4 64$ **b.** $\log_5 0.2$ **c.** $\log_{1/5} 125$ **d.** $\log_{36} 6$

SOLUTION

To help you find the value of $\log_b y$, ask yourself what power of b gives you y.

a. What power of 4 gives you 64? $4^3 = 64$, so $\log_4 64 = 3$.

b. What power of 5 gives you 0.2? $5^{-1} = 0.2$, so $\log_5 0.2 = -1$.

c. What power of $\frac{1}{5}$ gives you 125? $\left(\frac{1}{5}\right)^{-3} = 125$, so $\log_{1/5} 125 = -3$.

d. What power of 36 gives you 6? $36^{1/2} = 6$, so $\log_{36} 6 = \frac{1}{2}$.

A **common logarithm** is a logarithm with base 10. It is denoted by $\log_{10}$ or simply by log. A **natural logarithm** is a logarithm with base e. It can be denoted by $\log_e$ but is usually denoted by ln.

Common Logarithm	**Natural Logarithm**
$\log_{10} x = \log x$	$\log_e x = \ln x$

EXAMPLE 4 Evaluating Common and Natural Logarithms

Evaluate (a) log 8 and (b) ln 0.3 using a calculator. Round your answer to three decimal places.

SOLUTION

Most calculators have keys for evaluating common and natural logarithms.

```
log(8)
         .903089987
ln(0.3)
       -1.203972804
```

a. $\log 8 \approx 0.903$

b. $\ln 0.3 \approx -1.204$

Check your answers by rewriting each logarithm in exponential form and evaluating.

Check

```
10^(0.903)
         7.99834255
e^(-1.204)
        .2999918414
```

Monitoring Progress  Help in English and Spanish at *BigIdeasMath.com*

Rewrite the equation in exponential form.

1. $\log_3 81 = 4$ **2.** $\log_7 7 = 1$ **3.** $\log_{14} 1 = 0$ **4.** $\log_{1/2} 32 = -5$

Rewrite the equation in logarithmic form.

5. $7^2 = 49$ **6.** $50^0 = 1$ **7.** $4^{-1} = \frac{1}{4}$ **8.** $256^{1/8} = 2$

Evaluate the logarithm. If necessary, use a calculator and round your answer to three decimal places.

9. $\log_2 32$ **10.** $\log_{27} 3$ **11.** $\log 12$ **12.** $\ln 0.75$

Using Inverse Properties

By the definition of a logarithm, it follows that the logarithmic function $g(x) = \log_b x$ is the inverse of the exponential function $f(x) = b^x$. This means that

$$g(f(x)) = \log_b b^x = x \qquad \text{and} \qquad f(g(x)) = b^{\log_b x} = x.$$

In other words, exponential functions and logarithmic functions "undo" each other.

EXAMPLE 5 Using Inverse Properties

Simplify (a) $10^{\log 4}$ and (b) $\log_5 25^x$.

SOLUTION

a. $10^{\log 4} = 4$ — $b^{\log_b x} = x$

b. $\log_5 25^x = \log_5 (5^2)^x$ — Express 25 as a power with base 5.

$= \log_5 5^{2x}$ — Power of a Power Property

$= 2x$ — $\log_b b^x = x$

EXAMPLE 6 Finding Inverse Functions

Find the inverse of each function.

a. $f(x) = 6^x$

b. $y = \ln(x + 3)$

SOLUTION

a. From the definition of logarithm, the inverse of $f(x) = 6^x$ is $g(x) = \log_6 x$.

b.

$y = \ln(x + 3)$ — Write original function.

$x = \ln(y + 3)$ — Switch x and y.

$e^x = y + 3$ — Write in exponential form.

$e^x - 3 = y$ — Subtract 3 from each side.

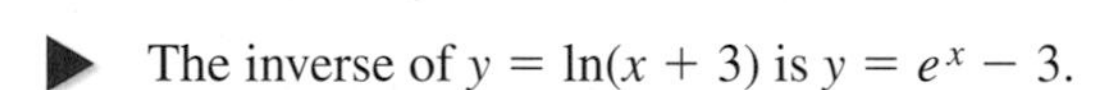

▶ The inverse of $y = \ln(x + 3)$ is $y = e^x - 3$.

Check

a. $f(g(x)) = 6^{\log_6 x} = x$ ✓

$g(f(x)) = \log_6 6^x = x$ ✓

b.

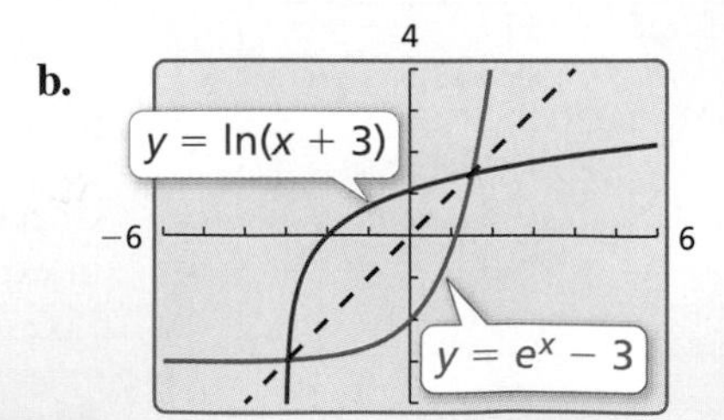

The graphs appear to be reflections of each other in the line $y = x$. ✓

Monitoring Progress

Help in English and Spanish at *BigIdeasMath.com*

Simplify the expression.

13. $8^{\log_8 x}$

14. $\log_7 7^{-3x}$

15. $\log_2 64^x$

16. $e^{\ln 20}$

17. Find the inverse of $y = 4^x$.

18. Find the inverse of $y = \ln(x - 5)$.

Graphing Logarithmic Functions

You can use the inverse relationship between exponential and logarithmic functions to graph logarithmic functions.

Core Concept

Parent Graphs for Logarithmic Functions

The graph of $f(x) = \log_b x$ is shown below for $b > 1$ and for $0 < b < 1$. Because $f(x) = \log_b x$ and $g(x) = b^x$ are inverse functions, the graph of $f(x) = \log_b x$ is the reflection of the graph of $g(x) = b^x$ in the line $y = x$.

Graph of $f(x) = \log_b x$ for $b > 1$

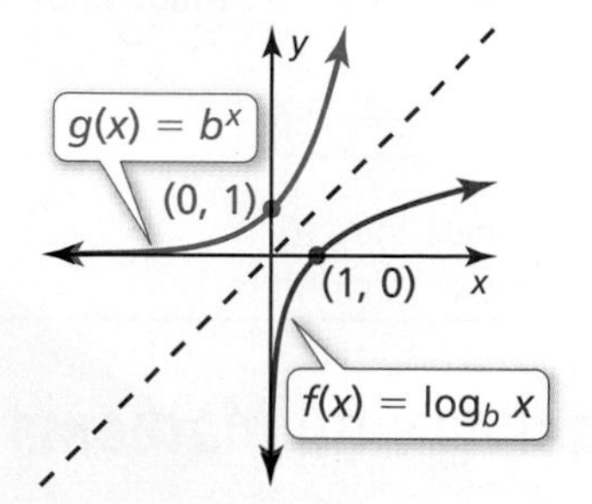

Graph of $f(x) = \log_b x$ for $0 < b < 1$

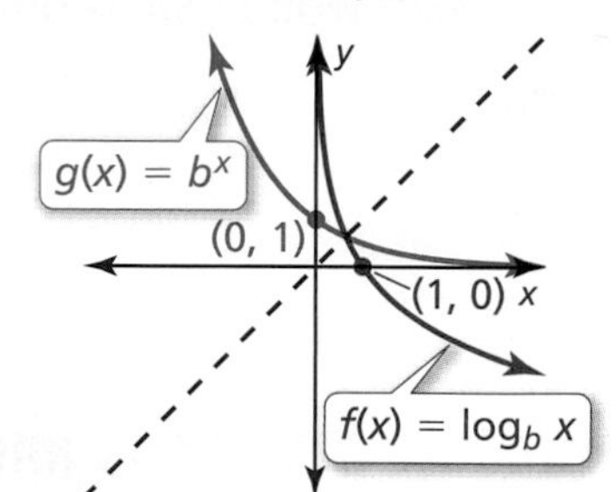

Note that the y-axis is a vertical asymptote of the graph of $f(x) = \log_b x$. The domain of $f(x) = \log_b x$ is $x > 0$, and the range is all real numbers.

EXAMPLE 7 **Graphing a Logarithmic Function**

Graph $f(x) = \log_3 x$.

SOLUTION

Step 1 Find the inverse of f. From the definition of logarithm, the inverse of $f(x) = \log_3 x$ is $g(x) = 3^x$.

Step 2 Make a table of values for $g(x) = 3^x$.

x	-2	-1	0	1	2
$g(x)$	$\frac{1}{9}$	$\frac{1}{3}$	1	3	9

Step 3 Plot the points from the table and connect them with a smooth curve.

Step 4 Because $f(x) = \log_3 x$ and $g(x) = 3^x$ are inverse functions, the graph of f is obtained by reflecting the graph of g in the line $y = x$. To do this, reverse the coordinates of the points on g and plot these new points on the graph of f.

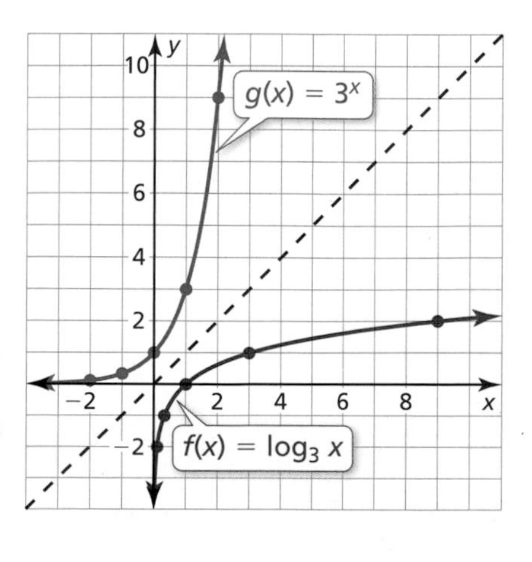

Monitoring Progress

Help in English and Spanish at *BigIdeasMath.com*

Graph the function.

19. $y = \log_2 x$

20. $f(x) = \log_5 x$

21. $y = \log_{1/2} x$

6.3 Exercises

Dynamic Solutions available at *BigIdeasMath.com*

Vocabulary and Core Concept Check

1. **COMPLETE THE SENTENCE** A logarithm with base 10 is called a(n) __________ logarithm.
2. **COMPLETE THE SENTENCE** The expression $\log_3 9$ is read as __________.
3. **WRITING** Describe the relationship between $y = 7^x$ and $y = \log_7 x$.
4. **DIFFERENT WORDS, SAME QUESTION** Which is different? Find "both" answers.

What power of 4 gives you 16?	What is log base 4 of 16?
Evaluate 4^2.	Evaluate $\log_4 16$.

Monitoring Progress and Modeling with Mathematics

In Exercises 5–10, rewrite the equation in exponential form. *(See Example 1.)*

5. $\log_3 9 = 2$
6. $\log_4 4 = 1$
7. $\log_6 1 = 0$
8. $\log_7 343 = 3$
9. $\log_{1/2} 16 = -4$
10. $\log_3 \frac{1}{3} = -1$

In Exercises 11–16, rewrite the equation in logarithmic form. *(See Example 2.)*

11. $6^2 = 36$
12. $12^0 = 1$
13. $16^{-1} = \frac{1}{16}$
14. $5^{-2} = \frac{1}{25}$
15. $125^{2/3} = 25$
16. $49^{1/2} = 7$

In Exercises 17–24, evaluate the logarithm. *(See Example 3.)*

17. $\log_3 81$
18. $\log_7 49$
19. $\log_3 3$
20. $\log_{1/2} 1$
21. $\log_5 \frac{1}{625}$
22. $\log_8 \frac{1}{512}$
23. $\log_4 0.25$
24. $\log_{10} 0.001$

25. **NUMBER SENSE** Order the logarithms from least value to greatest value.

$\log_5 23$	$\log_6 38$	$\log_7 8$	$\log_2 10$

26. **WRITING** Explain why the expressions $\log_2(-1)$ and $\log_1 1$ are not defined.

In Exercises 27–32, evaluate the logarithm using a calculator. Round your answer to three decimal places. *(See Example 4.)*

27. $\log 6$
28. $\ln 12$
29. $\ln \frac{1}{3}$
30. $\log \frac{2}{7}$
31. $3 \ln 0.5$
32. $\log 0.6 + 1$

33. **MODELING WITH MATHEMATICS** Skydivers use an instrument called an *altimeter* to track their altitude as they fall. The altimeter determines altitude by measuring air pressure. The altitude h (in meters) above sea level is related to the air pressure P (in pascals) by the function shown in the diagram. What is the altitude above sea level when the air pressure is 57,000 pascals?

34. **MODELING WITH MATHEMATICS** The pH value for a substance measures how acidic or alkaline the substance is. It is given by the formula $\text{pH} = -\log[\text{H}^+]$, where H^+ is the hydrogen ion concentration (in moles per liter). Find the pH of each substance.

    a. baking soda: $[\text{H}^+] = 10^{-8}$ moles per liter

    b. vinegar: $[\text{H}^+] = 10^{-3}$ moles per liter

In Exercises 35–40, simplify the expression. *(See Example 5.)*

35. $7^{\log_7 x}$

36. $3^{\log_3 5x}$

37. $e^{\ln 4}$

38. $10^{\log 15}$

39. $\log_3 3^{2x}$

40. $\ln e^{x+1}$

41. ERROR ANALYSIS Describe and correct the error in rewriting $4^{-3} = \frac{1}{64}$ in logarithmic form.

$\log_4(-3) = \frac{1}{64}$

42. ERROR ANALYSIS Describe and correct the error in simplifying the expression $\log_4 64^x$.

✗

$$\begin{aligned} \log_4 64^x &= \log_4(16 \cdot 4^x) \\ &= \log_4(4^2 \cdot 4^x) \\ &= \log_4 4^{2+x} \\ &= 2 + x \end{aligned}$$

In Exercises 43–52, find the inverse of the function. *(See Example 6.)*

43. $y = 0.3^x$

44. $y = 11^x$

45. $y = \log_2 x$

46. $y = \log_{1/5} x$

47. $y = \ln(x - 1)$

48. $y = \ln 2x$

49. $y = e^{3x}$

50. $y = e^{x-4}$

51. $y = 5^x - 9$

52. $y = 13 + \log x$

53. PROBLEM SOLVING The wind speed s (in miles per hour) near the center of a tornado can be modeled by $s = 93 \log d + 65$, where d is the distance (in miles) that the tornado travels.

a. In 1925, a tornado traveled 220 miles through three states. Estimate the wind speed near the center of the tornado.

b. Find the inverse of the given function. Describe what the inverse represents.

54. MODELING WITH MATHEMATICS The energy magnitude M of an earthquake can be modeled by $M = \frac{2}{3} \log E - 9.9$, where E is the amount of energy released (in ergs).

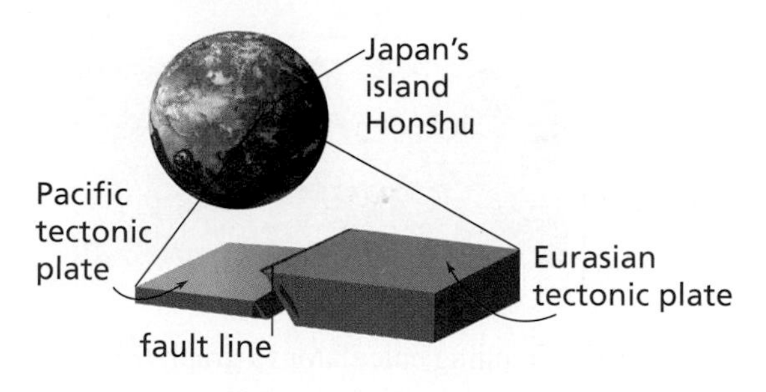

a. In 2011, a powerful earthquake in Japan, caused by the slippage of two tectonic plates along a fault, released 2.24×10^{28} ergs. What was the energy magnitude of the earthquake?

b. Find the inverse of the given function. Describe what the inverse represents.

In Exercises 55–60, graph the function. *(See Example 7.)*

55. $y = \log_4 x$

56. $y = \log_6 x$

57. $y = \log_{1/3} x$

58. $y = \log_{1/4} x$

59. $y = \log_2 x - 1$

60. $y = \log_3(x + 2)$

USING TOOLS In Exercises 61–64, use a graphing calculator to graph the function. Determine the domain, range, and asymptote of the function.

61. $y = \log(x + 2)$

62. $y = -\ln x$

63. $y = \ln(-x)$

64. $y = 3 - \log x$

65. MAKING AN ARGUMENT Your friend states that every logarithmic function will pass through the point (1, 0). Is your friend correct? Explain your reasoning.

66. ANALYZING RELATIONSHIPS Rank the functions in order from the least average rate of change to the greatest average rate of change over the interval $1 \le x \le 10$.

a. $y = \log_6 x$

b. $y = \log_{3/5} x$

c.

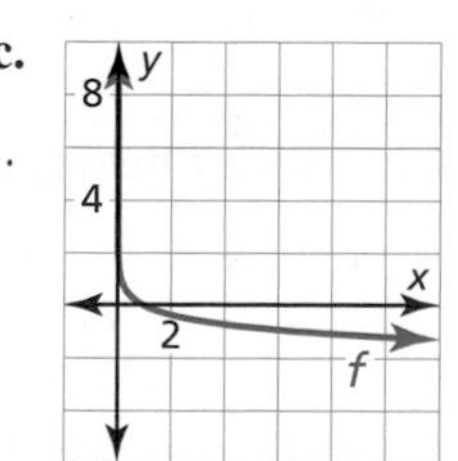

d.

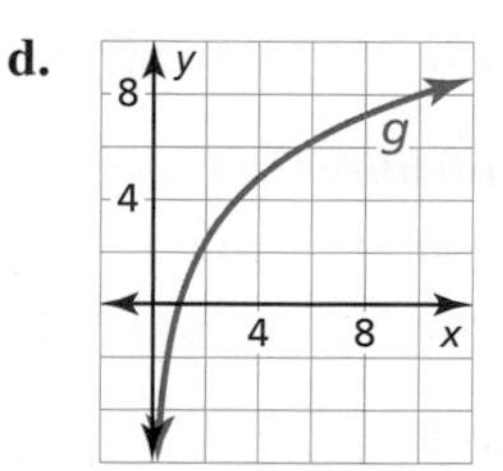

67. **PROBLEM SOLVING** Biologists have found that the length ℓ (in inches) of an alligator and its weight w (in pounds) are related by the function $\ell = 27.1 \ln w - 32.8$.

a. Use a graphing calculator to graph the function.

b. Use your graph to estimate the weight of an alligator that is 10 feet long.

c. Use the *zero* feature to find the x-intercept of the graph of the function. Does this x-value make sense in the context of the situation? Explain.

68. **HOW DO YOU SEE IT?** The figure shows the graphs of the two functions f and g.

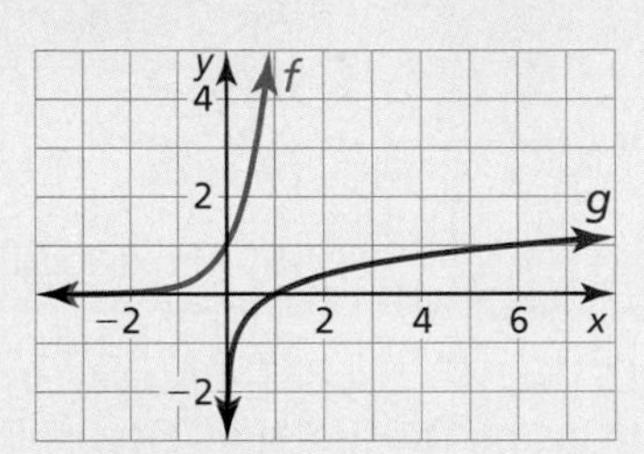

a. Compare the end behavior of the logarithmic function g to that of the exponential function f.

b. Determine whether the functions are inverse functions. Explain.

c. What is the base of each function? Explain.

69. **PROBLEM SOLVING** A study in Florida found that the number s of fish species in a pool or lake can be modeled by the function

$$s = 30.6 - 20.5 \log A + 3.8(\log A)^2$$

where A is the area (in square meters) of the pool or lake.

a. Use a graphing calculator to graph the function on the domain $200 \leq A \leq 35{,}000$.

b. Use your graph to estimate the number of species in a lake with an area of 30,000 square meters.

c. Use your graph to estimate the area of a lake that contains six species of fish.

d. Describe what happens to the number of fish species as the area of a pool or lake increases. Explain why your answer makes sense.

70. **THOUGHT PROVOKING** Write a logarithmic function that has an output of -4. Then sketch the graph of your function.

71. **CRITICAL THINKING** Evaluate each logarithm. (*Hint*: For each logarithm $\log_b x$, rewrite b and x as powers of the same base.)

a. $\log_{125} 25$ b. $\log_8 32$

c. $\log_{27} 81$ d. $\log_4 128$

Maintaining Mathematical Proficiency

Reviewing what you learned in previous grades and lessons

Let $f(x) = \sqrt[3]{x}$. Write a rule for g that represents the indicated transformation of the graph of f. *(Section 5.3)*

72. $g(x) = -f(x)$

73. $g(x) = f\left(\frac{1}{2}x\right)$

74. $g(x) = f(-x) + 3$

75. $g(x) = f(x + 2)$

Identify the function family to which f belongs. Compare the graph of f to the graph of its parent function. *(Section 1.1)*

76.

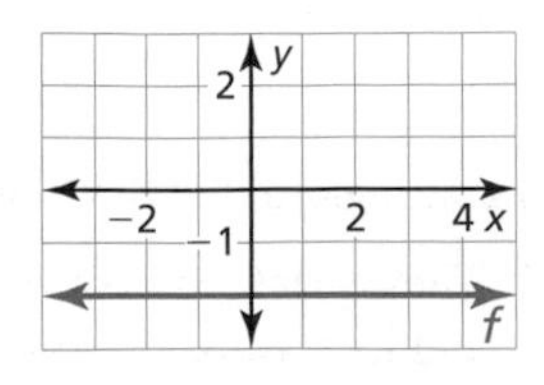

77.

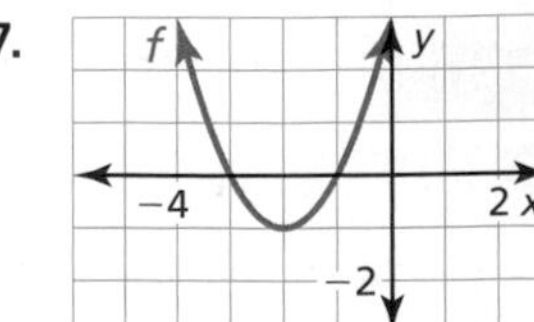

78.

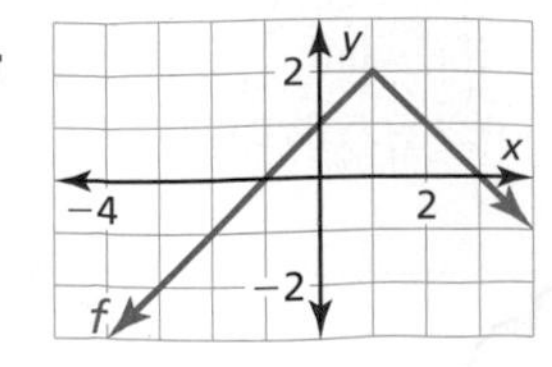

6.4 Transformations of Exponential and Logarithmic Functions

Essential Question

How can you transform the graphs of exponential and logarithmic functions?

EXPLORATION 1 Identifying Transformations

Work with a partner. Each graph shown is a transformation of the parent function

$f(x) = e^x$ or $f(x) = \ln x$.

Match each function with its graph. Explain your reasoning. Then describe the transformation of f represented by g.

a. $g(x) = e^{x+2} - 3$ **b.** $g(x) = -e^{x+2} + 1$ **c.** $g(x) = e^{x-2} - 1$

d. $g(x) = \ln(x + 2)$ **e.** $g(x) = 2 + \ln x$ **f.** $g(x) = 2 + \ln(-x)$

A.

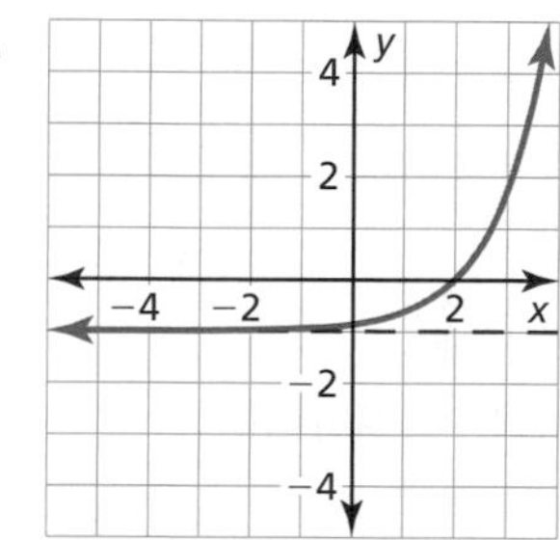

B.

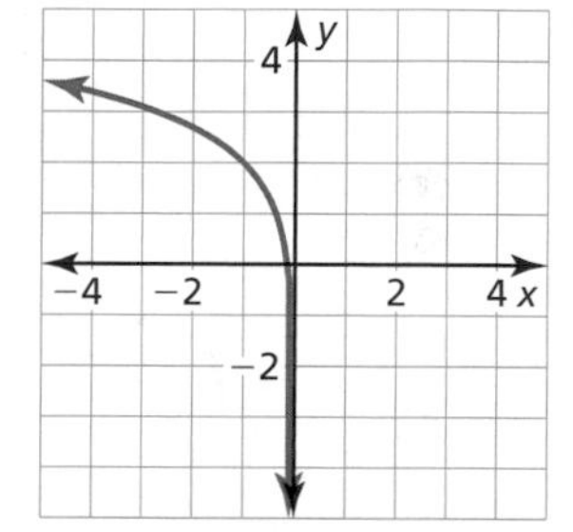

C.

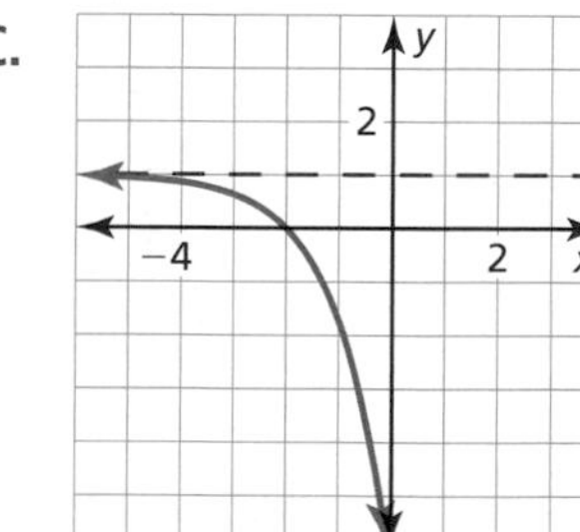

D.

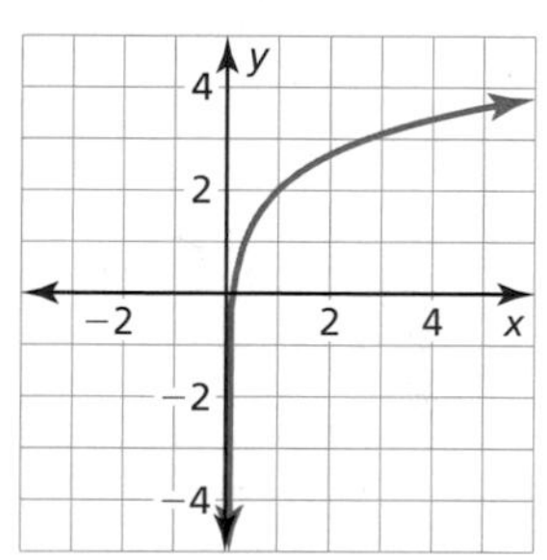

E.

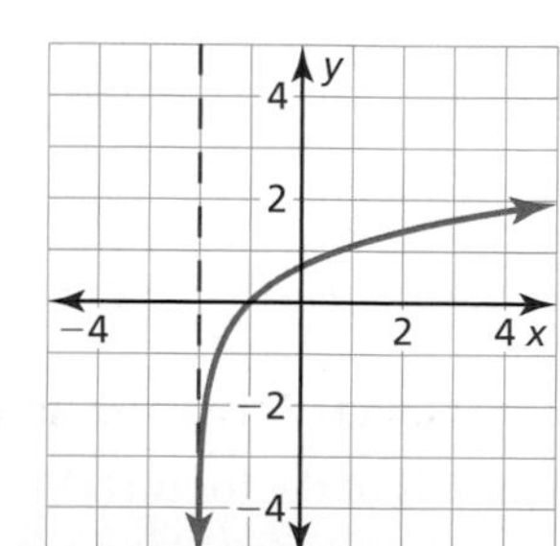

F.

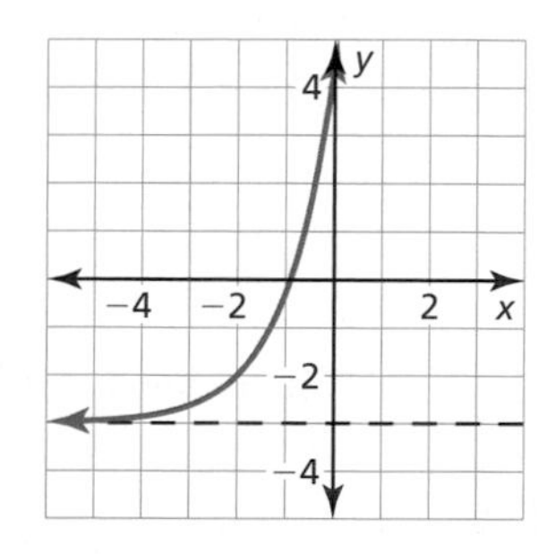

EXPLORATION 2 Characteristics of Graphs

Work with a partner. Determine the domain, range, and asymptote of each function in Exploration 1. Justify your answers.

REASONING QUANTITATIVELY

To be proficient in math, you need to make sense of quantities and their relationships in problem situations.

Communicate Your Answer

3. How can you transform the graphs of exponential and logarithmic functions?

4. Find the inverse of each function in Exploration 1. Then check your answer by using a graphing calculator to graph each function and its inverse in the same viewing window.

6.4 Lesson

Core Vocabulary

Previous
exponential function
logarithmic function
transformations

What You Will Learn

- Transform graphs of exponential functions.
- Transform graphs of logarithmic functions.
- Write transformations of graphs of exponential and logarithmic functions.

Transforming Graphs of Exponential Functions

You can transform graphs of exponential and logarithmic functions in the same way you transformed graphs of functions in previous chapters. Examples of transformations of the graph of $f(x) = 4^x$ are shown below.

Core Concept

Transformation	$f(x)$ Notation	Examples	
Horizontal Translation Graph shifts left or right.	$f(x - h)$	$g(x) = 4^{x-3}$ $g(x) = 4^{x+2}$	3 units right 2 units left
Vertical Translation Graph shifts up or down.	$f(x) + k$	$g(x) = 4^x + 5$ $g(x) = 4^x - 1$	5 units up 1 unit down
Reflection Graph flips over x- or y-axis.	$f(-x)$ $-f(x)$	$g(x) = 4^{-x}$ $g(x) = -4^x$	in the y-axis in the x-axis
Horizontal Stretch or Shrink Graph stretches away from or shrinks toward y-axis.	$f(ax)$	$g(x) = 4^{2x}$ $g(x) = 4^{x/2}$	shrink by a factor of $\frac{1}{2}$ stretch by a factor of 2
Vertical Stretch or Shrink Graph stretches away from or shrinks toward x-axis.	$a \cdot f(x)$	$g(x) = 3(4^x)$ $g(x) = \frac{1}{4}(4^x)$	stretch by a factor of 3 shrink by a factor of $\frac{1}{4}$

EXAMPLE 1 Translating an Exponential Function

Describe the transformation of $f(x) = \left(\frac{1}{2}\right)^x$ represented by $g(x) = \left(\frac{1}{2}\right)^x - 4$. Then graph each function.

SOLUTION

Notice that the function is of the form $g(x) = \left(\frac{1}{2}\right)^x + k$.

Rewrite the function to identify k.

$$g(x) = \left(\frac{1}{2}\right)^x + \underset{\substack{\uparrow \\ k}}{(-4)}$$

STUDY TIP

Notice in the graph that the vertical translation also shifted the asymptote 4 units down, so the range of g is $y > -4$.

- Because $k = -4$, the graph of g is a translation 4 units down of the graph of f.

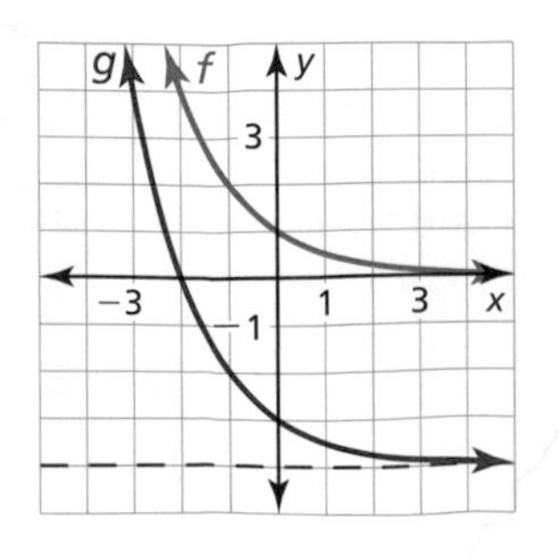

EXAMPLE 2 Translating a Natural Base Exponential Function

Describe the transformation of $f(x) = e^x$ represented by $g(x) = e^{x+3} + 2$. Then graph each function.

SOLUTION

Notice that the function is of the form $g(x) = e^{x-h} + k$. Rewrite the function to identify h and k.

$$g(x) = e^{x-(-3)} + 2$$

h ↑ (−3), k ↑ (2)

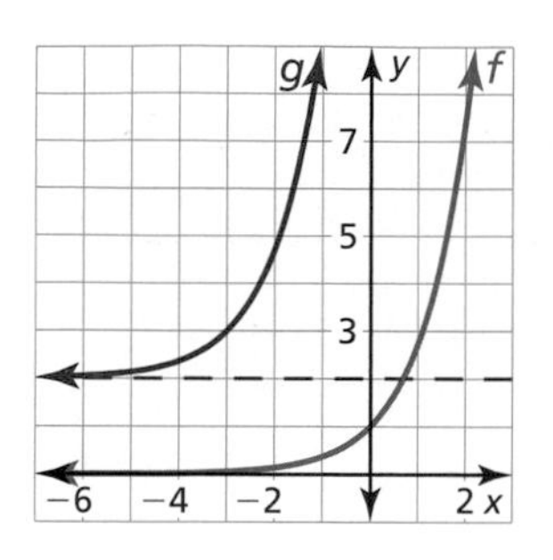

STUDY TIP

Notice in the graph that the vertical translation also shifted the asymptote 2 units up, so the range of g is $y > 2$.

▶ Because $h = -3$ and $k = 2$, the graph of g is a translation 3 units left and 2 units up of the graph of f.

EXAMPLE 3 Transforming Exponential Functions

Describe the transformation of f represented by g. Then graph each function.

a. $f(x) = 3^x$, $g(x) = 3^{3x-5}$

b. $f(x) = e^{-x}$, $g(x) = -\frac{1}{8}e^{-x}$

LOOKING FOR STRUCTURE

In Example 3(a), the horizontal shrink follows the translation. In the function $h(x) = 3^{3(x-5)}$, the translation 5 units right follows the horizontal shrink by a factor of $\frac{1}{3}$.

SOLUTION

a. Notice that the function is of the form $g(x) = 3^{ax-h}$, where $a = 3$ and $h = 5$.

▶ So, the graph of g is a translation 5 units right, followed by a horizontal shrink by a factor of $\frac{1}{3}$ of the graph of f.

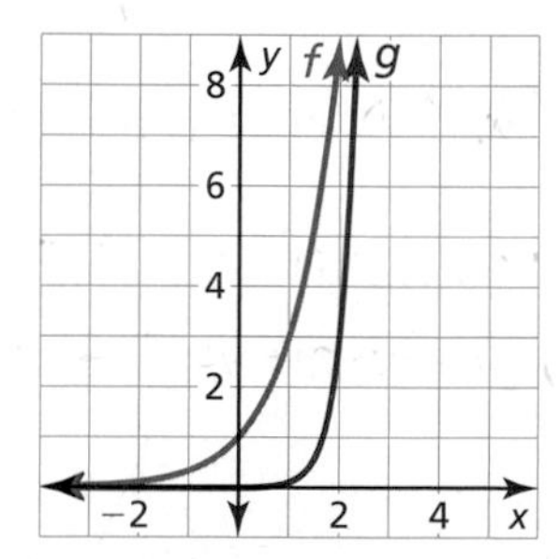

b. Notice that the function is of the form $g(x) = ae^{-x}$, where $a = -\frac{1}{8}$.

▶ So, the graph of g is a reflection in the x-axis and a vertical shrink by a factor of $\frac{1}{8}$ of the graph of f.

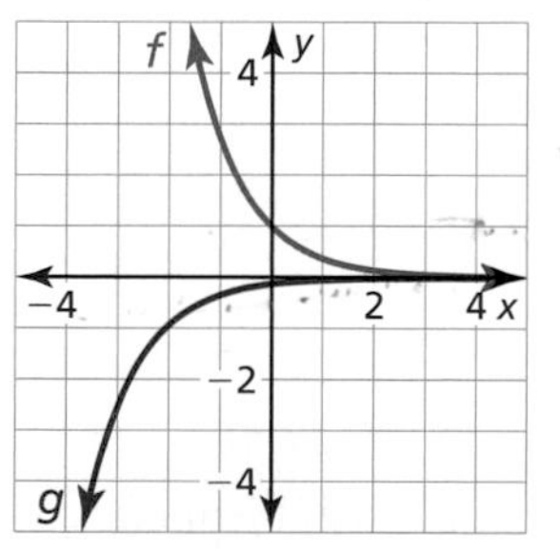

Monitoring Progress

Help in English and Spanish at *BigIdeasMath.com*

Describe the transformation of f represented by g. Then graph each function.

1. $f(x) = 2^x$, $g(x) = 2^{x-3} + 1$
2. $f(x) = e^{-x}$, $g(x) = e^{-x} - 5$
3. $f(x) = 0.4^x$, $g(x) = 0.4^{-2x}$
4. $f(x) = e^x$, $g(x) = -e^{x+6}$

Transforming Graphs of Logarithmic Functions

Examples of transformations of the graph of $f(x) = \log x$ are shown below.

Core Concept

Transformation	$f(x)$ Notation	Examples	
Horizontal Translation Graph shifts left or right.	$f(x - h)$	$g(x) = \log(x - 4)$ $g(x) = \log(x + 7)$	4 units right 7 units left
Vertical Translation Graph shifts up or down.	$f(x) + k$	$g(x) = \log x + 3$ $g(x) = \log x - 1$	3 units up 1 unit down
Reflection Graph flips over x- or y-axis.	$f(-x)$ $-f(x)$	$g(x) = \log(-x)$ $g(x) = -\log x$	in the y-axis in the x-axis
Horizontal Stretch or Shrink Graph stretches away from or shrinks toward y-axis.	$f(ax)$	$g(x) = \log(4x)$ $g(x) = \log\left(\frac{1}{3}x\right)$	shrink by a factor of $\frac{1}{4}$ stretch by a factor of 3
Vertical Stretch or Shrink Graph stretches away from or shrinks toward x-axis.	$a \cdot f(x)$	$g(x) = 5 \log x$ $g(x) = \frac{2}{3} \log x$	stretch by a factor of 5 shrink by a factor of $\frac{2}{3}$

EXAMPLE 4 Transforming Logarithmic Functions

Describe the transformation of f represented by g. Then graph each function.

a. $f(x) = \log x$, $g(x) = \log\left(-\frac{1}{2}x\right)$ **b.** $f(x) = \log_{1/2} x$, $g(x) = 2 \log_{1/2}(x + 4)$

SOLUTION

a. Notice that the function is of the form $g(x) = \log(ax)$, where $a = -\frac{1}{2}$.

▶ So, the graph of g is a reflection in the y-axis and a horizontal stretch by a factor of 2 of the graph of f.

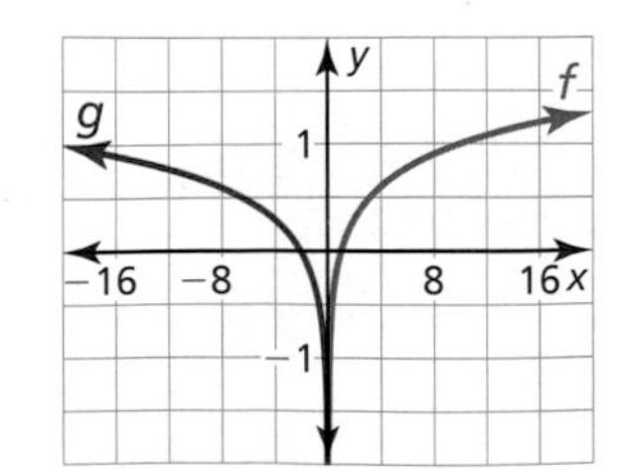

STUDY TIP

In Example 4(b), notice in the graph that the horizontal translation also shifted the asymptote 4 units left, so the domain of g is $x > -4$.

b. Notice that the function is of the form $g(x) = a \log_{1/2}(x - h)$, where $a = 2$ and $h = -4$.

▶ So, the graph of g is a horizontal translation 4 units left and a vertical stretch by a factor of 2 of the graph of f.

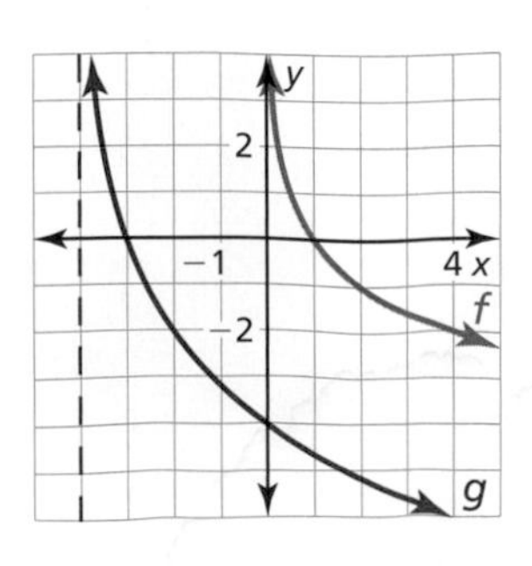

Monitoring Progress  Help in English and Spanish at BigIdeasMath.com

Describe the transformation of *f* represented by *g*. Then graph each function.

5. $f(x) = \log_2 x$, $g(x) = -3 \log_2 x$

6. $f(x) = \log_{1/4} x$, $g(x) = \log_{1/4}(4x) - 5$

Writing Transformations of Graphs of Functions

EXAMPLE 5 Writing a Transformed Exponential Function

Let the graph of g be a reflection in the x-axis followed by a translation 4 units right of the graph of $f(x) = 2^x$. Write a rule for g.

SOLUTION

Check

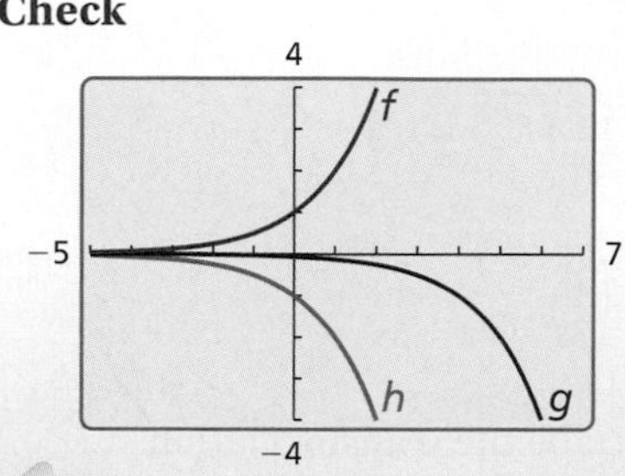

Step 1 First write a function h that represents the reflection of f.

$h(x) = -f(x)$ Multiply the output by −1.

$= -2^x$ Substitute 2^x for $f(x)$.

Step 2 Then write a function g that represents the translation of h.

$g(x) = h(x - 4)$ Subtract 4 from the input.

$= -2^{x-4}$ Replace x with $x - 4$ in $h(x)$.

▶ The transformed function is $g(x) = -2^{x-4}$.

EXAMPLE 6 Writing a Transformed Logarithmic Function

Let the graph of g be a translation 2 units up followed by a vertical stretch by a factor of 2 of the graph of $f(x) = \log_{1/3} x$. Write a rule for g.

SOLUTION

Check

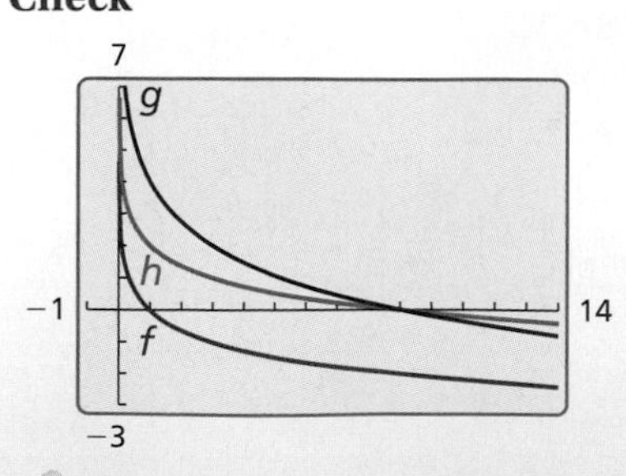

Step 1 First write a function h that represents the translation of f.

$h(x) = f(x) + 2$ Add 2 to the output.

$= \log_{1/3} x + 2$ Substitute $\log_{1/3} x$ for $f(x)$.

Step 2 Then write a function g that represents the vertical stretch of h.

$g(x) = 2 \cdot h(x)$ Multiply the output by 2.

$= 2 \cdot (\log_{1/3} x + 2)$ Substitute $\log_{1/3} x + 2$ for $h(x)$.

$= 2 \log_{1/3} x + 4$ Distributive Property

▶ The transformed function is $g(x) = 2 \log_{1/3} x + 4$.

Monitoring Progress  Help in English and Spanish at BigIdeasMath.com

7. Let the graph of g be a horizontal stretch by a factor of 3, followed by a translation 2 units up of the graph of $f(x) = e^{-x}$. Write a rule for g.

8. Let the graph of g be a reflection in the y-axis, followed by a translation 4 units to the left of the graph of $f(x) = \log x$. Write a rule for g.

6.4 Exercises

Dynamic Solutions available at *BigIdeasMath.com*

Vocabulary and Core Concept Check

1. **WRITING** Given the function $f(x) = ab^{x-h} + k$, describe the effects of a, h, and k on the graph of the function.

2. **COMPLETE THE SENTENCE** The graph of $g(x) = \log_4(-x)$ is a reflection in the __________ of the graph of $f(x) = \log_4 x$.

Monitoring Progress and Modeling with Mathematics

In Exercises 3–6, match the function with its graph. Explain your reasoning.

3. $f(x) = 2^{x+2} - 2$

4. $g(x) = 2^{x+2} + 2$

5. $h(x) = 2^{x-2} - 2$

6. $k(x) = 2^{x-2} + 2$

A.

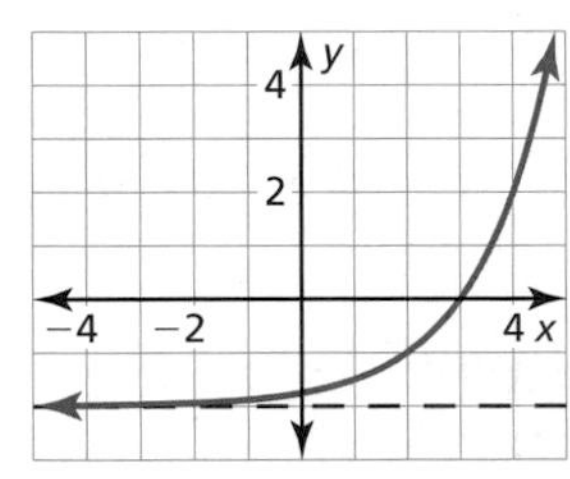

B.

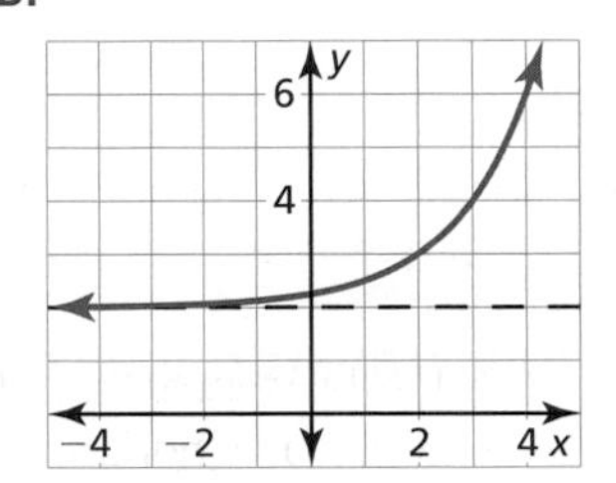

C.

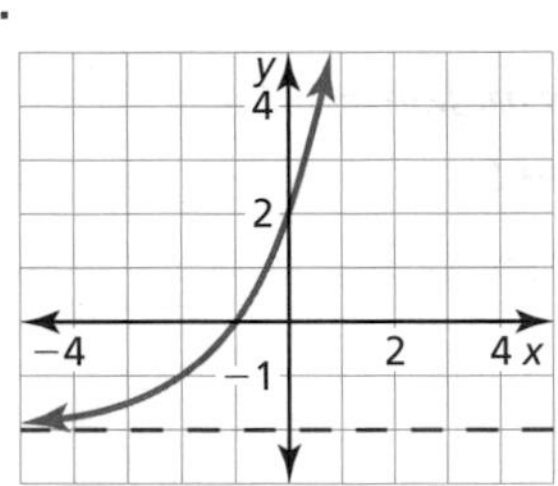

D.

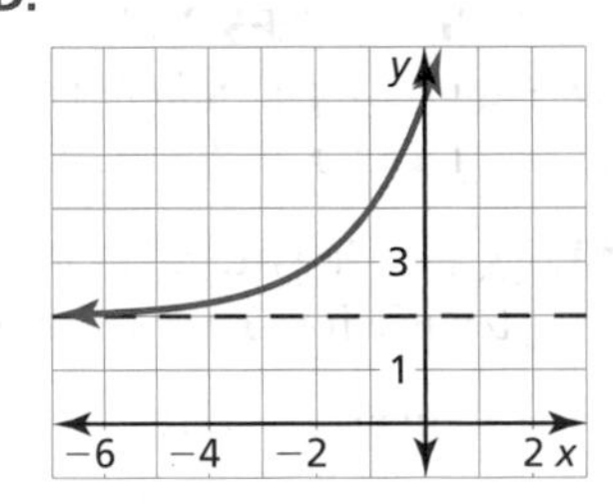

In Exercises 7–16, describe the transformation of f represented by g. Then graph each function. *(See Examples 1 and 2.)*

7. $f(x) = 3^x$, $g(x) = 3^x + 5$

8. $f(x) = 4^x$, $g(x) = 4^x - 8$

9. $f(x) = e^x$, $g(x) = e^x - 1$

10. $f(x) = e^x$, $g(x) = e^x + 4$

11. $f(x) = 2^x$, $g(x) = 2^{x-7}$

12. $f(x) = 5^x$, $g(x) = 5^{x+1}$

13. $f(x) = e^{-x}$, $g(x) = e^{-x} + 6$

14. $f(x) = e^{-x}$, $g(x) = e^{-x} - 9$

15. $f(x) = \left(\frac{1}{4}\right)^x$, $g(x) = \left(\frac{1}{4}\right)^{x-3} + 12$

16. $f(x) = \left(\frac{1}{3}\right)^x$, $g(x) = \left(\frac{1}{3}\right)^{x+2} - \frac{2}{3}$

In Exercises 17–24, describe the transformation of f represented by g. Then graph each function. *(See Example 3.)*

17. $f(x) = e^x$, $g(x) = e^{2x}$

18. $f(x) = e^x$, $g(x) = \frac{4}{3}e^x$

19. $f(x) = 2^x$, $g(x) = -2^{x-3}$

20. $f(x) = 4^x$, $g(x) = 4^{0.5x-5}$

21. $f(x) = e^{-x}$, $g(x) = 3e^{-6x}$

22. $f(x) = e^{-x}$, $g(x) = e^{-5x} + 2$

23. $f(x) = \left(\frac{1}{2}\right)^x$, $g(x) = 6\left(\frac{1}{2}\right)^{x+5} - 2$

24. $f(x) = \left(\frac{3}{4}\right)^x$, $g(x) = -\left(\frac{3}{4}\right)^{x-7} + 1$

ERROR ANALYSIS In Exercises 25 and 26, describe and correct the error in graphing the function.

25. $f(x) = 2^x + 3$

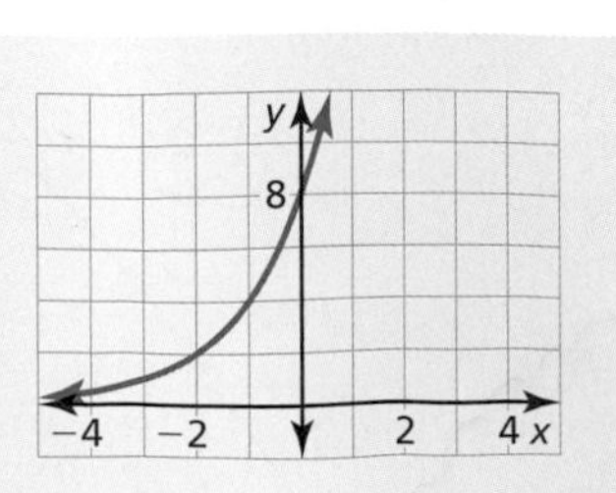

26. $f(x) = 3^{-x}$

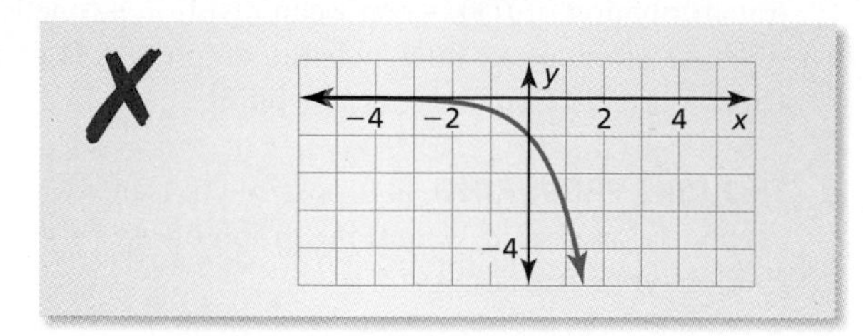

In Exercises 27–30, describe the transformation of *f* represented by *g*. Then graph each function. *(See Example 4.)*

27. $f(x) = \log_4 x,\ g(x) = 3\log_4 x - 5$

28. $f(x) = \log_{1/3} x,\ g(x) = \log_{1/3}(-x) + 6$

29. $f(x) = \log_{1/5} x,\ g(x) = -\log_{1/5}(x - 7)$

30. $f(x) = \log_2 x,\ g(x) = \log_2(x + 2) - 3$

ANALYZING RELATIONSHIPS In Exercises 31–34, match the function with the correct transformation of the graph of *f*. Explain your reasoning.

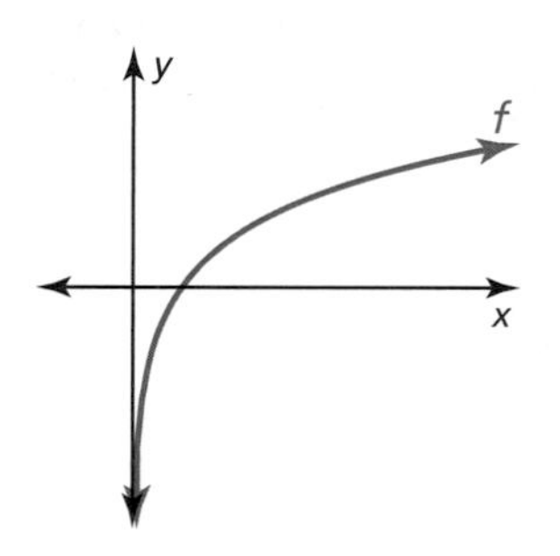

31. $y = f(x - 2)$

32. $y = f(x + 2)$

33. $y = 2f(x)$

34. $y = f(2x)$

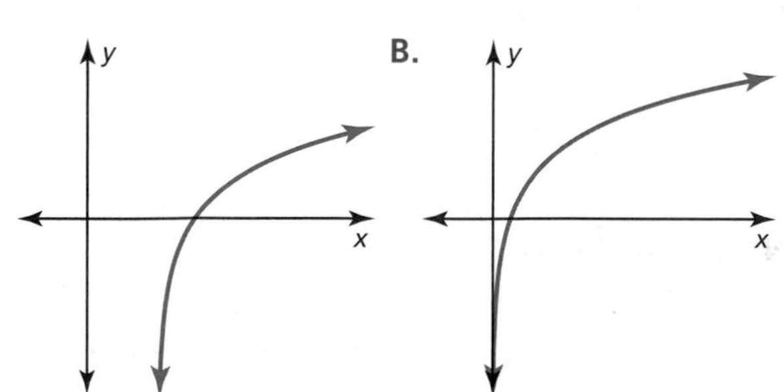

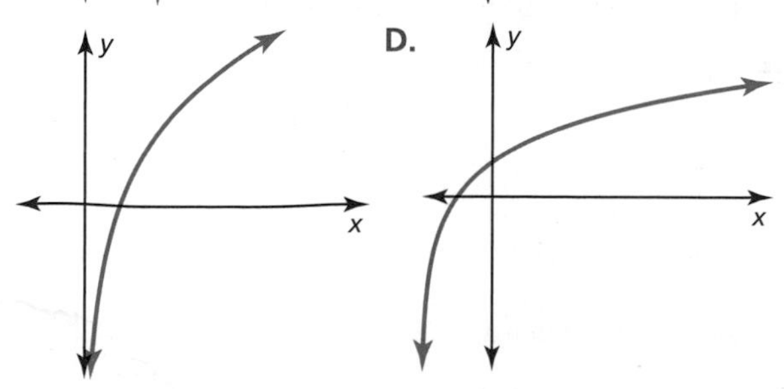

In Exercises 35–38, write a rule for *g* that represents the indicated transformations of the graph of *f*. *(See Example 5.)*

35. $f(x) = 5^x$; translation 2 units down, followed by a reflection in the y-axis

36. $f(x) = \left(\frac{2}{3}\right)^x$; reflection in the x-axis, followed by a vertical stretch by a factor of 6 and a translation 4 units left

37. $f(x) = e^x$; horizontal shrink by a factor of $\frac{1}{2}$, followed by a translation 5 units up

38. $f(x) = e^{-x}$; translation 4 units right and 1 unit down, followed by a vertical shrink by a factor of $\frac{1}{3}$

In Exercises 39–42, write a rule for *g* that represents the indicated transformation of the graph of *f*. *(See Example 6.)*

39. $f(x) = \log_6 x$; vertical stretch by a factor of 6, followed by a translation 5 units down

40. $f(x) = \log_5 x$; reflection in the x-axis, followed by a translation 9 units left

41. $f(x) = \log_{1/2} x$; translation 3 units left and 2 units up, followed by a reflection in the y-axis

42. $f(x) = \ln x$; translation 3 units right and 1 unit up, followed by a horizontal stretch by a factor of 8

JUSTIFYING STEPS In Exercises 43 and 44, justify each step in writing a rule for *g* that represents the indicated transformations of the graph of *f*.

43. $f(x) = \log_7 x$; reflection in the x-axis, followed by a translation 6 units down

$$h(x) = -f(x)$$
$$= -\log_7 x$$
$$g(x) = h(x) - 6$$
$$= -\log_7 x - 6$$

44. $f(x) = 8^x$; vertical stretch by a factor of 4, followed by a translation 1 unit up and 3 units left

$$h(x) = 4 \cdot f(x)$$
$$= 4 \cdot 8^x$$
$$g(x) = h(x + 3) + 1$$
$$= 4 \cdot 8^{x+3} + 1$$

USING STRUCTURE In Exercises 45–48, describe the transformation of the graph of f represented by the graph of g. Then give an equation of the asymptote.

45. $f(x) = e^x$, $g(x) = e^x + 4$

46. $f(x) = 3^x$, $g(x) = 3^{x-9}$

47. $f(x) = \ln x$, $g(x) = \ln(x + 6)$

48. $f(x) = \log_{1/5} x$, $g(x) = \log_{1/5} x + 13$

49. MODELING WITH MATHEMATICS The slope S of a beach is related to the average diameter d (in millimeters) of the sand particles on the beach by the equation $S = 0.159 + 0.118 \log d$. Describe the transformation of $f(d) = \log d$ represented by S. Then use the function to determine the slope of a beach for each sand type below.

Sand particle	Diameter (mm), d
fine sand	0.125
medium sand	0.25
coarse sand	0.5
very coarse sand	1

50. HOW DO YOU SEE IT? The graphs of $f(x) = b^x$ and $g(x) = \left(\frac{1}{b}\right)^x$ are shown for $b = 2$.

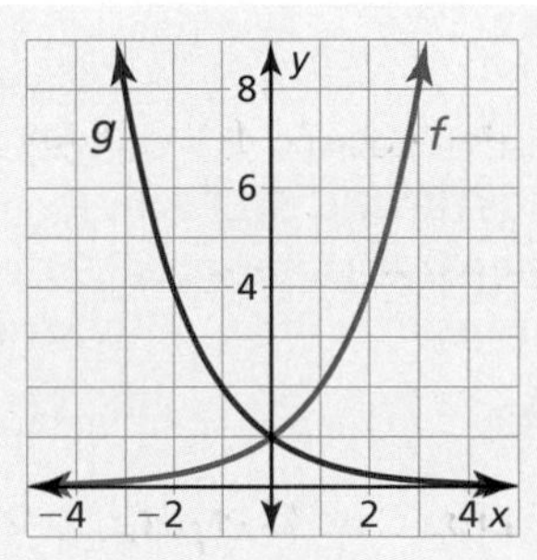

a. Use the graph to describe a transformation of the graph of f that results in the graph of g.

b. Does your answer in part (a) change when $0 < b < 1$? Explain.

51. MAKING AN ARGUMENT Your friend claims a single transformation of $f(x) = \log x$ can result in a function g whose graph never intersects the graph of f. Is your friend correct? Explain your reasoning.

52. THOUGHT PROVOKING Is it possible to transform the graph of $f(x) = e^x$ to obtain the graph of $g(x) = \ln x$? Explain your reasoning.

53. ABSTRACT REASONING Determine whether each statement is *always*, *sometimes*, or *never* true. Explain your reasoning.

a. A vertical translation of the graph of $f(x) = \log x$ changes the equation of the asymptote.

b. A vertical translation of the graph of $f(x) = e^x$ changes the equation of the asymptote.

c. A horizontal shrink of the graph of $f(x) = \log x$ does not change the domain.

d. The graph of $g(x) = ab^{x-h} + k$ does not intersect the x-axis.

54. PROBLEM SOLVING The amount P (in grams) of 100 grams of plutonium-239 that remains after t years can be modeled by $P = 100(0.99997)^t$.

a. Describe the domain and range of the function.

b. How much plutonium-239 is present after 12,000 years?

c. Describe the transformation of the function if the initial amount of plutonium were 550 grams.

d. Does the transformation in part (c) affect the domain and range of the function? Explain your reasoning.

55. CRITICAL THINKING Consider the graph of the function $h(x) = e^{-x-2}$. Describe the transformation of the graph of $f(x) = e^{-x}$ represented by the graph of h. Then describe the transformation of the graph of $g(x) = e^x$ represented by the graph of h. Justify your answers.

56. OPEN-ENDED Write a function of the form $y = ab^{x-h} + k$ whose graph has a y-intercept of 5 and an asymptote of $y = 2$.

Maintaining Mathematical Proficiency

Reviewing what you learned in previous grades and lessons

Perform the indicated operation. *(Section 5.5)*

57. Let $f(x) = x^4$ and $g(x) = x^2$. Find $(fg)(x)$. Then evaluate the product when $x = 3$.

58. Let $f(x) = 4x^6$ and $g(x) = 2x^3$. Find $\left(\frac{f}{g}\right)(x)$. Then evaluate the quotient when $x = 5$.

59. Let $f(x) = 6x^3$ and $g(x) = 8x^3$. Find $(f + g)(x)$. Then evaluate the sum when $x = 2$.

60. Let $f(x) = 2x^2$ and $g(x) = 3x^2$. Find $(f - g)(x)$. Then evaluate the difference when $x = 6$.

6.1–6.4 What Did You Learn?

Core Vocabulary

exponential function, *p. 296*
exponential growth function, *p. 296*
growth factor, *p. 296*
asymptote, *p. 296*
exponential decay function, *p. 296*
decay factor, *p. 296*
natural base e, *p. 304*
logarithm of y with base b, *p. 310*
common logarithm, *p. 311*
natural logarithm, *p. 311*

Core Concepts

Section 6.1

Parent Function for Exponential Growth Functions, *p. 296*
Parent Function for Exponential Decay Functions, *p. 296*
Exponential Growth and Decay Models, *p. 297*
Compound Interest, *p. 299*

Section 6.2

The Natural Base e, *p. 304*
Natural Base Functions, *p. 305*
Continuously Compounded Interest, *p. 306*

Section 6.3

Definition of Logarithm with Base b, *p. 310*
Parent Graphs for Logarithmic Functions, *p. 313*

Section 6.4

Transforming Graphs of Exponential Functions, *p. 318*
Transforming Graphs of Logarithmic Functions, *p. 320*

Mathematical Practices

1. How did you check to make sure your answer was reasonable in Exercise 23 on page 300?
2. How can you justify your conclusions in Exercises 23–26 on page 307?
3. How did you monitor and evaluate your progress in Exercise 66 on page 315?

Study Skills

Forming a Weekly Study Group

- Select students who are just as dedicated to doing well in the math class as you are.
- Find a regular meeting place that has minimal distractions.
- Compare schedules and plan at least one time a week to meet, allowing at least 1.5 hours for study time.

6.1–6.4 Quiz

Tell whether the function represents *exponential growth* or *exponential decay*. Explain your reasoning. *(Sections 6.1 and 6.2)*

1. $f(x) = (4.25)^x$
2. $y = \left(\frac{3}{8}\right)^x$
3. $y = e^{0.6x}$
4. $f(x) = 5e^{-2x}$

Simplify the expression. *(Sections 6.2 and 6.3)*

5. $e^8 \cdot e^4$
6. $\frac{15e^3}{3e}$
7. $(5e^{4x})^3$
8. $e^{\ln 9}$
9. $\log_7 49^x$
10. $\log_3 81^{-2x}$

Rewrite the expression in exponential or logarithmic form. *(Section 6.3)*

11. $\log_4 1024 = 5$
12. $\log_{1/3} 27 = -3$
13. $7^4 = 2401$
14. $4^{-2} = 0.0625$

Evaluate the logarithm. If necessary, use a calculator and round your answer to three decimal places. *(Section 6.3)*

15. $\log 45$
16. $\ln 1.4$
17. $\log_2 32$

Graph the function and its inverse. *(Section 6.3)*

18. $f(x) = \left(\frac{1}{9}\right)^x$
19. $y = \ln(x - 7)$
20. $f(x) = \log_5(x + 1)$

The graph of g is a transformation of the graph of f. Write a rule for g. *(Section 6.4)*

21. $f(x) = \log_3 x$

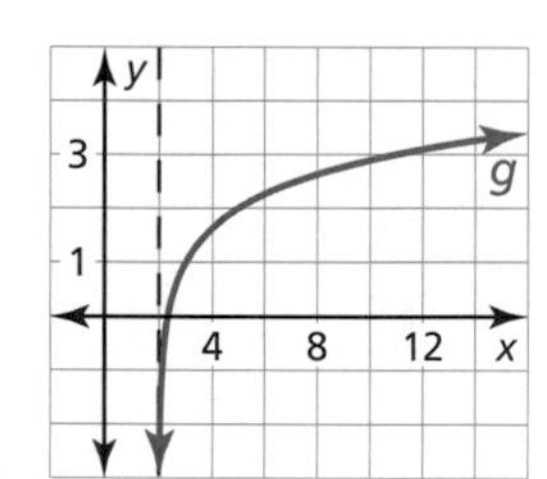

22. $f(x) = 3^x$

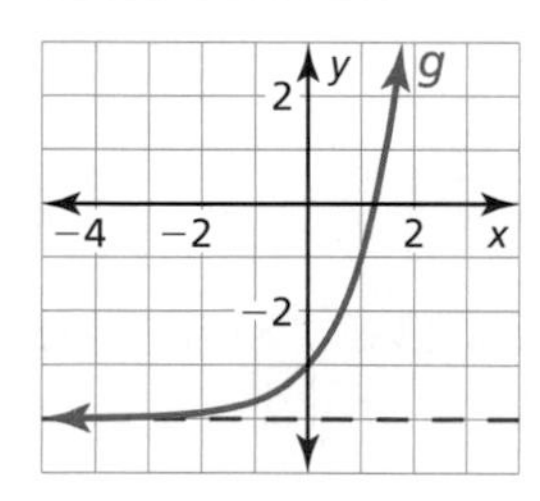

23. $f(x) = \log_{1/2} x$

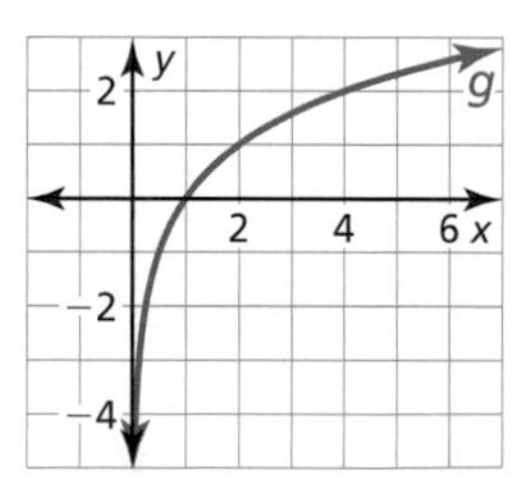

24. You purchase an antique lamp for $150. The value of the lamp increases by 2.15% each year. Write an exponential model that gives the value y (in dollars) of the lamp t years after you purchased it. *(Section 6.1)*

25. A local bank advertises two certificate of deposit (CD) accounts that you can use to save money and earn interest. The interest is compounded monthly for both accounts. *(Section 6.1)*

    a. You deposit the minimum required amounts in each CD account. How much money is in each account at the end of its term? How much interest does each account earn? Justify your answers.

    b. Describe the benefits and drawbacks of each account.

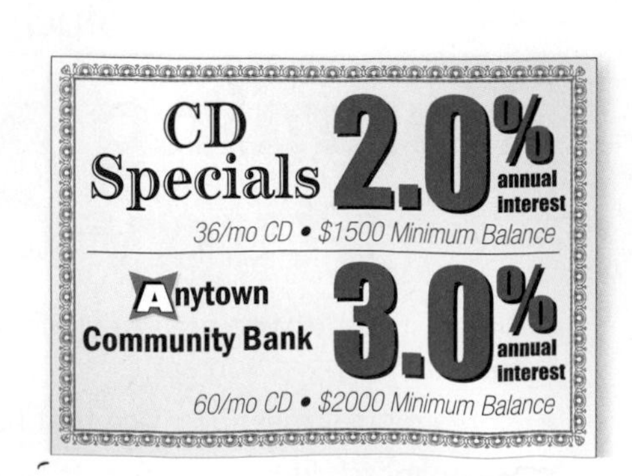

26. The Richter scale is used for measuring the magnitude of an earthquake. The Richter magnitude R is given by $R = 0.67 \ln E + 1.17$, where E is the energy (in kilowatt-hours) released by the earthquake. Graph the model. What is the Richter magnitude for an earthquake that releases 23,000 kilowatt-hours of energy? *(Section 6.4)*

6.5 Properties of Logarithms

Essential Question

How can you use properties of exponents to derive properties of logarithms?

Let

$x = \log_b m$ and $y = \log_b n$.

The corresponding exponential forms of these two equations are

$b^x = m$ and $b^y = n$.

CONSTRUCTING VIABLE ARGUMENTS

To be proficient in math, you need to understand and use stated assumptions, definitions, and previously established results.

EXPLORATION 1 Product Property of Logarithms

Work with a partner. To derive the Product Property, multiply m and n to obtain

$mn = b^x b^y = b^{x+y}$.

The corresponding logarithmic form of $mn = b^{x+y}$ is $\log_b mn = x + y$. So,

$\log_b mn = $ ______. Product Property of Logarithms

EXPLORATION 2 Quotient Property of Logarithms

Work with a partner. To derive the Quotient Property, divide m by n to obtain

$$\frac{m}{n} = \frac{b^x}{b^y} = b^{x-y}.$$

The corresponding logarithmic form of $\frac{m}{n} = b^{x-y}$ is $\log_b \frac{m}{n} = x - y$. So,

$\log_b \frac{m}{n} = $ ______. Quotient Property of Logarithms

EXPLORATION 3 Power Property of Logarithms

Work with a partner. To derive the Power Property, substitute b^x for m in the expression $\log_b m^n$, as follows.

$\log_b m^n = \log_b(b^x)^n$	Substitute b^x for m.
$= \log_b b^{nx}$	Power of a Power Property of Exponents
$= nx$	Inverse Property of Logarithms

So, substituting $\log_b m$ for x, you have

$\log_b m^n = $ ______. Power Property of Logarithms

Communicate Your Answer

4. How can you use properties of exponents to derive properties of logarithms?

5. Use the properties of logarithms that you derived in Explorations 1–3 to evaluate each logarithmic expression.

a. $\log_4 16^3$

b. $\log_3 81^{-3}$

c. $\ln e^2 + \ln e^5$

d. $2 \ln e^6 - \ln e^5$

e. $\log_5 75 - \log_5 3$

f. $\log_4 2 + \log_4 32$

6.5 Lesson

Core Vocabulary

Previous
base
properties of exponents

What You Will Learn

- Use the properties of logarithms to evaluate logarithms.
- Use the properties of logarithms to expand or condense logarithmic expressions.
- Use the change-of-base formula to evaluate logarithms.

Properties of Logarithms

You know that the logarithmic function with base b is the inverse function of the exponential function with base b. Because of this relationship, it makes sense that logarithms have properties similar to properties of exponents.

Core Concept

Properties of Logarithms

Let b, m, and n be positive real numbers with $b \neq 1$.

Product Property $\log_b mn = \log_b m + \log_b n$

Quotient Property $\log_b \dfrac{m}{n} = \log_b m - \log_b n$

Power Property $\log_b m^n = n \log_b m$

STUDY TIP

These three properties of logarithms correspond to these three properties of exponents.

$a^m a^n = a^{m+n}$

$\dfrac{a^m}{a^n} = a^{m-n}$

$(a^m)^n = a^{mn}$

EXAMPLE 1 Using Properties of Logarithms

Use $\log_2 3 \approx 1.585$ and $\log_2 7 \approx 2.807$ to evaluate each logarithm.

a. $\log_2 \frac{3}{7}$ **b.** $\log_2 21$ **c.** $\log_2 49$

SOLUTION

a. $\log_2 \frac{3}{7} = \log_2 3 - \log_2 7$ Quotient Property

$\approx 1.585 - 2.807$ Use the given values of $\log_2 3$ and $\log_2 7$.

$= -1.222$ Subtract.

b. $\log_2 21 = \log_2(3 \cdot 7)$ Write 21 as $3 \cdot 7$.

$= \log_2 3 + \log_2 7$ Product Property

$\approx 1.585 + 2.807$ Use the given values of $\log_2 3$ and $\log_2 7$.

$= 4.392$ Add.

c. $\log_2 49 = \log_2 7^2$ Write 49 as 7^2.

$= 2 \log_2 7$ Power Property

$\approx 2(2.807)$ Use the given value $\log_2 7$.

$= 5.614$ Multiply.

COMMON ERROR

Note that in general $\log_b \dfrac{m}{n} \neq \dfrac{\log_b m}{\log_b n}$ and $\log_b mn \neq (\log_b m)(\log_b n)$.

Monitoring Progress

Help in English and Spanish at *BigIdeasMath.com*

Use $\log_6 5 \approx 0.898$ and $\log_6 8 \approx 1.161$ to evaluate the logarithm.

1. $\log_6 \dfrac{5}{8}$ **2.** $\log_6 40$ **3.** $\log_6 64$ **4.** $\log_6 125$

Rewriting Logarithmic Expressions

You can use the properties of logarithms to expand and condense logarithmic expressions.

EXAMPLE 2 Expanding a Logarithmic Expression

Expand $\ln \frac{5x^7}{y}$.

> **STUDY TIP**
> When you are expanding or condensing an expression involving logarithms, you can assume that any variables are positive.

SOLUTION

$$\ln \frac{5x^7}{y} = \ln 5x^7 - \ln y \qquad \text{Quotient Property}$$

$$= \ln 5 + \ln x^7 - \ln y \qquad \text{Product Property}$$

$$= \ln 5 + 7 \ln x - \ln y \qquad \text{Power Property}$$

EXAMPLE 3 Condensing a Logarithmic Expression

Condense $\log 9 + 3 \log 2 - \log 3$.

SOLUTION

$$\log 9 + 3 \log 2 - \log 3 = \log 9 + \log 2^3 - \log 3 \qquad \text{Power Property}$$

$$= \log(9 \cdot 2^3) - \log 3 \qquad \text{Product Property}$$

$$= \log \frac{9 \cdot 2^3}{3} \qquad \text{Quotient Property}$$

$$= \log 24 \qquad \text{Simplify.}$$

Monitoring Progress

Help in English and Spanish at *BigIdeasMath.com*

Expand the logarithmic expression.

5. $\log_6 3x^4$

6. $\ln \frac{5}{12x}$

Condense the logarithmic expression.

7. $\log x - \log 9$

8. $\ln 4 + 3 \ln 3 - \ln 12$

Change-of-Base Formula

Logarithms with any base other than 10 or e can be written in terms of common or natural logarithms using the *change-of-base formula*. This allows you to evaluate any logarithm using a calculator.

Core Concept

Change-of-Base Formula

If a, b, and c are positive real numbers with $b \neq 1$ and $c \neq 1$, then

$$\log_c a = \frac{\log_b a}{\log_b c}.$$

In particular, $\log_c a = \frac{\log a}{\log c}$ and $\log_c a = \frac{\ln a}{\ln c}$.

EXAMPLE 4 Changing a Base Using Common Logarithms

Evaluate $\log_3 8$ using common logarithms.

SOLUTION

$$\log_3 8 = \frac{\log 8}{\log 3} \qquad \log_c a = \frac{\log a}{\log c}$$

$$\approx \frac{0.9031}{0.4771} \approx 1.893 \qquad \text{Use a calculator. Then divide.}$$

ANOTHER WAY

In Example 4, $\log_3 8$ can be evaluated using natural logarithms.

$\log_3 8 = \frac{\ln 8}{\ln 3} \approx 1.893$

Notice that you get the same answer whether you use natural logarithms or common logarithms in the change-of-base formula.

EXAMPLE 5 Changing a Base Using Natural Logarithms

Evaluate $\log_6 24$ using natural logarithms.

SOLUTION

$$\log_6 24 = \frac{\ln 24}{\ln 6} \qquad \log_c a = \frac{\ln a}{\ln c}$$

$$\approx \frac{3.1781}{1.7918} \approx 1.774 \qquad \text{Use a calculator. Then divide.}$$

EXAMPLE 6 Solving a Real-Life Problem

For a sound with intensity I (in watts per square meter), the loudness $L(I)$ of the sound (in decibels) is given by the function

$$L(I) = 10 \log \frac{I}{I_0}$$

where I_0 is the intensity of a barely audible sound (about 10^{-12} watts per square meter). An artist in a recording studio turns up the volume of a track so that the intensity of the sound doubles. By how many decibels does the loudness increase?

SOLUTION

Let I be the original intensity, so that $2I$ is the doubled intensity.

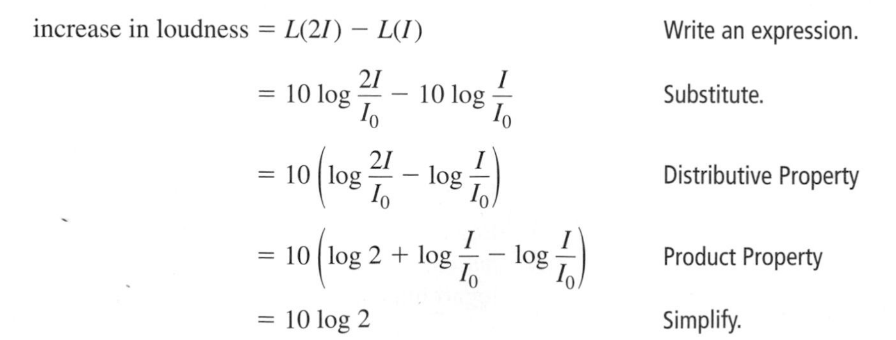

$$\text{increase in loudness} = L(2I) - L(I) \qquad \text{Write an expression.}$$

$$= 10 \log \frac{2I}{I_0} - 10 \log \frac{I}{I_0} \qquad \text{Substitute.}$$

$$= 10\left(\log \frac{2I}{I_0} - \log \frac{I}{I_0}\right) \qquad \text{Distributive Property}$$

$$= 10\left(\log 2 + \log \frac{I}{I_0} - \log \frac{I}{I_0}\right) \qquad \text{Product Property}$$

$$= 10 \log 2 \qquad \text{Simplify.}$$

▶ The loudness increases by 10 log 2 decibels, or about 3 decibels.

Monitoring Progress

Help in English and Spanish at *BigIdeasMath.com*

Use the change-of-base formula to evaluate the logarithm.

9. $\log_5 8$ **10.** $\log_8 14$ **11.** $\log_{26} 9$ **12.** $\log_{12} 30$

13. WHAT IF? In Example 6, the artist turns up the volume so that the intensity of the sound triples. By how many decibels does the loudness increase?

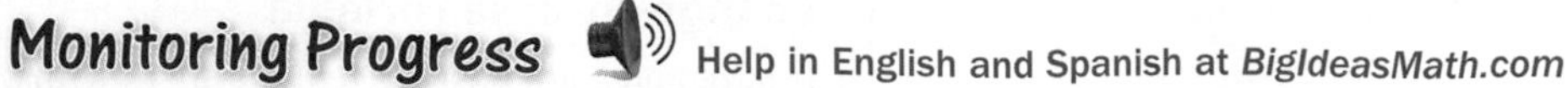

6.5 Exercises

Dynamic Solutions available at *BigIdeasMath.com*

Vocabulary and Core Concept Check

1. **COMPLETE THE SENTENCE** To condense the expression $\log_3 2x + \log_3 y$, you need to use the __________ Property of Logarithms.

2. **WRITING** Describe two ways to evaluate $\log_7 12$ using a calculator.

Monitoring Progress and Modeling with Mathematics

In Exercises 3–8, use $\log_7 4 \approx 0.712$ and $\log_7 12 \approx 1.277$ to evaluate the logarithm. *(See Example 1.)*

3. $\log_7 3$
4. $\log_7 48$
5. $\log_7 16$
6. $\log_7 64$
7. $\log_7 \frac{1}{4}$
8. $\log_7 \frac{1}{3}$

In Exercises 9–12, match the expression with the logarithm that has the same value. Justify your answer.

9. $\log_3 6 - \log_3 2$
10. $2 \log_3 6$
11. $6 \log_3 2$
12. $\log_3 6 + \log_3 2$

A. $\log_3 64$

B. $\log_3 3$

C. $\log_3 12$

D. $\log_3 36$

In Exercises 13–20, expand the logarithmic expression. *(See Example 2.)*

13. $\log_3 4x$
14. $\log_8 3x$
15. $\log 10x^5$
16. $\ln 3x^4$
17. $\ln \frac{x}{3y}$
18. $\ln \frac{6x^2}{y^4}$
19. $\log_7 5\sqrt{x}$
20. $\log_5 \sqrt[3]{x^2 y}$

ERROR ANALYSIS In Exercises 21 and 22, describe and correct the error in expanding the logarithmic expression.

21.

$\log_2 5x = (\log_2 5)(\log_2 x)$

22.

$\ln 8x^3 = 3 \ln 8 + \ln x$

In Exercises 23–30, condense the logarithmic expression. *(See Example 3.)*

23. $\log_4 7 - \log_4 10$
24. $\ln 12 - \ln 4$
25. $6 \ln x + 4 \ln y$
26. $2 \log x + \log 11$
27. $\log_5 4 + \frac{1}{3} \log_5 x$
28. $6 \ln 2 - 4 \ln y$
29. $5 \ln 2 + 7 \ln x + 4 \ln y$
30. $\log_3 4 + 2 \log_3 \frac{1}{2} + \log_3 x$

31. **REASONING** Which of the following is *not* equivalent to $\log_5 \frac{y^4}{3x}$? Justify your answer.

Ⓐ $4 \log_5 y - \log_5 3x$

Ⓑ $4 \log_5 y - \log_5 3 + \log_5 x$

Ⓒ $4 \log_5 y - \log_5 3 - \log_5 x$

Ⓓ $\log_5 y^4 - \log_5 3 - \log_5 x$

32. **REASONING** Which of the following equations is correct? Justify your answer.

Ⓐ $\log_7 x + 2 \log_7 y = \log_7(x + y^2)$

Ⓑ $9 \log x - 2 \log y = \log \frac{x^9}{y^2}$

Ⓒ $5 \log_4 x + 7 \log_2 y = \log_6 x^5 y^7$

Ⓓ $\log_9 x - 5 \log_9 y = \log_9 \frac{x}{5y}$

In Exercises 33–40, use the change-of-base formula to evaluate the logarithm. *(See Examples 4 and 5.)*

33. $\log_4 7$

34. $\log_5 13$

35. $\log_9 15$

36. $\log_8 22$

37. $\log_6 17$

38. $\log_2 28$

39. $\log_7 \frac{3}{16}$

40. $\log_3 \frac{9}{40}$

41. MAKING AN ARGUMENT Your friend claims you can use the change-of-base formula to graph $y = \log_3 x$ using a graphing calculator. Is your friend correct? Explain your reasoning.

42. HOW DO YOU SEE IT? Use the graph to determine the value of $\frac{\log 8}{\log 2}$.

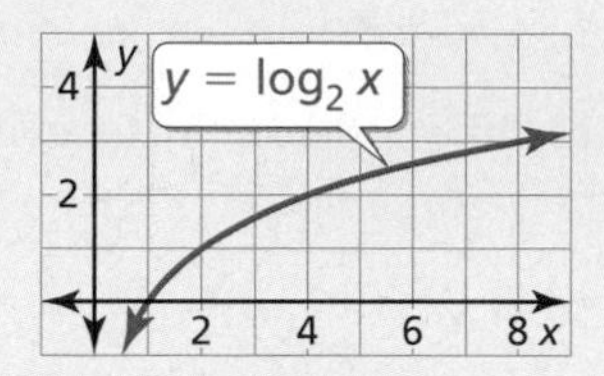

MODELING WITH MATHEMATICS In Exercises 43 and 44, use the function $L(I)$ given in Example 6.

43. The blue whale can produce sound with an intensity that is 1 million times greater than the intensity of the loudest sound a human can make. Find the difference in the decibel levels of the sounds made by a blue whale and a human. *(See Example 6.)*

44. The intensity of the sound of a certain television advertisement is 10 times greater than the intensity of the television program. By how many decibels does the loudness increase?

Intensity of Television Sound

During show:
Intensity = I

During ad:
Intensity = $10I$

45. REWRITING A FORMULA Under certain conditions, the wind speed s (in knots) at an altitude of h meters above a grassy plain can be modeled by the function

$$s(h) = 2 \ln 100h.$$

a. By what amount does the wind speed increase when the altitude doubles?

b. Show that the given function can be written in terms of common logarithms as

$$s(h) = \frac{2}{\log e}(\log h + 2).$$

46. THOUGHT PROVOKING Determine whether the formula

$$\log_b(M + N) = \log_b M + \log_b N$$

is true for all positive, real values of M, N, and b (with $b \neq 1$). Justify your answer.

47. USING STRUCTURE Use the properties of exponents to prove the change-of-base formula. (*Hint:* Let $x = \log_b a$, $y = \log_b c$, and $z = \log_c a$.)

48. CRITICAL THINKING Describe *three* ways to transform the graph of $f(x) = \log x$ to obtain the graph of $g(x) = \log 100x - 1$. Justify your answers.

Maintaining Mathematical Proficiency

Reviewing what you learned in previous grades and lessons

Solve the inequality by graphing. *(Section 3.6)*

49. $x^2 - 4 > 0$

50. $2(x - 6)^2 - 5 \geq 37$

51. $x^2 + 13x + 42 < 0$

52. $-x^2 - 4x + 6 \leq -6$

Solve the equation by graphing the related system of equations. *(Section 3.5)*

53. $4x^2 - 3x - 6 = -x^2 + 5x + 3$

54. $-(x + 3)(x - 2) = x^2 - 6x$

55. $2x^2 - 4x - 5 = -(x + 3)^2 + 10$

56. $-(x + 7)^2 + 5 = (x + 10)^2 - 3$

6.6 Solving Exponential and Logarithmic Equations

Essential Question How can you solve exponential and logarithmic equations?

EXPLORATION 1 **Solving Exponential and Logarithmic Equations**

Work with a partner. Match each equation with the graph of its related system of equations. Explain your reasoning. Then use the graph to solve the equation.

a. $e^x = 2$ **b.** $\ln x = -1$

c. $2^x = 3^{-x}$ **d.** $\log_4 x = 1$

e. $\log_5 x = \frac{1}{2}$ **f.** $4^x = 2$

A.
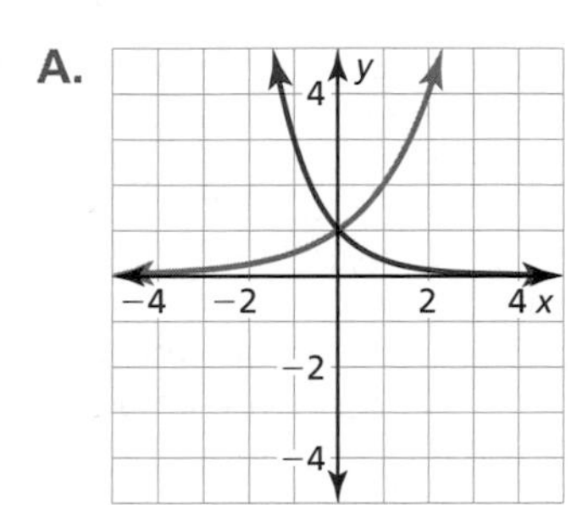

B.
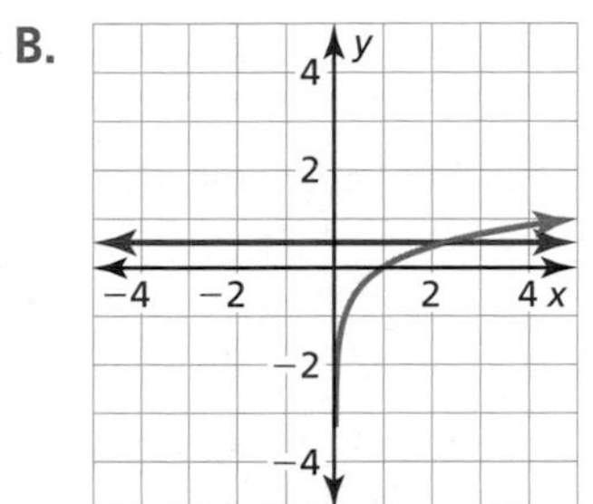

C.
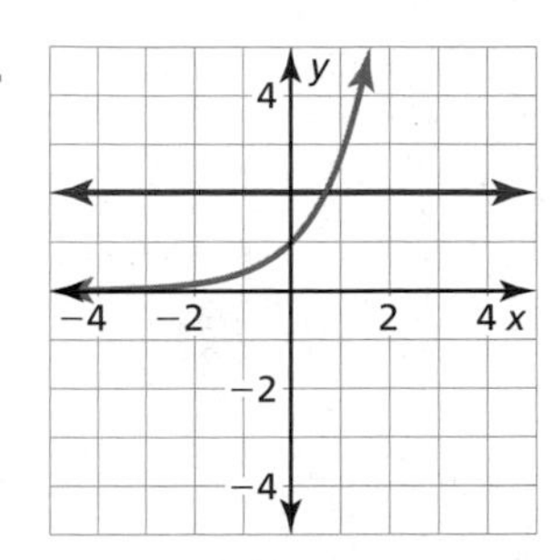

D.
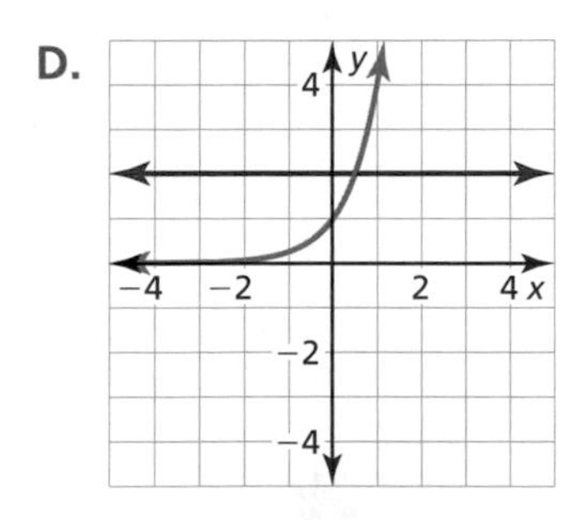

E.
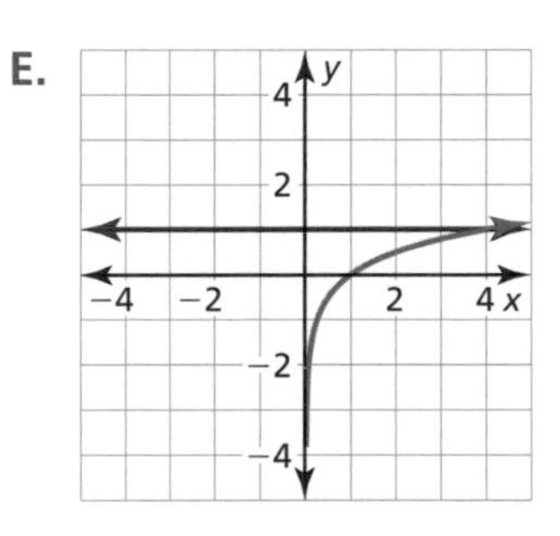

F.
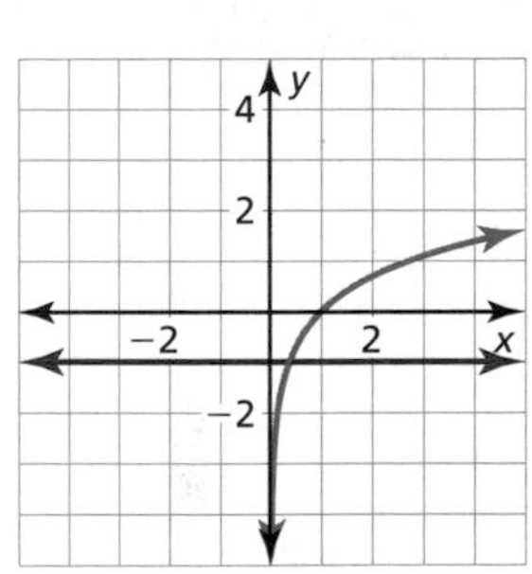

MAKING SENSE OF PROBLEMS

To be proficient in math, you need to plan a solution pathway rather than simply jumping into a solution attempt.

EXPLORATION 2 **Solving Exponential and Logarithmic Equations**

Work with a partner. Look back at the equations in Explorations 1(a) and 1(b). Suppose you want a more accurate way to solve the equations than using a graphical approach.

a. Show how you could use a *numerical approach* by creating a table. For instance, you might use a spreadsheet to solve the equations.

b. Show how you could use an *analytical approach*. For instance, you might try solving the equations by using the inverse properties of exponents and logarithms.

Communicate Your Answer

3. How can you solve exponential and logarithmic equations?

4. Solve each equation using any method. Explain your choice of method.

a. $16^x = 2$ **b.** $2^x = 4^{2x+1}$

c. $2^x = 3^{x+1}$ **d.** $\log x = \frac{1}{2}$

e. $\ln x = 2$ **f.** $\log_3 x = \frac{3}{2}$

6.6 Lesson

Core Vocabulary

exponential equations, *p. 334*
logarithmic equations, *p. 335*

Previous
extraneous solution
inequality

What You Will Learn

- Solve exponential equations.
- Solve logarithmic equations.
- Solve exponential and logarithmic inequalities.

Solving Exponential Equations

Exponential equations are equations in which variable expressions occur as exponents. The result below is useful for solving certain exponential equations.

Core Concept

Property of Equality for Exponential Equations

Algebra If b is a positive real number other than 1, then $b^x = b^y$ if and only if $x = y$.

Example If $3^x = 3^5$, then $x = 5$. If $x = 5$, then $3^x = 3^5$.

The preceding property is useful for solving an exponential equation when each side of the equation uses the same base (or can be rewritten to use the same base). When it is not convenient to write each side of an exponential equation using the same base, you can try to solve the equation by taking a logarithm of each side.

EXAMPLE 1 Solving Exponential Equations

Solve each equation.

a. $100^x = \left(\frac{1}{10}\right)^{x-3}$ **b.** $2^x = 7$

SOLUTION

a.

$100^x = \left(\frac{1}{10}\right)^{x-3}$	Write original equation.
$(10^2)^x = (10^{-1})^{x-3}$	Rewrite 100 and $\frac{1}{10}$ as powers with base 10.
$10^{2x} = 10^{-x+3}$	Power of a Power Property
$2x = -x + 3$	Property of Equality for Exponential Equations
$x = 1$	Solve for x.

Check

$$100^1 \stackrel{?}{=} \left(\frac{1}{10}\right)^{1-3}$$

$$100 \stackrel{?}{=} \left(\frac{1}{10}\right)^{-2}$$

$$100 = 100 \checkmark$$

b.

$2^x = 7$	Write original equation.
$\log_2 2^x = \log_2 7$	Take $\log_2$ of each side.
$x = \log_2 7$	$\log_b b^x = x$
$x \approx 2.807$	Use a calculator.

Check

Enter $y = 2^x$ and $y = 7$ in a graphing calculator. Use the *intersect* feature to find the intersection point of the graphs. The graphs intersect at about (2.807, 7). So, the solution of $2^x = 7$ is about 2.807. ✓

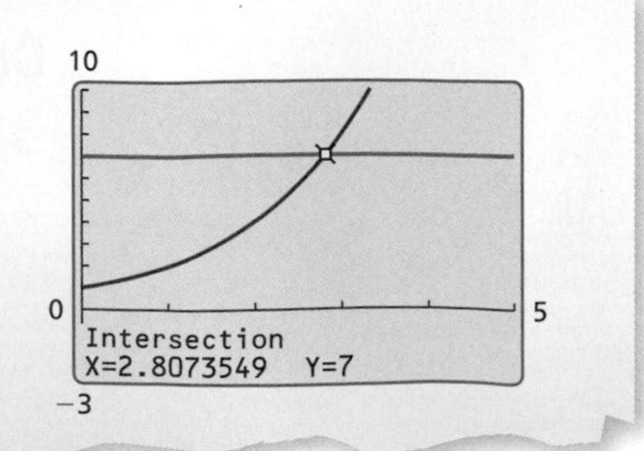

An important application of exponential equations is *Newton's Law of Cooling*. This law states that for a cooling substance with initial temperature T_0, the temperature T after t minutes can be modeled by

$$T = (T_0 - T_R)e^{-rt} + T_R$$

where T_R is the surrounding temperature and r is the cooling rate of the substance.

LOOKING FOR STRUCTURE

Notice that Newton's Law of Cooling models the temperature of a cooling body by adding a constant function, T_R, to a decaying exponential function, $(T_0 - T_R)e^{-rt}$.

EXAMPLE 2 Solving a Real-Life Problem

You are cooking *aleecha*, an Ethiopian stew. When you take it off the stove, its temperature is 212°F. The room temperature is 70°F, and the cooling rate of the stew is $r = 0.046$. How long will it take to cool the stew to a serving temperature of 100°F?

SOLUTION

Use Newton's Law of Cooling with $T = 100$, $T_0 = 212$, $T_R = 70$, and $r = 0.046$.

$T = (T_0 - T_R)e^{-rt} + T_R$	Newton's Law of Cooling
$100 = (212 - 70)e^{-0.046t} + 70$	Substitute for T, T_0, T_R, and r.
$30 = 142e^{-0.046t}$	Subtract 70 from each side.
$0.211 \approx e^{-0.046t}$	Divide each side by 142.
$\ln 0.211 \approx \ln e^{-0.046t}$	Take natural log of each side.
$-1.556 \approx -0.046t$	$\ln e^x = \log_e e^x = x$
$33.8 \approx t$	Divide each side by -0.046.

▶ You should wait about 34 minutes before serving the stew.

Monitoring Progress  Help in English and Spanish at *BigIdeasMath.com*

Solve the equation.

1. $2^x = 5$
2. $7^{9x} = 15$
3. $4e^{-0.3x} - 7 = 13$
4. **WHAT IF?** In Example 2, how long will it take to cool the stew to 100°F when the room temperature is 75°F?

Solving Logarithmic Equations

Logarithmic equations are equations that involve logarithms of variable expressions. You can use the next property to solve some types of logarithmic equations.

Core Concept

Property of Equality for Logarithmic Equations

Algebra If b, x, and y are positive real numbers with $b \neq 1$, then $\log_b x = \log_b y$ if and only if $x = y$.

Example If $\log_2 x = \log_2 7$, then $x = 7$. If $x = 7$, then $\log_2 x = \log_2 7$.

The preceding property implies that if you are given an equation $x = y$, then you can exponentiate each side to obtain an equation of the form $b^x = b^y$. This technique is useful for solving some logarithmic equations.

EXAMPLE 3 Solving Logarithmic Equations

Solve (a) $\ln(4x - 7) = \ln(x + 5)$ and (b) $\log_2(5x - 17) = 3$.

SOLUTION

a. $\ln(4x - 7) = \ln(x + 5)$ — Write original equation.

$4x - 7 = x + 5$ — Property of Equality for Logarithmic Equations

$3x - 7 = 5$ — Subtract x from each side.

$3x = 12$ — Add 7 to each side.

$x = 4$ — Divide each side by 3.

Check

$\ln(4 \cdot 4 - 7) \stackrel{?}{=} \ln(4 + 5)$

$\ln(16 - 7) \stackrel{?}{=} \ln 9$

$\ln 9 = \ln 9$ ✓

b. $\log_2(5x - 17) = 3$ — Write original equation.

$2^{\log_2(5x - 17)} = 2^3$ — Exponentiate each side using base 2.

$5x - 17 = 8$ — $b^{\log_b x} = x$

$5x = 25$ — Add 17 to each side.

$x = 5$ — Divide each side by 5.

Check

$\log_2(5 \cdot 5 - 17) \stackrel{?}{=} 3$

$\log_2(25 - 17) \stackrel{?}{=} 3$

$\log_2 8 \stackrel{?}{=} 3$

Because $2^3 = 8$, $\log_2 8 = 3$. ✓

Because the domain of a logarithmic function generally does not include all real numbers, be sure to check for extraneous solutions of logarithmic equations. You can do this algebraically or graphically.

EXAMPLE 4 Solving a Logarithmic Equation

Solve $\log 2x + \log(x - 5) = 2$.

SOLUTION

$\log 2x + \log(x - 5) = 2$ — Write original equation.

$\log[2x(x - 5)] = 2$ — Product Property of Logarithms

$10^{\log[2x(x - 5)]} = 10^2$ — Exponentiate each side using base 10.

$2x(x - 5) = 100$ — $b^{\log_b x} = x$

$2x^2 - 10x = 100$ — Distributive Property

$2x^2 - 10x - 100 = 0$ — Write in standard form.

$x^2 - 5x - 50 = 0$ — Divide each side by 2.

$(x - 10)(x + 5) = 0$ — Factor.

$x = 10$ or $x = -5$ — Zero-Product Property

▶ The apparent solution $x = -5$ is extraneous. So, the only solution is $x = 10$.

Check

$\log(2 \cdot 10) + \log(10 - 5) \stackrel{?}{=} 2$

$\log 20 + \log 5 \stackrel{?}{=} 2$

$\log 100 \stackrel{?}{=} 2$

$2 = 2$ ✓

$\log[2 \cdot (-5)] + \log(-5 - 5) \stackrel{?}{=} 2$

$\log(-10) + \log(-10) \stackrel{?}{=} 2$

Because $\log(-10)$ is not defined, -5 is not a solution. ✗

Monitoring Progress

Help in English and Spanish at *BigIdeasMath.com*

Solve the equation. Check for extraneous solutions.

5. $\ln(7x - 4) = \ln(2x + 11)$

6. $\log_2(x - 6) = 5$

7. $\log 5x + \log(x - 1) = 2$

8. $\log_4(x + 12) + \log_4 x = 3$

Solving Exponential and Logarithmic Inequalities

Exponential inequalities are inequalities in which variable expressions occur as exponents, and *logarithmic inequalities* are inequalities that involve logarithms of variable expressions. To solve exponential and logarithmic inequalities algebraically, use these properties. Note that the properties are true for $\leq$ and $\geq$.

STUDY TIP

Be sure you understand that these properties of inequality are only true for values of $b > 1$.

Exponential Property of Inequality: If b is a positive real number greater than 1, then $b^x > b^y$ if and only if $x > y$, and $b^x < b^y$ if and only if $x < y$.

Logarithmic Property of Inequality: If b, x, and y are positive real numbers with $b > 1$, then $\log_b x > \log_b y$ if and only if $x > y$, and $\log_b x < \log_b y$ if and only if $x < y$.

You can also solve an inequality by taking a logarithm of each side or by exponentiating.

EXAMPLE 5 Solving an Exponential Inequality

Solve $3^x < 20$.

SOLUTION

$3^x < 20$	Write original inequality.
$\log_3 3^x < \log_3 20$	Take $\log_3$ of each side.
$x < \log_3 20$	$\log_b b^x = x$

▶ The solution is $x < \log_3 20$. Because $\log_3 20 \approx 2.727$, the approximate solution is $x < 2.727$.

EXAMPLE 6 Solving a Logarithmic Inequality

Solve $\log x \leq 2$.

SOLUTION

Method 1 Use an algebraic approach.

$\log x \leq 2$	Write original inequality.
$10^{\log_{10} x} \leq 10^2$	Exponentiate each side using base 10.
$x \leq 100$	$b^{\log_b x} = x$

▶ Because $\log x$ is only defined when $x > 0$, the solution is $0 < x \leq 100$.

Method 2 Use a graphical approach.

Graph $y = \log x$ and $y = 2$ in the same viewing window. Use the *intersect* feature to determine that the graphs intersect when $x = 100$. The graph of $y = \log x$ is on or below the graph of $y = 2$ when $0 < x \leq 100$.

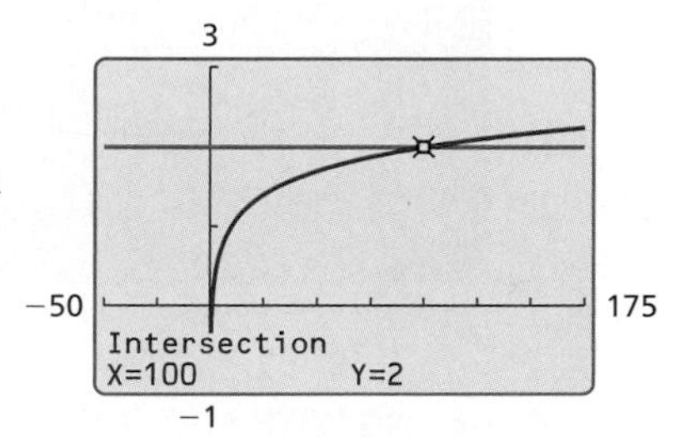

▶ The solution is $0 < x \leq 100$.

Monitoring Progress

Help in English and Spanish at *BigIdeasMath.com*

Solve the inequality.

9. $e^x < 2$ **10.** $10^{2x-6} > 3$ **11.** $\log x + 9 < 45$ **12.** $2 \ln x - 1 > 4$

6.6 Exercises

Dynamic Solutions available at *BigIdeasMath.com*

Vocabulary and Core Concept Check

1. **COMPLETE THE SENTENCE** The equation $3^{x-1} = 34$ is an example of a(n) ________ equation.
2. **WRITING** Compare the methods for solving exponential and logarithmic equations.
3. **WRITING** When do logarithmic equations have extraneous solutions?
4. **COMPLETE THE SENTENCE** If b is a positive real number other than 1, then $b^x = b^y$ if and only if ________.

Monitoring Progress and Modeling with Mathematics

In Exercises 5–16, solve the equation. *(See Example 1.)*

5. $7^{3x+5} = 7^{1-x}$
6. $e^{2x} = e^{3x-1}$
7. $5^{x-3} = 25^{x-5}$
8. $6^{2x-6} = 36^{3x-5}$
9. $3^x = 7$
10. $5^x = 33$
11. $49^{5x+2} = \left(\frac{1}{7}\right)^{11-x}$
12. $512^{5x-1} = \left(\frac{1}{8}\right)^{-4-x}$
13. $7^{5x} = 12$
14. $11^{6x} = 38$
15. $3e^{4x} + 9 = 15$
16. $2e^{2x} - 7 = 5$

17. **MODELING WITH MATHEMATICS** The length ℓ (in centimeters) of a scalloped hammerhead shark can be modeled by the function

$$\ell = 266 - 219e^{-0.05t}$$

where t is the age (in years) of the shark. How old is a shark that is 175 centimeters long?

18. **MODELING WITH MATHEMATICS** One hundred grams of radium are stored in a container. The amount R (in grams) of radium present after t years can be modeled by $R = 100e^{-0.00043t}$. After how many years will only 5 grams of radium be present?

In Exercises 19 and 20, use Newton's Law of Cooling to solve the problem. *(See Example 2.)*

19. You are driving on a hot day when your car overheats and stops running. The car overheats at 280°F and can be driven again at 230°F. When it is 80°F outside, the cooling rate of the car is $r = 0.0058$. How long do you have to wait until you can continue driving?

20. You cook a turkey until the internal temperature reaches 180°F. The turkey is placed on the table until the internal temperature reaches 100°F and it can be carved. When the room temperature is 72°F, the cooling rate of the turkey is $r = 0.067$. How long do you have to wait until you can carve the turkey?

In Exercises 21–32, solve the equation. *(See Example 3.)*

21. $\ln(4x - 7) = \ln(x + 11)$
22. $\ln(2x - 4) = \ln(x + 6)$
23. $\log_2(3x - 4) = \log_2 5$
24. $\log(7x + 3) = \log 38$
25. $\log_2(4x + 8) = 5$
26. $\log_3(2x + 1) = 2$
27. $\log_7(4x + 9) = 2$
28. $\log_5(5x + 10) = 4$
29. $\log(12x - 9) = \log 3x$
30. $\log_6(5x + 9) = \log_6 6x$
31. $\log_2(x^2 - x - 6) = 2$
32. $\log_3(x^2 + 9x + 27) = 2$

In Exercises 33–40, solve the equation. Check for extraneous solutions. *(See Example 4.)*

33. $\log_2 x + \log_2(x - 2) = 3$

34. $\log_6 3x + \log_6(x - 1) = 3$

35. $\ln x + \ln(x + 3) = 4$

36. $\ln x + \ln(x - 2) = 5$

37. $\log_3 3x^2 + \log_3 3 = 2$

38. $\log_4(-x) + \log_4(x + 10) = 2$

39. $\log_3(x - 9) + \log_3(x - 3) = 2$

40. $\log_5(x + 4) + \log_5(x + 1) = 2$

ERROR ANALYSIS In Exercises 41 and 42, describe and correct the error in solving the equation.

41. ✗

$$\log_3(5x - 1) = 4$$
$$3^{\log_3(5x - 1)} = 4^3$$
$$5x - 1 = 64$$
$$5x = 65$$
$$x = 13$$

42. ✗

$$\log_4(x + 12) + \log_4 x = 3$$
$$\log_4[(x + 12)(x)] = 3$$
$$4^{\log_4[(x + 12)(x)]} = 4^3$$
$$(x + 12)(x) = 64$$
$$x^2 + 12x - 64 = 0$$
$$(x + 16)(x - 4) = 0$$
$$x = -16 \quad \text{or} \quad x = 4$$

43. PROBLEM SOLVING You deposit \$100 in an account that pays 6% annual interest. How long will it take for the balance to reach \$1000 for each frequency of compounding?

a. annual **b.** quarterly

c. daily **d.** continuously

44. MODELING WITH MATHEMATICS The *apparent magnitude* of a star is a measure of the brightness of the star as it appears to observers on Earth. The apparent magnitude M of the dimmest star that can be seen with a telescope is $M = 5 \log D + 2$, where D is the diameter (in millimeters) of the telescope's objective lens. What is the diameter of the objective lens of a telescope that can reveal stars with a magnitude of 12?

45. ANALYZING RELATIONSHIPS Approximate the solution of each equation using the graph.

a. $1 - 5^{5-x} = -9$ **b.** $\log_2 5x = 2$

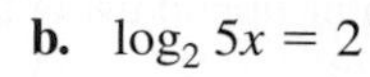

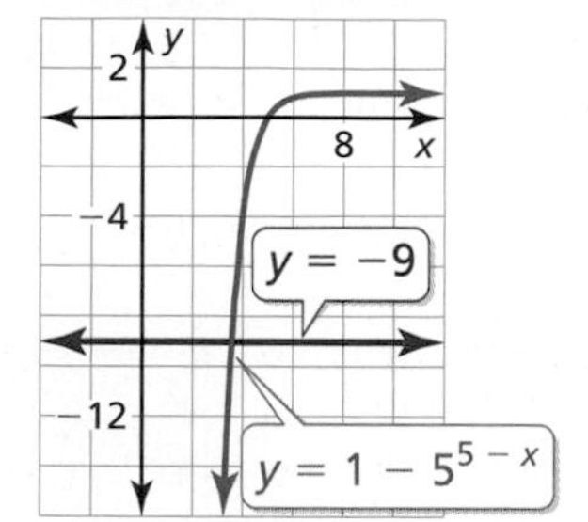

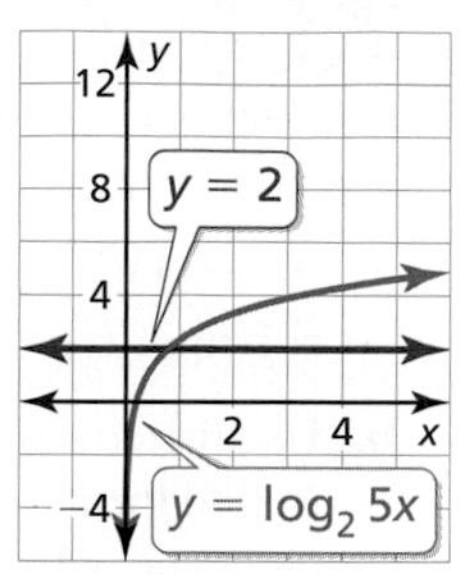

46. MAKING AN ARGUMENT Your friend states that a logarithmic equation cannot have a negative solution because logarithmic functions are not defined for negative numbers. Is your friend correct? Justify your answer.

In Exercises 47–54, solve the inequality. *(See Examples 5 and 6.)*

47. $9^x > 54$ **48.** $4^x \leq 36$

49. $\ln x \geq 3$ **50.** $\log_4 x < 4$

51. $3^{4x-5} < 8$ **52.** $e^{3x+4} > 11$

53. $-3 \log_5 x + 6 \leq 9$ **54.** $-4 \log_5 x - 5 \geq 3$

55. COMPARING METHODS Solve $\log_5 x < 2$ algebraically and graphically. Which method do you prefer? Explain your reasoning.

56. PROBLEM SOLVING You deposit \$1000 in an account that pays 3.5% annual interest compounded monthly. When is your balance at least \$1200? \$3500?

57. PROBLEM SOLVING An investment that earns a rate of return r doubles in value in t years, where $t = \frac{\ln 2}{\ln(1 + r)}$ and r is expressed as a decimal. What rates of return will double the value of an investment in less than 10 years?

58. PROBLEM SOLVING Your family purchases a new car for \$20,000. Its value decreases by 15% each year. During what interval does the car's value exceed \$10,000?

USING TOOLS In Exercises 59–62, use a graphing calculator to solve the equation.

59. $\ln 2x = 3^{-x+2}$ **60.** $\log x = 7^{-x}$

61. $\log x = 3^{x-3}$ **62.** $\ln 2x = e^{x-3}$

63. **REWRITING A FORMULA** A biologist can estimate the age of an African elephant by measuring the length of its footprint and using the equation $\ell = 45 - 25.7e^{-0.09a}$, where ℓ is the length (in centimeters) of the footprint and a is the age (in years).

a. Rewrite the equation, solving for a in terms of ℓ.

b. Use the equation in part (a) to find the ages of the elephants whose footprints are shown.

64. **HOW DO YOU SEE IT?** Use the graph to solve the inequality $4 \ln x + 6 > 9$. Explain your reasoning.

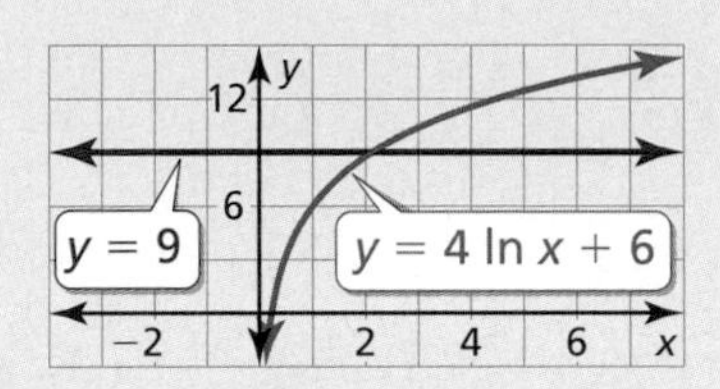

65. **OPEN-ENDED** Write an exponential equation that has a solution of $x = 4$. Then write a logarithmic equation that has a solution of $x = -3$.

66. **THOUGHT PROVOKING** Give examples of logarithmic or exponential equations that have one solution, two solutions, and no solutions.

CRITICAL THINKING In Exercises 67–72, solve the equation.

67. $2^{x+3} = 5^{3x-1}$

68. $10^{3x-8} = 2^{5-x}$

69. $\log_3(x - 6) = \log_9 2x$

70. $\log_4 x = \log_8 4x$

71. $2^{2x} - 12 \cdot 2^x + 32 = 0$

72. $5^{2x} + 20 \cdot 5^x - 125 = 0$

73. **WRITING** In Exercises 67–70, you solved exponential and logarithmic equations with different bases. Describe general methods for solving such equations.

74. **PROBLEM SOLVING** When X-rays of a fixed wavelength strike a material x centimeters thick, the intensity $I(x)$ of the X-rays transmitted through the material is given by $I(x) = I_0e^{-\mu x}$, where I_0 is the initial intensity and μ is a value that depends on the type of material and the wavelength of the X-rays. The table shows the values of μ for various materials and X-rays of medium wavelength.

Material	Aluminum	Copper	Lead
Value of μ	0.43	3.2	43

a. Find the thickness of aluminum shielding that reduces the intensity of X-rays to 30% of their initial intensity. (*Hint*: Find the value of x for which $I(x) = 0.3I_0$.)

b. Repeat part (a) for the copper shielding.

c. Repeat part (a) for the lead shielding.

d. Your dentist puts a lead apron on you before taking X-rays of your teeth to protect you from harmful radiation. Based on your results from parts (a)–(c), explain why lead is a better material to use than aluminum or copper.

Maintaining Mathematical Proficiency

Reviewing what you learned in previous grades and lessons

Write an equation in point-slope form of the line that passes through the given point and has the given slope. *(Skills Review Handbook)*

75. $(1, -2);\ m = 4$

76. $(3, 2);\ m = -2$

77. $(3, -8);\ m = -\frac{1}{3}$

78. $(2, 5);\ m = 2$

Use finite differences to determine the degree of the polynomial function that fits the data. Then use technology to find the polynomial function. *(Section 4.9)*

79. $(-3, -50), (-2, -13), (-1, 0), (0, 1), (1, 2), (2, 15), (3, 52), (4, 125)$

80. $(-3, 139), (-2, 32), (-1, 1), (0, -2), (1, -1), (2, 4), (3, 37), (4, 146)$

81. $(-3, -327), (-2, -84), (-1, -17), (0, -6), (1, -3), (2, -32), (3, -189), (4, -642)$

6.7 Modeling with Exponential and Logarithmic Functions

Essential Question How can you recognize polynomial, exponential, and logarithmic models?

EXPLORATION 1 **Recognizing Different Types of Models**

Work with a partner. Match each type of model with the appropriate scatter plot. Use a regression program to find a model that fits the scatter plot.

a. linear (positive slope) **b.** linear (negative slope) **c.** quadratic

d. cubic **e.** exponential **f.** logarithmic

A.

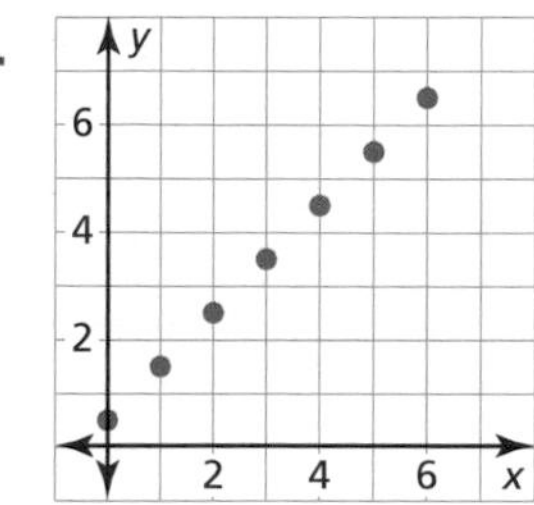

B.

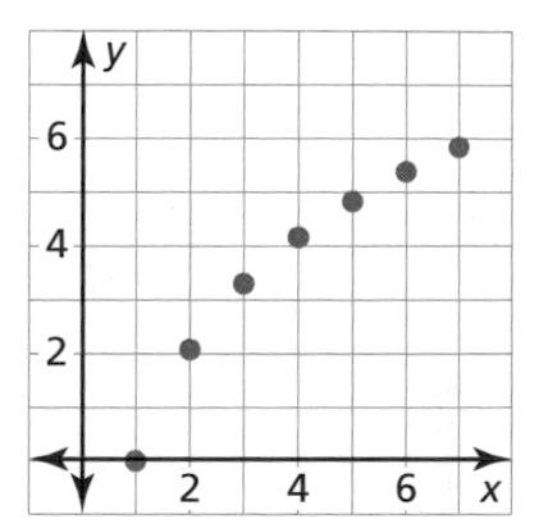

C.

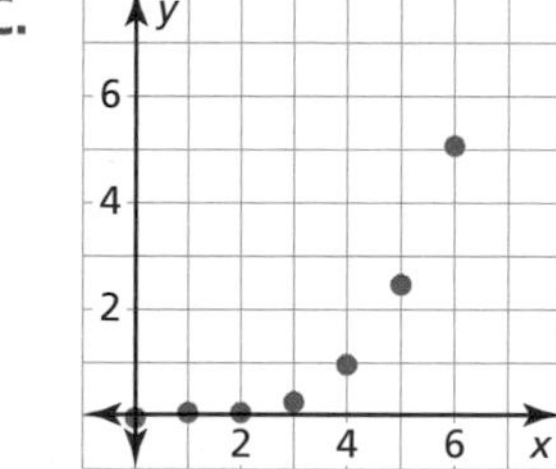

D.

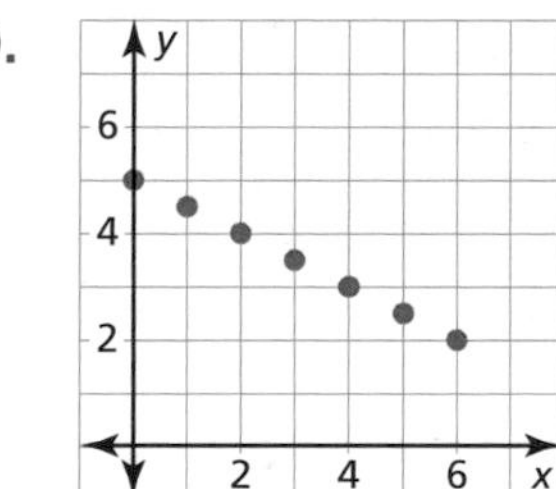

E.

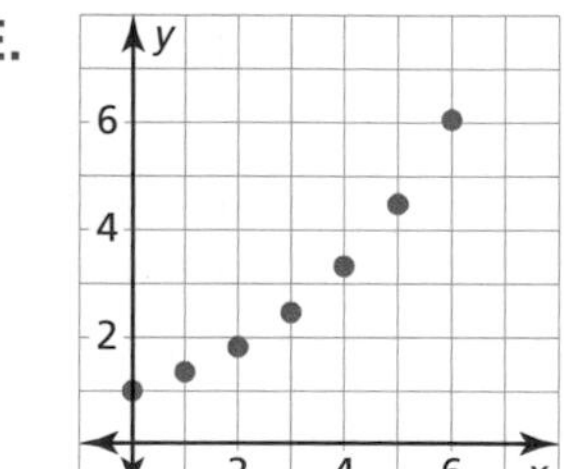

F.

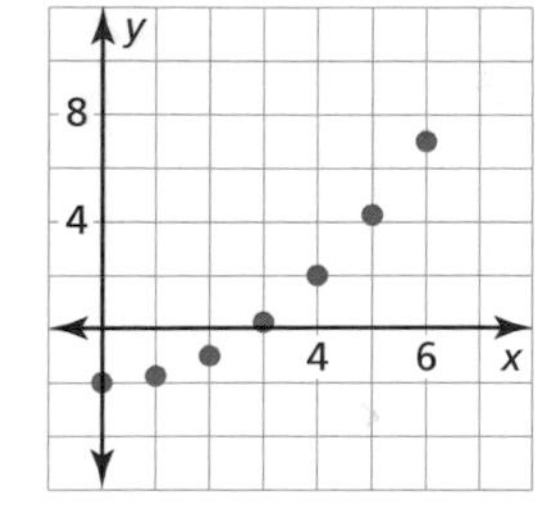

USING TOOLS STRATEGICALLY

To be proficient in math, you need to use technological tools to explore and deepen your understanding of concepts.

EXPLORATION 2 **Exploring Gaussian and Logistic Models**

Work with a partner. Two common types of functions that are related to exponential functions are given. Use a graphing calculator to graph each function. Then determine the domain, range, intercept, and asymptote(s) of the function.

a. Gaussian Function: $f(x) = e^{-x^2}$ **b.** Logistic Function: $f(x) = \dfrac{1}{1 + e^{-x}}$

Communicate Your Answer

3. How can you recognize polynomial, exponential, and logarithmic models?

4. Use the Internet or some other reference to find real-life data that can be modeled using one of the types given in Exploration 1. Create a table and a scatter plot of the data. Then use a regression program to find a model that fits the data.

6.7 Lesson

Core Vocabulary

Previous
finite differences
common ratio
point-slope form

What You Will Learn

- Classify data sets.
- Write exponential functions.
- Use technology to find exponential and logarithmic models.

Classifying Data

You have analyzed *finite differences* of data with equally-spaced inputs to determine what type of polynomial function can be used to model the data. For exponential data with equally-spaced inputs, the outputs are multiplied by a constant factor. So, consecutive outputs form a constant ratio.

EXAMPLE 1 Classifying Data Sets

Determine the type of function represented by each table.

a.

x	−2	−1	0	1	2	3	4
y	0.5	1	2	4	8	16	32

b.

x	−2	0	2	4	6	8	10
y	2	0	2	8	18	32	50

SOLUTION

a. The inputs are equally spaced. Look for a pattern in the outputs.

x	−2	−1	0	1	2	3	4
y	0.5	1	2	4	8	16	32

×2 ×2 ×2 ×2 ×2 ×2

▶ As x increases by 1, y is multiplied by 2. So, the common ratio is 2, and the data in the table represent an exponential function.

b. The inputs are equally spaced. The outputs do not have a common ratio. So, analyze the finite differences.

REMEMBER
First differences of linear functions are constant, second differences of quadratic functions are constant, and so on.

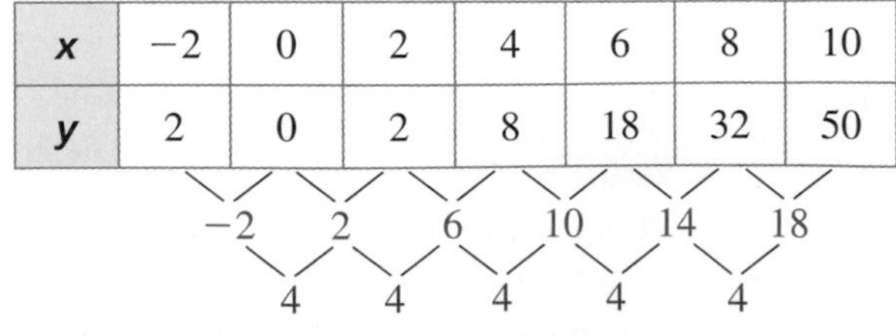

x	−2	0	2	4	6	8	10
y	2	0	2	8	18	32	50

▶ The second differences are constant. So, the data in the table represent a quadratic function.

Monitoring Progress

Help in English and Spanish at *BigIdeasMath.com*

Determine the type of function represented by the table. Explain your reasoning.

1.

x	0	10	20	30
y	15	12	9	6

2.

x	0	2	4	6
y	27	9	3	1

Writing Exponential Functions

You know that two points determine a line. Similarly, two points determine an exponential curve.

EXAMPLE 2 Writing an Exponential Function Using Two Points

Write an exponential function $y = ab^x$ whose graph passes through (1, 6) and (3, 54).

SOLUTION

Step 1 Substitute the coordinates of the two given points into $y = ab^x$.

$6 = ab^1$ Equation 1: Substitute 6 for y and 1 for x.

$54 = ab^3$ Equation 2: Substitute 54 for y and 3 for x.

Step 2 Solve for a in Equation 1 to obtain $a = \frac{6}{b}$ and substitute this expression for a in Equation 2.

$54 = \left(\frac{6}{b}\right)b^3$ Substitute $\frac{6}{b}$ for a in Equation 2.

$54 = 6b^2$ Simplify.

$9 = b^2$ Divide each side by 6.

$3 = b$ Take the positive square root because $b > 0$.

REMEMBER

You know that b must be positive by the definition of an exponential function.

Step 3 Determine that $a = \frac{6}{b} = \frac{6}{3} = 2$.

▶ So, the exponential function is $y = 2(3^x)$.

Data do not always show an *exact* exponential relationship. When the data in a scatter plot show an *approximately* exponential relationship, you can model the data with an exponential function.

EXAMPLE 3 Finding an Exponential Model

A store sells trampolines. The table shows the numbers y of trampolines sold during the xth year that the store has been open. Write a function that models the data.

Year, x	Number of trampolines, y
1	12
2	16
3	25
4	36
5	50
6	67
7	96

SOLUTION

Step 1 Make a scatter plot of the data. The data appear exponential.

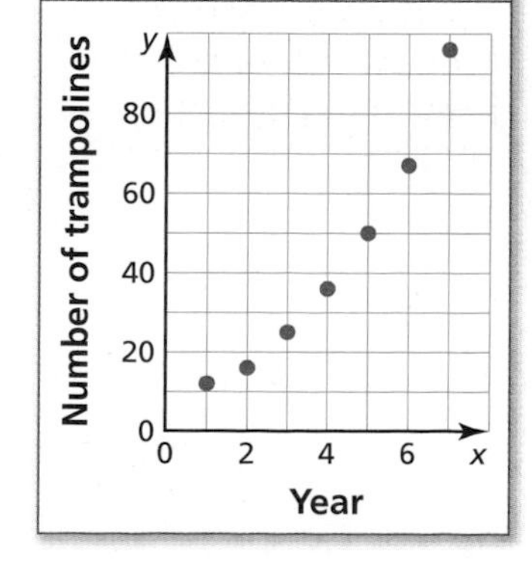

Step 2 Choose any two points to write a model, such as (1, 12) and (4, 36). Substitute the coordinates of these two points into $y = ab^x$.

$12 = ab^1$

$36 = ab^4$

Solve for a in the first equation to obtain $a = \frac{12}{b}$. Substitute to obtain $b = \sqrt[3]{3} \approx 1.44$ and $a = \frac{12}{\sqrt[3]{3}} \approx 8.32$.

▶ So, an exponential function that models the data is $y = 8.32(1.44)^x$.

A set of more than two points (x, y) fits an exponential pattern if and only if the set of transformed points $(x, \ln y)$ fits a linear pattern.

Graph of points (x, y)

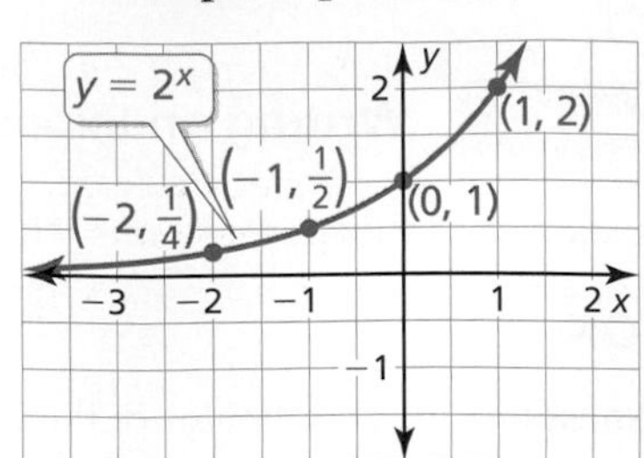

The graph is an exponential curve.

Graph of points $(x, \ln y)$

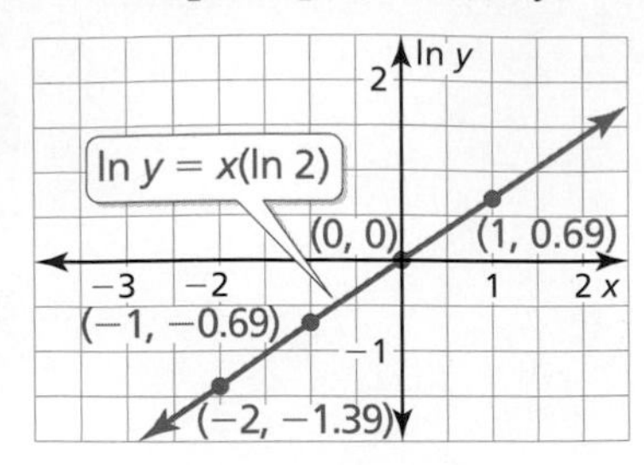

The graph is a line.

EXAMPLE 4 Writing a Model Using Transformed Points

Use the data from Example 3. Create a scatter plot of the data pairs $(x, \ln y)$ to show that an exponential model should be a good fit for the original data pairs (x, y). Then write an exponential model for the original data.

SOLUTION

Step 1 Create a table of data pairs $(x, \ln y)$.

x	1	2	3	4	5	6	7
$\ln y$	2.48	2.77	3.22	3.58	3.91	4.20	4.56

Step 2 Plot the transformed points as shown. The points lie close to a line, so an exponential model should be a good fit for the original data.

Step 3 Find an exponential model $y = ab^x$ by choosing any two points on the line, such as (1, 2.48) and (7, 4.56). Use these points to write an equation of the line. Then solve for y.

LOOKING FOR STRUCTURE

Because the axes are x and $\ln y$, the point-slope form is rewritten as $\ln y - \ln y_1 = m(x - x_1)$. The slope of the line through (1, 2.48) and (7, 4.56) is

$$\frac{4.56 - 2.48}{7 - 1} \approx 0.35.$$

$\ln y - 2.48 = 0.35(x - 1)$ — Equation of line

$\ln y = 0.35x + 2.13$ — Simplify.

$y = e^{0.35x + 2.13}$ — Exponentiate each side using base e.

$y = e^{0.35x}(e^{2.13})$ — Use properties of exponents.

$y = 8.41(1.42)^x$ — Simplify.

▶ So, an exponential function that models the data is $y = 8.41(1.42)^x$.

Monitoring Progress  Help in English and Spanish at *BigIdeasMath.com*

Write an exponential function $y = ab^x$ whose graph passes through the given points.

3. (2, 12), (3, 24) **4.** (1, 2), (3, 32) **5.** (2, 16), (5, 2)

6. WHAT IF? Repeat Examples 3 and 4 using the sales data from another store.

Year, x	1	2	3	4	5	6	7
Number of trampolines, y	15	23	40	52	80	105	140

Using Technology

You can use technology to find best-fit models for exponential and logarithmic data.

EXAMPLE 5 Finding an Exponential Model

Use a graphing calculator to find an exponential model for the data in Example 3. Then use this model and the models in Examples 3 and 4 to predict the number of trampolines sold in the eighth year. Compare the predictions.

SOLUTION

Enter the data into a graphing calculator and perform an exponential regression. The model is $y = 8.46(1.42)^x$.

```
ExpReg
 y=a*b^x
 a=8.457377971
 b=1.418848603
 r2=.9972445053
 r=.9986213023
```

Substitute $x = 8$ into each model to predict the number of trampolines sold in the eighth year.

Example 3: $y = 8.32(1.44)^8 \approx 154$

Example 4: $y = 8.41(1.42)^8 \approx 139$

Regression model: $y = 8.46(1.42)^8 \approx 140$

▶ The predictions are close for the regression model and the model in Example 4 that used transformed points. These predictions are less than the prediction for the model in Example 3.

EXAMPLE 6 Finding a Logarithmic Model

Weather balloons carry instruments that send back information such as wind speed, temperature, and air pressure.

The atmospheric pressure decreases with increasing altitude. At sea level, the average air pressure is 1 atmosphere (1.033227 kilograms per square centimeter). The table shows the pressures p (in atmospheres) at selected altitudes h (in kilometers). Use a graphing calculator to find a logarithmic model of the form $h = a + b \ln p$ that represents the data. Estimate the altitude when the pressure is 0.75 atmosphere.

Air pressure, *p*	1	0.55	0.25	0.12	0.06	0.02
Altitude, *h*	0	5	10	15	20	25

SOLUTION

Enter the data into a graphing calculator and perform a logarithmic regression. The model is $h = 0.86 - 6.45 \ln p$.

```
LnReg
 y=a+blnx
 a=.8626578705
 b=-6.447382985
 r2=.9925582287
 r=-.996272166
```

Substitute $p = 0.75$ into the model to obtain

$h = 0.86 - 6.45 \ln 0.75 \approx 2.7.$

▶ So, when the air pressure is 0.75 atmosphere, the altitude is about 2.7 kilometers.

Monitoring Progress  Help in English and Spanish at *BigIdeasMath.com*

7. Use a graphing calculator to find an exponential model for the data in Monitoring Progress Question 6.

8. Use a graphing calculator to find a logarithmic model of the form $p = a + b \ln h$ for the data in Example 6. Explain why the result is an error message.

6.7 Exercises

Dynamic Solutions available at *BigIdeasMath.com*

Vocabulary and Core Concept Check

1. **COMPLETE THE SENTENCE** Given a set of more than two data pairs (x, y), you can decide whether a(n) __________ function fits the data well by making a scatter plot of the points $(x, \ln y)$.

2. **WRITING** Given a table of values, explain how you can determine whether an exponential function is a good model for a set of data pairs (x, y).

Monitoring Progress and Modeling with Mathematics

In Exercises 3–6, determine the type of function represented by the table. Explain your reasoning. *(See Example 1.)*

3.

x	0	3	6	9	12	15
y	0.25	1	4	16	64	256

4.

x	−4	−3	−2	−1	0	1	2
y	16	8	4	2	1	$\frac{1}{2}$	$\frac{1}{4}$

5.

x	5	10	15	20	25	30
y	4	3	7	16	30	49

6.

x	−3	1	5	9	13
y	8	−3	−14	−25	−36

In Exercises 7–16, write an exponential function $y = ab^x$ whose graph passes through the given points. *(See Example 2.)*

7. (1, 3), (2, 12)

8. (2, 24), (3, 144)

9. (3, 1), (5, 4)

10. (3, 27), (5, 243)

11. (1, 2), (3, 50)

12. (1, 40), (3, 640)

13. (−1, 10), (4, 0.31)

14. (2, 6.4), (5, 409.6)

15.

(4, 4)
(1, 0.5)

16.

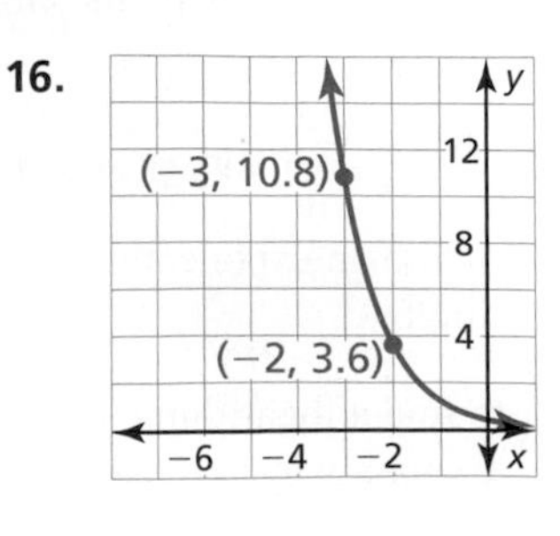

ERROR ANALYSIS In Exercises 17 and 18, describe and correct the error in determining the type of function represented by the data.

17.

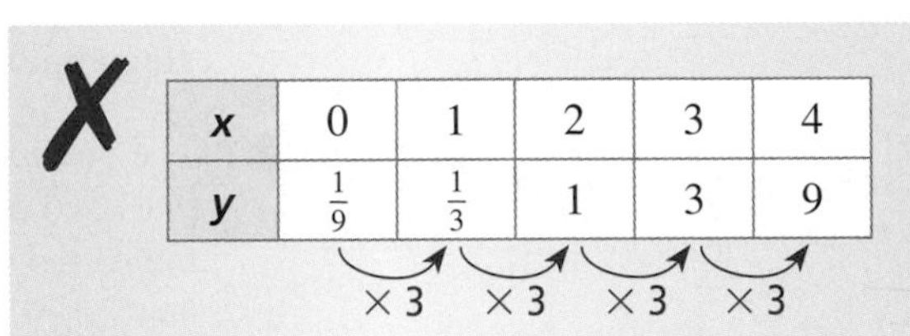

x	0	1	2	3	4
y	$\frac{1}{9}$	$\frac{1}{3}$	1	3	9

×3 ×3 ×3 ×3

The outputs have a common ratio of 3, so the data represent a linear function.

18.

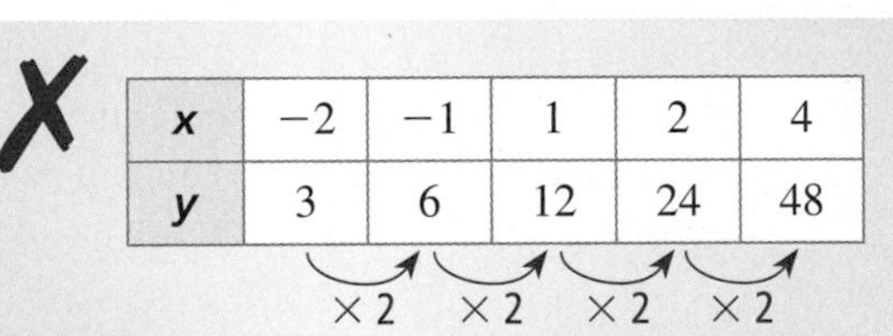

x	−2	−1	1	2	4
y	3	6	12	24	48

×2 ×2 ×2 ×2

The outputs have a common ratio of 2, so the data represent an exponential function.

19. **MODELING WITH MATHEMATICS** A store sells motorized scooters. The table shows the numbers y of scooters sold during the xth year that the store has been open. Write a function that models the data. *(See Example 3.)*

x	y
1	9
2	14
3	19
4	25
5	37
6	53
7	71

20. MODELING WITH MATHEMATICS The table shows the numbers y of visits to a website during the xth month. Write a function that models the data. Then use your model to predict the number of visits after 1 year.

x	1	2	3	4	5	6	7
y	22	39	70	126	227	408	735

In Exercises 21–24, determine whether the data show an exponential relationship. Then write a function that models the data.

21.

x	1	6	11	16	21
y	12	28	76	190	450

22.

x	−3	−1	1	3	5
y	2	7	24	68	194

23.

x	0	10	20	30	40	50	60
y	66	58	48	42	31	26	21

24.

x	−20	−13	−6	1	8	15
y	25	19	14	11	8	6

25. MODELING WITH MATHEMATICS Your visual near point is the closest point at which your eyes can see an object distinctly. The diagram shows the near point y (in centimeters) at age x (in years). Create a scatter plot of the data pairs $(x, \ln y)$ to show that an exponential model should be a good fit for the original data pairs (x, y). Then write an exponential model for the original data. *(See Example 4.)*

Visual Near Point Distances

Age 20
12 cm

Age 30
15 cm

Age 40
25 cm

Age 50
40 cm

Age 60
100 cm

26. MODELING WITH MATHEMATICS Use the data from Exercise 19. Create a scatter plot of the data pairs $(x, \ln y)$ to show that an exponential model should be a good fit for the original data pairs (x, y). Then write an exponential model for the original data.

In Exercises 27–30, create a scatter plot of the points $(x, \ln y)$ to determine whether an exponential model fits the data. If so, find an exponential model for the data.

27.

x	1	2	3	4	5
y	18	36	72	144	288

28.

x	1	4	7	10	13
y	3.3	10.1	30.6	92.7	280.9

29.

x	−13	−6	1	8	15
y	9.8	12.2	15.2	19	23.8

30.

x	−8	−5	−2	1	4
y	1.4	1.67	5.32	6.41	7.97

31. USING TOOLS Use a graphing calculator to find an exponential model for the data in Exercise 19. Then use the model to predict the number of motorized scooters sold in the tenth year. *(See Example 5.)*

32. USING TOOLS A doctor measures an astronaut's pulse rate y (in beats per minute) at various times x (in minutes) after the astronaut has finished exercising. The results are shown in the table. Use a graphing calculator to find an exponential model for the data. Then use the model to predict the astronaut's pulse rate after 16 minutes.

x	y
0	172
2	132
4	110
6	92
8	84
10	78
12	75

33. USING TOOLS An object at a temperature of 160°C is removed from a furnace and placed in a room at 20°C. The table shows the temperatures d (in degrees Celsius) at selected times t (in hours) after the object was removed from the furnace. Use a graphing calculator to find a logarithmic model of the form $t = a + b \ln d$ that represents the data. Estimate how long it takes for the object to cool to 50°C. *(See Example 6.)*

d	160	90	56	38	29	24
t	0	1	2	3	4	5

34. USING TOOLS The f-stops on a camera control the amount of light that enters the camera. Let s be a measure of the amount of light that strikes the film and let f be the f-stop. The table shows several f-stops on a 35-millimeter camera. Use a graphing calculator to find a logarithmic model of the form $s = a + b \ln f$ that represents the data. Estimate the amount of light that strikes the film when $f = 5.657$.

f	s
1.414	1
2.000	2
2.828	3
4.000	4
11.314	7

35. DRAWING CONCLUSIONS The table shows the average weight (in kilograms) of an Atlantic cod that is x years old from the Gulf of Maine.

Age, x	1	2	3	4	5
Weight, y	0.751	1.079	1.702	2.198	3.438

a. Show that an exponential model fits the data. Then find an exponential model for the data.

b. By what percent does the weight of an Atlantic cod increase each year in this period of time? Explain.

36. HOW DO YOU SEE IT? The graph shows a set of data points $(x, \ln y)$. Do the data pairs (x, y) fit an exponential pattern? Explain your reasoning.

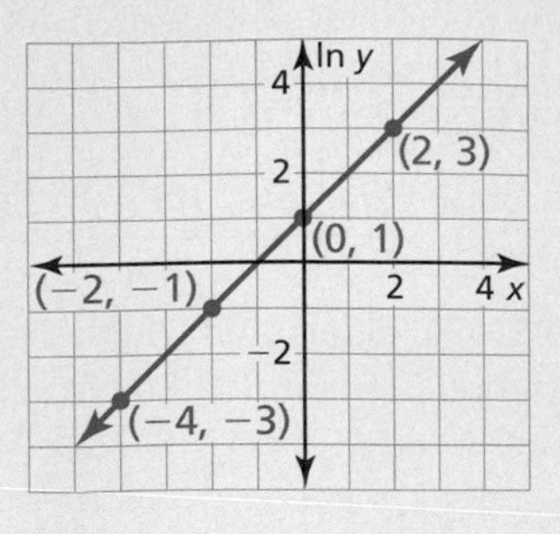

37. MAKING AN ARGUMENT Your friend says it is possible to find a logarithmic model of the form $d = a + b \ln t$ for the data in Exercise 33. Is your friend correct? Explain.

38. THOUGHT PROVOKING Is it possible to write y as an exponential function of x? Explain your reasoning. (Assume p is positive.)

x	y
1	p
2	$2p$
3	$4p$
4	$8p$
5	$16p$

39. CRITICAL THINKING You plant a sunflower seedling in your garden. The height h (in centimeters) of the seedling after t weeks can be modeled by the *logistic function*

$$h(t) = \frac{256}{1 + 13e^{-0.65t}}.$$

a. Find the time it takes the sunflower seedling to reach a height of 200 centimeters.

b. Use a graphing calculator to graph the function. Interpret the meaning of the asymptote in the context of this situation.

Maintaining Mathematical Proficiency

Reviewing what you learned in previous grades and lessons

Tell whether x and y are in a proportional relationship. Explain your reasoning. *(Skills Review Handbook)*

40. $y = \frac{x}{2}$ **41.** $y = 3x - 12$ **42.** $y = \frac{5}{x}$ **43.** $y = -2x$

Identify the focus, directrix, and axis of symmetry of the parabola. Then graph the equation. *(Section 2.3)*

44. $x = \frac{1}{8}y^2$ **45.** $y = 4x^2$ **46.** $x^2 = 3y$ **47.** $y^2 = \frac{2}{5}x$

6.5–6.7 What Did You Learn?

Core Vocabulary

exponential equations, *p. 334*
logarithmic equations, *p. 335*

Core Concepts

Section 6.5

Properties of Logarithms, *p. 328*
Change-of-Base Formula, *p. 329*

Section 6.6

Property of Equality for Exponential Equations, *p. 334*
Property of Equality for Logarithmic Equations, *p. 335*

Section 6.7

Classifying Data, *p. 342*
Writing Exponential Functions, *p. 343*
Using Exponential and Logarithmic Regression, *p. 345*

Mathematical Practices

1. Explain how you used properties of logarithms to rewrite the function in part (b) of Exercise 45 on page 332.
2. How can you use cases to analyze the argument given in Exercise 46 on page 339?

Performance Task

Measuring Natural Disasters

In 2005, an earthquake measuring 4.1 on the Richter scale barely shook the city of Ocotillo, California, leaving virtually no damage. But in 1906, an earthquake with an estimated 8.2 on the same scale devastated the city of San Francisco. Does twice the measurement on the Richter scale mean twice the intensity of the earthquake?

To explore the answers to these questions and more, go to ***BigIdeasMath.com***.

6 Chapter Review

Dynamic Solutions available at *BigIdeasMath.com*

6.1 Exponential Growth and Decay Functions (pp. 295–302)

Tell whether the function $y = 3^x$ represents *exponential growth* or *exponential decay*. Then graph the function.

Step 1 Identify the value of the base. The base, 3, is greater than 1, so the function represents exponential growth.

Step 2 Make a table of values.

x	−2	−1	0	1	2
y	$\frac{1}{9}$	$\frac{1}{3}$	1	3	9

Step 3 Plot the points from the table.

Step 4 Draw, from *left to right*, a smooth curve that begins just above the x-axis, passes through the plotted points, and moves up to the right.

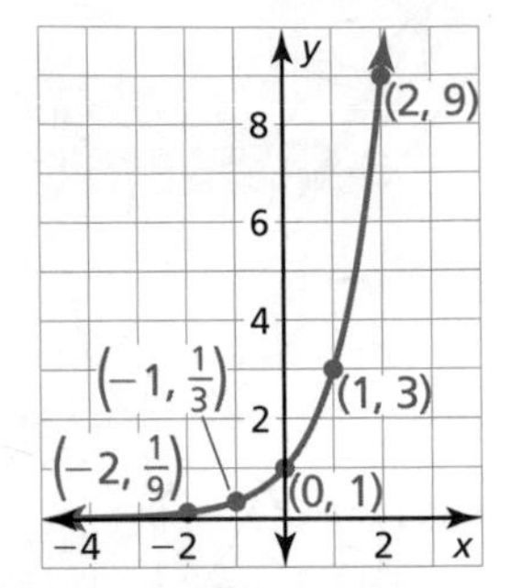

Tell whether the function represents exponential growth or exponential decay. Identify the percent increase or decrease. Then graph the function.

1. $f(x) = \left(\frac{1}{3}\right)^x$ **2.** $y = 5^x$ **3.** $f(x) = (0.2)^x$

4. You deposit \$1500 in an account that pays 7% annual interest. Find the balance after 2 years when the interest is compounded daily.

6.2 The Natural Base e (pp. 303–308)

Simplify each expression.

a. $\frac{18e^{13}}{2e^7} = 9e^{13-7} = 9e^6$ **b.** $(2e^{3x})^3 = 2^3(e^{3x})^3 = 8e^{9x}$

Simplify the expression.

5. $e^4 \cdot e^{11}$ **6.** $\frac{20e^3}{10e^6}$ **7.** $(-3e^{-5x})^2$

Tell whether the function represents *exponential growth* or *exponential decay*. Then graph the function.

8. $f(x) = \frac{1}{3}e^x$ **9.** $y = 6e^{-x}$ **10.** $y = 3e^{-0.75x}$

6.3 Logarithms and Logarithmic Functions (pp. 309–316)

Find the inverse of the function $y = \ln(x - 2)$.

$y = \ln(x - 2)$	Write original function.
$x = \ln(y - 2)$	Switch x and y.
$e^x = y - 2$	Write in exponential form.
$e^x + 2 = y$	Add 2 to each side.

▶ The inverse of $y = \ln(x - 2)$ is $y = e^x + 2$.

Check

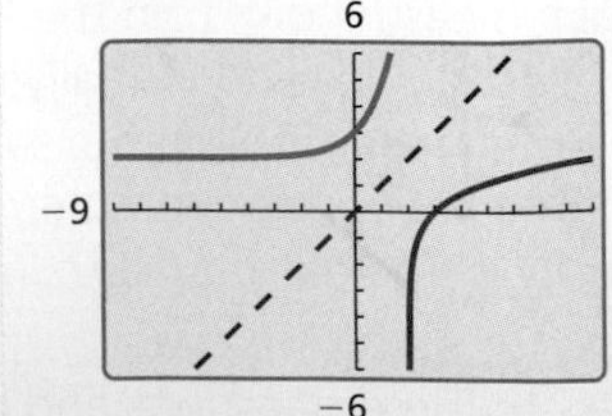

The graphs appear to be reflections of each other in the line $y = x$. ✓

Evaluate the logarithm.

11. $\log_2 8$

12. $\log_6 \frac{1}{36}$

13. $\log_5 1$

Find the inverse of the function.

14. $f(x) = 8^x$

15. $y = \ln(x - 4)$

16. $y = \log(x + 9)$

17. Graph $y = \log_{1/5} x$.

6.4 Transformations of Exponential and Logarithmic Functions *(pp. 317–324)*

Describe the transformation of $f(x) = \left(\frac{1}{3}\right)^x$ represented by $g(x) = \left(\frac{1}{3}\right)^{x-1} + 3$. Then graph each function.

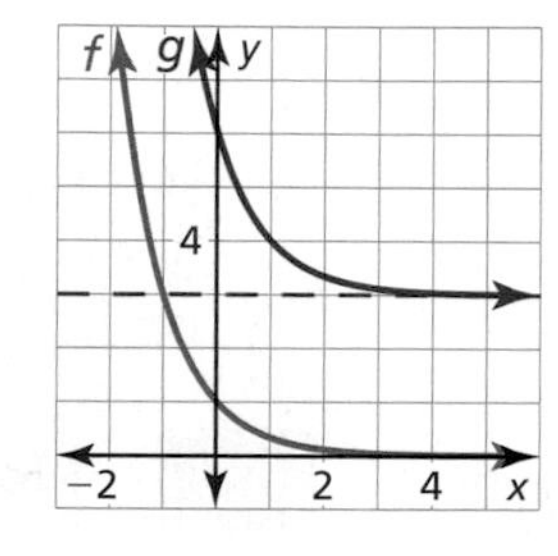

Notice that the function is of the form $g(x) = \left(\frac{1}{3}\right)^{x-h} + k$, where $h = 1$ and $k = 3$.

▶ So, the graph of g is a translation 1 unit right and 3 units up of the graph of f.

Describe the transformation of f represented by g. Then graph each function.

18. $f(x) = e^{-x}$, $g(x) = e^{-5x} - 8$

19. $f(x) = \log_4 x$, $g(x) = \frac{1}{2}\log_4(x + 5)$

Write a rule for g.

20. Let the graph of g be a vertical stretch by a factor of 3, followed by a translation 6 units left and 3 units up of the graph of $f(x) = e^x$.

21. Let the graph of g be a translation 2 units down, followed by a reflection in the y-axis of the graph of $f(x) = \log x$.

6.5 Properties of Logarithms *(pp. 327–332)*

Expand $\ln \frac{12x^5}{y}$.

$\ln \frac{12x^5}{y} = \ln 12x^5 - \ln y$	Quotient Property
$= \ln 12 + \ln x^5 - \ln y$	Product Property
$= \ln 12 + 5 \ln x - \ln y$	Power Property

Expand or condense the logarithmic expression.

22. $\log_8 3xy$

23. $\log 10x^3y$

24. $\ln \frac{3y}{x^5}$

25. $3 \log_7 4 + \log_7 6$

26. $\log_2 12 - 2 \log_2 x$

27. $2 \ln x + 5 \ln 2 - \ln 8$

Use the change-of-base formula to evaluate the logarithm.

28. $\log_2 10$

29. $\log_7 9$

30. $\log_{23} 42$

6.6 Solving Exponential and Logarithmic Equations (pp. 333–340)

Solve $\ln(3x - 9) = \ln(2x + 6)$.

$\ln(3x - 9) = \ln(2x + 6)$	Write original equation.
$3x - 9 = 2x + 6$	Property of Equality for Logarithmic Equations
$x - 9 = 6$	Subtract 2x from each side.
$x = 15$	Add 9 to each side.

Check

$\ln(3 \cdot 15 - 9) \stackrel{?}{=} \ln(2 \cdot 15 + 6)$

$\ln(45 - 9) \stackrel{?}{=} \ln(30 + 6)$

$\ln 36 = \ln 36$ ✓

Solve the equation. Check for extraneous solutions.

31. $5^x = 8$ **32.** $\log_3(2x - 5) = 2$ **33.** $\ln x + \ln(x + 2) = 3$

Solve the inequality.

34. $6^x > 12$ **35.** $\ln x \leq 9$ **36.** $e^{4x - 2} \geq 16$

6.7 Modeling with Exponential and Logarithmic Functions (pp. 341–348)

Write an exponential function whose graph passes through (1, 3) and (4, 24).

Step 1 Substitute the coordinates of the two given points into $y = ab^x$.

$3 = ab^1$	Equation 1: Substitute 3 for y and 1 for x.
$24 = ab^4$	Equation 2: Substitute 24 for y and 4 for x.

Step 2 Solve for a in Equation 1 to obtain $a = \frac{3}{b}$ and substitute this expression for a in Equation 2.

$24 = \left(\frac{3}{b}\right)b^4$	Substitute $\frac{3}{b}$ for a in Equation 2.
$24 = 3b^3$	Simplify.
$8 = b^3$	Divide each side by 3.
$2 = b$	Take cube root of each side.

Step 3 Determine that $a = \frac{3}{b} = \frac{3}{2}$.

▶ So, the exponential function is $y = \frac{3}{2}(2^x)$.

Write an exponential model for the data pairs (x, y).

37. (3, 8), (5, 2)

38.

x	1	2	3	4
$\ln y$	1.64	2.00	2.36	2.72

39. A shoe store sells a new type of basketball shoe. The table shows the pairs sold s over time t (in weeks). Use a graphing calculator to find a logarithmic model of the form $s = a + b \ln t$ that represents the data. Estimate how many pairs of shoes are sold after 6 weeks.

Week, t	1	3	5	7	9
Pairs sold, s	5	32	48	58	65

6 Chapter Test

Graph the equation. State the domain, range, and asymptote.

1. $y = \left(\frac{1}{2}\right)^x$

2. $y = \log_{1/5} x$

3. $y = 4e^{-2x}$

Describe the transformation of f represented by g. Then write a rule for g.

4. $f(x) = \log x$

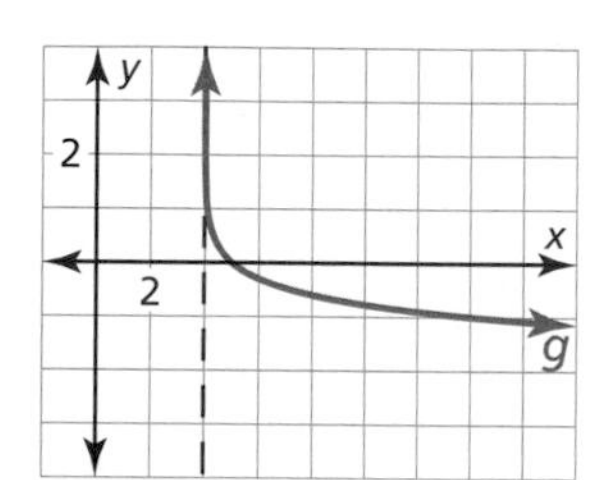

5. $f(x) = e^x$

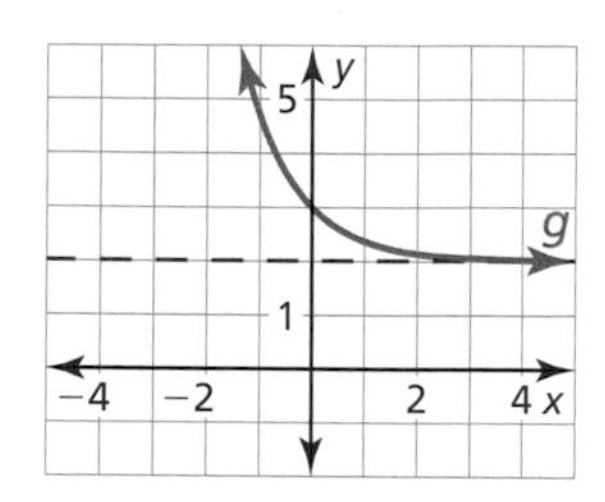

6. $f(x) = \left(\frac{1}{4}\right)^x$

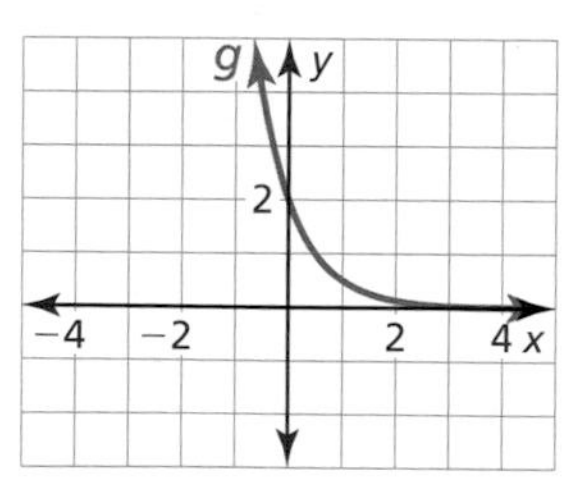

Evaluate the logarithm. Use $\log_3 4 \approx 1.262$ and $\log_3 13 \approx 2.335$, if necessary.

7. $\log_3 52$

8. $\log_3 \frac{13}{9}$

9. $\log_3 16$

10. $\log_3 8 + \log_3 \frac{1}{2}$

11. Describe the similarities and differences in solving the equations $4^{5x-2} = 16$ and $\log_4(10x + 6) = 1$. Then solve each equation.

12. Without calculating, determine whether $\log_5 11$, $\frac{\log 11}{\log 5}$, and $\frac{\ln 11}{\ln 5}$ are equivalent expressions. Explain your reasoning.

13. The amount y of oil collected by a petroleum company drilling on the U.S. continental shelf can be modeled by $y = 12.263 \ln x - 45.381$, where y is measured in billions of barrels and x is the number of wells drilled. About how many barrels of oil would you expect to collect after drilling 1000 wells? Find the inverse function and describe the information you obtain from finding the inverse.

14. The percent L of surface light that filters down through bodies of water can be modeled by the exponential function $L(x) = 100e^{kx}$, where k is a measure of the murkiness of the water and x is the depth (in meters) below the surface.

    a. A recreational submersible is traveling in clear water with a k-value of about -0.02. Write a function that gives the percent of surface light that filters down through clear water as a function of depth.

    b. Tell whether your function in part (a) represents exponential growth or exponential decay. Explain your reasoning.

    c. Estimate the percent of surface light available at a depth of 40 meters.

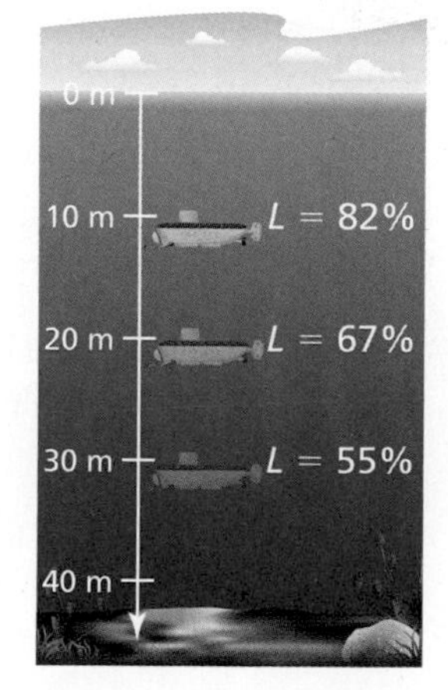

15. The table shows the values y (in dollars) of a new snowmobile after x years of ownership. Describe three different ways to find an exponential model that represents the data. Then write and use a model to find the year when the snowmobile is worth \$2500.

Year, x	0	1	2	3	4
Value, y	4200	3780	3402	3061.80	2755.60

6 Cumulative Assessment

1. Select every value of b for the equation $y = b^x$ that could result in the graph shown.

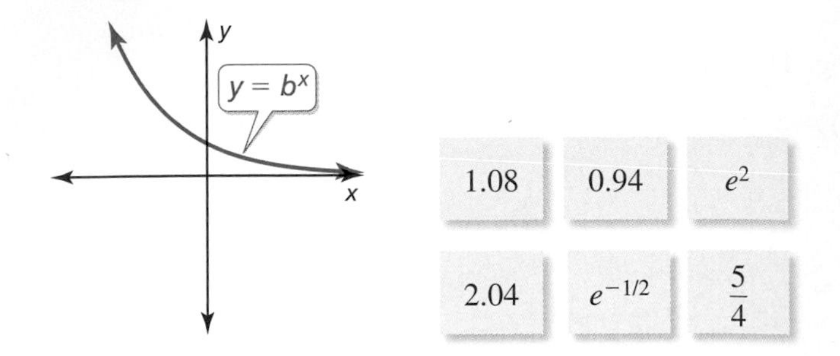

2. Your friend claims more interest is earned when an account pays interest compounded continuously than when it pays interest compounded daily. Do you agree with your friend? Justify your answer.

3. You are designing a rectangular picnic cooler with a length four times its width and height twice its width. The cooler has insulation that is 1 inch thick on each of the four sides and 2 inches thick on the top and bottom.

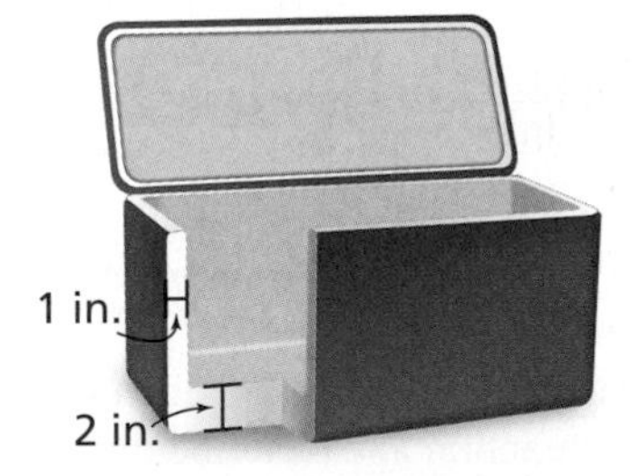

   a. Let x represent the width of the cooler. Write a polynomial function T that gives the volume of the rectangular prism formed by the outer surfaces of the cooler.

   b. Write a polynomial function C for the volume of the inside of the cooler.

   c. Let I be a polynomial function that represents the volume of the insulation. How is I related to T and C?

   d. Write I in standard form. What is the volume of the insulation when the width of the cooler is 8 inches?

4. What is the solution to the logarithmic inequality $-4 \log_2 x \geq -20$?

   Ⓐ $x \leq 32$

   Ⓑ $0 \leq x \leq 32$

   Ⓒ $0 < x \leq 32$

   Ⓓ $x \geq 32$

5. Describe the transformation of $f(x) = \log_2 x$ represented by the graph of g.

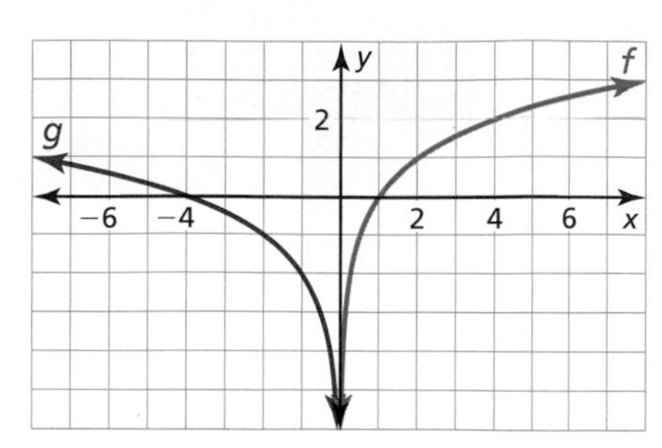

6. Let $f(x) = 2x^3 - 4x^2 + 8x - 1$, $g(x) = 2x - 3x^4 - 6x^3 + 5$, and $h(x) = -7 + x^2 + x$. Order the following functions from least degree to greatest degree.

A. $(f + g)(x)$ **B.** $(hg)(x)$

C. $(h - f)(x)$ **D.** $(fh)(x)$

7. Write an exponential model that represents each data set. Compare the two models.

a.

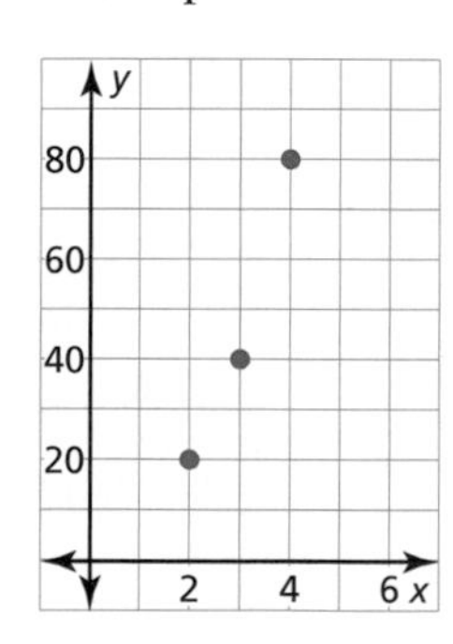

b.

x	2	3	4	5	6
y	4.5	13.5	40.5	121.5	364.5

8. Choose a method to solve each quadratic equation. Explain your choice of method.

a. $x^2 + 4x = 10$ **b.** $x^2 = -12$

c. $4(x - 1)^2 = 6x + 2$ **d.** $x^2 - 3x - 18 = 0$

9. At the annual pumpkin-tossing contest, contestants compete to see whose catapult will send pumpkins the longest distance. The table shows the horizontal distances y (in feet) a pumpkin travels when launched at different angles x (in degrees). Create a scatter plot of the data. Do the data show a linear, quadratic, or exponential relationship? Use technology to find a model for the data. Find the angle(s) at which a launched pumpkin travels 500 feet.

Angle (degrees), x	20	30	40	50	60	70
Distance (feet), y	372	462	509	501	437	323

7 Rational Functions

Cost of Fuel ***(p. 397)***

Galapagos Penguin ***(p. 382)***

Lightning Strike ***(p. 371)***

3-D Printer ***(p. 369)***

Volunteer Project ***(p. 362)***

Maintaining Mathematical Proficiency

Adding and Subtracting Rational Numbers

Example 1 **Find the sum $-\frac{3}{4} + \frac{1}{3}$.**

$$-\frac{3}{4} + \frac{1}{3} = -\frac{9}{12} + \frac{4}{12}$$ Rewrite using the LCD (least common denominator).

$$= \frac{-9 + 4}{12}$$ Write the sum of the numerators over the common denominator.

$$= -\frac{5}{12}$$ Add.

Example 2 **Find the difference $\frac{7}{8} - \left(-\frac{5}{8}\right)$.**

$$\frac{7}{8} - \left(-\frac{5}{8}\right) = \frac{7}{8} + \frac{5}{8}$$ Add the opposite of $-\frac{5}{8}$.

$$= \frac{7 + 5}{8}$$ Write the sum of the numerators over the common denominator.

$$= \frac{12}{8}$$ Add.

$$= \frac{3}{2}, \text{ or } 1\frac{1}{2}$$ Simplify.

Evaluate.

1. $\frac{3}{5} + \frac{2}{3}$

2. $-\frac{4}{7} + \frac{1}{6}$

3. $\frac{7}{9} - \frac{4}{9}$

4. $\frac{5}{12} - \left(-\frac{1}{2}\right)$

5. $\frac{2}{7} + \frac{1}{7} - \frac{6}{7}$

6. $\frac{3}{10} - \frac{3}{4} + \frac{2}{5}$

Simplifying Complex Fractions

Example 3 **Simplify $\frac{\frac{1}{2}}{\frac{4}{5}}$.**

$$\frac{\frac{1}{2}}{\frac{4}{5}} = \frac{1}{2} \div \frac{4}{5}$$ Rewrite the quotient.

$$= \frac{1}{2} \cdot \frac{5}{4}$$ Multiply by the reciprocal of $\frac{4}{5}$.

$$= \frac{1 \cdot 5}{2 \cdot 4}$$ Multiply the numerators and denominators.

$$= \frac{5}{8}$$ Simplify.

Simplify.

7. $\frac{\frac{3}{8}}{\frac{5}{6}}$

8. $\frac{\frac{1}{4}}{-\frac{5}{7}}$

9. $\frac{\frac{2}{3}}{\frac{2}{3} + \frac{1}{4}}$

10. ABSTRACT REASONING For what value of x is the expression $\frac{1}{x}$ undefined? Explain your reasoning.

Mathematical Practices

Mathematically proficient students are careful about specifying units of measure and clarifying the relationship between quantities in a problem.

Specifying Units of Measure

Core Concept

Converting Units of Measure

To convert from one unit of measure to another unit of measure, you can begin by writing the new units. Then multiply the old units by the appropriate conversion factors. For example, you can convert 60 miles per hour to feet per second as follows.

old units → $\frac{60 \text{ mi}}{1 \text{ h}} = \frac{? \text{ ft}}{1 \text{ sec}}$ ← new units

$$\frac{60 \not{\text{mi}}}{1 \not{\text{h}}} \cdot \frac{1 \not{\text{h}}}{60 \not{\text{min}}} \cdot \frac{1 \not{\text{min}}}{60 \text{ sec}} \cdot \frac{5280 \text{ ft}}{1 \not{\text{mi}}} = \frac{5280 \text{ ft}}{60 \text{ sec}}$$

$$= \frac{88 \text{ ft}}{1 \text{ sec}}$$

EXAMPLE 1 Converting Units of Measure

You are given two job offers. Which has the greater annual income?

- \$45,000 per year
- \$22 per hour

SOLUTION

One way to answer this question is to convert \$22 per hour to dollars per year and then compare the two annual salaries. Assume there are 40 hours in a work week.

$$\frac{22 \text{ dollars}}{1 \text{ h}} = \frac{? \text{ dollars}}{1 \text{ yr}}$$ Write new units.

$$\frac{22 \text{ dollars}}{1 \not{\text{h}}} \cdot \frac{40 \not{\text{h}}}{1 \not{\text{week}}} \cdot \frac{52 \not{\text{weeks}}}{1 \text{ yr}} = \frac{45{,}760 \text{ dollars}}{1 \text{ yr}}$$ Multiply by conversion factors.

▶ The second offer has the greater annual salary.

Monitoring Progress

1. You drive a car at a speed of 60 miles per hour. What is the speed in meters per second?
2. A hose carries a pressure of 200 pounds per square inch. What is the pressure in kilograms per square centimeter?
3. A concrete truck pours concrete at the rate of 1 cubic yard per minute. What is the rate in cubic feet per hour?
4. Water in a pipe flows at a rate of 10 gallons per minute. What is the rate in liters per second?

7.1 Inverse Variation

Essential Question How can you recognize when two quantities vary directly or inversely?

EXPLORATION 1 **Recognizing Direct Variation**

Work with a partner. You hang different weights from the same spring.

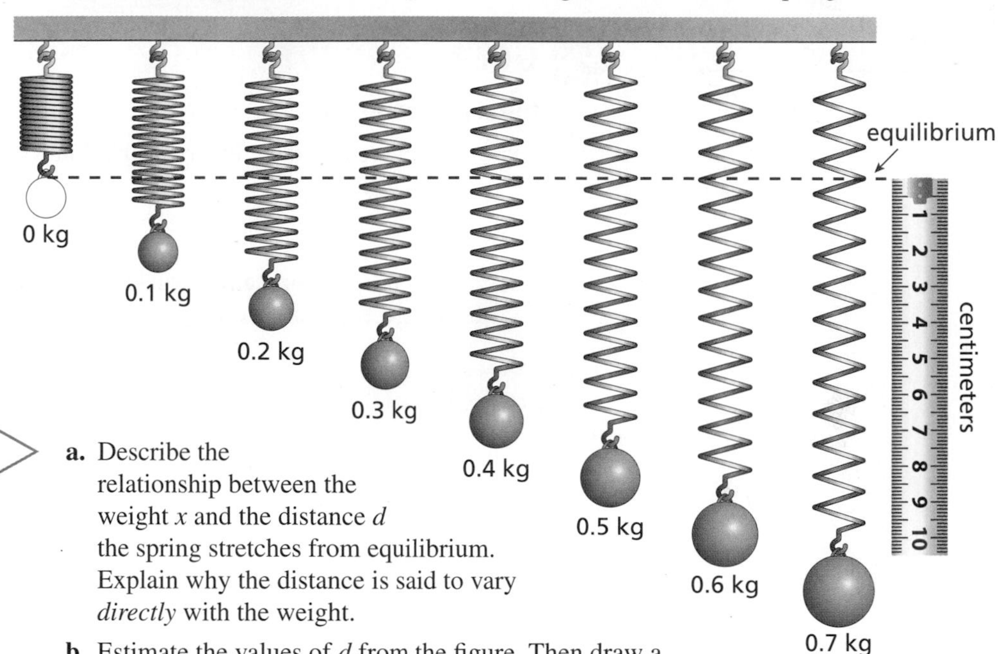

REASONING QUANTITATIVELY

To be proficient in math, you need to make sense of quantities and their relationships in problem situations.

a. Describe the relationship between the weight x and the distance d the spring stretches from equilibrium. Explain why the distance is said to vary *directly* with the weight.

b. Estimate the values of d from the figure. Then draw a scatter plot of the data. What are the characteristics of the graph?

c. Write an equation that represents d as a function of x.

d. In physics, the relationship between d and x is described by *Hooke's Law*. How would you describe Hooke's Law?

EXPLORATION 2 **Recognizing Inverse Variation**

Work with a partner. The table shows the length x (in inches) and the width y (in inches) of a rectangle. The area of each rectangle is 64 square inches.

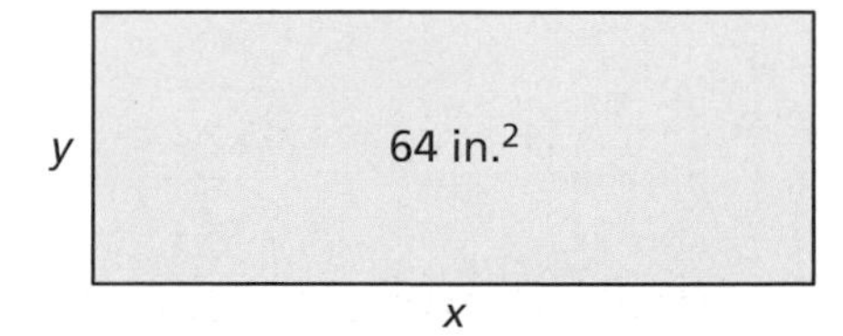

x	y
1	
2	
4	
8	
16	
32	
64	

a. Copy and complete the table.

b. Describe the relationship between x and y. Explain why y is said to vary *inversely* with x.

c. Draw a scatter plot of the data. What are the characteristics of the graph?

d. Write an equation that represents y as a function of x.

Communicate Your Answer

3. How can you recognize when two quantities vary directly or inversely?

4. Does the flapping rate of the wings of a bird vary directly or inversely with the length of its wings? Explain your reasoning.

7.1 Lesson

Core Vocabulary

inverse variation, *p. 360*
constant of variation, *p. 360*

Previous
direct variation
ratios

What You Will Learn

- Classify direct and inverse variation.
- Write inverse variation equations.

Classifying Direct and Inverse Variation

You have learned that two variables x and y show direct variation when $y = ax$ for some nonzero constant a. Another type of variation is called *inverse variation*.

Core Concept

Inverse Variation

Two variables x and y show **inverse variation** when they are related as follows:

$$y = \frac{a}{x}, a \neq 0$$

The constant a is the **constant of variation**, and y is said to *vary inversely* with x.

EXAMPLE 1 Classifying Equations

Tell whether x and y show *direct variation*, *inverse variation*, or *neither*.

a. $xy = 5$

b. $y = x - 4$

c. $\frac{y}{2} = x$

STUDY TIP

The equation in part (b) does not show direct variation because $y = x - 4$ is not of the form $y = ax$.

SOLUTION

Given Equation	Solved for y	Type of Variation
a. $xy = 5$	$y = \frac{5}{x}$	inverse
b. $y = x - 4$	$y = x - 4$	neither
c. $\frac{y}{2} = x$	$y = 2x$	direct

Monitoring Progress

Help in English and Spanish at *BigIdeasMath.com*

Tell whether x and y show *direct variation*, *inverse variation*, or *neither*.

1. $6x = y$
2. $xy = -0.25$
3. $y + x = 10$

The general equation $y = ax$ for direct variation can be rewritten as $\frac{y}{x} = a$. So, a set of data pairs (x, y) shows direct variation when the ratios $\frac{y}{x}$ are constant.

The general equation $y = \frac{a}{x}$ for inverse variation can be rewritten as $xy = a$. So, a set of data pairs (x, y) shows inverse variation when the products xy are constant.

EXAMPLE 2 Classifying Data

Tell whether x and y show *direct variation*, *inverse variation*, or *neither*.

a.

x	2	4	6	8
y	−12	−6	−4	−3

b.

x	1	2	3	4
y	2	4	8	16

SOLUTION

a. Find the products xy and ratios $\frac{y}{x}$.

xy	−24	−24	−24	−24
$\frac{y}{x}$	$\frac{-12}{2} = -6$	$\frac{-6}{4} = -\frac{3}{2}$	$\frac{-4}{6} = -\frac{2}{3}$	$-\frac{3}{8}$

The products are constant.

The ratios are not constant.

▶ So, x and y show inverse variation.

b. Find the products xy and ratios $\frac{y}{x}$.

xy	2	8	24	64
$\frac{y}{x}$	$\frac{2}{1} = 2$	$\frac{4}{2} = 2$	$\frac{8}{3}$	$\frac{16}{4} = 4$

The products are not constant.

The ratios are not constant.

▶ So, x and y show neither direct nor inverse variation.

ANALYZING RELATIONSHIPS

In Example 2(b), notice in the original table that as x increases by 1, y is multiplied by 2. So, the data in the table represent an exponential function.

Monitoring Progress

Help in English and Spanish at *BigIdeasMath.com*

Tell whether x and y show *direct variation*, *inverse variation*, or *neither*.

4.

x	−4	−3	−2	−1
y	20	15	10	5

5.

x	1	2	3	4
y	60	30	20	15

Writing Inverse Variation Equations

EXAMPLE 3 Writing an Inverse Variation Equation

The variables x and y vary inversely, and $y = 4$ when $x = 3$. Write an equation that relates x and y. Then find y when $x = -2$.

SOLUTION

$y = \frac{a}{x}$ — Write general equation for inverse variation.

$4 = \frac{a}{3}$ — Substitute 4 for y and 3 for x.

$12 = a$ — Multiply each side by 3.

▶ The inverse variation equation is $y = \frac{12}{x}$. When $x = -2$, $y = \frac{12}{-2} = -6$.

ANOTHER WAY

Because x and y vary inversely, you also know that the products xy are constant. This product equals the constant of variation a. So, you can quickly determine that $a = xy = 3(4) = 12$.

EXAMPLE 4 Modeling with Mathematics

The time t (in hours) that it takes a group of volunteers to build a playground varies inversely with the number n of volunteers. It takes a group of 10 volunteers 8 hours to build the playground.

- Make a table showing the time that it would take to build the playground when the number of volunteers is 15, 20, 25, and 30.
- What happens to the time it takes to build the playground as the number of volunteers increases?

SOLUTION

1. **Understand the Problem** You are given a description of two quantities that vary inversely and one pair of data values. You are asked to create a table that gives additional data pairs.

2. **Make a Plan** Use the time that it takes 10 volunteers to build the playground to find the constant of variation. Then write an inverse variation equation and substitute for the different numbers of volunteers to find the corresponding times.

3. **Solve the Problem**

$t = \frac{a}{n}$ Write general equation for inverse variation.

$8 = \frac{a}{10}$ Substitute 8 for t and 10 for n.

$80 = a$ Multiply each side by 10.

The inverse variation equation is $t = \frac{80}{n}$. Make a table of values.

n	15	20	25	30
t	$\frac{80}{15}$ = 5 h 20 min	$\frac{80}{20}$ = 4 h	$\frac{80}{25}$ = 3 h 12 min	$\frac{80}{30}$ = 2 h 40 min

▶ As the number of volunteers increases, the time it takes to build the playground decreases.

4. **Look Back** Because the time decreases as the number of volunteers increases, the time for 5 volunteers to build the playground should be greater than 8 hours.

$$t = \frac{80}{5} = 16 \text{ hours } \checkmark$$

LOOKING FOR A PATTERN

Notice that as the number of volunteers increases by 5, the time decreases by a lesser and lesser amount.

From $n = 15$ to $n = 20$, t decreases by 1 hour 20 minutes.

From $n = 20$ to $n = 25$, t decreases by 48 minutes.

From $n = 25$ to $n = 30$, t decreases by 32 minutes.

Monitoring Progress

Help in English and Spanish at *BigIdeasMath.com*

The variables x and y vary inversely. Use the given values to write an equation relating x and y. Then find y when $x = 2$.

6. $x = 4, y = 5$ **7.** $x = 6, y = -1$ **8.** $x = \frac{1}{2}, y = 16$

9. WHAT IF? In Example 4, it takes a group of 10 volunteers 12 hours to build the playground. How long would it take a group of 15 volunteers?

7.1 Exercises

Dynamic Solutions available at *BigIdeasMath.com*

Vocabulary and Core Concept Check

1. **VOCABULARY** Explain how direct variation equations and inverse variation equations are different.

2. **DIFFERENT WORDS, SAME QUESTION** Which is different? Find "both" answers.

What is an inverse variation equation relating x and y with $a = 4$?	What is an equation for which the ratios $\frac{y}{x}$ are constant and $a = 4$?
What is an equation for which y varies inversely with x and $a = 4$?	What is an equation for which the products xy are constant and $a = 4$?

Monitoring Progress and Modeling with Mathematics

In Exercises 3–10, tell whether x and y show *direct variation, inverse variation,* or *neither*. *(See Example 1.)*

3. $y = \frac{2}{x}$

4. $xy = 12$

5. $\frac{y}{x} = 8$

6. $4x = y$

7. $y = x + 4$

8. $x + y = 6$

9. $8y = x$

10. $xy = \frac{1}{5}$

In Exercises 11–14, tell whether x and y show *direct variation, inverse variation,* or *neither*. *(See Example 2.)*

11.

x	12	18	23	29	34
y	132	198	253	319	374

12.

x	1.5	2.5	4	7.5	10
y	13.5	22.5	36	67.5	90

13.

x	4	6	8	8.4	12
y	21	14	10.5	10	7

14.

x	4	5	6.2	7	11
y	16	11	10	9	6

In Exercises 15–22, the variables x and y vary inversely. Use the given values to write an equation relating x and y. Then find y when $x = 3$. *(See Example 3.)*

15. $x = 5, y = -4$

16. $x = 1, y = 9$

17. $x = -3, y = 8$

18. $x = 7, y = 2$

19. $x = \frac{3}{4}, y = 28$

20. $x = -4, y = -\frac{5}{4}$

21. $x = -12, y = -\frac{1}{6}$

22. $x = \frac{5}{3}, y = -7$

ERROR ANALYSIS **In Exercises 23 and 24, the variables x and y vary inversely. Describe and correct the error in writing an equation relating x and y.**

23. $x = 8, y = 5$

✗

$y = ax$

$5 = a(8)$

$\frac{5}{8} = a$

So, $y = \frac{5}{8}x$.

24. $x = 5, y = 2$

✗

$xy = a$

$5 \cdot 2 = a$

$10 = a$

So, $y = 10x$.

25. MODELING WITH MATHEMATICS The number y of songs that can be stored on an MP3 player varies inversely with the average size x of a song. A certain MP3 player can store 2500 songs when the average size of a song is 4 megabytes (MB). *(See Example 4.)*

a. Make a table showing the numbers of songs that will fit on the MP3 player when the average size of a song is 2 MB, 2.5 MB, 3 MB, and 5 MB.

b. What happens to the number of songs as the average song size increases?

26. MODELING WITH MATHEMATICS When you stand on snow, the average pressure P (in pounds per square inch) that you exert on the snow varies inversely with the total area A (in square inches) of the soles of your footwear. Suppose the pressure is 0.43 pound per square inch when you wear the snowshoes shown. Write an equation that gives P as a function of A. Then find the pressure when you wear the boots shown.

27. PROBLEM SOLVING Computer chips are etched onto silicon wafers. The table compares the area A (in square millimeters) of a computer chip with the number c of chips that can be obtained from a silicon wafer. Write a model that gives c as a function of A. Then predict the number of chips per wafer when the area of a chip is 81 square millimeters.

Area (mm²), A	58	62	66	70
Number of chips, c	448	424	392	376

28. HOW DO YOU SEE IT? Does the graph of f represent inverse variation or direct variation? Explain your reasoning.

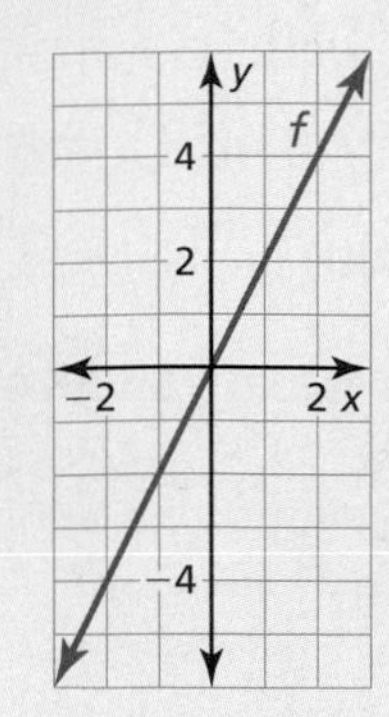

29. MAKING AN ARGUMENT You have enough money to buy 5 hats for \$10 each or 10 hats for \$5 each. Your friend says this situation represents inverse variation. Is your friend correct? Explain your reasoning.

30. THOUGHT PROVOKING The weight w (in pounds) of an object varies inversely with the square of the distance d (in miles) of the object from the center of Earth. At sea level (3978 miles from the center of the Earth), an astronaut weighs 210 pounds. How much does the astronaut weigh 200 miles above sea level?

31. OPEN-ENDED Describe a real-life situation that can be modeled by an inverse variation equation.

32. CRITICAL THINKING Suppose x varies inversely with y and y varies inversely with z. How does x vary with z? Justify your answer.

33. USING STRUCTURE To balance the board in the diagram, the distance (in feet) of each animal from the center of the board must vary inversely with its weight (in pounds). What is the distance of each animal from the fulcrum? Justify your answer.

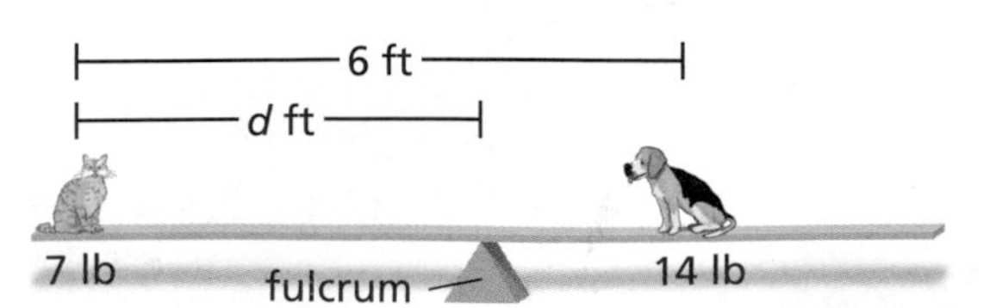

Maintaining Mathematical Proficiency

Reviewing what you learned in previous grades and lessons

Divide. *(Section 4.3)*

34. $(x^2 + 2x - 99) \div (x + 11)$

35. $(3x^4 - 13x^2 - x^3 + 6x - 30) \div (3x^2 - x + 5)$

Graph the function. Then state the domain and range. *(Section 6.4)*

36. $f(x) = 5^x + 4$

37. $g(x) = e^{x-1}$

38. $y = \ln 3x - 6$

39. $h(x) = 2 \ln (x + 9)$

7.2 Graphing Rational Functions

Essential Question

What are some of the characteristics of the graph of a rational function?

The parent function for rational functions with a linear numerator and a linear denominator is

$f(x) = \frac{1}{x}$.   Parent function

The graph of this function, shown at the right, is a *hyperbola*.

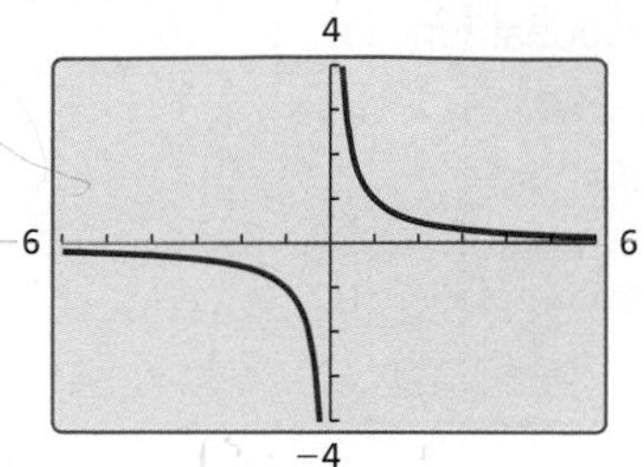

EXPLORATION 1 Identifying Graphs of Rational Functions

Work with a partner. Each function is a transformation of the graph of the parent function $f(x) = \frac{1}{x}$. Match the function with its graph. Explain your reasoning. Then describe the transformation.

a. $g(x) = \frac{1}{x - 1}$ **b.** $g(x) = \frac{-1}{x - 1}$ **c.** $g(x) = \frac{x + 1}{x - 1}$

d. $g(x) = \frac{x - 2}{x + 1}$ **e.** $g(x) = \frac{x}{x + 2}$ **f.** $g(x) = \frac{-x}{x + 2}$

A.

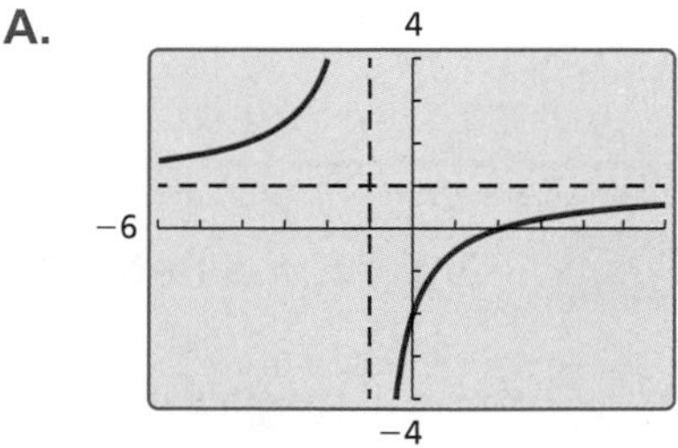

B.

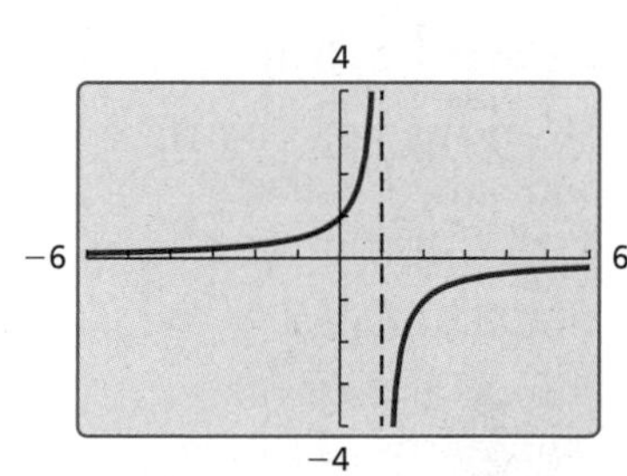

C.

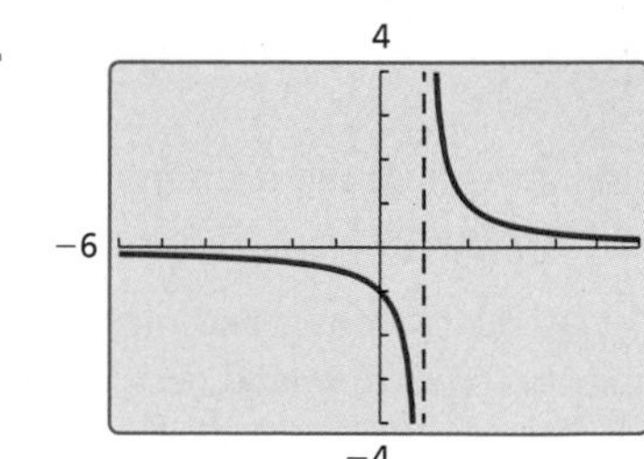

D.

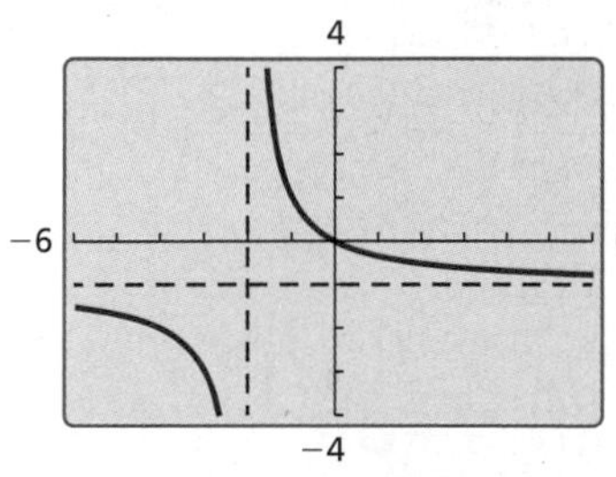

E.

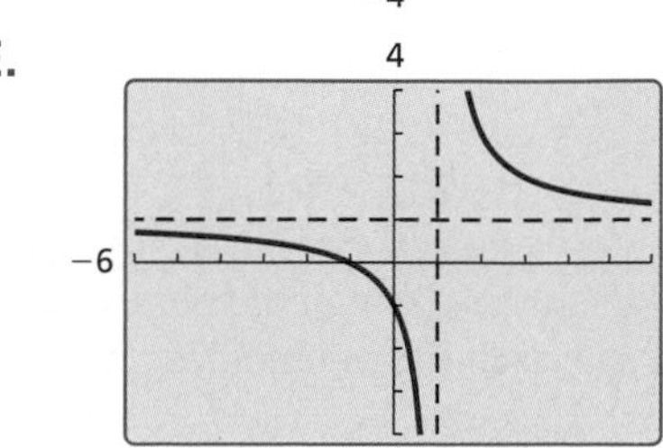

F.

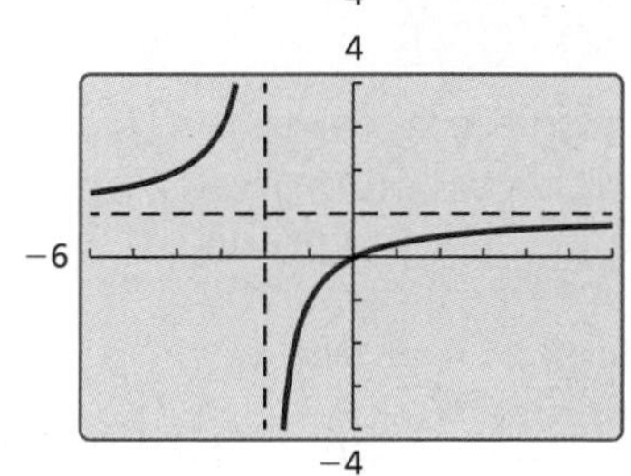

LOOKING FOR STRUCTURE

To be proficient in math, you need to look closely to discern a pattern or structure.

Communicate Your Answer

2. What are some of the characteristics of the graph of a rational function?

3. Determine the intercepts, asymptotes, domain, and range of the rational function $g(x) = \frac{x - a}{x - b}$.

7.2 Lesson

Core Vocabulary

rational function, *p. 366*

Previous
domain
range
asymptote
long division

What You Will Learn

- Graph simple rational functions.
- Translate simple rational functions.
- Graph other rational functions.

Graphing Simple Rational Functions

A **rational function** has the form $f(x) = \frac{p(x)}{q(x)}$, where $p(x)$ and $q(x)$ are polynomials and $q(x) \neq 0$. The inverse variation function $f(x) = \frac{a}{x}$ is a rational function. The graph of this function when $a = 1$ is shown below.

Core Concept

Parent Function for Simple Rational Functions

The graph of the parent function $f(x) = \frac{1}{x}$ is a *hyperbola*, which consists of two symmetrical parts called branches. The domain and range are all nonzero real numbers.

Any function of the form $g(x) = \frac{a}{x}$ $(a \neq 0)$ has the same asymptotes, domain, and range as the function $f(x) = \frac{1}{x}$.

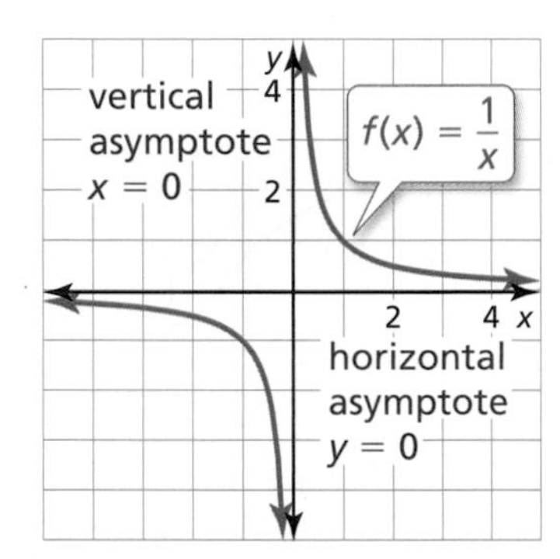

STUDY TIP

Notice that $\frac{1}{x} \to 0$ as $x \to \infty$ and as $x \to -\infty$. This explains why $y = 0$ is a horizontal asymptote of the graph of $f(x) = \frac{1}{x}$. You can also analyze y-values as x approaches 0 to see why $x = 0$ is a vertical asymptote.

EXAMPLE 1 Graphing a Rational Function of the Form $y = \frac{a}{x}$

Graph $g(x) = \frac{4}{x}$. Compare the graph with the graph of $f(x) = \frac{1}{x}$.

SOLUTION

Step 1 The function is of the form $g(x) = \frac{a}{x}$, so the asymptotes are $x = 0$ and $y = 0$. Draw the asymptotes.

Step 2 Make a table of values and plot the points. Include both positive and negative values of x.

x	-3	-2	-1	1	2	3
y	$-\frac{4}{3}$	-2	-4	4	2	$\frac{4}{3}$

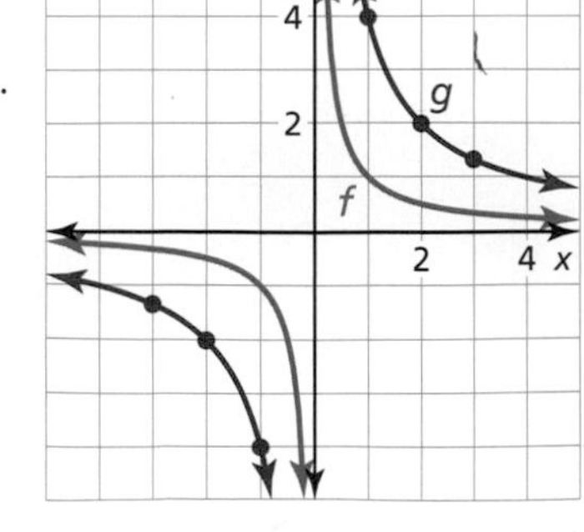

Step 3 Draw the two branches of the hyperbola so that they pass through the plotted points and approach the asymptotes.

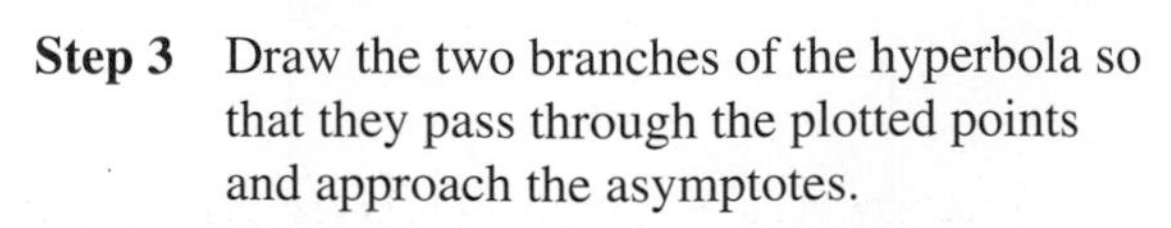

▶ The graph of g lies farther from the axes than the graph of f. Both graphs lie in the first and third quadrants and have the same asymptotes, domain, and range.

LOOKING FOR STRUCTURE

Because the function is of the form $g(x) = a \cdot f(x)$, where $a = 4$, the graph of g is a vertical stretch by a factor of 4 of the graph of f.

Monitoring Progress

Help in English and Spanish at *BigIdeasMath.com*

1. Graph $g(x) = \frac{-6}{x}$. Compare the graph with the graph of $f(x) = \frac{1}{x}$.

Translating Simple Rational Functions

Core Concept

Graphing Translations of Simple Rational Functions

To graph a rational function of the form $y = \frac{a}{x - h} + k$, follow these steps:

Step 1 Draw the asymptotes $x = h$ and $y = k$.

Step 2 Plot points to the left and to the right of the vertical asymptote.

Step 3 Draw the two branches of the hyperbola so that they pass through the plotted points and approach the asymptotes.

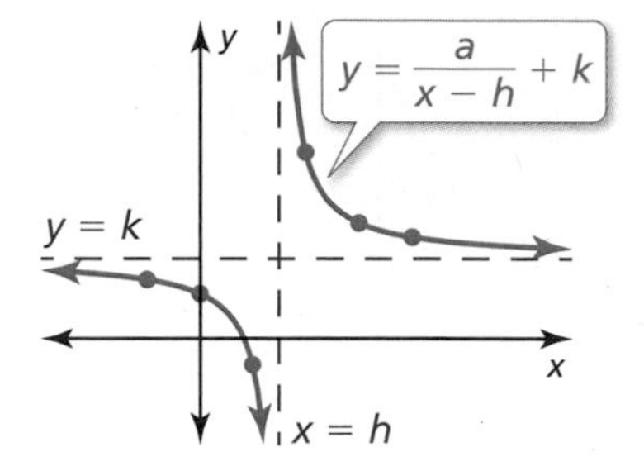

EXAMPLE 2 Graphing a Translation of a Rational Function

Graph $g(x) = \frac{-4}{x + 2} - 1$. State the domain and range.

SOLUTION

Step 1 Draw the asymptotes $x = -2$ and $y = -1$.

Step 2 Plot points to the left of the vertical asymptote, such as $(-3, 3)$, $(-4, 1)$, and $(-6, 0)$. Plot points to the right of the vertical asymptote, such as $(-1, -5)$, $(0, -3)$, and $(2, -2)$.

Step 3 Draw the two branches of the hyperbola so that they pass through the plotted points and approach the asymptotes.

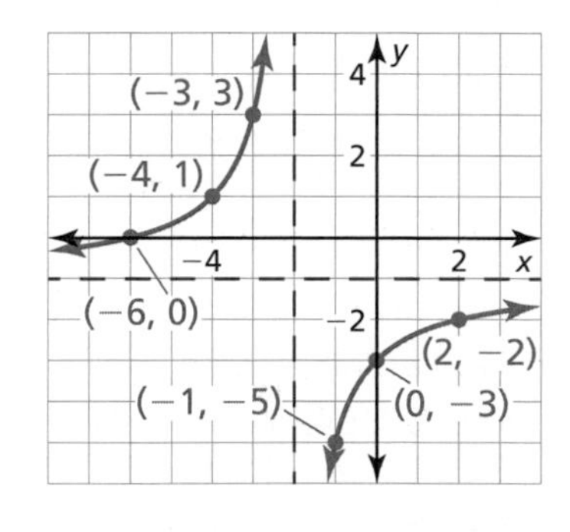

▶ The domain is all real numbers except -2 and the range is all real numbers except -1.

LOOKING FOR STRUCTURE

Let $f(x) = \frac{-4}{x}$. Notice that g is of the form $g(x) = f(x - h) + k$, where $h = -2$ and $k = -1$. So, the graph of g is a translation 2 units left and 1 unit down of the graph of f.

Monitoring Progress

Help in English and Spanish at *BigIdeasMath.com*

Graph the function. State the domain and range.

2. $y = \frac{3}{x} - 2$ **3.** $y = \frac{-1}{x + 4}$ **4.** $y = \frac{1}{x - 1} + 5$

Graphing Other Rational Functions

All rational functions of the form $y = \frac{ax + b}{cx + d}$ also have graphs that are hyperbolas.

- The vertical asymptote of the graph is the line $x = -\frac{d}{c}$ because the function is undefined when the denominator $cx + d$ is zero.
- The horizontal asymptote is the line $y = \frac{a}{c}$.

EXAMPLE 3 Graphing a Rational Function of the Form $y = \dfrac{ax + b}{cx + d}$

Graph $f(x) = \dfrac{2x + 1}{x - 3}$. State the domain and range.

SOLUTION

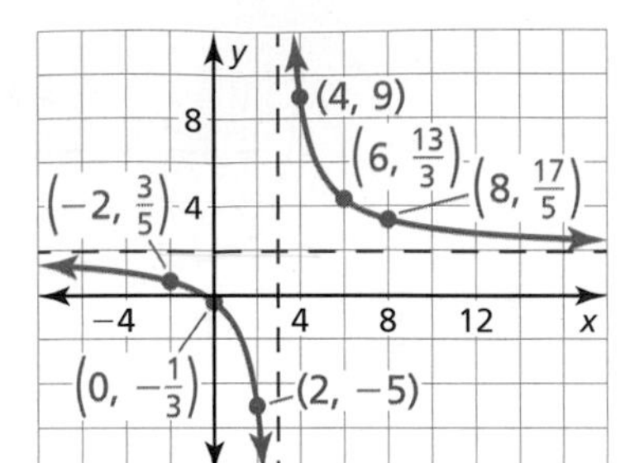

Step 1 Draw the asymptotes. Solve $x - 3 = 0$ for x to find the vertical asymptote $x = 3$. The horizontal asymptote is the line $y = \frac{a}{c} = \frac{2}{1} = 2$.

Step 2 Plot points to the left of the vertical asymptote, such as $(2, -5)$, $\left(0, -\frac{1}{3}\right)$, and $\left(-2, \frac{3}{5}\right)$. Plot points to the right of the vertical asymptote, such as $(4, 9)$, $\left(6, \frac{13}{3}\right)$, and $\left(8, \frac{17}{5}\right)$.

Step 3 Draw the two branches of the hyperbola so that they pass through the plotted points and approach the asymptotes.

▶ The domain is all real numbers except 3 and the range is all real numbers except 2.

Rewriting a rational function may reveal properties of the function and its graph. For example, rewriting a rational function in the form $y = \dfrac{a}{x - h} + k$ reveals that it is a translation of $y = \dfrac{a}{x}$ with vertical asymptote $x = h$ and horizontal asymptote $y = k$.

EXAMPLE 4 Rewriting and Graphing a Rational Function

Rewrite $g(x) = \dfrac{3x + 5}{x + 1}$ in the form $g(x) = \dfrac{a}{x - h} + k$. Graph the function. Describe the graph of g as a transformation of the graph of $f(x) = \dfrac{a}{x}$.

ANOTHER WAY

You will use a different method to rewrite g in Example 5 of Lesson 7.4.

SOLUTION

Rewrite the function by using long division:

$$\begin{array}{r} 3 \\ x + 1 \overline{)\, 3x + 5} \\ \underline{3x + 3} \\ 2 \end{array}$$

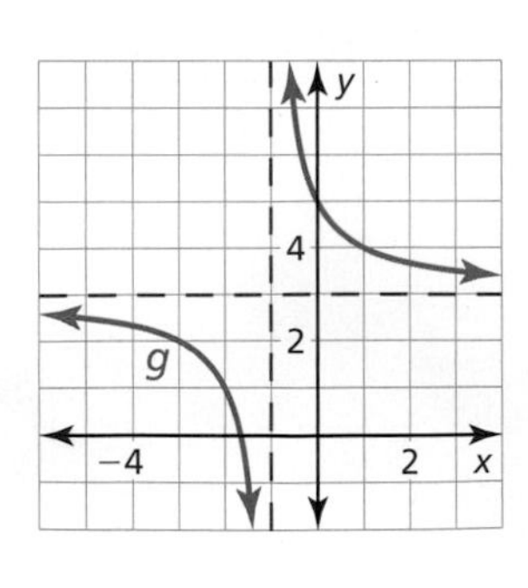

▶ The rewritten function is $g(x) = \dfrac{2}{x + 1} + 3$. The graph of g is a translation 1 unit left and 3 units up of the graph of $f(x) = \dfrac{2}{x}$.

Monitoring Progress

Help in English and Spanish at *BigIdeasMath.com*

Graph the function. State the domain and range.

5. $f(x) = \dfrac{x - 1}{x + 3}$ **6.** $f(x) = \dfrac{2x + 1}{4x - 2}$ **7.** $f(x) = \dfrac{-3x + 2}{-x - 1}$

8. Rewrite $g(x) = \dfrac{2x + 3}{x + 1}$ in the form $g(x) = \dfrac{a}{x - h} + k$. Graph the function. Describe the graph of g as a transformation of the graph of $f(x) = \dfrac{a}{x}$.

EXAMPLE 5 Modeling with Mathematics

A 3-D printer builds up layers of materials to make three-dimensional models. Each deposited layer bonds to the layer below it. A company decides to make small display models of engine components using a 3-D printer. The printer costs \$1000. The material for each model costs \$50.

- Estimate how many models must be printed for the average cost per model to fall to \$90.
- What happens to the average cost as more models are printed?

SOLUTION

1. **Understand the Problem** You are given the cost of a printer and the cost to create a model using the printer. You are asked to find the number of models for which the average cost falls to \$90.

2. **Make a Plan** Write an equation that represents the average cost. Use a graphing calculator to estimate the number of models for which the average cost is about \$90. Then analyze the horizontal asymptote of the graph to determine what happens to the average cost as more models are printed.

USING A GRAPHING CALCULATOR

Because the number of models and average cost cannot be negative, choose a viewing window in the first quadrant.

3. **Solve the Problem** Let c be the average cost (in dollars) and m be the number of models printed.

$$c = \frac{\text{(Unit cost)(Number printed)} + \text{(Cost of printer)}}{\text{Number printed}} = \frac{50m + 1000}{m}$$

Use a graphing calculator to graph the function.

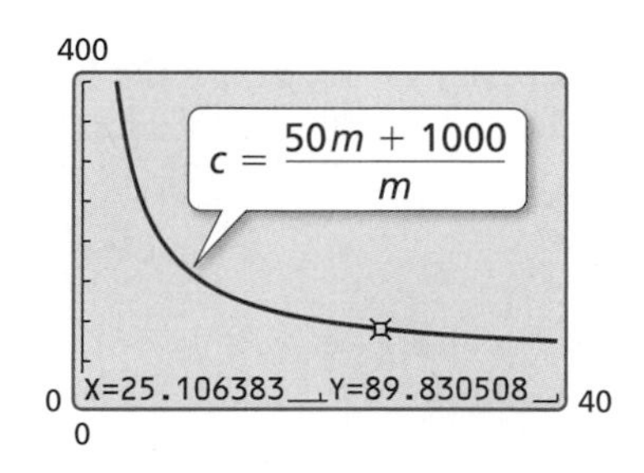

▶ Using the *trace* feature, the average cost falls to \$90 per model after about 25 models are printed. Because the horizontal asymptote is $c = 50$, the average cost approaches \$50 as more models are printed.

4. **Look Back** Use a graphing calculator to create tables of values for large values of m. The tables show that the average cost approaches \$50 as more models are printed.

X	Y1
0	ERROR
50	70
100	60
150	56.667
200	55
250	54
300	53.333
X=0	

X	Y1
0	ERROR
10000	50.1
20000	50.05
30000	50.033
40000	50.025
50000	50.02
60000	50.017
X=0	

Monitoring Progress  Help in English and Spanish at *BigIdeasMath.com*

9. **WHAT IF?** How do the answers in Example 5 change when the cost of the 3-D printer is \$800?

7.2 Exercises

Dynamic Solutions available at *BigIdeasMath.com*

Vocabulary and Core Concept Check

1. **COMPLETE THE SENTENCE** The function $y = \frac{7}{x+4} + 3$ has a(n) _________ of all real numbers except 3 and a(n) _________ of all real numbers except -4.

2. **WRITING** Is $f(x) = \frac{-3x+5}{2^x+1}$ a rational function? Explain your reasoning.

Monitoring Progress and Modeling with Mathematics

In Exercises 3–10, graph the function. Compare the graph with the graph of $f(x) = \frac{1}{x}$. *(See Example 1.)*

3. $g(x) = \frac{3}{x}$

4. $g(x) = \frac{10}{x}$

5. $g(x) = \frac{-5}{x}$

6. $g(x) = \frac{-9}{x}$

7. $g(x) = \frac{15}{x}$

8. $g(x) = \frac{-12}{x}$

9. $g(x) = \frac{-0.5}{x}$

10. $g(x) = \frac{0.1}{x}$

In Exercises 11–18, graph the function. State the domain and range. *(See Example 2.)*

11. $g(x) = \frac{4}{x} + 3$

12. $y = \frac{2}{x} - 3$

13. $h(x) = \frac{6}{x-1}$

14. $y = \frac{1}{x+2}$

15. $h(x) = \frac{-3}{x+2}$

16. $f(x) = \frac{-2}{x-7}$

17. $g(x) = \frac{-3}{x-4} - 1$

18. $y = \frac{10}{x+7} - 5$

ERROR ANALYSIS In Exercises 19 and 20, describe and correct the error in graphing the rational function.

19. $y = \frac{-8}{x}$

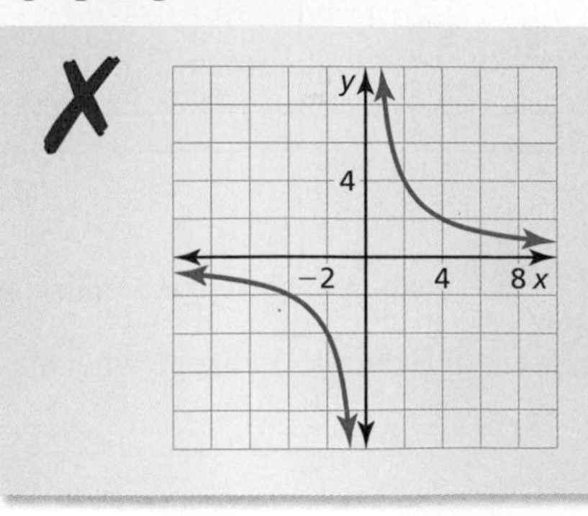

20. $y = \frac{2}{x-1} - 2$

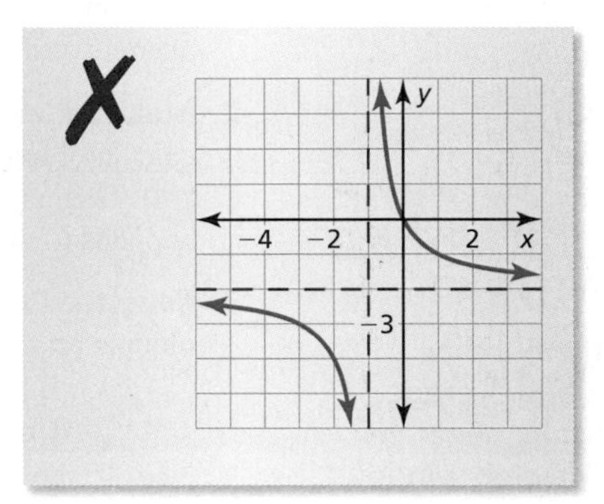

ANALYZING RELATIONSHIPS In Exercises 21–24, match the function with its graph. Explain your reasoning.

21. $g(x) = \frac{2}{x-3} + 1$

22. $h(x) = \frac{2}{x+3} + 1$

23. $f(x) = \frac{2}{x-3} - 1$

24. $y = \frac{2}{x+3} - 1$

A.

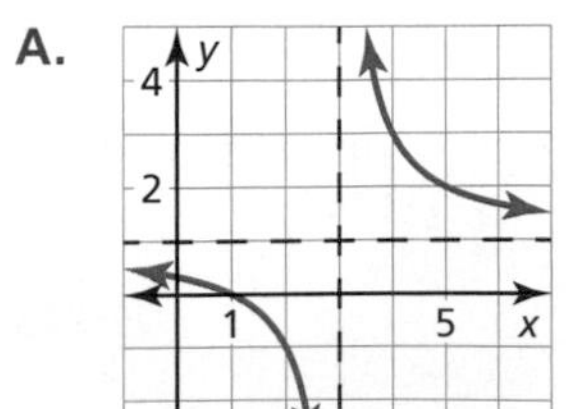

B.

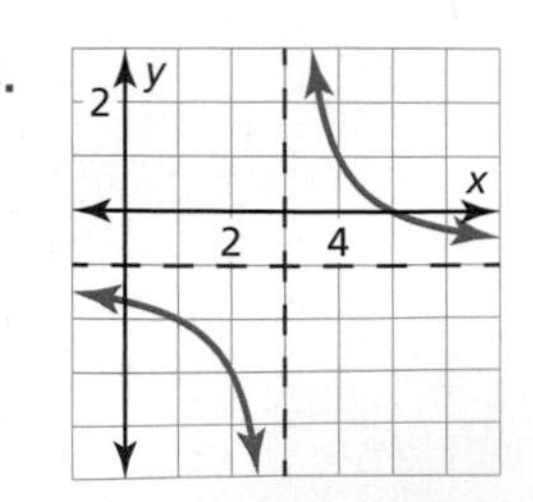

C.

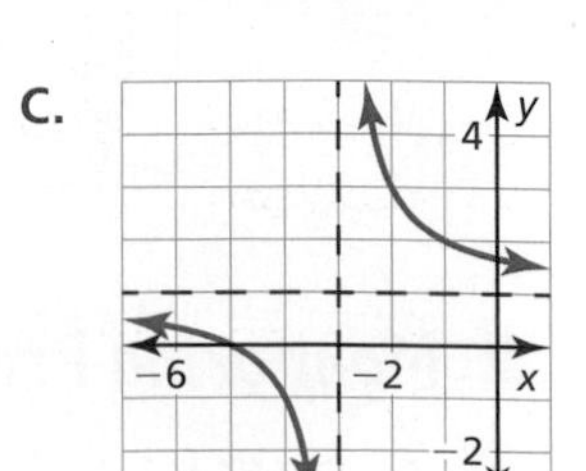

D.

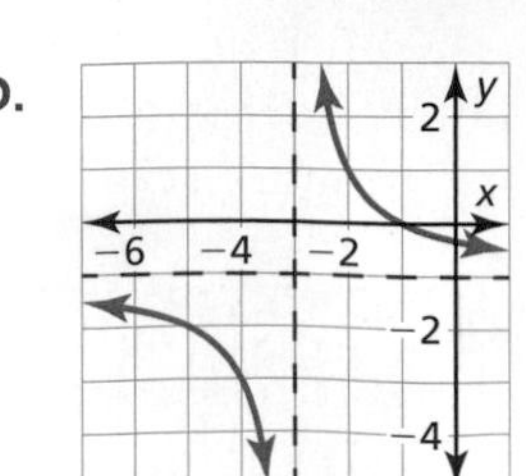

In Exercises 25–32, graph the function. State the domain and range. *(See Example 3.)*

25. $f(x) = \frac{x + 4}{x - 3}$

26. $y = \frac{x - 1}{x + 5}$

27. $y = \frac{x + 6}{4x - 8}$

28. $h(x) = \frac{8x + 3}{2x - 6}$

29. $f(x) = \frac{-5x + 2}{4x + 5}$

30. $g(x) = \frac{6x - 1}{3x - 1}$

31. $h(x) = \frac{-5x}{-2x - 3}$

32. $y = \frac{-2x + 3}{-x + 10}$

In Exercises 33–40, rewrite the function in the form $g(x) = \frac{a}{x - h} + k$. Graph the function. Describe the graph of g as a transformation of the graph of $f(x) = \frac{a}{x}$. *(See Example 4.)*

33. $g(x) = \frac{5x + 6}{x + 1}$

34. $g(x) = \frac{7x + 4}{x - 3}$

35. $g(x) = \frac{2x - 4}{x - 5}$

36. $g(x) = \frac{4x - 11}{x - 2}$

37. $g(x) = \frac{x + 18}{x - 6}$

38. $g(x) = \frac{x + 2}{x - 8}$

39. $g(x) = \frac{7x - 20}{x + 13}$

40. $g(x) = \frac{9x - 3}{x + 7}$

41. PROBLEM SOLVING Your school purchases a math software program. The program has an initial cost of \$500 plus \$20 for each student that uses the program. *(See Example 5.)*

a. Estimate how many students must use the program for the average cost per student to fall to \$30.

b. What happens to the average cost as more students use the program?

42. PROBLEM SOLVING To join a rock climbing gym, you must pay an initial fee of \$100 and a monthly fee of \$59.

a. Estimate how many months you must purchase a membership for the average cost per month to fall to \$69.

b. What happens to the average cost as the number of months that you are a member increases?

43. USING STRUCTURE What is the vertical asymptote of the graph of the function $y = \frac{2}{x + 4} + 7$?

Ⓐ $x = -7$
Ⓑ $x = -4$
Ⓒ $x = 4$
Ⓓ $x = 7$

44. REASONING What are the x-intercept(s) of the graph of the function $y = \frac{x - 5}{x^2 - 1}$?

Ⓐ $1, -1$
Ⓑ 5
Ⓒ 1
Ⓓ -5

45. USING TOOLS The time t (in seconds) it takes for sound to travel 1 kilometer can be modeled by

$$t = \frac{1000}{0.6T + 331}$$

where T is the air temperature (in degrees Celsius).

a. You are 1 kilometer from a lightning strike. You hear the thunder 2.9 seconds later. Use a graph to find the approximate air temperature.

b. Find the average rate of change in the time it takes sound to travel 1 kilometer as the air temperature increases from 0°C to 10°C.

46. MODELING WITH MATHEMATICS A business is studying the cost to remove a pollutant from the ground at its site. The function $y = \frac{15x}{1.1 - x}$ models the estimated cost y (in thousands of dollars) to remove x percent (expressed as a decimal) of the pollutant.

a. Graph the function. Describe a reasonable domain and range.

b. How much does it cost to remove 20% of the pollutant? 40% of the pollutant? 80% of the pollutant? Does doubling the percentage of the pollutant removed double the cost? Explain.

USING TOOLS In Exercises 47–50, use a graphing calculator to graph the function. Then determine whether the function is *even*, *odd*, or *neither*.

47. $h(x) = \frac{6}{x^2 + 1}$

48. $f(x) = \frac{2x^2}{x^2 - 9}$

49. $y = \frac{x^3}{3x^2 + x^4}$

50. $f(x) = \frac{4x^2}{2x^3 - x}$

51. **MAKING AN ARGUMENT** Your friend claims it is possible for a rational function to have two vertical asymptotes. Is your friend correct? Justify your answer.

52. **HOW DO YOU SEE IT?** Use the graph of f to determine the equations of the asymptotes. Explain.

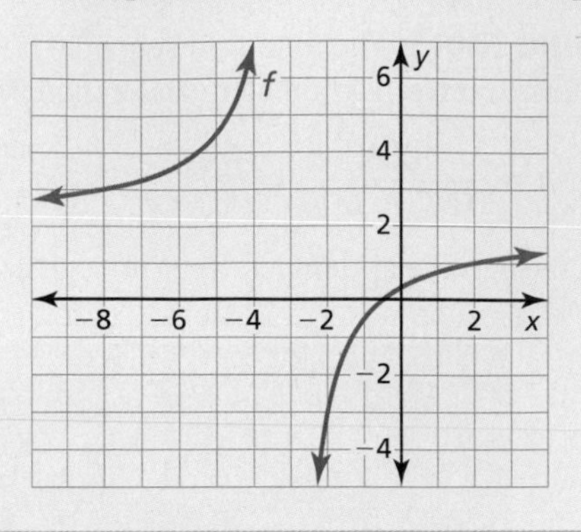

53. **DRAWING CONCLUSIONS** In what line(s) is the graph of $y = \frac{1}{x}$ symmetric? What does this symmetry tell you about the inverse of the function $f(x) = \frac{1}{x}$?

54. **THOUGHT PROVOKING** There are four basic types of conic sections: parabola, circle, ellipse, and hyperbola. Each of these can be represented by the intersection of a double-napped cone and a plane. The intersections for a parabola, circle, and ellipse are shown below. Sketch the intersection for a hyperbola.

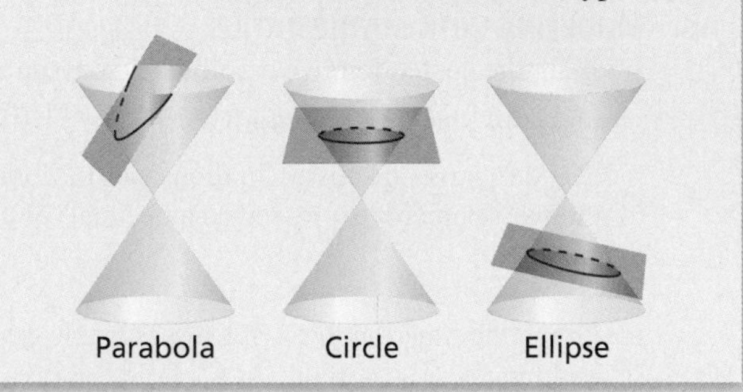

55. **REASONING** The graph of the rational function f is a hyperbola. The asymptotes of the graph of f intersect at (3, 2). The point (2, 1) is on the graph. Find another point on the graph. Explain your reasoning.

56. **ABSTRACT REASONING** Describe the intervals where the graph of $y = \frac{a}{x}$ is increasing or decreasing when (a) $a > 0$ and (b) $a < 0$. Explain your reasoning.

57. **PROBLEM SOLVING** An Internet service provider charges a \$50 installation fee and a monthly fee of \$43. The table shows the average monthly costs y of a competing provider for x months of service. Under what conditions would a person choose one provider over the other? Explain your reasoning.

Months, x	Average monthly cost (dollars), y
6	\$49.83
12	\$46.92
18	\$45.94
24	\$45.45

58. **MODELING WITH MATHEMATICS** The Doppler effect occurs when the source of a sound is moving relative to a listener, so that the frequency f_ℓ (in hertz) heard by the listener is different from the frequency f_s (in hertz) at the source. In both equations below, r is the speed (in miles per hour) of the sound source.

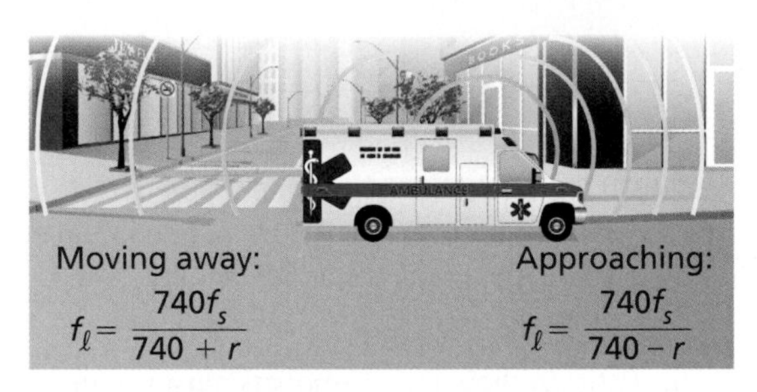

a. An ambulance siren has a frequency of 2000 hertz. Write two equations modeling the frequencies heard when the ambulance is approaching and when the ambulance is moving away.

b. Graph the equations in part (a) using the domain $0 \le r \le 60$.

c. For any speed r, how does the frequency heard for an approaching sound source compare with the frequency heard when the source moves away?

Maintaining Mathematical Proficiency

Reviewing what you learned in previous grades and lessons

Factor the polynomial. *(Skills Review Handbook)*

59. $4x^2 - 4x - 80$

60. $3x^2 - 3x - 6$

61. $2x^2 - 2x - 12$

62. $10x^2 + 31x - 14$

Simplify the expression. *(Section 5.2)*

63. $3^2 \cdot 3^4$

64. $2^{1/2} \cdot 2^{3/5}$

65. $\frac{6^{5/6}}{6^{1/6}}$

66. $\frac{6^8}{6^{10}}$

7.1–7.2 What Did You Learn?

Core Vocabulary

inverse variation, *p. 360*
constant of variation, *p. 360*
rational function, *p. 366*

Core Concepts

Section 7.1

Inverse Variation, *p. 360*
Writing Inverse Variation Equations, *p. 361*

Section 7.2

Parent Function for Simple Rational Functions, *p. 366*
Graphing Translations of Simple Rational Functions, *p. 367*

Mathematical Practices

1. Explain the meaning of the given information in Exercise 25 on page 364.
2. How are you able to recognize whether the logic used in Exercise 29 on page 364 is correct or flawed?
3. How can you evaluate the reasonableness of your answer in part (b) of Exercise 41 on page 371?
4. How did the context allow you to determine a reasonable domain and range for the function in Exercise 46 on page 371?

Study Skills

Analyzing Your Errors

Study Errors

What Happens: You do not study the right material or you do not learn it well enough to remember it on a test without resources such as notes.

How to Avoid This Error: Take a practice test. Work with a study group. Discuss the topics on the test with your teacher. Do not try to learn a whole chapter's worth of material in one night.

7.1–7.2 Quiz

Tell whether x and y show *direct variation*, *inverse variation*, or *neither*. Explain your reasoning. *(Section 7.1)*

1. $x + y = 7$

2. $\frac{2}{5}x = y$

3. $xy = 0.45$

4.

x	3	6	9	12
y	9	18	27	36

5.

x	1	2	3	4
y	−24	−12	−8	−6

6.

x	2	4	6	8
y	72	36	18	9

7. The variables x and y vary inversely, and $y = 10$ when $x = 5$. Write an equation that relates x and y. Then find y when $x = -2$. *(Section 7.1)*

Match the equation with the correct graph. Explain your reasoning. *(Section 7.2)*

8. $f(x) = \frac{3}{x} + 2$

9. $y = \frac{-2}{x+3} - 2$

10. $h(x) = \frac{2x+2}{3x+1}$

A.

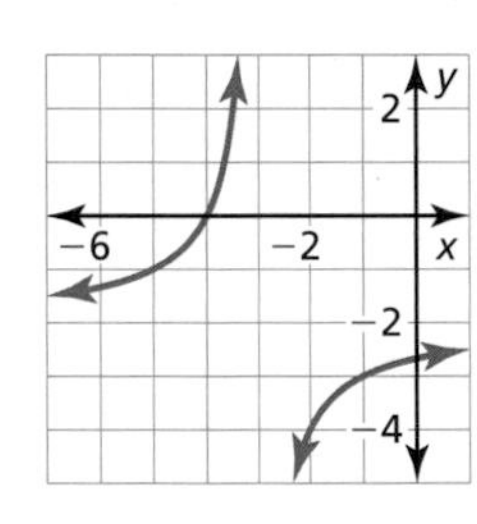

B.

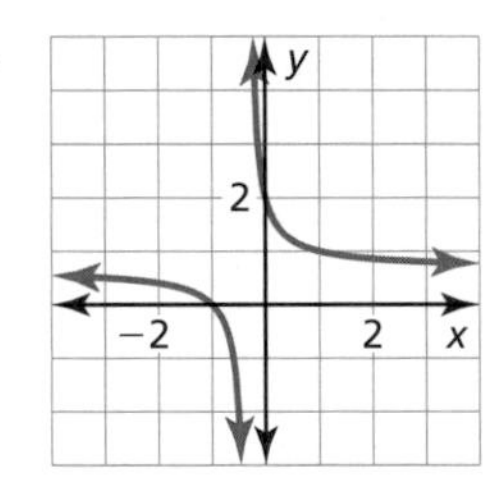

C.

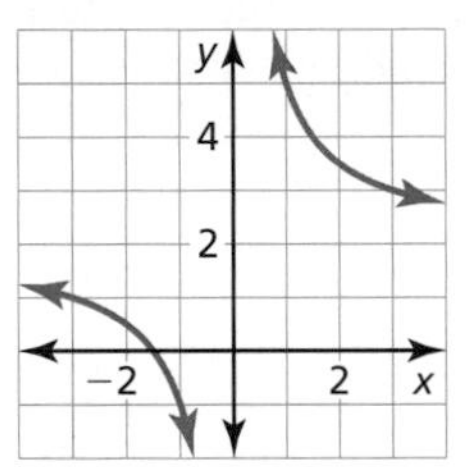

11. Rewrite $g(x) = \frac{2x+9}{x+8}$ in the form $g(x) = \frac{a}{x-h} + k$. Graph the function. Describe the graph of g as a transformation of the graph of $f(x) = \frac{a}{x}$. *(Section 7.2)*

12. The time t (in minutes) required to empty a tank varies inversely with the pumping rate r (in gallons per minute). The rate of a certain pump is 70 gallons per minute. It takes the pump 20 minutes to empty the tank. Complete the table for the times it takes the pump to empty a tank for the given pumping rates. *(Section 7.1)*

Pumping rate (gal/min)	Time (min)
50	
60	
65	
70	

13. A pitcher throws 16 strikes in the first 38 pitches. The table shows how a pitcher's strike percentage changes when the pitcher throws x consecutive strikes after the first 38 pitches. Write a rational function for the strike percentage in terms of x. Graph the function. How many consecutive strikes must the pitcher throw to reach a strike percentage of 0.60? *(Section 7.2)*

x	Total strikes	Total pitches	Strike percentage
0	16	38	0.42
5	21	43	0.49
10	26	48	0.54
x	$x + 16$	$x + 38$	

7.3 Multiplying and Dividing Rational Expressions

Essential Question How can you determine the excluded values in a product or quotient of two rational expressions?

You can multiply and divide rational expressions in much the same way that you multiply and divide fractions. Values that make the denominator of an expression zero are *excluded values*.

$$\frac{1}{\cancel{x}} \cdot \frac{\cancel{x}}{x+1} = \frac{1}{x+1}, \; x \neq 0 \qquad \text{Product of rational expressions}$$

$$\frac{1}{x} \div \frac{x}{x+1} = \frac{1}{x} \cdot \frac{x+1}{x} = \frac{x+1}{x^2}, \; x \neq -1 \qquad \text{Quotient of rational expressions}$$

EXPLORATION 1 Multiplying and Dividing Rational Expressions

Work with a partner. Find the product or quotient of the two rational expressions. Then match the product or quotient with its excluded values. Explain your reasoning.

Product or Quotient	Excluded Values
a. $\frac{1}{x-1} \cdot \frac{x-2}{x+1} =$	**A.** -1, 0, and 2
b. $\frac{1}{x-1} \cdot \frac{-1}{x-1} =$	**B.** -2 and 1
c. $\frac{1}{x-2} \cdot \frac{x-2}{x+1} =$	**C.** -2, 0, and 1
d. $\frac{x+2}{x-1} \cdot \frac{-x}{x+2} =$	**D.** -1 and 2
e. $\frac{x}{x+2} \div \frac{x+1}{x+2} =$	**E.** -1, 0, and 1
f. $\frac{x}{x-2} \div \frac{x+1}{x} =$	**F.** -1 and 1
g. $\frac{x}{x+2} \div \frac{x}{x-1} =$	**G.** -2 and -1
h. $\frac{x+2}{x} \div \frac{x+1}{x-1} =$	**H.** 1

REASONING ABSTRACTLY

To be proficient in math, you need to know and flexibly use different properties of operations and objects.

EXPLORATION 2 Writing a Product or Quotient

Work with a partner. Write a product or quotient of rational expressions that has the given excluded values. Justify your answer.

a. -1 **b.** -1 and 3 **c.** -1, 0, and 3

Communicate Your Answer

3. How can you determine the excluded values in a product or quotient of two rational expressions?

4. Is it possible for the product or quotient of two rational expressions to have *no* excluded values? Explain your reasoning. If it is possible, give an example.

7.3 Lesson

What You Will Learn

- Simplify rational expressions.
- Multiply rational expressions.
- Divide rational expressions.

Core Vocabulary

rational expression, *p. 376*
simplified form of a rational expression, *p. 376*

Previous
fractions
polynomials
domain
equivalent expressions
reciprocal

Simplifying Rational Expressions

A **rational expression** is a fraction whose numerator and denominator are nonzero polynomials. The *domain* of a rational expression excludes values that make the denominator zero. A rational expression is in **simplified form** when its numerator and denominator have no common factors (other than ± 1).

Core Concept

Simplifying Rational Expressions

Let a, b, and c be expressions with $b \neq 0$ and $c \neq 0$.

Property $\quad \frac{a\cancel{c}}{b\cancel{c}} = \frac{a}{b}$ $\quad$ Divide out common factor c.

Examples $\quad \frac{15}{65} = \frac{3 \cdot \cancel{5}}{13 \cdot \cancel{5}} = \frac{3}{13}$ $\quad$ Divide out common factor 5.

$\frac{4\cancel{(x+3)}}{(x+3)\cancel{(x+3)}} = \frac{4}{x+3}$ $\quad$ Divide out common factor $x + 3$.

STUDY TIP

Notice that you can divide out common factors in the second expression at the right. You cannot, however, divide out like terms in the third expression.

Simplifying a rational expression usually requires two steps. First, factor the numerator and denominator. Then, divide out any factors that are common to both the numerator and denominator. Here is an example:

$$\frac{x^2 + 7x}{x^2} = \frac{x(x+7)}{x \cdot x} = \frac{x+7}{x}$$

EXAMPLE 1 Simplifying a Rational Expression

Simplify $\frac{x^2 - 4x - 12}{x^2 - 4}$.

SOLUTION

$$\frac{x^2 - 4x - 12}{x^2 - 4} = \frac{(x+2)(x-6)}{(x+2)(x-2)} \quad \text{Factor numerator and denominator.}$$

$$= \frac{\cancel{(x+2)}(x-6)}{\cancel{(x+2)}(x-2)} \quad \text{Divide out common factor.}$$

$$= \frac{x-6}{x-2},\quad x \neq -2 \quad \text{Simplified form}$$

COMMON ERROR

Do not divide out variable terms that are not factors.

$$\frac{x-6}{x-2} \neq \frac{-6}{-2}$$

The original expression is undefined when $x = -2$. To make the original and simplified expressions equivalent, restrict the domain of the simplified expression by excluding $x = -2$. Both expressions are undefined when $x = 2$, so it is not necessary to list it.

Monitoring Progress  Help in English and Spanish at *BigIdeasMath.com*

Simplify the rational expression, if possible.

1. $\frac{2(x+1)}{(x+1)(x+3)}$
2. $\frac{x+4}{x^2-16}$
3. $\frac{4}{x(x+2)}$
4. $\frac{x^2-2x-3}{x^2-x-6}$

Multiplying Rational Expressions

The rule for multiplying rational expressions is the same as the rule for multiplying numerical fractions: multiply numerators, multiply denominators, and write the new fraction in simplified form. Similar to rational numbers, rational expressions are closed under multiplication.

Core Concept

Multiplying Rational Expressions

Let a, b, c, and d be expressions with $b \neq 0$ and $d \neq 0$.

Property $\frac{a}{b} \cdot \frac{c}{d} = \frac{ac}{bd}$ Simplify $\frac{ac}{bd}$ if possible.

Example $\frac{5x^2}{2xy^2} \cdot \frac{6xy^3}{10y} = \frac{30x^3y^3}{20xy^3} = \frac{\cancel{10} \cdot 3 \cdot \cancel{x} \cdot x^2 \cdot \cancel{y^3}}{\cancel{10} \cdot 2 \cdot \cancel{x} \cdot \cancel{y^3}} = \frac{3x^2}{2}, \quad x \neq 0, y \neq 0$

ANOTHER WAY

In Example 2, you can first simplify each rational expression, then multiply, and finally simplify the result.

$$\frac{8x^3y}{2xy^2} \cdot \frac{7x^4y^3}{4y}$$

$$= \frac{4x^2}{y} \cdot \frac{7x^4y^2}{4}$$

$$= \frac{\cancel{4} \cdot 7 \cdot x^6 \cdot \cancel{y} \cdot y}{\cancel{4} \cdot \cancel{y}}$$

$$= 7x^6y, \quad x \neq 0, y \neq 0$$

EXAMPLE 2 Multiplying Rational Expressions

Find the product $\frac{8x^3y}{2xy^2} \cdot \frac{7x^4y^3}{4y}$.

SOLUTION

$\frac{8x^3y}{2xy^2} \cdot \frac{7x^4y^3}{4y} = \frac{56x^7y^4}{8xy^3}$ Multiply numerators and denominators.

$= \frac{\cancel{8} \cdot 7 \cdot \cancel{x} \cdot x^6 \cdot \cancel{y^3} \cdot y}{\cancel{8} \cdot \cancel{x} \cdot \cancel{y^3}}$ Factor and divide out common factors.

$= 7x^6y, \quad x \neq 0, y \neq 0$ Simplified form

EXAMPLE 3 Multiplying Rational Expressions

Find the product $\frac{3x - 3x^2}{x^2 + 4x - 5} \cdot \frac{x^2 + x - 20}{3x}$.

SOLUTION

$\frac{3x - 3x^2}{x^2 + 4x - 5} \cdot \frac{x^2 + x - 20}{3x} = \frac{3x(1 - x)}{(x - 1)(x + 5)} \cdot \frac{(x + 5)(x - 4)}{3x}$ Factor numerators and denominators.

$= \frac{3x(1 - x)(x + 5)(x - 4)}{(x - 1)(x + 5)(3x)}$ Multiply numerators and denominators.

$= \frac{3x(-1)(x - 1)(x + 5)(x - 4)}{(x - 1)(x + 5)(3x)}$ Rewrite $1 - x$ as $(-1)(x - 1)$.

$= \frac{\cancel{3x}(-1)\cancel{(x - 1)}\cancel{(x + 5)}(x - 4)}{\cancel{(x - 1)}\cancel{(x + 5)}\cancel{(3x)}}$ Divide out common factors.

$= -x + 4, \quad x \neq -5, x \neq 0, x \neq 1$ Simplified form

Check

X	Y1	Y2
-5	ERROR	9
-4	8	8
-3	7	7
-2	6	6
-1	5	5
0	ERROR	4
1	ERROR	3
X=-4		

Check the simplified expression. Enter the original expression as y_1 and the simplified expression as y_2 in a graphing calculator. Then use the *table* feature to compare the values of the two expressions. The values of y_1 and y_2 are the same, except when $x = -5$, $x = 0$, and $x = 1$. So, when these values are excluded from the domain of the simplified expression, it is equivalent to the original expression.

EXAMPLE 4 Multiplying a Rational Expression by a Polynomial

Find the product $\frac{x+2}{x^3-27} \cdot (x^2+3x+9)$.

STUDY TIP

Notice that $x^2 + 3x + 9$ does not equal zero for any real value of x. So, no values must be excluded from the domain to make the simplified form equivalent to the original.

SOLUTION

$$\frac{x+2}{x^3-27} \cdot (x^2+3x+9) = \frac{x+2}{x^3-27} \cdot \frac{x^2+3x+9}{1}$$ Write polynomial as a rational expression.

$$= \frac{(x+2)(x^2+3x+9)}{(x-3)(x^2+3x+9)}$$ Multiply. Factor denominator.

$$= \frac{(x+2)\cancel{(x^2+3x+9)}}{(x-3)\cancel{(x^2+3x+9)}}$$ Divide out common factors.

$$= \frac{x+2}{x-3}$$ Simplified form

Monitoring Progress

Help in English and Spanish at *BigIdeasMath.com*

Find the product.

5. $\frac{3x^5y^2}{8xy} \cdot \frac{6xy^2}{9x^3y}$

6. $\frac{2x^2-10x}{x^2-25} \cdot \frac{x+3}{2x^2}$

7. $\frac{x+5}{x^3-1} \cdot (x^2+x+1)$

Dividing Rational Expressions

To divide one rational expression by another, multiply the first rational expression by the reciprocal of the second rational expression. Rational expressions are closed under nonzero division.

Core Concept

Dividing Rational Expressions

Let a, b, c, and d be expressions with $b \neq 0$, $c \neq 0$, and $d \neq 0$.

Property $\frac{a}{b} \div \frac{c}{d} = \frac{a}{b} \cdot \frac{d}{c} = \frac{ad}{bc}$ Simplify $\frac{ad}{bc}$ if possible.

Example $\frac{7}{x+1} \div \frac{x+2}{2x-3} = \frac{7}{x+1} \cdot \frac{2x-3}{x+2} = \frac{7(2x-3)}{(x+1)(x+2)}, x \neq \frac{3}{2}$

EXAMPLE 5 Dividing Rational Expressions

Find the quotient $\frac{7x}{2x-10} \div \frac{x^2-6x}{x^2-11x+30}$.

SOLUTION

$$\frac{7x}{2x-10} \div \frac{x^2-6x}{x^2-11x+30} = \frac{7x}{2x-10} \cdot \frac{x^2-11x+30}{x^2-6x}$$ Multiply by reciprocal.

$$= \frac{7x}{2(x-5)} \cdot \frac{(x-5)(x-6)}{x(x-6)}$$ Factor.

$$= \frac{7\cancel{x}\cancel{(x-5)}\cancel{(x-6)}}{2\cancel{(x-5)}\cancel{(x)}\cancel{(x-6)}}$$ Multiply. Divide out common factors.

$$= \frac{7}{2}, \quad x \neq 0, x \neq 5, x \neq 6$$ Simplified form

EXAMPLE 6 Dividing a Rational Expression by a Polynomial

Find the quotient $\frac{6x^2 + x - 15}{4x^2} \div (3x^2 + 5x)$.

SOLUTION

$$\frac{6x^2 + x - 15}{4x^2} \div (3x^2 + 5x) = \frac{6x^2 + x - 15}{4x^2} \cdot \frac{1}{3x^2 + 5x}$$ Multiply by reciprocal.

$$= \frac{(3x + 5)(2x - 3)}{4x^2} \cdot \frac{1}{x(3x + 5)}$$ Factor.

$$= \frac{\cancel{(3x + 5)}(2x - 3)}{4x^2(x)\cancel{(3x + 5)}}$$ Divide out common factors.

$$= \frac{2x - 3}{4x^3}, \quad x \neq -\frac{5}{3}$$ Simplified form

EXAMPLE 7 Solving a Real-Life Problem

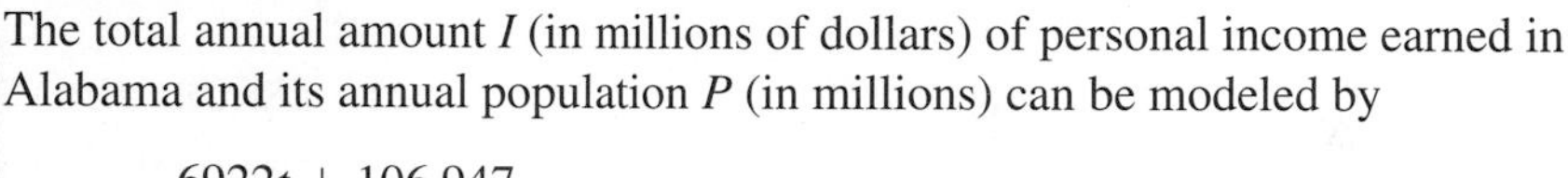

The total annual amount I (in millions of dollars) of personal income earned in Alabama and its annual population P (in millions) can be modeled by

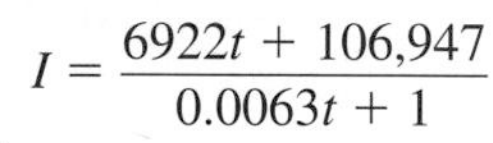

$$I = \frac{6922t + 106{,}947}{0.0063t + 1}$$

and

$$P = 0.0343t + 4.432$$

where t represents the year, with $t = 1$ corresponding to 2001. Find a model M for the annual per capita income. (Per capita means per person.) Estimate the per capita income in 2010. (Assume $t > 0$.)

SOLUTION

To find a model M for the annual per capita income, divide the total amount I by the population P.

$$M = \frac{6922t + 106{,}947}{0.0063t + 1} \div (0.0343t + 4.432)$$ Divide I by P.

$$= \frac{6922t + 106{,}947}{0.0063t + 1} \cdot \frac{1}{0.0343t + 4.432}$$ Multiply by reciprocal.

$$= \frac{6922t + 106{,}947}{(0.0063t + 1)(0.0343t + 4.432)}$$ Multiply.

To estimate Alabama's per capita income in 2010, let $t = 10$ in the model.

$$M = \frac{6922 \cdot 10 + 106{,}947}{(0.0063 \cdot 10 + 1)(0.0343 \cdot 10 + 4.432)}$$ Substitute 10 for t.

$$\approx 34{,}707$$ Use a calculator.

▶ In 2010, the per capita income in Alabama was about $34,707.

Monitoring Progress

Help in English and Spanish at *BigIdeasMath.com*

Find the quotient.

8. $\frac{4x}{5x - 20} \div \frac{x^2 - 2x}{x^2 - 6x + 8}$

9. $\frac{2x^2 + 3x - 5}{6x} \div (2x^2 + 5x)$

7.3 Exercises

Dynamic Solutions available at *BigIdeasMath.com*

Vocabulary and Core Concept Check

1. **WRITING** Describe how to multiply and divide two rational expressions.

2. **WHICH ONE DOESN'T BELONG?** Which rational expression does *not* belong with the other three? Explain your reasoning.

$\frac{x-4}{x^2}$	$\frac{x^2+4x-12}{x^2+6x}$	$\frac{9+x}{3x^2}$	$\frac{x^2-x-12}{x^2-6x}$

Monitoring Progress and Modeling with Mathematics

In Exercises 3–10, simplify the expression, if possible. *(See Example 1.)*

3. $\frac{2x^2}{3x^2-4x}$

4. $\frac{7x^3-x^2}{2x^3}$

5. $\frac{x^2-3x-18}{x^2-7x+6}$

6. $\frac{x^2+13x+36}{x^2-7x+10}$

7. $\frac{x^2+11x+18}{x^3+8}$

8. $\frac{x^2-7x+12}{x^3-27}$

9. $\frac{32x^4-50}{4x^3-12x^2-5x+15}$

10. $\frac{3x^3-3x^2+7x-7}{27x^4-147}$

In Exercises 11–20, find the product. *(See Examples 2, 3, and 4.)*

11. $\frac{4xy^3}{x^2y} \cdot \frac{y}{8x}$

12. $\frac{48x^5y^3}{y^4} \cdot \frac{x^2y}{6x^3y^2}$

13. $\frac{x^2(x-4)}{x-3} \cdot \frac{(x-3)(x+6)}{x^3}$

14. $\frac{x^3(x+5)}{x-9} \cdot \frac{(x-9)(x+8)}{3x^3}$

15. $\frac{x^2-3x}{x-2} \cdot \frac{x^2+x-6}{x}$

16. $\frac{x^2-4x}{x-1} \cdot \frac{x^2+3x-4}{2x}$

17. $\frac{x^2+3x-4}{x^2+4x+4} \cdot \frac{2x^2+4x}{x^2-4x+3}$

18. $\frac{x^2-x-6}{4x^3} \cdot \frac{2x^2+2x}{x^2+5x+6}$

19. $\frac{x^2+5x-36}{x^2-49} \cdot (x^2-11x+28)$

20. $\frac{x^2-x-12}{x^2-16} \cdot (x^2+2x-8)$

21. **ERROR ANALYSIS** Describe and correct the error in simplifying the rational expression.

✗ $\frac{x^2+\cancel{16}^{\,2}x+\cancel{48}^{\,3}}{x^2+\cancel{8}_{\,1}x+\cancel{16}_{\,1}} = \frac{x^2+2x+3}{x^2+x+1}$

22. **ERROR ANALYSIS** Describe and correct the error in finding the product.

✗ $\frac{x^2-25}{3-x} \cdot \frac{x-3}{x+5} = \frac{(x+5)(x-5)}{3-x} \cdot \frac{x-3}{x+5}$

$= \frac{\cancel{(x+5)}(x-5)\cancel{(x-3)}}{\cancel{(3-x)}\cancel{(x+5)}}$

$= x-5, x \neq 3, x \neq -5$

23. **USING STRUCTURE** Which rational expression is in simplified form?

Ⓐ $\frac{x^2-x-6}{x^2+3x+2}$

Ⓑ $\frac{x^2+6x+8}{x^2+2x-3}$

Ⓒ $\frac{x^2-6x+9}{x^2-2x-3}$

Ⓓ $\frac{x^2+3x-4}{x^2+x-2}$

24. **COMPARING METHODS** Find the product below by multiplying the numerators and denominators, then simplifying. Then find the product by simplifying each expression, then multiplying. Which method do you prefer? Explain.

$\frac{4x^2y}{2x^3} \cdot \frac{12y^4}{24x^2}$

25. WRITING Compare the function $f(x) = \frac{(3x - 7)(x + 6)}{(3x - 7)}$ to the function $g(x) = x + 6$.

26. MODELING WITH MATHEMATICS Write a model in terms of x for the total area of the base of the building.

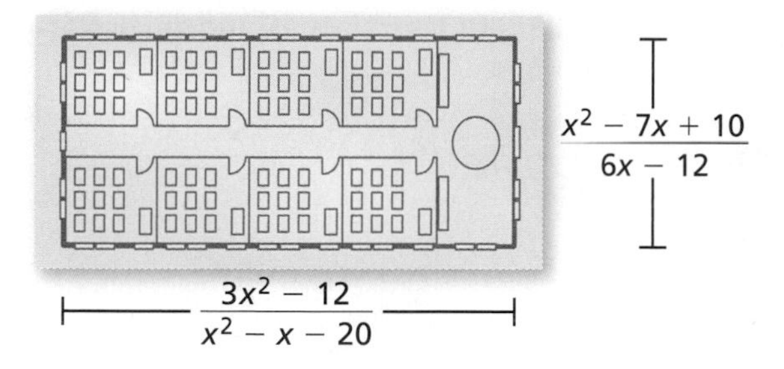

In Exercises 27–34, find the quotient. *(See Examples 5 and 6.)*

27. $\frac{32x^3y}{y^8} \div \frac{y^7}{8x^4}$

28. $\frac{2xyz}{x^3z^3} \div \frac{6y^4}{2x^2z^2}$

29. $\frac{x^2 - x - 6}{2x^4 - 6x^3} \div \frac{x + 2}{4x^3}$

30. $\frac{2x^2 - 12x}{x^2 - 7x + 6} \div \frac{2x}{3x - 3}$

31. $\frac{x^2 - x - 6}{x + 4} \div (x^2 - 6x + 9)$

32. $\frac{x^2 - 5x - 36}{x + 2} \div (x^2 - 18x + 81)$

33. $\frac{x^2 + 9x + 18}{x^2 + 6x + 8} \div \frac{x^2 - 3x - 18}{x^2 + 2x - 8}$

34. $\frac{x^2 - 3x - 40}{x^2 + 8x - 20} \div \frac{x^2 + 13x + 40}{x^2 + 12x + 20}$

In Exercises 35 and 36, use the following information.

Manufacturers often package products in a way that uses the least amount of material. One measure of the efficiency of a package is the ratio of its surface area S to its volume V. The smaller the ratio, the more efficient the packaging.

35. You are examining three cylindrical containers.

a. Write an expression for the efficiency ratio $\frac{S}{V}$ of a cylinder.

b. Find the efficiency ratio for each cylindrical can listed in the table. Rank the three cans according to efficiency.

	Soup	Coffee	Paint
Height, h	10.2 cm	15.9 cm	19.4 cm
Radius, r	3.4 cm	7.8 cm	8.4 cm

36. PROBLEM SOLVING A popcorn company is designing a new tin with the same square base and twice the height of the old tin.

a. Write an expression for the efficiency ratio $\frac{S}{V}$ of each tin.

b. Did the company make a good decision by creating the new tin? Explain.

37. MODELING WITH MATHEMATICS The total amount I (in millions of dollars) of healthcare expenditures and the residential population P (in millions) in the United States can be modeled by

$$I = \frac{171{,}000t + 1{,}361{,}000}{1 + 0.018t} \quad \text{and}$$

$$P = 2.96t + 278.649$$

where t is the number of years since 2000. Find a model M for the annual healthcare expenditures per resident. Estimate the annual healthcare expenditures per resident in 2010. *(See Example 7.)*

38. MODELING WITH MATHEMATICS The total amount I (in millions of dollars) of school expenditures from prekindergarten to a college level and the enrollment P (in millions) in prekindergarten through college in the United States can be modeled by

$$I = \frac{17{,}913t + 709{,}569}{1 - 0.028t} \text{ and } P = 0.5906t + 70.219$$

where t is the number of years since 2001. Find a model M for the annual education expenditures per student. Estimate the annual education expenditures per student in 2009.

39. USING EQUATIONS Refer to the population model P in Exercise 37.

a. Interpret the meaning of the coefficient of t.

b. Interpret the meaning of the constant term.

40. HOW DO YOU SEE IT? Use the graphs of f and g to determine the excluded values of the functions $h(x) = (fg)(x)$ and $k(x) = \left(\frac{f}{g}\right)(x)$. Explain your reasoning.

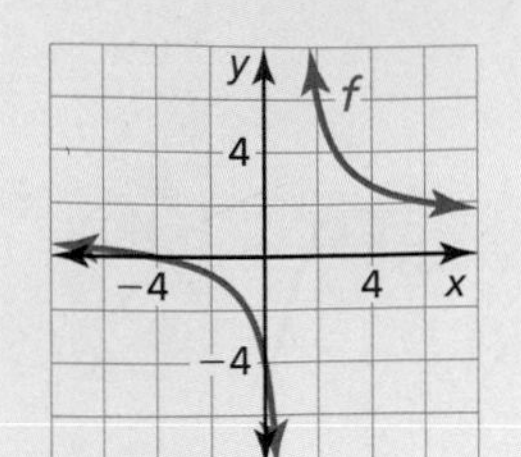

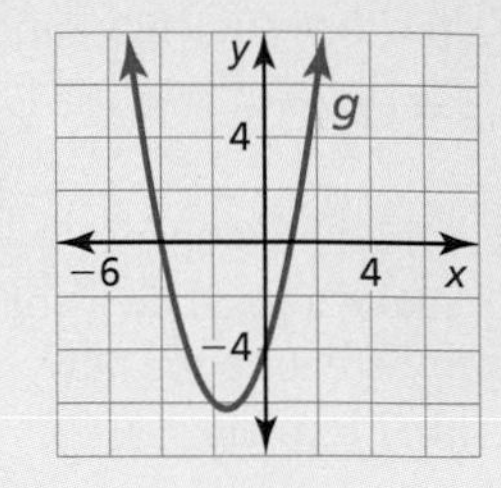

41. DRAWING CONCLUSIONS Complete the table for the function $y = \frac{x + 4}{x^2 - 16}$. Then use the *trace* feature of a graphing calculator to explain the behavior of the function at $x = -4$.

x	y
−3.5	
−3.8	
−3.9	
−4.1	
−4.2	

42. MAKING AN ARGUMENT You and your friend are asked to state the domain of the expression below.

$$\frac{x^2 + 6x - 27}{x^2 + 4x - 45}$$

Your friend claims the domain is all real numbers except 5. You claim the domain is all real numbers except −9 and 5. Who is correct? Explain.

43. MATHEMATICAL CONNECTIONS Find the ratio of the perimeter to the area of the triangle shown.

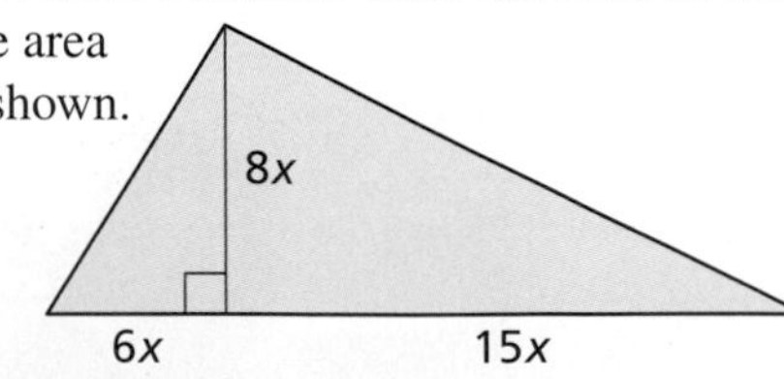

44. CRITICAL THINKING Find the expression that makes the following statement true. Assume $x \neq -2$ and $x \neq 5$.

$$\frac{x - 5}{x^2 + 2x - 35} \div \frac{\square}{x^2 - 3x - 10} = \frac{x + 2}{x + 7}$$

USING STRUCTURE In Exercises 45 and 46, perform the indicated operations.

45. $\frac{2x^2 + x - 15}{2x^2 - 11x - 21} \cdot (6x + 9) \div \frac{2x - 5}{3x - 21}$

46. $(x^3 + 8) \cdot \frac{x - 2}{x^2 - 2x + 4} \div \frac{x^2 - 4}{x - 6}$

47. REASONING Animals that live in temperatures several degrees colder than their bodies must avoid losing heat to survive. Animals can better conserve body heat as their surface area to volume ratios decrease. Find the surface area to volume ratio of each penguin shown by using cylinders to approximate their shapes. Which penguin is better equipped to live in a colder environment? Explain your reasoning.

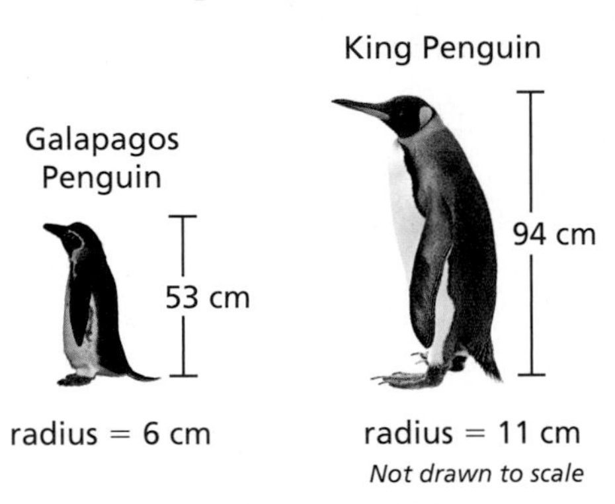

48. THOUGHT PROVOKING Is it possible to write two radical functions whose product when graphed is a parabola and whose quotient when graphed is a hyperbola? Justify your answer.

49. REASONING Find two rational functions f and g that have the stated product and quotient.

$$(fg)(x) = x^2, \left(\frac{f}{g}\right)(x) = \frac{(x - 1)^2}{(x + 2)^2}$$

Maintaining Mathematical Proficiency

Reviewing what you learned in previous grades and lessons

Solve the equation. Check your solution. *(Skills Review Handbook)*

50. $\frac{1}{2}x + 4 = \frac{3}{2}x + 5$ **51.** $\frac{1}{3}x - 2 = \frac{3}{4}x$ **52.** $\frac{1}{4}x - \frac{3}{5} = \frac{9}{2}x - \frac{4}{5}$ **53.** $\frac{1}{2}x + \frac{1}{3} = \frac{3}{4}x - \frac{1}{5}$

Write the prime factorization of the number. If the number is prime, then write *prime*.
(Skills Review Handbook)

54. 42 **55.** 91 **56.** 72 **57.** 79

7.4 Adding and Subtracting Rational Expressions

Essential Question

How can you determine the domain of the sum or difference of two rational expressions?

You can add and subtract rational expressions in much the same way that you add and subtract fractions.

$$\frac{x}{x+1} + \frac{2}{x+1} = \frac{x+2}{x+1}$$ Sum of rational expressions

$$\frac{1}{x} - \frac{1}{2x} = \frac{2}{2x} - \frac{1}{2x} = \frac{1}{2x}$$ Difference of rational expressions

EXPLORATION 1 Adding and Subtracting Rational Expressions

Work with a partner. Find the sum or difference of the two rational expressions. Then match the sum or difference with its domain. Explain your reasoning.

Sum or Difference	Domain
a. $\frac{1}{x-1} + \frac{3}{x-1} =$	**A.** all real numbers except -2
b. $\frac{1}{x-1} + \frac{1}{x} =$	**B.** all real numbers except -1 and 1
c. $\frac{1}{x-2} + \frac{1}{2-x} =$	**C.** all real numbers except 1
d. $\frac{1}{x-1} + \frac{-1}{x+1} =$	**D.** all real numbers except 0
e. $\frac{x}{x+2} - \frac{x+1}{2+x} =$	**E.** all real numbers except -2 and 1
f. $\frac{x}{x-2} - \frac{x+1}{x} =$	**F.** all real numbers except 0 and 1
g. $\frac{x}{x+2} - \frac{x}{x-1} =$	**G.** all real numbers except 2
h. $\frac{x+2}{x} - \frac{x+1}{x} =$	**H.** all real numbers except 0 and 2

CONSTRUCTING VIABLE ARGUMENTS

To be proficient in math, you need to justify your conclusions and communicate them to others.

EXPLORATION 2 Writing a Sum or Difference

Work with a partner. Write a sum or difference of rational expressions that has the given domain. Justify your answer.

a. all real numbers except -1

b. all real numbers except -1 and 3

c. all real numbers except -1, 0, and 3

Communicate Your Answer

3. How can you determine the domain of the sum or difference of two rational expressions?

4. Your friend found a sum as follows. Describe and correct the error(s).

$$\frac{x}{x+4} + \frac{3}{x-4} = \frac{x+3}{2x}$$

7.4 Lesson

Core Vocabulary

complex fraction, p. 387

Previous
rational numbers
reciprocal

What You Will Learn

- Add or subtract rational expressions.
- Rewrite rational expressions and graph the related function.
- Simplify complex fractions.

Adding or Subtracting Rational Expressions

As with numerical fractions, the procedure used to add (or subtract) two rational expressions depends upon whether the expressions have like or unlike denominators. To add (or subtract) rational expressions with like denominators, simply add (or subtract) their numerators. Then place the result over the common denominator.

Core Concept

Adding or Subtracting with Like Denominators

Let a, b, and c be expressions with $c \neq 0$.

Addition

$$\frac{a}{c} + \frac{b}{c} = \frac{a+b}{c}$$

Subtraction

$$\frac{a}{c} - \frac{b}{c} = \frac{a-b}{c}$$

EXAMPLE 1 Adding or Subtracting with Like Denominators

a. $\dfrac{7}{4x} + \dfrac{3}{4x} = \dfrac{7+3}{4x} = \dfrac{10}{4x} = \dfrac{5}{2x}$ Add numerators and simplify.

b. $\dfrac{2x}{x+6} - \dfrac{5}{x+6} = \dfrac{2x-5}{x+6}$ Subtract numerators.

Monitoring Progress  Help in English and Spanish at *BigIdeasMath.com*

Find the sum or difference.

1. $\dfrac{8}{12x} - \dfrac{5}{12x}$ **2.** $\dfrac{2}{3x^2} + \dfrac{1}{3x^2}$ **3.** $\dfrac{4x}{x-2} - \dfrac{x}{x-2}$ **4.** $\dfrac{2x^2}{x^2+1} + \dfrac{2}{x^2+1}$

To add (or subtract) two rational expressions with *unlike* denominators, find a common denominator. Rewrite each rational expression using the common denominator. Then add (or subtract).

Core Concept

Adding or Subtracting with Unlike Denominators

Let a, b, c, and d be expressions with $c \neq 0$ and $d \neq 0$.

Addition

$$\frac{a}{c} + \frac{b}{d} = \frac{ad}{cd} + \frac{bc}{cd} = \frac{ad+bc}{cd}$$

Subtraction

$$\frac{a}{c} - \frac{b}{d} = \frac{ad}{cd} - \frac{bc}{cd} = \frac{ad-bc}{cd}$$

You can always find a common denominator of two rational expressions by multiplying the denominators, as shown above. However, when you use the least common denominator (LCD), which is the least common multiple (LCM) of the denominators, simplifying your answer may take fewer steps.

To find the LCM of two (or more) expressions, factor the expressions completely. The LCM is the product of the highest power of each factor that appears in any of the expressions.

EXAMPLE 2 Finding a Least Common Multiple (LCM)

Find the least common multiple of $4x^2 - 16$ and $6x^2 - 24x + 24$.

SOLUTION

Step 1 Factor each polynomial. Write numerical factors as products of primes.

$$4x^2 - 16 = 4(x^2 - 4) = (2^2)(x + 2)(x - 2)$$

$$6x^2 - 24x + 24 = 6(x^2 - 4x + 4) = (2)(3)(x - 2)^2$$

Step 2 The LCM is the product of the highest power of each factor that appears in either polynomial.

$$\text{LCM} = (2^2)(3)(x + 2)(x - 2)^2 = 12(x + 2)(x - 2)^2$$

EXAMPLE 3 Adding with Unlike Denominators

Find the sum $\dfrac{7}{9x^2} + \dfrac{x}{3x^2 + 3x}$.

SOLUTION

Method 1 Use the definition for adding rational expressions with unlike denominators.

$$\frac{7}{9x^2} + \frac{x}{3x^2 + 3x} = \frac{7(3x^2 + 3x) + x(9x^2)}{9x^2(3x^2 + 3x)} \qquad \frac{a}{c} + \frac{b}{d} = \frac{ad + bc}{cd}$$

$$= \frac{21x^2 + 21x + 9x^3}{9x^2(3x^2 + 3x)} \qquad \text{Distributive Property}$$

$$= \frac{\cancel{3x}(3x^2 + 7x + 7)}{9x^2(x + 1)(\cancel{3x})} \qquad \text{Factor. Divide out common factors.}$$

$$= \frac{3x^2 + 7x + 7}{9x^2(x + 1)} \qquad \text{Simplify.}$$

Method 2 Find the LCD and then add. To find the LCD, factor each denominator and write each factor to the highest power that appears in either denominator. Note that $9x^2 = 3^2x^2$ and $3x^2 + 3x = 3x(x + 1)$, so the LCD is $9x^2(x + 1)$.

$$\frac{7}{9x^2} + \frac{x}{3x^2 + 3x} = \frac{7}{9x^2} + \frac{x}{3x(x + 1)} \qquad \text{Factor second denominator.}$$

$$= \frac{7}{9x^2} \cdot \frac{x + 1}{x + 1} + \frac{x}{3x(x + 1)} \cdot \frac{3x}{3x} \qquad \text{LCD is } 9x^2(x + 1).$$

$$= \frac{7x + 7}{9x^2(x + 1)} + \frac{3x^2}{9x^2(x + 1)} \qquad \text{Multiply.}$$

$$= \frac{3x^2 + 7x + 7}{9x^2(x + 1)} \qquad \text{Add numerators.}$$

Note in Examples 1 and 3 that when adding or subtracting rational expressions, the result is a rational expression. In general, similar to rational numbers, rational expressions are closed under addition and subtraction.

EXAMPLE 4 Subtracting with Unlike Denominators

Find the difference $\dfrac{x+2}{2x-2} - \dfrac{-2x-1}{x^2-4x+3}$.

SOLUTION

COMMON ERROR

When subtracting rational expressions, remember to distribute the negative sign to all the terms in the quantity that is being subtracted.

$$\frac{x+2}{2x-2} - \frac{-2x-1}{x^2-4x+3} = \frac{x+2}{2(x-1)} - \frac{-2x-1}{(x-1)(x-3)}$$ Factor each denominator.

$$= \frac{x+2}{2(x-1)} \cdot \frac{x-3}{x-3} - \frac{-2x-1}{(x-1)(x-3)} \cdot \frac{2}{2}$$ LCD is $2(x-1)(x-3)$.

$$= \frac{x^2-x-6}{2(x-1)(x-3)} - \frac{-4x-2}{2(x-1)(x-3)}$$ Multiply.

$$= \frac{x^2-x-6-(-4x-2)}{2(x-1)(x-3)}$$ Subtract numerators.

$$= \frac{x^2+3x-4}{2(x-1)(x-3)}$$ Simplify numerator.

$$= \frac{\cancel{(x-1)}(x+4)}{2\cancel{(x-1)}(x-3)}$$ Factor numerator. Divide out common factors.

$$= \frac{x+4}{2(x-3)},\ x \neq -1$$ Simplify.

Monitoring Progress

Help in English and Spanish at *BigIdeasMath.com*

5. Find the least common multiple of $5x^3$ and $10x^2 - 15x$.

Find the sum or difference.

6. $\dfrac{3}{4x} - \dfrac{1}{7}$ **7.** $\dfrac{1}{3x^2} + \dfrac{x}{9x^2-12}$ **8.** $\dfrac{x}{x^2-x-12} + \dfrac{5}{12x-48}$

Rewriting Rational Functions

Rewriting a rational expression may reveal properties of the related function and its graph. In Example 4 of Section 7.2, you used long division to rewrite a rational expression. In the next example, you will use inspection.

EXAMPLE 5 Rewriting and Graphing a Rational Function

Rewrite the function $g(x) = \dfrac{3x+5}{x+1}$ in the form $g(x) = \dfrac{a}{x-h} + k$. Graph the function. Describe the graph of g as a transformation of the graph of $f(x) = \dfrac{a}{x}$.

SOLUTION

Rewrite by inspection:

$$\frac{3x+5}{x+1} = \frac{3x+3+2}{x+1} = \frac{3(x+1)+2}{x+1} = \frac{3\cancel{(x+1)}}{\cancel{x+1}} + \frac{2}{x+1} = 3 + \frac{2}{x+1}$$

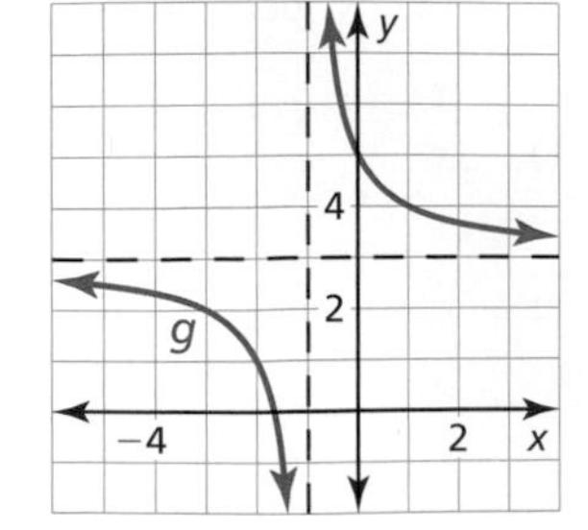

The rewritten function is $g(x) = \dfrac{2}{x+1} + 3$. The graph of g is a translation 1 unit left and 3 units up of the graph of $f(x) = \dfrac{2}{x}$.

Monitoring Progress

Help in English and Spanish at *BigIdeasMath.com*

9. Rewrite $g(x) = \dfrac{2x-4}{x-3}$ in the form $g(x) = \dfrac{a}{x-h} + k$. Graph the function. Describe the graph of g as a transformation of the graph of $f(x) = \dfrac{a}{x}$.

Complex Fractions

A **complex fraction** is a fraction that contains a fraction in its numerator or denominator. A complex fraction can be simplified using either of the methods below.

Core Concept

Simplifying Complex Fractions

Method 1 If necessary, simplify the numerator and denominator by writing each as a single fraction. Then divide by multiplying the numerator by the reciprocal of the denominator.

Method 2 Multiply the numerator and the denominator by the LCD of *every* fraction in the numerator and denominator. Then simplify.

EXAMPLE 6 Simplifying a Complex Fraction

Simplify $\dfrac{\dfrac{5}{x+4}}{\dfrac{1}{x+4}+\dfrac{2}{x}}$.

SOLUTION

Method 1

$$\frac{\frac{5}{x+4}}{\frac{1}{x+4}+\frac{2}{x}} = \frac{\frac{5}{x+4}}{\frac{3x+8}{x(x+4)}}$$ Add fractions in denominator.

$$= \frac{5}{x+4} \cdot \frac{x(x+4)}{3x+8}$$ Multiply by reciprocal.

$$= \frac{5x\cancel{(x+4)}}{\cancel{(x+4)}(3x+8)}$$ Divide out common factors.

$$= \frac{5x}{3x+8},\ x \neq -4,\ x \neq 0$$ Simplify.

Method 2 The LCD of all the fractions in the numerator and denominator is $x(x+4)$.

$$\frac{\frac{5}{x+4}}{\frac{1}{x+4}+\frac{2}{x}} = \frac{\frac{5}{x+4}}{\frac{1}{x+4}+\frac{2}{x}} \cdot \frac{x(x+4)}{x(x+4)}$$ Multiply numerator and denominator by the LCD.

$$= \frac{\frac{5}{\cancel{x+4}} \cdot x\cancel{(x+4)}}{\frac{1}{\cancel{x+4}} \cdot x\cancel{(x+4)} + \frac{2}{\cancel{x}} \cdot \cancel{x}(x+4)}$$ Divide out common factors.

$$= \frac{5x}{x+2(x+4)}$$ Simplify.

$$= \frac{5x}{3x+8},\ x \neq -4,\ x \neq 0$$ Simplify.

Monitoring Progress  Help in English and Spanish at *BigIdeasMath.com*

Simplify the complex fraction.

10. $\dfrac{\frac{x}{6}-\frac{x}{3}}{\frac{x}{5}-\frac{7}{10}}$

11. $\dfrac{\frac{2}{x}-4}{\frac{2}{x}+3}$

12. $\dfrac{\frac{3}{x+5}}{\frac{2}{x-3}+\frac{1}{x+5}}$

7.4 Exercises

Dynamic Solutions available at *BigIdeasMath.com*

Vocabulary and Core Concept Check

1. **COMPLETE THE SENTENCE** A fraction that contains a fraction in its numerator or denominator is called a(n) ________.

2. **WRITING** Explain how adding and subtracting rational expressions is similar to adding and subtracting numerical fractions.

Monitoring Progress and Modeling with Mathematics

In Exercises 3–8, find the sum or difference. *(See Example 1.)*

3. $\frac{15}{4x} + \frac{5}{4x}$

4. $\frac{x}{16x^2} - \frac{4}{16x^2}$

5. $\frac{9}{x+1} - \frac{2x}{x+1}$

6. $\frac{3x^2}{x-8} + \frac{6x}{x-8}$

7. $\frac{5x}{x+3} + \frac{15}{x+3}$

8. $\frac{4x^2}{2x-1} - \frac{1}{2x-1}$

In Exercises 9–16, find the least common multiple of the expressions. *(See Example 2.)*

9. $3x, 3(x-2)$

10. $2x^2, 4x + 12$

11. $2x, 2x(x-5)$

12. $24x^2, 8x^2 - 16x$

13. $x^2 - 25, x - 5$

14. $9x^2 - 16, 3x^2 + x - 4$

15. $x^2 + 3x - 40, x - 8$

16. $x^2 - 2x - 63, x + 7$

ERROR ANALYSIS In Exercises 17 and 18, describe and correct the error in finding the sum.

17. ✗ $\frac{2}{5x} + \frac{4}{x^2} = \frac{2+4}{5x+x^2} = \frac{6}{x(5+x)}$

18. ✗ $\frac{x}{x+2} + \frac{4}{x-5} = \frac{x+4}{(x+2)(x-5)}$

In Exercises 19–26, find the sum or difference. *(See Examples 3 and 4.)*

19. $\frac{12}{5x} - \frac{7}{6x}$

20. $\frac{8}{3x^2} + \frac{5}{4x}$

21. $\frac{3}{x+4} - \frac{1}{x+6}$

22. $\frac{9}{x-3} + \frac{2x}{x+1}$

23. $\frac{12}{x^2 + 5x - 24} + \frac{3}{x-3}$

24. $\frac{x^2 - 5}{x^2 + 5x - 14} - \frac{x+3}{x+7}$

25. $\frac{x+2}{x-4} + \frac{2}{x} + \frac{5x}{3x-1}$

26. $\frac{x+3}{x^2 - 25} - \frac{x-1}{x-5} + \frac{3}{x+3}$

REASONING In Exercises 27 and 28, tell whether the statement is *always*, *sometimes*, or *never* true. Explain.

27. The LCD of two rational expressions is the product of the denominators.

28. The LCD of two rational expressions will have a degree greater than or equal to that of the denominator with the higher degree.

29. **ANALYZING EQUATIONS** How would you begin to rewrite the function $g(x) = \frac{4x+1}{x+2}$ to obtain the form $g(x) = \frac{a}{x-h} + k$?

   Ⓐ $g(x) = \frac{4(x+2) - 7}{x+2}$

   Ⓑ $g(x) = \frac{4(x+2) + 1}{x+2}$

   Ⓒ $g(x) = \frac{(x+2) + (3x-1)}{x+2}$

   Ⓓ $g(x) = \frac{4x + 2 - 1}{x+2}$

30. **ANALYZING EQUATIONS** How would you begin to rewrite the function $g(x) = \frac{x}{x-5}$ to obtain the form $g(x) = \frac{a}{x-h} + k$?

   Ⓐ $g(x) = \frac{x(x+5)(x-5)}{x-5}$

   Ⓑ $g(x) = \frac{x-5+5}{x-5}$

   Ⓒ $g(x) = \frac{x}{x-5+5}$

   Ⓓ $g(x) = \frac{x}{x} - \frac{x}{5}$

In Exercises 31–38, rewrite the function g in the form $g(x) = \frac{a}{x-h} + k$. Graph the function. Describe the graph of g as a transformation of the graph of $f(x) = \frac{a}{x}$. *(See Example 5.)*

31. $g(x) = \dfrac{5x-7}{x-1}$ **32.** $g(x) = \dfrac{6x+4}{x+5}$

33. $g(x) = \dfrac{12x}{x-5}$ **34.** $g(x) = \dfrac{8x}{x+13}$

35. $g(x) = \dfrac{2x+3}{x}$ **36.** $g(x) = \dfrac{4x-6}{x}$

37. $g(x) = \dfrac{3x+11}{x-3}$ **38.** $g(x) = \dfrac{7x-9}{x+10}$

In Exercises 39–44, simplify the complex fraction. *(See Example 6.)*

39. $\dfrac{\frac{x}{3} - 6}{10 + \frac{4}{x}}$ **40.** $\dfrac{15 - \frac{2}{x}}{\frac{x}{5} + 4}$

41. $\dfrac{\frac{1}{2x-5} - \frac{7}{8x-20}}{\frac{x}{2x-5}}$ **42.** $\dfrac{\frac{16}{x-2}}{\frac{4}{x+1} + \frac{6}{x}}$

43. $\dfrac{\frac{1}{3x^2-3}}{\frac{5}{x+1} - \frac{x+4}{x^2-3x-4}}$ **44.** $\dfrac{\frac{3}{x-2} - \frac{6}{x^2-4}}{\frac{3}{x+2} + \frac{1}{x-2}}$

45. PROBLEM SOLVING The total time T (in hours) needed to fly from New York to Los Angeles and back can be modeled by the equation below, where d is the distance (in miles) each way, a is the average airplane speed (in miles per hour), and j is the average speed (in miles per hour) of the jet stream. Simplify the equation. Then find the total time it takes to fly 2468 miles when $a = 510$ miles per hour and $j = 115$ miles per hour.

$$T = \frac{d}{a-j} + \frac{d}{a+j}$$

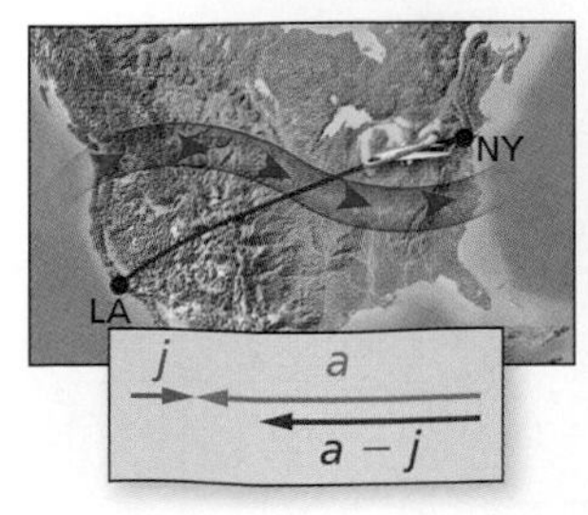

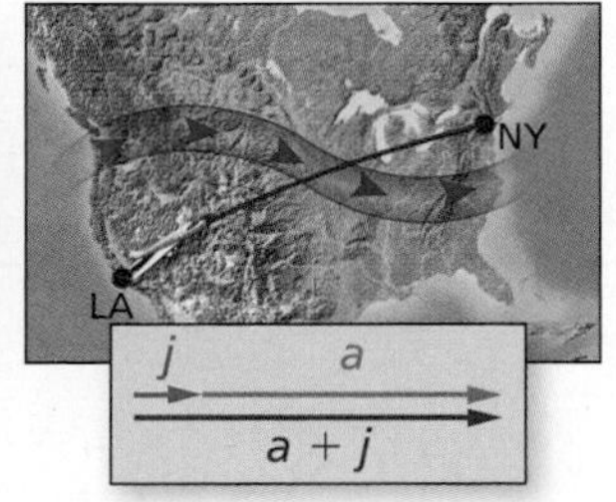

46. REWRITING A FORMULA The total resistance R_t of two resistors in a parallel circuit with resistances R_1 and R_2 (in ohms) is given by the equation shown. Simplify the complex fraction. Then find the total resistance when $R_1 = 2000$ ohms and $R_2 = 5600$ ohms.

$$R_t = \frac{1}{\frac{1}{R_1} + \frac{1}{R_2}}$$

47. PROBLEM SOLVING You plan a trip that involves a 40-mile bus ride and a train ride. The entire trip is 140 miles. The time (in hours) the bus travels is $y_1 = \frac{40}{x}$, where x is the average speed (in miles per hour) of the bus. The time (in hours) the train travels is $y_2 = \frac{100}{x+30}$. Write and simplify a model that shows the total time y of the trip.

48. PROBLEM SOLVING You participate in a sprint triathlon that involves swimming, bicycling, and running. The table shows the distances (in miles) and your average speed for each portion of the race.

	Distance (miles)	Speed (miles per hour)
Swimming	0.5	r
Bicycling	22	$15r$
Running	6	$r + 5$

a. Write a model in simplified form for the total time (in hours) it takes to complete the race.

b. How long does it take to complete the race if you can swim at an average speed of 2 miles per hour? Justify your answer.

49. MAKING AN ARGUMENT Your friend claims that the least common multiple of two numbers is always greater than each of the numbers. Is your friend correct? Justify your answer.

50. HOW DO YOU SEE IT? Use the graph of the function $f(x) = \frac{a}{x - h} + k$ to determine the values of h and k.

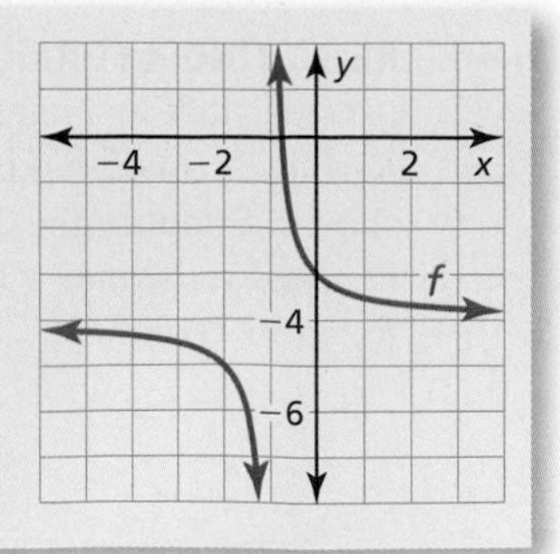

51. REWRITING A FORMULA You borrow P dollars to buy a car and agree to repay the loan over t years at a monthly interest rate of i (expressed as a decimal). Your monthly payment M is given by either formula below.

$$M = \frac{Pi}{1 - \left(\frac{1}{1 + i}\right)^{12t}} \quad \text{or} \quad M = \frac{Pi(1 + i)^{12t}}{(1 + i)^{12t} - 1}$$

a. Show that the formulas are equivalent by simplifying the first formula.

b. Find your monthly payment when you borrow \$15,500 at a monthly interest rate of 0.5% and repay the loan over 4 years.

52. THOUGHT PROVOKING Is it possible to write two rational functions whose sum is a quadratic function? Justify your answer.

53. USING TOOLS Use technology to rewrite the function $g(x) = \frac{(97.6)(0.024) + x(0.003)}{12.2 + x}$ in the form $g(x) = \frac{a}{x - h} + k$. Describe the graph of g as a transformation of the graph of $f(x) = \frac{a}{x}$.

54. MATHEMATICAL CONNECTIONS Find an expression for the surface area of the box.

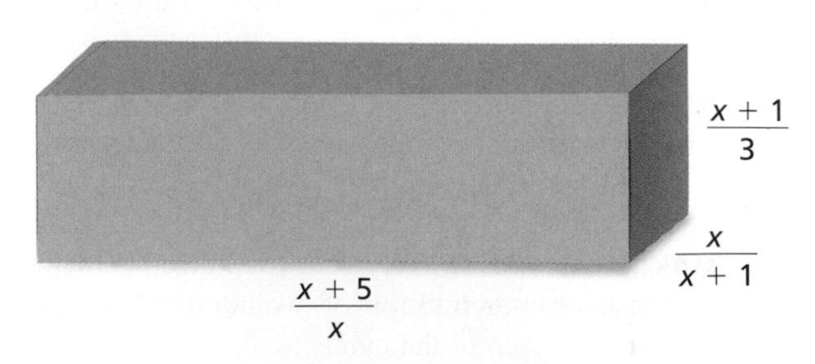

55. PROBLEM SOLVING You are hired to wash the new cars at a car dealership with two other employees. You take an average of 40 minutes to wash a car (R_1 = 1/40 car per minute). The second employee washes a car in x minutes. The third employee washes a car in $x + 10$ minutes.

a. Write expressions for the rates that each employee can wash a car.

b. Write a single expression R for the combined rate of cars washed per minute by the group.

c. Evaluate your expression in part (b) when the second employee washes a car in 35 minutes. How many cars per hour does this represent? Explain your reasoning.

56. MODELING WITH MATHEMATICS The amount A (in milligrams) of aspirin in a person's bloodstream can be modeled by

$$A = \frac{391t^2 + 0.112}{0.218t^4 + 0.991t^2 + 1}$$

where t is the time (in hours) after one dose is taken.

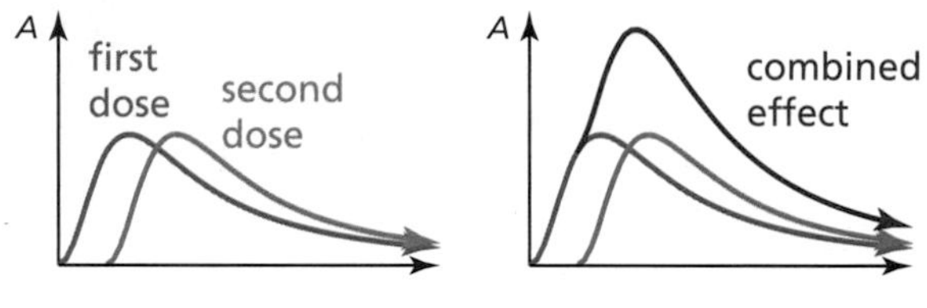

a. A second dose is taken 1 hour after the first dose. Write an equation to model the amount of the second dose in the bloodstream.

b. Write a model for the *total* amount of aspirin in the bloodstream after the second dose is taken.

57. FINDING A PATTERN Find the next two expressions in the pattern shown. Then simplify all five expressions. What value do the expressions approach?

$$1 + \frac{1}{2 + \frac{1}{2}},\ 1 + \frac{1}{2 + \frac{1}{2 + \frac{1}{2}}},\ 1 + \frac{1}{2 + \frac{1}{2 + \frac{1}{2 + \frac{1}{2}}}},\ \ldots$$

Maintaining Mathematical Proficiency

Reviewing what you learned in previous grades and lessons

Solve the system by graphing. *(Section 3.5)*

58. $y = x^2 + 6$
$y = 3x + 4$

59. $2x^2 - 3x - y = 0$
$\frac{5}{2}x - y = \frac{9}{4}$

60. $3 = y - x^2 - x$
$y = -x^2 - 3x - 5$

61. $y = (x + 2)^2 - 3$
$y = x^2 + 4x + 5$

7.5 Solving Rational Equations

Essential Question

How can you solve a rational equation?

EXPLORATION 1 Solving Rational Equations

Work with a partner. Match each equation with the graph of its related system of equations. Explain your reasoning. Then use the graph to solve the equation.

a. $\dfrac{2}{x-1} = 1$ **b.** $\dfrac{2}{x-2} = 2$ **c.** $\dfrac{-x-1}{x-3} = x + 1$

d. $\dfrac{2}{x-1} = x$ **e.** $\dfrac{1}{x} = \dfrac{-1}{x-2}$ **f.** $\dfrac{1}{x} = x^2$

A.

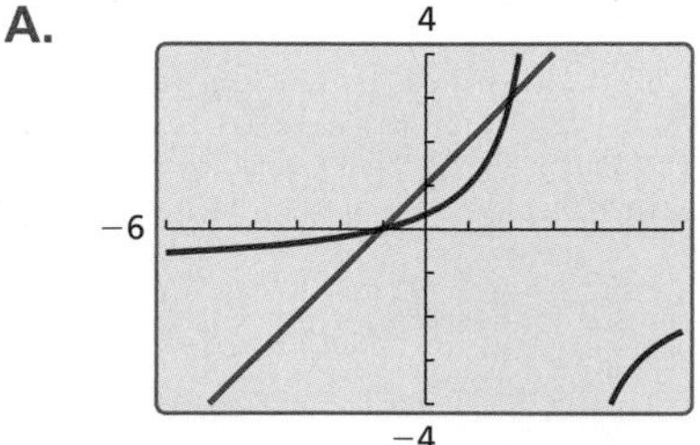

B.

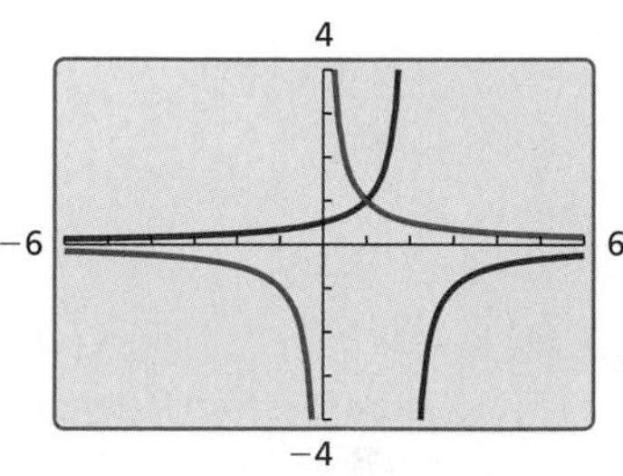

C.

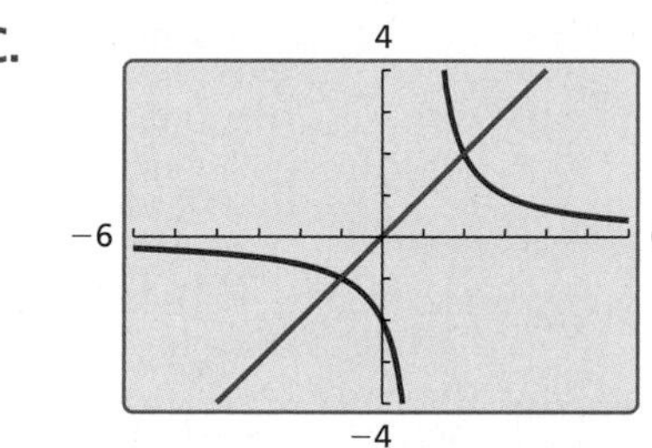

D.

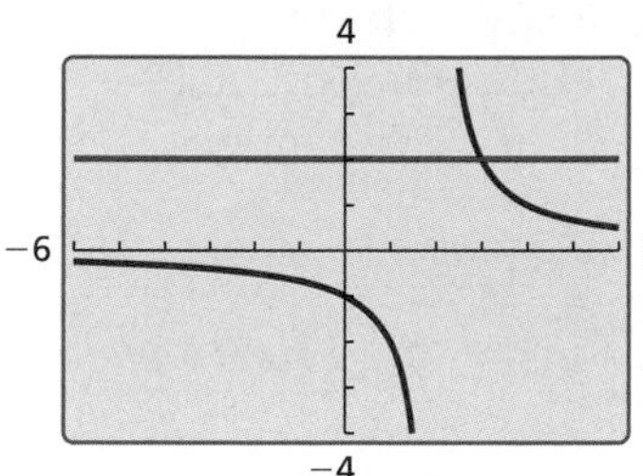

E.

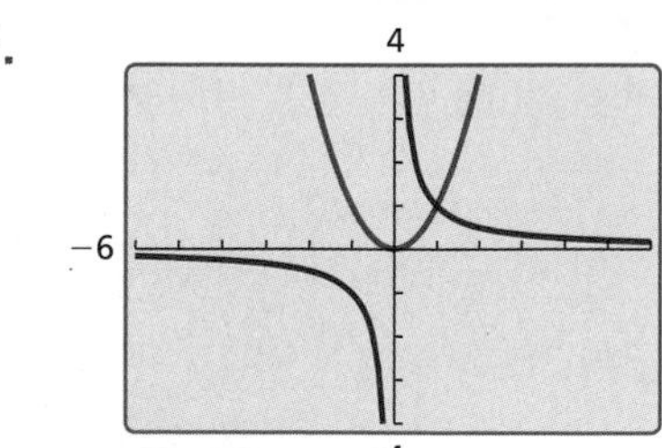

F.

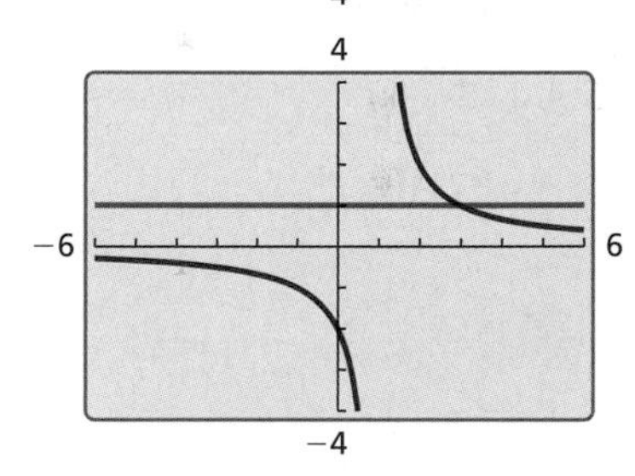

EXPLORATION 2 Solving Rational Equations

Work with a partner. Look back at the equations in Explorations 1(d) and 1(e). Suppose you want a more accurate way to solve the equations than using a graphical approach.

a. Show how you could use a *numerical approach* by creating a table. For instance, you might use a spreadsheet to solve the equations.

b. Show how you could use an *analytical approach*. For instance, you might use the method you used to solve proportions.

MAKING SENSE OF PROBLEMS

To be proficient in math, you need to plan a solution pathway rather than simply jumping into a solution attempt.

Communicate Your Answer

3. How can you solve a rational equation?

4. Use the method in either Exploration 1 or 2 to solve each equation.

a. $\dfrac{x+1}{x-1} = \dfrac{x-1}{x+1}$ **b.** $\dfrac{1}{x+1} = \dfrac{1}{x^2+1}$ **c.** $\dfrac{1}{x^2-1} = \dfrac{1}{x-1}$

7.5 Lesson

Core Vocabulary

cross multiplying, *p. 392*

Previous
proportion
extraneous solution
inverse of a function

What You Will Learn

- Solve rational equations by cross multiplying.
- Solve rational equations by using the least common denominator.
- Use inverses of functions.

Solving by Cross Multiplying

You can use **cross multiplying** to solve a rational equation when each side of the equation is a single rational expression.

EXAMPLE 1 Solving a Rational Equation by Cross Multiplying

Solve $\frac{3}{x+1} = \frac{9}{4x+5}$.

SOLUTION

$\frac{3}{x+1} = \frac{9}{4x+5}$	Write original equation.
$3(4x+5) = 9(x+1)$	Cross multiply.
$12x + 15 = 9x + 9$	Distributive Property
$3x + 15 = 9$	Subtract $9x$ from each side.
$3x = -6$	Subtract 15 from each side.
$x = -2$	Divide each side by 3.

▶ The solution is $x = -2$. Check this in the original equation.

Check

$$\frac{3}{-2+1} \stackrel{?}{=} \frac{9}{4(-2)+5}$$

$$\frac{3}{-1} \stackrel{?}{=} \frac{9}{-3}$$

$$-3 = -3 \checkmark$$

EXAMPLE 2 Writing and Using a Rational Model

An *alloy* is formed by mixing two or more metals. Sterling silver is an alloy composed of 92.5% silver and 7.5% copper by weight. You have 15 ounces of 800 grade silver, which is 80% silver and 20% copper by weight. How much pure silver should you mix with the 800 grade silver to make sterling silver?

SOLUTION

$$\text{percent of copper in mixture} = \frac{\text{weight of copper in mixture}}{\text{total weight of mixture}}$$

$\frac{7.5}{100} = \frac{(0.2)(15)}{15+x}$	x is the amount of silver added.
$7.5(15+x) = 100(0.2)(15)$	Cross multiply.
$112.5 + 7.5x = 300$	Simplify.
$7.5x = 187.5$	Subtract 112.5 from each side.
$x = 25$	Divide each side by 7.5.

▶ You should mix 25 ounces of pure silver with the 15 ounces of 800 grade silver.

Monitoring Progress

Help in English and Spanish at *BigIdeasMath.com*

Solve the equation by cross multiplying. Check your solution(s).

1. $\frac{3}{5x} = \frac{2}{x-7}$ **2.** $\frac{-4}{x+3} = \frac{5}{x-3}$ **3.** $\frac{1}{2x+5} = \frac{x}{11x+8}$

Solving by Using the Least Common Denominator

When a rational equation is not expressed as a proportion, you can solve it by multiplying each side of the equation by the least common denominator of the rational expressions.

EXAMPLE 3 Solving Rational Equations by Using the LCD

Solve each equation.

a. $\frac{5}{x} + \frac{7}{4} = -\frac{9}{x}$ **b.** $1 - \frac{8}{x-5} = \frac{3}{x}$

SOLUTION

a.

$\frac{5}{x} + \frac{7}{4} = -\frac{9}{x}$	Write original equation.
$4x\left(\frac{5}{x} + \frac{7}{4}\right) = 4x\left(-\frac{9}{x}\right)$	Multiply each side by the LCD, $4x$.
$20 + 7x = -36$	Simplify.
$7x = -56$	Subtract 20 from each side.
$x = -8$	Divide each side by 7.

The solution is $x = -8$. Check this in the original equation.

Check

$\frac{5}{-8} + \frac{7}{4} \stackrel{?}{=} -\frac{9}{-8}$

$-\frac{5}{8} + \frac{14}{8} \stackrel{?}{=} \frac{9}{8}$

$\frac{9}{8} = \frac{9}{8}$ ✓

b.

$1 - \frac{8}{x-5} = \frac{3}{x}$	Write original equation.
$x(x-5)\left(1 - \frac{8}{x-5}\right) = x(x-5) \cdot \frac{3}{x}$	Multiply each side by the LCD, $x(x-5)$.
$x(x-5) - 8x = 3(x-5)$	Simplify.
$x^2 - 5x - 8x = 3x - 15$	Distributive Property
$x^2 - 16x + 15 = 0$	Write in standard form.
$(x-1)(x-15) = 0$	Factor.
$x = 1$ or $x = 15$	Zero-Product Property

The solutions are $x = 1$ and $x = 15$. Check these in the original equation.

Check

$1 - \frac{8}{1-5} \stackrel{?}{=} \frac{3}{1}$	Substitute for x.	$1 - \frac{8}{15-5} \stackrel{?}{=} \frac{3}{15}$
$1 + 2 \stackrel{?}{=} 3$	Simplify.	$1 - \frac{4}{5} \stackrel{?}{=} \frac{1}{5}$
$3 = 3$ ✓		$\frac{1}{5} = \frac{1}{5}$ ✓

Monitoring Progress  Help in English and Spanish at *BigIdeasMath.com*

Solve the equation by using the LCD. Check your solution(s).

4. $\frac{15}{x} + \frac{4}{5} = \frac{7}{x}$ **5.** $\frac{3x}{x+1} - \frac{5}{2x} = \frac{3}{2x}$ **6.** $\frac{4x+1}{x+1} = \frac{12}{x^2-1} + 3$

When solving a rational equation, you may obtain solutions that are extraneous. Be sure to check for extraneous solutions by checking your solutions in the *original* equation.

EXAMPLE 4 Solving an Equation with an Extraneous Solution

Solve $\dfrac{6}{x-3} = \dfrac{8x^2}{x^2-9} - \dfrac{4x}{x+3}$.

SOLUTION

Write each denominator in factored form. The LCD is $(x+3)(x-3)$.

$$\frac{6}{x-3} = \frac{8x^2}{(x+3)(x-3)} - \frac{4x}{x+3}$$

$$(x+3)(x-3) \cdot \frac{6}{x-3} = (x+3)(x-3) \cdot \frac{8x^2}{(x+3)(x-3)} - (x+3)(x-3) \cdot \frac{4x}{x+3}$$

$$6(x+3) = 8x^2 - 4x(x-3)$$

$$6x + 18 = 8x^2 - 4x^2 + 12x$$

$$0 = 4x^2 + 6x - 18$$

$$0 = 2x^2 + 3x - 9$$

$$0 = (2x-3)(x+3)$$

$$2x - 3 = 0 \quad \text{or} \quad x + 3 = 0$$

$$x = \frac{3}{2} \quad \text{or} \quad x = -3$$

Check

Check $x = \frac{3}{2}$:

$$\frac{6}{\frac{3}{2}-3} \stackrel{?}{=} \frac{8\left(\frac{3}{2}\right)^2}{\left(\frac{3}{2}\right)^2-9} - \frac{4\left(\frac{3}{2}\right)}{\frac{3}{2}+3}$$

$$\frac{6}{-\frac{3}{2}} \stackrel{?}{=} \frac{18}{-\frac{27}{4}} - \frac{6}{\frac{9}{2}}$$

$$-4 \stackrel{?}{=} -\frac{8}{3} - \frac{4}{3}$$

$$-4 = -4 \checkmark$$

Check $x = -3$:

$$\frac{6}{-3-3} \stackrel{?}{=} \frac{8(-3)^2}{(-3)^2-9} - \frac{4(-3)}{-3+3}$$

$$\frac{6}{-6} \stackrel{?}{=} \frac{72}{0} - \frac{-12}{0} \; ✗$$

Division by zero is undefined.

ANOTHER WAY

You can also graph each side of the equation and find the x-value where the graphs intersect.

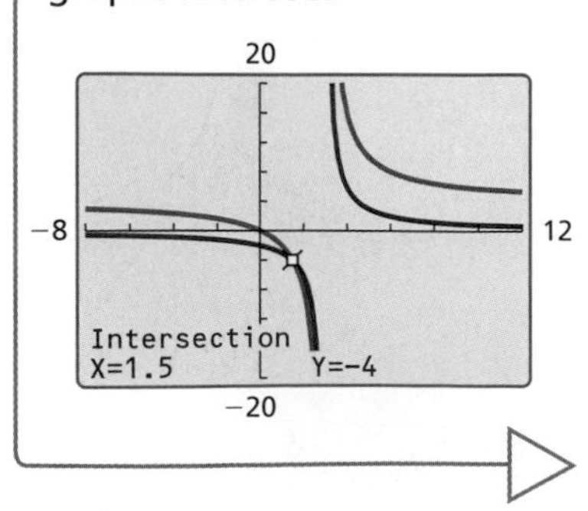

▶ The apparent solution $x = -3$ is extraneous. So, the only solution is $x = \frac{3}{2}$.

Monitoring Progress  Help in English and Spanish at *BigIdeasMath.com*

Solve the equation. Check your solution(s).

7. $\dfrac{9}{x-2} + \dfrac{6x}{x+2} = \dfrac{9x^2}{x^2-4}$

8. $\dfrac{7}{x-1} - 5 = \dfrac{6}{x^2-1}$

Using Inverses of Functions

EXAMPLE 5 Finding the Inverse of a Rational Function

Consider the function $f(x) = \frac{2}{x+3}$. Determine whether the inverse of f is a function. Then find the inverse.

SOLUTION

Graph the function f. Notice that no horizontal line intersects the graph more than once. So, the inverse of f is a function. Find the inverse.

$y = \frac{2}{x+3}$	Set y equal to $f(x)$.
$x = \frac{2}{y+3}$	Switch x and y.
$x(y+3) = 2$	Cross multiply.
$y + 3 = \frac{2}{x}$	Divide each side by x.
$y = \frac{2}{x} - 3$	Subtract 3 from each side.

$f(x) = \frac{2}{x+3}$

▶ So, the inverse of f is $g(x) = \frac{2}{x} - 3$.

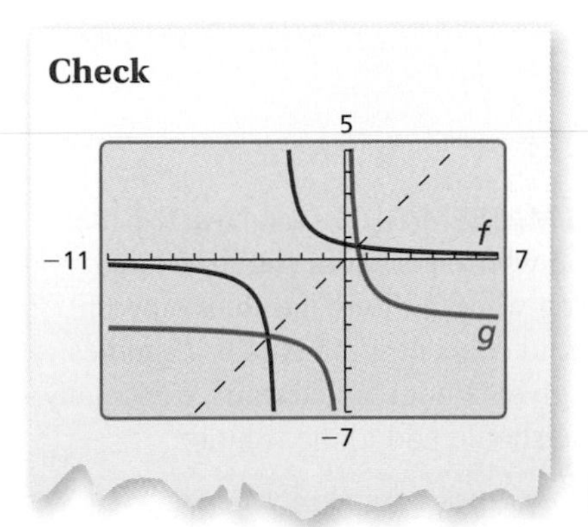

EXAMPLE 6 Solving a Real-Life Problem

In Section 7.2 Example 5, you wrote the function $c = \frac{50m + 1000}{m}$, which represents the average cost c (in dollars) of making m models using a 3-D printer. Find how many models must be printed for the average cost per model to fall to \$90 by (a) solving an equation, and (b) using the inverse of the function.

REMEMBER

In part (b), the variables are meaningful. Switching them to find the inverse would create confusion. So, solve for m without switching variables.

SOLUTION

a. Substitute 90 for c and solve by cross multiplying.

$$90 = \frac{50m + 1000}{m}$$

$$90m = 50m + 1000$$

$$40m = 1000$$

$$m = 25$$

b. Solve the equation for m.

$$c = \frac{50m + 1000}{m}$$

$$c = 50 + \frac{1000}{m}$$

$$c - 50 = \frac{1000}{m}$$

$$m = \frac{1000}{c - 50}$$

When $c = 90$, $m = \frac{1000}{90 - 50} = 25$.

▶ So, the average cost falls to \$90 per model after 25 models are printed.

Monitoring Progress

Help in English and Spanish at *BigIdeasMath.com*

9. Consider the function $f(x) = \frac{1}{x} - 2$. Determine whether the inverse of f is a function. Then find the inverse.

10. WHAT IF? How do the answers in Example 6 change when $c = \frac{50m + 800}{m}$?

7.5 Exercises

Dynamic Solutions available at *BigIdeasMath.com*

Vocabulary and Core Concept Check

1. **WRITING** When can you solve a rational equation by cross multiplying? Explain.

2. **WRITING** A student solves the equation $\frac{4}{x-3} = \frac{x}{x-3}$ and obtains the solutions 3 and 4. Are either of these extraneous solutions? Explain.

Monitoring Progress and Modeling with Mathematics

In Exercises 3–10, solve the equation by cross multiplying. Check your solution(s). *(See Example 1.)*

3. $\frac{4}{2x} = \frac{5}{x+6}$

4. $\frac{9}{3x} = \frac{4}{x+2}$

5. $\frac{6}{x-1} = \frac{9}{x+1}$

6. $\frac{8}{3x-2} = \frac{2}{x-1}$

7. $\frac{x}{2x+7} = \frac{x-5}{x-1}$

8. $\frac{-2}{x-1} = \frac{x-8}{x+1}$

9. $\frac{x^2-3}{x+2} = \frac{x-3}{2}$

10. $\frac{-1}{x-3} = \frac{x-4}{x^2-27}$

11. **USING EQUATIONS** So far in your volleyball practice, you have put into play 37 of the 44 serves you have attempted. Solve the equation $\frac{90}{100} = \frac{37+x}{44+x}$ to find the number of consecutive serves you need to put into play in order to raise your serve percentage to 90%.

12. **USING EQUATIONS** So far this baseball season, you have 12 hits out of 60 times at-bat. Solve the equation $0.360 = \frac{12+x}{60+x}$ to find the number of consecutive hits you need to raise your batting average to 0.360.

13. **MODELING WITH MATHEMATICS** Brass is an alloy composed of 55% copper and 45% zinc by weight. You have 25 ounces of copper. How many ounces of zinc do you need to make brass? *(See Example 2.)*

14. **MODELING WITH MATHEMATICS** You have 0.2 liter of an acid solution whose acid concentration is 16 moles per liter. You want to dilute the solution with water so that its acid concentration is only 12 moles per liter. Use the given model to determine how many liters of water you should add to the solution.

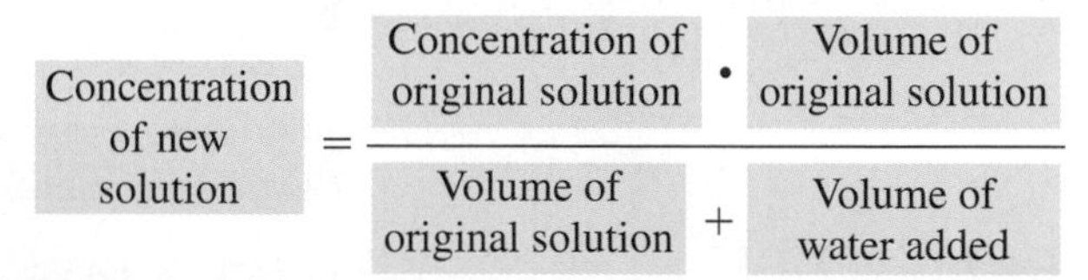

Concentration of new solution
=
Concentration of original solution
•
Volume of original solution
Volume of original solution
+
Volume of water added

USING STRUCTURE In Exercises 15–18, identify the LCD of the rational expressions in the equation.

15. $\frac{x}{x+3} + \frac{1}{x} = \frac{3}{x}$

16. $\frac{5x}{x-1} - \frac{7}{x} = \frac{9}{x}$

17. $\frac{2}{x+1} + \frac{x}{x+4} = \frac{1}{2}$

18. $\frac{4}{x+9} + \frac{3x}{2x-1} = \frac{10}{3}$

In Exercises 19–30, solve the equation by using the LCD. Check your solution(s). *(See Examples 3 and 4.)*

19. $\frac{3}{2} + \frac{1}{x} = 2$

20. $\frac{2}{3x} + \frac{1}{6} = \frac{4}{3x}$

21. $\frac{x-3}{x-4} + 4 = \frac{3x}{x}$

22. $\frac{2}{x-3} + \frac{1}{x} = \frac{x-1}{x-3}$

23. $\frac{6x}{x+4} + 4 = \frac{2x+2}{x-1}$

24. $\frac{10}{x} + 3 = \frac{x+9}{x-4}$

25. $\frac{18}{x^2-3x} - \frac{6}{x-3} = \frac{5}{x}$

26. $\frac{10}{x^2-2x} + \frac{4}{x} = \frac{5}{x-2}$

27. $\frac{x+1}{x+6} + \frac{1}{x} = \frac{2x+1}{x+6}$

28. $\frac{x+3}{x-3} + \frac{x}{x-5} = \frac{x+5}{x-5}$

29. $\frac{5}{x} - 2 = \frac{2}{x+3}$

30. $\frac{5}{x^2+x-6} = 2 + \frac{x-3}{x-2}$

ERROR ANALYSIS **In Exercises 31 and 32, describe and correct the error in the first step of solving the equation.**

31. ✗

$$\frac{5}{3x} + \frac{2}{x^2} = 1$$

$$3x^3 \cdot \frac{5}{3x} + 3x^3 \cdot \frac{2}{x^2} = 1$$

32. ✗

$$\frac{7x+1}{2x+5} + 4 = \frac{10x-3}{3x}$$

$$(2x+5)3x \cdot \frac{7x+1}{2x+5} + 4 = \frac{10x-3}{3x} \cdot (2x+5)3x$$

33. PROBLEM SOLVING You can paint a room in 8 hours. Working together, you and your friend can paint the room in just 5 hours.

a. Let t be the time (in hours) your friend would take to paint the room when working alone. Copy and complete the table.
(*Hint*: (Work done) = (Work rate) × (Time))

	Work rate	Time	Work done
You	$\frac{1 \text{ room}}{8 \text{ hours}}$	5 hours	
Friend		5 hours	

b. Explain what the sum of the expressions represents in the last column. Write and solve an equation to find how long your friend would take to paint the room when working alone.

34. PROBLEM SOLVING You can clean a park in 2 hours. Working together, you and your friend can clean the park in just 1.2 hours.

a. Let t be the time (in hours) your friend would take to clean the park when working alone. Copy and complete the table.
(*Hint*: (Work done) = (Work rate) × (Time))

	Work rate	Time	Work done
You	$\frac{1 \text{ park}}{2 \text{ hours}}$	1.2 hours	
Friend		1.2 hours	

b. Explain what the sum of the expressions represents in the last column. Write and solve an equation to find how long your friend would take to clean the park when working alone.

35. OPEN-ENDED Give an example of a rational equation that you would solve using cross multiplication and one that you would solve using the LCD. Explain your reasoning.

36. OPEN-ENDED Describe a real-life situation that can be modeled by a rational equation. Justify your answer.

In Exercises 37–44, determine whether the inverse of f is a function. Then find the inverse. *(See Example 5.)*

37. $f(x) = \frac{2}{x-4}$

38. $f(x) = \frac{7}{x+6}$

39. $f(x) = \frac{3}{x} - 2$

40. $f(x) = \frac{5}{x} - 6$

41. $f(x) = \frac{4}{11-2x}$

42. $f(x) = \frac{8}{9+5x}$

43. $f(x) = \frac{1}{x^2} + 4$

44. $f(x) = \frac{1}{x^4} - 7$

45. PROBLEM SOLVING The cost of fueling your car for 1 year can be calculated using this equation:

$$\text{Fuel cost for 1 year} = \frac{\text{Miles driven} \cdot \text{Price per gallon of fuel}}{\text{Fuel-efficiency rate}}$$

Last year you drove 9000 miles, paid \$3.24 per gallon of gasoline, and spent a total of \$1389 on gasoline. Find the fuel-efficiency rate of your car by (a) solving an equation, and (b) using the inverse of the function. *(See Example 6.)*

46. PROBLEM SOLVING The recommended percent p (in decimal form) of nitrogen (by volume) in the air that a diver breathes is given by $p = \frac{105.07}{d+33}$, where d is the depth (in feet) of the diver. Find the depth when the air contains 47% recommended nitrogen by (a) solving an equation, and (b) using the inverse of the function.

USING TOOLS In Exercises 47–50, use a graphing calculator to solve the equation $f(x) = g(x)$.

47. $f(x) = \frac{2}{3x}$, $g(x) = x$

48. $f(x) = -\frac{3}{5x}$, $g(x) = -x$

49. $f(x) = \frac{1}{x} + 1$, $g(x) = x^2$

50. $f(x) = \frac{2}{x} + 1$, $g(x) = x^2 + 1$

51. MATHEMATICAL CONNECTIONS *Golden rectangles* are rectangles for which the ratio of the width w to the length ℓ is equal to the ratio of ℓ to $\ell + w$. The ratio of the length to the width for these rectangles is called the golden ratio. Find the value of the golden ratio using a rectangle with a width of 1 unit.

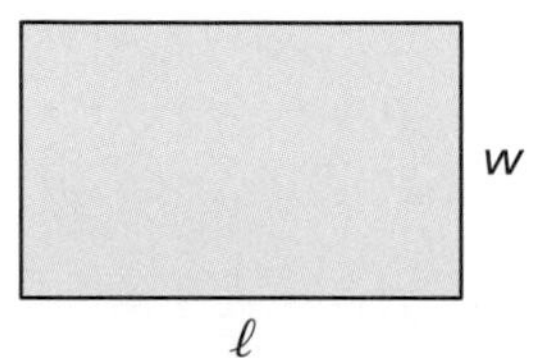

52. HOW DO YOU SEE IT? Use the graph to identify the solution(s) of the rational equation $\frac{4(x-1)}{x-1} = \frac{2x-2}{x+1}$. Explain your reasoning.

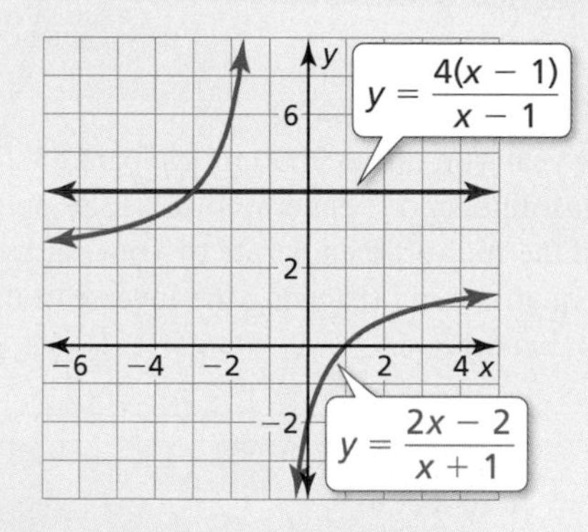

USING STRUCTURE In Exercises 53 and 54, find the inverse of the function. (*Hint:* Try rewriting the function by using either inspection or long division.)

53. $f(x) = \frac{3x+1}{x-4}$

54. $f(x) = \frac{4x-7}{2x+3}$

55. ABSTRACT REASONING Find the inverse of rational functions of the form $y = \frac{ax+b}{cx+d}$. Verify your answer is correct by using it to find the inverses in Exercises 53 and 54.

56. THOUGHT PROVOKING Is it possible to write a rational equation that has the following number of solutions? Justify your answers.

a. no solution

b. exactly one solution

c. exactly two solutions

d. infinitely many solutions

57. CRITICAL THINKING Let a be a nonzero real number. Tell whether each statement is *always true*, *sometimes true*, or *never true*. Explain your reasoning.

a. For the equation $\frac{1}{x-a} = \frac{x}{x-a}$, $x = a$ is an extraneous solution.

b. The equation $\frac{3}{x-a} = \frac{x}{x-a}$ has exactly one solution.

c. The equation $\frac{1}{x-a} = \frac{2}{x+a} + \frac{2a}{x^2-a^2}$ has no solution.

58. MAKING AN ARGUMENT Your friend claims that it is not possible for a rational equation of the form $\frac{x-a}{b} = \frac{x-c}{d}$, where $b \neq 0$ and $d \neq 0$, to have extraneous solutions. Is your friend correct? Explain your reasoning.

Maintaining Mathematical Proficiency

Reviewing what you learned in previous grades and lessons

Is the domain discrete or continuous? Explain. Graph the function using its domain. *(Skills Review Handbook)*

59. The linear function $y = 0.25x$ represents the amount of money y (in dollars) of x quarters in your pocket. You have a maximum of eight quarters in your pocket.

60. A store sells broccoli for \$2 per pound. The total cost t of the broccoli is a function of the number of pounds p you buy.

Evaluate the function for the given value of x. *(Section 4.1)*

61. $f(x) = x^3 - 2x + 7$; $x = -2$

62. $g(x) = -2x^4 + 7x^3 + x - 2$; $x = 3$

63. $h(x) = -x^3 + 3x^2 + 5x$; $x = 3$

64. $k(x) = -2x^3 - 4x^2 + 12x - 5$; $x = -5$

7.3–7.5 What Did You Learn?

Core Vocabulary

rational expression, *p. 376*
simplified form of a rational expression, *p. 376*
complex fraction, *p. 387*
cross multiplying, *p. 392*

Core Concepts

Section 7.3

Simplifying Rational Expressions, *p. 376*
Multiplying Rational Expressions, *p. 377*
Dividing Rational Expressions, *p. 378*

Section 7.4

Adding or Subtracting with Like Denominators, *p. 384*
Adding or Subtracting with Unlike Denominators, *p. 384*
Simplifying Complex Fractions, *p. 387*

Section 7.5

Solving Rational Equations by Cross Multiplying, *p. 392*
Solving Rational Equations by Using the Least Common Denominator, *p. 393*
Using Inverses of Functions, *p. 395*

Mathematical Practices

1. In Exercise 37 on page 381, what type of equation did you expect to get as your solution? Explain why this type of equation is appropriate in the context of this situation.
2. Write a simpler problem that is similar to Exercise 44 on page 382. Describe how to use the simpler problem to gain insight into the solution of the more complicated problem in Exercise 44.
3. In Exercise 57 on page 390, what conjecture did you make about the value the given expressions were approaching? What logical progression led you to determine whether your conjecture was correct?
4. Compare the methods for solving Exercise 45 on page 397. Be sure to discuss the similarities and differences between the methods as precisely as possible.

Performance Task

Circuit Design

A thermistor is a resistor whose resistance varies with temperature. Thermistors are an engineer's dream because they are inexpensive, small, rugged, and accurate. The one problem with thermistors is their responses to temperature are not linear. How would you design a circuit that corrects this problem?

To explore the answers to these questions and more, go to ***BigIdeasMath.com***.

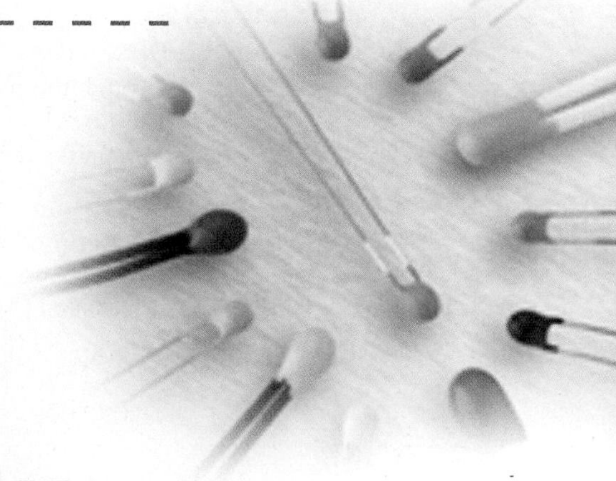

7 Chapter Review

Dynamic Solutions available at *BigIdeasMath.com*

7.1 Inverse Variation (pp. 359–364)

The variables x and y vary inversely, and $y = 12$ when $x = 3$. Write an equation that relates x and y. Then find y when $x = -4$.

$y = \frac{a}{x}$ Write general equation for inverse variation.

$12 = \frac{a}{3}$ Substitute 12 for y and 3 for x.

$36 = a$ Multiply each side by 3.

▶ The inverse variation equation is $y = \frac{36}{x}$. When $x = -4$, $y = \frac{36}{-4} = -9$.

Tell whether x and y show *direct variation*, *inverse variation*, or *neither*.

1. $xy = 5$

2. $5y = 6x$

3. $15 = \frac{x}{y}$

4. $y - 3 = 2x$

5.

x	7	11	15	20
y	35	55	75	100

6.

x	5	8	10	20
y	6.4	4	3.2	1.6

The variables x and y vary inversely. Use the given values to write an equation relating x and y. Then find y when $x = -3$.

7. $x = 1, y = 5$

8. $x = -4, y = -6$

9. $x = \frac{5}{2}, y = 18$

10. $x = -12, y = \frac{2}{3}$

7.2 Graphing Rational Functions (pp. 365–372)

Graph $y = \frac{2x + 5}{x - 1}$. State the domain and range.

Step 1 Draw the asymptotes. Solve $x - 1 = 0$ for x to find the vertical asymptote $x = 1$. The horizontal asymptote is the line $y = \frac{a}{c} = \frac{2}{1} = 2$.

Step 2 Plot points to the left of the vertical asymptote, such as $\left(-2, -\frac{1}{3}\right)$, $\left(-1, -\frac{3}{2}\right)$, and $(0, -5)$. Plot points to the right of the vertical asymptote, such as $\left(3, \frac{11}{2}\right)$, $\left(5, \frac{15}{4}\right)$, and $\left(7, \frac{19}{6}\right)$.

Step 3 Draw the two branches of the hyperbola so that they pass through the plotted points and approach the asymptotes.

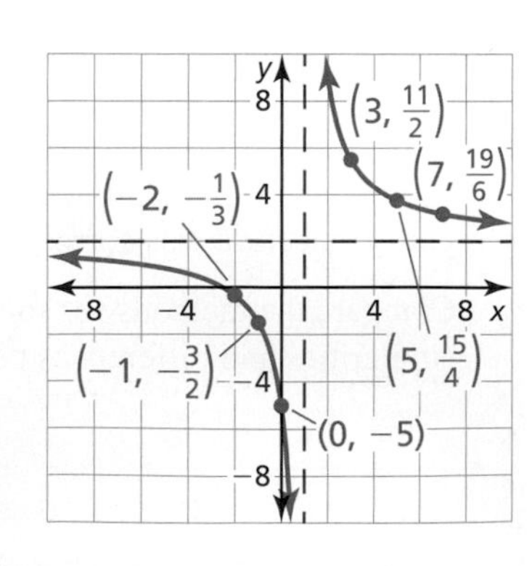

▶ The domain is all real numbers except 1 and the range is all real numbers except 2.

Graph the function. State the domain and range.

11. $y = \frac{4}{x - 3}$

12. $y = \frac{1}{x + 5} + 2$

13. $f(x) = \frac{3x - 2}{x - 4}$

7.3 Multiplying and Dividing Rational Expressions (pp. 375–382)

Find the quotient $\frac{3x+27}{6x-48} \div \frac{x^2+9x}{x^2-4x-32}$.

$$\frac{3x+27}{6x-48} \div \frac{x^2+9x}{x^2-4x-32} = \frac{3x+27}{6x-48} \cdot \frac{x^2-4x-32}{x^2+9x}$$ Multiply by reciprocal.

$$= \frac{3(x+9)}{6(x-8)} \cdot \frac{(x+4)(x-8)}{x(x+9)}$$ Factor.

$$= \frac{\cancel{3}\cancel{(x+9)}(x+4)\cancel{(x-8)}}{2(\cancel{3})\cancel{(x-8)}(x)\cancel{(x+9)}}$$ Multiply. Divide out common factors.

$$= \frac{x+4}{2x},\ x \neq 8,\ x \neq -9,\ x \neq -4$$ Simplified form

Find the product or quotient.

14. $\frac{80x^4}{y^3} \cdot \frac{xy}{5x^2}$

15. $\frac{x-3}{2x-8} \cdot \frac{6x^2-96}{x^2-9}$

16. $\frac{16x^2-8x+1}{x^3-7x^2+12x} \div \frac{20x^2-5x}{15x^3}$

17. $\frac{x^2-13x+40}{x^2-2x-15} \div (x^2-5x-24)$

7.4 Adding and Subtracting Rational Expressions (pp. 383–390)

Find the sum $\frac{x}{6x+24} + \frac{x+2}{x^2+9x+20}$.

$$\frac{x}{6x+24} + \frac{x+2}{x^2+9x+20} = \frac{x}{6(x+4)} + \frac{x+2}{(x+4)(x+5)}$$ Factor each denominator.

$$= \frac{x}{6(x+4)} \cdot \frac{x+5}{x+5} + \frac{x+2}{(x+4)(x+5)} \cdot \frac{6}{6}$$ LCD is $6(x+4)(x+5)$.

$$= \frac{x^2+5x}{6(x+4)(x+5)} + \frac{6x+12}{6(x+4)(x+5)}$$ Multiply.

$$= \frac{x^2+11x+12}{6(x+4)(x+5)}$$ Add numerators.

Find the sum or difference.

18. $\frac{5}{6(x+3)} + \frac{x+4}{2x}$

19. $\frac{5x}{x+8} + \frac{4x-9}{x^2+5x-24}$

20. $\frac{x+2}{x^2+4x+3} - \frac{5x}{x^2-9}$

Rewrite the function in the form $g(x) = \frac{a}{x-h} + k$. Graph the function. Describe the graph of g as a transformation of the graph of $f(x) = \frac{a}{x}$.

21. $g(x) = \frac{5x+1}{x-3}$

22. $g(x) = \frac{4x+2}{x+7}$

23. $g(x) = \frac{9x-10}{x-1}$

24. Let f be the focal length of a thin camera lens, p be the distance between the lens and an object being photographed, and q be the distance between the lens and the film. For the photograph to be in focus, the variables should satisfy the lens equation to the right. Simplify the complex fraction.

$$f = \frac{1}{\frac{1}{p} + \frac{1}{q}}$$

7.5 Solving Rational Equations (pp. 391–398)

Solve $\frac{-4}{x+3} = \frac{x-1}{x+3} + \frac{x}{x-4}$.

The LCD is $(x+3)(x-4)$.

$$\frac{-4}{x+3} = \frac{x-1}{x+3} + \frac{x}{x-4}$$

$$(x+3)(x-4) \cdot \frac{-4}{x+3} = (x+3)(x-4) \cdot \frac{x-1}{x+3} + (x+3)(x-4) \cdot \frac{x}{x-4}$$

$$-4(x-4) = (x-1)(x-4) + x(x+3)$$

$$-4x + 16 = x^2 - 5x + 4 + x^2 + 3x$$

$$0 = 2x^2 + 2x - 12$$

$$0 = x^2 + x - 6$$

$$0 = (x+3)(x-2)$$

$$x + 3 = 0 \quad \text{or} \quad x - 2 = 0$$

$$x = -3 \quad \text{or} \quad x = 2$$

Check

Check $x = -3$:

$$\frac{-4}{-3+3} \stackrel{?}{=} \frac{-3-1}{-3+3} + \frac{-3}{-3-4}$$

$$\frac{-4}{0} \stackrel{?}{=} \frac{-4}{0} + \frac{-3}{-7} \quad ✗$$

Division by zero is undefined.

Check $x = 2$:

$$\frac{-4}{2+3} \stackrel{?}{=} \frac{2-1}{2+3} + \frac{2}{2-4}$$

$$\frac{-4}{5} \stackrel{?}{=} \frac{1}{5} + \frac{2}{-2}$$

$$\frac{-4}{5} = \frac{-4}{5} \quad ✓$$

▶ The apparent solution $x = -3$ is extraneous. So, the only solution is $x = 2$.

Solve the equation. Check your solution(s).

25. $\frac{5}{x} = \frac{7}{x+2}$ **26.** $\frac{8(x-1)}{x^2-4} = \frac{4}{x+2}$ **27.** $\frac{2(x+7)}{x+4} - 2 = \frac{2x+20}{2x+8}$

Determine whether the inverse of *f* is a function. Then find the inverse.

28. $f(x) = \frac{3}{x+6}$ **29.** $f(x) = \frac{10}{x-7}$ **30.** $f(x) = \frac{1}{x} + 8$

31. At a bowling alley, shoe rentals cost $3 and each game costs $4. The average cost c (in dollars) of bowling n games is given by $c = \frac{4n+3}{n}$. Find how many games you must bowl for the average cost to fall to $4.75 by (a) solving an equation, and (b) using the inverse of a function.

7 Chapter Test

The variables x and y vary inversely. Use the given values to write an equation relating x and y. Then find y when $x = 4$.

1. $x = 5, y = 2$

2. $x = -4, y = \frac{7}{2}$

3. $x = \frac{3}{4}, y = \frac{5}{8}$

The graph shows the function $y = \dfrac{1}{x - h} + k$. Determine whether the value of each constant h and k is *positive*, *negative*, or *zero*. Explain your reasoning.

4.

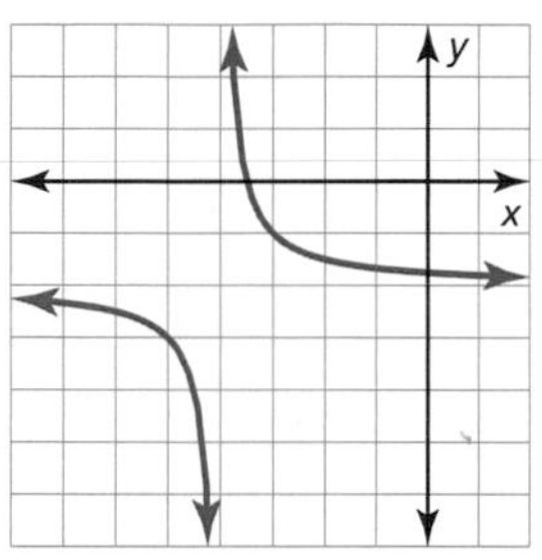

5.

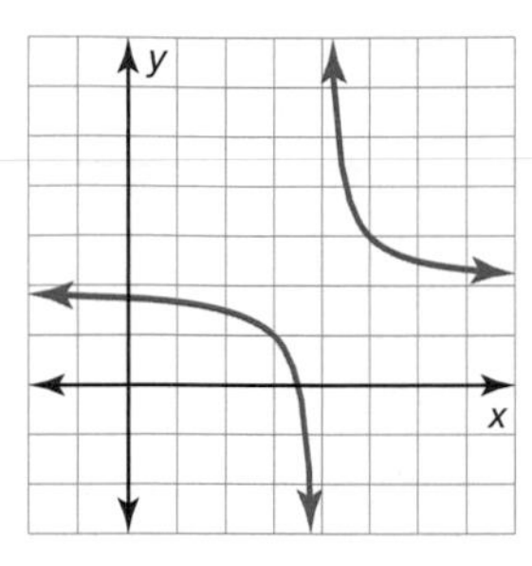

6.

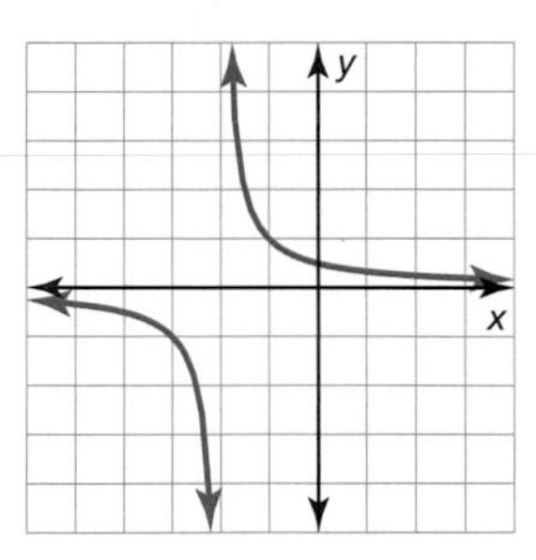

Perform the indicated operation.

7. $\dfrac{3x^2y}{4x^3y^5} \div \dfrac{6y^2}{2xy^3}$

8. $\dfrac{3x}{x^2 + x - 12} - \dfrac{6}{x + 4}$

9. $\dfrac{x^2 - 3x - 4}{x^2 - 3x - 18} \bullet \dfrac{x - 6}{x + 1}$

10. $\dfrac{4}{x + 5} + \dfrac{2x}{x^2 - 25}$

11. Let $g(x) = \dfrac{(x + 3)(x - 2)}{x + 3}$. Simplify $g(x)$. Determine whether the graph of $f(x) = x - 2$ and the graph of g are different. Explain your reasoning.

12. You start a small beekeeping business. Your initial costs are \$500 for equipment and bees. You estimate it will cost \$1.25 per pound to collect, clean, bottle, and label the honey. How many pounds of honey must you produce before your average cost per pound is \$1.79? Justify your answer.

13. You can use a simple lever to lift a 300-pound rock. The force F (in foot-pounds) needed to lift the rock is inversely related to the distance d (in feet) from the pivot point of the lever. To lift the rock, you need 60 pounds of force applied to a lever with a distance of 10 feet from the pivot point. What force is needed when you increase the distance to 15 feet from the pivot point? Justify your answer.

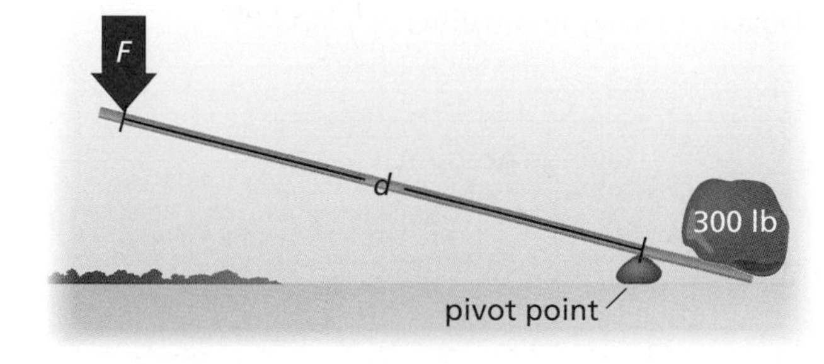

14. Three tennis balls fit tightly in a can as shown.

a. Write an expression for the height h of the can in terms of its radius r. Then rewrite the formula for the volume of a cylinder in terms of r only.

b. Find the percent of the can's volume that is *not* occupied by tennis balls.

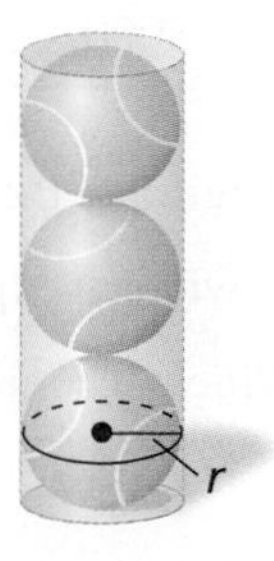

7 Cumulative Assessment

1. Which of the following functions are shown in the graph? Select all that apply. Justify your answers.

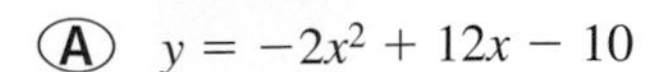

Ⓐ $y = -2x^2 + 12x - 10$

Ⓑ $y = x^2 - 6x + 13$

Ⓒ $y = -2(x - 3)^2 + 8$

Ⓓ $y = -(x - 1)(x - 5)$

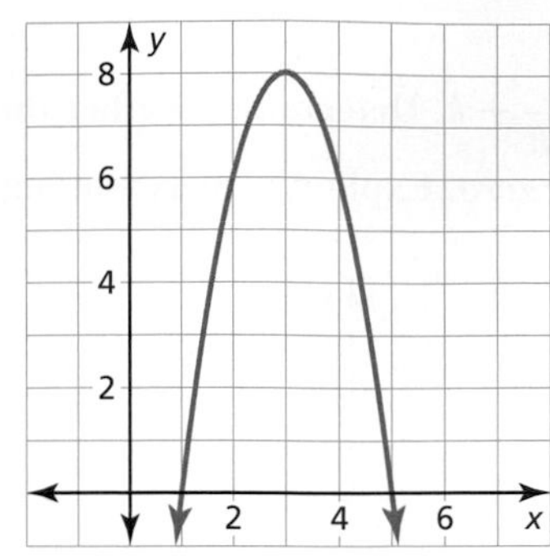

2. You step onto an escalator and begin descending. After riding for 12 feet, you realize that you dropped your keys on the upper floor and walk back up the escalator to retrieve them. The total time T of your trip down and up the escalator is given by

$$T = \frac{12}{s} + \frac{12}{w - s}$$

where s is the speed of the escalator and w is your walking speed. The trip took 9 seconds, and you walk at a speed of 6 feet per second. Find two possible speeds of the escalator.

3. The graph of a rational function has asymptotes that intersect at the point (4, 3). Choose the correct values to complete the equation of the function. Then graph the function.

$$y = \frac{\square x + 6}{\square x + \square}$$

12	−3
9	−6
3	−12

4. The tables below give the amounts A (in dollars) of money in two different bank accounts over time t (in years).

Checking Account				
t	1	2	3	4
A	5000	5110	5220	5330

Savings Account				
t	1	2	3	4
A	5000	5100	5202	5306.04

a. Determine the type of function represented by the data in each table.

b. Provide an explanation for the type of growth of each function.

c. Which account has a greater value after 10 years? after 15 years? Justify your answers.

5. Order the expressions from least to greatest. Justify your answer.

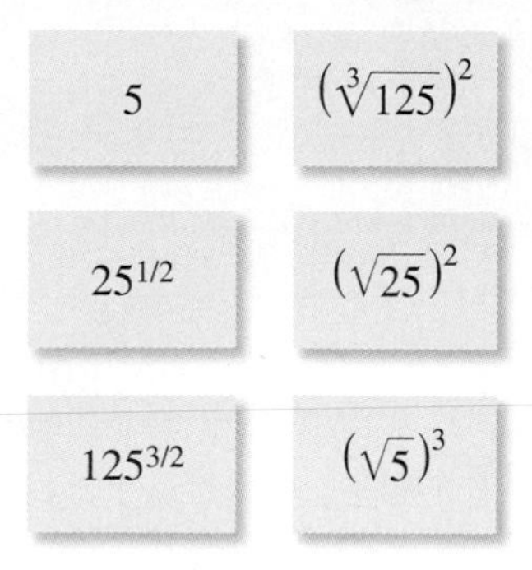

6. A movie grosses $37 million after the first week of release. The weekly gross sales y decreases by 30% each week. Write an exponential decay function that represents the weekly gross sales in week x. What is a reasonable domain and range in this situation? Explain your reasoning.

7. Choose the correct relationship among the variables in the table. Justify your answer by writing an equation that relates p, q, and r.

p	q	r
−12	20	16
3	1	−10
30	−82	−8
−1.5	4	0.5

Ⓐ The variable p varies directly with the difference of q and r.

Ⓑ The variable r varies inversely with the difference of p and q.

Ⓒ The variable q varies inversely with the sum of p and r.

Ⓓ The variable p varies directly with the sum of q and r.

8. You have taken five quizzes in your history class, and your average score is 83 points. You think you can score 95 points on each remaining quiz. How many quizzes do you need to take to raise your average quiz score to 90 points? Justify your answer.

8 Sequences and Series

Tree Farm *(p. 449)*

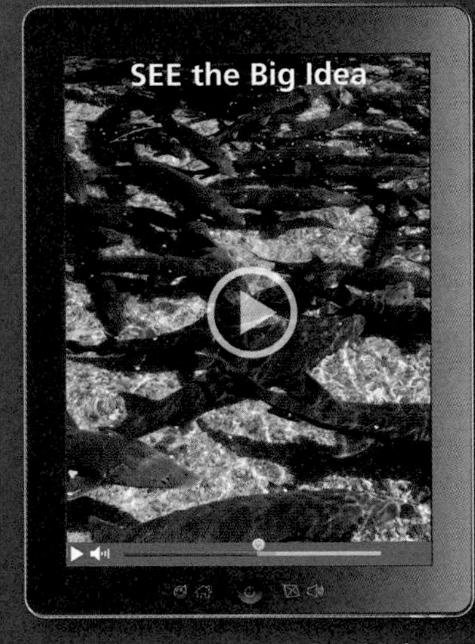

Fish Population *(p. 445)*

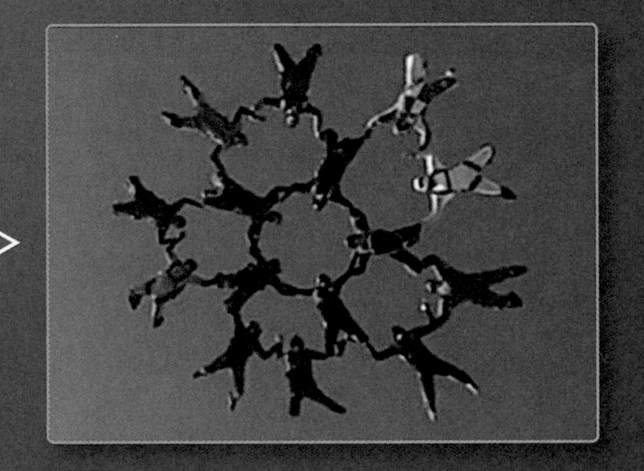
Skydiving *(p. 431)*

Marching Band *(p. 423)*

Museum Skylight *(p. 416)*

Maintaining Mathematical Proficiency

Evaluating Functions

Example 1 Evaluate the function $y = 2x^2 - 10$ for the values $x = 0, 1, 2, 3$, and 4.

Input, x	$2x^2 - 10$	Output, y
0	$2(0)^2 - 10$	-10
1	$2(1)^2 - 10$	-8
2	$2(2)^2 - 10$	-2
3	$2(3)^2 - 10$	8
4	$2(4)^2 - 10$	22

Copy and complete the table to evaluate the function.

1. $y = 3 - 2^x$

x	y
1	
2	
3	

2. $y = 5x^2 + 1$

x	y
2	
3	
4	

3. $y = -4x + 24$

x	y
5	
10	
15	

Solving Equations

Example 2 Solve the equation $45 = 5(3)^x$.

$45 = 5(3)^x$ Write original equation.

$\frac{45}{5} = \frac{5(3)^x}{5}$ Divide each side by 5.

$9 = 3^x$ Simplify.

$\log_3 9 = \log_3 3^x$ Take $\log_3$ of each side.

$2 = x$ Simplify.

Solve the equation. Check your solution(s).

4. $7x + 3 = 31$

5. $\frac{1}{16} = 4\left(\frac{1}{2}\right)^x$

6. $216 = 3(x + 6)$

7. $2^x + 16 = 144$

8. $\frac{1}{4}x - 8 = 17$

9. $8\left(\frac{3}{4}\right)^x = \frac{27}{8}$

10. ABSTRACT REASONING The graph of the exponential decay function $f(x) = b^x$ has an asymptote $y = 0$. How is the graph of f different from a scatter plot consisting of the points $(1, b^1)$, $(2, b^1 + b^2)$, $(3, b^1 + b^2 + b^3)$, . . .? How is the graph of f similar?

Mathematical Practices

Mathematically proficient students consider the available tools when solving a mathematical problem.

Using Appropriate Tools Strategically

Core Concept

Using a Spreadsheet

To use a spreadsheet, it is common to write one cell as a function of another cell. For instance, in the spreadsheet shown, the cells in column A starting with cell A2 contain functions of the cell in the preceding row. Also, the cells in column B contain functions of the cells in the same row in column A.

	A	B
1	1	0
2	2	2
3	3	4
4	4	6
5	5	8
6	6	10
7	7	12
8	8	14

B1 = 2*A1−2

A2 = A1+1

EXAMPLE 1 **Using a Spreadsheet**

You deposit $1000 in stocks that earn 15% interest compounded annually. Use a spreadsheet to find the balance at the end of each year for 8 years. Describe the type of growth.

SOLUTION

You can enter the given information into a spreadsheet and generate the graph shown. From the formula in the spreadsheet, you can see that the growth pattern is exponential. The graph also appears to be exponential.

	A	B
1	Year	Balance
2	0	$1000.00
3	1	$1150.00
4	2	$1322.50
5	3	$1520.88
6	4	$1749.01
7	5	$2011.36
8	6	$2313.06
9	7	$2660.02
10	8	$3059.02

B3 = B2*1.15

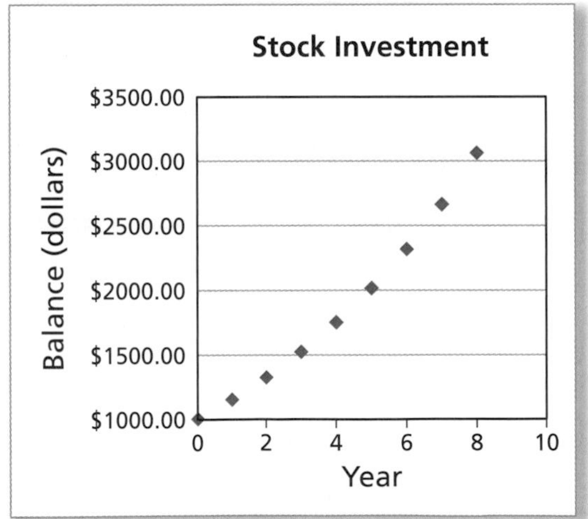

Monitoring Progress

Use a spreadsheet to help you answer the question.

1. A pilot flies a plane at a speed of 500 miles per hour for 4 hours. Find the total distance flown at 30-minute intervals. Describe the pattern.
2. A population of 60 rabbits increases by 25% each year for 8 years. Find the population at the end of each year. Describe the type of growth.
3. An endangered population has 500 members. The population declines by 10% each decade for 80 years. Find the population at the end of each decade. Describe the type of decline.
4. The top eight runners finishing a race receive cash prizes. First place receives $200, second place receives $175, third place receives $150, and so on. Find the fifth through eighth place prizes. Describe the type of decline.

8.1 Defining and Using Sequences and Series

Essential Question

How can you write a rule for the *n*th term of a sequence?

A **sequence** is an ordered list of numbers. There can be a limited number or an infinite number of *terms* of a sequence.

$a_1, a_2, a_3, a_4, \ldots, a_n, \ldots$ Terms of a sequence

Here is an example.

$1, 4, 7, 10, \ldots, 3n - 2, \ldots$

CONSTRUCTING VIABLE ARGUMENTS

To be proficient in math, you need to reason inductively about data.

EXPLORATION 1 Writing Rules for Sequences

Work with a partner. Match each sequence with its graph. The horizontal axes represent *n*, the position of each term in the sequence. Then write a rule for the *n*th term of the sequence, and use the rule to find a_{10}.

a. $1, 2.5, 4, 5.5, 7, \ldots$

b. $8, 6.5, 5, 3.5, 2, \ldots$

c. $\frac{1}{4}, \frac{4}{4}, \frac{9}{4}, \frac{16}{4}, \frac{25}{4}, \ldots$

d. $\frac{25}{4}, \frac{16}{4}, \frac{9}{4}, \frac{4}{4}, \frac{1}{4}, \ldots$

e. $\frac{1}{2}, 1, 2, 4, 8, \ldots$

f. $8, 4, 2, 1, \frac{1}{2}, \ldots$

A.

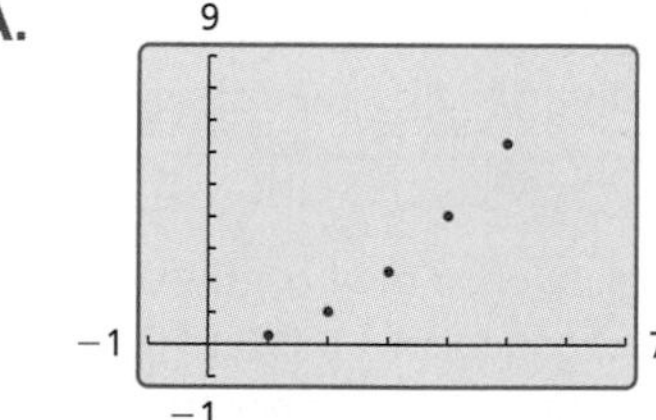

B.

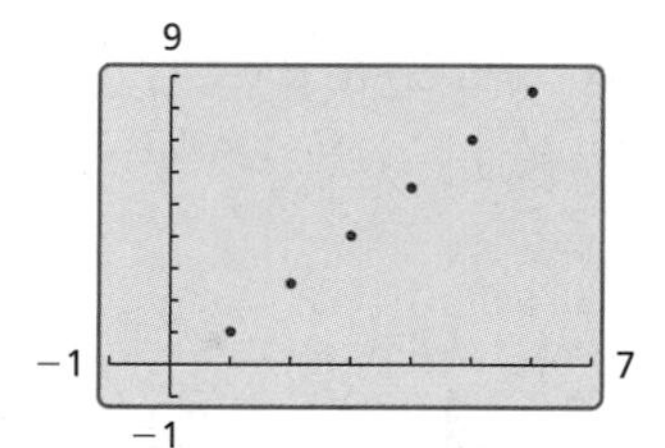

C.

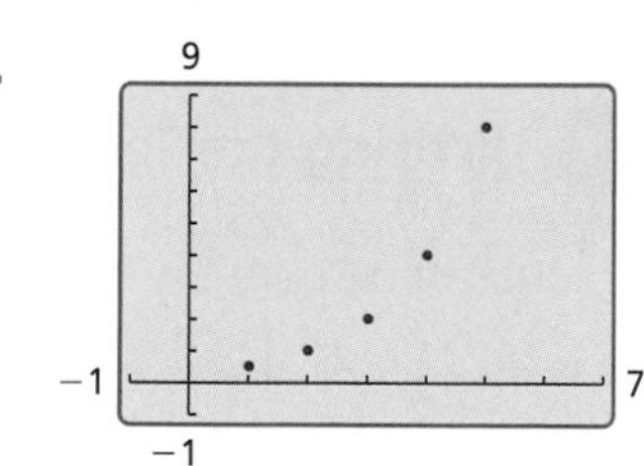

D.

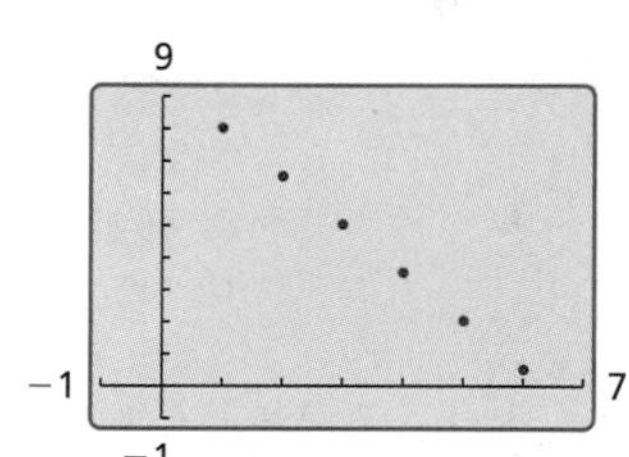

E.

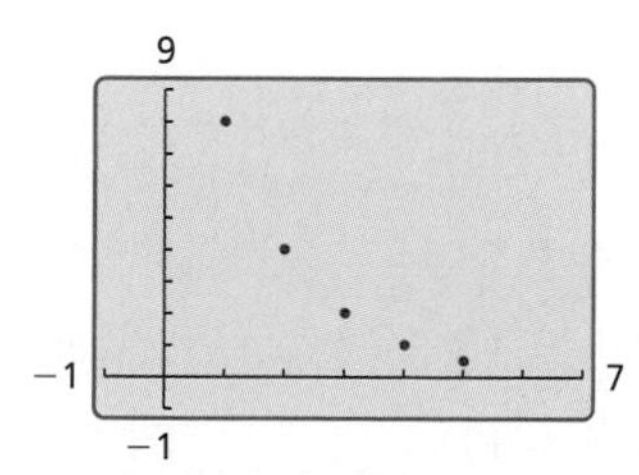

F.

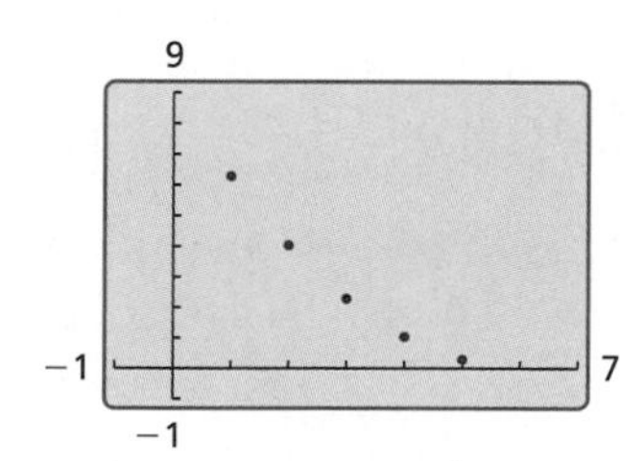

Communicate Your Answer

2. How can you write a rule for the *n*th term of a sequence?

3. What do you notice about the relationship between the terms in (a) an arithmetic sequence and (b) a geometric sequence? Justify your answers.

8.1 Lesson

Core Vocabulary

sequence, *p. 410*
terms of a sequence, *p. 410*
series, *p. 412*
summation notation, *p. 412*
sigma notation, *p. 412*

Previous
domain
range

What You Will Learn

- Use sequence notation to write terms of sequences.
- Write a rule for the *n*th term of a sequence.
- Sum the terms of a sequence to obtain a series and use summation notation.

Writing Terms of Sequences

Core Concept

Sequences

A **sequence** is an ordered list of numbers. A *finite sequence* is a function that has a limited number of terms and whose domain is the finite set $\{1, 2, 3, \ldots, n\}$. The values in the range are called the **terms** of the sequence.

Domain: $1 \quad 2 \quad 3 \quad 4 \quad \ldots \quad n$ — Relative position of each term

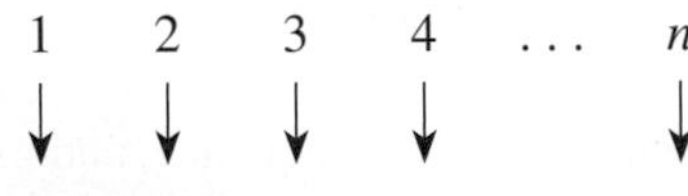

Range: $a_1 \quad a_2 \quad a_3 \quad a_4 \quad \ldots \quad a_n$ — Terms of the sequence

An *infinite sequence* is a function that continues without stopping and whose domain is the set of positive integers. Here are examples of a finite sequence and an infinite sequence.

Finite sequence: 2, 4, 6, 8 **Infinite sequence:** 2, 4, 6, 8, . . .

A sequence can be specified by an equation, or *rule*. For example, both sequences above can be described by the rule $a_n = 2n$ or $f(n) = 2n$.

The domain of a sequence may begin with 0 instead of 1. When this is the case, the domain of a finite sequence is the set $\{0, 1, 2, 3, \ldots, n\}$ and the domain of an infinite sequence becomes the set of nonnegative integers. Unless otherwise indicated, assume the domain of a sequence begins with 1.

EXAMPLE 1 Writing the Terms of Sequences

Write the first six terms of (a) $a_n = 2n + 5$ and (b) $f(n) = (-3)^{n-1}$.

SOLUTION

a.		b.
$a_1 = 2(1) + 5 = 7$	1st term	$f(1) = (-3)^{1-1} = 1$
$a_2 = 2(2) + 5 = 9$	2nd term	$f(2) = (-3)^{2-1} = -3$
$a_3 = 2(3) + 5 = 11$	3rd term	$f(3) = (-3)^{3-1} = 9$
$a_4 = 2(4) + 5 = 13$	4th term	$f(4) = (-3)^{4-1} = -27$
$a_5 = 2(5) + 5 = 15$	5th term	$f(5) = (-3)^{5-1} = 81$
$a_6 = 2(6) + 5 = 17$	6th term	$f(6) = (-3)^{6-1} = -243$

Monitoring Progress

Help in English and Spanish at *BigIdeasMath.com*

Write the first six terms of the sequence.

1. $a_n = n + 4$ **2.** $f(n) = (-2)^{n-1}$ **3.** $a_n = \dfrac{n}{n+1}$

Writing Rules for Sequences

When the terms of a sequence have a recognizable pattern, you may be able to write a rule for the *n*th term of the sequence.

STUDY TIP

When you are given only the first several terms of a sequence, there may be more than one rule for the *n*th term. For instance, the sequence 2, 4, 8, . . . can be given by $a_n = 2^n$ or $a_n = n^2 - n + 2$.

EXAMPLE 2 Writing Rules for Sequences

Describe the pattern, write the next term, and write a rule for the *n*th term of the sequences (a) $-1, -8, -27, -64, \ldots$ and (b) $0, 2, 6, 12, \ldots$.

SOLUTION

a. You can write the terms as $(-1)^3, (-2)^3, (-3)^3, (-4)^3, \ldots$. The next term is $a_5 = (-5)^3 = -125$. A rule for the *n*th term is $a_n = (-n)^3$.

b. You can write the terms as $0(1), 1(2), 2(3), 3(4), \ldots$. The next term is $f(5) = 4(5) = 20$. A rule for the *n*th term is $f(n) = (n - 1)n$.

To graph a sequence, let the horizontal axis represent the position numbers (the domain) and the vertical axis represent the terms (the range).

EXAMPLE 3 Solving a Real-Life Problem

You work in a grocery store and are stacking apples in the shape of a square pyramid with seven layers. Write a rule for the number of apples in each layer. Then graph the sequence.

SOLUTION

Step 1 Make a table showing the number of fruit in the first three layers. Let a_n represent the number of apples in layer *n*.

Layer, *n*	1	2	3
Number of apples, a_n	$1 = 1^2$	$4 = 2^2$	$9 = 3^2$

COMMON ERROR

Although the plotted points in Example 3 follow a curve, do not draw the curve because the sequence is defined only for integer values of *n*, specifically $n = 1, 2, 3, 4, 5, 6,$ and 7.

Step 2 Write a rule for the number of apples in each layer. From the table, you can see that $a_n = n^2$.

Step 3 Plot the points (1, 1), (2, 4), (3, 9), (4, 16), (5, 25), (6, 36), and (7, 49). The graph is shown at the right.

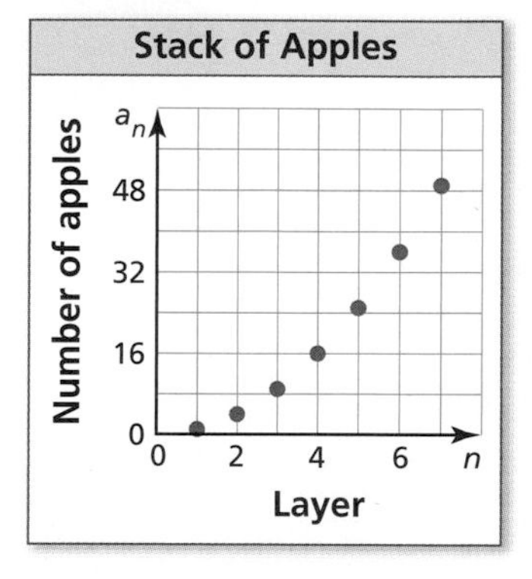

Monitoring Progress

Help in English and Spanish at *BigIdeasMath.com*

Describe the pattern, write the next term, graph the first five terms, and write a rule for the *n*th term of the sequence.

4. $3, 5, 7, 9, \ldots$

5. $3, 8, 15, 24, \ldots$

6. $1, -2, 4, -8, \ldots$

7. $2, 5, 10, 17, \ldots$

8. **WHAT IF?** In Example 3, suppose there are nine layers of apples. How many apples are in the ninth layer?

Writing Rules for Series

Core Concept

Series and Summation Notation

When the terms of a sequence are added together, the resulting expression is a **series**. A series can be finite or infinite.

Finite series: $2 + 4 + 6 + 8$

Infinite series: $2 + 4 + 6 + 8 + \cdots$

You can use **summation notation** to write a series. For example, the two series above can be written in summation notation as follows:

Finite series: $2 + 4 + 6 + 8 = \sum_{i=1}^{4} 2i$

READING

When written in summation notation, this series is read as "the sum of $2i$ for values of i from 1 to 4."

Infinite series: $2 + 4 + 6 + 8 + \cdots = \sum_{i=1}^{\infty} 2i$

For both series, the *index of summation* is i and the *lower limit of summation* is 1. The *upper limit of summation* is 4 for the finite series and ∞ (infinity) for the infinite series. Summation notation is also called **sigma notation** because it uses the uppercase Greek letter *sigma*, written Σ.

EXAMPLE 4 Writing Series Using Summation Notation

Write each series using summation notation.

a. $25 + 50 + 75 + \cdots + 250$

b. $\frac{1}{2} + \frac{2}{3} + \frac{3}{4} + \frac{4}{5} + \cdots$

SOLUTION

a. Notice that the first term is 25(1), the second is 25(2), the third is 25(3), and the last is 25(10). So, the terms of the series can be written as:

$$a_i = 25i, \text{ where } i = 1, 2, 3, \ldots, 10$$

The lower limit of summation is 1 and the upper limit of summation is 10.

▶ The summation notation for the series is $\sum_{i=1}^{10} 25i$.

b. Notice that for each term, the denominator of the fraction is 1 more than the numerator. So, the terms of the series can be written as:

$$a_i = \frac{i}{i + 1}, \text{ where } i = 1, 2, 3, 4, \ldots$$

The lower limit of summation is 1 and the upper limit of summation is infinity.

▶ The summation notation for the series is $\sum_{i=1}^{\infty} \frac{i}{i + 1}$.

Monitoring Progress

Help in English and Spanish at *BigIdeasMath.com*

Write the series using summation notation.

9. $5 + 10 + 15 + \cdots + 100$

10. $\frac{1}{2} + \frac{4}{5} + \frac{9}{10} + \frac{16}{17} + \cdots$

11. $6 + 36 + 216 + 1296 + \cdots$

12. $5 + 6 + 7 + \cdots + 12$

The index of summation for a series does not have to be i—any letter can be used. Also, the index does not have to begin at 1. For instance, the index begins at 4 in the next example.

COMMON ERROR

Be sure to use the correct lower and upper limits of summation when finding the sum of a series.

EXAMPLE 5 **Finding the Sum of a Series**

Find the sum $\sum_{k=4}^{8}(3 + k^2)$.

SOLUTION

$$\sum_{k=4}^{8}(3 + k^2) = (3 + 4^2) + (3 + 5^2) + (3 + 6^2) + (3 + 7^2) + (3 + 8^2)$$
$$= 19 + 28 + 39 + 52 + 67$$
$$= 205$$

For series with many terms, finding the sum by adding the terms can be tedious. Below are formulas you can use to find the sums of three special types of series.

Core Concept

Formulas for Special Series

Sum of n terms of 1: $\sum_{i=1}^{n} 1 = n$

Sum of first n positive integers: $\sum_{i=1}^{n} i = \frac{n(n + 1)}{2}$

Sum of squares of first n positive integers: $\sum_{i=1}^{n} i^2 = \frac{n(n + 1)(2n + 1)}{6}$

EXAMPLE 6 **Using a Formula for a Sum**

How many apples are in the stack in Example 3?

SOLUTION

From Example 3, you know that the ith term of the series is given by $a_i = i^2$, where $i = 1, 2, 3, \ldots, 7$. Using summation notation and the third formula listed above, you can find the total number of apples as follows:

$$1^2 + 2^2 + \cdots + 7^2 = \sum_{i=1}^{7} i^2 = \frac{7(7 + 1)(2 \cdot 7 + 1)}{6} = \frac{7(8)(15)}{6} = 140$$

▶ There are 140 apples in the stack. Check this by adding the number of apples in each of the seven layers.

Monitoring Progress

Help in English and Spanish at *BigIdeasMath.com*

Find the sum.

13. $\sum_{i=1}^{5} 8i$

14. $\sum_{k=3}^{7}(k^2 - 1)$

15. $\sum_{i=1}^{34} 1$

16. $\sum_{k=1}^{6} k$

17. WHAT IF? Suppose there are nine layers in the apple stack in Example 3. How many apples are in the stack?

8.1 Exercises

Dynamic Solutions available at *BigIdeasMath.com*

Vocabulary and Core Concept Check

1. **VOCABULARY** What is another name for summation notation?

2. **COMPLETE THE SENTENCE** In a sequence, the numbers are called __________ of the sequence.

3. **WRITING** Compare sequences and series.

4. **WHICH ONE DOESN'T BELONG?** Which does *not* belong with the other three? Explain your reasoning.

$\sum_{i=1}^{6} i^2$	91	$1 + 4 + 9 + 16 + 25 + 36$	$\sum_{i=0}^{5} i^2$

Monitoring Progress and Modeling with Mathematics

In Exercises 5–14, write the first six terms of the sequence. *(See Example 1.)*

5. $a_n = n + 2$

6. $a_n = 6 - n$

7. $a_n = n^2$

8. $f(n) = n^3 + 2$

9. $f(n) = 4^{n-1}$

10. $a_n = -n^2$

11. $a_n = n^2 - 5$

12. $a_n = (n + 3)^2$

13. $f(n) = \frac{2n}{n + 2}$

14. $f(n) = \frac{n}{2n - 1}$

In Exercises 15–26, describe the pattern, write the next term, and write a rule for the *n*th term of the sequence. *(See Example 2.)*

15. 1, 6, 11, 16, . . .

16. 1, 2, 4, 8, . . .

17. 3.1, 3.8, 4.5, 5.2, . . .

18. 9, 16.8, 24.6, 32.4, . . .

19. 5.8, 4.2, 2.6, 1, −0.6 . . .

20. −4, 8, −12, 16, . . .

21. $\frac{1}{4}, \frac{2}{4}, \frac{3}{4}, \frac{4}{4}, \ldots$

22. $\frac{1}{10}, \frac{3}{20}, \frac{5}{30}, \frac{7}{40}, \ldots$

23. $\frac{2}{3}, \frac{2}{6}, \frac{2}{9}, \frac{2}{12}, \ldots$

24. $\frac{2}{3}, \frac{4}{4}, \frac{6}{5}, \frac{8}{6}, \ldots$

25. 2, 9, 28, 65, . . .

26. 1.2, 4.2, 9.2, 16.2, . . .

27. **FINDING A PATTERN** Which rule gives the total number of squares in the *n*th figure of the pattern shown? Justify your answer.

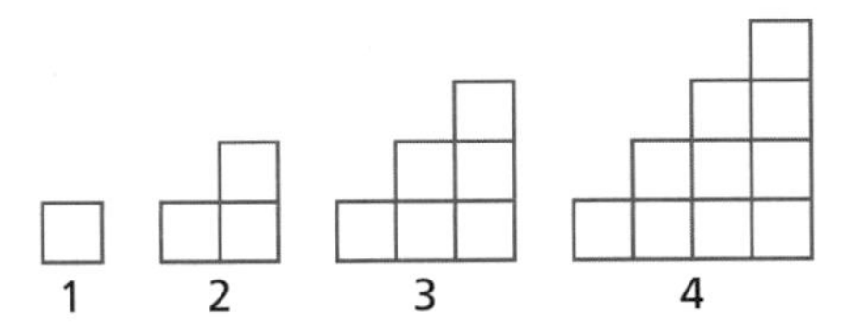

Ⓐ $a_n = 3n - 3$

Ⓑ $a_n = 4n - 5$

Ⓒ $a_n = n$

Ⓓ $a_n = \frac{n(n + 1)}{2}$

28. **FINDING A PATTERN** Which rule gives the total number of green squares in the *n*th figure of the pattern shown? Justify your answer.

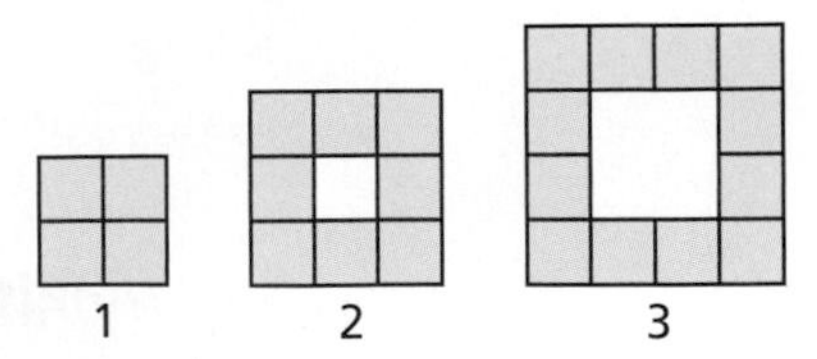

Ⓐ $a_n = n^2 - 1$

Ⓑ $a_n = \frac{n^2}{2}$

Ⓒ $a_n = 4n$

Ⓓ $a_n = 2n + 1$

29. **MODELING WITH MATHEMATICS** Rectangular tables are placed together along their short edges, as shown in the diagram. Write a rule for the number of people that can be seated around n tables arranged in this manner. Then graph the sequence. *(See Example 3.)*

30. **MODELING WITH MATHEMATICS** An employee at a construction company earns \$33,000 for the first year of employment. Employees at the company receive raises of \$2400 each year. Write a rule for the salary of the employee each year. Then graph the sequence.

In Exercises 31–38, write the series using summation notation. *(See Example 4.)*

31. $7 + 10 + 13 + 16 + 19$

32. $5 + 11 + 17 + 23 + 29$

33. $4 + 7 + 12 + 19 + \cdots$

34. $-1 + 2 + 7 + 14 + \cdots$

35. $\frac{1}{3} + \frac{1}{9} + \frac{1}{27} + \frac{1}{81} + \cdots$

36. $\frac{1}{4} + \frac{2}{5} + \frac{3}{6} + \frac{4}{7} + \cdots$

37. $-3 + 4 - 5 + 6 - 7$

38. $-2 + 4 - 8 + 16 - 32$

In Exercises 39–50, find the sum. *(See Examples 5 and 6.)*

39. $\sum_{i=1}^{6} 2i$

40. $\sum_{i=1}^{5} 7i$

41. $\sum_{n=0}^{4} n^3$

42. $\sum_{k=1}^{4} 3k^2$

43. $\sum_{k=3}^{6} (5k - 2)$

44. $\sum_{n=1}^{5} (n^2 - 1)$

45. $\sum_{i=2}^{8} \frac{2}{i}$

46. $\sum_{k=4}^{6} \frac{k}{k+1}$

47. $\sum_{i=1}^{35} 1$

48. $\sum_{n=1}^{16} n$

49. $\sum_{i=10}^{25} i$

50. $\sum_{n=1}^{18} n^2$

ERROR ANALYSIS In Exercises 51 and 52, describe and correct the error in finding the sum of the series.

51. ✗

$$\sum_{n=1}^{10} (3n - 5) = -2 + 1 + 4 + 7 + 10$$
$$= 20$$

52. ✗

$$\sum_{i=2}^{4} i^2 = \frac{4(4+1)(2 \cdot 4 + 1)}{6}$$
$$= \frac{180}{6}$$
$$= 30$$

53. **PROBLEM SOLVING** You want to save \$500 for a school trip. You begin by saving a penny on the first day. You save an additional penny each day after that. For example, you will save two pennies on the second day, three pennies on the third day, and so on.

a. How much money will you have saved after 100 days?

b. Use a series to determine how many days it takes you to save \$500.

54. **MODELING WITH MATHEMATICS** You begin an exercise program. The first week you do 25 push-ups. Each week you do 10 more push-ups than the previous week. How many push-ups will you do in the ninth week? Justify your answer.

55. **MODELING WITH MATHEMATICS** For a display at a sports store, you are stacking soccer balls in a pyramid whose base is an equilateral triangle with five layers. Write a rule for the number of soccer balls in each layer. Then graph the sequence.

56. HOW DO YOU SEE IT? Use the diagram to determine the sum of the series. Explain your reasoning.

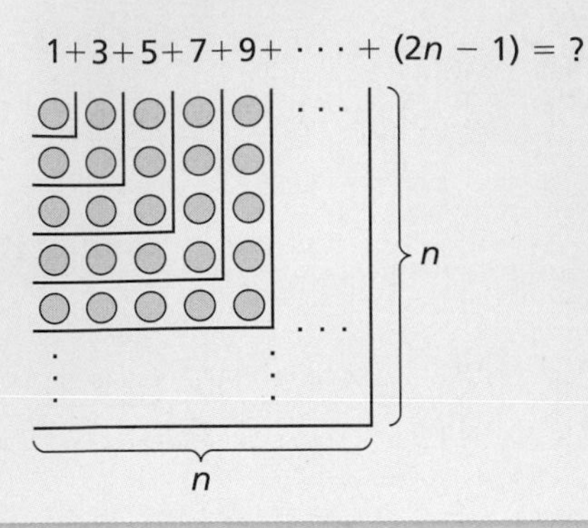

57. MAKING AN ARGUMENT You use a calculator to evaluate $\sum_{i=3}^{1659} i$ because the lower limit of summation is 3, not 1. Your friend claims there is a way to use the formula for the sum of the first n positive integers. Is your friend correct? Explain.

58. MATHEMATICAL CONNECTIONS A *regular* polygon has equal angle measures and equal side lengths. For a regular n-sided polygon ($n \geq 3$), the measure a_n of an interior angle is given by $a_n = \frac{180(n-2)}{n}$.

a. Write the first five terms of the sequence.

b. Write a rule for the sequence giving the sum T_n of the measures of the interior angles in each regular n-sided polygon.

c. Use your rule in part (b) to find the sum of the interior angle measures in the Guggenheim Museum skylight, which is a regular dodecagon.

Guggenheim Museum Skylight

59. USING STRUCTURE Determine whether each statement is true. If so, provide a proof. If not, provide a counterexample.

a. $\sum_{i=1}^{n} ca_i = c\sum_{i=1}^{n} a_i$

b. $\sum_{i=1}^{n} (a_i + b_i) = \sum_{i=1}^{n} a_i + \sum_{i=1}^{n} b_i$

c. $\sum_{i=1}^{n} a_i b_i = \sum_{i=1}^{n} a_i \sum_{i=1}^{n} b_i$

d. $\sum_{i=1}^{n} (a_i)^c = \left(\sum_{i=1}^{n} a_i\right)^c$

60. THOUGHT PROVOKING In this section, you learned the following formulas.

$$\sum_{i=1}^{n} 1 = n$$

$$\sum_{i=1}^{n} i = \frac{n(n+1)}{2}$$

$$\sum_{i=1}^{n} i^2 = \frac{n(n+1)(2n+1)}{6}$$

Write a formula for the sum of the cubes of the first n positive integers.

61. MODELING WITH MATHEMATICS In the puzzle called the Tower of Hanoi, the object is to use a series of moves to take the rings from one peg and stack them in order on another peg. A move consists of moving exactly one ring, and no ring may be placed on top of a smaller ring. The minimum number a_n of moves required to move n rings is 1 for 1 ring, 3 for 2 rings, 7 for 3 rings, 15 for 4 rings, and 31 for 5 rings.

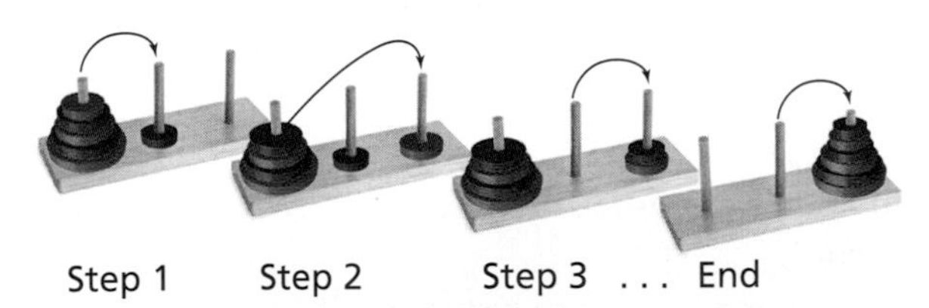

Step 1 Step 2 Step 3 . . . End

a. Write a rule for the sequence.

b. What is the minimum number of moves required to move 6 rings? 7 rings? 8 rings?

Maintaining Mathematical Proficiency

Reviewing what you learned in previous grades and lessons

Solve the system. Check your solution. *(Section 1.4)*

62. $2x - y - 3z = 6$
$x + y + 4z = -1$
$3x - 2z = 8$

63. $2x - 2y + z = 5$
$-2x + 3y + 2z = -1$
$x - 4y + 5z = 4$

64. $2x - 3y + z = 4$
$x - 2z = 1$
$y + z = 2$

8.2 Analyzing Arithmetic Sequences and Series

Essential Question How can you recognize an arithmetic sequence from its graph?

In an **arithmetic sequence**, the difference of consecutive terms, called the *common difference*, is constant. For example, in the arithmetic sequence 1, 4, 7, 10, . . . , the common difference is 3.

EXPLORATION 1 Recognizing Graphs of Arithmetic Sequences

Work with a partner. Determine whether each graph shows an arithmetic sequence. If it does, then write a rule for the nth term of the sequence, and use a spreadsheet to find the sum of the first 20 terms. What do you notice about the graph of an arithmetic sequence?

a.

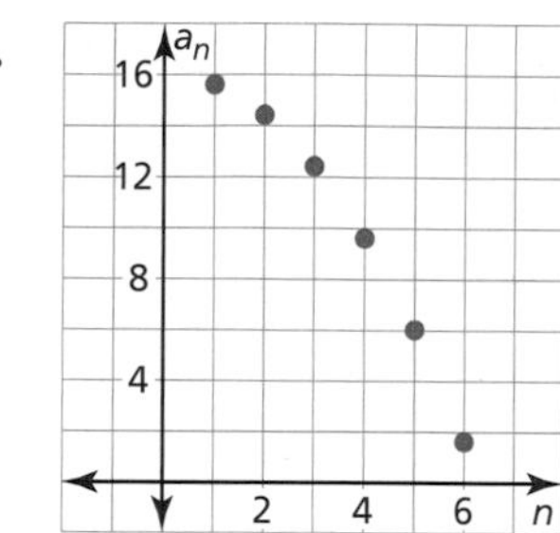

b.

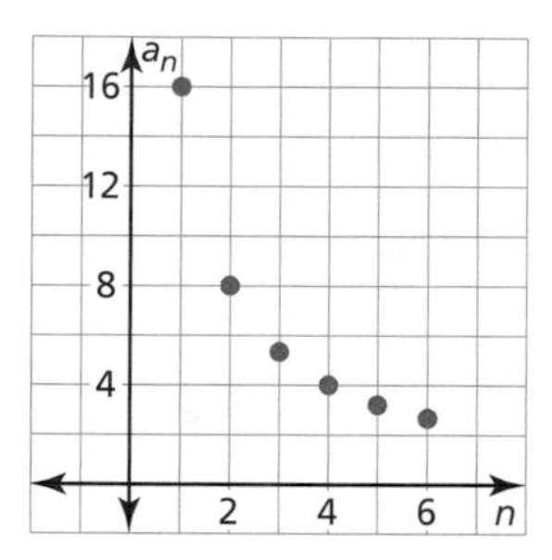

c.

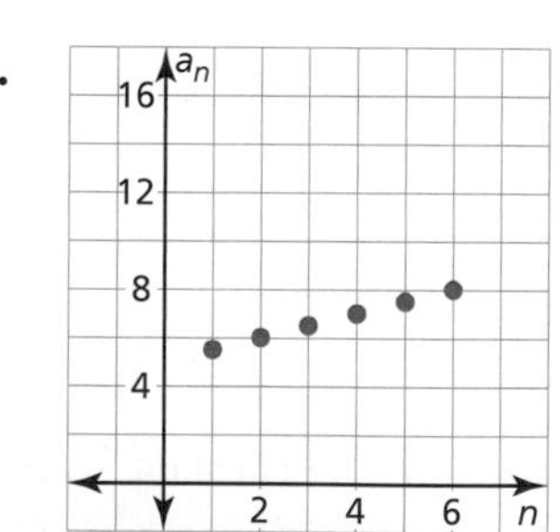

d.

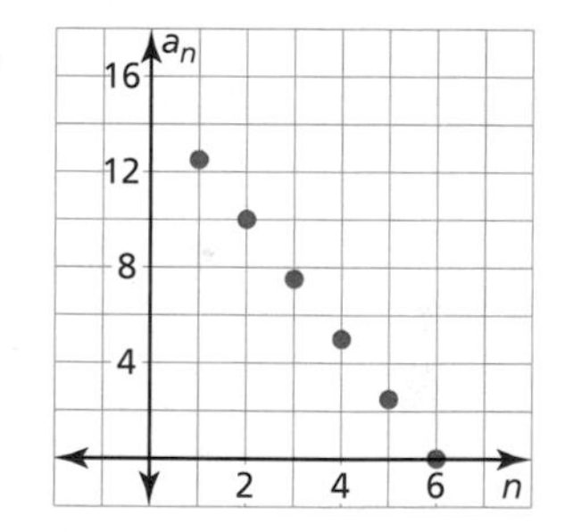

EXPLORATION 2 Finding the Sum of an Arithmetic Sequence

Work with a partner. A teacher of German mathematician Carl Friedrich Gauss (1777–1855) asked him to find the sum of all the whole numbers from 1 through 100. To the astonishment of his teacher, Gauss came up with the answer after only a few moments. Here is what Gauss did:

$$\begin{array}{rcrcrcccr} 1 & + & 2 & + & 3 & + & \cdots & + & 100 \\ 100 & + & 99 & + & 98 & + & \cdots & + & 1 \\ \hline 101 & + & 101 & + & 101 & + & \cdots & + & 101 \end{array} \qquad \frac{100 \times 101}{2} = 5050$$

REASONING ABSTRACTLY

To be proficient in math, you need to make sense of quantities and their relationships in problem situations.

Explain Gauss's thought process. Then write a formula for the sum S_n of the first n terms of an arithmetic sequence. Verify your formula by finding the sums of the first 20 terms of the arithmetic sequences in Exploration 1. Compare your answers to those you obtained using a spreadsheet.

Communicate Your Answer

3. How can you recognize an arithmetic sequence from its graph?

4. Find the sum of the terms of each arithmetic sequence.

a. 1, 4, 7, 10, . . . , 301 **b.** 1, 2, 3, 4, . . . , 1000 **c.** 2, 4, 6, 8, . . . , 800

8.2 Lesson

Core Vocabulary

arithmetic sequence, *p. 418*
common difference, *p. 418*
arithmetic series, *p. 420*

Previous
linear function
mean

What You Will Learn

- Identify arithmetic sequences.
- Write rules for arithmetic sequences.
- Find sums of finite arithmetic series.

Identifying Arithmetic Sequences

In an **arithmetic sequence**, the difference of consecutive terms is constant. This constant difference is called the **common difference** and is denoted by d.

EXAMPLE 1 Identifying Arithmetic Sequences

Tell whether each sequence is arithmetic.

a. $-9, -2, 5, 12, 19, \ldots$

b. $23, 15, 9, 5, 3, \ldots$

SOLUTION

Find the differences of consecutive terms.

a. $a_2 - a_1 = -2 - (-9) = 7$

$a_3 - a_2 = 5 - (-2) = 7$

$a_4 - a_3 = 12 - 5 = 7$

$a_5 - a_4 = 19 - 12 = 7$

▶ Each difference is 7, so the sequence is arithmetic.

b. $a_2 - a_1 = 15 - 23 = -8$

$a_3 - a_2 = 9 - 15 = -6$

$a_4 - a_3 = 5 - 9 = -4$

$a_5 - a_4 = 3 - 5 = -2$

▶ The differences are not constant, so the sequence is not arithmetic.

Monitoring Progress  Help in English and Spanish at *BigIdeasMath.com*

Tell whether the sequence is arithmetic. Explain your reasoning.

1. $2, 5, 8, 11, 14, \ldots$

2. $15, 9, 3, -3, -9, \ldots$

3. $8, 4, 2, 1, \frac{1}{2}, \ldots$

Writing Rules for Arithmetic Sequences

Core Concept

Rule for an Arithmetic Sequence

Algebra The nth term of an arithmetic sequence with first term a_1 and common difference d is given by:

$$a_n = a_1 + (n - 1)d$$

Example The nth term of an arithmetic sequence with a first term of 3 and a common difference of 2 is given by:

$$a_n = 3 + (n - 1)2, \text{ or } a_n = 2n + 1$$

EXAMPLE 2 Writing a Rule for the *n*th Term

Write a rule for the *n*th term of each sequence. Then find a_{15}.

a. 3, 8, 13, 18, . . . **b.** 55, 47, 39, 31, . . .

SOLUTION

a. The sequence is arithmetic with first term $a_1 = 3$, and common difference $d = 8 - 3 = 5$. So, a rule for the *n*th term is

$a_n = a_1 + (n - 1)d$ Write general rule.

$= 3 + (n - 1)5$ Substitute 3 for a_1 and 5 for d.

$= 5n - 2.$ Simplify.

▶ A rule is $a_n = 5n - 2$, and the 15th term is $a_{15} = 5(15) - 2 = 73$.

b. The sequence is arithmetic with first term $a_1 = 55$, and common difference $d = 47 - 55 = -8$. So, a rule for the *n*th term is

$a_n = a_1 + (n - 1)d$ Write general rule.

$= 55 + (n - 1)(-8)$ Substitute 55 for a_1 and -8 for d.

$= -8n + 63.$ Simplify.

▶ A rule is $a_n = -8n + 63$, and the 15th term is $a_{15} = -8(15) + 63 = -57$.

COMMON ERROR

In the general rule for an arithmetic sequence, note that the common difference *d* is multiplied by *n* − 1, not *n*.

Monitoring Progress  Help in English and Spanish at *BigIdeasMath.com*

4. Write a rule for the *n*th term of the sequence 7, 11, 15, 19, ... Then find a_{15}.

EXAMPLE 3 Writing a Rule Given a Term and Common Difference

One term of an arithmetic sequence is $a_{19} = -45$. The common difference is $d = -3$. Write a rule for the *n*th term. Then graph the first six terms of the sequence.

SOLUTION

Step 1 Use the general rule to find the first term.

$a_n = a_1 + (n - 1)d$ Write general rule.

$a_{19} = a_1 + (19 - 1)d$ Substitute 19 for *n*.

$-45 = a_1 + 18(-3)$ Substitute −45 for a_{19} and −3 for *d*.

$9 = a_1$ Solve for a_1.

Step 2 Write a rule for the *n*th term.

$a_n = a_1 + (n - 1)d$ Write general rule.

$= 9 + (n - 1)(-3)$ Substitute 9 for a_1 and −3 for *d*.

$= -3n + 12$ Simplify.

Step 3 Use the rule to create a table of values for the sequence. Then plot the points.

n	1	2	3	4	5	6
a_n	9	6	3	0	−3	−6

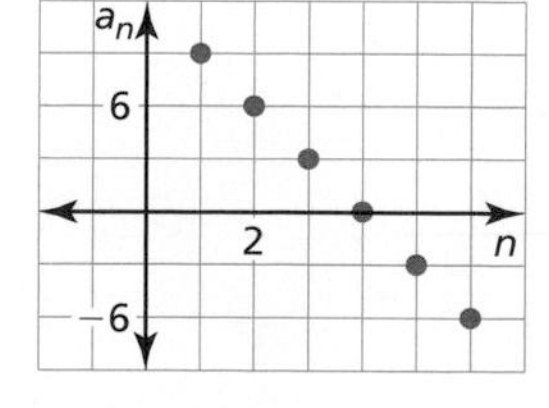

ANALYZING RELATIONSHIPS

Notice that the points lie on a line. This is true for any arithmetic sequence. So, an arithmetic sequence is a linear function whose domain is a subset of the integers. You can also use function notation to write sequences:

$f(n) = -3n + 12.$

EXAMPLE 4 Writing a Rule Given Two Terms

Two terms of an arithmetic sequence are $a_7 = 17$ and $a_{26} = 93$. Write a rule for the nth term.

SOLUTION

Step 1 Write a system of equations using $a_n = a_1 + (n - 1)d$. Substitute 26 for n to write Equation 1. Substitute 7 for n to write Equation 2.

$a_{26} = a_1 + (26 - 1)d$ ➡ $93 = a_1 + 25d$ Equation 1

$a_7 = a_1 + (7 - 1)d$ ➡ $17 = a_1 + 6d$ Equation 2

Step 2 Solve the system.

$76 = 19d$ Subtract.

$4 = d$ Solve for d.

$93 = a_1 + 25(4)$ Substitute for d in Equation 1.

$-7 = a_1$ Solve for a_1.

Step 3 Write a rule for a_n.

$a_n = a_1 + (n - 1)d$ Write general rule.

$= -7 + (n - 1)4$ Substitute for a_1 and d.

$= 4n - 11$ Simplify.

Check

Use the rule to verify that the 7th term is 17 and the 26th term is 93.

$a_7 = 4(7) - 11 = 17$ ✓

$a_{26} = 4(26) - 11 = 93$ ✓

Monitoring Progress  Help in English and Spanish at *BigIdeasMath.com*

Write a rule for the nth term of the sequence. Then graph the first six terms of the sequence.

5. $a_{11} = 50, d = 7$

6. $a_7 = 71, a_{16} = 26$

Finding Sums of Finite Arithmetic Series

The expression formed by adding the terms of an arithmetic sequence is called an **arithmetic series**. The sum of the first n terms of an arithmetic series is denoted by S_n. To find a rule for S_n, you can write S_n in two different ways and add the results.

$$S_n = a_1 + (a_1 + d) + (a_1 + 2d) + \cdots + a_n$$

$$S_n = a_n + (a_n - d) + (a_n - 2d) + \cdots + a_1$$

$$2S_n = \underbrace{(a_1 + a_n) + (a_1 + a_n) + (a_1 + a_n) + \cdots + (a_1 + a_n)}_{(a_1 + a_n) \text{ is added } n \text{ times.}}$$

You can conclude that $2S_n = n(a_1 + a_n)$, which leads to the following result.

Core Concept

The Sum of a Finite Arithmetic Series

The sum of the first n terms of an arithmetic series is

$$S_n = n\left(\frac{a_1 + a_n}{2}\right).$$

In words, S_n is the mean of the first and nth terms, multiplied by the number of terms.

EXAMPLE 5 Finding the Sum of an Arithmetic Series

Find the sum $\sum_{i=1}^{20}(3i + 7)$.

SOLUTION

Step 1 Find the first and last terms.

$a_1 = 3(1) + 7 = 10$ Identify first term.

$a_{20} = 3(20) + 7 = 67$ Identify last term.

Step 2 Find the sum.

$S_{20} = 20\left(\frac{a_1 + a_{20}}{2}\right)$ Write rule for S_{20}.

$= 20\left(\frac{10 + 67}{2}\right)$ Substitute 10 for a_1 and 67 for a_{20}.

$= 770$ Simplify.

STUDY TIP

This sum is actually a *partial* sum. You cannot find the complete sum of an infinite arithmetic series because its terms continue indefinitely.

EXAMPLE 6 Solving a Real-Life Problem

You are making a house of cards similar to the one shown.

a. Write a rule for the number of cards in the nth row when the top row is row 1.

b. How many cards do you need to make a house of cards with 12 rows?

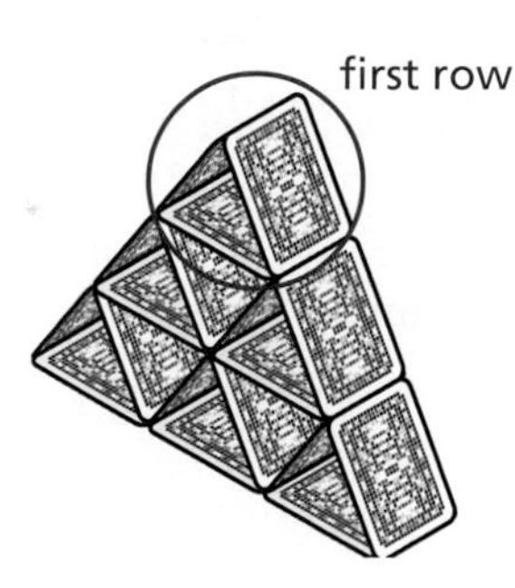

SOLUTION

a. Starting with the top row, the number of cards in the rows are 3, 6, 9, 12, ... These numbers form an arithmetic sequence with a first term of 3 and a common difference of 3. So, a rule for the sequence is:

$a_n = a_1 + (n - 1)d$ Write general rule.

$= 3 + (n - 1)(3)$ Substitute 3 for a_1 and 3 for d.

$= 3n$ Simplify.

b. Find the sum of an arithmetic series with first term $a_1 = 3$ and last term $a_{12} = 3(12) = 36$.

$$S_{12} = 12\left(\frac{a_1 + a_{12}}{2}\right) = 12\left(\frac{3 + 36}{2}\right) = 234$$

▶ So, you need 234 cards to make a house of cards with 12 rows.

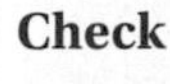

Check

Use a graphing calculator to check the sum.

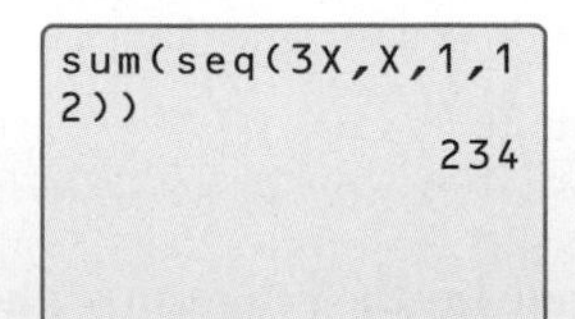

Monitoring Progress

Help in English and Spanish at *BigIdeasMath.com*

Find the sum.

7. $\sum_{i=1}^{10} 9i$ **8.** $\sum_{k=1}^{12}(7k + 2)$ **9.** $\sum_{n=1}^{20}(-4n + 6)$

10. WHAT IF? In Example 6, how many cards do you need to make a house of cards with eight rows?

8.2 Exercises

Dynamic Solutions available at *BigIdeasMath.com*

Vocabulary and Core Concept Check

1. **COMPLETE THE SENTENCE** The constant difference between consecutive terms of an arithmetic sequence is called the ________________.

2. **DIFFERENT WORDS, SAME QUESTION** Which is different? Find "both" answers.

What sequence consists of all the positive odd numbers?

What sequence starts with 1 and has a common difference of 2?

What sequence has an nth term of $a_n = 1 + (n - 1)2$?

What sequence has an nth term of $a_n = 2n + 1$?

Monitoring Progress and Modeling with Mathematics

In Exercises 3–10, tell whether the sequence is arithmetic. Explain your reasoning. *(See Example 1.)*

3. $1, -1, -3, -5, -7, \ldots$

4. $12, 6, 0, -6, -12, \ldots$

5. $5, 8, 13, 20, 29, \ldots$

6. $3, 5, 9, 15, 23, \ldots$

7. $36, 18, 9, \frac{9}{2}, \frac{9}{4}, \ldots$

8. $81, 27, 9, 3, 1, \ldots$

9. $\frac{1}{2}, \frac{3}{4}, 1, \frac{5}{4}, \frac{3}{2}, \ldots$

10. $\frac{1}{6}, \frac{1}{2}, \frac{5}{6}, \frac{7}{6}, \frac{3}{2}, \ldots$

11. **WRITING EQUATIONS** Write a rule for the arithmetic sequence with the given description.
    a. The first term is -3 and each term is 6 less than the previous term.
    b. The first term is 7 and each term is 5 more than the previous term.

12. **WRITING** Compare the terms of an arithmetic sequence when $d > 0$ to when $d < 0$.

In Exercises 13–20, write a rule for the nth term of the sequence. Then find a_{20}. *(See Example 2.)*

13. $12, 20, 28, 36, \ldots$

14. $7, 12, 17, 22, \ldots$

15. $51, 48, 45, 42, \ldots$

16. $86, 79, 72, 65, \ldots$

17. $-1, -\frac{1}{3}, \frac{1}{3}, 1, \ldots$

18. $-2, -\frac{5}{4}, -\frac{1}{2}, \frac{1}{4}, \ldots$

19. $2.3, 1.5, 0.7, -0.1, \ldots$

20. $11.7, 10.8, 9.9, 9, \ldots$

ERROR ANALYSIS In Exercises 21 and 22, describe and correct the error in writing a rule for the nth term of the arithmetic sequence $22, 9, -4, -17, -30, \ldots$.

21.

Use $a_1 = 22$ and $d = -13$.

$a_n = a_1 + nd$

$a_n = 22 + n(-13)$

$a_n = 22 - 13n$

22.

The first term is 22 and the common difference is -13.

$a_n = -13 + (n - 1)(22)$

$a_n = -35 + 22n$

In Exercises 23–28, write a rule for the nth term of the sequence. Then graph the first six terms of the sequence. *(See Example 3.)*

23. $a_{11} = 43, d = 5$

24. $a_{13} = 42, d = 4$

25. $a_{20} = -27, d = -2$

26. $a_{15} = -35, d = -3$

27. $a_{17} = -5, d = -\frac{1}{2}$

28. $a_{21} = -25, d = -\frac{3}{2}$

29. **USING EQUATIONS** One term of an arithmetic sequence is $a_8 = -13$. The common difference is -8. What is a rule for the nth term of the sequence?

    Ⓐ $a_n = 51 + 8n$

    Ⓑ $a_n = 35 + 8n$

    Ⓒ $a_n = 51 - 8n$

    Ⓓ $a_n = 35 - 8n$

30. FINDING A PATTERN One term of an arithmetic sequence is $a_{12} = 43$. The common difference is 6. What is another term of the sequence?

Ⓐ $a_3 = -11$ Ⓑ $a_4 = -53$

Ⓒ $a_5 = 13$ Ⓓ $a_6 = -47$

In Exercises 31–38, write a rule for the *n*th term of the arithmetic sequence. *(See Example 4.)*

31. $a_5 = 41, a_{10} = 96$

32. $a_7 = 58, a_{11} = 94$

33. $a_6 = -8, a_{15} = -62$

34. $a_8 = -15, a_{17} = -78$

35. $a_{18} = -59, a_{21} = -71$

36. $a_{12} = -38, a_{19} = -73$

37. $a_8 = 12, a_{16} = 22$

38. $a_{12} = 9, a_{27} = 15$

WRITING EQUATIONS In Exercises 39–44, write a rule for the sequence with the given terms.

39.

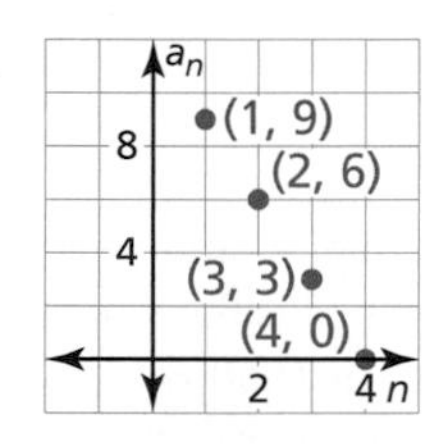

40.

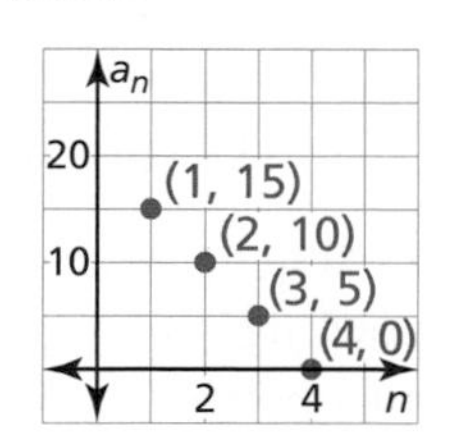

41.

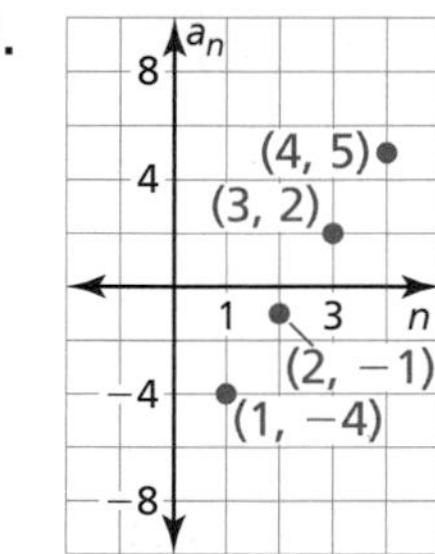

42.

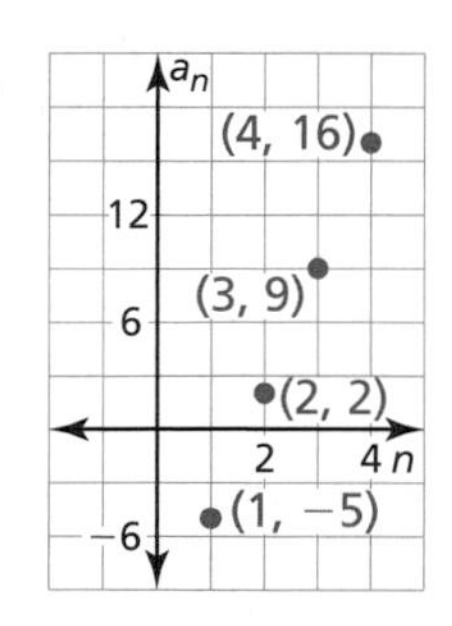

43.

n	4	5	6	7	8
a_n	25	29	33	37	41

44.

n	4	5	6	7	8
a_n	31	39	47	55	63

45. WRITING Compare the graph of $a_n = 3n + 1$, where n is a positive integer, with the graph of $f(x) = 3x + 1$, where x is a real number.

46. DRAWING CONCLUSIONS Describe how doubling each term in an arithmetic sequence changes the common difference of the sequence. Justify your answer.

In Exercises 47–52, find the sum. *(See Example 5.)*

47. $\sum_{i=1}^{20} (2i - 3)$ **48.** $\sum_{i=1}^{26} (4i + 7)$

49. $\sum_{i=1}^{33} (6 - 2i)$ **50.** $\sum_{i=1}^{31} (-3 - 4i)$

51. $\sum_{i=1}^{41} (-2.3 + 0.1i)$ **52.** $\sum_{i=1}^{39} (-4.1 + 0.4i)$

NUMBER SENSE In Exercises 53 and 54, find the sum of the arithmetic sequence.

53. The first 19 terms of the sequence 9, 2, −5, −12, ...

54. The first 22 terms of the sequence 17, 9, 1, −7, ...

55. MODELING WITH MATHEMATICS A marching band is arranged in rows. The first row has three band members, and each row after the first has two more band members than the row before it. *(See Example 6.)*

a. Write a rule for the number of band members in the nth row.

b. How many band members are in a formation with seven rows?

56. MODELING WITH MATHEMATICS Domestic bees make their honeycomb by starting with a single hexagonal cell, then forming ring after ring of hexagonal cells around the initial cell, as shown. The number of cells in successive rings forms an arithmetic sequence.

Initial cell

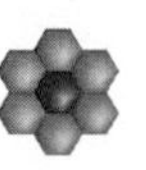

1 ring

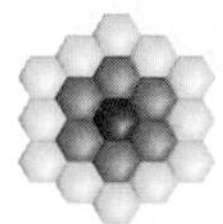

2 rings

a. Write a rule for the number of cells in the nth ring.

b. How many cells are in the honeycomb after the ninth ring is formed?

57. MATHEMATICAL CONNECTIONS A quilt is made up of strips of cloth, starting with an inner square surrounded by rectangles to form successively larger squares. The inner square and all rectangles have a width of 1 foot. Write an expression using summation notation that gives the sum of the areas of all the strips of cloth used to make the quilt shown. Then evaluate the expression.

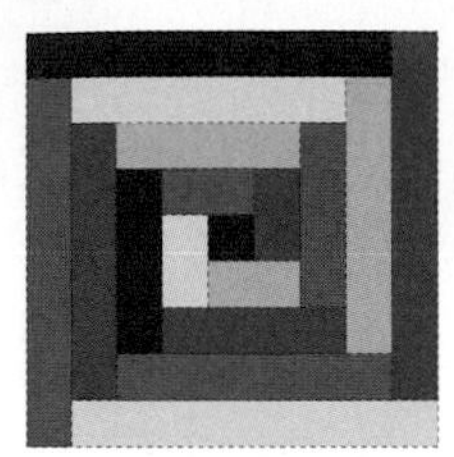

58. HOW DO YOU SEE IT? Which graph(s) represents an arithmetic sequence? Explain your reasoning.

a.

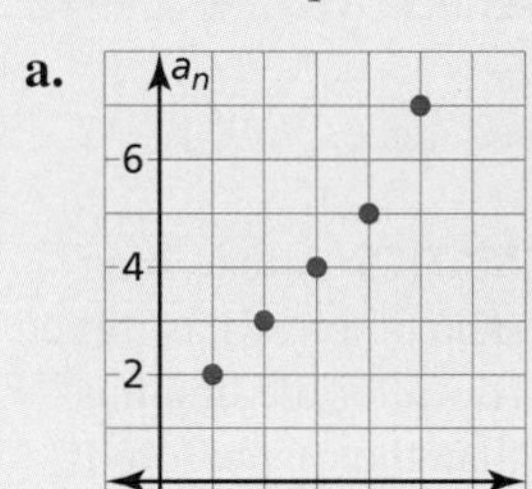

b.

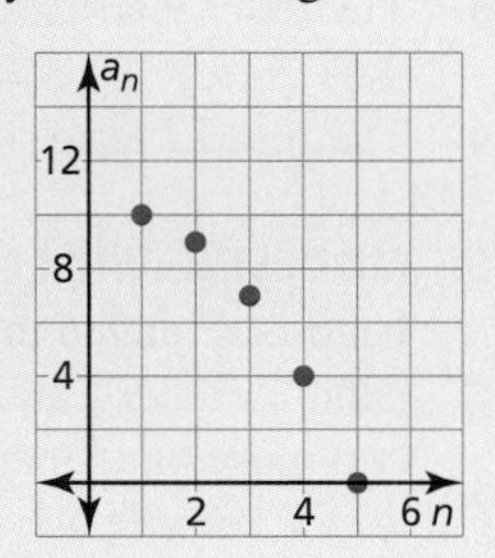

c.

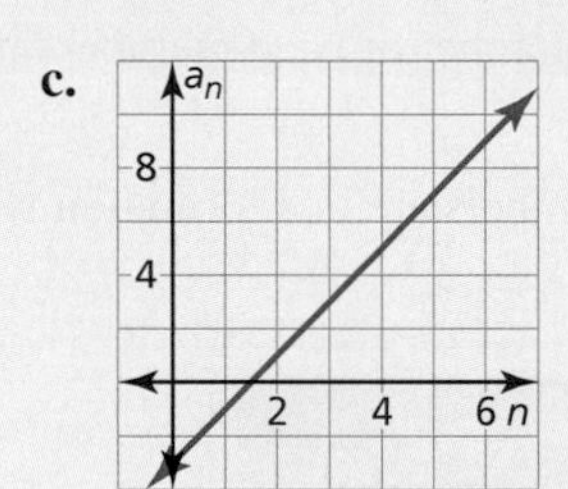

d.

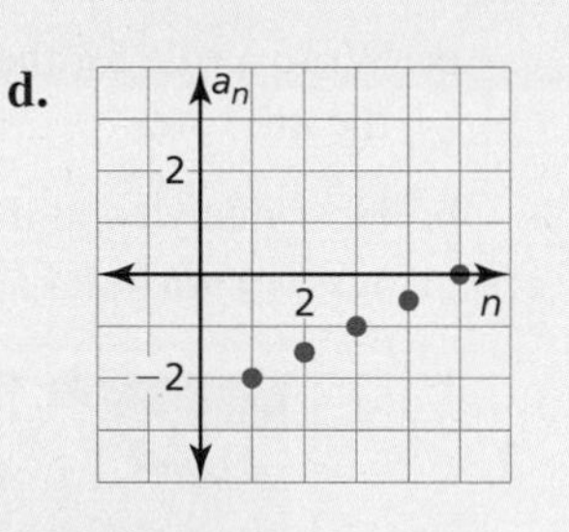

59. MAKING AN ARGUMENT Your friend believes the sum of a series doubles when the common difference of an arithmetic series is doubled and the first term and number of terms in the series remain unchanged. Is your friend correct? Explain your reasoning.

60. THOUGHT PROVOKING In number theory, the *Dirichlet Prime Number Theorem* states that if a and b are relatively prime, then the arithmetic sequence

$$a, a + b, a + 2b, a + 3b, \ldots$$

contains infinitely many prime numbers. Find the first 10 primes in the sequence when $a = 3$ and $b = 4$.

61. REASONING Find the sum of the positive odd integers less than 300. Explain your reasoning.

62. USING EQUATIONS Find the value of n.

a. $\sum_{i=1}^{n}(3i + 5) = 544$ **b.** $\sum_{i=1}^{n}(-4i - 1) = -1127$

c. $\sum_{i=5}^{n}(7 + 12i) = 455$ **d.** $\sum_{i=3}^{n}(-3 - 4i) = -507$

63. ABSTRACT REASONING A theater has n rows of seats, and each row has d more seats than the row in front of it. There are x seats in the last (nth) row and a total of y seats in the entire theater. How many seats are in the front row of the theater? Write your answer in terms of n, x, and y.

64. CRITICAL THINKING The expressions $3 - x$, x, and $1 - 3x$ are the first three terms in an arithmetic sequence. Find the value of x and the next term in the sequence.

65. CRITICAL THINKING One of the major sources of our knowledge of Egyptian mathematics is the Ahmes papyrus, which is a scroll copied in 1650 B.C. by an Egyptian scribe. The following problem is from the Ahmes papyrus.

Divide 10 hekats of barley among 10 men so that the common difference is $\frac{1}{8}$ of a hekat of barley.

Use what you know about arithmetic sequences and series to determine what portion of a hekat each man should receive.

Maintaining Mathematical Proficiency

Reviewing what you learned in previous grades and lessons

Simplify the expression. *(Section 5.2)*

66. $\dfrac{7}{7^{1/3}}$

67. $\dfrac{3^{-2}}{3^{-4}}$

68. $\left(\dfrac{9}{49}\right)^{1/2}$

69. $(5^{1/2} \cdot 5^{1/4})$

Tell whether the function represents *exponential growth* or *exponential decay*. Then graph the function. *(Section 6.2)*

70. $y = 2e^x$

71. $y = e^{-3x}$

72. $y = 3e^{-x}$

73. $y = e^{0.25x}$

8.3 Analyzing Geometric Sequences and Series

Essential Question How can you recognize a geometric sequence from its graph?

In a **geometric sequence**, the ratio of any term to the previous term, called the *common ratio*, is constant. For example, in the geometric sequence 1, 2, 4, 8, . . . , the common ratio is 2.

EXPLORATION 1 Recognizing Graphs of Geometric Sequences

Work with a partner. Determine whether each graph shows a geometric sequence. If it does, then write a rule for the *n*th term of the sequence and use a spreadsheet to find the sum of the first 20 terms. What do you notice about the graph of a geometric sequence?

a.

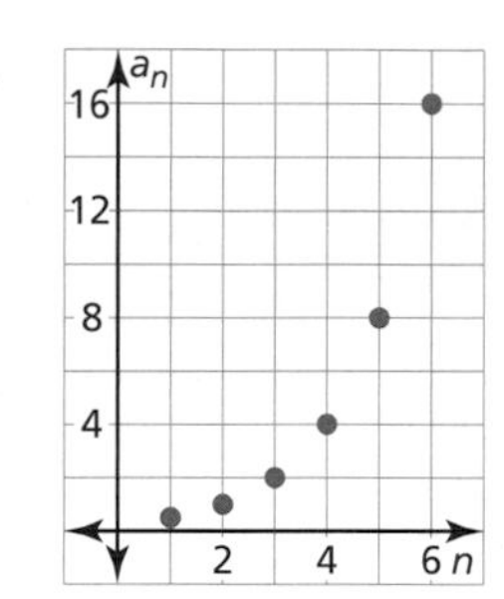

b.

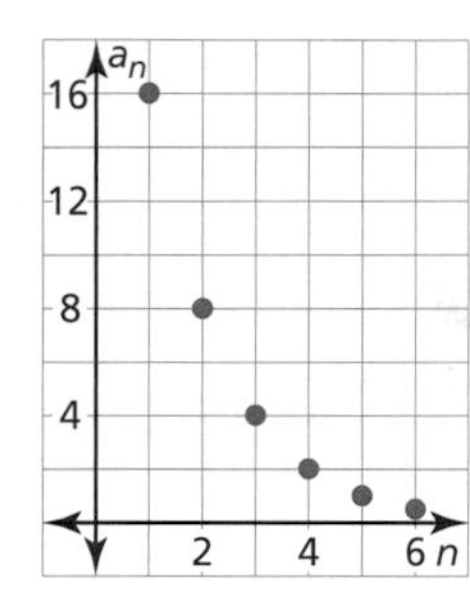

c.

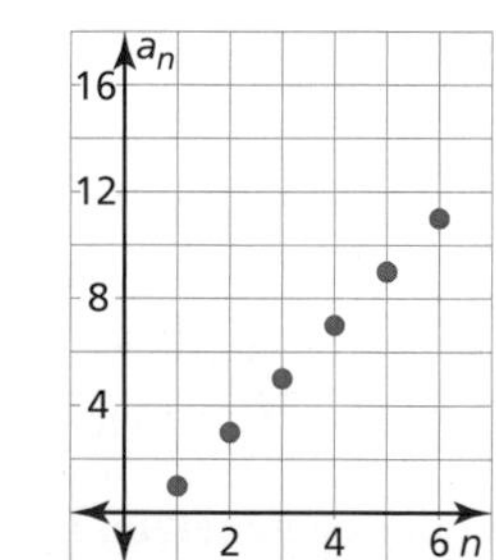

d.

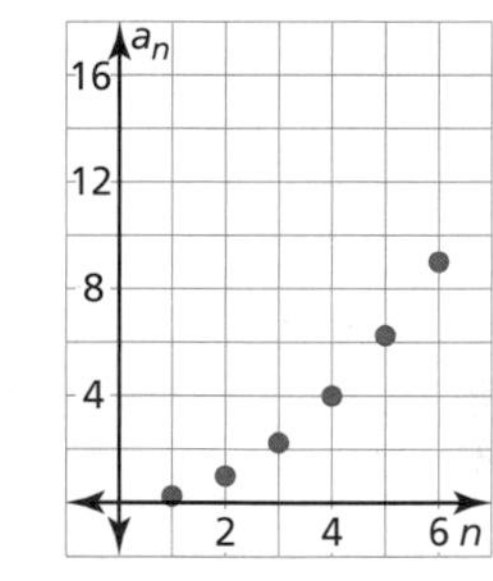

EXPLORATION 2 Finding the Sum of a Geometric Sequence

Work with a partner. You can write the *n*th term of a geometric sequence with first term a_1 and common ratio r as

$$a_n = a_1 r^{n-1}.$$

So, you can write the sum S_n of the first n terms of a geometric sequence as

$$S_n = a_1 + a_1 r + a_1 r^2 + a_1 r^3 + \cdots + a_1 r^{n-1}.$$

Rewrite this formula by finding the difference $S_n - rS_n$ and solving for S_n. Then verify your rewritten formula by finding the sums of the first 20 terms of the geometric sequences in Exploration 1. Compare your answers to those you obtained using a spreadsheet.

LOOKING FOR REGULARITY IN REPEATED REASONING

To be proficient in math, you need to notice when calculations are repeated, and look both for general methods and for shortcuts.

Communicate Your Answer

3. How can you recognize a geometric sequence from its graph?

4. Find the sum of the terms of each geometric sequence.

a. 1, 2, 4, 8, . . . , 8192 **b.** 0.1, 0.01, 0.001, 0.0001, . . . , 10^{-10}

8.3 Lesson

Core Vocabulary

geometric sequence, *p. 426*
common ratio, *p. 426*
geometric series, *p. 428*

Previous
exponential function
properties of exponents

What You Will Learn

- Identify geometric sequences.
- Write rules for geometric sequences.
- Find sums of finite geometric series.

Identifying Geometric Sequences

In a **geometric sequence**, the ratio of any term to the previous term is constant. This constant ratio is called the **common ratio** and is denoted by r.

EXAMPLE 1 Identifying Geometric Sequences

Tell whether each sequence is geometric.

a. 6, 12, 20, 30, 42, . . .

b. 256, 64, 16, 4, 1, . . .

SOLUTION

Find the ratios of consecutive terms.

a. $\frac{a_2}{a_1} = \frac{12}{6} = 2$ $\quad \frac{a_3}{a_2} = \frac{20}{12} = \frac{5}{3}$ $\quad \frac{a_4}{a_3} = \frac{30}{20} = \frac{3}{2}$ $\quad \frac{a_5}{a_4} = \frac{42}{30} = \frac{7}{5}$

▶ The ratios are not constant, so the sequence is not geometric.

b. $\frac{a_2}{a_1} = \frac{64}{256} = \frac{1}{4}$ $\quad \frac{a_3}{a_2} = \frac{16}{64} = \frac{1}{4}$ $\quad \frac{a_4}{a_3} = \frac{4}{16} = \frac{1}{4}$ $\quad \frac{a_5}{a_4} = \frac{1}{4}$

▶ Each ratio is $\frac{1}{4}$, so the sequence is geometric.

Monitoring Progress  Help in English and Spanish at *BigIdeasMath.com*

Tell whether the sequence is geometric. Explain your reasoning.

1. 27, 9, 3, 1, $\frac{1}{3}$, . . . **2.** 2, 6, 24, 120, 720, . . . **3.** −1, 2, −4, 8, −16, . . .

Writing Rules for Geometric Sequences

Core Concept

Rule for a Geometric Sequence

Algebra The nth term of a geometric sequence with first term a_1 and common ratio r is given by:

$$a_n = a_1 r^{n-1}$$

Example The nth term of a geometric sequence with a first term of 2 and a common ratio of 3 is given by:

$$a_n = 2(3)^{n-1}$$

EXAMPLE 2 Writing a Rule for the *n*th Term

Write a rule for the *n*th term of each sequence. Then find a_8.

a. 5, 15, 45, 135, . . .

b. 88, −44, 22, −11, . . .

COMMON ERROR

In the general rule for a geometric sequence, note that the exponent is $n - 1$, not n.

SOLUTION

a. The sequence is geometric with first term $a_1 = 5$ and common ratio $r = \frac{15}{5} = 3$. So, a rule for the *n*th term is

$$a_n = a_1 r^{n-1} \quad \text{Write general rule.}$$

$$= 5(3)^{n-1}. \quad \text{Substitute 5 for } a_1 \text{ and 3 for } r.$$

▶ A rule is $a_n = 5(3)^{n-1}$, and the 8th term is $a_8 = 5(3)^{8-1} = 10{,}935$.

b. The sequence is geometric with first term $a_1 = 88$ and common ratio $r = \frac{-44}{88} = -\frac{1}{2}$. So, a rule for the *n*th term is

$$a_n = a_1 r^{n-1} \quad \text{Write general rule.}$$

$$= 88\left(-\frac{1}{2}\right)^{n-1}. \quad \text{Substitute 88 for } a_1 \text{ and } -\frac{1}{2} \text{ for } r.$$

▶ A rule is $a_n = 88\left(-\frac{1}{2}\right)^{n-1}$, and the 8th term is $a_8 = 88\left(-\frac{1}{2}\right)^{8-1} = -\frac{11}{16}$.

Monitoring Progress

Help in English and Spanish at *BigIdeasMath.com*

4. Write a rule for the *n*th term of the sequence 3, 15, 75, 375, ... Then find a_9.

EXAMPLE 3 Writing a Rule Given a Term and Common Ratio

One term of a geometric sequence is $a_4 = 12$. The common ratio is $r = 2$. Write a rule for the *n*th term. Then graph the first six terms of the sequence.

SOLUTION

Step 1 Use the general rule to find the first term.

$$a_n = a_1 r^{n-1} \quad \text{Write general rule.}$$

$$a_4 = a_1 r^{4-1} \quad \text{Substitute 4 for } n.$$

$$12 = a_1(2)^3 \quad \text{Substitute 12 for } a_4 \text{ and 2 for } r.$$

$$1.5 = a_1 \quad \text{Solve for } a_1.$$

Step 2 Write a rule for the *n*th term.

$$a_n = a_1 r^{n-1} \quad \text{Write general rule.}$$

$$= 1.5(2)^{n-1} \quad \text{Substitute 1.5 for } a_1 \text{ and 2 for } r.$$

Step 3 Use the rule to create a table of values for the sequence. Then plot the points.

n	1	2	3	4	5	6
a_n	1.5	3	6	12	24	48

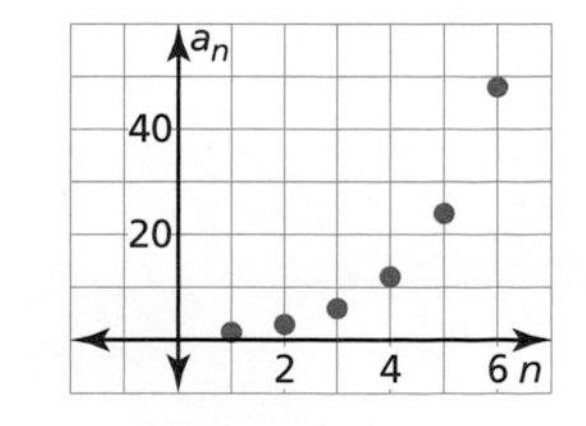

ANALYZING RELATIONSHIPS

Notice that the points lie on an exponential curve because consecutive terms change by equal factors. So, a geometric sequence in which $r > 0$ and $r \neq 1$ is an exponential function whose domain is a subset of the integers.

EXAMPLE 4 Writing a Rule Given Two Terms

Two terms of a geometric sequence are $a_2 = 12$ and $a_5 = -768$. Write a rule for the nth term.

SOLUTION

Step 1 Write a system of equations using $a_n = a_1r^{n-1}$. Substitute 2 for n to write Equation 1. Substitute 5 for n to write Equation 2.

$a_2 = a_1r^{2-1}$ ➡ $12 = a_1r$ Equation 1

$a_5 = a_1r^{5-1}$ ➡ $-768 = a_1r^4$ Equation 2

Step 2 Solve the system.

$\frac{12}{r} = a_1$ Solve Equation 1 for a_1.

$-768 = \frac{12}{r}(r^4)$ Substitute for a_1 in Equation 2.

$-768 = 12r^3$ Simplify.

$-4 = r$ Solve for r.

$12 = a_1(-4)$ Substitute for r in Equation 1.

$-3 = a_1$ Solve for a_1.

Step 3 Write a rule for a_n.

$a_n = a_1r^{n-1}$ Write general rule.

$= -3(-4)^{n-1}$ Substitute for a_1 and r.

Check

Use the rule to verify that the 2nd term is 12 and the 5th term is -768.

$a_2 = -3(-4)^{2-1}$

$= -3(-4) = 12$ ✓

$a_5 = -3(-4)^{5-1}$

$= -3(256) = -768$ ✓

Monitoring Progress

Help in English and Spanish at *BigIdeasMath.com*

Write a rule for the *n*th term of the sequence. Then graph the first six terms of the sequence.

5. $a_6 = -96, r = -2$

6. $a_2 = 12, a_4 = 3$

Finding Sums of Finite Geometric Series

The expression formed by adding the terms of a geometric sequence is called a **geometric series**. The sum of the first n terms of a geometric series is denoted by S_n. You can develop a rule for S_n as follows.

$$
\begin{aligned}
S_n &= a_1 + a_1r + a_1r^2 + a_1r^3 + \cdots + a_1r^{n-1} \\
-rS_n &= \quad - a_1r - a_1r^2 - a_1r^3 - \cdots - a_1r^{n-1} - a_1r^n \\
\hline
S_n - rS_n &= a_1 + 0 + 0 + 0 + \cdots + 0 \quad - a_1r^n \\
S_n(1 - r) &= a_1(1 - r^n)
\end{aligned}
$$

When $r \neq 1$, you can divide each side of this equation by $1 - r$ to obtain the following rule for S_n.

Core Concept

The Sum of a Finite Geometric Series

The sum of the first n terms of a geometric series with common ratio $r \neq 1$ is

$$S_n = a_1\left(\frac{1 - r^n}{1 - r}\right).$$

EXAMPLE 5 Finding the Sum of a Geometric Series

Find the sum $\sum_{k=1}^{10} 4(3)^{k-1}$.

SOLUTION

Check

Use a graphing calculator to check the sum.

```
sum(seq(4*3^(X-1
),X,1,10))
                118096
```

Step 1 Find the first term and the common ratio.

$a_1 = 4(3)^{1-1} = 4$ Identify first term.

$r = 3$ Identify common ratio.

Step 2 Find the sum.

$S_{10} = a_1\left(\frac{1 - r^{10}}{1 - r}\right)$ Write rule for S_{10}.

$= 4\left(\frac{1 - 3^{10}}{1 - 3}\right)$ Substitute 4 for a_1 and 3 for r.

$= 118{,}096$ Simplify.

EXAMPLE 6 Solving a Real-Life Problem

You can calculate the monthly payment M (in dollars) for a loan using the formula

$$M = \frac{L}{\sum_{k=1}^{t}\left(\frac{1}{1+i}\right)^k}$$

where L is the loan amount (in dollars), i is the monthly interest rate (in decimal form), and t is the term (in months). Calculate the monthly payment on a 5-year loan for \$20,000 with an annual interest rate of 6%.

SOLUTION

USING TECHNOLOGY

Storing the value of $\frac{1}{1.005}$ helps minimize mistakes and also assures an accurate answer. Rounding this value to 0.995 results in a monthly payment of \$386.94.

Step 1 Substitute for L, i, and t. The loan amount is $L = 20{,}000$, the monthly interest rate is $i = \frac{0.06}{12} = 0.005$, and the term is $t = 5(12) = 60$.

$$M = \frac{20{,}000}{\sum_{k=1}^{60}\left(\frac{1}{1+0.005}\right)^k}$$

Step 2 Notice that the denominator is a geometric series with first term $\frac{1}{1.005}$ and common ratio $\frac{1}{1.005}$. Use a calculator to find the monthly payment.

```
1/1.005→R
          .9950248756
R((1-R^60)/(1-R)
)
          51.72556075
20000/Ans
          386.6560306
```

▶ So, the monthly payment is \$386.66.

Monitoring Progress

Help in English and Spanish at *BigIdeasMath.com*

Find the sum.

7. $\sum_{k=1}^{8} 5^{k-1}$

8. $\sum_{i=1}^{12} 6(-2)^{i-1}$

9. $\sum_{t=1}^{7} -16(0.5)^{t-1}$

10. WHAT IF? In Example 6, how does the monthly payment change when the annual interest rate is 5%?

8.3 Exercises

Dynamic Solutions available at *BigIdeasMath.com*

Vocabulary and Core Concept Check

1. **COMPLETE THE SENTENCE** The constant ratio of consecutive terms in a geometric sequence is called the __________.

2. **WRITING** How can you determine whether a sequence is geometric from its graph?

3. **COMPLETE THE SENTENCE** The nth term of a geometric sequence has the form $a_n =$ __________.

4. **VOCABULARY** State the rule for the sum of the first n terms of a geometric series.

Monitoring Progress and Modeling with Mathematics

In Exercises 5–12, tell whether the sequence is geometric. Explain your reasoning. *(See Example 1.)*

5. 96, 48, 24, 12, 6, . . .

6. 729, 243, 81, 27, 9, . . .

7. 2, 4, 6, 8, 10, . . .

8. 5, 20, 35, 50, 65, . . .

9. 0.2, 3.2, −12.8, 51.2, −204.8, . . .

10. 0.3, −1.5, 7.5, −37.5, 187.5, . . .

11. $\frac{1}{2}, \frac{1}{6}, \frac{1}{18}, \frac{1}{54}, \frac{1}{162}, \ldots$

12. $\frac{1}{4}, \frac{1}{16}, \frac{1}{64}, \frac{1}{256}, \frac{1}{1024}, \ldots$

13. **WRITING EQUATIONS** Write a rule for the geometric sequence with the given description.

    a. The first term is −3, and each term is 5 times the previous term.

    b. The first term is 72, and each term is $\frac{1}{3}$ times the previous term.

14. **WRITING** Compare the terms of a geometric sequence when $r > 1$ to when $0 < r < 1$.

In Exercises 15–22, write a rule for the nth term of the sequence. Then find a_7. *(See Example 2.)*

15. 4, 20, 100, 500, . . .

16. 6, 24, 96, 384, . . .

17. 112, 56, 28, 14, . . .

18. 375, 75, 15, 3, . . .

19. $4, 6, 9, \frac{27}{2}, \ldots$

20. $2, \frac{3}{2}, \frac{9}{8}, \frac{27}{32}, \ldots$

21. 1.3, −3.9, 11.7, −35.1, . . .

22. 1.5, −7.5, 37.5, −187.5, . . .

In Exercises 23–30, write a rule for the nth term of the sequence. Then graph the first six terms of the sequence. *(See Example 3.)*

23. $a_3 = 4, r = 2$

24. $a_3 = 27, r = 3$

25. $a_2 = 30, r = \frac{1}{2}$

26. $a_2 = 64, r = \frac{1}{4}$

27. $a_4 = -192, r = 4$

28. $a_4 = -500, r = 5$

29. $a_5 = 3, r = -\frac{1}{3}$

30. $a_5 = 1, r = -\frac{1}{5}$

ERROR ANALYSIS In Exercises 31 and 32, describe and correct the error in writing a rule for the nth term of the geometric sequence for which $a_2 = 48$ and $r = 6$.

31.

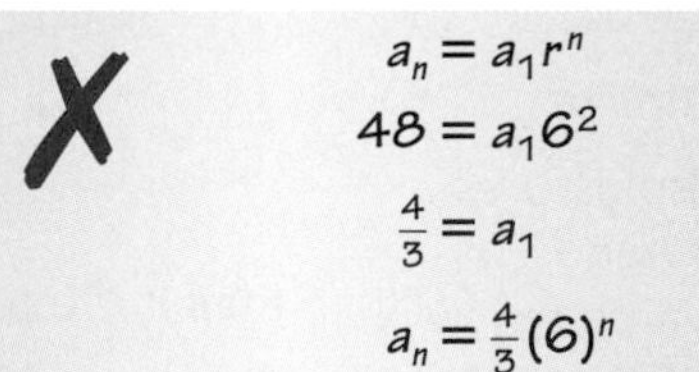

$a_n = a_1r^n$
$48 = a_16^2$
$\frac{4}{3} = a_1$
$a_n = \frac{4}{3}(6)^n$

32.

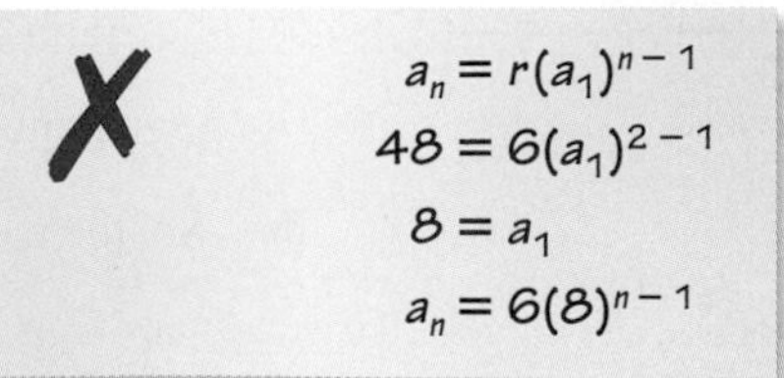

$a_n = r(a_1)^{n-1}$
$48 = 6(a_1)^{2-1}$
$8 = a_1$
$a_n = 6(8)^{n-1}$

In Exercises 33–40, write a rule for the nth term of the geometric sequence. *(See Example 4.)*

33. $a_2 = 28, a_5 = 1792$

34. $a_1 = 11, a_4 = 88$

35. $a_1 = -6, a_5 = -486$

36. $a_2 = -10, a_6 = -6250$

37. $a_2 = 64, a_4 = 1$

38. $a_1 = 1, a_2 = 49$

39. $a_2 = -72, a_6 = -\frac{1}{18}$

40. $a_2 = -48, a_5 = \frac{3}{4}$

WRITING EQUATIONS **In Exercises 41–46, write a rule for the sequence with the given terms.**

41.

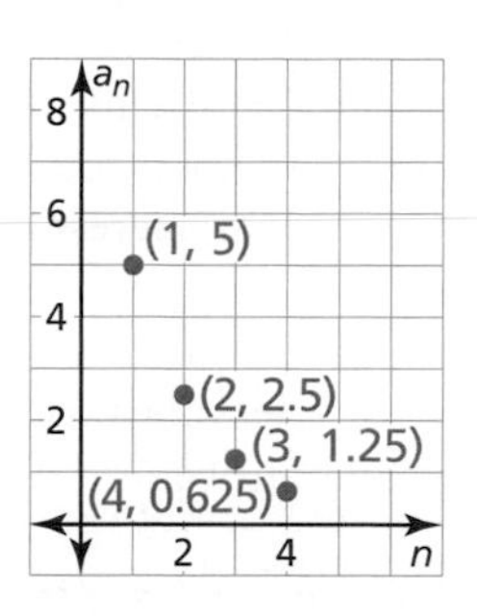

42.

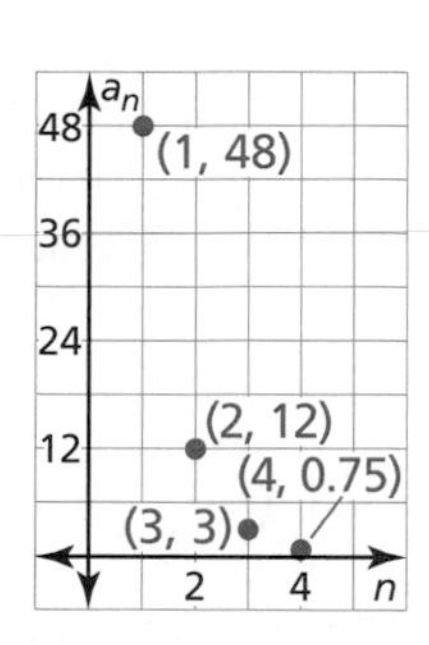

43.

(1, 5)
(2, 2.5)
(3, 1.25)
(4, 0.625)

44.

(1, 48)
(2, 12)
(3, 3)
(4, 0.75)

45.

n	2	3	4	5	6
a_n	−12	24	−48	96	−192

46.

n	2	3	4	5	6
a_n	−21	63	−189	567	−1701

In Exercises 47–52, find the sum. *(See Example 5.)*

47. $\sum_{i=1}^{9} 6(7)^{i-1}$

48. $\sum_{i=1}^{10} 7(4)^{i-1}$

49. $\sum_{i=1}^{10} 4\left(\frac{3}{4}\right)^{i-1}$

50. $\sum_{i=1}^{8} 5\left(\frac{1}{3}\right)^{i-1}$

51. $\sum_{i=0}^{8} 8\left(-\frac{2}{3}\right)^{i}$

52. $\sum_{i=0}^{9} 9\left(-\frac{3}{4}\right)^{i}$

NUMBER SENSE **In Exercises 53 and 54, find the sum.**

53. The first 8 terms of the geometric sequence $-12, -48, -192, -768, \ldots$.

54. The first 9 terms of the geometric sequence $-14, -42, -126, -378, \ldots$.

55. WRITING Compare the graph of $a_n = 5(3)^{n-1}$, where n is a positive integer, to the graph of $f(x) = 5 \bullet 3^{x-1}$, where x is a real number.

56. ABSTRACT REASONING Use the rule for the sum of a finite geometric series to write each polynomial as a rational expression.

a. $1 + x + x^2 + x^3 + x^4$

b. $3x + 6x^3 + 12x^5 + 24x^7$

MODELING WITH MATHEMATICS **In Exercises 57 and 58, use the monthly payment formula given in Example 6.**

57. You are buying a new car. You take out a 5-year loan for \$15,000. The annual interest rate of the loan is 4%. Calculate the monthly payment. *(See Example 6.)*

58. You are buying a new house. You take out a 30-year mortgage for \$200,000. The annual interest rate of the loan is 4.5%. Calculate the monthly payment.

59. MODELING WITH MATHEMATICS A regional soccer tournament has 64 participating teams. In the first round of the tournament, 32 games are played. In each successive round, the number of games decreases by a factor of $\frac{1}{2}$.

a. Write a rule for the number of games played in the nth round. For what values of n does the rule make sense? Explain.

b. Find the total number of games played in the regional soccer tournament.

60. MODELING WITH MATHEMATICS In a skydiving formation with R rings, each ring after the first has twice as many skydivers as the preceding ring. The formation for $R = 2$ is shown.

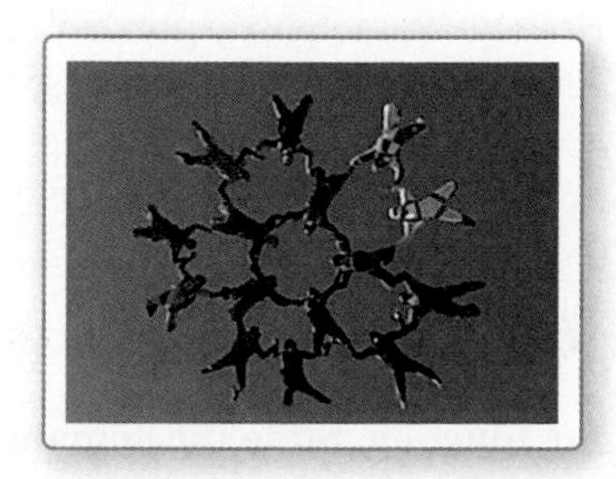

a. Let a_n be the number of skydivers in the nth ring. Write a rule for a_n.

b. Find the total number of skydivers when there are four rings.

61. PROBLEM SOLVING The *Sierpinski carpet* is a fractal created using squares. The process involves removing smaller squares from larger squares. First, divide a large square into nine congruent squares. Then remove the center square. Repeat these steps for each smaller square, as shown below. Assume that each side of the initial square is 1 unit long.

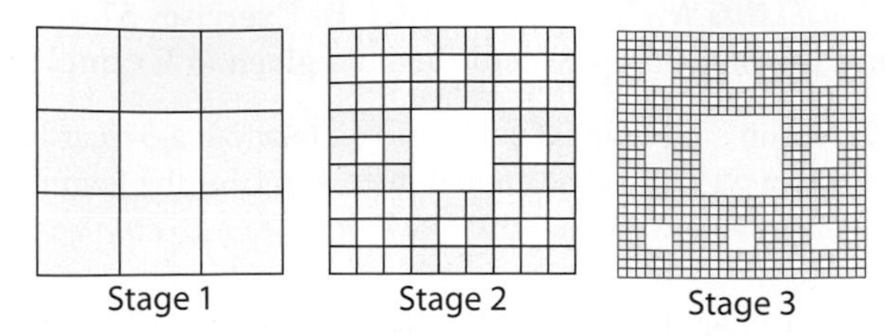

a. Let a_n be the total number of squares removed at the nth stage. Write a rule for a_n. Then find the total number of squares removed through Stage 8.

b. Let b_n be the remaining area of the original square after the nth stage. Write a rule for b_n. Then find the remaining area of the original square after Stage 12.

62. HOW DO YOU SEE IT? Match each sequence with its graph. Explain your reasoning.

a. $a_n = 10\left(\frac{1}{2}\right)^{n-1}$ b. $a_n = 10(2)^{n-1}$

A.
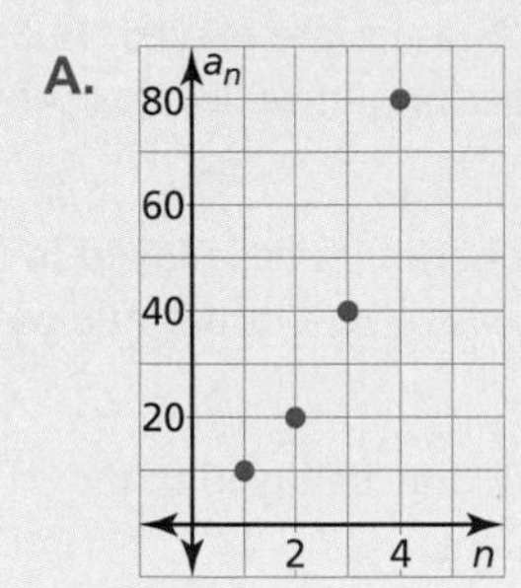

B.
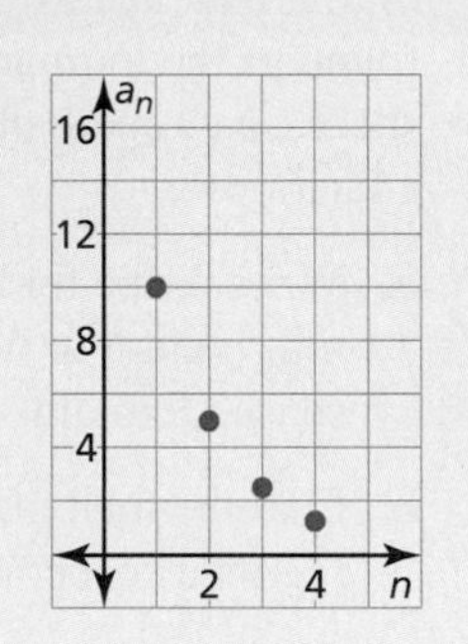

63. CRITICAL THINKING On January 1, you deposit $2000 in a retirement account that pays 5% annual interest. You make this deposit each January 1 for the next 30 years. How much money do you have in your account immediately after you make your last deposit?

64. THOUGHT PROVOKING The first four iterations of the fractal called the *Koch snowflake* are shown below. Find the perimeter and area of each iteration. Do the perimeters and areas form geometric sequences? Explain your reasoning.

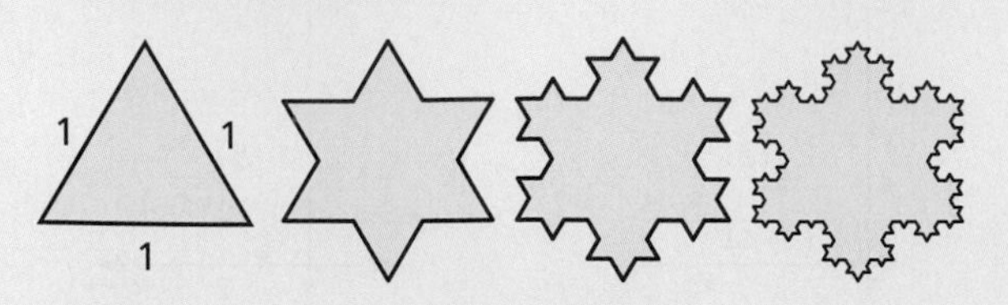

65. MAKING AN ARGUMENT You and your friend are comparing two loan options for a $165,000 house. Loan 1 is a 15-year loan with an annual interest rate of 3%. Loan 2 is a 30-year loan with an annual interest rate of 4%. Your friend claims the total amount repaid over the loan will be less for Loan 2. Is your friend correct? Justify your answer.

66. CRITICAL THINKING Let L be the amount of a loan (in dollars), i be the monthly interest rate (in decimal form), t be the term (in months), and M be the monthly payment (in dollars).

a. When making monthly payments, you are paying the loan amount plus the interest the loan gathers each month. For a 1-month loan, $t = 1$, the equation for repayment is $L(1 + i) - M = 0$. For a 2-month loan, $t = 2$, the equation is $[L(1 + i) - M](1 + i) - M = 0$. Solve both of these repayment equations for L.

b. Use the pattern in the equations you solved in part (a) to write a repayment equation for a t-month loan. (*Hint*: L is equal to M times a geometric series.) Then solve the equation for M.

c. Use the rule for the sum of a finite geometric series to show that the formula in part (b) is equivalent to

$$M = L\left(\frac{i}{1 - (1 + i)^{-t}}\right).$$

Use this formula to check your answers in Exercises 57 and 58.

Maintaining Mathematical Proficiency

Reviewing what you learned in previous grades and lessons

Graph the function. State the domain and range. *(Section 7.2)*

67. $f(x) = \frac{1}{x - 3}$

68. $g(x) = \frac{2}{x} + 3$

69. $h(x) = \frac{1}{x - 2} + 1$

70. $p(x) = \frac{3}{x + 1} - 2$

8.1–8.3 What Did You Learn?

Core Vocabulary

sequence, *p. 410*
terms of a sequence, *p. 410*
series, *p. 412*
summation notation, *p. 412*
sigma notation, *p. 412*
arithmetic sequence, *p. 418*
common difference, *p. 418*
arithmetic series, *p. 420*
geometric sequence, *p. 426*
common ratio, *p. 426*
geometric series, *p. 428*

Core Concepts

Section 8.1

Sequences, *p. 410*
Series and Summation Notation, *p. 412*
Formulas for Special Series, *p. 413*

Section 8.2

Rule for an Arithmetic Sequence, *p. 418*
The Sum of a Finite Arithmetic Series, *p. 420*

Section 8.3

Rule for a Geometric Sequence, *p. 426*
The Sum of a Finite Geometric Series, *p. 428*

Mathematical Practices

1. Explain how viewing each arrangement as individual tables can be helpful in Exercise 29 on page 415.
2. How can you use tools to find the sum of the arithmetic series in Exercises 53 and 54 on page 423?
3. How did understanding the domain of each function help you to compare the graphs in Exercise 55 on page 431?

Study Skills

Keeping Your Mind Focused

- Before doing homework, review the concept boxes and examples. Talk through the examples out loud.
- Complete homework as though you were also preparing for a quiz. Memorize the different types of problems, formulas, rules, and so on.

8.1–8.3 Quiz

Describe the pattern, write the next term, and write a rule for the *n*th term of the sequence. *(Section 8.1)*

1. $1, 7, 13, 19, \ldots$

2. $-5, 10, -15, 20, \ldots$

3. $\frac{1}{20}, \frac{2}{30}, \frac{3}{40}, \frac{4}{50}, \ldots$

Write the series using summation notation. Then find the sum of the series. *(Section 8.1)*

4. $1 + 2 + 3 + 4 + \cdots + 15$

5. $0 + \frac{1}{2} + \frac{2}{3} + \frac{3}{4} + \cdots + \frac{7}{8}$

6. $9 + 16 + 25 + \cdots + 100$

Write a rule for the *n*th term of the sequence. *(Sections 8.2 and 8.3)*

7.

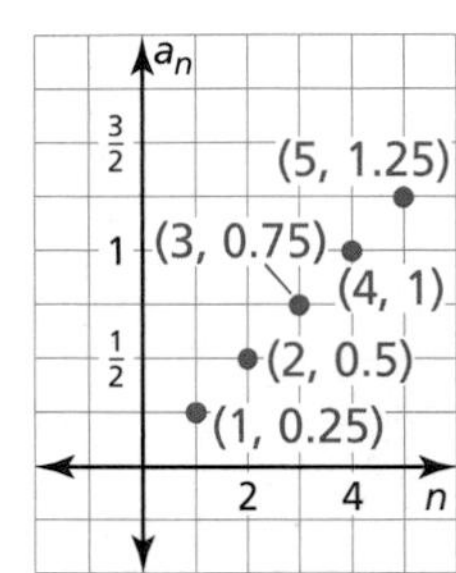

8.

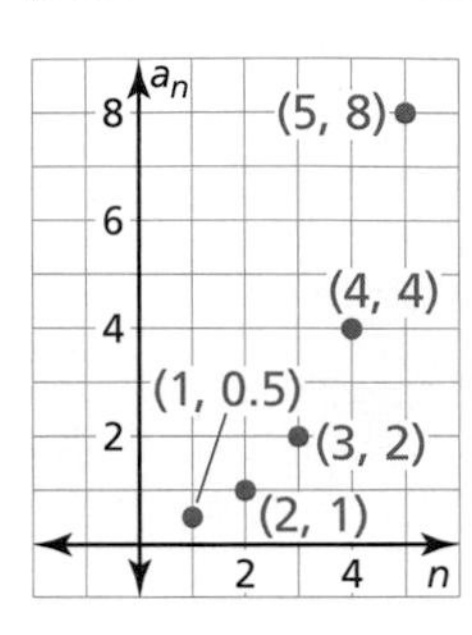

9.

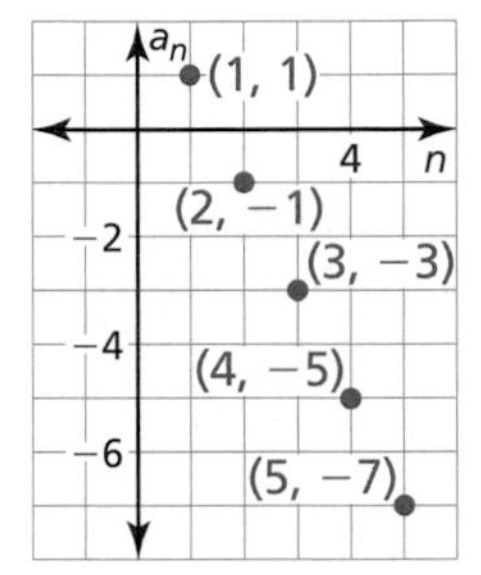

Tell whether the sequence is *arithmetic*, *geometric*, or *neither*. Write a rule for the *n*th term of the sequence. Then find a_9. *(Sections 8.2 and 8.3)*

10. $13, 6, -1, -8, \ldots$

11. $\frac{1}{2}, \frac{1}{3}, \frac{1}{4}, \frac{1}{5}, \ldots$

12. $1, -3, 9, -27, \ldots$

13. One term of an arithmetic sequence is $a_{12} = 19$. The common difference is $d = 7$. Write a rule for the *n*th term. Then graph the first six terms of the sequence. *(Section 8.2)*

14. Two terms of a geometric sequence are $a_6 = -50$ and $a_9 = -6250$. Write a rule for the *n*th term. *(Section 8.3)*

Find the sum. *(Sections 8.2 and 8.3)*

15. $\sum_{n=1}^{9} (3n + 5)$

16. $\sum_{k=1}^{5} 11(-3)^{k-2}$

17. $\sum_{i=1}^{12} -4\left(\frac{1}{2}\right)^{i+3}$

18. Pieces of chalk are stacked in a pile. Part of the pile is shown. The bottom row has 15 pieces of chalk, and the top row has 6 pieces of chalk. Each row has one less piece of chalk than the row below it. How many pieces of chalk are in the pile? *(Section 8.2)*

19. You accept a job as an environmental engineer that pays a salary of $45,000 in the first year. After the first year, your salary increases by 3.5% per year. *(Section 8.3)*

a. Write a rule giving your salary a_n for your *n*th year of employment.

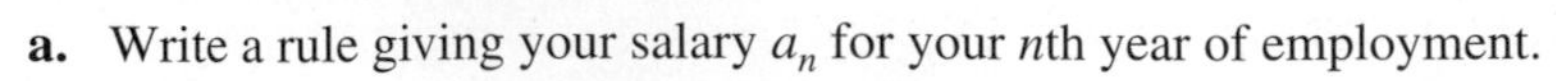

b. What will your salary be during your fifth year of employment?

c. You work 10 years for the company. What are your total earnings? Justify your answer.

8.4 Finding Sums of Infinite Geometric Series

Essential Question How can you find the sum of an infinite geometric series?

EXPLORATION 1 **Finding Sums of Infinite Geometric Series**

USING TOOLS STRATEGICALLY

To be proficient in math, you need to use technological tools, such as a spreadsheet, to explore and deepen your understanding of concepts.

Work with a partner. Enter each geometric series in a spreadsheet. Then use the spreadsheet to determine whether the infinite geometric series has a finite sum. If it does, find the sum. Explain your reasoning. (The figure shows a partially completed spreadsheet for part (a).)

a. $1 + \frac{1}{2} + \frac{1}{4} + \frac{1}{8} + \frac{1}{16} + \cdots$

b. $1 + \frac{1}{3} + \frac{1}{9} + \frac{1}{27} + \frac{1}{81} + \cdots$

c. $1 + \frac{3}{2} + \frac{9}{4} + \frac{27}{8} + \frac{81}{16} + \cdots$

d. $1 + \frac{5}{4} + \frac{25}{16} + \frac{125}{64} + \frac{625}{256} + \cdots$

e. $1 + \frac{4}{5} + \frac{16}{25} + \frac{64}{125} + \frac{256}{625} + \cdots$

f. $1 + \frac{9}{10} + \frac{81}{100} + \frac{729}{1000} + \frac{6561}{10{,}000} + \cdots$

	A	B
1	1	1
2	2	0.5
3	3	0.25
4	4	0.125
5	5	0.0625
6	6	0.03125
7	7	
8	8	
9	9	
10	10	
11	11	
12	12	
13	13	
14	14	
15	15	
16	Sum	

EXPLORATION 2 **Writing a Conjecture**

Work with a partner. Look back at the infinite geometric series in Exploration 1. Write a conjecture about how you can determine whether the infinite geometric series

$$a_1 + a_1r + a_1r^2 + a_1r^3 + \cdots$$

has a finite sum.

EXPLORATION 3 **Writing a Formula**

Work with a partner. In Lesson 8.3, you learned that the sum of the first n terms of a geometric series with first term a_1 and common ratio $r \neq 1$ is

$$S_n = a_1\left(\frac{1 - r^n}{1 - r}\right).$$

When an infinite geometric series has a finite sum, what happens to r^n as n increases? Explain your reasoning. Write a formula to find the sum of an infinite geometric series. Then verify your formula by checking the sums you obtained in Exploration 1.

Communicate Your Answer

4. How can you find the sum of an infinite geometric series?
5. Find the sum of each infinite geometric series, if it exists.
   a. $1 + 0.1 + 0.01 + 0.001 + 0.0001 + \cdots$
   b. $2 + \frac{4}{3} + \frac{8}{9} + \frac{16}{27} + \frac{32}{81} + \cdots$

8.4 Lesson

Core Vocabulary

partial sum, *p. 436*

Previous
repeating decimal
fraction in simplest form
rational number

What You Will Learn

- Find partial sums of infinite geometric series.
- Find sums of infinite geometric series.

Partial Sums of Infinite Geometric Series

The sum S_n of the first n terms of an infinite series is called a **partial sum**. The partial sums of an infinite geometric series may approach a limiting value.

EXAMPLE 1 Finding Partial Sums

Consider the infinite geometric series

$$\frac{1}{2} + \frac{1}{4} + \frac{1}{8} + \frac{1}{16} + \frac{1}{32} + \cdots.$$

Find and graph the partial sums S_n for $n = 1, 2, 3, 4$, and 5. Then describe what happens to S_n as n increases.

SOLUTION

Step 1 Find the partial sums.

$$S_1 = \frac{1}{2} = 0.5$$

$$S_2 = \frac{1}{2} + \frac{1}{4} = 0.75$$

$$S_3 = \frac{1}{2} + \frac{1}{4} + \frac{1}{8} \approx 0.88$$

$$S_4 = \frac{1}{2} + \frac{1}{4} + \frac{1}{8} + \frac{1}{16} \approx 0.94$$

$$S_5 = \frac{1}{2} + \frac{1}{4} + \frac{1}{8} + \frac{1}{16} + \frac{1}{32} \approx 0.97$$

Step 2 Plot the points (1, 0.5), (2, 0.75), (3, 0.88), (4, 0.94), and (5, 0.97). The graph is shown at the right.

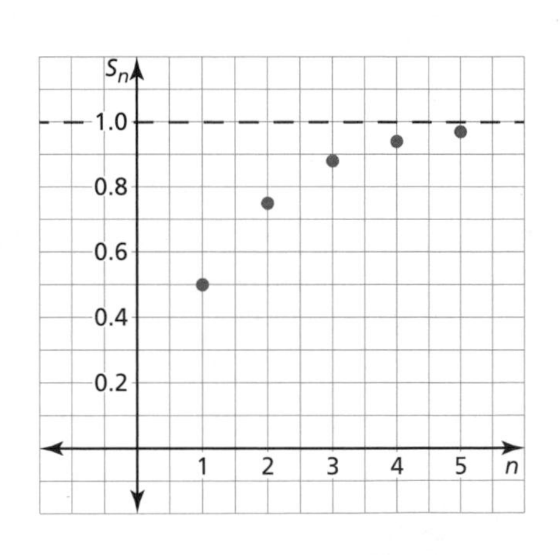

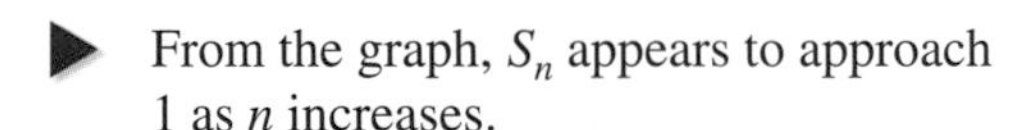

- From the graph, S_n appears to approach 1 as n increases.

Sums of Infinite Geometric Series

In Example 1, you can understand why S_n approaches 1 as n increases by considering the rule for the sum of a finite geometric series.

$$S_n = a_1\left(\frac{1 - r^n}{1 - r}\right) = \frac{1}{2}\left(\frac{1 - \left(\frac{1}{2}\right)^n}{1 - \frac{1}{2}}\right) = 1 - \left(\frac{1}{2}\right)^n$$

As n increases, $\left(\frac{1}{2}\right)^n$ approaches 0, so S_n approaches 1. Therefore, 1 is defined to be the sum of the infinite geometric series in Example 1. More generally, as n increases for *any* infinite geometric series with common ratio r between -1 and 1, the value of S_n approaches

$$S_n = a_1\left(\frac{1 - r^n}{1 - r}\right) \approx a_1\left(\frac{1 - 0}{1 - r}\right) = \frac{a_1}{1 - r}.$$

Core Concept

The Sum of an Infinite Geometric Series

The sum of an infinite geometric series with first term a_1 and common ratio r is given by

$$S = \frac{a_1}{1 - r}$$

provided $|r| < 1$. If $|r| \geq 1$, then the series has no sum.

UNDERSTANDING MATHEMATICAL TERMS

Even though a geometric series with a common ratio of $|r| < 1$ has *infinitely* many terms, the series has a *finite* sum.

EXAMPLE 2 Finding Sums of Infinite Geometric Series

Find the sum of each infinite geometric series.

a. $\sum_{i=1}^{\infty} 3(0.7)^{i-1}$ **b.** $1 + 3 + 9 + 27 + \cdots$ **c.** $1 - \frac{3}{4} + \frac{9}{16} - \frac{27}{64} + \cdots$

SOLUTION

a. For this series, $a_1 = 3(0.7)^{1-1} = 3$ and $r = 0.7$. The sum of the series is

$S = \frac{a_1}{1 - r}$ Formula for sum of an infinite geometric series

$= \frac{3}{1 - 0.7}$ Substitute 3 for a_1 and 0.7 for r.

$= 10.$ Simplify.

b. For this series, $a_1 = 1$ and $a_2 = 3$. So, the common ratio is $r = \frac{3}{1} = 3$.

Because $|3| \geq 1$, the sum does not exist.

STUDY TIP

For the geometric series in part (b), the graph of the partial sums S_n for $n = 1, 2, 3, 4, 5$, and 6 are shown. From the graph, it appears that as n increases, the partial sums do not approach a fixed number.

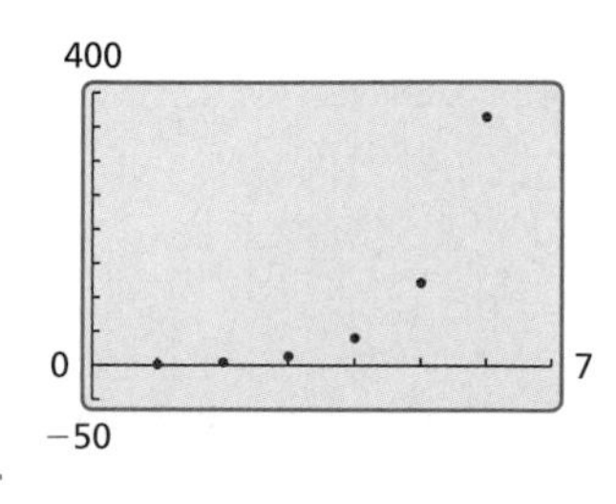

c. For this series, $a_1 = 1$ and $a_2 = -\frac{3}{4}$. So, the common ratio is

$$r = \frac{-\frac{3}{4}}{1} = -\frac{3}{4}.$$

The sum of the series is

$S = \frac{a_1}{1 - r}$ Formula for sum of an infinite geometric series

$= \frac{1}{1 - \left(-\frac{3}{4}\right)}$ Substitute 1 for a_1 and $-\frac{3}{4}$ for r.

$= \frac{4}{7}.$ Simplify.

Monitoring Progress

Help in English and Spanish at *BigIdeasMath.com*

1. Consider the infinite geometric series

$$\frac{2}{5} + \frac{4}{25} + \frac{8}{125} + \frac{16}{1625} + \frac{32}{3125} + \cdots.$$

Find and graph the partial sums S_n for $n = 1, 2, 3, 4$, and 5. Then describe what happens to S_n as n increases.

Find the sum of the infinite geometric series, if it exists.

2. $\sum_{n=1}^{\infty} \left(-\frac{1}{2}\right)^{n-1}$ **3.** $\sum_{n=1}^{\infty} 3\left(\frac{5}{4}\right)^{n-1}$ **4.** $3 + \frac{3}{4} + \frac{3}{16} + \frac{3}{64} + \cdots$

EXAMPLE 3 Solving a Real-Life Problem

A pendulum that is released to swing freely travels 18 inches on the first swing. On each successive swing, the pendulum travels 80% of the distance of the previous swing. What is the total distance the pendulum swings?

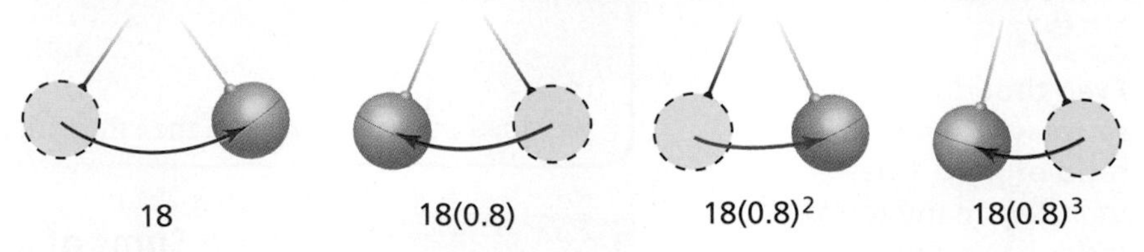

SOLUTION

The total distance traveled by the pendulum is given by the infinite geometric series

$$18 + 18(0.8) + 18(0.8)^2 + 18(0.8)^3 + \cdots.$$

For this series, $a_1 = 18$ and $r = 0.8$. The sum of the series is

$S = \frac{a_1}{1 - r}$ Formula for sum of an infinite geometric series

$= \frac{18}{1 - 0.8}$ Substitute 18 for a_1 and 0.8 for r.

$= 90.$ Simplify.

▶ The pendulum travels a total distance of 90 inches, or 7.5 feet.

REMEMBER

Because a repeating decimal is a rational number, it can be written as $\frac{a}{b}$, where a and b are integers and $b \neq 0$.

EXAMPLE 4 Writing a Repeating Decimal as a Fraction

Write 0.242424 . . . as a fraction in simplest form.

SOLUTION

Write the repeating decimal as an infinite geometric series.

$$0.242424\ldots = 0.24 + 0.0024 + 0.000024 + 0.00000024 + \cdots$$

For this series, $a_1 = 0.24$ and $r = \frac{0.0024}{0.24} = 0.01$. Next, write the sum of the series.

$S = \frac{a_1}{1 - r}$ Formula for sum of an infinite geometric series

$= \frac{0.24}{1 - 0.01}$ Substitute 0.24 for a_1 and 0.01 for r.

$= \frac{0.24}{0.99}$ Simplify.

$= \frac{24}{99}$ Write as a quotient of integers.

$= \frac{8}{33}$ Simplify.

Monitoring Progress  Help in English and Spanish at *BigIdeasMath.com*

5. **WHAT IF?** In Example 3, suppose the pendulum travels 10 inches on its first swing. What is the total distance the pendulum swings?

Write the repeating decimal as a fraction in simplest form.

6. 0.555 . . .
7. 0.727272 . . .
8. 0.131313 . . .

8.4 Exercises

Dynamic Solutions available at *BigIdeasMath.com*

Vocabulary and Core Concept Check

1. **COMPLETE THE SENTENCE** The sum S_n of the first n terms of an infinite series is called a(n) ________.

2. **WRITING** Explain how to tell whether the series $\sum_{i=1}^{\infty} a_1 r^{i-1}$ has a sum.

Monitoring Progress and Modeling with Mathematics

In Exercises 3–6, consider the infinite geometric series. Find and graph the partial sums S_n for $n = 1, 2, 3, 4,$ and 5. Then describe what happens to S_n as n increases. *(See Example 1.)*

3. $\frac{1}{2} + \frac{1}{6} + \frac{1}{18} + \frac{1}{54} + \frac{1}{162} + \cdots$

4. $\frac{2}{3} + \frac{1}{3} + \frac{1}{6} + \frac{1}{12} + \frac{1}{24} + \cdots$

5. $4 + \frac{12}{5} + \frac{36}{25} + \frac{108}{125} + \frac{324}{625} + \cdots$

6. $2 + \frac{2}{6} + \frac{2}{36} + \frac{2}{216} + \frac{2}{1296} + \cdots$

In Exercises 7–14, find the sum of the infinite geometric series, if it exists. *(See Example 2.)*

7. $\sum_{n=1}^{\infty} 8\left(\frac{1}{5}\right)^{n-1}$

8. $\sum_{k=1}^{\infty} -6\left(\frac{3}{2}\right)^{k-1}$

9. $\sum_{k=1}^{\infty} \frac{11}{3}\left(\frac{3}{8}\right)^{k-1}$

10. $\sum_{i=1}^{\infty} \frac{2}{5}\left(\frac{5}{3}\right)^{i-1}$

11. $2 + \frac{6}{4} + \frac{18}{16} + \frac{54}{64} + \cdots$

12. $-5 - 2 - \frac{4}{5} - \frac{8}{25} - \cdots$

13. $3 + \frac{5}{2} + \frac{25}{12} + \frac{125}{72} + \cdots$

14. $\frac{1}{2} - \frac{5}{3} + \frac{50}{9} - \frac{500}{27} + \cdots$

ERROR ANALYSIS In Exercises 15 and 16, describe and correct the error in finding the sum of the infinite geometric series.

15. $\sum_{n=1}^{\infty} \left(\frac{7}{2}\right)^{n-1}$

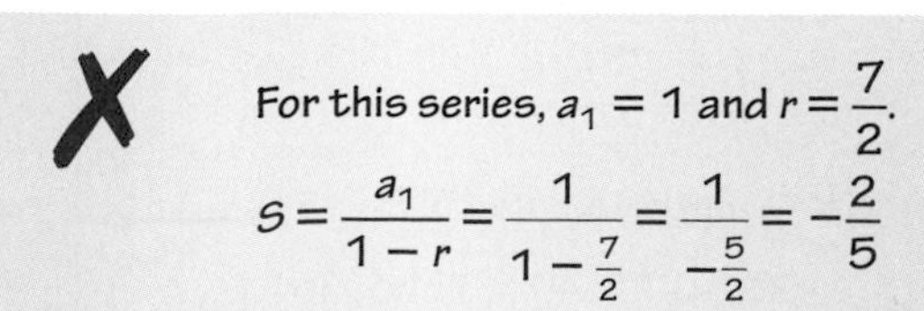

For this series, $a_1 = 1$ and $r = \frac{7}{2}$.
$S = \frac{a_1}{1-r} = \frac{1}{1-\frac{7}{2}} = \frac{1}{-\frac{5}{2}} = -\frac{2}{5}$

16. $4 + \frac{8}{3} + \frac{16}{9} + \frac{32}{27} + \cdots$

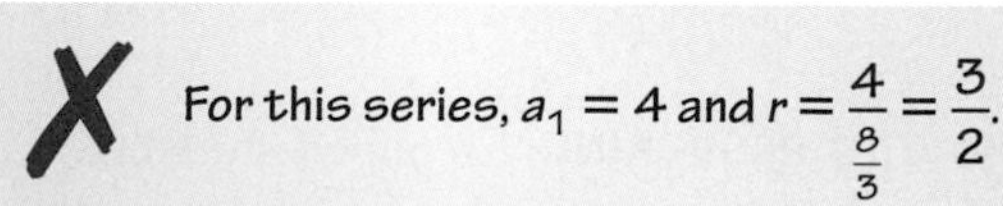

For this series, $a_1 = 4$ and $r = \frac{4}{\frac{8}{3}} = \frac{3}{2}$.
Because $\left|\frac{3}{2}\right| > 1$, the series has no sum.

17. **MODELING WITH MATHEMATICS** You push your younger cousin on a tire swing one time and then allow your cousin to swing freely. On the first swing, your cousin travels a distance of 14 feet. On each successive swing, your cousin travels 75% of the distance of the previous swing. What is the total distance your cousin swings? *(See Example 3.)*

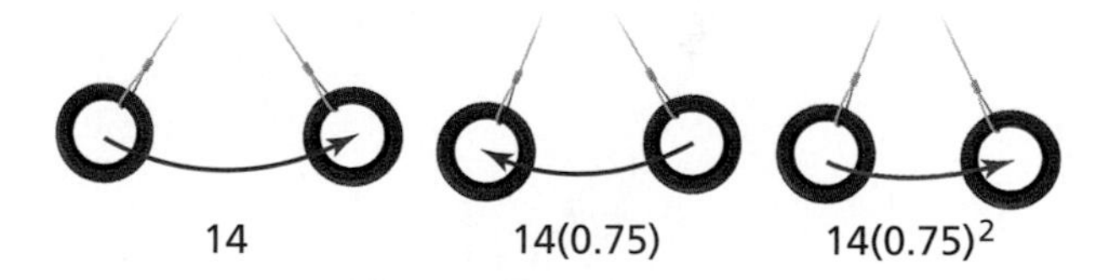

18. **MODELING WITH MATHEMATICS** A company had a profit of $350,000 in its first year. Since then, the company's profit has decreased by 12% per year. Assuming this trend continues, what is the total profit the company can make over the course of its lifetime? Justify your answer.

In Exercises 19–24, write the repeating decimal as a fraction in simplest form. *(See Example 4.)*

19. 0.222 . . .

20. 0.444 . . .

21. 0.161616 . . .

22. 0.625625625 . . .

23. 32.323232 . . .

24. 130.130130130 . . .

25. **PROBLEM SOLVING** Find two infinite geometric series whose sums are each 6. Justify your answers.

26. HOW DO YOU SEE IT? The graph shows the partial sums of the geometric series $a_1 + a_2 + a_3 + a_4 + \cdots$.

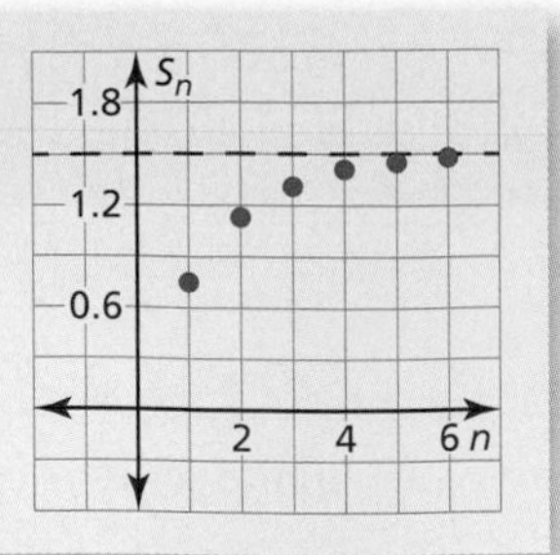

What is the value of $\sum_{n=1}^{\infty} a_n$? Explain.

27. MODELING WITH MATHEMATICS A radio station has a daily contest in which a random listener is asked a trivia question. On the first day, the station gives \$500 to the first listener who answers correctly. On each successive day, the winner receives 90% of the winnings from the previous day. What is the total amount of prize money the radio station gives away during the contest?

28. THOUGHT PROVOKING Archimedes used the sum of a geometric series to compute the area enclosed by a parabola and a straight line. In "Quadrature of the Parabola," he proved that the area of the region is $\frac{4}{3}$ the area of the inscribed triangle. The first term of the series for the parabola below is represented by the area of the blue triangle and the second term is represented by the area of the red triangles. Use Archimedes' result to find the area of the region. Then write the area as the sum of an infinite geometric series.

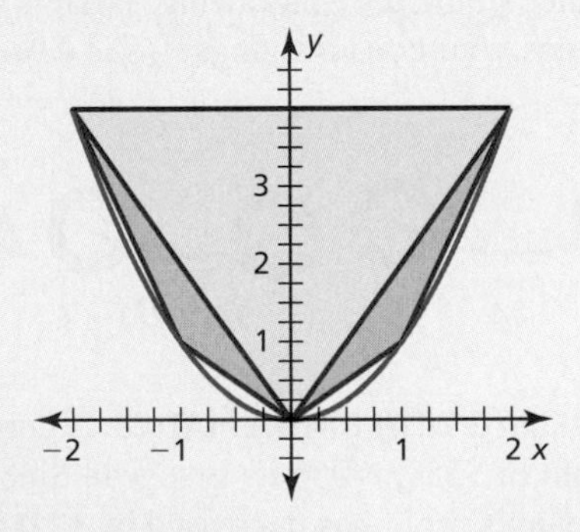

29. DRAWING CONCLUSIONS Can a person running at 20 feet per second ever catch up to a tortoise that runs 10 feet per second when the tortoise has a 20-foot head start? The Greek mathematician Zeno said no. He reasoned as follows:

The person will keep halving the distance but will never catch up to the tortoise.
20 ft
10 ft

Looking at the race as Zeno did, the distances and the times it takes the person to run those distances both form infinite geometric series. Using the table, show that both series have finite sums. Does the person catch up to the tortoise? Justify your answer.

Distance (ft)	20	10	5	2.5	. . .
Time (sec)	1	0.5	0.25	0.125	. . .

30. MAKING AN ARGUMENT Your friend claims that 0.999 . . . is equal to 1. Is your friend correct? Justify your answer.

31. CRITICAL THINKING The *Sierpinski triangle* is a fractal created using equilateral triangles. The process involves removing smaller triangles from larger triangles by joining the midpoints of the sides of the larger triangles as shown. Assume that the initial triangle has an area of 1 square foot.

a. Let a_n be the total area of all the triangles that are removed at Stage n. Write a rule for a_n.

b. Find $\sum_{n=1}^{\infty} a_n$. Interpret your answer in the context of this situation.

Maintaining Mathematical Proficiency

Reviewing what you learned in previous grades and lessons

Determine the type of function represented by the table. *(Section 6.7)*

32.

x	−3	−2	−1	0	1
y	0.5	1.5	4.5	13.5	40.5

33.

x	0	4	8	12	16
y	−7	−1	2	2	−1

Determine whether the sequence is *arithmetic*, *geometric*, or *neither*. *(Sections 8.2 and 8.3)*

34. −7, −1, 5, 11, 17, . . .

35. 0, −1, −3, −7, −15, . . .

36. 13.5, 40.5, 121.5, 364.5, . . .

8.5 Using Recursive Rules with Sequences

Essential Question

How can you define a sequence recursively?

A **recursive rule** gives the beginning term(s) of a sequence and a *recursive equation* that tells how a_n is related to one or more preceding terms.

EXPLORATION 1 Evaluating a Recursive Rule

Work with a partner. Use each recursive rule and a spreadsheet to write the first six terms of the sequence. Classify the sequence as arithmetic, geometric, or neither. Explain your reasoning. (The figure shows a partially completed spreadsheet for part (a).)

a. $a_1 = 7, a_n = a_{n-1} + 3$

b. $a_1 = 5, a_n = a_{n-1} - 2$

c. $a_1 = 1, a_n = 2a_{n-1}$

d. $a_1 = 1, a_n = \frac{1}{2}(a_{n-1})^2$

e. $a_1 = 3, a_n = a_{n-1} + 1$

f. $a_1 = 4, a_n = \frac{1}{2}a_{n-1} - 1$

g. $a_1 = 4, a_n = \frac{1}{2}a_{n-1}$

h. $a_1 = 4, a_2 = 5, a_n = a_{n-1} + a_{n-2}$

	A	B
1	n	nth Term
2	1	7
3	2	10
4	3	
5	4	
6	5	
7	6	

B2+3

ATTENDING TO PRECISION

To be proficient in math, you need to communicate precisely to others.

EXPLORATION 2 Writing a Recursive Rule

Work with a partner. Write a recursive rule for the sequence. Explain your reasoning.

a. 3, 6, 9, 12, 15, 18, . . .

b. 18, 14, 10, 6, 2, −2, . . .

c. 3, 6, 12, 24, 48, 96, . . .

d. 128, 64, 32, 16, 8, 4, . . .

e. 5, 5, 5, 5, 5, 5, . . .

f. 1, 1, 2, 3, 5, 8, . . .

EXPLORATION 3 Writing a Recursive Rule

Work with a partner. Write a recursive rule for the sequence whose graph is shown.

a.

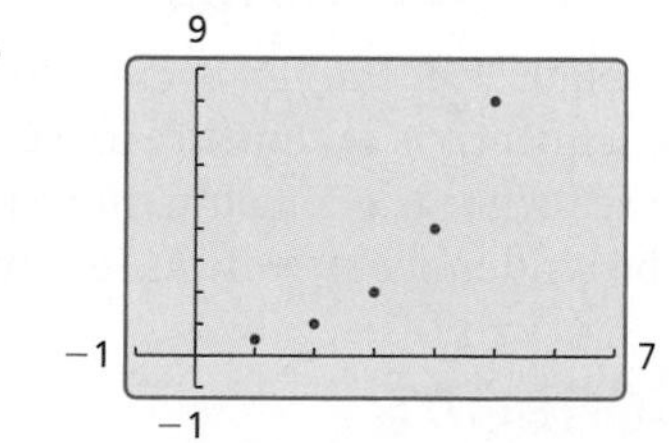

b.

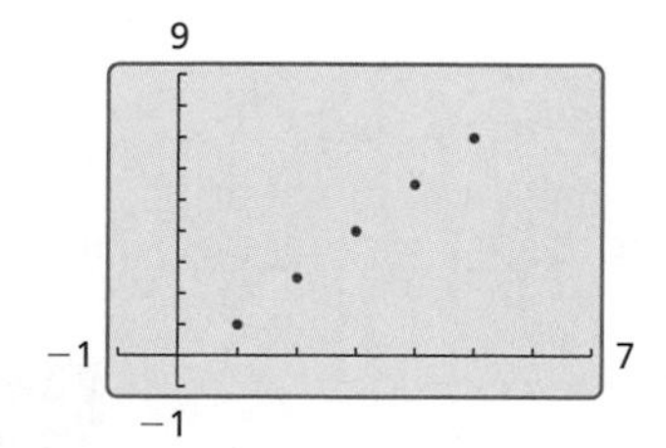

Communicate Your Answer

4. How can you define a sequence recursively?

5. Write a recursive rule that is different from those in Explorations 1–3. Write the first six terms of the sequence. Then graph the sequence and classify it as arithmetic, geometric, or neither.

8.5 Lesson

Core Vocabulary

explicit rule, *p. 442*
recursive rule, *p. 442*

What You Will Learn

- Evaluate recursive rules for sequences.
- Write recursive rules for sequences.
- Translate between recursive and explicit rules for sequences.
- Use recursive rules to solve real-life problems.

Evaluating Recursive Rules

So far in this chapter, you have worked with explicit rules for the nth term of a sequence, such as $a_n = 3n - 2$ and $a_n = 7(0.5)^n$. An **explicit rule** gives a_n as a function of the term's position number n in the sequence.

In this section, you will learn another way to define a sequence—by a *recursive rule*. A **recursive rule** gives the beginning term(s) of a sequence and a *recursive equation* that tells how a_n is related to one or more preceding terms.

EXAMPLE 1 Evaluating Recursive Rules

Write the first six terms of each sequence.

a. $a_0 = 1,\ a_n = a_{n-1} + 4$

b. $f(1) = 1, f(n) = 3 \cdot f(n-1)$

SOLUTION

a.		**b.**
$a_0 = 1$	1st term	$f(1) = 1$
$a_1 = a_0 + 4 = 1 + 4 = 5$	2nd term	$f(2) = 3 \cdot f(1) = 3(1) = 3$
$a_2 = a_1 + 4 = 5 + 4 = 9$	3rd term	$f(3) = 3 \cdot f(2) = 3(3) = 9$
$a_3 = a_2 + 4 = 9 + 4 = 13$	4th term	$f(4) = 3 \cdot f(3) = 3(9) = 27$
$a_4 = a_3 + 4 = 13 + 4 = 17$	5th term	$f(5) = 3 \cdot f(4) = 3(27) = 81$
$a_5 = a_4 + 4 = 17 + 4 = 21$	6th term	$f(6) = 3 \cdot f(5) = 3(81) = 243$

Monitoring Progress

Help in English and Spanish at *BigIdeasMath.com*

Write the first six terms of the sequence.

1. $a_1 = 3,\ a_n = a_{n-1} - 7$

2. $a_0 = 162,\ a_n = 0.5a_{n-1}$

3. $f(0) = 1, f(n) = f(n-1) + n$

4. $a_1 = 4,\ a_n = 2a_{n-1} - 1$

Writing Recursive Rules

In part (a) of Example 1, the *differences* of consecutive terms of the sequence are constant, so the sequence is arithmetic. In part (b), the *ratios* of consecutive terms are constant, so the sequence is geometric. In general, rules for arithmetic and geometric sequences can be written recursively as follows.

Core Concept

Recursive Equations for Arithmetic and Geometric Sequences

Arithmetic Sequence

$a_n = a_{n-1} + d$, where d is the common difference

Geometric Sequence

$a_n = r \cdot a_{n-1}$, where r is the common ratio

EXAMPLE 2 Writing Recursive Rules

Write a recursive rule for (a) 3, 13, 23, 33, 43, . . . and (b) 16, 40, 100, 250, 625, ...

SOLUTION

Use a table to organize the terms and find the pattern.

> **COMMON ERROR**
>
> A recursive *equation* for a sequence does not include the initial term. To write a recursive *rule* for a sequence, the initial term(s) must be included.

a.

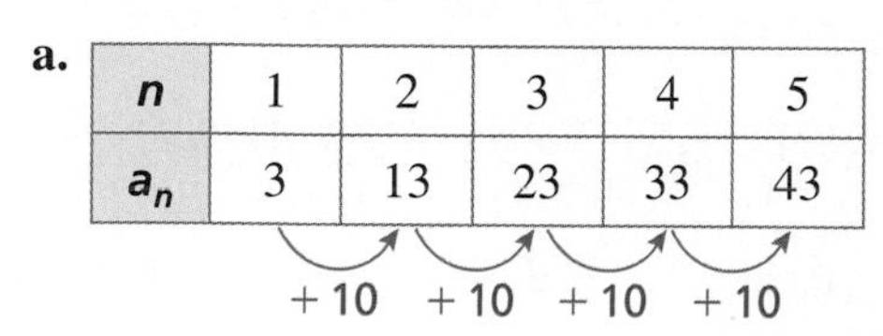

n	1	2	3	4	5
a_n	3	13	23	33	43

The sequence is arithmetic with first term $a_1 = 3$ and common difference $d = 10$.

$a_n = a_{n-1} + d$ Recursive equation for arithmetic sequence

$= a_{n-1} + 10$ Substitute 10 for d.

▶ A recursive rule for the sequence is $a_1 = 3$, $a_n = a_{n-1} + 10$.

b.

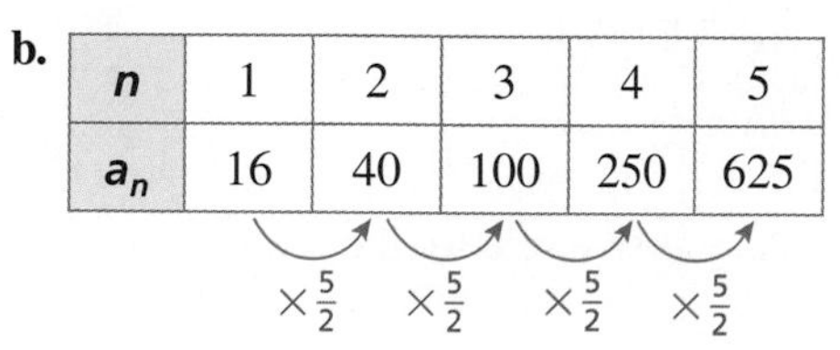

n	1	2	3	4	5
a_n	16	40	100	250	625

The sequence is geometric with first term $a_1 = 16$ and common ratio $r = \frac{5}{2}$.

$a_n = r \cdot a_{n-1}$ Recursive equation for geometric sequence

$= \frac{5}{2}a_{n-1}$ Substitute $\frac{5}{2}$ for r.

▶ A recursive rule for the sequence is $a_1 = 16$, $a_n = \frac{5}{2}a_{n-1}$.

> **STUDY TIP**
>
> The sequence in part (a) of Example 3 is called the *Fibonacci sequence*. The sequence in part (b) lists *factorial numbers*. You will learn more about *factorials* in Chapter 10.

EXAMPLE 3 Writing Recursive Rules

Write a recursive rule for each sequence.

a. 1, 1, 2, 3, 5, . . .

b. 1, 1, 2, 6, 24, . . .

SOLUTION

a. The terms have neither a common difference nor a common ratio. Beginning with the third term in the sequence, each term is the sum of the two previous terms.

▶ A recursive rule for the sequence is $a_1 = 1$, $a_2 = 1$, $a_n = a_{n-2} + a_{n-1}$.

b. The terms have neither a common difference nor a common ratio. Denote the first term by $a_0 = 1$. Note that $a_1 = 1 = 1 \cdot a_0$, $a_2 = 2 = 2 \cdot a_1$, $a_3 = 6 = 3 \cdot a_2$, and so on.

▶ A recursive rule for the sequence is $a_0 = 1$, $a_n = n \cdot a_{n-1}$.

Monitoring Progress  Help in English and Spanish at *BigIdeasMath.com*

Write a recursive rule for the sequence.

5. 2, 14, 98, 686, 4802, . . .

6. 19, 13, 7, 1, −5, . . .

7. 11, 22, 33, 44, 55, . . .

8. 1, 2, 2, 4, 8, 32, . . .

Translating Between Recursive and Explicit Rules

EXAMPLE 4 Translating from Explicit Rules to Recursive Rules

Write a recursive rule for (a) $a_n = -6 + 8n$ and (b) $a_n = -3\left(\frac{1}{2}\right)^{n-1}$.

SOLUTION

a. The explicit rule represents an arithmetic sequence with first term $a_1 = -6 + 8(1) = 2$ and common difference $d = 8$.

$a_n = a_{n-1} + d$ — Recursive equation for arithmetic sequence

$a_n = a_{n-1} + 8$ — Substitute 8 for d.

▶ A recursive rule for the sequence is $a_1 = 2$, $a_n = a_{n-1} + 8$.

b. The explicit rule represents a geometric sequence with first term $a_1 = -3\left(\frac{1}{2}\right)^0 = -3$ and common ratio $r = \frac{1}{2}$.

$a_n = r \cdot a_{n-1}$ — Recursive equation for geometric sequence

$a_n = \frac{1}{2}a_{n-1}$ — Substitute $\frac{1}{2}$ for r.

▶ A recursive rule for the sequence is $a_1 = -3$, $a_n = \frac{1}{2}a_{n-1}$.

EXAMPLE 5 Translating from Recursive Rules to Explicit Rules

Write an explicit rule for each sequence.

a. $a_1 = -5$, $a_n = a_{n-1} - 2$

b. $a_1 = 10$, $a_n = 2a_{n-1}$

SOLUTION

a. The recursive rule represents an arithmetic sequence with first term $a_1 = -5$ and common difference $d = -2$.

$a_n = a_1 + (n - 1)d$ — Explicit rule for arithmetic sequence

$a_n = -5 + (n - 1)(-2)$ — Substitute -5 for a_1 and -2 for d.

$a_n = -3 - 2n$ — Simplify.

▶ An explicit rule for the sequence is $a_n = -3 - 2n$.

b. The recursive rule represents a geometric sequence with first term $a_1 = 10$ and common ratio $r = 2$.

$a_n = a_1r^{n-1}$ — Explicit rule for geometric sequence

$a_n = 10(2)^{n-1}$ — Substitute 10 for a_1 and 2 for r.

▶ An explicit rule for the sequence is $a_n = 10(2)^{n-1}$.

Monitoring Progress  Help in English and Spanish at *BigIdeasMath.com*

Write a recursive rule for the sequence.

9. $a_n = 17 - 4n$

10. $a_n = 16(3)^{n-1}$

Write an explicit rule for the sequence.

11. $a_1 = -12$, $a_n = a_{n-1} + 16$

12. $a_1 = 2$, $a_n = -6a_{n-1}$

Solving Real-Life Problems

EXAMPLE 6 **Solving a Real-Life Problem**

A lake initially contains 5200 fish. Each year, the population declines 30% due to fishing and other causes, so the lake is restocked with 400 fish.

a. Write a recursive rule for the number a_n of fish at the start of the nth year.

b. Find the number of fish at the start of the fifth year.

c. Describe what happens to the population of fish over time.

SOLUTION

a. Write a recursive rule. The initial value is 5200. Because the population declines 30% each year, 70% of the fish remain in the lake from one year to the next. Also, 400 fish are added each year. Here is a verbal model for the recursive equation.

Fish at start of year n	$= 0.7 \cdot$	Fish at start of year $n - 1$	$+$	New fish added
⬇		⬇		⬇
a_n	$= 0.7 \cdot$	a_{n-1}	$+$	400

▶ A recursive rule is $a_1 = 5200$, $a_n = (0.7)a_{n-1} + 400$.

b. Find the number of fish at the start of the fifth year. Enter 5200 (the value of a_1) in a graphing calculator. Then enter the rule

$.7 \times \text{Ans} + 400$

to find a_2. Press the enter button three more times to find $a_5 \approx 2262$.

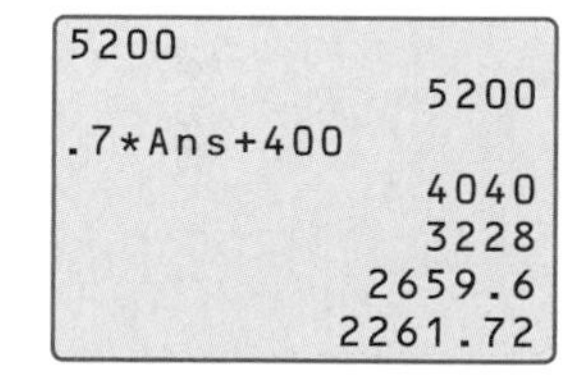

▶ There are about 2262 fish in the lake at the start of the fifth year.

c. Describe what happens to the population of fish over time. Continue pressing enter on the calculator. The screen at the right shows the fish populations for years 44 to 50. Observe that the population of fish approaches 1333.

1333.334178
1333.333924
1333.333747
1333.333623
1333.333536
1333.333475
1333.333433

▶ Over time, the population of fish in the lake stabilizes at about 1333 fish.

Check

Set a graphing calculator to *sequence* and *dot* modes. Graph the sequence and use the *trace* feature. From the graph, it appears the sequence approaches 1333.

u=.7*u(n−1)+400
n=75
X=75 Y=1333.3333

Monitoring Progress  Help in English and Spanish at *BigIdeasMath.com*

13. WHAT IF? In Example 6, suppose 75% of the fish remain each year. What happens to the population of fish over time?

EXAMPLE 7 Modeling with Mathematics

> **REMEMBER**
> In Section 8.3, you used a formula involving a geometric series to calculate the monthly payment for a similar loan.

You borrow \$150,000 at 6% annual interest compounded monthly for 30 years. The monthly payment is \$899.33.

- Find the balance after the third payment.
- Due to rounding in the calculations, the last payment is often different from the original payment. Find the amount of the last payment.

SOLUTION

1. **Understand the Problem** You are given the conditions of a loan. You are asked to find the balance after the third payment and the amount of the last payment.

2. **Make a Plan** Because the balance after each payment depends on the balance after the previous payment, write a recursive rule that gives the balance after each payment. Then use a spreadsheet to find the balance after each payment, rounded to the nearest cent.

3. **Solve the Problem** Because the monthly interest rate is $\frac{0.06}{12} = 0.005$, the balance increases by a factor of 1.005 each month, and then the payment of \$899.33 is subtracted.

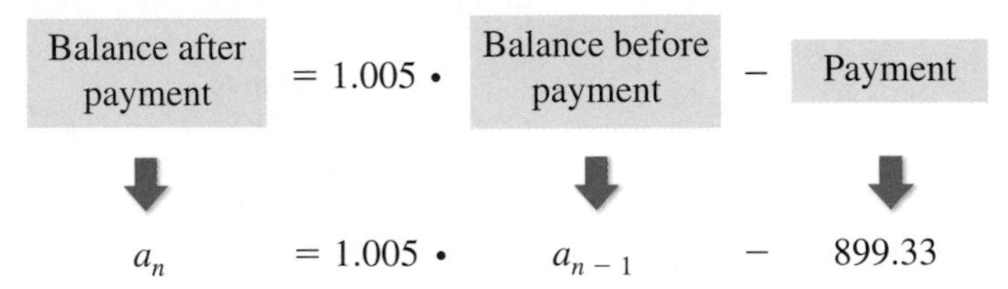

$$a_n = 1.005 \cdot a_{n-1} - 899.33$$

Use a spreadsheet and the recursive rule to find the balance after the third payment and after the 359th payment.

	A	B	
1	Payment number	Balance after payment	
2	1	149850.67	B2 =Round(1.005*150000−899.33, 2)
3	2	149700.59	B3 =Round(1.005*B2−899.33, 2)
4	3	149549.76	⋮
358	357	2667.38	
359	358	1781.39	
360	359	890.97	B360 =Round(1.005*B359−899.33, 2)

▶ The balance after the third payment is \$149,549.76. The balance after the 359th payment is \$890.97, so the final payment is 1.005(890.97) = \$895.42.

4. **Look Back** By continuing the spreadsheet for the 360th payment using the original monthly payment of \$899.33, the balance is −3.91.

361	360	−3.91	B361 =Round(1.005*B360−899.33, 2)

This shows an overpayment of \$3.91. So, it is reasonable that the last payment is \$899.33 − \$3.91 = \$895.42.

Monitoring Progress  Help in English and Spanish at *BigIdeasMath.com*

14. WHAT IF? How do the answers in Example 7 change when the annual interest rate is 7.5% and the monthly payment is \$1048.82?

8.5 Exercises

Dynamic Solutions available at *BigIdeasMath.com*

Vocabulary and Core Concept Check

1. **COMPLETE THE SENTENCE** A recursive __________ tells how the nth term of a sequence is related to one or more preceding terms.

2. **WRITING** Explain the difference between an explicit rule and a recursive rule for a sequence.

Monitoring Progress and Modeling with Mathematics

In Exercises 3–10, write the first six terms of the sequence. *(See Example 1.)*

3. $a_1 = 1$
   $a_n = a_{n-1} + 3$

4. $a_1 = 1$
   $a_n = a_{n-1} - 5$

5. $f(0) = 4$
   $f(n) = 2f(n-1)$

6. $f(0) = 10$
   $f(n) = \frac{1}{2}f(n-1)$

7. $a_1 = 2$
   $a_n = (a_{n-1})^2 + 1$

8. $a_1 = 1$
   $a_n = (a_{n-1})^2 - 10$

9. $f(0) = 2, f(1) = 4$
   $f(n) = f(n-1) - f(n-2)$

10. $f(1) = 2, f(2) = 3$
    $f(n) = f(n-1) \cdot f(n-2)$

In Exercises 11–22, write a recursive rule for the sequence. *(See Examples 2 and 3.)*

11. $21, 14, 7, 0, -7, \ldots$

12. $54, 43, 32, 21, 10, \ldots$

13. $3, 12, 48, 192, 768, \ldots$

14. $4, -12, 36, -108, \ldots$

15. $44, 11, \frac{11}{4}, \frac{11}{16}, \frac{11}{64}, \ldots$

16. $1, 8, 15, 22, 29, \ldots$

17. $2, 5, 10, 50, 500, \ldots$

18. $3, 5, 15, 75, 1125, \ldots$

19. $1, 4, 5, 9, 14, \ldots$

20. $16, 9, 7, 2, 5, \ldots$

21. $6, 12, 36, 144, 720, \ldots$

22. $-3, -1, 2, 6, 11, \ldots$

In Exercises 23–26, write a recursive rule for the sequence shown in the graph.

23.
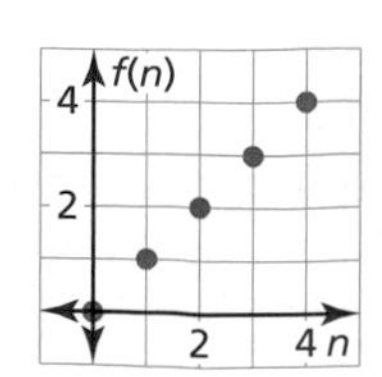

24.
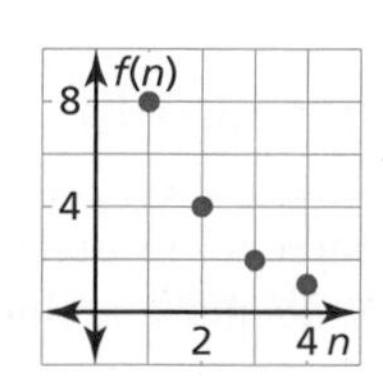

25.
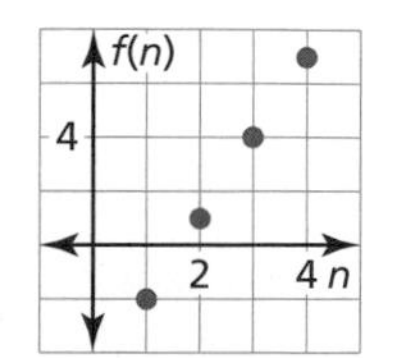

26.
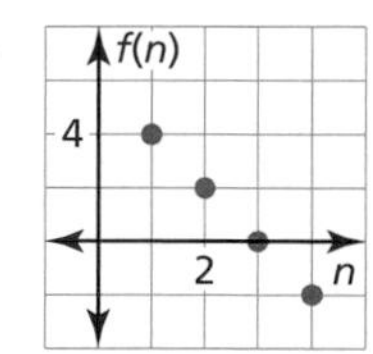

ERROR ANALYSIS In Exercises 27 and 28, describe and correct the error in writing a recursive rule for the sequence 5, 2, 3, −1, 4, ...

27.

Beginning with the third term in the sequence, each term a_n equals $a_{n-2} - a_{n-1}$. So, a recursive rule is given by

$$a_n = a_{n-2} - a_{n-1}.$$

28.

Beginning with the second term in the sequence, each term a_n equals $a_{n-1} - 3$. So, a recursive rule is given by

$$a_1 = 5, a_n = a_{n-1} - 3.$$

In Exercises 29–38, write a recursive rule for the sequence. *(See Example 4.)*

29. $a_n = 3 + 4n$

30. $a_n = -2 - 8n$

31. $a_n = 12 - 10n$

32. $a_n = 9 - 5n$

33. $a_n = 12(11)^{n-1}$

34. $a_n = -7(6)^{n-1}$

35. $a_n = 2.5 - 0.6n$

36. $a_n = -1.4 + 0.5n$

37. $a_n = -\frac{1}{2}\left(\frac{1}{4}\right)^{n-1}$

38. $a_n = \frac{1}{4}(5)^{n-1}$

39. REWRITING A FORMULA You have saved \$82 to buy a bicycle. You save an additional \$30 each month. The explicit rule $a_n = 30n + 82$ gives the amount saved after n months. Write a recursive rule for the amount you have saved n months from now.

40. REWRITING A FORMULA Your salary is given by the explicit rule $a_n = 35{,}000(1.04)^{n-1}$, where n is the number of years you have worked. Write a recursive rule for your salary.

In Exercises 41–48, write an explicit rule for the sequence. *(See Example 5.)*

41. $a_1 = 3, a_n = a_{n-1} - 6$ **42.** $a_1 = 16, a_n = a_{n-1} + 7$

43. $a_1 = -2, a_n = 3a_{n-1}$ **44.** $a_1 = 13, a_n = 4a_{n-1}$

45. $a_1 = -12, a_n = a_{n-1} + 9.1$

46. $a_1 = -4, a_n = 0.65a_{n-1}$

47. $a_1 = 5, a_n = a_{n-1} - \frac{1}{3}$ **48.** $a_1 = -5, a_n = \frac{1}{4}a_{n-1}$

49. REWRITING A FORMULA A grocery store arranges cans in a pyramid-shaped display with 20 cans in the bottom row and two fewer cans in each subsequent row going up. The number of cans in each row is represented by the recursive rule $a_1 = 20$, $a_n = a_{n-1} - 2$. Write an explicit rule for the number of cans in row n.

50. REWRITING A FORMULA The value of a car is given by the recursive rule $a_1 = 25{,}600$, $a_n = 0.86a_{n-1}$, where n is the number of years since the car was new. Write an explicit rule for the value of the car after n years.

51. USING STRUCTURE What is the 1000th term of the sequence whose first term is $a_1 = 4$ and whose nth term is $a_n = a_{n-1} + 6$? Justify your answer.

Ⓐ 4006 Ⓑ 5998

Ⓒ 1010 Ⓓ 10,000

52. USING STRUCTURE What is the 873rd term of the sequence whose first term is $a_1 = 0.01$ and whose nth term is $a_n = 1.01a_{n-1}$? Justify your answer.

Ⓐ 58.65 Ⓑ 8.73

Ⓒ 1.08 Ⓓ 586,459.38

53. PROBLEM SOLVING An online music service initially has 50,000 members. Each year, the company loses 20% of its current members and gains 5000 new members. *(See Example 6.)*

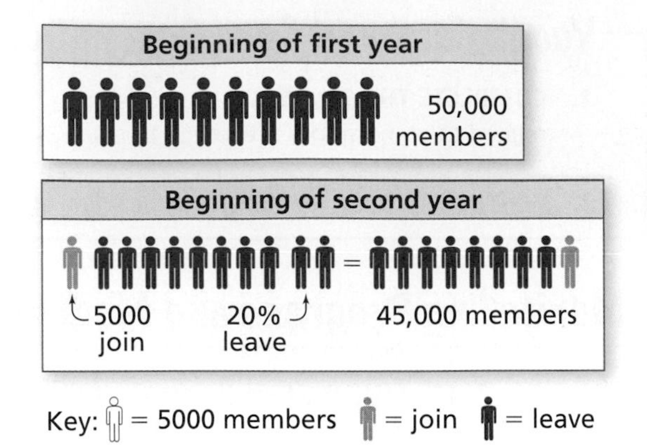

a. Write a recursive rule for the number a_n of members at the start of the nth year.

b. Find the number of members at the start of the fifth year.

c. Describe what happens to the number of members over time.

54. PROBLEM SOLVING You add chlorine to a swimming pool. You add 34 ounces of chlorine the first week and 16 ounces every week thereafter. Each week, 40% of the chlorine in the pool evaporates.

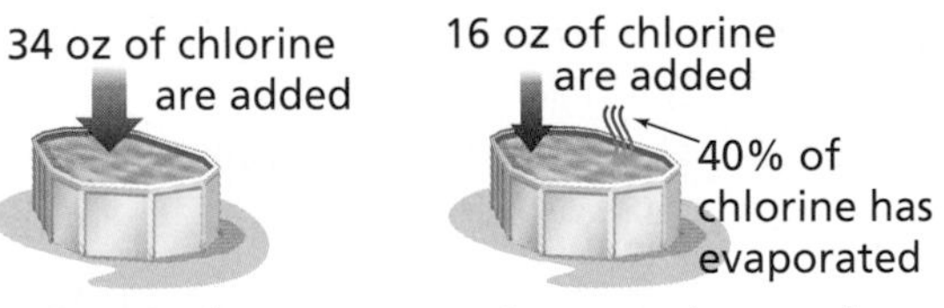

a. Write a recursive rule for the amount of chlorine in the pool at the start of the nth week.

b. Find the amount of chlorine in the pool at the start of the third week.

c. Describe what happens to the amount of chlorine in the pool over time.

55. OPEN-ENDED Give an example of a real-life situation which you can represent with a recursive rule that does not approach a limit. Write a recursive rule that represents the situation.

56. OPEN-ENDED Give an example of a sequence in which each term after the third term is a function of the three terms preceding it. Write a recursive rule for the sequence and find its first eight terms.

57. MODELING WITH MATHEMATICS You borrow \$2000 at 9% annual interest compounded monthly for 2 years. The monthly payment is \$91.37. *(See Example 7.)*

a. Find the balance after the fifth payment.

b. Find the amount of the last payment.

58. MODELING WITH MATHEMATICS You borrow \$10,000 to build an extra bedroom onto your house. The loan is secured for 7 years at an annual interest rate of 11.5%. The monthly payment is \$173.86.

a. Find the balance after the fourth payment.

b. Find the amount of the last payment.

59. COMPARING METHODS In 1202, the mathematician Leonardo Fibonacci wrote *Liber Abaci,* in which he proposed the following rabbit problem:

Begin with a pair of newborn rabbits. When a pair of rabbits is two months old, the rabbits begin producing a new pair of rabbits each month. Assume none of the rabbits die.

Month	1	2	3	4	5	6
Pairs at start of month	1	1	2	3	5	8

This problem produces a sequence called the Fibonacci sequence, which has both a recursive formula and an explicit formula as follows.

Recursive: $a_1 = 1, a_2 = 1, a_n = a_{n-2} + a_{n-1}$

Explicit: $f_n = \frac{1}{\sqrt{5}}\left(\frac{1+\sqrt{5}}{2}\right)^n - \frac{1}{\sqrt{5}}\left(\frac{1-\sqrt{5}}{2}\right)^n, n \geq 1$

Use each formula to determine how many rabbits there will be after one year. Justify your answers.

60. USING TOOLS A town library initially has 54,000 books in its collection. Each year, 2% of the books are lost or discarded. The library can afford to purchase 1150 new books each year.

a. Write a recursive rule for the number a_n of books in the library at the beginning of the nth year.

b. Use the *sequence* mode and the *dot* mode of a graphing calculator to graph the sequence. What happens to the number of books in the library over time? Explain.

61. DRAWING CONCLUSIONS A tree farm initially has 9000 trees. Each year, 10% of the trees are harvested and 800 seedlings are planted.

a. Write a recursive rule for the number of trees on the tree farm at the beginning of the nth year.

b. What happens to the number of trees after an extended period of time?

62. DRAWING CONCLUSIONS You sprain your ankle and your doctor prescribes 325 milligrams of an anti-inflammatory drug every 8 hours for 10 days. Sixty percent of the drug is removed from the bloodstream every 8 hours.

a. Write a recursive rule for the amount of the drug in the bloodstream after n doses.

b. The value that a drug level approaches after an extended period of time is called the *maintenance level*. What is the maintenance level of this drug given the prescribed dosage?

c. How does doubling the dosage affect the maintenance level of the drug? Justify your answer.

63. FINDING A PATTERN A fractal tree starts with a single branch (the trunk). At each stage, each new branch from the previous stage grows two more branches, as shown.

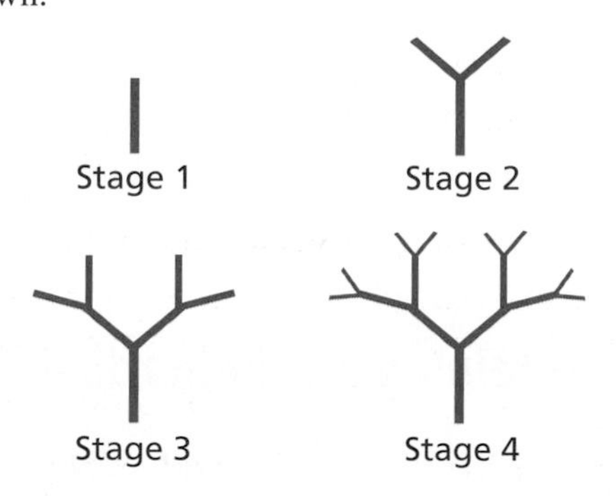

a. List the number of new branches in each of the first seven stages. What type of sequence do these numbers form?

b. Write an explicit rule and a recursive rule for the sequence in part (a).

64. **THOUGHT PROVOKING** Let $a_1 = 34$. Then write the terms of the sequence until you discover a pattern.

$$a_{n+1} = \begin{cases} \frac{1}{2}a_n, & \text{if } a_n \text{ is even} \\ 3a_n + 1, & \text{if } a_n \text{ is odd} \end{cases}$$

Do the same for $a_1 = 25$. What can you conclude?

65. **MODELING WITH MATHEMATICS** You make a \$500 down payment on a \$3500 diamond ring. You borrow the remaining balance at 10% annual interest compounded monthly. The monthly payment is \$213.59. How long does it take to pay back the loan? What is the amount of the last payment? Justify your answers.

66. **HOW DO YOU SEE IT?** The graph shows the first six terms of the sequence $a_1 = p$, $a_n = ra_{n-1}$.

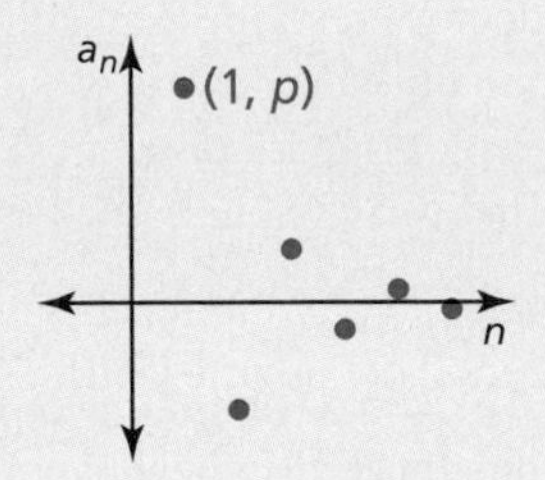

a. Describe what happens to the values in the sequence as n increases.

b. Describe the set of possible values for r. Explain your reasoning.

67. **REASONING** The rule for a recursive sequence is as follows.

$$f(1) = 3, f(2) = 10$$
$$f(n) = 4 + 2f(n-1) - f(n-2)$$

a. Write the first five terms of the sequence.

b. Use finite differences to find a pattern. What type of relationship do the terms of the sequence show?

c. Write an explicit rule for the sequence.

68. **MAKING AN ARGUMENT** Your friend says it is impossible to write a recursive rule for a sequence that is neither arithmetic nor geometric. Is your friend correct? Justify your answer.

69. **CRITICAL THINKING** The first four triangular numbers T_n and the first four square numbers S_n are represented by the points in each diagram.

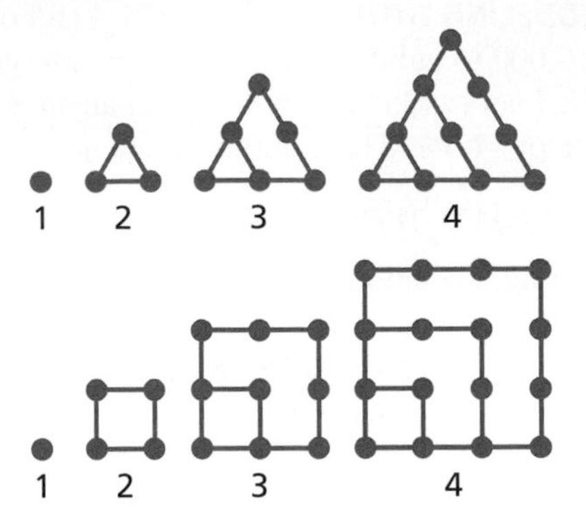

a. Write an explicit rule for each sequence.

b. Write a recursive rule for each sequence.

c. Write a rule for the square numbers in terms of the triangular numbers. Draw diagrams to explain why this rule is true.

70. **CRITICAL THINKING** You are saving money for retirement. You plan to withdraw \$30,000 at the beginning of each year for 20 years after you retire. Based on the type of investment you are making, you can expect to earn an annual return of 8% on your savings after you retire.

a. Let a_n be your balance n years after retiring. Write a recursive equation that shows how a_n is related to a_{n-1}.

b. Solve the equation from part (a) for a_{n-1}. Find a_0, the minimum amount of money you should have in your account when you retire. (*Hint:* Let $a_{20} = 0$.)

Maintaining Mathematical Proficiency

Reviewing what you learned in previous grades and lessons

Solve the equation. Check your solution. *(Section 5.4)*

71. $\sqrt{x} + 2 = 7$

72. $2\sqrt{x} - 5 = 15$

73. $\sqrt[3]{x} + 16 = 19$

74. $2\sqrt[3]{x} - 13 = -5$

The variables x and y vary inversely. Use the given values to write an equation relating x and y. Then find y when $x = 4$. *(Section 7.1)*

75. $x = 2, y = 9$

76. $x = -4, y = 3$

77. $x = 10, y = 32$

8.4–8.5 What Did You Learn?

Core Vocabulary

partial sum, *p. 436*
explicit rule, *p. 442*
recursive rule, *p. 442*

Core Concepts

Section 8.4

Partial Sums of Infinite Geometric Series, *p. 436*
The Sum of an Infinite Geometric Series, *p. 437*

Section 8.5

Evaluating Recursive Rules, *p. 442*
Recursive Equations for Arithmetic and Geometric Sequences, *p. 442*
Translating Between Recursive and Explicit Rules, *p. 444*

Mathematical Practices

1. Describe how labeling the axes in Exercises 3–6 on page 439 clarifies the relationship between the quantities in the problems.
2. What logical progression of arguments can you use to determine whether the statement in Exercise 30 on page 440 is true?
3. Describe how the structure of the equation presented in Exercise 40 on page 448 allows you to determine the starting salary and the raise you receive each year.
4. Does the recursive rule in Exercise 61 on page 449 make sense when $n = 5$? Explain your reasoning.

Performance Task

Integrated Circuits and Moore's Law

In April of 1965, an engineer named Gordon Moore noticed how quickly the size of electronics was shrinking. He predicted how the number of transistors that could fit on a 1-inch diameter circuit would increase over time. In 1965, only 50 transistors fit on the circuit. A decade later, about 65,000 transistors could fit on the circuit. Moore's prediction was accurate and is now known as Moore's Law. What was his prediction? How many transistors will be able to fit on a 1-inch circuit when you graduate from high school?

To explore the answers to this question and more, go to ***BigIdeasMath.com***.

8 Chapter Review

Dynamic Solutions available at *BigIdeasMath.com*

8.1 Defining and Using Sequences and Series (pp. 409–416)

Find the sum $\sum_{i=1}^{4} (i^2 - 3)$.

$$\sum_{i=1}^{4} (i^2 - 3) = (1^2 - 3) + (2^2 - 3) + (3^2 - 3) + (4^2 - 3)$$

$$= -2 + 1 + 6 + 13$$

$$= 18$$

1. Describe the pattern shown in the figure. Then write a rule for the nth layer of the figure, where $n = 1$ represents the top layer.

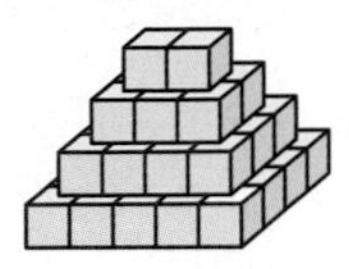

Write the series using summation notation.

2. $7 + 10 + 13 + \cdots + 40$

3. $0 + 2 + 6 + 12 + \cdots$

Find the sum.

4. $\sum_{i=2}^{7} (9 - i^3)$

5. $\sum_{i=1}^{46} i$

6. $\sum_{i=1}^{12} i^2$

7. $\sum_{i=1}^{5} \frac{3 + i}{2}$

8.2 Analyzing Arithmetic Sequences and Series (pp. 417–424)

Write a rule for the nth term of the sequence 9, 14, 19, 24, ... Then find a_{14}.

The sequence is arithmetic with first term $a_1 = 9$ and common difference $d = 14 - 9 = 5$. So, a rule for the nth term is

$a_n = a_1 + (n - 1)d$	Write general rule.
$= 9 + (n - 1)5$	Substitute 9 for a_1 and 5 for d.
$= 5n + 4.$	Simplify.

▶ A rule is $a_n = 5n + 4$, and the 14th term is $a_{14} = 5(14) + 4 = 74$.

8. Tell whether the sequence $12, 4, -4, -12, -20, \ldots$ is arithmetic. Explain your reasoning.

Write a rule for the nth term of the arithmetic sequence. Then graph the first six terms of the sequence.

9. 2, 8, 14, 20, . . .

10. $a_{14} = 42, d = 3$

11. $a_6 = -12, a_{12} = -36$

12. Find the sum $\sum_{i=1}^{36} (2 + 3i)$.

13. You take a job with a starting salary of $37,000. Your employer offers you an annual raise of $1500 for the next 6 years. Write a rule for your salary in the nth year. What are your total earnings in 6 years?

8.3 Analyzing Geometric Sequences and Series *(pp. 425–432)*

Find the sum $\sum_{i=1}^{8} 6(3)^{i-1}$.

Step 1 Find the first term and the common ratio.

$a_1 = 6(3)^{1-1} = 6$ Identify first term.

$r = 3$ Identify common ratio.

Step 2 Find the sum.

$$S_8 = a_1\left(\frac{1 - r^8}{1 - r}\right)$$ Write rule for S_8.

$$= 6\left(\frac{1 - 3^8}{1 - 3}\right)$$ Substitute 6 for a_1 and 3 for r.

$$= 19{,}680$$ Simplify.

14. Tell whether the sequence 7, 14, 28, 56, 112, . . . is geometric. Explain your reasoning.

Write a rule for the *n*th term of the geometric sequence. Then graph the first six terms of the sequence.

15. $25, 10, 4, \frac{8}{5}, \ldots$

16. $a_5 = 162, r = -3$

17. $a_3 = 16, a_5 = 256$

18. Find the sum $\sum_{i=1}^{9} 5(-2)^{i-1}$.

8.4 Finding Sums of Infinite Geometric Series *(pp. 435–440)*

Find the sum of the series $\sum_{i=1}^{\infty} \left(\frac{4}{5}\right)^{i-1}$, if it exists.

For this series, $a_1 = 1$ and $r = \frac{4}{5}$. Because $\left|\frac{4}{5}\right| < 1$, the sum of the series exists.

The sum of the series is

$$S = \frac{a_1}{1 - r}$$ Formula for the sum of an infinite geometric series

$$= \frac{1}{1 - \frac{4}{5}}$$ Substitute 1 for a_1 and $\frac{4}{5}$ for r.

$$= 5.$$ Simplify.

19. Consider the infinite geometric series $1, -\frac{1}{4}, \frac{1}{16}, -\frac{1}{64}, \frac{1}{256}, \ldots$. Find and graph the partial sums S_n for $n = 1, 2, 3, 4$, and 5. Then describe what happens to S_n as n increases.

20. Find the sum of the infinite geometric series $-2 + \frac{1}{2} - \frac{1}{8} + \frac{1}{32} + \cdots$, if it exists.

21. Write the repeating decimal 0.1212 . . . as a fraction in simplest form.

8.5 Using Recursive Rules with Sequences (pp. 441–450)

a. Write the first six terms of the sequence $a_0 = 46, a_n = a_{n-1} - 8$.

$a_0 = 46$	1st term
$a_1 = a_0 - 8 = 46 - 8 = 38$	2nd term
$a_2 = a_1 - 8 = 38 - 8 = 30$	3rd term
$a_3 = a_2 - 8 = 30 - 8 = 22$	4th term
$a_4 = a_3 - 8 = 22 - 8 = 14$	5th term
$a_5 = a_4 - 8 = 14 - 8 = 6$	6th term

b. Write a recursive rule for the sequence 6, 10, 14, 18, 22, ...

Use a table to organize the terms and find the pattern.

n	1	2	3	4	5
a_n	6	10	14	18	22

+4 +4 +4 +4

The sequence is arithmetic with the first term $a_1 = 6$ and common difference $d = 4$.

$a_n = a_{n-1} + d$ Recursive equation for arithmetic sequence

$= a_{n-1} + 4$ Substitute 4 for d.

▶ A recursive rule for the sequence is $a_1 = 6, a_n = a_{n-1} + 4$.

Write the first six terms of the sequence.

22. $a_1 = 7, a_n = a_{n-1} + 11$ **23.** $a_1 = 6, a_n = 4a_{n-1}$ **24.** $f(0) = 4, f(n) = f(n-1) + 2n$

Write a recursive rule for the sequence.

25. $9, 6, 4, \frac{8}{3}, \frac{16}{9}, \ldots$ **26.** 2, 2, 4, 12, 48, . . . **27.** 7, 3, 4, −1, 5, . . .

28. Write a recursive rule for $a_n = 105\left(\frac{3}{5}\right)^{n-1}$.

Write an explicit rule for the sequence.

29. $a_1 = -4, a_n = a_{n-1} + 26$ **30.** $a_1 = 8, a_n = -5a_{n-1}$ **31.** $a_1 = 26, a_n = \frac{2}{5}a_{n-1}$

32. A town's population increases at a rate of about 4% per year. In 2010, the town had a population of 11,120. Write a recursive rule for the population P_n of the town in year n. Let $n = 1$ represent 2010.

33. The numbers 1, 6, 15, 28, . . . are called hexagonal numbers because they represent the number of dots used to make hexagons, as shown. Write a recursive rule for the nth hexagonal number.

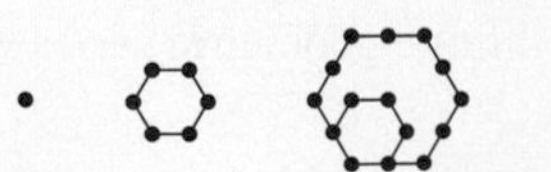
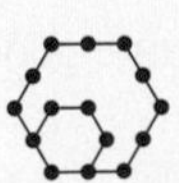
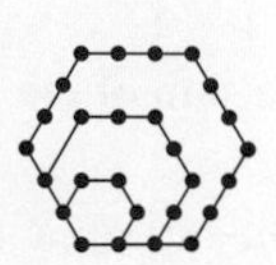

8 Chapter Test

Find the sum.

1. $\sum_{i=1}^{24}(6i - 13)$

2. $\sum_{n=1}^{16} n^2$

3. $\sum_{k=1}^{\infty} 2(0.8)^{k-1}$

4. $\sum_{i=1}^{6} 4(-3)^{i-1}$

Determine whether the graph represents an arithmetic sequence, geometric sequence, or neither. Explain your reasoning. Then write a rule for the *n*th term.

5.

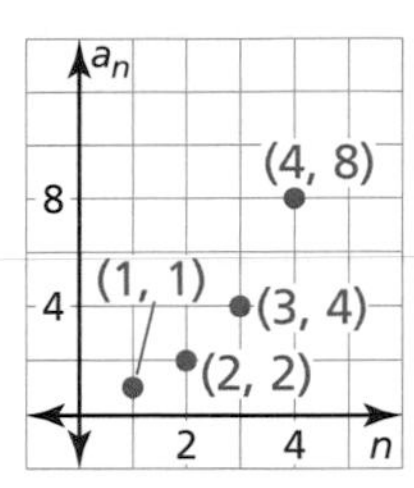

6.

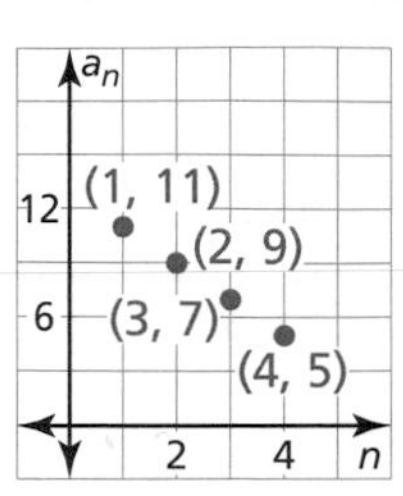

7.

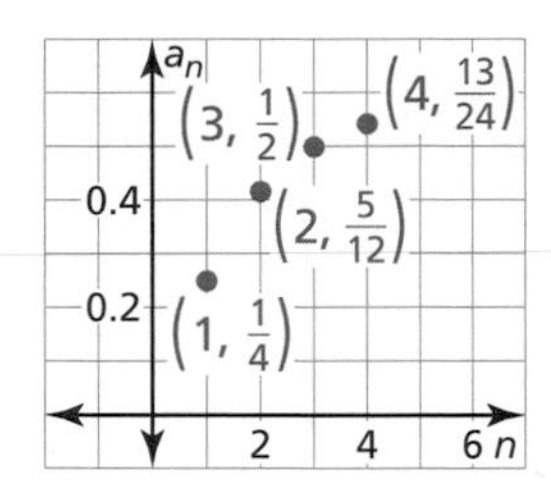

Write a recursive rule for the sequence. Then find a_9.

8. $a_1 = 32, r = \frac{1}{2}$

9. $a_n = 2 + 7n$

10. 2, 0, −3, −7, −12, . . .

11. Write a recursive rule for the sequence 5, −20, 80, −320, 1280, ... Then write an explicit rule for the sequence using your recursive rule.

12. The numbers a, b, and c are the first three terms of an arithmetic sequence. Is b half of the sum of a and c? Explain your reasoning.

13. Use the pattern of checkerboard quilts shown.

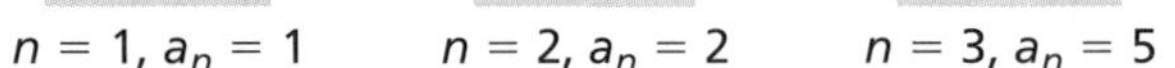

a. What does n represent for each quilt? What does a_n represent?

b. Make a table that shows n and a_n for $n = 1, 2, 3, 4, 5, 6, 7$, and 8.

c. Use the rule $a_n = \frac{n^2}{2} + \frac{1}{4}[1 - (-1)^n]$ to find a_n for $n = 1, 2, 3, 4, 5, 6, 7$, and 8. Compare these values to those in your table in part (b). What can you conclude? Explain.

14. During a baseball season, a company pledges to donate \$5000 to a charity plus \$100 for each home run hit by the local team. Does this situation represent a sequence or a series? Explain your reasoning.

15. The length ℓ_1 of the first loop of a spring is 16 inches. The length ℓ_2 of the second loop is 0.9 times the length of the first loop. The length ℓ_3 of the third loop is 0.9 times the length of the second loop, and so on. Suppose the spring has infinitely many loops, would its length be finite or infinite? Explain. Find the length of the spring, if possible.

$\ell_1 = 16$ in.

$\ell_2 = 16(0.9)$ in.

$\ell_3 = 16(0.9)^2$ in.

8 Cumulative Assessment

1. The frequencies (in hertz) of the notes on a piano form a geometric sequence. The frequencies of G (labeled 8) and A (labeled 10) are shown in the diagram. What is the approximate frequency of E flat (labeled 4)?

Ⓐ 247 Hz

Ⓑ 311 Hz

Ⓒ 330 Hz

Ⓓ 554 Hz

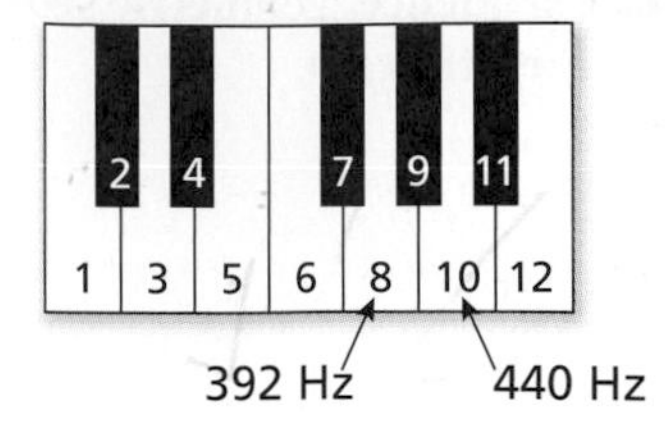

2. You take out a loan for \$16,000 with an interest rate of 0.75% per month. At the end of each month, you make a payment of \$300.

a. Write a recursive rule for the balance a_n of the loan at the beginning of the nth month.

b. How much do you owe at the beginning of the 18th month?

c. How long will it take to pay off the loan?

d. If you pay \$350 instead of \$300 each month, how long will it take to pay off the loan? How much money will you save? Explain.

3. The table shows that the force F (in pounds) needed to loosen a certain bolt with a wrench depends on the length ℓ (in inches) of the wrench's handle. Write an equation that relates ℓ and F. Describe the relationship.

Length, ℓ	4	6	10	12
Force, F	375	250	150	125

4. Order the functions from the least average rate of change to the greatest average rate of change on the interval $1 \le x \le 4$. Justify your answers.

A. $f(x) = 4\sqrt{x} + 2$

B. x and y vary inversely, and $y = 2$ when $x = 5$.

C.

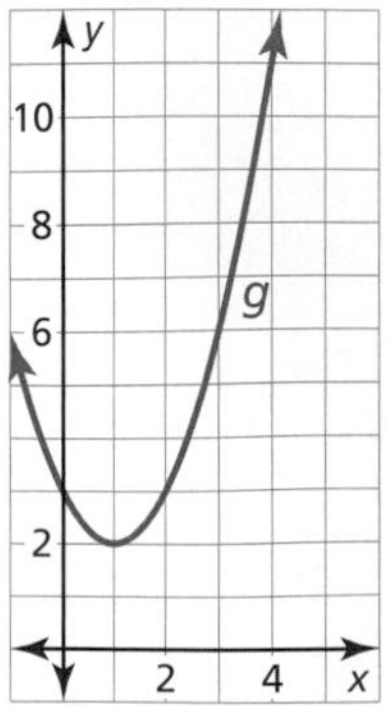

D.

x	y
1	−4
2	−1
3	2
4	5

5. A running track is shaped like a rectangle with two semicircular ends, as shown. The track has 8 lanes that are each 1.22 meters wide. The lanes are numbered from 1 to 8 starting from the inside lane. The distance from the center of a semicircle to the inside of a lane is called the curve radius of that lane. The curve radius of lane 1 is 36.5 meters, as shown in the figure.

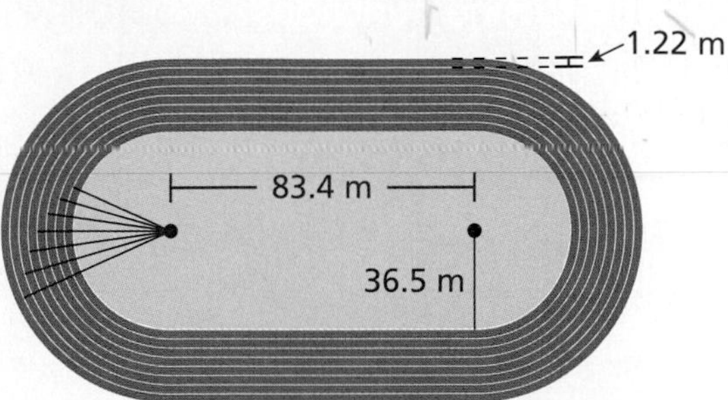

a. Is the sequence formed by the curve radii arithmetic, geometric, or neither? Explain.

b. Write a rule for the sequence formed by the curve radii.

c. World records must be set on tracks that have a curve radius of at most 50 meters in the outside lane. Does the track shown meet the requirement? Explain.

6. The diagram shows the bounce heights of a basketball and a baseball dropped from a height of 10 feet. On each bounce, the basketball bounces to 36% of its previous height, and the baseball bounces to 30% of its previous height. About how much greater is the total distance traveled by the basketball than the total distance traveled by the baseball?

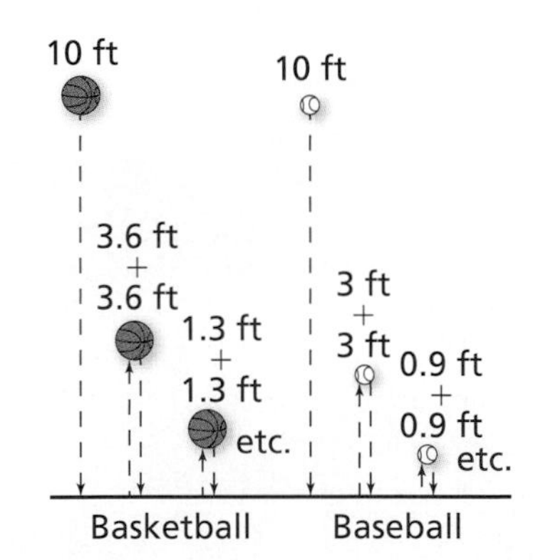

Ⓐ 1.34 feet   Ⓑ 2.00 feet

Ⓒ 2.68 feet   Ⓓ 5.63 feet

7. Classify the solution(s) of each equation as real numbers, imaginary numbers, or pure imaginary numbers. Justify your answers.

a. $x + \sqrt{-16} = 0$ b. $(11 - 2i) - (-3i + 6) = 8 + x$ c. $3x^2 - 14 = -20$

d. $x^2 + 2x = -3$ e. $x^2 = 16$ f. $x^2 - 5x - 8 = 0$

9 Trigonometric Ratios and Functions

Sundial *(p. 518)*

Tuning Fork *(p. 510)*

Ferris Wheel *(p. 494)*

Terminator *(p. 476)*

Parasailing *(p. 465)*

Maintaining Mathematical Proficiency

Absolute Value

Example 1 **Order the expressions by value from least to greatest:** $|6|, |-3|, \frac{2}{|-4|}, |10-6|$

$|6| = 6$ $\qquad$ $|-3| = 3$ ← The absolute value of a negative number is positive.

$\frac{2}{|-4|} = \frac{2}{4} = \frac{1}{2}$ $\qquad$ $|10-6| = |4| = 4$

▶ So, the order is $\frac{2}{|-4|}$, $|-3|$, $|10-6|$, and $|6|$.

Order the expressions by value from least to greatest.

1. $|4|, |2-9|, |6+4|, -|7|$

2. $|9-3|, |0|, |-4|, \frac{|-5|}{|2|}$

3. $|-8^3|, |-2 \cdot 8|, |9-1|, |9| + |-2| - |1|$

4. $|-4+20|, -|4^2|, |5| - |3 \cdot 2|, |-15|$

Pythagorean Theorem

Example 2 **Find the missing side length of the triangle.**

10 cm, 26 cm, b

$a^2 + b^2 = c^2$	Write the Pythagorean Theorem.
$10^2 + b^2 = 26^2$	Substitute 10 for a and 26 for c.
$100 + b^2 = 676$	Evaluate powers.
$b^2 = 576$	Subtract 100 from each side.
$b = 24$	Take positive square root of each side.

▶ So, the length is 24 centimeters.

Find the missing side length of the triangle.

5.

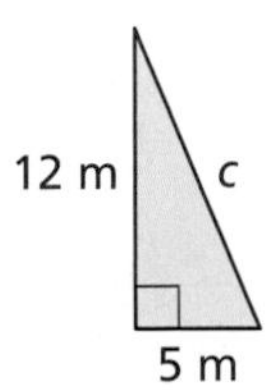

6.

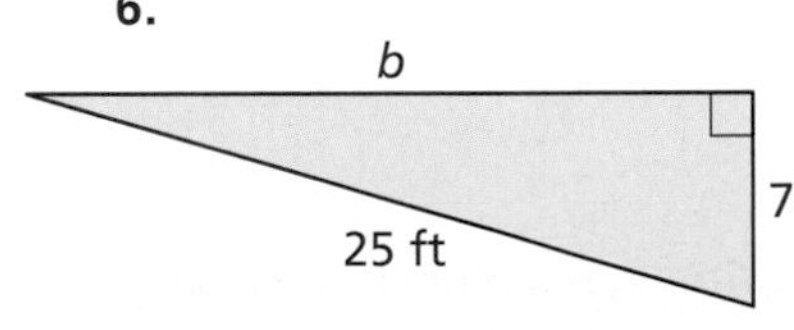

7.

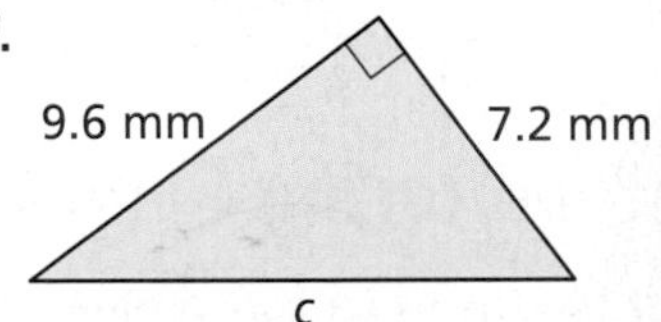

8.

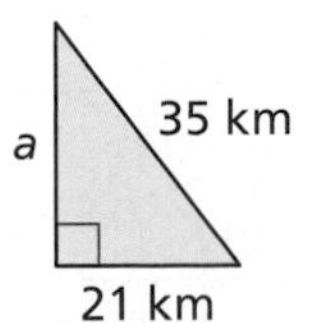

9.

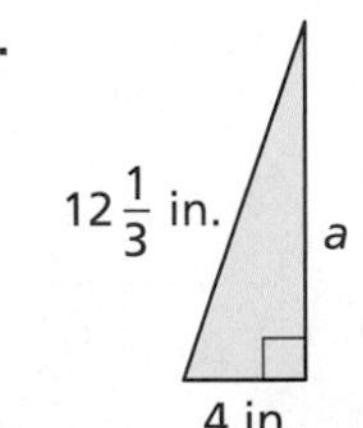

10.

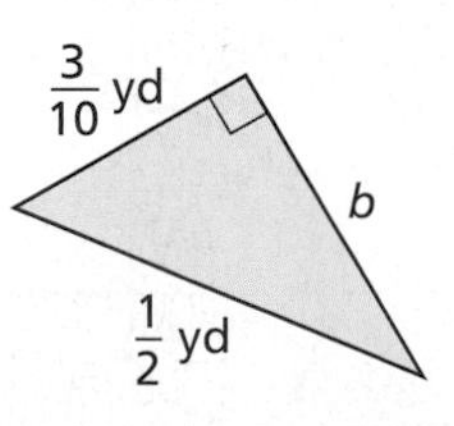

11. ABSTRACT REASONING The line segments connecting the points (x_1, y_1), (x_2, y_1), and (x_2, y_2) form a triangle. Is the triangle a right triangle? Justify your answer.

Mathematical Practices

Mathematically proficient students reason quantitatively by creating valid representations of problems.

Reasoning Abstractly and Quantitatively

Core Concept

The Unit Circle

The *unit circle* is a circle in the coordinate plane. Its center is at the origin, and it has a radius of 1 unit. The equation of the unit circle is

$x^2 + y^2 = 1.$ Equation of unit circle

As the point (x, y) starts at $(1, 0)$ and moves counterclockwise around the unit circle, the angle θ (the Greek letter *theta*) moves from 0° through 360°.

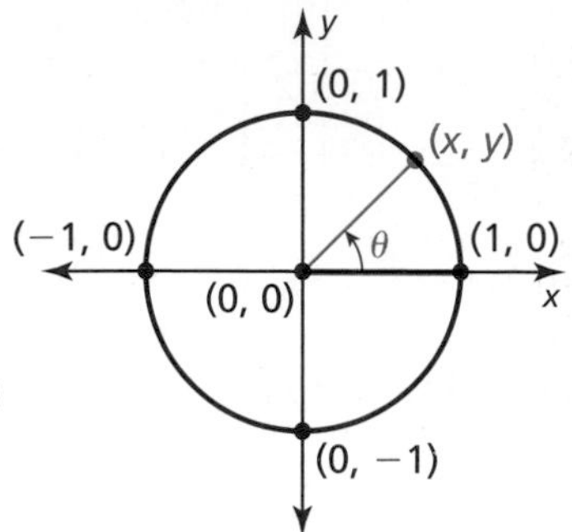

EXAMPLE 1 **Finding Coordinates of a Point on the Unit Circle**

Find the exact coordinates of the point (x, y) on the unit circle.

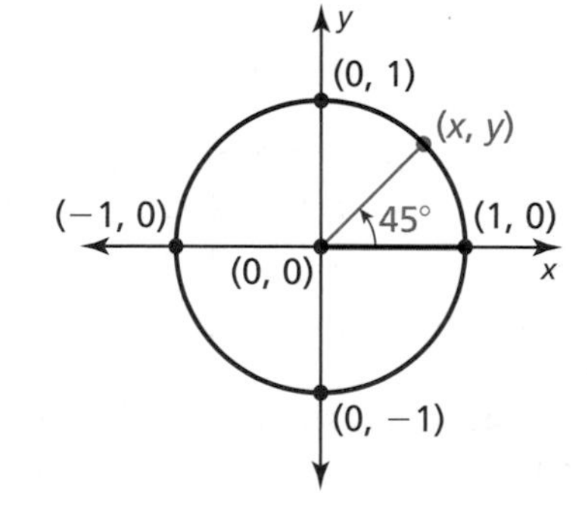

SOLUTION

Because $\theta = 45^\circ$, (x, y) lies on the line $y = x$.

$x^2 + y^2 = 1$ Write equation of unit circle.

$x^2 + x^2 = 1$ Substitute x for y.

$2x^2 = 1$ Add like terms.

$x^2 = \frac{1}{2}$ Divide each side by 2.

$x = \frac{1}{\sqrt{2}}$ Take positive square root of each side.

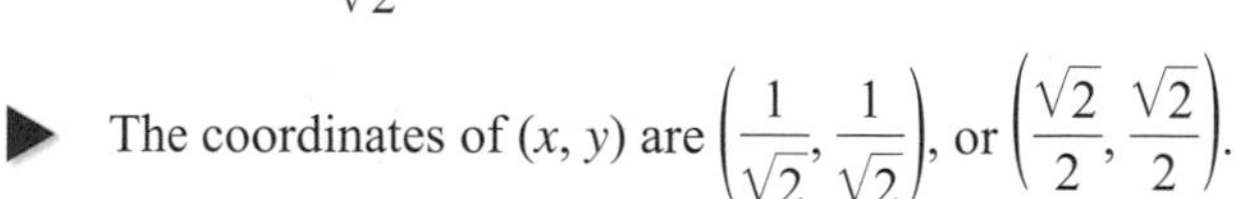

▶ The coordinates of (x, y) are $\left(\frac{1}{\sqrt{2}}, \frac{1}{\sqrt{2}}\right)$, or $\left(\frac{\sqrt{2}}{2}, \frac{\sqrt{2}}{2}\right)$.

Monitoring Progress

Find the exact coordinates of the point (x, y) on the unit circle.

1.

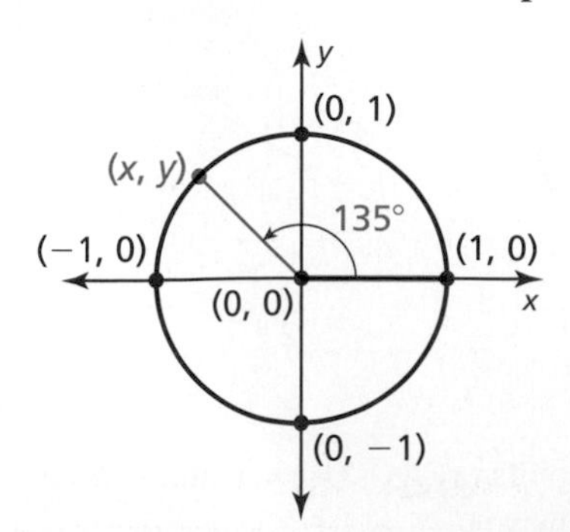

2.

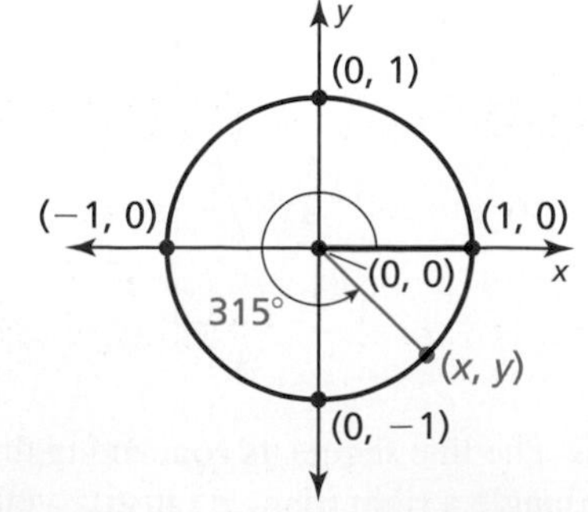

3.

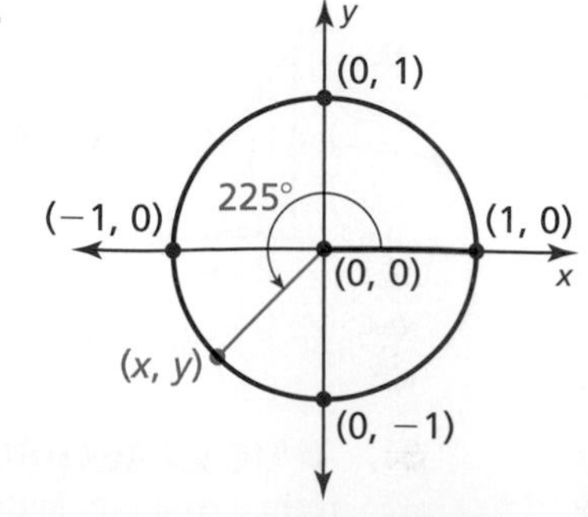

9.1 Right Triangle Trigonometry

Essential Question

How can you find a trigonometric function of an acute angle θ?

Consider one of the acute angles θ of a right triangle. Ratios of a right triangle's side lengths are used to define the six *trigonometric functions*, as shown.

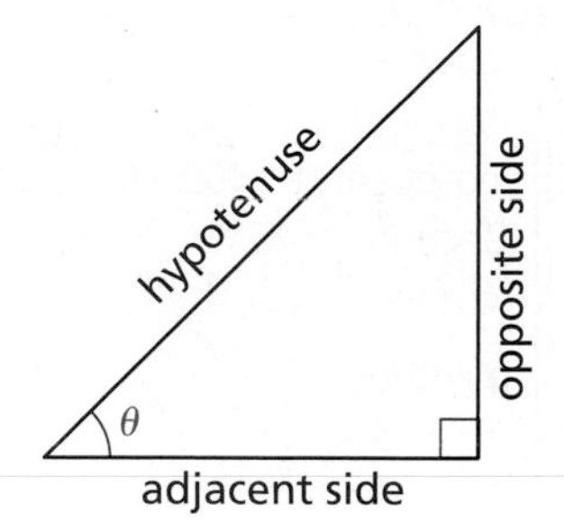

Sine $\sin \theta = \dfrac{\text{opp.}}{\text{hyp.}}$ **Cosine** $\cos \theta = \dfrac{\text{adj.}}{\text{hyp.}}$

Tangent $\tan \theta = \dfrac{\text{opp.}}{\text{adj.}}$ **Cotangent** $\cot \theta = \dfrac{\text{adj.}}{\text{opp.}}$

Secant $\sec \theta = \dfrac{\text{hyp.}}{\text{adj.}}$ **Cosecant** $\csc \theta = \dfrac{\text{hyp.}}{\text{opp.}}$

EXPLORATION 1 Trigonometric Functions of Special Angles

Work with a partner. Find the exact values of the sine, cosine, and tangent functions for the angles 30°, 45°, and 60° in the right triangles shown.

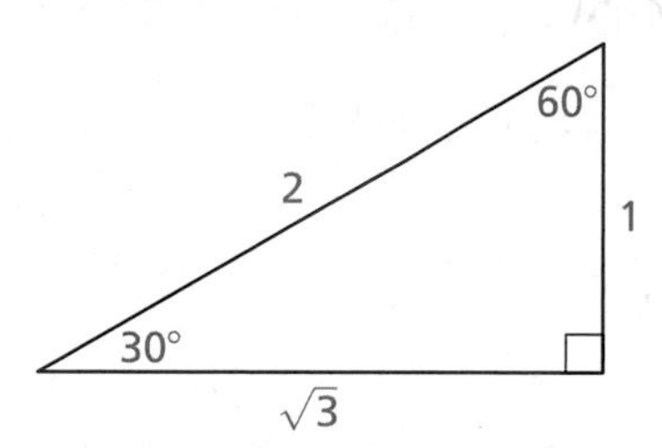

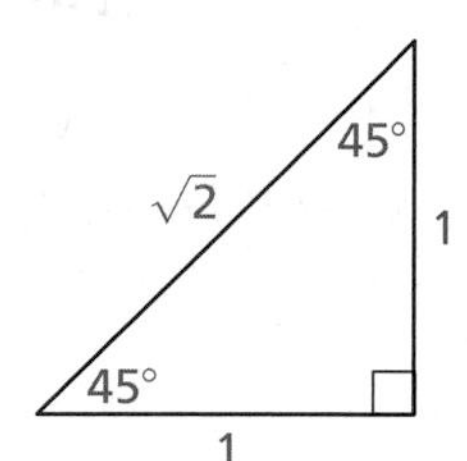

CONSTRUCTING VIABLE ARGUMENTS

To be proficient in math, you need to understand and use stated assumptions, definitions, and previously established results in constructing arguments.

EXPLORATION 2 Exploring Trigonometric Identities

Work with a partner.

Use the definitions of the trigonometric functions to explain why each *trigonometric identity* is true.

a. $\sin \theta = \cos(90° - \theta)$

b. $\cos \theta = \sin(90° - \theta)$

c. $\sin \theta = \dfrac{1}{\csc \theta}$

d. $\tan \theta = \dfrac{1}{\cot \theta}$

Use the definitions of the trigonometric functions to complete each trigonometric identity.

e. $(\sin \theta)^2 + (\cos \theta)^2 =$ ______

f. $(\sec \theta)^2 - (\tan \theta)^2 =$ ______

Communicate Your Answer

3. How can you find a trigonometric function of an acute angle θ?

4. Use a calculator to find the lengths x and y of the legs of the right triangle shown.

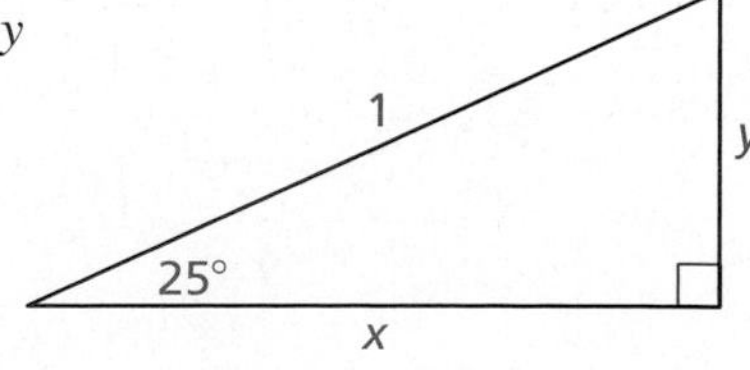

9.1 Lesson

What You Will Learn

- Evaluate trigonometric functions of acute angles.
- Find unknown side lengths and angle measures of right triangles.
- Use trigonometric functions to solve real-life problems.

Core Vocabulary

sine, *p. 462*
cosine, *p. 462*
tangent, *p. 462*
cosecant, *p. 462*
secant, *p. 462*
cotangent, *p. 462*

Previous
right triangle
hypotenuse
acute angle
Pythagorean Theorem
reciprocal
complementary angles

The Six Trigonometric Functions

Consider a right triangle that has an acute angle θ (the Greek letter *theta*). The three sides of the triangle are the *hypotenuse*, the side *opposite* θ, and the side *adjacent* to θ.

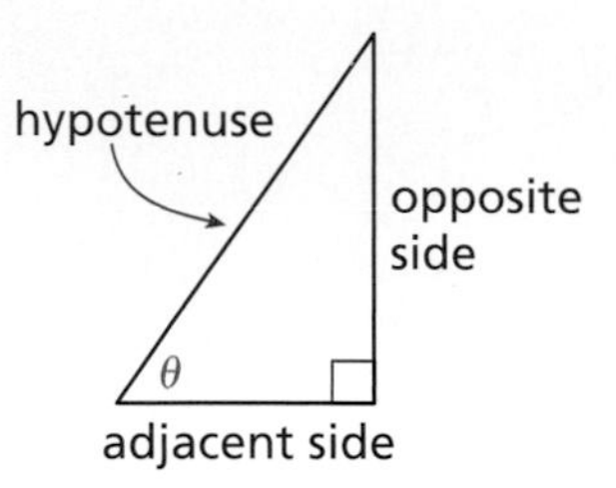

Ratios of a right triangle's side lengths are used to define the six trigonometric functions: **sine**, **cosine**, **tangent**, **cosecant**, **secant**, and **cotangent**. These six functions are abbreviated sin, cos, tan, csc, sec, and cot, respectively.

Core Concept

Right Triangle Definitions of Trigonometric Functions

Let θ be an acute angle of a right triangle. The six trigonometric functions of θ are defined as shown.

$$\sin\theta = \frac{\text{opposite}}{\text{hypotenuse}} \qquad \cos\theta = \frac{\text{adjacent}}{\text{hypotenuse}} \qquad \tan\theta = \frac{\text{opposite}}{\text{adjacent}}$$

$$\csc\theta = \frac{\text{hypotenuse}}{\text{opposite}} \qquad \sec\theta = \frac{\text{hypotenuse}}{\text{adjacent}} \qquad \cot\theta = \frac{\text{adjacent}}{\text{opposite}}$$

The abbreviations *opp.*, *adj.*, and *hyp.* are often used to represent the side lengths of the right triangle. Note that the ratios in the second row are reciprocals of the ratios in the first row.

$$\csc\theta = \frac{1}{\sin\theta} \qquad \sec\theta = \frac{1}{\cos\theta} \qquad \cot\theta = \frac{1}{\tan\theta}$$

REMEMBER

The Pythagorean Theorem states that $a^2 + b^2 = c^2$ for a right triangle with hypotenuse of length c and legs of lengths a and b.

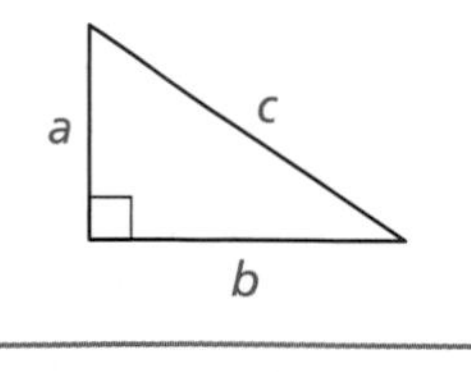

EXAMPLE 1 Evaluating Trigonometric Functions

Evaluate the six trigonometric functions of the angle θ.

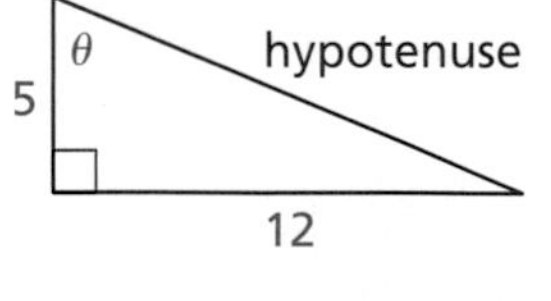

SOLUTION

From the Pythagorean Theorem, the length of the hypotenuse is

$$\begin{aligned}\text{hyp.} &= \sqrt{5^2 + 12^2}\\ &= \sqrt{169}\\ &= 13.\end{aligned}$$

Using adj. $= 5$, opp. $= 12$, and hyp. $= 13$, the values of the six trigonometric functions of θ are:

$$\sin\theta = \frac{\text{opp.}}{\text{hyp.}} = \frac{12}{13} \qquad \cos\theta = \frac{\text{adj.}}{\text{hyp.}} = \frac{5}{13} \qquad \tan\theta = \frac{\text{opp.}}{\text{adj.}} = \frac{12}{5}$$

$$\csc\theta = \frac{\text{hyp.}}{\text{opp.}} = \frac{13}{12} \qquad \sec\theta = \frac{\text{hyp.}}{\text{adj.}} = \frac{13}{5} \qquad \cot\theta = \frac{\text{adj.}}{\text{opp.}} = \frac{5}{12}$$

EXAMPLE 2 Evaluating Trigonometric Functions

In a right triangle, θ is an acute angle and $\sin \theta = \frac{4}{7}$. Evaluate the other five trigonometric functions of θ.

SOLUTION

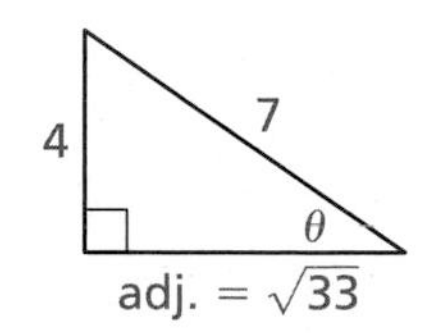

Step 1 Draw a right triangle with acute angle θ such that the leg opposite θ has length 4 and the hypotenuse has length 7.

Step 2 Find the length of the adjacent side. By the Pythagorean Theorem, the length of the other leg is

$$\text{adj.} = \sqrt{7^2 - 4^2} = \sqrt{33}.$$

Step 3 Find the values of the remaining five trigonometric functions.

Because $\sin \theta = \frac{4}{7}$, $\csc \theta = \frac{\text{hyp.}}{\text{opp.}} = \frac{7}{4}$. The other values are:

$$\cos \theta = \frac{\text{adj.}}{\text{hyp.}} = \frac{\sqrt{33}}{7} \qquad \tan \theta = \frac{\text{opp.}}{\text{adj.}} = \frac{4}{\sqrt{33}} = \frac{4\sqrt{33}}{33}$$

$$\sec \theta = \frac{\text{hyp.}}{\text{adj.}} = \frac{7}{\sqrt{33}} = \frac{7\sqrt{33}}{33} \qquad \cot \theta = \frac{\text{adj.}}{\text{opp.}} = \frac{\sqrt{33}}{4}$$

Monitoring Progress  Help in English and Spanish at *BigIdeasMath.com*

Evaluate the six trigonometric functions of the angle θ.

1.

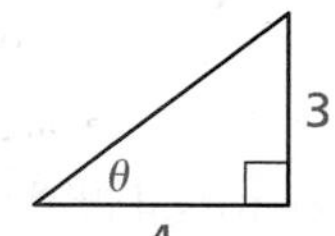

2.

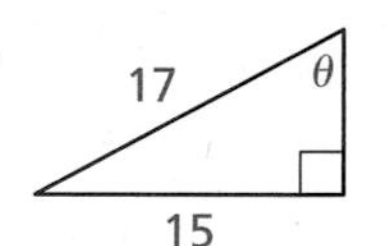

3.

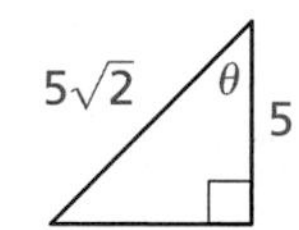

4. In a right triangle, θ is an acute angle and $\cos \theta = \frac{7}{10}$. Evaluate the other five trigonometric functions of θ.

The angles 30°, 45°, and 60° occur frequently in trigonometry. You can use the trigonometric values for these angles to find unknown side lengths in special right triangles.

Core Concept

Trigonometric Values for Special Angles

The table gives the values of the six trigonometric functions for the angles 30°, 45°, and 60°. You can obtain these values from the triangles shown.

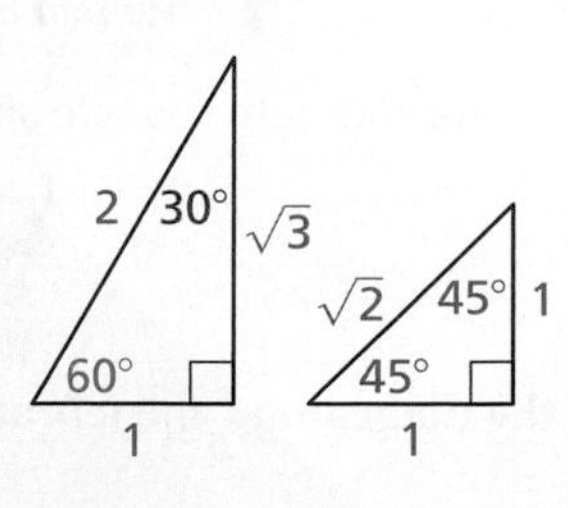

θ	$\sin \theta$	$\cos \theta$	$\tan \theta$	$\csc \theta$	$\sec \theta$	$\cot \theta$
30°	$\frac{1}{2}$	$\frac{\sqrt{3}}{2}$	$\frac{\sqrt{3}}{3}$	2	$\frac{2\sqrt{3}}{3}$	$\sqrt{3}$
45°	$\frac{\sqrt{2}}{2}$	$\frac{\sqrt{2}}{2}$	1	$\sqrt{2}$	$\sqrt{2}$	1
60°	$\frac{\sqrt{3}}{2}$	$\frac{1}{2}$	$\sqrt{3}$	$\frac{2\sqrt{3}}{3}$	2	$\frac{\sqrt{3}}{3}$

Finding Side Lengths and Angle Measures

EXAMPLE 3 Finding an Unknown Side Length

Find the value of x for the right triangle.

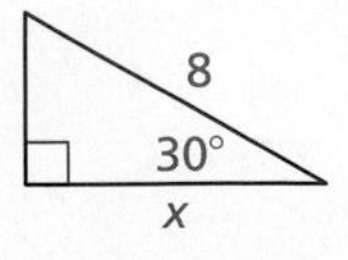

SOLUTION

Write an equation using a trigonometric function that involves the ratio of x and 8. Solve the equation for x.

$\cos 30° = \frac{\text{adj.}}{\text{hyp.}}$ Write trigonometric equation.

$\frac{\sqrt{3}}{2} = \frac{x}{8}$ Substitute.

$4\sqrt{3} = x$ Multiply each side by 8.

▶ The length of the side is $x = 4\sqrt{3} \approx 6.93$.

READING

Throughout this chapter, a capital letter is used to denote both an angle of a triangle and its measure. The same letter in lowercase is used to denote the length of the side opposite that angle.

Finding all unknown side lengths and angle measures of a triangle is called *solving the triangle*. Solving right triangles that have acute angles other than 30°, 45°, and 60° may require the use of a calculator. Be sure the calculator is set in *degree* mode.

EXAMPLE 4 Using a Calculator to Solve a Right Triangle

Solve $\triangle ABC$.

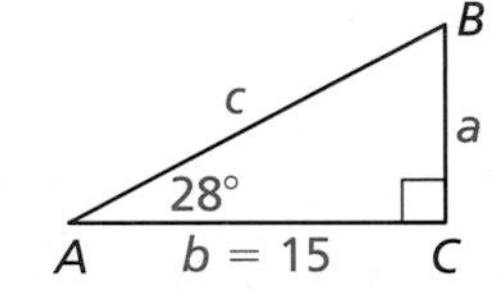

SOLUTION

Because the triangle is a right triangle, A and B are complementary angles. So, $B = 90° - 28° = 62°$.

Next, write two equations using trigonometric functions, one that involves the ratio of a and 15, and one that involves c and 15. Solve the first equation for a and the second equation for c.

$\tan 28° = \frac{\text{opp.}}{\text{adj.}}$	Write trigonometric equation.	$\sec 28° = \frac{\text{hyp.}}{\text{adj.}}$
$\tan 28° = \frac{a}{15}$	Substitute.	$\sec 28° = \frac{c}{15}$
$15(\tan 28°) = a$	Solve for the variable.	$15\left(\frac{1}{\cos 28°}\right) = c$
$7.98 \approx a$	Use a calculator.	$16.99 \approx c$

▶ So, $B = 62°$, $a \approx 7.98$, and $c \approx 16.99$.

Monitoring Progress

Help in English and Spanish at *BigIdeasMath.com*

5. Find the value of x for the right triangle shown.

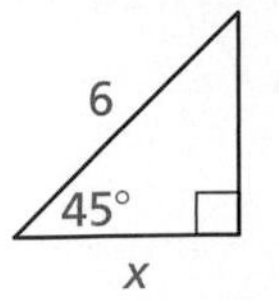

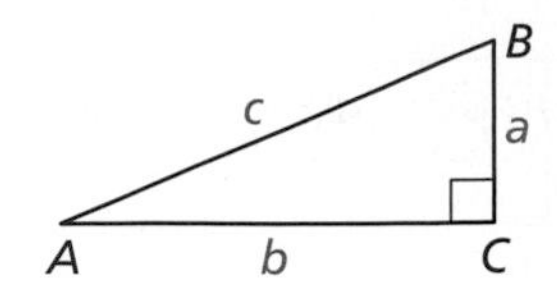

Solve $\triangle ABC$ using the diagram at the left and the given measurements.

6. $B = 45°$, $c = 5$

7. $A = 32°$, $b = 10$

8. $A = 71°$, $c = 20$

9. $B = 60°$, $a = 7$

Solving Real-Life Problems

EXAMPLE 5 Using Indirect Measurement

You are hiking near a canyon. While standing at A, you measure an angle of 90° between B and C, as shown. You then walk to B and measure an angle of 76° between A and C. The distance between A and B is about 2 miles. How wide is the canyon between A and C?

FINDING AN ENTRY POINT

The tangent function is used to find the unknown distance because it involves the ratio of x and 2.

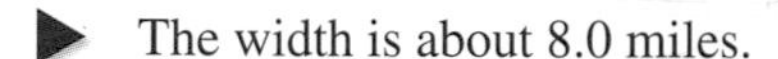

SOLUTION

$$\tan 76^\circ = \frac{x}{2}$$ Write trigonometric equation.

$$2(\tan 76^\circ) = x$$ Multiply each side by 2.

$$8.0 \approx x$$ Use a calculator.

▶ The width is about 8.0 miles.

If you look at a point above you, such as the top of a building, the angle that your line of sight makes with a line parallel to the ground is called the *angle of elevation*. At the top of the building, the angle between a line parallel to the ground and your line of sight is called the *angle of depression*. These two angles have the same measure.

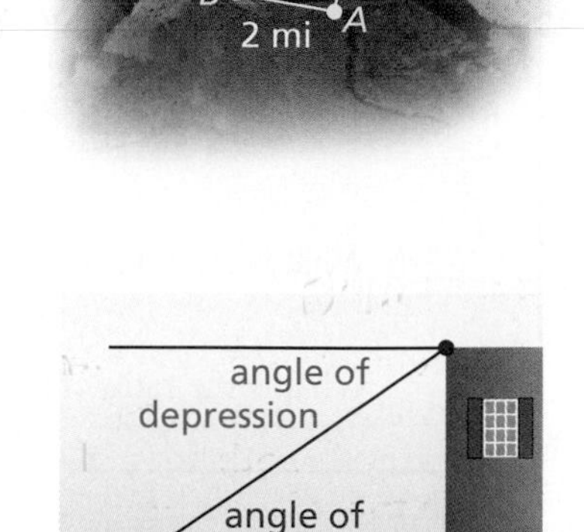

EXAMPLE 6 Using an Angle of Elevation

A parasailer is attached to a boat with a rope 72 feet long. The angle of elevation from the boat to the parasailer is 28°. Estimate the parasailer's height above the boat.

SOLUTION

Step 1 Draw a diagram that represents the situation.

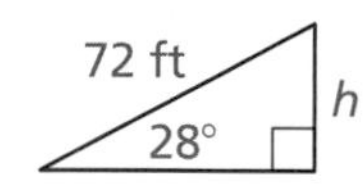

Step 2 Write and solve an equation to find the height h.

$$\sin 28^\circ = \frac{h}{72}$$ Write trigonometric equation.

$$72(\sin 28^\circ) = h$$ Multiply each side by 72.

$$33.8 \approx h$$ Use a calculator.

▶ The height of the parasailer above the boat is about 33.8 feet.

Monitoring Progress  Help in English and Spanish at *BigIdeasMath.com*

10. In Example 5, find the distance between B and C.

11. WHAT IF? In Example 6, estimate the height of the parasailer above the boat when the angle of elevation is 38°.

9.1 Exercises

Dynamic Solutions available at *BigIdeasMath.com*

Vocabulary and Core Concept Check

1. **COMPLETE THE SENTENCE** In a right triangle, the two trigonometric functions of θ that are defined using the lengths of the hypotenuse and the side adjacent to θ are __________ and __________.

2. **VOCABULARY** Compare an angle of elevation to an angle of depression.

3. **WRITING** Explain what it means to solve a right triangle.

4. **DIFFERENT WORDS, SAME QUESTION** Which is different? Find "both" answers.

What is the cosecant of θ?

What is $\dfrac{1}{\sin \theta}$?

What is the ratio of the side opposite θ to the hypotenuse?

What is the ratio of the hypotenuse to the side opposite θ?

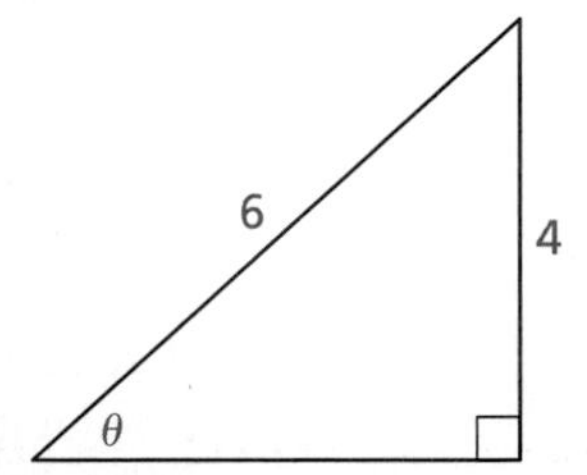

Monitoring Progress and Modeling with Mathematics

In Exercises 5–10, evaluate the six trigonometric functions of the angle θ. *(See Example 1.)*

5. 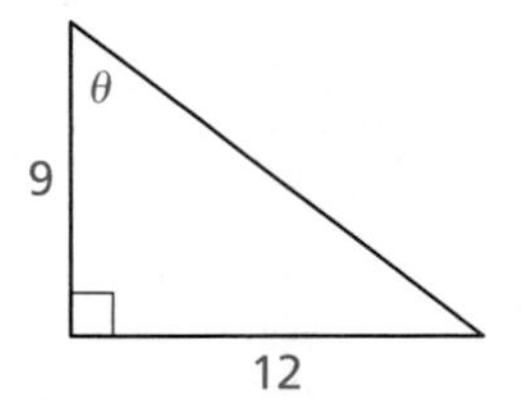

6. 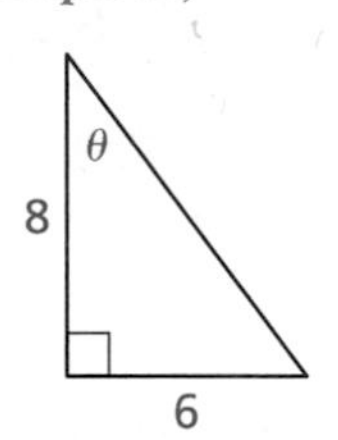

7. 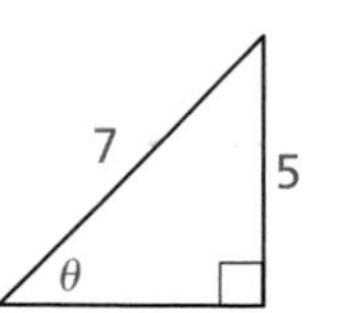

8. 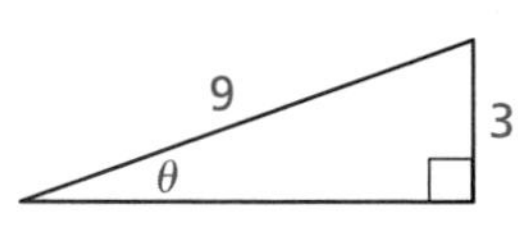

9. 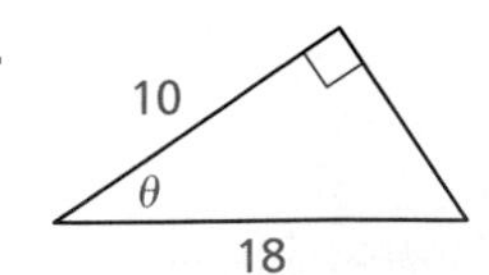

10. 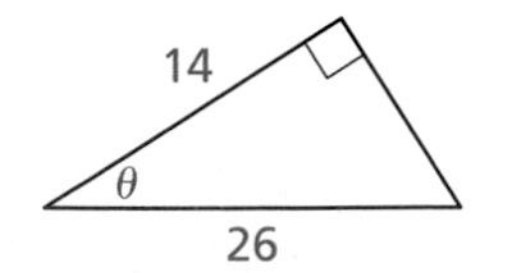

11. **REASONING** Let θ be an acute angle of a right triangle. Use the two trigonometric functions $\tan \theta = \dfrac{4}{9}$ and $\sec \theta = \dfrac{\sqrt{97}}{9}$ to sketch and label the right triangle. Then evaluate the other four trigonometric functions of θ.

12. **ANALYZING RELATIONSHIPS** Evaluate the six trigonometric functions of the $90° - \theta$ angle in Exercises 5–10. Describe the relationships you notice.

In Exercises 13–18, let θ be an acute angle of a right triangle. Evaluate the other five trigonometric functions of θ. *(See Example 2.)*

13. $\sin \theta = \frac{7}{11}$

14. $\cos \theta = \frac{5}{12}$

15. $\tan \theta = \frac{7}{6}$

16. $\csc \theta = \frac{15}{8}$

17. $\sec \theta = \frac{14}{9}$

18. $\cot \theta = \frac{16}{11}$

19. **ERROR ANALYSIS** Describe and correct the error in finding $\sin \theta$ of the triangle below.

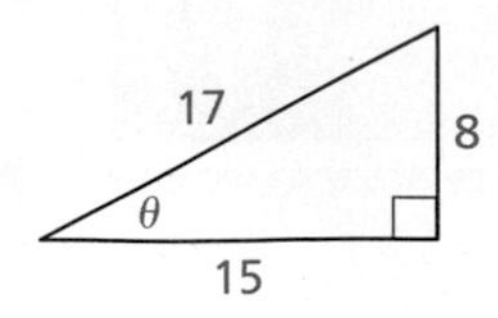

$\sin \theta = \dfrac{\text{opp.}}{\text{hyp.}} = \dfrac{15}{17}$

20. **ERROR ANALYSIS** Describe and correct the error in finding $\csc \theta$, given that θ is an acute angle of a right triangle and $\cos \theta = \frac{7}{11}$.

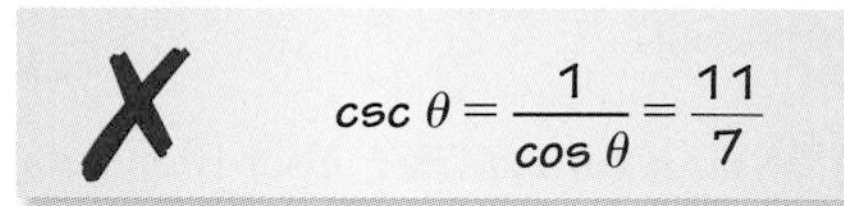

In Exercises 21–26, find the value of x for the right triangle. *(See Example 3.)*

21. 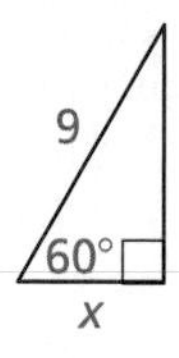

22. 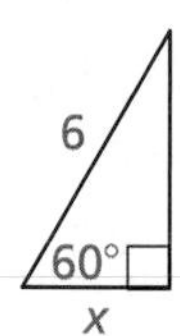

23. 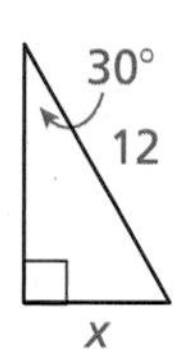

24. 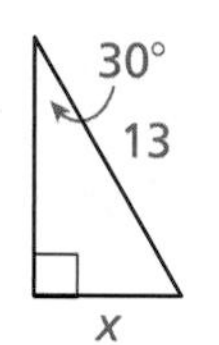

25. 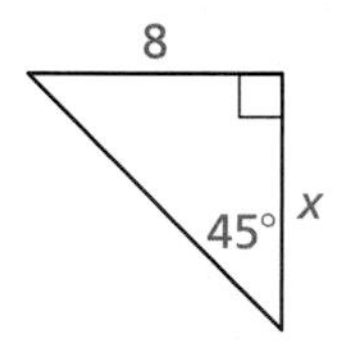

26. 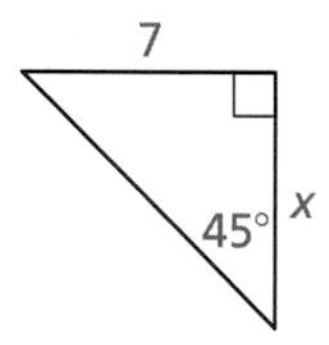

USING TOOLS In Exercises 27–32, evaluate the trigonometric function using a calculator. Round your answer to four decimal places.

27. $\cos 14°$
28. $\tan 31°$
29. $\csc 59°$
30. $\sin 23°$
31. $\cot 6°$
32. $\sec 11°$

In Exercises 33–40, solve $\triangle ABC$ using the diagram and the given measurements. *(See Example 4.)*

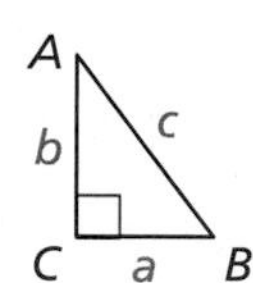

33. $B = 36°, a = 23$
34. $A = 27°, b = 9$
35. $A = 55°, a = 17$
36. $B = 16°, b = 14$
37. $A = 43°, b = 31$
38. $B = 31°, a = 23$
39. $B = 72°, c = 12.8$
40. $A = 64°, a = 7.4$

41. **MODELING WITH MATHEMATICS** To measure the width of a river, you plant a stake on one side of the river, directly across from a boulder. You then walk 100 meters to the right of the stake and measure a 79° angle between the stake and the boulder. What is the width w of the river? *(See Example 5.)*

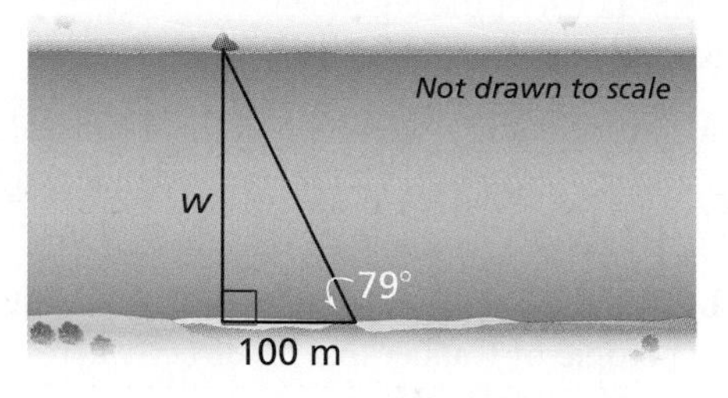

42. **MODELING WITH MATHEMATICS** Katoomba Scenic Railway in Australia is the steepest railway in the world. The railway makes an angle of about 52° with the ground. The railway extends horizontally about 458 feet. What is the height of the railway?

43. **MODELING WITH MATHEMATICS** A person whose eye level is 1.5 meters above the ground is standing 75 meters from the base of the Jin Mao Building in Shanghai, China. The person estimates the angle of elevation to the top of the building is about 80°. What is the approximate height of the building? *(See Example 6.)*

44. **MODELING WITH MATHEMATICS** The Duquesne Incline in Pittsburgh, Pennsylvania, has an angle of elevation of 30°. The track has a length of about 800 feet. Find the height of the incline.

45. **MODELING WITH MATHEMATICS** You are standing on the Grand View Terrace viewing platform at Mount Rushmore, 1000 feet from the base of the monument.

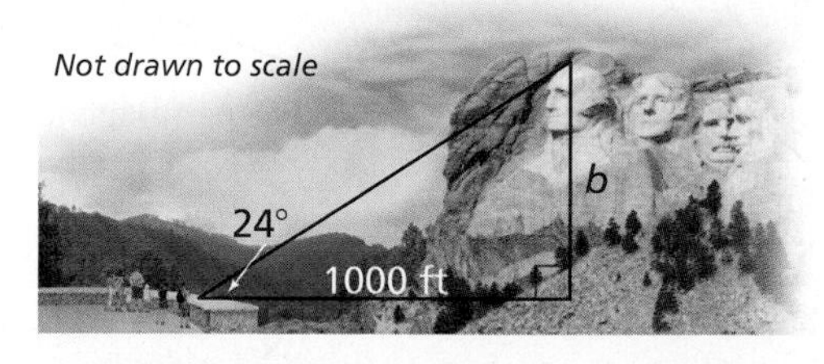

   a. You look up at the top of Mount Rushmore at an angle of 24°. How high is the top of the monument from where you are standing? Assume your eye level is 5.5 feet above the platform.

   b. The elevation of the Grand View Terrace is 5280 feet. Use your answer in part (a) to find the elevation of the top of Mount Rushmore.

46. **WRITING** Write a real-life problem that can be solved using a right triangle. Then solve your problem.

47. MATHEMATICAL CONNECTIONS The Tropic of Cancer is the circle of latitude farthest north of the equator where the Sun can appear directly overhead. It lies 23.5° north of the equator, as shown.

North Pole
Tropic of Cancer
23.5°
equator
South Pole

a. Find the circumference of the Tropic of Cancer using 3960 miles as the approximate radius of Earth.

b. What is the distance between two points on the Tropic of Cancer that lie directly across from each other?

48. HOW DO YOU SEE IT? Use the figure to answer each question.

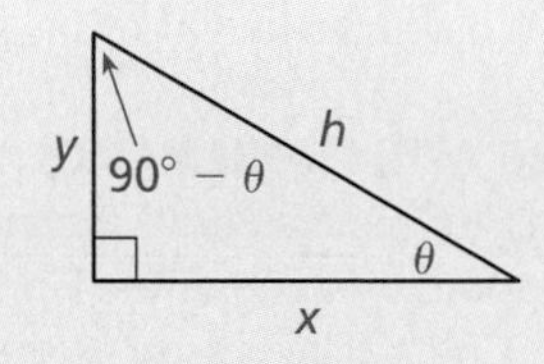

a. Which side is adjacent to θ?

b. Which side is opposite of θ?

c. Does $\cos \theta = \sin(90^\circ - \theta)$? Explain.

49. PROBLEM SOLVING A passenger in an airplane sees two towns directly to the left of the plane.

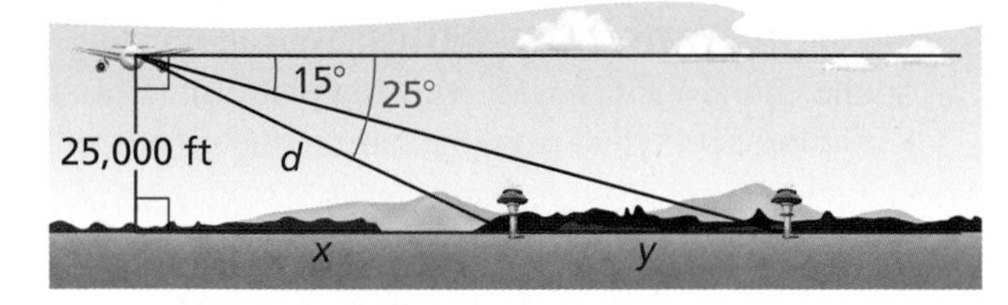

a. What is the distance d from the airplane to the first town?

b. What is the horizontal distance x from the airplane to the first town?

c. What is the distance y between the two towns? Explain the process you used to find your answer.

50. PROBLEM SOLVING You measure the angle of elevation from the ground to the top of a building as 32°. When you move 50 meters closer to the building, the angle of elevation is 53°. What is the height of the building?

51. MAKING AN ARGUMENT Your friend claims it is possible to draw a right triangle so the values of the cosine function of the acute angles are equal. Is your friend correct? Explain your reasoning.

52. THOUGHT PROVOKING Consider a semicircle with a radius of 1 unit, as shown below. Write the values of the six trigonometric functions of the angle θ. Explain your reasoning.

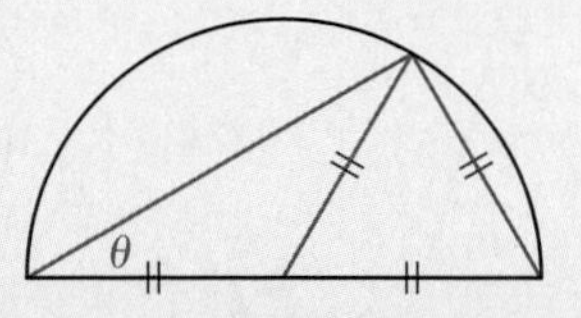

53. CRITICAL THINKING A procedure for approximating π based on the work of Archimedes is to inscribe a regular hexagon in a circle.

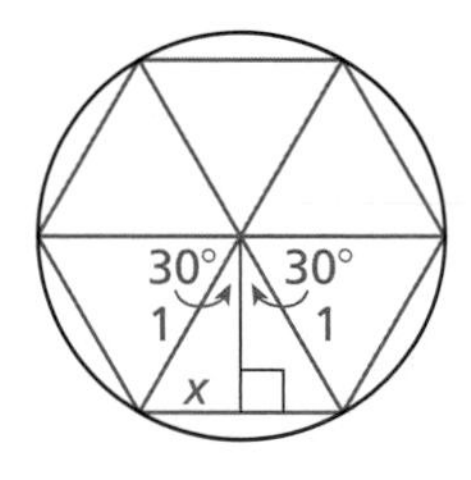

a. Use the diagram to solve for x. What is the perimeter of the hexagon?

b. Show that a regular n-sided polygon inscribed in a circle of radius 1 has a perimeter of $2n \cdot \sin\left(\frac{180}{n}\right)^\circ$.

c. Use the result from part (b) to find an expression in terms of n that approximates π. Then evaluate the expression when $n = 50$.

Maintaining Mathematical Proficiency

Reviewing what you learned in previous grades and lessons

Perform the indicated conversion. *(Skills Review Handbook)*

54. 5 years to seconds

55. 12 pints to gallons

56. 5.6 meters to millimeters

Find the circumference and area of the circle with the given radius or diameter. *(Skills Review Handbook)*

57. $r = 6$ centimeters

58. $r = 11$ inches

59. $d = 14$ feet

9.2 Angles and Radian Measure

Essential Question How can you find the measure of an angle in radians?

Let the vertex of an angle be at the origin, with one side of the angle on the positive x-axis. The *radian measure* of the angle is a measure of the intercepted arc length on a circle of radius 1. To convert between degree and radian measure, use the fact that

$$\frac{\pi \text{ radians}}{180^\circ} = 1.$$

EXPLORATION 1 Writing Radian Measures of Angles

Work with a partner. Write the radian measure of each angle with the given degree measure. Explain your reasoning.

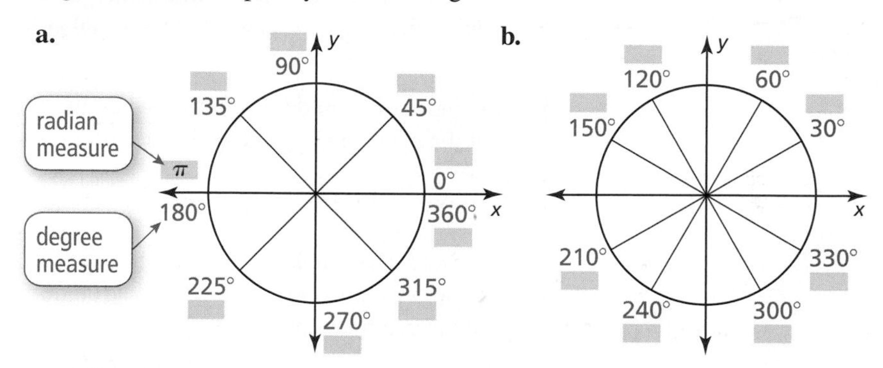

EXPLORATION 2 Writing Degree Measures of Angles

Work with a partner. Write the degree measure of each angle with the given radian measure. Explain your reasoning.

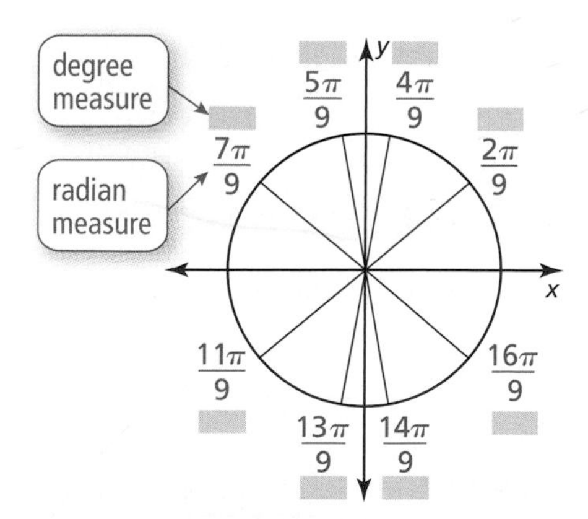

Communicate Your Answer

REASONING ABSTRACTLY

To be proficient in math, you need to make sense of quantities and their relationships in problem situations.

3. How can you find the measure of an angle in radians?

4. The figure shows an angle whose measure is 30 radians. What is the measure of the angle in degrees? How many times greater is 30 radians than 30 degrees? Justify your answers.

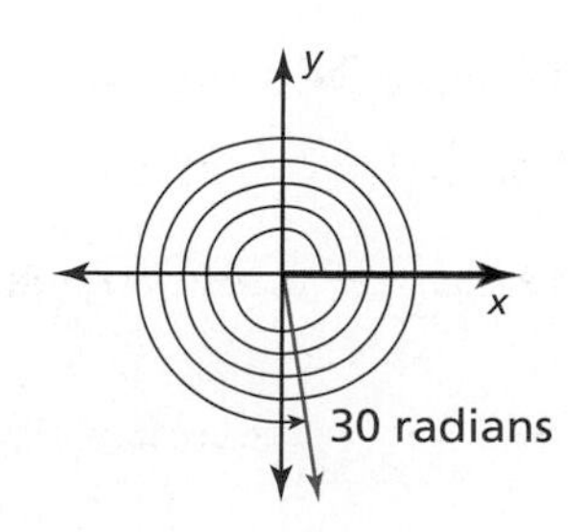

9.2 Lesson

What You Will Learn

- ▶ Draw angles in standard position.
- ▶ Find coterminal angles.
- ▶ Use radian measure.

Core Vocabulary

initial side, *p. 470*
terminal side, *p. 470*
standard position, *p. 470*
coterminal, *p. 471*
radian, *p. 471*
sector, *p. 472*
central angle, *p. 472*

Previous
radius of a circle
circumference of a circle

Drawing Angles in Standard Position

In this lesson, you will expand your study of angles to include angles with measures that can be any real numbers.

Core Concept

Angles in Standard Position

In a coordinate plane, an angle can be formed by fixing one ray, called the **initial side**, and rotating the other ray, called the **terminal side**, about the vertex.

An angle is in **standard position** when its vertex is at the origin and its initial side lies on the positive x-axis.

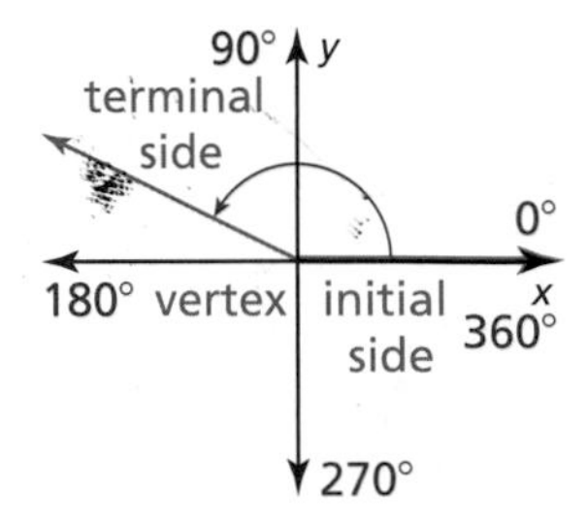

The measure of an angle is positive when the rotation of its terminal side is counterclockwise and negative when the rotation is clockwise. The terminal side of an angle can rotate more than 360°.

EXAMPLE 1 Drawing Angles in Standard Position

Draw an angle with the given measure in standard position.

a. 240° **b.** 500° **c.** −50°

SOLUTION

a. Because 240° is 60° more than 180°, the terminal side is 60° counterclockwise past the negative x-axis.

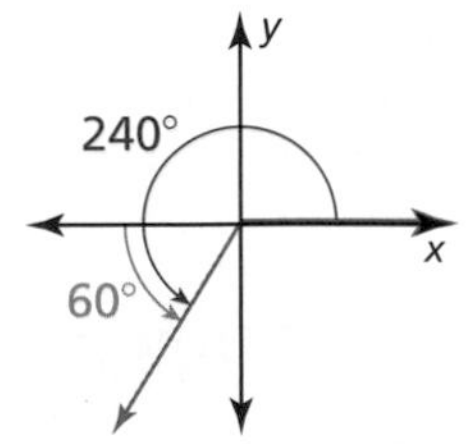

b. Because 500° is 140° more than 360°, the terminal side makes one complete rotation 360° counterclockwise plus 140° more.

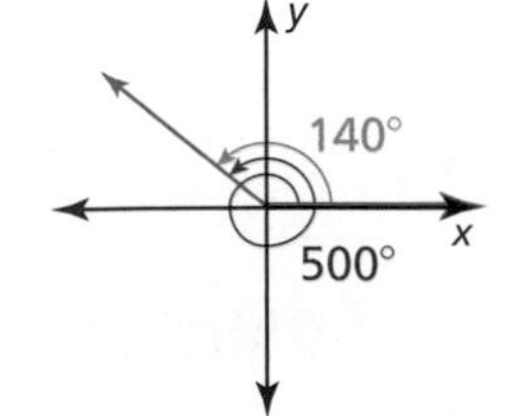

c. Because −50° is negative, the terminal side is 50° clockwise from the positive x-axis.

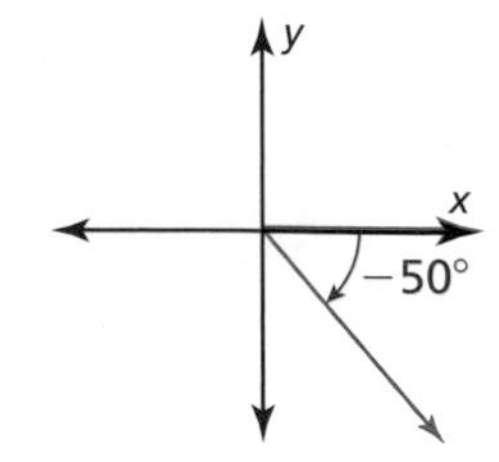

Monitoring Progress

Help in English and Spanish at *BigIdeasMath.com*

Draw an angle with the given measure in standard position.

1. 65° **2.** 300° **3.** −120° **4.** −450°

Finding Coterminal Angles

In Example 1(b), the angles 500° and 140° are **coterminal** because their terminal sides coincide. An angle coterminal with a given angle can be found by adding or subtracting multiples of 360°.

STUDY TIP

If two angles differ by a multiple of 360°, then the angles are coterminal.

EXAMPLE 2 **Finding Coterminal Angles**

Find one positive angle and one negative angle that are coterminal with (a) −45° and (b) 395°.

SOLUTION

There are many such angles, depending on what multiple of 360° is added or subtracted.

a. $-45° + 360° = 315°$
$-45° - 360° = -405°$

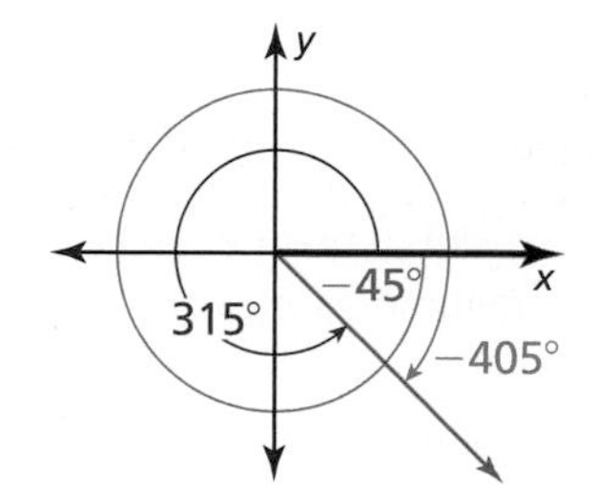

b. $395° - 360° = 35°$
$395° - 2(360°) = -325°$

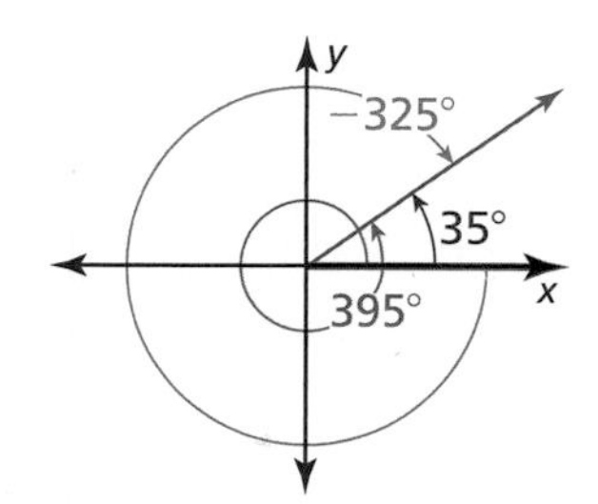

Monitoring Progress  Help in English and Spanish at *BigIdeasMath.com*

Find one positive angle and one negative angle that are coterminal with the given angle.

5. 80° **6.** 230° **7.** 740° **8.** −135°

Using Radian Measure

Angles can also be measured in *radians*. To define a radian, consider a circle with radius r centered at the origin, as shown. One **radian** is the measure of an angle in standard position whose terminal side intercepts an arc of length r.

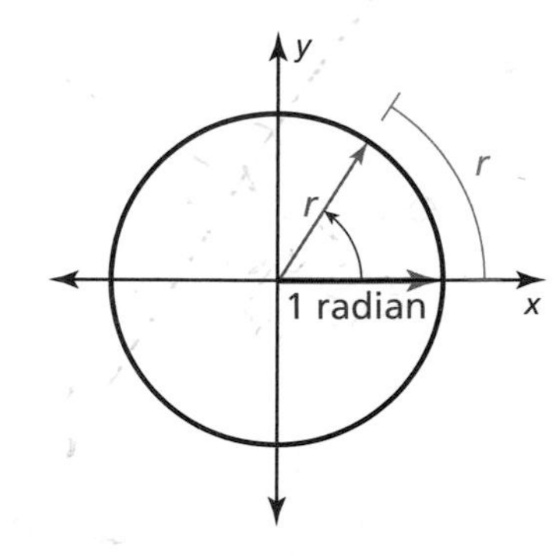

Because the circumference of a circle is $2\pi r$, there are 2π radians in a full circle. So, degree measure and radian measure are related by the equation $360° = 2\pi$ radians, or $180° = \pi$ radians.

STUDY TIP

Notice that 1 radian is approximately equal to 57.3°.

$180° = \pi$ radians

$\frac{180°}{\pi} = 1$ radian

$57.3° \approx 1$ radian

Core Concept

Converting Between Degrees and Radians

Degrees to radians

Multiply degree measure by

$$\frac{\pi \text{ radians}}{180°}.$$

Radians to degrees

Multiply radian measure by

$$\frac{180°}{\pi \text{ radians}}.$$

EXAMPLE 3 Convert Between Degrees and Radians

Convert the degree measure to radians or the radian measure to degrees.

a. $120°$ **b.** $-\frac{\pi}{12}$

READING

The unit "radians" is often omitted. For instance, the measure $-\frac{\pi}{12}$ radians may be written simply as $-\frac{\pi}{12}$.

SOLUTION

a. $120° = 120 \cancel{\text{degrees}}\left(\frac{\pi \text{ radians}}{180 \cancel{\text{degrees}}}\right)$

$= \frac{2\pi}{3}$

b. $-\frac{\pi}{12} = \left(-\frac{\pi}{12} \cancel{\text{radians}}\right)\left(\frac{180°}{\pi \cancel{\text{radians}}}\right)$

$= -15°$

Concept Summary

Degree and Radian Measures of Special Angles

The diagram shows equivalent degree and radian measures for special angles from $0°$ to $360°$ (0 radians to 2π radians).

You may find it helpful to memorize the equivalent degree and radian measures of special angles in the first quadrant and for $90° = \frac{\pi}{2}$ radians. All other special angles shown are multiples of these angles.

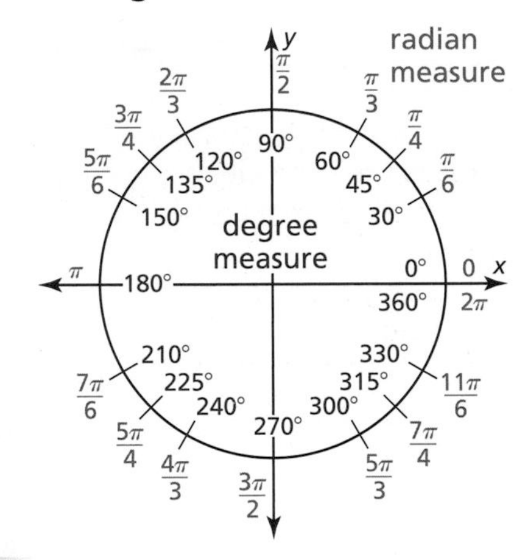

Monitoring Progress

Help in English and Spanish at *BigIdeasMath.com*

Convert the degree measure to radians or the radian measure to degrees.

9. $135°$ **10.** $-40°$ **11.** $\frac{5\pi}{4}$ **12.** -6.28

A **sector** is a region of a circle that is bounded by two radii and an arc of the circle. The **central angle** θ of a sector is the angle formed by the two radii. There are simple formulas for the arc length and area of a sector when the central angle is measured in radians.

Core Concept

Arc Length and Area of a Sector

The arc length s and area A of a sector with radius r and central angle θ (measured in radians) are as follows.

Arc length: $s = r\theta$

Area: $A = \frac{1}{2}r^2\theta$

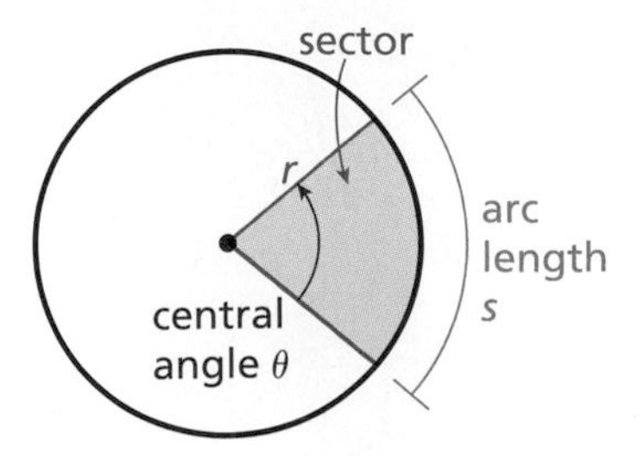

EXAMPLE 4 Modeling with Mathematics

A softball field forms a sector with the dimensions shown. Find the length of the outfield fence and the area of the field.

SOLUTION

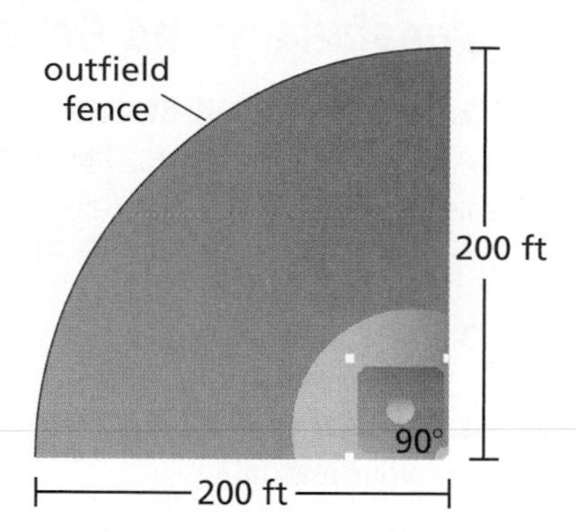

1. **Understand the Problem** You are given the dimensions of a softball field. You are asked to find the length of the outfield fence and the area of the field.

2. **Make a Plan** Find the measure of the central angle in radians. Then use the arc length and area of a sector formulas.

3. **Solve the Problem**

Step 1 Convert the measure of the central angle to radians.

$$90^\circ = 90 \text{ \sout{degrees}} \left(\frac{\pi \text{ radians}}{180 \text{ \sout{degrees}}} \right)$$

$$= \frac{\pi}{2} \text{ radians}$$

COMMON ERROR

You must write the measure of an angle in radians when using these formulas for the arc length and area of a sector.

Step 2 Find the arc length and the area of the sector.

Arc length: $s = r\theta$

$$= 200\left(\frac{\pi}{2}\right)$$

$$= 100\pi$$

$$\approx 314$$

Area: $A = \frac{1}{2}r^2\theta$

$$= \frac{1}{2}(200)^2\left(\frac{\pi}{2}\right)$$

$$= 10{,}000\pi$$

$$\approx 31{,}416$$

ANOTHER WAY

Because the central angle is 90°, the sector represents $\frac{1}{4}$ of a circle with a radius of 200 feet. So,

$$s = \frac{1}{4} \cdot 2\pi r = \frac{1}{4} \cdot 2\pi(200)$$

$$= 100\pi$$

and

$$A = \frac{1}{4} \cdot \pi r^2 = \frac{1}{4} \cdot \pi(200)^2$$

$$= 10{,}000\pi.$$

▶ The length of the outfield fence is about 314 feet. The area of the field is about 31,416 square feet.

4. **Look Back** To check the area of the field, consider the square formed using the two 200-foot sides.

By drawing the diagonal, you can see that the area of the field is less than the area of the square but greater than one-half of the area of the square.

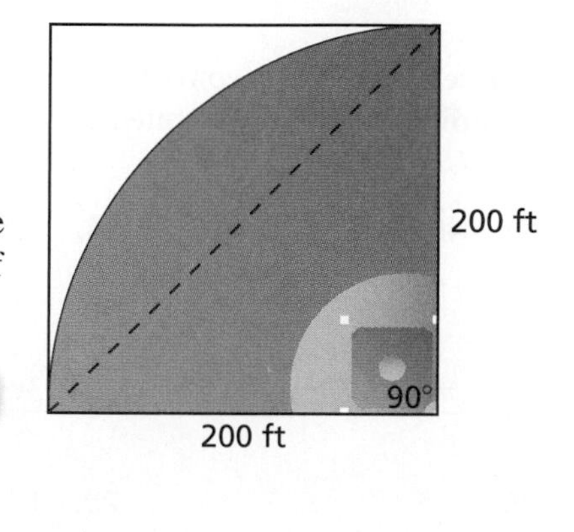

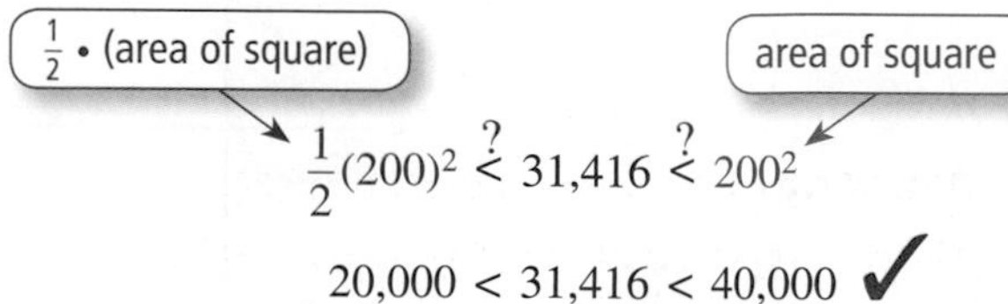

$$\frac{1}{2}(200)^2 \stackrel{?}{<} 31{,}416 \stackrel{?}{<} 200^2$$

$$20{,}000 < 31{,}416 < 40{,}000 \checkmark$$

Monitoring Progress

Help in English and Spanish at *BigIdeasMath.com*

13. **WHAT IF?** In Example 4, the outfield fence is 220 feet from home plate. Estimate the length of the outfield fence and the area of the field.

9.2 Exercises

Dynamic Solutions available at *BigIdeasMath.com*

Vocabulary and Core Concept Check

1. **COMPLETE THE SENTENCE** An angle is in standard position when its vertex is at the __________ and its __________ lies on the positive x-axis.

2. **WRITING** Explain how the sign of an angle measure determines its direction of rotation.

3. **VOCABULARY** In your own words, define a radian.

4. **WHICH ONE DOESN'T BELONG?** Which angle does *not* belong with the other three? Explain your reasoning.

$-90°$	$450°$	$90°$	$-270°$

Monitoring Progress and Modeling with Mathematics

In Exercises 5–8, draw an angle with the given measure in standard position. *(See Example 1.)*

5. $110°$
6. $450°$
7. $-900°$
8. $-10°$

In Exercises 9–12, find one positive angle and one negative angle that are coterminal with the given angle. *(See Example 2.)*

9. $70°$
10. $255°$
11. $-125°$
12. $-800°$

In Exercises 13–20, convert the degree measure to radians or the radian measure to degrees. *(See Example 3.)*

13. $40°$
14. $315°$
15. $-260°$
16. $-500°$
17. $\frac{\pi}{9}$
18. $\frac{3\pi}{4}$
19. -5
20. 12

21. **WRITING** The terminal side of an angle in standard position rotates one-sixth of a revolution counterclockwise from the positive x-axis. Describe how to find the measure of the angle in both degree and radian measures.

22. **OPEN-ENDED** Using radian measure, give one positive angle and one negative angle that are coterminal with the angle shown. Justify your answers.

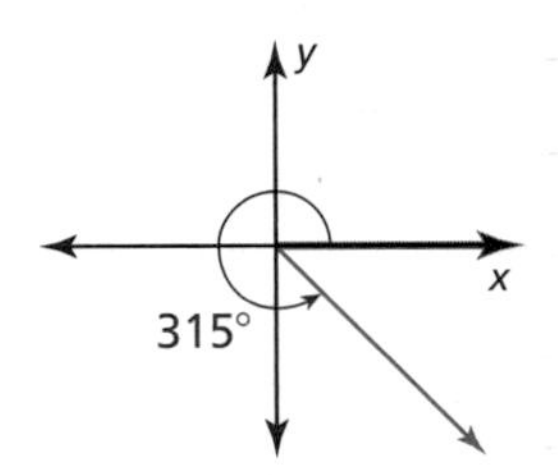

ANALYZING RELATIONSHIPS In Exercises 23–26, match the angle measure with the angle.

23. $600°$
24. $-\frac{9\pi}{4}$
25. $\frac{5\pi}{6}$
26. $-240°$

A.

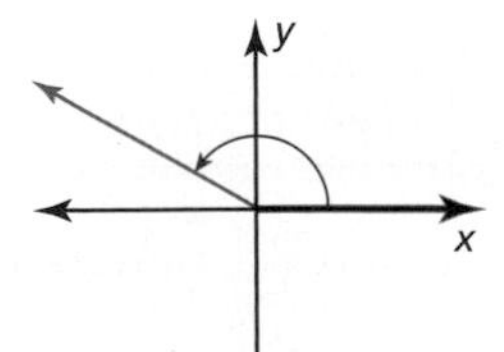

B.

C.

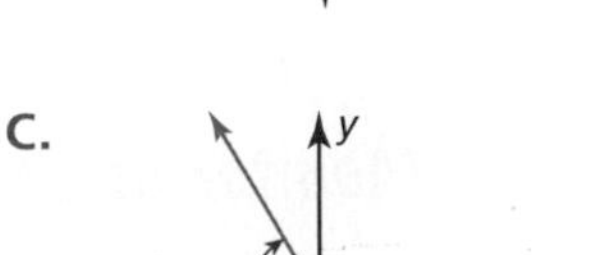

D.

27. **MODELING WITH MATHEMATICS** The observation deck of a building forms a sector with the dimensions shown. Find the length of the safety rail and the area of the deck. *(See Example 4.)*

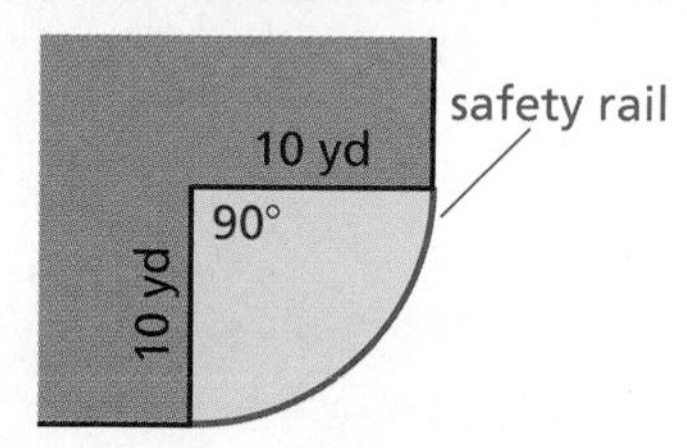

28. **MODELING WITH MATHEMATICS** In the men's shot put event at the 2012 Summer Olympic Games, the length of the winning shot was 21.89 meters. A shot put must land within a sector having a central angle of 34.92° to be considered fair.

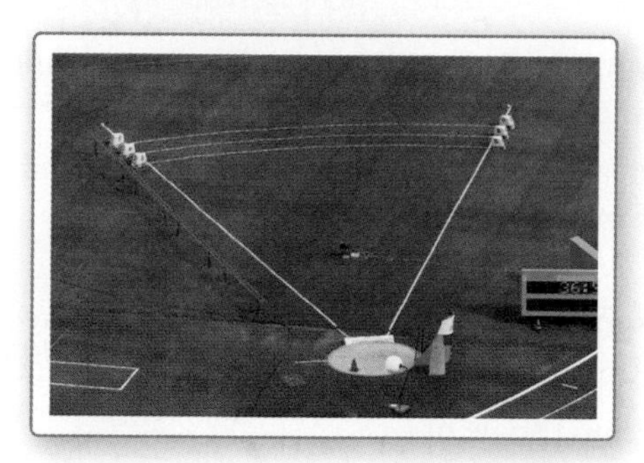

a. The officials draw an arc across the fair landing area, marking the farthest throw. Find the length of the arc.

b. All fair throws in the 2012 Olympics landed within a sector bounded by the arc in part (a). What is the area of this sector?

29. **ERROR ANALYSIS** Describe and correct the error in converting the degree measure to radians.

✗

$$24° = 24 \text{ ~~degrees~~} \left(\frac{180 \text{ ~~degrees~~}}{\pi \text{ radians}}\right)$$

$$= \frac{4320}{\pi} \text{ radians}$$

$$\approx 1375.1 \text{ radians}$$

30. **ERROR ANALYSIS** Describe and correct the error in finding the area of a sector with a radius of 6 centimeters and a central angle of 40°.

✗

$$A = \frac{1}{2}(6)^2(40) = 720 \text{ cm}^2$$

31. **PROBLEM SOLVING** When a CD player reads information from the outer edge of a CD, the CD spins about 200 revolutions per minute. At that speed, through what angle does a point on the CD spin in one minute? Give your answer in both degree and radian measures.

32. **PROBLEM SOLVING** You work every Saturday from 9:00 A.M. to 5:00 P.M. Draw a diagram that shows the rotation completed by the hour hand of a clock during this time. Find the measure of the angle generated by the hour hand in both degrees and radians. Compare this angle with the angle generated by the minute hand from 9:00 A.M. to 5:00 P.M.

USING TOOLS In Exercises 33–38, use a calculator to evaluate the trigonometric function.

33. $\cos \frac{4\pi}{3}$

34. $\sin \frac{7\pi}{8}$

35. $\csc \frac{10\pi}{11}$

36. $\cot\left(-\frac{6\pi}{5}\right)$

37. $\cot(-14)$

38. $\cos 6$

39. **MODELING WITH MATHEMATICS** The rear windshield wiper of a car rotates 120°, as shown. Find the area cleared by the wiper.

40. **MODELING WITH MATHEMATICS** A scientist performed an experiment to study the effects of gravitational force on humans. In order for humans to experience twice Earth's gravity, they were placed in a centrifuge 58 feet long and spun at a rate of about 15 revolutions per minute.

a. Through how many radians did the people rotate each second?

b. Find the length of the arc through which the people rotated each second.

41. **REASONING** In astronomy, the *terminator* is the day-night line on a planet that divides the planet into daytime and nighttime regions. The terminator moves across the surface of a planet as the planet rotates. It takes about 4 hours for Earth's terminator to move across the continental United States. Through what angle has Earth rotated during this time? Give your answer in both degree and radian measures.

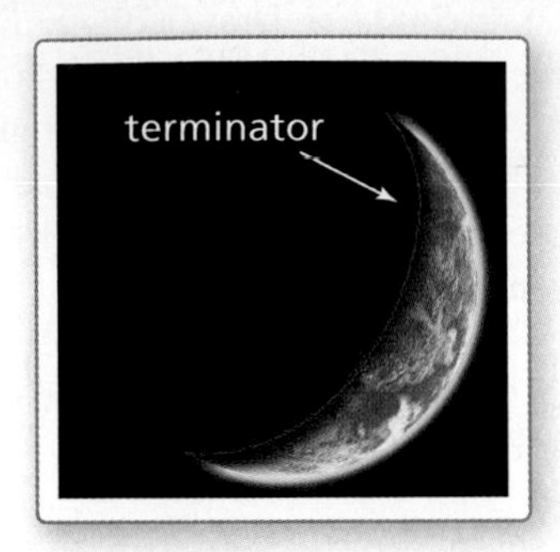

42. **HOW DO YOU SEE IT?** Use the graph to find the measure of θ. Explain your reasoning.

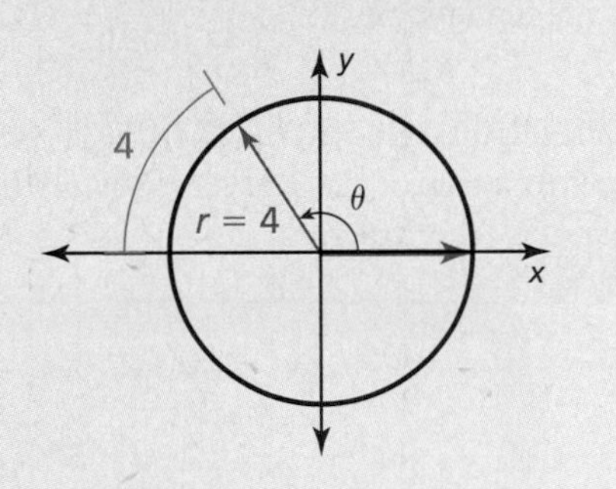

43. **MODELING WITH MATHEMATICS** A dartboard is divided into 20 sectors. Each sector is worth a point value from 1 to 20 and has shaded regions that double or triple this value. A sector is shown below. Find the areas of the entire sector, the double region, and the triple region.

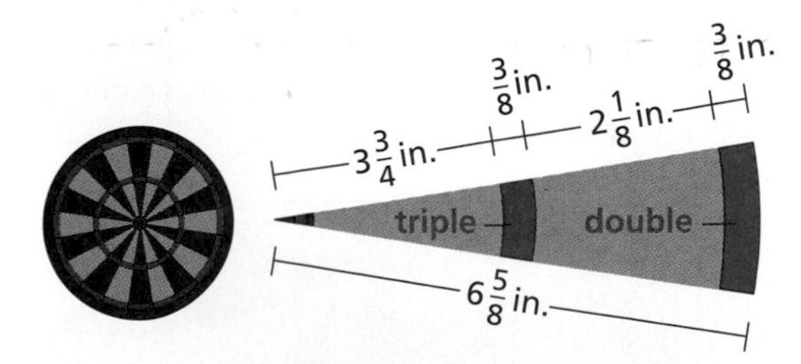

44. **THOUGHT PROVOKING** π is an irrational number, which means that it cannot be written as the ratio of two whole numbers. π can, however, be written exactly as a *continued fraction*, as follows.

$$3 + \cfrac{1}{7 + \cfrac{1}{15 + \cfrac{1}{1 + \cfrac{1}{292 + \cfrac{1}{1 + \cfrac{1}{1 + \cfrac{1}{1 + \cdots}}}}}}}$$

Show how to use this continued fraction to obtain a decimal approximation for π.

45. **MAKING AN ARGUMENT** Your friend claims that when the arc length of a sector equals the radius, the area can be given by $A = \frac{s^2}{2}$. Is your friend correct? Explain.

46. **PROBLEM SOLVING** A spiral staircase has 15 steps. Each step is a sector with a radius of 42 inches and a central angle of $\frac{\pi}{8}$.

   a. What is the length of the arc formed by the outer edge of a step?

   b. Through what angle would you rotate by climbing the stairs?

   c. How many square inches of carpeting would you need to cover the 15 steps?

47. **MULTIPLE REPRESENTATIONS** There are 60 *minutes* in 1 degree of arc, and 60 *seconds* in 1 minute of arc. The notation $50° \, 30' \, 10''$ represents an angle with a measure of 50 degrees, 30 minutes, and 10 seconds.

   a. Write the angle measure 70.55° using the notation above.

   b. Write the angle measure $110° \, 45' \, 30''$ to the nearest hundredth of a degree. Justify your answer.

Maintaining Mathematical Proficiency

Reviewing what you learned in previous grades and lessons

Find the distance between the two points. *(Skills Review Handbook)*

48. $(1, 4), (3, 6)$

49. $(-7, -13), (10, 8)$

50. $(-3, 9), (-3, 16)$

51. $(2, 12), (8, -5)$

52. $(-14, -22), (-20, -32)$

53. $(4, 16), (-1, 34)$

9.3 Trigonometric Functions of Any Angle

Essential Question

How can you use the unit circle to define the trigonometric functions of any angle?

Let θ be an angle in standard position with (x, y) a point on the terminal side of θ and $r = \sqrt{x^2 + y^2} \neq 0$. The six trigonometric functions of θ are defined as shown.

$\sin \theta = \frac{y}{r}$ $\qquad \csc \theta = \frac{r}{y}, y \neq 0$

$\cos \theta = \frac{x}{r}$ $\qquad \sec \theta = \frac{r}{x}, x \neq 0$

$\tan \theta = \frac{y}{x}, x \neq 0$ $\qquad \cot \theta = \frac{x}{y}, y \neq 0$

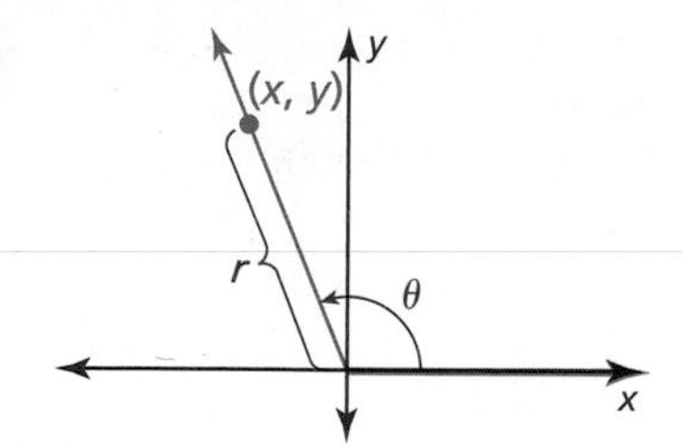

EXPLORATION 1 Writing Trigonometric Functions

Work with a partner. Find the sine, cosine, and tangent of the angle θ in standard position whose terminal side intersects the unit circle at the point (x, y) shown.

a.

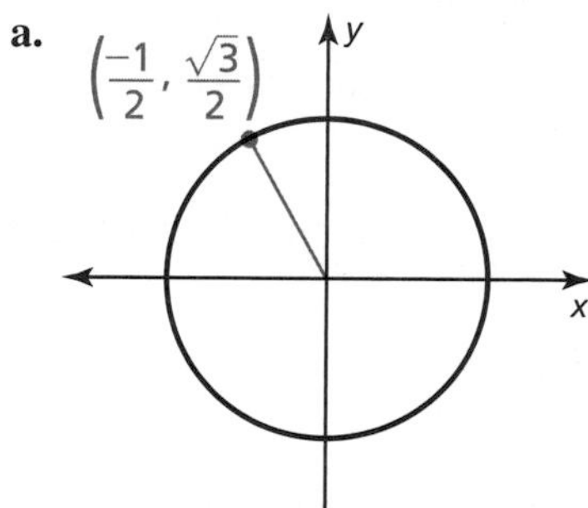

b.

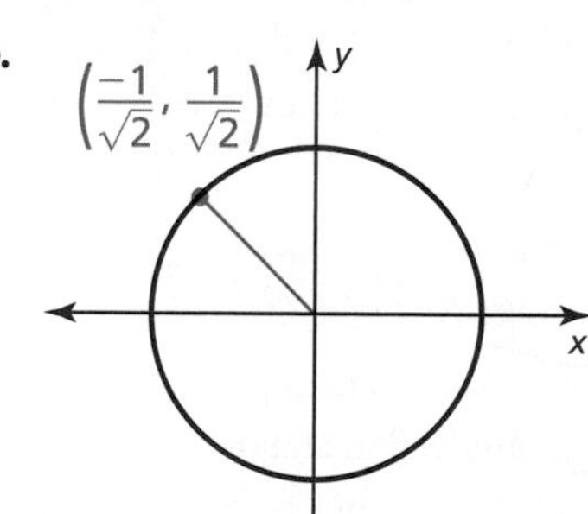

c.

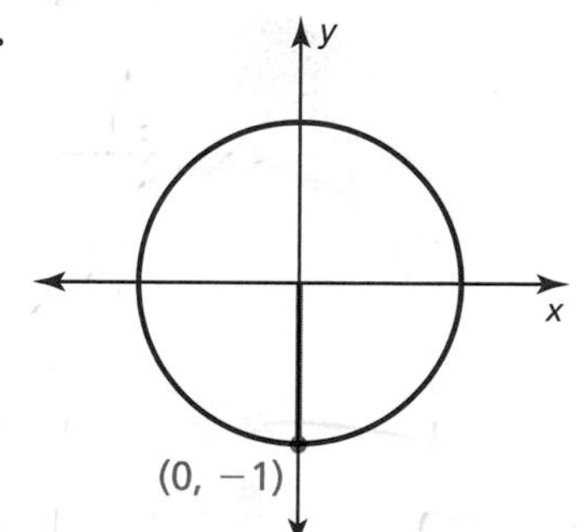

d.

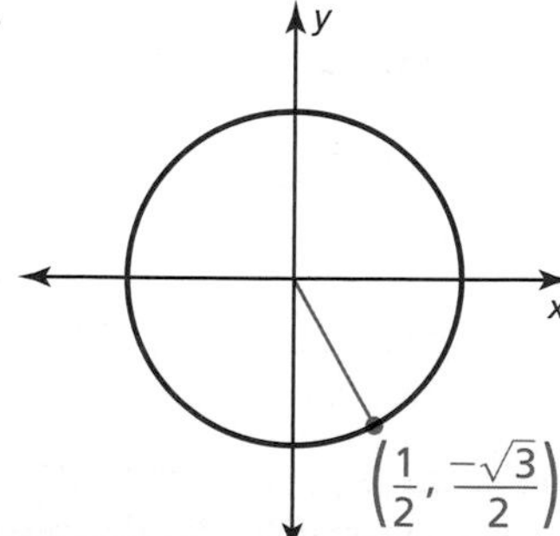

e.

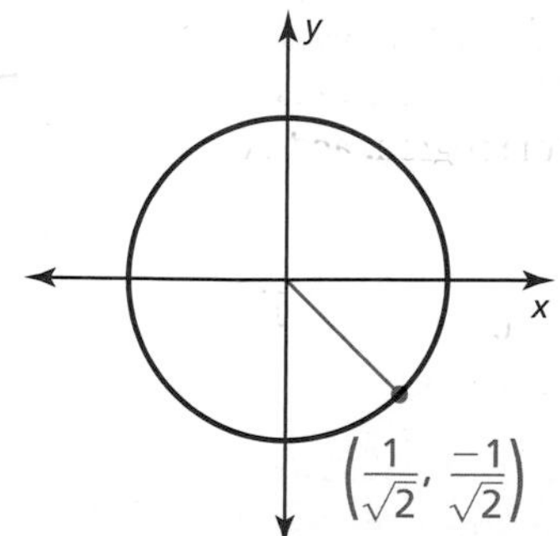

f.

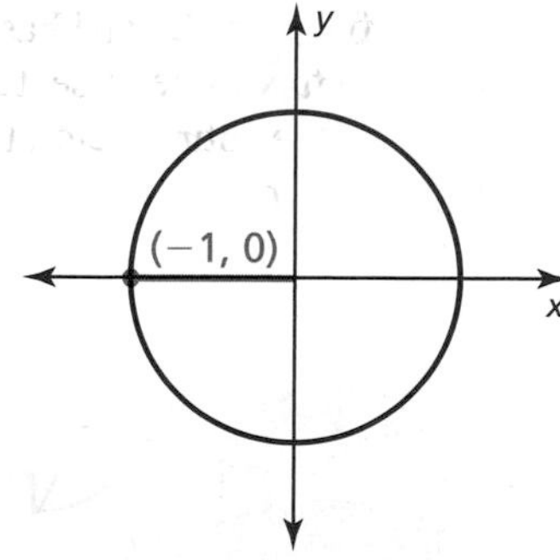

CONSTRUCTING VIABLE ARGUMENTS

To be proficient in math, you need to understand and use stated assumptions, definitions, and previously established results.

Communicate Your Answer

2. How can you use the unit circle to define the trigonometric functions of any angle?

3. For which angles are each function undefined? Explain your reasoning.

a. tangent **b.** cotangent **c.** secant **d.** cosecant

9.3 Lesson

Core Vocabulary

unit circle, *p. 479*
quadrantal angle, *p. 479*
reference angle, *p. 480*

Previous
circle
radius
Pythagorean Theorem

What You Will Learn

- Evaluate trigonometric functions of any angle.
- Find and use reference angles to evaluate trigonometric functions.

Trigonometric Functions of Any Angle

You can generalize the right-triangle definitions of trigonometric functions so that they apply to any angle in standard position.

Core Concept

General Definitions of Trigonometric Functions

Let θ be an angle in standard position, and let (x, y) be the point where the terminal side of θ intersects the circle $x^2 + y^2 = r^2$. The six trigonometric functions of θ are defined as shown.

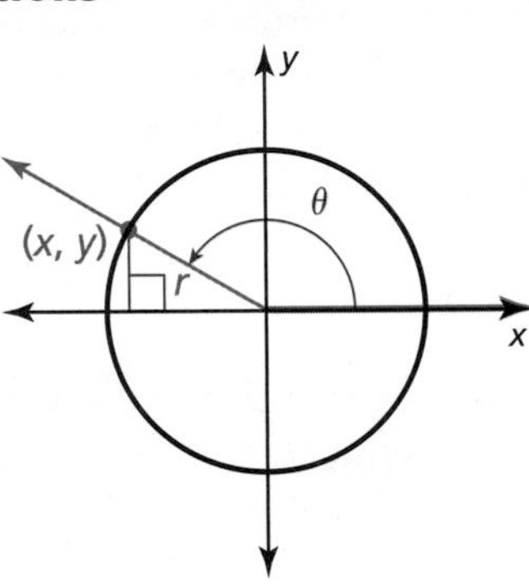

$\sin \theta = \frac{y}{r}$ $\qquad$ $\csc \theta = \frac{r}{y}, y \neq 0$

$\cos \theta = \frac{x}{r}$ $\qquad$ $\sec \theta = \frac{r}{x}, x \neq 0$

$\tan \theta = \frac{y}{x}, x \neq 0$ $\qquad$ $\cot \theta = \frac{x}{y}, y \neq 0$

These functions are sometimes called *circular functions.*

EXAMPLE 1 Evaluating Trigonometric Functions Given a Point

Let $(-4, 3)$ be a point on the terminal side of an angle θ in standard position. Evaluate the six trigonometric functions of θ.

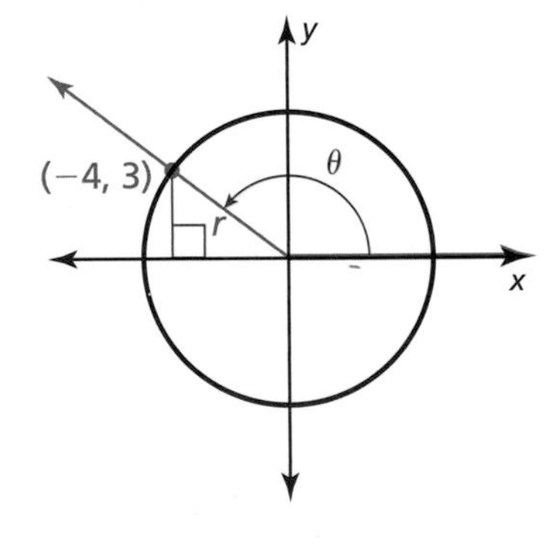

SOLUTION

Use the Pythagorean Theorem to find the length of r.

$$\begin{aligned} r &= \sqrt{x^2 + y^2} \\ &= \sqrt{(-4)^2 + 3^2} \\ &= \sqrt{25} \\ &= 5 \end{aligned}$$

Using $x = -4$, $y = 3$, and $r = 5$, the values of the six trigonometric functions of θ are:

$\sin \theta = \frac{y}{r} = \frac{3}{5}$ $\qquad$ $\csc \theta = \frac{r}{y} = \frac{5}{3}$

$\cos \theta = \frac{x}{r} = -\frac{4}{5}$ $\qquad$ $\sec \theta = \frac{r}{x} = -\frac{5}{4}$

$\tan \theta = \frac{y}{x} = -\frac{3}{4}$ $\qquad$ $\cot \theta = \frac{x}{y} = -\frac{4}{3}$

Core Concept

The Unit Circle

The circle $x^2 + y^2 = 1$, which has center (0, 0) and radius 1, is called the **unit circle**. The values of $\sin \theta$ and $\cos \theta$ are simply the y-coordinate and x-coordinate, respectively, of the point where the terminal side of θ intersects the unit circle.

$$\sin \theta = \frac{y}{r} = \frac{y}{1} = y$$

$$\cos \theta = \frac{x}{r} = \frac{x}{1} = x$$

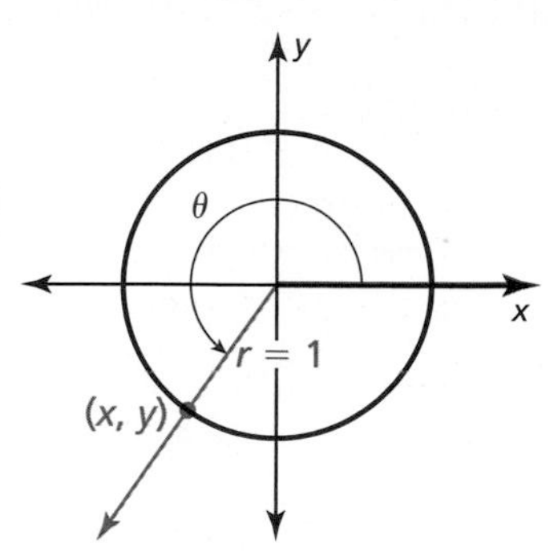

ANOTHER WAY

The general circle $x^2 + y^2 = r^2$ can also be used to find the six trigonometric functions of θ. The terminal side of θ intersects the circle at $(0, -r)$. So,

$$\sin \theta = \frac{y}{r} = \frac{-r}{r} = -1.$$

The other functions can be evaluated similarly.

It is convenient to use the unit circle to find trigonometric functions of **quadrantal angles**. A quadrantal angle is an angle in standard position whose terminal side lies on an axis. The measure of a quadrantal angle is always a multiple of 90°, or $\frac{\pi}{2}$ radians.

EXAMPLE 2 Using the Unit Circle

Use the unit circle to evaluate the six trigonometric functions of $\theta = 270°$.

SOLUTION

Step 1 Draw a unit circle with the angle $\theta = 270°$ in standard position.

Step 2 Identify the point where the terminal side of θ intersects the unit circle. The terminal side of θ intersects the unit circle at $(0, -1)$.

Step 3 Find the values of the six trigonometric functions. Let $x = 0$ and $y = -1$ to evaluate the trigonometric functions.

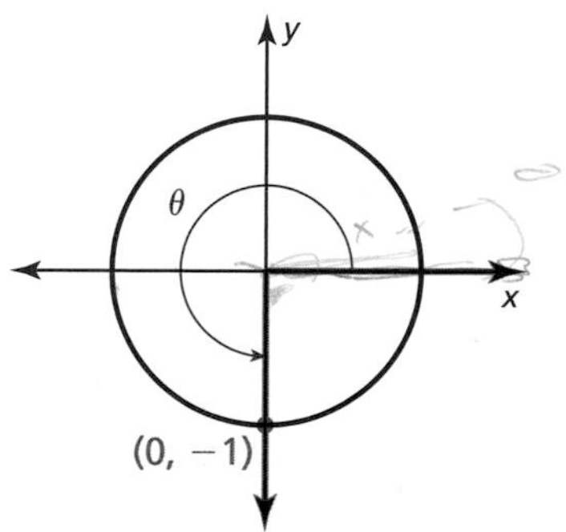

$$\sin \theta = \frac{y}{r} = \frac{-1}{1} = -1 \qquad \csc \theta = \frac{r}{y} = \frac{1}{-1} = -1$$

$$\cos \theta = \frac{x}{r} = \frac{0}{1} = 0 \qquad \sec \theta = \frac{r}{x} = \frac{1}{0} \quad \text{undefined}$$

$$\tan \theta = \frac{y}{x} = \frac{-1}{0} \quad \text{undefined} \qquad \cot \theta = \frac{x}{y} = \frac{0}{-1} = 0$$

Monitoring Progress  Help in English and Spanish at *BigIdeasMath.com*

Evaluate the six trigonometric functions of θ.

1. 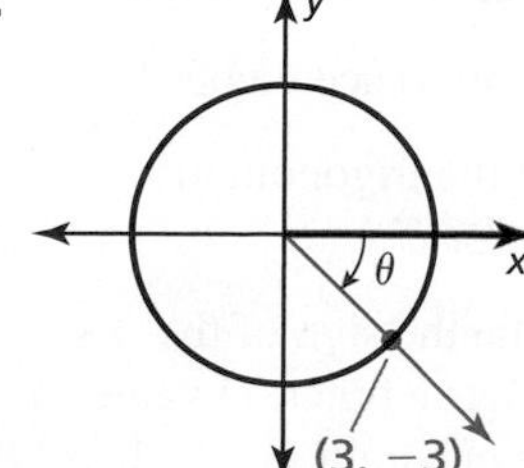

2. 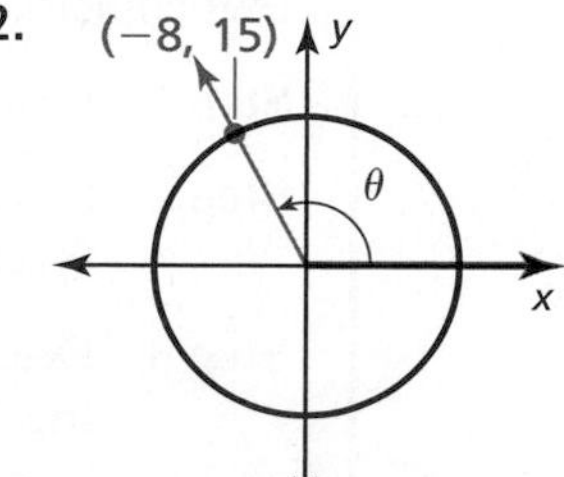

3. 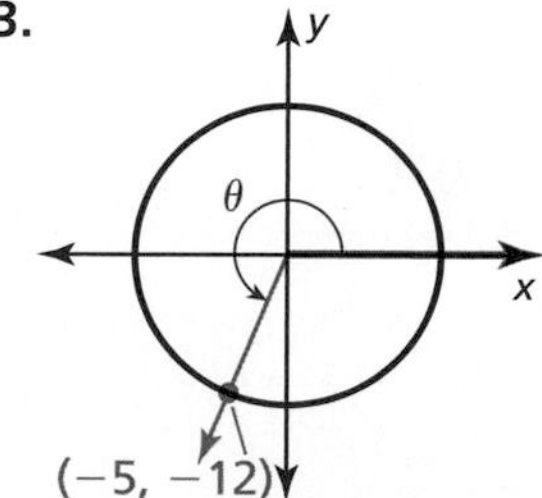

4. Use the unit circle to evaluate the six trigonometric functions of $\theta = 180°$.

Reference Angles

Core Concept

Reference Angle Relationships

Let θ be an angle in standard position. The **reference angle** for θ is the acute angle θ' formed by the terminal side of θ and the x-axis. The relationship between θ and θ' is shown below for nonquadrantal angles θ such that $90° < \theta < 360°$ or, in radians, $\frac{\pi}{2} < \theta < 2\pi$.

READING
The symbol θ' is read as "theta prime."

Quadrant II

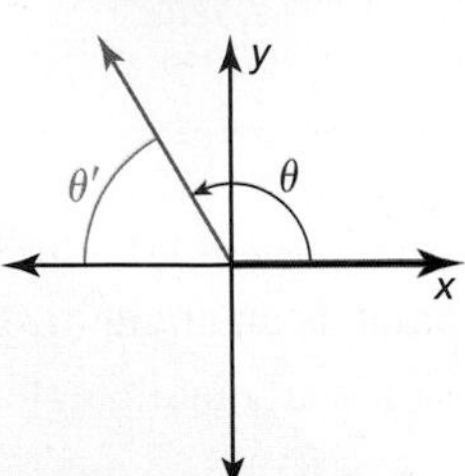

Degrees: $\theta' = 180° - \theta$
Radians: $\theta' = \pi - \theta$

Quadrant III

Degrees: $\theta' = \theta - 180°$
Radians: $\theta' = \theta - \pi$

Quadrant IV

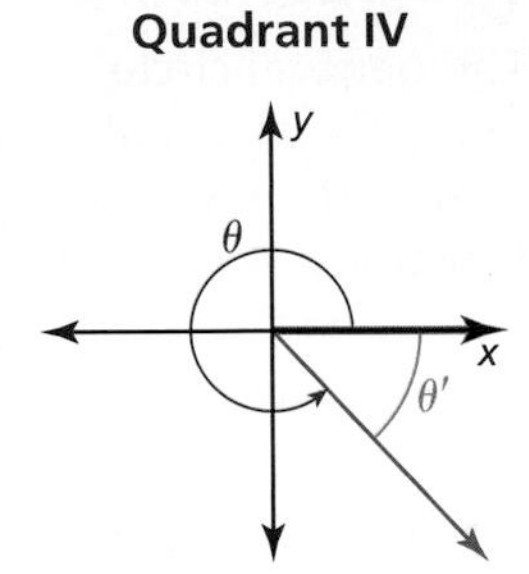

Degrees: $\theta' = 360° - \theta$
Radians: $\theta' = 2\pi - \theta$

EXAMPLE 3 Finding Reference Angles

Find the reference angle θ' for (a) $\theta = \frac{5\pi}{3}$ and (b) $\theta = -130°$.

SOLUTION

a. The terminal side of θ lies in Quadrant IV. So, $\theta' = 2\pi - \frac{5\pi}{3} = \frac{\pi}{3}$. The figure at the right shows $\theta = \frac{5\pi}{3}$ and $\theta' = \frac{\pi}{3}$.

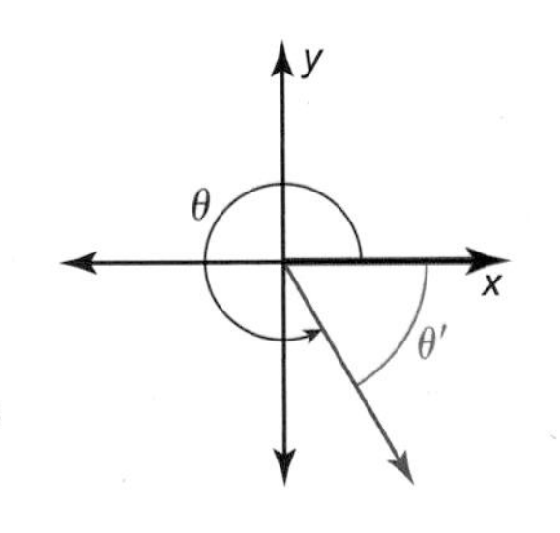

b. Note that θ is coterminal with 230°, whose terminal side lies in Quadrant III. So, $\theta' = 230° - 180° = 50°$. The figure at the left shows $\theta = -130°$ and $\theta' = 50°$.

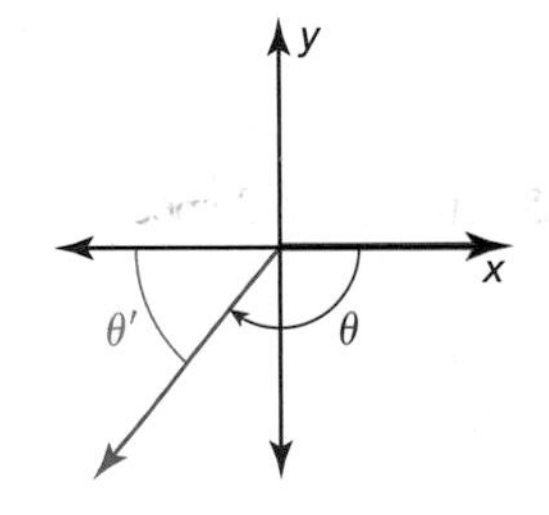

Reference angles allow you to evaluate a trigonometric function for any angle θ. The sign of the trigonometric function value depends on the quadrant in which θ lies.

Core Concept

Evaluating Trigonometric Functions

Use these steps to evaluate a trigonometric function for any angle θ:

Step 1 Find the reference angle θ'.

Step 2 Evaluate the trigonometric function for θ'.

Step 3 Determine the sign of the trigonometric function value from the quadrant in which θ lies.

Signs of Function Values

Quadrant II	Quadrant I
$\sin\theta, \csc\theta$: +	$\sin\theta, \csc\theta$: +
$\cos\theta, \sec\theta$: −	$\cos\theta, \sec\theta$: +
$\tan\theta, \cot\theta$: −	$\tan\theta, \cot\theta$: +
Quadrant III	**Quadrant IV**
$\sin\theta, \csc\theta$: −	$\sin\theta, \csc\theta$: −
$\cos\theta, \sec\theta$: −	$\cos\theta, \sec\theta$: +
$\tan\theta, \cot\theta$: +	$\tan\theta, \cot\theta$: −

EXAMPLE 4 Using Reference Angles to Evaluate Functions

Evaluate (a) $\tan(-240°)$ and (b) $\csc \frac{17\pi}{6}$.

SOLUTION

a. The angle $-240°$ is coterminal with $120°$. The reference angle is $\theta' = 180° - 120° = 60°$. The tangent function is negative in Quadrant II, so

$$\tan(-240°) = -\tan 60° = -\sqrt{3}.$$

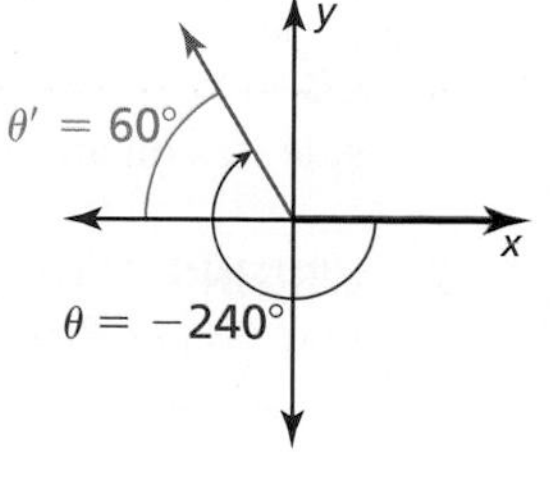

b. The angle $\frac{17\pi}{6}$ is coterminal with $\frac{5\pi}{6}$. The reference angle is

$$\theta' = \pi - \frac{5\pi}{6} = \frac{\pi}{6}.$$

The cosecant function is positive in Quadrant II, so

$$\csc \frac{17\pi}{6} = \csc \frac{\pi}{6} = 2.$$

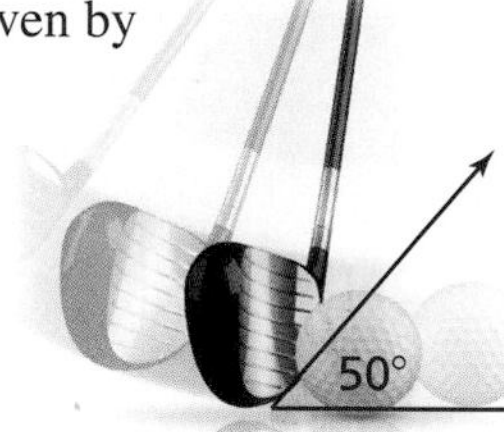

EXAMPLE 5 Solving a Real-Life Problem

The horizontal distance d (in feet) traveled by a projectile launched at an angle θ and with an initial speed v (in feet per second) is given by

$$d = \frac{v^2}{32} \sin 2\theta. \quad \text{Model for horizontal distance}$$

INTERPRETING MODELS

This model neglects air resistance and assumes that the projectile's starting and ending heights are the same.

Estimate the horizontal distance traveled by a golf ball that is hit at an angle of 50° with an initial speed of 105 feet per second.

SOLUTION

Note that the golf ball is launched at an angle of $\theta = 50°$ with initial speed of $v = 105$ feet per second.

$$d = \frac{v^2}{32} \sin 2\theta \quad \text{Write model for horizontal distance.}$$

$$= \frac{105^2}{32} \sin(2 \cdot 50°) \quad \text{Substitute 105 for } v \text{ and } 50° \text{ for } \theta.$$

$$\approx 339 \quad \text{Use a calculator.}$$

▶ The golf ball travels a horizontal distance of about 339 feet.

Monitoring Progress  Help in English and Spanish at BigIdeasMath.com

Sketch the angle. Then find its reference angle.

5. $210°$ **6.** $-260°$ **7.** $\frac{-7\pi}{9}$ **8.** $\frac{15\pi}{4}$

Evaluate the function without using a calculator.

9. $\cos(-210°)$ **10.** $\sec \frac{11\pi}{4}$

11. Use the model given in Example 5 to estimate the horizontal distance traveled by a track and field long jumper who jumps at an angle of 20° and with an initial speed of 27 feet per second.

9.3 Exercises

Dynamic Solutions available at *BigIdeasMath.com*

Vocabulary and Core Concept Check

1. **COMPLETE THE SENTENCE** A(n) __________ is an angle in standard position whose terminal side lies on an axis.

2. **WRITING** Given an angle θ in standard position with its terminal side in Quadrant III, explain how you can use a reference angle to find $\cos \theta$.

Monitoring Progress and Modeling with Mathematics

In Exercises 3–8, evaluate the six trigonometric functions of θ. *(See Example 1.)*

3.

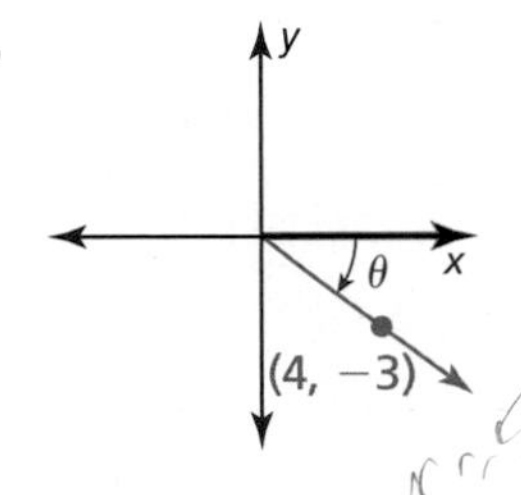

4.

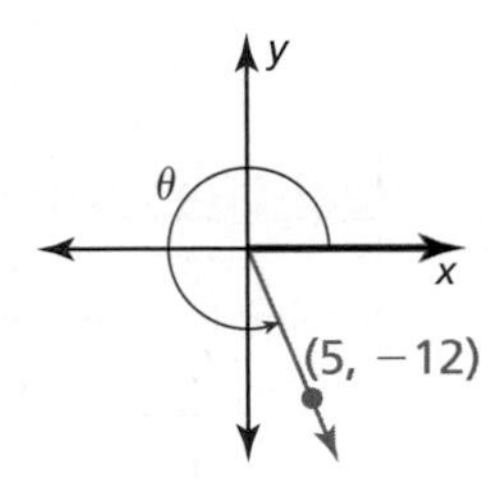

5.

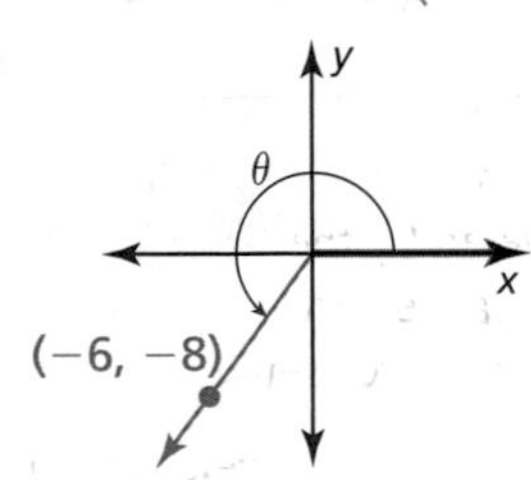

6.

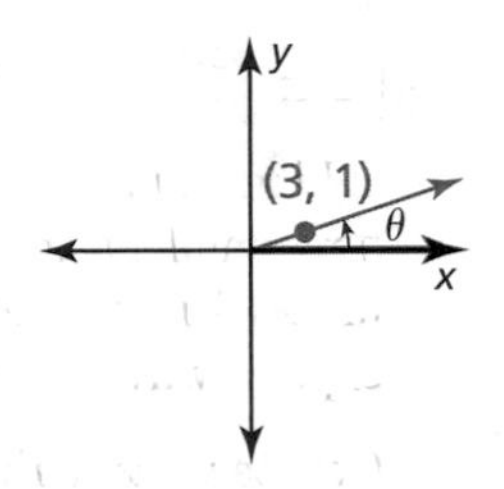

7.

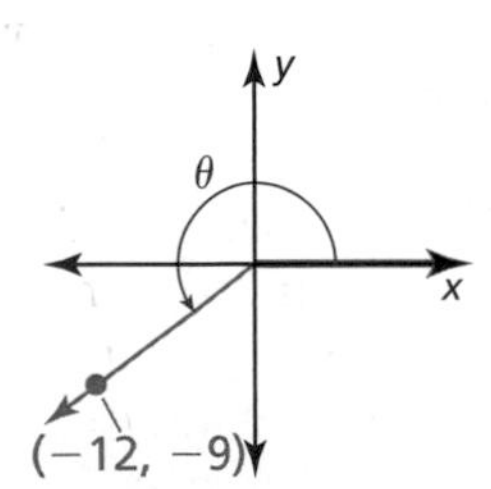

8.

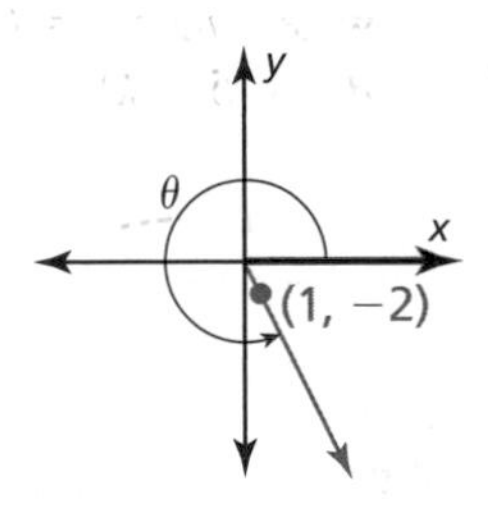

In Exercises 9–14, use the unit circle to evaluate the six trigonometric functions of θ. *(See Example 2.)*

9. $\theta = 0°$

10. $\theta = 540°$

11. $\theta = \frac{\pi}{2}$

12. $\theta = \frac{7\pi}{2}$

13. $\theta = -270°$

14. $\theta = -2\pi$

In Exercises 15–22, sketch the angle. Then find its reference angle. *(See Example 3.)*

15. $-100°$

16. $150°$

17. $320°$

18. $-370°$

19. $\frac{15\pi}{4}$

20. $\frac{8\pi}{3}$

21. $-\frac{5\pi}{6}$

22. $-\frac{13\pi}{6}$

23. ERROR ANALYSIS Let $(-3, 2)$ be a point on the terminal side of an angle θ in standard position. Describe and correct the error in finding $\tan \theta$.

$\tan \theta = \frac{x}{y} = -\frac{3}{2}$

24. ERROR ANALYSIS Describe and correct the error in finding a reference angle θ' for $\theta = 650°$.

θ is coterminal with 290°, whose terminal side lies in Quadrant IV.

So, $\theta' = 290° - 270° = 20°$.

In Exercises 25–32, evaluate the function without using a calculator. *(See Example 4.)*

25. $\sec 135°$

26. $\tan 240°$

27. $\sin(-150°)$

28. $\csc(-420°)$

29. $\tan\left(-\frac{3\pi}{4}\right)$

30. $\cot\left(\frac{-8\pi}{3}\right)$

31. $\cos \frac{7\pi}{4}$

32. $\sec \frac{11\pi}{6}$

In Exercises 33–36, use the model for horizontal distance given in Example 5.

33. You kick a football at an angle of 60° with an initial speed of 49 feet per second. Estimate the horizontal distance traveled by the football. *(See Example 5.)*

34. The "frogbot" is a robot designed for exploring rough terrain on other planets. It can jump at a 45° angle with an initial speed of 14 feet per second. Estimate the horizontal distance the frogbot can jump on Earth.

35. At what speed must the in-line skater launch himself off the ramp in order to land on the other side of the ramp?

36. To win a javelin throwing competition, your last throw must travel a horizontal distance of at least 100 feet. You release the javelin at a 40° angle with an initial speed of 71 feet per second. Do you win the competition? Justify your answer.

37. MODELING WITH MATHEMATICS A rock climber is using a rock climbing treadmill that is 10 feet long. The climber begins by lying horizontally on the treadmill, which is then rotated about its midpoint by 110° so that the rock climber is climbing toward the top. If the midpoint of the treadmill is 6 feet above the ground, how high above the ground is the top of the treadmill?

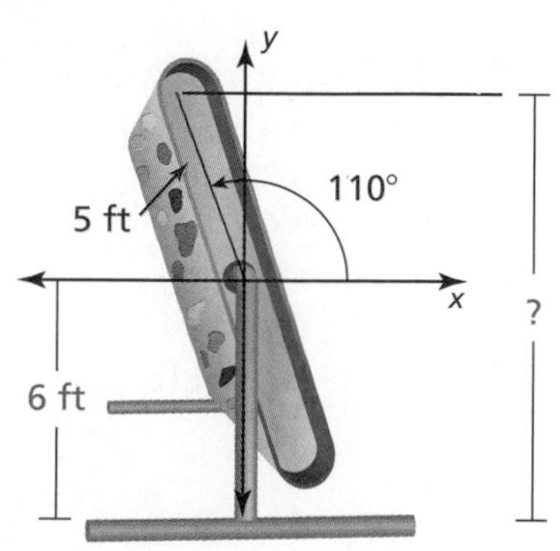

38. REASONING A Ferris wheel has a radius of 75 feet. You board a car at the bottom of the Ferris wheel, which is 10 feet above the ground, and rotate 255° counterclockwise before the ride temporarily stops. How high above the ground are you when the ride stops? If the radius of the Ferris wheel is doubled, is your height above the ground doubled? Explain your reasoning.

39. DRAWING CONCLUSIONS A sprinkler at ground level is used to water a garden. The water leaving the sprinkler has an initial speed of 25 feet per second.

a. Use the model for horizontal distance given in Example 5 to complete the table.

Angle of sprinkler, θ	Horizontal distance water travels, d
30°	
35°	
40°	
45°	
50°	
55°	
60°	

b. Which value of θ appears to maximize the horizontal distance traveled by the water? Use the model for horizontal distance and the unit circle to explain why your answer makes sense.

c. Compare the horizontal distance traveled by the water when $\theta = (45 - k)°$ with the distance when $\theta = (45 + k)°$, for $0 < k < 45$.

40. MODELING WITH MATHEMATICS Your school's marching band is performing at halftime during a football game. In the last formation, the band members form a circle 100 feet wide in the center of the field. You start at a point on the circle 100 feet from the goal line, march 300° around the circle, and then walk toward the goal line to exit the field. How far from the goal line are you at the point where you leave the circle?

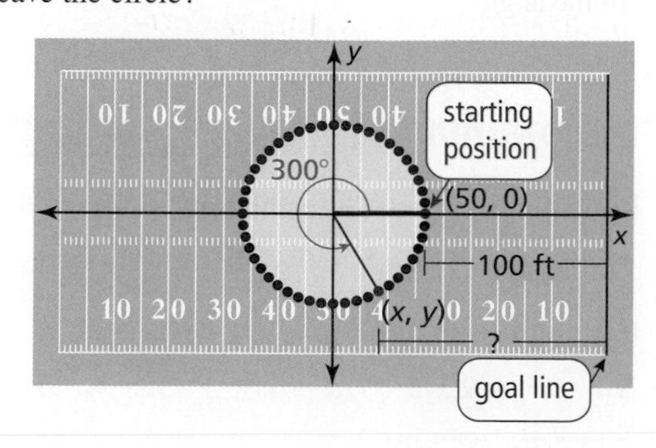

41. ANALYZING RELATIONSHIPS Use symmetry and the given information to label the coordinates of the other points corresponding to special angles on the unit circle.

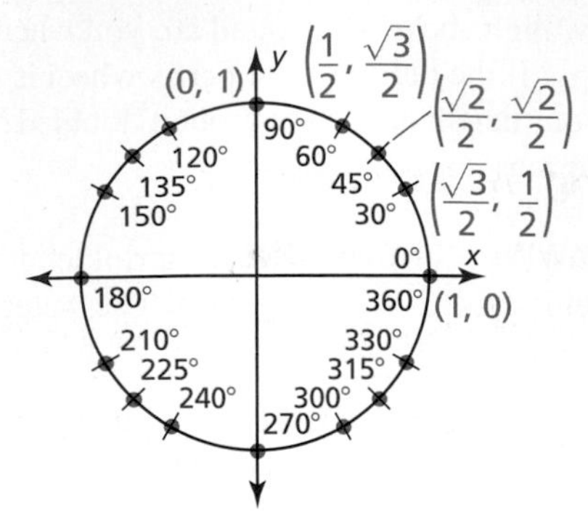

42. THOUGHT PROVOKING Use the interactive unit circle tool at *BigIdeasMath.com* to describe all values of θ for each situation.

a. $\sin \theta > 0$, $\cos \theta < 0$, and $\tan \theta > 0$

b. $\sin \theta > 0$, $\cos \theta < 0$, and $\tan \theta < 0$

43. CRITICAL THINKING Write $\tan \theta$ as the ratio of two other trigonometric functions. Use this ratio to explain why $\tan 90°$ is undefined but $\cot 90° = 0$.

44. HOW DO YOU SEE IT? Determine whether each of the six trigonometric functions of θ is *positive*, *negative*, or *zero*. Explain your reasoning.

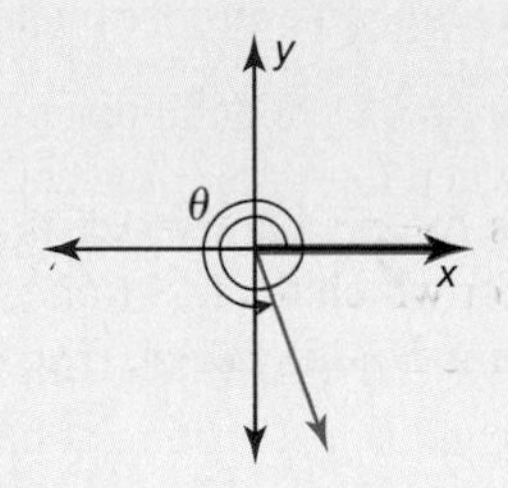

45. USING STRUCTURE A line with slope m passes through the origin. An angle θ in standard position has a terminal side that coincides with the line. Use a trigonometric function to relate the slope of the line to the angle.

46. MAKING AN ARGUMENT Your friend claims that the only solution to the trigonometric equation $\tan \theta = \sqrt{3}$ is $\theta = 60°$. Is your friend correct? Explain your reasoning.

47. PROBLEM SOLVING When two atoms in a molecule are bonded to a common atom, chemists are interested in both the bond angle and the lengths of the bonds. An ozone molecule is made up of two oxygen atoms bonded to a third oxygen atom, as shown.

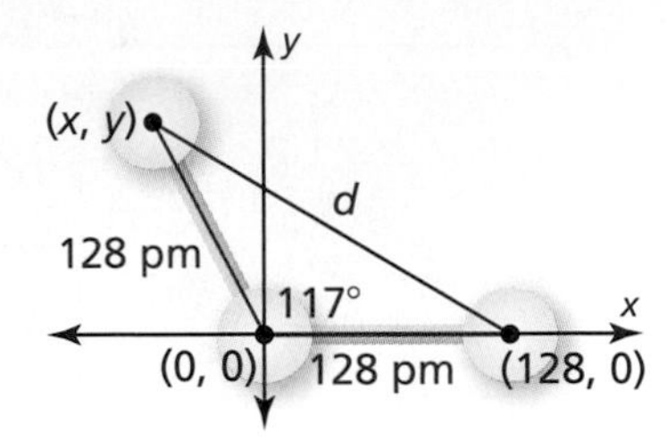

a. In the diagram, coordinates are given in picometers (pm). (*Note*: 1 pm $= 10^{-12}$ m) Find the coordinates (x, y) of the center of the oxygen atom in Quadrant II.

b. Find the distance d (in picometers) between the centers of the two unbonded oxygen atoms.

48. MATHEMATICAL CONNECTIONS The latitude of a point on Earth is the degree measure of the shortest arc from that point to the equator. For example, the latitude of point P in the diagram equals the degree measure of arc PE. At what latitude θ is the circumference of the circle of latitude at P half the distance around the equator?

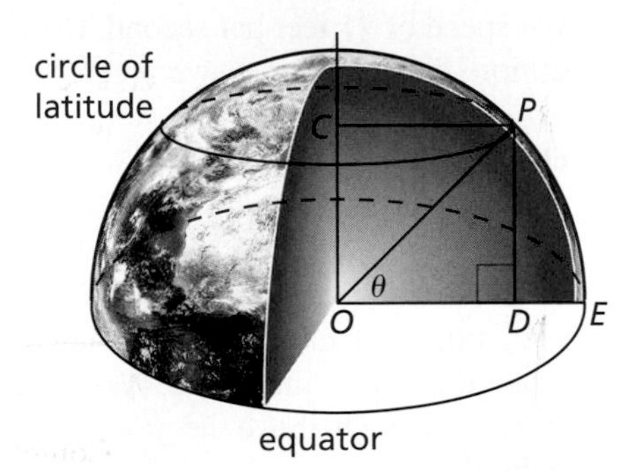

Maintaining Mathematical Proficiency

Reviewing what you learned in previous grades and lessons

Find all real zeros of the polynomial function. *(Section 4.6)*

49. $f(x) = x^4 + 2x^3 + x^2 + 8x - 12$

50. $f(x) = x^5 + 4x^4 - 14x^3 - 14x^2 - 15x - 18$

Graph the function. *(Section 4.8)*

51. $f(x) = 2(x + 3)^2(x - 1)$

52. $f(x) = \frac{1}{3}(x - 4)(x + 5)(x + 9)$

53. $f(x) = x^2(x + 1)^3(x - 2)$

9.4 Graphing Sine and Cosine Functions

Essential Question What are the characteristics of the graphs of the sine and cosine functions?

EXPLORATION 1 Graphing the Sine Function

Work with a partner.

a. Complete the table for $y = \sin x$, where x is an angle measure in radians.

x	-2π	$-\frac{7\pi}{4}$	$-\frac{3\pi}{2}$	$-\frac{5\pi}{4}$	$-\pi$	$-\frac{3\pi}{4}$	$-\frac{\pi}{2}$	$-\frac{\pi}{4}$	0
$y = \sin x$									
x	$\frac{\pi}{4}$	$\frac{\pi}{2}$	$\frac{3\pi}{4}$	π	$\frac{5\pi}{4}$	$\frac{3\pi}{2}$	$\frac{7\pi}{4}$	2π	$\frac{9\pi}{4}$
$y = \sin x$									

b. Plot the points (x, y) from part (a). Draw a smooth curve through the points to sketch the graph of $y = \sin x$.

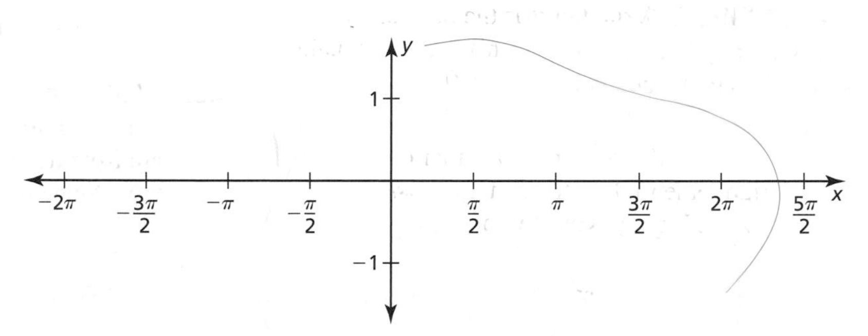

c. Use the graph to identify the x-intercepts, the x-values where the local maximums and minimums occur, and the intervals for which the function is increasing or decreasing over $-2\pi \le x \le 2\pi$. Is the sine function *even*, *odd*, or *neither*?

EXPLORATION 2 Graphing the Cosine Function

Work with a partner.

a. Complete a table for $y = \cos x$ using the same values of x as those used in Exploration 1.

b. Plot the points (x, y) from part (a) and sketch the graph of $y = \cos x$.

c. Use the graph to identify the x-intercepts, the x-values where the local maximums and minimums occur, and the intervals for which the function is increasing or decreasing over $-2\pi \le x \le 2\pi$. Is the cosine function *even*, *odd*, or *neither*?

LOOKING FOR STRUCTURE

To be proficient in math, you need to look closely to discern a pattern or structure.

Communicate Your Answer

3. What are the characteristics of the graphs of the sine and cosine functions?

4. Describe the end behavior of the graph of $y = \sin x$.

9.4 Lesson

Core Vocabulary

amplitude, *p. 486*
periodic function, *p. 486*
cycle, *p. 486*
period, *p. 486*
phase shift, *p. 488*
midline, *p. 488*

Previous
transformations
x-intercept

What You Will Learn

- Explore characteristics of sine and cosine functions.
- Stretch and shrink graphs of sine and cosine functions.
- Translate graphs of sine and cosine functions.
- Reflect graphs of sine and cosine functions.

Exploring Characteristics of Sine and Cosine Functions

In this lesson, you will learn to graph sine and cosine functions. The graphs of sine and cosine functions are related to the graphs of the parent functions $y = \sin x$ and $y = \cos x$, which are shown below.

x	-2π	$-\frac{3\pi}{2}$	$-\pi$	$-\frac{\pi}{2}$	0	$\frac{\pi}{2}$	π	$\frac{3\pi}{2}$	2π
$y = \sin x$	0	1	0	-1	0	1	0	-1	0
$y = \cos x$	1	0	-1	0	1	0	-1	0	1

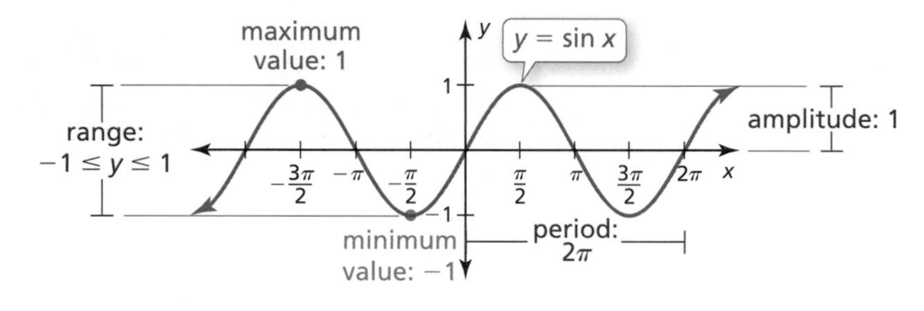

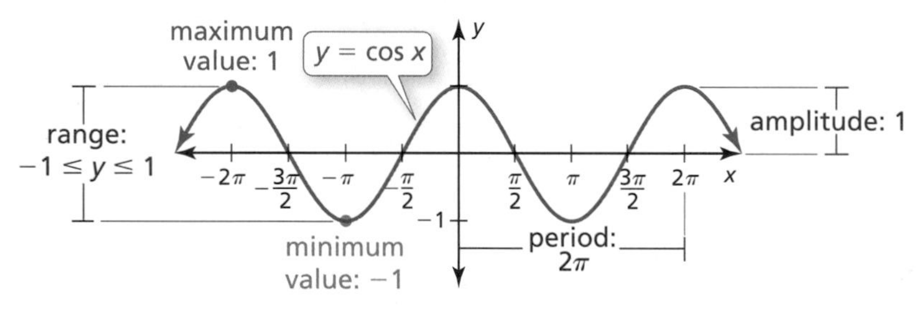

Core Concept

Characteristics of $y = \sin x$ and $y = \cos x$

- The domain of each function is all real numbers.
- The range of each function is $-1 \le y \le 1$. So, the minimum value of each function is -1 and the maximum value is 1.
- The **amplitude** of the graph of each function is one-half of the difference of the maximum value and the minimum value, or $\frac{1}{2}[1 - (-1)] = 1$.
- Each function is **periodic**, which means that its graph has a repeating pattern. The shortest repeating portion of the graph is called a **cycle**. The horizontal length of each cycle is called the **period**. Each graph shown above has a period of 2π.
- The x-intercepts for $y = \sin x$ occur when $x = 0, \pm\pi, \pm 2\pi, \pm 3\pi, \ldots$.
- The x-intercepts for $y = \cos x$ occur when $x = \pm\frac{\pi}{2}, \pm\frac{3\pi}{2}, \pm\frac{5\pi}{2}, \pm\frac{7\pi}{2}, \ldots$.

Stretching and Shrinking Sine and Cosine Functions

The graphs of $y = a \sin bx$ and $y = a \cos bx$ represent transformations of their parent functions. The value of a indicates a vertical stretch ($a > 1$) or a vertical shrink ($0 < a < 1$) and changes the amplitude of the graph. The value of b indicates a horizontal stretch ($0 < b < 1$) or a horizontal shrink ($b > 1$) and changes the period of the graph.

> **REMEMBER**
>
> The graph of $y = a \cdot f(x)$ is a vertical stretch or shrink of the graph of $y = f(x)$ by a factor of a.
>
> The graph of $y = f(bx)$ is a horizontal stretch or shrink of the graph of $y = f(x)$ by a factor of $\frac{1}{b}$.

$$y = a \sin bx$$

$$y = a \cos bx$$

vertical stretch or shrink by a factor of a (points to a) — horizontal stretch or shrink by a factor of $\frac{1}{b}$ (points to b)

Core Concept

Amplitude and Period

The amplitude and period of the graphs of $y = a \sin bx$ and $y = a \cos bx$, where a and b are nonzero real numbers, are as follows:

$$\text{Amplitude} = |a| \qquad \text{Period} = \frac{2\pi}{|b|}$$

Each graph below shows five key points that partition the interval $0 \le x \le \frac{2\pi}{b}$ into four equal parts. You can use these points to sketch the graphs of $y = a \sin bx$ and $y = a \cos bx$. The x-intercepts, maximum, and minimum occur at these points.

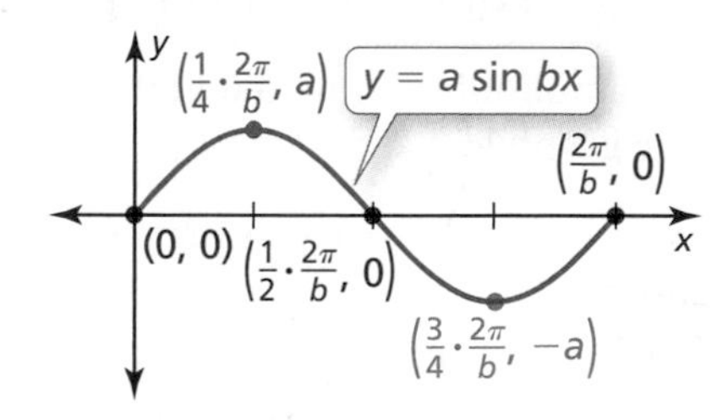

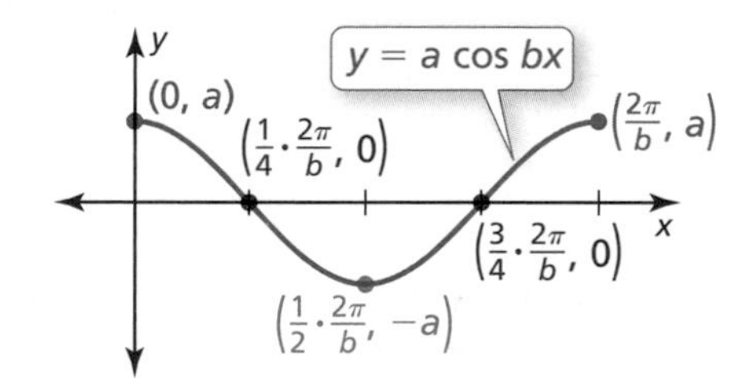

EXAMPLE 1 **Graphing a Sine Function**

Identify the amplitude and period of $g(x) = 4 \sin x$. Then graph the function and describe the graph of g as a transformation of the graph of $f(x) = \sin x$.

> **REMEMBER**
>
> A vertical stretch of a graph does not change its x-intercept(s). So, it makes sense that the x-intercepts of $g(x) = 4 \sin x$ and $f(x) = \sin x$ are the same.

SOLUTION

The function is of the form $g(x) = a \sin bx$ where $a = 4$ and $b = 1$. So, the amplitude is $a = 4$ and the period is $\frac{2\pi}{b} = \frac{2\pi}{1} = 2\pi$.

Intercepts: $(0, 0)$; $\left(\frac{1}{2} \cdot 2\pi, 0\right) = (\pi, 0)$; $(2\pi, 0)$

Maximum: $\left(\frac{1}{4} \cdot 2\pi, 4\right) = \left(\frac{\pi}{2}, 4\right)$

Minimum: $\left(\frac{3}{4} \cdot 2\pi, -4\right) = \left(\frac{3\pi}{2}, -4\right)$

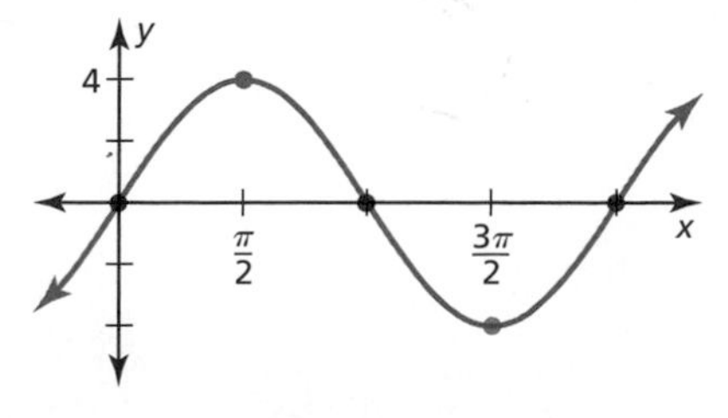

▶ The graph of g is a vertical stretch by a factor of 4 of the graph of f.

EXAMPLE 2 Graphing a Cosine Function

Identify the amplitude and period of $g(x) = \frac{1}{2}\cos 2\pi x$. Then graph the function and describe the graph of g as a transformation of the graph of $f(x) = \cos x$.

SOLUTION

The function is of the form $g(x) = a \cos bx$ where $a = \frac{1}{2}$ and $b = 2\pi$. So, the amplitude is $a = \frac{1}{2}$ and the period is $\frac{2\pi}{b} = \frac{2\pi}{2\pi} = 1$.

STUDY TIP

After you have drawn one complete cycle of the graph in Example 2 on the interval $0 \leq x \leq 1$, you can extend the graph by repeating the cycle as many times as desired to the left and right of $0 \leq x \leq 1$.

Intercepts: $\left(\frac{1}{4} \cdot 1, 0\right) = \left(\frac{1}{4}, 0\right)$; $\left(\frac{3}{4} \cdot 1, 0\right) = \left(\frac{3}{4}, 0\right)$

Maximums: $\left(0, \frac{1}{2}\right)$; $\left(1, \frac{1}{2}\right)$

Minimum: $\left(\frac{1}{2} \cdot 1, -\frac{1}{2}\right) = \left(\frac{1}{2}, -\frac{1}{2}\right)$

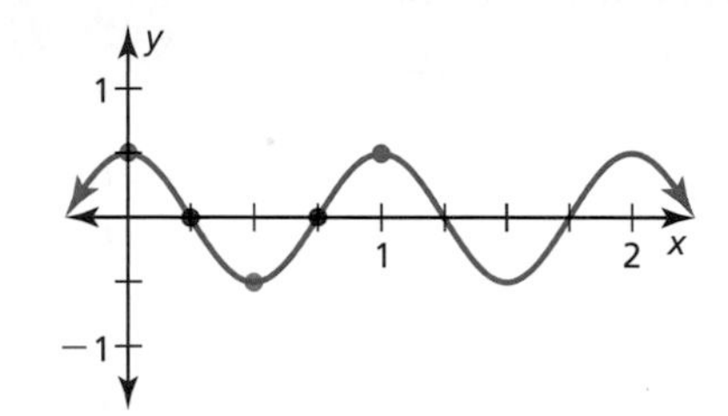

▶ The graph of g is a vertical shrink by a factor of $\frac{1}{2}$ and a horizontal shrink by a factor of $\frac{1}{2\pi}$ of the graph of f.

Monitoring Progress

Help in English and Spanish at *BigIdeasMath.com*

Identify the amplitude and period of the function. Then graph the function and describe the graph of g as a transformation of the graph of its parent function.

1. $g(x) = \frac{1}{4}\sin x$ **2.** $g(x) = \cos 2x$ **3.** $g(x) = 2 \sin \pi x$ **4.** $g(x) = \frac{1}{3}\cos \frac{1}{2}x$

REMEMBER

The graph of $y = f(x) + k$ is a vertical translation of the graph of $y = f(x)$.

The graph of $y = f(x - h)$ is a horizontal translation of the graph of $y = f(x)$.

Translating Sine and Cosine Functions

The graphs of $y = a \sin b(x - h) + k$ and $y = a \cos b(x - h) + k$ represent translations of $y = a \sin bx$ and $y = a \cos bx$. The value of k indicates a translation up $(k > 0)$ or down $(k < 0)$. The value of h indicates a translation left $(h < 0)$ or right $(h > 0)$. A horizontal translation of a periodic function is called a **phase shift**.

Core Concept

Graphing $y = a \sin b(x - h) + k$ and $y = a \cos b(x - h) + k$

To graph $y = a \sin b(x - h) + k$ or $y = a \cos b(x - h) + k$ where $a > 0$ and $b > 0$, follow these steps:

Step 1 Identify the amplitude a, the period $\frac{2\pi}{b}$, the horizontal shift h, and the vertical shift k of the graph.

Step 2 Draw the horizontal line $y = k$, called the **midline** of the graph.

Step 3 Find the five key points by translating the key points of $y = a \sin bx$ or $y = a \cos bx$ horizontally h units and vertically k units.

Step 4 Draw the graph through the five translated key points.

EXAMPLE 3 Graphing a Vertical Translation

Graph $g(x) = 2 \sin 4x + 3$.

LOOKING FOR STRUCTURE

The graph of g is a translation 3 units up of the graph of $f(x) = 2 \sin 4x$. So, add 3 to the y-coordinates of the five key points of f.

SOLUTION

Step 1 Identify the amplitude, period, horizontal shift, and vertical shift.

Amplitude: $a = 2$ Horizontal shift: $h = 0$

Period: $\frac{2\pi}{b} = \frac{2\pi}{4} = \frac{\pi}{2}$ Vertical shift: $k = 3$

Step 2 Draw the midline of the graph, $y = 3$.

Step 3 Find the five key points.

On $y = k$: $(0, 0 + 3) = (0, 3)$; $\left(\frac{\pi}{4}, 0 + 3\right) = \left(\frac{\pi}{4}, 3\right)$; $\left(\frac{\pi}{2}, 0 + 3\right) = \left(\frac{\pi}{2}, 3\right)$

Maximum: $\left(\frac{\pi}{8}, 2 + 3\right) = \left(\frac{\pi}{8}, 5\right)$

Minimum: $\left(\frac{3\pi}{8}, -2 + 3\right) = \left(\frac{3\pi}{8}, 1\right)$

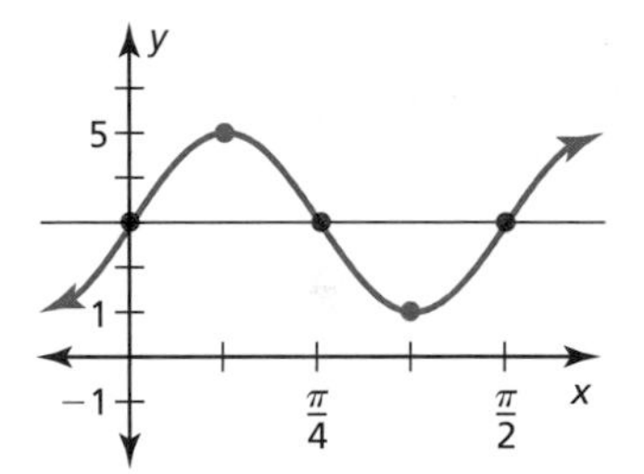

Step 4 Draw the graph through the key points.

EXAMPLE 4 Graphing a Horizontal Translation

Graph $g(x) = 5 \cos \frac{1}{2}(x - 3\pi)$.

LOOKING FOR STRUCTURE

The graph of g is a translation 3π units right of the graph of $f(x) = 5 \cos \frac{1}{2}x$. So, add 3π to the x-coordinates of the five key points of f.

SOLUTION

Step 1 Identify the amplitude, period, horizontal shift, and vertical shift.

Amplitude: $a = 5$ Horizontal shift: $h = 3\pi$

Period: $\frac{2\pi}{b} = \frac{2\pi}{\frac{1}{2}} = 4\pi$ Vertical shift: $k = 0$

Step 2 Draw the midline of the graph. Because $k = 0$, the midline is the x-axis.

Step 3 Find the five key points.

On $y = k$: $(\pi + 3\pi, 0) = (4\pi, 0)$; $(3\pi + 3\pi, 0) = (6\pi, 0)$

Maximums: $(0 + 3\pi, 5) = (3\pi, 5)$; $(4\pi + 3\pi, 5) = (7\pi, 5)$

Minimum: $(2\pi + 3\pi, -5) = (5\pi, -5)$

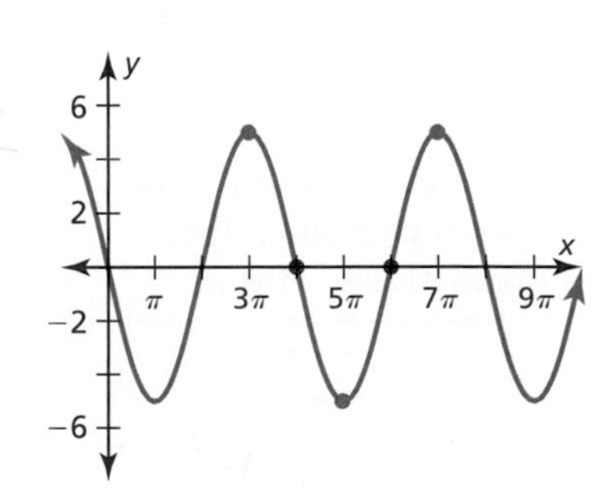

Step 4 Draw the graph through the key points.

Monitoring Progress

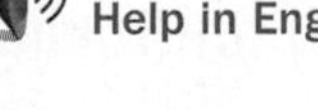

Help in English and Spanish at *BigIdeasMath.com*

Graph the function.

5. $g(x) = \cos x + 4$ **6.** $g(x) = \frac{1}{2} \sin\left(x - \frac{\pi}{2}\right)$ **7.** $g(x) = \sin(x + \pi) - 1$

Reflecting Sine and Cosine Functions

You have graphed functions of the form $y = a \sin b(x - h) + k$ and $y = a \cos b(x - h) + k$, where $a > 0$ and $b > 0$. To see what happens when $a < 0$, consider the graphs of $y = -\sin x$ and $y = -\cos x$.

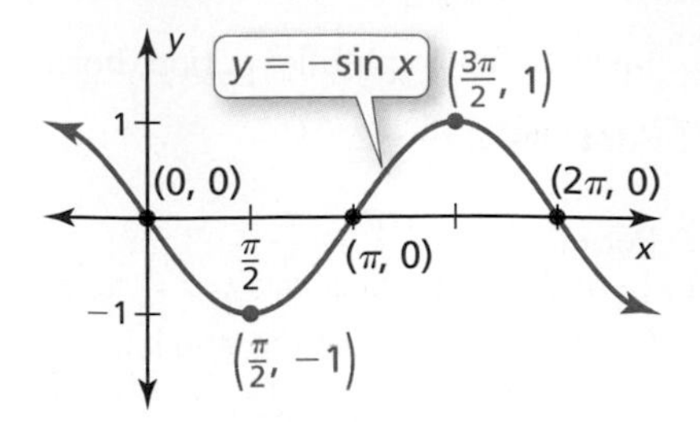

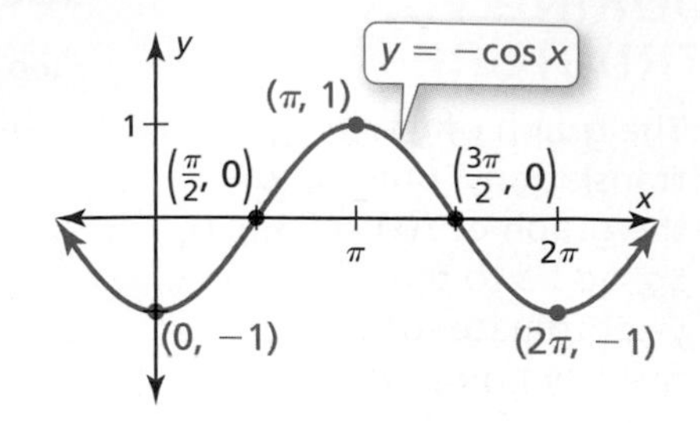

REMEMBER

This result makes sense because the graph of $y = -f(x)$ is a reflection in the x-axis of the graph of $y = f(x)$.

The graphs are reflections of the graphs of $y = \sin x$ and $y = \cos x$ in the x-axis. In general, when $a < 0$, the graphs of $y = a \sin b(x - h) + k$ and $y = a \cos b(x - h) + k$ are reflections of the graphs of $y = |a| \sin b(x - h) + k$ and $y = |a| \cos b(x - h) + k$, respectively, in the midline $y = k$.

EXAMPLE 5 Graphing a Reflection

Graph $g(x) = -2 \sin \frac{2}{3}\left(x - \frac{\pi}{2}\right)$.

SOLUTION

Step 1 Identify the amplitude, period, horizontal shift, and vertical shift.

Amplitude: $|a| = |-2| = 2$ Horizontal shift: $h = \frac{\pi}{2}$

Period: $\frac{2\pi}{b} = \frac{2\pi}{\frac{2}{3}} = 3\pi$ Vertical shift: $k = 0$

Step 2 Draw the midline of the graph. Because $k = 0$, the midline is the x-axis.

Step 3 Find the five key points of $f(x) = |-2| \sin \frac{2}{3}\left(x - \frac{\pi}{2}\right)$.

On $y = k$: $\left(0 + \frac{\pi}{2}, 0\right) = \left(\frac{\pi}{2}, 0\right)$; $\left(\frac{3\pi}{2} + \frac{\pi}{2}, 0\right) = (2\pi, 0)$; $\left(3\pi + \frac{\pi}{2}, 0\right) = \left(\frac{7\pi}{2}, 0\right)$

Maximum: $\left(\frac{3\pi}{4} + \frac{\pi}{2}, 2\right) = \left(\frac{5\pi}{4}, 2\right)$ Minimum: $\left(\frac{9\pi}{4} + \frac{\pi}{2}, -2\right) = \left(\frac{11\pi}{4}, -2\right)$

STUDY TIP

In Example 5, the maximum value and minimum value of f are the minimum value and maximum value, respectively, of g.

Step 4 Reflect the graph. Because $a < 0$, the graph is reflected in the midline $y = 0$. So, $\left(\frac{5\pi}{4}, 2\right)$ becomes $\left(\frac{5\pi}{4}, -2\right)$ and $\left(\frac{11\pi}{4}, -2\right)$ becomes $\left(\frac{11\pi}{4}, 2\right)$.

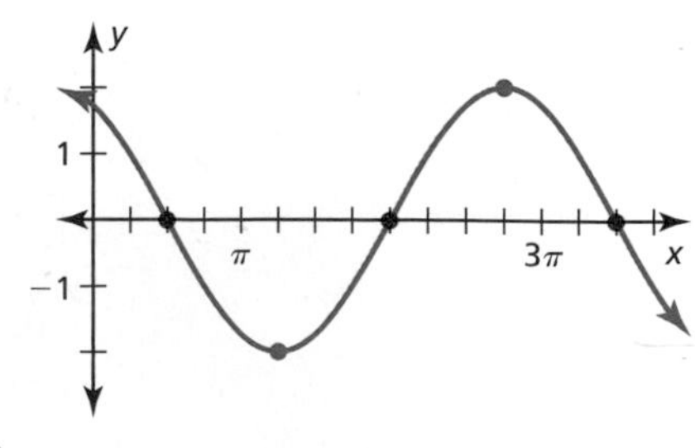

Step 5 Draw the graph through the key points.

Monitoring Progress  Help in English and Spanish at *BigIdeasMath.com*

Graph the function.

8. $g(x) = -\cos\left(x + \frac{\pi}{2}\right)$ **9.** $g(x) = -3 \sin \frac{1}{2}x + 2$ **10.** $g(x) = -2 \cos 4x - 1$

9.4 Exercises

Dynamic Solutions available at *BigIdeasMath.com*

Vocabulary and Core Concept Check

1. **COMPLETE THE SENTENCE** The shortest repeating portion of the graph of a periodic function is called a(n) _________.

2. **WRITING** Compare the amplitudes and periods of the functions $y = \frac{1}{2}\cos x$ and $y = 3 \cos 2x$.

3. **VOCABULARY** What is a phase shift? Give an example of a sine function that has a phase shift.

4. **VOCABULARY** What is the midline of the graph of the function $y = 2 \sin 3(x + 1) - 2$?

Monitoring Progress and Modeling with Mathematics

USING STRUCTURE In Exercises 5–8, determine whether the graph represents a periodic function. If so, identify the period.

5.

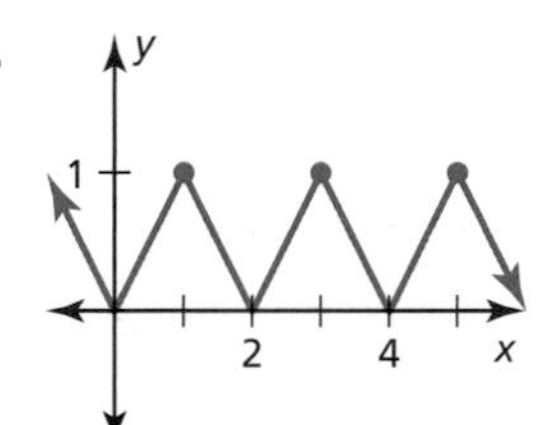

6.

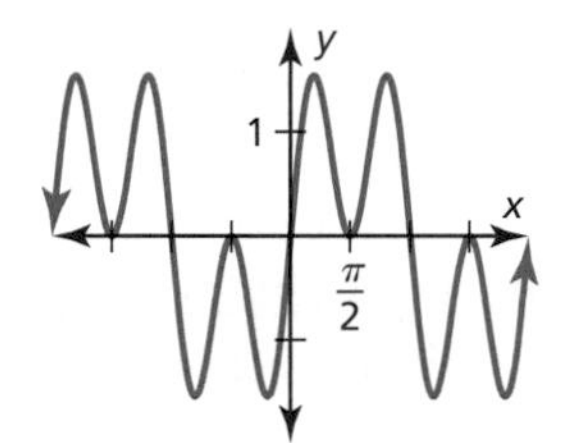

7.

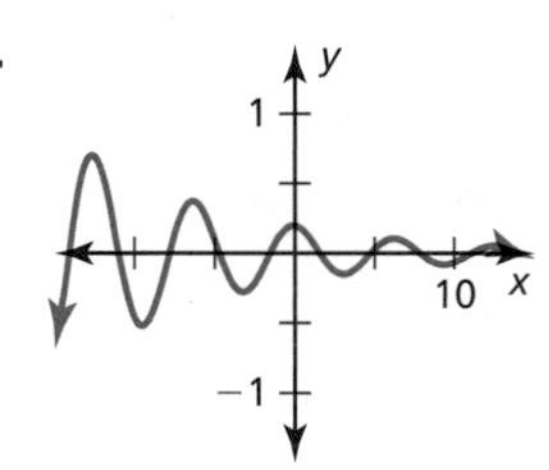

8.

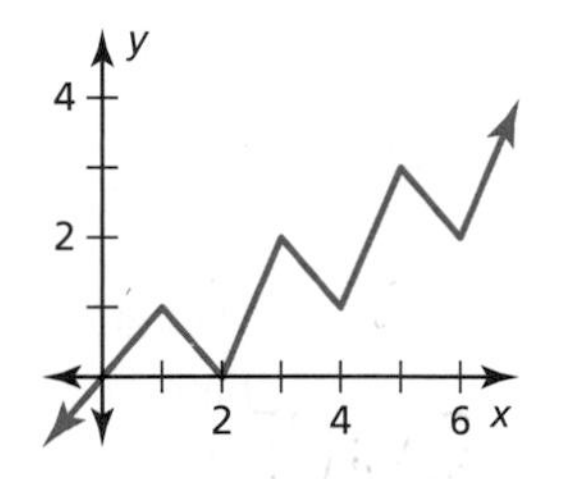

In Exercises 9–12, identify the amplitude and period of the graph of the function.

9.

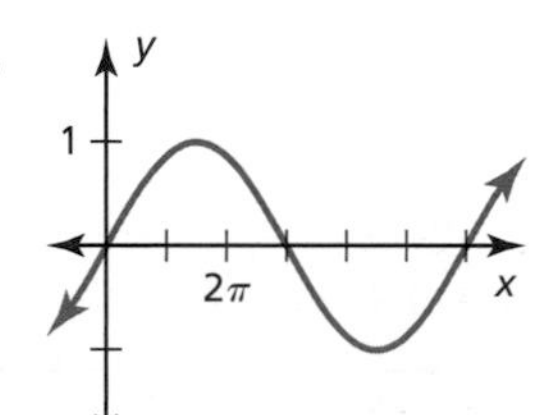

10.

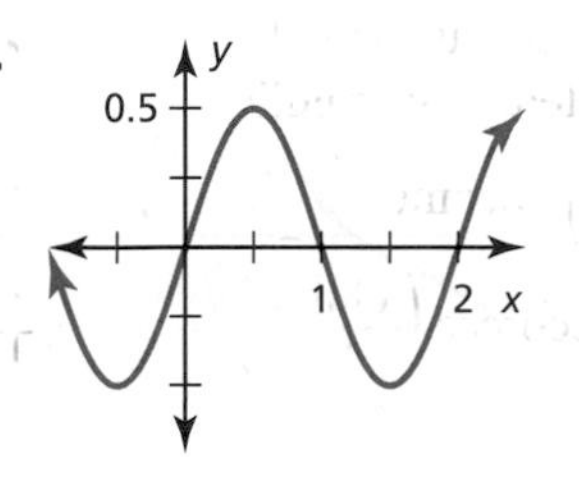

11.

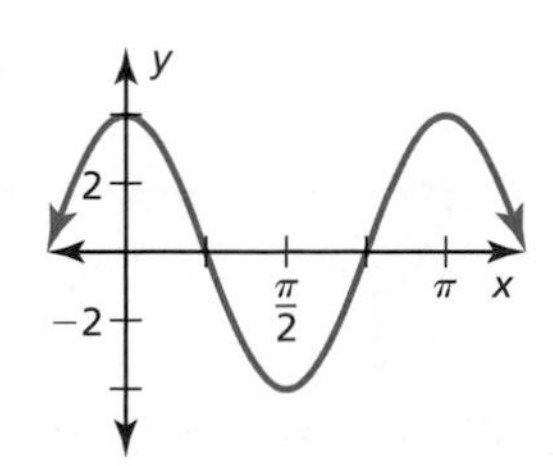

12.

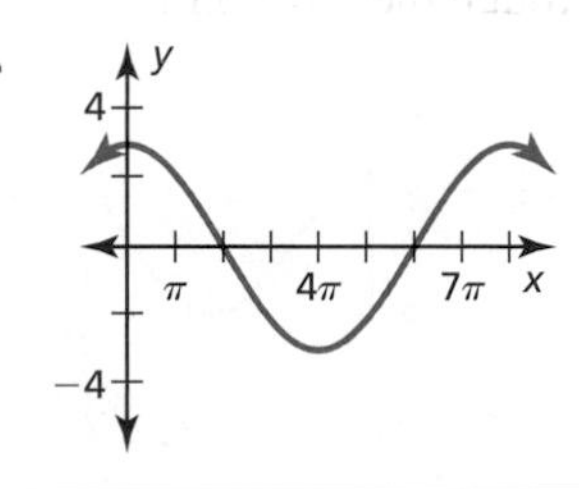

In Exercises 13–20, identify the amplitude and period of the function. Then graph the function and describe the graph of *g* as a transformation of the graph of its parent function. *(See Examples 1 and 2.)*

13. $g(x) = 3 \sin x$

14. $g(x) = 2 \sin x$

15. $g(x) = \cos 3x$

16. $g(x) = \cos 4x$

17. $g(x) = \sin 2\pi x$

18. $g(x) = 3 \sin 2x$

19. $g(x) = \frac{1}{3} \cos 4x$

20. $g(x) = \frac{1}{2} \cos 4\pi x$

21. **ANALYZING EQUATIONS** Which functions have an amplitude of 4 and a period of 2?

    (A) $y = 4 \cos 2x$

    (B) $y = -4 \sin \pi x$

    (C) $y = 2 \sin 4x$

    (D) $y = 4 \cos \pi x$

22. **WRITING EQUATIONS** Write an equation of the form $y = a \sin bx$, where $a > 0$ and $b > 0$, so that the graph has the given amplitude and period.

    a. amplitude: 1
    period: 5

    b. amplitude: 10
    period: 4

    c. amplitude: 2
    period: 2π

    d. amplitude: $\frac{1}{2}$
    period: 3π

23. **MODELING WITH MATHEMATICS** The motion of a pendulum can be modeled by the function $d = 4 \cos 8\pi t$, where d is the horizontal displacement (in inches) of the pendulum relative to its position at rest and t is the time (in seconds). Find and interpret the period and amplitude in the context of this situation. Then graph the function.

24. **MODELING WITH MATHEMATICS** A buoy bobs up and down as waves go past. The vertical displacement y (in feet) of the buoy with respect to sea level can be modeled by $y = 1.75 \cos \frac{\pi}{3}t$, where t is the time (in seconds). Find and interpret the period and amplitude in the context of the problem. Then graph the function.

In Exercises 25–34, graph the function. *(See Examples 3 and 4.)*

25. $g(x) = \sin x + 2$
26. $g(x) = \cos x - 4$
27. $g(x) = \cos\left(x - \frac{\pi}{2}\right)$
28. $g(x) = \sin\left(x + \frac{\pi}{4}\right)$
29. $g(x) = 2 \cos x - 1$
30. $g(x) = 3 \sin x + 1$
31. $g(x) = \sin 2(x + \pi)$
32. $g(x) = \cos 2(x - \pi)$
33. $g(x) = \sin \frac{1}{2}(x + 2\pi) + 3$
34. $g(x) = \cos \frac{1}{2}(x - 3\pi) - 5$

35. **ERROR ANALYSIS** Describe and correct the error in finding the period of the function $y = \sin \frac{2}{3}x$.

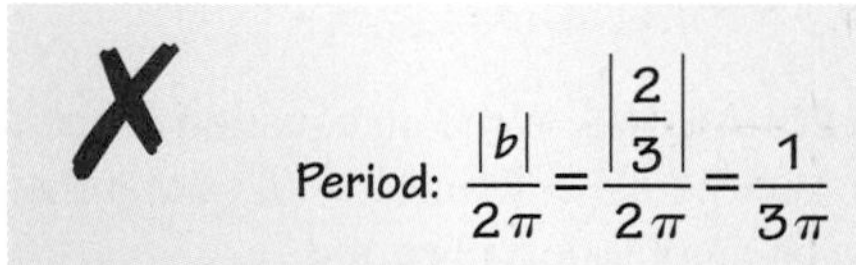

36. **ERROR ANALYSIS** Describe and correct the error in determining the point where the maximum value of the function $y = 2 \sin\left(x - \frac{\pi}{2}\right)$ occurs.

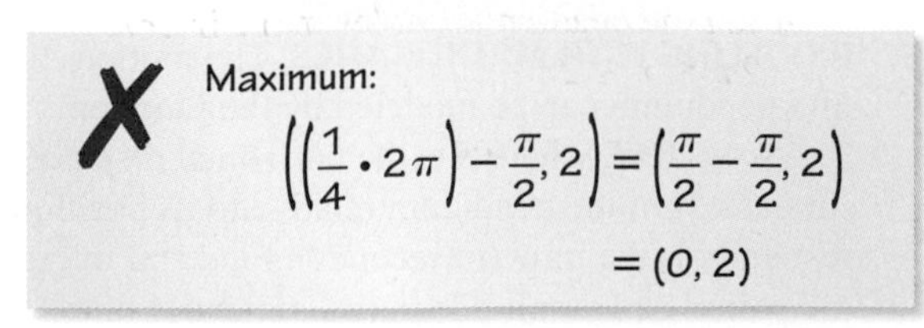

USING STRUCTURE In Exercises 37–40, describe the transformation of the graph of f represented by the function g.

37. $f(x) = \cos x$, $g(x) = 2 \cos\left(x - \frac{\pi}{2}\right) + 1$
38. $f(x) = \sin x$, $g(x) = 3 \sin\left(x + \frac{\pi}{4}\right) - 2$
39. $f(x) = \sin x$, $g(x) = \sin 3(x + 3\pi) - 5$
40. $f(x) = \cos x$, $g(x) = \cos 6(x - \pi) + 9$

In Exercises 41–48, graph the function. *(See Example 5.)*

41. $g(x) = -\cos x + 3$
42. $g(x) = -\sin x - 5$
43. $g(x) = -\sin \frac{1}{2}x - 2$
44. $g(x) = -\cos 2x + 1$
45. $g(x) = -\sin(x - \pi) + 4$
46. $g(x) = -\cos(x + \pi) - 2$
47. $g(x) = -4 \cos\left(x + \frac{\pi}{4}\right) - 1$
48. $g(x) = -5 \sin\left(x - \frac{\pi}{2}\right) + 3$

49. **USING EQUATIONS** Which of the following is a point where the maximum value of the graph of $y = -4 \cos\left(x - \frac{\pi}{2}\right)$ occurs?

Ⓐ $\left(-\frac{\pi}{2}, 4\right)$
Ⓑ $\left(\frac{\pi}{2}, 4\right)$
Ⓒ $(0, 4)$
Ⓓ $(\pi, 4)$

50. **ANALYZING RELATIONSHIPS** Match each function with its graph. Explain your reasoning.

a. $y = 3 + \sin x$
b. $y = -3 + \cos x$
c. $y = \sin 2\left(x - \frac{\pi}{2}\right)$
d. $y = \cos 2\left(x - \frac{\pi}{2}\right)$

A.
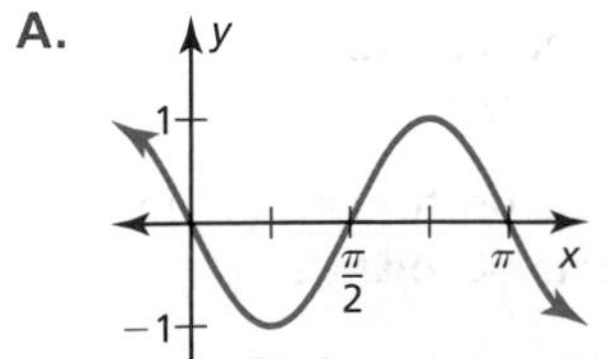

B.
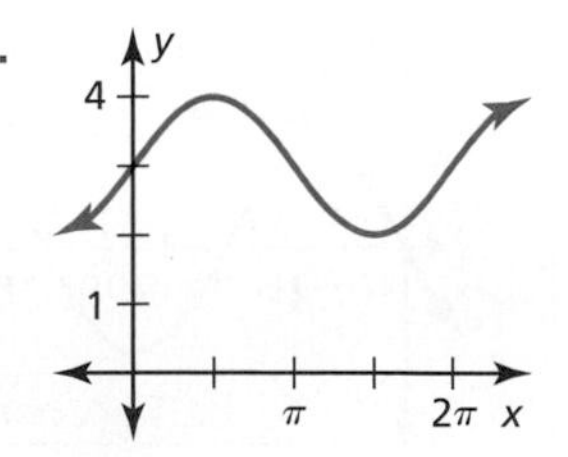

C.
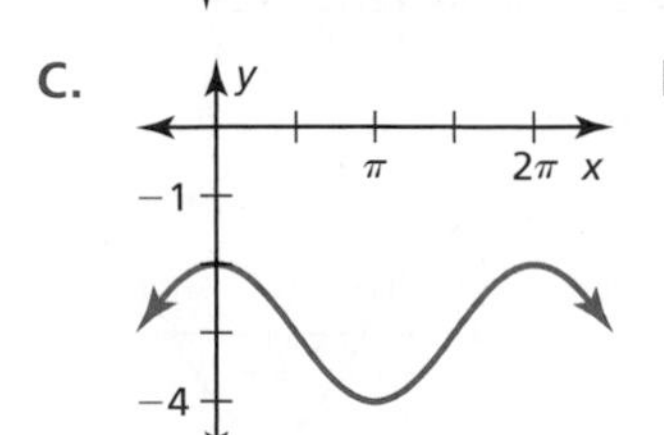

D.
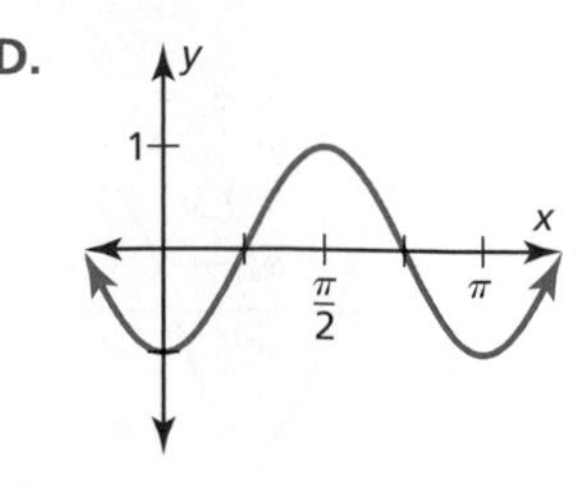

WRITING EQUATIONS In Exercises 51–54, write a rule for g that represents the indicated transformations of the graph of f.

51. $f(x) = 3 \sin x$; translation 2 units up and π units right

52. $f(x) = \cos 2\pi x$; translation 4 units down and 3 units left

53. $f(x) = \frac{1}{3} \cos \pi x$; translation 1 unit down, followed by a reflection in the line $y = -1$

54. $f(x) = \frac{1}{2} \sin 6x$; translation $\frac{3}{2}$ units down and 1 unit right, followed by a reflection in the line $y = -\frac{3}{2}$

55. MODELING WITH MATHEMATICS The height h (in feet) of a swing above the ground can be modeled by the function $h = -8 \cos \theta + 10$, where the pivot is 10 feet above the ground, the rope is 8 feet long, and θ is the angle that the rope makes with the vertical. Graph the function. What is the height of the swing when θ is 45°?

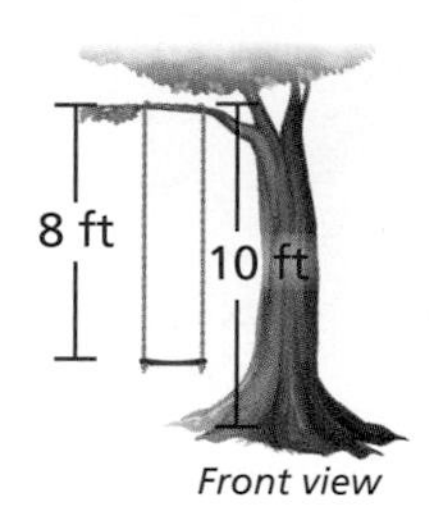

Front view

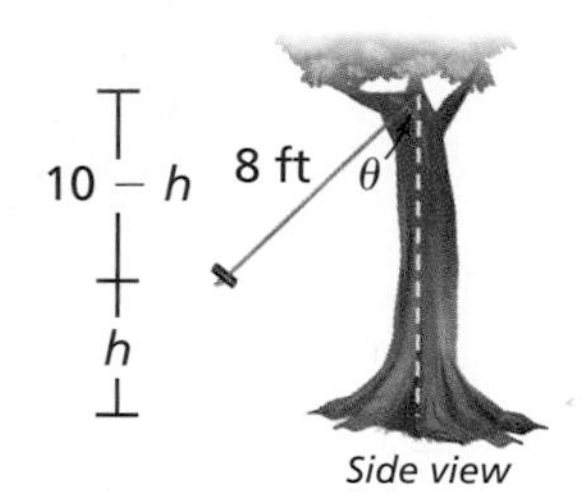

Side view

56. DRAWING A CONCLUSION In a particular region, the population L (in thousands) of lynx (the predator) and the population H (in thousands) of hares (the prey) can be modeled by the equations

$$L = 11.5 + 6.5 \sin \frac{\pi}{5} t$$

$$H = 27.5 + 17.5 \cos \frac{\pi}{5} t$$

where t is the time in years.

a. Determine the ratio of hares to lynx when $t = 0$, 2.5, 5, and 7.5 years.

b. Use the figure to explain how the changes in the two populations appear to be related.

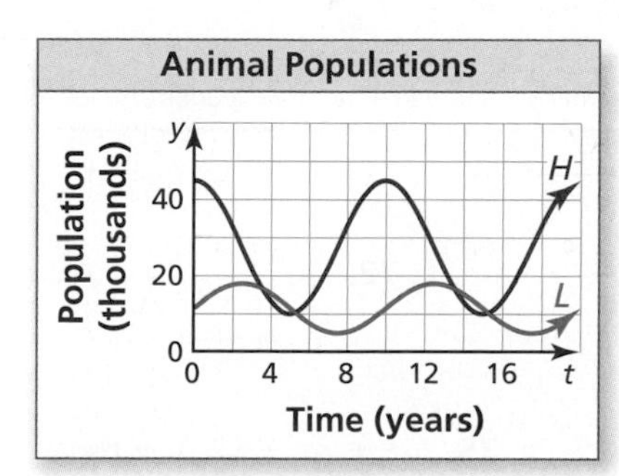

57. USING TOOLS The average wind speed s (in miles per hour) in the Boston Harbor can be approximated by

$$s = 3.38 \sin \frac{\pi}{180}(t + 3) + 11.6$$

where t is the time in days and $t = 0$ represents January 1. Use a graphing calculator to graph the function. On which days of the year is the average wind speed 10 miles per hour? Explain your reasoning.

58. USING TOOLS The water depth d (in feet) for the Bay of Fundy can be modeled by $d = 35 - 28 \cos \frac{\pi}{6.2} t$, where t is the time in hours and $t = 0$ represents midnight. Use a graphing calculator to graph the function. At what time(s) is the water depth 7 feet? Explain.

high tide

low tide

59. MULTIPLE REPRESENTATIONS Find the average rate of change of each function over the interval $0 < x < \pi$.

a. $y = 2 \cos x$

b.

x	0	$\frac{\pi}{2}$	π	$\frac{3\pi}{2}$	2π
$f(x) = -\cos x$	-1	0	1	0	-1

c.

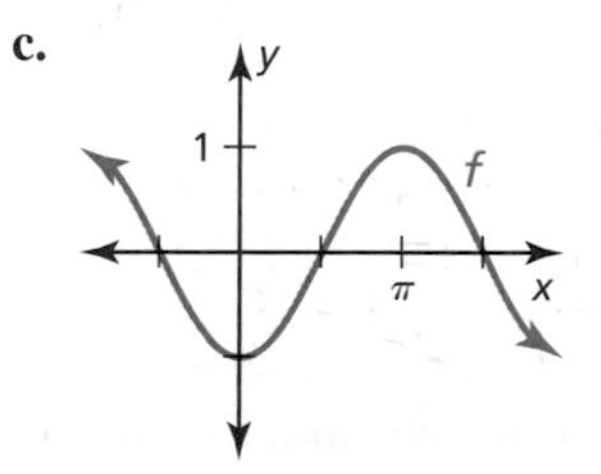

60. REASONING Consider the functions $y = \sin(-x)$ and $y = \cos(-x)$.

a. Construct a table of values for each equation using the quadrantal angles in the interval $-2\pi \leq x \leq 2\pi$.

b. Graph each function.

c. Describe the transformations of the graphs of the parent functions.

61. **MODELING WITH MATHEMATICS** You are riding a Ferris wheel that turns for 180 seconds. Your height h (in feet) above the ground at any time t (in seconds) can be modeled by the equation

$h = 85 \sin \frac{\pi}{20}(t - 10) + 90.$

a. Graph the function.

b. How many cycles does the Ferris wheel make in 180 seconds?

c. What are your maximum and minimum heights?

62. **HOW DO YOU SEE IT?** Use the graph to answer each question.

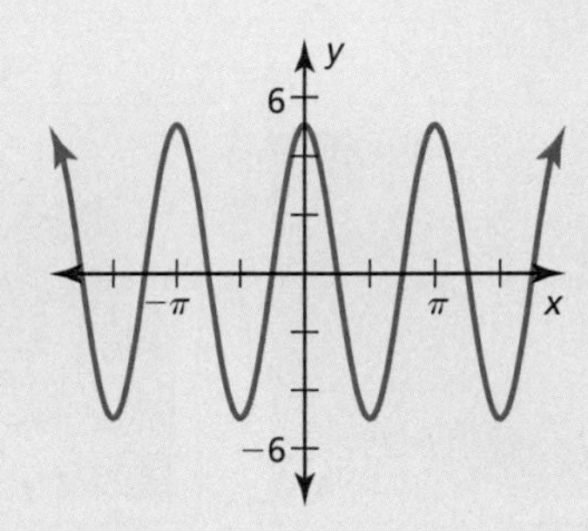

a. Does the graph represent a function of the form $f(x) = a \sin bx$ or $f(x) = a \cos bx$? Explain.

b. Identify the maximum value, minimum value, period, and amplitude of the function.

63. **FINDING A PATTERN** Write an expression in terms of the integer n that represents all the x-intercepts of the graph of the function $y = \cos 2x$. Justify your answer.

64. **MAKING AN ARGUMENT** Your friend states that for functions of the form $y = a \sin bx$ and $y = a \cos bx$, the values of a and b affect the x-intercepts of the graph of the function. Is your friend correct? Explain.

65. **CRITICAL THINKING** Describe a transformation of the graph of $f(x) = \sin x$ that results in the graph of $g(x) = \cos x$.

66. **THOUGHT PROVOKING** Use a graphing calculator to find a function of the form $y = \sin b_1x + \cos b_2x$ whose graph matches that shown below.

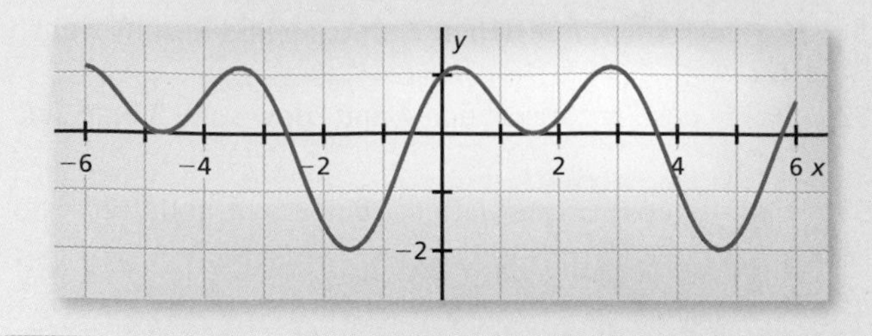

67. **PROBLEM SOLVING** For a person at rest, the blood pressure P (in millimeters of mercury) at time t (in seconds) is given by the function

$$P = 100 - 20 \cos \frac{8\pi}{3}t.$$

Graph the function. One cycle is equivalent to one heartbeat. What is the pulse rate (in heartbeats per minute) of the person?

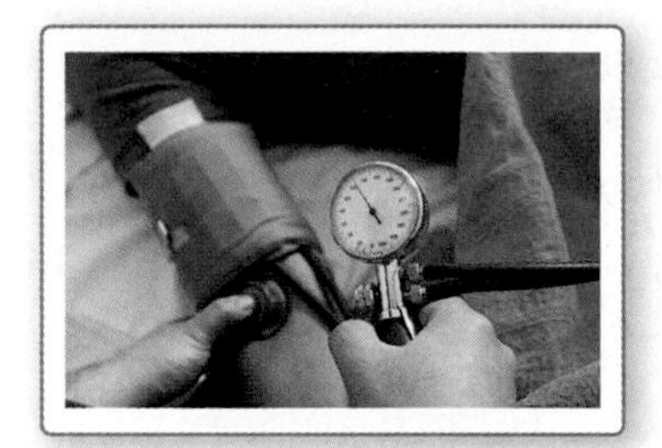

68. **PROBLEM SOLVING** The motion of a spring can be modeled by $y = A \cos kt$, where y is the vertical displacement (in feet) of the spring relative to its position at rest, A is the initial displacement (in feet), k is a constant that measures the elasticity of the spring, and t is the time (in seconds).

a. You have a spring whose motion can be modeled by the function $y = 0.2 \cos 6t$. Find the initial displacement and the period of the spring. Then graph the function.

b. When a damping force is applied to the spring, the motion of the spring can be modeled by the function $y = 0.2e^{-4.5t} \cos 4t$. Graph this function. What effect does damping have on the motion?

Maintaining Mathematical Proficiency

Reviewing what you learned in previous grades and lessons

Simplify the rational expression, if possible. *(Section 7.3)*

69. $\frac{x^2 + x - 6}{x + 3}$

70. $\frac{x^3 - 2x^2 - 24x}{x^2 - 2x - 24}$

71. $\frac{x^2 - 4x - 5}{x^2 + 4x - 5}$

72. $\frac{x^2 - 16}{x^2 + x - 20}$

Find the least common multiple of the expressions. *(Section 7.4)*

73. $2x, 2(x - 5)$

74. $x^2 - 4, x + 2$

75. $x^2 + 8x + 12, x + 6$

9.1–9.4 What Did You Learn?

Core Vocabulary

sine, *p. 462*
cosine, *p. 462*
tangent, *p. 462*
cosecant, *p. 462*
secant, *p. 462*
cotangent, *p. 462*
initial side, *p. 470*
terminal side, *p. 470*
standard position, *p. 470*
coterminal, *p. 471*
radian, *p. 471*
sector, *p. 472*
central angle, *p. 472*
unit circle, *p. 479*
quadrantal angle, *p. 479*
reference angle, *p. 480*
amplitude, *p. 486*
periodic function, *p. 486*
cycle, *p. 486*
period, *p. 486*
phase shift, *p. 488*
midline, *p. 488*

Core Concepts

Section 9.1

Right Triangle Definitions of Trigonometric Functions, *p. 462*
Trigonometric Values for Special Angles, *p. 463*

Section 9.2

Angles in Standard Position, *p. 470*
Converting Between Degrees and Radians, *p. 471*
Degree and Radian Measures of Special Angles, *p. 472*
Arc Length and Area of a Sector, *p. 472*

Section 9.3

General Definitions of Trigonometric Functions, *p. 478*
The Unit Circle, *p. 479*
Reference Angle Relationships, *p. 480*
Evaluating Trigonometric Functions, *p. 480*

Section 9.4

Characteristics of $y = \sin x$ and $y = \cos x$, *p. 486*
Amplitude and Period, *p. 487*
Graphing $y = a \sin b(x - h) + k$ and $y = a \cos b(x - h) + k$, *p. 488*

Mathematical Practices

1. Make a conjecture about the horizontal distances traveled in part (c) of Exercise 39 on page 483.
2. Explain why the quantities in part (a) of Exercise 56 on page 493 make sense in the context of the situation.

Study Skills

Form a Final Exam Study Group

Form a study group several weeks before the final exam. The intent of this group is to review what you have already learned while continuing to learn new material.

9.1–9.4 Quiz

1. In a right triangle, θ is an acute angle and $\sin \theta = \frac{2}{7}$. Evaluate the other five trigonometric functions of θ. *(Section 9.1)*

Find the value of x for the right triangle. *(Section 9.1)*

2. 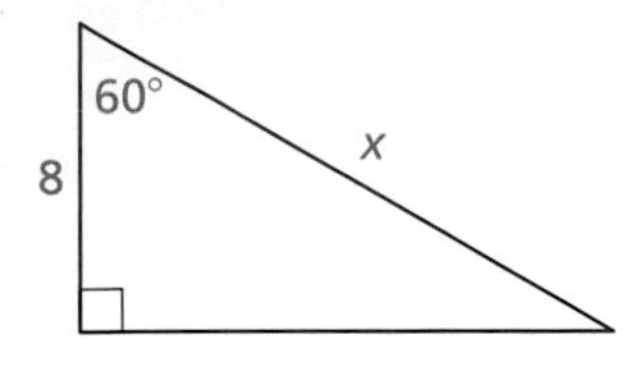

3. 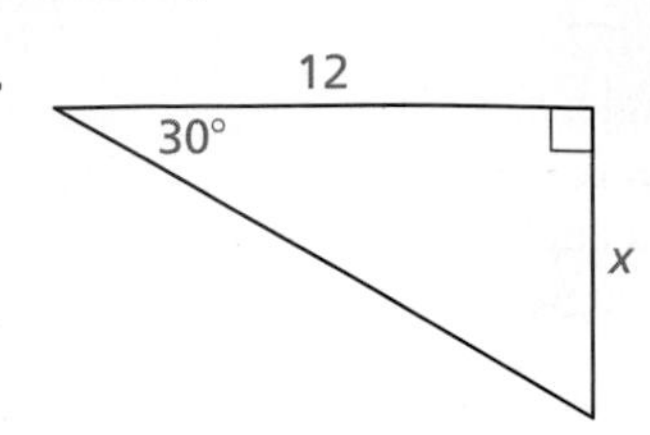

4. 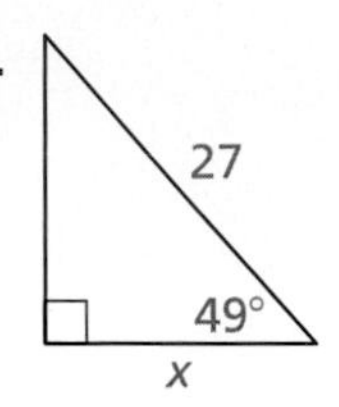

Draw an angle with the given measure in standard position. Then find one positive angle and one negative angle that are coterminal with the given angle. *(Section 9.2)*

5. 40°

6. $\frac{5\pi}{6}$

7. -960°

Convert the degree measure to radians or the radian measure to degrees. *(Section 9.2)*

8. $\frac{3\pi}{10}$

9. -60°

10. 72°

Evaluate the six trigonometric functions of θ. *(Section 9.3)*

11. 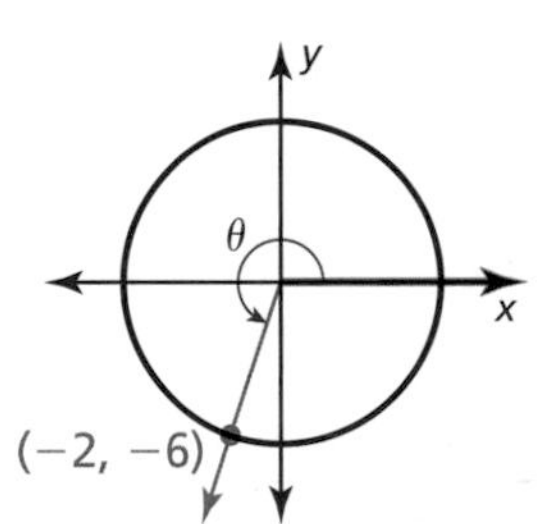

12. 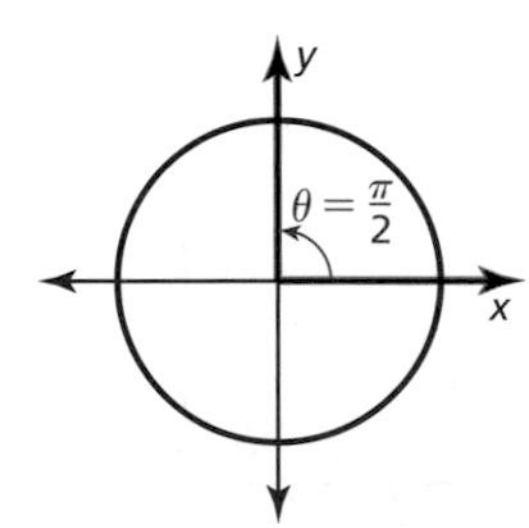

13. 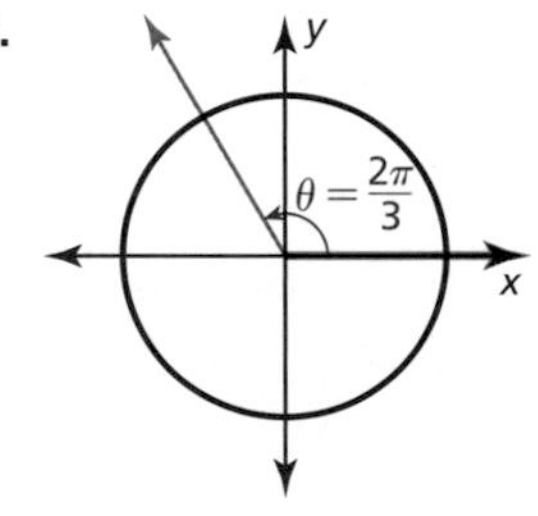

14. Identify the amplitude and period of $g(x) = 3 \sin x$. Then graph the function and describe the graph of g as a transformation of the graph of $f(x) = \sin x$. *(Section 9.4)*

15. Identify the amplitude and period of $g(x) = \cos 5\pi x + 3$. Then graph the function and describe the graph of g as a transformation of the graph of $f(x) = \cos x$. *(Section 9.4)*

16. You are flying a kite at an angle of 70°. You have let out a total of 400 feet of string and are holding the reel steady 4 feet above the ground. *(Section 9.1)*

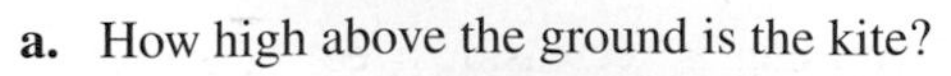

a. How high above the ground is the kite?

b. A friend watching the kite estimates that the angle of elevation to the kite is 85°. How far from your friend are you standing?

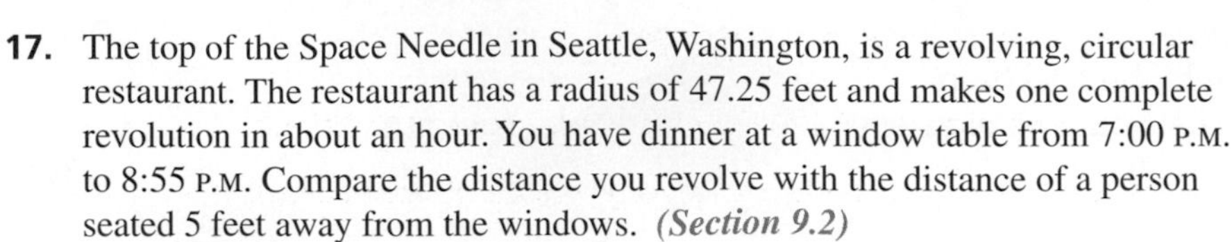

17. The top of the Space Needle in Seattle, Washington, is a revolving, circular restaurant. The restaurant has a radius of 47.25 feet and makes one complete revolution in about an hour. You have dinner at a window table from 7:00 P.M. to 8:55 P.M. Compare the distance you revolve with the distance of a person seated 5 feet away from the windows. *(Section 9.2)*

9.5 Graphing Other Trigonometric Functions

Essential Question What are the characteristics of the graph of the tangent function?

EXPLORATION 1 Graphing the Tangent Function

Work with a partner.

a. Complete the table for $y = \tan x$, where x is an angle measure in radians.

x	$-\frac{\pi}{2}$	$-\frac{\pi}{3}$	$-\frac{\pi}{4}$	$-\frac{\pi}{6}$	0	$\frac{\pi}{6}$	$\frac{\pi}{4}$	$\frac{\pi}{3}$	$\frac{\pi}{2}$
$y = \tan x$									
x	$\frac{2\pi}{3}$	$\frac{3\pi}{4}$	$\frac{5\pi}{6}$	π	$\frac{7\pi}{6}$	$\frac{5\pi}{4}$	$\frac{4\pi}{3}$	$\frac{3\pi}{2}$	$\frac{5\pi}{3}$
$y = \tan x$									

b. The graph of $y = \tan x$ has vertical asymptotes at x-values where $\tan x$ is undefined. Plot the points (x, y) from part (a). Then use the asymptotes to sketch the graph of $y = \tan x$.

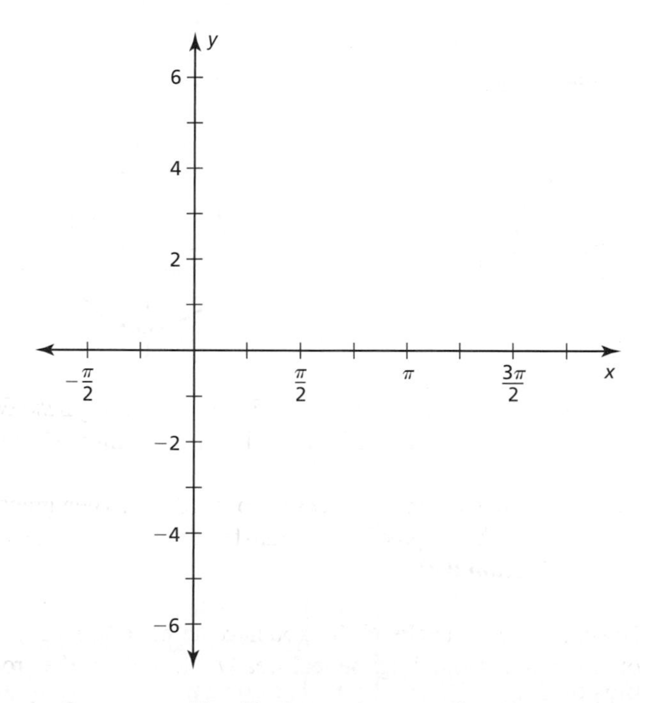

c. For the graph of $y = \tan x$, identify the asymptotes, the x-intercepts, and the intervals for which the function is increasing or decreasing over $-\frac{\pi}{2} \le x \le \frac{3\pi}{2}$. Is the tangent function *even*, *odd*, or *neither*?

MAKING SENSE OF PROBLEMS

To be proficient in math, you need to consider analogous problems and try special cases of the original problem in order to gain insight into its solution.

Communicate Your Answer

2. What are the characteristics of the graph of the tangent function?

3. Describe the asymptotes of the graph of $y = \cot x$ on the interval $-\frac{\pi}{2} < x < \frac{3\pi}{2}$.

9.5 Lesson

Core Vocabulary

Previous
asymptote
period
amplitude
x-intercept
transformations

What You Will Learn

- Explore characteristics of tangent and cotangent functions.
- Graph tangent and cotangent functions.
- Graph secant and cosecant functions.

Exploring Tangent and Cotangent Functions

The graphs of tangent and cotangent functions are related to the graphs of the parent functions $y = \tan x$ and $y = \cot x$, which are graphed below.

← x approaches $-\frac{\pi}{2}$ | x approaches $\frac{\pi}{2}$ →

x	$-\frac{\pi}{2}$	-1.57	-1.5	$-\frac{\pi}{4}$	0	$\frac{\pi}{4}$	1.5	1.57	$\frac{\pi}{2}$
$y = \tan x$	Undef.	-1256	-14.10	-1	0	1	14.10	1256	Undef.

← $\tan x$ approaches $-\infty$ | $\tan x$ approaches ∞ →

Because $\tan x = \frac{\sin x}{\cos x}$, $\tan x$ is undefined for x-values at which $\cos x = 0$, such as

$x = \pm\frac{\pi}{2} \approx \pm 1.571.$

The table indicates that the graph has asymptotes at these values. The table represents one cycle of the graph, so the period of the graph is π.

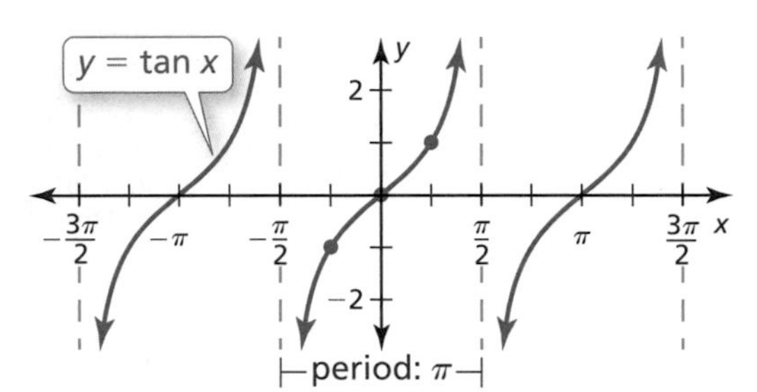

You can use a similar approach to graph $y = \cot x$. Because $\cot x = \frac{\cos x}{\sin x}$, $\cot x$ is undefined for x-values at which $\sin x = 0$, which are multiples of π. The graph has asymptotes at these values. The period of the graph is also π.

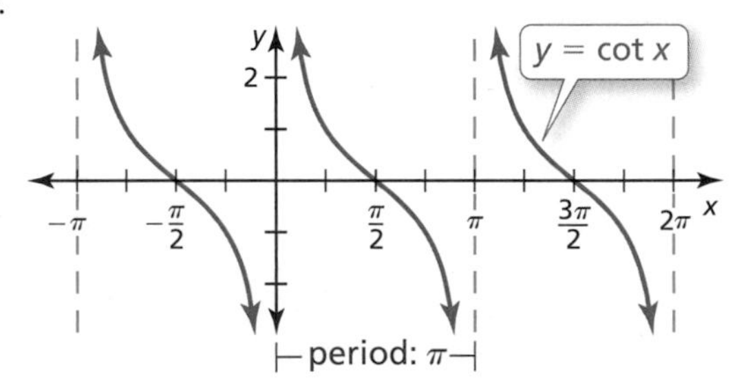

Core Concept

Characteristics of $y = \tan x$ and $y = \cot x$

The functions $y = \tan x$ and $y = \cot x$ have the following characteristics.

- The domain of $y = \tan x$ is all real numbers except odd multiples of $\frac{\pi}{2}$. At these x-values, the graph has vertical asymptotes.
- The domain of $y = \cot x$ is all real numbers except multiples of π. At these x-values, the graph has vertical asymptotes.
- The range of each function is all real numbers. So, the functions do not have maximum or minimum values, and the graphs do not have an amplitude.
- The period of each graph is π.
- The x-intercepts for $y = \tan x$ occur when $x = 0, \pm\pi, \pm 2\pi, \pm 3\pi, \ldots$.
- The x-intercepts for $y = \cot x$ occur when $x = \pm\frac{\pi}{2}, \pm\frac{3\pi}{2}, \pm\frac{5\pi}{2}, \pm\frac{7\pi}{2}, \ldots$.

STUDY TIP

Odd multiples of $\frac{\pi}{2}$ are values such as these:

$\pm 1 \cdot \frac{\pi}{2} = \pm\frac{\pi}{2}$

$\pm 3 \cdot \frac{\pi}{2} = \pm\frac{3\pi}{2}$

$\pm 5 \cdot \frac{\pi}{2} = \pm\frac{5\pi}{2}$

Graphing Tangent and Cotangent Functions

The graphs of $y = a \tan bx$ and $y = a \cot bx$ represent transformations of their parent functions. The value of a indicates a vertical stretch ($a > 1$) or a vertical shrink ($0 < a < 1$). The value of b indicates a horizontal stretch ($0 < b < 1$) or a horizontal shrink ($b > 1$) and changes the period of the graph.

Core Concept

Period and Vertical Asymptotes of $y = a \tan bx$ and $y = a \cot bx$

The period and vertical asymptotes of the graphs of $y = a \tan bx$ and $y = a \cot bx$, where a and b are nonzero real numbers, are as follows.

- The period of the graph of each function is $\frac{\pi}{|b|}$.
- The vertical asymptotes for $y = a \tan bx$ are at odd multiples of $\frac{\pi}{2|b|}$.
- The vertical asymptotes for $y = a \cot bx$ are at multiples of $\frac{\pi}{|b|}$.

Each graph below shows five key x-values that you can use to sketch the graphs of $y = a \tan bx$ and $y = a \cot bx$ for $a > 0$ and $b > 0$. These are the x-intercept, the x-values where the asymptotes occur, and the x-values halfway between the x-intercept and the asymptotes. At each halfway point, the value of the function is either a or $-a$.

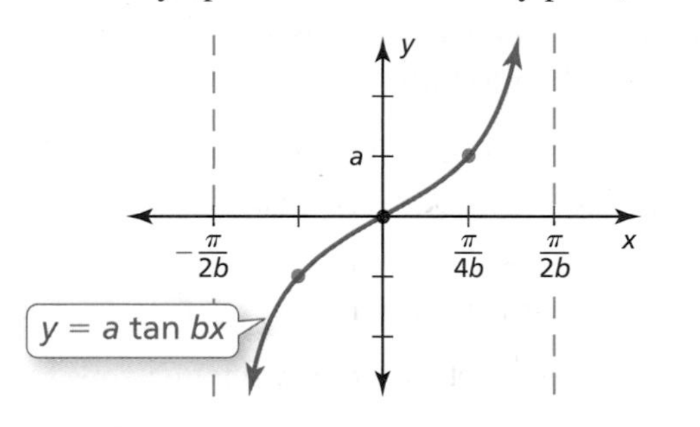

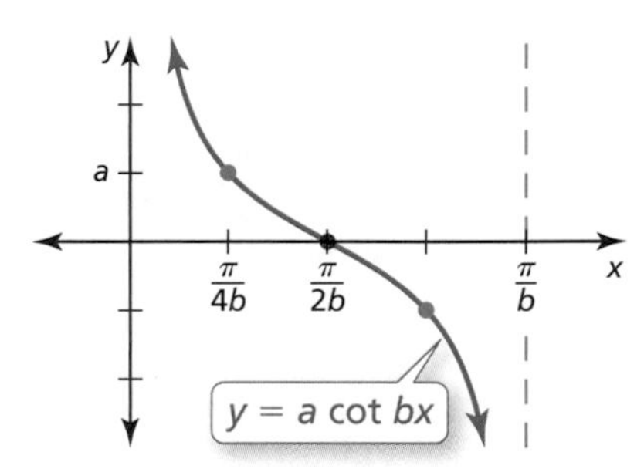

EXAMPLE 1 **Graphing a Tangent Function**

Graph one period of $g(x) = 2 \tan 3x$. Describe the graph of g as a transformation of the graph of $f(x) = \tan x$.

SOLUTION

The function is of the form $g(x) = a \tan bx$ where $a = 2$ and $b = 3$. So, the period is $\frac{\pi}{|b|} = \frac{\pi}{3}$.

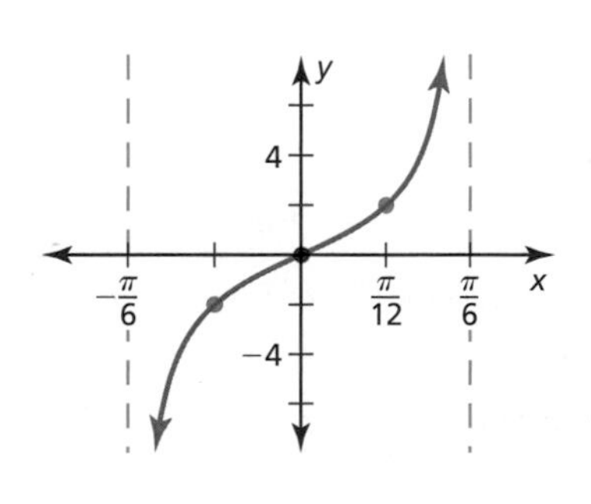

Intercept: $(0, 0)$

Asymptotes: $x = \frac{\pi}{2|b|} = \frac{\pi}{2(3)}$, or $x = \frac{\pi}{6}$; $\quad x = -\frac{\pi}{2|b|} = -\frac{\pi}{2(3)}$, or $x = -\frac{\pi}{6}$

Halfway points: $\left(\frac{\pi}{4b}, a\right) = \left(\frac{\pi}{4(3)}, 2\right) = \left(\frac{\pi}{12}, 2\right)$;

$$\left(-\frac{\pi}{4b}, -a\right) = \left(-\frac{\pi}{4(3)}, -2\right) = \left(-\frac{\pi}{12}, -2\right)$$

▶ The graph of g is a vertical stretch by a factor of 2 and a horizontal shrink by a factor of $\frac{1}{3}$ of the graph of f.

EXAMPLE 2 Graphing a Cotangent Function

Graph one period of $g(x) = \cot \frac{1}{2}x$. Describe the graph of g as a transformation of the graph of $f(x) = \cot x$.

SOLUTION

The function is of the form $g(x) = a \cot bx$ where $a = 1$ and $b = \frac{1}{2}$. So, the period is $\frac{\pi}{|b|} = \frac{\pi}{\frac{1}{2}} = 2\pi$.

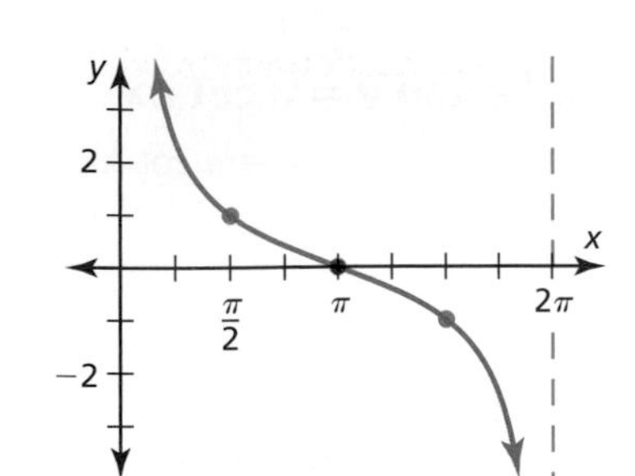

Intercept: $\left(\frac{\pi}{2b}, 0\right) = \left(\frac{\pi}{2\left(\frac{1}{2}\right)}, 0\right) = (\pi, 0)$

Asymptotes: $x = 0$; $x = \frac{\pi}{|b|} = \frac{\pi}{\frac{1}{2}}$, or $x = 2\pi$

Halfway points: $\left(\frac{\pi}{4b}, a\right) = \left(\frac{\pi}{4\left(\frac{1}{2}\right)}, 1\right) = \left(\frac{\pi}{2}, 1\right)$; $\left(\frac{3\pi}{4b}, -a\right) = \left(\frac{3\pi}{4\left(\frac{1}{2}\right)}, -1\right) = \left(\frac{3\pi}{2}, -1\right)$

▶ The graph of g is a horizontal stretch by a factor of 2 of the graph of f.

Monitoring Progress

Help in English and Spanish at *BigIdeasMath.com*

Graph one period of the function. Describe the graph of g as a transformation of the graph of its parent function.

1. $g(x) = \tan 2x$
2. $g(x) = \frac{1}{3} \cot x$
3. $g(x) = 2 \cot 4x$
4. $g(x) = 5 \tan \pi x$

Graphing Secant and Cosecant Functions

The graphs of secant and cosecant functions are related to the graphs of the parent functions $y = \sec x$ and $y = \csc x$, which are shown below.

STUDY TIP

Because $\sec x = \frac{1}{\cos x}$, $\sec x$ is undefined for x-values at which $\cos x = 0$. The graph of $y = \sec x$ has vertical asymptotes at these x-values. You can use similar reasoning to understand the vertical asymptotes of the graph of $y = \csc x$.

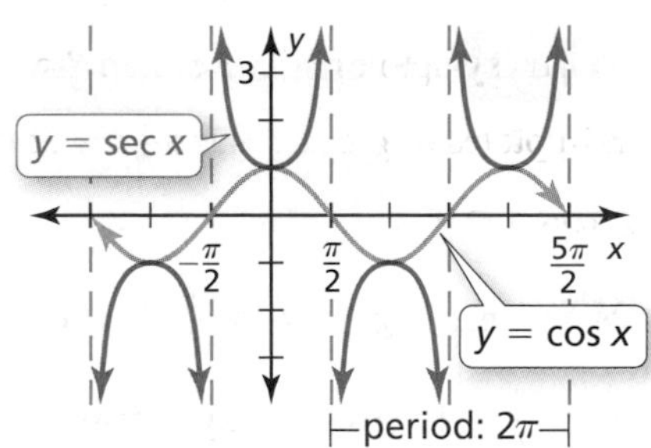

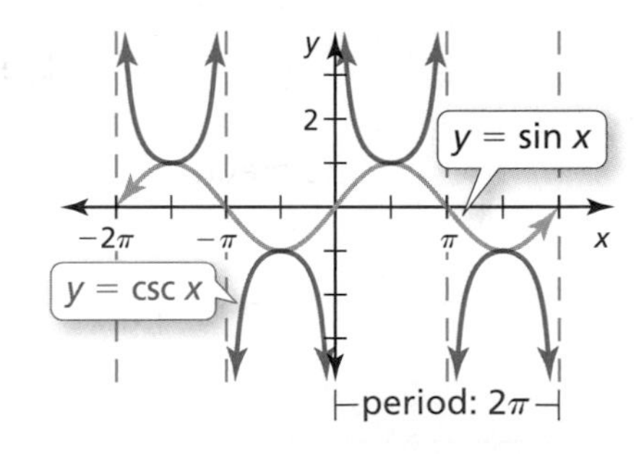

Core Concept

Characteristics of $y = \sec x$ and $y = \csc x$

The functions $y = \sec x$ and $y = \csc x$ have the following characteristics.

- The domain of $y = \sec x$ is all real numbers except odd multiples of $\frac{\pi}{2}$. At these x-values, the graph has vertical asymptotes.
- The domain of $y = \csc x$ is all real numbers except multiples of π. At these x-values, the graph has vertical asymptotes.
- The range of each function is $y \leq -1$ and $y \geq 1$. So, the graphs do not have an amplitude.
- The period of each graph is 2π.

To graph $y = a \sec bx$ or $y = a \csc bx$, first graph the function $y = a \cos bx$ or $y = a \sin bx$, respectively. Then use the asymptotes and several points to sketch a graph of the function. Notice that the value of b represents a horizontal stretch or shrink by a factor of $\frac{1}{b}$, so the period of $y = a \sec bx$ and $y = a \csc bx$ is $\frac{2\pi}{|b|}$.

EXAMPLE 3 Graphing a Secant Function

Graph one period of $g(x) = 2 \sec x$. Describe the graph of g as a transformation of the graph of $f(x) = \sec x$.

SOLUTION

Step 1 Graph the function $y = 2 \cos x$.

The period is $\frac{2\pi}{1} = 2\pi$.

Step 2 Graph asymptotes of g. Because the asymptotes of g occur when $2 \cos x = 0$, graph $x = -\frac{\pi}{2}$, $x = \frac{\pi}{2}$, and $x = \frac{3\pi}{2}$.

Step 3 Plot points on g, such as $(0, 2)$ and $(\pi, -2)$. Then use the asymptotes to sketch the curve.

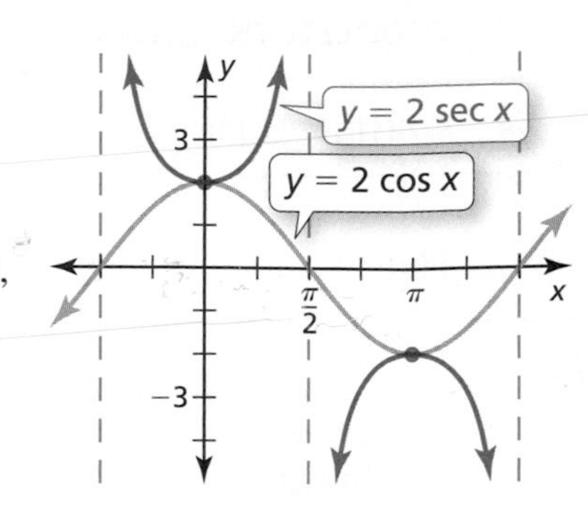

▶ The graph of g is a vertical stretch by a factor of 2 of the graph of f.

EXAMPLE 4 Graphing a Cosecant Function

Graph one period of $g(x) = \frac{1}{2} \csc \pi x$. Describe the graph of g as a transformation of the graph of $f(x) = \csc x$.

SOLUTION

Step 1 Graph the function $y = \frac{1}{2} \sin \pi x$. The period is $\frac{2\pi}{\pi} = 2$.

Step 2 Graph asymptotes of g. Because the asymptotes of g occur when $\frac{1}{2} \sin \pi x = 0$, graph $x = 0$, $x = 1$, and $x = 2$.

Step 3 Plot points on g, such as $\left(\frac{1}{2}, \frac{1}{2}\right)$ and $\left(\frac{3}{2}, -\frac{1}{2}\right)$. Then use the asymptotes to sketch the curve.

y
1
1
2
x
$y = \frac{1}{2} \sin \pi x$
$y = \frac{1}{2} \csc \pi x$

LOOKING FOR A PATTERN

In Examples 3 and 4, notice that the plotted points are on both graphs. Also, these points represent a local maximum on one graph and a local minimum on the other graph.

▶ The graph of g is a vertical shrink by a factor of $\frac{1}{2}$ and a horizontal shrink by a factor of $\frac{1}{\pi}$ of the graph of f.

Monitoring Progress

Help in English and Spanish at *BigIdeasMath.com*

Graph one period of the function. Describe the graph of g as a transformation of the graph of its parent function.

5. $g(x) = \csc 3x$ **6.** $g(x) = \frac{1}{2} \sec x$ **7.** $g(x) = 2 \csc 2x$ **8.** $g(x) = 2 \sec \pi x$

9.5 Exercises

Dynamic Solutions available at *BigIdeasMath.com*

Vocabulary and Core Concept Check

1. **WRITING** Explain why the graphs of the tangent, cotangent, secant, and cosecant functions do not have an amplitude.

2. **COMPLETE THE SENTENCE** The ________ and ________ functions are undefined for x-values at which $\sin x = 0$.

3. **COMPLETE THE SENTENCE** The period of the function $y = \sec x$ is ______, and the period of $y = \cot x$ is ______.

4. **WRITING** Explain how to graph a function of the form $y = a \sec bx$.

Monitoring Progress and Modeling with Mathematics

In Exercises 5–12, graph one period of the function. Describe the graph of g as a transformation of the graph of its parent function. *(See Examples 1 and 2.)*

5. $g(x) = 2 \tan x$
6. $g(x) = 3 \tan x$
7. $g(x) = \cot 3x$
8. $g(x) = \cot 2x$
9. $g(x) = 3 \cot \frac{1}{4}x$
10. $g(x) = 4 \cot \frac{1}{2}x$
11. $g(x) = \frac{1}{2} \tan \pi x$
12. $g(x) = \frac{1}{3} \tan 2\pi x$

13. **ERROR ANALYSIS** Describe and correct the error in finding the period of the function $y = \cot 3x$.

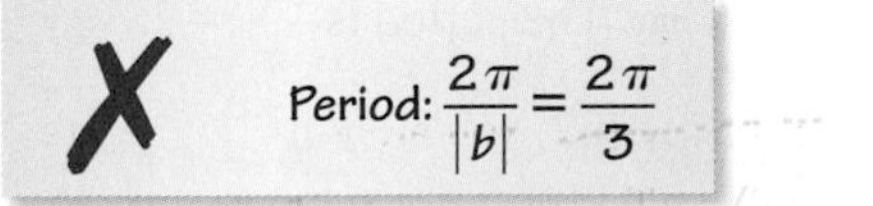

14. **ERROR ANALYSIS** Describe and correct the error in describing the transformation of $f(x) = \tan x$ represented by $g(x) = 2 \tan 5x$.

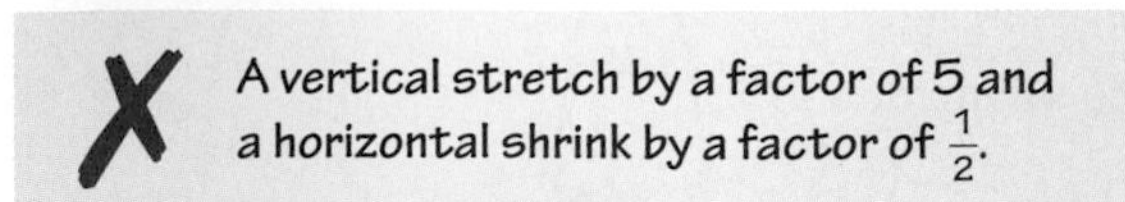

15. **ANALYZING RELATIONSHIPS** Use the given graph to graph each function.

 a. $f(x) = 3 \sec 2x$ b. $f(x) = 4 \csc 3x$

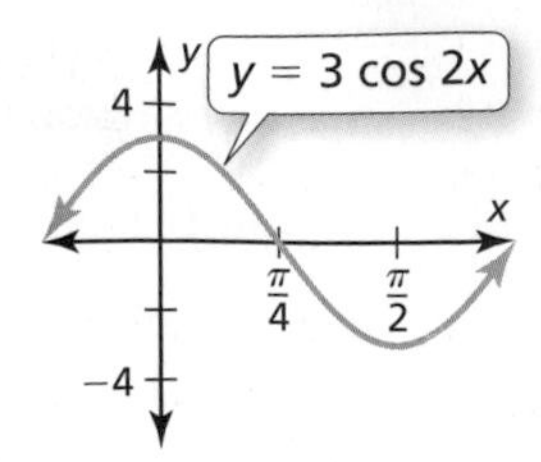

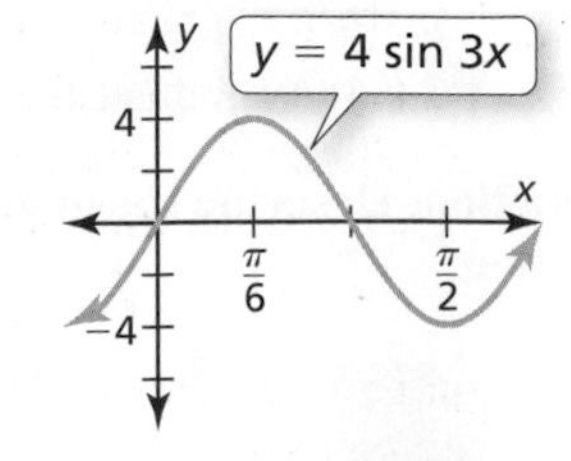

16. **USING EQUATIONS** Which of the following are asymptotes of the graph of $y = 3 \tan 4x$?

 Ⓐ $x = \frac{\pi}{8}$ Ⓑ $x = \frac{\pi}{4}$

 Ⓒ $x = 0$ Ⓓ $x = -\frac{5\pi}{8}$

In Exercises 17–24, graph one period of the function. Describe the graph of g as a transformation of the graph of its parent function. *(See Examples 3 and 4.)*

17. $g(x) = 3 \csc x$
18. $g(x) = 2 \csc x$
19. $g(x) = \sec 4x$
20. $g(x) = \sec 3x$
21. $g(x) = \frac{1}{2} \sec \pi x$
22. $g(x) = \frac{1}{4} \sec 2\pi x$
23. $g(x) = \csc \frac{\pi}{2}x$
24. $g(x) = \csc \frac{\pi}{4}x$

ATTENDING TO PRECISION **In Exercises 25–28, use the graph to write a function of the form $y = a \tan bx$.**

25.

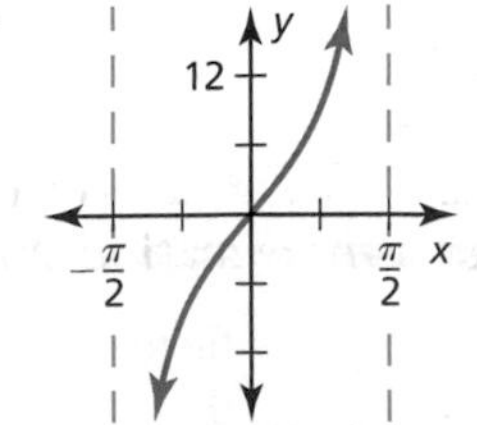

26.

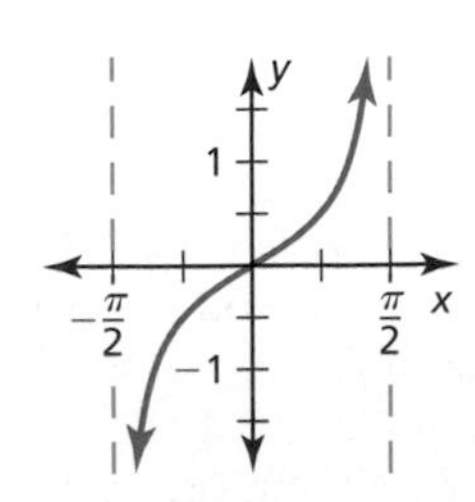

27.

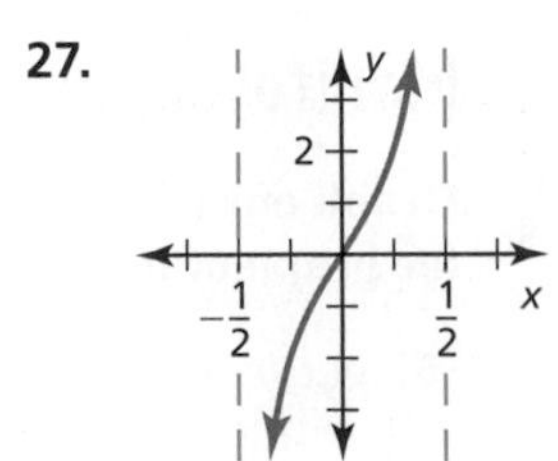

28.

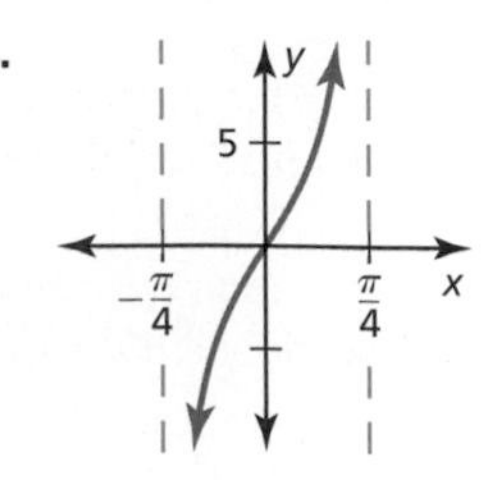

USING STRUCTURE In Exercises 29–34, match the equation with the correct graph. Explain your reasoning.

29. $g(x) = 4 \tan x$

30. $g(x) = 4 \cot x$

31. $g(x) = 4 \csc \pi x$

32. $g(x) = 4 \sec \pi x$

33. $g(x) = \sec 2x$

34. $g(x) = \csc 2x$

A.

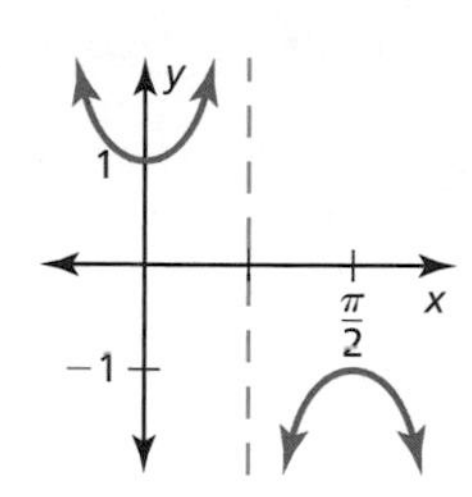

B.

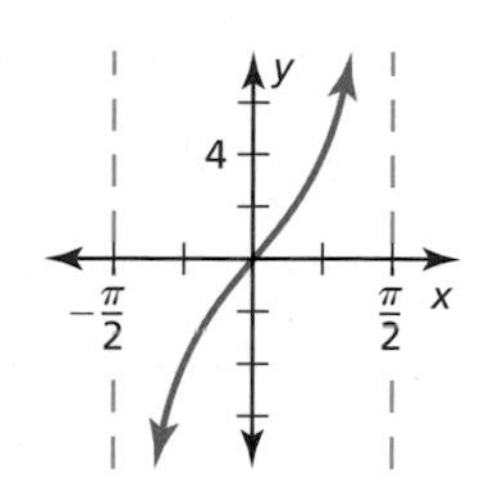

C.

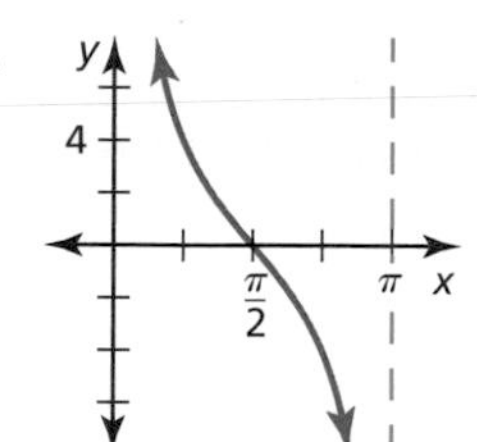

D.

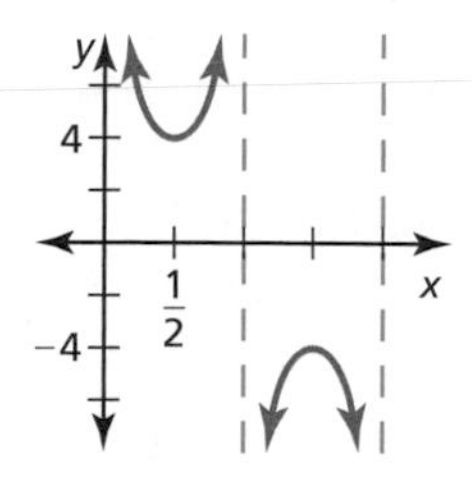

E.

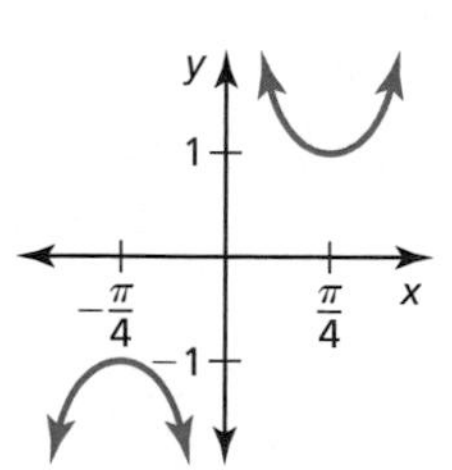

F.

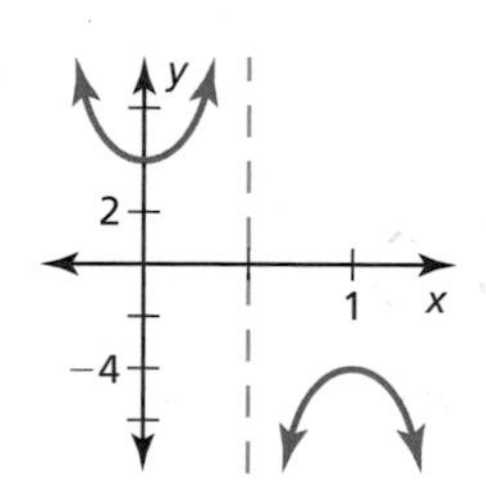

35. WRITING Explain why there is more than one tangent function whose graph passes through the origin and has asymptotes at $x = -\pi$ and $x = \pi$.

36. USING EQUATIONS Graph one period of each function. Describe the transformation of the graph of its parent function.

a. $g(x) = \sec x + 3$ **b.** $g(x) = \csc x - 2$

c. $g(x) = \cot(x - \pi)$ **d.** $g(x) = -\tan x$

WRITING EQUATIONS In Exercises 37–40, write a rule for g that represents the indicated transformation of the graph of f.

37. $f(x) = \cot 2x$; translation 3 units up and $\frac{\pi}{2}$ units left

38. $f(x) = 2 \tan x$; translation π units right, followed by a horizontal shrink by a factor of $\frac{1}{3}$

39. $f(x) = 5 \sec (x - \pi)$; translation 2 units down, followed by a reflection in the x-axis

40. $f(x) = 4 \csc x$; vertical stretch by a factor of 2 and a reflection in the x-axis

41. MULTIPLE REPRESENTATIONS Which function has a greater local maximum value? Which has a greater local minimum value? Explain.

A. $f(x) = \frac{1}{4} \csc \pi x$ 

B.

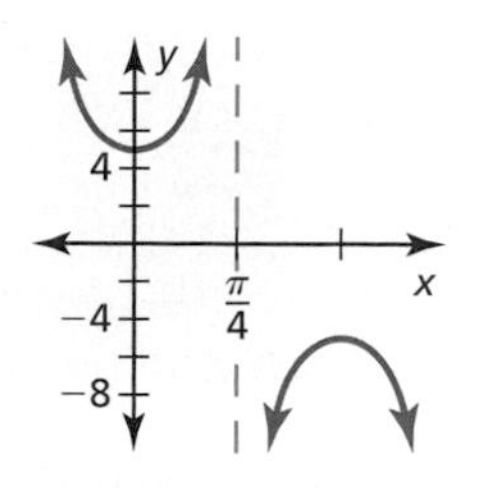

42. ANALYZING RELATIONSHIPS Order the functions from the least average rate of change to the greatest average rate of change over the interval $-\frac{\pi}{4} < x < \frac{\pi}{4}$.

A.

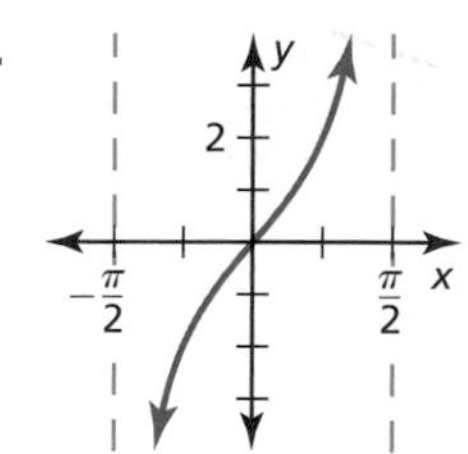

B.

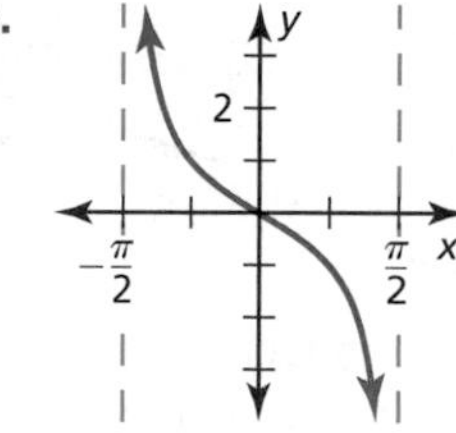

C.

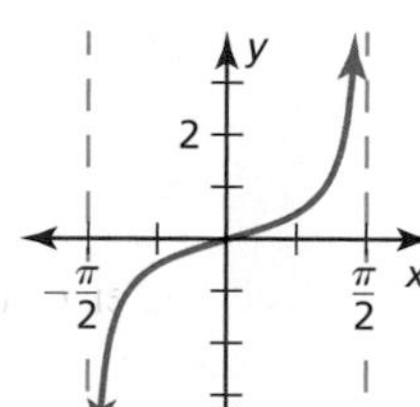

D.

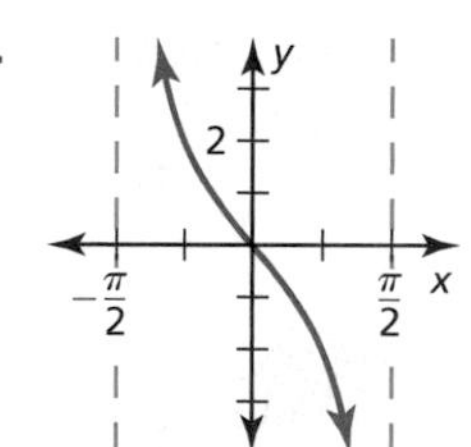

43. REASONING You are standing on a bridge 140 feet above the ground. You look down at a car traveling away from the underpass. The distance d (in feet) the car is from the base of the bridge can be modeled by $d = 140 \tan \theta$. Graph the function. Describe what happens to θ as d increases.

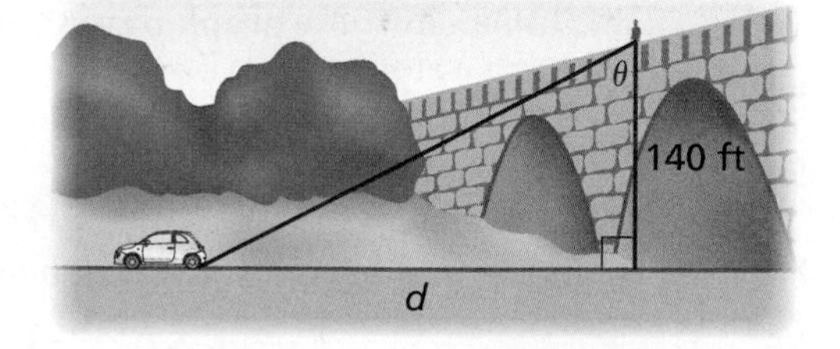

44. USING TOOLS You use a video camera to pan up the Statue of Liberty. The height h (in feet) of the part of the Statue of Liberty that can be seen through your video camera after time t (in seconds) can be modeled by $h = 100 \tan \frac{\pi}{36}t$. Graph the function using a graphing calculator. What viewing window did you use? Explain.

45. MODELING WITH MATHEMATICS You are standing 120 feet from the base of a 260-foot building. You watch your friend go down the side of the building in a glass elevator.

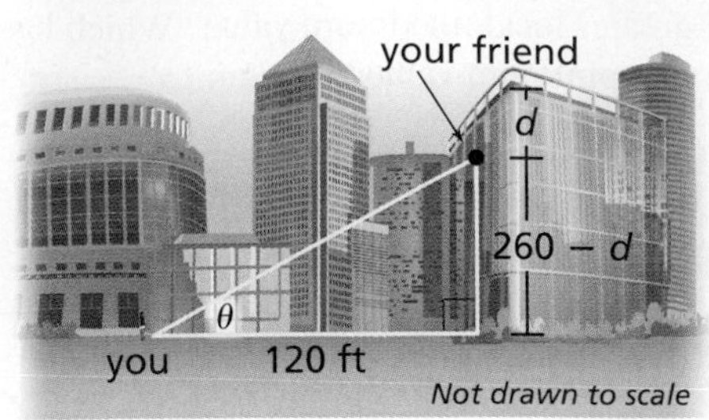

a. Write an equation that gives the distance d (in feet) your friend is from the top of the building as a function of the angle of elevation θ.

b. Graph the function found in part (a). Explain how the graph relates to this situation.

46. MODELING WITH MATHEMATICS You are standing 300 feet from the base of a 200-foot cliff. Your friend is rappelling down the cliff.

a. Write an equation that gives the distance d (in feet) your friend is from the top of the cliff as a function of the angle of elevation θ.

b. Graph the function found in part (a).

c. Use a graphing calculator to determine the angle of elevation when your friend has rappelled halfway down the cliff.

47. MAKING AN ARGUMENT Your friend states that it is not possible to write a cosecant function that has the same graph as $y = \sec x$. Is your friend correct? Explain your reasoning.

48. HOW DO YOU SEE IT? Use the graph to answer each question.

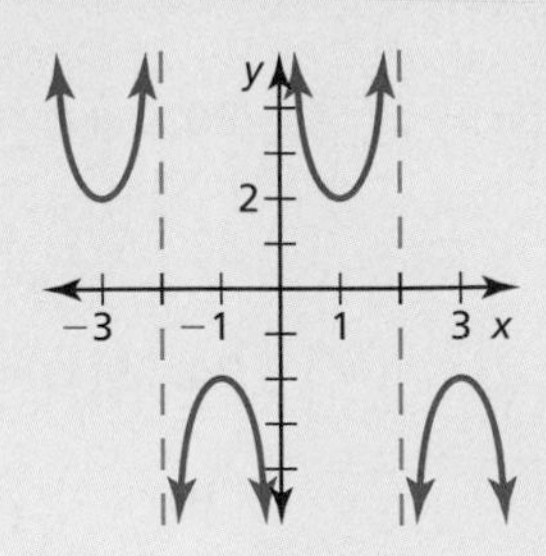

a. What is the period of the graph?

b. What is the range of the function?

c. Is the function of the form $f(x) = a \csc bx$ or $f(x) = a \sec bx$? Explain.

49. ABSTRACT REASONING Rewrite $a \sec bx$ in terms of $\cos bx$. Use your results to explain the relationship between the local maximums and minimums of the cosine and secant functions.

50. THOUGHT PROVOKING A trigonometric equation that is true for all values of the variable for which both sides of the equation are defined is called a *trigonometric identity*. Use a graphing calculator to graph the function

$$y = \frac{1}{2}\left(\tan \frac{x}{2} + \cot \frac{x}{2}\right).$$

Use your graph to write a trigonometric identity involving this function. Explain your reasoning.

51. CRITICAL THINKING Find a tangent function whose graph intersects the graph of $y = 2 + 2 \sin x$ only at minimum points of the sine function.

Maintaining Mathematical Proficiency

Reviewing what you learned in previous grades and lessons

Write a cubic function whose graph passes through the given points. *(Section 4.9)*

52. $(-1, 0), (1, 0), (3, 0), (0, 3)$

53. $(-2, 0), (1, 0), (3, 0), (0, -6)$

54. $(-1, 0), (2, 0), (3, 0), (1, -2)$

55. $(-3, 0), (-1, 0), (3, 0), (-2, 1)$

Find the amplitude and period of the graph of the function. *(Section 9.4)*

56.

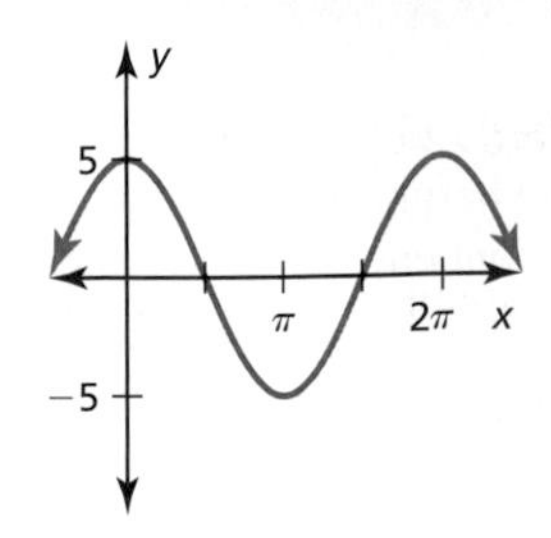

57.

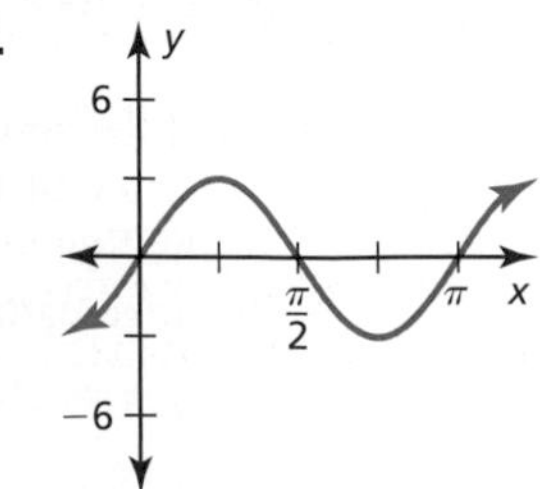

58.

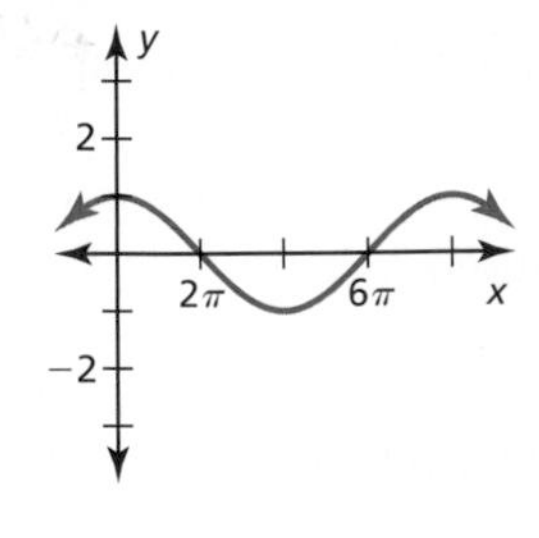

9.6 Modeling with Trigonometric Functions

Essential Question What are the characteristics of the real-life problems that can be modeled by trigonometric functions?

EXPLORATION 1 **Modeling Electric Currents**

MODELING WITH MATHEMATICS

To be proficient in math, you need to apply the mathematics you know to solve problems arising in everyday life.

Work with a partner. Find a sine function that models the electric current shown in each oscilloscope screen. State the amplitude and period of the graph.

a.

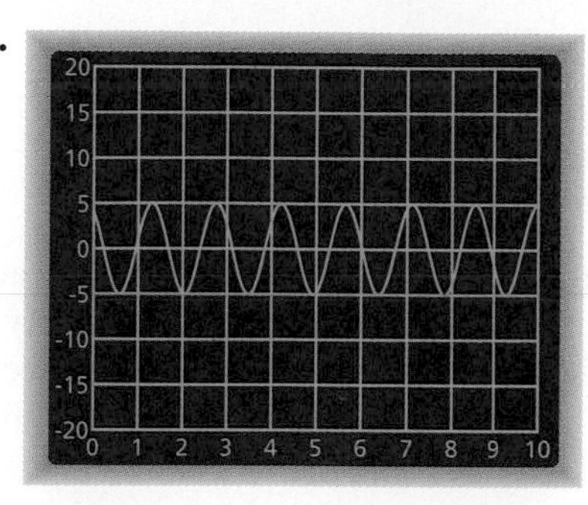

b.

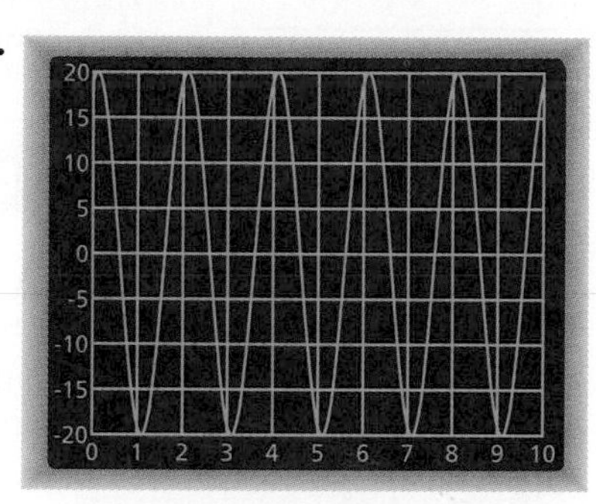

c.

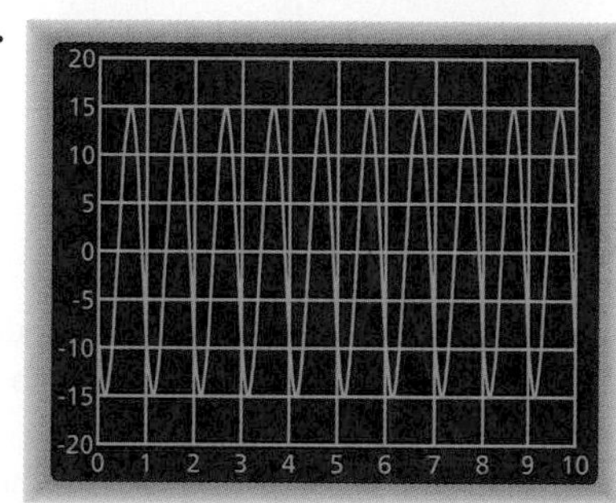

d.

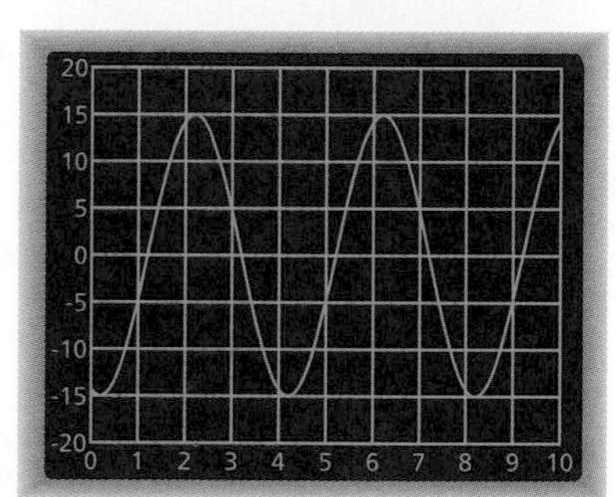

e.

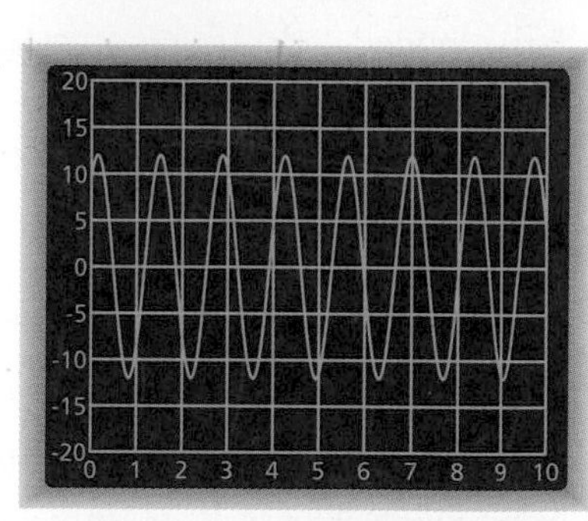

f.

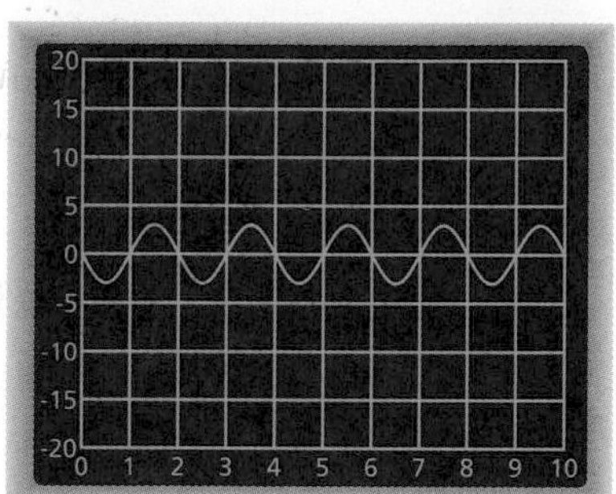

Communicate Your Answer

2. What are the characteristics of the real-life problems that can be modeled by trigonometric functions?

3. Use the Internet or some other reference to find examples of real-life situations that can be modeled by trigonometric functions.

9.6 Lesson

Core Vocabulary

frequency, *p. 506*
sinusoid, *p. 507*

Previous
amplitude
period
midline

What You Will Learn

- Interpret and use frequency.
- Write trigonometric functions.
- Use technology to find trigonometric models.

Frequency

The periodic nature of trigonometric functions makes them useful for modeling *oscillating* motions or repeating patterns that occur in real life. Some examples are sound waves, the motion of a pendulum, and seasons of the year. In such applications, the reciprocal of the period is called the **frequency**, which gives the number of cycles per unit of time.

EXAMPLE 1 Using Frequency

A sound consisting of a single frequency is called a *pure tone*. An audiometer produces pure tones to test a person's auditory functions. An audiometer produces a pure tone with a frequency f of 2000 hertz (cycles per second). The maximum pressure P produced from the pure tone is 2 millipascals. Write and graph a sine model that gives the pressure P as a function of the time t (in seconds).

SOLUTION

Step 1 Find the values of a and b in the model $P = a \sin bt$. The maximum pressure is 2, so $a = 2$. Use the frequency f to find b.

$$\text{frequency} = \frac{1}{\text{period}}$$ Write relationship involving frequency and period.

$$2000 = \frac{b}{2\pi}$$ Substitute.

$$4000\pi = b$$ Multiply each side by 2π.

The pressure P as a function of time t is given by $P = 2 \sin 4000\pi t$.

Step 2 Graph the model. The amplitude is $a = 2$ and the period is

$$\frac{1}{f} = \frac{1}{2000}.$$

The key points are:

Intercepts: $(0, 0)$; $\left(\frac{1}{2} \cdot \frac{1}{2000}, 0\right) = \left(\frac{1}{4000}, 0\right)$; $\left(\frac{1}{2000}, 0\right)$

Maximum: $\left(\frac{1}{4} \cdot \frac{1}{2000}, 2\right) = \left(\frac{1}{8000}, 2\right)$

Minimum: $\left(\frac{3}{4} \cdot \frac{1}{2000}, -2\right) = \left(\frac{3}{8000}, -2\right)$

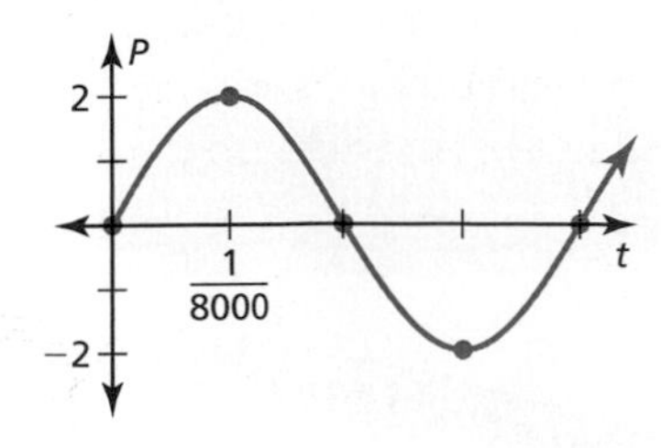

▶ The graph of $P = 2 \sin 4000\pi t$ is shown at the left.

Monitoring Progress  Help in English and Spanish at *BigIdeasMath.com*

1. **WHAT IF?** In Example 1, how would the function change when the audiometer produced a pure tone with a frequency of 1000 hertz?

Writing Trigonometric Functions

Graphs of sine and cosine functions are called **sinusoids**. One method to write a sine or cosine function that models a sinusoid is to find the values of a, b, h, and k for

$$y = a \sin b(x - h) + k \quad \text{or} \quad y = a \cos b(x - h) + k$$

where $|a|$ is the amplitude, $\frac{2\pi}{b}$ is the period ($b > 0$), h is the horizontal shift, and k is the vertical shift.

EXAMPLE 2 Writing a Trigonometric Function

Write a function for the sinusoid shown.

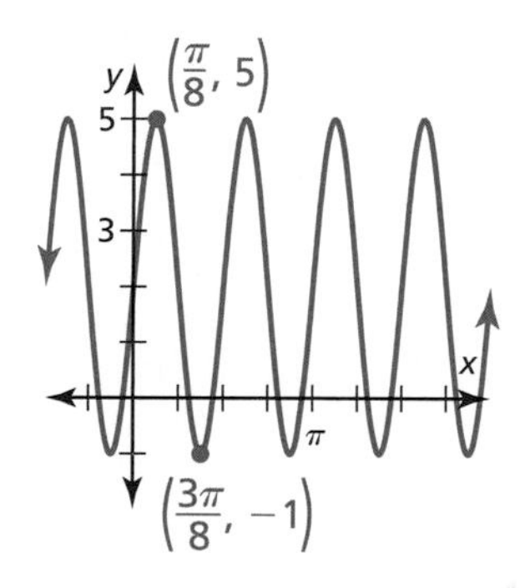

SOLUTION

Step 1 Find the maximum and minimum values. From the graph, the maximum value is 5 and the minimum value is -1.

Step 2 Identify the vertical shift, k. The value of k is the mean of the maximum and minimum values.

$$k = \frac{(\text{maximum value}) + (\text{minimum value})}{2} = \frac{5 + (-1)}{2} = \frac{4}{2} = 2$$

STUDY TIP

Because the graph repeats every $\frac{\pi}{2}$ units, the period is $\frac{\pi}{2}$.

Step 3 Decide whether the graph should be modeled by a sine or cosine function. Because the graph crosses the midline $y = 2$ on the y-axis, the graph is a sine curve with no horizontal shift. So, $h = 0$.

Step 4 Find the amplitude and period. The period is

$$\frac{\pi}{2} = \frac{2\pi}{b} \quad \Rightarrow \quad b = 4.$$

The amplitude is

$$|a| = \frac{(\text{maximum value}) - (\text{minimum value})}{2} = \frac{5 - (-1)}{2} = \frac{6}{2} = 3.$$

The graph is not a reflection, so $a > 0$. Therefore, $a = 3$.

▶ The function is $y = 3 \sin 4x + 2$. Check this by graphing the function on a graphing calculator.

Check

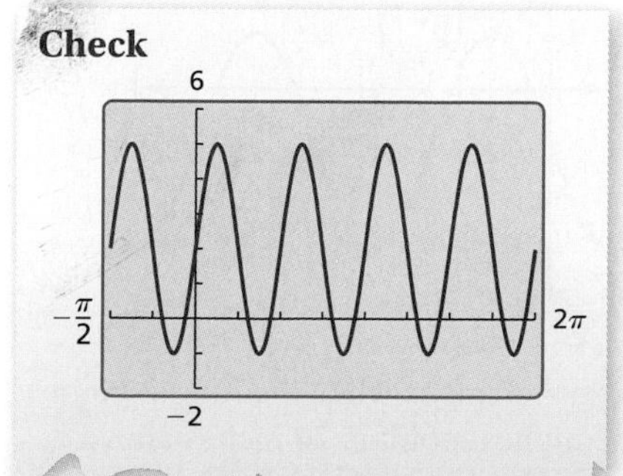

EXAMPLE 3 Modeling Circular Motion

Two people swing jump ropes, as shown in the diagram. The highest point of the middle of each rope is 75 inches above the ground, and the lowest point is 3 inches. The rope makes 2 revolutions per second. Write a model for the height h (in inches) of a rope as a function of the time t (in seconds) given that the rope is at its lowest point when $t = 0$.

SOLUTION

A rope oscillates between 3 inches and 75 inches above the ground. So, a sine or cosine function may be an appropriate model for the height over time.

Step 1 Identify the maximum and minimum values. The maximum height of a rope is 75 inches. The minimum height is 3 inches.

Step 2 Identify the vertical shift, k.

$$k = \frac{(\text{maximum value}) + (\text{minimum value})}{2} = \frac{75 + 3}{2} = 39$$

Step 3 Decide whether the height should be modeled by a sine or cosine function. When $t = 0$, the height is at its minimum. So, use a cosine function whose graph is a reflection in the x-axis with no horizontal shift ($h = 0$).

Step 4 Find the amplitude and period.

The amplitude is $|a| = \dfrac{(\text{maximum value}) - (\text{minimum value})}{2} = \dfrac{75 - 3}{2} = 36.$

Because the graph is a reflection in the x-axis, $a < 0$. So, $a = -36$. Because a rope is rotating at a rate of 2 revolutions per second, one revolution is completed in 0.5 second. So, the period is $\dfrac{2\pi}{b} = 0.5$, and $b = 4\pi$.

▶ A model for the height of a rope is $h(t) = -36 \cos 4\pi t + 39$.

Check

Use the *table* feature of a graphing calculator to check your model.

X	Y1	
0	3	2 revolutions
.25	75	
.5	3	
.75	75	
1	3	
1.25	75	
1.5	3	
X=0		

Monitoring Progress

Help in English and Spanish at *BigIdeasMath.com*

Write a function for the sinusoid.

2.

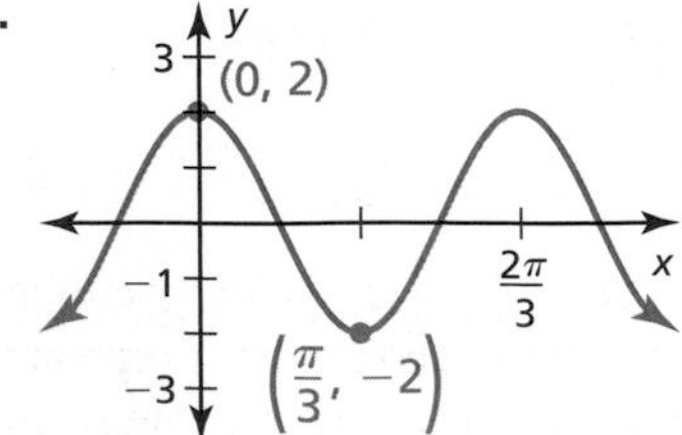

3.

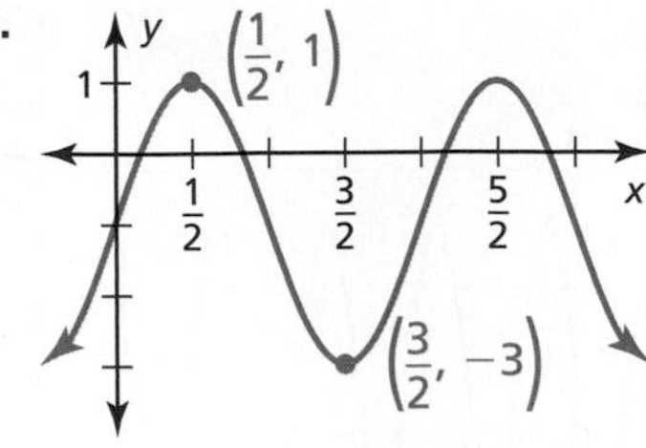

4. WHAT IF? Describe how the model in Example 3 changes when the lowest point of a rope is 5 inches above the ground and the highest point is 70 inches above the ground.

Using Technology to Find Trigonometric Models

Another way to model sinusoids is to use a graphing calculator that has a sinusoidal regression feature.

Using Sinusoidal Regression

The table shows the numbers N of hours of daylight in Denver, Colorado, on the 15th day of each month, where $t = 1$ represents January. Write a model that gives N as a function of t and interpret the period of its graph.

t	1	2	3	4	5	6
N	9.68	10.75	11.93	13.27	14.38	14.98

t	7	8	9	10	11	12
N	14.70	13.73	12.45	11.17	9.98	9.38

SOLUTION

Step 1 Enter the data in a graphing calculator.

L1	L2	L3 1
1	9.68	------
2	10.75	
3	11.93	
4	13.27	
5	14.38	
6	14.98	
7	14.7	

L1(1)=1

Step 2 Make a scatter plot.

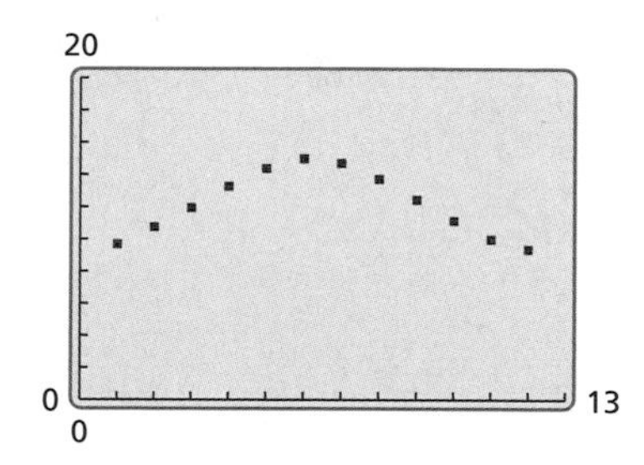

Step 3 The scatter plot appears sinusoidal. So, perform a sinusoidal regression.

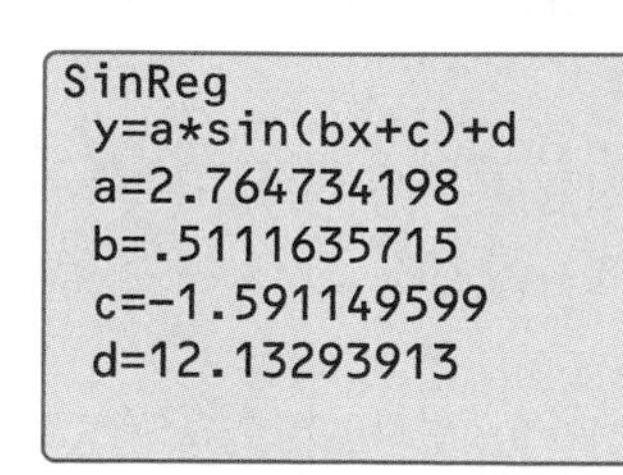

Step 4 Graph the data and the model in the same viewing window.

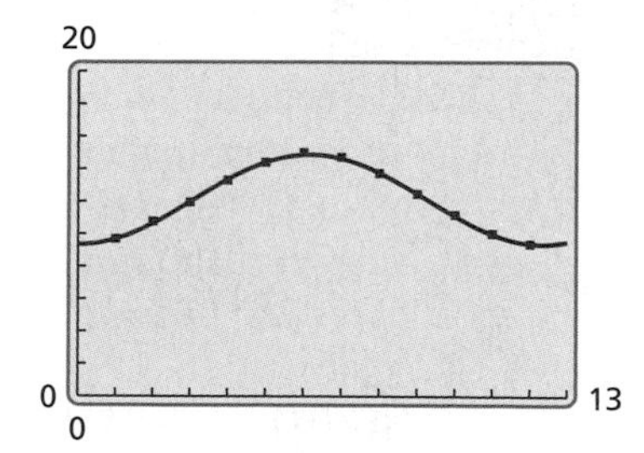

STUDY TIP

Notice that the *sinusoidal regression* feature finds a model of the form $y = a \sin(bx + c) + d$. This function has a period of $\frac{2\pi}{b}$ because it can be written as $y = a \sin b\left(x + \frac{c}{b}\right) + d$.

▶ The model appears to be a good fit. So, a model for the data is $N = 2.76 \sin(0.511t - 1.59) + 12.1$. The period, $\frac{2\pi}{0.511} \approx 12$, makes sense because there are 12 months in a year and you would expect this pattern to continue in following years.

Monitoring Progress

Help in English and Spanish at *BigIdeasMath.com*

5. The table shows the average daily temperature T (in degrees Fahrenheit) for a city each month, where $m = 1$ represents January. Write a model that gives T as a function of m and interpret the period of its graph.

m	1	2	3	4	5	6	7	8	9	10	11	12
T	29	32	39	48	59	68	74	72	65	54	45	35

9.6 Exercises

Dynamic Solutions available at *BigIdeasMath.com*

Vocabulary and Core Concept Check

1. **COMPLETE THE SENTENCE** Graphs of sine and cosine functions are called ________.

2. **WRITING** Describe how to find the frequency of the function whose graph is shown.

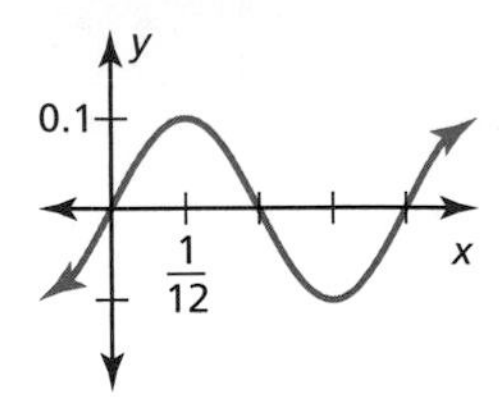

Monitoring Progress and Modeling with Mathematics

In Exercises 3–10, find the frequency of the function.

3. $y = \sin x$

4. $y = \sin 3x$

5. $y = \cos 4x + 2$

6. $y = -\cos 2x$

7. $y = \sin 3\pi x$

8. $y = \cos \frac{\pi x}{4}$

9. $y = \frac{1}{2}\cos 0.75x - 8$

10. $y = 3 \sin 0.2x + 6$

11. **MODELING WITH MATHEMATICS** The lowest frequency of sounds that can be heard by humans is 20 hertz. The maximum pressure P produced from a sound with a frequency of 20 hertz is 0.02 millipascal. Write and graph a sine model that gives the pressure P as a function of the time t (in seconds). *(See Example 1.)*

12. **MODELING WITH MATHEMATICS** A middle-A tuning fork vibrates with a frequency f of 440 hertz (cycles per second). You strike a middle-A tuning fork with a force that produces a maximum pressure of 5 pascals. Write and graph a sine model that gives the pressure P as a function of the time t (in seconds).

In Exercises 13–16, write a function for the sinusoid. *(See Example 2.)*

13.

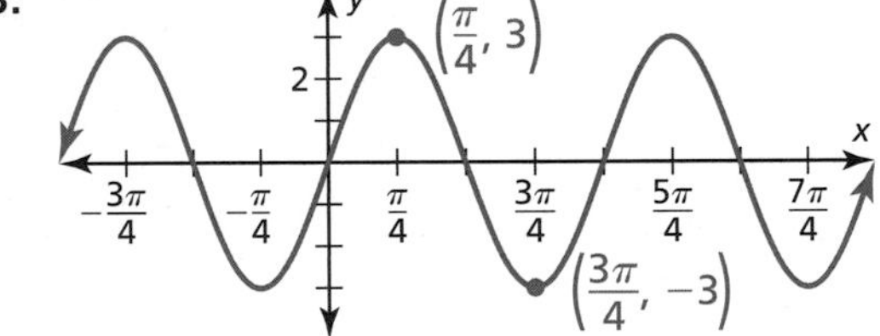

14.

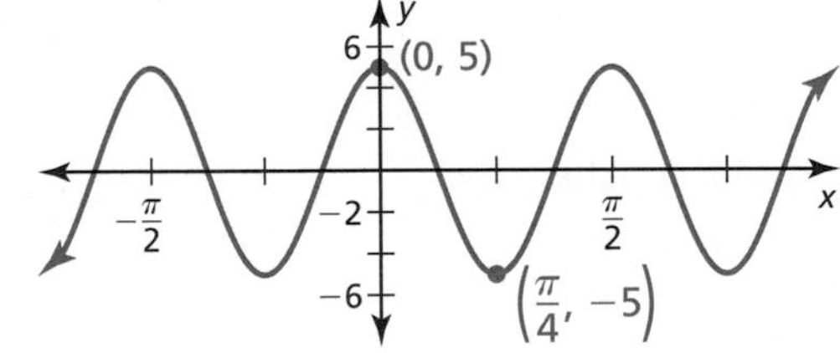

15.

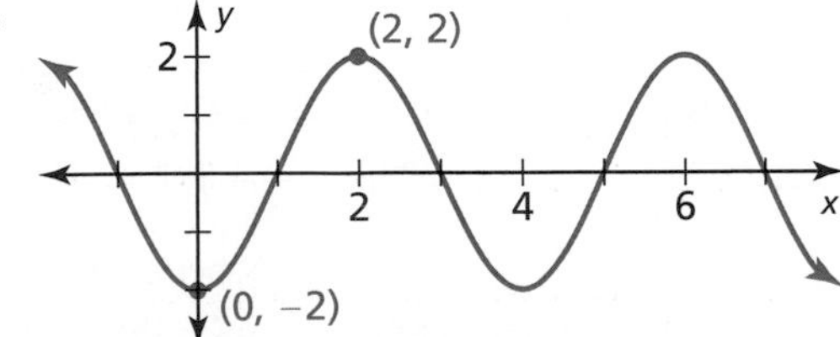

16.

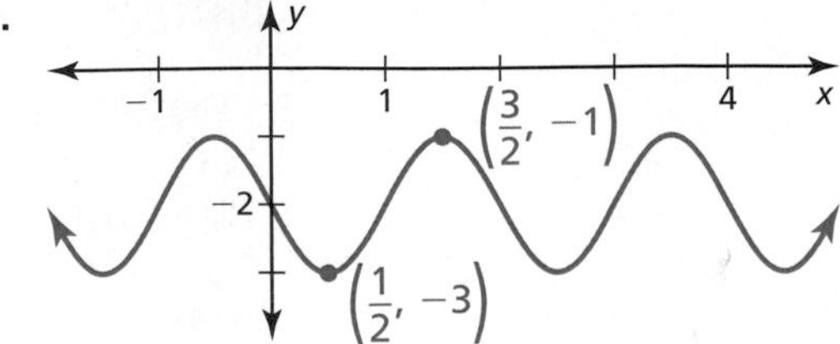

17. **ERROR ANALYSIS** Describe and correct the error in finding the amplitude of a sinusoid with a maximum point at (2, 10) and a minimum point at (4, −6).

> ✗ $|a| = \dfrac{(\text{maximum value}) + (\text{minimum value})}{2}$
>
> $= \dfrac{10 - 6}{2}$
>
> $= 2$

18. **ERROR ANALYSIS** Describe and correct the error in finding the vertical shift of a sinusoid with a maximum point at (3, −2) and a minimum point at (7, −8).

> ✗ $k = \dfrac{(\text{maximum value}) + (\text{minimum value})}{2}$
>
> $= \dfrac{7 + 3}{2}$
>
> $= 5$

19. **MODELING WITH MATHEMATICS** One of the largest sewing machines in the world has a *flywheel* (which turns as the machine sews) that is 5 feet in diameter. The highest point of the handle at the edge of the flywheel is 9 feet above the ground, and the lowest point is 4 feet. The wheel makes a complete turn every 2 seconds. Write a model for the height h (in feet) of the handle as a function of the time t (in seconds) given that the handle is at its lowest point when $t = 0$. *(See Example 3.)*

20. **MODELING WITH MATHEMATICS** The Great Laxey Wheel, located on the Isle of Man, is the largest working water wheel in the world. The highest point of a bucket on the wheel is 70.5 feet above the viewing platform, and the lowest point is 2 feet below the viewing platform. The wheel makes a complete turn every 24 seconds. Write a model for the height h (in feet) of the bucket as a function of time t (in seconds) given that the bucket is at its lowest point when $t = 0$.

USING TOOLS In Exercises 21 and 22, the time t is measured in months, where $t = 1$ represents January. Write a model that gives the average monthly high temperature D as a function of t and interpret the period of the graph. *(See Example 4.)*

21. **Air Temperatures in Apple Valley, CA**

t	1	2	3	4	5	6
D	60	63	69	75	85	94
t	7	8	9	10	11	12
D	99	99	93	81	69	60

22. **Water Temperatures at Miami Beach, FL**

t	1	2	3	4	5	6
D	71	73	75	78	81	85
t	7	8	9	10	11	12
D	86	85	84	81	76	73

23. **MODELING WITH MATHEMATICS** A circuit has an alternating voltage of 100 volts that peaks every 0.5 second. Write a sinusoidal model for the voltage V as a function of the time t (in seconds).

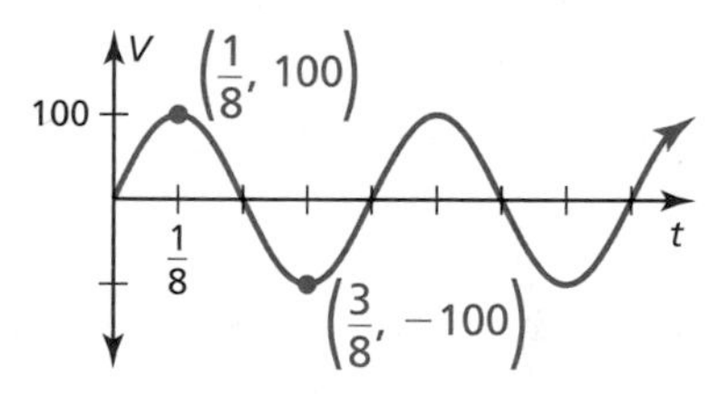

24. **MULTIPLE REPRESENTATIONS** The graph shows the average daily temperature of Lexington, Kentucky. The average daily temperature of Louisville, Kentucky, is modeled by $y = -22 \cos \frac{\pi}{6}t + 57$, where y is the temperature (in degrees Fahrenheit) and t is the number of months since January 1. Which city has the greater average daily temperature? Explain.

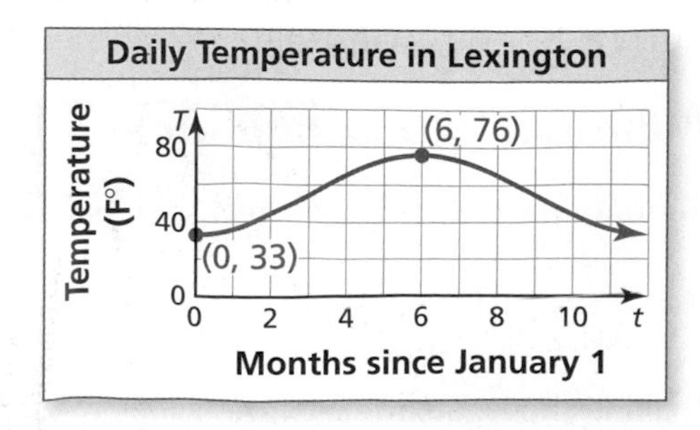

25. USING TOOLS The table shows the numbers of employees N (in thousands) at a sporting goods company each year for 11 years. The time t is measured in years, with $t = 1$ representing the first year.

t	1	2	3	4	5	6
N	20.8	22.7	24.6	23.2	20	17.5

t	7	8	9	10	11	12
N	16.7	17.8	21	22	24.1	

a. Use sinusoidal regression to find a model that gives N as a function of t.

b. Predict the number of employees at the company in the 12th year.

26. THOUGHT PROVOKING The figure shows a tangent line drawn to the graph of the function $y = \sin x$. At several points on the graph, draw a tangent line to the graph and estimate its slope. Then plot the points (x, m), where m is the slope of the tangent line. What can you conclude?

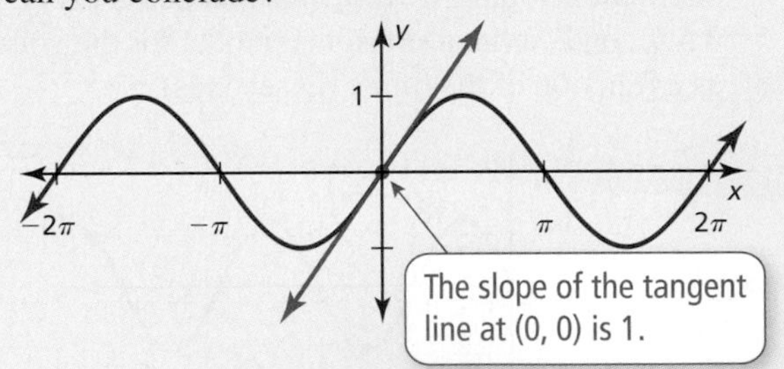

27. REASONING Determine whether you would use a sine or cosine function to model each sinusoid with the y-intercept described. Explain your reasoning.

a. The y-intercept occurs at the maximum value of the function.

b. The y-intercept occurs at the minimum value of the function.

c. The y-intercept occurs halfway between the maximum and minimum values of the function.

28. HOW DO YOU SEE IT? What is the frequency of the function whose graph is shown? Explain.

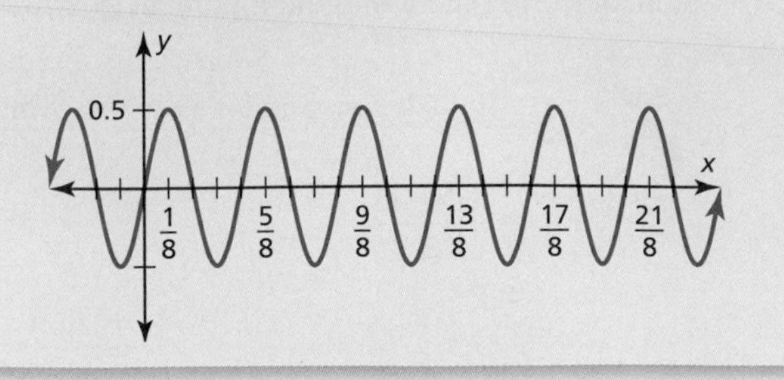

29. USING STRUCTURE During one cycle, a sinusoid has a minimum at $\left(\frac{\pi}{2}, 3\right)$ and a maximum at $\left(\frac{\pi}{4}, 8\right)$. Write a sine function *and* a cosine function for the sinusoid. Use a graphing calculator to verify that your answers are correct.

30. MAKING AN ARGUMENT Your friend claims that a function with a frequency of 2 has a greater period than a function with a frequency of $\frac{1}{2}$. Is your friend correct? Explain your reasoning.

31. PROBLEM SOLVING The low tide at a port is 3.5 feet and occurs at midnight. After 6 hours, the port is at high tide, which is 16.5 feet.

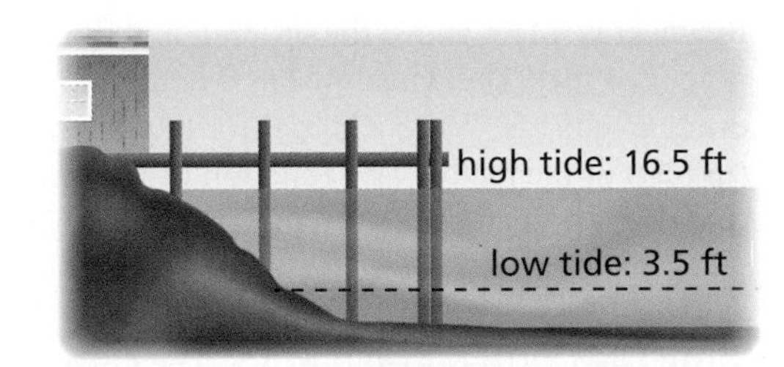

a. Write a sinusoidal model that gives the tide depth d (in feet) as a function of the time t (in hours). Let $t = 0$ represent midnight.

b. Find all the times when low and high tides occur in a 24-hour period.

c. Explain how the graph of the function you wrote in part (a) is related to a graph that shows the tide depth d at the port t hours after 3:00 A.M.

Maintaining Mathematical Proficiency

Reviewing what you learned in previous grades and lessons

Simplify the expression. *(Section 5.2)*

32. $\frac{17}{\sqrt{2}}$ **33.** $\frac{3}{\sqrt{6} - 2}$ **34.** $\frac{8}{\sqrt{10} + 3}$ **35.** $\frac{13}{\sqrt{3} + \sqrt{11}}$

Expand the logarithmic expression. *(Section 6.5)*

36. $\log_8 \frac{x}{7}$ **37.** $\ln 2x$ **38.** $\log_3 5x^3$ **39.** $\ln \frac{4x^6}{y}$

9.7 Using Trigonometric Identities

Essential Question

How can you verify a trigonometric identity?

EXPLORATION 1 Writing a Trigonometric Identity

Work with a partner. In the figure, the point (x, y) is on a circle of radius c with center at the origin.

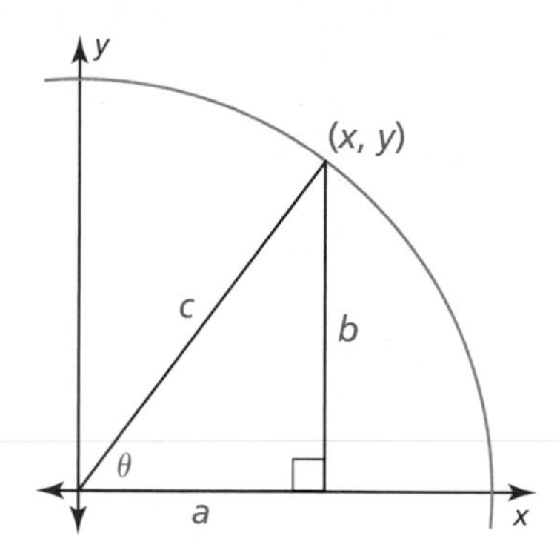

a. Write an equation that relates a, b, and c.

b. Write expressions for the sine and cosine ratios of angle θ.

c. Use the results from parts (a) and (b) to find the sum of $\sin^2\theta$ and $\cos^2\theta$. What do you observe?

d. Complete the table to verify that the identity you wrote in part (c) is valid for angles (of your choice) in each of the four quadrants.

	θ	$\sin^2 \theta$	$\cos^2 \theta$	$\sin^2 \theta + \cos^2 \theta$
QI				
QII				
QIII				
QIV				

EXPLORATION 2 Writing Other Trigonometric Identities

Work with a partner. The trigonometric identity you derived in Exploration 1 is called a Pythagorean identity. There are two other Pythagorean identities. To derive them, recall the four relationships:

$$\tan \theta = \frac{\sin \theta}{\cos \theta} \qquad \cot \theta = \frac{\cos \theta}{\sin \theta}$$

$$\sec \theta = \frac{1}{\cos \theta} \qquad \csc \theta = \frac{1}{\sin \theta}$$

REASONING ABSTRACTLY

To be proficient in math, you need to know and flexibly use different properties of operations and objects.

a. Divide each side of the Pythagorean identity you derived in Exploration 1 by $\cos^2\theta$ and simplify. What do you observe?

b. Divide each side of the Pythagorean identity you derived in Exploration 1 by $\sin^2\theta$ and simplify. What do you observe?

Communicate Your Answer

3. How can you verify a trigonometric identity?

4. Is $\sin \theta = \cos \theta$ a trigonometric identity? Explain your reasoning.

5. Give some examples of trigonometric identities that are different than those in Explorations 1 and 2.

9.7 Lesson

Core Vocabulary

trigonometric identity, *p. 514*

Previous
unit circle

What You Will Learn

- Use trigonometric identities to evaluate trigonometric functions and simplify trigonometric expressions.
- Verify trigonometric identities.

Using Trigonometric Identities

Recall that when an angle θ is in standard position with its terminal side intersecting the unit circle at (x, y), then $x = \cos\theta$ and $y = \sin\theta$. Because (x, y) is on a circle centered at the origin with radius 1, it follows that

$$x^2 + y^2 = 1$$

and

$$\cos^2\theta + \sin^2\theta = 1.$$

(cos θ, sin θ) = (x, y)
r = 1
θ
x
y

STUDY TIP

Note that $\sin^2\theta$ represents $(\sin\theta)^2$ and $\cos^2\theta$ represents $(\cos\theta)^2$.

The equation $\cos^2\theta + \sin^2\theta = 1$ is true for any value of θ. A trigonometric equation that is true for all values of the variable for which both sides of the equation are defined is called a **trigonometric identity**. In Section 9.1, you used reciprocal identities to find the values of the cosecant, secant, and cotangent functions. These and other fundamental trigonometric identities are listed below.

Core Concept

Fundamental Trigonometric Identities

Reciprocal Identities

$$\csc\theta = \frac{1}{\sin\theta} \qquad \sec\theta = \frac{1}{\cos\theta} \qquad \cot\theta = \frac{1}{\tan\theta}$$

Tangent and Cotangent Identities

$$\tan\theta = \frac{\sin\theta}{\cos\theta} \qquad \cot\theta = \frac{\cos\theta}{\sin\theta}$$

Pythagorean Identities

$$\sin^2\theta + \cos^2\theta = 1 \qquad 1 + \tan^2\theta = \sec^2\theta \qquad 1 + \cot^2\theta = \csc^2\theta$$

Cofunction Identities

$$\sin\left(\frac{\pi}{2} - \theta\right) = \cos\theta \qquad \cos\left(\frac{\pi}{2} - \theta\right) = \sin\theta \qquad \tan\left(\frac{\pi}{2} - \theta\right) = \cot\theta$$

Negative Angle Identities

$$\sin(-\theta) = -\sin\theta \qquad \cos(-\theta) = \cos\theta \qquad \tan(-\theta) = -\tan\theta$$

In this section, you will use trigonometric identities to do the following.

- Evaluate trigonometric functions.
- Simplify trigonometric expressions.
- Verify other trigonometric identities.

EXAMPLE 1 Finding Trigonometric Values

Given that $\sin\theta = \frac{4}{5}$ and $\frac{\pi}{2} < \theta < \pi$, find the values of the other five trigonometric functions of θ.

SOLUTION

Step 1 Find $\cos\theta$.

$\sin^2\theta + \cos^2\theta = 1$ Write Pythagorean identity.

$\left(\frac{4}{5}\right)^2 + \cos^2\theta = 1$ Substitute $\frac{4}{5}$ for $\sin\theta$.

$\cos^2\theta = 1 - \left(\frac{4}{5}\right)^2$ Subtract $\left(\frac{4}{5}\right)^2$ from each side.

$\cos^2\theta = \frac{9}{25}$ Simplify.

$\cos\theta = \pm\frac{3}{5}$ Take square root of each side.

$\cos\theta = -\frac{3}{5}$ Because θ is in Quadrant II, $\cos\theta$ is negative.

Step 2 Find the values of the other four trigonometric functions of θ using the values of $\sin\theta$ and $\cos\theta$.

$$\tan\theta = \frac{\sin\theta}{\cos\theta} = \frac{\frac{4}{5}}{-\frac{3}{5}} = -\frac{4}{3} \qquad \cot\theta = \frac{\cos\theta}{\sin\theta} = \frac{-\frac{3}{5}}{\frac{4}{5}} = -\frac{3}{4}$$

$$\csc\theta = \frac{1}{\sin\theta} = \frac{1}{\frac{4}{5}} = \frac{5}{4} \qquad \sec\theta = \frac{1}{\cos\theta} = \frac{1}{-\frac{3}{5}} = -\frac{5}{3}$$

EXAMPLE 2 Simplifying Trigonometric Expressions

Simplify (a) $\tan\left(\frac{\pi}{2} - \theta\right)\sin\theta$ and (b) $\sec\theta\tan^2\theta + \sec\theta$.

SOLUTION

a. $\tan\left(\frac{\pi}{2} - \theta\right)\sin\theta = \cot\theta\sin\theta$ Cofunction identity

$= \left(\frac{\cos\theta}{\sin\theta}\right)(\sin\theta)$ Cotangent identity

$= \cos\theta$ Simplify.

b. $\sec\theta\tan^2\theta + \sec\theta = \sec\theta(\sec^2\theta - 1) + \sec\theta$ Pythagorean identity

$= \sec^3\theta - \sec\theta + \sec\theta$ Distributive Property

$= \sec^3\theta$ Simplify.

Monitoring Progress

Help in English and Spanish at *BigIdeasMath.com*

1. Given that $\cos\theta = \frac{1}{6}$ and $0 < \theta < \frac{\pi}{2}$, find the values of the other five trigonometric functions of θ.

Simplify the expression.

2. $\sin x \cot x \sec x$

3. $\cos\theta - \cos\theta\sin^2\theta$

4. $\frac{\tan x \csc x}{\sec x}$

Verifying Trigonometric Identities

You can use the fundamental identities from this chapter to verify new trigonometric identities. When verifying an identity, begin with the expression on one side. Use algebra and trigonometric properties to manipulate the expression until it is identical to the other side.

EXAMPLE 3 Verifying a Trigonometric Identity

Verify the identity $\dfrac{\sec^2 \theta - 1}{\sec^2 \theta} = \sin^2 \theta$.

SOLUTION

$$\frac{\sec^2 \theta - 1}{\sec^2 \theta} = \frac{\sec^2 \theta}{\sec^2 \theta} - \frac{1}{\sec^2 \theta}$$ Write as separate fractions.

$$= 1 - \left(\frac{1}{\sec \theta}\right)^2$$ Simplify.

$$= 1 - \cos^2 \theta$$ Reciprocal identity

$$= \sin^2 \theta$$ Pythagorean identity

Notice that verifying an identity is not the same as solving an equation. When verifying an identity, you cannot assume that the two sides of the equation are equal because you are trying to verify that they are equal. So, you cannot use any properties of equality, such as adding the same quantity to each side of the equation.

EXAMPLE 4 Verifying a Trigonometric Identity

Verify the identity $\sec x + \tan x = \dfrac{\cos x}{1 - \sin x}$.

SOLUTION

LOOKING FOR STRUCTURE

To verify the identity, you must introduce $1 - \sin x$ into the denominator. Multiply the numerator and the denominator by $1 - \sin x$ so you get an equivalent expression.

$$\sec x + \tan x = \frac{1}{\cos x} + \tan x$$ Reciprocal identity

$$= \frac{1}{\cos x} + \frac{\sin x}{\cos x}$$ Tangent identity

$$= \frac{1 + \sin x}{\cos x}$$ Add fractions.

$$= \frac{1 + \sin x}{\cos x} \cdot \frac{1 - \sin x}{1 - \sin x}$$ Multiply by $\dfrac{1 - \sin x}{1 - \sin x}$.

$$= \frac{1 - \sin^2 x}{\cos x(1 - \sin x)}$$ Simplify numerator.

$$= \frac{\cos^2 x}{\cos x(1 - \sin x)}$$ Pythagorean identity

$$= \frac{\cos x}{1 - \sin x}$$ Simplify.

Monitoring Progress  Help in English and Spanish at *BigIdeasMath.com*

Verify the identity.

5. $\cot(-\theta) = -\cot \theta$

6. $\csc^2 x(1 - \sin^2 x) = \cot^2 x$

7. $\cos x \csc x \tan x = 1$

8. $(\tan^2 x + 1)(\cos^2 x - 1) = -\tan^2 x$

9.7 Exercises

Dynamic Solutions available at *BigIdeasMath.com*

Vocabulary and Core Concept Check

1. **WRITING** Describe the difference between a trigonometric identity and a trigonometric equation.

2. **WRITING** Explain how to use trigonometric identities to determine whether $\sec(-\theta) = \sec\theta$ or $\sec(-\theta) = -\sec\theta$.

Monitoring Progress and Modeling with Mathematics

In Exercises 3–10, find the values of the other five trigonometric functions of θ. *(See Example 1.)*

3. $\sin\theta = \frac{1}{3}, 0 < \theta < \frac{\pi}{2}$

4. $\sin\theta = -\frac{7}{10}, \pi < \theta < \frac{3\pi}{2}$

5. $\tan\theta = -\frac{3}{7}, \frac{\pi}{2} < \theta < \pi$

6. $\cot\theta = -\frac{2}{5}, \frac{\pi}{2} < \theta < \pi$

7. $\cos\theta = -\frac{5}{6}, \pi < \theta < \frac{3\pi}{2}$

8. $\sec\theta = \frac{9}{4}, \frac{3\pi}{2} < \theta < 2\pi$

9. $\cot\theta = -3, \frac{3\pi}{2} < \theta < 2\pi$

10. $\csc\theta = -\frac{5}{3}, \pi < \theta < \frac{3\pi}{2}$

In Exercises 11–20, simplify the expression. *(See Example 2.)*

11. $\sin x \cot x$

12. $\cos\theta(1 + \tan^2\theta)$

13. $\frac{\sin(-\theta)}{\cos(-\theta)}$

14. $\frac{\cos^2 x}{\cot^2 x}$

15. $\frac{\cos\left(\frac{\pi}{2} - x\right)}{\csc x}$

16. $\sin\left(\frac{\pi}{2} - \theta\right)\sec\theta$

17. $\frac{\csc^2 x - \cot^2 x}{\sin(-x)\cot x}$

18. $\frac{\cos^2 x \tan^2(-x) - 1}{\cos^2 x}$

19. $\frac{\cos\left(\frac{\pi}{2} - \theta\right)}{\csc\theta} + \cos^2\theta$

20. $\frac{\sec x \sin x + \cos\left(\frac{\pi}{2} - x\right)}{1 + \sec x}$

ERROR ANALYSIS In Exercises 21 and 22, describe and correct the error in simplifying the expression.

21.

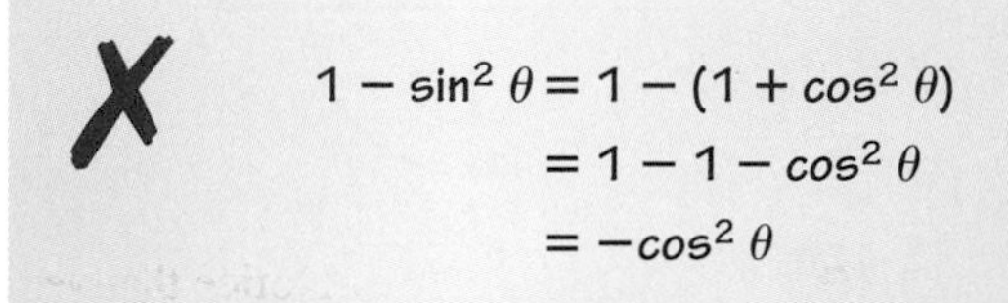

$1 - \sin^2\theta = 1 - (1 + \cos^2\theta)$
$= 1 - 1 - \cos^2\theta$
$= -\cos^2\theta$

22.

$$\tan x \csc x = \frac{\cos x}{\sin x} \cdot \frac{1}{\sin x} = \frac{\cos x}{\sin^2 x}$$

In Exercises 23–30, verify the identity. *(See Examples 3 and 4.)*

23. $\sin x \csc x = 1$

24. $\tan\theta \csc\theta \cos\theta = 1$

25. $\cos\left(\frac{\pi}{2} - x\right)\cot x = \cos x$

26. $\sin\left(\frac{\pi}{2} - x\right)\tan x = \sin x$

27. $\frac{\cos\left(\frac{\pi}{2} - \theta\right) + 1}{1 - \sin(-\theta)} = 1$

28. $\frac{\sin^2(-x)}{\tan^2 x} = \cos^2 x$

29. $\frac{1 + \cos x}{\sin x} + \frac{\sin x}{1 + \cos x} = 2\csc x$

30. $\frac{\sin x}{1 - \cos(-x)} = \csc x + \cot x$

31. **USING STRUCTURE** A function f is *odd* when $f(-x) = -f(x)$. A function f is *even* when $f(-x) = f(x)$. Which of the six trigonometric functions are odd? Which are even? Justify your answers using identities and graphs.

32. **ANALYZING RELATIONSHIPS** As the value of $\cos\theta$ increases, what happens to the value of $\sec\theta$? Explain your reasoning.

33. **MAKING AN ARGUMENT** Your friend simplifies an expression and obtains $\sec x \tan x - \sin x$. You simplify the same expression and obtain $\sin x \tan^2 x$. Are your answers equivalent? Justify your answer.

34. **HOW DO YOU SEE IT?** The figure shows the unit circle and the angle θ.

    a. Is $\sin \theta$ positive or negative? $\cos \theta$? $\tan \theta$?

    b. In what quadrant does the terminal side of $-\theta$ lie?

    c. Is $\sin(-\theta)$ positive or negative? $\cos(-\theta)$? $\tan(-\theta)$?

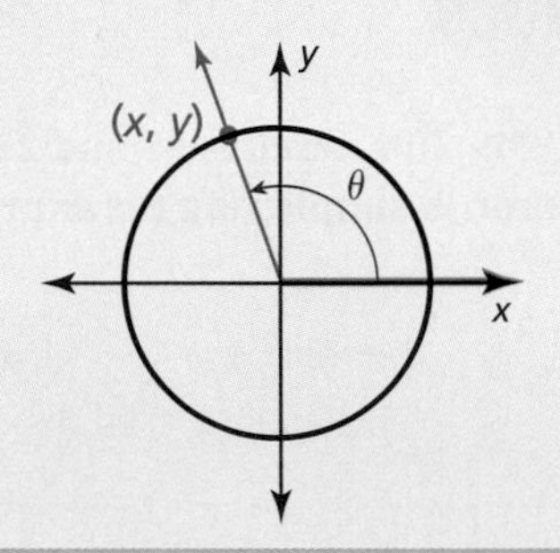

35. **MODELING WITH MATHEMATICS** A vertical *gnomon* (the part of a sundial that projects a shadow) has height h. The length s of the shadow cast by the gnomon when the angle of the Sun above the horizon is θ can be modeled by the equation below. Show that the equation below is equivalent to $s = h \cot \theta$.

$$s = \frac{h \sin(90° - \theta)}{\sin \theta}$$

36. **THOUGHT PROVOKING** Explain how you can use a trigonometric identity to find all the values of x for which $\sin x = \cos x$.

37. **DRAWING CONCLUSIONS** *Static friction* is the amount of force necessary to keep a stationary object on a flat surface from moving. Suppose a book weighing W pounds is lying on a ramp inclined at an angle θ. The coefficient of static friction u for the book can be found using the equation $uW \cos \theta = W \sin \theta$.

    a. Solve the equation for u and simplify the result.

    b. Use the equation from part (a) to determine what happens to the value of u as the angle θ increases from 0° to 90°.

38. **PROBLEM SOLVING** When light traveling in a medium (such as air) strikes the surface of a second medium (such as water) at an angle θ_1, the light begins to travel at a different angle θ_2. This change of direction is defined by Snell's law, $n_1 \sin \theta_1 = n_2 \sin \theta_2$, where n_1 and n_2 are the *indices of refraction* for the two mediums. Snell's law can be derived from the equation

$$\frac{n_1}{\sqrt{\cot^2 \theta_1 + 1}} = \frac{n_2}{\sqrt{\cot^2 \theta_2 + 1}}.$$

air: n_1
water: n_2
θ_1
θ_2

    a. Simplify the equation to derive Snell's law.

    b. What is the value of n_1 when $\theta_1 = 55°$, $\theta_2 = 35°$, and $n_2 = 2$?

    c. If $\theta_1 = \theta_2$, then what must be true about the values of n_1 and n_2? Explain when this situation would occur.

39. **WRITING** Explain how transformations of the graph of the parent function $f(x) = \sin x$ support the cofunction identity $\sin\left(\frac{\pi}{2} - \theta\right) = \cos \theta$.

40. **USING STRUCTURE** Verify each identity.

    a. $\ln|\sec \theta| = -\ln|\cos \theta|$

    b. $\ln|\tan \theta| = \ln|\sin \theta| - \ln|\cos \theta|$

Maintaining Mathematical Proficiency

Reviewing what you learned in previous grades and lessons

Find the value of x for the right triangle. *(Section 9.1)*

41.
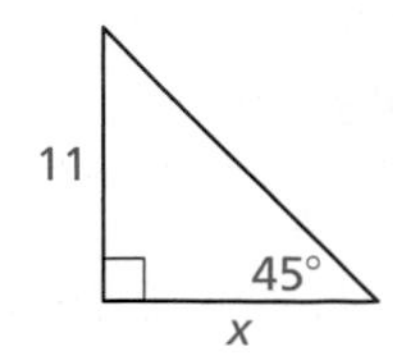

42.
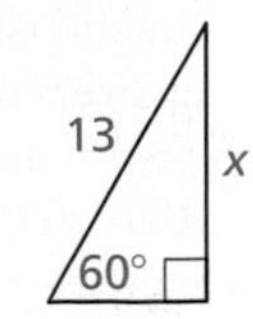

43.
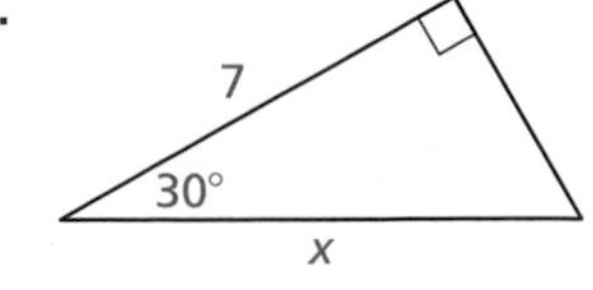

9.8 Using Sum and Difference Formulas

Essential Question How can you evaluate trigonometric functions of the sum or difference of two angles?

EXPLORATION 1 **Deriving a Difference Formula**

Work with a partner.

a. Explain why the two triangles shown are congruent.

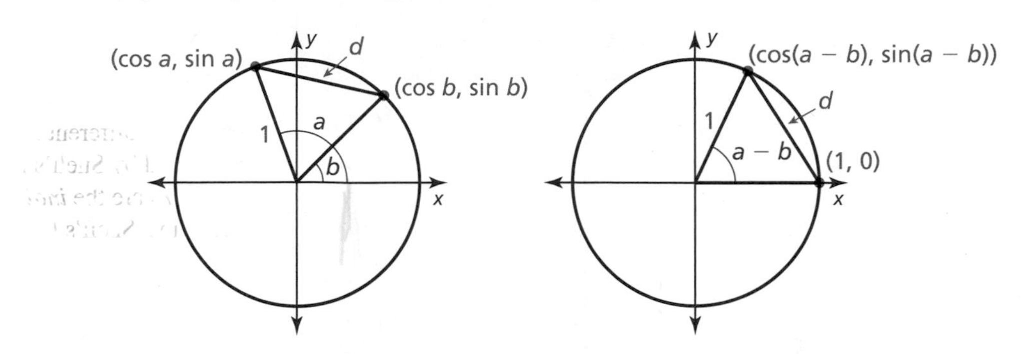

b. Use the Distance Formula to write an expression for d in the first unit circle.

c. Use the Distance Formula to write an expression for d in the second unit circle.

d. Write an equation that relates the expressions in parts (b) and (c). Then simplify this equation to obtain a formula for $\cos(a - b)$.

CONSTRUCTING VIABLE ARGUMENTS

To be proficient in math, you need to understand and use stated assumptions, definitions, and previously established results.

EXPLORATION 2 **Deriving a Sum Formula**

Work with a partner. Use the difference formula you derived in Exploration 1 to write a formula for $\cos(a + b)$ in terms of sine and cosine of a and b. *Hint*: Use the fact that

$$\cos(a + b) = \cos[a - (-b)].$$

EXPLORATION 3 **Deriving Difference and Sum Formulas**

Work with a partner. Use the formulas you derived in Explorations 1 and 2 to write formulas for $\sin(a - b)$ and $\sin(a + b)$ in terms of sine and cosine of a and b. *Hint*: Use the cofunction identities

$$\sin\left(\frac{\pi}{2} - a\right) = \cos a \text{ and } \cos\left(\frac{\pi}{2} - a\right) = \sin a$$

and the fact that

$$\cos\left[\left(\frac{\pi}{2} - a\right) + b\right] = \sin(a - b) \text{ and } \sin(a + b) = \sin[a - (-b)].$$

Communicate Your Answer

4. How can you evaluate trigonometric functions of the sum or difference of two angles?

5. a. Find the exact values of sin 75° and cos 75° using sum formulas. Explain your reasoning.

b. Find the exact values of sin 75° and cos 75° using difference formulas. Compare your answers to those in part (a).

9.8 Lesson

Core Vocabulary

Previous
ratio

What You Will Learn

- Use sum and difference formulas to evaluate and simplify trigonometric expressions.
- Use sum and difference formulas to solve trigonometric equations and rewrite real-life formulas.

Using Sum and Difference Formulas

In this lesson, you will study formulas that allow you to evaluate trigonometric functions of the sum or difference of two angles.

Core Concept

Sum and Difference Formulas

Sum Formulas	Difference Formulas
$\sin(a+b)=\sin a\cos b+\cos a\sin b$	$\sin(a-b)=\sin a\cos b-\cos a\sin b$
$\cos(a+b)=\cos a\cos b-\sin a\sin b$	$\cos(a-b)=\cos a\cos b+\sin a\sin b$
$\tan(a+b)=\dfrac{\tan a+\tan b}{1-\tan a\tan b}$	$\tan(a-b)=\dfrac{\tan a-\tan b}{1+\tan a\tan b}$

In general, $\sin(a+b)\neq\sin a+\sin b$. Similar statements can be made for the other trigonometric functions of sums and differences.

EXAMPLE 1 Evaluating Trigonometric Expressions

Find the exact value of (a) $\sin 15°$ and (b) $\tan\frac{7\pi}{12}$.

SOLUTION

a. $\sin 15° = \sin(60° - 45°)$ — Substitute 60° − 45° for 15°.

$= \sin 60°\cos 45° - \cos 60°\sin 45°$ — Difference formula for sine

$= \frac{\sqrt{3}}{2}\left(\frac{\sqrt{2}}{2}\right) - \frac{1}{2}\left(\frac{\sqrt{2}}{2}\right)$ — Evaluate.

$= \frac{\sqrt{6}-\sqrt{2}}{4}$ — Simplify.

▶ The exact value of $\sin 15°$ is $\frac{\sqrt{6}-\sqrt{2}}{4}$. Check this with a calculator.

Check

```
sin(15°)
        .2588190451
(√(6)-√(2))/4
        .2588190451
```

b. $\tan\frac{7\pi}{12} = \tan\left(\frac{\pi}{3}+\frac{\pi}{4}\right)$ — Substitute $\frac{\pi}{3}+\frac{\pi}{4}$ for $\frac{7\pi}{12}$.

$= \dfrac{\tan\frac{\pi}{3}+\tan\frac{\pi}{4}}{1-\tan\frac{\pi}{3}\tan\frac{\pi}{4}}$ — Sum formula for tangent

$= \dfrac{\sqrt{3}+1}{1-\sqrt{3}\cdot 1}$ — Evaluate.

$= -2-\sqrt{3}$ — Simplify.

▶ The exact value of $\tan\frac{7\pi}{12}$ is $-2-\sqrt{3}$. Check this with a calculator.

Check

```
tan(7π/12)
        -3.732050808
-2-√(3)
        -3.732050808
```

EXAMPLE 2 Using a Difference Formula

ANOTHER WAY

You can also use a Pythagorean identity and quadrant signs to find sin a and cos b.

Find $\cos(a - b)$ given that $\cos a = -\frac{4}{5}$ with $\pi < a < \frac{3\pi}{2}$ and $\sin b = \frac{5}{13}$ with $0 < b < \frac{\pi}{2}$.

SOLUTION

Step 1 Find sin a and cos b.

Because $\cos a = -\frac{4}{5}$ and a is in Quadrant III, $\sin a = -\frac{3}{5}$, as shown in the figure.

Because $\sin b = \frac{5}{13}$ and b is in Quadrant I, $\cos b = \frac{12}{13}$, as shown in the figure.

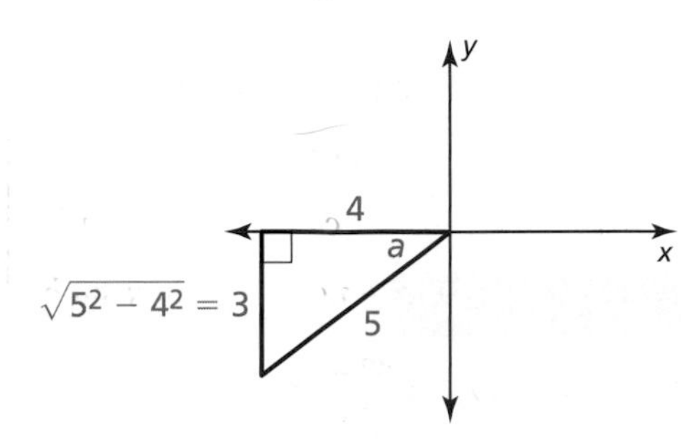

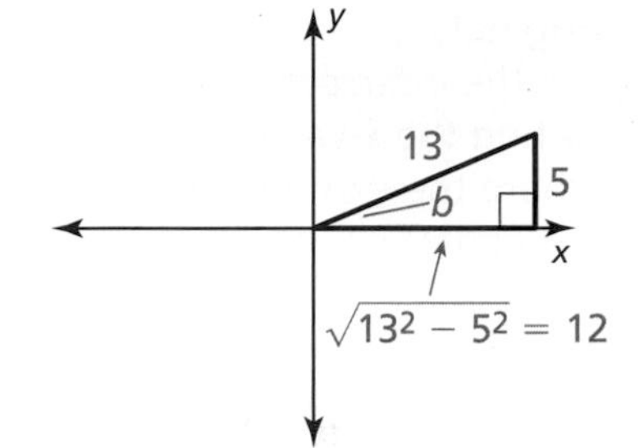

Step 2 Use the difference formula for cosine to find $\cos(a - b)$.

$$\cos(a - b) = \cos a \cos b + \sin a \sin b \qquad \text{Difference formula for cosine}$$

$$= -\frac{4}{5}\left(\frac{12}{13}\right) + \left(-\frac{3}{5}\right)\left(\frac{5}{13}\right) \qquad \text{Evaluate.}$$

$$= -\frac{63}{65} \qquad \text{Simplify.}$$

▶ The value of $\cos(a - b)$ is $-\frac{63}{65}$.

EXAMPLE 3 Simplifying an Expression

Simplify the expression $\cos(x + \pi)$.

SOLUTION

$$\cos(x + \pi) = \cos x \cos \pi - \sin x \sin \pi \qquad \text{Sum formula for cosine}$$

$$= (\cos x)(-1) - (\sin x)(0) \qquad \text{Evaluate.}$$

$$= -\cos x \qquad \text{Simplify.}$$

Monitoring Progress

Help in English and Spanish at *BigIdeasMath.com*

Find the exact value of the expression.

1. $\sin 105°$ **2.** $\cos 15°$ **3.** $\tan \frac{5\pi}{12}$ **4.** $\cos \frac{\pi}{12}$

5. Find $\sin(a - b)$ given that $\sin a = \frac{8}{17}$ with $0 < a < \frac{\pi}{2}$ and $\cos b = -\frac{24}{25}$ with $\pi < b < \frac{3\pi}{2}$.

Simplify the expression.

6. $\sin(x + \pi)$ **7.** $\cos(x - 2\pi)$ **8.** $\tan(x - \pi)$

Solving Equations and Rewriting Formulas

EXAMPLE 4 Solving a Trigonometric Equation

Solve $\sin\left(x + \frac{\pi}{3}\right) + \sin\left(x - \frac{\pi}{3}\right) = 1$ for $0 \le x < 2\pi$.

SOLUTION

> **ANOTHER WAY**
> You can also solve the equation by using a graphing calculator. First, graph each side of the original equation. Then use the *intersect* feature to find the x-value(s) where the expressions are equal.

$$\sin\left(x + \frac{\pi}{3}\right) + \sin\left(x - \frac{\pi}{3}\right) = 1 \quad \text{Write equation.}$$

$$\sin x \cos \frac{\pi}{3} + \cos x \sin \frac{\pi}{3} + \sin x \cos \frac{\pi}{3} - \cos x \sin \frac{\pi}{3} = 1 \quad \text{Use formulas.}$$

$$\frac{1}{2}\sin x + \frac{\sqrt{3}}{2}\cos x + \frac{1}{2}\sin x - \frac{\sqrt{3}}{2}\cos x = 1 \quad \text{Evaluate.}$$

$$\sin x = 1 \quad \text{Simplify.}$$

▶ In the interval $0 \le x < 2\pi$, the solution is $x = \frac{\pi}{2}$.

EXAMPLE 5 Rewriting a Real-Life Formula

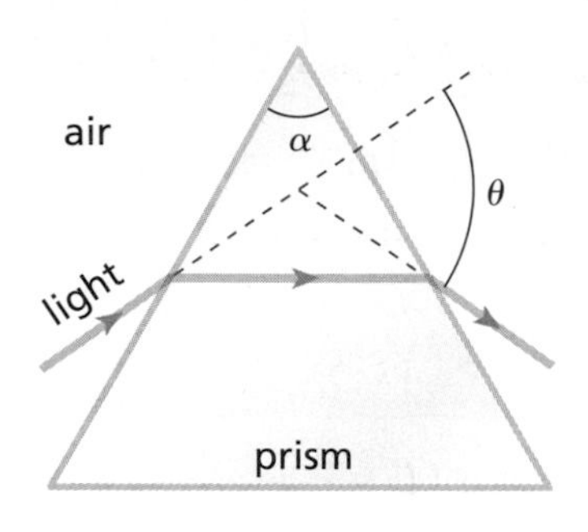

The *index of refraction* of a transparent material is the ratio of the speed of light in a vacuum to the speed of light in the material. A triangular prism, like the one shown, can be used to measure the index of refraction using the formula

$$n = \frac{\sin\left(\frac{\theta}{2} + \frac{\alpha}{2}\right)}{\sin \frac{\theta}{2}}.$$

For $\alpha = 60°$, show that the formula can be rewritten as $n = \frac{\sqrt{3}}{2} + \frac{1}{2}\cot \frac{\theta}{2}$.

SOLUTION

$$n = \frac{\sin\left(\frac{\theta}{2} + 30°\right)}{\sin \frac{\theta}{2}} \quad \text{Write formula with } \frac{\alpha}{2} = \frac{60°}{2} = 30°.$$

$$= \frac{\sin \frac{\theta}{2} \cos 30° + \cos \frac{\theta}{2} \sin 30°}{\sin \frac{\theta}{2}} \quad \text{Sum formula for sine}$$

$$= \frac{\left(\sin \frac{\theta}{2}\right)\left(\frac{\sqrt{3}}{2}\right) + \left(\cos \frac{\theta}{2}\right)\left(\frac{1}{2}\right)}{\sin \frac{\theta}{2}} \quad \text{Evaluate.}$$

$$= \frac{\frac{\sqrt{3}}{2}\sin \frac{\theta}{2}}{\sin \frac{\theta}{2}} + \frac{\frac{1}{2}\cos \frac{\theta}{2}}{\sin \frac{\theta}{2}} \quad \text{Write as separate fractions.}$$

$$= \frac{\sqrt{3}}{2} + \frac{1}{2}\cot \frac{\theta}{2} \quad \text{Simplify.}$$

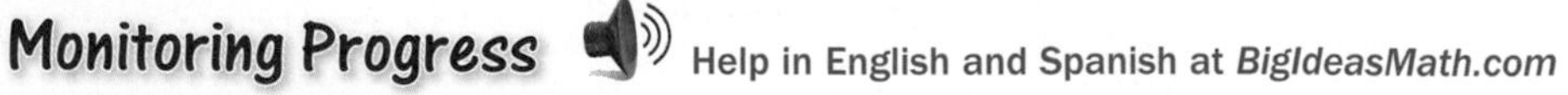

Monitoring Progress Help in English and Spanish at *BigIdeasMath.com*

9. Solve $\sin\left(\frac{\pi}{4} - x\right) - \sin\left(x + \frac{\pi}{4}\right) = 1$ for $0 \le x < 2\pi$.

9.8 Exercises

Dynamic Solutions available at *BigIdeasMath.com*

Vocabulary and Core Concept Check

1. **COMPLETE THE SENTENCE** Write the expression $\cos 130° \cos 40° - \sin 130° \sin 40°$ as the cosine of an angle.

2. **WRITING** Explain how to evaluate $\tan 75°$ using either the sum or difference formula for tangent.

Monitoring Progress and Modeling with Mathematics

In Exercises 3–10, find the exact value of the expression. *(See Example 1.)*

3. $\tan(-15°)$
4. $\tan 195°$
5. $\sin \frac{23\pi}{12}$
6. $\sin(-165°)$
7. $\cos 105°$
8. $\cos \frac{11\pi}{12}$
9. $\tan \frac{17\pi}{12}$
10. $\sin\left(-\frac{7\pi}{12}\right)$

In Exercises 11–16, evaluate the expression given that $\cos a = \frac{4}{5}$ with $0 < a < \frac{\pi}{2}$ and $\sin b = -\frac{15}{17}$ with $\frac{3\pi}{2} < b < 2\pi$. *(See Example 2.)*

11. $\sin(a + b)$
12. $\sin(a - b)$
13. $\cos(a - b)$
14. $\cos(a + b)$
15. $\tan(a + b)$
16. $\tan(a - b)$

In Exercises 17–22, simplify the expression. *(See Example 3.)*

17. $\tan(x + \pi)$
18. $\cos\left(x - \frac{\pi}{2}\right)$
19. $\cos(x + 2\pi)$
20. $\tan(x - 2\pi)$
21. $\sin\left(x - \frac{3\pi}{2}\right)$
22. $\tan\left(x + \frac{\pi}{2}\right)$

ERROR ANALYSIS In Exercises 23 and 24, describe and correct the error in simplifying the expression.

23. ✗

$$\tan\left(x + \frac{\pi}{4}\right) = \frac{\tan x + \tan\frac{\pi}{4}}{1 + \tan x \tan\frac{\pi}{4}} = \frac{\tan x + 1}{1 + \tan x} = 1$$

24. ✗

$$\sin\left(x - \frac{\pi}{4}\right) = \sin\frac{\pi}{4}\cos x - \cos\frac{\pi}{4}\sin x = \frac{\sqrt{2}}{2}\cos x - \frac{\sqrt{2}}{2}\sin x = \frac{\sqrt{2}}{2}(\cos x - \sin x)$$

25. What are the solutions of the equation $2 \sin x - 1 = 0$ for $0 \le x < 2\pi$?

    Ⓐ $\frac{\pi}{3}$ Ⓑ $\frac{\pi}{6}$ Ⓒ $\frac{2\pi}{3}$ Ⓓ $\frac{5\pi}{6}$

26. What are the solutions of the equation $\tan x + 1 = 0$ for $0 \le x < 2\pi$?

    Ⓐ $\frac{\pi}{4}$ Ⓑ $\frac{3\pi}{4}$ Ⓒ $\frac{5\pi}{4}$ Ⓓ $\frac{7\pi}{4}$

In Exercises 27–32, solve the equation for $0 \le x < 2\pi$. *(See Example 4.)*

27. $\sin\left(x + \frac{\pi}{2}\right) = \frac{1}{2}$
28. $\tan\left(x - \frac{\pi}{4}\right) = 0$
29. $\cos\left(x + \frac{\pi}{6}\right) - \cos\left(x - \frac{\pi}{6}\right) = 1$
30. $\sin\left(x + \frac{\pi}{4}\right) + \sin\left(x - \frac{\pi}{4}\right) = 0$
31. $\tan(x + \pi) - \tan(\pi - x) = 0$
32. $\sin(x + \pi) + \cos(x + \pi) = 0$

33. **USING EQUATIONS** Derive the cofunction identity $\sin\left(\frac{\pi}{2} - \theta\right) = \cos\theta$ using the difference formula for sine.

34. MAKING AN ARGUMENT Your friend claims it is possible to use the difference formula for tangent to derive the cofunction identity $\tan\left(\frac{\pi}{2} - \theta\right) = \cot \theta$. Is your friend correct? Explain your reasoning.

35. MODELING WITH MATHEMATICS A photographer is at a height h taking aerial photographs with a 35-millimeter camera. The ratio of the image length WQ to the length NA of the actual object is given by the formula

$$\frac{WQ}{NA} = \frac{35 \tan(\theta - t) + 35 \tan t}{h \tan \theta}$$

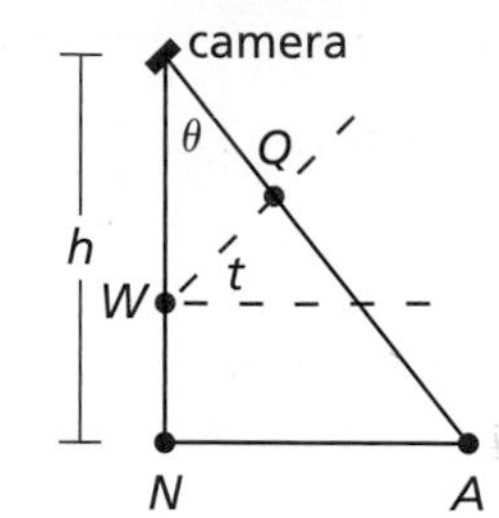

where θ is the angle between the vertical line perpendicular to the ground and the line from the camera to point A and t is the tilt angle of the film. When $t = 45°$, show that the formula can be rewritten as $\frac{WQ}{NA} = \frac{70}{h(1 + \tan \theta)}$. *(See Example 5.)*

36. MODELING WITH MATHEMATICS When a wave travels through a taut string, the displacement y of each point on the string depends on the time t and the point's position x. The equation of a *standing wave* can be obtained by adding the displacements of two waves traveling in opposite directions. Suppose a standing wave can be modeled by the formula

$$y = A \cos\left(\frac{2\pi t}{3} - \frac{2\pi x}{5}\right) + A \cos\left(\frac{2\pi t}{3} + \frac{2\pi x}{5}\right).$$

When $t = 1$, show that the formula can be rewritten as $y = -A \cos \frac{2\pi x}{5}$.

37. MODELING WITH MATHEMATICS The busy signal on a touch-tone phone is a combination of two tones with frequencies of 480 hertz and 620 hertz. The individual tones can be modeled by the equations:

480 hertz: $y_1 = \cos 960\pi t$

620 hertz: $y_2 = \cos 1240\pi t$

The sound of the busy signal can be modeled by $y_1 + y_2$. Show that $y_1 + y_2 = 2 \cos 1100\pi t \cos 140\pi t$.

38. HOW DO YOU SEE IT? Explain how to use the figure to solve the equation $\sin\left(x + \frac{\pi}{4}\right) - \sin\left(\frac{\pi}{4} - x\right) = 0$ for $0 \le x < 2\pi$.

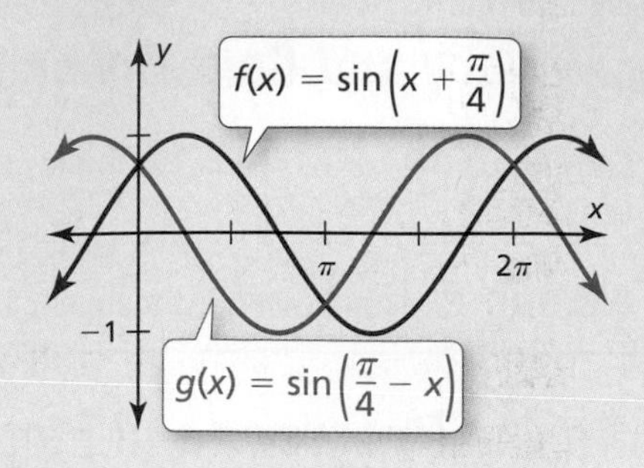

39. MATHEMATICAL CONNECTIONS The figure shows the acute angle of intersection, $\theta_2 - \theta_1$, of two lines with slopes m_1 and m_2.

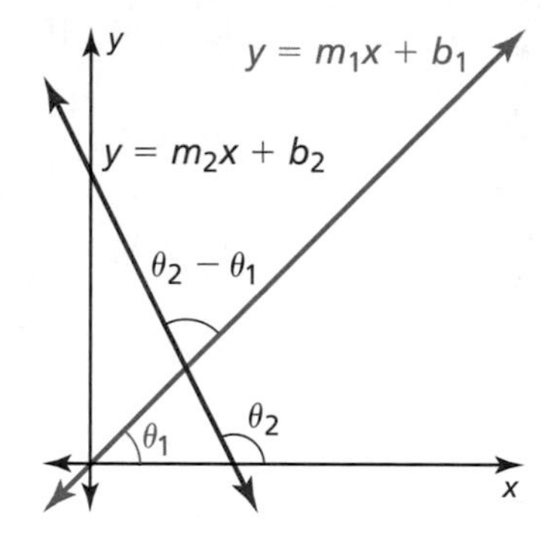

a. Use the difference formula for tangent to write an equation for $\tan(\theta_2 - \theta_1)$ in terms of m_1 and m_2.

b. Use the equation from part (a) to find the acute angle of intersection of the lines $y = x - 1$ and $y = \left(\frac{1}{\sqrt{3} - 2}\right)x + \frac{4 - \sqrt{3}}{2 - \sqrt{3}}$.

40. THOUGHT PROVOKING Rewrite each function. Justify your answers.

a. Write $\sin 3x$ as a function of $\sin x$.

b. Write $\cos 3x$ as a function of $\cos x$.

c. Write $\tan 3x$ as a function of $\tan x$.

Maintaining Mathematical Proficiency

Reviewing what you learned in previous grades and lessons

Solve the equation. Check your solution(s). *(Section 7.5)*

41. $1 - \frac{9}{x - 2} = -\frac{7}{2}$

42. $\frac{12}{x} + \frac{3}{4} = \frac{8}{x}$

43. $\frac{2x - 3}{x + 1} = \frac{10}{x^2 - 1} + 5$

9.5–9.8 What Did You Learn?

Core Vocabulary

frequency, *p. 506*
sinusoid, *p. 507*
trigonometric identity, *p. 514*

Core Concepts

Section 9.5

Characteristics of $y = \tan x$ and $y = \cot x$, *p. 498*
Period and Vertical Asymptotes of $y = a \tan bx$ and $y = a \cot bx$, *p. 499*
Characteristics of $y = \sec x$ and $y = \csc x$, *p. 500*

Section 9.6

Frequency, *p. 506*
Writing Trigonometric Functions, *p. 507*
Using Technology to Find Trigonometric Models, *p. 509*

Section 9.7

Fundamental Trigonometric Identities, *p. 514*

Section 9.8

Sum and Difference Formulas, *p. 520*
Trigonometric Equations and Real-Life Formulas, *p. 522*

Mathematical Practices

1. Explain why the relationship between θ and d makes sense in the context of the situation in Exercise 43 on page 503.
2. How can you use definitions to relate the slope of a line with the tangent of an angle in Exercise 39 on page 524?

Performance Task

Lightening the Load

You need to move a heavy table across the room. What is the easiest way to move it? Should you push it? Should you tie a rope around one leg of the table and pull it? How can trigonometry help you make the right decision?

To explore the answers to these questions and more, go to ***BigIdeasMath.com***.

9 Chapter Review

Dynamic Solutions available at *BigIdeasMath.com*

9.1 Right Triangle Trigonometry *(pp. 461–468)*

Evaluate the six trigonometric functions of the angle θ.

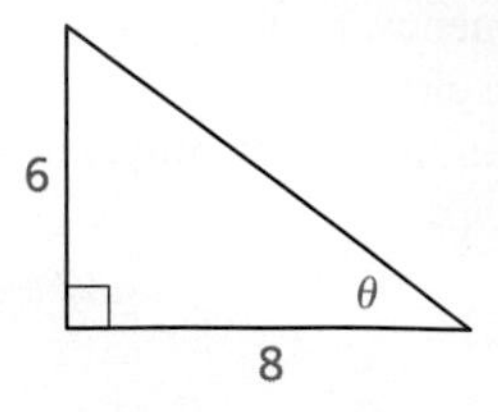

From the Pythagorean Theorem, the length of the hypotenuse is

$$\text{hyp.} = \sqrt{6^2 + 8^2} = \sqrt{100} = 10.$$

Using adj. = 8, opp. = 6, and hyp. = 10, the values of the six trigonometric functions of θ are:

$\sin\theta = \dfrac{\text{opp.}}{\text{hyp.}} = \dfrac{6}{10} = \dfrac{3}{5}$ $\quad\cos\theta = \dfrac{\text{adj.}}{\text{hyp.}} = \dfrac{8}{10} = \dfrac{4}{5}$ $\quad\tan\theta = \dfrac{\text{opp.}}{\text{adj.}} = \dfrac{6}{8} = \dfrac{3}{4}$

$\csc\theta = \dfrac{\text{hyp.}}{\text{opp.}} = \dfrac{10}{6} = \dfrac{5}{3}$ $\quad\sec\theta = \dfrac{\text{hyp.}}{\text{adj.}} = \dfrac{10}{8} = \dfrac{5}{4}$ $\quad\cot\theta = \dfrac{\text{adj.}}{\text{opp.}} = \dfrac{8}{6} = \dfrac{4}{3}$

1. In a right triangle, θ is an acute angle and $\cos\theta = \frac{6}{11}$. Evaluate the other five trigonometric functions of θ.

2. The shadow of a tree measures 25 feet from its base. The angle of elevation to the Sun is 31°. How tall is the tree?

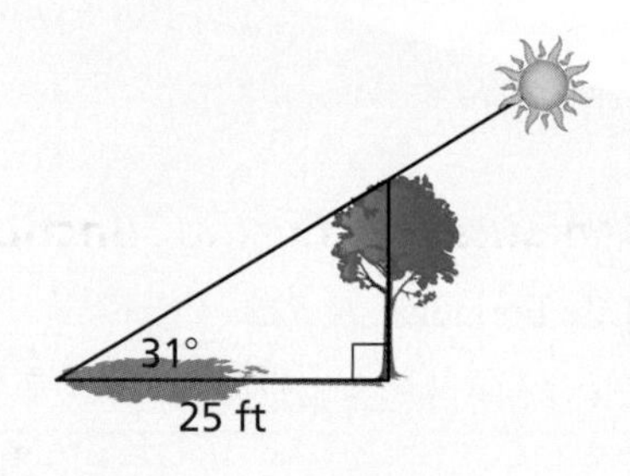

9.2 Angles and Radian Measure *(pp. 469–476)*

Convert the degree measure to radians or the radian measure to degrees.

a. 110°

$$110° = 110 \text{ degrees}\left(\frac{\pi \text{ radians}}{180 \text{ degrees}}\right) = \frac{11\pi}{18}$$

b. $\dfrac{7\pi}{12}$

$$\frac{7\pi}{12} = \frac{7\pi}{12} \text{ radians}\left(\frac{180°}{\pi \text{ radians}}\right) = 105°$$

3. Find one positive angle and one negative angle that are coterminal with 382°.

Convert the degree measure to radians or the radian measure to degrees.

4. 30° **5.** 225° **6.** $\dfrac{3\pi}{4}$ **7.** $\dfrac{5\pi}{3}$

8. A sprinkler system on a farm rotates 140° and sprays water up to 35 meters. Draw a diagram that shows the region that can be irrigated with the sprinkler. Then find the area of the region.

9.3 Trigonometric Functions of Any Angle (pp. 477–484)

Evaluate csc 210°.

The reference angle is $\theta' = 210° - 180° = 30°$. The cosecant function is negative in Quadrant III, so $\csc 210° = -\csc 30° = -2$.

Evaluate the six trigonometric functions of θ.

9.

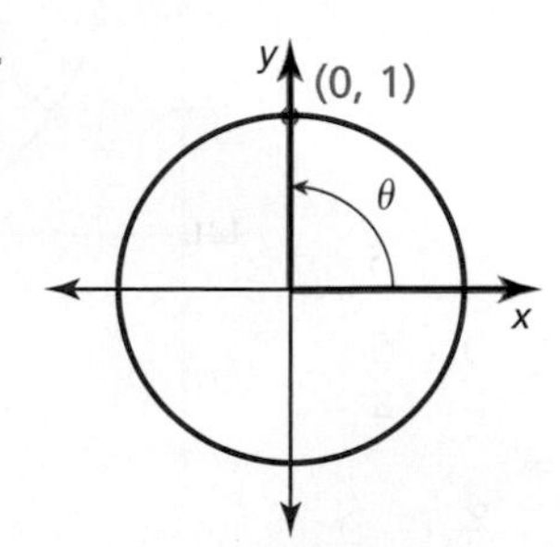

10.

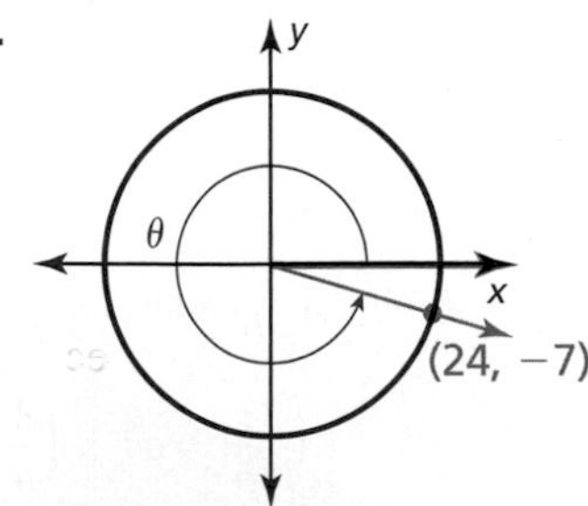

11.

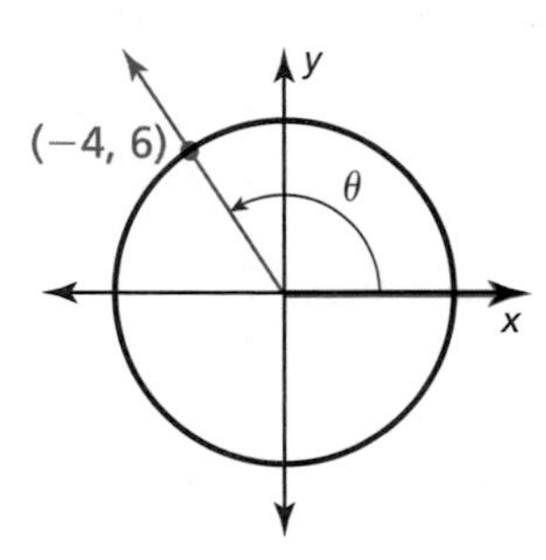

Evaluate the function without using a calculator.

12. $\tan 330°$ **13.** $\sec(-405°)$ **14.** $\sin \frac{13\pi}{6}$ **15.** $\sec \frac{11\pi}{3}$

9.4 Graphing Sine and Cosine Functions (pp. 485–494)

Identify the amplitude and period of $g(x) = \frac{1}{2} \sin 2x$. Then graph the function and describe the graph of g as a transformation of the graph of $f(x) = \sin x$.

The function is of the form $g(x) = a \sin bx$, where $a = \frac{1}{2}$ and $b = 2$. So, the amplitude is $a = \frac{1}{2}$ and the period is $\frac{2\pi}{b} = \frac{2\pi}{2} = \pi$.

Intercepts: $(0, 0)$; $\left(\frac{1}{2} \bullet \pi, 0\right) = \left(\frac{\pi}{2}, 0\right)$; $(\pi, 0)$

Maximum: $\left(\frac{1}{4} \bullet \pi, \frac{1}{2}\right) = \left(\frac{\pi}{4}, \frac{1}{2}\right)$

Minimum: $\left(\frac{3}{4} \bullet \pi, -\frac{1}{2}\right) = \left(\frac{3\pi}{4}, -\frac{1}{2}\right)$

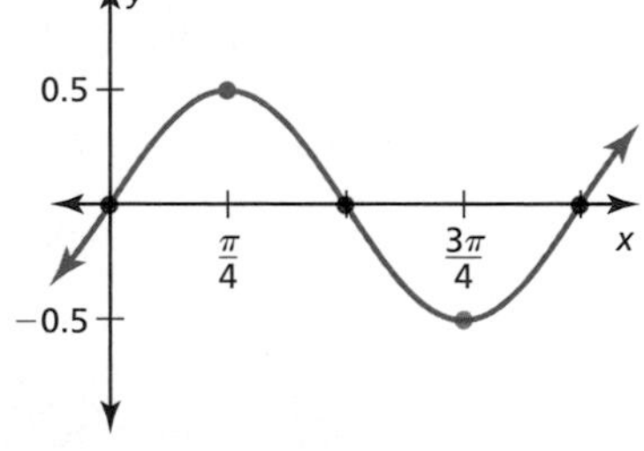

▶ The graph of g is a vertical shrink by a factor of $\frac{1}{2}$ and a horizontal shrink by a factor of $\frac{1}{2}$ of the graph of f.

Identify the amplitude and period of the function. Then graph the function and describe the graph of g as a transformation of the graph of the parent function.

16. $g(x) = 8 \cos x$ **17.** $g(x) = 6 \sin \pi x$ **18.** $g(x) = \frac{1}{4} \cos 4x$

Graph the function.

19. $g(x) = \cos(x + \pi) + 2$ **20.** $g(x) = -\sin x - 4$ **21.** $g(x) = 2 \sin\left(x + \frac{\pi}{2}\right)$

9.5 Graphing Other Trigonometric Functions (pp. 497–504)

a. Graph one period of $g(x) = 7 \cot \pi x$. Describe the graph of g as a transformation of the graph of $f(x) = \cot x$.

The function is of the form $g(x) = a \cot bx$, where $a = 7$ and $b = \pi$. So, the period is $\frac{\pi}{|b|} = \frac{\pi}{\pi} = 1$.

Intercepts: $\left(\frac{\pi}{2b}, 0\right) = \left(\frac{\pi}{2\pi}, 0\right) = \left(\frac{1}{2}, 0\right)$

Asymptotes: $x = 0$; $x = \frac{\pi}{|b|} = \frac{\pi}{\pi}$, or $x = 1$

Halfway points: $\left(\frac{\pi}{4b}, a\right) = \left(\frac{\pi}{4\pi}, 7\right) = \left(\frac{1}{4}, 7\right)$;

$\left(\frac{3\pi}{4b}, -a\right) = \left(\frac{3\pi}{4\pi}, -7\right) = \left(\frac{3}{4}, -7\right)$

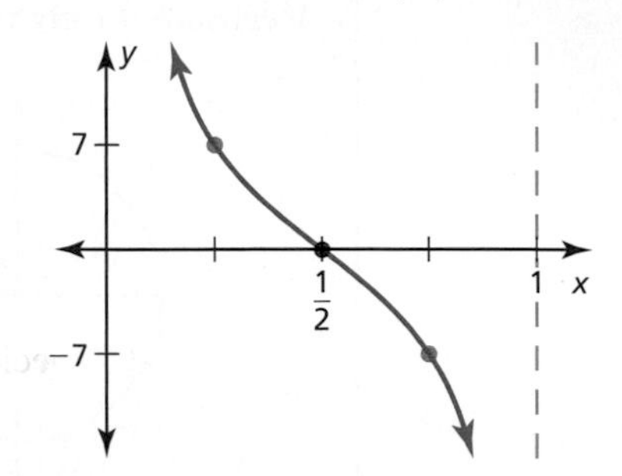

▶ The graph of g is a vertical stretch by a factor of 7 and a horizontal shrink by a factor of $\frac{1}{\pi}$ of the graph of f.

b. Graph one period of $g(x) = 9 \sec x$. Describe the graph of g as a transformation of the graph of $f(x) = \sec x$.

Step 1 Graph the function $y = 9 \cos x$.

The period is $\frac{2\pi}{1} = 2\pi$.

Step 2 Graph asymptotes of g. Because the asymptotes of g occur when $9 \cos x = 0$, graph $x = -\frac{\pi}{2}$, $x = \frac{\pi}{2}$, and $x = \frac{3\pi}{2}$.

Step 3 Plot the points on g, such as $(0, 9)$ and $(\pi, -9)$. Then use the asymptotes to sketch the curve.

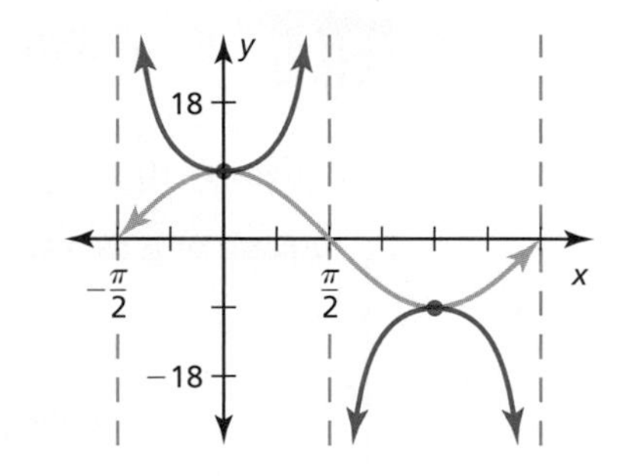

▶ The graph of g is a vertical stretch by a factor of 9 of the graph of f.

Graph one period of the function. Describe the graph of g as a transformation of the graph of its parent function.

22. $g(x) = \tan \frac{1}{2}x$

23. $g(x) = 2 \cot x$

24. $g(x) = 4 \tan 3\pi x$

Graph the function.

25. $g(x) = 5 \csc x$

26. $g(x) = \sec \frac{1}{2}x$

27. $g(x) = 5 \sec \pi x$

28. $g(x) = \frac{1}{2} \csc \frac{\pi}{4}x$

9.6 Modeling with Trigonometric Functions (pp. 505–512)

Write a function for the sinusoid shown.

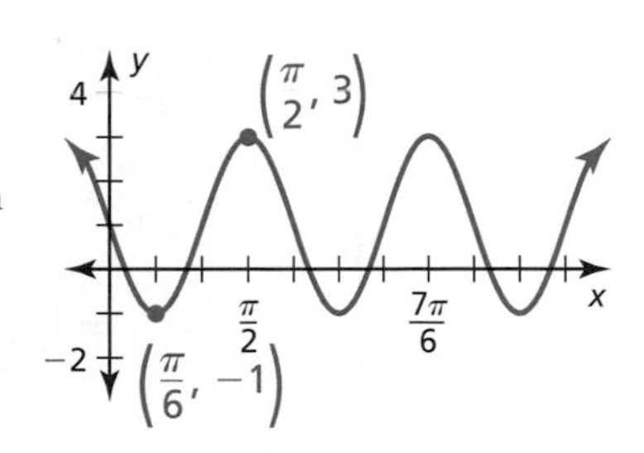

Step 1 Find the maximum and minimum values. From the graph, the maximum value is 3 and the minimum value is −1.

Step 2 Identify the vertical shift, k. The value of k is the mean of the maximum and minimum values.

$$k = \frac{(\text{maximum value}) + (\text{minimum value})}{2} = \frac{3 + (-1)}{2} = \frac{2}{2} = 1$$

Step 3 Decide whether the graph should be modeled by a sine or cosine function. Because the graph crosses the midline $y = 1$ on the y-axis and then decreases to its minimum value, the graph is a sine curve with a reflection in the x-axis and no horizontal shift. So, $h = 0$.

Step 4 Find the amplitude and period.

The period is $\frac{2\pi}{3} = \frac{2\pi}{b}$. So, $b = 3$.

The amplitude is

$$|a| = \frac{(\text{maximum value}) - (\text{minimum value})}{2} = \frac{3 - (-1)}{2} = \frac{4}{2} = 2.$$

Because the graph is a reflection in the x-axis, $a < 0$. So, $a = -2$.

▶ The function is $y = -2 \sin 3x + 1$.

Write a function for the sinusoid.

29.

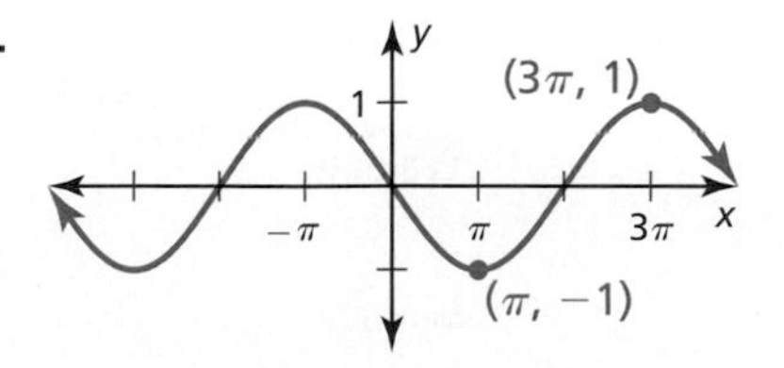

30.

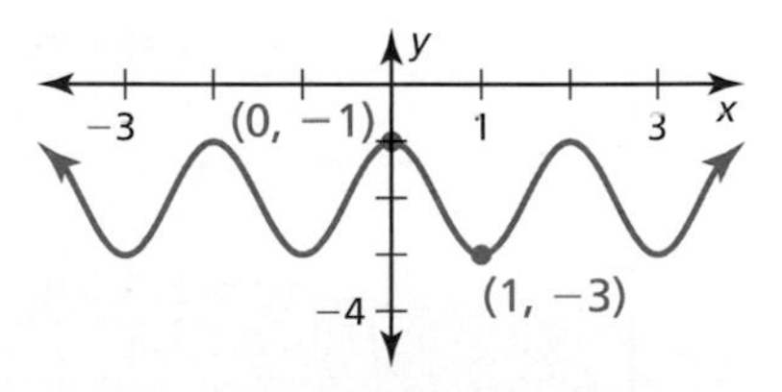

31. You put a reflector on a spoke of your bicycle wheel. The highest point of the reflector is 25 inches above the ground, and the lowest point is 2 inches. The reflector makes 1 revolution per second. Write a model for the height h (in inches) of a reflector as a function of time t (in seconds) given that the reflector is at its lowest point when $t = 0$.

32. The table shows the monthly precipitation P (in inches) for Bismarck, North Dakota, where $t = 1$ represents January. Write a model that gives P as a function of t and interpret the period of its graph.

t	1	2	3	4	5	6	7	8	9	10	11	12
P	0.5	0.5	0.9	1.5	2.2	2.6	2.6	2.2	1.6	1.3	0.7	0.4

9.7 Using Trigonometric Identities (pp. 513–518)

Verify the identity $\frac{\cot^2 \theta}{\csc \theta} = \csc \theta - \sin \theta$.

$\frac{\cot^2 \theta}{\csc \theta} = \frac{\csc^2 \theta - 1}{\csc \theta}$ — Pythagorean identity

$= \frac{\csc^2 \theta}{\csc \theta} - \frac{1}{\csc \theta}$ — Write as separate fractions.

$= \csc \theta - \frac{1}{\csc \theta}$ — Simplify.

$= \csc \theta - \sin \theta$ — Reciprocal identity

Simplify the expression.

33. $\cot^2 x - \cot^2 x \cos^2 x$

34. $\frac{(\sec x + 1)(\sec x - 1)}{\tan x}$

35. $\sin\left(\frac{\pi}{2} - x\right) \tan x$

Verify the identity.

36. $\frac{\cos x \sec x}{1 + \tan^2 x} = \cos^2 x$

37. $\tan\left(\frac{\pi}{2} - x\right) \cot x = \csc^2 x - 1$

9.8 Using Sum and Difference Formulas (pp. 519–524)

Find the exact value of sin 105°.

$\sin 105° = \sin(45° + 60°)$ — Substitute 45° + 60° for 105°.

$= \sin 45° \cos 60° + \cos 45° \sin 60°$ — Sum formula for sine

$= \frac{\sqrt{2}}{2} \cdot \frac{1}{2} + \frac{\sqrt{2}}{2} \cdot \frac{\sqrt{3}}{2}$ — Evaluate.

$= \frac{\sqrt{2} + \sqrt{6}}{4}$ — Simplify.

▶ The exact value of sin 105° is $\frac{\sqrt{2} + \sqrt{6}}{4}$.

Find the exact value of the expression.

38. $\sin 75°$

39. $\tan(-15°)$

40. $\cos \frac{\pi}{12}$

41. Find $\tan(a + b)$, given that $\tan a = \frac{1}{4}$ with $\pi < a < \frac{3\pi}{2}$ and $\tan b = \frac{3}{7}$ with $0 < b < \frac{\pi}{2}$.

Solve the equation for $0 \le x < 2\pi$.

42. $\cos\left(x + \frac{3\pi}{4}\right) + \cos\left(x - \frac{3\pi}{4}\right) = 1$

43. $\tan(x + \pi) + \cos\left(x + \frac{\pi}{2}\right) = 0$

9 Chapter Test

Verify the identity.

1. $\dfrac{\cos^2 x + \sin^2 x}{1 + \tan^2 x} = \cos^2 x$

2. $\dfrac{1 + \sin x}{\cos x} + \dfrac{\cos x}{1 + \sin x} = 2 \sec x$

3. $\cos\left(x + \dfrac{3\pi}{2}\right) = \sin x$

4. Evaluate $\sec(-300°)$ without using a calculator.

Write a function for the sinusoid.

5.

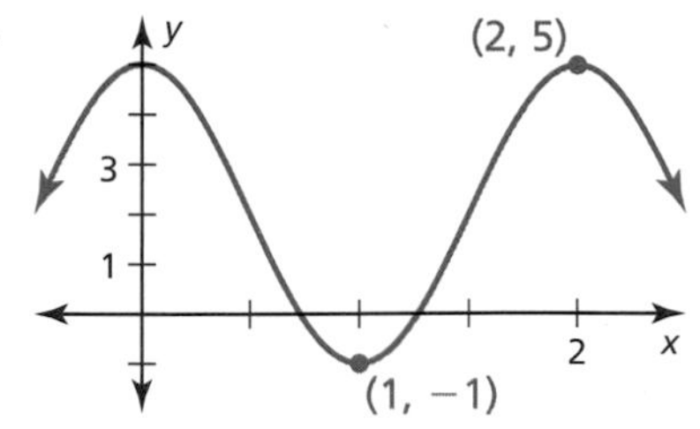

6.

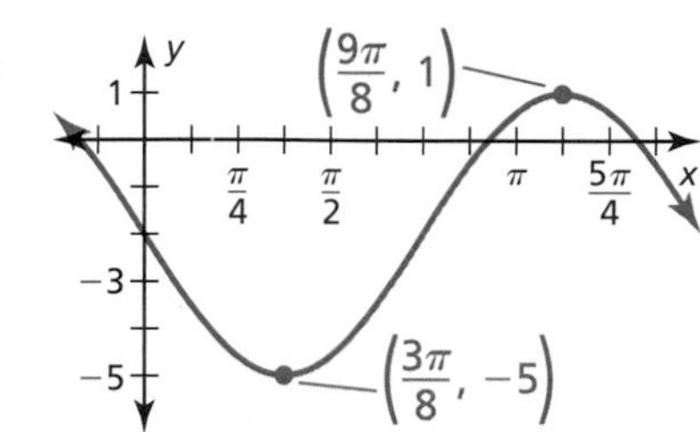

Graph the function. Then describe the graph of g as a transformation of the graph of its parent function.

7. $g(x) = -4 \tan 2x$

8. $g(x) = -2 \cos \frac{1}{3}x + 3$

9. $g(x) = 3 \csc \pi x$

Convert the degree measure to radians or the radian measure to degrees. Then find one positive angle and one negative angle that are coterminal with the given angle.

10. $-50°$

11. $\dfrac{4\pi}{5}$

12. $\dfrac{8\pi}{3}$

13. Find the arc length and area of a sector with radius $r = 13$ inches and central angle $\theta = 40°$.

Evaluate the six trigonometric functions of the angle θ.

14.

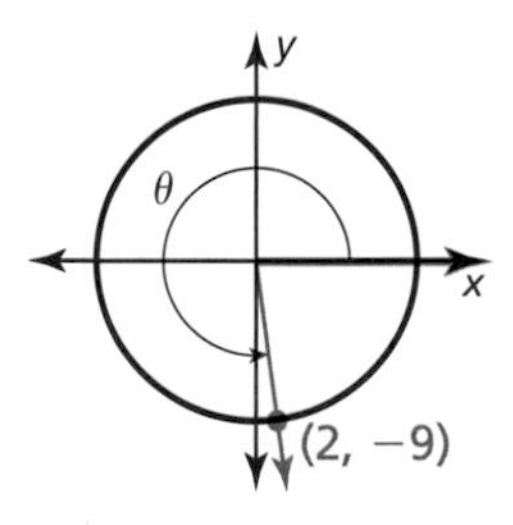

15.

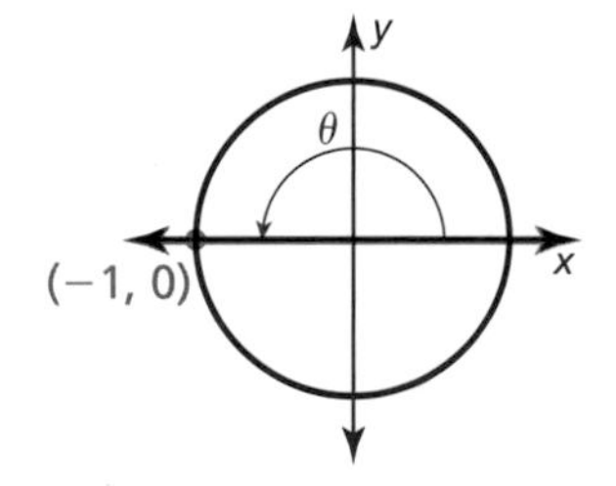

16. In which quadrant does the terminal side of θ lie when $\cos \theta < 0$ and $\tan \theta > 0$? Explain.

17. How tall is the building? Justify your answer.

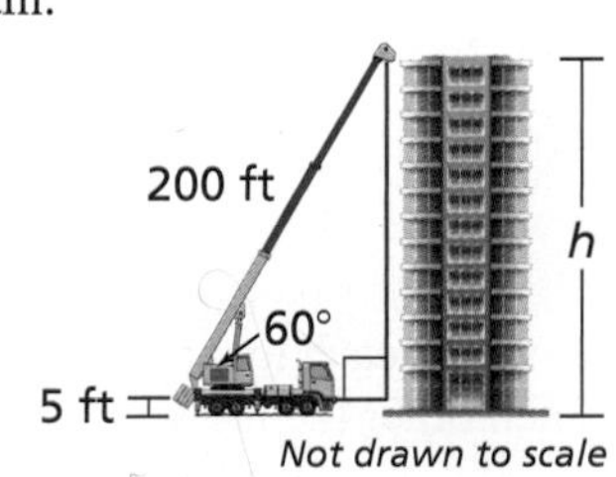

18. The table shows the average daily high temperatures T (in degrees Fahrenheit) in Baltimore, Maryland, where $m = 1$ represents January. Write a model that gives T as a function of m and interpret the period of its graph.

m	1	2	3	4	5	6	7	8	9	10	11	12
T	41	45	54	65	74	83	87	85	78	67	56	45

9 Cumulative Assessment

1. Which expressions are equivalent to 1?

$\tan x \sec x \cos x$	$\sin^2 x + \cos^2 x$	$\dfrac{\cos^2(-x) \tan^2 x}{\sin^2(-x)}$	$\cos\left(\dfrac{\pi}{2} - x\right) \csc x$

2. Which rational expression represents the ratio of the perimeter to the area of the playground shown in the diagram?

Ⓐ $\dfrac{9}{7x}$

Ⓑ $\dfrac{11}{14x}$

Ⓒ $\dfrac{1}{x}$

Ⓓ $\dfrac{1}{2x}$

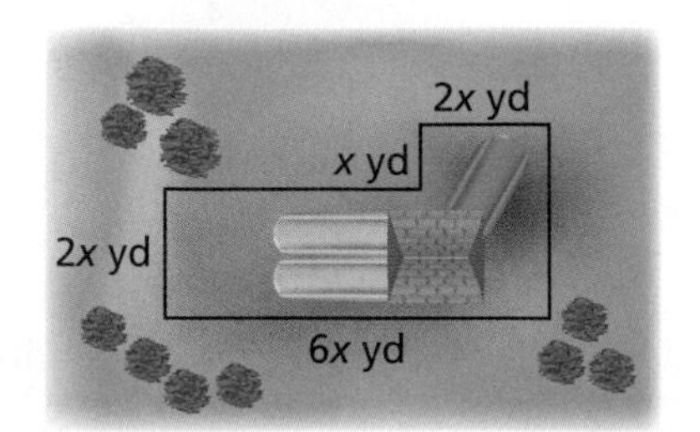

3. The chart shows the average monthly temperatures (in degrees Fahrenheit) and the gas usages (in cubic feet) of a household for 12 months.

a. Use a graphing calculator to find trigonometric models for the average temperature y_1 as a function of time and the gas usage y_2 (in thousands of cubic feet) as a function of time. Let $t = 1$ represent January.

b. Graph the two regression equations in the same coordinate plane on your graphing calculator. Describe the relationship between the graphs.

January	February	March	April
32°F	21°F	15°F	22°F
20,000 ft³	27,000 ft³	23,000 ft³	22,000 ft³
May	**June**	**July**	**August**
35°F	49°F	62°F	78°F
21,000 ft³	14,000 ft³	8,000 ft³	9,000 ft³
September	**October**	**November**	**December**
71°F	63°F	55°F	40°F
13,000 ft³	15,000 ft³	19,000 ft³	23,000 ft³

4. Evaluate each logarithm using $\log_2 5 \approx 2.322$ and $\log_2 3 \approx 1.585$, if necessary. Then order the logarithms by value from least to greatest.

a. $\log 1000$

b. $\log_2 15$

c. $\ln e$

d. $\log_2 9$

e. $\log_2 \frac{5}{3}$

f. $\log_2 1$

5. Which function is *not* represented by the graph?

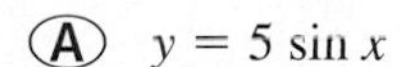

Ⓐ $y = 5 \sin x$

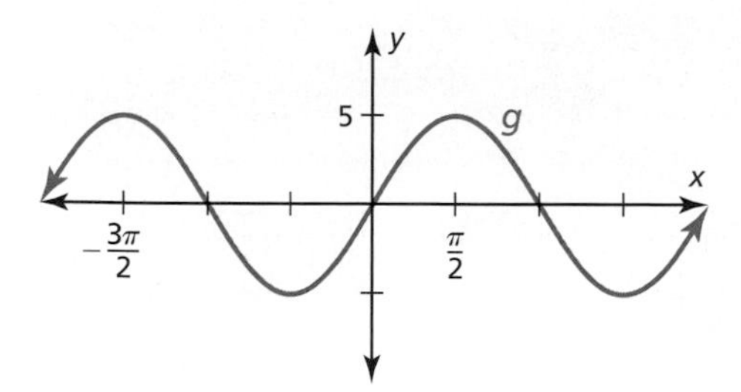

Ⓑ $y = 5 \cos\left(\frac{\pi}{2} - x\right)$

Ⓒ $y = 5 \cos\left(x + \frac{\pi}{2}\right)$

Ⓓ $y = -5 \sin(x + \pi)$

6. Complete each statement with < or > so that each statement is true.

a. θ ▢ 3 radians

b. $\tan \theta$ ▢ 0

c. θ' ▢ 45°

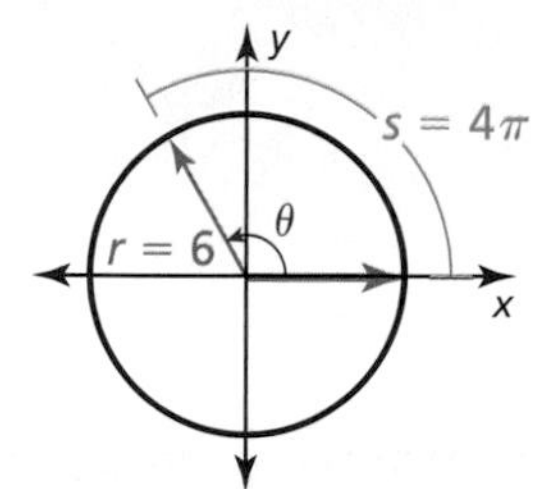

7. Use the Rational Root Theorem and the graph to find all the real zeros of the function $f(x) = 2x^3 - x^2 - 13x - 6$.

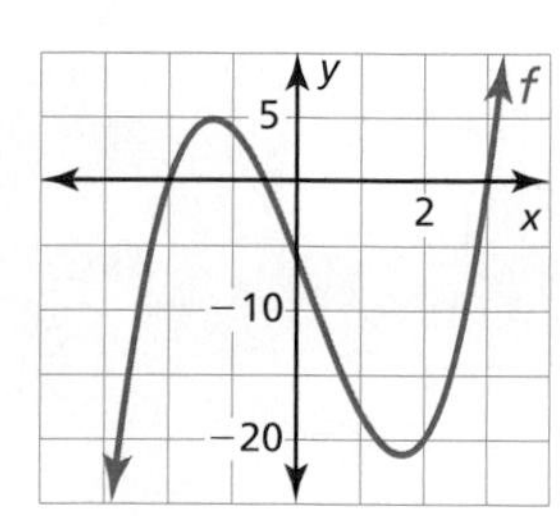

8. Your friend claims -210° is coterminal with the angle $\frac{5\pi}{6}$. Is your friend correct? Explain your reasoning.

9. Company A and Company B offer the same starting annual salary of $20,000. Company A gives a $1000 raise each year. Company B gives a 4% raise each year.

a. Write rules giving the salaries a_n and b_n for your nth year of employment at Company A and Company B, respectively. Tell whether the sequence represented by each rule is *arithmetic*, *geometric*, or *neither*.

b. Graph each sequence in the same coordinate plane.

c. Under what conditions would you choose to work for Company B?

d. After 20 years of employment, compare your total earnings.

10 Probability

Class Ring *(p. 583)*

Horse Racing *(p. 571)*

Tree Growth *(p. 568)*

Jogging *(p. 557)*

Coaching *(p. 552)*

Maintaining Mathematical Proficiency

Finding a Percent

Example 1 **What percent of 12 is 9?**

$$\frac{a}{w} = \frac{p}{100}$$ Write the percent proportion.

$$\frac{9}{12} = \frac{p}{100}$$ Substitute 9 for *a* and 12 for *w*.

$$100 \cdot \frac{9}{12} = 100 \cdot \frac{p}{100}$$ Multiplication Property of Equality.

$$75 = p$$ Simplify.

▶ So, 9 is 75% of 12.

Write and solve a proportion to answer the question.

1. What percent of 30 is 6? **2.** What number is 68% of 25? **3.** 34.4 is what percent of 86?

Making a Histogram

Example 2 **The frequency table shows the ages of people at a gym. Display the data in a histogram.**

Age	Frequency
10–19	7
20–29	12
30–39	6
40–49	4
50–59	0
60–69	3

Step 1 Draw and label the axes.

Step 2 Draw a bar to represent the frequency of each interval.

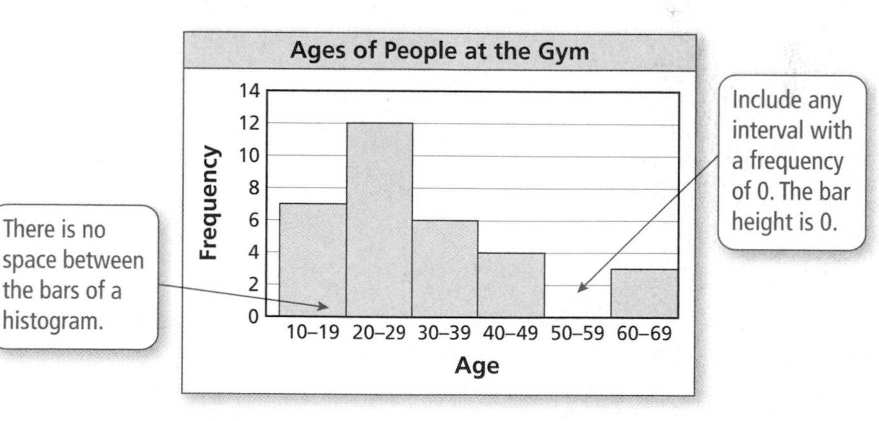

Display the data in a histogram.

4.

	Movies Watched per Week		
Movies	0–1	2–3	4–5
Frequency	35	11	6

5. ABSTRACT REASONING You want to purchase either a sofa or an arm chair at a furniture store. Each item has the same retail price. The sofa is 20% off. The arm chair is 10% off, and you have a coupon to get an additional 10% off the discounted price of the chair. Are the items equally priced after the discounts are applied? Explain.

Mathematical Practices

Mathematically proficient students apply the mathematics they know to solve real-life problems.

Modeling with Mathematics

Core Concept

Likelihoods and Probabilities

The **probability of an event** is a measure of the likelihood that the event will occur. Probability is a number from 0 to 1, including 0 and 1. The diagram relates *likelihoods* (described in words) and probabilities.

Words	Impossible	Unlikely	Equally likely to happen or not happen	Likely	Certain
Fraction	0	$\frac{1}{4}$	$\frac{1}{2}$	$\frac{3}{4}$	1
Decimal	0	0.25	0.5	0.75	1
Percent	0%	25%	50%	75%	100%

EXAMPLE 1 Describing Likelihoods

Describe the likelihood of each event.

Probability of an Asteroid or a Meteoroid Hitting Earth			
Name	Diameter	Probability of impact	Flyby date
a. Meteoroid	6 in.	0.75	Any day
b. Apophis	886 ft	0	2029
c. 2000 SG344	121 ft	$\frac{1}{435}$	2068–2110

SOLUTION

a. On any given day, it is *likely* that a meteoroid of this size will enter Earth's atmosphere. If you have ever seen a "shooting star," then you have seen a meteoroid.

b. A probability of 0 means this event is *impossible.*

c. With a probability of $\frac{1}{435} \approx 0.23\%$, this event is very *unlikely*. Of 435 identical asteroids, you would expect only one of them to hit Earth.

Monitoring Progress

In Exercises 1 and 2, describe the event as unlikely, equally likely to happen or not happen, or likely. Explain your reasoning.

1. The oldest child in a family is a girl.
2. The two oldest children in a family with three children are girls.
3. Give an example of an event that is certain to occur.

10.1 Sample Spaces and Probability

Essential Question How can you list the possible outcomes in the sample space of an experiment?

The **sample space** of an experiment is the set of all possible outcomes for that experiment.

EXPLORATION 1 Finding the Sample Space of an Experiment

Work with a partner. In an experiment, three coins are flipped. List the possible outcomes in the sample space of the experiment.

EXPLORATION 2 Finding the Sample Space of an Experiment

Work with a partner. List the possible outcomes in the sample space of the experiment.

a. One six-sided die is rolled.

b. Two six-sided dice are rolled.

EXPLORATION 3 Finding the Sample Space of an Experiment

Work with a partner. In an experiment, a spinner is spun.

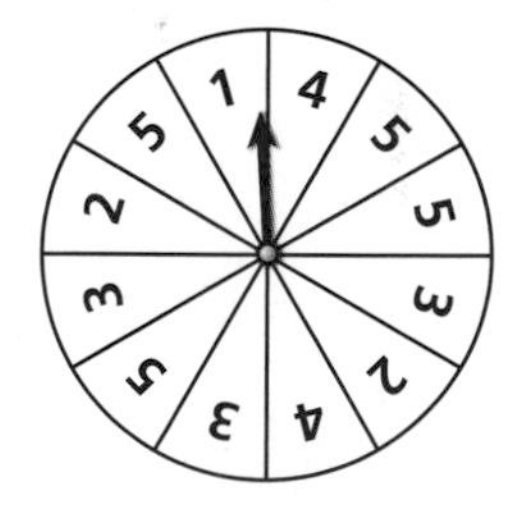

a. How many ways can you spin a 1? 2? 3? 4? 5?

b. List the sample space.

c. What is the total number of outcomes?

EXPLORATION 4 Finding the Sample Space of an Experiment

Work with a partner. In an experiment, a bag contains 2 blue marbles and 5 red marbles. Two marbles are drawn from the bag.

a. How many ways can you choose two blue? a red then blue? a blue then red? two red?

b. List the sample space.

c. What is the total number of outcomes?

LOOKING FOR A PATTERN

To be proficient in math, you need to look closely to discern a pattern or structure.

Communicate Your Answer

5. How can you list the possible outcomes in the sample space of an experiment?

6. For Exploration 3, find the ratio of the number of each possible outcome to the total number of outcomes. Then find the sum of these ratios. Repeat for Exploration 4. What do you observe?

10.1 Lesson

Core Vocabulary

probability experiment, *p. 538*
outcome, *p. 538*
event, *p. 538*
sample space, *p. 538*
probability of an event, *p. 538*
theoretical probability, *p. 539*
geometric probability, *p. 540*
experimental probability, *p. 541*

Previous
tree diagram

What You Will Learn

- Find sample spaces.
- Find theoretical probabilities.
- Find experimental probabilities.

Sample Spaces

A **probability experiment** is an action, or trial, that has varying results. The possible results of a probability experiment are **outcomes**. For instance, when you roll a six-sided die, there are 6 possible outcomes: 1, 2, 3, 4, 5, or 6. A collection of one or more outcomes is an **event**, such as rolling an odd number. The set of all possible outcomes is called a **sample space**.

EXAMPLE 1 Finding a Sample Space

You flip a coin and roll a six-sided die. How many possible outcomes are in the sample space? List the possible outcomes.

SOLUTION

Use a tree diagram to find the outcomes in the sample space.

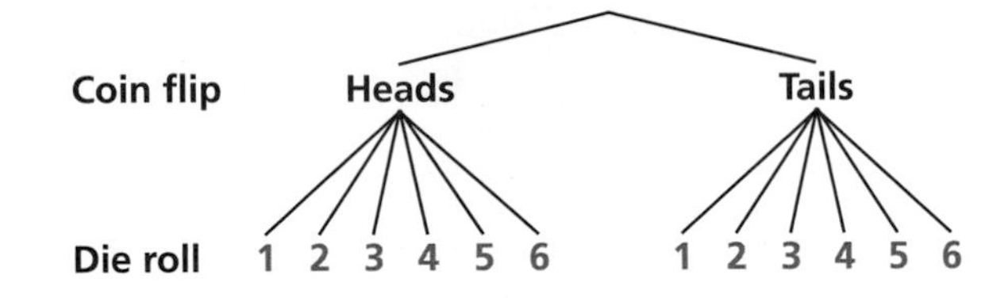

ANOTHER WAY

Using H for "heads" and T for "tails," you can list the outcomes as shown below.

H1	H2	H3	H4	H5	H6
T1	T2	T3	T4	T5	T6

▶ The sample space has 12 possible outcomes. They are listed below.

Heads, 1	Heads, 2	Heads, 3	Heads, 4	Heads, 5	Heads, 6
Tails, 1	Tails, 2	Tails, 3	Tails, 4	Tails, 5	Tails, 6

Monitoring Progress  Help in English and Spanish at *BigIdeasMath.com*

Find the number of possible outcomes in the sample space. Then list the possible outcomes.

1. You flip two coins.
2. You flip two coins and roll a six-sided die.

Theoretical Probabilities

The **probability of an event** is a measure of the likelihood, or chance, that the event will occur. Probability is a number from 0 to 1, including 0 and 1, and can be expressed as a decimal, fraction, or percent.

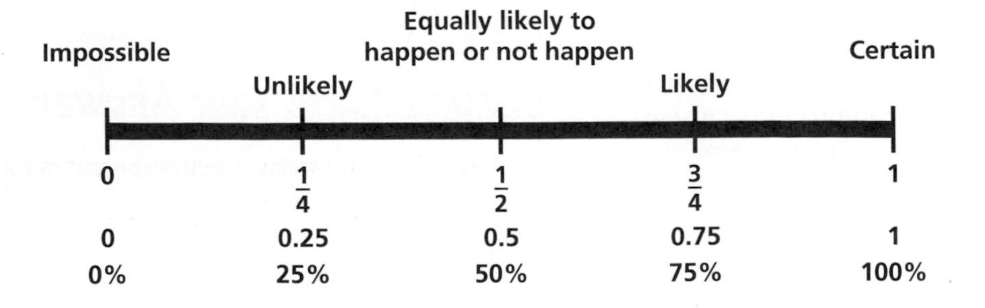

The outcomes for a specified event are called *favorable outcomes*. When all outcomes are equally likely, the **theoretical probability** of the event can be found using the following.

$$\text{Theoretical probability} = \frac{\text{Number of favorable outcomes}}{\text{Total number of outcomes}}$$

The probability of event A is written as $P(A)$.

ATTENDING TO PRECISION

Notice that the question uses the phrase "exactly two answers." This phrase is more precise than saying "two answers," which may be interpreted as "at least two" or as "exactly two."

EXAMPLE 2 Finding a Theoretical Probability

A student taking a quiz randomly guesses the answers to four true-false questions. What is the probability of the student guessing exactly two correct answers?

SOLUTION

Step 1 Find the outcomes in the sample space. Let C represent a correct answer and I represent an incorrect answer. The possible outcomes are:

Number correct	Outcome
0	IIII
1	CIII ICII IICI IIIC
2	IICC ICIC ICCI CIIC CICI CCII
3	ICCC CICC CCIC CCCI
4	CCCC

exactly two correct → 2

Step 2 Identify the number of favorable outcomes and the total number of outcomes. There are 6 favorable outcomes with exactly two correct answers and the total number of outcomes is 16.

Step 3 Find the probability of the student guessing exactly two correct answers. Because the student is randomly guessing, the outcomes should be equally likely. So, use the theoretical probability formula.

$$P(\text{exactly two correct answers}) = \frac{\text{Number of favorable outcomes}}{\text{Total number of outcomes}}$$

$$= \frac{6}{16}$$

$$= \frac{3}{8}$$

▶ The probability of the student guessing exactly two correct answers is $\frac{3}{8}$, or 37.5%.

The sum of the probabilities of all outcomes in a sample space is 1. So, when you know the probability of event A, you can find the probability of the *complement* of event A. The *complement* of event A consists of all outcomes that are not in A and is denoted by $\overline{A}$. The notation $\overline{A}$ is read as "A bar." You can use the following formula to find $P(\overline{A})$.

Core Concept

Probability of the Complement of an Event

The probability of the complement of event A is

$$P(\overline{A}) = 1 - P(A).$$

EXAMPLE 3 Finding Probabilities of Complements

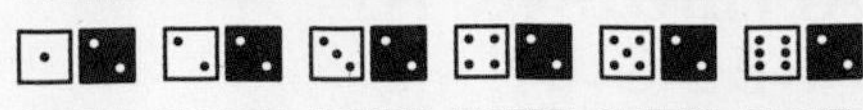

When two six-sided dice are rolled, there are 36 possible outcomes, as shown. Find the probability of each event.

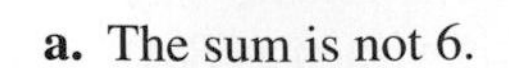

a. The sum is not 6.

b. The sum is less than or equal to 9.

SOLUTION

a. $P(\text{sum is not } 6) = 1 - P(\text{sum is } 6) = 1 - \frac{5}{36} = \frac{31}{36} \approx 0.861$

b. $P(\text{sum} \le 9) = 1 - P(\text{sum} > 9) = 1 - \frac{6}{36} = \frac{30}{36} = \frac{5}{6} \approx 0.833$

Some probabilities are found by calculating a ratio of two lengths, areas, or volumes. Such probabilities are called **geometric probabilities**.

EXAMPLE 4 Using Area to Find Probability

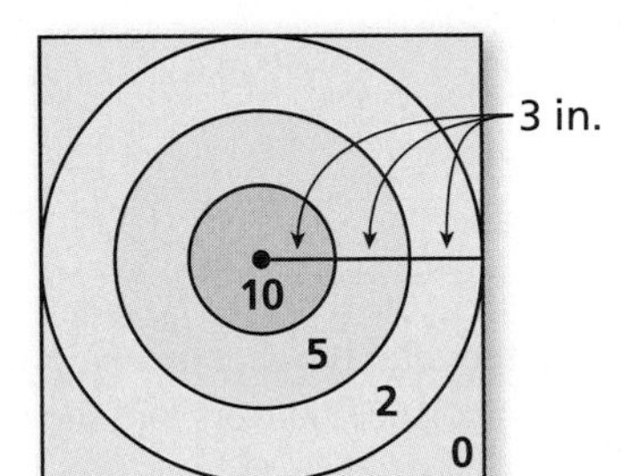

You throw a dart at the board shown. Your dart is equally likely to hit any point inside the square board. Are you more likely to get 10 points or 0 points?

SOLUTION

The probability of getting 10 points is

$$P(10 \text{ points}) = \frac{\text{Area of smallest circle}}{\text{Area of entire board}} = \frac{\pi \cdot 3^2}{18^2} = \frac{9\pi}{324} = \frac{\pi}{36} \approx 0.0873.$$

The probability of getting 0 points is

$$P(0 \text{ points}) = \frac{\text{Area outside largest circle}}{\text{Area of entire board}}$$
$$= \frac{18^2 - (\pi \cdot 9^2)}{18^2}$$
$$= \frac{324 - 81\pi}{324}$$
$$= \frac{4 - \pi}{4}$$
$$\approx 0.215.$$

▶ You are more likely to get 0 points.

Monitoring Progress

Help in English and Spanish at *BigIdeasMath.com*

3. You flip a coin and roll a six-sided die. What is the probability that the coin shows tails and the die shows 4?

Find $P(\overline{A})$.

4. $P(A) = 0.45$

5. $P(A) = \frac{1}{4}$

6. $P(A) = 1$

7. $P(A) = 0.03$

8. In Example 4, are you more likely to get 10 points or 5 points?

9. In Example 4, are you more likely to score points (10, 5, or 2) or get 0 points?

Experimental Probabilities

An **experimental probability** is based on repeated *trials* of a probability experiment. The number of trials is the number of times the probability experiment is performed. Each trial in which a favorable outcome occurs is called a *success*. The experimental probability can be found using the following.

$$\text{Experimental probability} = \frac{\text{Number of successes}}{\text{Number of trials}}$$

EXAMPLE 5 Finding an Experimental Probability

Spinner Results			
red	**green**	**blue**	**yellow**
5	9	3	3

Each section of the spinner shown has the same area. The spinner was spun 20 times. The table shows the results. For which color is the experimental probability of stopping on the color the same as the theoretical probability?

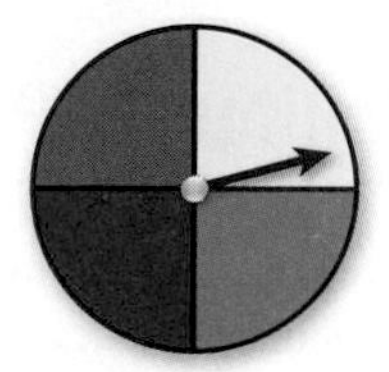

SOLUTION

The theoretical probability of stopping on each of the four colors is $\frac{1}{4}$. Use the outcomes in the table to find the experimental probabilities.

$P(\text{red}) = \frac{5}{20} = \frac{1}{4}$ $\qquad P(\text{green}) = \frac{9}{20}$

$P(\text{blue}) = \frac{3}{20}$ $\qquad P(\text{yellow}) = \frac{3}{20}$

▶ The experimental probability of stopping on red is the same as the theoretical probability.

EXAMPLE 6 Solving a Real-Life Problem

In the United States, a survey of 2184 adults ages 18 and over found that 1328 of them have at least one pet. The types of pets these adults have are shown in the figure. What is the probability that a pet-owning adult chosen at random has a dog?

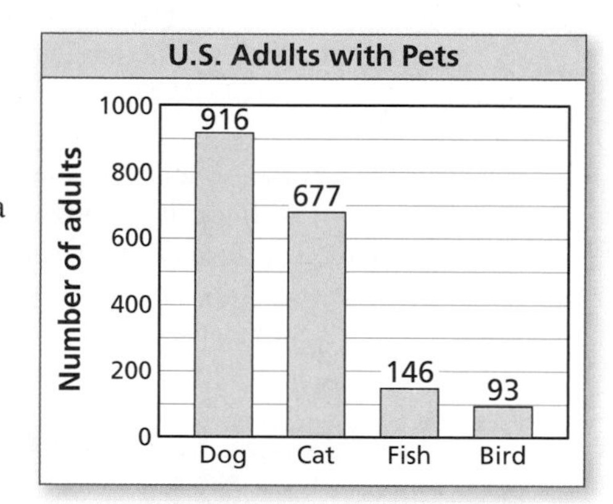

SOLUTION

The number of trials is the number of pet-owning adults, 1328. A success is a pet-owning adult who has a dog. From the graph, there are 916 adults who said that they have a dog.

$$P(\text{pet-owning adult has a dog}) = \frac{916}{1328} = \frac{229}{332} \approx 0.690$$

▶ The probability that a pet-owning adult chosen at random has a dog is about 69%.

Monitoring Progress  Help in English and Spanish at *BigIdeasMath.com*

10. In Example 5, for which color is the experimental probability of stopping on the color greater than the theoretical probability?

11. In Example 6, what is the probability that a pet-owning adult chosen at random owns a fish?

10.1 Exercises

Dynamic Solutions available at *BigIdeasMath.com*

Vocabulary and Core Concept Check

1. **COMPLETE THE SENTENCE** A number that describes the likelihood of an event is the __________ of the event.

2. **WRITING** Describe the difference between theoretical probability and experimental probability.

Monitoring Progress and Modeling with Mathematics

In Exercises 3–6, find the number of possible outcomes in the sample space. Then list the possible outcomes. *(See Example 1.)*

3. You roll a die and flip three coins.

4. You flip a coin and draw a marble at random from a bag containing two purple marbles and one white marble.

5. A bag contains four red cards numbered 1 through 4, four white cards numbered 1 through 4, and four black cards numbered 1 through 4. You choose a card at random.

6. You draw two marbles without replacement from a bag containing three green marbles and four black marbles.

7. **PROBLEM SOLVING** A game show airs on television five days per week. Each day, a prize is randomly placed behind one of two doors. The contestant wins the prize by selecting the correct door. What is the probability that exactly two of the five contestants win a prize during a week? *(See Example 2.)*

8. **PROBLEM SOLVING** Your friend has two standard decks of 52 playing cards and asks you to randomly draw one card from each deck. What is the probability that you will draw two spades?

9. **PROBLEM SOLVING** When two six-sided dice are rolled, there are 36 possible outcomes. Find the probability that (a) the sum is not 4 and (b) the sum is greater than 5. *(See Example 3.)*

10. **PROBLEM SOLVING** The age distribution of a population is shown. Find the probability of each event.

Age Distribution

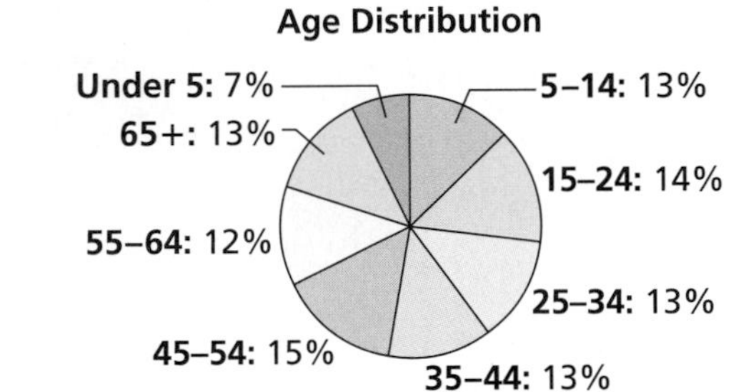

a. A person chosen at random is at least 15 years old.

b. A person chosen at random is from 25 to 44 years old.

11. **ERROR ANALYSIS** A student randomly guesses the answers to two true-false questions. Describe and correct the error in finding the probability of the student guessing both answers correctly.

The student can either guess two incorrect answers, two correct answers, or one of each. So the probability of guessing both answers correctly is $\frac{1}{3}$.

12. **ERROR ANALYSIS** A student randomly draws a number between 1 and 30. Describe and correct the error in finding the probability that the number drawn is greater than 4.

The probability that the number is less than 4 is $\frac{3}{30}$, or $\frac{1}{10}$. So, the probability that the number is greater than 4 is $1 - \frac{1}{10}$, or $\frac{9}{10}$.

13. **MATHEMATICAL CONNECTIONS** You throw a dart at the board shown. Your dart is equally likely to hit any point inside the square board. What is the probability your dart lands in the yellow region? *(See Example 4.)*

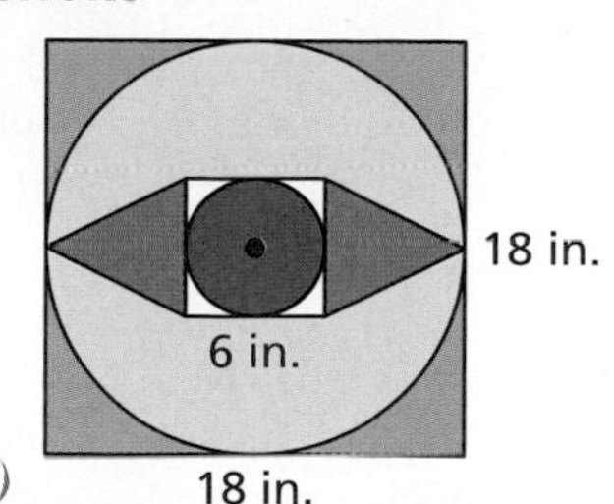

14. **MATHEMATICAL CONNECTIONS** The map shows the length (in miles) of shoreline along the Gulf of Mexico for each state that borders the body of water. What is the probability that a ship coming ashore at a random point in the Gulf of Mexico lands in the given state?

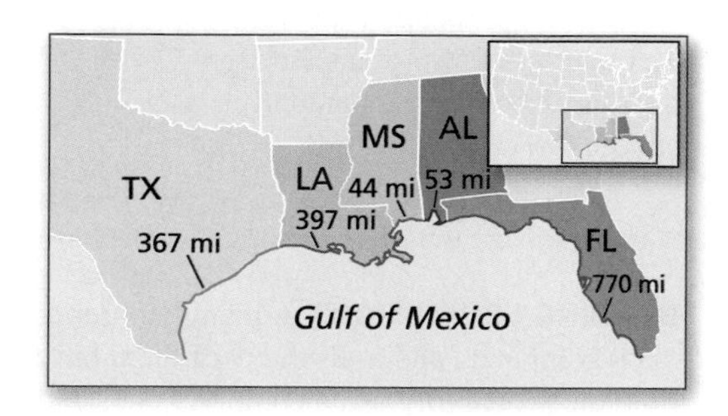

a. Texas
b. Alabama
c. Florida
d. Louisiana

15. **DRAWING CONCLUSIONS** You roll a six-sided die 60 times. The table shows the results. For which number is the experimental probability of rolling the number the same as the theoretical probability? *(See Example 5.)*

Six-sided Die Results					
⚀	⚁	⚂	⚃	⚄	⚅
11	14	7	10	6	12

16. **DRAWING CONCLUSIONS** A bag contains 5 marbles that are each a different color. A marble is drawn, its color is recorded, and then the marble is placed back in the bag. This process is repeated until 30 marbles have been drawn. The table shows the results. For which marble is the experimental probability of drawing the marble the same as the theoretical probability?

Drawing Results				
white	black	red	green	blue
5	6	8	2	9

17. **REASONING** Refer to the spinner shown. The spinner is divided into sections with the same area.

a. What is the theoretical probability that the spinner stops on a multiple of 3?

b. You spin the spinner 30 times. It stops on a multiple of 3 twenty times. What is the experimental probability of stopping on a multiple of 3?

c. Explain why the probability you found in part (b) is different than the probability you found in part (a).

18. **OPEN-ENDED** Describe a real-life event that has a probability of 0. Then describe a real-life event that has a probability of 1.

19. **DRAWING CONCLUSIONS** A survey of 2237 adults ages 18 and over asked which sport is their favorite. The results are shown in the figure. What is the probability that an adult chosen at random prefers auto racing? *(See Example 6.)*

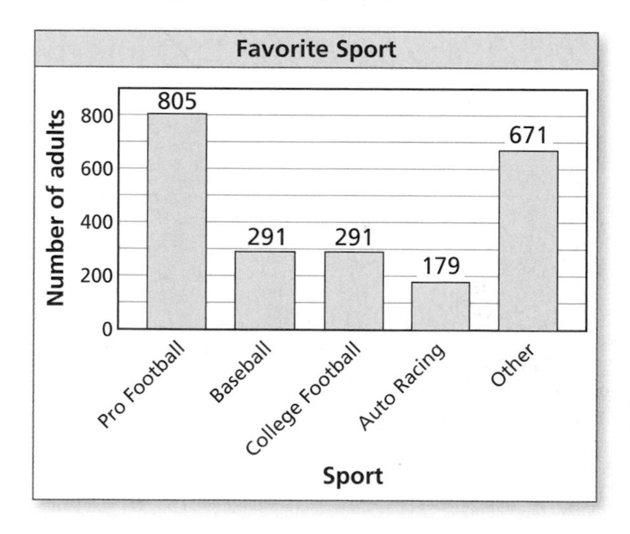

20. **DRAWING CONCLUSIONS** A survey of 2392 adults ages 18 and over asked what type of food they would be most likely to choose at a restaurant. The results are shown in the figure. What is the probability that an adult chosen at random prefers Italian food?

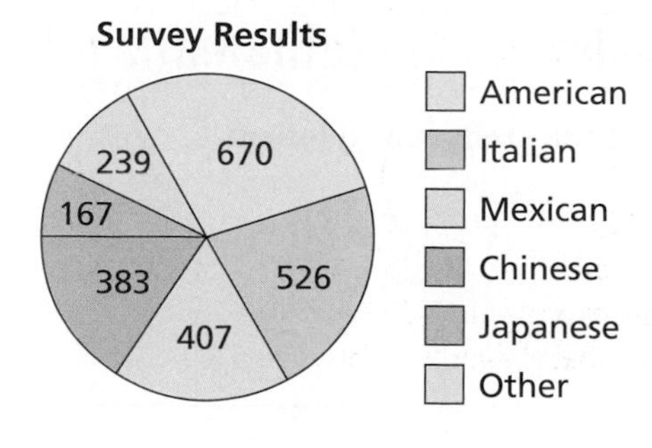

21. **ANALYZING RELATIONSHIPS** Refer to the board in Exercise 13. Order the likelihoods that the dart lands in the given region from least likely to most likely.

A. green **B.** not blue

C. red **D.** not yellow

22. **ANALYZING RELATIONSHIPS** Refer to the chart below. Order the following events from least likely to most likely.

Four-Day Forecast			
Friday	Saturday	Sunday	Monday
Chance of Rain **5%**	Chance of Rain **30%**	Chance of Rain **80%**	Chance of Rain **90%**

A. It rains on Sunday.

B. It does not rain on Saturday.

C. It rains on Monday.

D. It does not rain on Friday.

23. **USING TOOLS** Use the figure in Example 3 to answer each question.

a. List the possible sums that result from rolling two six-sided dice.

b. Find the theoretical probability of rolling each sum.

c. The table below shows a simulation of rolling two six-sided dice three times. Use a random number generator to simulate rolling two six-sided dice 50 times. Compare the experimental probabilities of rolling each sum with the theoretical probabilities.

	A	B	C
1	First Die	Second Die	Sum
2	4	6	10
3	3	5	8
4	1	6	7
5			

24. **MAKING AN ARGUMENT** You flip a coin three times. It lands on heads twice and on tails once. Your friend concludes that the theoretical probability of the coin landing heads up is $P(\text{heads up}) = \frac{2}{3}$. Is your friend correct? Explain your reasoning.

25. **MATHEMATICAL CONNECTIONS** A sphere fits inside a cube so that it touches each side, as shown. What is the probability a point chosen at random inside the cube is also inside the sphere?

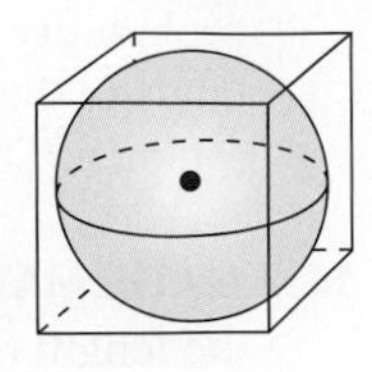

26. **HOW DO YOU SEE IT?** Consider the graph of f shown. What is the probability that the graph of $y = f(x) + c$ intersects the x-axis when c is a randomly chosen integer from 1 to 6? Explain.

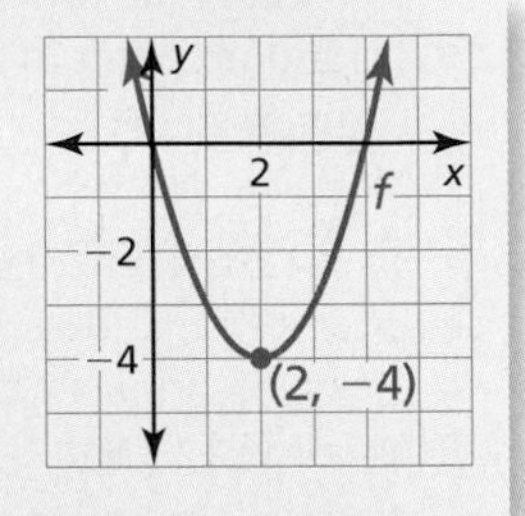

27. **DRAWING CONCLUSIONS** A manufacturer tests 1200 computers and finds that 9 of them have defects. Find the probability that a computer chosen at random has a defect. Predict the number of computers with defects in a shipment of 15,000 computers. Explain your reasoning.

28. **THOUGHT PROVOKING** The tree diagram shows a sample space. Write a probability problem that can be represented by the sample space. Then write the answer(s) to the problem.

Box A	Box B	Outcomes	Sum	Product
1	1	(1, 1)	2	1
	2	(1, 2)	3	2
2	1	(2, 1)	3	2
	2	(2, 2)	4	4
3	1	(3, 1)	4	3
	2	(3, 2)	5	6

Maintaining Mathematical Proficiency

Reviewing what you learned in previous grades and lessons

Find the product or quotient. *(Section 7.3)*

29. $\frac{3x}{y} \cdot \frac{2x^3}{y^2}$

30. $\frac{4x^9y}{3x^3} \cdot \frac{2xy}{8y^2}$

31. $\frac{x+3}{x^4-2} \cdot (x^2 - 7x + 6)$

32. $\frac{2y}{5x} \div \frac{y}{6x}$

33. $\frac{3x}{12x-11} \div \frac{x+1}{5x}$

34. $\frac{3x^2 + 2x - 13}{x^4} \div (x^2 + 9)$

10.2 Independent and Dependent Events

Essential Question

Essential Question How can you determine whether two events are independent or dependent?

Two events are **independent events** when the occurrence of one event does not affect the occurrence of the other event. Two events are **dependent events** when the occurrence of one event *does* affect the occurrence of the other event.

EXPLORATION 1 Identifying Independent and Dependent Events

Work with a partner. Determine whether the events are independent or dependent. Explain your reasoning.

REASONING ABSTRACTLY

To be proficient in math, you need to make sense of quantities and their relationships in problem situations.

a. Two six-sided dice are rolled.

b. Six pieces of paper, numbered 1 through 6, are in a bag. Two pieces of paper are selected one at a time without replacement.

EXPLORATION 2 Finding Experimental Probabilities

Work with a partner.

a. In Exploration 1(a), experimentally estimate the probability that the sum of the two numbers rolled is 7. Describe your experiment.

b. In Exploration 1(b), experimentally estimate the probability that the sum of the two numbers selected is 7. Describe your experiment.

EXPLORATION 3 Finding Theoretical Probabilities

Work with a partner.

a. In Exploration 1(a), find the theoretical probability that the sum of the two numbers rolled is 7. Then compare your answer with the experimental probability you found in Exploration 2(a).

b. In Exploration 1(b), find the theoretical probability that the sum of the two numbers selected is 7. Then compare your answer with the experimental probability you found in Exploration 2(b).

c. Compare the probabilities you obtained in parts (a) and (b).

Communicate Your Answer

4. How can you determine whether two events are independent or dependent?

5. Determine whether the events are independent or dependent. Explain your reasoning.

a. You roll a 4 on a six-sided die and spin red on a spinner.

b. Your teacher chooses a student to lead a group, chooses another student to lead a second group, and chooses a third student to lead a third group.

10.2 Lesson

Core Vocabulary

independent events, *p. 546*
dependent events, *p. 547*
conditional probability, *p. 547*

Previous
probability
sample space

What You Will Learn

- Determine whether events are independent events.
- Find probabilities of independent and dependent events.
- Find conditional probabilities.

Determining Whether Events Are Independent

Two events are **independent events** when the occurrence of one event does not affect the occurrence of the other event.

Core Concept

Probability of Independent Events

Words Two events A and B are independent events if and only if the probability that both events occur is the product of the probabilities of the events.

Symbols $P(A \text{ and } B) = P(A) \cdot P(B)$

EXAMPLE 1 Determining Whether Events Are Independent

A student taking a quiz randomly guesses the answers to four true-false questions. Use a sample space to determine whether guessing Question 1 correctly and guessing Question 2 correctly are independent events.

SOLUTION

Using the sample space in Example 2 on page 539:

$P(\text{correct on Question 1}) = \frac{8}{16} = \frac{1}{2}$ $\qquad$ $P(\text{correct on Question 2}) = \frac{8}{16} = \frac{1}{2}$

$P(\text{correct on Question 1 and correct on Question 2}) = \frac{4}{16} = \frac{1}{4}$

▶ Because $\frac{1}{2} \cdot \frac{1}{2} = \frac{1}{4}$, the events are independent.

EXAMPLE 2 Determining Whether Events Are Independent

A group of four students includes one boy and three girls. The teacher randomly selects one of the students to be the speaker and a different student to be the recorder. Use a sample space to determine whether randomly selecting a girl first and randomly selecting a girl second are independent events.

SOLUTION

Let B represent the boy. Let G_1, G_2, and G_3 represent the three girls. Use a table to list the outcomes in the sample space.

Number of girls	Outcome	
1	G_1B	BG_1
1	G_2B	BG_2
1	G_3B	BG_3
2	G_1G_2	G_2G_1
2	G_1G_3	G_3G_1
2	G_2G_3	G_3G_2

Using the sample space:

$P(\text{girl first}) = \frac{9}{12} = \frac{3}{4}$ $\qquad$ $P(\text{girl second}) = \frac{9}{12} = \frac{3}{4}$

$P(\text{girl first and girl second}) = \frac{6}{12} = \frac{1}{2}$

▶ Because $\frac{3}{4} \cdot \frac{3}{4} \neq \frac{1}{2}$, the events are not independent.

Monitoring Progress

Help in English and Spanish at *BigIdeasMath.com*

1. In Example 1, determine whether guessing Question 1 incorrectly and guessing Question 2 correctly are independent events.
2. In Example 2, determine whether randomly selecting a girl first and randomly selecting a boy second are independent events.

Finding Probabilities of Events

In Example 1, it makes sense that the events are independent because the second guess should not be affected by the first guess. In Example 2, however, the selection of the second person *depends* on the selection of the first person because the same person cannot be selected twice. These events are *dependent*. Two events are **dependent events** when the occurrence of one event *does* affect the occurrence of the other event.

The probability that event B occurs given that event A has occurred is called the **conditional probability** of B given A and is written as $P(B \mid A)$.

MAKING SENSE OF PROBLEMS

One way that you can find $P(\text{girl second} \mid \text{girl first})$ is to list the 9 outcomes in which a girl is chosen first and then find the fraction of these outcomes in which a girl is chosen second:

$G_1B \quad G_2B \quad G_3B$

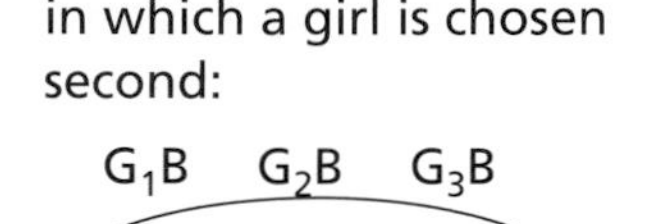

Core Concept

Probability of Dependent Events

Words If two events A and B are dependent events, then the probability that both events occur is the product of the probability of the first event and the conditional probability of the second event given the first event.

Symbols $P(A \text{ and } B) = P(A) \cdot P(B \mid A)$

Example Using the information in Example 2:

$$P(\text{girl first and girl second}) = P(\text{girl first}) \cdot P(\text{girl second} \mid \text{girl first})$$

$$= \frac{9}{12} \cdot \frac{6}{9} = \frac{1}{2}$$

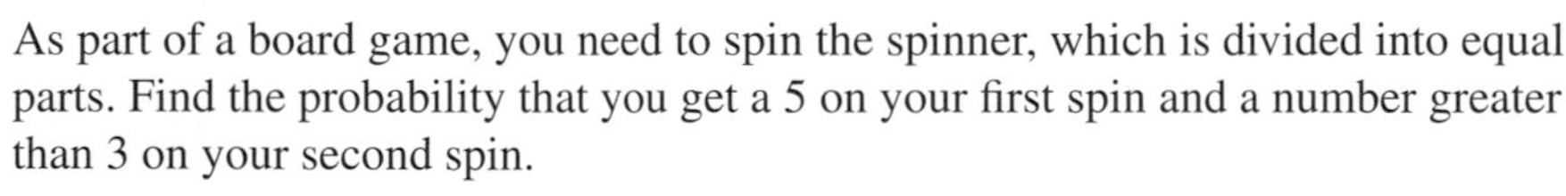

EXAMPLE 3 Finding the Probability of Independent Events

As part of a board game, you need to spin the spinner, which is divided into equal parts. Find the probability that you get a 5 on your first spin and a number greater than 3 on your second spin.

SOLUTION

Let event A be "5 on first spin" and let event B be "greater than 3 on second spin."

The events are independent because the outcome of your second spin is not affected by the outcome of your first spin. Find the probability of each event and then multiply the probabilities.

$P(A) = \frac{1}{8}$ 1 of the 8 sections is a "5."

$P(B) = \frac{5}{8}$ 5 of the 8 sections (4, 5, 6, 7, 8) are greater than 3.

$$P(A \text{ and } B) = P(A) \cdot P(B) = \frac{1}{8} \cdot \frac{5}{8} = \frac{5}{64} \approx 0.078$$

▶ So, the probability that you get a 5 on your first spin and a number greater than 3 on your second spin is about 7.8%.

EXAMPLE 4 Finding the Probability of Dependent Events

A bag contains twenty \$1 bills and five \$100 bills. You randomly draw a bill from the bag, set it aside, and then randomly draw another bill from the bag. Find the probability that both events A and B will occur.

Event A: The first bill is \$100. **Event B:** The second bill is \$100.

SOLUTION

The events are dependent because there is one less bill in the bag on your second draw than on your first draw. Find $P(A)$ and $P(B|A)$. Then multiply the probabilities.

$P(A) = \frac{5}{25}$ 5 of the 25 bills are \$100 bills.

$P(B|A) = \frac{4}{24}$ 4 of the remaining 24 bills are \$100 bills.

$$P(A \text{ and } B) = P(A) \cdot P(B|A) = \frac{5}{25} \cdot \frac{4}{24} = \frac{1}{5} \cdot \frac{1}{6} = \frac{1}{30} \approx 0.033.$$

▶ So, the probability that you draw two \$100 bills is about 3.3%.

EXAMPLE 5 Comparing Independent and Dependent Events

You randomly select 3 cards from a standard deck of 52 playing cards. What is the probability that all 3 cards are hearts when (a) you replace each card before selecting the next card, and (b) you do not replace each card before selecting the next card? Compare the probabilities.

STUDY TIP

The formulas for finding probabilities of independent and dependent events can be extended to three or more events.

SOLUTION

Let event A be "first card is a heart," event B be "second card is a heart," and event C be "third card is a heart."

a. Because you replace each card before you select the next card, the events are independent. So, the probability is

$$P(A \text{ and } B \text{ and } C) = P(A) \cdot P(B) \cdot P(C) = \frac{13}{52} \cdot \frac{13}{52} \cdot \frac{13}{52} = \frac{1}{64} \approx 0.016.$$

b. Because you do not replace each card before you select the next card, the events are dependent. So, the probability is

$$P(A \text{ and } B \text{ and } C) = P(A) \cdot P(B|A) \cdot P(C|A \text{ and } B)$$

$$= \frac{13}{52} \cdot \frac{12}{51} \cdot \frac{11}{50} = \frac{11}{850} \approx 0.013.$$

▶ So, you are $\frac{1}{64} \div \frac{11}{850} \approx 1.2$ times more likely to select 3 hearts when you replace each card before you select the next card.

Monitoring Progress  Help in English and Spanish at *BigIdeasMath.com*

3. In Example 3, what is the probability that you spin an even number and then an odd number?

4. In Example 4, what is the probability that both bills are \$1 bills?

5. In Example 5, what is the probability that none of the cards drawn are hearts when (a) you replace each card, and (b) you do not replace each card? Compare the probabilities.

Finding Conditional Probabilities

EXAMPLE 6 Using a Table to Find Conditional Probabilities

	Pass	Fail
Defective	3	36
Non-defective	450	11

A quality-control inspector checks for defective parts. The table shows the results of the inspector's work. Find (a) the probability that a defective part "passes," and (b) the probability that a non-defective part "fails."

SOLUTION

a. $P(\text{pass} \mid \text{defective}) = \dfrac{\text{Number of defective parts "passed"}}{\text{Total number of defective parts}}$

$$= \frac{3}{3 + 36} = \frac{3}{39} = \frac{1}{13} \approx 0.077, \text{ or about } 7.7\%$$

b. $P(\text{fail} \mid \text{non-defective}) = \dfrac{\text{Number of non-defective parts "failed"}}{\text{Total number of non-defective parts}}$

$$= \frac{11}{450 + 11} = \frac{11}{461} \approx 0.024, \text{ or about } 2.4\%$$

STUDY TIP

Note that when A and B are independent, this rule still applies because $P(B) = P(B \mid A)$.

You can rewrite the formula for the probability of dependent events to write a rule for finding conditional probabilities.

$P(A) \cdot P(B \mid A) = P(A \text{ and } B)$ Write formula.

$P(B \mid A) = \dfrac{P(A \text{ and } B)}{P(A)}$ Divide each side by $P(A)$.

EXAMPLE 7 Finding a Conditional Probability

At a school, 60% of students buy a school lunch. Only 10% of students buy lunch and dessert. What is the probability that a student who buys lunch also buys dessert?

SOLUTION

Let event A be "buys lunch" and let event B be "buys dessert." You are given $P(A) = 0.6$ and $P(A \text{ and } B) = 0.1$. Use the formula to find $P(B \mid A)$.

$P(B \mid A) = \dfrac{P(A \text{ and } B)}{P(A)}$ Write formula for conditional probability.

$= \dfrac{0.1}{0.6}$ Substitute 0.1 for $P(A \text{ and } B)$ and 0.6 for $P(A)$.

$= \dfrac{1}{6} \approx 0.167$ Simplify.

▶ So, the probability that a student who buys lunch also buys dessert is about 16.7%.

Monitoring Progress

Help in English and Spanish at *BigIdeasMath.com*

6. In Example 6, find (a) the probability that a non-defective part "passes," and (b) the probability that a defective part "fails."
7. At a coffee shop, 80% of customers order coffee. Only 15% of customers order coffee and a bagel. What is the probability that a customer who orders coffee also orders a bagel?

10.2 Exercises

Dynamic Solutions available at *BigIdeasMath.com*

Vocabulary and Core Concept Check

1. **WRITING** Explain the difference between dependent events and independent events, and give an example of each.

2. **COMPLETE THE SENTENCE** The probability that event B will occur given that event A has occurred is called the __________ of B given A and is written as __________.

Monitoring Progress and Modeling with Mathematics

In Exercises 3–6, tell whether the events are independent or dependent. Explain your reasoning.

3. A box of granola bars contains an assortment of flavors. You randomly choose a granola bar and eat it. Then you randomly choose another bar.

   Event *A*: You choose a coconut almond bar first.

   Event *B*: You choose a cranberry almond bar second.

4. You roll a six-sided die and flip a coin.

   Event *A*: You get a 4 when rolling the die.

   Event *B*: You get tails when flipping the coin.

5. Your MP3 player contains hip-hop and rock songs. You randomly choose a song. Then you randomly choose another song without repeating song choices.

   Event *A*: You choose a hip-hop song first.

   Event *B*: You choose a rock song second.

6. There are 22 novels of various genres on a shelf. You randomly choose a novel and put it back. Then you randomly choose another novel.

   Event *A*: You choose a mystery novel.

   Event *B*: You choose a science fiction novel.

In Exercises 7–10, determine whether the events are independent. *(See Examples 1 and 2.)*

7. You play a game that involves spinning a wheel. Each section of the wheel shown has the same area. Use a sample space to determine whether randomly spinning blue and then green are independent events.

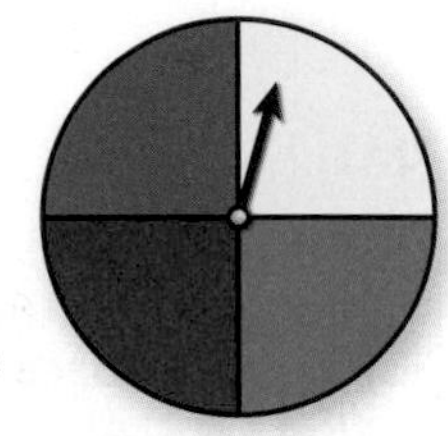

8. You have one red apple and three green apples in a bowl. You randomly select one apple to eat now and another apple for your lunch. Use a sample space to determine whether randomly selecting a green apple first and randomly selecting a green apple second are independent events.

9. A student is taking a multiple-choice test where each question has four choices. The student randomly guesses the answers to the five-question test. Use a sample space to determine whether guessing Question 1 correctly and Question 2 correctly are independent events.

10. A vase contains four white roses and one red rose. You randomly select two roses to take home. Use a sample space to determine whether randomly selecting a white rose first and randomly selecting a white rose second are independent events.

11. **PROBLEM SOLVING** You play a game that involves spinning the money wheel shown. You spin the wheel twice. Find the probability that you get more than \$500 on your first spin and then go bankrupt on your second spin. *(See Example 3.)*

12. **PROBLEM SOLVING** You play a game that involves drawing two numbers from a hat. There are 25 pieces of paper numbered from 1 to 25 in the hat. Each number is replaced after it is drawn. Find the probability that you will draw the 3 on your first draw and a number greater than 10 on your second draw.

13. **PROBLEM SOLVING** A drawer contains 12 white socks and 8 black socks. You randomly choose 1 sock and do not replace it. Then you randomly choose another sock. Find the probability that both events A and B will occur. *(See Example 4.)*

    Event A: The first sock is white.

    Event B: The second sock is white.

14. **PROBLEM SOLVING** A word game has 100 tiles, 98 of which are letters and 2 of which are blank. The numbers of tiles of each letter are shown. You randomly draw 1 tile, set it aside, and then randomly draw another tile. Find the probability that both events A and B will occur.

    Event A: The first tile is a consonant.

    Event B: The second tile is a vowel.

A – 9	H – 2	O – 8	V – 2
B – 2	I – 9	P – 2	W – 2
C – 2	J – 1	Q – 1	X – 1
D – 4	K – 1	R – 6	Y – 2
E – 12	L – 4	S – 4	Z – 1
F – 2	M – 2	T – 6	– 2 Blank
G – 3	N – 6	U – 4	

15. **ERROR ANALYSIS** Events A and B are independent. Describe and correct the error in finding $P(A \text{ and } B)$.

    ✗ $P(A) = 0.6$ $P(B) = 0.2$

    $P(A \text{ and } B) = 0.6 + 0.2 = 0.8$

16. **ERROR ANALYSIS** A shelf contains 3 fashion magazines and 4 health magazines. You randomly choose one to read, set it aside, and randomly choose another for your friend to read. Describe and correct the error in finding the probability that both events A and B occur.

    Event A: The first magazine is fashion.

    Event B: The second magazine is health.

    ✗ $P(A) = \frac{3}{7}$ $P(B|A) = \frac{4}{7}$

    $P(A \text{ and } B) = \frac{3}{7} \cdot \frac{4}{7} = \frac{12}{49} \approx 0.245$

17. **NUMBER SENSE** Events A and B are independent. Suppose $P(B) = 0.4$ and $P(A \text{ and } B) = 0.13$. Find $P(A)$.

18. **NUMBER SENSE** Events A and B are dependent. Suppose $P(B|A) = 0.6$ and $P(A \text{ and } B) = 0.15$. Find $P(A)$.

19. **ANALYZING RELATIONSHIPS** You randomly select three cards from a standard deck of 52 playing cards. What is the probability that all three cards are face cards when (a) you replace each card before selecting the next card, and (b) you do not replace each card before selecting the next card? Compare the probabilities. *(See Example 5.)*

20. **ANALYZING RELATIONSHIPS** A bag contains 9 red marbles, 4 blue marbles, and 7 yellow marbles. You randomly select three marbles from the bag. What is the probability that all three marbles are red when (a) you replace each marble before selecting the next marble, and (b) you do not replace each marble before selecting the next marble? Compare the probabilities.

21. **ATTEND TO PRECISION** The table shows the number of species in the United States listed as endangered and threatened. Find (a) the probability that a randomly selected endangered species is a bird, and (b) the probability that a randomly selected mammal is endangered. *(See Example 6.)*

	Endangered	Threatened
Mammals	70	16
Birds	80	16
Other	318	142

22. **ATTEND TO PRECISION** The table shows the number of tropical cyclones that formed during the hurricane seasons over a 12-year period. Find (a) the probability to predict whether a future tropical cyclone in the Northern Hemisphere is a hurricane, and (b) the probability to predict whether a hurricane is in the Southern Hemisphere.

Type of Tropical Cyclone	Northern Hemisphere	Southern Hemisphere
tropical depression	100	107
tropical storm	342	487
hurricane	379	525

23. **PROBLEM SOLVING** At a school, 43% of students attend the homecoming football game. Only 23% of students go to the game and the homecoming dance. What is the probability that a student who attends the football game also attends the dance? *(See Example 7.)*

24. **PROBLEM SOLVING** At a gas station, 84% of customers buy gasoline. Only 5% of customers buy gasoline and a beverage. What is the probability that a customer who buys gasoline also buys a beverage?

25. **PROBLEM SOLVING** You and 19 other students volunteer to present the "Best Teacher" award at a school banquet. One student volunteer will be chosen to present the award. Each student worked at least 1 hour in preparation for the banquet. You worked for 4 hours, and the group worked a combined total of 45 hours. For each situation, describe a process that gives you a "fair" chance to be chosen, and find the probability that you are chosen.

   a. "Fair" means equally likely.

   b. "Fair" means proportional to the number of hours each student worked in preparation.

26. **HOW DO YOU SEE IT?** A bag contains one red marble and one blue marble. The diagrams show the possible outcomes of randomly choosing two marbles using different methods. For each method, determine whether the marbles were selected with or without replacement.

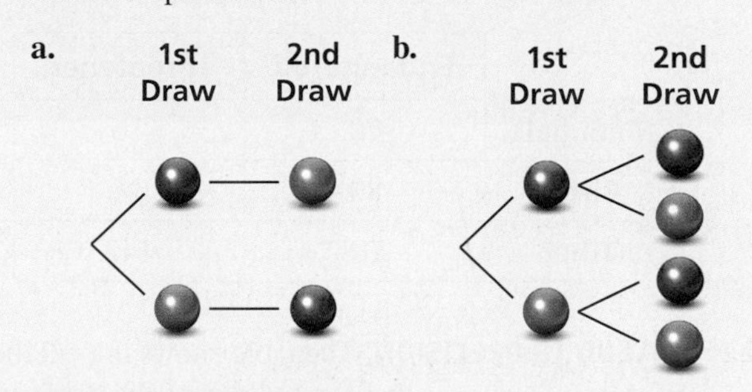

27. **MAKING AN ARGUMENT** A meteorologist claims that there is a 70% chance of rain. When it rains, there is a 75% chance that your softball game will be rescheduled. Your friend believes the game is more likely to be rescheduled than played. Is your friend correct? Explain your reasoning.

28. **THOUGHT PROVOKING** Two six-sided dice are rolled once. Events A and B are represented by the diagram. Describe each event. Are the two events dependent or independent? Justify your reasoning.

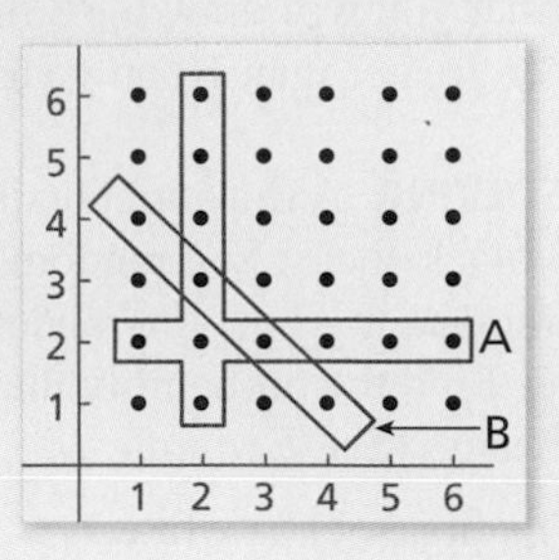

29. **MODELING WITH MATHEMATICS** A football team is losing by 14 points near the end of a game. The team scores two touchdowns (worth 6 points each) before the end of the game. After each touchdown, the coach must decide whether to go for 1 point with a kick (which is successful 99% of the time) or 2 points with a run or pass (which is successful 45% of the time).

   a. If the team goes for 1 point after each touchdown, what is the probability that the team wins? loses? ties?

   b. If the team goes for 2 points after each touchdown, what is the probability that the team wins? loses? ties?

   c. Can you develop a strategy so that the coach's team has a probability of winning the game that is greater than the probability of losing? If so, explain your strategy and calculate the probabilities of winning and losing the game.

30. **ABSTRACT REASONING** Assume that A and B are independent events.

   a. Explain why $P(B) = P(B \mid A)$ and $P(A) = P(A \mid B)$.

   b. Can $P(A \text{ and } B)$ also be defined as $P(B) \cdot P(A \mid B)$? Justify your reasoning.

Maintaining Mathematical Proficiency

Reviewing what you learned in previous grades and lessons

Solve the equation. Check your solution. *(Skills Review Handbook)*

31. $\frac{9}{10}x = 0.18$

32. $\frac{1}{4}x + 0.5x = 1.5$

33. $0.3x - \frac{3}{5}x + 1.6 = 1.555$

10.3 Two-Way Tables and Probability

Essential Question How can you construct and interpret a two-way table?

EXPLORATION 1 Completing and Using a Two-Way Table

Work with a partner. A *two-way table* displays the same information as a Venn diagram. In a two-way table, one category is represented by the rows and the other category is represented by the columns.

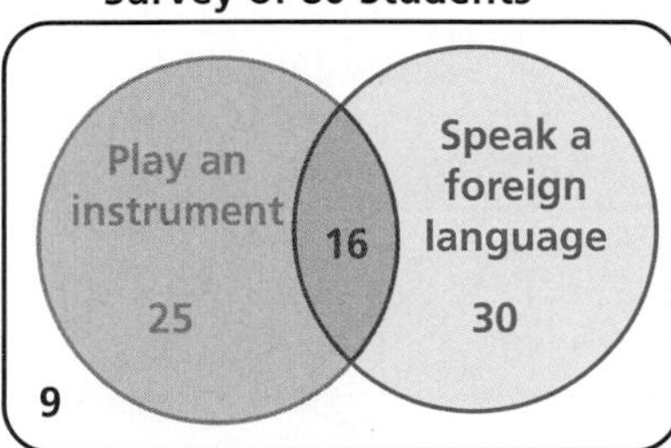

The Venn diagram shows the results of a survey in which 80 students were asked whether they play a musical instrument and whether they speak a foreign language. Use the Venn diagram to complete the two-way table. Then use the two-way table to answer each question.

	Play an Instrument	Do Not Play an Instrument	Total
Speak a Foreign Language			
Do Not Speak a Foreign Language			
Total			

a. How many students play an instrument?

b. How many students speak a foreign language?

c. How many students play an instrument and speak a foreign language?

d. How many students do not play an instrument and do not speak a foreign language?

e. How many students play an instrument and do not speak a foreign language?

EXPLORATION 2 Two-Way Tables and Probability

Work with a partner. In Exploration 1, one student is selected at random from the 80 students who took the survey. Find the probability that the student

a. plays an instrument.

b. speaks a foreign language.

c. plays an instrument and speaks a foreign language.

d. does not play an instrument and does not speak a foreign language.

e. plays an instrument and does not speak a foreign language.

MODELING WITH MATHEMATICS

To be proficient in math, you need to identify important quantities in a practical situation and map their relationships using such tools as diagrams and two-way tables.

EXPLORATION 3 Conducting a Survey

Work with your class. Conduct a survey of the students in your class. Choose two categories that are different from those given in Explorations 1 and 2. Then summarize the results in both a Venn diagram and a two-way table. Discuss the results.

Communicate Your Answer

4. How can you construct and interpret a two-way table?

5. How can you use a two-way table to determine probabilities?

10.3 Lesson

Core Vocabulary

two-way table, *p. 554*
joint frequency, *p. 554*
marginal frequency, *p. 554*
joint relative frequency, *p. 555*
marginal relative frequency, *p. 555*
conditional relative frequency, *p. 555*

Previous
conditional probability

READING

A two-way table is also called a *contingency table*, or a *two-way frequency table*.

What You Will Learn

- Make two-way tables.
- Find relative and conditional relative frequencies.
- Use conditional relative frequencies to find conditional probabilities.

Making Two-Way Tables

A **two-way table** is a frequency table that displays data collected from one source that belong to two different categories. One category of data is represented by rows and the other is represented by columns. Suppose you randomly survey freshmen and sophomores about whether they are attending a school concert. A two-way table is one way to organize your results.

Each entry in the table is called a **joint frequency**. The sums of the rows and columns are called **marginal frequencies**, which you will find in Example 1.

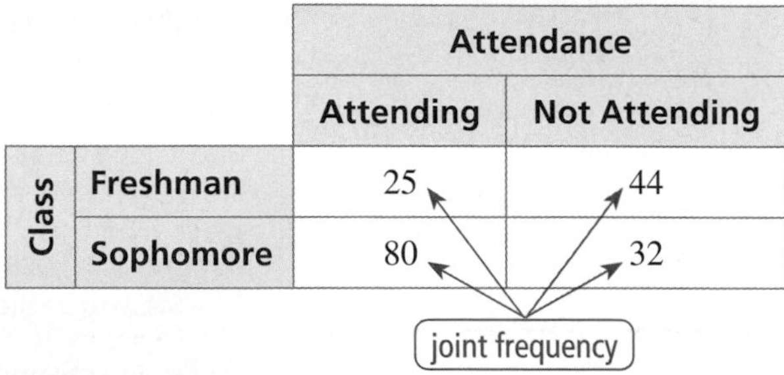

		Attendance	
		Attending	Not Attending
Class	Freshman	25	44
	Sophomore	80	32

EXAMPLE 1 Making a Two-Way Table

In another survey similar to the one above, 106 juniors and 114 seniors respond. Of those, 42 juniors and 77 seniors plan on attending. Organize these results in a two-way table. Then find and interpret the marginal frequencies.

SOLUTION

Step 1 Find the joint frequencies. Because 42 of the 106 juniors are attending, $106 - 42 = 64$ juniors are not attending. Because 77 of the 114 seniors are attending, $114 - 77 = 37$ seniors are not attending. Place each joint frequency in its corresponding cell.

Step 2 Find the marginal frequencies. Create a new column and row for the sums. Then add the entries and interpret the results.

		Attendance		
		Attending	Not Attending	Total
Class	Junior	42	64	106
	Senior	77	37	114
	Total	119	101	220

106 juniors responded.
114 seniors responded.
220 students were surveyed.
119 students are attending.
101 students are not attending.

Step 3 Find the sums of the marginal frequencies. Notice the sums $106 + 114 = 220$ and $119 + 101 = 220$ are equal. Place this value at the bottom right.

Monitoring Progress

Help in English and Spanish at *BigIdeasMath.com*

1. You randomly survey students about whether they are in favor of planting a community garden at school. Of 96 boys surveyed, 61 are in favor. Of 88 girls surveyed, 17 are against. Organize the results in a two-way table. Then find and interpret the marginal frequencies.

Finding Relative and Conditional Relative Frequencies

You can display values in a two-way table as frequency counts (as in Example 1) or as *relative frequencies*.

Core Concept

Relative and Conditional Relative Frequencies

A **joint relative frequency** is the ratio of a frequency that is not in the total row or the total column to the total number of values or observations.

A **marginal relative frequency** is the sum of the joint relative frequencies in a row or a column.

A **conditional relative frequency** is the ratio of a joint relative frequency to the marginal relative frequency. You can find a conditional relative frequency using a row total or a column total of a two-way table.

STUDY TIP

Two-way tables can display relative frequencies based on the total number of observations, the row totals, or the column totals.

EXAMPLE 2 Finding Joint and Marginal Relative Frequencies

Use the survey results in Example 1 to make a two-way table that shows the joint and marginal relative frequencies.

SOLUTION

To find the joint relative frequencies, divide each frequency by the total number of students in the survey. Then find the sum of each row and each column to find the marginal relative frequencies.

		Attendance		
		Attending	Not Attending	Total
Class	Junior	$\frac{42}{220} \approx 0.191$	$\frac{64}{220} \approx 0.291$	0.482
	Senior	$\frac{77}{220} = 0.35$	$\frac{37}{220} \approx 0.168$	0.518
	Total	0.541	0.459	1

About 29.1% of the students in the survey are juniors and are *not* attending the concert.

About 51.8% of the students in the survey are seniors.

INTERPRETING MATHEMATICAL RESULTS

Relative frequencies can be interpreted as probabilities. The probability that a randomly selected student is a junior and is *not* attending the concert is 29.1%.

EXAMPLE 3 Finding Conditional Relative Frequencies

Use the survey results in Example 1 to make a two-way table that shows the conditional relative frequencies based on the row totals.

SOLUTION

Use the marginal relative frequency of each *row* to calculate the conditional relative frequencies.

		Attendance	
		Attending	Not Attending
Class	Junior	$\frac{0.191}{0.482} \approx 0.396$	$\frac{0.291}{0.482} \approx 0.604$
	Senior	$\frac{0.35}{0.518} \approx 0.676$	$\frac{0.168}{0.518} \approx 0.324$

Given that a student is a senior, the conditional relative frequency that he or she is *not* attending the concert is about 32.4%.

Monitoring Progress  Help in English and Spanish at *BigIdeasMath.com*

2. Use the survey results in Monitoring Progress Question 1 to make a two-way table that shows the joint and marginal relative frequencies.

3. Use the survey results in Example 1 to make a two-way table that shows the conditional relative frequencies based on the column totals. Interpret the conditional relative frequencies in the context of the problem.

4. Use the survey results in Monitoring Progress Question 1 to make a two-way table that shows the conditional relative frequencies based on the row totals. Interpret the conditional relative frequencies in the context of the problem.

Finding Conditional Probabilities

You can use conditional relative frequencies to find conditional probabilities.

EXAMPLE 4 Finding Conditional Probabilities

A satellite TV provider surveys customers in three cities. The survey asks whether they would recommend the TV provider to a friend. The results, given as joint relative frequencies, are shown in the two-way table.

		Location		
		Glendale	**Santa Monica**	**Long Beach**
Response	**Yes**	0.29	0.27	0.32
	No	0.05	0.03	0.04

a. What is the probability that a randomly selected customer who is located in Glendale will recommend the provider?

b. What is the probability that a randomly selected customer who will not recommend the provider is located in Long Beach?

c. Determine whether recommending the provider to a friend and living in Long Beach are independent events.

SOLUTION

a. $P(\text{yes} \mid \text{Glendale}) = \dfrac{P(\text{Glendale and yes})}{P(\text{Glendale})} = \dfrac{0.29}{0.29 + 0.05} \approx 0.853$

▶ So, the probability that a customer who is located in Glendale will recommend the provider is about 85.3%.

INTERPRETING MATHEMATICAL RESULTS

The probability 0.853 is a conditional relative frequency based on a column total. The condition is that the customer lives in Glendale.

b. $P(\text{Long Beach} \mid \text{no}) = \dfrac{P(\text{no and Long Beach})}{P(\text{no})} = \dfrac{0.04}{0.05 + 0.03 + 0.04} \approx 0.333$

▶ So, the probability that a customer who will not recommend the provider is located in Long Beach is about 33.3%.

c. Use the formula $P(B) = P(B \mid A)$ and compare $P(\text{Long Beach})$ and $P(\text{Long Beach} \mid \text{yes})$.

$P(\text{Long Beach}) = 0.32 + 0.04 = 0.36$

$P(\text{Long Beach} \mid \text{yes}) = \dfrac{P(\text{Yes and Long Beach})}{P(\text{yes})} = \dfrac{0.32}{0.29 + 0.27 + 0.32} \approx 0.36$

▶ Because $P(\text{Long Beach}) \approx P(\text{Long Beach} \mid \text{yes})$, the two events are independent.

Monitoring Progress  Help in English and Spanish at *BigIdeasMath.com*

5. In Example 4, what is the probability that a randomly selected customer who is located in Santa Monica will not recommend the provider to a friend?

6. In Example 4, determine whether recommending the provider to a friend and living in Santa Monica are independent events. Explain your reasoning.

EXAMPLE 5 Comparing Conditional Probabilities

A jogger wants to burn a certain number of calories during his workout. He maps out three possible jogging routes. Before each workout, he randomly selects a route, and then determines the number of calories he burns and whether he reaches his goal. The table shows his findings. Which route should he use?

	Reaches Goal	Does Not Reach Goal
Route A	𝍸 𝍸 I	𝍸 I
Route B	𝍸 𝍸 I	IIII
Route C	𝍸 𝍸 II	𝍸 I

SOLUTION

Step 1 Use the findings to make a two-way table that shows the joint and marginal relative frequencies. There are a total of 50 observations in the table.

		Result		
		Reaches Goal	Does Not Reach Goal	Total
Route	A	0.22	0.12	0.34
	B	0.22	0.08	0.30
	C	0.24	0.12	0.36
	Total	0.68	0.32	1

Step 2 Find the conditional probabilities by dividing each joint relative frequency in the "Reaches Goal" column by the marginal relative frequency in its corresponding row.

$$P(\text{reaches goal} \mid \text{Route A}) = \frac{P(\text{Route A and reaches goal})}{P(\text{Route A})} = \frac{0.22}{0.34} \approx 0.647$$

$$P(\text{reaches goal} \mid \text{Route B}) = \frac{P(\text{Route B and reaches goal})}{P(\text{Route B})} = \frac{0.22}{0.30} \approx 0.733$$

$$P(\text{reaches goal} \mid \text{Route C}) = \frac{P(\text{Route C and reaches goal})}{P(\text{Route C})} = \frac{0.24}{0.36} \approx 0.667$$

▶ Based on the sample, the probability that he reaches his goal is greatest when he uses Route B. So, he should use Route B.

Monitoring Progress  Help in English and Spanish at *BigIdeasMath.com*

7. A manager is assessing three employees in order to offer one of them a promotion. Over a period of time, the manager records whether the employees meet or exceed expectations on their assigned tasks. The table shows the manager's results. Which employee should be offered the promotion? Explain.

	Exceed Expectations	Meet Expectations
Joy	𝍸 IIII	𝍸 I
Elena	𝍸 𝍸 II	𝍸 III
Sam	𝍸 𝍸 I	𝍸 II

10.3 Exercises

Dynamic Solutions available at *BigIdeasMath.com*

Vocabulary and Core Concept Check

1. **COMPLETE THE SENTENCE** A(n) ____________ displays data collected from the same source that belongs to two different categories.

2. **WRITING** Compare the definitions of joint relative frequency, marginal relative frequency, and conditional relative frequency.

Monitoring Progress and Modeling with Mathematics

In Exercises 3 and 4, complete the two-way table.

3.

		Preparation		
		Studied	Did Not Study	Total
Grade	Pass		6	
	Fail			10
	Total	38		50

4.

		Response		
		Yes	No	Total
Role	Student	56		
	Teacher		7	10
	Total		49	

5. **MODELING WITH MATHEMATICS** You survey 171 males and 180 females at Grand Central Station in New York City. Of those, 132 males and 151 females wash their hands after using the public rest rooms. Organize these results in a two-way table. Then find and interpret the marginal frequencies. *(See Example 1.)*

6. **MODELING WITH MATHEMATICS** A survey asks 60 teachers and 48 parents whether school uniforms reduce distractions in school. Of those, 49 teachers and 18 parents say uniforms reduce distractions in school. Organize these results in a two-way table. Then find and interpret the marginal frequencies.

USING STRUCTURE In Exercises 7 and 8, use the two-way table to create a two-way table that shows the joint and marginal relative frequencies.

7.

		Dominant Hand		
		Left	Right	Total
Gender	Female	11	104	115
	Male	24	92	116
	Total	35	196	231

8.

		Gender		
		Male	Female	Total
Experience	Expert	62	6	68
	Average	275	24	299
	Novice	40	3	43
	Total	377	33	410

9. **MODELING WITH MATHEMATICS** Use the survey results from Exercise 5 to make a two-way table that shows the joint and marginal relative frequencies. *(See Example 2.)*

10. **MODELING WITH MATHEMATICS** In a survey, 49 people received a flu vaccine before the flu season and 63 people did not receive the vaccine. Of those who receive the flu vaccine, 16 people got the flu. Of those who did not receive the vaccine, 17 got the flu. Make a two-way table that shows the joint and marginal relative frequencies.

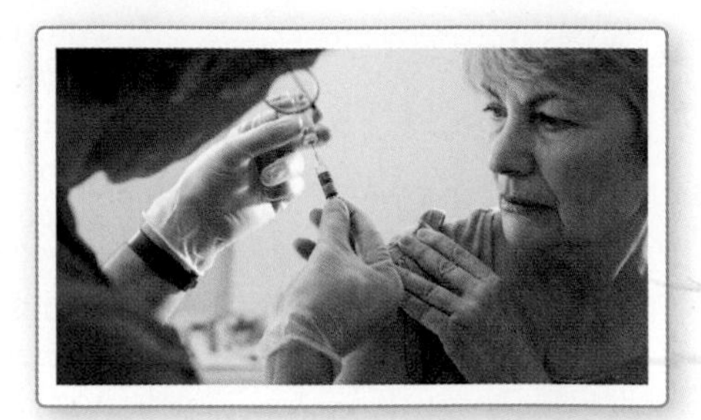

11. **MODELING WITH MATHEMATICS** A survey finds that 110 people ate breakfast and 30 people skipped breakfast. Of those who ate breakfast, 10 people felt tired. Of those who skipped breakfast, 10 people felt tired. Make a two-way table that shows the conditional relative frequencies based on the breakfast totals. *(See Example 3.)*

12. **MODELING WITH MATHEMATICS** Use the survey results from Exercise 10 to make a two-way table that shows the conditional relative frequencies based on the flu vaccine totals.

13. **PROBLEM SOLVING** Three different local hospitals in New York surveyed their patients. The survey asked whether the patient's physician communicated efficiently. The results, given as joint relative frequencies, are shown in the two-way table. *(See Example 4.)*

		Location		
		Glens Falls	Saratoga	Albany
Response	Yes	0.123	0.288	0.338
	No	0.042	0.077	0.131

a. What is the probability that a randomly selected patient located in Saratoga was satisfied with the communication of the physician?

b. What is the probability that a randomly selected patient who was not satisfied with the physician's communication is located in Glens Falls?

c. Determine whether being satisfied with the communication of the physician and living in Saratoga are independent events.

14. **PROBLEM SOLVING** A researcher surveys a random sample of high school students in seven states. The survey asks whether students plan to stay in their home state after graduation. The results, given as joint relative frequencies, are shown in the two-way table.

		Location		
		Nebraska	North Carolina	Other States
Response	Yes	0.044	0.051	0.056
	No	0.400	0.193	0.256

a. What is the probability that a randomly selected student who lives in Nebraska plans to stay in his or her home state after graduation?

b. What is the probability that a randomly selected student who does not plan to stay in his or her home state after graduation lives in North Carolina?

c. Determine whether planning to stay in their home state and living in Nebraska are independent events.

ERROR ANALYSIS In Exercises 15 and 16, describe and correct the error in finding the given conditional probability.

		City			
		Tokyo	London	Washington, D.C.	Total
Response	Yes	0.049	0.136	0.171	0.356
	No	0.341	0.112	0.191	0.644
	Total	0.39	0.248	0.362	1

15. $P(\text{yes} \mid \text{Tokyo})$

✗

$$P(\text{yes} \mid \text{Tokyo}) = \frac{P(\text{Tokyo and yes})}{P(\text{Tokyo})} = \frac{0.049}{0.356} \approx 0.138$$

16. $P(\text{London} \mid \text{no})$

✗

$$P(\text{London} \mid \text{no}) = \frac{P(\text{no and London})}{P(\text{London})} = \frac{0.112}{0.248} \approx 0.452$$

17. **PROBLEM SOLVING** You want to find the quickest route to school. You map out three routes. Before school, you randomly select a route and record whether you are late or on time. The table shows your findings. Assuming you leave at the same time each morning, which route should you use? Explain. *(See Example 5.)*

	On Time	Late
Route A	𝍸 II	IIII
Route B	𝍸 𝍸 I	III
Route C	𝍸 𝍸 II	IIII

18. **PROBLEM SOLVING** A teacher is assessing three groups of students in order to offer one group a prize. Over a period of time, the teacher records whether the groups meet or exceed expectations on their assigned tasks. The table shows the teacher's results. Which group should be awarded the prize? Explain.

	Exceed Expectations	Meet Expectations
Group 1	𝍸 𝍸 II	IIII
Group 2	𝍸 III	𝍸
Group 3	𝍸 IIII	𝍸 I

19. **OPEN-ENDED** Create and conduct a survey in your class. Organize the results in a two-way table. Then create a two-way table that shows the joint and marginal frequencies.

20. **HOW DO YOU SEE IT?** A research group surveys parents and coaches of high school students about whether competitive sports are important in school. The two-way table shows the results of the survey.

		Role		
		Parent	Coach	Total
Important	Yes	880	456	1336
	No	120	45	165
	Total	1000	501	1501

a. What does 120 represent?

b. What does 1336 represent?

c. What does 1501 represent?

21. **MAKING AN ARGUMENT** Your friend uses the table below to determine which workout routine is the best. Your friend decides that Routine B is the best option because it has the fewest tally marks in the "Does Not Reach Goal" column. Is your friend correct? Explain your reasoning.

	Reached Goal	Does Not Reach Goal
Routine A	𝍸	III
Routine B	IIII	II
Routine C	𝍸 II	IIII

22. **MODELING WITH MATHEMATICS** A survey asks students whether they prefer math class or science class. Of the 150 male students surveyed, 62% prefer math class over science class. Of the female students surveyed, 74% prefer math. Construct a two-way table to show the number of students in each category if 350 students were surveyed.

23. **MULTIPLE REPRESENTATIONS** Use the Venn diagram to construct a two-way table. Then use your table to answer the questions.

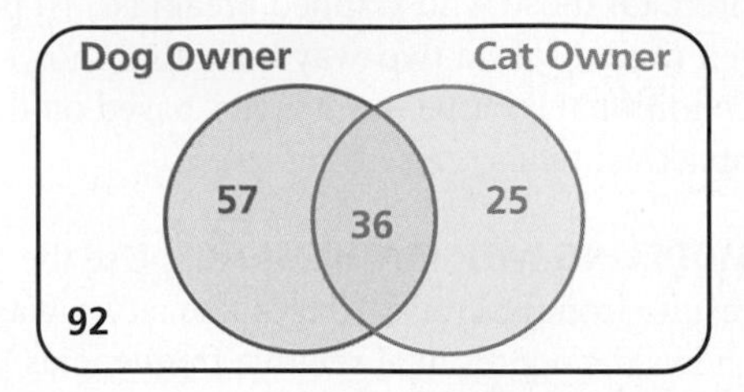

a. What is the probability that a randomly selected person does not own either pet?

b. What is the probability that a randomly selected person who owns a dog also owns a cat?

24. **WRITING** Compare two-way tables and Venn diagrams. Then describe the advantages and disadvantages of each.

25. **PROBLEM SOLVING** A company creates a new snack, N, and tests it against its current leader, L. The table shows the results.

	Prefer L	Prefer N
Current L Consumer	72	46
Not Current L Consumer	52	114

The company is deciding whether it should try to improve the snack before marketing it, and to whom the snack should be marketed. Use probability to explain the decisions the company should make when the total size of the snack's market is expected to (a) change very little, and (b) expand very rapidly.

26. **THOUGHT PROVOKING** Bayes' Theorem is given by

$$P(A|B) = \frac{P(B|A) \cdot P(A)}{P(B)}.$$

Use a two-way table to write an example of Bayes' Theorem.

Maintaining Mathematical Proficiency

Reviewing what you learned in previous grades and lessons

Draw a Venn diagram of the sets described. *(Skills Review Handbook)*

27. Of the positive integers less than 15, set A consists of the factors of 15 and set B consists of all odd numbers.

28. Of the positive integers less than 14, set A consists of all prime numbers and set B consists of all even numbers.

29. Of the positive integers less than 24, set A consists of the multiples of 2 and set B consists of all the multiples of 3.

10.1–10.3 What Did You Learn?

Core Vocabulary

probability experiment, *p. 538*
outcome, *p. 538*
event, *p. 538*
sample space, *p. 538*
probability of an event, *p. 538*
theoretical probability, *p. 539*
geometric probability, *p. 540*
experimental probability, *p. 541*
independent events, *p. 546*
dependent events, *p. 547*
conditional probability, *p. 547*
two-way table, *p. 554*
joint frequency, *p. 554*
marginal frequency, *p. 554*
joint relative frequency, *p. 555*
marginal relative frequency, *p. 555*
conditional relative frequency, *p. 555*

Core Concepts

Section 10.1

Theoretical Probabilities, *p. 538*
Probability of the Complement of an Event, *p. 539*
Experimental Probabilities, *p. 541*

Section 10.2

Probability of Independent Events, *p. 546*
Probability of Dependent Events, *p. 547*
Finding Conditional Probabilities, *p. 549*

Section 10.3

Making Two-Way Tables, *p. 554*
Relative and Conditional Relative Frequencies, *p. 555*

Mathematical Practices

1. How can you use a number line to analyze the error in Exercise 12 on page 542?
2. Explain how you used probability to correct the flawed logic of your friend in Exercise 21 on page 560.

Study Skills

Making a Mental Cheat Sheet

- Write down important information on note cards.
- Memorize the information on the note cards, placing the ones containing information you know in one stack and the ones containing information you do not know in another stack. Keep working on the information you do not know.

10.1–10.3 Quiz

1. You randomly draw a marble out of a bag containing 8 green marbles, 4 blue marbles, 12 yellow marbles, and 10 red marbles. Find the probability of drawing a marble that is not yellow. *(Section 10.1)*

Find $P(\overline{A})$. *(Section 10.1)*

2. $P(A) = 0.32$

3. $P(A) = \frac{8}{9}$

4. $P(A) = 0.01$

5. You roll a six-sided die 30 times. A 5 is rolled 8 times. What is the theoretical probability of rolling a 5? What is the experimental probability of rolling a 5? *(Section 10.1)*

6. Events A and B are independent. Find the missing probability. *(Section 10.2)*

$P(A) = 0.25$

$P(B) =$ _____

$P(A \text{ and } B) = 0.05$

7. Events A and B are dependent. Find the missing probability. *(Section 10.2)*

$P(A) = 0.6$

$P(B|A) = 0.2$

$P(A \text{ and } B) =$ _____

8. Find the probability that a dart thrown at the circular target shown will hit the given region. Assume the dart is equally likely to hit any point inside the target. *(Section 10.1)*

a. the center circle

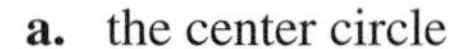

b. outside the square

c. inside the square but outside the center circle

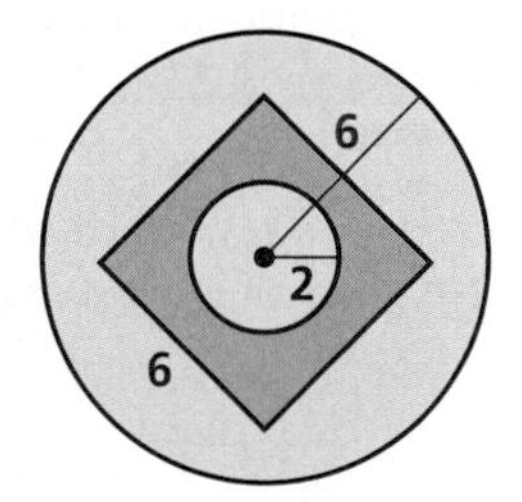

9. A survey asks 13-year-old and 15-year-old students about their eating habits. Four hundred students are surveyed, 100 male students and 100 female students from each age group. The bar graph shows the number of students who said they eat fruit every day. *(Section 10.2)*

a. Find the probability that a female student, chosen at random from the students surveyed, eats fruit every day.

b. Find the probability that a 15-year-old student, chosen at random from the students surveyed, eats fruit every day.

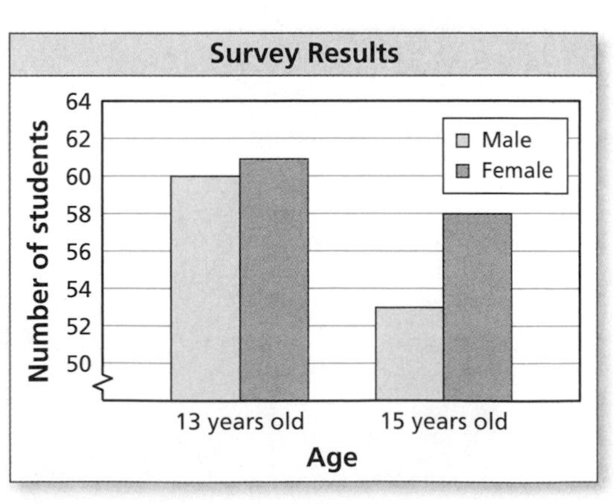

10. There are 14 boys and 18 girls in a class. The teacher allows the students to vote whether they want to take a test on Friday or on Monday. A total of 6 boys and 10 girls vote to take the test on Friday. Organize the information in a two-way table. Then find and interpret the marginal frequencies. *(Section 10.3)*

11. Three schools compete in a cross country invitational. Of the 15 athletes on your team, 9 achieve their goal times. Of the 20 athletes on the home team, 6 achieve their goal times. On your rival's team, 8 of the 13 athletes achieve their goal times. Organize the information in a two-way table. Then determine the probability that a randomly selected runner who achieves his or her goal time is from your school. *(Section 10.3)*

10.4 Probability of Disjoint and Overlapping Events

Essential Question

How can you find probabilities of disjoint and overlapping events?

Two events are **disjoint**, or **mutually exclusive**, when they have no outcomes in common. Two events are **overlapping** when they have one or more outcomes in common.

MODELING WITH MATHEMATICS

To be proficient in math, you need to map the relationships between important quantities in a practical situation using such tools as diagrams.

EXPLORATION 1 Disjoint Events and Overlapping Events

Work with a partner. A six-sided die is rolled. Draw a Venn diagram that relates the two events. Then decide whether the events are disjoint or overlapping.

a. Event *A*: The result is an even number.
Event *B*: The result is a prime number.

b. Event *A*: The result is 2 or 4.
Event *B*: The result is an odd number.

EXPLORATION 2 Finding the Probability that Two Events Occur

Work with a partner. A six-sided die is rolled. For each pair of events, find (a) $P(A)$, (b) $P(B)$, (c) $P(A \text{ and } B)$, and (d) $P(A \text{ or } B)$.

a. Event *A*: The result is an even number.
Event *B*: The result is a prime number.

b. Event *A*: The result is 2 or 4.
Event *B*: The result is an odd number.

EXPLORATION 3 Discovering Probability Formulas

Work with a partner.

a. In general, if event *A* and event *B* are disjoint, then what is the probability that event *A* or event *B* will occur? Use a Venn diagram to justify your conclusion.

b. In general, if event *A* and event *B* are overlapping, then what is the probability that event *A* or event *B* will occur? Use a Venn diagram to justify your conclusion.

c. Conduct an experiment using a six-sided die. Roll the die 50 times and record the results. Then use the results to find the probabilities described in Exploration 2. How closely do your experimental probabilities compare to the theoretical probabilities you found in Exploration 2?

Communicate Your Answer

4. How can you find probabilities of disjoint and overlapping events?

5. Give examples of disjoint events and overlapping events that do not involve dice.

10.4 Lesson

Core Vocabulary

compound event, *p. 564*
overlapping events, *p. 564*
disjoint or mutually exclusive events, *p. 564*

Previous
Venn diagram

What You Will Learn

- ▶ Find probabilities of compound events.
- ▶ Use more than one probability rule to solve real-life problems.

Compound Events

When you consider all the outcomes for either of two events A and B, you form the *union* of A and B, as shown in the first diagram. When you consider only the outcomes shared by both A and B, you form the *intersection* of A and B, as shown in the second diagram. The union or intersection of two events is called a **compound event**.

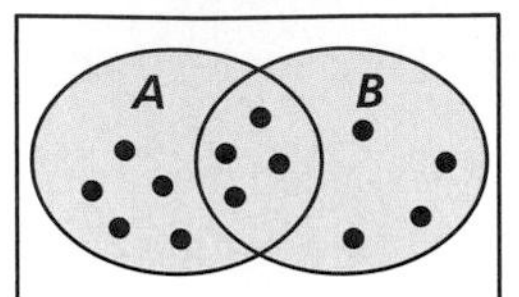

Union of A and B

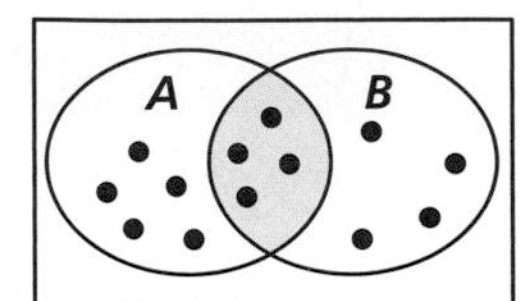

Intersection of A and B

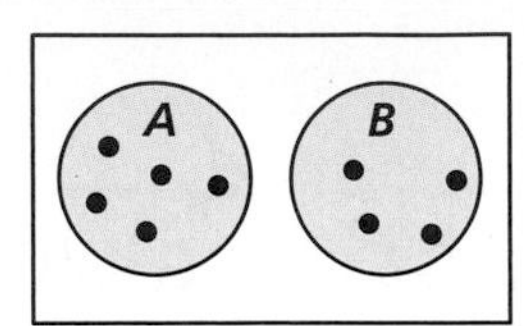

Intersection of A and B is empty.

To find $P(A \text{ or } B)$ you must consider what outcomes, if any, are in the intersection of A and B. Two events are **overlapping** when they have one or more outcomes in common, as shown in the first two diagrams. Two events are **disjoint**, or **mutually exclusive**, when they have no outcomes in common, as shown in the third diagram.

STUDY TIP

If two events A and B are overlapping, then the outcomes in the intersection of A and B are counted *twice* when $P(A)$ and $P(B)$ are added. So, $P(A \text{ and } B)$ must be subtracted from the sum.

Core Concept

Probability of Compound Events

If A and B are any two events, then the probability of A or B is

$$P(A \text{ or } B) = P(A) + P(B) - P(A \text{ and } B).$$

If A and B are disjoint events, then the probability of A or B is

$$P(A \text{ or } B) = P(A) + P(B).$$

EXAMPLE 1 Finding the Probability of Disjoint Events

A card is randomly selected from a standard deck of 52 playing cards. What is the probability that it is a 10 *or* a face card?

SOLUTION

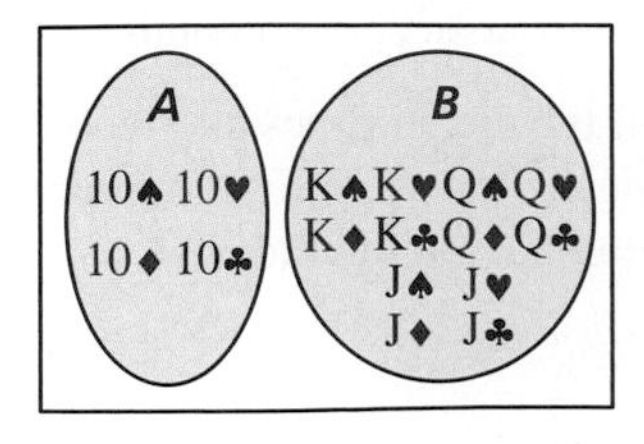

Let event A be selecting a 10 and event B be selecting a face card. From the diagram, A has 4 outcomes and B has 12 outcomes. Because A and B are disjoint, the probability is

$P(A \text{ or } B) = P(A) + P(B)$	Write disjoint probability formula.
$= \frac{4}{52} + \frac{12}{52}$	Substitute known probabilities.
$= \frac{16}{52}$	Add.
$= \frac{4}{13}$	Simplify.
$\approx 0.308.$	Use a calculator.

COMMON ERROR

When two events A and B overlap, as in Example 2, $P(A \text{ or } B)$ does not equal $P(A) + P(B)$.

EXAMPLE 2 Finding the Probability of Overlapping Events

A card is randomly selected from a standard deck of 52 playing cards. What is the probability that it is a face card *or* a spade?

SOLUTION

Let event A be selecting a face card and event B be selecting a spade. From the diagram, A has 12 outcomes and B has 13 outcomes. Of these, 3 outcomes are common to A and B. So, the probability of selecting a face card or a spade is

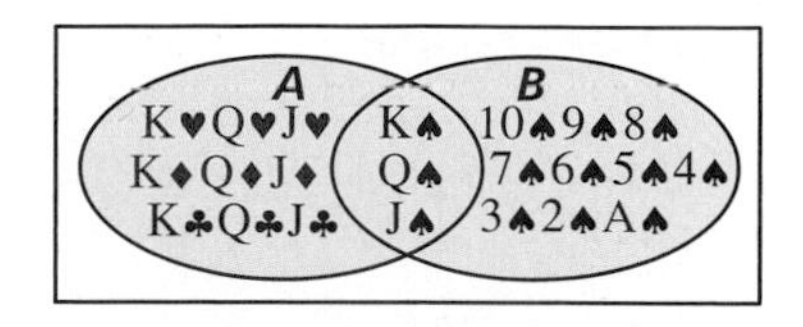

$$P(A \text{ or } B) = P(A) + P(B) - P(A \text{ and } B)$$ Write general formula.

$$= \frac{12}{52} + \frac{13}{52} - \frac{3}{52}$$ Substitute known probabilities.

$$= \frac{22}{52}$$ Add.

$$= \frac{11}{26}$$ Simplify.

$$\approx 0.423.$$ Use a calculator.

EXAMPLE 3 Using a Formula to Find $P(A \text{ and } B)$

Out of 200 students in a senior class, 113 students are either varsity athletes or on the honor roll. There are 74 seniors who are varsity athletes and 51 seniors who are on the honor roll. What is the probability that a randomly selected senior is both a varsity athlete *and* on the honor roll?

SOLUTION

Let event A be selecting a senior who is a varsity athlete and event B be selecting a senior on the honor roll. From the given information, you know that $P(A) = \frac{74}{200}$, $P(B) = \frac{51}{200}$, and $P(A \text{ or } B) = \frac{113}{200}$. The probability that a randomly selected senior is both a varsity athlete *and* on the honor roll is $P(A \text{ and } B)$.

$$P(A \text{ or } B) = P(A) + P(B) - P(A \text{ and } B)$$ Write general formula.

$$\frac{113}{200} = \frac{74}{200} + \frac{51}{200} - P(A \text{ and } B)$$ Substitute known probabilities.

$$P(A \text{ and } B) = \frac{74}{200} + \frac{51}{200} - \frac{113}{200}$$ Solve for $P(A \text{ and } B)$.

$$P(A \text{ and } B) = \frac{12}{200}$$ Simplify.

$$P(A \text{ and } B) = \frac{3}{50}, \text{ or } 0.06$$ Simplify.

Monitoring Progress  Help in English and Spanish at *BigIdeasMath.com*

A card is randomly selected from a standard deck of 52 playing cards. Find the probability of the event.

1. selecting an ace *or* an 8
2. selecting a 10 *or* a diamond
3. **WHAT IF?** In Example 3, suppose 32 seniors are in the band and 64 seniors are in the band or on the honor roll. What is the probability that a randomly selected senior is both in the band and on the honor roll?

Using More Than One Probability Rule

In the first four sections of this chapter, you have learned several probability rules. The solution to some real-life problems may require the use of two or more of these probability rules, as shown in the next example.

EXAMPLE 4 Solving a Real-Life Problem

The American Diabetes Association estimates that 8.3% of people in the United States have diabetes. Suppose that a medical lab has developed a simple diagnostic test for diabetes that is 98% accurate for people who have the disease and 95% accurate for people who do not have it. The medical lab gives the test to a randomly selected person. What is the probability that the diagnosis is correct?

SOLUTION

Let event A be "person has diabetes" and event B be "correct diagnosis." Notice that the probability of B depends on the occurrence of A, so the events are dependent. When A occurs, $P(B) = 0.98$. When A does not occur, $P(B) = 0.95$.

A probability tree diagram, where the probabilities are given along the branches, can help you see the different ways to obtain a correct diagnosis. Use the complements of events A and B to complete the diagram, where $\overline{A}$ is "person does not have diabetes" and $\overline{B}$ is "incorrect diagnosis." Notice that the probabilities for all branches from the same point must sum to 1.

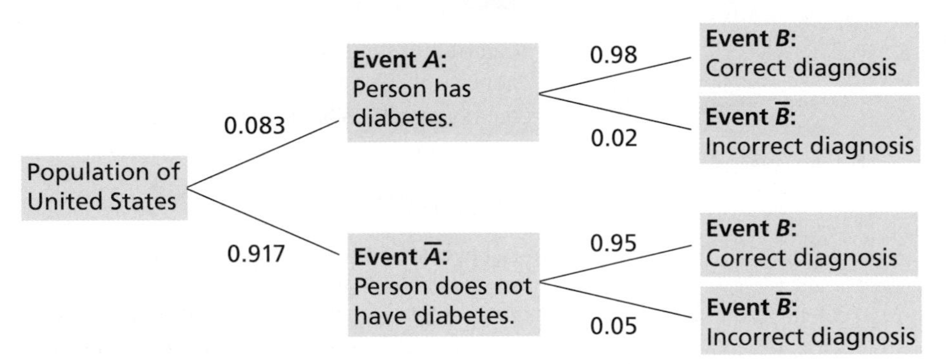

To find the probability that the diagnosis is correct, follow the branches leading to event B.

$P(B) = P(A \text{ and } B) + P(\overline{A} \text{ and } B)$	Use tree diagram.
$= P(A) \cdot P(B \mid A) + P(\overline{A}) \cdot P(B \mid \overline{A})$	Probability of dependent events
$= (0.083)(0.98) + (0.917)(0.95)$	Substitute.
≈ 0.952	Use a calculator.

▶ The probability that the diagnosis is correct is about 0.952, or 95.2%.

Monitoring Progress  Help in English and Spanish at *BigIdeasMath.com*

4. In Example 4, what is the probability that the diagnosis is *incorrect*?

5. A high school basketball team leads at halftime in 60% of the games in a season. The team wins 80% of the time when they have the halftime lead, but only 10% of the time when they do not. What is the probability that the team wins a particular game during the season?

10.4 Exercises

Dynamic Solutions available at *BigIdeasMath.com*

Vocabulary and Core Concept Check

1. **WRITING** Are the events A and $\overline{A}$ disjoint? Explain. Then give an example of a real-life event and its complement.

2. **DIFFERENT WORDS, SAME QUESTION** Which is different? Find "both" answers.

How many outcomes are in the intersection of A and B?

How many outcomes are shared by both A and B?

How many outcomes are in the union of A and B?

How many outcomes in B are also in A?

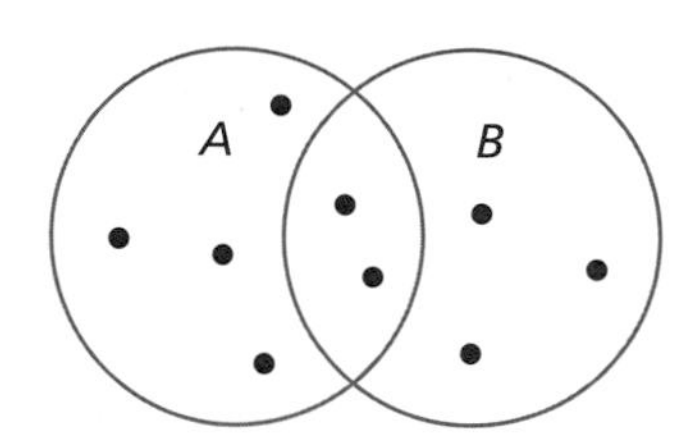

Monitoring Progress and Modeling with Mathematics

In Exercises 3–6, events A and B are disjoint. Find $P(A \text{ or } B)$.

3. $P(A) = 0.3$, $P(B) = 0.1$

4. $P(A) = 0.55$, $P(B) = 0.2$

5. $P(A) = \frac{1}{3}$, $P(B) = \frac{1}{4}$

6. $P(A) = \frac{2}{3}$, $P(B) = \frac{1}{5}$

7. **PROBLEM SOLVING** Your dart is equally likely to hit any point inside the board shown. You throw a dart and pop a balloon. What is the probability that the balloon is red or blue? *(See Example 1.)*

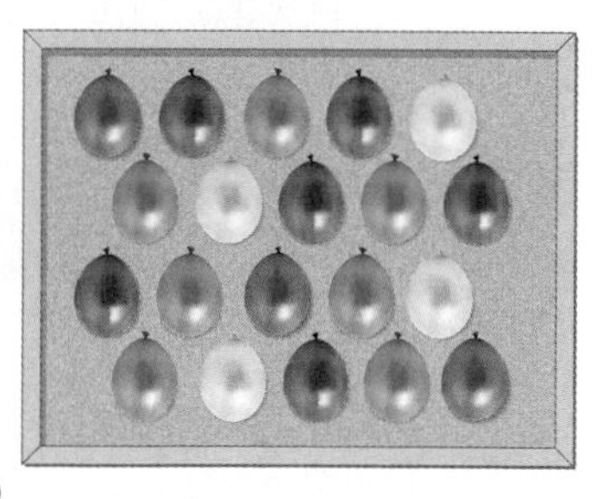

8. **PROBLEM SOLVING** You and your friend are among several candidates running for class president. You estimate that there is a 45% chance you will win and a 25% chance your friend will win. What is the probability that you or your friend win the election?

9. **PROBLEM SOLVING** You are performing an experiment to determine how well plants grow under different light sources. Of the 30 plants in the experiment, 12 receive visible light, 15 receive ultraviolet light, and 6 receive both visible and ultraviolet light. What is the probability that a plant in the experiment receives visible or ultraviolet light? *(See Example 2.)*

10. **PROBLEM SOLVING** Of 162 students honored at an academic awards banquet, 48 won awards for mathematics and 78 won awards for English. There are 14 students who won awards for both mathematics and English. A newspaper chooses a student at random for an interview. What is the probability that the student interviewed won an award for English or mathematics?

ERROR ANALYSIS In Exercises 11 and 12, describe and correct the error in finding the probability of randomly drawing the given card from a standard deck of 52 playing cards.

11. ✗

$$P(\text{heart or face card}) = P(\text{heart}) + P(\text{face card}) = \frac{13}{52} + \frac{12}{52} = \frac{25}{52}$$

12. ✗

$$P(\text{club or } 9) = P(\text{club}) + P(9) + P(\text{club and } 9) = \frac{13}{52} + \frac{4}{52} + \frac{1}{52} = \frac{9}{26}$$

In Exercises 13 and 14, you roll a six-sided die. Find $P(A \text{ or } B)$.

13. Event A: Roll a 6.
Event B: Roll a prime number.

14. Event A: Roll an odd number.
Event B: Roll a number less than 5.

15. **DRAWING CONCLUSIONS** A group of 40 trees in a forest are not growing properly. A botanist determines that 34 of the trees have a disease or are being damaged by insects, with 18 trees having a disease and 20 being damaged by insects. What is the probability that a randomly selected tree has both a disease and is being damaged by insects? *(See Example 3.)*

16. **DRAWING CONCLUSIONS** A company paid overtime wages or hired temporary help during 9 months of the year. Overtime wages were paid during 7 months, and temporary help was hired during 4 months. At the end of the year, an auditor examines the accounting records and randomly selects one month to check the payroll. What is the probability that the auditor will select a month in which the company paid overtime wages and hired temporary help?

17. **DRAWING CONCLUSIONS** A company is focus testing a new type of fruit drink. The focus group is 47% male. Of the responses, 40% of the males and 54% of the females said they would buy the fruit drink. What is the probability that a randomly selected person would buy the fruit drink? *(See Example 4.)*

18. **DRAWING CONCLUSIONS** The Redbirds trail the Bluebirds by one goal with 1 minute left in the hockey game. The Redbirds' coach must decide whether to remove the goalie and add a frontline player. The probabilities of each team scoring are shown in the table.

	Goalie	No goalie
Redbirds score	0.1	0.3
Bluebirds score	0.1	0.6

a. Find the probability that the Redbirds score and the Bluebirds do not score when the coach leaves the goalie in.

b. Find the probability that the Redbirds score and the Bluebirds do not score when the coach takes the goalie out.

c. Based on parts (a) and (b), what should the coach do?

19. **PROBLEM SOLVING** You can win concert tickets from a radio station if you are the first person to call when the song of the day is played, or if you are the first person to correctly answer the trivia question. The song of the day is announced at a random time between 7:00 and 7:30 A.M. The trivia question is asked at a random time between 7:15 and 7:45 A.M. You begin listening to the radio station at 7:20. Find the probability that you miss the announcement of the song of the day or the trivia question.

20. **HOW DO YOU SEE IT?** Are events A and B disjoint events? Explain your reasoning.

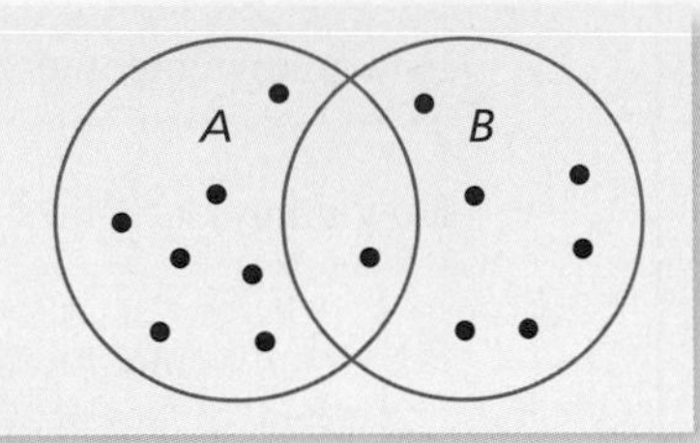

21. **PROBLEM SOLVING** You take a bus from your neighborhood to your school. The express bus arrives at your neighborhood at a random time between 7:30 and 7:36 A.M. The local bus arrives at your neighborhood at a random time between 7:30 and 7:40 A.M. You arrive at the bus stop at 7:33 A.M. Find the probability that you missed both the express bus and the local bus.

22. **THOUGHT PROVOKING** Write a general rule for finding $P(A \text{ or } B \text{ or } C)$ for (a) disjoint and (b) overlapping events A, B, and C.

23. **MAKING AN ARGUMENT** A bag contains 40 cards numbered 1 through 40 that are either red or blue. A card is drawn at random and placed back in the bag. This is done four times. Two red cards are drawn, numbered 31 and 19, and two blue cards are drawn, numbered 22 and 7. Your friend concludes that red cards and even numbers must be mutually exclusive. Is your friend correct? Explain.

Maintaining Mathematical Proficiency

Reviewing what you learned in previous grades and lessons

Write the first six terms of the sequence. *(Section 8.5)*

24. $a_1 = 4,\ a_n = 2a_{n-1} + 3$

25. $a_1 = 1,\ a_n = \dfrac{n(n-1)}{a_{n-1}}$

26. $a_1 = 2,\ a_2 = 6,\ a_n = \dfrac{(n+1)a_{n-1}}{a_{n-2}}$

10.5 Permutations and Combinations

Essential Question How can a tree diagram help you visualize the number of ways in which two or more events can occur?

EXPLORATION 1 Reading a Tree Diagram

Work with a partner. Two coins are flipped and the spinner is spun. The tree diagram shows the possible outcomes.

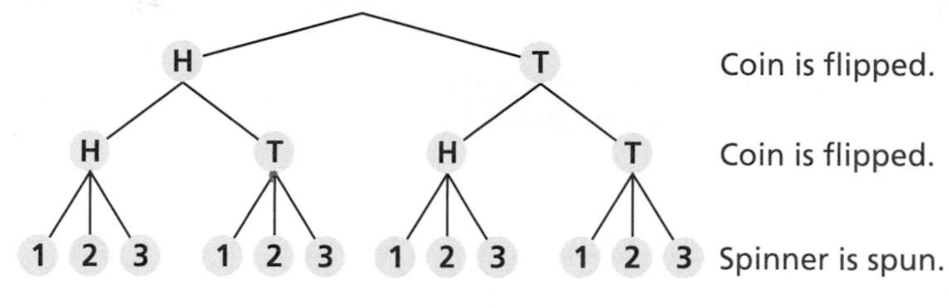

a. How many outcomes are possible?

b. List the possible outcomes.

EXPLORATION 2 Reading a Tree Diagram

Work with a partner. Consider the tree diagram below.

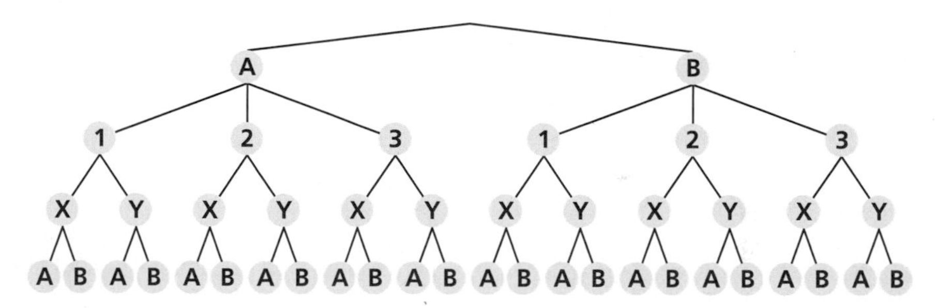

a. How many events are shown?

b. What outcomes are possible for each event?

c. How many outcomes are possible?

d. List the possible outcomes.

CONSTRUCTING VIABLE ARGUMENTS

To be proficient in math, you need to make conjectures and build a logical progression of statements to explore the truth of your conjectures.

EXPLORATION 3 Writing a Conjecture

Work with a partner.

a. Consider the following general problem: Event 1 can occur in m ways and event 2 can occur in n ways. Write a conjecture about the number of ways the two events can occur. Explain your reasoning.

b. Use the conjecture you wrote in part (a) to write a conjecture about the number of ways *more than* two events can occur. Explain your reasoning.

c. Use the results of Explorations 1(a) and 2(c) to verify your conjectures.

Communicate Your Answer

4. How can a tree diagram help you visualize the number of ways in which two or more events can occur?

5. In Exploration 1, the spinner is spun a second time. How many outcomes are possible?

10.5 Lesson

What You Will Learn

- Use the formula for the number of permutations.
- Use the formula for the number of combinations.
- Use combinations and the Binomial Theorem to expand binomials.

Core Vocabulary

permutation, *p. 570*
n factorial, *p. 570*
combination, *p. 572*
Binomial Theorem, *p. 574*

Previous
Fundamental Counting Principle
Pascal's Triangle

Permutations

A **permutation** is an arrangement of objects in which order is important. For instance, the 6 possible permutations of the letters A, B, and C are shown.

ABC ACB BAC BCA CAB CBA

EXAMPLE 1 Counting Permutations

Consider the number of permutations of the letters in the word JULY. In how many ways can you arrange (a) all of the letters and (b) 2 of the letters?

SOLUTION

a. Use the Fundamental Counting Principle to find the number of permutations of the letters in the word JULY.

$$\text{Number of permutations} = \left(\begin{array}{l}\text{Choices for}\\ \text{1st letter}\end{array}\right)\left(\begin{array}{l}\text{Choices for}\\ \text{2nd letter}\end{array}\right)\left(\begin{array}{l}\text{Choices for}\\ \text{3rd letter}\end{array}\right)\left(\begin{array}{l}\text{Choices for}\\ \text{4th letter}\end{array}\right)$$

$$= 4 \cdot 3 \cdot 2 \cdot 1$$

$$= 24$$

▶ There are 24 ways you can arrange all of the letters in the word JULY.

b. When arranging 2 letters of the word JULY, you have 4 choices for the first letter and 3 choices for the second letter.

$$\text{Number of permutations} = \left(\begin{array}{l}\text{Choices for}\\ \text{1st letter}\end{array}\right)\left(\begin{array}{l}\text{Choices for}\\ \text{2nd letter}\end{array}\right)$$

$$= 4 \cdot 3$$

$$= 12$$

▶ There are 12 ways you can arrange 2 of the letters in the word JULY.

REMEMBER

Fundamental Counting Principle: If one event can occur in *m* ways and another event can occur in *n* ways, then the number of ways that both events can occur is $m \cdot n$. The Fundamental Counting Principle can be extended to three or more events.

Monitoring Progress

Help in English and Spanish at *BigIdeasMath.com*

1. In how many ways can you arrange the letters in the word HOUSE?

2. In how many ways can you arrange 3 of the letters in the word MARCH?

In Example 1(a), you evaluated the expression $4 \cdot 3 \cdot 2 \cdot 1$. This expression can be written as 4! and is read "4 *factorial*." For any positive integer *n*, the product of the integers from 1 to *n* is called ***n* factorial** and is written as

$$n! = n \cdot (n - 1) \cdot (n - 2) \cdot \cdots \cdot 3 \cdot 2 \cdot 1.$$

As a special case, the value of 0! is defined to be 1.

In Example 1(b), you found the permutations of 4 objects taken 2 at a time. You can find the number of permutations using the formulas on the next page.

Core Concept

Permutations

Formulas	Examples
The number of permutations of n objects is given by $${}_nP_n = n!.$$	The number of permutations of 4 objects is $${}_4P_4 = 4! = 4 \cdot 3 \cdot 2 \cdot 1 = 24.$$
The number of permutations of n objects taken r at a time, where $r \le n$, is given by $${}_nP_r = \frac{n!}{(n-r)!}.$$	The number of permutations of 4 objects taken 2 at a time is $${}_4P_2 = \frac{4!}{(4-2)!} = \frac{4 \cdot 3 \cdot \cancel{2!}}{\cancel{2!}} = 12.$$

USING A GRAPHING CALCULATOR

Most graphing calculators can calculate permutations.

```
4 nPr 4
                24
4 nPr 2
                12
```

EXAMPLE 2 Using a Permutations Formula

Ten horses are running in a race. In how many different ways can the horses finish first, second, and third? (Assume there are no ties.)

SOLUTION

To find the number of permutations of 3 horses chosen from 10, find ${}_{10}P_3$.

$${}_{10}P_3 = \frac{10!}{(10-3)!}$$ Permutations formula

$$= \frac{10!}{7!}$$ Subtract.

$$= \frac{10 \cdot 9 \cdot 8 \cdot \cancel{7!}}{\cancel{7!}}$$ Expand factorial. Divide out common factor, 7!.

$$= 720$$ Simplify.

▶ There are 720 ways for the horses to finish first, second, and third.

STUDY TIP

When you divide out common factors, remember that 7! is a factor of 10!.

EXAMPLE 3 Finding a Probability Using Permutations

For a town parade, you will ride on a float with your soccer team. There are 12 floats in the parade, and their order is chosen at random. Find the probability that your float is first and the float with the school chorus is second.

SOLUTION

Step 1 Write the number of possible outcomes as the number of permutations of the 12 floats in the parade. This is ${}_{12}P_{12} = 12!$.

Step 2 Write the number of favorable outcomes as the number of permutations of the other floats, given that the soccer team is first and the chorus is second. This is ${}_{10}P_{10} = 10!$.

Step 3 Find the probability.

$$P(\text{soccer team is 1st, chorus is 2nd}) = \frac{10!}{12!}$$ Form a ratio of favorable to possible outcomes.

$$= \frac{\cancel{10!}}{12 \cdot 11 \cdot \cancel{10!}}$$ Expand factorial. Divide out common factor, 10!.

$$= \frac{1}{132}$$ Simplify.

Monitoring Progress  Help in English and Spanish at *BigIdeasMath.com*

3. **WHAT IF?** In Example 2, suppose there are 8 horses in the race. In how many different ways can the horses finish first, second, and third? (Assume there are no ties.)

4. **WHAT IF?** In Example 3, suppose there are 14 floats in the parade. Find the probability that the soccer team is first and the chorus is second.

Combinations

A **combination** is a selection of objects in which order is *not* important. For instance, in a drawing for 3 identical prizes, you would use combinations, because the order of the winners would not matter. If the prizes were different, then you would use permutations, because the order would matter.

EXAMPLE 4 Counting Combinations

Count the possible combinations of 2 letters chosen from the list A, B, C, D.

SOLUTION

List all of the permutations of 2 letters from the list A, B, C, D. Because order is not important in a combination, cross out any duplicate pairs.

AB	AC	AD	~~BA~~	BC	BD
~~CA~~	~~CB~~	CD	~~DA~~	~~DB~~	~~DC~~

BD and DB are the same pair.

▶ There are 6 possible combinations of 2 letters from the list A, B, C, D.

Monitoring Progress  Help in English and Spanish at *BigIdeasMath.com*

5. Count the possible combinations of 3 letters chosen from the list A, B, C, D, E.

In Example 4, you found the number of combinations of objects by making an organized list. You can also find the number of combinations using the following formula.

USING A GRAPHING CALCULATOR

Most graphing calculators can calculate combinations.

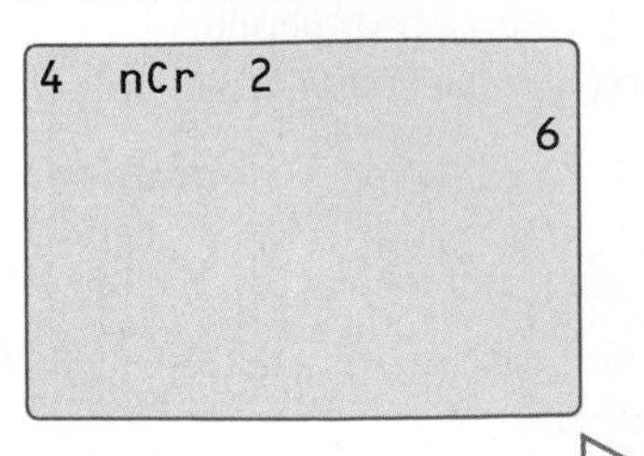

Core Concept

Combinations

Formula The number of combinations of n objects taken r at a time, where $r \leq n$, is given by

$$_nC_r = \frac{n!}{(n-r)! \cdot r!}.$$

Example The number of combinations of 4 objects taken 2 at a time is

$$_4C_2 = \frac{4!}{(4-2)! \cdot 2!} = \frac{4 \cdot 3 \cdot \cancel{2!}}{\cancel{2!} \cdot (2 \cdot 1)} = 6.$$

EXAMPLE 5 Using the Combinations Formula

You order a sandwich at a restaurant. You can choose 2 side dishes from a list of 8. How many combinations of side dishes are possible?

SOLUTION

The order in which you choose the side dishes is not important. So, to find the number of combinations of 8 side dishes taken 2 at a time, find ${}_8C_2$.

Check

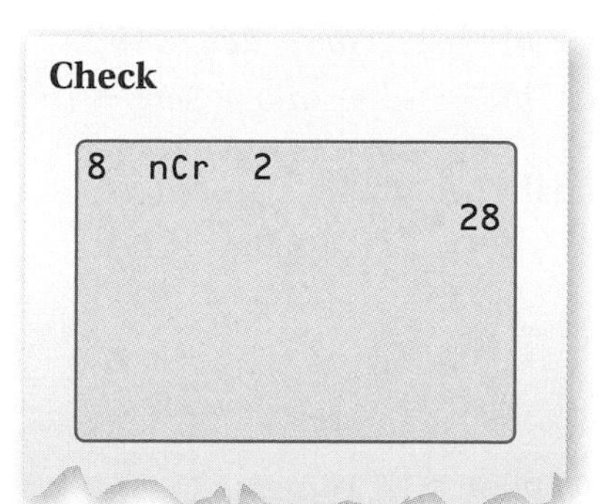

$${}_8C_2 = \frac{8!}{(8-2)! \cdot 2!}$$ Combinations formula

$$= \frac{8!}{6! \cdot 2!}$$ Subtract.

$$= \frac{8 \cdot 7 \cdot \cancel{6!}}{\cancel{6!} \cdot (2 \cdot 1)}$$ Expand factorials. Divide out common factor, 6!.

$$= 28$$ Multiply.

▶ There are 28 different combinations of side dishes you can order.

EXAMPLE 6 Finding a Probability Using Combinations

A yearbook editor has selected 14 photos, including one of you and one of your friend, to use in a collage for the yearbook. The photos are placed at random. There is room for 2 photos at the top of the page. What is the probability that your photo and your friend's photo are the 2 placed at the top of the page?

SOLUTION

Step 1 Write the number of possible outcomes as the number of combinations of 14 photos taken 2 at a time, or ${}_{14}C_2$, because the order in which the photos are chosen is not important.

$${}_{14}C_2 = \frac{14!}{(14-2)! \cdot 2!}$$ Combinations formula

$$= \frac{14!}{12! \cdot 2!}$$ Subtract.

$$= \frac{14 \cdot 13 \cdot \cancel{12!}}{\cancel{12!} \cdot (2 \cdot 1)}$$ Expand factorials. Divide out common factor, 12!.

$$= 91$$ Multiply.

Step 2 Find the number of favorable outcomes. Only one of the possible combinations includes your photo and your friend's photo.

Step 3 Find the probability.

$$P(\text{your photo and your friend's photos are chosen}) = \frac{1}{91}$$

6. **WHAT IF?** In Example 5, suppose you can choose 3 side dishes out of the list of 8 side dishes. How many combinations are possible?

7. **WHAT IF?** In Example 6, suppose there are 20 photos in the collage. Find the probability that your photo and your friend's photo are the 2 placed at the top of the page.

Binomial Expansions

In Section 4.2, you used Pascal's Triangle to find binomial expansions. The table shows that the coefficients in the expansion of $(a + b)^n$ correspond to combinations.

	n	Pascal's Triangle as Numbers	Pascal's Triangle as Combinations	Binomial Expansion
0th row	0	1	${}_0C_0$	$(a + b)^0 = 1$
1st row	1	1 1	${}_1C_0 \quad {}_1C_1$	$(a + b)^1 = 1a + 1b$
2nd row	2	1 2 1	${}_2C_0 \quad {}_2C_1 \quad {}_2C_2$	$(a + b)^2 = 1a^2 + 2ab + 1b^2$
3rd row	3	1 3 3 1	${}_3C_0 \quad {}_3C_1 \quad {}_3C_2 \quad {}_3C_3$	$(a + b)^3 = 1a^3 + 3a^2b + 3ab^2 + 1b^3$

The results in the table are generalized in the **Binomial Theorem**.

Core Concept

The Binomial Theorem

For any positive integer n, the binomial expansion of $(a + b)^n$ is

$$(a + b)^n = {}_nC_0\, a^n b^0 + {}_nC_1\, a^{n-1} b^1 + {}_nC_2\, a^{n-2} b^2 + \cdots + {}_nC_n\, a^0 b^n.$$

Notice that each term in the expansion of $(a + b)^n$ has the form ${}_nC_r\, a^{n-r} b^r$, where r is an integer from 0 to n.

EXAMPLE 7 Using the Binomial Theorem

a. Use the Binomial Theorem to write the expansion of $(x^2 + y)^3$.

b. Find the coefficient of x^4 in the expansion of $(3x + 2)^{10}$.

SOLUTION

a. $(x^2 + y)^3 = {}_3C_0(x^2)^3y^0 + {}_3C_1(x^2)^2y^1 + {}_3C_2(x^2)^1y^2 + {}_3C_3(x^2)^0y^3$

$= (1)(x^6)(1) + (3)(x^4)(y^1) + (3)(x^2)(y^2) + (1)(1)(y^3)$

$= x^6 + 3x^4y + 3x^2y^2 + y^3$

b. From the Binomial Theorem, you know

$$(3x + 2)^{10} = {}_{10}C_0(3x)^{10}(2)^0 + {}_{10}C_1(3x)^9(2)^1 + \cdots + {}_{10}C_{10}(3x)^0(2)^{10}.$$

Each term in the expansion has the form ${}_{10}C_r(3x)^{10-r}(2)^r$. The term containing x^4 occurs when $r = 6$.

$${}_{10}C_6(3x)^4(2)^6 = (210)(81x^4)(64) = 1{,}088{,}640x^4$$

▶ The coefficient of x^4 is 1,088,640.

Monitoring Progress

Help in English and Spanish at *BigIdeasMath.com*

8. Use the Binomial Theorem to write the expansion of (a) $(x + 3)^5$ and (b) $(2p - q)^4$.

9. Find the coefficient of x^5 in the expansion of $(x - 3)^7$.

10. Find the coefficient of x^3 in the expansion of $(2x + 5)^8$.

10.5 Exercises

Dynamic Solutions available at *BigIdeasMath.com*

Vocabulary and Core Concept Check

1. **COMPLETE THE SENTENCE** An arrangement of objects in which order is important is called a(n) ________.

2. **WHICH ONE DOESN'T BELONG?** Which expression does *not* belong with the other three? Explain your reasoning.

$\dfrac{7!}{2! \cdot 5!}$	$_7C_5$	$_7C_2$	$\dfrac{7!}{(7-2)!}$

Monitoring Progress and Modeling with Mathematics

In Exercises 3–8, find the number of ways you can arrange (a) all of the letters and (b) 2 of the letters in the given word. *(See Example 1.)*

3. AT
4. TRY
5. ROCK
6. WATER
7. FAMILY
8. FLOWERS

In Exercises 9–16, evaluate the expression.

9. $_5P_2$
10. $_7P_3$
11. $_9P_1$
12. $_6P_5$
13. $_8P_6$
14. $_{12}P_0$
15. $_{30}P_2$
16. $_{25}P_5$

17. **PROBLEM SOLVING** Eleven students are competing in an art contest. In how many different ways can the students finish first, second, and third? *(See Example 2.)*

18. **PROBLEM SOLVING** Six friends go to a movie theater. In how many different ways can they sit together in a row of 6 empty seats?

19. **PROBLEM SOLVING** You and your friend are 2 of 8 servers working a shift in a restaurant. At the beginning of the shift, the manager randomly assigns one section to each server. Find the probability that you are assigned Section 1 and your friend is assigned Section 2. *(See Example 3.)*

20. **PROBLEM SOLVING** You make 6 posters to hold up at a basketball game. Each poster has a letter of the word TIGERS. You and 5 friends sit next to each other in a row. The posters are distributed at random. Find the probability that TIGERS is spelled correctly when you hold up the posters.

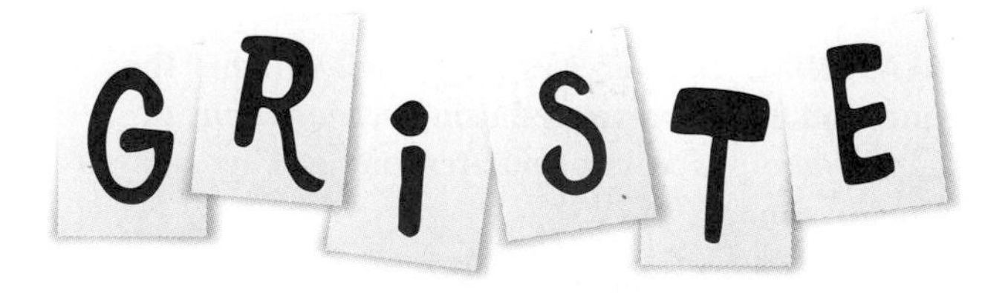

In Exercises 21–24, count the possible combinations of *r* letters chosen from the given list. *(See Example 4.)*

21. A, B, C, D; $r = 3$
22. L, M, N, O; $r = 2$
23. U, V, W, X, Y, Z; $r = 3$
24. D, E, F, G, H; $r = 4$

In Exercises 25–32, evaluate the expression.

25. $_5C_1$
26. $_8C_5$
27. $_9C_9$
28. $_8C_6$
29. $_{12}C_3$
30. $_{11}C_4$
31. $_{15}C_8$
32. $_{20}C_5$

33. PROBLEM SOLVING Each year, 64 golfers participate in a golf tournament. The golfers play in groups of 4. How many groups of 4 golfers are possible? *(See Example 5.)*

34. PROBLEM SOLVING You want to purchase vegetable dip for a party. A grocery store sells 7 different flavors of vegetable dip. You have enough money to purchase 2 flavors. How many combinations of 2 flavors of vegetable dip are possible?

ERROR ANALYSIS In Exercises 35 and 36, describe and correct the error in evaluating the expression.

35.

✗ $_{11}P_7 = \dfrac{11!}{(11-7)} = \dfrac{11!}{4} = 9{,}979{,}200$

36.

✗ $_9C_4 = \dfrac{9!}{(9-4)!} = \dfrac{9!}{5!} = 3024$

REASONING In Exercises 37–40, tell whether the question can be answered using *permutations* or *combinations*. Explain your reasoning. Then answer the question.

37. To complete an exam, you must answer 8 questions from a list of 10 questions. In how many ways can you complete the exam?

38. Ten students are auditioning for 3 different roles in a play. In how many ways can the 3 roles be filled?

39. Fifty-two athletes are competing in a bicycle race. In how many orders can the bicyclists finish first, second, and third? (Assume there are no ties.)

40. An employee at a pet store needs to catch 5 tetras in an aquarium containing 27 tetras. In how many groupings can the employee capture 5 tetras?

41. CRITICAL THINKING Compare the quantities $_{50}C_9$ and $_{50}C_{41}$ without performing any calculations. Explain your reasoning.

42. CRITICAL THINKING Show that each identity is true for any whole numbers r and n, where $0 \le r \le n$.

a. $_nC_n = 1$

b. $_nC_r = {}_nC_{n-r}$

c. $_{n+1}C_r = {}_nC_r + {}_nC_{r-1}$

43. REASONING Consider a set of 4 objects.

a. Are there more permutations of all 4 of the objects or of 3 of the objects? Explain your reasoning.

b. Are there more combinations of all 4 of the objects or of 3 of the objects? Explain your reasoning.

c. Compare your answers to parts (a) and (b).

44. OPEN-ENDED Describe a real-life situation where the number of possibilities is given by $_5P_2$. Then describe a real-life situation that can be modeled by $_5C_2$.

45. REASONING Complete the table for each given value of r. Then write an inequality relating $_nP_r$ and $_nC_r$. Explain your reasoning.

	$r = 0$	$r = 1$	$r = 2$	$r = 3$
$_3P_r$				
$_3C_r$				

46. REASONING Write an equation that relates $_nP_r$ and $_nC_r$. Then use your equation to find and interpret the value of $\dfrac{_{182}P_4}{_{182}C_4}$.

47. PROBLEM SOLVING You and your friend are in the studio audience on a television game show. From an audience of 300 people, 2 people are randomly selected as contestants. What is the probability that you and your friend are chosen? *(See Example 6.)*

48. **PROBLEM SOLVING** You work 5 evenings each week at a bookstore. Your supervisor assigns you 5 evenings at random from the 7 possibilities. What is the probability that your schedule does not include working on the weekend?

REASONING In Exercises 49 and 50, find the probability of winning a lottery using the given rules. Assume that lottery numbers are selected at random.

49. You must correctly select 6 numbers, each an integer from 0 to 49. The order is not important.

50. You must correctly select 4 numbers, each an integer from 0 to 9. The order is important.

In Exercises 51–58, use the Binomial Theorem to write the binomial expansion. *(See Example 7a.)*

51. $(x + 2)^3$ **52.** $(c - 4)^5$

53. $(a + 3b)^4$ **54.** $(4p - q)^6$

55. $(w^3 - 3)^4$ **56.** $(2s^4 + 5)^5$

57. $(3u + v^2)^6$ **58.** $(x^3 - y^2)^4$

In Exercises 59–66, use the given value of n to find the coefficient of x^n in the expansion of the binomial. *(See Example 7b.)*

59. $(x - 2)^{10}, n = 5$ **60.** $(x - 3)^7, n = 4$

61. $(x^2 - 3)^8, n = 6$ **62.** $(3x + 2)^5, n = 3$

63. $(2x + 5)^{12}, n = 7$ **64.** $(3x - 1)^9, n = 2$

65. $\left(\frac{1}{2}x - 4\right)^{11}, n = 4$ **66.** $\left(\frac{1}{4}x + 6\right)^6, n = 3$

67. **REASONING** Write the eighth row of Pascal's Triangle as combinations and as numbers.

68. **PROBLEM SOLVING** The first four triangular numbers are 1, 3, 6, and 10.

a. Use Pascal's Triangle to write the first four triangular numbers as combinations.

1
1 1
1 2 1
1 3 3 1
1 4 6 4 1
1 5 10 10 5 1

b. Use your result from part (a) to write an explicit rule for the nth triangular number T_n.

69. **MATHEMATICAL CONNECTIONS** A polygon is convex when no line that contains a side of the polygon contains a point in the interior of the polygon. Consider a convex polygon with n sides.

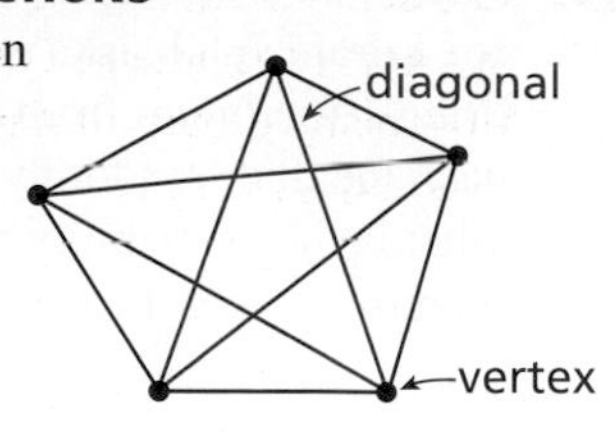

a. Use the combinations formula to write an expression for the number of diagonals in an n-sided polygon.

b. Use your result from part (a) to write a formula for the number of diagonals of an n-sided convex polygon.

70. **PROBLEM SOLVING** You are ordering a burrito with 2 main ingredients and 3 toppings. The menu below shows the possible choices. How many different burritos are possible?

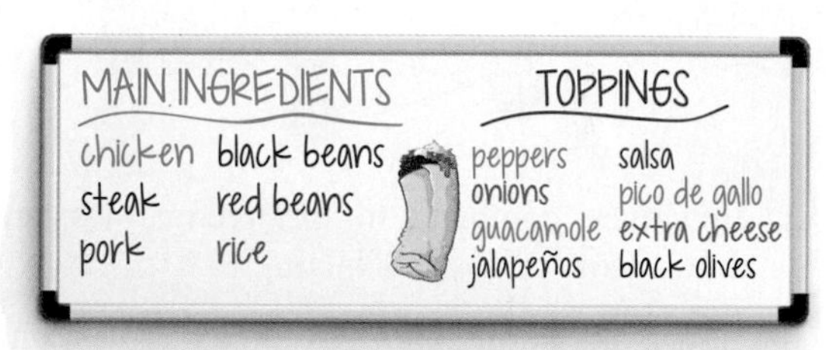

71. **PROBLEM SOLVING** You want to purchase 2 different types of contemporary music CDs and 1 classical music CD from the music collection shown. How many different sets of music types can you choose for your purchase?

72. **PROBLEM SOLVING** Every student in your history class is required to present a project in front of the class. Each day, 4 students make their presentations in an order chosen at random by the teacher. You make your presentation on the first day.

a. What is the probability that you are chosen to be the first or second presenter on the first day?

b. What is the probability that you are chosen to be the second or third presenter on the first day? Compare your answer with that in part (a).

73. PROBLEM SOLVING The organizer of a cast party for a drama club asks each of the 6 cast members to bring 1 food item from a list of 10 items. Assuming each member randomly chooses a food item to bring, what is the probability that at least 2 of the 6 cast members bring the same item?

74. HOW DO YOU SEE IT? A bag contains one green marble, one red marble, and one blue marble. The diagram shows the possible outcomes of randomly drawing three marbles from the bag without replacement.

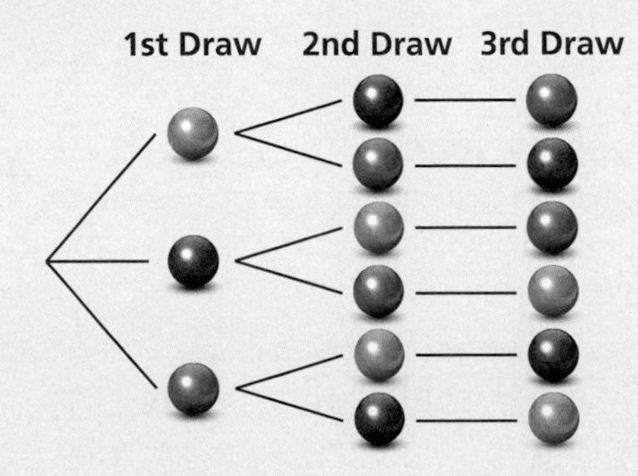

a. How many combinations of three marbles can be drawn from the bag? Explain.

b. How many permutations of three marbles can be drawn from the bag? Explain.

75. PROBLEM SOLVING You are one of 10 students performing in a school talent show. The order of the performances is determined at random. The first 5 performers go on stage before the intermission.

a. What is the probability that you are the last performer before the intermission and your rival performs immediately before you?

b. What is the probability that you are *not* the first performer?

76. THOUGHT PROVOKING How many integers, greater than 999 but not greater than 4000, can be formed with the digits 0, 1, 2, 3, and 4? Repetition of digits is allowed.

77. PROBLEM SOLVING Consider a standard deck of 52 playing cards. The order in which the cards are dealt for a "hand" does not matter.

a. How many different 5-card hands are possible?

b. How many different 5-card hands have all 5 cards of a single suit?

78. PROBLEM SOLVING There are 30 students in your class. Your science teacher chooses 5 students at random to complete a group project. Find the probability that you and your 2 best friends in the science class are chosen to work in the group. Explain how you found your answer.

79. PROBLEM SOLVING Follow the steps below to explore a famous probability problem called the *birthday problem*. (Assume there are 365 equally likely birthdays possible.)

a. What is the probability that at least 2 people share the same birthday in a group of 6 randomly chosen people? in a group of 10 randomly chosen people?

b. Generalize the results from part (a) by writing a formula for the probability $P(n)$ that at least 2 people in a group of n people share the same birthday. (*Hint*: Use ${}_nP_r$ notation in your formula.)

c. Enter the formula from part (b) into a graphing calculator. Use the *table* feature to make a table of values. For what group size does the probability that at least 2 people share the same birthday first exceed 50%?

Maintaining Mathematical Proficiency

Reviewing what you learned in previous grades and lessons

80. A bag contains 12 white marbles and 3 black marbles. You pick 1 marble at random. What is the probability that you pick a black marble? *(Section 10.1)*

81. The table shows the result of flipping two coins 12 times. For what outcome is the experimental probability the same as the theoretical probability? *(Section 10.1)*

HH	HT	TH	TT
2	6	3	1

10.6 Binomial Distributions

Essential Question How can you determine the frequency of each outcome of an event?

EXPLORATION 1 **Analyzing Histograms**

Work with a partner. The histograms show the results when n coins are flipped.

STUDY TIP

When 4 coins are flipped ($n = 4$), the possible outcomes are

TTTT TTTH TTHT TTHH

THTT THTH THHT THHH

HTTT HTTH HTHT HTHH

HHTT HHTH HHHT HHHH.

The histogram shows the numbers of outcomes having 0, 1, 2, 3, and 4 heads.

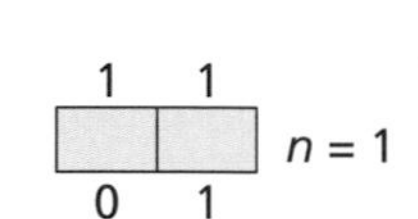

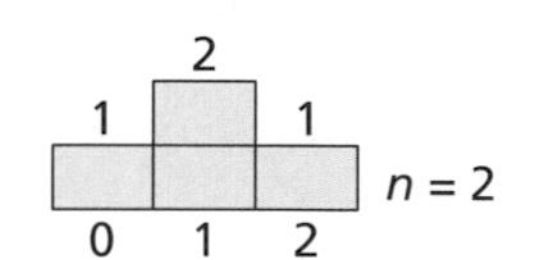

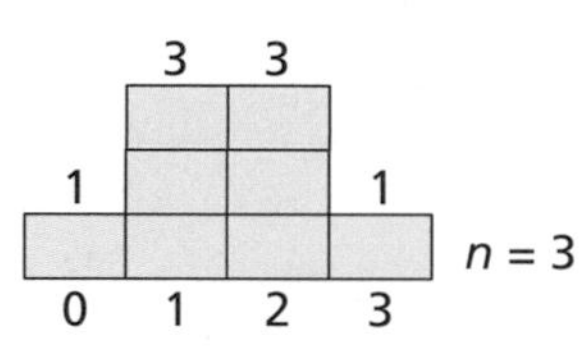

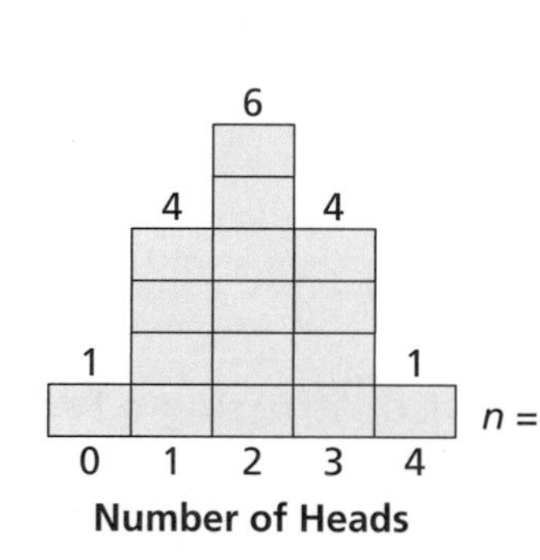

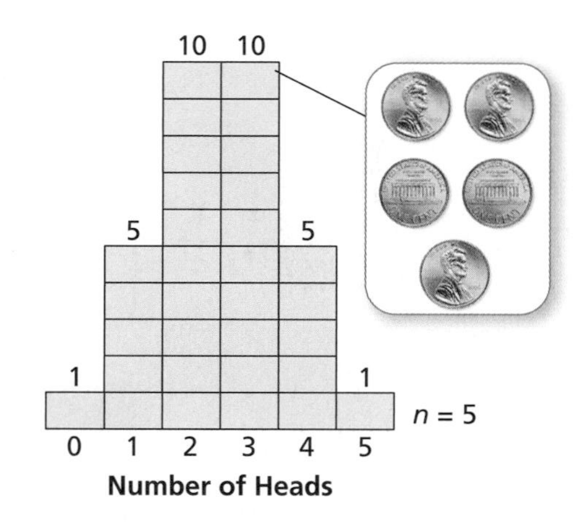

a. In how many ways can 3 heads occur when 5 coins are flipped?

b. Draw a histogram that shows the numbers of heads that can occur when 6 coins are flipped.

c. In how many ways can 3 heads occur when 6 coins are flipped?

EXPLORATION 2 **Determining the Number of Occurrences**

Work with a partner.

LOOKING FOR A PATTERN

To be proficient in math, you need to look closely to discern a pattern or structure.

a. Complete the table showing the numbers of ways in which 2 heads can occur when n coins are flipped.

n	3	4	5	6	7
Occurrences of 2 heads					

b. Determine the pattern shown in the table. Use your result to find the number of ways in which 2 heads can occur when 8 coins are flipped.

Communicate Your Answer

3. How can you determine the frequency of each outcome of an event?

4. How can you use a histogram to find the probability of an event?

10.6 Lesson

Core Vocabulary

random variable, *p. 580*
probability distribution, *p. 580*
binomial distribution, *p. 581*
binomial experiment, *p. 581*

Previous
histogram

What You Will Learn

- Construct and interpret probability distributions.
- Construct and interpret binomial distributions.

Probability Distributions

A **random variable** is a variable whose value is determined by the outcomes of a probability experiment. For example, when you roll a six-sided die, you can define a random variable x that represents the number showing on the die. So, the possible values of x are 1, 2, 3, 4, 5, and 6. For every random variable, a *probability distribution* can be defined.

Core Concept

Probability Distributions

A **probability distribution** is a function that gives the probability of each possible value of a random variable. The sum of all the probabilities in a probability distribution must equal 1.

Probability Distribution for Rolling a Six-Sided Die						
x	1	2	3	4	5	6
$P(x)$	$\frac{1}{6}$	$\frac{1}{6}$	$\frac{1}{6}$	$\frac{1}{6}$	$\frac{1}{6}$	$\frac{1}{6}$

EXAMPLE 1 Constructing a Probability Distribution

Let x be a random variable that represents the sum when two six-sided dice are rolled. Make a table and draw a histogram showing the probability distribution for x.

SOLUTION

Step 1 Make a table. The possible values of x are the integers from 2 to 12. The table shows how many outcomes of rolling two dice produce each value of x. Divide the number of outcomes for x by 36 to find $P(x)$.

STUDY TIP
Recall that there are 36 possible outcomes when rolling two six-sided dice. These are listed in Example 3 on page 540.

x (sum)	2	3	4	5	6	7	8	9	10	11	12
Outcomes	1	2	3	4	5	6	5	4	3	2	1
$P(x)$	$\frac{1}{36}$	$\frac{1}{18}$	$\frac{1}{12}$	$\frac{1}{9}$	$\frac{5}{36}$	$\frac{1}{6}$	$\frac{5}{36}$	$\frac{1}{9}$	$\frac{1}{12}$	$\frac{1}{18}$	$\frac{1}{36}$

Step 2 Draw a histogram where the intervals are given by x and the frequencies are given by $P(x)$.

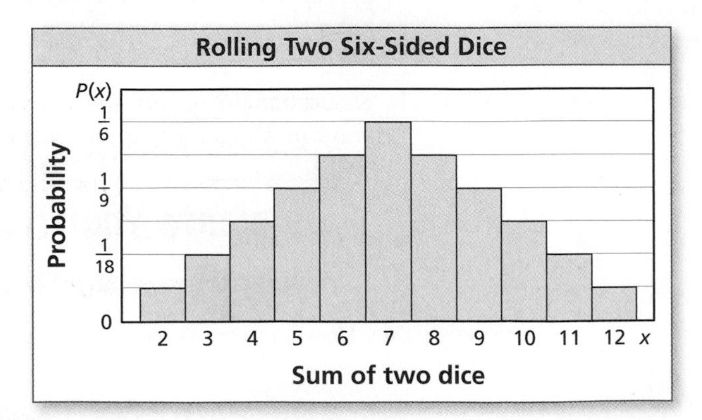

EXAMPLE 2 Interpreting a Probability Distribution

Use the probability distribution in Example 1 to answer each question.

a. What is the most likely sum when rolling two six-sided dice?

b. What is the probability that the sum of the two dice is at least 10?

SOLUTION

a. The most likely sum when rolling two six-sided dice is the value of x for which $P(x)$ is greatest. This probability is greatest for $x = 7$. So, when rolling the two dice, the most likely sum is 7.

b. The probability that the sum of the two dice is at least 10 is

$$\begin{aligned} P(x \geq 10) &= P(x = 10) + P(x = 11) + P(x = 12) \\ &= \frac{3}{36} + \frac{2}{36} + \frac{1}{36} \\ &= \frac{6}{36} \\ &= \frac{1}{6} \\ &\approx 0.167. \end{aligned}$$

▶ The probability is about 16.7%.

Monitoring Progress  Help in English and Spanish at *BigIdeasMath.com*

An octahedral die has eight sides numbered 1 through 8. Let x be a random variable that represents the sum when two such dice are rolled.

1. Make a table and draw a histogram showing the probability distribution for x.

2. What is the most likely sum when rolling the two dice?

3. What is the probability that the sum of the two dice is at most 3?

Binomial Distributions

One type of probability distribution is a **binomial distribution**. A binomial distribution shows the probabilities of the outcomes of a *binomial experiment*.

Core Concept

Binomial Experiments

A **binomial experiment** meets the following conditions.

- There are n independent trials.
- Each trial has only two possible outcomes: success and failure.
- The probability of success is the same for each trial. This probability is denoted by p. The probability of failure is $1 - p$.

For a binomial experiment, the probability of exactly k successes in n trials is

$$P(k \text{ successes}) = {}_nC_k p^k(1 - p)^{n-k}.$$

EXAMPLE 3 Constructing a Binomial Distribution

According to a survey, about 33% of people ages 16 and older in the U.S. own an electronic book reading device, or e-reader. You ask 6 randomly chosen people (ages 16 and older) whether they own an e-reader. Draw a histogram of the binomial distribution for your survey.

ATTENDING TO PRECISION

When probabilities are rounded, the sum of the probabilities may differ slightly from 1.

SOLUTION

The probability that a randomly selected person has an e-reader is $p = 0.33$. Because you survey 6 people, $n = 6$.

$P(k = 0) = {}_6C_0(0.33)^0(0.67)^6 \approx 0.090$

$P(k = 1) = {}_6C_1(0.33)^1(0.67)^5 \approx 0.267$

$P(k = 2) = {}_6C_2(0.33)^2(0.67)^4 \approx 0.329$

$P(k = 3) = {}_6C_3(0.33)^3(0.67)^3 \approx 0.216$

$P(k = 4) = {}_6C_4(0.33)^4(0.67)^2 \approx 0.080$

$P(k = 5) = {}_6C_5(0.33)^5(0.67)^1 \approx 0.016$

$P(k = 6) = {}_6C_6(0.33)^6(0.67)^0 \approx 0.001$

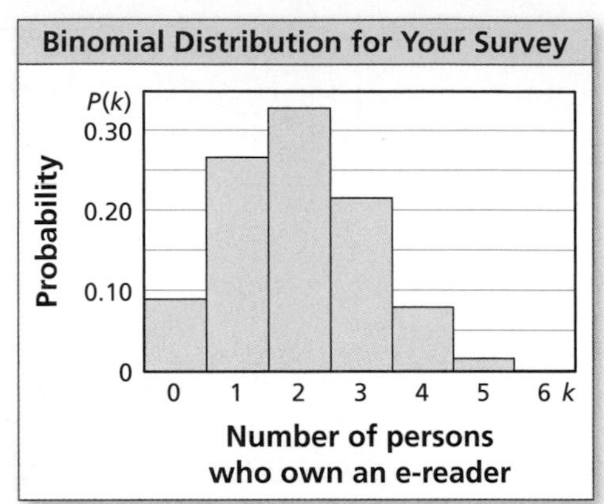

A histogram of the distribution is shown.

EXAMPLE 4 Interpreting a Binomial Distribution

Use the binomial distribution in Example 3 to answer each question.

a. What is the most likely outcome of the survey?

b. What is the probability that at most 2 people have an e-reader?

COMMON ERROR

Because a person may not have an e-reader, be sure you include $P(k = 0)$ when finding the probability that at most 2 people have an e-reader.

SOLUTION

a. The most likely outcome of the survey is the value of k for which $P(k)$ is greatest. This probability is greatest for $k = 2$. The most likely outcome is that 2 of the 6 people own an e-reader.

b. The probability that at most 2 people have an e-reader is

$$P(k \leq 2) = P(k = 0) + P(k = 1) + P(k = 2)$$
$$\approx 0.090 + 0.267 + 0.329$$
$$\approx 0.686.$$

▶ The probability is about 68.6%.

Monitoring Progress

Help in English and Spanish at *BigIdeasMath.com*

According to a survey, about 85% of people ages 18 and older in the U.S. use the Internet or e-mail. You ask 4 randomly chosen people (ages 18 and older) whether they use the Internet or e-mail.

4. Draw a histogram of the binomial distribution for your survey.

5. What is the most likely outcome of your survey?

6. What is the probability that at most 2 people you survey use the Internet or e-mail?

10.6 Exercises

Dynamic Solutions available at *BigIdeasMath.com*

Vocabulary and Core Concept Check

1. **VOCABULARY** What is a random variable?

2. **WRITING** Give an example of a binomial experiment and describe how it meets the conditions of a binomial experiment.

Monitoring Progress and Modeling with Mathematics

In Exercises 3–6, make a table and draw a histogram showing the probability distribution for the random variable. *(See Example 1.)*

3. $x =$ the number on a table tennis ball randomly chosen from a bag that contains 5 balls labeled "1," 3 balls labeled "2," and 2 balls labeled "3."

4. $c = 1$ when a randomly chosen card out of a standard deck of 52 playing cards is a heart and $c = 2$ otherwise.

5. $w = 1$ when a randomly chosen letter from the English alphabet is a vowel and $w = 2$ otherwise.

6. $n =$ the number of digits in a random integer from 0 through 999.

In Exercises 7 and 8, use the probability distribution to determine (a) the number that is most likely to be spun on a spinner, and (b) the probability of spinning an even number. *(See Example 2.)*

7.

Spinner Results

$P(x)$

Probability: 0, $\frac{1}{4}$, $\frac{1}{2}$

Number on spinner: 1, 2, 3, 4, x

8.

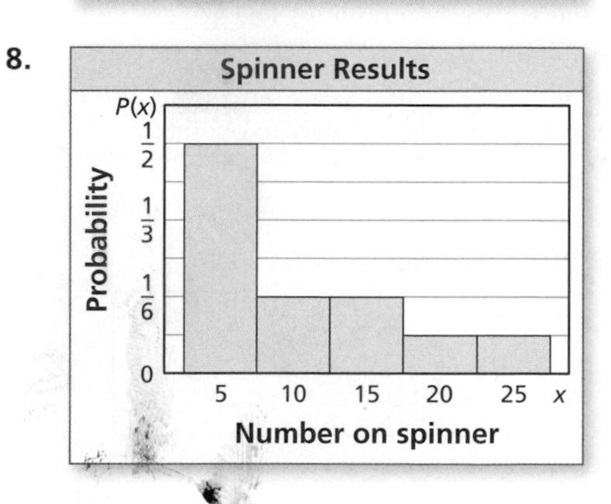

USING EQUATIONS In Exercises 9–12, calculate the probability of flipping a coin 20 times and getting the given number of heads.

9. 1

10. 4

11. 18

12. 20

13. **MODELING WITH MATHEMATICS** According to a survey, 27% of high school students in the United States buy a class ring. You ask 6 randomly chosen high school students whether they own a class ring. *(See Examples 3 and 4.)*

  a. Draw a histogram of the binomial distribution for your survey.

  b. What is the most likely outcome of your survey?

  c. What is the probability that at most 2 people have a class ring?

14. **MODELING WITH MATHEMATICS** According to a survey, 48% of adults in the United States believe that Unidentified Flying Objects (UFOs) are observing our planet. You ask 8 randomly chosen adults whether they believe UFOs are watching Earth.

  a. Draw a histogram of the binomial distribution for your survey.

  b. What is the most likely outcome of your survey?

  c. What is the probability that at most 3 people believe UFOs are watching Earth?

ERROR ANALYSIS In Exercises 15 and 16, describe and correct the error in calculating the probability of rolling a 1 exactly 3 times in 5 rolls of a six-sided die.

15. ✗ $P(k = 3) = {}_5C_3\left(\frac{1}{6}\right)^{5-3}\left(\frac{5}{6}\right)^3$

≈ 0.161

16. ✗ $P(k = 3) = \left(\frac{1}{6}\right)^3\left(\frac{5}{6}\right)^{5-3}$

≈ 0.003

17. MATHEMATICAL CONNECTIONS At most 7 gopher holes appear each week on the farm shown. Let x represent how many of the gopher holes appear in the carrot patch. Assume that a gopher hole has an equal chance of appearing at any point on the farm.

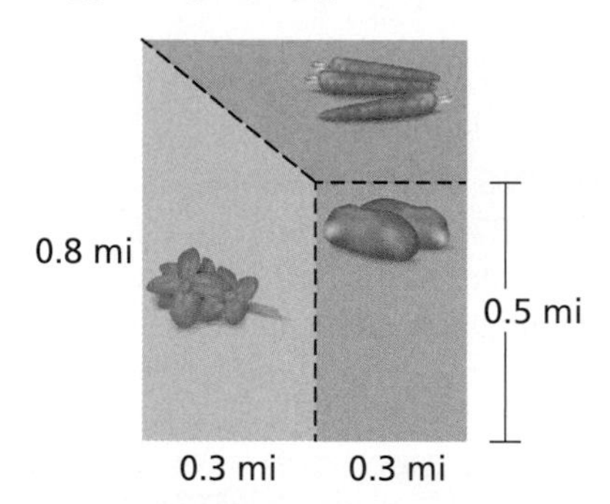

a. Find $P(x)$ for $x = 0, 1, 2, \ldots, 7$.

b. Make a table showing the probability distribution for x.

c. Make a histogram showing the probability distribution for x.

18. HOW DO YOU SEE IT? Complete the probability distribution for the random variable x. What is the probability the value of x is greater than 2?

x	1	2	3	4
$P(x)$	0.1	0.3	0.4	

19. MAKING AN ARGUMENT The binomial distribution shows the results of a binomial experiment. Your friend claims that the probability p of a success must be greater than the probability $1 - p$ of a failure. Is your friend correct? Explain your reasoning.

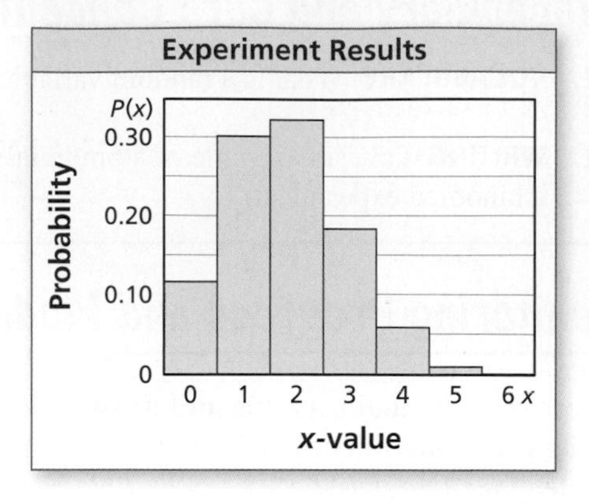

20. THOUGHT PROVOKING There are 100 coins in a bag. Only one of them has a date of 2010. You choose a coin at random, check the date, and then put the coin back in the bag. You repeat this 100 times. Are you certain of choosing the 2010 coin at least once? Explain your reasoning.

21. MODELING WITH MATHEMATICS Assume that having a male and having a female child are independent events, and that the probability of each is 0.5.

a. A couple has 4 male children. Evaluate the validity of this statement: "The first 4 kids were all boys, so the next one will probably be a girl."

b. What is the probability of having 4 male children and then a female child?

c. Let x be a random variable that represents the number of children a couple already has when they have their first female child. Draw a histogram of the distribution of $P(x)$ for $0 \leq x \leq 10$. Describe the shape of the histogram.

22. CRITICAL THINKING An entertainment system has n speakers. Each speaker will function properly with probability p, independent of whether the other speakers are functioning. The system will operate effectively when at least 50% of its speakers are functioning. For what values of p is a 5-speaker system more likely to operate than a 3-speaker system?

Maintaining Mathematical Proficiency

Reviewing what you learned in previous grades and lessons

List the possible outcomes for the situation. *(Section 10.1)*

23. guessing the gender of three children

24. picking one of two doors and one of three curtains

10.4–10.6 What Did You Learn?

Core Vocabulary

compound event, *p. 564*
overlapping events, *p. 564*
disjoint events, *p. 564*
mutually exclusive events, *p. 564*
permutation, *p. 570*
n factorial, *p. 570*
combination, *p. 572*
Binomial Theorem, *p. 574*
random variable, *p. 580*
probability distribution, *p. 580*
binomial distribution, *p. 581*
binomial experiment, *p. 581*

Core Concepts

Section 10.4

Probability of Compound Events, *p. 564*

Section 10.5

Permutations, *p. 571*
Combinations, *p. 572*
The Binomial Theorem, *p. 574*

Section 10.6

Probability Distributions, *p. 580*
Binomial Experiments, *p. 581*

Mathematical Practices

1. How can you use diagrams to understand the situation in Exercise 22 on page 568?
2. Describe a relationship between the results in part (a) and part (b) in Exercise 74 on page 578.
3. Explain how you were able to break the situation into cases to evaluate the validity of the statement in part (a) of Exercise 21 on page 584.

Performance Task

A New Dartboard

You are a graphic artist working for a company on a new design for the board in the game of darts. You are eager to begin the project, but the team cannot decide on the terms of the game. Everyone agrees that the board should have four colors. But some want the probabilities of hitting each color to be equal, while others want them to be different. You offer to design two boards, one for each group. How do you get started? How creative can you be with your designs?

To explore the answers to these questions and more, go to ***BigIdeasMath.com***.

10 Chapter Review

Dynamic Solutions available at *BigIdeasMath.com*

10.1 Sample Spaces and Probability *(pp. 537–544)*

Each section of the spinner shown has the same area. The spinner was spun 30 times. The table shows the results. For which color is the experimental probability of stopping on the color the same as the theoretical probability?

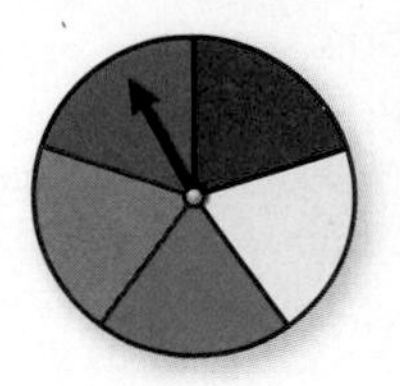

Spinner Results	
green	4
orange	6
red	9
blue	8
yellow	3

SOLUTION

The theoretical probability of stopping on each of the five colors is $\frac{1}{5}$. Use the outcomes in the table to find the experimental probabilities.

$P(\text{green}) = \frac{4}{30} = \frac{2}{15}$ $\quad P(\text{orange}) = \frac{6}{30} = \frac{1}{5}$ $\quad P(\text{red}) = \frac{9}{30} = \frac{3}{10}$ $\quad P(\text{blue}) = \frac{8}{30} = \frac{4}{15}$ $\quad P(\text{yellow}) = \frac{3}{30} = \frac{1}{10}$

▶ The experimental probability of stopping on orange is the same as the theoretical probability.

1. A bag contains 9 tiles, one for each letter in the word HAPPINESS. You choose a tile at random. What is the probability that you choose a tile with the letter S? What is the probability that you choose a tile with a letter other than P?
2. You throw a dart at the board shown. Your dart is equally likely to hit any point inside the square board. Are you most likely to get 5 points, 10 points, or 20 points?

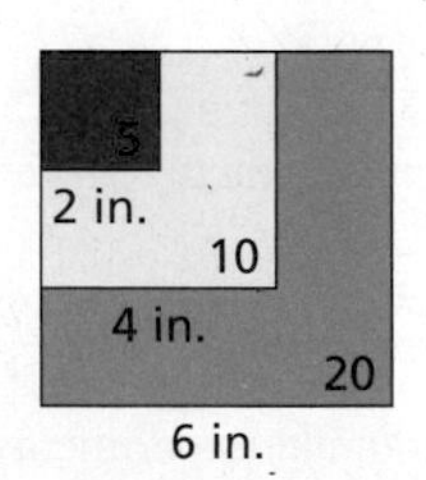

10.2 Independent and Dependent Events *(pp. 545–552)*

You randomly select 2 cards from a standard deck of 52 playing cards. What is the probability that both cards are jacks when (a) you replace the first card before selecting the second, and (b) you do not replace the first card. Compare the probabilities.

SOLUTION

Let event A be "first card is a jack" and event B be "second card is a jack."

a. Because you replace the first card before you select the second card, the events are independent. So, the probability is

$$P(A \text{ and } B) = P(A) \cdot P(B) = \frac{4}{52} \cdot \frac{4}{52} = \frac{16}{2704} = \frac{1}{169} \approx 0.006.$$

b. Because you do not replace the first card before you select the second card, the events are dependent. So, the probability is

$$P(A \text{ and } B) = P(A) \cdot P(B|A) = \frac{4}{52} \cdot \frac{3}{51} = \frac{12}{2652} = \frac{1}{221} \approx 0.005.$$

▶ So, you are $\frac{1}{169} \div \frac{1}{221} \approx 1.3$ times more likely to select 2 jacks when you replace the first card before you select the second card.

Find the probability of randomly selecting the given marbles from a bag of 5 red, 8 green, and 3 blue marbles when (a) you replace the first marble before drawing the second, and (b) you do not replace the first marble. Compare the probabilities.

3. red, then green
4. blue, then red
5. green, then green

10.3 Two-Way Tables and Probability (pp. 553–560)

A survey asks residents of the east and west sides of a city whether they support the construction of a bridge. The results, given as joint relative frequencies, are shown in the two-way table. What is the probability that a randomly selected resident from the east side will support the project?

		Location	
		East Side	West Side
Response	Yes	0.47	0.36
	No	0.08	0.09

SOLUTION

Find the joint and marginal relative frequencies. Then use these values to find the conditional probability.

$$P(\text{yes} \mid \text{east side}) = \frac{P(\text{east side and yes})}{P(\text{east side})} = \frac{0.47}{0.47 + 0.08} \approx 0.855$$

▶ So, the probability that a resident of the east side of the city will support the project is about 85.5%.

6. What is the probability that a randomly selected resident who does not support the project in the example above is from the west side?

7. After a conference, 220 men and 270 women respond to a survey. Of those, 200 men and 230 women say the conference was impactful. Organize these results in a two-way table. Then find and interpret the marginal frequencies.

10.4 Probability of Disjoint and Overlapping Events (pp. 563–568)

Let A and B be events such that $P(A) = \frac{2}{3}$, $P(B) = \frac{1}{2}$, and $P(A \text{ and } B) = \frac{1}{3}$. Find $P(A \text{ or } B)$.

SOLUTION

$P(A \text{ or } B) = P(A) + P(B) - P(A \text{ and } B)$	Write general formula.
$= \frac{2}{3} + \frac{1}{2} - \frac{1}{3}$	Substitute known probabilities.
$= \frac{5}{6}$	Simplify.
≈ 0.833	Use a calculator.

8. Let A and B be events such that $P(A) = 0.32$, $P(B) = 0.48$, and $P(A \text{ and } B) = 0.12$. Find $P(A \text{ or } B)$.

9. Out of 100 employees at a company, 92 employees either work part time or work 5 days each week. There are 14 employees who work part time and 80 employees who work 5 days each week. What is the probability that a randomly selected employee works both part time and 5 days each week?

10.5 Permutations and Combinations (pp. 569–578)

A 5-digit code consists of 5 different integers from 0 to 9. How many different codes are possible?

SOLUTION

To find the number of permutations of 5 integers chosen from 10, find ${}_{10}P_5$.

$${}_{10}P_5 = \frac{10!}{(10-5)!}$$ Permutations formula

$$= \frac{10!}{5!}$$ Subtract.

$$= \frac{10 \cdot 9 \cdot 8 \cdot 7 \cdot 6 \cdot \cancel{5!}}{\cancel{5!}}$$ Expand factorials. Divide out common factor, 5!.

$$= 30{,}240$$ Simplify.

▶ There are 30,240 possible codes.

Evaluate the expression.

10. ${}_7P_6$ **11.** ${}_{13}P_{10}$ **12.** ${}_6C_2$ **13.** ${}_8C_4$

14. Use the Binomial Theorem to write the expansion of $(2x + y^2)^4$.

15. A random drawing will determine which 3 people in a group of 9 will win concert tickets. What is the probability that you and your 2 friends will win the tickets?

10.6 Binomial Distributions (pp. 579–584)

According to a survey, about 21% of adults in the U.S. visited an art museum last year. You ask 4 randomly chosen adults whether they visited an art museum last year. Draw a histogram of the binomial distribution for your survey.

SOLUTION

The probability that a randomly selected person visited an art museum is $p = 0.21$. Because you survey 4 people, $n = 4$.

$$P(k = 0) = {}_4C_0(0.21)^0(0.79)^4 \approx 0.390$$

$$P(k = 1) = {}_4C_1(0.21)^1(0.79)^3 \approx 0.414$$

$$P(k = 2) = {}_4C_2(0.21)^2(0.79)^2 \approx 0.165$$

$$P(k = 3) = {}_4C_3(0.21)^3(0.79)^1 \approx 0.029$$

$$P(k = 4) = {}_4C_4(0.21)^4(0.79)^0 \approx 0.002$$

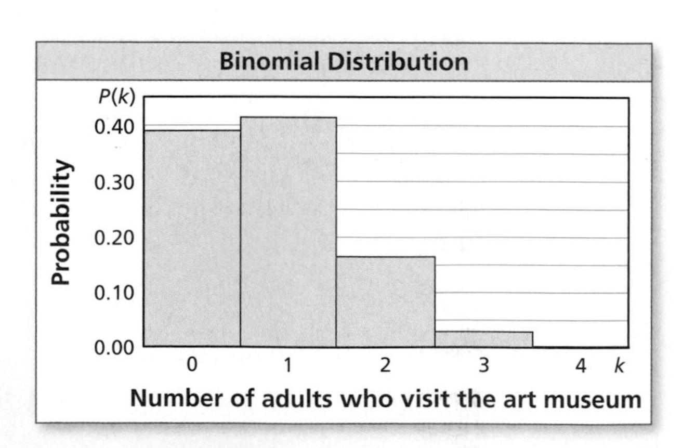

16. Find the probability of flipping a coin 12 times and getting exactly 4 heads.

17. A basketball player makes a free throw 82.6% of the time. The player attempts 5 free throws. Draw a histogram of the binomial distribution of the number of successful free throws. What is the most likely outcome?

10 Chapter Test

You roll a six-sided die. Find the probability of the event described. Explain your reasoning.

1. You roll a number less than 5.

2. You roll a multiple of 3.

Evaluate the expression.

3. ${}_7P_2$

4. ${}_8P_3$

5. ${}_6C_3$

6. ${}_{12}C_7$

7. Use the Binomial Theorem to write the binomial expansion of $(x + y^2)^5$.

8. You find the probability $P(A \text{ or } B)$ by using the equation $P(A \text{ or } B) = P(A) + P(B) - P(A \text{ and } B)$. Describe why it is necessary to subtract $P(A \text{ and } B)$ when the events A and B are overlapping. Then describe why it is *not* necessary to subtract $P(A \text{ and } B)$ when the events A and B are disjoint.

9. Is it possible to use the formula $P(A \text{ and } B) = P(A) \cdot P(B \mid A)$ when events A and B are independent? Explain your reasoning.

10. According to a survey, about 58% of families sit down for a family dinner at least four times per week. You ask 5 randomly chosen families whether they have a family dinner at least four times per week.

a. Draw a histogram of the binomial distribution for the survey.

b. What is the most likely outcome of the survey?

c. What is the probability that at least 3 families have a family dinner four times per week?

11. You are choosing a cell phone company to sign with for the next 2 years. The three plans you consider are equally priced. You ask several of your neighbors whether they are satisfied with their current cell phone company. The table shows the results. According to this survey, which company should you choose?

	Satisfied	Not Satisfied
Company A	IIII	II
Company B	IIII	III
Company C	~~IIII~~ I	~~IIII~~

12. The surface area of Earth is about 196.9 million square miles. The land area is about 57.5 million square miles and the rest is water. What is the probability that a meteorite that reaches the surface of Earth will hit land? What is the probability that it will hit water?

13. Consider a bag that contains all the chess pieces in a set, as shown in the diagram.

	King	Queen	Bishop	Rook	Knight	Pawn	
Black	1	1	2	2	2	8	
White	1	1	2	2	2	8	

a. You choose one piece at random. Find the probability that you choose a black piece or a queen.

b. You choose one piece at random, do not replace it, then choose a second piece at random. Find the probability that you choose a king, then a pawn.

14. Three volunteers are chosen at random from a group of 12 to help at a summer camp.

a. What is the probability that you, your brother, and your friend are chosen?

b. The first person chosen will be a counselor, the second will be a lifeguard, and the third will be a cook. What is the probability that you are the cook, your brother is the lifeguard, and your friend is the counselor?

10 Cumulative Assessment

1. According to a survey, 63% of Americans consider themselves sports fans. You randomly select 14 Americans to survey.

   a. Draw a histogram of the binomial distribution of your survey.

   b. What is the most likely number of Americans who consider themselves sports fans?

   c. What is the probability at least 7 Americans consider themselves sports fans?

2. Order the acute angles from smallest to largest. Explain your reasoning.

$\tan \theta_1 = 1$	$\tan \theta_2 = \frac{1}{2}$
$\tan \theta_3 = \frac{\sqrt{3}}{3}$	$\tan \theta_4 = \frac{23}{4}$
$\tan \theta_5 = \frac{38}{5}$	$\tan \theta_6 = \sqrt{3}$

3. You order a fruit smoothie made with 2 liquid ingredients and 3 fruit ingredients from the menu shown. How many different fruit smoothies can you order?

Liquids
• Water • Tea
• Coconut Water
• Almond Milk
• Apple Juice
• Orange Juice
Fruits
• Orange
• Banana
• Pineapple
• Cantaloupe
• Strawberry
• Watermelon
• Kiwi
• Peach
• Blueberry
• Pomegranate

4. Which statements describe the transformation of the graph of $f(x) = x^3 - x$ represented by $g(x) = 4(x - 2)^3 - 4(x - 2)$?

   Ⓐ a vertical stretch by a factor of 4

   Ⓑ a vertical shrink by a factor of $\frac{1}{4}$

   Ⓒ a horizontal shrink by a factor of $\frac{1}{4}$

   Ⓓ a horizontal stretch by a factor of 4

   Ⓔ a horizontal translation 2 units to the right

   Ⓕ a horizontal translation 2 units to the left

5. Use the diagram to explain why the equation is true.

$$P(A) + P(B) = P(A \text{ or } B) + P(A \text{ and } B)$$

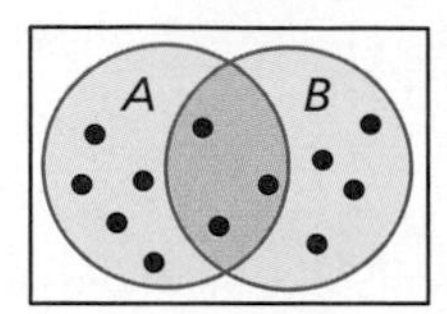

6. For the sequence $-\frac{1}{2}, -\frac{1}{4}, -\frac{1}{6}, -\frac{1}{8}, \ldots$, describe the pattern, write the next term, graph the first five terms, and write a rule for the nth term.

7. A survey asked male and female students about whether they prefer to take gym class or choir. The table shows the results of the survey.

		Class		
		Gym	Choir	Total
Gender	Male			50
	Female	23		
	Total		49	106

a. Complete the two-way table.

b. What is the probability that a randomly selected student is female and prefers choir?

c. What is the probability that a randomly selected male student prefers gym class?

8. The owner of a lawn-mowing business has three mowers. As long as one of the mowers is working, the owner can stay productive. One of the mowers is unusable 10% of the time, one is unusable 8% of the time, and one is unusable 18% of the time.

a. Find the probability that all three mowers are unusable on a given day.

b. Find the probability that at least one of the mowers is unusable on a given day.

c. Suppose the least-reliable mower stops working completely. How does this affect the probability that the lawn-mowing business can be productive on a given day?

9. Write a system of quadratic inequalities whose solution is represented in the graph.

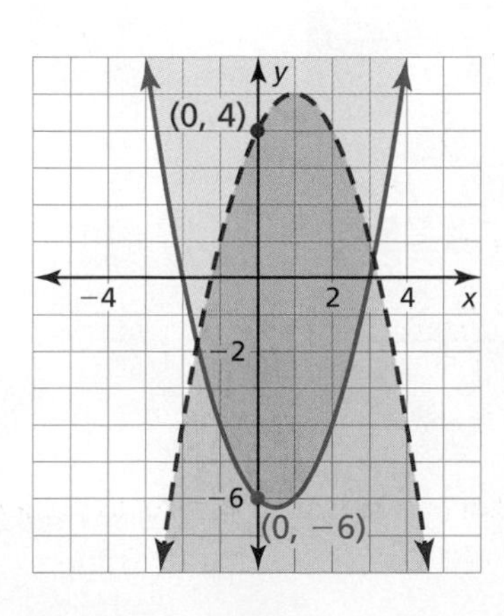

11 Data Analysis and Statistics

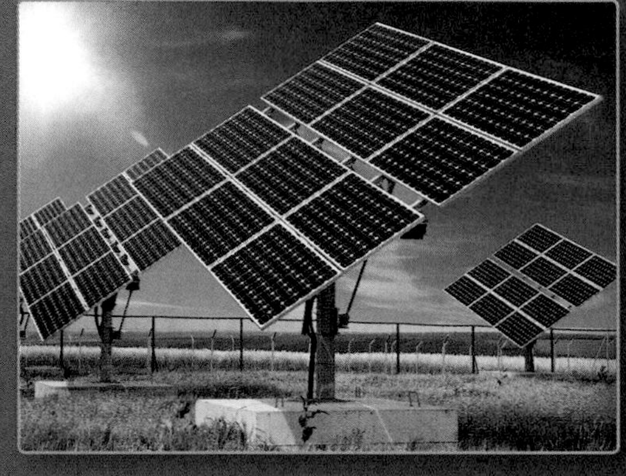

Solar Power *(p. 631)*

Reading *(p. 624)*

Volcano Damage *(p. 615)*

SAT Scores *(p. 605)*

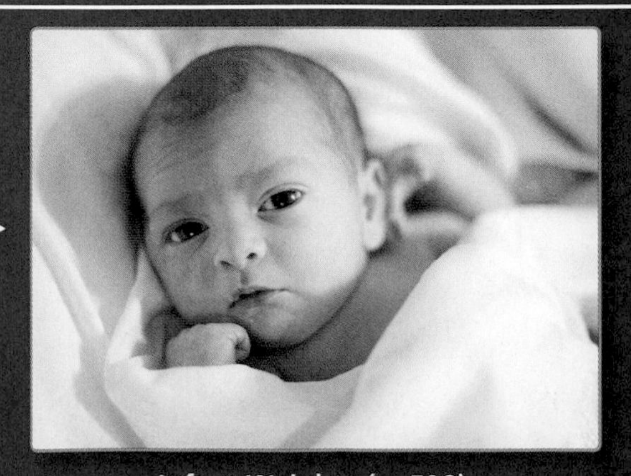

Infant Weights *(p. 598)*

Maintaining Mathematical Proficiency

Comparing Measures of Center

Example 1 **Find the mean, median, and mode of the data set 4, 11, 16, 8, 9, 40, 4, 12, 13, 5, and 10. Then determine which measure of center best represents the data. Explain.**

Mean $\bar{x} = \dfrac{4 + 11 + 16 + 8 + 9 + 40 + 4 + 12 + 13 + 5 + 10}{11} = 12$

Median 4, 4, 5, 8, 9, 10, 11, 12, 13, 16, 40 Order the data. The middle value is 10.

Mode 4, 4, 5, 8, 9, 10, 11, 12, 13, 16, 40 4 occurs most often.

▶ The mean is 12, the median is 10, and the mode is 4. The median best represents the data. The mode is less than most of the data, and the mean is greater than most of the data.

Find the mean, median, and mode of the data set. Then determine which measure of center best represents the data. Explain.

1. 36, 82, 94, 83, 86, 82 **2.** 74, 89, 71, 70, 68, 70 **3.** 1, 18, 12, 16, 11, 15, 17, 44, 44

Finding a Standard Deviation

Example 2 **Find and interpret the standard deviation of the data set 10, 2, 6, 8, 12, 15, 18, and 25. Use a table to organize your work.**

x	$\bar{x}$	$x - \bar{x}$	$(x - \bar{x})^2$
10	12	−2	4
2	12	−10	100
6	12	−6	36
8	12	−4	16
12	12	0	0
15	12	3	9
18	12	6	36
25	12	13	169

Step 1 Find the mean, $\bar{x}$.

$$\bar{x} = \frac{96}{8} = 12$$

Step 2 Find the deviation of each data value, $x - \bar{x}$, as shown in the table.

Step 3 Square each deviation, $(x - \bar{x})^2$, as shown in the table.

Step 4 Find the mean of the squared deviations.

$$\frac{(x_1 - \bar{x})^2 + (x_2 - \bar{x})^2 + \cdots + (x_n - \bar{x})^2}{n} = \frac{4 + 100 + \cdots + 169}{8} = \frac{370}{8} = 46.25$$

Step 5 Use a calculator to take the square root of the mean of the squared deviations.

$$\sqrt{\frac{(x_1 - \bar{x})^2 + (x_2 - \bar{x})^2 + \cdots + (x_n - \bar{x})^2}{n}} = \sqrt{\frac{370}{8}} = \sqrt{46.25} \approx 6.80$$

▶ The standard deviation is about 6.80. This means that the typical data value differs from the mean by about 6.80 units.

Find and interpret the standard deviation of the data set.

4. 43, 48, 41, 51, 42 **5.** 28, 26, 21, 44, 29, 32 **6.** 65, 56, 49, 66, 62, 52, 53, 49

7. **ABSTRACT REASONING** Describe a data set that has a standard deviation of zero. Can a standard deviation be negative? Explain your reasoning.

Mathematical Practices

Mathematically proficient students use diagrams and graphs to show relationships between data. They also analyze data to draw conclusions.

Modeling with Mathematics

Information Design

Information design is the designing of data and information so it can be understood and used. Throughout this book, you have seen several types of information design. In the modern study of statistics, many types of designs require technology to analyze the data and organize the graphical design.

EXAMPLE 1 **Comparing Age Pyramids**

You can use an *age pyramid* to compare the ages of males and females in the population of a country. Compare the mean, median, and mode of each age pyramid.

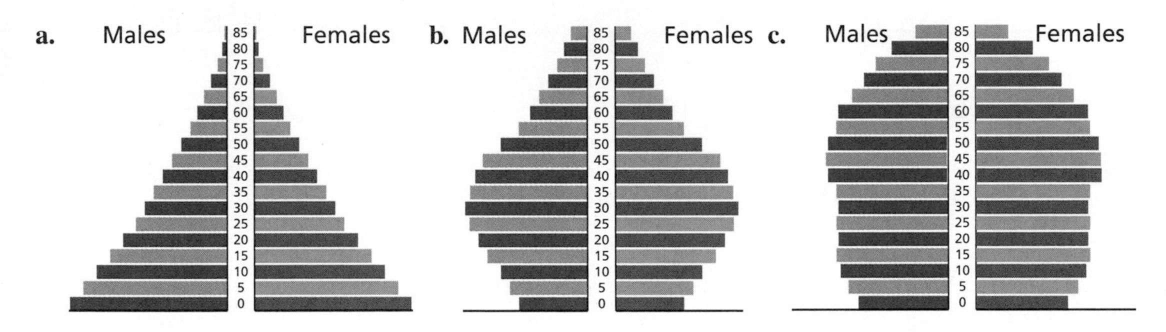

SOLUTION

a. The relative frequency of each successive age group (from 0–4 to 85+) is less than the preceding age group. The mean is roughly 25 years, the median is roughly 20 years, and the mode is the youngest age group, 0–4 years.

b. The mean, median, and mode are all roughly 32 years.

c. The mean, median, and mode are all roughly middle age, around 40 or 45 years.

Monitoring Progress

Use the Internet or some other reference to determine which age pyramid is that of Canada, Japan, and Mexico. Compare the mean, median, and mode of the three age pyramids.

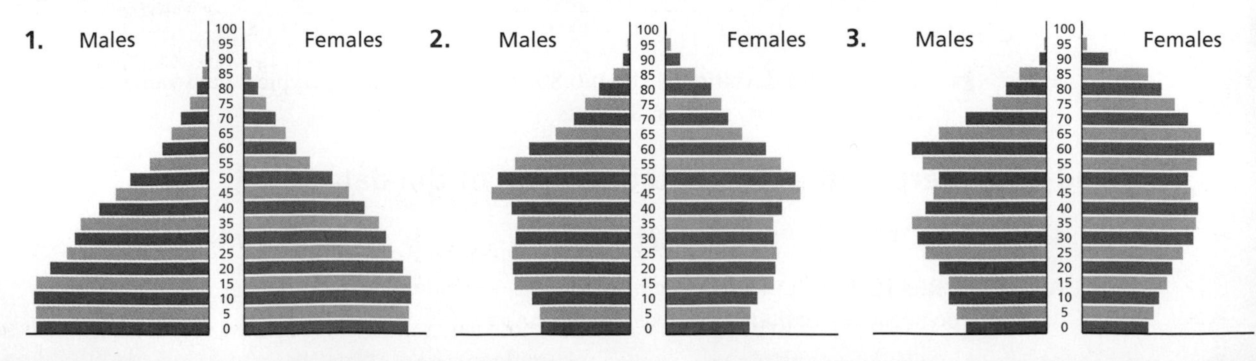

11.1 Using Normal Distributions

Essential Question

In a normal distribution, about what percent of the data lies within one, two, and three standard deviations of the mean?

Recall that the standard deviation σ of a numerical data set is given by

$$\sigma = \sqrt{\frac{(x_1 - \mu)^2 + (x_2 - \mu)^2 + \cdots + (x_n - \mu)^2}{n}}$$

where n is the number of values in the data set and μ is the mean of the data set.

EXPLORATION 1 Analyzing a Normal Distribution

Work with a partner. In many naturally occurring data sets, the histogram of the data is bell-shaped. In statistics, such data sets are said to have a *normal distribution*. For the normal distribution shown below, estimate the percent of the data that lies within one, two, and three standard deviations of the mean. Each square on the grid represents 1%.

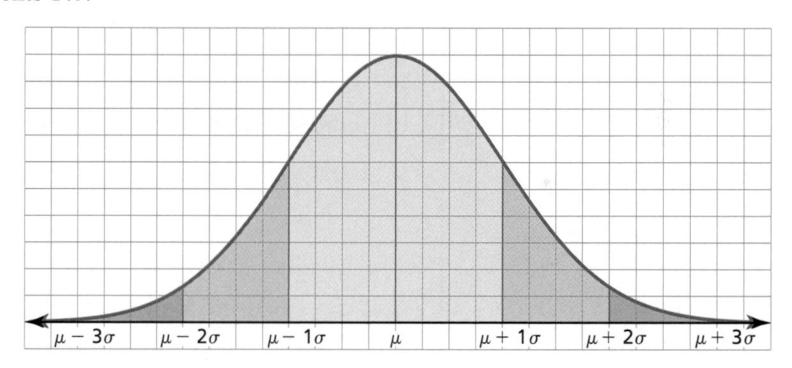

EXPLORATION 2 Analyzing a Data Set

Work with a partner. A famous data set was collected in Scotland in the mid-1800s. It contains the chest sizes (in inches) of 5738 men in the Scottish Militia. Do the data fit a normal distribution? Explain.

MODELING WITH MATHEMATICS

To be proficient in math, you need to analyze relationships mathematically to draw conclusions.

Chest size	Number of men
33	3
34	18
35	81
36	185
37	420
38	749
39	1073
40	1079
41	934
42	658
43	370
44	92
45	50
46	21
47	4
48	1

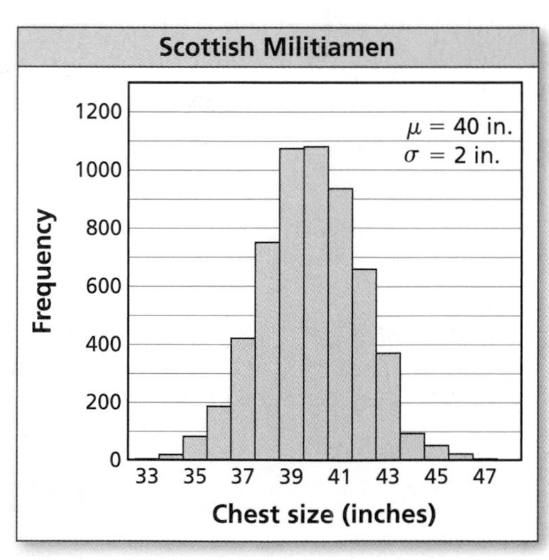

Communicate Your Answer

3. In a normal distribution, about what percent of the data lies within one, two, and three standard deviations of the mean?

4. Use the Internet or some other reference to find another data set that is normally distributed. Display your data in a histogram.

11.1 Lesson

Core Vocabulary

normal distribution, *p. 596*
normal curve, *p. 596*
standard normal distribution, *p. 597*
z-score, *p. 597*

Previous
probability distribution
symmetric
mean
standard deviation
skewed
median

What You Will Learn

- Calculate probabilities using normal distributions.
- Use *z*-scores and the standard normal table to find probabilities.
- Recognize data sets that are normal.

Normal Distributions

You have studied probability distributions. One type of probability distribution is a *normal distribution*. The graph of a **normal distribution** is a bell-shaped curve called a **normal curve** that is symmetric about the mean.

Core Concept

Areas Under a Normal Curve

A normal distribution with mean μ (the Greek letter *mu*) and standard deviation σ (the Greek letter *sigma*) has these properties.

- The total area under the related normal curve is 1.
- About 68% of the area lies within 1 standard deviation of the mean.
- About 95% of the area lies within 2 standard deviations of the mean.
- About 99.7% of the area lies within 3 standard deviations of the mean.

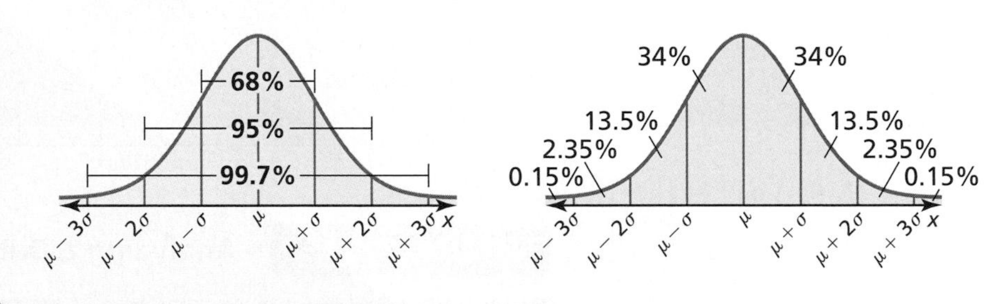

USING A GRAPHING CALCULATOR

A graphing calculator can be used to find areas under normal curves. For example, the normal distribution shown below has mean 0 and standard deviation 1. The graphing calculator screen shows that the area within 1 standard deviation of the mean is about 0.68, or 68%.

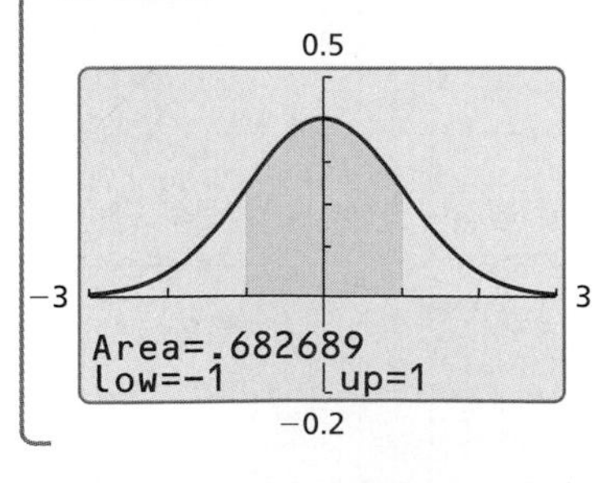

From the second bulleted statement above and the symmetry of a normal curve, you can deduce that 34% of the area lies within 1 standard deviation to the left of the mean, and 34% of the area lies within 1 standard deviation to the right of the mean. The second diagram above shows other partial areas based on the properties of a normal curve.

The areas under a normal curve can be interpreted as probabilities in a normal distribution. So, in a normal distribution, the probability that a randomly chosen *x*-value is between *a* and *b* is given by the area under the normal curve between *a* and *b*.

EXAMPLE 1 Finding a Normal Probability

A normal distribution has mean μ and standard deviation σ. An *x*-value is randomly selected from the distribution. Find $P(\mu - 2\sigma \leq x \leq \mu)$.

SOLUTION

The probability that a randomly selected *x*-value lies between $\mu - 2\sigma$ and μ is the shaded area under the normal curve shown.

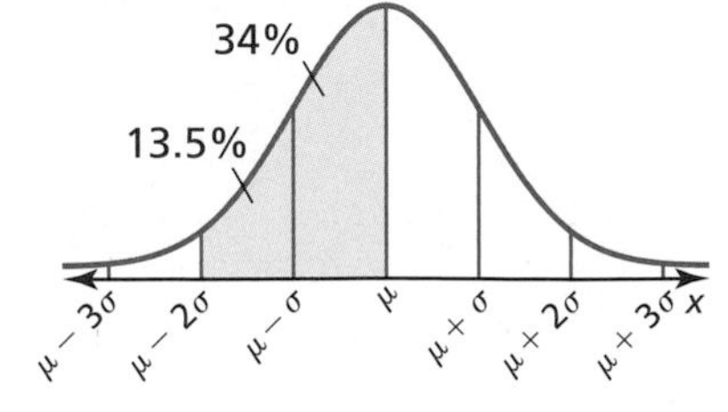

$P(\mu - 2\sigma \leq x \leq \mu) = 0.135 + 0.34 = 0.475$

EXAMPLE 2 Interpreting Normally Distributed Data

The scores for a state's peace officer standards and training test are normally distributed with a mean of 55 and a standard deviation of 12. The test scores range from 0 to 100.

a. About what percent of the people taking the test have scores between 43 and 67?

b. An agency in the state will only hire applicants with test scores of 67 or greater. About what percent of the people have test scores that make them eligible to be hired by the agency?

SOLUTION

a. The scores of 43 and 67 represent one standard deviation on either side of the mean, as shown. So, about 68% of the people taking the test have scores between 43 and 67.

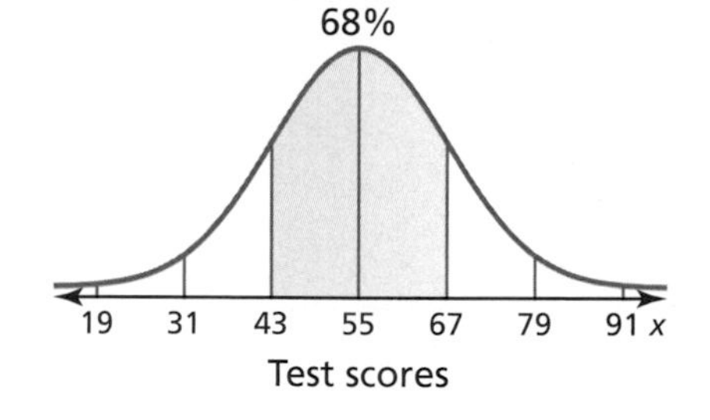

b. A score of 67 is one standard deviation to the right of the mean, as shown. So, the percent of the people who have test scores that make them eligible to be hired by the agency is about 13.5% + 2.35% + 0.15%, or 16%.

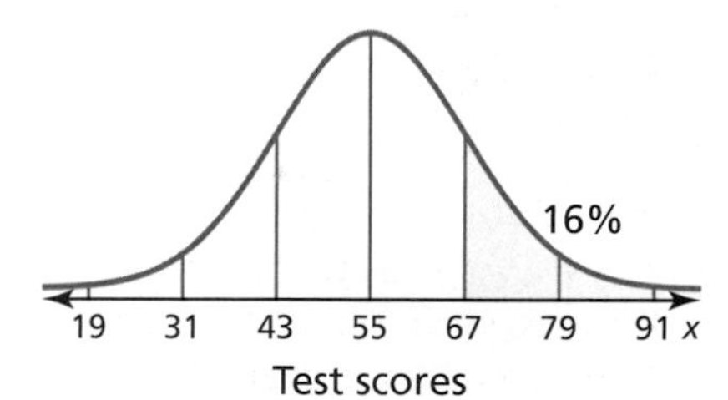

Check

a.

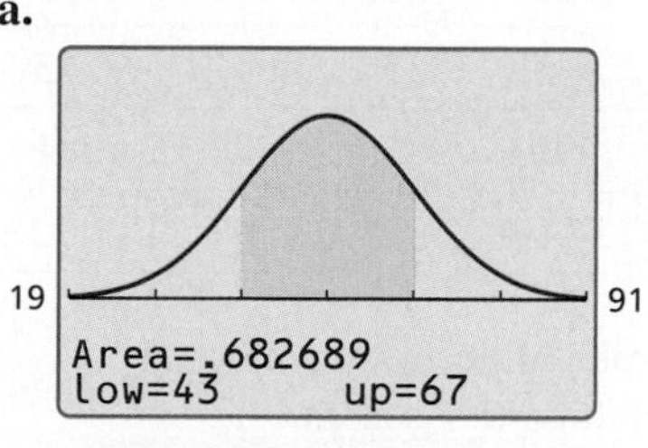

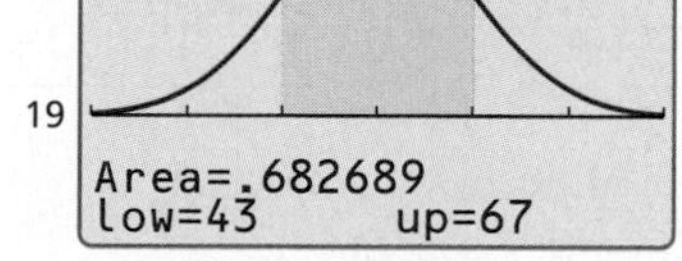

b.

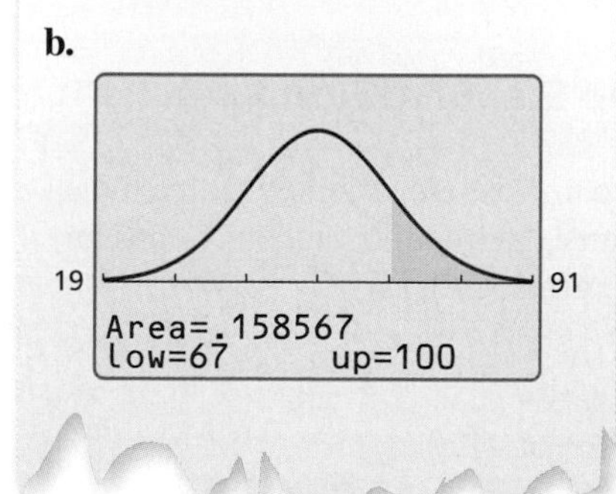

Monitoring Progress  Help in English and Spanish at *BigIdeasMath.com*

A normal distribution has mean μ and standard deviation σ. Find the indicated probability for a randomly selected x-value from the distribution.

1. $P(x \leq \mu)$
2. $P(x \geq \mu)$
3. $P(\mu \leq x \leq \mu + 2\sigma)$
4. $P(\mu - \sigma \leq x \leq \mu)$
5. $P(x \leq \mu - 3\sigma)$
6. $P(x \geq \mu + \sigma)$
7. **WHAT IF?** In Example 2, about what percent of the people taking the test have scores between 43 and 79?

The Standard Normal Distribution

The **standard normal distribution** is the normal distribution with mean 0 and standard deviation 1. The formula below can be used to transform x-values from a normal distribution with mean μ and standard deviation σ into z-values having a standard normal distribution.

Formula $z = \dfrac{x - \mu}{\sigma}$ ← Subtract the mean from the given x-value, then divide by the standard deviation.

$z = -3$ $z = -2$ $z = -1$ $z = 0$ $z = 1$ $z = 2$ $z = 3$ x

The z-value for a particular x-value is called the **z-score** for the x-value and is the number of standard deviations the x-value lies above or below the mean μ.

READING

In the table, the value .0000+ means "slightly more than 0" and the value 1.0000− means "slightly less than 1."

For a randomly selected z-value from a standard normal distribution, you can use the table below to find the probability that z is less than or equal to a given value. For example, the table shows that $P(z \leq -0.4) = 0.3446$. You can find the value of $P(z \leq -0.4)$ in the table by finding the value where row −0 and column .4 intersect.

Standard Normal Table

z	.0	.1	.2	.3	.4	.5	.6	.7	.8	.9
−3	.0013	.0010	.0007	.0005	.0003	.0002	.0002	.0001	.0001	.0000+
−2	.0228	.0179	.0139	.0107	.0082	.0062	.0047	.0035	.0026	.0019
−1	.1587	.1357	.1151	.0968	.0808	.0668	.0548	.0446	.0359	.0287
−0	.5000	.4602	.4207	.3821	.3446	.3085	.2743	.2420	.2119	.1841
0	.5000	.5398	.5793	.6179	.6554	.6915	.7257	.7580	.7881	.8159
1	.8413	.8643	.8849	.9032	.9192	.9332	.9452	.9554	.9641	.9713
2	.9772	.9821	.9861	.9893	.9918	.9938	.9953	.9965	.9974	.9981
3	.9987	.9990	.9993	.9995	.9997	.9998	.9998	.9999	.9999	1.0000−

You can also use the standard normal table to find probabilities for any normal distribution by first converting values from the distribution to z-scores.

EXAMPLE 3 Using a z-Score and the Standard Normal Table

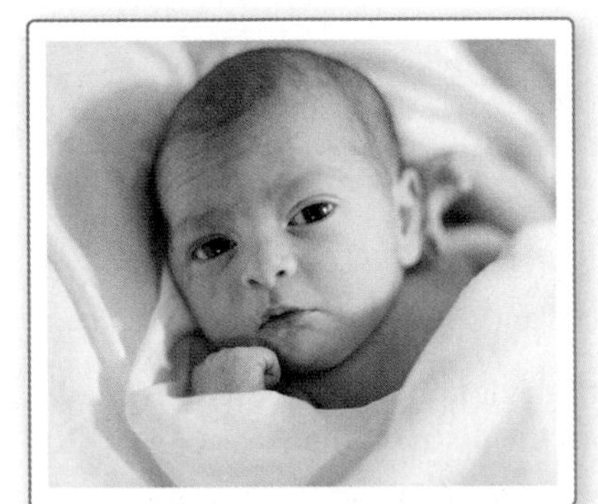

A study finds that the weights of infants at birth are normally distributed with a mean of 3270 grams and a standard deviation of 600 grams. An infant is randomly chosen. What is the probability that the infant weighs 4170 grams or less?

SOLUTION

Step 1 Find the z-score corresponding to an x-value of 4170.

$$z = \frac{x - \mu}{\sigma} = \frac{4170 - 3270}{600} = 1.5$$

Step 2 Use the table to find $P(z \leq 1.5)$. The table shows that $P(z \leq 1.5) = 0.9332$.

Standard Normal Table

z	.0	.1	.2	.3	.4	.5	.6	.7	.8	.9
−3	.0013	.0010	.0007	.0005	.0003	.0002	.0002	.0001	.0001	.0000+
−2	.0228	.0179	.0139	.0107	.0082	.0062	.0047	.0035	.0026	.0019
−1	.1587	.1357	.1151	.0968	.0808	.0668	.0548	.0446	.0359	.0287
−0	.5000	.4602	.4207	.3821	.3446	.3085	.2743	.2420	.2119	.1841
0	.5000	.5398	.5793	.6179	.6554	.6915	.7257	.7580	.7881	.8159
1	.8413	.8643	.8849	.9032	.9192	.9332	.9452	.9554	.9641	.9713

STUDY TIP

When n% of the data are less than or equal to a certain value, that value is called the nth *percentile*. In Example 3, a weight of 4170 grams is the 93rd percentile.

▶ So, the probability that the infant weighs 4170 grams or less is about 0.9332.

Monitoring Progress  Help in English and Spanish at *BigIdeasMath.com*

8. **WHAT IF?** In Example 3, what is the probability that the infant weighs 3990 grams or more?
9. Explain why it makes sense that $P(z \leq 0) = 0.5$.

Recognizing Normal Distributions

Not all distributions are normal. For instance, consider the histograms shown below. The first histogram has a normal distribution. Notice that it is bell-shaped and symmetric. Recall that a distribution is symmetric when you can draw a vertical line that divides the histogram into two parts that are mirror images. Some distributions are skewed. The second histogram is *skewed left* and the third histogram is *skewed right*. The second and third histograms do *not* have normal distributions.

UNDERSTANDING MATHEMATICAL TERMS

Be sure you understand that you cannot use a normal distribution to interpret skewed distributions. The areas under a normal curve do not correspond to the areas of a skewed distribution.

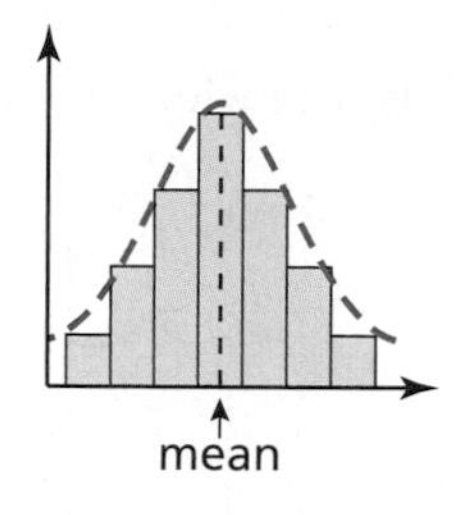

Bell-shaped and symmetric
- histogram has a normal distribution
- mean = median

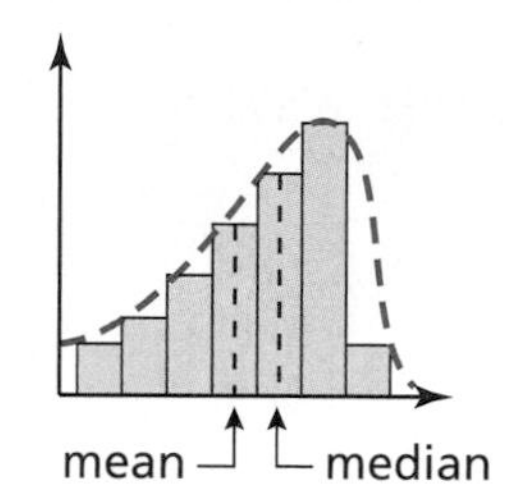

Skewed left
- histogram does not have a normal distribution
- mean < median

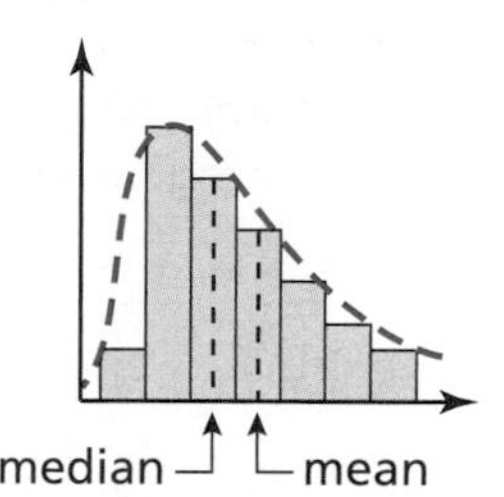

Skewed right
- histogram does not have a normal distribution
- mean > median

EXAMPLE 4 Recognizing Normal Distributions

Determine whether each histogram has a normal distribution.

a.

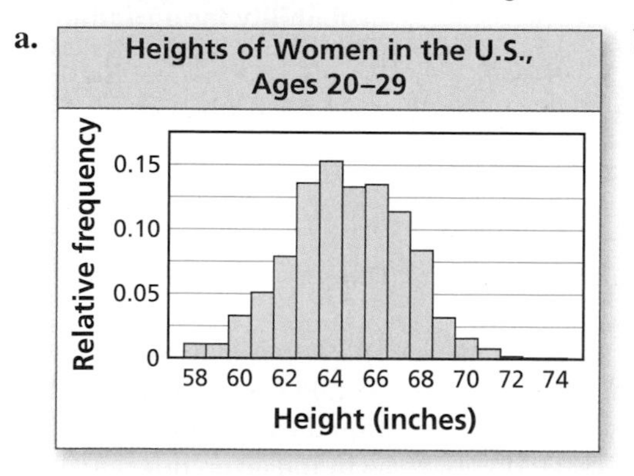

b.

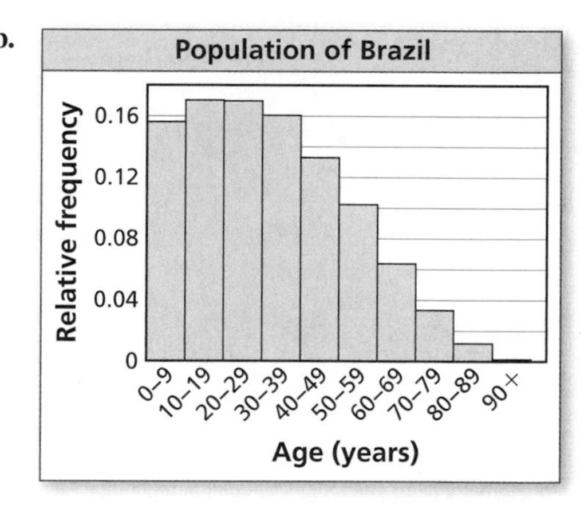

SOLUTION

a. The histogram is bell-shaped and fairly symmetric. So, the histogram has an approximately normal distribution.

b. The histogram is skewed right. So, the histogram does not have a normal distribution, and you cannot use a normal distribution to interpret the histogram.

Monitoring Progress  Help in English and Spanish at *BigIdeasMath.com*

10. Determine whether the histogram has a normal distribution.

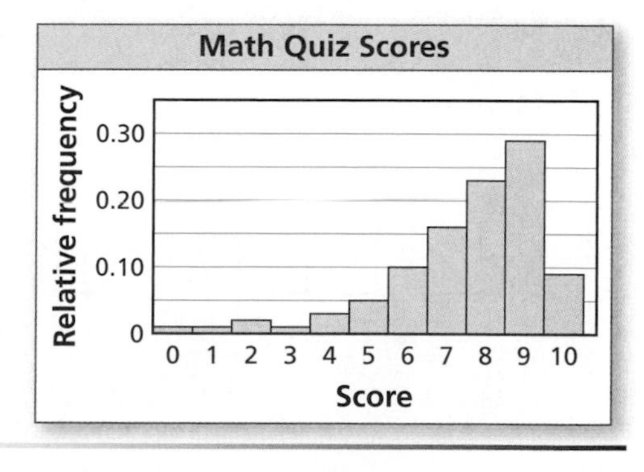

11.1 Exercises

Dynamic Solutions available at *BigIdeasMath.com*

Vocabulary and Core Concept Check

1. **WRITING** Describe how to use the standard normal table to find $P(z \leq 1.4)$.

2. **WHICH ONE DOESN'T BELONG?** Which histogram does *not* belong with the other three? Explain your reasoning.

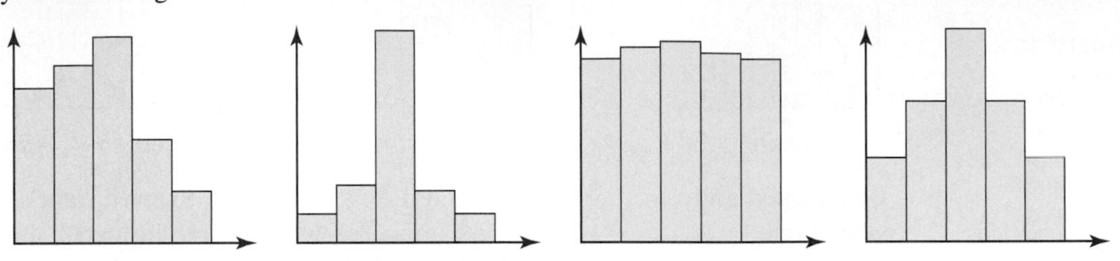

Monitoring Progress and Modeling with Mathematics

ATTENDING TO PRECISION **In Exercises 3–6, give the percent of the area under the normal curve represented by the shaded region(s).**

3.

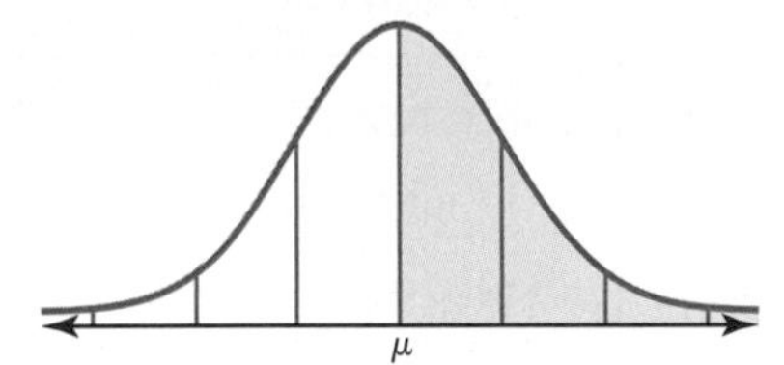

4.

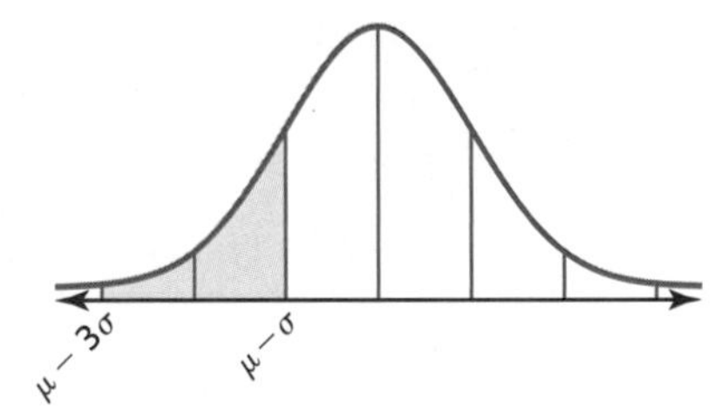

5.

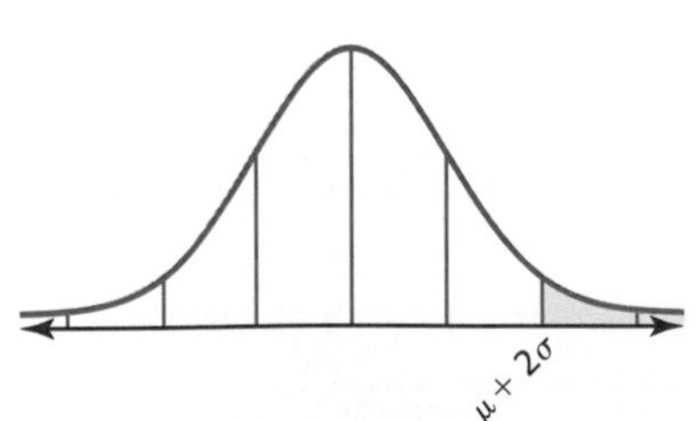

6.

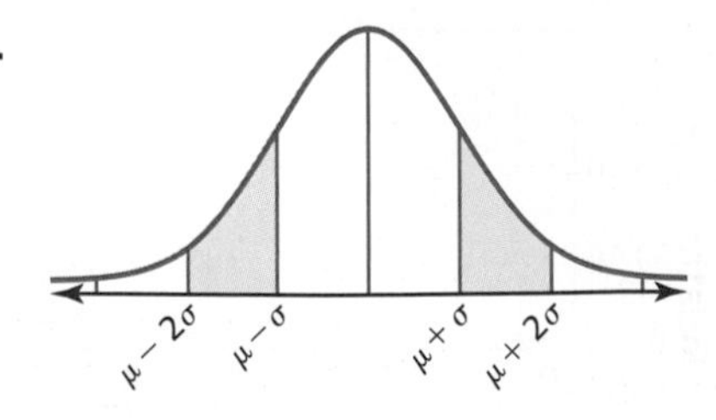

In Exercises 7–12, a normal distribution has mean μ and standard deviation σ. Find the indicated probability for a randomly selected x-value from the distribution. *(See Example 1.)*

7. $P(x \leq \mu - \sigma)$

8. $P(x \geq \mu - \sigma)$

9. $P(x \geq \mu + 2\sigma)$

10. $P(x \leq \mu + \sigma)$

11. $P(\mu - \sigma \leq x \leq \mu + \sigma)$

12. $P(\mu - 3\sigma \leq x \leq \mu)$

In Exercises 13–18, a normal distribution has a mean of 33 and a standard deviation of 4. Find the probability that a randomly selected x-value from the distribution is in the given interval.

13. between 29 and 37

14. between 33 and 45

15. at least 25

16. at least 29

17. at most 37

18. at most 21

19. **PROBLEM SOLVING** The wing lengths of houseflies are normally distributed with a mean of 4.6 millimeters and a standard deviation of 0.4 millimeter. *(See Example 2.)*

wing length

   a. About what percent of houseflies have wing lengths between 3.8 millimeters and 5.0 millimeters?

   b. About what percent of houseflies have wing lengths longer than 5.8 millimeters?

20. **PROBLEM SOLVING** The times a fire department takes to arrive at the scene of an emergency are normally distributed with a mean of 6 minutes and a standard deviation of 1 minute.

a. For about what percent of emergencies does the fire department arrive at the scene in 8 minutes or less?

b. The goal of the fire department is to reach the scene of an emergency in 5 minutes or less. About what percent of the time does the fire department achieve its goal?

ERROR ANALYSIS In Exercises 21 and 22, a normal distribution has a mean of 25 and a standard deviation of 2. Describe and correct the error in finding the probability that a randomly selected x-value is in the given interval.

21. between 23 and 27

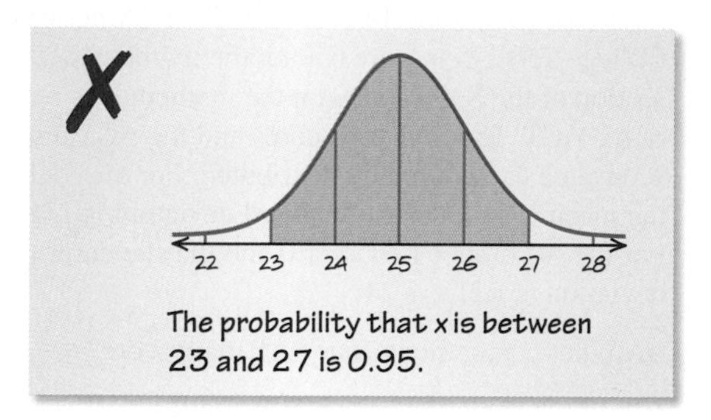

22. at least 21

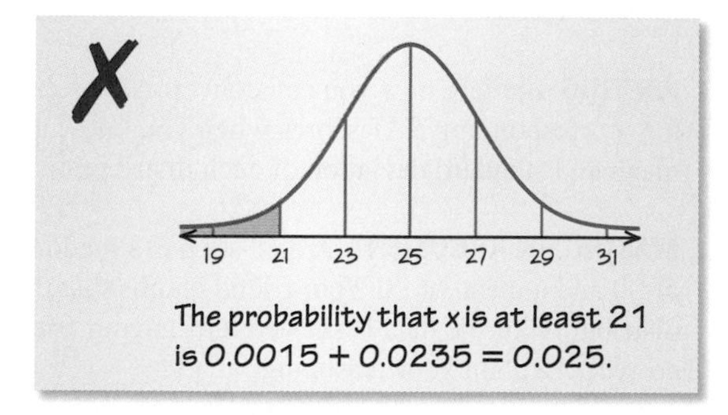

19 21 23 25 27 29 31
The probability that x is at least 21 is 0.0015 + 0.0235 = 0.025.

23. **PROBLEM SOLVING** A busy time to visit a bank is during its Friday evening rush hours. For these hours, the waiting times at the drive-through window are normally distributed with a mean of 8 minutes and a standard deviation of 2 minutes. You have no more than 11 minutes to do your banking and still make it to your meeting on time. What is the probability that you will be late for the meeting? *(See Example 3.)*

24. **PROBLEM SOLVING** Scientists conducted aerial surveys of a seal sanctuary and recorded the number x of seals they observed during each survey. The numbers of seals observed were normally distributed with a mean of 73 seals and a standard deviation of 14.1 seals. Find the probability that at most 50 seals were observed during a randomly chosen survey.

In Exercises 25 and 26, determine whether the histogram has a normal distribution. *(See Example 4.)*

25.

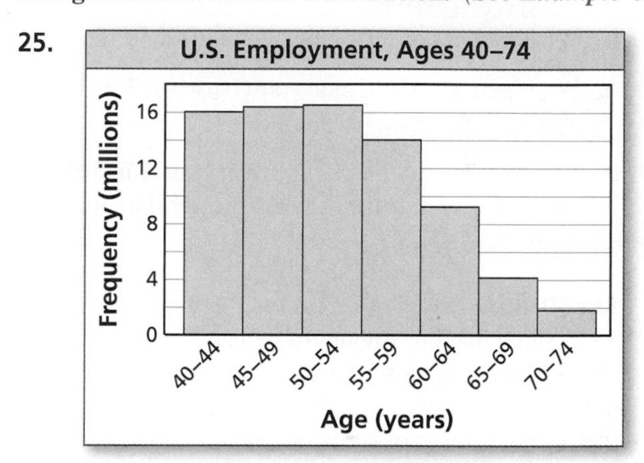

26.

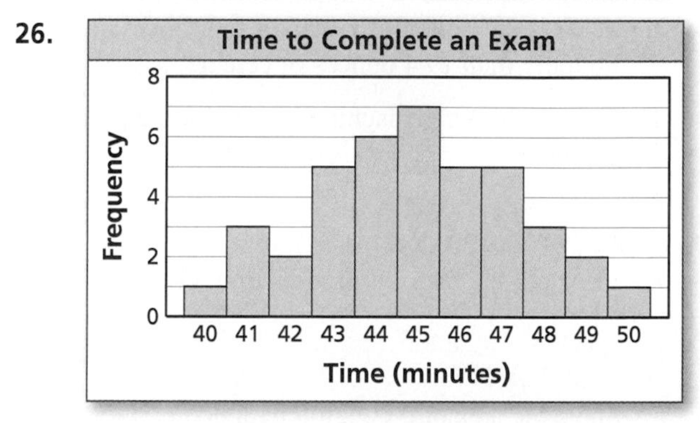

27. **ANALYZING RELATIONSHIPS** The table shows the numbers of tickets that are sold for various baseball games in a league over an entire season. Display the data in a histogram. Do the data fit a normal distribution? Explain.

Tickets sold	Frequency
150–189	1
190–229	2
230–269	4
270–309	8
310–349	8
350–389	7

28. **PROBLEM SOLVING** The guayule plant, which grows in the southwestern United States and in Mexico, is one of several plants that can be used as a source of rubber. In a large group of guayule plants, the heights of the plants are normally distributed with a mean of 12 inches and a standard deviation of 2 inches.

a. What percent of the plants are taller than 16 inches?

b. What percent of the plants are at most 13 inches?

c. What percent of the plants are between 7 inches and 14 inches?

d. What percent of the plants are at least 3 inches taller than or at least 3 inches shorter than the mean height?

29. **REASONING** Boxes of cereal are filled by a machine. Tests show that the amount of cereal in each box varies. The weights are normally distributed with a mean of 20 ounces and a standard deviation of 0.25 ounce. Four boxes of cereal are randomly chosen.

a. What is the probability that all four boxes contain no more than 19.4 ounces of cereal?

b. Do you think the machine is functioning properly? Explain.

30. **THOUGHT PROVOKING** Sketch the graph of the standard normal distribution function, given by

$$f(x) = \frac{1}{\sqrt{2\pi}}e^{-x^2/2}.$$

Estimate the area of the region bounded by the x-axis, the graph of f, and the vertical lines $x = -3$ and $x = 3$.

31. **REASONING** For normally distributed data, describe the value that represents the 84th percentile in terms of the mean and standard deviation.

32. **HOW DO YOU SEE IT?** In the figure, the shaded region represents 47.5% of the area under a normal curve. What are the mean and standard deviation of the normal distribution?

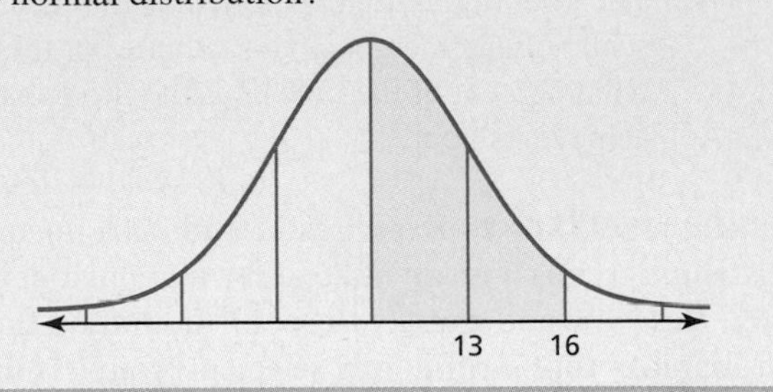

33. **DRAWING CONCLUSIONS** You take both the SAT (Scholastic Aptitude Test) and the ACT (American College Test). You score 650 on the mathematics section of the SAT and 29 on the mathematics section of the ACT. The SAT test scores and the ACT test scores are each normally distributed. For the SAT, the mean is 514 and the standard deviation is 118. For the ACT, the mean is 21.0 and the standard deviation is 5.3.

a. What percentile is your SAT math score?

b. What percentile is your ACT math score?

c. On which test did you perform better? Explain your reasoning.

34. **WRITING** Explain how you can convert ACT scores into corresponding SAT scores when you know the mean and standard deviation of each distribution.

35. **MAKING AN ARGUMENT** A data set has a median of 80 and a mean of 90. Your friend claims that the distribution of the data is skewed left. Is your friend correct? Explain your reasoning.

36. **CRITICAL THINKING** The average scores on a statistics test are normally distributed with a mean of 75 and a standard deviation of 10. You randomly select a test score x. Find $P(|x - \mu| \geq 15)$.

Maintaining Mathematical Proficiency

Reviewing what you learned in previous grades and lessons

Graph the function. Identify the x-intercepts and the points where the local maximums and local minimums occur. Determine the intervals for which the function is increasing or decreasing. *(Section 4.8)*

37. $f(x) = x^3 - 4x^2 + 5$

38. $g(x) = \frac{1}{4}x^4 - 2x^2 - x - 3$

39. $h(x) = -0.5x^2 + 3x + 7$

40. $f(x) = -x^4 + 6x^2 - 13$

11.2 Populations, Samples, and Hypotheses

Essential Question How can you test theoretical probability using sample data?

EXPLORATION 1 Using Sample Data

Work with a partner.

a. When two six-sided dice are rolled, what is the theoretical probability that you roll the same number on both dice?

b. Conduct an experiment to check your answer in part (a). What sample size did you use? Explain your reasoning.

c. Use the dice rolling simulator at *BigIdeasMath.com* to complete the table and check your answer to part (a). What happens as you increase the sample size?

Number of Rolls	Number of Times Same Number Appears	Experimental Probability
100		
500		
1000		
5000		
10,000		

USING TOOLS STRATEGICALLY

To be proficient in math, you need to use technology to visualize the results of varying assumptions, explore consequences, and compare predictions with data.

EXPLORATION 2 Using Sample Data

Work with a partner.

a. When three six-sided dice are rolled, what is the theoretical probability that you roll the same number on all three dice?

b. Compare the theoretical probability you found in part (a) with the theoretical probability you found in Exploration 1(a).

c. Conduct an experiment to check your answer in part (a). How does adding a die affect the sample size that you use? Explain your reasoning.

d. Use the dice rolling simulator at *BigIdeasMath.com* to check your answer to part (a). What happens as you increase the sample size?

Communicate Your Answer

3. How can you test theoretical probability using sample data?

4. Conduct an experiment to determine the probability of rolling a sum of 7 when two six-sided dice are rolled. Then find the theoretical probability and compare your answers.

11.2 Lesson

Core Vocabulary

population, *p. 604*
sample, *p. 604*
parameter, *p. 605*
statistic, *p. 605*
hypothesis, *p. 605*

Previous
Venn diagram
proportion

What You Will Learn

- Distinguish between populations and samples.
- Analyze hypotheses.

Populations and Samples

A **population** is the collection of all data, such as responses, measurements, or counts, that you want information about. A **sample** is a subset of a population.

A *census* consists of data from an entire population. But, unless a population is small, it is usually impractical to obtain all the population data. In most studies, information must be obtained from a *random sample*. (You will learn more about random sampling and data collection in the next section.)

It is important for a sample to be representative of a population so that sample data can be used to draw conclusions about the population. When the sample is not representative of the population, the conclusions may not be valid. Drawing conclusions about populations is an important use of *statistics*. Recall that statistics is the science of collecting, organizing, and interpreting data.

EXAMPLE 1 Distinguishing Between Populations and Samples

Identify the population and the sample. Describe the sample.

a. In the United States, a survey of 2184 adults ages 18 and over found that 1328 of them own at least one pet.

b. To estimate the gasoline mileage of new cars sold in the United States, a consumer advocacy group tests 845 new cars and finds they have an average of 25.1 miles per gallon.

SOLUTION

a. The population consists of the responses of all adults ages 18 and over in the United States, and the sample consists of the responses of the 2184 adults in the survey. Notice in the diagram that the sample is a subset of the responses of all adults in the United States. The sample consists of 1328 adults who said they own at least one pet and 856 adults who said they do not own any pets.

Population: responses of all adults ages 18 and over in the United States

Sample: 2184 responses of adults in survey

b. The population consists of the gasoline mileages of all new cars sold in the United States, and the sample consists of the gasoline mileages of the 845 new cars tested by the group. Notice in the diagram that the sample is a subset of the gasoline mileages of all new cars in the United States. The sample consists of 845 new cars with an average of 25.1 miles per gallon.

Population: gasoline mileages of all new cars sold in the United States

Sample: gasoline mileages of 845 new cars in test

A numerical description of a population characteristic is called a **parameter**. A numerical description of a sample characteristic is called a **statistic**. Because some populations are too large to measure, a statistic, such as the sample mean, is used to estimate the parameter, such as the population mean. It is important that you are able to distinguish between a parameter and a statistic.

EXAMPLE 2 Distinguishing Between Parameters and Statistics

a. For all students taking the SAT in a recent year, the mean mathematics score was 514. Is the mean score a parameter or a statistic? Explain your reasoning.

b. A survey of 1060 women, ages 20–29 in the United States, found that the standard deviation of their heights is about 2.6 inches. Is the standard deviation of the heights a parameter or a statistic? Explain your reasoning.

SOLUTION

a. Because the mean score of 514 is based on all students who took the SAT in a recent year, it is a parameter.

b. Because there are more than 1060 women ages 20–29 in the United States, the survey is based on a subset of the population (all women ages 20–29 in the United States). So, the standard deviation of the heights is a statistic. Note that if the sample is representative of the population, then you can estimate that the standard deviation of the heights of all women ages 20–29 in the United States is about 2.6 inches.

Monitoring Progress  Help in English and Spanish at *BigIdeasMath.com*

In Monitoring Progress Questions 1 and 2, identify the population and the sample.

1. To estimate the retail prices for three grades of gasoline sold in the United States, the Energy Information Association calls 800 retail gasoline outlets, records the prices, and then determines the average price for each grade.
2. A survey of 4464 shoppers in the United States found that they spent an average of $407.02 from Thursday through Sunday during a recent Thanksgiving holiday.
3. A survey found that the median salary of 1068 statisticians is about $72,800. Is the median salary a parameter or a statistic? Explain your reasoning.
4. The mean age of U.S. representatives at the start of the 113th Congress was about 57 years. Is the mean age a parameter or a statistic? Explain your reasoning.

UNDERSTANDING MATHEMATICAL TERMS

A *population proportion* is the ratio of members of a population with a particular characteristic to the total members of the population. A *sample proportion* is the ratio of members of a sample of the population with a particular characteristic to the total members of the sample.

Analyzing Hypotheses

In statistics, a **hypothesis** is a claim about a characteristic of a population. Here are some examples.

1. A drug company claims that patients using its weight-loss drug lose an average of 24 pounds in the first 3 months.
2. A medical researcher claims that the proportion of U.S. adults living with one or more chronic conditions, such as high blood pressure, is 0.45, or 45%.

To analyze a hypothesis, you need to distinguish between results that can easily occur by chance and results that are highly unlikely to occur by chance. One way to analyze a hypothesis is to perform a *simulation*. When the results are highly unlikely to occur, the hypothesis is probably false.

EXAMPLE 3 Analyzing a Hypothesis

You roll a six-sided die 5 times and do not get an even number. The probability of this happening is $\left(\frac{1}{2}\right)^5 = 0.03125$, so you suspect this die favors odd numbers. The die maker claims the die does not favor odd numbers or even numbers. What should you conclude when you roll the actual die 50 times and get (a) 26 odd numbers and (b) 35 odd numbers?

INTERPRETING MATHEMATICAL RESULTS

Results of other simulations may have histograms different from the one shown, but the shape should be similar. Note that the histogram is fairly bell-shaped and symmetric, which means the distribution is approximately normal. By increasing the number of samples or the sample sizes (or both), you should get a histogram that more closely resembles a normal distribution.

SOLUTION

The maker's claim, or hypothesis, is "the die does not favor odd numbers or even numbers." This is the same as saying that the proportion of odd numbers rolled, in the long run, is 0.50. So, assume the probability of rolling an odd number is 0.50. Simulate the rolling of the die by repeatedly drawing 200 random samples of size 50 from a population of 50% ones and 50% zeros. Let the population of ones represent the event of rolling an odd number and make a histogram of the distribution of the sample proportions.

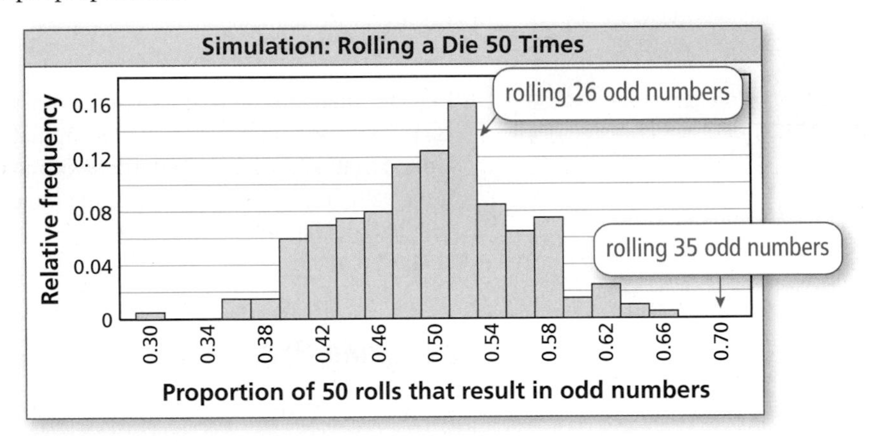

a. Getting 26 odd numbers in 50 rolls corresponds to a proportion of $\frac{26}{50} = 0.52$. In the simulation, this result had a relative frequency of 0.16. In fact, most of the results are close to 0.50. Because this result can easily occur by chance, you can conclude that the maker's claim is most likely true.

b. Getting 35 odd numbers in 50 rolls corresponds to a proportion of $\frac{35}{50} = 0.70$. In the simulation, this result did not occur. Because getting 35 odd numbers is highly unlikely to occur by chance, you can conclude that the maker's claim is most likely false.

JUSTIFYING CONCLUSIONS

In Example 3(b), the theoretical probability of getting 35 odd numbers in 50 rolls is about 0.002. So, while unlikely, it is possible that you incorrectly concluded that the die maker's claim is false.

Monitoring Progress  Help in English and Spanish at *BigIdeasMath.com*

5. WHAT IF? In Example 3, what should you conclude when you roll the actual die 50 times and get (a) 24 odd numbers and (b) 31 odd numbers?

In Example 3(b), you concluded the maker's claim is probably false. In general, such conclusions may or may not be correct. The table summarizes the incorrect and correct decisions that can be made about a hypothesis.

		Truth of Hypothesis	
		Hypothesis is true.	**Hypothesis is false.**
Decision	**You decide that the hypothesis is true.**	correct decision	incorrect decision
	You decide that the hypothesis is false.	incorrect decision	correct decision

11.2 Exercises

Dynamic Solutions available at *BigIdeasMath.com*

Vocabulary and Core Concept Check

1. **COMPLETE THE SENTENCE** A portion of a population that can be studied in order to make predictions about the entire population is a(n) ___________.

2. **WRITING** Describe the difference between a parameter and a statistic. Give an example of each.

3. **VOCABULARY** What is a hypothesis in statistics?

4. **WRITING** Describe two ways you can make an incorrect decision when analyzing a hypothesis.

Monitoring Progress and Modeling with Mathematics

In Exercises 5–8, determine whether the data are collected from a population or a sample. Explain your reasoning.

5. the number of high school students in the United States

6. the color of every third car that passes your house

7. a survey of 100 spectators at a sporting event with 1800 spectators

8. the age of each dentist in the United States

In Exercises 9–12, identify the population and sample. Describe the sample. *(See Example 1.)*

9. In the United States, a survey of 1152 adults ages 18 and over found that 403 of them pretend to use their smartphones to avoid talking to someone.

10. In the United States, a survey of 1777 adults ages 18 and over found that 1279 of them do some kind of spring cleaning every year.

11. In a school district, a survey of 1300 high school students found that 1001 of them like the new, healthy cafeteria food choices.

12. In the United States, a survey of 2000 households with at least one child found that 1280 of them eat dinner together every night.

In Exercises 13–16, determine whether the numerical value is a parameter or a statistic. Explain your reasoning. *(See Example 2.)*

13. The average annual salary of some physical therapists in a state is $76,210.

14. In a recent year, 53% of the senators in the United States Senate were Democrats.

15. Seventy-three percent of all the students in a school would prefer to have school dances on Saturday.

16. A survey of U.S. adults found that 10% believe a cleaning product they use is not safe for the environment.

17. **ERROR ANALYSIS** A survey of 1270 high school students found that 965 students felt added stress because of their workload. Describe and correct the error in identifying the population and the sample.

The population consists of all the students in the high school. The sample consists of the 965 students who felt added stress.

18. **ERROR ANALYSIS** Of all the players on a National Football League team, the mean age is 26 years. Describe and correct the error in determining whether the mean age represents a parameter or statistic.

Because the mean age of 26 is based only on one football team, it is a statistic.

19. MODELING WITH MATHEMATICS You flip a coin 4 times and do not get a tails. You suspect this coin favors heads. The coin maker claims that the coin does not favor heads or tails. You simulate flipping the coin 50 times by repeatedly drawing 200 random samples of size 50. The histogram shows the results. What should you conclude when you flip the actual coin 50 times and get (a) 27 heads and (b) 33 heads? *(See Example 3.)*

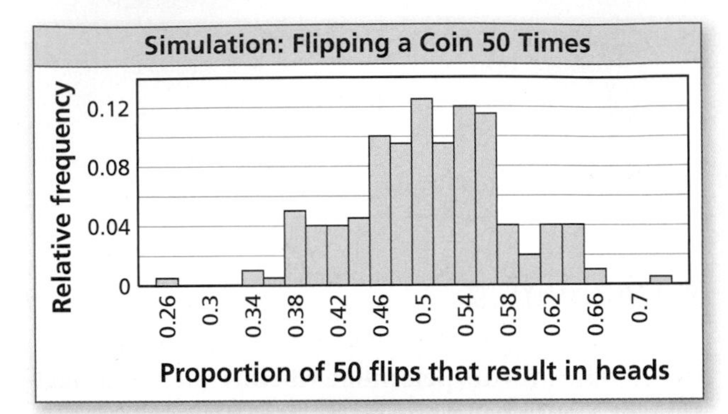

20. MODELING WITH MATHEMATICS Use the histogram in Exercise 19 to determine what you should conclude when you flip the actual coin 50 times and get (a) 17 heads and (b) 23 heads.

21. MAKING AN ARGUMENT A random sample of five people at a movie theater from a population of 200 people gave the film 4 out of 4 stars. Your friend concludes that everyone in the movie theater would give the film 4 stars. Is your friend correct? Explain your reasoning.

22. HOW DO YOU SEE IT? Use the Venn diagram to identify the population and sample. Explain your reasoning.

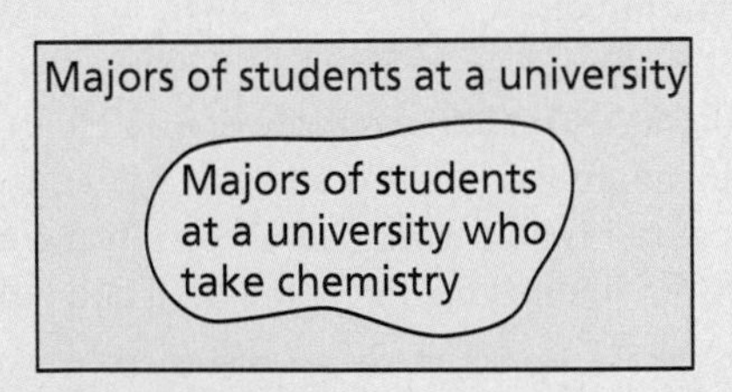

23. OPEN-ENDED Find a newspaper or magazine article that describes a survey. Identify the population and sample. Describe the sample.

24. THOUGHT PROVOKING You choose a random sample of 200 from a population of 2000. Each person in the sample is asked how many hours of sleep he or she gets each night. The mean of your sample is 8 hours. Is it possible that the mean of the entire population is only 7.5 hours of sleep each night? Explain.

25. DRAWING CONCLUSIONS You perform two simulations of repeatedly selecting a marble out of a bag with replacement that contains three red marbles and three blue marbles. The first simulation uses 20 random samples of size 10, and the second uses 400 random samples of size 10. The histograms show the results. Which simulation should you use to accurately analyze a hypothesis? Explain.

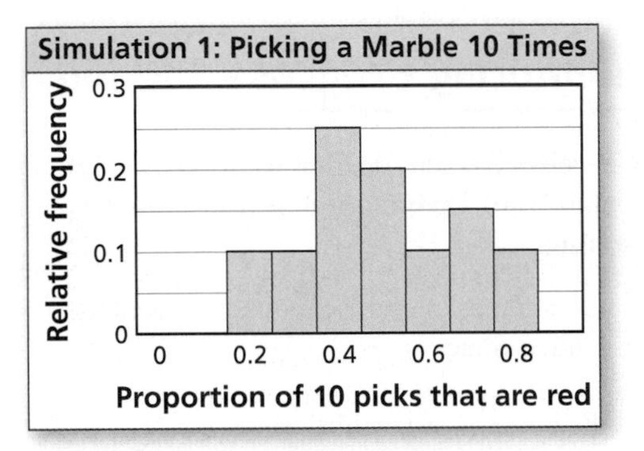

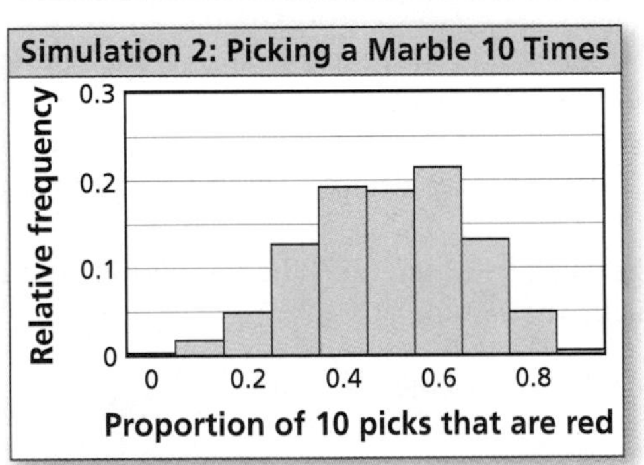

26. PROBLEM SOLVING You roll an eight-sided die five times and get a four every time. You suspect that the die favors the number four. The die maker claims that the die does not favor any number.

a. Perform a simulation involving 50 trials of rolling the actual die and getting a four to test the die maker's claim. Display the results in a histogram.

b. What should you conclude when you roll the actual die 50 times and get 20 fours? 7 fours?

Maintaining Mathematical Proficiency

Reviewing what you learned in previous grades and lessons

Solve the equation by completing the square. *(Section 3.3)*

27. $x^2 - 10x - 4 = 0$ **28.** $3t^2 + 6t = 18$ **29.** $s^2 + 10s + 8 = 0$

Solve the equation using the Quadratic Formula. *(Section 3.4)*

30. $n^2 + 2n + 2 = 0$ **31.** $4z^2 + 28z = 15$ **32.** $5w - w^2 = -11$

11.3 Collecting Data

Essential Question What are some considerations when undertaking a statistical study?

The goal of any statistical study is to collect data and then use the data to make a decision. Any decision you make using the results of a statistical study is only as reliable as the process used to obtain the data. If the process is flawed, then the resulting decision is questionable.

EXPLORATION 1 Analyzing Sampling Techniques

Work with a partner. Determine whether each sample is representative of the population. Explain your reasoning.

JUSTIFYING CONCLUSIONS

To be proficient in math, you need to justify your conclusions and communicate them to others.

a. To determine the number of hours people exercise during a week, researchers use random-digit dialing and call 1500 people.

b. To determine how many text messages high school students send in a week, researchers post a survey on a website and receive 750 responses.

c. To determine how much money college students spend on clothes each semester, a researcher surveys 450 college students as they leave the university library.

d. To determine the quality of service customers receive, an airline sends an e-mail survey to each customer after the completion of a flight.

EXPLORATION 2 Analyzing Survey Questions

Work with a partner. Determine whether each survey question is biased. Explain your reasoning. If so, suggest an unbiased rewording of the question.

a. Does eating nutritious, whole-grain foods improve your health?

b. Do you ever attempt the dangerous activity of texting while driving?

c. How many hours do you sleep each night?

d. How can the mayor of your city improve his or her public image?

EXPLORATION 3 Analyzing Survey Randomness and Truthfulness

Work with a partner. Discuss each potential problem in obtaining a random survey of a population. Include suggestions for overcoming the problem.

a. The people selected might not be a random sample of the population.

b. The people selected might not be willing to participate in the survey.

c. The people selected might not be truthful when answering the question.

d. The people selected might not understand the survey question.

Communicate Your Answer

4. What are some considerations when undertaking a statistical study?

5. Find a real-life example of a biased survey question. Then suggest an unbiased rewording of the question.

11.3 Lesson

Core Vocabulary

random sample, *p. 610*
self-selected sample, *p. 610*
systematic sample, *p. 610*
stratified sample, *p. 610*
cluster sample, *p. 610*
convenience sample, *p. 610*
bias, *p. 611*
unbiased sample, *p. 611*
biased sample, *p. 611*
experiment, *p. 612*
observational study, *p. 612*
survey, *p. 612*
simulation, *p. 612*
biased question, *p. 613*

Previous
population
sample

What You Will Learn

- Identify types of sampling methods in statistical studies.
- Recognize bias in sampling.
- Analyze methods of collecting data.
- Recognize bias in survey questions.

Identifying Sampling Methods in Statistical Studies

The steps in a typical statistical study are shown below.

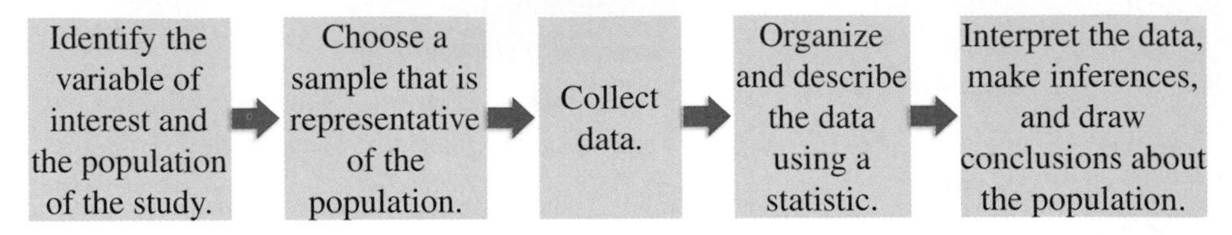
Identify the variable of interest and the population of the study.
Choose a sample that is representative of the population.
Collect data.
Organize and describe the data using a statistic.
Interpret the data, make inferences, and draw conclusions about the population.

There are many different ways of sampling a population, but a *random sample* is preferred because it is most likely to be representative of a population. In a **random sample**, each member of a population has an equal chance of being selected.

The other types of samples given below are defined by the methods used to select members. Each sampling method has its advantages and disadvantages.

Core Concept

Types of Samples

For a **self-selected sample**, members of a population can volunteer to be in the sample.

For a **systematic sample**, a rule is used to select members of a population. For instance, selecting every other person.

STUDY TIP
A stratified sample ensures that every segment of a population is represented.

For a **stratified sample**, a population is divided into smaller groups that share a similar characteristic. A sample is then randomly selected from each group.

STUDY TIP
With cluster sampling, a member of a population cannot belong to more than one cluster.

For a **cluster sample**, a population is divided into groups, called *clusters*. All of the members in one or more of the clusters are selected.

For a **convenience sample**, only members of a population who are easy to reach are selected.

EXAMPLE 1 Identifying Types of Samples

You want to determine whether students in your school like the new design of the school's website. Identify the type of sample described.

a. You list all of the students alphabetically and choose every sixth student.

b. You mail questionnaires and use only the questionnaires that are returned.

c. You ask all of the students in your algebra class.

d. You randomly select two students from each classroom.

SOLUTION

a. You are using a rule to select students. So, the sample is a *systematic* sample.

b. The students can choose whether to respond. So, the sample is a *self-selected* sample.

c. You are selecting students who are readily available. So, the sample is a *convenience* sample.

d. The students are divided into similar groups by their classrooms, and two students are selected at random from each group. So, the sample is a *stratified* sample.

Monitoring Progress  Help in English and Spanish at *BigIdeasMath.com*

1. **WHAT IF?** In Example 1, you divide the students in your school according to their zip codes, then select all of the students that live in one zip code. What type of sample are you using?
2. Describe another method you can use to obtain a stratified sample in Example 1.

Recognizing Bias in Sampling

STUDY TIP
All good sampling methods rely on random sampling.

A **bias** is an error that results in a misrepresentation of a population. In order to obtain reliable information and draw accurate conclusions about a population, it is important to select an *unbiased sample*. An **unbiased sample** is representative of the population that you want information about. A sample that overrepresents or under-represents part of the population is a **biased sample**. When a sample is biased, the data are invalid. A *random sample* can help reduce the possibility of a biased sample.

EXAMPLE 2 Identifying Bias in Samples

Identify the type of sample and explain why the sample is biased.

a. A news organization asks its viewers to participate in an online poll about bullying.

b. A computer science teacher wants to know how students at a school most often access the Internet. The teacher asks students in one of the computer science classes.

SOLUTION

a. The viewers can choose whether to participate in the poll. So, the sample is a *self-selected* sample. The sample is biased because people who go online and respond to the poll most likely have a strong opinion on the subject of bullying.

b. The teacher selects students who are readily available. So, the sample is a *convenience* sample. The sample is biased because other students in the school do not have an opportunity to be chosen.

EXAMPLE 3 Selecting an Unbiased Sample

You are a member of your school's yearbook committee. You want to poll members of the senior class to find out what the theme of the yearbook should be. There are 246 students in the senior class. Describe a method for selecting a random sample of 50 seniors to poll.

STUDY TIP

When you obtain a duplicate integer during the generation, ignore it and generate a new, unique integer as a replacement.

SOLUTION

Step 1 Make a list of all 246 seniors. Assign each senior a different integer from 1 to 246.

Step 2 Generate 50 unique random integers from 1 to 246 using the *randInt* feature of a graphing calculator.

Step 3 Choose the 50 students who correspond to the 50 integers you generated in Step 2.

```
randInt(1,246)
                84
               245
                50
               197
               235
                55
```

Monitoring Progress

Help in English and Spanish at *BigIdeasMath.com*

3. The manager of a concert hall wants to know how often people in the community attend concerts. The manager asks 45 people standing in line for a rock concert how many concerts they attend per year. Identify the type of sample the manager is using and explain why the sample is biased.
4. In Example 3, what is another method you can use to generate a random sample of 50 students? Explain why your sampling method is random.

Analyzing Methods of Data Collection

There are several ways to collect data for a statistical study. The objective of the study often dictates the best method for collecting the data.

READING

A *census* is a survey that obtains data from every member of a population. Often, a census is not practical because of its cost or the time required to gather the data. The U.S. population census is conducted every 10 years.

Core Concept

Methods of Collecting Data

An **experiment** imposes a treatment on individuals in order to collect data on their response to the treatment. The treatment may be a medical treatment, or it can be any action that might affect a variable in the experiment, such as adding methanol to gasoline and then measuring its effect on fuel efficiency.

An **observational study** observes individuals and measures variables without controlling the individuals or their environment. This type of study is used when it is difficult to control or isolate the variable being studied, or when it may be unethical to subject people to a certain treatment or to withhold it from them.

A **survey** is an investigation of one or more characteristics of a population. In a survey, every member of a sample is asked one or more questions.

A **simulation** uses a model to reproduce the conditions of a situation or process so that the simulated outcomes closely match the real-world outcomes. Simulations allow you to study situations that are impractical or dangerous to create in real life.

EXAMPLE 4 Identifying Methods of Data Collection

Identify the method of data collection each situation describes.

a. A researcher records whether people at a gas station use hand sanitizer.

b. A landscaper fertilizes 20 lawns with a regular fertilizer mix and 20 lawns with a new organic fertilizer. The landscaper then compares the lawns after 10 weeks and determines which fertilizer is better.

SOLUTION

a. The researcher is gathering data without controlling the individuals or applying a treatment. So, this situation is an *observational study*.

b. A treatment (organic fertilizer) is being applied to some of the individuals (lawns) in the study. So, this situation is an *experiment*.

Monitoring Progress  Help in English and Spanish at *BigIdeasMath.com*

Identify the method of data collection the situation describes.

5. Members of a student council at your school ask every eighth student who enters the cafeteria whether they like the snacks in the school's vending machines.

6. A park ranger measures and records the heights of trees in a park as they grow.

7. A researcher uses a computer program to help determine how fast an influenza virus might spread within a city.

STUDY TIP

Bias may also be introduced in survey questioning in other ways, such as by the order in which questions are asked or by respondents giving answers they believe will please the questioner.

Recognizing Bias in Survey Questions

When designing a survey, it is important to word survey questions so they do not lead to biased results. Answers to poorly worded questions may not accurately reflect the opinions or actions of those being surveyed. Questions that are flawed in a way that leads to inaccurate results are called **biased questions**. Avoid questions that:

- encourage a particular response
- do not provide enough information to give an accurate opinion
- are too sensitive to answer truthfully
- address more than one issue

EXAMPLE 5 Identify and Correct Bias in Survey Questioning

A dentist surveys his patients by asking, "Do you brush your teeth at least twice per day and floss every day?" Explain why the question may be biased or otherwise introduce bias into the survey. Then describe a way to correct the flaw.

SOLUTION

Patients who brush less than twice per day or do not floss daily may be afraid to admit this because the dentist is asking the question. One improvement may be to have patients answer questions about dental hygiene on paper and then put the paper anonymously into a box.

Monitoring Progress  Help in English and Spanish at *BigIdeasMath.com*

8. Explain why the survey question below may be biased or otherwise introduce bias into the survey. Then describe a way to correct the flaw.

"Do you agree that our school cafeteria should switch to a healthier menu?"

11.3 Exercises

Dynamic Solutions available at *BigIdeasMath.com*

Vocabulary and Core Concept Check

1. **VOCABULARY** Describe the difference between a stratified sample and a cluster sample.

2. **COMPLETE THE SENTENCE** A sample for which each member of a population has an equal chance of being selected is a(n) __________ sample.

3. **WRITING** Describe a situation in which you would use a simulation to collect data.

4. **WRITING** Describe the difference between an unbiased sample and a biased sample. Give one example of each.

Monitoring Progress and Modeling with Mathematics

In Exercises 5–8, identify the type of sample described. *(See Example 1.)*

5. The owners of a chain of 260 retail stores want to assess employee job satisfaction. Employees from 12 stores near the headquarters are surveyed.

6. Each employee in a company writes their name on a card and places it in a hat. The employees whose names are on the first two cards drawn each win a gift card.

7. A taxicab company wants to know whether its customers are satisfied with the service. Drivers survey every tenth customer during the day.

8. The owner of a community pool wants to ask patrons whether they think the water should be colder. Patrons are divided into four age groups, and a sample is randomly surveyed from each age group.

In Exercises 9–12, identify the type of sample and explain why the sample is biased. *(See Example 2.)*

9. A town council wants to know whether residents support having an off-leash area for dogs in the town park. Eighty dog owners are surveyed at the park.

10. A sportswriter wants to determine whether baseball coaches think wooden bats should be mandatory in collegiate baseball. The sportswriter mails surveys to all collegiate coaches and uses the surveys that are returned.

11. You want to find out whether booth holders at a convention were pleased with their booth locations. You divide the convention center into six sections and survey every booth holder in the fifth section.

12. Every tenth employee who arrives at a company health fair answers a survey that asks for opinions about new health-related programs.

13. **ERROR ANALYSIS** Surveys are mailed to every other household in a neighborhood. Each survey that is returned is used. Describe and correct the error in identifying the type of sample that is used.

Because the surveys were mailed to every other household, the sample is a systematic sample.

14. **ERROR ANALYSIS** A researcher wants to know whether the U.S. workforce supports raising the minimum wage. Fifty high school students chosen at random are surveyed. Describe and correct the error in determining whether the sample is biased.

Because the students were chosen at random, the sample is not biased.

In Exercises 15–18, determine whether the sample is biased. Explain your reasoning.

15. Every third person who enters an athletic event is asked whether he or she supports the use of instant replay in officiating the event.

16. A governor wants to know whether voters in the state support building a highway that will pass through a state forest. Business owners in a town near the proposed highway are randomly surveyed.

17. To assess customers' experiences making purchases online, a rating company e-mails purchasers and asks that they click on a link and complete a survey.

18. Your school principal randomly selects five students from each grade to complete a survey about classroom participation.

19. **WRITING** The staff of a student newsletter wants to conduct a survey of the students' favorite television shows. There are 1225 students in the school. Describe a method for selecting a random sample of 250 students to survey. *(See Example 3.)*

20. **WRITING** A national collegiate athletic association wants to survey 15 of the 120 head football coaches in a division about a proposed rules change. Describe a method for selecting a random sample of coaches to survey.

In Exercises 21–24, identify the method of data collection the situation describes. *(See Example 4.)*

21. A researcher uses technology to estimate the damage that will be done if a volcano erupts.

22. The owner of a restaurant asks 20 customers whether they are satisfied with the quality of their meals.

23. A researcher compares incomes of people who live in rural areas with those who live in large urban areas.

24. A researcher places bacteria samples in two different climates. The researcher then measures the bacteria growth in each sample after 3 days.

In Exercises 25–28, explain why the survey question may be biased or otherwise introduce bias into the survey. Then describe a way to correct the flaw. *(See Example 5.)*

25. "Do you agree that the budget of our city should be cut?"

26. "Would you rather watch the latest award-winning movie or just read some book?"

27. "The tap water coming from our western water supply contains twice the level of arsenic of water from our eastern supply. Do you think the government should address this health problem?"

28. A child asks, "Do you support the construction of a new children's hospital?"

In Exercises 29–32, determine whether the survey question may be biased or otherwise introduce bias into the survey. Explain your reasoning.

29. "Do you favor government funding to help prevent acid rain?"

30. "Do you think that renovating the old town hall would be a mistake?"

31. A police officer asks mall visitors, "Do you wear your seat belt regularly?"

32. "Do you agree with the amendments to the Clean Air Act?"

33. **REASONING** A researcher studies the effect of fiber supplements on heart disease. The researcher identified 175 people who take fiber supplements and 175 people who do not take fiber supplements. The study found that those who took the supplements had 19.6% fewer heart attacks. The researcher concludes that taking fiber supplements reduces the chance of heart attacks.

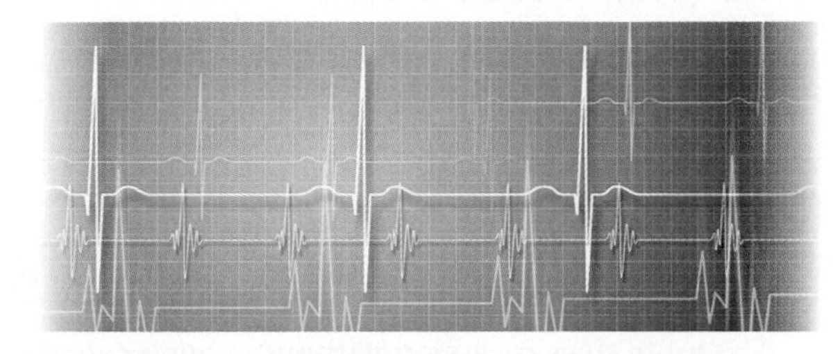

a. Explain why the researcher's conclusion may not be valid.

b. Describe how the researcher could have conducted the study differently to produce valid results.

34. HOW DO YOU SEE IT? A poll is conducted to predict the results of a statewide election in New Mexico before all the votes are counted. Fifty voters in each of the state's 33 counties are asked how they voted as they leave the polls.

a. Identify the type of sample described.

b. Explain how the diagram shows that the polling method could result in a biased sample.

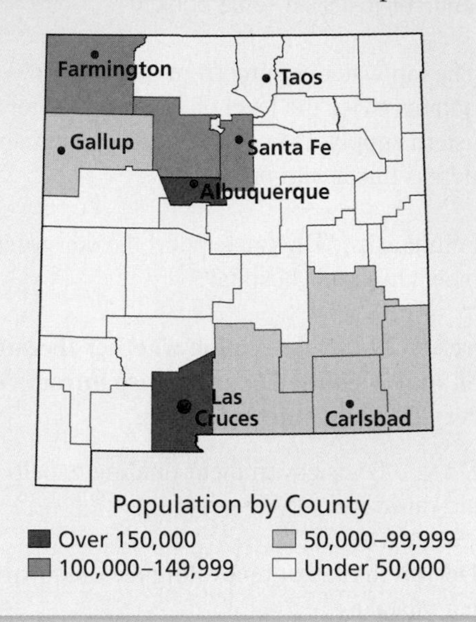

35. WRITING Consider each type of sample listed on page 610. Which of the samples are most likely to lead to biased results? Explain.

36. THOUGHT PROVOKING What is the difference between a "blind experiment" and a "double-blind experiment?" Describe a possible advantage of the second type of experiment over the first.

37. WRITING A college wants to survey its graduating seniors to find out how many have already found jobs in their field of study after graduation.

a. What is the objective of the survey?

b. Describe the population for the survey.

c. Write two unbiased questions for the survey.

38. REASONING About 3.2% of U.S. adults follow a vegetarian-based diet. Two randomly selected groups of people were asked whether they follow such a diet. The first sample consists of 20 people and the second sample consists of 200 people. Which sample proportion is more likely to be representative of the national percentage? Explain.

39. MAKING AN ARGUMENT The U.S. Census is taken every 10 years to gather data from the population. Your friend claims that the sample cannot be biased. Is your friend correct? Explain.

40. OPEN-ENDED An airline wants to know whether travelers have enough leg room on its planes.

a. What method of data collection is appropriate for this situation?

b. Describe a sampling method that is likely to give biased results. Explain.

c. Describe a sampling method that is *not* likely to give biased results. Explain.

d. Write one biased question and one unbiased question for this situation.

41. REASONING A website contains a link to a survey that asks how much time each person spends on the Internet each week.

a. What type of sampling method is used in this situation?

b. Which population is likely to respond to the survey? What can you conclude?

Maintaining Mathematical Proficiency

Reviewing what you learned in previous grades and lessons

Evaluate the expression without using a calculator. *(Section 5.1)*

42. $4^{5/2}$

43. $27^{2/3}$

44. $-64^{1/3}$

45. $8^{-2/3}$

Simplify the expression. *(Section 5.2)*

46. $(4^{3/2} \cdot 4^{1/4})^4$

47. $(6^{1/3} \cdot 3^{1/3})^{-2}$

48. $\sqrt[3]{4} \cdot \sqrt[3]{16}$

49. $\dfrac{\sqrt[4]{405}}{\sqrt[4]{5}}$

11.1–11.3 What Did You Learn?

Core Vocabulary

normal distribution, *p. 596*
normal curve, *p. 596*
standard normal distribution, *p. 597*
z-score, *p. 597*
population, *p. 604*
sample, *p. 604*
parameter, *p. 605*
statistic, *p. 605*
hypothesis, *p. 605*
random sample, *p. 610*
self-selected sample, *p. 610*
systematic sample, *p. 610*
stratified sample, *p. 610*
cluster sample, *p. 610*
convenience sample, *p. 610*
bias, *p. 611*
unbiased sample, *p. 611*
biased sample, *p. 611*
experiment, *p. 612*
observational study, *p. 612*
survey, *p. 612*
simulation, *p. 612*
biased question, *p. 613*

Core Concepts

Section 11.1

Areas Under a Normal Curve, *p. 596*
Using *z*-Scores and the Standard Normal Table, *p. 597*
Recognizing Normal Distributions, *p. 599*

Section 11.2

Distinguishing Between Populations and Samples, *p. 604*
Analyzing Hypotheses, *p. 606*

Section 11.3

Types of Samples, *p. 610*
Methods of Collecting Data, *p. 612*

Mathematical Practices

1. What previously established results, if any, did you use to solve Exercise 31 on page 602?
2. What external resources, if any, did you use to answer Exercise 36 on page 616?

Study Skills

Reworking Your Notes

It's almost impossible to write down in your notes all the detailed information you are taught in class. A good way to reinforce the concepts and put them into your long-term memory is to rework your notes. When you take notes, leave extra space on the pages. You can go back after class and fill in:

- important definitions and rules
- additional examples
- questions you have about the material

11.1–11.3 Quiz

A normal distribution has a mean of 32 and a standard deviation of 4. Find the probability that a randomly selected x-value from the distribution is in the given interval. *(Section 11.1)*

1. at least 28 **2.** between 20 and 32 **3.** at most 26 **4.** at most 35

Determine whether the histogram has a normal distribution. *(Section 11.1)*

5.

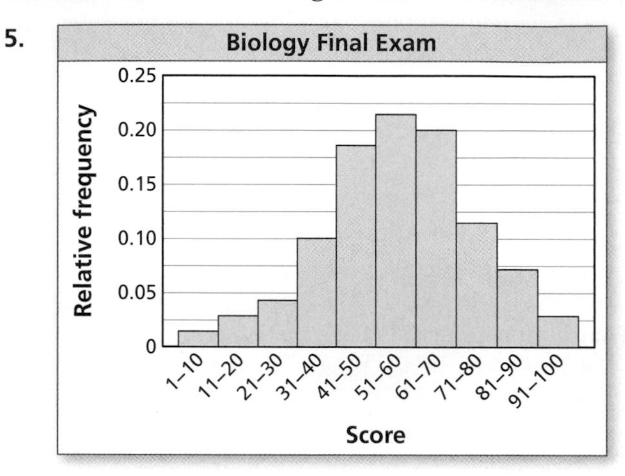

6.

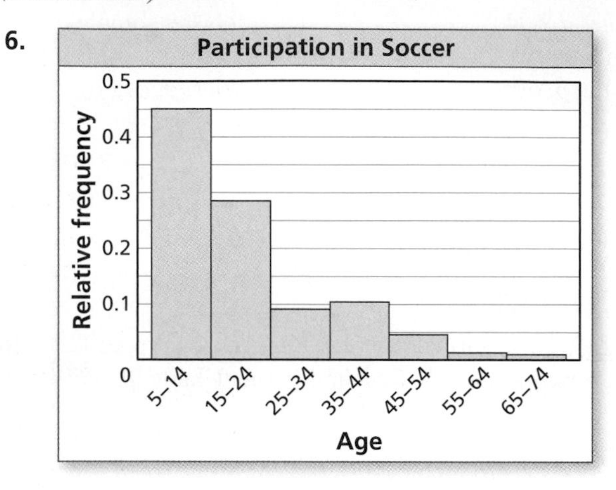

7. A survey of 1654 high school seniors determined that 1125 plan to attend college. Identify the population and the sample. Describe the sample. *(Section 11.2)*

8. A survey of all employees at a company found that the mean one-way daily commute to work of the employees is 25.5 minutes. Is the mean time a parameter or a statistic? Explain your reasoning. *(Section 11.2)*

9. A researcher records the number of bacteria present in several samples in a laboratory. Identify the method of data collection. *(Section 11.3)*

10. You spin a five-color spinner, which is divided into equal parts, five times and every time the spinner lands on red. You suspect the spinner favors red. The maker of the spinner claims that the spinner does not favor any color. You simulate spinning the spinner 50 times by repeatedly drawing 200 random samples of size 50. The histogram shows the results. Use the histogram to determine what you should conclude when you spin the actual spinner 50 times and the spinner lands on red (a) 9 times and (b) 19 times. *(Section 11.2)*

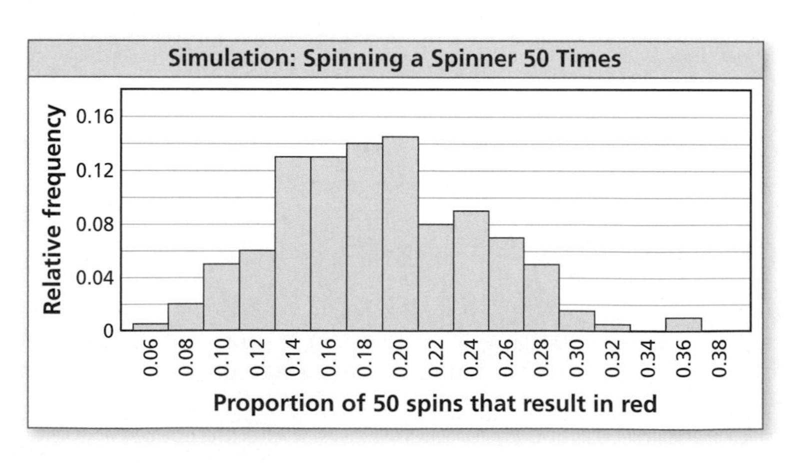

11. A local television station wants to find the number of hours per week people in the viewing area watch sporting events on television. The station surveys people at a nearby sports stadium. *(Section 11.3)*

a. Identify the type of sample described. **b.** Is the sample biased? Explain your reasoning.

c. Describe a method for selecting a random sample of 200 people to survey.

11.4 Experimental Design

Essential Question How can you use an experiment to test a conjecture?

EXPLORATION 1 Using an Experiment

Work with a partner. Standard white playing dice are manufactured with black dots that are indentations, as shown. So, the side with six indentations is the lightest side and the side with one indentation is the heaviest side.

You make a conjecture that when you roll a standard playing die, the number 6 will come up more often than the number 1 because 6 is the lightest side. To test your conjecture, roll a standard playing die 25 times. Record the results in the table. Does the experiment confirm your conjecture? Explain your reasoning.

Number						
Rolls						

EXPLORATION 2 Analyzing an Experiment

Work with a partner. To overcome the imbalance of standard playing dice, one of the authors of this book invented and patented 12-sided dice, on which each number from 1 through 6 appears twice (on opposing sides). See *BigIdeasMath.com*.

As part of the patent process, a standard playing die was rolled 27,090 times. The results are shown below.

Number	1	2	3	4	5	6
Rolls	4293	4524	4492	4397	4623	4761

What can you conclude from the results of this experiment? Explain your reasoning.

CONSTRUCTING VIABLE ARGUMENTS

To be proficient in math, you need to make conjectures and perform experiments to explore the truth of your conjectures.

Communicate Your Answer

3. How can you use an experiment to test a conjecture?

4. Exploration 2 shows the results of rolling a standard playing die 27,090 times to test the conjecture in Exploration 1. Why do you think the number of trials was so large?

5. Make a conjecture about the outcomes of rolling the 12-sided die in Exploration 2. Then design an experiment that could be used to test your conjecture. Be sure that your experiment is practical to complete and includes enough trials to give meaningful results.

11.4 Lesson

Core Vocabulary

controlled experiment, *p. 620*
control group, *p. 620*
treatment group, *p. 620*
randomization, *p. 620*
randomized comparative experiment, *p. 620*
placebo, *p. 620*
replication, *p. 622*

Previous
sample size

What You Will Learn

- Describe experiments.
- Recognize how randomization applies to experiments and observational studies.
- Analyze experimental designs.

Describing Experiments

In a **controlled experiment**, two groups are studied under identical conditions with the exception of one variable. The group under ordinary conditions that is subjected to no treatment is the **control group**. The group that is subjected to the treatment is the **treatment group**.

Randomization is a process of randomly assigning subjects to different treatment groups. In a **randomized comparative experiment**, subjects are randomly assigned to the control group or the treatment group. In some cases, subjects in the control group are given a **placebo**, which is a harmless, unmedicated treatment that resembles the actual treatment. The comparison of the control group and the treatment group makes it possible to determine any effects of the treatment.

Randomization minimizes bias and produces groups of individuals who are theoretically similar in all ways before the treatment is applied. Conclusions drawn from an experiment that is not a randomized comparative experiment may not be valid.

EXAMPLE 1 Evaluating Published Reports

Determine whether each study is a randomized comparative experiment. If it is, describe the treatment, the treatment group, and the control group. If it is not, explain why not and discuss whether the conclusions drawn from the study are valid.

a.

Health Watch

Vitamin C Lowers Cholesterol

At a health clinic, patients were given the choice of whether to take a dietary supplement of 500 milligrams of vitamin C each day. Fifty patients who took the supplement were monitored for one year, as were 50 patients who did not take the supplement. At the end of one year, patients who took the supplement had 15% lower cholesterol levels than patients in the other group.

b.

Supermarket Checkout

Check Out Even Faster

To test the new design of its self checkout, a grocer gathered 142 customers and randomly divided them into two groups. One group used the new self checkout and one group used the old self checkout to buy the same groceries. Users of the new self checkout were able to complete their purchases 16% faster.

STUDY TIP

The study in part (a) is an *observational study* because the treatment is not being imposed.

SOLUTION

a. The study is not a randomized comparative experiment because the individuals were not randomly assigned to a control group and a treatment group. The conclusion that vitamin C lowers cholesterol may or may not be valid. There may be other reasons why patients who took the supplement had lower cholesterol levels. For instance, patients who voluntarily take the supplement may be more likely to have other healthy eating or lifestyle habits that could affect their cholesterol levels.

b. The study is a randomized comparative experiment. The treatment is the use of the new self checkout. The treatment group is the individuals who use the new self checkout. The control group is the individuals who use the old self checkout.

Monitoring Progress  Help in English and Spanish at *BigIdeasMath.com*

Motorist News

Early Birds Make Better Drivers

A recent study shows that adults who rise before 6:30 A.M. are better drivers than other adults. The study monitored the driving records of 140 volunteers who always wake up before 6:30 and 140 volunteers who never wake up before 6:30. The early risers had 12% fewer accidents.

1. Determine whether the study is a randomized comparative experiment. If it is, describe the treatment, the treatment group, and the control group. If it is not, explain why not and discuss whether the conclusions drawn from the study are valid.

Randomization in Experiments and Observational Studies

You have already learned about random sampling and its usefulness in surveys. Randomization applies to experiments and observational studies as shown below.

Experiment	Observational study
Individuals are assigned at random to the treatment group or the control group.	When possible, random samples can be selected for the groups being studied.

Good experiments and observational studies are designed to compare data from two or more groups and to show any relationship between variables. Only a well-designed *experiment*, however, can determine a cause-and-effect relationship.

Core Concept

Comparative Studies and Causality

- A rigorous randomized comparative experiment, by eliminating sources of variation other than the controlled variable, can make valid cause-and-effect conclusions possible.
- An observational study can identify *correlation* between variables, but not *causality*. Variables, other than what is being measured, may be affecting the results.

EXAMPLE 2 Designing an Experiment or Observational Study

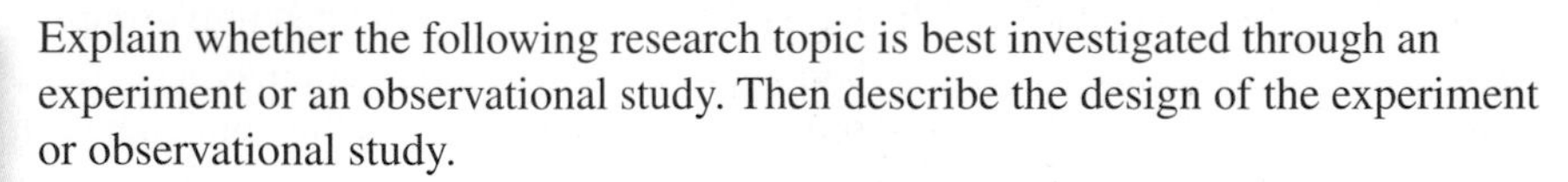

Explain whether the following research topic is best investigated through an experiment or an observational study. Then describe the design of the experiment or observational study.

You want to know whether vigorous exercise in older people results in longer life.

SOLUTION

The treatment, vigorous exercise, is not possible for those people who are already unhealthy, so it is not ethical to assign individuals to a control or treatment group. Use an observational study. Randomly choose one group of individuals who already exercise vigorously. Then randomly choose one group of individuals who do not exercise vigorously. Monitor the ages of the individuals in both groups at regular intervals. Note that because you are using an observational study, you should be able to identify a *correlation* between vigorous exercise in older people and longevity, but not *causality*.

Monitoring Progress  Help in English and Spanish at *BigIdeasMath.com*

2. Determine whether the following research topic is best investigated through an experiment or an observational study. Then describe the design of the experiment or observational study.

   You want to know whether flowers sprayed twice per day with a mist of water stay fresh longer than flowers that are not sprayed.

Analyzing Experimental Designs

An important part of experimental design is *sample size*, or the number of subjects in the experiment. To improve the validity of the experiment, **replication** is required, which is repetition of the experiment under the same or similar conditions.

UNDERSTANDING MATHEMATICAL TERMS

The *validity* of an experiment refers to the reliability of the results. The results of a valid experiment are more likely to be accepted.

EXAMPLE 3 Analyzing Experimental Designs

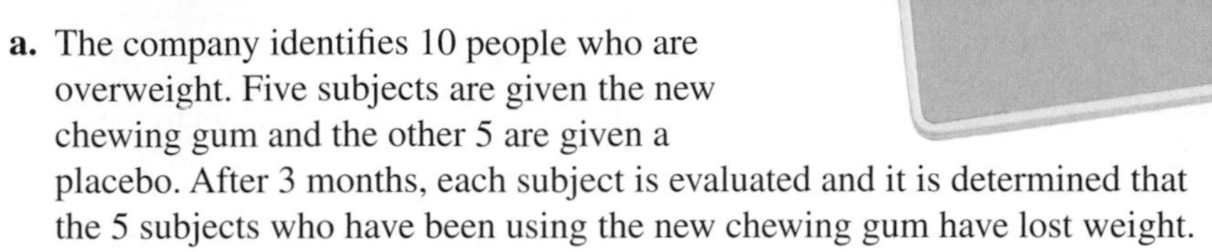

A pharmaceutical company wants to test the effectiveness of a new chewing gum designed to help people lose weight. Identify a potential problem, if any, with each experimental design. Then describe how you can improve it.

a. The company identifies 10 people who are overweight. Five subjects are given the new chewing gum and the other 5 are given a placebo. After 3 months, each subject is evaluated and it is determined that the 5 subjects who have been using the new chewing gum have lost weight.

b. The company identifies 10,000 people who are overweight. The subjects are divided into groups according to gender. Females receive the new chewing gum and males receive the placebo. After 3 months, a significantly large number of the female subjects have lost weight.

c. The company identifies 10,000 people who are overweight. The subjects are divided into groups according to age. Within each age group, subjects are randomly assigned to receive the new chewing gum or the placebo. After 3 months, a significantly large number of the subjects who received the new chewing gum have lost weight.

STUDY TIP

The experimental design described in part (c) is an example of *randomized block design.*

SOLUTION

a. The sample size is not large enough to produce valid results. To improve the validity of the experiment, the sample size must be larger and the experiment must be replicated.

b. Because the subjects are divided into groups according to gender, the groups are not similar. The new chewing gum may have more of an effect on women than on men, or more of an effect on men than on women. It is not possible to see such an effect with the experiment the way it is designed. The subjects can be divided into groups according to gender, but within each group, they must be randomly assigned to the treatment group or the control group.

c. The subjects are divided into groups according to a similar characteristic (age). Because subjects within each age group are randomly assigned to receive the new chewing gum or the placebo, replication is possible.

Monitoring Progress

Help in English and Spanish at *BigIdeasMath.com*

3. In Example 3, the company identifies 250 people who are overweight. The subjects are randomly assigned to a treatment group or a control group. In addition, each subject is given a DVD that documents the dangers of obesity. After 3 months, most of the subjects placed in the treatment group have lost weight. Identify a potential problem with the experimental design. Then describe how you can improve it.

4. You design an experiment to test the effectiveness of a vaccine against a strain of influenza. In the experiment, 100,000 people receive the vaccine and another 100,000 people receive a placebo. Identify a potential problem with the experimental design. Then describe how you can improve it.

11.4 Exercises

Dynamic Solutions available at *BigIdeasMath.com*

Vocabulary and Core Concept Check

1. **COMPLETE THE SENTENCE** Repetition of an experiment under the same or similar conditions is called _________.

2. **WRITING** Describe the difference between the control group and the treatment group in a controlled experiment.

Monitoring Progress and Modeling with Mathematics

In Exercises 3 and 4, determine whether the study is a randomized comparative experiment. If it is, describe the treatment, the treatment group, and the control group. If it is not, explain why not and discuss whether the conclusions drawn from the study are valid. *(See Example 1.)*

3. ***Insomnia***

   New Drug Improves Sleep

   To test a new drug for insomnia, a pharmaceutical company randomly divided 200 adult volunteers into two groups. One group received the drug and one group received a placebo. After one month, the adults who took the drug slept 18% longer, while those who took the placebo experienced no significant change.

4. ***Dental Health***

   Milk Fights Cavities

   At a middle school, students can choose to drink milk or other beverages at lunch. Seventy-five students who chose milk were monitored for one year, as were 75 students who chose other beverages. At the end of the year, students in the "milk" group had 25% fewer cavities than students in the other group.

ERROR ANALYSIS In Exercises 5 and 6, describe and correct the error in describing the study.

A company's researchers want to study the effects of adding shea butter to their existing hair conditioner. They monitor the hair quality of 30 randomly selected customers using the regular conditioner and 30 randomly selected customers using the new shea butter conditioner.

5. 

   The control group is individuals who do not use either of the conditioners.

6. 

   The study is an observational study.

In Exercises 7–10, explain whether the research topic is best investigated through an experiment or an observational study. Then describe the design of the experiment or observational study. *(See Example 2.)*

7. A researcher wants to compare the body mass index of smokers and nonsmokers.

8. A restaurant chef wants to know which pasta sauce recipe is preferred by more diners.

9. A farmer wants to know whether a new fertilizer affects the weight of the fruit produced by strawberry plants.

10. You want to know whether homes that are close to parks or schools have higher property values.

11. **DRAWING CONCLUSIONS** A company wants to test whether a nutritional supplement has an adverse effect on an athlete's heart rate while exercising. Identify a potential problem, if any, with each experimental design. Then describe how you can improve it. *(See Example 3.)*

    a. The company randomly selects 250 athletes. Half of the athletes receive the supplement and their heart rates are monitored while they run on a treadmill. The other half of the athletes are given a placebo and their heart rates are monitored while they lift weights. The heart rates of the athletes who took the supplement significantly increased while exercising.

    b. The company selects 1000 athletes. The athletes are divided into two groups based on age. Within each age group, the athletes are randomly assigned to receive the supplement or the placebo. The athletes' heart rates are monitored while they run on a treadmill. There was no significant difference in the increases in heart rates between the two groups.

12. **DRAWING CONCLUSIONS** A researcher wants to test the effectiveness of reading novels on raising intelligence quotient (IQ) scores. Identify a potential problem, if any, with each experimental design. Then describe how you can improve it.

 a. The researcher selects 500 adults and randomly divides them into two groups. One group reads novels daily and one group does not read novels. At the end of 1 year, each adult is evaluated and it is determined that neither group had an increase in IQ scores.

 b. Fifty adults volunteer to spend time reading novels every day for 1 year. Fifty other adults volunteer to refrain from reading novels for 1 year. Each adult is evaluated and it is determined that the adults who read novels raised their IQ scores by 3 points more than the other group.

13. **DRAWING CONCLUSIONS** A fitness company claims that its workout program will increase vertical jump heights in 6 weeks. To test the workout program, 10 athletes are divided into two groups. The double bar graph shows the results of the experiment. Identify the potential problems with the experimental design. Then describe how you can improve it.

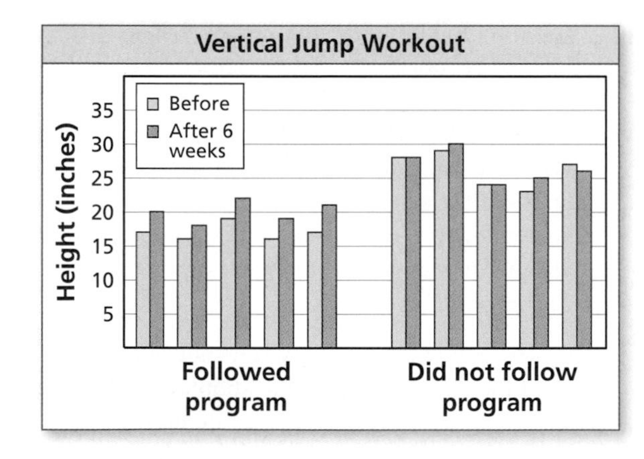

14. **WRITING** Explain why observational studies, rather than experiments, are usually used in astronomy.

15. **MAKING AN ARGUMENT** Your friend wants to determine whether the number of siblings has an effect on a student's grades. Your friend claims to be able to show causality between the number of siblings and grades. Is your friend correct? Explain.

16. **HOW DO YOU SEE IT?** To test the effect political advertisements have on voter preferences, a researcher selects 400 potential voters and randomly divides them into two groups. The circle graphs show the results of the study.

Survey Results

Watching 30 Minutes of TV with No Ads

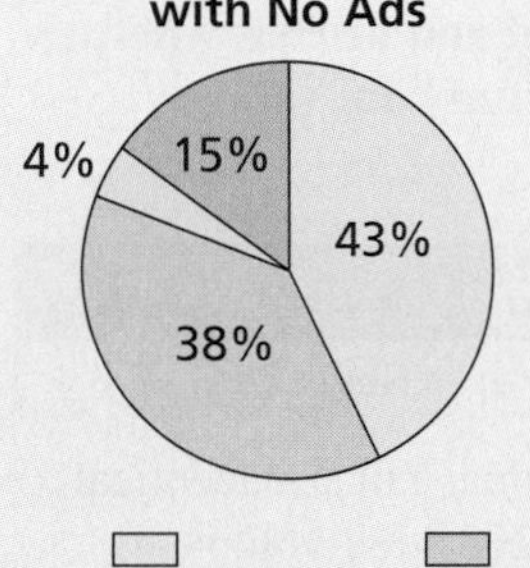

Watching 30 Minutes of TV with Ads for Candidate B

3%
8%
45%
44%

Candidate A Candidate B Candidate C Undecided

 a. Is the study a randomized comparative experiment? Explain.

 b. Describe the treatment.

 c. Can you conclude that the political advertisements were effective? Explain.

17. **WRITING** Describe the *placebo effect* and how it affects the results of an experiment. Explain how a researcher can minimize the placebo effect.

18. **THOUGHT PROVOKING** Make a hypothesis about something that interests you. Design an experiment that could show that your hypothesis is probably true.

19. **REASONING** Will replicating an experiment on many individuals produce data that are more likely to accurately represent a population than performing the experiment only once? Explain.

Maintaining Mathematical Proficiency

Reviewing what you learned in previous grades and lessons

Draw a dot plot that represents the data. Identify the shape of the distribution. *(Skills Review Handbook)*

20. Ages: 24, 21, 22, 26, 22, 23, 25, 23, 23, 24, 20, 25

21. Golf strokes: 4, 3, 4, 3, 3, 2, 7, 5, 3, 4

Tell whether the function represents *exponential growth* or *exponential decay*. Then graph the function. *(Section 6.1)*

22. $y = 4^x$

23. $y = (0.95)^x$

24. $y = (0.2)^x$

25. $y = (1.25)^x$

11.5 Making Inferences from Sample Surveys

Essential Question How can you use a sample survey to infer a conclusion about a population?

EXPLORATION 1 Making an Inference from a Sample

Work with a partner. You conduct a study to determine what percent of the high school students in your city would prefer an upgraded model of their current cell phone. Based on your intuition and talking with a few acquaintances, you think that 50% of high school students would prefer an upgrade. You survey 50 randomly chosen high school students and find that 20 of them prefer an upgraded model.

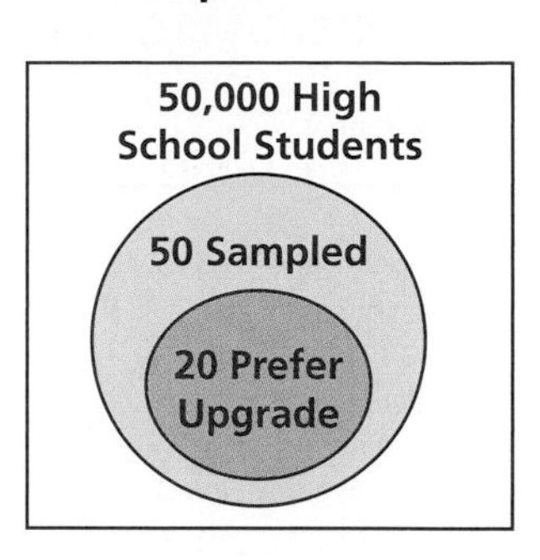

MODELING WITH MATHEMATICS

To be proficient in math, you need to apply the mathematics you know to solve problems arising in everyday life.

a. Based on your sample survey, what percent of the high school students in your city would prefer an upgraded model? Explain your reasoning.

b. In spite of your sample survey, is it still possible that 50% of the high school students in your city prefer an upgraded model? Explain your reasoning.

c. To investigate the likelihood that you could have selected a sample of 50 from a population in which 50% of the population does prefer an upgraded model, you create a binomial distribution as shown below. From the distribution, estimate the probability that exactly 20 students surveyed prefer an upgraded model. Is this event likely to occur? Explain your reasoning.

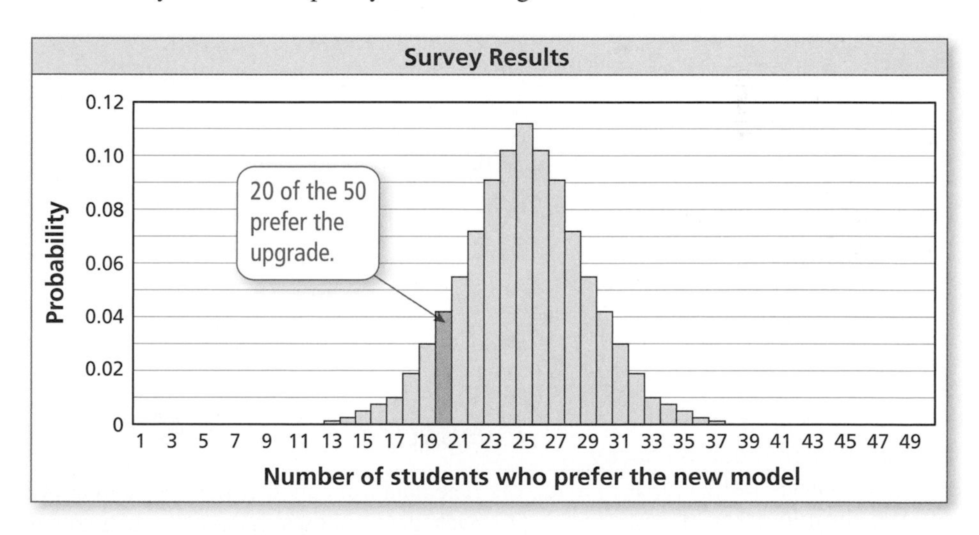

d. When making inferences from sample surveys, the sample must be random. In the situation described above, describe how you could design and conduct a survey using a random sample of 50 high school students who live in a large city.

Communicate Your Answer

2. How can you use a sample survey to infer a conclusion about a population?

3. In Exploration 1(c), what is the probability that exactly 25 students you survey prefer an upgraded model?

11.5 Lesson

Core Vocabulary

descriptive statistics, *p. 626*
inferential statistics, *p. 626*
margin of error, *p. 629*

Previous
statistic
parameter

What You Will Learn

- Estimate population parameters.
- Analyze estimated population parameters.
- Find margins of error for surveys.

Estimating Population Parameters

The study of statistics has two major branches: *descriptive statistics* and *inferential statistics*. **Descriptive statistics** involves the organization, summarization, and display of data. So far, you have been using descriptive statistics in your studies of data analysis and statistics. **Inferential statistics** involves using a sample to draw conclusions about a population. You can use statistics to make reasonable predictions, or *inferences*, about an entire population when the sample is representative of the population.

EXAMPLE 1 Estimating a Population Mean

The numbers of friends for a random sample of 40 teen users of a social networking website are shown in the table. Estimate the population mean μ.

Number of Friends				
281	342	229	384	320
247	298	248	312	445
385	286	314	260	186
287	342	225	308	343
262	220	320	310	150
274	291	300	410	255
279	351	370	257	350
369	215	325	338	278

REMEMBER

Recall that $\overline{x}$ denotes the sample mean. It is read as "*x* bar."

SOLUTION

To estimate the unknown population mean μ, find the sample mean $\overline{x}$.

$$\overline{x} = \frac{\Sigma x}{n} = \frac{11,966}{40} = 299.15$$

STUDY TIP

The probability that the population mean is *exactly* 299.15 is virtually 0, but the sample mean is a good estimate of μ.

▶ So, the mean number of friends for all teen users of the website is about 299.

Monitoring Progress  Help in English and Spanish at *BigIdeasMath.com*

1. The data from another random sample of 30 teen users of the social networking website are shown in the table. Estimate the population mean μ.

Number of Friends				
305	237	261	374	341
257	243	352	330	189
297	418	275	288	307
295	288	341	322	271
209	164	363	228	390
313	315	263	299	285

Not every random sample results in the same estimate of a population parameter; there will be some sampling variability. Larger sample sizes, however, tend to produce more accurate estimates.

REMEMBER

A *population proportion* is the ratio of members of a population with a particular characteristic to the total members of the population. A *sample proportion* is the ratio of members of a sample of the population with a particular characteristic to the total members of the sample.

STUDY TIP

Statistics and probability provide information that you can use to weigh evidence and make decisions.

EXAMPLE 2 Estimating Population Proportions

A student newspaper wants to predict the winner of a city's mayoral election. Two candidates, A and B, are running for office. Eight staff members conduct surveys of randomly selected residents. The residents are asked whether they will vote for Candidate A. The results are shown in the table.

Sample Size	Number of Votes for Candidate A in the Sample	Percent of Votes for Candidate A in the Sample
5	2	40%
12	4	33.3%
20	12	60%
30	17	56.7%
50	29	58%
125	73	58.4%
150	88	58.7%
200	118	59%

a. Based on the results of the first two sample surveys, do you think Candidate A will win the election? Explain.

b. Based on the results in the table, do you think Candidate A will win the election? Explain.

SOLUTION

a. The results of the first two surveys (sizes 5 and 12) show that fewer than 50% of the residents will vote for Candidate A. Because there are only two candidates, one candidate needs more than 50% of the votes to win.

▶ Based on these surveys, you can predict Candidate A will not win the election.

b. As the sample sizes increase, the estimated percent of votes approaches 59%. You can predict that 59% of the city residents will vote for Candidate A.

▶ Because 59% of the votes are more than the 50% needed to win, you should feel confident that Candidate A will win the election.

Monitoring Progress

Help in English and Spanish at *BigIdeasMath.com*

2. Two candidates are running for class president. The table shows the results of four surveys of random students in the class. The students were asked whether they will vote for the incumbent. Do you think the incumbent will be reelected? Explain.

Sample Size	Number of "Yes" Responses	Percent of Votes for Incumbent
10	7	70%
20	11	55%
30	13	43.3%
40	17	42.5%

Analyzing Estimated Population Parameters

An estimated population parameter is a hypothesis. You learned in Section 11.2 that one way to analyze a hypothesis is to perform a simulation.

EXAMPLE 3 Analyzing an Estimated Population Proportion

A national polling company claims 34% of U.S. adults say mathematics is the most valuable school subject in their lives. You survey a random sample of 50 adults.

a. What can you conclude about the accuracy of the claim that the population proportion is 0.34 when 15 adults in your survey say mathematics is the most valuable subject?

b. What can you conclude about the accuracy of the claim when 25 adults in your survey say mathematics is the most valuable subject?

c. Assume that the true population proportion is 0.34. Estimate the variation among sample proportions using samples of size 50.

SOLUTION

```
randInt(0,99,50)
{76 10 27 54 41...
```

The polling company's claim (hypothesis) is that the population proportion of U.S. adults who say mathematics is the most valuable school subject is 0.34. To analyze this claim, simulate choosing 80 random samples of size 50 using a random number generator on a graphing calculator. Generate 50 random numbers from 0 to 99 for each sample. Let numbers 1 through 34 represent adults who say math. Find the sample proportions and make a dot plot showing the distribution of the sample proportions.

Simulation: Polling 50 Adults

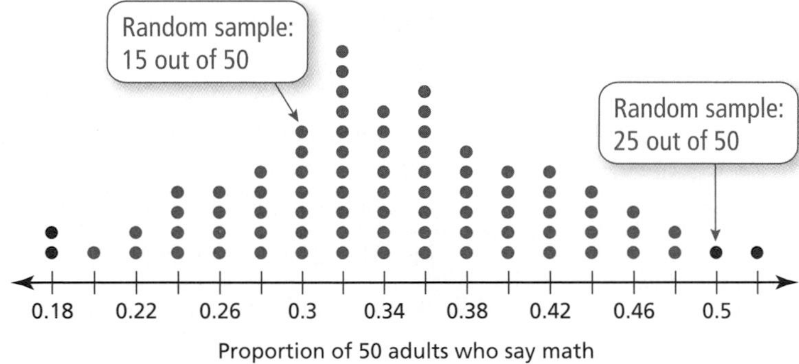

STUDY TIP

The dot plot shows the results of one simulation. Results of other simulations may give slightly different results but the shape should be similar.

a. Note that 15 out of 50 corresponds to a sample proportion of $\frac{15}{50} = 0.3$. In the simulation, this result occurred in 7 of the 80 random samples. It is *likely* that 15 adults out of 50 would say math is the most valuable subject when the true population percentage is 34%. So, you can conclude the company's claim is probably accurate.

b. Note that 25 out of 50 corresponds to a sample proportion of $\frac{25}{50} = 0.5$. In the simulation, this result occurred in only 1 of the 80 random samples. So, it is *unlikely* that 25 adults out of 50 would say math is the most valuable subject when the true population percentage is 34%. So, you can conclude the company's claim is probably *not* accurate.

c. Note that the dot plot is fairly bell-shaped and symmetric, so the distribution is approximately normal. In a normal distribution, you know that about 95% of the possible sample proportions will lie within two standard deviations of 0.34. Excluding the two least and two greatest sample proportions, represented by red dots ● in the dot plot, leaves 76 of 80, or 95%, of the sample proportions. These 76 proportions range from 0.2 to 0.48. So, 95% of the time, a sample proportion should lie in the interval from 0.2 to 0.48.

INTERPRETING MATHEMATICAL RESULTS

Note that the sample proportion 0.3 in part (a) lies in this interval, while the sample proportion 0.5 in part (b) falls outside this interval.

Monitoring Progress

Help in English and Spanish at *BigIdeasMath.com*

3. **WHAT IF?** In Example 3, what can you conclude about the accuracy of the claim that the population proportion is 0.34 when 21 adults in your random sample say mathematics is the most valuable subject?

Finding Margins of Error for Surveys

When conducting a survey, you need to make the size of your sample large enough so that it accurately represents the population. As the sample size increases, the *margin of error* decreases.

The **margin of error** gives a limit on how much the responses of the sample would differ from the responses of the population. For example, if 40% of the people in a poll favor a new tax law, and the margin of error is ±4%, then it is likely that between 36% and 44% of the entire population favor a new tax law.

Core Concept

Margin of Error Formula

When a random sample of size n is taken from a large population, the margin of error is approximated by

$$\text{Margin of error} = \pm\frac{1}{\sqrt{n}}.$$

This means that if the percent of the sample responding a certain way is p (expressed as a decimal), then the percent of the population who would respond the same way is likely to be between $p - \frac{1}{\sqrt{n}}$ and $p + \frac{1}{\sqrt{n}}$.

EXAMPLE 4 Finding a Margin of Error

In a survey of 2048 people in the U.S., 55% said that television is their main source of news. (a) What is the margin of error for the survey? (b) Give an interval that is likely to contain the exact percent of all people who use television as their main source of news.

Americans' Main News Source

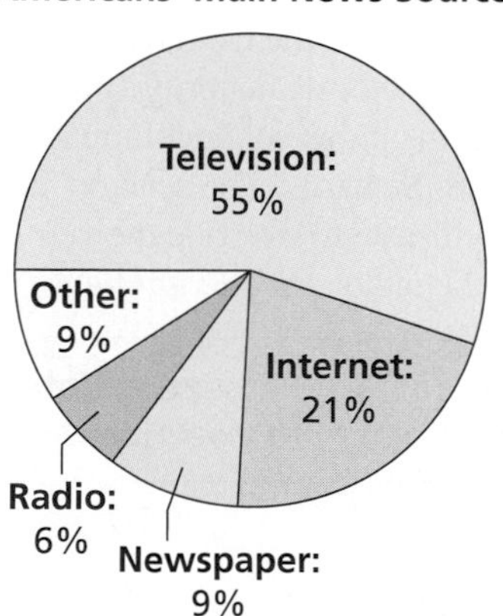

SOLUTION

a. Use the margin of error formula.

$$\text{Margin of error} = \pm\frac{1}{\sqrt{n}} = \pm\frac{1}{\sqrt{2048}} \approx \pm 0.022$$

▶ The margin of error for the survey is about ±2.2%.

b. To find the interval, subtract and add 2.2% to the percent of people surveyed who said television is their main source of news (55%).

$$55\% - 2.2\% = 52.8\% \qquad 55\% + 2.2\% = 57.2\%$$

▶ It is likely that the exact percent of all people in the U.S. who use television as their main source of news is between 52.8% and 57.2%.

Monitoring Progress

Help in English and Spanish at *BigIdeasMath.com*

4. In a survey of 1028 people in the U.S., 87% reported using the Internet. Give an interval that is likely to contain the exact percent of all people in the U.S. who use the Internet.

11.5 Exercises

Dynamic Solutions available at *BigIdeasMath.com*

Vocabulary and Core Concept Check

1. **COMPLETE THE SENTENCE** The __________ gives a limit on how much the responses of the sample would differ from the responses of the population.

2. **WRITING** What is the difference between descriptive and inferential statistics?

Monitoring Progress and Modeling with Mathematics

3. **PROBLEM SOLVING** The numbers of text messages sent each day by a random sample of 30 teen cellphone users are shown in the table. Estimate the population mean μ. *(See Example 1.)*

Number of Text Messages				
30	60	59	83	41
37	66	63	60	92
53	42	47	32	79
53	80	41	51	85
73	71	69	31	69
57	60	70	91	67

4. **PROBLEM SOLVING** The incomes for a random sample of 35 U.S. households are shown in the table. Estimate the population mean μ.

Income of U.S. Households				
14,300	52,100	74,800	51,000	91,500
72,800	50,500	15,000	37,600	22,100
40,000	65,400	50,000	81,100	99,800
43,300	32,500	76,300	83,400	24,600
30,800	62,100	32,800	21,900	64,400
73,100	20,000	49,700	71,000	45,900
53,200	45,500	55,300	19,100	63,100

5. **PROBLEM SOLVING** Use the data in Exercise 3 to answer each question.

   a. Estimate the population proportion p of teen cellphone users who send more than 70 text messages each day.

   b. Estimate the population proportion p of teen cellphone users who send fewer than 50 text messages each day.

6. **WRITING** A survey asks a random sample of U.S. teenagers how many hours of television they watch each night. The survey reveals that the sample mean is 3 hours per night. How confident are you that the average of all U.S. teenagers is exactly 3 hours per night? Explain your reasoning.

7. **DRAWING CONCLUSIONS** When the President of the United States vetoes a bill, the Congress can override the veto by a two-thirds majority vote in each House. Five news organizations conduct individual random surveys of U.S. Senators. The senators are asked whether they will vote to override the veto. The results are shown in the table. *(See Example 2.)*

Sample Size	Number of Votes to Override Veto	Percent of Votes to Override Veto
7	6	85.7%
22	16	72.7%
28	21	75%
31	17	54.8%
49	27	55.1%

   a. Based on the results of the first two surveys, do you think the Senate will vote to override the veto? Explain.

   b. Based on the results in the table, do you think the Senate will vote to override the veto? Explain.

8. **DRAWING CONCLUSIONS** Your teacher lets the students decide whether to have their test on Friday or Monday. The table shows the results from four surveys of randomly selected students in your grade who are taking the same class. The students are asked whether they want to have the test on Friday.

Sample Size	Number of "Yes" Responses	Percent of Votes
10	8	80%
20	12	60%
30	16	53.3%
40	18	45%

a. Based on the results of the first two surveys, do you think the test will be on Friday? Explain.

b. Based on the results in the table, do you think the test will be on Friday? Explain.

9. **MODELING WITH MATHEMATICS** A national polling company claims that 54% of U.S. adults are married. You survey a random sample of 50 adults. *(See Example 3.)*

a. What can you conclude about the accuracy of the claim that the population proportion is 0.54 when 31 adults in your survey are married?

b. What can you conclude about the accuracy of the claim that the population proportion is 0.54 when 19 adults in your survey are married?

c. Assume that the true population proportion is 0.54. Estimate the variation among sample proportions for samples of size 50.

10. **MODELING WITH MATHEMATICS** Employee engagement is the level of commitment and involvement an employee has toward the company and its values. A national polling company claims that only 29% of U.S. employees feel engaged at work. You survey a random sample of 50 U.S. employees.

a. What can you conclude about the accuracy of the claim that the population proportion is 0.29 when 16 employees feel engaged at work?

b. What can you conclude about the accuracy of the claim that the population proportion is 0.29 when 23 employees feel engaged at work?

c. Assume that the true population proportion is 0.29. Estimate the variation among sample proportions for samples of size 50.

In Exercises 11–16, find the margin of error for a survey that has the given sample size. Round your answer to the nearest tenth of a percent.

11. 260
12. 1000
13. 2024
14. 6400
15. 3275
16. 750

17. **ATTENDING TO PRECISION** In a survey of 1020 U.S. adults, 41% said that their top priority for saving is retirement. *(See Example 4.)*

a. What is the margin of error for the survey?

b. Give an interval that is likely to contain the exact percent of all U.S. adults whose top priority for saving is retirement.

18. **ATTENDING TO PRECISION** In a survey of 1022 U.S. adults, 76% said that more emphasis should be placed on producing domestic energy from solar power.

a. What is the margin of error for the survey?

b. Give an interval that is likely to contain the exact percent of all U.S. adults who think more emphasis should be placed on producing domestic energy from solar power.

19. **ERROR ANALYSIS** In a survey, 8% of adult Internet users said they participate in sports fantasy leagues online. The margin of error is ±4%. Describe and correct the error in calculating the sample size.

✗

$$\pm 0.08 = \pm \frac{1}{\sqrt{n}}$$

$$0.0064 = \frac{1}{n}$$

$$n \approx 156$$

20. **ERROR ANALYSIS** In a random sample of 2500 consumers, 61% prefer Game A over Game B. Describe and correct the error in giving an interval that is likely to contain the exact percent of all consumers who prefer Game A over Game B.

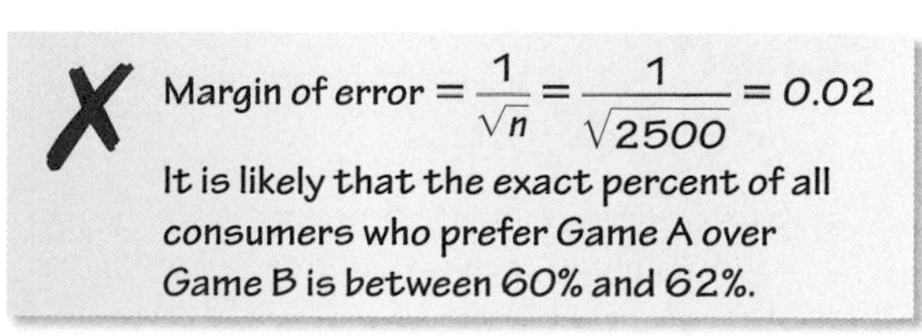

Margin of error = 1/√n = 1/√2500 = 0.02
It is likely that the exact percent of all consumers who prefer Game A over Game B is between 60% and 62%.

21. **MAKING AN ARGUMENT** Your friend states that it is possible to have a margin of error between 0 and 100 percent, not including 0 or 100 percent. Is your friend correct? Explain your reasoning.

22. **HOW DO YOU SEE IT?** The figure shows the distribution of the sample proportions from three simulations using different sample sizes. Which simulation has the least margin of error? the greatest? Explain your reasoning.

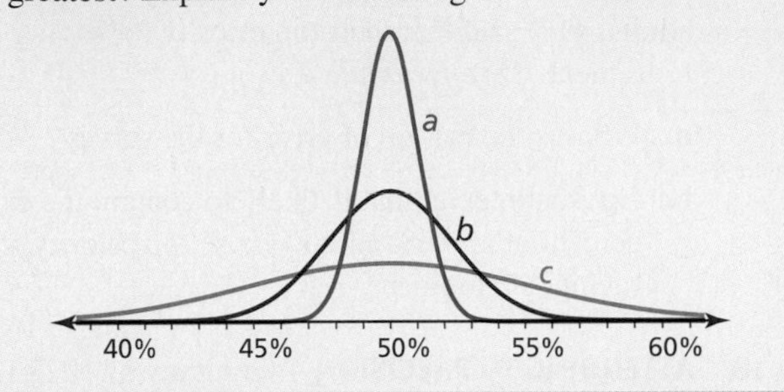

23. **REASONING** A developer claims that the percent of city residents who favor building a new football stadium is likely between 52.3% and 61.7%. How many residents were surveyed?

24. **ABSTRACT REASONING** Suppose a random sample of size n is required to produce a margin of error of $\pm E$. Write an expression in terms of n for the sample size needed to reduce the margin of error to $\pm \frac{1}{2}E$. How many times must the sample size be increased to cut the margin of error in half? Explain.

25. **PROBLEM SOLVING** A survey reported that 47% of the voters surveyed, or about 235 voters, said they voted for Candidate A and the remainder said they voted for Candidate B.

a. How many voters were surveyed?

b. What is the margin of error for the survey?

c. For each candidate, find an interval that is likely to contain the exact percent of all voters who voted for the candidate.

d. Based on your intervals in part (c), can you be confident that Candidate B won? If not, how many people in the sample would need to vote for Candidate B for you to be confident that Candidate B won? (*Hint:* Find the least number of voters for Candidate B so that the intervals do not overlap.)

26. **THOUGHT PROVOKING** Consider a large population in which p percent (in decimal form) have a certain characteristic. To be reasonably sure that you are choosing a sample that is representative of a population, you should choose a random sample of n people where

$$n > 9\left(\frac{1-p}{p}\right).$$

a. Suppose $p = 0.5$. How large does n need to be?

b. Suppose $p = 0.01$. How large does n need to be?

c. What can you conclude from parts (a) and (b)?

27. **CRITICAL THINKING** In a survey, 52% of the respondents said they prefer sports drink X and 48% said they prefer sports drink Y. How many people would have to be surveyed for you to be confident that sports drink X is truly preferred by more than half the population? Explain.

Maintaining Mathematical Proficiency

Reviewing what you learned in previous grades and lessons

Find the inverse of the function. *(Section 6.3)*

28. $y = 10^{x-3}$

29. $y = 2^x - 5$

30. $y = \ln(x + 5)$

31. $y = \log_6 x - 1$

Determine whether the graph represents an arithmetic sequence or a geometric sequence. Then write a rule for the nth term. *(Section 8.2 and Section 8.3)*

32.

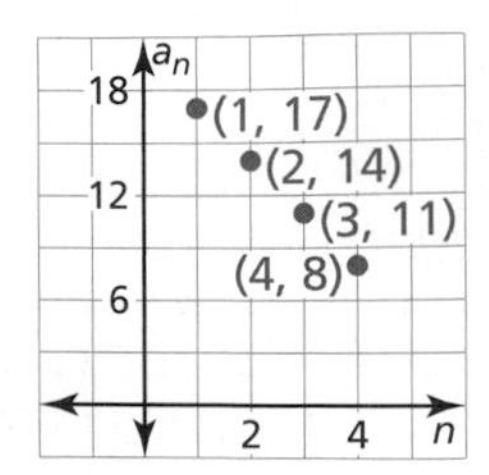

33.

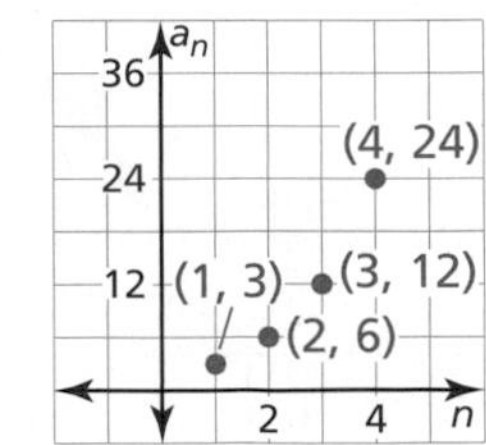

34.

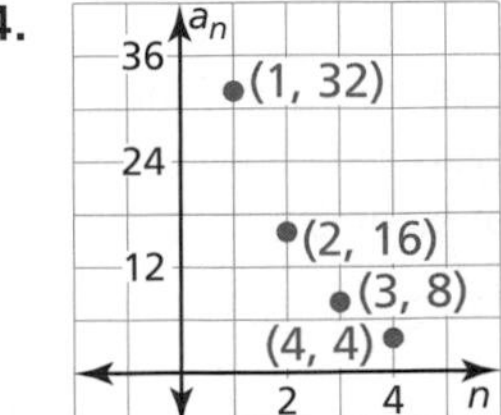

11.6 Making Inferences from Experiments

Essential Question How can you test a hypothesis about an experiment?

EXPLORATION 1 **Resampling Data**

Work with a partner. A randomized comparative experiment tests whether water with dissolved calcium affects the yields of yellow squash plants. The table shows the results.

Yield (kilograms)	
Control Group	**Treatment Group**
1.0	1.1
1.2	1.3
1.5	1.4
0.9	1.2
1.1	1.0
1.4	1.7
0.8	1.8
0.9	1.1
1.3	1.1
1.6	1.8

a. Find the mean yield of the control group and the mean yield of the treatment group. Then find the difference of the two means. Record the results.

b. Write each yield measurement from the table on an equal-sized piece of paper. Place the pieces of paper in a bag, shake, and randomly choose 10 pieces of paper. Call this the "control" group, and call the 10 pieces in the bag the "treatment" group. Then repeat part (a) and return the pieces to the bag. Perform this resampling experiment five times.

c. How does the difference in the means of the control and treatment groups compare with the differences resulting from chance?

EXPLORATION 2 **Evaluating Results**

Work as a class. To conclude that the treatment is responsible for the difference in yield, you need strong evidence to reject the hypothesis:

Water dissolved in calcium has no effect on the yields of yellow squash plants.

To evaluate this hypothesis, compare the experimental difference of means with the resampling differences.

a. Collect all the resampling differences of means found in Exploration 1(b) for the whole class and display these values in a histogram.

b. Draw a vertical line on your class histogram to represent the experimental difference of means found in Exploration 1(a).

c. Where on the histogram should the experimental difference of means lie to give evidence for rejecting the hypothesis?

d. Is your class able to reject the hypothesis? Explain your reasoning.

MODELING WITH MATHEMATICS

To be proficient in math, you need to identify important quantities in a practical situation, map their relationships using such tools as diagrams and graphs, and analyze those relationships mathematically to draw conclusions.

Communicate Your Answer

3. How can you test a hypothesis about an experiment?

4. The randomized comparative experiment described in Exploration 1 is replicated and the results are shown in the table. Repeat Explorations 1 and 2 using this data set. Explain any differences in your answers.

	Yield (kilograms)									
Control Group	0.9	0.9	1.4	0.6	1.0	1.1	0.7	0.6	1.2	1.3
Treatment Group	1.0	1.2	1.2	1.3	1.0	1.8	1.7	1.2	1.0	1.9

11.6 Lesson

Core Vocabulary

Previous
randomized comparative experiment
control group
treatment group
mean
dot plot
outlier
simulation
hypothesis

What You Will Learn

- Organize data from an experiment with two samples.
- Resample data using a simulation to analyze a hypothesis.
- Make inferences about a treatment.

Experiments with Two Samples

In this lesson, you will compare data from two samples in an experiment to make inferences about a treatment using a method called *resampling*. Before learning about this method, consider the experiment described in Example 1.

EXAMPLE 1 Organizing Data from an Experiment

A randomized comparative experiment tests whether a soil supplement affects the total yield (in kilograms) of cherry tomato plants. The control group has 10 plants and the treatment group, which receives the soil supplement, has 10 plants. The table shows the results.

	Total Yield of Tomato Plants (kilograms)									
Control Group	1.2	1.3	0.9	1.4	2.0	1.2	0.7	1.9	1.4	1.7
Treatment Group	1.4	0.9	1.5	1.8	1.6	1.8	2.4	1.9	1.9	1.7

a. Find the mean yield of the control group, $\overline{x}_{\text{control}}$.

b. Find the mean yield of the treatment group, $\overline{x}_{\text{treatment}}$.

c. Find the experimental difference of the means, $\overline{x}_{\text{treatment}} - \overline{x}_{\text{control}}$.

d. Display the data in a double dot plot.

e. What can you conclude?

SOLUTION

a. $\overline{x}_{\text{control}} = \dfrac{1.2 + 1.3 + 0.9 + 1.4 + 2.0 + 1.2 + 0.7 + 1.9 + 1.4 + 1.7}{10} = \dfrac{13.7}{10} = 1.37$

▶ The mean yield of the control group is 1.37 kilograms.

b. $\overline{x}_{\text{treatment}} = \dfrac{1.4 + 0.9 + 1.5 + 1.8 + 1.6 + 1.8 + 2.4 + 1.9 + 1.9 + 1.7}{10} = \dfrac{16.9}{10} = 1.69$

▶ The mean yield of the treatment group is 1.69 kilograms.

c. $\overline{x}_{\text{treatment}} - \overline{x}_{\text{control}} = 1.69 - 1.37 = 0.32$

▶ The experimental difference of the means is 0.32 kilogram.

d.

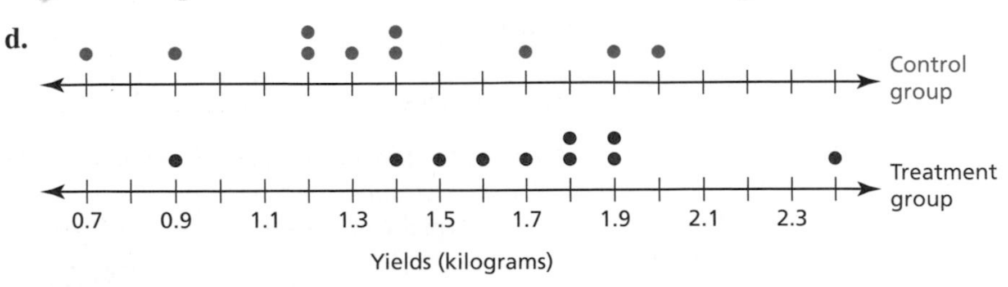

e. The plot of the data shows that the two data sets tend to be fairly symmetric and have no extreme values (outliers). So, the mean is a suitable measure of center. The mean yield of the treatment group is 0.32 kilogram more than the control group. It appears that the soil supplement might be slightly effective, but the sample size is small and the difference could be due to chance.

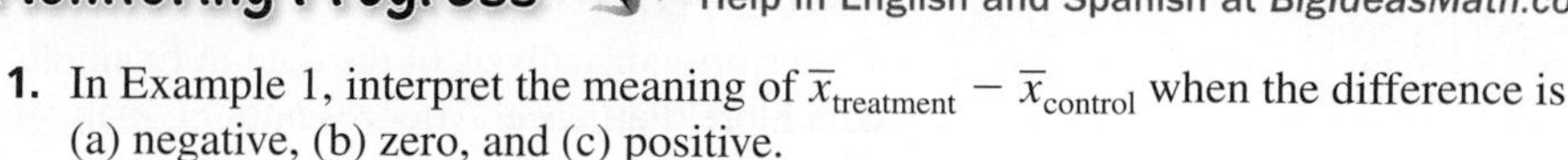

1. In Example 1, interpret the meaning of $\overline{x}_{\text{treatment}} - \overline{x}_{\text{control}}$ when the difference is (a) negative, (b) zero, and (c) positive.

Resampling Data Using a Simulation

The samples in Example 1 are too small to make inferences about the treatment. Statisticians have developed a method called resampling to overcome this problem. Here is one way to resample: combine the measurements from both groups, and repeatedly create new "control" and "treatment" groups at random from the measurements without repeats. Example 2 shows one resampling of the data in Example 1.

EXAMPLE 2 Resampling Data Using a Simulation

Resample the data in Example 1 using a simulation. Use the mean yields of the new control and treatment groups to calculate the difference of the means.

SOLUTION

Step 1 Combine the measurements from both groups and assign a number to each value. Let the numbers 1 through 10 represent the data in the original control group, and let the numbers 11 through 20 represent the data in the original treatment group, as shown.

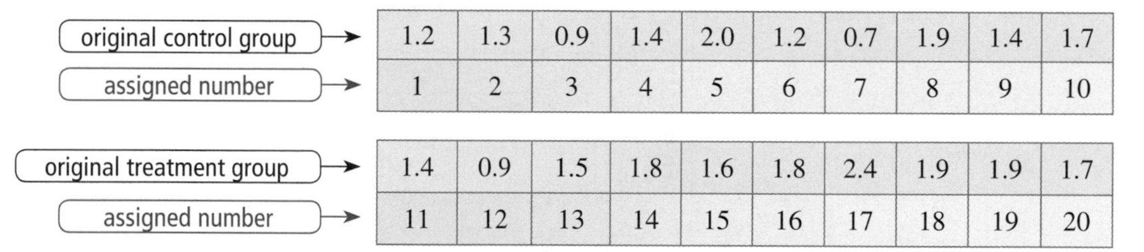

original control group →	1.2	1.3	0.9	1.4	2.0	1.2	0.7	1.9	1.4	1.7
assigned number →	1	2	3	4	5	6	7	8	9	10

original treatment group →	1.4	0.9	1.5	1.8	1.6	1.8	2.4	1.9	1.9	1.7
assigned number →	11	12	13	14	15	16	17	18	19	20

Step 2 Use a random number generator. Randomly generate 20 numbers from 1 through 20 *without repeating a number*. The table shows the results.

```
randIntNoRep(1,20)
{14 19 4 3 18 9...
```

14	19	4	3	18	9	5	15	2	7
1	17	20	16	6	8	13	12	11	10

Use the first 10 numbers to make the new control group, and the next 10 to make the new treatment group. The results are shown in the next table.

	Resample of Tomato Plant Yields (kilograms)									
New Control Group	1.8	1.9	1.4	0.9	1.9	1.4	2.0	1.6	1.3	0.7
New Treatment Group	1.2	2.4	1.7	1.8	1.2	1.9	1.5	0.9	1.4	1.7

Step 3 Find the mean yields of the new control and treatment groups.

$$\overline{x}_{\text{new control}} = \frac{1.8 + 1.9 + 1.4 + 0.9 + 1.9 + 1.4 + 2.0 + 1.6 + 1.3 + 0.7}{10} = \frac{14.9}{10} = 1.49$$

$$\overline{x}_{\text{new treatment}} = \frac{1.2 + 2.4 + 1.7 + 1.8 + 1.2 + 1.9 + 1.5 + 0.9 + 1.4 + 1.7}{10} = \frac{15.7}{10} = 1.57$$

▶ So, $\overline{x}_{\text{new treatment}} - \overline{x}_{\text{new control}} = 1.57 - 1.49 = 0.08$. This is less than the experimental difference found in Example 1.

Making Inferences About a Treatment

To perform an analysis of the data in Example 1, you will need to resample the data more than once. After resampling many times, you can see how often you get differences between the new groups that are at least as large as the one you measured.

EXAMPLE 3 Making Inferences About a Treatment

To conclude that the treatment in Example 1 is responsible for the difference in yield, you need to analyze this hypothesis:

The soil nutrient has no effect on the yield of the cherry tomato plants.

Simulate 200 resamplings of the data in Example 1. Compare the experimental difference of 0.32 from Example 1 with the resampling differences. What can you conclude about the hypothesis? Does the soil nutrient have an effect on the yield?

SOLUTION

The histogram shows the results of the simulation. The histogram is approximately bell-shaped and fairly symmetric, so the differences have an approximately normal distribution.

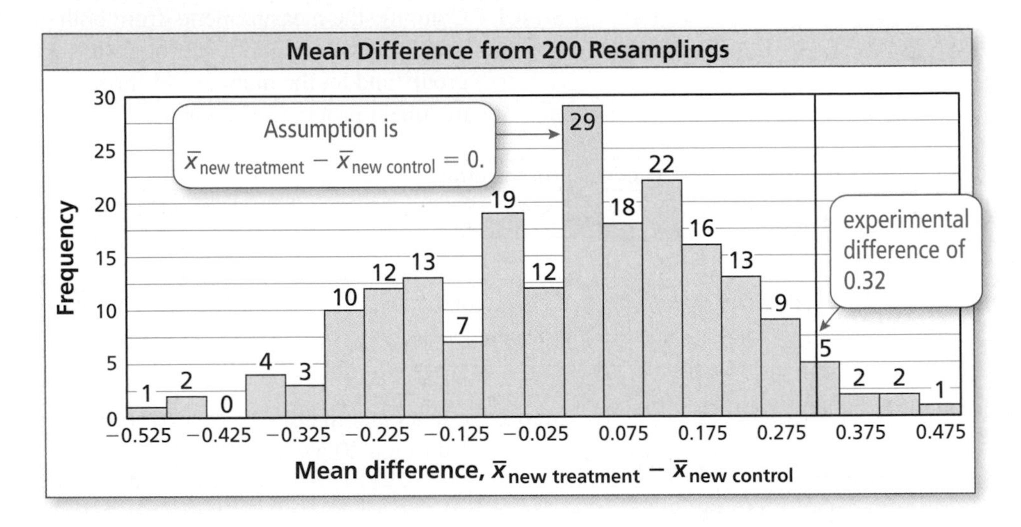

Note that the hypothesis assumes that the difference of the mean yields is 0. The experimental difference of 0.32, however, lies close to the right tail. From the graph, there are about 5 to 10 values out of 200 that are greater than 0.32, which is at most 5% of the values. Also, the experimental difference falls outside the middle 90% of the resampling differences. (The middle 90% is the area of the bars from −0.275 to 0.275, which contains 180 of the 200 values, or 90%.) This means it is unlikely to get a difference this large when you assume that the difference is 0, suggesting the control group and the treatment group differ.

INTERPRETING MATHEMATICAL RESULTS

With this conclusion, you can be 90% confident that the soil supplement does have an effect.

▶ You can conclude that the hypothesis is most likely false. So, the soil nutrient *does* have an effect on the yield of cherry tomato plants. Because the mean difference is positive, the treatment *increases* the yield.

Monitoring Progress  Help in English and Spanish at *BigIdeasMath.com*

2. In Example 3, what are the consequences of concluding that the hypothesis is false when it is actually true?

11.6 Exercises

Dynamic Solutions available at *BigIdeasMath.com*

Vocabulary and Core Concept Check

1. **COMPLETE THE SENTENCE** A method in which new samples are repeatedly drawn from the data set is called ____________.

2. **DIFFERENT WORDS, SAME QUESTION** Which is different? Find "both" answers.

What is the experimental difference of the means?

What is $\bar{x}_{\text{treatment}} - \bar{x}_{\text{control}}$?

	Weight of Tumor (grams)					
Control Group	3.3	3.2	3.7	3.5	3.3	3.4
Treatment Group	0.4	0.6	0.5	0.6	0.7	0.5

What is the square root of the average of the squared differences from −2.85?

What is the difference between the mean of the treatment group and the mean of the control group?

Monitoring Progress and Modeling with Mathematics

3. **PROBLEM SOLVING** A randomized comparative experiment tests whether music therapy affects the depression scores of college students. The depression scores range from 20 to 80, with scores greater than 50 being associated with depression. The control group has eight students and the treatment group, which receives the music therapy, has eight students. The table shows the results. *(See Example 1.)*

	Depression Score			
Control Group	49	45	43	47
Treatment Group	39	40	39	37
Control Group	46	45	47	46
Treatment Group	41	40	42	43

a. Find the mean score of the control group.

b. Find the mean score of the treatment group.

c. Find the experimental difference of the means.

d. Display the data in a double dot plot.

e. What can you conclude?

4. **PROBLEM SOLVING** A randomized comparative experiment tests whether low-level laser therapy affects the waist circumference of adults. The control group has eight adults and the treatment group, which receives the low-level laser therapy, has eight adults. The table shows the results.

	Circumference (inches)			
Control Group	34.6	35.4	33	34.6
Treatment Group	31.4	33	32.4	32.6
Control Group	35.2	35.2	36.2	35
Treatment Group	33.4	33.4	34.8	33

a. Find the mean circumference of the control group.

b. Find the mean circumference of the treatment group.

c. Find the experimental difference of the means.

d. Display the data in a double dot plot.

e. What can you conclude?

5. **ERROR ANALYSIS** In a randomized comparative experiment, the mean score of the treatment group is 11 and the mean score of the control group is 16. Describe and correct the error in interpreting the experimental difference of the means.

$\bar{x}_{\text{control}} - \bar{x}_{\text{treatment}} = 16 - 11 = 5$

So, you can conclude the treatment increases the score.

6. **REASONING** In Exercise 4, interpret the meaning of $\bar{x}_{\text{treatment}} - \bar{x}_{\text{control}}$ when the difference is positive, negative, and zero.

7. **MODELING WITH MATHEMATICS** Resample the data in Exercise 3 using a simulation. Use the means of the new control and treatment groups to calculate the difference of the means. *(See Example 2.)*

8. **MODELING WITH MATHEMATICS** Resample the data in Exercise 4 using a simulation. Use the means of the new control and treatment groups to calculate the difference of the means.

9. **DRAWING CONCLUSIONS** To analyze the hypothesis below, use the histogram which shows the results from 200 resamplings of the data in Exercise 3.

   Music therapy has no effect on the depression score.

   Compare the experimental difference in Exercise 3 with the resampling differences. What can you conclude about the hypothesis? Does music therapy have an effect on the depression score? *(See Example 3.)*

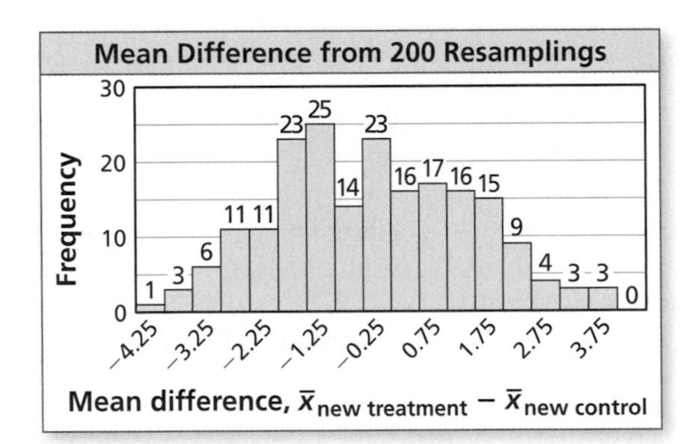

10. **DRAWING CONCLUSIONS** Suppose the experimental difference of the means in Exercise 3 had been -0.75. Compare this experimental difference of means with the resampling differences in the histogram in Exercise 9. What can you conclude about the hypothesis? Does music therapy have an effect on the depression score?

11. **WRITING** Compare the histogram in Exercise 9 to the histogram below. Determine which one provides stronger evidence against the hypothesis, *Music therapy has no effect on the depression score*. Explain.

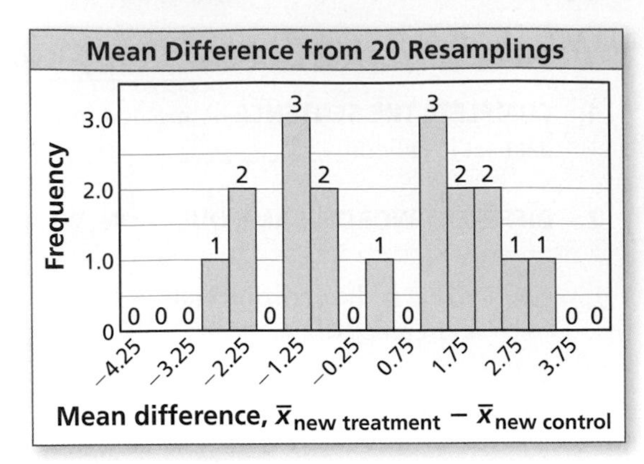

12. **HOW DO YOU SEE IT?** Without calculating, determine whether the experimental difference, $\bar{x}_{\text{treatment}} - \bar{x}_{\text{control}}$, is positive, negative, or zero. What can you conclude about the effect of the treatment? Explain.

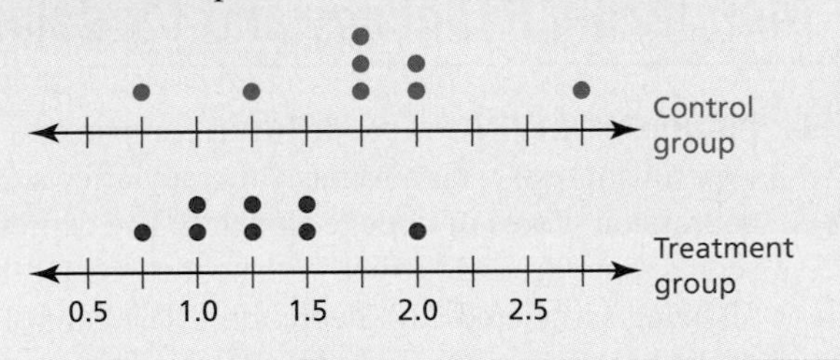

13. **MAKING AN ARGUMENT** Your friend states that the mean of the resampling differences of the means should be close to 0 as the number of resamplings increase. Is your friend correct? Explain your reasoning.

14. **THOUGHT PROVOKING** Describe an example of an observation that can be made from an experiment. Then give four possible inferences that could be made from the observation.

15. **CRITICAL THINKING** In Exercise 4, how many resamplings of the treatment and control groups are theoretically possible? Explain.

Maintaining Mathematical Proficiency

Reviewing what you learned in previous grades and lessons

Factor the polynomial completely. *(Section 4.4)*

16. $5x^3 - 15x^2$
17. $y^3 - 8$
18. $z^3 + 5z^2 - 9z - 45$
19. $81w^4 - 16$

Determine whether the inverse of f is a function. Then find the inverse. *(Section 7.5)*

20. $f(x) = \dfrac{3}{x + 5}$
21. $f(x) = \dfrac{1}{2x - 1}$
22. $f(x) = \dfrac{2}{x} - 4$
23. $f(x) = \dfrac{3}{x^2} + 1$

11.4–11.6 What Did You Learn?

Core Vocabulary

controlled experiment, *p. 620*
control group, *p. 620*
treatment group, *p. 620*
randomization, *p. 620*
randomized comparative experiment, *p. 620*
placebo, *p. 620*
replication, *p. 622*
descriptive statistics, *p. 626*
inferential statistics, *p. 626*
margin of error, *p. 629*

Core Concepts

Section 11.4

Randomization in Experiments and Observational Studies, *p. 621*
Comparative Studies and Causality, *p. 621*
Analyzing Experimental Designs, *p. 622*

Section 11.5

Estimating Population Parameters, *p. 626*
Analyzing Estimated Population Parameters, *p. 628*

Section 11.6

Experiments with Two Samples, *p. 634*
Resampling Data Using Simulations, *p. 635*
Making Inferences About Treatments, *p. 636*

Mathematical Practices

1. In Exercise 7 on page 623, find a partner and discuss your answers. What questions should you ask your partner to determine whether an observational study or an experiment is more appropriate?
2. In Exercise 23 on page 632, how did you use the given interval to find the sample size?

Performance Task

Curving the Test

Test scores are sometimes curved for different reasons using different techniques. Curving began with the assumption that a good test would result in scores that were normally distributed about a C average. Is this assumption valid? Are test scores in your class normally distributed? If not, how are they distributed? Which curving algorithms preserve the distribution and which algorithms change it?

To explore the answers to these questions and more, go to ***BigIdeasMath.com***.

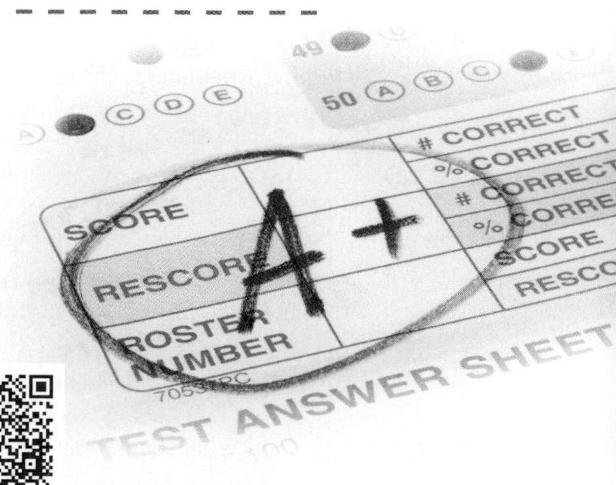

11 Chapter Review

Dynamic Solutions available at *BigIdeasMath.com*

11.1 Using Normal Distributions *(pp. 595–602)*

A normal distribution has mean μ and standard deviation σ. An x-value is randomly selected from the distribution. Find $P(\mu - 2\sigma \le x \le \mu + 3\sigma)$.

The probability that a randomly selected x-value lies between $\mu - 2\sigma$ and $\mu + 3\sigma$ is the shaded area under the normal curve shown.

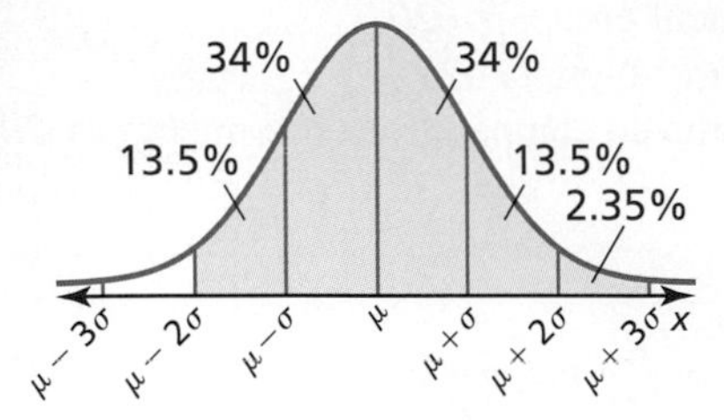

▶ $P(\mu - 2\sigma \le x \le \mu + 3\sigma) = 0.135 + 0.34 + 0.34 + 0.135 + 0.0235 = 0.9735$

1. A normal distribution has mean μ and standard deviation σ. An x-value is randomly selected from the distribution. Find $P(x \le \mu - 3\sigma)$.
2. The scores received by juniors on the math portion of the PSAT are normally distributed with a mean of 48.6 and a standard deviation of 11.4. What is the probability that a randomly selected score is at least 76?

11.2 Populations, Samples, and Hypotheses *(pp. 603–608)*

You suspect a die favors the number six. The die maker claims the die does not favor any number. What should you conclude when you roll the actual die 50 times and get a six 13 times?

The maker's claim, or hypothesis, is "the die does not favor any number." This is the same as saying that the proportion of sixes rolled, in the long run, is $\frac{1}{6}$. So, assume the probability of rolling a six is $\frac{1}{6}$. Simulate the rolling of the die by repeatedly drawing 200 random samples of size 50 from a population of numbers from one through six. Make a histogram of the distribution of the sample proportions.

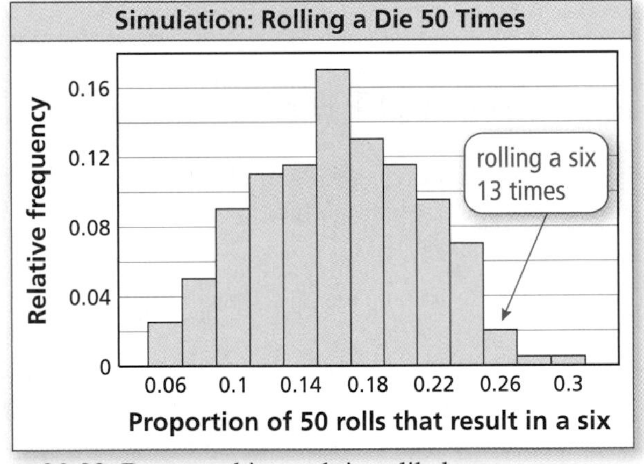

▶ Getting a six 13 times corresponds to a proportion of $\frac{13}{50} = 0.26$. In the simulation, this result had a relative frequency of 0.02. Because this result is unlikely to occur by chance, you can conclude that the maker's claim is most likely false.

3. To estimate the average number of miles driven by U.S. motorists each year, a researcher conducts a survey of 1000 drivers, records the number of miles they drive in a year, and then determines the average. Identify the population and the sample.
4. A pitcher throws 40 fastballs in a game. A baseball analyst records the speeds of 10 fastballs and finds that the mean speed is 92.4 miles per hour. Is the mean speed a parameter or a statistic? Explain.
5. A prize on a game show is placed behind either Door A or Door B. You suspect the prize is more often behind Door A. The show host claims the prize is randomly placed behind either door. What should you conclude when the prize is behind Door A for 32 out of 50 contestants?

11.3 Collecting Data (pp. 609–616)

You want to determine how many people in the senior class plan to study mathematics after high school. You survey every senior in your calculus class. Identify the type of sample described and determine whether the sample is biased.

▶ You select students who are readily available. So, the sample is a *convenience* sample. The sample is biased because students in a calculus class are more likely to study mathematics after high school.

6. A researcher wants to determine how many people in a city support the construction of a new road connecting the high school to the north side of the city. Fifty residents from each side of the city are surveyed. Identify the type of sample described and determine whether the sample is biased.

7. A researcher records the number of people who use a coupon when they dine at a certain restaurant. Identify the method of data collection.

8. Explain why the survey question below may be biased or otherwise introduce bias into the survey. Then describe a way to correct the flaw.

"Do you think the city should replace the outdated police cars it is using?"

11.4 Experimental Design (pp. 619–624)

Determine whether the study is a randomized comparative experiment. If it is, describe the treatment, the treatment group, and the control group. If it is not, explain why not and discuss whether the conclusions drawn from the study are valid.

Headphones Hurt Hearing
A study of 100 college and high school students compared their times spent listening to music using headphones with hearing loss. Twelve percent of people who listened to headphones more than one hour per day were found to have measurable hearing loss over the course of the three-year study.

▶ The study is not a randomized comparative experiment because the individuals were not randomly assigned to a control group and a treatment group. The conclusion that headphone use impairs hearing ability may or may not be valid. For instance, people who listen to more than an hour of music per day may be more likely to attend loud concerts that are known to affect hearing.

9. A restaurant manager wants to know which type of sandwich bread attracts the most repeat customers. Is the topic best investigated through an experiment or an observational study? Describe how you would design the experiment or observational study.

10. A researcher wants to test the effectiveness of a sleeping pill. Identify a potential problem, if any, with the experimental design below. Then describe how you can improve it.

The researcher asks for 16 volunteers who have insomnia. Eight volunteers are given the sleeping pill and the other 8 volunteers are given a placebo. Results are recorded for 1 month.

11. Determine whether the study is a randomized comparative experiment. If it is, describe the treatment, the treatment group, and the control group. If it is not, explain why not and discuss whether the conclusions drawn from the study are valid.

Cleaner Cars in Less Time!
To test the new design of a car wash, an engineer gathered 80 customers and randomly divided them into two groups. One group used the old design to wash their cars and one group used the new design to wash their cars. Users of the new car wash design were able to wash their cars 30% faster.

11.5 Making Inferences from Sample Surveys (pp. 625–632)

Before the Thanksgiving holiday, in a survey of 2368 people, 85% said they are thankful for the health of their family. What is the margin of error for the survey?

Use the margin of error formula.

$$\text{Margin of error} = \pm\frac{1}{\sqrt{n}} = \pm\frac{1}{\sqrt{2368}} \approx \pm 0.021$$

▶ The margin of error for the survey is about ±2.1%.

12. In a survey of 1017 U.S. adults, 62% said that they prefer saving money over spending it. Give an interval that is likely to contain the exact percent of all U.S. adults who prefer saving money over spending it.

13. There are two candidates for homecoming king. The table shows the results from four random surveys of the students in the school. The students were asked whether they will vote for Candidate A. Do you think Candidate A will be the homecoming king? Explain.

Sample Size	Number of "Yes" Responses	Percent of Votes
8	6	75%
22	14	63.6%
34	16	47.1%
62	29	46.8%

11.6 Making Inferences from Experiments (pp. 633–638)

A randomized comparative experiment tests whether a new fertilizer affects the length (in inches) of grass after one week. The control group has 10 sections of land and the treatment group, which is fertilized, has 10 sections of land. The table shows the results.

	Grass Length (inches)									
Control Group	4.5	4.5	4.8	4.4	4.4	4.7	4.3	4.5	4.1	4.2
Treatment Group	4.6	4.8	5.0	4.8	4.7	4.6	4.9	4.9	4.8	4.4

a. Find the experimental difference of the means, $\bar{x}_{\text{treatment}} - \bar{x}_{\text{control}}$.

$\bar{x}_{\text{treatment}} - \bar{x}_{\text{control}} = 4.75 - 4.44 = 0.31$

▶ The experimental difference of the means is 0.31 inch.

b. What can you conclude?

▶ The two data sets tend to be fairly symmetric and have no extreme values. So, the mean is a suitable measure of center. The mean length of the treatment group is 0.31 inch longer than the control group. It appears that the fertilizer might be slightly effective, but the sample size is small and the difference could be due to chance.

14. Describe how to use a simulation to resample the data in the example above. Explain how this allows you to make inferences about the data when the sample size is small.

11 Chapter Test

1. Market researchers want to know whether more men or women buy their product. Explain whether this research topic is best investigated through an experiment or an observational study. Then describe the design of the experiment or observational study.

2. You want to survey 100 of the 2774 four-year colleges in the United States about their tuition cost. Describe a method for selecting a random sample of colleges to survey.

3. The grade point averages of all the students in a high school are normally distributed with a mean of 2.95 and a standard deviation of 0.72. Are these numerical values parameters or statistics? Explain.

A normal distribution has a mean of 72 and a standard deviation of 5. Find the probability that a randomly selected x-value from the distribution is in the given interval.

4. between 67 and 77
5. at least 75
6. at most 82

7. A researcher wants to test the effectiveness of a new medication designed to lower blood pressure. Identify a potential problem, if any, with the experimental design. Then describe how you can improve it.

   The researcher identifies 30 people with high blood pressure. Fifteen people with the highest blood pressures are given the medication and the other 15 are given a placebo. After 1 month, the subjects are evaluated.

8. A randomized comparative experiment tests whether a vitamin supplement increases human bone density (in grams per square centimeter). The control group has eight people and the treatment group, which receives the vitamin supplement, has eight people. The table shows the results.

	Bone Density (g/cm²)							
Control Group	0.9	1.2	1.0	0.8	1.3	1.1	0.9	1.0
Treatment Group	1.2	1.0	0.9	1.3	1.2	0.9	1.3	1.2

   a. Find the mean yields of the control group, $\overline{x}_{\text{control}}$, and the treatment group, $\overline{x}_{\text{treatment}}$.

   b. Find the experimental difference of the means, $\overline{x}_{\text{treatment}} - \overline{x}_{\text{control}}$.

   c. Display the data in a double dot plot. What can you conclude?

   d. Five hundred resamplings of the data are simulated. Out of the 500 resampling differences, 231 are greater than the experimental difference in part (b). What can you conclude about the hypothesis, *The vitamin supplement has no effect on human bone density*? Explain your reasoning.

9. In a recent survey of 1600 randomly selected U.S. adults, 81% said they have purchased a product online.

   a. Identify the population and the sample. Describe the sample.

   b. Find the margin of error for the survey.

   c. Give an interval that is likely to contain the exact percent of all U.S. adults who have purchased a product online.

   d. You survey 75 teachers at your school. The results are shown in the graph. Would you use the recent survey or your survey to estimate the percent of U.S. adults who have purchased a product online? Explain.

Have You Purchased a Product Online?

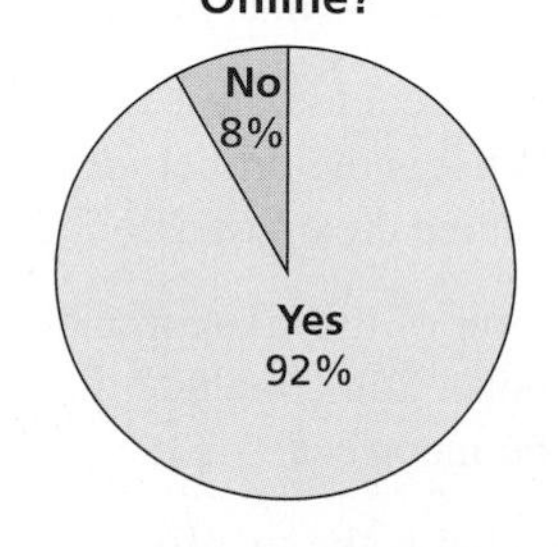

11 Cumulative Assessment

1. Your friend claims any system formed by three of the following equations will have exactly one solution.

$3x + y + 3z = 6$	$x + y + z = 2$	$4x - 2y + 4z = 8$
$x - y + z = 2$	$2x + y + z = 4$	$3x + y + 9z = 12$

   a. Write a linear system that would support your friend's claim.

   b. Write a linear system that shows your friend's claim is incorrect.

2. Which of the following samples are biased? If the sample is biased, explain why it is biased.

   (A) A restaurant asks customers to participate in a survey about the food sold at the restaurant. The restaurant uses the surveys that are returned.

   (B) You want to know the favorite sport of students at your school. You randomly select athletes to survey at the winter sports banquet.

   (C) The owner of a store wants to know whether the store should stay open 1 hour later each night. Each cashier surveys every fifth customer.

   (D) The owner of a movie theater wants to know whether the volume of its movies is too loud. Patrons under the age of 18 are randomly surveyed.

3. A survey asks adults about their favorite way to eat ice cream. The results of the survey are displayed in the table shown.

Survey Results	
Cup	45%
Cone	29%
Sundae	18%
Other	8%
(*margin of error* ±2.11%)	

   a. How many people were surveyed?

   b. Why might the conclusion, "Adults generally do not prefer to eat their ice cream in a cone" be inaccurate to draw from this data?

   c. You decide to test the results of the poll by surveying adults chosen at random. What is the probability that at least three out of the six people you survey prefer to eat ice cream in a cone?

   d. Four of the six respondents in your study said they prefer to eat their ice cream in a cone. You conclude that the other survey is inaccurate. Why might this conclusion be incorrect?

   e. What is the margin of error for your survey?

4. You are making a lampshade out of fabric for the lamp shown. The pattern for the lampshade is shown in the diagram on the left.

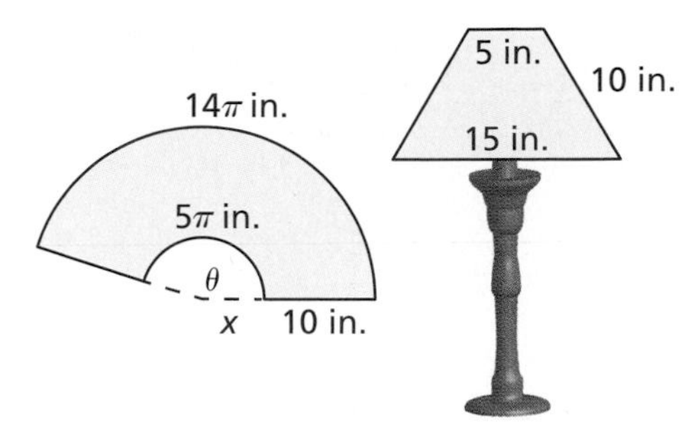

a. Use the smaller sector to write an equation that relates θ and x.

b. Use the larger sector to write an equation that relates θ and $x + 10$.

c. Solve the system of equations from parts (a) and (b) for x and θ.

d. Find the amount of fabric (in square inches) that you will use to make the lampshade.

5. For all students taking the Medical College Admission Test over a period of 3 years, the mean score was 25.1. During the same 3 years, a group of 1000 students who took the test had a mean score of 25.3. Classify each mean as a parameter or a statistic. Explain.

6. Complete the table for the four equations. Explain your reasoning.

Equation	Is the inverse a function?		Is the function its own inverse?	
	Yes	No	Yes	No
$y = -x$				
$y = 3 \ln x + 2$				
$y = \left(\frac{1}{x}\right)^2$				
$y = \frac{x}{x-1}$				

7. The normal distribution shown has mean 63 and standard deviation 8. Find the percent of the area under the normal curve that is represented by the shaded region. Then describe another interval under the normal curve that has the same area.

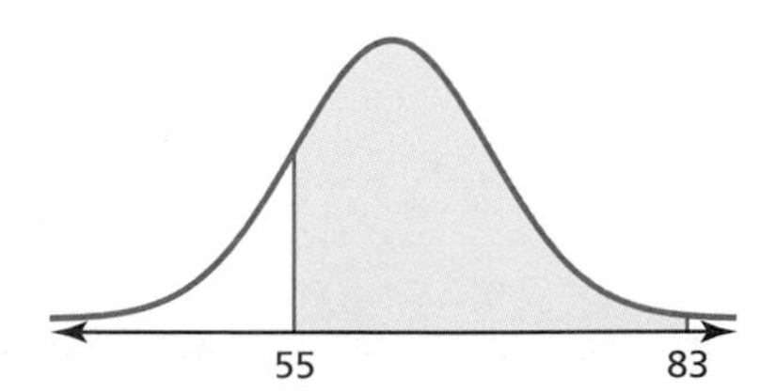

8. Which of the rational expressions *cannot* be simplified?

Ⓐ $\dfrac{2x^2 + 5x - 3}{x^2 - 7x + 12}$

Ⓑ $\dfrac{3x^3 + 21x^2 + 30x}{x^2 - 25}$

Ⓒ $\dfrac{x^3 + 27}{x^2 - 3x + 9}$

Ⓓ $\dfrac{x^3 + 2x^2 - 8x - 16}{2x^2 - 21x + 55}$

Selected Answers

Chapter 1

Chapter 1 Maintaining Mathematical Proficiency *(p. 1)*

1. 47 **2.** −46 **3.** $3\frac{3}{5}$ **4.** 4 **5.** 13 **6.** 0

7.

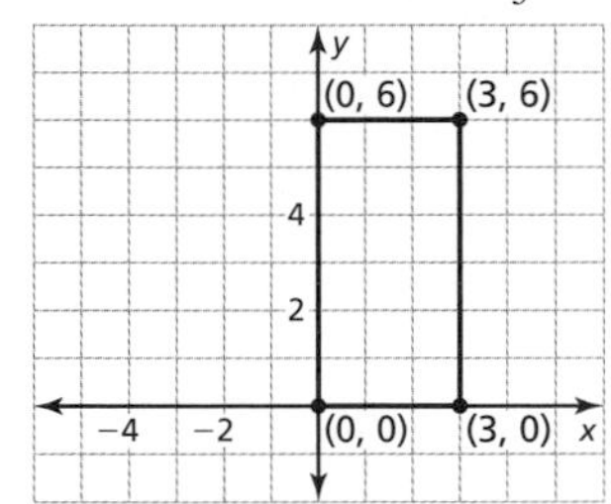

8.

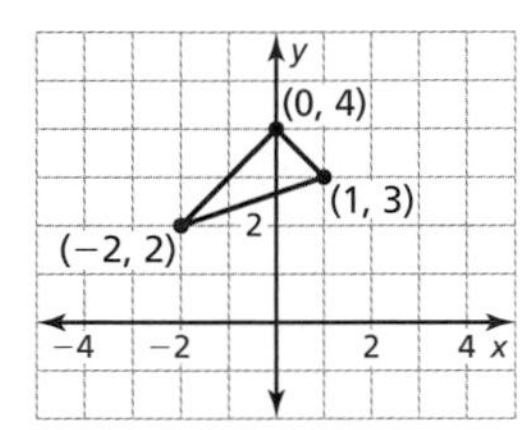

9.

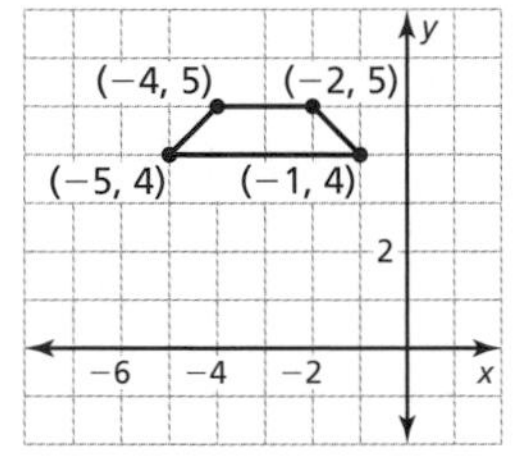

10. *Sample answer:* 12 + 18 ÷ 3 equals 18 when division is performed first and 10 when addition is performed first; yes; If the point (3, 2) is translated up 3 units then reflected across the x-axis, the new coordinate is (3, −5). If it is reflected across the x-axis first then translated up 3, the new coordinate is (3, 1).

1.1 Vocabulary and Core Concept Check *(p. 8)*

1. parent function

1.1 Monitoring Progress and Modeling with Mathematics *(pp. 8–10)*

3. absolute value; The graph is a vertical stretch with a translation 2 units left and 8 units down; The domain of each function is all real numbers, but the range of f is $y \geq -8$, and the range of the parent function is $y \geq 0$.

5. linear; The graph is a vertical stretch and a translation 2 units down; The domain and range of each function is all real numbers.

7.

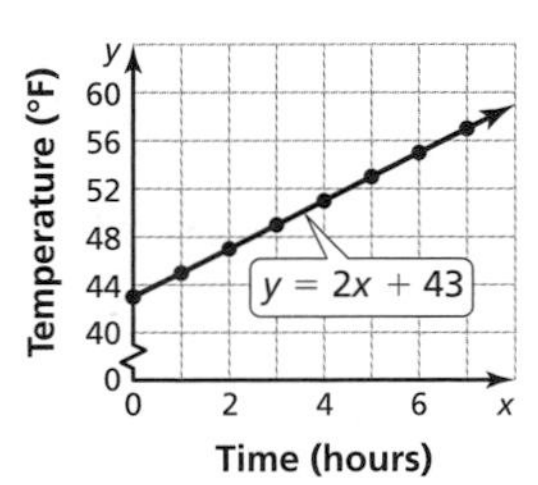

linear; The temperature is increasing by the same amount at each interval.

9.

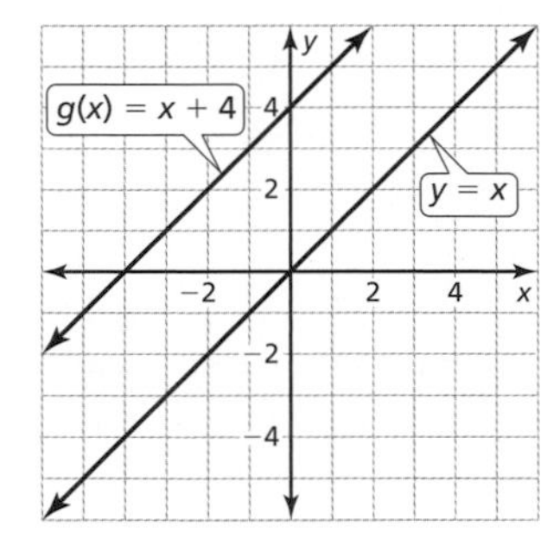

The graph of g is a vertical translation 4 units up of the parent linear function.

11.

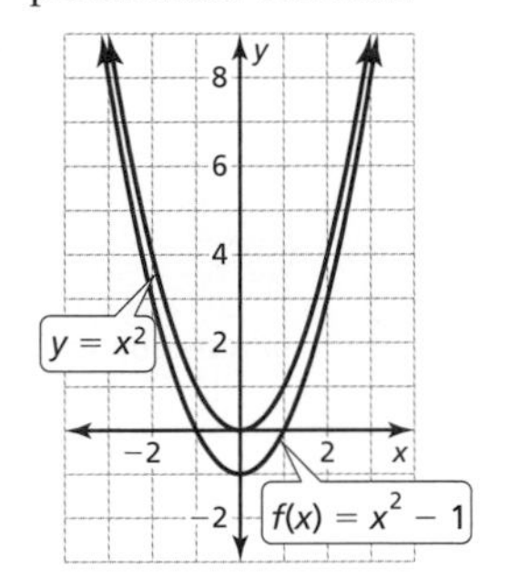

The graph of f is a vertical translation 1 unit down of the parent quadratic function.

13.

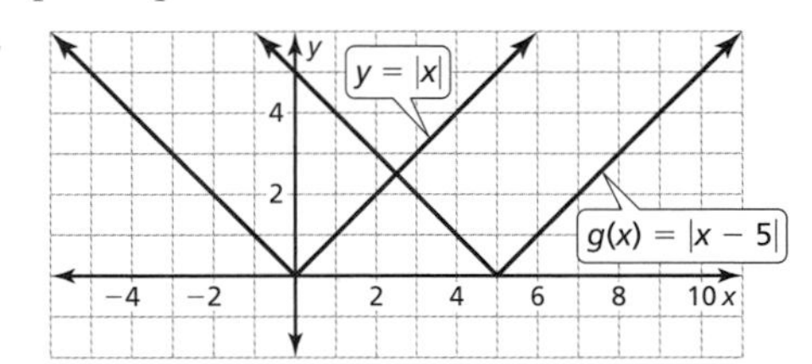

The graph of g is a horizontal translation 5 units right of the parent absolute value function.

15.

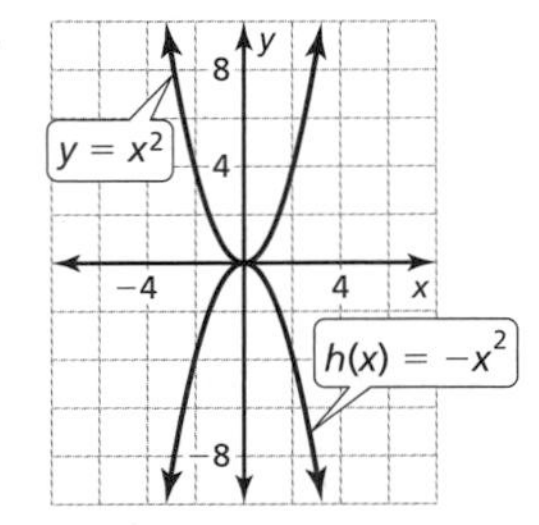

The graph of h is a reflection in the x-axis of the parent quadratic function.

17.

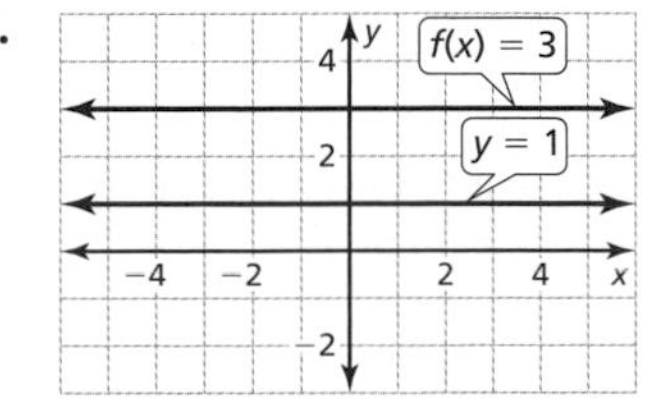

The graph of f is a vertical translation 2 units up of the parent constant function.

19.

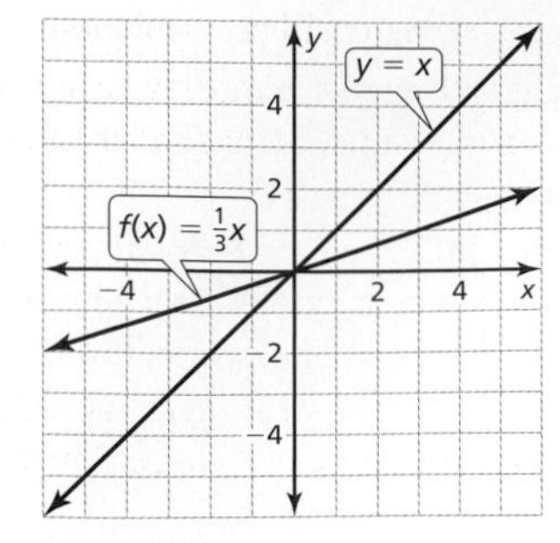

The graph of f is a vertical shrink of the parent linear function.

21.

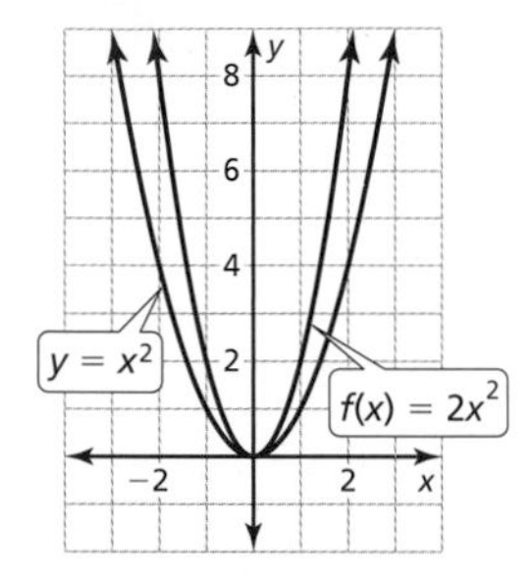

The graph of f is a vertical stretch of the parent quadratic function.

23.

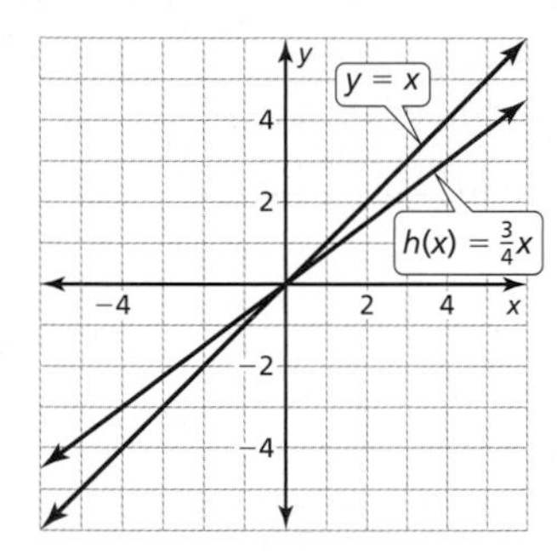

The graph of h is a vertical shrink of the parent linear function.

25.

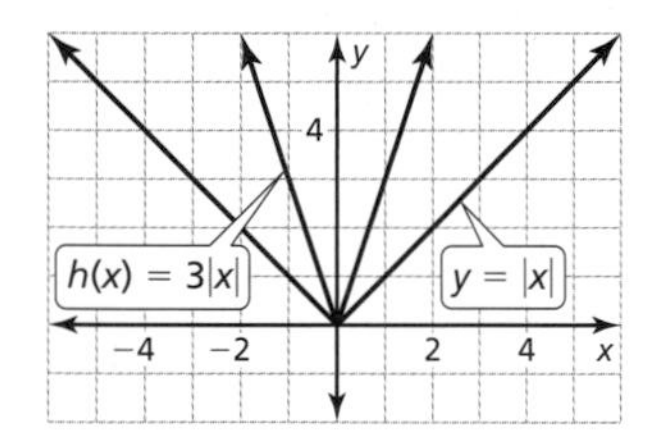

The graph of h is a vertical stretch of the parent absolute value function.

27.

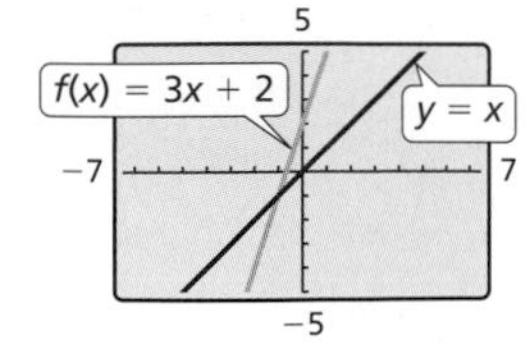

The graph of f is a vertical stretch followed by a translation 2 units up of the parent linear function.

29.

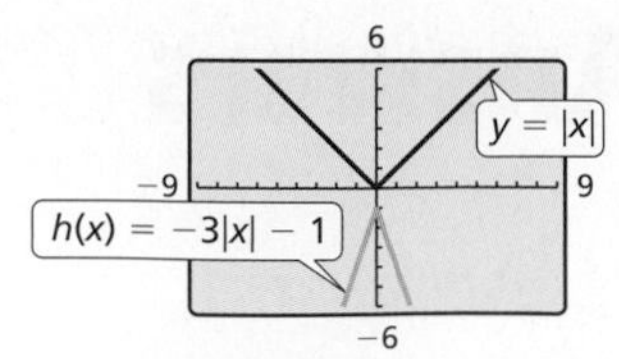

The graph of h is a vertical stretch and a reflection in the x-axis followed by a translation 1 unit down of the parent absolute value function.

31.

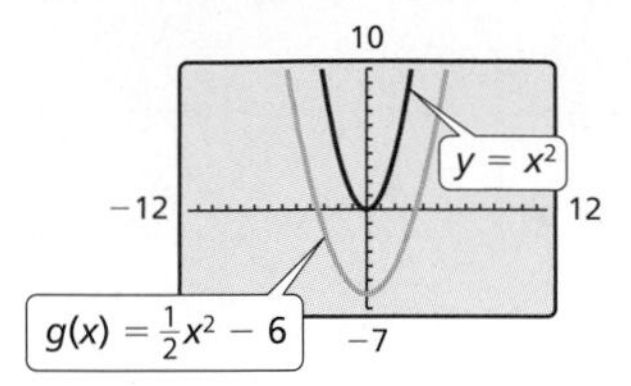

The graph of g is a vertical shrink followed by a translation 6 units down of the parent quadratic function.

33.

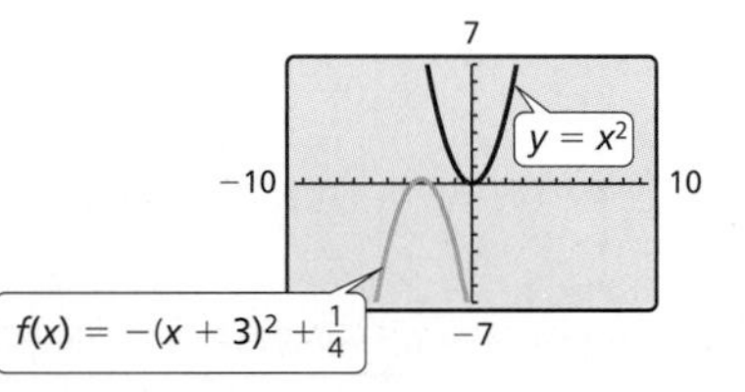

The graph of f is a reflection in the x-axis followed by a translation 3 units left and $\frac{1}{4}$ unit up of the parent quadratic function.

35. It is a vertical stretch, not shrink. The graph is a reflection in the x-axis followed by a vertical stretch of the parent quadratic function.

37. $(2, -1), (-1, -4), (2, -5)$

39. absolute value; domain is all real numbers; range is $y \geq -1$

41. linear; domain is all real numbers; range is all real numbers

43. quadratic; domain is all real numbers; range is $y \geq -2$

45. absolute value; 8 mi/h

47. no; f is shifted right and g is shifted down.

49. yes; Shifting the parent linear function down 2 units will create the same graph as shifting it 2 units right.

51. **a.** quadratic

b. 0; At the moment the ball is released, 0 seconds have passed.

c. 5.2; Because $f(t)$ represents the height of the ball, find $f(0)$.

53. **a.** vertical translation; The graph will have a vertical stretch and will be shifted 3 units down.

b. horizontal translation; The graph will be shifted 8 units right.

c. both; The graph will be shifted 2 units left and 4 units up.

d. neither; The graph will have a vertical stretch.

1.1 Maintaining Mathematical Proficiency *(p. 10)*

55. no **57.** yes **59.** x-intercept: 0; y-intercept: 0

61. x-intercept: $\frac{1}{3}$; y-intercept: 1

1.2 Vocabulary and Core Concept Check *(p. 16)*

1. shrink

1.2 Monitoring Progress and Modeling with Mathematics (pp. 16–18)

3. $g(x) = x - 1$ **5.** $g(x) = |4x + 3|$

7. $g(x) = 4 - |x - 2|$

9. f could be translated 3 units up or 3 units right.

11. $g(x) = 5x - 2$ **13.** $g(x) = |6x| - 2$

15. $g(x) = -3 + |-x - 11|$ **17.** $g(x) = 5x + 10$

19. $g(x) = |4x| + 4$ **21.** $g(x) = -|x - 4| + 1$

23. C; The graph has been translated left.

25. D; The graph has been translated up. **27.** $g(x) = 2x + 1$

29. $g(x) = \left|\frac{1}{2}x - 1\right|$ **31.** $g(x) = -|x| - 8$

33. Translating a graph to the right requires subtraction, not addition; $g(x) = |x - 3| + 2$

35. no; Suppose a graph contains the point (3, 2) and is translated up 3 units then reflected in the x-axis. The new coordinate is (3, −5). If it is reflected in the x-axis first then translated up 3, the new coordinate is (3, 1).

37. The graph has been translated 6 units left; $A = 9$

39. The graph has been reflected in the x-axis; $A = 16$

41. **a.** $f(x) + (c - b)$ **b.** $f\left(x + \frac{c - b}{m}\right)$

43. vertical stretch, translation, reflection; *Sample answer:* $-(4|x| - 2) = -4|x| + 2$

45. $a = -2$, $b = 1$, and $c = 0$, $g(x) = -2|x - 1|$ represents the transformation of $f(x)$.

1.2 Maintaining Mathematical Proficiency (p. 18)

47. −5 **49.** 0

51.

1.3 Vocabulary and Core Concept Check (p. 26)

1. slope-intercept

1.3 Monitoring Progress and Modeling with Mathematics (pp. 26–28)

3. $y = \frac{1}{5}x$; The tip increases \$0.20 for each dollar spent on the meal.

5. $y = 50x + 100$; The balance increases \$50 each week.

7. $y = 55x$; The number of words increases by 55 each minute.

9. Greenville Journal; 5 lines

11. The original balance of \$100 should have been included; After 10 years, the increase in balance will be \$70, resulting in a new balance of \$170.

13. yes; *Sample answer:* $y = 4.25x + 1.75$; $y = 65.5$; After 15 minutes, you have burned 65.5 calories.

15. yes; *Sample answer:* $y = -4.6x + 96$; $y = 27$; After 15 hours, the battery will have 27% of life remaining.

17. $y = 380.03x + 11{,}290$; \$16,990.45; The annual tuition increases about \$380 each year and the cost of tuition in 2005 is about \$11,290.

19. $y = 0.42x + 1.44$; $r = 0.61$; weak positive correlation

21. $y = -0.45x + 4.26$; $r = -0.67$; weak negative correlation

23. $y = 0.61x + 0.10$; $r = 0.95$; strong positive correlation

25. **a.** *Sample answer:* height and weight; temperature and ice cream sales; Correlation is positive because as the first goes up, so does the second.

b. *Sample answer:* miles driven and gas remaining; hours used and battery life remaining; Correlation is negative because as the first goes up, the second goes down.

c. *Sample answer:* age and length of hair; typing speed and shoe size; There is no relationship between the first and second.

27. no; Because r is close to 0, the points do not lie close to the line.

29. It is negative; As x increases, y increases, so z decreases.

31. about 2.2 mi

1.3 Maintaining Mathematical Proficiency (p. 28)

33. (16, −41) **35.** $\left(1, \frac{1}{2}\right)$ **37.** $\left(\frac{16}{17}, \frac{15}{17}\right)$

1.4 Vocabulary and Core Concept Check (p. 34)

1. ordered triple

1.4 Monitoring Progress and Modeling with Mathematics (pp. 34–36)

3. (1, 2, −1) **5.** (3, −1, −4) **7.** $\left(\frac{151}{64}, \frac{9}{8}, -\frac{51}{32}\right)$

9. The entire second equation should be multiplied by 4, not just the x-term.

$$\begin{array}{r} 4x - y + 2z = -18 \\ -4x + 8y + 4z = 44 \\ \hline 7y + 6z = 26 \end{array}$$

11. no solution **13.** $(z - 1, 1, z)$ **15.** no solution

17. A small pizza costs \$5, a liter of soda costs \$1, and a salad costs \$3.

19. (4, −3, 2) **21.** no solution **23.** (7, 3, 5)

25. (3, 2, 1) **27.** $\left(\frac{-3z + 3}{5}, \frac{-13z + 13}{5}, z\right)$ **29.** 1%

31. *Sample answer:* When one variable has the same coefficient or its opposite in each equation. The system

$3x + 2y - 4z = -5$

$2x + 2y + 3z = 8$

$5x - 2y - 7z = -9$

can be solved by eliminating y first.

33. $\ell + m + n = 65$, $n = \ell + m - 15$, $\ell = \frac{1}{3}m$; $\ell = 10$ ft, $m = 30$ ft, $n = 25$ ft

35. **a.** *Sample answer:* $a = -1$, $b = -1$, $c = -1$; Use elimination on equations 1 and 2.

b. *Sample answer:* $a = 4$, $b = 4$, $c = 5$; The solution is $\left(\frac{2}{3}, -\frac{2}{3}, 2\right)$.

c. *Sample answer*: $a = 5$, $b = 5$, $c = 5$; Use elimination on equations 1 and 2.

37. 350 ft^2

39. **a.** $r + \ell + i = 12$, $2.50r + 4l + 2i = 32$, $r = 2l + 2i$

b. 8 roses, 2 lilies, 2 irises

c. no; *Sample answer:* 8 roses, 4 lilies, 0 irises; 8 roses, 0 lilies, 4 irises; 8 roses, 3 lilies, 1 iris

41. $a = 12$, $b = -4$, $c = 10$; These values are the only ones which can satisfy the linear system at (−1, 2, −3).

43. $t + a = g$, $t + b = a$, $2g = 3b$; 5 tangerines

Selected Answers

1.4 Maintaining Mathematical Proficiency *(p. 36)*

45. $9m^2 + 6m + 1$ **47.** $16 - 8y + y^2$

49. $g(x) = -|x| + 5$ **51.** $g(x) = 3|x| - 15$

Chapter 1 Review *(pp. 38–40)*

1.

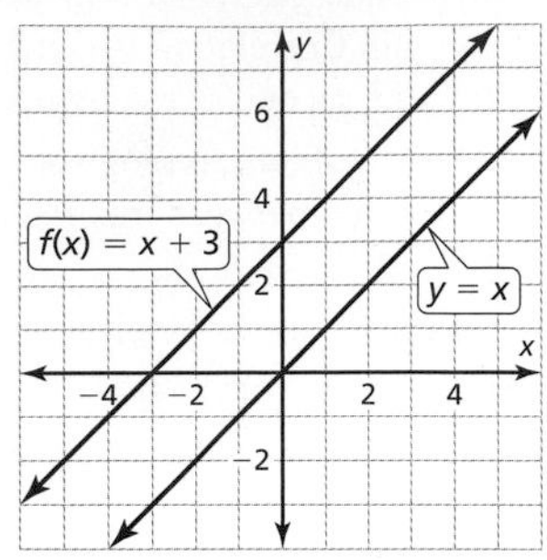

The graph of f is a translation 3 units up of the parent linear function.

2.

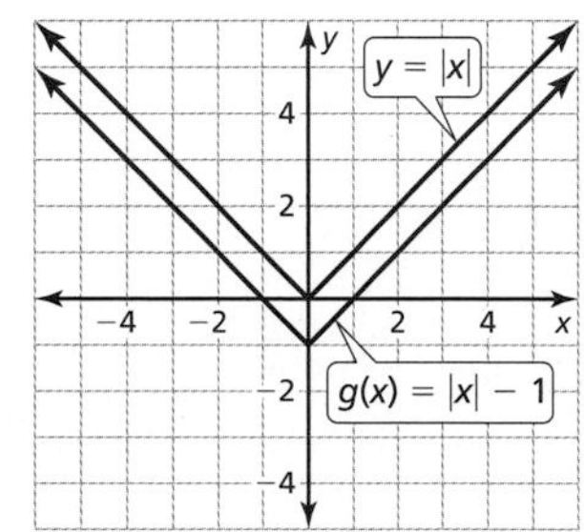

The graph of g is a translation 1 unit down of the parent absolute value function.

3.

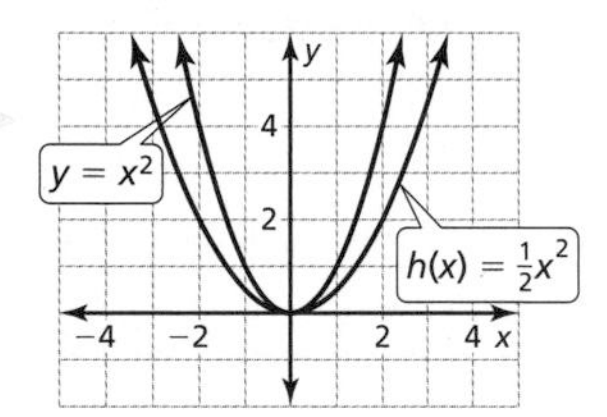

The graph of h is a vertical shrink by a factor of $\frac{1}{2}$ of the parent quadratic function.

4.

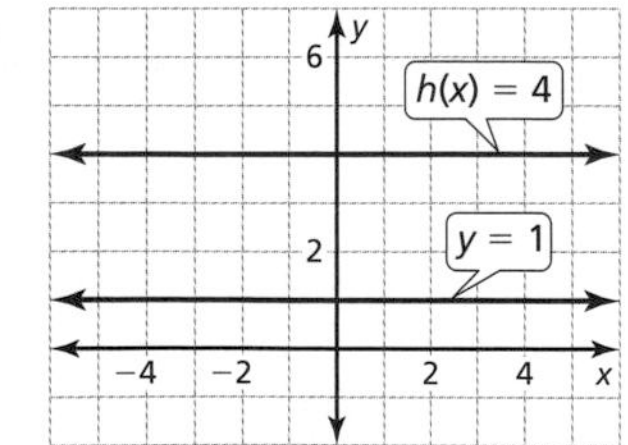

The graph of h is a translation 3 units up of the parent constant function.

5.

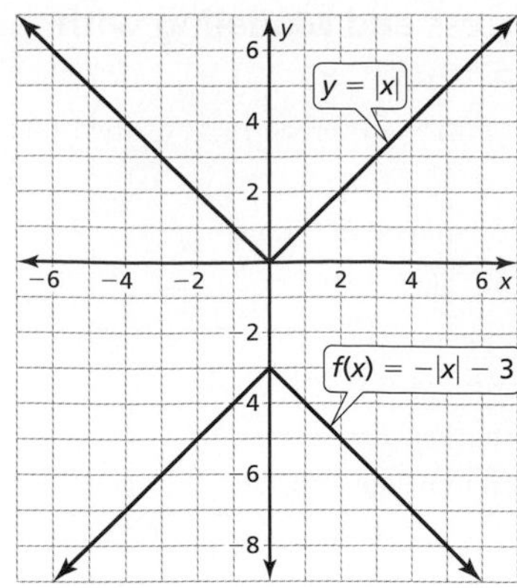

The graph of f is a reflection in the x-axis followed by a translation 3 units down of the parent absolute value function.

6.

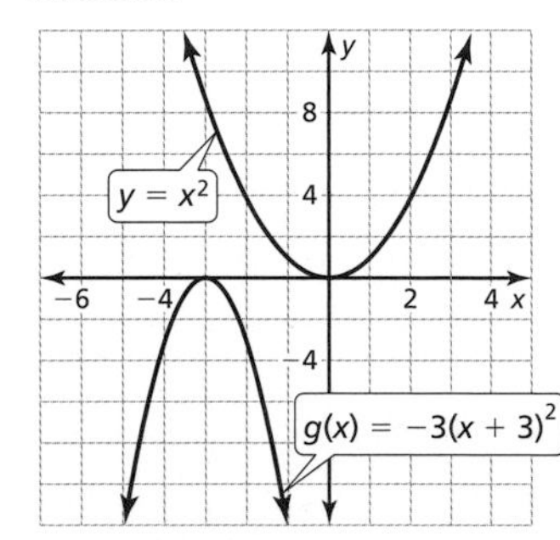

The graph of g is a vertical stretch by a factor of 3 followed by a reflection in the x-axis and a translation 3 units left

7. $g(x) = -|x + 4|$ **8.** $g(x) = \frac{1}{2}|x| + 2$

9. $g(x) = -x - 3$ **10.** $y = 0.03x + 1.23$

11. $y = 0.35x$; 15.75 mi **12.** $(4, -2, 1)$

13. $\left(-\frac{4}{3}, -\frac{17}{3}, \frac{26}{3}\right)$ **14.** $(9 + 4y, y, -7 - 5y)$

15. no solution **16.** $(-11, -8, 3)$ **17.** $(-16, 12, 10)$

18. 200 student tickets, 350 adult tickets, and 50 children under 12 tickets

Chapter 2

Chapter 2 Maintaining Mathematical Proficiency *(p. 45)*

1. $-\frac{7}{2}$ **2.** $\frac{4}{3}$ **3.** -3.6 **4.** 5 **5.** -10

6. 8 **7.** about 6.32 **8.** about 8.06 **9.** about 2.24

10. about 12.65 **11.** 10 **12.** about 15.30

13. $d = |b - a|$; yes; Find the distance between the two y-coordinates by subtraction. Take the absolute value of the result, because distance is always positive. This is possible because when $x_1 = x_2$, the distance formula simplifies as shown.

$$d = \sqrt{(x_2 - x_1)^2 + (y_2 - y_1)^2}$$

$$d = \sqrt{(y_2 - y_1)^2}$$

$$d = |y_2 - y_1|$$

2.1 Vocabulary and Core Concept Check *(p. 52)*

1. parabola

2.1 Monitoring Progress and Modeling with Mathematics *(pp. 52–54)*

3. The graph of g is a translation 3 units down of the graph of f.

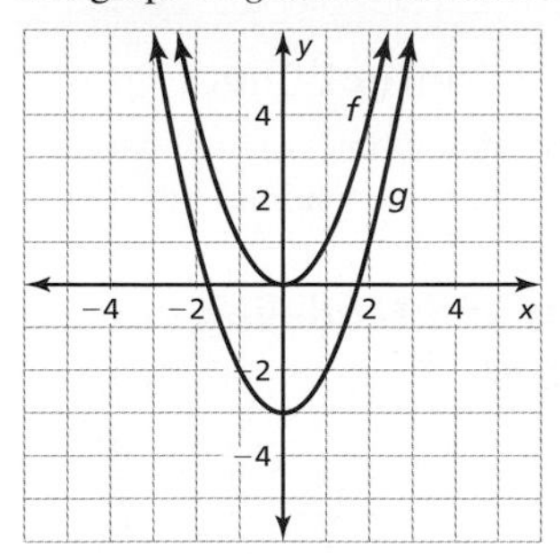

5. The graph of g is a translation 2 units left of the graph of f.

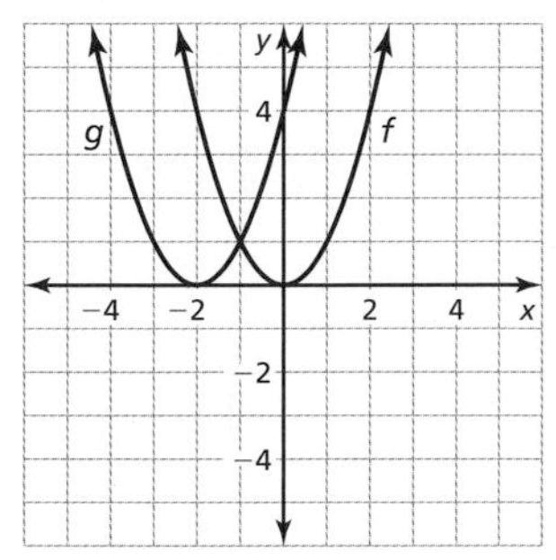

7. The graph of g is a translation 1 unit right of the graph of f.

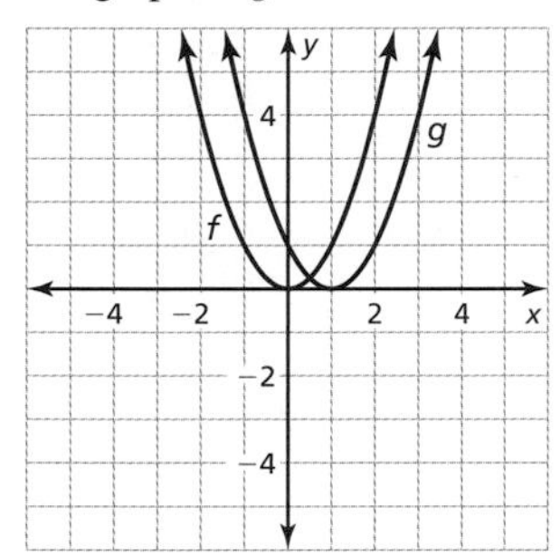

9. The graph of g is a translation 6 units left and 2 units down of the graph of f.

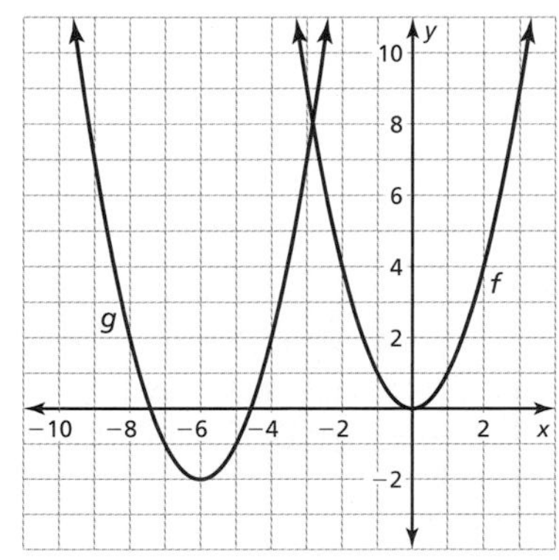

11. The graph of g is a translation 7 units right and 1 unit up of the graph of f.

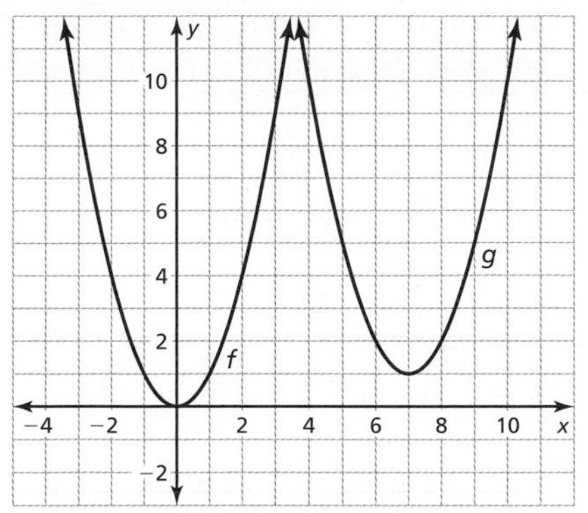

13. A; The graph has been translated 1 unit right.

15. C; The graph has been translated 1 unit right and 1 unit up.

17. The graph of g is a reflection in the x-axis of the graph of f.

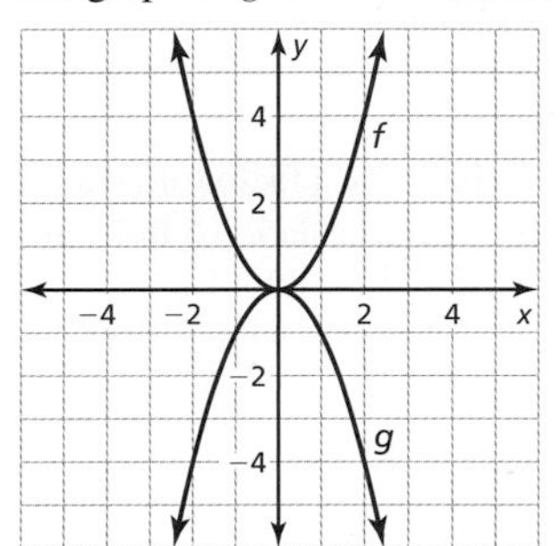

19. The graph of g is a vertical stretch by a factor of 3 of the graph of f.

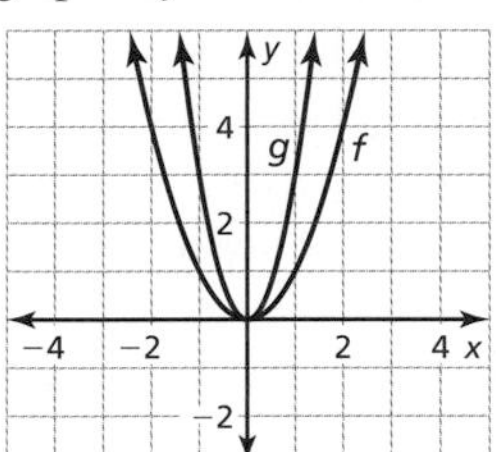

21. The graph of g is a horizontal shrink by a factor of $\frac{1}{2}$ of the graph of f.

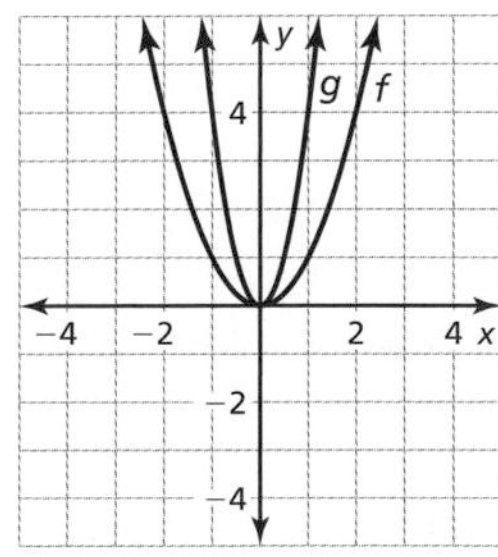

23. The graph of g is a vertical shrink by a factor of $\frac{1}{5}$ followed by a translation 4 units down.

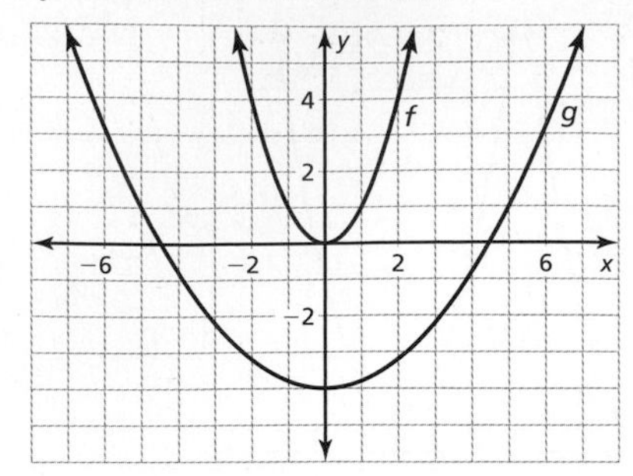

25. The graph is a reflection in the x-axis, not y-axis; The graph is a reflection in the x-axis and a vertical stretch by a factor of 6, followed by a translation 4 units up of the graph of the parent quadratic function.

27. The graph of f is a vertical stretch by a factor of 3 followed by a translation 2 units left and 1 unit up of the graph of the parent quadratic function; $(-2, 1)$

29. The graph of f is a vertical stretch by a factor of 2 followed by a reflection in the x-axis and a translation 5 units up of the graph of the parent quadratic function; $(0, 5)$

31. $g(x) = -4x^2 + 2$; $(0, 2)$

33. $g(x) = 8\left(\frac{1}{2}x\right)^2 - 4$; $(0, -4)$

35. C; The graph is a vertical stretch by a factor of 2 followed by a translation 1 unit right and 2 units down of the parent quadratic function.

37. D; The graph is a vertical stretch by a factor of 2 and a reflection in the x-axis, followed by a translation 1 unit right and 2 units up of the parent quadratic function.

39. F; The graph is a vertical stretch by a factor of 2 and a reflection in the x-axis followed by a translation 1 unit left and 2 units down of the parent quadratic function.

41. Subtract 6 from the output; Substitute $2x^2 + 6x$ for $f(x)$; Multiply the output by -1; Substitute $2x^2 + 6x - 6$ for $h(x)$; Simplify.

43. $h(x) = -0.03(x - 14)^2 + 10.99$

45. **a.** $y = \frac{-5}{1089}(x - 33)^2 + 5$

b. The domain is $0 \le x \le 66$ and the range is $0 \le y \le 5$; The domain represents the time the fish was in the air and the range represents the height of the fish.

c. yes; The value changes to $-\frac{1}{225}$; The vertex has changed but it still goes through the point $(0, 0)$, so there has been a horizontal stretch or shrink which changes the value of a.

47. **a.** $a = 2, h = 1, k = 6$; $g(x) = 2(x - 1)^2 + 6$

b. $g(x) = 2f(x - 1) + 6$; For each function, a, h, and k are the same but the second function does not indicate the type of function that is being translated.

c. $a = 2, h = 1, k = 3$; $g(x) = 2(x - 1)^2 + 3$; $g(x) = 2f(x - 1) + 3$; For each function, a, h, and k are the same, but the answer in part (b) does not indicate the type of function that is being translated.

d. *Sample answer:* vertex form; Writing a transformed function using function notation requires an extra step of substituting $f(x)$ into the newly transformed function.

49. a vertical shrink by a factor of $\frac{7}{16}$

2.1 Maintaining Mathematical Proficiency *(p. 54)*

51. $(4, 4)$

2.2 Vocabulary and Core Concept Check *(p. 61)*

1. If a is positive, then the quadratic function will have a minimum. If a is negative, then the quadratic function will have a maximum.

2.2 Monitoring Progress and Modeling with Mathematics *(pp. 61–64)*

3.

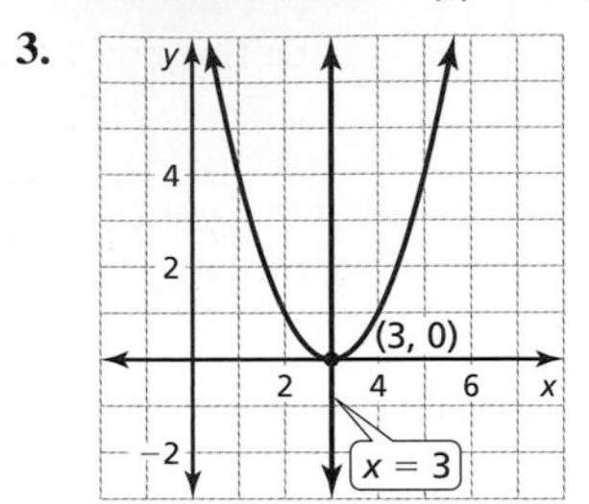

5.

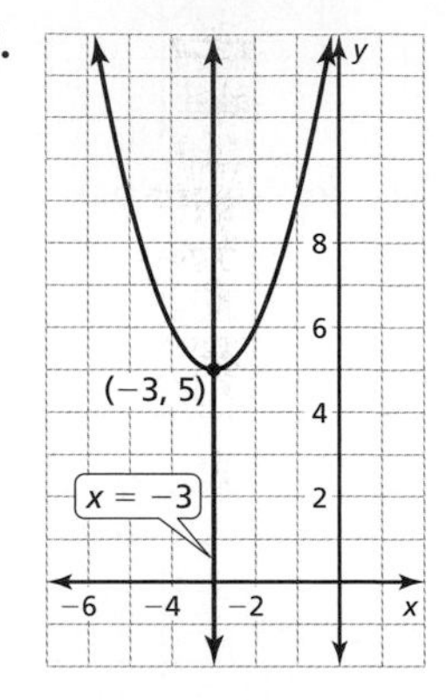

7.

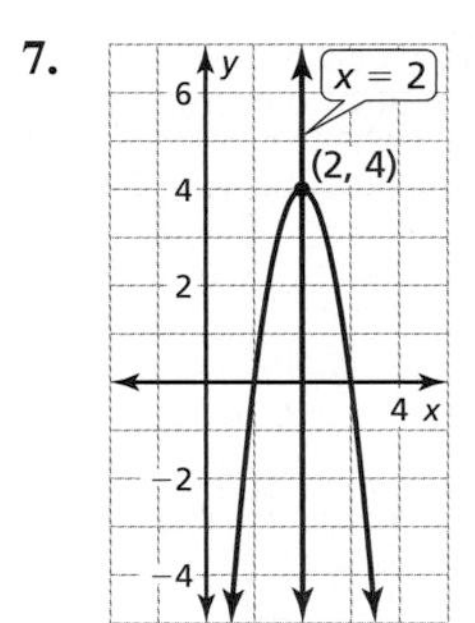

9.

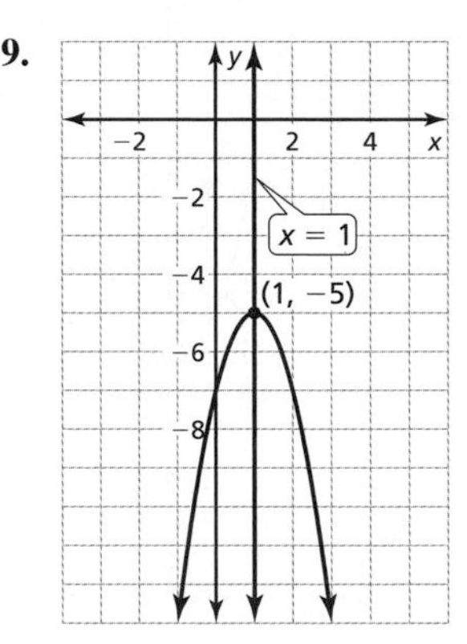

11.

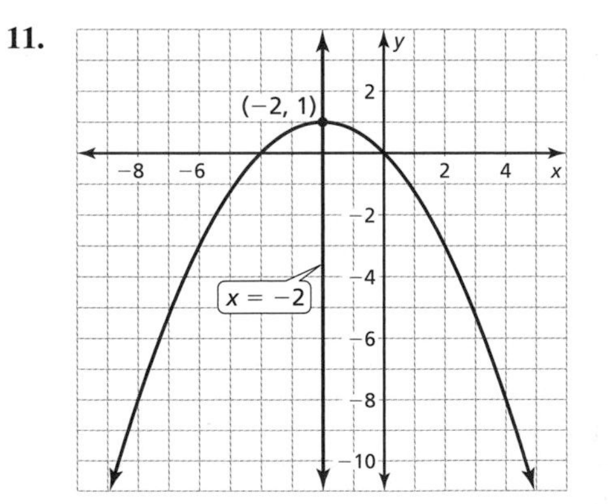

13.

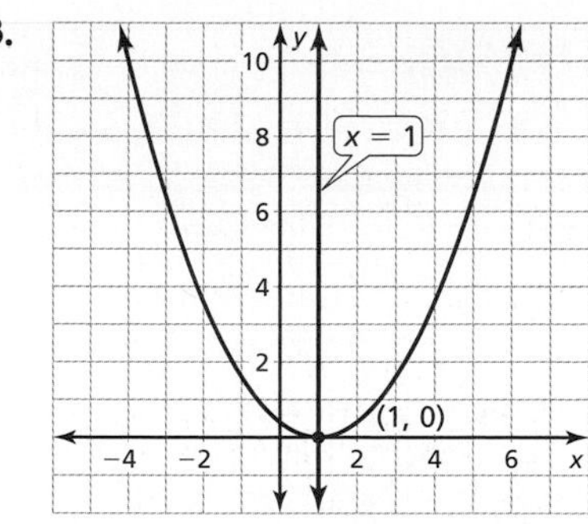

15. C **17.** B

19.

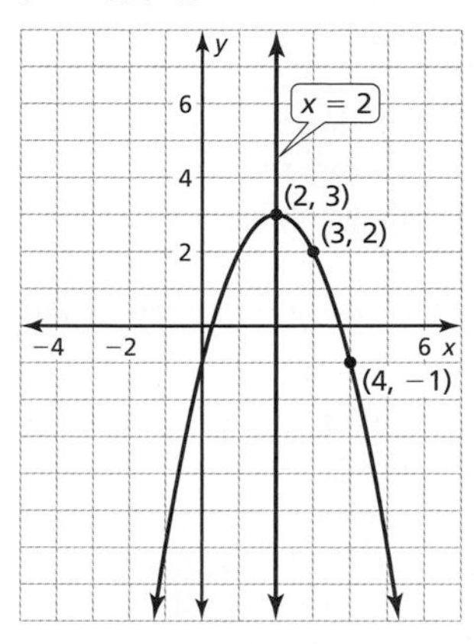

21.

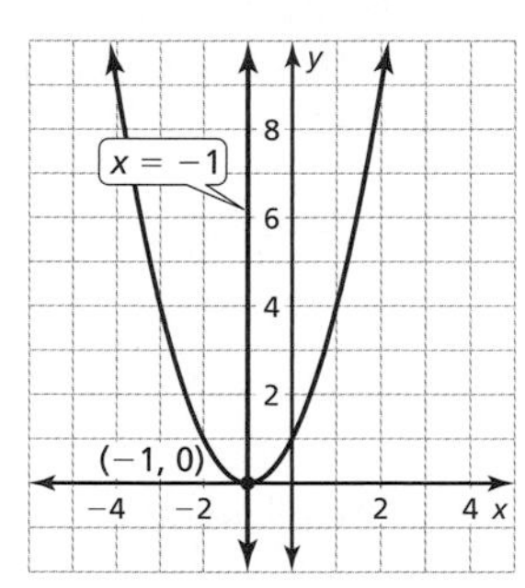

23.

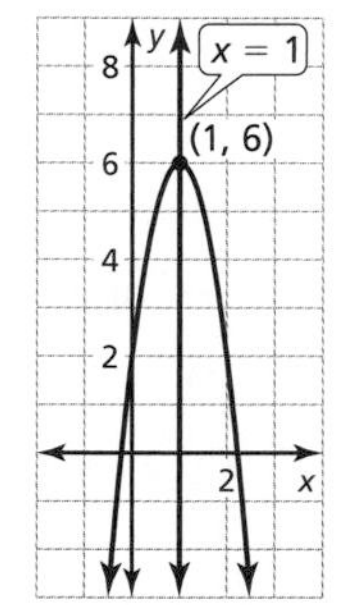

25.

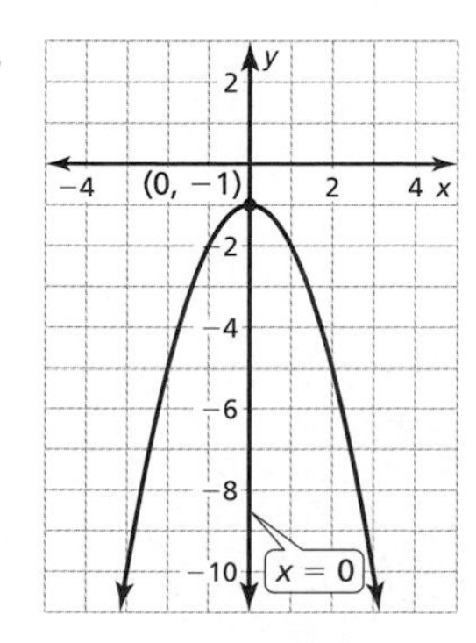

27.

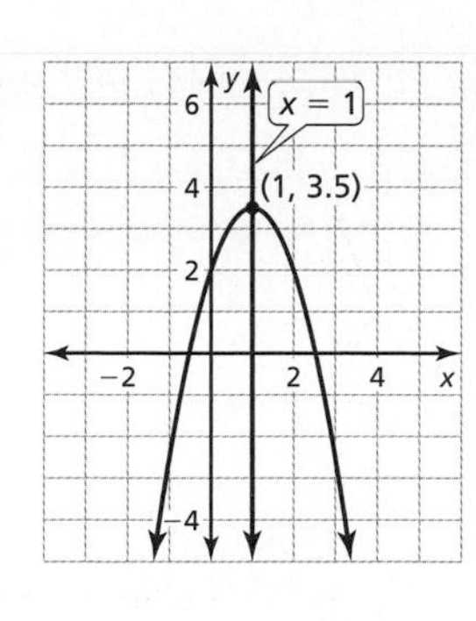

29.

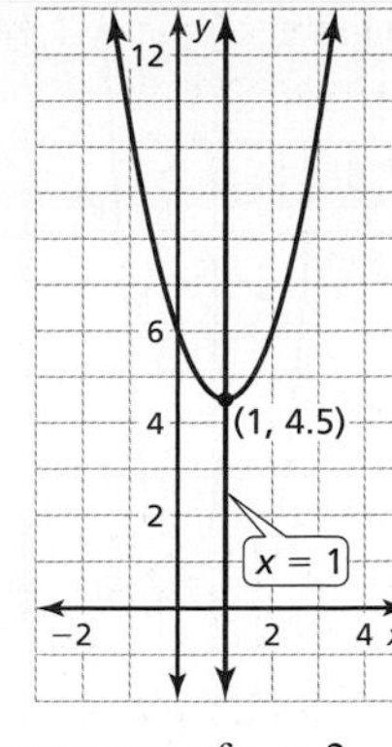

31. Both functions have an axis of symmetry of $x = 2$.

33. The formula is missing the negative sign; The x-coordinate of the vertex is

$$x = -\frac{b}{2a} = -\frac{24}{2(4)} = -3.$$

35. (25, 18.5); When the basketball is at its highest point, it is 25 feet from its starting point and 18.5 feet off the ground.

37. B

39. The minimum value is -1. The domain is all real numbers and the range is $y \geq -1$. The function is decreasing to the left of $x = 0$ and increasing to the right of $x = 0$.

41. The maximum value is 2. The domain is all real numbers and the range is $y \leq 2$. The function is increasing to the left of $x = -2$ and decreasing to the right of $x = -2$.

43. The maximum value is 15. The domain is all real numbers and the range is $y \leq 15$. The function is increasing to the left of $x = 2$ and decreasing to the right of $x = 2$.

45. The minimum value is -18. The domain is all real numbers and the range is $y \geq -18$. The function is decreasing to the left of $x = 3$ and increasing to the right of $x = 3$.

47. The minimum value is -7. The domain is all real numbers and the range is $y \geq -7$. The function is decreasing to the left of $x = 6$ and increasing to the right of $x = 6$.

49. **a.** 1 m

b. 3.25 m

c. The diver is ascending from 0 meters to 0.5 meter and descending from 0.5 meter until hitting the water after approximately 1.1 meters.

51. $A = w(20 - w) = -w^2 + 20w$; The maximum area is 100 square units.

53.

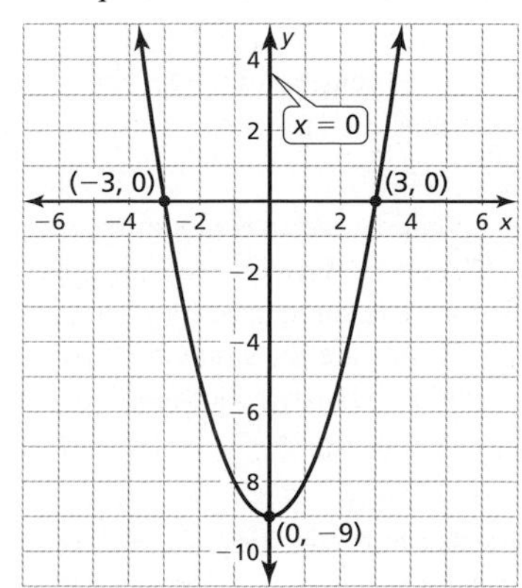

55.

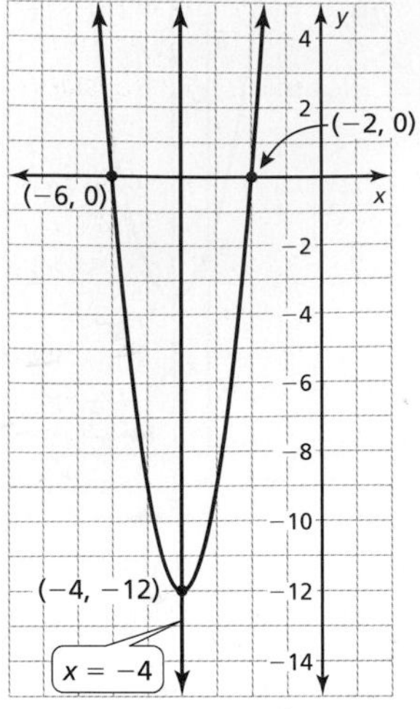

57.

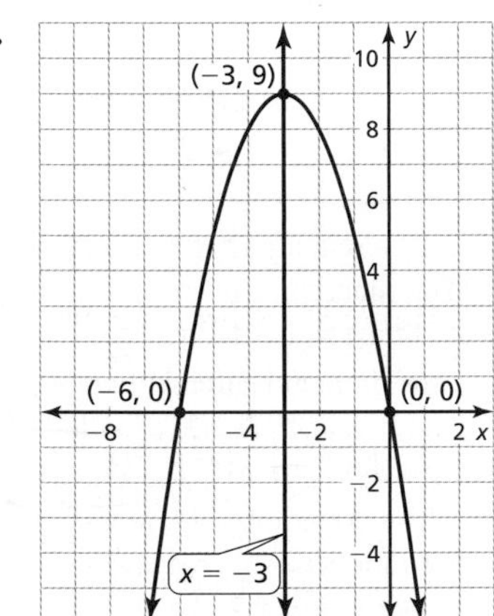

59.

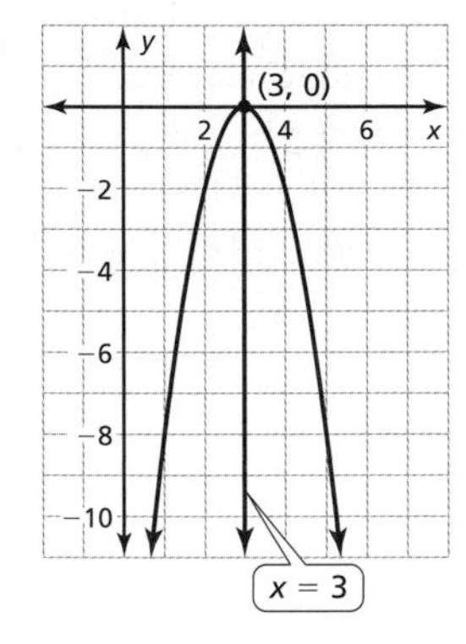

61. $p = 2$, $q = -6$; The graph is decreasing to the left of $x = -2$ and increasing to the right of $x = -2$.

63. $p = 4$, $q = 2$; The graph is increasing to the left of $x = 3$ and decreasing to the right of $x = 3$.

65. the second kick; the first kick

67. no; Either of the points could be the axis of symmetry, or neither of the points could be the axis of symmetry. You can only determine the axis of symmetry if the y-coordinates of the two points are the same, because the axis of symmetry would lie halfway between the two points.

69. $1.75

71. All three graphs are the same; $f(x) = x^2 + 4x + 3$, $g(x) = x^2 + 4x + 3$

73. no; The vertex of the graph is (3.25, 2.1125), which means the mouse cannot jump over a fence that is higher than 2.1125 feet.

75.

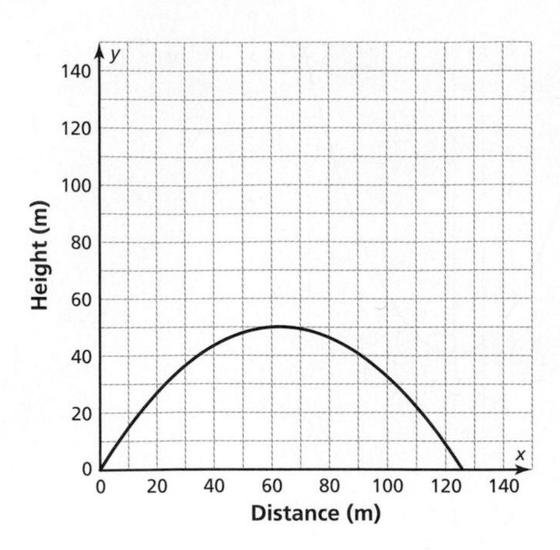

The domain is $0 \leq x \leq 126$ and the range is $0 \leq y \leq 50$; The domain represents the distance from the start of the bridge on one side of the river, and the range represents the height of the bridge.

77. no; The vertex must lie on the axis of symmetry, and (0, 5) does not lie on $x = -1$.

79. **a.** about 14.1%; about 55.5 cm^3/g

b. about 13.6%; about 44.1 cm^3/g

c. The domain for hot-air popping is $5.52 \leq x \leq 22.6$, and the range is $0 \leq y \leq 55.5$. The domain for hot-oil popping is $5.35 \leq x \leq 21.8$, and the range is $0 \leq y \leq 44.1$. This means that the moisture content for the kernels can range from 5.52% to 22.6% and 5.35% to 21.8%, while the popping volume can range from 0 to 55.5 cubic centimeters per gram and 0 to 44.11 cubic centimeters per gram.

2.2 Maintaining Mathematical Proficiency *(p. 64)*

81. 4 **83.** no solution **85.** 2 **87.** −12

2.3 Vocabulary and Core Concept Check *(p. 72)*

1. focus; directrix

2.3 Monitoring Progress and Modeling with Mathematics *(pp. 72–74)*

3. $y = \frac{1}{4}x^2$ **5.** $y = -\frac{1}{8}x^2$ **7.** $y = \frac{1}{24}x^2$

9. $y = -\frac{1}{40}x^2$

11. A, B and D; Each has a value for p that is negative. Substituting in a negative value for p in $y = \frac{1}{4p}x^2$ results in a parabola that has been reflected across the x-axis.

13. The focus is (0, 2). The directrix is $y = -2$. The axis of symmetry is the y-axis.

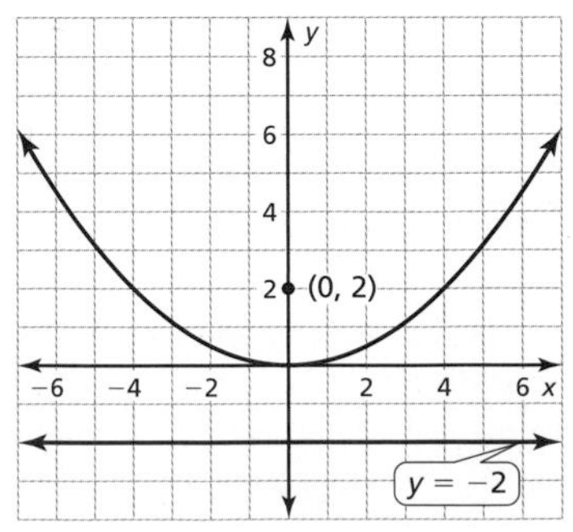

15. The focus is $(-5, 0)$. The directrix is $x = 5$. The axis of symmetry is the x-axis.

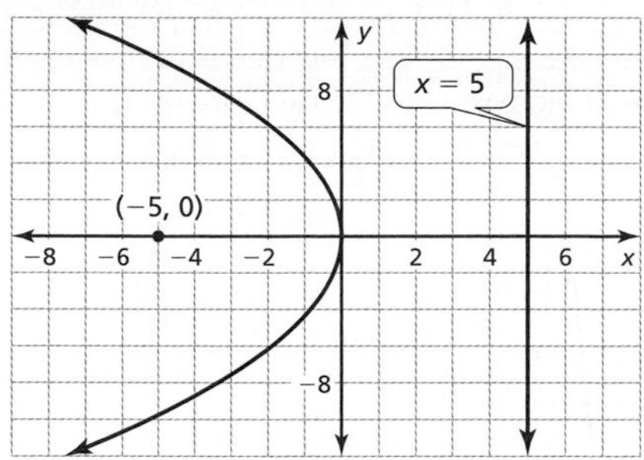

17. The focus is $(4, 0)$. The directrix is $x = -4$. The axis of symmetry is the x-axis.

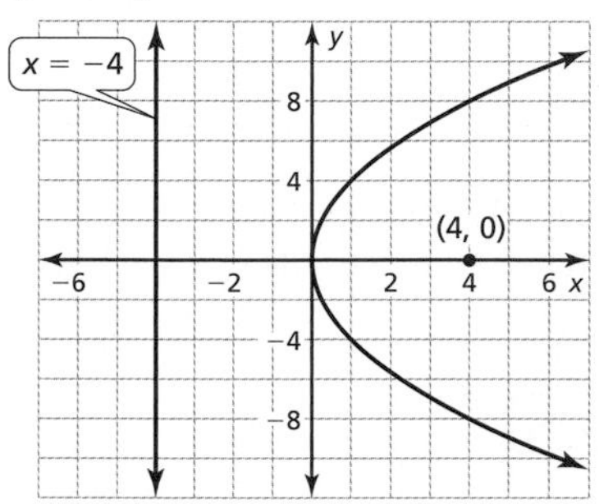

19. The focus is $\left(0, -\frac{1}{8}\right)$. The directrix is $y = \frac{1}{8}$. The axis of symmetry is the y-axis.

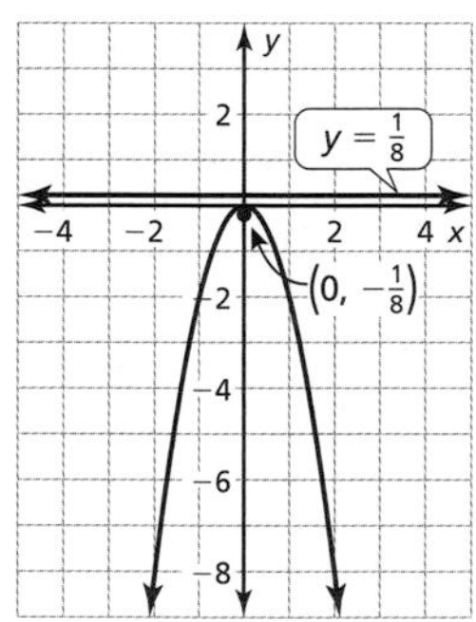

21. Instead of a vertical axis of symmetry, the graph should have a horizontal axis of symmetry.

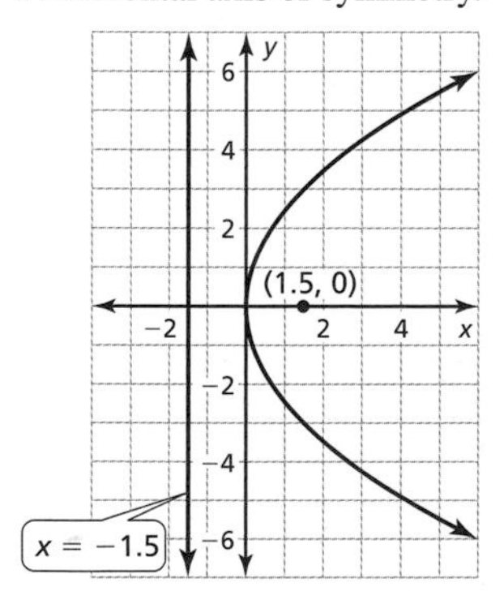

23. 9.5 in.; The receiver should be placed at the focus. The distance from the vertex to the focus is $p = \frac{38}{4} = 9.5$ in.

25. $y = \frac{1}{32}x^2$ **27.** $x = -\frac{1}{10}y^2$ **29.** $x = \frac{1}{12}y^2$

31. $x = \frac{1}{40}y^2$ **33.** $y = -\frac{3}{20}x^2$ **35.** $y = \frac{7}{24}x^2$

37. $x = -\frac{1}{16}y^2 - 4$ **39.** $y = \frac{1}{6}x^2 + 1$

41. The vertex is $(3, 2)$. The focus is $(3, 4)$. The directrix is $y = 0$. The axis of symmetry is $x = 3$. The graph is a vertical shrink by a factor of $\frac{1}{2}$ followed by a translation 3 units right and 2 units up.

43. The vertex is $(1, 3)$. The focus is $(5, 3)$. The directrix is $x = -3$. The axis of symmetry is $y = 3$. The graph is a horizontal shrink by a factor of $\frac{1}{4}$ followed by a translation 1 unit right and 3 units up.

45. The vertex is $(2, -4)$. The focus is $\left(\frac{23}{12}, -4\right)$. The directrix is $x = \frac{25}{12}$. The axis of symmetry is $y = -4$. The graph is a horizontal stretch by a factor of 12 followed by a reflection in the y-axis and a translation 2 units right and 4 units down.

47. $x = \frac{1}{5.2}y^2$; about 3.08 in.

49. As $|p|$ increases, the graph gets wider; As $|p|$ increases, the constant in the function gets smaller which results in a vertical shrink, making the graph wider.

51. $y = \frac{1}{4}x^2$ **53.** $x = \frac{1}{4p}y^2$

2.3 Maintaining Mathematical Proficiency *(p. 74)*

55. $y = 3x - 7$ **57.** $y = -\frac{1}{2}x + \frac{5}{2}$

59. $y = 3.98x + 0.92$

2.4 Vocabulary and Core Concept Check *(p. 80)*

1. A quadratic model is appropriate when the second differences are constant.

2.4 Monitoring Progress and Modeling with Mathematics *(pp. 80–82)*

3. $y = -3(x + 2)^2 + 6$ **5.** $y = 0.06(x - 3)^2 + 2$

7. $y = -\frac{1}{3}(x + 6)^2 - 12$ **9.** $y = -4(x - 2)(x - 4)$

11. $y = \frac{1}{10}(x - 12)(x + 6)$ **13.** $y = 2.25(x + 16)(x + 2)$

15. If given the x-intercepts, it is easier to write the equation in intercept form. If given the vertex, it is easier to write the equation in vertex form.

17. $y = -16(x - 3)^2 + 150$ **19.** $y = -0.75x^2 + 3x$

21. The x-intercepts were substituted incorrectly.

$y = a(x - p)(x - q)$

$4 = a(3 + 1)(3 - 2)$

$a = 1$

$y = (x + 1)(x - 2)$

23. $S(C) = 180C^2$; 18,000 lbs

25. intercept form; The three points can be substituted into the intercept form of a quadratic equation to solve for a, and then the equation can be written. This method is much shorter than writing and solving a system of three equations, although it can only be used when given the intercepts.

27. **a.** parabola; not a constant rate of change

b. $h = -16t^2 + 280$ **c.** about 4.18 sec

d. The domain is $0 \leq t \leq 4.18$ and represents the time the sponge was in the air. The range is $0 \leq h \leq 280$ and represents the height of the sponge.

29. quadratic; The second differences are constant; $y = -2x^2 + 42x + 470$

31. neither; The first and second differences are not constant.

33. **a.** The vertex indicates that on the 6th day, 19 people were absent, more than any other day.

b. $y = -0.5(10 - 6)^2 + 19$; 11 students

c. From 0 to 6 days, the average rate of change was 3 students per day. From 6 to 11 days, the average rate of change was -2.5 students per day. The rate at which students were missing school was changing more rapidly as more became ill, in comparison to when the students were becoming well.

35. $y = -16x^2 + 6x + 22$; after about 1.24 sec; 1.375 sec

37. 155 tiles

2.4 Maintaining Mathematical Proficiency *(p. 82)*

39. $(x - 2)(x - 1)$ **41.** $5(x + 3)(x - 2)$

Chapter 2 Review *(pp. 84–86)*

1. The graph is a translation 4 units left of the parent quadratic function.

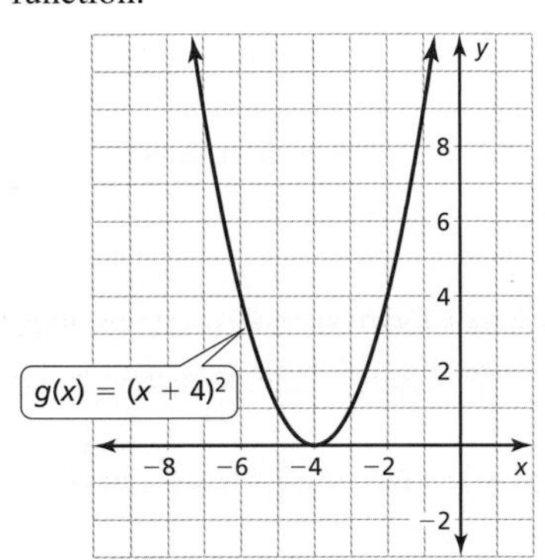

2. The graph is a translation 7 units right and 2 units up of the parent quadratic function.

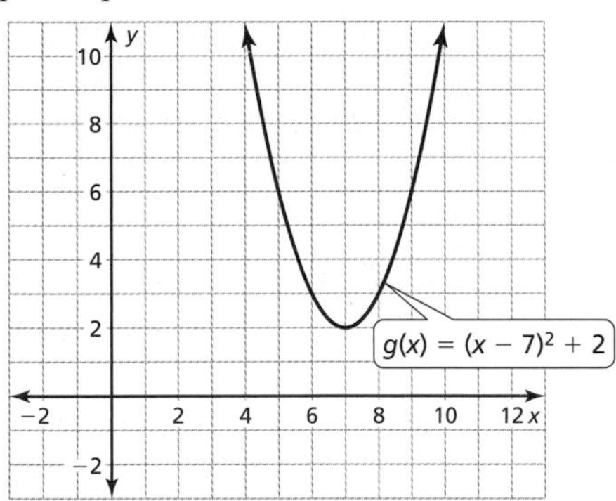

3. The graph is a vertical stretch by a factor of 3 followed by a reflection in the x-axis and a translation 2 units left and 1 unit down.

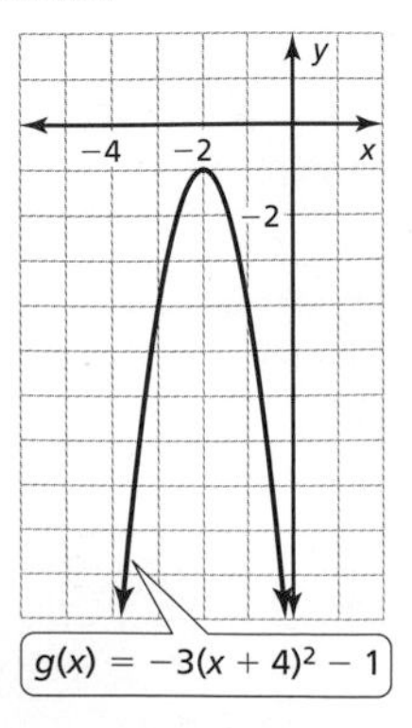

4. $g(x) = \frac{9}{4}(x + 5)^2 - 2$

5. $g(x) = (-x + 2)^2 - 2(-x + 2) + 3 = x^2 - 2x + 3$

6. The minimum value is -4; The function is decreasing to the left of $x = 1$ and increasing to the right of $x = 1$.

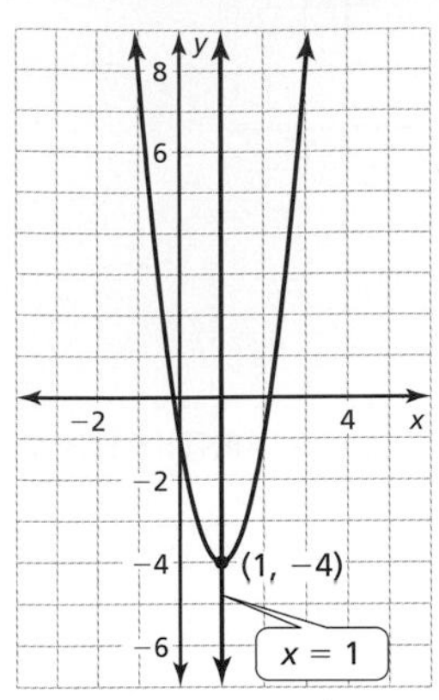

7. The maximum value is 35; The function is increasing to the left of $x = 4$ and decreasing to the right of $x = 4$.

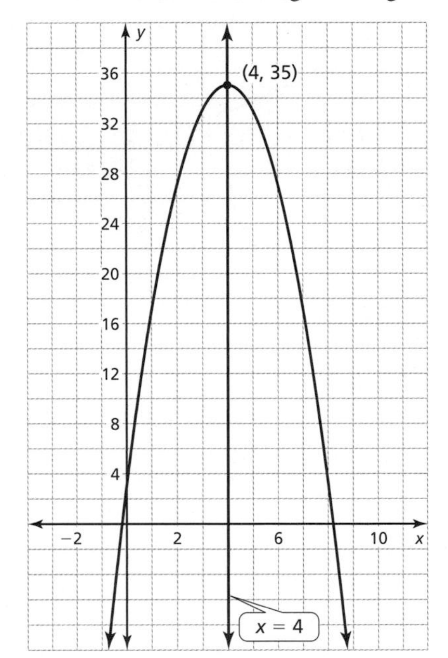

8. The minimum value is -25; The function is decreasing to the left of $x = -2$ and increasing to the right of $x = -2$.

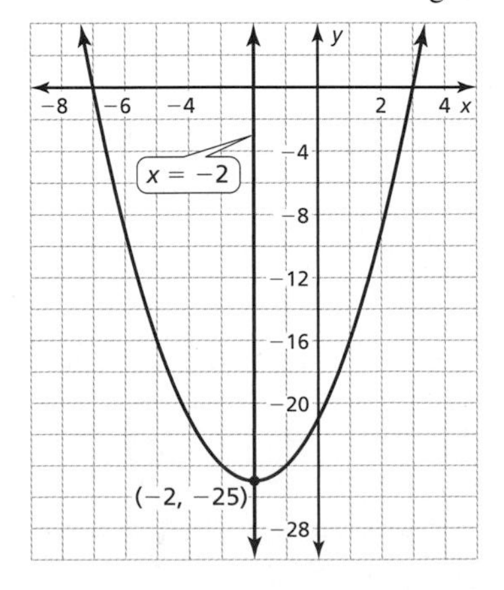

9. 2.25 in.

10. The focus is (0, 9), the directrix is $y = -9$, and the axis of symmetry is $x = 0$.

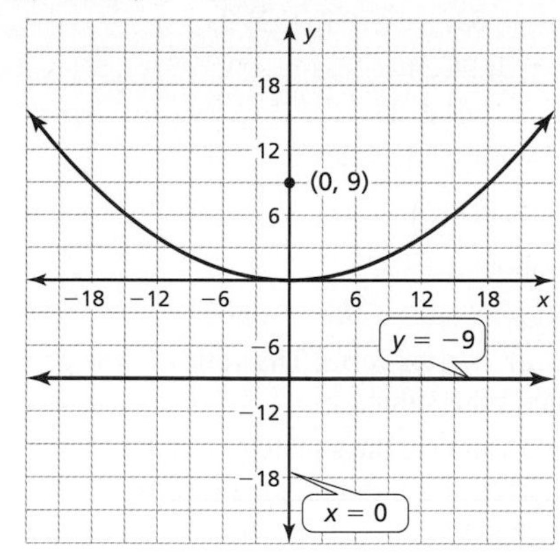

11. $x = -\frac{1}{8}y^2$ **12.** $y = -\frac{1}{16}(x - 2)^2 + 6$

13. $y = \frac{16}{81}(x - 10)^2 - 4$ **14.** $y = -\frac{3}{5}(x + 1)(x - 5)$

15. $y = 4x^2 + 5x + 1$

16. $y = -16x^2 + 150$; about 3.06 sec

Chapter 3

Chapter 3 Maintaining Mathematical Proficiency *(p. 91)*

1. $3\sqrt{3}$ **2.** $-4\sqrt{7}$ **3.** $\frac{\sqrt{11}}{8}$ **4.** $\frac{7\sqrt{3}}{10}$ **5.** $\frac{3\sqrt{2}}{7}$

6. $-\frac{\sqrt{65}}{11}$ **7.** $-4\sqrt{5}$ **8.** $4\sqrt{2}$ **9.** $(x - 6)(x + 6)$

10. $(x - 3)(x + 3)$ **11.** $(2x - 5)(2x + 5)$ **12.** $(x - 11)^2$

13. $(x + 14)^2$ **14.** $(7x + 15)^2$

15. $a = 16$ and $c = 1$, $a = 4$ and $c = 4$, $a = 1$ and $c = 16$; $2\sqrt{ac} = 8$

3.1 Vocabulary and Core Concept Check *(p. 99)*

1. Use the graph to find the x-intercepts of the function.

3.1 Monitoring Progress and Modeling with Mathematics *(pp. 99–102)*

3. $x = -1$ and $x = -2$ **5.** $x = 3$ and $x = -3$

7. $x = -1$ **9.** no real solution **11.** no real solution

13. $s = \pm 12$ **15.** $z = 1$ and $z = 11$ **17.** $x = 1 \pm \sqrt{2}$

19. no real solution **21.** A, B, and E

23. The ± was not used when taking the square root; $2(x + 1)^2 + 3 = 21$; $2(x + 1)^2 = 18$; $(x + 1)^2 = 9$; $x + 1 = \pm 3$; $x = 2$ and $x = -4$

25. **a.** *Sample answer:* $x^2 = 16$ **b.** $x^2 = 0$

c. *Sample answer:* $x^2 = -9$

27. $x = -3$ **29.** $x = 6$ and $x = 2$ **31.** $n = 0$ and $n = 6$

33. $w = 12$ and $w = 2$ **35.** $x = 4$ **37.** $x = 3$

39. $u = 0$ and $u = -9$; *Sample answer:* factoring because the equation can be factored

41. no real solution; *Sample answer:* square roots because the equation can be written in the form $u^2 = d$

43. $x = 6$ and $x = 2$; *Sample answer:* square roots because the equation can be written in the form $u^2 = d$

45. $x = -0.5$ and $x = -2.5$; *Sample answer:* factoring because the equation can be factored

47. $x = -2$ and $x = -4$ **49.** $x = 3$ and $x = -10$

51. $x = 3$ and $x = -2$ **53.** $x = -11$

55. $f(x) = x^2 - 19x + 88$ **57.** \$5.75; \$1983.75

59. **a.** $h(t) = -16t^2 + 188$; about 3.4 sec

b. 80 ft; The log fell 80 feet between 2 and 3 seconds.

61. 0.5 ft or 6 in.

63. The 20-foot wave requires a wind speed twice as great as the wind speed required for a 5-foot wave.

65. $x \approx 34.64$; about 207.84 ft, 277.12 ft, 173.20 ft, and 300 ft

67. the rock on Jupiter; Because the first term is negative, the height of the falling object will decrease faster as g gets larger.

69.

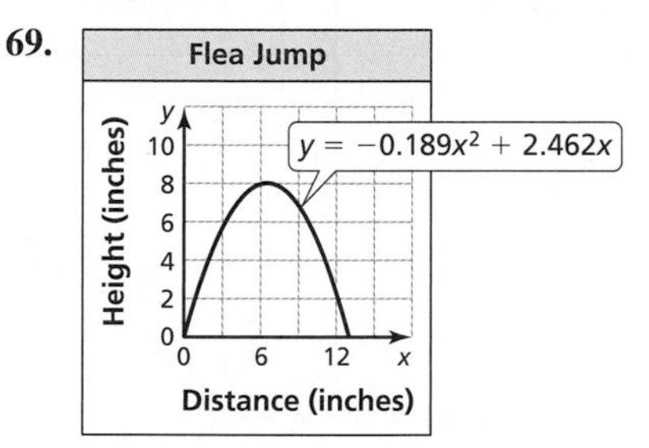

The vertex (6.5, 8.0) indicates that the flea's maximum jump is 6.5 inches away from and 8.0 inches above the starting point. The zeros $x = 13$ and $x = 0$ indicate when the flea is on the ground.

71. you; The function does not cross the x-axis.

73. **a.** $mn = a^2$ and $m + n = 0$

b. $m = \sqrt{-a^2} = a\sqrt{-1}$, $n = -\sqrt{-a^2} = -a\sqrt{-1}$; m and n are not real numbers.

75. 60 ft

3.1 Maintaining Mathematical Proficiency *(p. 102)*

77. $x^3 + 4x^2 + 6$ **79.** $-3x^3 + 7x^2 - 15x + 9$

81. $10x^3 - 2x^2 + 6x$ **83.** $-44x^3 + 33x^2 + 88x$

3.2 Vocabulary and Core Concept Check *(p. 108)*

1. $i = \sqrt{-1}$ and is used to write the square root of any negative number.

3. Add the real parts and the imaginary parts separately.

3.2 Monitoring Progress and Modeling with Mathematics *(pp. 108–110)*

5. $6i$ **7.** $3i\sqrt{2}$ **9.** $8i$ **11.** $-16i\sqrt{2}$

13. $x = 2$ and $y = 2$ **15.** $x = -2$ and $y = 4$

17. $x = 7$ and $y = -12$ **19.** $x = 6$ and $y = 28$

21. $13 + 2i$ **23.** $9 + 11i$ **25.** 19 **27.** $4 + 2i$

29. $-4 - 14i$ **31.** **a.** $-4 + 5i$ **b.** $2\sqrt{2} + 10i$

33. $(12 + 2i)$ ohms **35.** $(8 + i)$ ohms **37.** $-3 - 15i$

39. $14 - 5i$ **41.** 20 **43.** $-27 - 36i$

45. Distributive Property; Simplify; Definition of complex addition; Write in standard form.

47. $(6 - 7i) - (4 - 3i) = 2 - 4i$ **49.** $x = \pm 3i$

51. $x = \pm i\sqrt{7}$ **53.** $x = \pm 2i\sqrt{5}$ **55.** $x = \pm i\sqrt{2}$

57. $x = \pm 6i$ **59.** $x = \pm 3i\sqrt{3}$ **61.** $x = \pm 4i\sqrt{3}$

63. i^2 can be simplified; $15 - 3i + 10i - 2i^2 = 15 + 7i + 2 = 17 + 7i$

65. **a.** -8 **b.** $12 - 10i$ **c.** $21i$ **d.** $41 + 3i$ **e.** $-9i$
f. $-9 + 23i$ **g.** 14 **h.** $14i$

Real numbers	Imaginary numbers	Pure imaginary numbers
-8 14	$12 - 10i$ $41 + 3i$ $-9 + 23i$	$21i$ $-9i$ $14i$

67.

Powers of *i*	i^1	i^2	i^3	i^4	i^5	i^6	i^7	i^8	i^9	i^{10}	i^{11}	i^{12}
Simplified form	i	-1	$-i$	1	i	-1	$-i$	1	i	-1	$-i$	1

The results of i^n alternate in the pattern i, -1, $-i$, and 1.

69. $-28 + 27i$ **71.** $-15 - 25i$ **73.** $9 + 5i$

75. *Sample answer:* $3 + 2i$ and $3 - 2i$; The real parts are equal and the imaginary parts are opposites.

77. **a.** false; *Sample answer:* $(3 - 5i) + (4 + 5i) = 7$
b. true; *Sample answer:* $(3i)(2i) = 6i^2 = -6$
c. true; *Sample answer:* $3i = 0 + 3i$
d. false; *Sample answer:* $1 + 8i$

3.2 Maintaining Mathematical Proficiency *(p. 110)*

79. yes **81.** no **83.** $y = 2(x + 3)^2 - 3$

3.3 Vocabulary and Core Concept Check *(p. 116)*

1. $\left(\frac{b}{2}\right)^2$

3.3 Monitoring Progress and Modeling with Mathematics *(pp. 116–118)*

3. $x = 9$ and $x = -1$ **5.** $x = 9 \pm \sqrt{5}$

7. $y = 12 \pm 10i$ **9.** $w = \dfrac{-1 \pm 5\sqrt{3}}{2}$ **11.** $25; (x + 5)^2$

13. $36; (y - 6)^2$ **15.** $9; (x - 3)^2$ **17.** $\frac{25}{4}; \left(z - \frac{5}{2}\right)^2$

19. $\frac{169}{4}; \left(w + \frac{13}{2}\right)^2$ **21.** $4; x^2 + 4x + 4$

23. $36; x^2 + 12x + 36$ **25.** $x = -3 \pm \sqrt{6}$

27. $x = -2 \pm \sqrt{6}$ **29.** $z = \dfrac{-9 \pm \sqrt{85}}{2}$

31. $t = -2 \pm 2i$ **33.** $x = -3 \pm i$ **35.** $x = 5 \pm 2\sqrt{7}$

37. 36 should have been added to the right side of the equation instead of 9; $4x^2 + 24x - 11 = 0$; $4(x^2 + 6x) = 11$; $4(x^2 + 6x + 9) = 11 + 36$; $4(x + 3)^2 = 47$; $(x + 3)^2 = \dfrac{47}{4}$; $x + 3 = \dfrac{\pm\sqrt{47}}{2}$; $x = -3 \pm \dfrac{\sqrt{47}}{2}$; $x = \dfrac{-6 \pm \sqrt{47}}{2}$

39. yes; All of the steps would be the same as with two real solutions, with the exception of the constant being negative when you take the square root.

41. factoring; The equation can be factored; $x = 7$ and $x = -3$

43. square roots; The equation can be written in the form $u^2 = d$; $x = -8$ and $x = 0$

45. factoring; The equation can be factored; $x = -6$

47. completing the square; The equation cannot be factored or written in the form $u^2 = d$; $x = -1 \pm \dfrac{\sqrt{10}}{2}$

49. square roots; The equation can be written in the form $u^2 = d$; $x = \pm 10$

51. $x = -5 + 5\sqrt{3}$ **53.** $x = -2 + 2\sqrt{21}$

55. $f(x) = (x - 4)^2 + 3$; $(4, 3)$

57. $g(x) = (x + 6)^2 + 1$; $(-6, 1)$

59. $h(x) = (x + 1)^2 - 49$; $(-1, -49)$

61. $f(x) = \left(x - \frac{3}{2}\right)^2 + \frac{7}{4}$; $\left(\frac{3}{2}, \frac{7}{4}\right)$

63. **a.** 22 ft **b.** about 2.1 sec

65. **a.** \$3600 **b.** $y = -(x - 10)^2 + 3600$
c. *Sample answer:* vertex form; The vertex of the graph gives the maximum value.

67. *Sample answer:* Complete the square to find the vertex. Factor it into intercept form to find the two roots, find their average to obtain the time when the water reaches its maximum height, and then substitute the time into the function. Use the coefficients of the original function $y = f(x)$ to find the maximum height, $f\left(-\dfrac{b}{2a}\right)$; 125.44 ft

69. no; The problem cannot be solved by factoring because the answers are not rational.

71. $x = \dfrac{-b \pm \sqrt{b^2 - 4c}}{2}$ **73.** $x \approx 0.896$ cm

3.3 Maintaining Mathematical Proficiency *(p. 118)*

75. $y \leq -1$

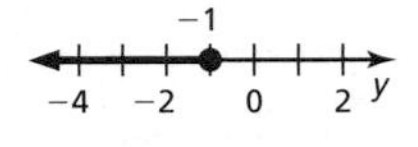

77. $s \geq -20$

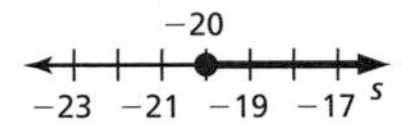

79.

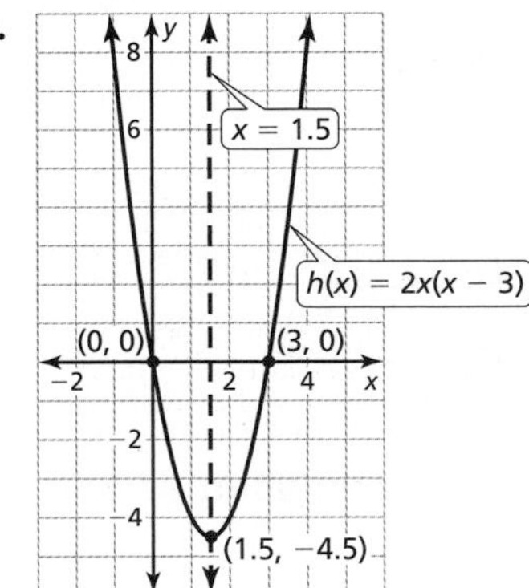

81.

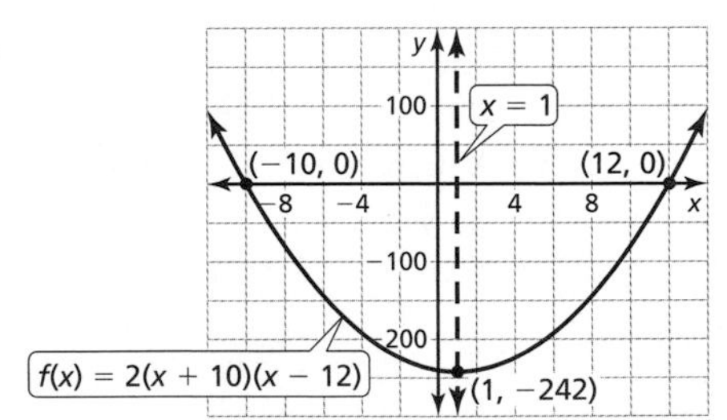

3.4 Vocabulary and Core Concept Check *(p. 127)*

1. $\dfrac{-b \pm \sqrt{b^2 - 4ac}}{2a}$

3. There will be two imaginary solutions.

3.4 Monitoring Progress and Modeling with Mathematics *(pp.127–130)*

5. $x = 3$ and $x = 1$ **7.** $x = -3 \pm i\sqrt{6}$ **9.** $x = 7$

11. $x = \dfrac{-1 \pm i\sqrt{14}}{3}$ **13.** $x = 5$ **15.** $x = \dfrac{3 \pm \sqrt{89}}{8}$

17. $z = 6 \pm \sqrt{30}$ **19.** 0; one real: $x = -6$

21. 400; two real: $n = 3$ and $n = -2$

23. -135; two imaginary: $x = \dfrac{5 \pm 3i\sqrt{15}}{8}$

25. 0; one real: $x = -4$ **27.** A

29. C; The discriminant is negative, so the graph has no x-intercepts.

31. A; The discriminant is positive, so the graph has two x-intercepts. The y-intercept is -9.

33. The i was left out after taking the square root;
$x = \dfrac{-10 \pm \sqrt{-196}}{2} = \dfrac{-10 \pm 14i}{2} = -5 \pm 7i$

35. *Sample answer:* $a = 1$ and $c = 5$; $x^2 + 4x + 5 = 0$

37. *Sample answer:* $a = 2$ and $c = 4$; $2x^2 - 8x + 4 = 0$

39. *Sample answer:* $a = 5$ and $c = -5$; $5x^2 + 10x + 5 = 0$

41. $-5x^2 + 8x - 12 = 0$ **43.** $-7x^2 + 4x - 5 = 0$

45. $3x^2 + 4x + 1 = 0$

47. $x = \pm 2\sqrt{2}$; *Sample answer:* square roots; The equation can be written in the form $u^2 = d$.

49. $x = 9$ and $x = -3$; *Sample answer:* factoring; The equation can be factored.

51. $x = 3$ and $x = 4$; *Sample answer:* factoring; The equation can be factored.

53. $x = 5 \pm i\sqrt{2}$; *Sample answer:* completing the square; Factor out 5, and $a = 1$ and b is an even number.

55. $x = \dfrac{-9 \pm \sqrt{33}}{8}$; *Sample answer:* Quadratic Formula; $a \neq 1$, b is not an even number, the equation cannot be factored, and it cannot be easily written in the form $u^2 = d$.

57. $x = \dfrac{-1 \pm \sqrt{5}}{2}$; *Sample answer:* Quadratic Formula; b is not an even number, the equation cannot be factored, and it cannot be easily written in the form $u^2 = d$.

59. $x = 6$ **61.** about 5.67 sec **63.** about 0.17 sec

65. a.

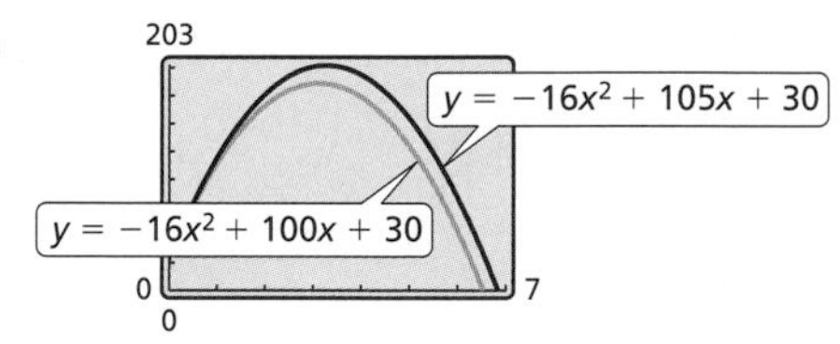

Both rockets start from the same height, but your friend's rocket does not go as high and lands about a half of a second earlier.

b. about 1 sec and 5.5625 sec; These are reasonable because $\dfrac{1 + 5.6}{2} = 3.3$ which is the axis of symmetry.

67. a. about 0.97 sec

b. the first bird; The second bird will reach the water after about 0.98 second.

69. 3.5 ft

71. a. $x = 6$, $x = -3$, $x = 5$, and $x = -2$ **b.** $x = \pm 3$

73. Add the solutions to get $\dfrac{-b}{a}$, then divide the result by 2 to get $-\dfrac{b}{2a}$; Because it is symmetric, the vertex of a parabola is in the middle of the two x-intercepts and the x-coordinate of the vertex is $-\dfrac{b}{2a}$.

75. If $x = 3i$ and $x = -2i$ are solutions, then the equation can be written as $a(x - 3i)(x + 2i) = ax^2 - aix + 6a$. a and ai cannot both be real numbers.

3.4 Maintaining Mathematical Proficiency *(p. 130)*

77. (4, 5) **79.** no solution

81.

83.

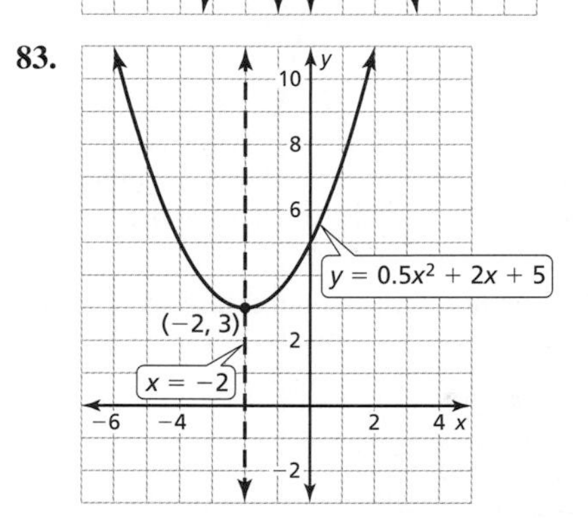

3.5 Vocabulary and Core Concept Check *(p. 136)*

1. There could be no solution, one solution, or two solutions.

3.5 Monitoring Progress and Modeling with Mathematics *(pp. 136–138)*

3. (0, 2) and (−2, 0) **5.** no solution **7.** (−4, 2)

9. (1, 1) and (3, 1) **11.** (−4, 1) **13.** (1, 4) and (9, 4)

15. (3, 8) and (−1, 4) **17.** (0, −8) **19.** no solution

21. (2, 3) and (−2, 3) **23.** no solution **25.** A and C

27. (2, 7) and (0, 5) **29.** no solution

31. about (−4.65, −4.71) and about (0.65, −15.29)

33. (−4, −4) and (−6, −4)

35. The terms that were added were not like terms; $0 = -2x^2 + 34x - 140$; $x = 7$ *or* $x = 10$

37. (0, −1); *Sample answer:* elimination because the equations are arranged with like terms in the same column

39. about (−11.31, 10) and about (5.31, 10); *Sample answer:* substitution because the second equation can be substituted into the first equation

41. (3, 3) and (5, 3); *Sample answer:* graphing because substitution and elimination would require more steps in this case

43. $x = 0$ **45.** $x \approx 0.63$ and $x \approx 2.37$

47. $x = 2$ and $x = 3$

49. The graphs intersect at the vertex of the quadratic function.

51. $d = 0.8t$; $d = 2.5t^2$; 0.32 min

53. no solution: $m = 1$; one solution: $m = 0$; two solutions: $m = -1$

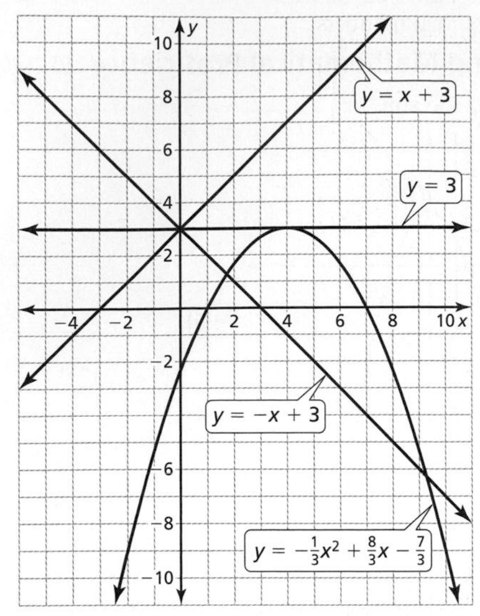

55. *Sample answer:* graphing and Quadratic Formula; graphing because it requires less time and steps than using the Quadratic Formula in this case

57. a. no solution, one solution, two solutions, three solutions, or four solutions

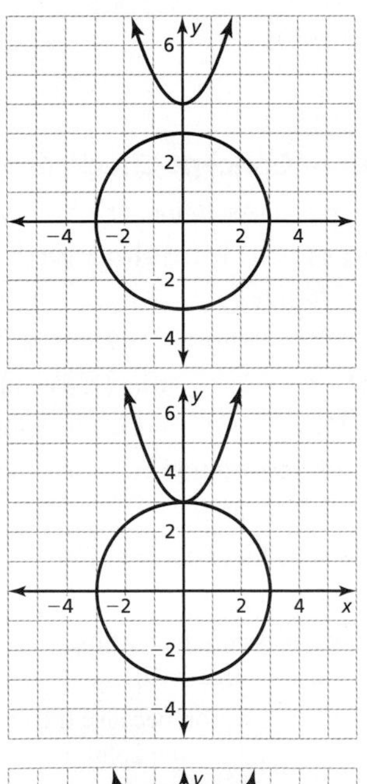

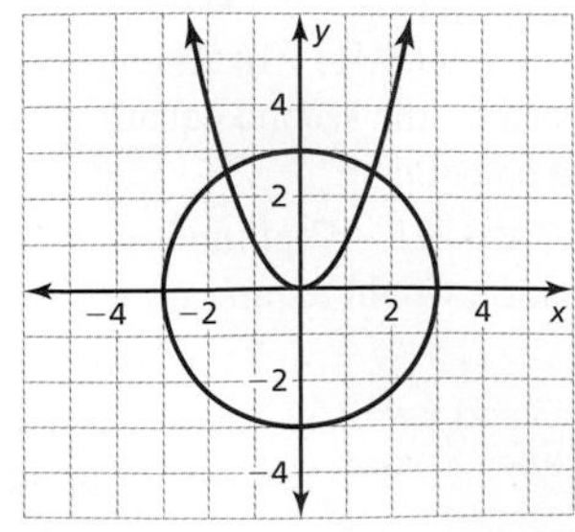

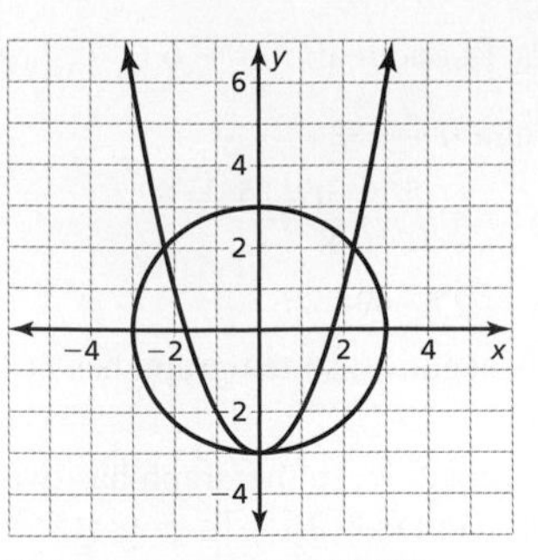

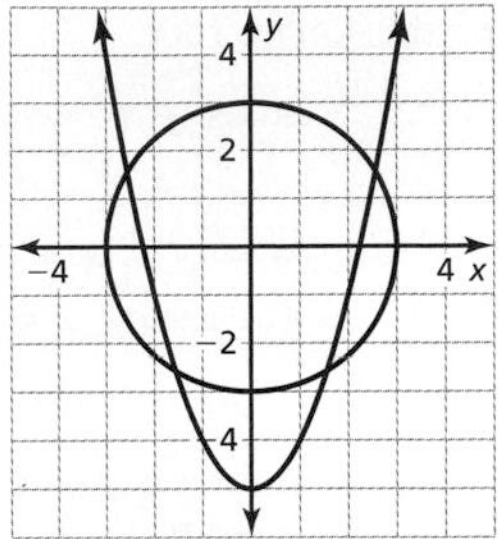

b. no solution, one solution, two solutions, or infinitely many solutions

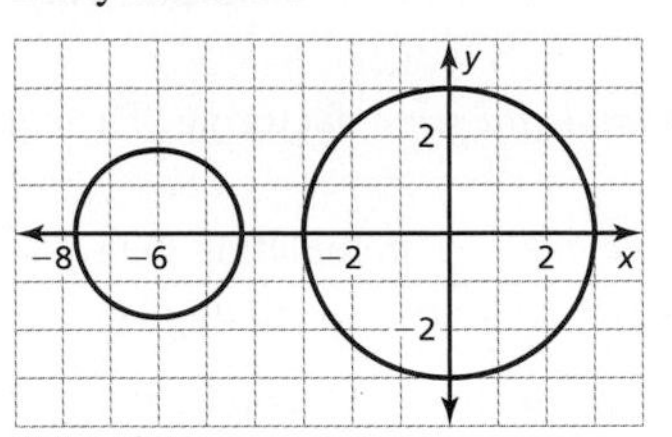

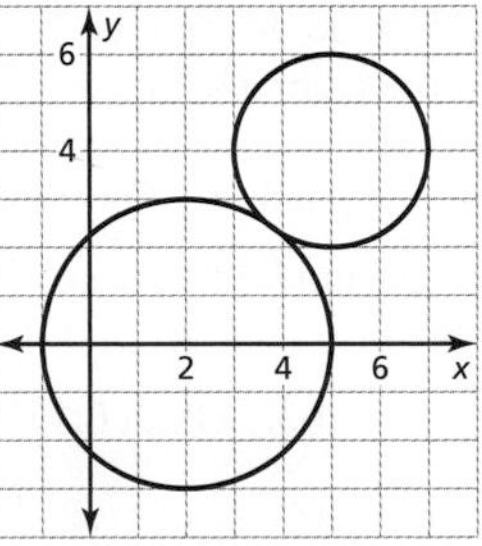

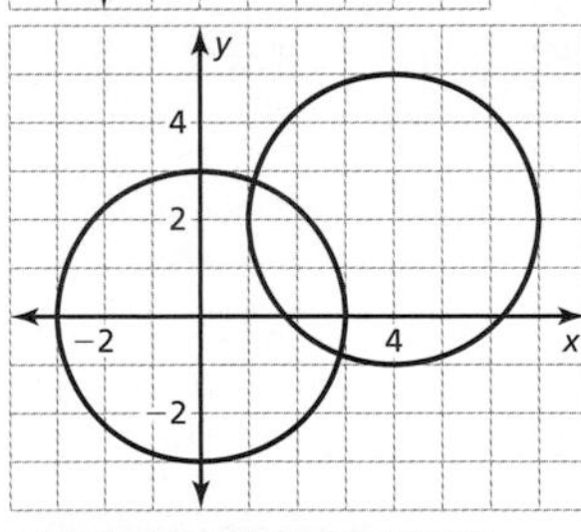

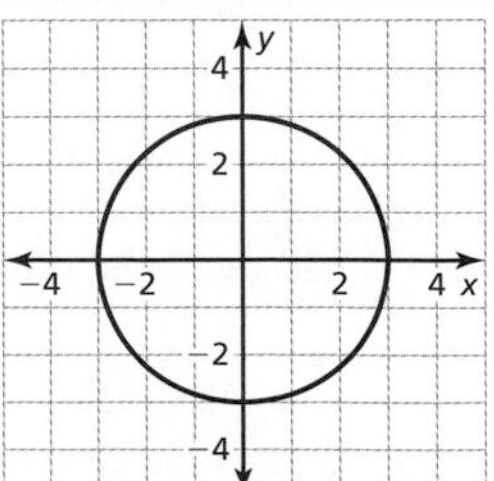

59. a. circle: $x^2 + y^2 = 1$, Oak Lane: $y = -\frac{1}{7}x + \frac{5}{7}$

b. $(-0.6, 0.8)$ and $(0.8, 0.6)$ **c.** about 1.41 mi

3.5 Maintaining Mathematical Proficiency *(p. 138)*

61. $x > 3$

−1 1 3 5 7

63. $x \leq -4$

65. $y < x - 2$

3.6 Vocabulary and Core Concept *(p. 144)*

1. The graph of a quadratic inequality in one variable consists of a number line, but the graph of a quadratic inequality in two variables consists of both the x- and y-axis.

3.6 Monitoring Progress and Modeling with Mathematics *(pp. 144–146)*

3. C; The x-intercepts are $x = -1$ and $x = -3$. The test point $(-2, 5)$ does not satisfy the inequality.

5. B; The x-intercepts are $x = 1$ and $x = 3$. The test point $(2, 5)$ does not satisfy the inequality.

7.

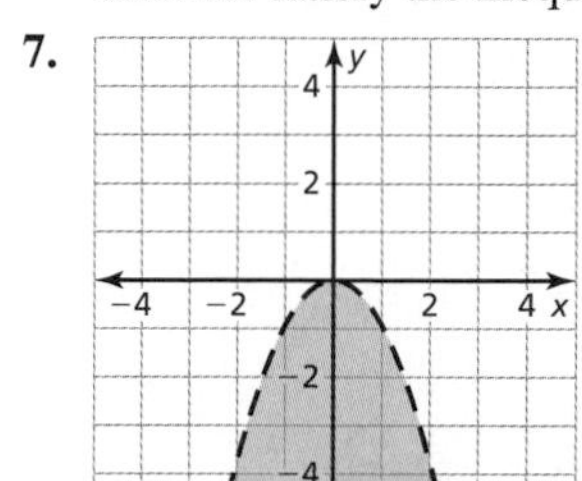

9.

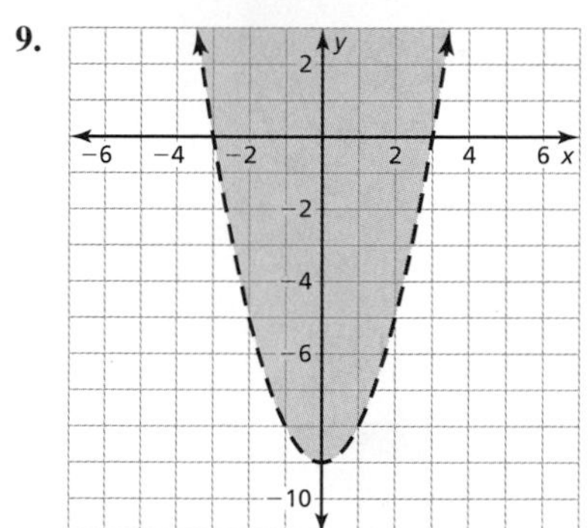

11.

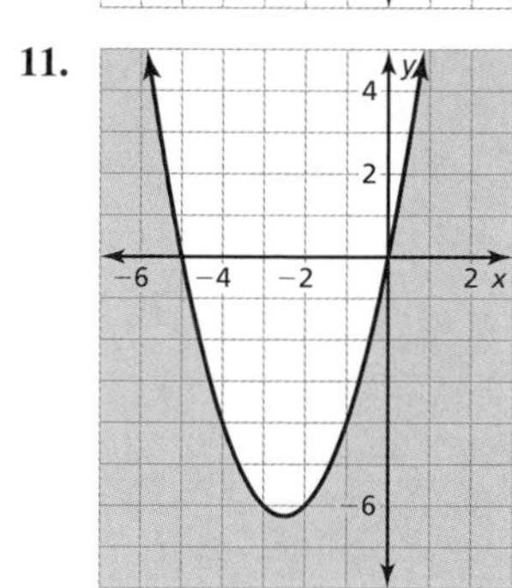

13.

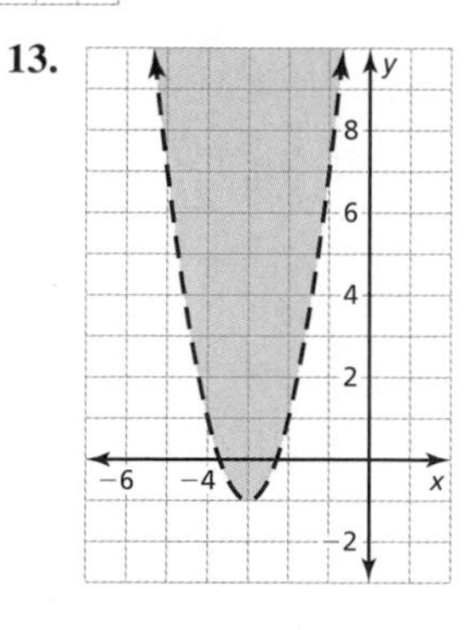

15. $y > f(x)$

17. The graph should be solid, not dashed.

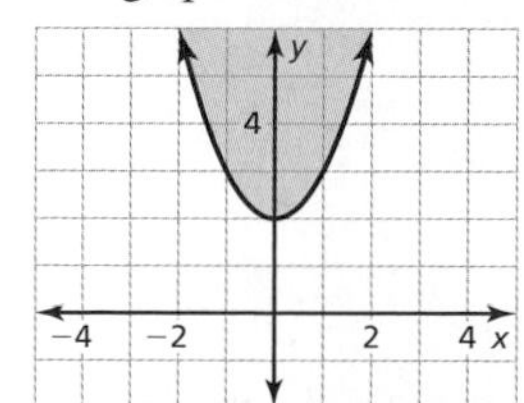

19. The solution represents weights that can be supported by shelves with various thicknesses.

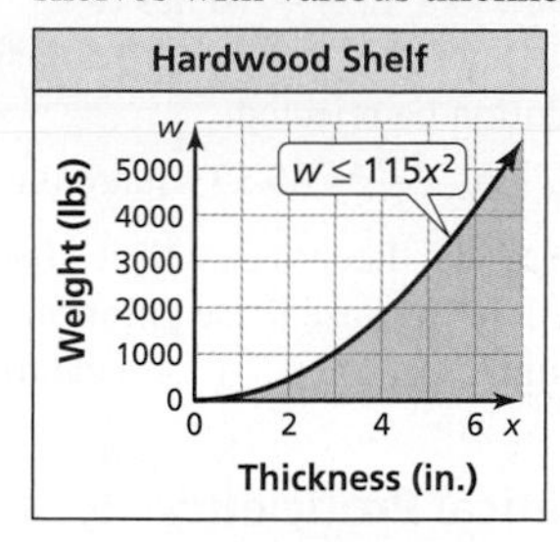

21.

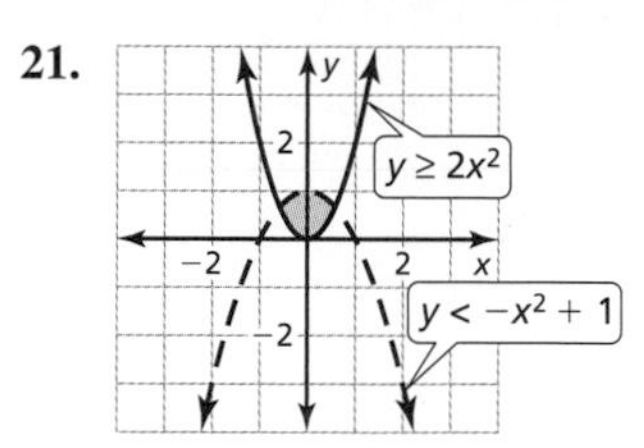

23.

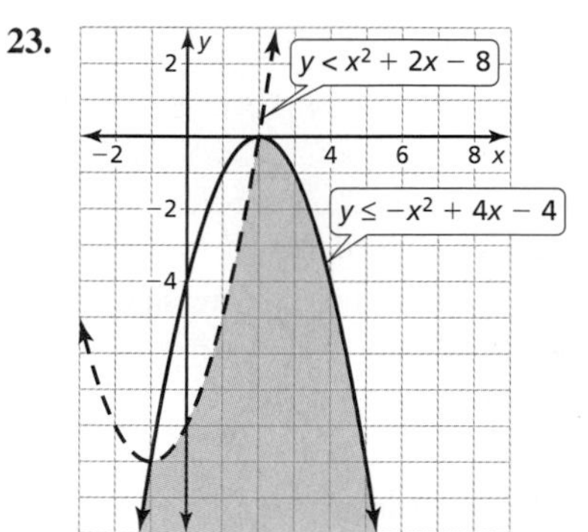

25.

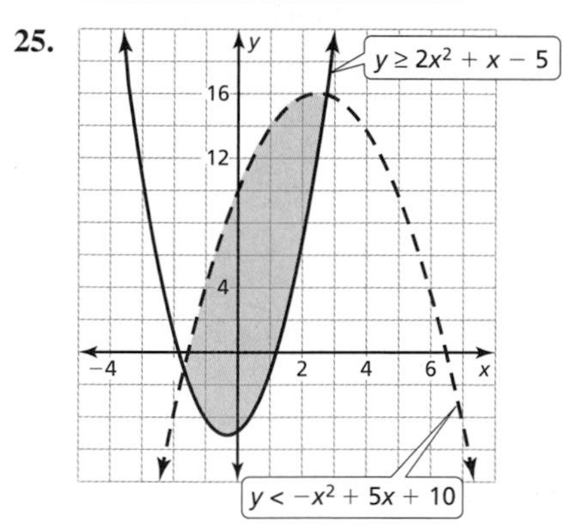

27. $-\frac{5}{2} < x < \frac{5}{2}$ **29.** $x \leq 4$ or $x \geq 7$ **31.** $-0.5 \leq x \leq 3$

33. $x < -2$ or $x > 4$ **35.** about $0.38 < x < 2.62$

37. $x < -7$ or $x > -1$ **39.** $-2 \leq x \leq \frac{4}{3}$

41. about $x \leq -6.87$ or $x \geq 0.87$

43. **a.** $x_1 < x < x_2$ **b.** $x < x_1$ or $x > x_2$ **c.** $x_1 < x < x_2$

45. about 55 m from the left pylon to about 447 m from the left pylon

47. **a.** $0.0051x^2 - 0.319x + 15 < 0.005x^2 - 0.23x + 22$, $16 \leq x \leq 70$

b. $A(x) < V(x)$ for $16 \leq x \leq 70$; Graph the inequalities only on $16 \leq x \leq 70$. $A(x)$ is always less than $V(x)$.

c. The driver would react more quickly to the siren of an approaching ambulance; The reaction time to audio stimuli is always less.

49. $0.00170x^2 + 0.145x + 2.35 > 10, 0 \le x \le 40$; after about 37 days; Because $L(x)$ is a parabola, $L(x) = 10$ has two solutions. Because the x-value must be positive, the domain requires that the negative solution be rejected.

51. **a.** $\frac{32}{3} \approx 10.67$ square units **b.** $\frac{256}{3} \approx 85.33$ square units

53. **a.** yes; The points on the parabola that are exactly 11 feet high are (6, 11) and (14, 11). Because these points are 8 feet apart, there is enough room for a 7-foot wide truck.

b. 8 ft **c.** about 11.2 ft

3.6 Maintaining Mathematical Proficiency *(p. 146)*

55.

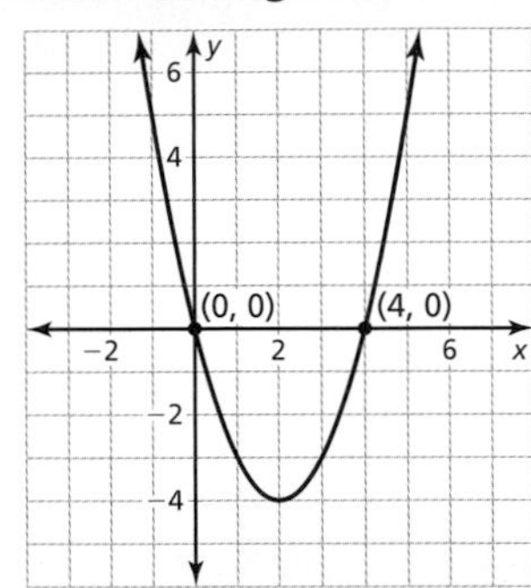

57. The maximum value is -1; The function is increasing to the left of $x = -3$ and decreasing to the right of $x = -3$.

59. The maximum value is 25; The function is increasing to the left of $x = -2$ and decreasing to the right of $x = -2$.

Chapter 3 Review *(pp. 148–150)*

1. $x = 4$ and $x = -2$ **2.** $x = \pm 2$

3. $x = 2$ and $x = -8$ **4.** $x = 6$ and $x = 2.5$

5. $(x + 18)(x + 35) = 1260$; $x = 10$; 28 ft by 45 ft

6. $x = 9$ and $y = -3$ **7.** $5 - 3i$ **8.** $11 + 10i$

9. $-62 + 11i$ **10.** $x = \pm i\sqrt{3}$ **11.** $x = \pm 4i$

12. 148 ft **13.** $x = -8 \pm \sqrt{47}$ **14.** $x = \dfrac{-4 \pm 3i}{2}$

15. $x = 3 \pm 3\sqrt{2}$ **16.** $y = (x - 1)^2 + 19$; (1, 19)

17. $x = \dfrac{5 \pm \sqrt{17}}{2}$ **18.** $x = 0.5$ and $x = -3$

19. $x = \dfrac{6 \pm i\sqrt{3}}{3}$ **20.** 0; one real solution: $x = -3$

21. 40; two real solutions: $x = 1 \pm \sqrt{10}$

22. 16; two real solutions: $x = -5$ and $x = -1$

23. $(-2, 6)$ and $(1, 0)$; *Sample answer:* substitution because both equations are already solved for y

24. (4, 5); *Sample answer:* elimination because adding the like terms eliminates y

25. about $(-0.32, 1.97)$ and $(0.92, -1.77)$; substitution because elimination is not a possibility with no like terms

26. $x \approx -0.14$ and $x \approx 1.77$

27.

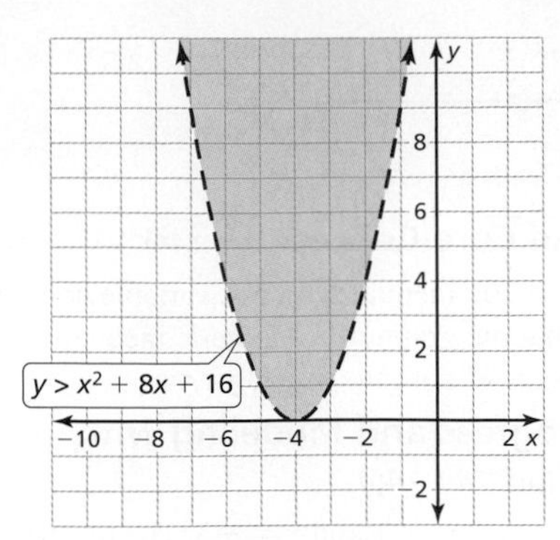

28.

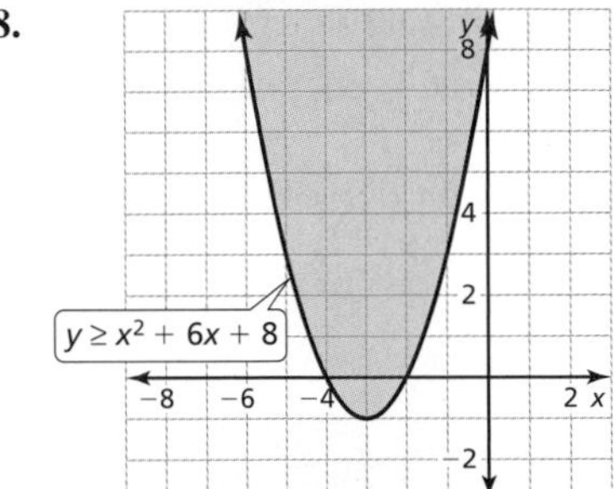

29.

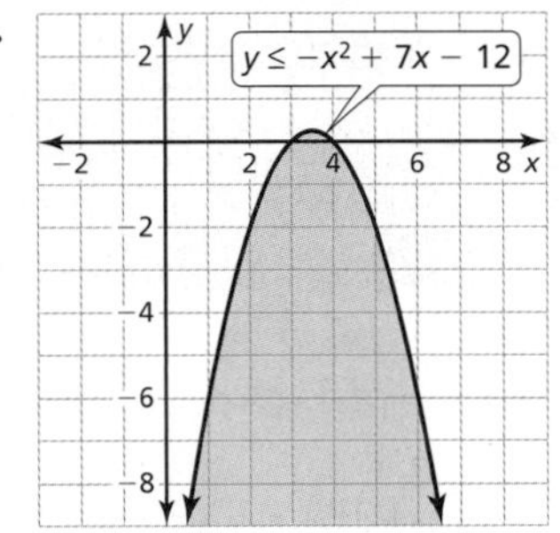

30.

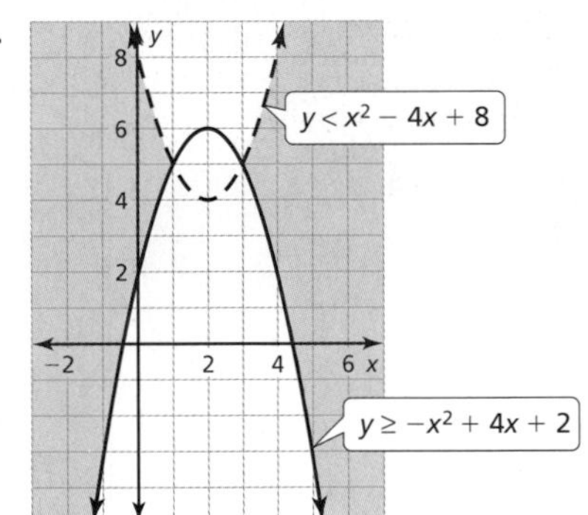

31.

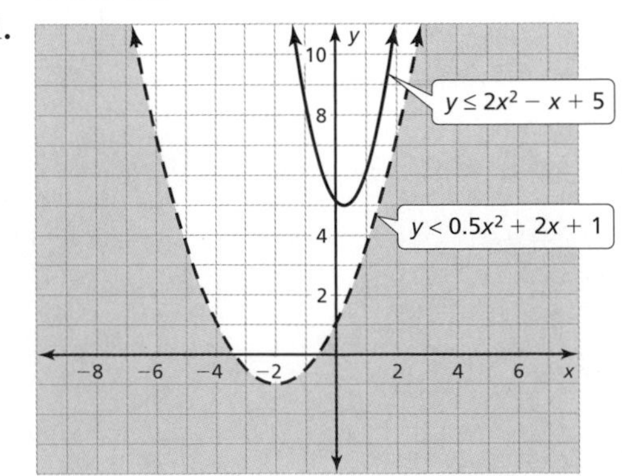

32.

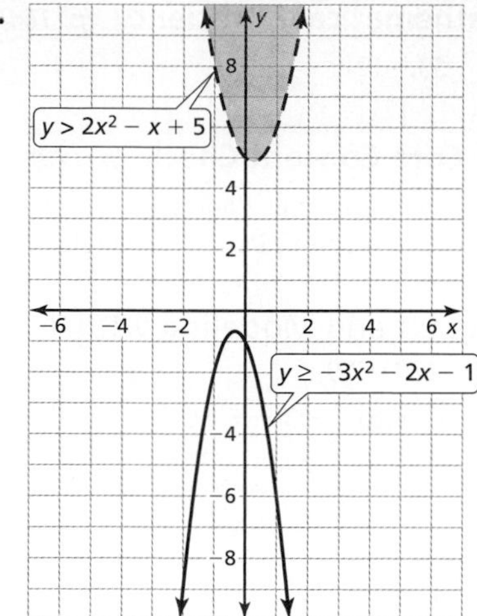

33. $x \leq -5$ or $x \geq 4$ **34.** $x < -7$ or $x > -3$

35. $\frac{2}{3} \leq x \leq 1$

Chapter 4

Chapter 4 Maintaining Mathematical Proficiency *(p. 155)*

1. $2x$ **2.** $4m + 3$ **3.** $-y + 6$ **4.** $x + 4$

5. $z - 4$ **6.** $5x$ **7.** 64 in.^3 **8.** $\frac{32\pi}{3} \text{ ft}^2 \approx 33.51 \text{ ft}^3$

9. 48 ft^3 **10.** $45\pi \text{ cm}^3 \approx 141.37 \text{ cm}^3$

11. no; If the volume of a cube is doubled, the side length is increased by a factor of $\sqrt[3]{2}$.

4.1 Vocabulary and Core Concept Check *(p. 162)*

1. The end behavior describes the behavior of a graph as x approaches positive infinity and negative infinity.

4.1 Monitoring Progress and Modeling with Mathematics *(pp. 162–164)*

3. polynomial function; $f(x) = 5x^3 - 6x^2 - 3x + 2$; degree: 3 (cubic), leading coefficient: 5

5. not a polynomial function

7. polynomial function; $h(x) = -\sqrt{7}x^4 + 8x^3 + \frac{5}{3}x^2 + x - \frac{1}{2}$; degree 4: (quartic), leading coefficient: $-\sqrt{7}$

9. The function is not in standard form so the wrong term was used to classify the function; f is a polynomial function. The degree is 4 and f is a quartic function. The leading coefficient is -7.

11. $h(-2) = -46$ **13.** $g(8) = -43$ **15.** $p\left(\frac{1}{2}\right) = \frac{45}{4}$

17. $h(x) \to -\infty$ as $x \to -\infty$ and $h(x) \to -\infty$ as $x \to \infty$

19. $f(x) \to \infty$ as $x \to -\infty$ and $f(x) \to \infty$ as $x \to \infty$

21. The degree of the function is odd and the leading coefficient is negative.

23. polynomial function; $f(x) = -4x^4 + \frac{5}{2}x^3 + \sqrt{2}x^2 + 4x - 6$; degree: 4 (quartic), leading coefficient: -4.

25.

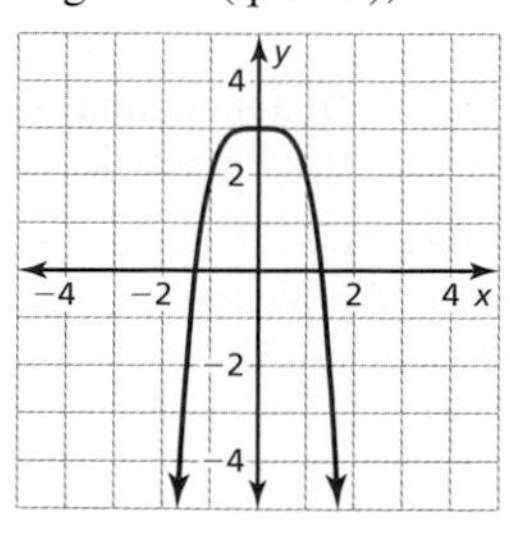

27.

29.

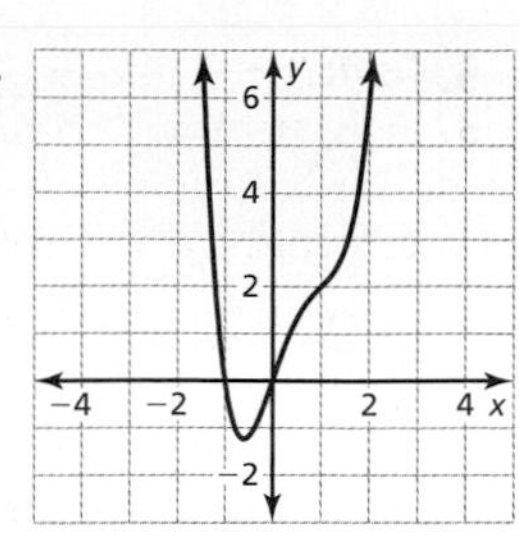

31.

33. a. The function is increasing when $x > 4$ and decreasing when $x < 4$.

b. $x < 3$ and $x > 5$

c. $3 < x < 5$

35. a. The function is increasing when $x < 0$ and $x > 2$ and decreasing when $0 < x < 2$.

b. $-1 < x < 2$ and $x > 2$

c. $x < -1$

37. The degree is even and the leading coefficient is positive.

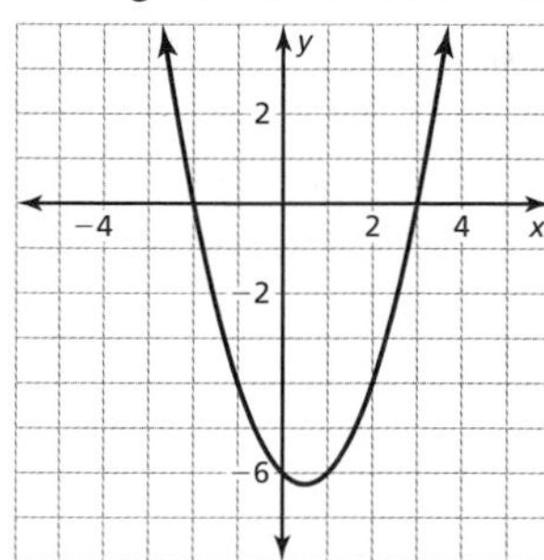

39. The degree is even and the leading coefficient is positive.

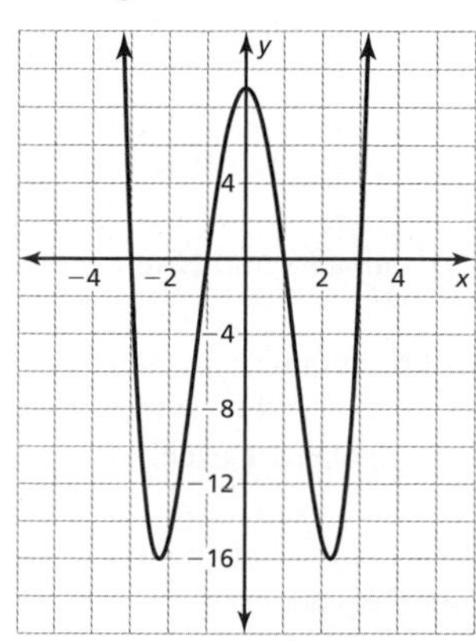

Selected Answers

41. a. 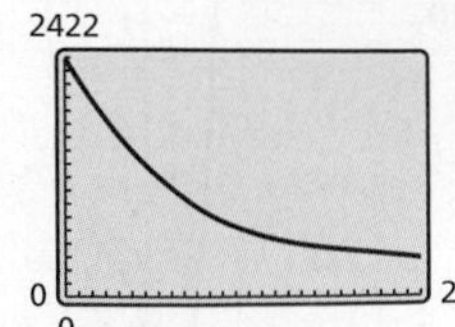

From 1980 to 2007 the number of open drive-in theaters decreased. Around the year 1995, the rate of decrease began to level off.

b. 1980 to 1995: about -119.6, 1995 to 2007: about -19.2; About 120 drive-in movie theaters closed each year on average from 1980 to 1995. From 1995 to 2007, drive-in movie theaters were closing at a much lower rate, with about 20 theaters closing each year.

c. Because the graph declines so sharply in the years leading up to 1980, it is most likely not accurate. The model may be valid for a few years before 1980, but in the long run, decline may not be reasonable. After 2007, the number of drive-in movie theaters declines sharply and soon becomes negative. Because negative values do not make sense given the context, the model cannot be used for years after 2007.

43. Because the graph of g is a reflection of the graph of f in the y-axis, the end behavior would be opposite; $g(x) \to -\infty$ as $x \to -\infty$ and $g(x) \to \infty$ as $x \to \infty$.

45. The viewing window is appropriate if it shows the end behavior of the graph as $x \to \infty$ and $x \to -\infty$.

47. a. 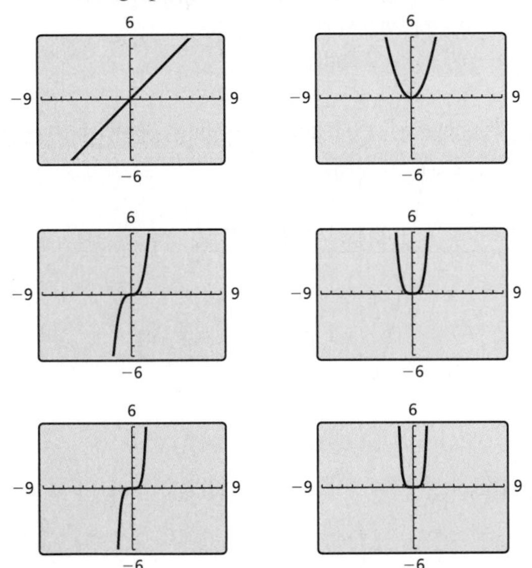

$y = x$, $y = x^3$, and $y = x^5$ are all symmetric with respect to the origin.

$y = x^2$, $y = x^4$, and $y = x^6$ are all symmetric with respect to the y-axis.

b. The graph of $y = x^{10}$ will be symmetric with respect to the y-axis. The graph of $y = x^{11}$ will be symmetric with respect to the origin; The exponent is even. The exponent is odd.

49. $f(-5) = -480$; Substituting the two given points into the function results in the system of equations $2 + b + c - 5 = 0$ and $16 + 4b + 2c - 5 = 3$. Solving for b and c gives $f(x) = 2x^3 - 7x^2 + 10x - 5$.

4.1 Maintaining Mathematical Proficiency *(p. 164)*

51. $-2x^2 + 3xy + y^2$ 53. $12kz - 4kw$

55. $-x^3y^2 + 3x^2y + 13xy - 12x + 9$

4.2 Vocabulary and Core Concept Check *(p. 170)*

1. The binomials could be multiplied in a horizontal format or a vertical format. The patterns from Pascal's Triangle could also be used.

4.2 Monitoring Progress and Modeling with Mathematics *(pp. 170–172)*

3. $x^2 + x + 1$ 5. $12x^5 + 5x^4 - 3x^3 + 6x - 4$

7. $7x^6 + 7x^5 + 8x^3 - 9x^2 + 11x - 5$

9. $-2x^3 - 14x^2 + 7x - 4$

11. $5x^6 - 7x^5 + 6x^4 + 9x^3 + 7$

13. $-x^5 + 7x^3 + 11x^2 + 10x - 4$

15. $P = 47.7t^2 + 678.5t + 17{,}667.4$; The constant term represents the total number of people attending degree-granting institutions at time $t = 0$.

17. $35x^5 + 21x^4 + 7x^3$ 19. $-10x^3 + 23x^2 - 24x + 18$

21. $x^4 - 5x^3 - 3x^2 + 22x + 20$

23. $3x^5 - 6x^4 - 6x^3 + 25x^2 - 23x + 7$

25. The negative was not distributed through the entire second set of parenthesis;

$$(x^2 - 3x + 4) - (x^3 + 7x - 2) = x^2 - 3x + 4 - x^3 - 7x + 2$$
$$= -x^3 + x^2 - 10x + 6$$

27. $x^3 + 3x^2 - 10x - 24$ 29. $12x^3 - 29x^2 + 7x + 6$

31. $-24x^3 + 86x^2 - 57x - 20$

33. $(a + b)(a - b) = a^2 - ab + ab - b^2 = a^2 - b^2$;

Sample answer: $24 \cdot 16 = (20 + 4)(20 - 4)$
$$= 20^2 - 4^2$$
$$= 400 - 16$$
$$= 384$$

35. $x^2 - 81$ 37. $9c^2 - 30c + 25$ 39. $49h^2 + 56h + 16$

41. $8k^3 + 72k^2 + 216k + 216$ 43. $8t^3 + 48t^2 + 96t + 64$

45. $16q^4 - 96q^3 + 216q^2 - 216q + 81$

47. $y^5z^5 + 5y^4z^4 + 10y^3z^3 + 10y^2z^2 + 5yz + 1$

49. $9a^8 + 66a^6b^2 + 97a^4b^4 - 88a^2b^6 + 16b^8$; *Sample answer:* Pascal's Triangle; Use Pascal's Triangle to expand the two binomials. Multiply the results vertically to find your final product.

51. $2x^3 + 10x^2 + 14x + 6$

53. a. $5000(1 + r)^3 + 1000(1 + r)^2 + 4000(1 + r)$

b. $7000r^3 + 25{,}000r^2 + 34{,}000r + 16{,}000$; 7000 is the total amount of money that gained interest for three years, 25,000 is the total amount of money that gained interest for two years, 34,000 is the total amount of money that gained interest for one year, and 16,000 is the total amount of money invested.

c. about $17,763.38

55. no; The sum of $(x + 3)$ and $(x - 3)$ is $2x$, a monomial. The product of $(x + 3)$ and $(x - 3)$ is $x^2 - 9$, a binomial.

57. equivalent; They produce the same graph.

59. not equivalent; Although they appear to produce the same graph, the table of values shows they are off by a constant of 1.

61.

1

1 1

1 2 1

1 3 3 1

1 4 6 4 1

1 5 10 10 5 1

1 6 15 20 15 6 1

1 7 21 35 35 21 7 1

1 8 28 56 70 56 28 8 1

1 9 36 84 126 126 84 36 9 1

1 10 45 120 210 252 210 120 45 10 1

$(x+3)^7 = x^7 + 21x^6 + 189x^5 + 945x^4 + 2835x^3 + 5103x^2 + 5103x + 2187$;

$(x-5)^9 = x^9 - 45x^8 + 900x^7 - 10{,}500x^6 + 78{,}750x^5 - 393{,}750x^4 + 1{,}312{,}500x^3 - 2{,}812{,}500x^2 + 3{,}515{,}625x - 1{,}953{,}125$

63. **a.** 5 **b.** 5 **c.** 9

d. $g(x) + h(x)$ has degree m. $g(x) - h(x)$ has degree m. $g(x) \cdot h(x)$ has degree $(m + n)$.

65. **a.** $(x^2 - y^2)^2 + (2xy)^2 = (x^2 + y^2)^2$

$(x^4 - 2x^2y^2 + y^4) + (4x^2y^2) = x^4 + 2x^2y^2 + y^4$

$x^4 + 2x^2y^2 + y^4 = x^4 + 2x^2y^2 + y^4$

b. The Pythagorean triple is 11, 60, and 61.

c. $121 + 3600 = 3721$

$3721 = 3721$

4.2 Maintaining Mathematical Proficiency *(p. 172)*

67. $5 + 11i$ **69.** $9 - 2i$

4.3 Vocabulary and Core Concept Check *(p. 177)*

1. To evaluate the function $f(x) = x^3 - 2x + 4$ when $x = 3$, synthetic division can be used to divide $f(x)$ by the factor $x - 3$. The remainder is the value of $f(3)$. So, $f(3) = 25$.

Sample answer: $\begin{array}{r|rrrr} 3 & 1 & 0 & -2 & 4 \\ & & 3 & 9 & 21 \\ \hline & 1 & 3 & 7 & 25 \end{array}$

3. $(x^3 - 2x^2 - 9x + 18) \div (x + 3) = x^2 - 5x + 6$

4.3 Monitoring Progress and Modeling with Mathematics *(pp. 177–178)*

5. $x + 5 + \dfrac{3}{x-4}$ **7.** $x + 1 + \dfrac{2x+3}{x^2-1}$

9. $5x^2 - 12x + 37 + \dfrac{-122x + 109}{x^2 + 2x - 4}$ **11.** $x + 12 + \dfrac{49}{x-4}$

13. $2x - 11 + \dfrac{62}{x+5}$ **15.** $x + 3 + \dfrac{18}{x-3}$

17. $x^3 + x^2 - 2x + 1 - \dfrac{6}{x-6}$

19. D; $(2)^2 + (2) - 3 = 3$ so the remainder must be 3.

21. C; $(2)^2 - (2) + 3 = 5$ so the remainder must be 5.

23. The quotient should be one degree less than the dividend.
$\dfrac{x^3 - 5x + 3}{x - 2} = x^2 + 2x - 1 + \dfrac{1}{x-2}$

25. $f(-1) = 37$ **27.** $f(2) = 11$

29. $f(6) = 181$ **31.** $f(3) = 115$

33. no; The Remainder Theorem states that $f(a) = 15$.

35. $\dfrac{A}{T} = \dfrac{-1.95x^3 + 70.1x^2 - 188x + 2150}{14.8x + 725}$

$= 0.13x^2 + 11.19x - 560.90 + \dfrac{408{,}563.25}{14.8x + 725}, \ 0 < x < 18$

37. A **39.** $2x + 5$

4.3 Maintaining Mathematical Proficiency *(p. 178)*

41. $x = 3$ **43.** $x = -7$

4.4 Vocabulary and Core Concept Check *(p. 184)*

1. quadratic; $3x^2$

3. It is written as a product of unfactorable polynomials with integer coefficients.

4.4 Monitoring Progress and Modeling with Mathematics *(pp. 184–186)*

5. $x(x-6)(x+4)$ **7.** $3p^3(p-8)(p+8)$

9. $q^2(2q-3)(q+6)$ **11.** $w^8(5w-2)(2w-3)$

13. $(x+4)(x^2-4x+16)$ **15.** $(g-7)(g^2+7g+49)$

17. $3h^6(h-4)(h^2+4h+16)$

19. $2t^4(2t+5)(4t^2-10t+125)$

21. $x^2 + 9$ is not a factorable binomial because it is not the difference of two squares; $3x^3 + 27x = 3x(x^2+9)$

23. $(y^2+6)(y-5)$ **25.** $(3a^2+8)(a+6)$

27. $(x-2)(x+2)(x-8)$ **29.** $(2q+3)(2q-3)(q-4)$

31. $(7k^2+3)(7k^2-3)$ **33.** $(c^2+5)(c^2+4)$

35. $(4z^2+9)(2z+3)(2z-3)$ **37.** $3r^2(r^3+5)(r^3-4)$

39. factor **41.** not a factor **43.** factor

45. $\begin{array}{r|rrrr} -4 & 1 & -1 & -20 & 0 \\ & & -4 & 20 & 0 \\ \hline & 1 & -5 & 0 & 0 \end{array}$

$g(x) = x(x+4)(x-5)$

47. $\begin{array}{r|rrrrr} 6 & 1 & -6 & 0 & -8 & 48 \\ & & 6 & 0 & 0 & -48 \\ \hline & 1 & 0 & 0 & -8 & 0 \end{array}$

$f(x) = (x-6)(x-2)(x^2+2x+4)$

49. $\begin{array}{r|rrrr} -7 & 1 & 0 & -37 & 84 \\ & & -7 & 49 & -84 \\ \hline & 1 & -7 & 12 & 0 \end{array}$

$r(x) = (x+7)(x-3)(x-4)$

51. D; The x-intercepts of the graph are 2, 3, and -1.

53. A; The x-intercepts of the graph are -2, -3, and 1.

55. The model makes sense for $x > 6.5$; When factored completely, the volume is $V = x(2x-13)(x-3)$. For all three dimensions of the box to have positive lengths, the value of x must be greater than 6.5.

57. $a^4(a+6)(a-5)$; A common monomial can be factored out to obtain a factorable trinomial in quadratic form.

59. $(z-3)(z+3)(z-7)$; Factoring by grouping can be used because the expression contains pairs of monomials that have a common factor. Difference of two squares can be used to factor one of the resulting binomials.

61. $(4r+9)(16r^2-36r+81)$; The sum of two cubes pattern can be used because the expression is of the form $a^3 + b^3$.

63. $(4n^2+1)(2n-1)(2n+1)$; The difference of two squares pattern can be used to factor the original expression and one of the resulting binomials.

65. **a.** no; $7z^4(2z + 3)(z - 2)$
b. no; $n(2 - n)(n + 6)(3n - 11)$ **c.** yes

67. 0.7 million

69. *Sample answer:* Factor Theorem and synthetic division; Calculations without a calculator are easier with this method because the values are lesser.

71. $k = 22$

7	2	−13	−22	105
		14	7	−105
	2	1	−15	0

73. **a.** $(c - d)(c + d)(7a + b)$ **b.** $(x^n - 1)(x^n - 1)$
c. $(a^3 - b^2)(ab + 1)^2$

75. **a.** $(x + 3)^2 + y^2 = 5^2$; The center of the circle is $(-3, 0)$ and the radius is 5.

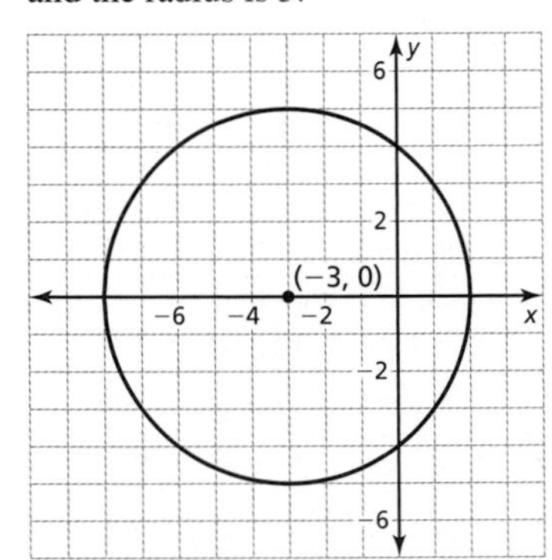

b. $(x - 2)^2 + y^2 = 3^2$; The center of the circle is $(2, 0)$ and the radius is 3.

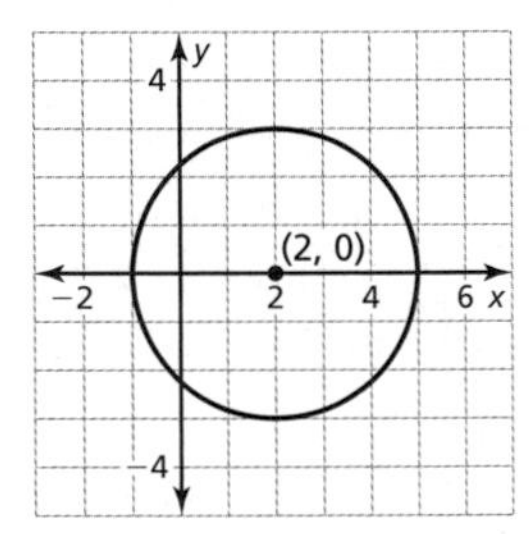

c. $(x - 4)^2 + (y + 1)^2 = 6^2$; The center of the circle is $(4, -1)$ and the radius is 6.

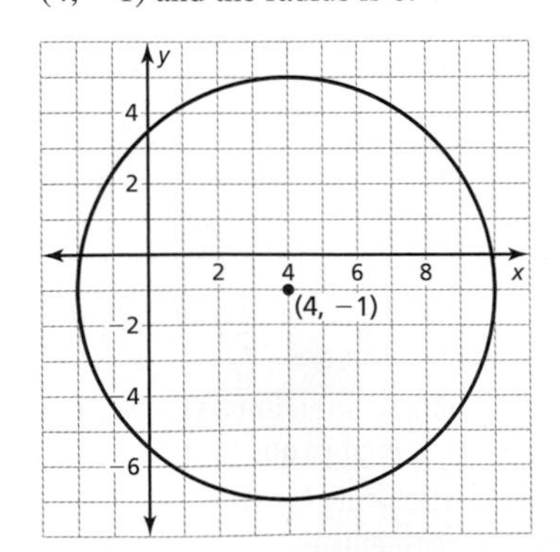

4.4 Maintaining Mathematical Proficiency *(p. 186)*

77. $x = 6$ and $x = -5$ **79.** $x = \frac{5}{3}$ and $x = 2$

81. $x = 18$ and $x = -6$ **83.** $x = -3$ and $x = -7$

4.5 Vocabulary and Core Concept Check *(p. 194)*

1. constant term; leading coefficient

4.5 Monitoring Progress and Modeling with Mathematics *(pp. 194–196)*

3. $z = -3$, $z = 0$, and $z = 4$ **5.** $x = 0$ and $x = 1$

7. $w = 0$ and $w = \pm\sqrt{10} \approx \pm 3.16$

9. $c = 0$, $c = 3$, and $c = \pm\sqrt{6} \approx \pm 2.45$ **11.** $n = -4$

13. $x = -3$, $x = 0$, and $x = 2$

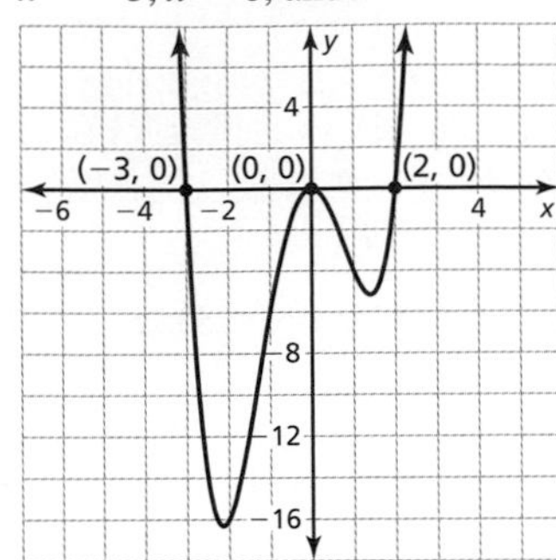

15. $x = 0$, $x = 5$, and $x = 6$

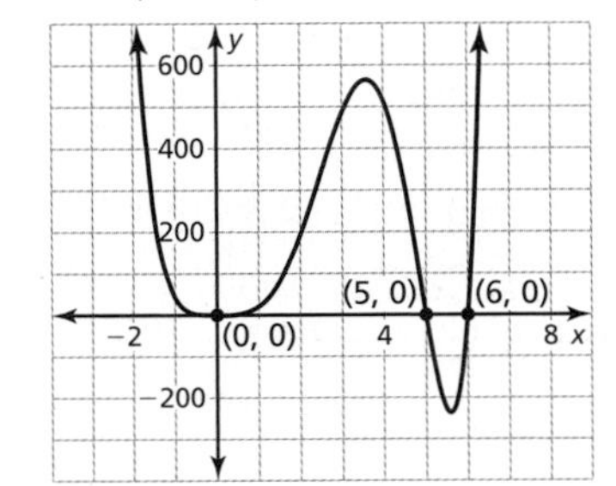

17. $x = -3$, $x = 0$, and $x = 5$

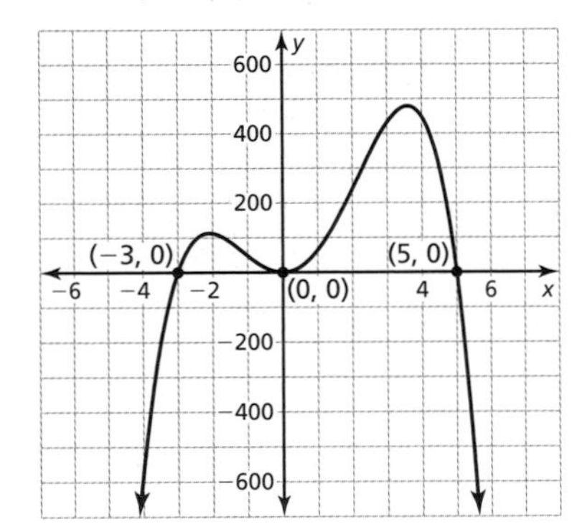

19. $x = -3$, $x = -1$, and $x = 3$

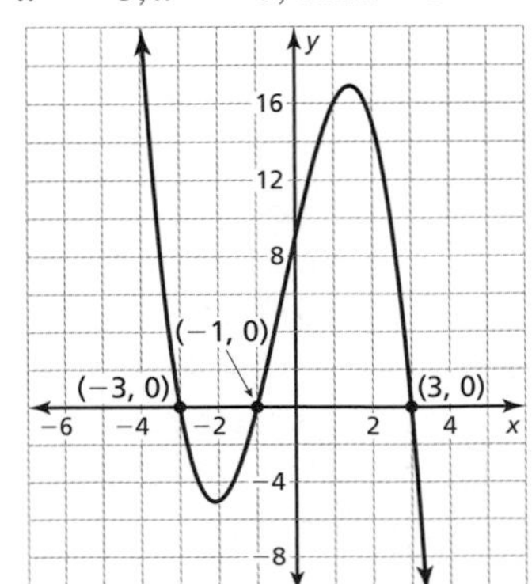

21. C

23. The $\pm$ was not included with each factor; ± 1, ± 3, ± 5, ± 9, ± 15, ± 45

25. $x = -5$, $x = 1$, and $x = 3$ **27.** $x = -1$, $x = 5$, and $x = 6$

29. $x = -3, x = 4,$ and $x = 5$

31. $x = -4, x = -0.5,$ and $x = 6$ 33. $-5, 3,$ and 4

35. $-5, -3,$ and -2 37. $-4, 1.5,$ and 3

39. $1, \frac{-1 + \sqrt{17}}{2} \approx 1.56,$ and $\frac{-1 - \sqrt{17}}{2} \approx -2.56$

41. $f(x) = x^3 - 7x^2 + 36$ 43. $f(x) = x^3 - 10x - 12$

45. $f(x) = x^4 - 32x^2 + 24x$

47. $x = -3, x = 3,$ and $x = 4$; *Sample answer:* graphing; The equation has three real solutions, all which can be found by graphing to find the x-intercepts.

49. 4 cm by 4 cm by 7 cm

51. The block is 3 meters high, 21 meters long, and 15 meters wide.

53. a. $-20t^3 + 252t^2 - 280t - 2400 = 0$
b. 1, 2, 3, 4, 5, 6, 8, 10 c. $t = 5$ years and $t = 10$ years

55. The length should be 8 feet, the width should be 4 feet, and the height should be 4 feet.

57. a. $k = 60$ b. $k = 33$ c. $k = 6$ 59. $x = 1$

61. $x = 2$

63. The height of each ramp is $\frac{5}{3}$ feet and the width of each ramp is 5 feet. The left ramp is to be 24 feet in length while the right ramp is to be 12 feet in length.

65. rs; Each factor of a_0 can be written as the numerator with each factor of a_n as the denominator, creating $r \times s$ factors.

4.5 Maintaining Mathematical Proficiency *(p. 196)*

67. not a polynomial function

69. not a polynomial function; The term $\sqrt[4]{x}$ has an exponent that is not a whole number.

71. $x = \pm 3i$ 73. $x = \pm\frac{\sqrt{2}}{4}$

4.6 Vocabulary and Core Concept Check *(p. 202)*

1. complex conjugates

4.6 Monitoring Progress and Modeling with Mathematics *(pp. 202–204)*

3. 4 5. 6 7. 7 9. $-1, 1, 2,$ and 4

11. $-2, -2, 1,$ and 3 13. $-3, -1, 2i,$ and $-2i$

15. $-4, -1, 2, i\sqrt{2},$ and $-i\sqrt{2}$

17. 2; The graph shows 2 real zeros, so the remaining zeros must be imaginary.

19. 2; The graph shows no real zeros, so all of the zeros must be imaginary.

21. $f(x) = x^3 + 4x^2 - 7x - 10$

23. $f(x) = x^3 - 11x^2 + 41x - 51$

25. $f(x) = x^3 - 4x^2 - 5x + 20$

27. $f(x) = x^5 - 8x^4 + 23x^3 - 32x^2 + 22x - 4$

29. The conjugate of the given imaginary zeros was not included.

$$\begin{aligned} f(x) &= (x-2)[x-(1+i)][x-(1-i)] \\ &= (x-2)[(x-1)-i][(x-1)+i] \\ &= (x-2)[(x-1)^2 - i^2] \\ &= (x-2)[(x^2-2x+1)-(-1)] \\ &= (x-2)(x^2-2x+2) \\ &= x^3 - 2x^2 + 2x - 2x^2 + 4x - 4 \\ &= x^3 - 4x^2 + 6x - 4 \end{aligned}$$

31. *Sample answer:* $y = x^6 - 4x^4 - x^2 + 4$;

$$\begin{aligned} y &= (x-1)(x+1)(x-2)(x+2)(x-i)(x+i) \\ &= (x^2-1)(x^2-4)(x^2+1) \\ &= (x^4-5x^2+4)(x^2+1) \\ &= x^6 + x^4 - 5x^4 - 5x^2 + 4x^2 + 4 \\ &= x^6 - 4x^4 - x^2 + 4 \end{aligned}$$

33.

Positive real zeros	Negative real zeros	Imaginary zeros	Total zeros
1	1	2	4

35.

Positive real zeros	Negative real zeros	Imaginary zeros	Total zeros
2	1	0	3
0	1	2	3

37.

Positive real zeros	Negative real zeros	Imaginary zeros	Total zeros
3	2	0	5
3	0	2	5
1	2	2	5
1	0	4	5

39.

Positive real zeros	Negative real zeros	Imaginary zeros	Total zeros
3	3	0	6
3	1	2	6
1	3	2	6
1	1	4	6

41. C; There are two sign changes in the coefficients of $f(-x)$. So, the number of negative real zeros is two or zero, not four.

43. in the year 1958 45. in the 3rd year and the 9th year

47. $x = 4.2577$

49. no; The Fundamental Theorem of Algebra applies to functions of degree greater than zero. Because the function $f(x) = 2$ is equivalent to $f(x) = 2x^0$, it has degree 0, and does not fall under the Fundamental Theorem of Algebra.

51.

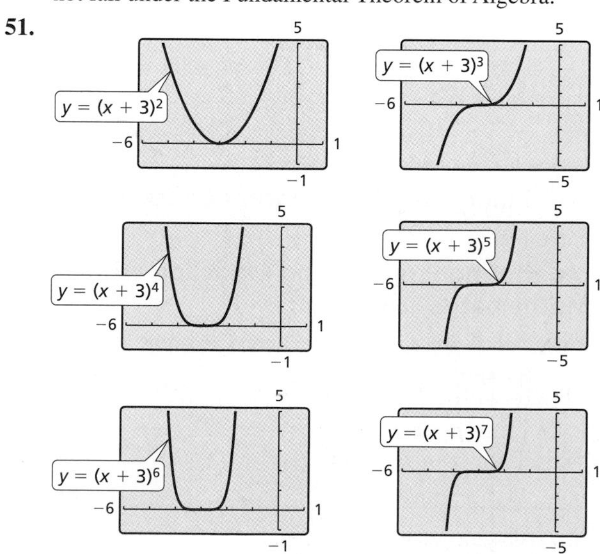

a. For all functions, $f(x) \to \infty$ as $x \to \infty$. When n is even, $f(x) \to \infty$ as $x \to -\infty$, but when n is odd, $f(x) \to -\infty$ as $x \to -\infty$.

b. As n increases, the graph becomes more flat near the zero $x = -3$.

c. The graph of g becomes more vertical and straight near $x = 4$.

53. a.

Deposit	Year 1	Year 2	Year 3	Year 4
1st Deposit	1000	$1000g$	$1000g^2$	$1000g^3$
2nd Deposit		1000	$1000g$	$1000g^2$
3rd Deposit			1000	$1000g$
4th Deposit				1000

b. $v = 1000g^3 + 1000g^2 + 1000g + 1000$

c. about 1.0484; about 4.84%

4.6 Maintaining Mathematical Proficiency *(p. 204)*

55. The function is a translation 4 units right and 6 units up of the parent quadratic function.

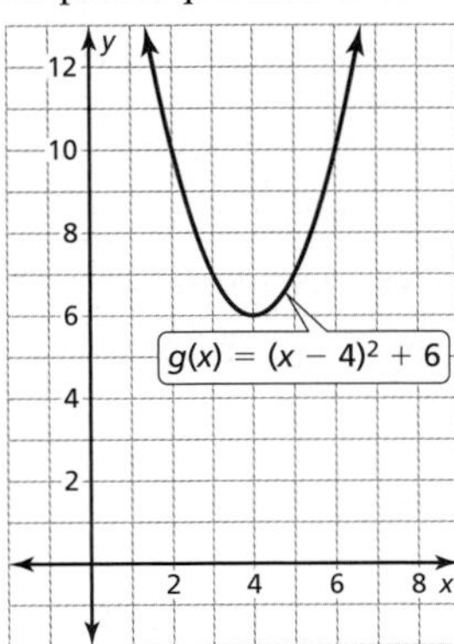

57. The function is a vertical stretch by a factor of 5 followed by a translation 4 units left of the parent quadratic function.

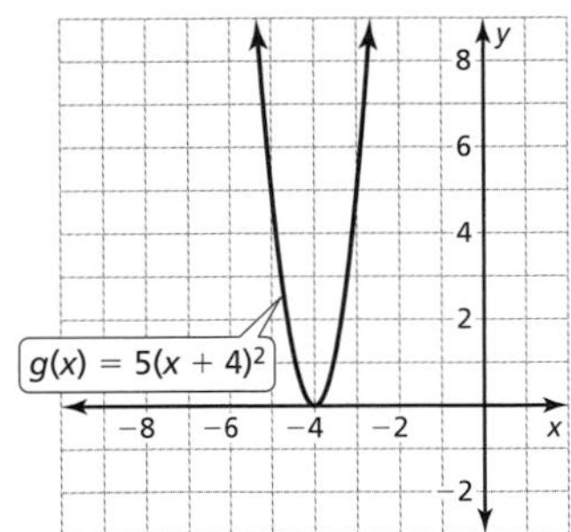

59. $g(x) = \left|\frac{1}{9}x + 1\right| - 3$

4.7 Vocabulary and Core Concept Check *(p. 209)*

1. horizontal

4.7 Monitoring Progress and Modeling with Mathematics *(pp. 209–210)*

3. The graph of g is a translation 3 units up of the graph of f.

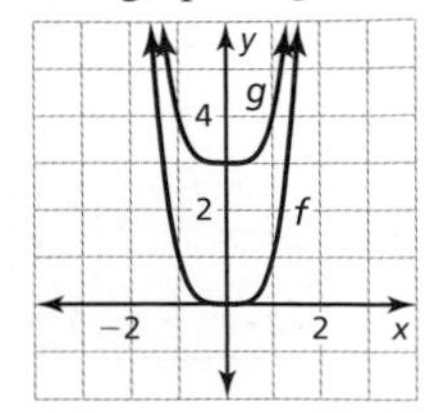

5. The graph of g is a translation 2 units right and 1 unit down of the graph of f.

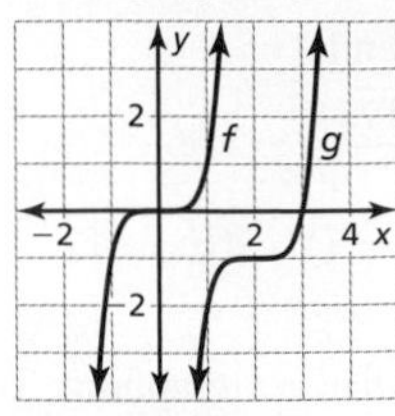

7. B; The graph has been translated 2 units right.

9. D; The graph has been translated 2 units right and 2 units up.

11. The graph of g is a vertical stretch by a factor of 2 followed by a reflection in the x-axis of the graph of f.

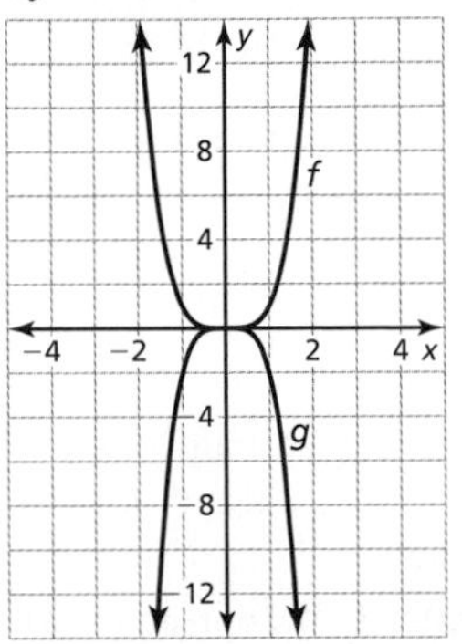

13. The graph of g is a vertical stretch by a factor of 5 followed by a translation 1 unit up of the graph of f.

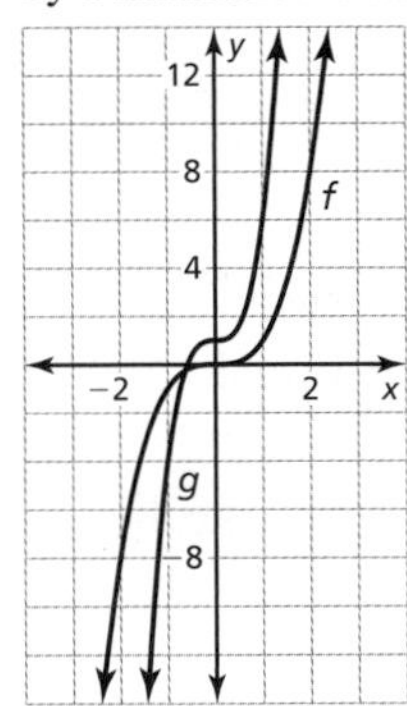

15. The graph of g is a vertical shrink by a factor of $\frac{3}{4}$ followed by a translation 4 units left of the graph of f.

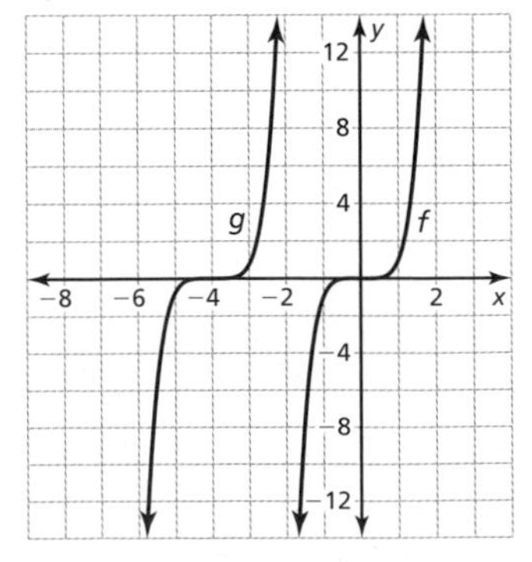

17. $g(x) = (x + 2)^4 + 1$;

The graph of g is a translation 2 units left of the graph of f.

19. $g(x) = -x^3 + x^2 - 3$;

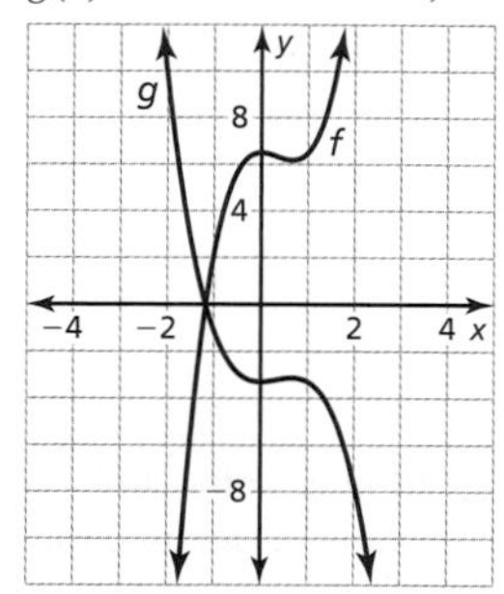

The graph of g is a vertical shrink by a factor of $\frac{1}{2}$ followed by a reflection in the x-axis of the graph of f.

21. The graph has been translated horizontally to the right 2 units instead of to the left 2 units.

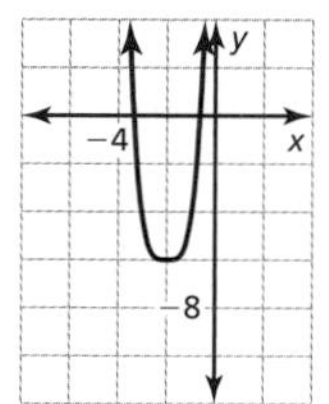

23. $g(x) = -x^3 + 9x^2 - 27x + 21$

25. $g(x) = -27x^3 - 18x^2 + 7$

27. $W(x) = 27x^3 - 12x$; $W(5) = 3315$; When x is 5 yards, the volume of the pyramid is 3315 cubic feet.

29. *Sample answer:* If the function is translated up and then reflected in the x-axis, the order is important; If the function is translated left and then reflected in the x-axis, the order is not important; Reflecting a graph in the x-axis does not affect its x-coordinate, but it does affect its y-coordinate. So, the order is only important if the translation is vertical.

31. **a.** 0 m, 4 m, and 7 m

b. $g(x) = -\frac{2}{5}(x - 2)(x - 6)^2(x - 9)$

33. $V(x) = 3\pi x^2(x + 3)$; $W(x) = \frac{\pi}{3}x^2\left(\frac{1}{3}x + 3\right)$;

$W(3) = 12\pi \approx 37.70$; When x is 3 feet, the volume of the cone is about 37.70 cubic yards.

4.7 Maintaining Mathematical Proficiency *(p. 210)*

35. The maximum value is 4; The domain is all real numbers and the range is $y \leq 4$. The function is increasing to the left of $x = 0$ and decreasing to the right of $x = 0$.

37. The maximum value is 9; The domain is all real numbers and the range is $y \leq 9$. The function is increasing to the left of $x = -5$ and decreasing to the right of $x = -5$.

39. The maximum value is 1; The domain is all real numbers and the range is $y \leq 1$. The function is increasing to the left of $x = 1$ and decreasing to the right of $x = 1$.

4.8 Vocabulary and Core Concept Check *(p. 216)*

1. turning

4.8 Monitoring Progress and Modeling with Mathematics *(pp. 216–218)*

3. A **5.** B

7.

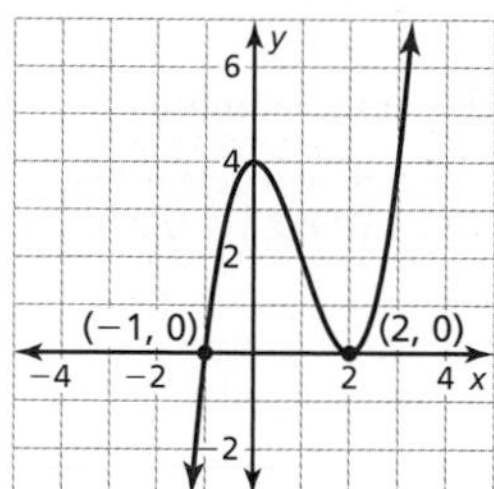

9.

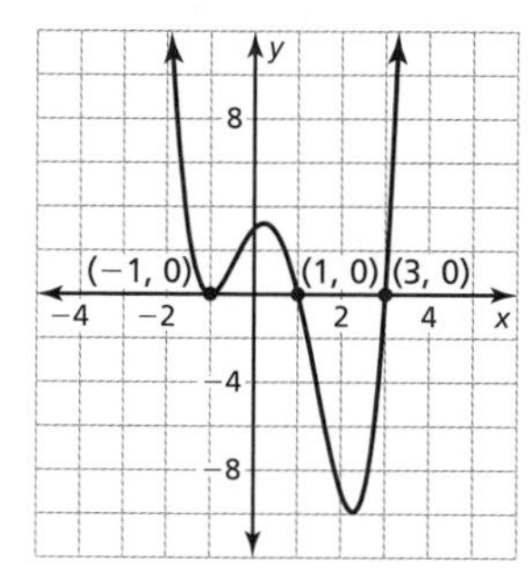

11.

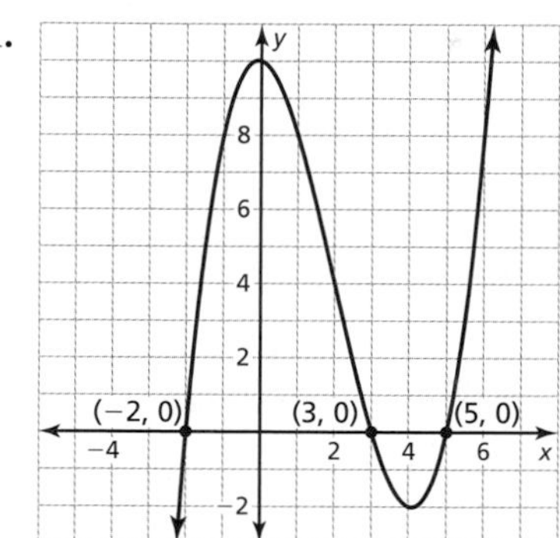

13.

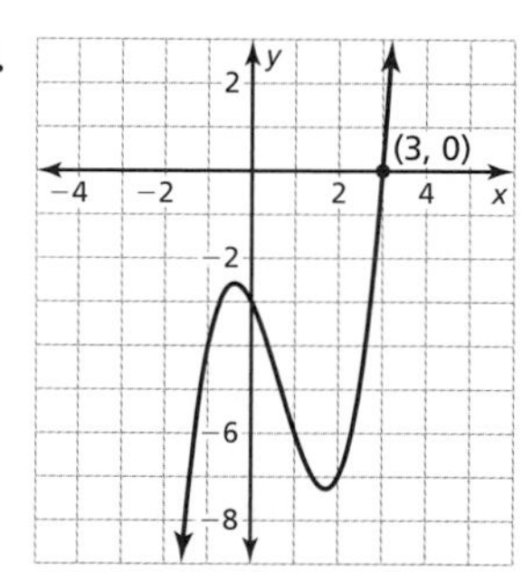

15. The x-intercepts should be -2 and 1.

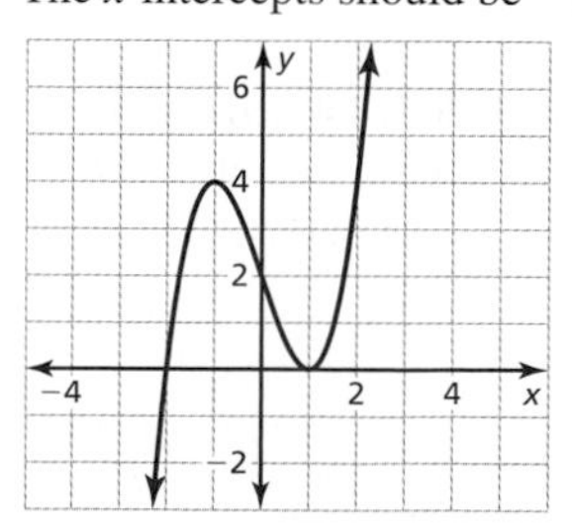

17. -1, 1, and 4 **19.** -4, $-\frac{1}{2}$, and 1 **21.** -4, $\frac{3}{4}$, and 3

23.

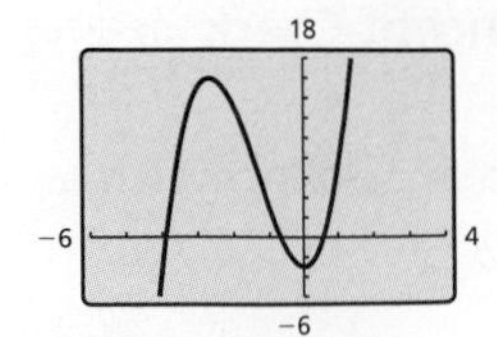

The x-intercepts of the graph are $x \approx -3.90$, $x \approx -0.67$, and $x \approx 0.57$. The function has a local maximum at $(-2.67, 15.96)$ and a local minimum at $(0, -3)$; The function is increasing when $x < -2.67$ and $x > 0$ and is decreasing when $-2.67 < x < 0$.

25.

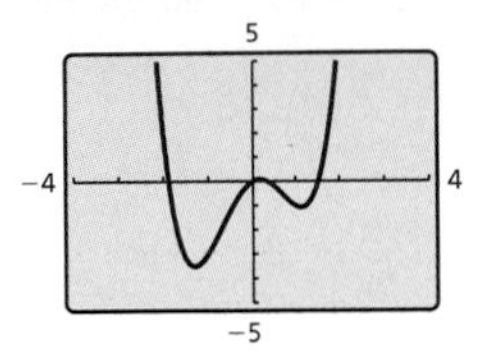

The x-intercepts of the graph are $x \approx -1.88$, $x = 0$, $x \approx 0.35$, and $x \approx 1.53$. The function has a local maximum at $(0.17, 0.08)$ and local minimums at $(-1.30, -3.51)$ and $(1.13, -1.07)$; The function is increasing when $-1.30 < x < 0.17$ and $x > 1.13$ and is decreasing when $x < -1.30$ and $0.17 < x < 1.13$.

27.

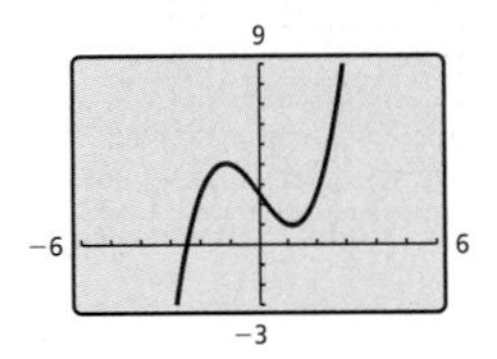

The x-intercept of the graph is $x \approx -2.46$. The function has a local maximum at $(-1.15, 4.04)$ and a local minimum at $(1.15, 0.96)$; The function is increasing when $x < -1.15$ and $x > 1.15$ and is decreasing when $-1.15 < x < 1.15$.

29.

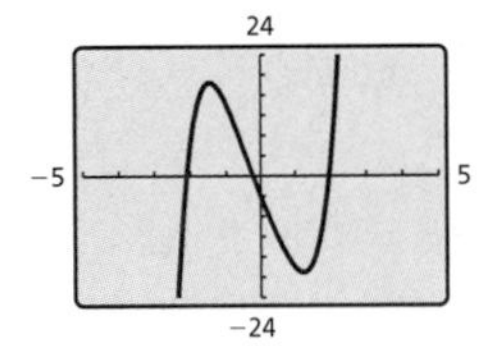

The x-intercepts of the graph are $x \approx -2.10$, $x \approx -0.23$, and $x \approx 1.97$. The function has a local maximum at $(-1.46, 18.45)$ and a local minimum at $(1.25, -19.07)$; The function is increasing when $x < -1.46$ and $x > 1.25$ and is decreasing when $-1.46 < x < 1.25$.

31. $(-0.29, 0.48)$ and $(0.29, -0.48)$; $(-0.29, 0.48)$ corresponds to a local maximum and $(0.29, -0.48)$ corresponds to a local minimum; The real zeros are -0.5, 0, and 0.5. The function is of at least degree 3.

33. $(1, 0)$, $(3, 0)$, and $(2, -2)$; $(1, 0)$ and $(3, 0)$ correspond to local maximums, and $(2, -2)$ corresponds to a local minimum; The real zeros are 1 and 3. The function is of at least degree 4.

35. $(-1.25, -10.65)$; $(-1.25, -10.65)$ corresponds to a local minimum; The real zeros are -2.07 and 1.78. The function is of at least degree 4.

37.

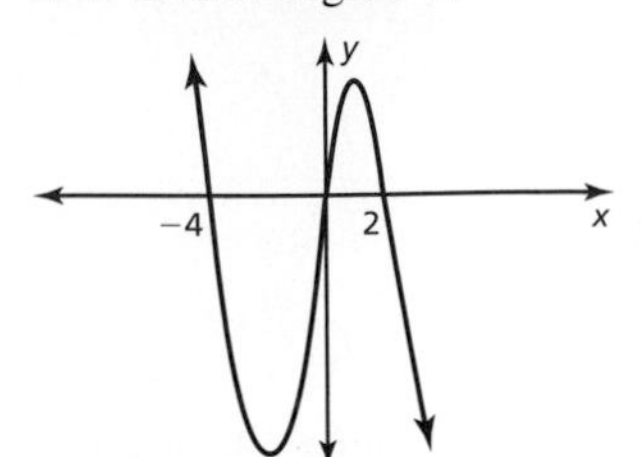

39. odd **41.** even **43.** neither **45.** even

47.

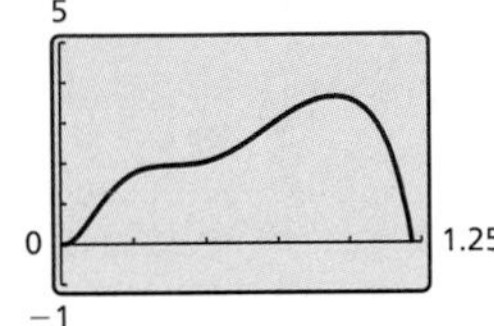

about 1 second into the stroke

49. A quadratic function only has one turning point, and it is always the maximum or minimum value of the function.

51. no; When multiplying two odd functions, the exponents of each term will be added, creating an even exponent. So, the product will not be an odd function.

53. **a.** $\dfrac{1100 - \pi r^2}{\pi r}$ **b.** $V = 550r - \dfrac{\pi}{2}r^3$ **c.** about 10.8 ft

55. $V(h) = 64\pi h - \dfrac{\pi}{4}h^3$; about 9.24 in.; about 1238.22 in.^{3}

4.8 Maintaining Mathematical Proficiency *(p. 218)*

57. quadratic; The second differences are constant.

4.9 Vocabulary and Core Concept Check *(p. 223)*

1. finite differences

4.9 Monitoring Progress and Modeling with Mathematics *(pp. 223–224)*

3. $f(x) = (x + 1)(x - 1)(x - 2)$

5. $f(x) = \frac{1}{7}(x + 5)(x - 1)(x - 4)$

7. $3; f(x) = \frac{2}{3}x^3 + 4x^2 - \frac{1}{3}x - 4$

9. $4; f(x) = -3x^4 - 5x^3 + 9x^2 + 3x - 1$

11. $4; f(x) = x^4 - 15x^3 + 81x^2 - 183x + 142$

13. The sign in each parentheses is wrong. The x-intercepts should have been subtracted from zero, not added.

$(-6, 0), (1, 0), (3, 0), (0, 54)$

$54 = a(0 + 6)(0 - 1)(0 - 3)$

$54 = 18a$

$a = 3$

$f(x) = 3(x + 6)(x - 1)(x - 3)$

15. *Sample answer:*
$y = (x - 3)(x - 4)(x + 1)$,
$y = 3(x - 3)(x - 4)(x - 1)$,
$y = \frac{1}{2}(x - 3)(x - 4)(x + 4)$;
$y = a(x - 3)(x - 4)(x - c)$
$6 = a(2 - 3)(2 - 4)(2 - c)$
$6 = 2a(2 - c)$
$3 = a(2 - c)$
$\frac{3}{2 - c} = a$
Any combination of a and c that fit the equation will contain these points.

17. $y = 0.002x^2 + 0.60x - 2.5$; about 15.9 mph

19. $d = \frac{1}{2}n^2 - \frac{3}{2}n$; 35

21. With real-life data sets, the numbers rarely fit a model perfectly. Because of this, the differences are rarely constant.

23. C, A, B, D

4.9 Maintaining Mathematical Proficiency *(p. 224)*

25. $x = \pm 6$ 27. $x = 3 \pm 2\sqrt{3}$ 29. $x = 1$ and $x = -2.5$

31. $x = \frac{-3 \pm \sqrt{29}}{10}$

Chapter 4 Review *(pp. 226–230)*

1. polynomial function; $h(x) = -15x^7 - x^3 + 2x^2$; It has degree 7 and has a leading coefficient of -15.

2. not a polynomial

3.

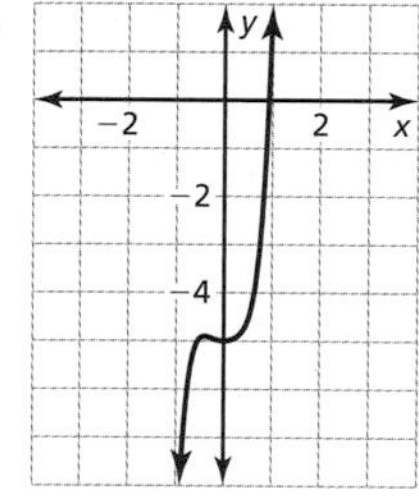

4.

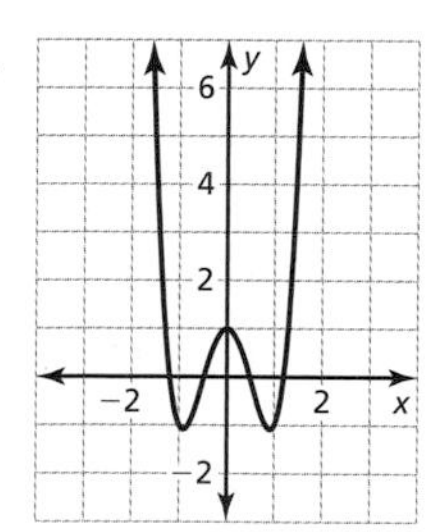

5.

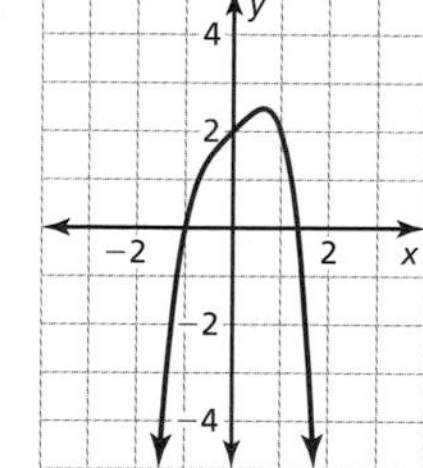

6. $4x^3 - 4x^2 - 4x - 8$ 7. $3x^4 + 3x^3 - x^2 - 3x + 15$

8. $2x^2 + 11x + 1$ 9. $2y^3 + 10y^2 + 5y - 21$

10. $8m^3 + 12m^2n + 6mn^2 + n^3$ 11. $s^3 + 3s^2 - 10s - 24$

12. $m^4 + 16m^3 + 96m^2 + 256m + 256$

13. $243s^5 + 810s^4 + 1080s^3 + 720s^2 + 240s + 32$

14. $z^6 + 6z^5 + 15z^4 + 20z^3 + 15z^2 + 6z + 1$

15. $x - 1 + \frac{4x - 3}{x^2 + 2x + 1}$ 16. $x^2 + 2x - 10 + \frac{7x + 43}{x^2 + x + 4}$

17. $x^3 - 4x^2 + 15x - 60 + \frac{233}{x + 4}$ 18. $g(5) = 546$

19. $8(2x - 1)(4x^2 + 2x + 1)$ 20. $2z(z^2 - 5)(z - 1)(z + 1)$

21. $(a - 2)(a + 2)(2a - 7)$

22.
$$\begin{array}{r|rrrrr} -2 & 1 & 2 & 0 & -27 & -54 \\ & & -2 & 0 & 0 & 54 \\ \hline & 1 & 0 & 0 & -27 & 0 \end{array}$$
$f(x) = (x + 2)(x - 3)(x^2 + 3x + 9)$

23. $x = -4$, $x = -2$, and $x = 3$

24. $x = -4$, $x = -3$, and $x = 2$

25. $f(x) = x^3 - 5x^2 + 5x - 1$

26. $f(x) = x^4 - 5x^3 + x^2 + 25x - 30$

27. $f(x) = x^4 - 9x^3 + 11x^2 + 51x - 30$

28. The length is 6 inches, the width is 2 inches, and the height is 20 inches; When $\ell(\ell - 4)(3\ell + 2) = 240$, $\ell = 6$.

29. $f(x) = x^3 - 5x^2 + 11x - 15$

30. $f(x) = x^4 - x^3 + 14x^2 - 16x - 32$

31. $f(x) = x^4 + 7x^3 + 6x^2 - 4x + 80$

32.

Positive real zeros	Negative real zeros	Imaginary zeros	Total zeros
2	0	2	4
0	0	4	4

33.

Positive real zeros	Negative real zeros	Imaginary zeros	Total zeros
1	3	0	4
1	1	2	4

34. The graph of g is a reflection in the y-axis followed by a translation 2 units up of the graph of f.

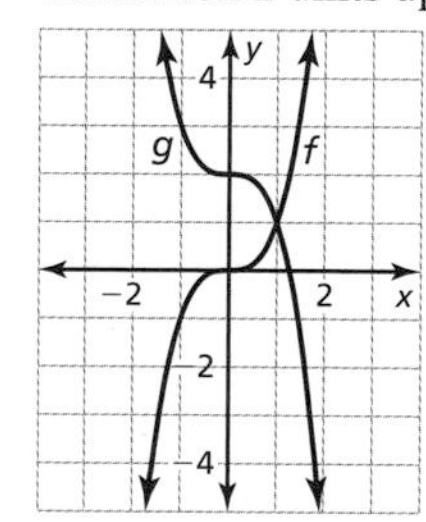

35. The graph of g is a reflection in the x-axis followed by a translation 9 units left of the graph of f.

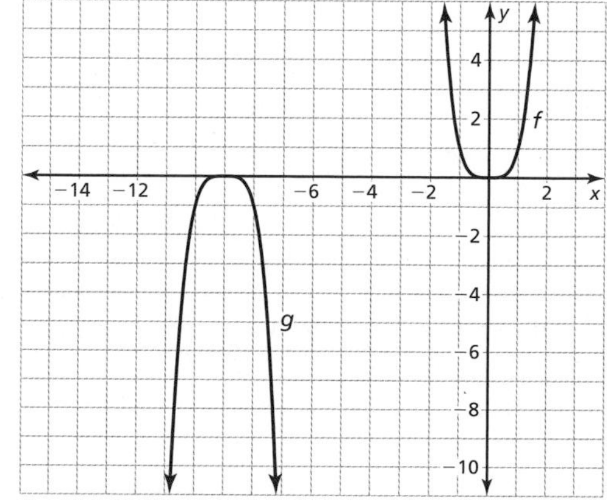

36. $g(x) = \frac{1}{1024}(x - 3)^5 + \frac{3}{4}(x - 3) - 5$

37. $g(x) = x^4 + 2x^3 - 7$

Selected Answers

38.

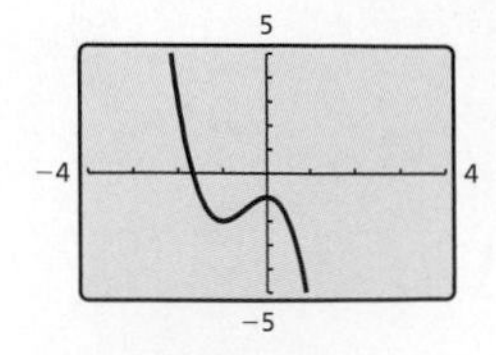

The x-intercept of the graph is $x \approx -1.68$. The function has a local maximum at $(0, -1)$ and a local minimum at $(-1, -2)$; The function is increasing when $-1 < x < 0$ and decreasing when $x < 1$ and $x > 0$.

39.

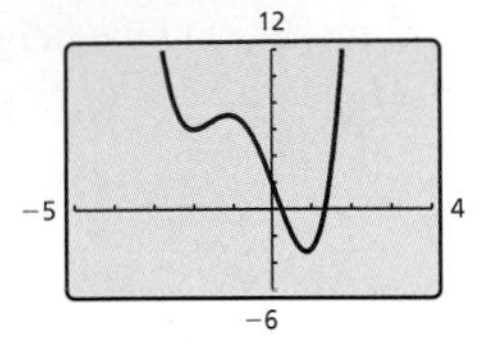

The x-intercepts of the graph are $x \approx 0.25$ and $x \approx 1.34$. The function has a local maximum at $(-1.13, 7.06)$ and local minimums at $(-2, 6)$ and $(0.88, -3.17)$; The function is increasing when $-2 < x < -1.13$ and $x > 0.88$ and is decreasing when $x < -2$ and $-1.13 < x < 0.88$.

40. odd **41.** even **42.** neither

43. $f(x) = \frac{3}{16}(x + 4)(x - 4)(x - 2)$

44. $3; f(x) = 2x^3 - 7x^2 - 6x$

Chapter 5

Chapter 5 Maintaining Mathematical Proficiency *(p. 235)*

1. y^7 **2.** n **3.** $\frac{1}{x^3}$ **4.** $3x^3$ **5.** $\frac{8w^9}{z^6}$ **6.** $\frac{m^{10}}{z^4}$

7. $y = 2 - 4x$ **8.** $y = 3 + 3x$ **9.** $y = \frac{13}{2}x + \frac{9}{2}$

10. $y = \frac{5}{x + 3}$ **11.** $y = \frac{8x - 3}{4x}$ **12.** $y = \frac{15 - 6x}{7x}$

13. sometimes; Order does not matter in Example 1 but does in Example 2.

5.1 Vocabulary and Core Concept Check *(p. 241)*

1. $\frac{1}{(\sqrt[t]{a})^s}$; t

3. When a is positive, it has two real fourth roots, $\pm\sqrt[4]{a}$, and one real fifth root $\sqrt[5]{a}$. When a is negative, it has no real fourth roots and one real fifth root, $\sqrt[5]{a}$.

5.1 Monitoring Progress and Modeling with Mathematics *(pp. 241–242)*

5. 2 **7.** 0 **9.** −2 **11.** 2 **13.** 125

15. −3 **17.** $\frac{1}{4}$

19. The cube root of 27 was calculated incorrectly; $27^{2/3} = (27^{1/3})^2 = 3^2 = 9$

21. B; The denominator of the exponent is 3 and the numerator is 4.

23. A; The denominator of the exponent is 4 and the exponent is negative.

25. 8 **27.** 0.34 **29.** 2840.40 **31.** 50.57

33. $r \approx 3.72$ ft **35.** $x = 5$ **37.** $x \approx -7.66$

39. $x \approx -2.17$ **41.** $x = \pm 2$ **43.** $x = \pm 3$

45. potatoes: 3.7%; ham: 2.4%; eggs: 1.7%

47. 3, 4; $\sqrt[4]{81} = 3$ and $\sqrt[4]{256} = 4$ **49.** about 753 ft^3/sec

5.1 Maintaining Mathematical Proficiency *(p. 242)*

51. 5^5 **53.** $\frac{1}{z^6}$ **55.** 5000 **57.** 0.82

5.2 Vocabulary and Core Concept Check *(p. 248)*

1. No radicands have perfect nth powers as factors other than 1, no radicands contain fractions, and no radicals appear in the denominator of a fraction.

5.2 Monitoring Progress and Modeling with Mathematics *(pp. 248–250)*

3. $9^{2/3}$ **5.** $6^{3/4}$ **7.** $\frac{5}{4}$ **9.** $3^{1/3}$ **11.** 4

13. 12 **15.** $2\sqrt[4]{3}$ **17.** 3 **19.** 6 **21.** $3\sqrt[4]{7}$

23. $\frac{\sqrt[3]{10}}{2}$ **25.** $\frac{\sqrt{6}}{4}$ **27.** $\frac{4\sqrt[3]{7}}{7}$ **29.** $\frac{1 - \sqrt{3}}{-2}$

31. $\frac{15 + 5\sqrt{2}}{7}$ **33.** $\frac{9\sqrt{3} - 9\sqrt{7}}{-4}$ **35.** $\frac{3\sqrt{2} + \sqrt{30}}{-2}$

37. $12\sqrt[3]{11}$ **39.** $12(11^{1/4})$ **41.** $-9\sqrt{3}$ **43.** $5\sqrt[5]{7}$

45. $6(3^{1/3})$

47. The radicand should not change when the expression is factored;
$3\sqrt[3]{12} + 5\sqrt[3]{12} = (3 + 5)\sqrt[3]{12} = 8\sqrt[3]{12}$

49. $3y^2$ **51.** $\frac{m^2}{n}$ **53.** $\frac{|g|}{|h|}$

55. Absolute value was not used to ensure that all variables are positive;
$\frac{\sqrt[6]{2^6(h^2)^6}}{\sqrt[6]{g^6}} = \frac{2h^2}{|g|}$

57. $9a^3b^6c^4\sqrt{ac}$ **59.** $\frac{2m\sqrt[5]{5mn^3}}{n^2}$ **61.** $\frac{\sqrt[6]{w^5}}{5w^6}$

63. $\frac{2v^{3/4}}{3w}$, $v \neq 0$ **65.** $21\sqrt[3]{y}$

67. $-2x^{7/2}$ **69.** $4w^2\sqrt{w}$

71. $P = 2x^3 + 4x^{2/3}$
$A = 2x^{11/3}$

73. about 0.45 mm

75. no; The second radical can be simplified to $18\sqrt{11}$. The difference is $-11\sqrt{11}$.

77. $10 + 6\sqrt{5}$

79. **a.** $r = \sqrt[3]{\frac{3V}{4\pi}}$

b. $S = 4\pi\left(\sqrt[3]{\frac{3V}{4\pi}}\right)^2$

$S = \frac{4\pi(3V)^{2/3}}{(4\pi)^{2/3}}$

$S = (4\pi)^{3/3 - 2/3}(3V)^{2/3}$

$S = (4\pi)^{1/3}(3V)^{2/3}$

c. The surface area of the larger balloon is $2^{2/3} \approx 1.59$ times as large as the surface area of the smaller balloon.

81. when n is even and $\frac{m}{n}$ is odd

5.2 Maintaining Mathematical Proficiency *(p. 250)*

83. The focus is $\left(-\frac{1}{4}, 0\right)$. The directrix is $x = \frac{1}{4}$. The axis of symmetry is $y = 0$.

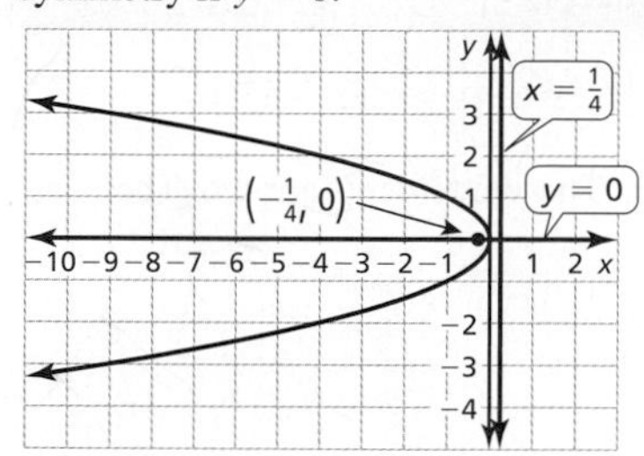

85. $g(x) = -x^4 + 3x^2 + 2x$; The graph of g is a reflection in the x-axis of the graph of f.

87. $g(x) = (x - 2)^3 - 4$; The graph of g is a translation 2 units right of the graph of f.

5.3 Vocabulary and Core Concept Check *(p.256)*

1. radical

5.3 Monitoring Progress and Modeling with Mathematics *(pp. 256–258)*

3. B **5.** F **7.** E

9.

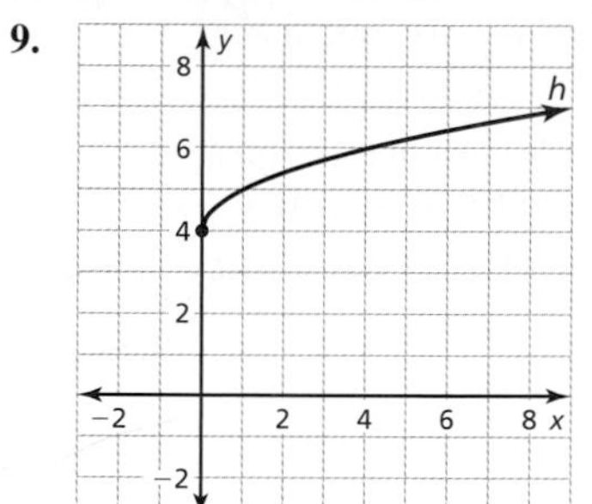

The domain is $x \geq 0$. The range is $y \geq 4$.

11.

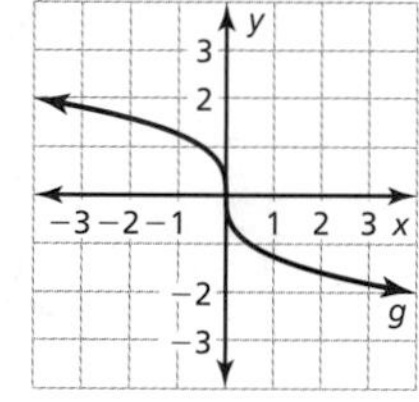

The domain and range are all real numbers.

13.

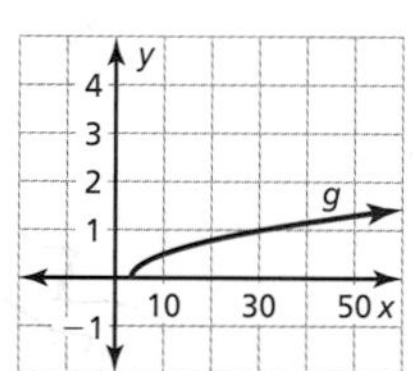

The domain is $x \geq 3$. The range is $y \geq 0$.

15.

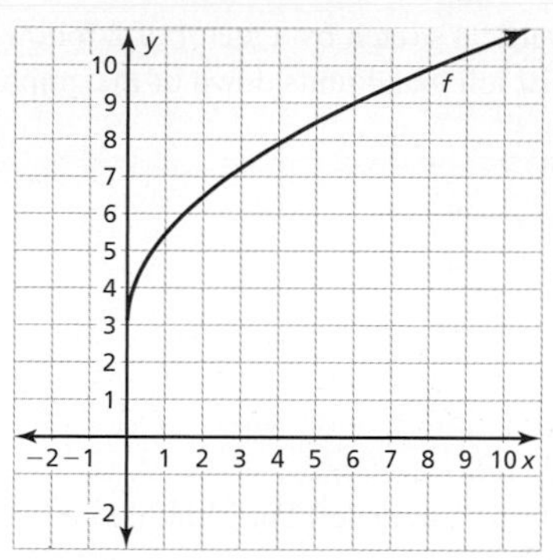

The domain is $x \geq 0$. The range is $y \geq 3$.

17.

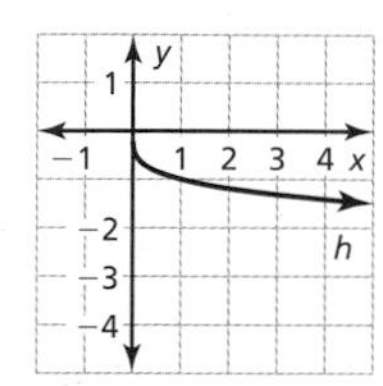

The domain is $x \geq 0$. The range is $y \leq 0$.

19. The graph of g is a translation 1 unit left and 8 units up of the graph of f.

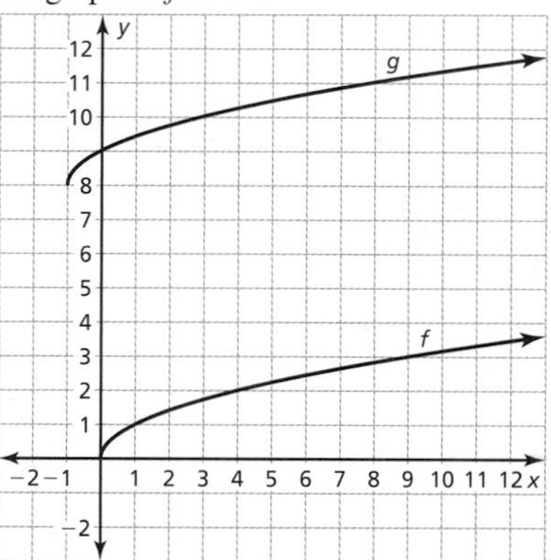

21. The graph of g is a reflection in the x-axis followed by a translation 1 unit down of the graph of f.

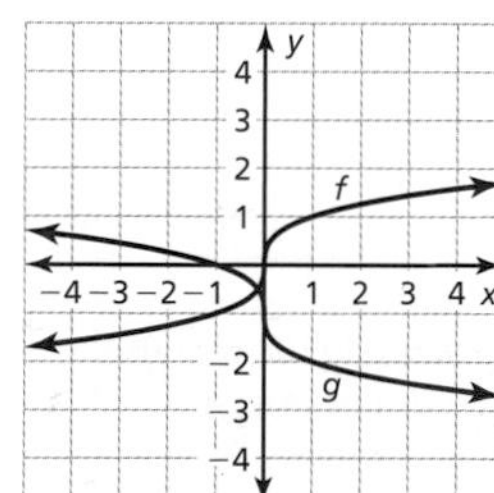

23. The graph of g is a vertical shrink by a factor of $\frac{1}{4}$ followed by a reflection in the y-axis of the graph of f.

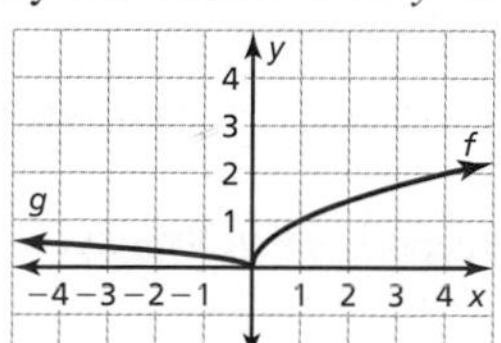

Selected Answers

25. The graph of g is a vertical stretch by a factor of 2 followed by a translation 5 units left and 4 units down of the graph of f.

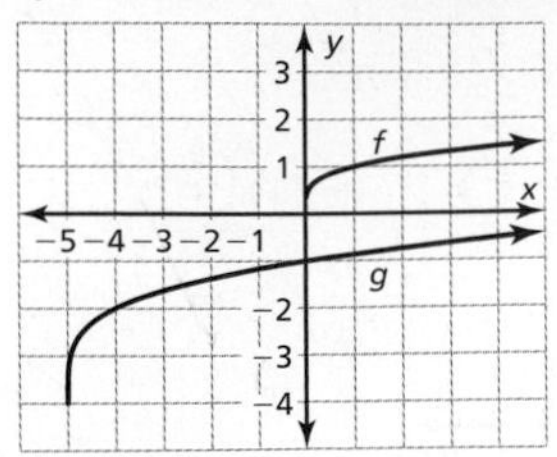

27. The graph was translated 2 units left but it should be translated 2 units right.

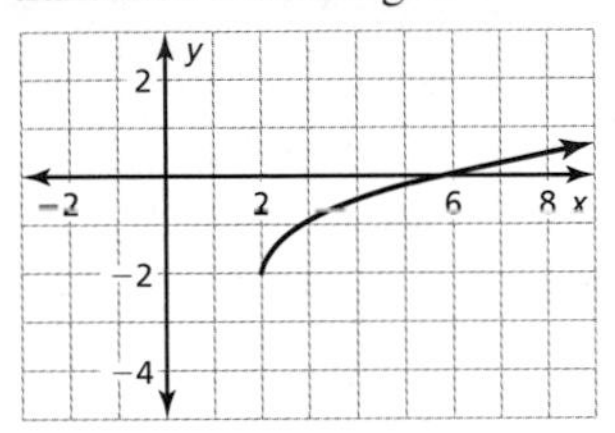

29. The domain is $x \leq -1$ and $x \geq 0$. The range is $y \geq 0$.

31. The domain is all real numbers. The range is $y \geq -\frac{\sqrt[3]{2}}{2}$.

33. The domain is all real numbers. The range is $y \geq \frac{\sqrt{14}}{4}$.

35. always **37.** always

39. $M(n) = 0.915\sqrt{n}$; about 91.5 mi **41.** $g(x) = 2\sqrt{x} + 8$

43. $g(x) = \sqrt{9x + 36}$ **45.** $g(x) = 2\sqrt{x + 1}$

47. $g(x) = 2\sqrt{x + 3}$ **49.** $g(x) = 2\sqrt{(x + 5)^2 - 2}$

51.

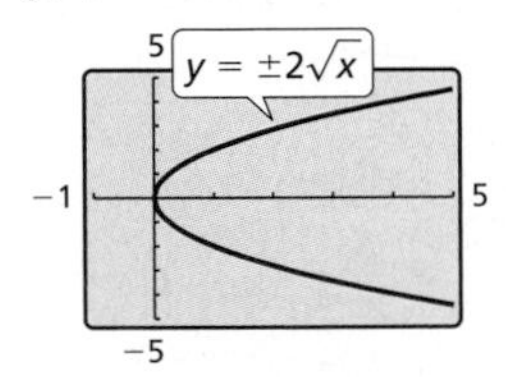

(0, 0), right

53.

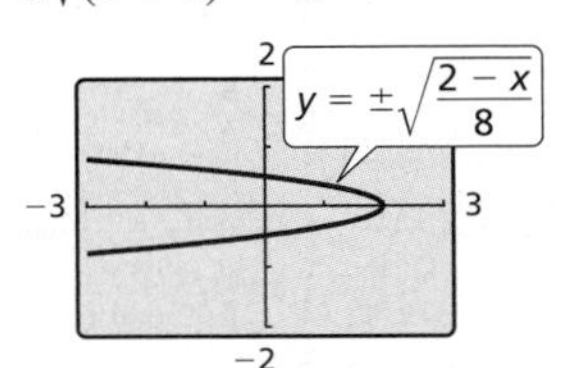

(2, 0), left

55.

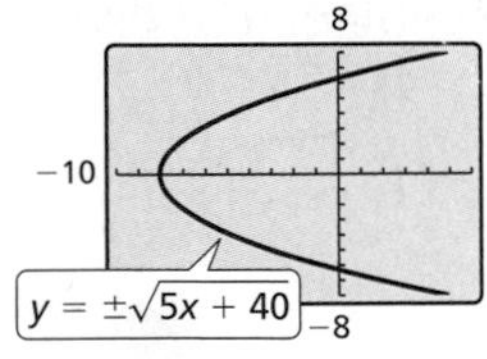

(−8, 0), right

57. $y = \pm\sqrt{9 - x^2}$

The radius is 3 units. The x-intercepts are ± 3. The y-intercepts are ± 3.

59.

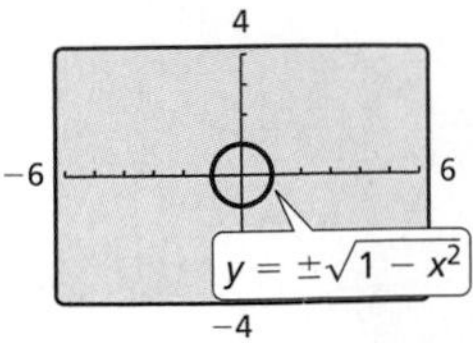

The radius is 1 unit. The x-intercepts are ± 1. The y-intercepts are ± 1.

61.

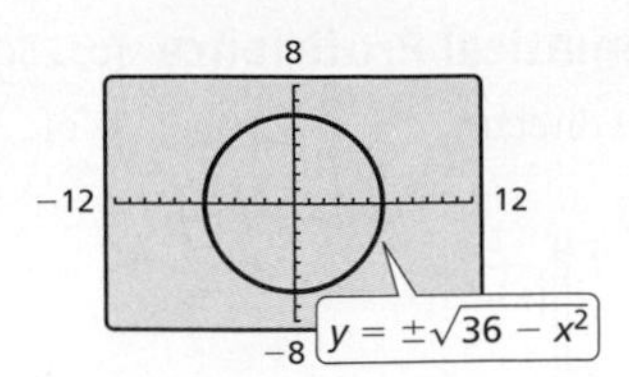

The radius is 6 units. The x-intercepts are ± 6. The y-intercepts are ± 6.

63.

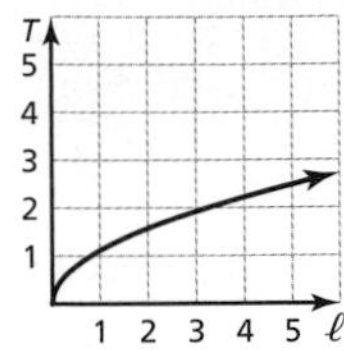

about 3 ft; *Sample answer:* Locate the T-value 2 on the graph and estimate the ℓ-value.

65.

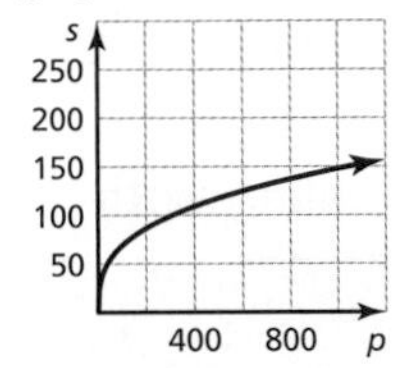

a. about 2468 hp **b.** about 0.04 mph/hp

67. a. the 165-lb skydiver

b. When $A = 1$, the diver is most likely vertical. When $A = 7$, the diver is most likely horizontal.

5.3 Maintaining Mathematical Proficiency *(p. 258)*

69. $x = 1$ and $x = -\frac{7}{3}$ **71.** $x = 2$ and $x = 6$

73. $-4 < x < -3$ **75.** $x < 0.5$ and $x > 6$

5.4 Vocabulary and Core Concept Check *(p. 266)*

1. no; The radicand does not contain a variable.

5.4 Monitoring Progress and Modeling with Mathematics *(pp. 266–268)*

3. $x = 7$ **5.** $x = 24$ **7.** $x = 6$ **9.** $x = -\frac{1000}{3}$

11. $x = 1024$ **13.** about 21.7 yr **15.** $x = 12$

17. $x = 14$ **19.** $x = 0$ and $x = \frac{1}{2}$ **21.** $x = 3$

23. $x = -1$ **25.** $x = 4$ **27.** $x = \pm 8$

29. no real solution **31.** $x = 3$ **33.** $x = 5$

35. Only one side of the equation was cubed;

$$\sqrt[3]{3x - 8} = 4$$
$$\left(\sqrt[3]{3x - 8}\right)^3 = 4^3$$
$$3x - 8 = 64$$
$$x = 24$$

37. $x \geq 64$ **39.** $x > 27$ **41.** $0 \leq x \leq \frac{25}{4}$

43. $x > -220$ **45.** about 0.15 in.

47. (3, 0) and (4, 1);

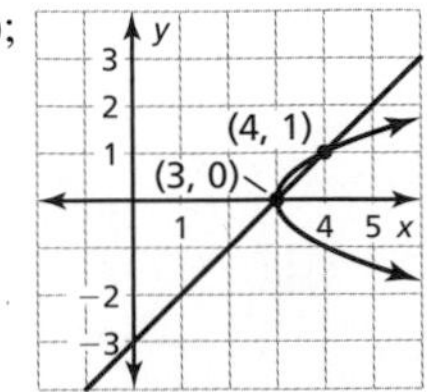

49. (0, −2) and (2, 0);

51. (0, −1);

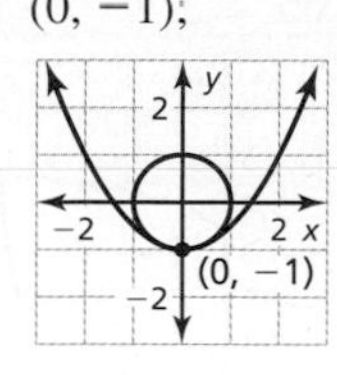

53. a. The greatest stopping distance is 450 feet on ice. On wet asphalt and snow, the stopping distance is 225 feet. The least stopping distance is 90 feet on dry asphalt.

b. about 272.2 ft; When $s = 35$ and $f = 0.15$, $d \approx 272.2$.

55. a. When solving the first equation, the solution is $x = 8$ with $x = 2$ as an extraneous solution. When solving the second equation, the solution is $x = 2$ with $x = 8$ as an extraneous solution.

b.

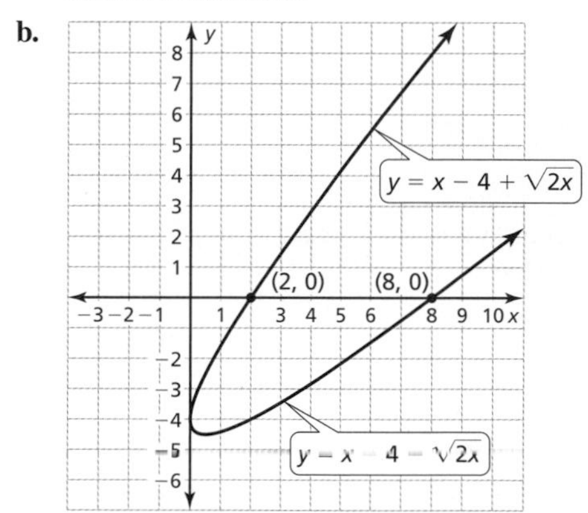

57. The square root of a quantity cannot be negative.

59. Raising the price would decrease demand.

61. $36\pi \approx 113.1$ ft^2

63. a. $h = h_0 - \dfrac{kt}{\pi r^2}$ **b.** about 5.75 in.

5.4 Maintaining Mathematical Proficiency *(p. 268)*

65. $x^5 + x^4 - 4x^2 + 3$ **67.** $x^3 + 11x - 8$

69. $g(x) = \frac{1}{2}x^3 - 2x^2$; The graph of g is a vertical shrink by a factor of $\frac{1}{2}$ followed by a translation 3 units down of the graph of f.

5.5 Vocabulary and Core Concept Check *(p. 273)*

1. You can add, subtract, multiply, or divide f and g.

5.5 Monitoring Progress and Modeling with Mathematics *(pp. 273–274)*

3. $(f + g)(x) = 14\sqrt[4]{x}$ and the domain is $x \geq 0$; $(f - g)(x) = -24\sqrt[4]{x}$ and the domain is $x \geq 0$; $(f + g)(16) = 28$; $(f - g)(16) = -48$

5. $(f + g)(x) = -7x^3 + 5x^2 + x$ and the domain is all real numbers; $(f - g)(x) = -7x^3 - 13x^2 + 11x$ and the domain is all real numbers; $(f + g)(-1) = 11$; $(f - g)(-1) = -17$

7. $(fg)(x) = 2x^{10/3}$ and the domain is all real numbers; $\left(\dfrac{f}{g}\right)(x) = 2x^{8/3}$ and the domain is $x \neq 0$; $(fg)(-27) = 118{,}098$; $\left(\dfrac{f}{g}\right)(-27) = 13{,}122$

9. $(fg)(x) = 36x^{3/2}$ and the domain is $x \geq 0$; $\left(\dfrac{f}{g}\right)(x) = \dfrac{4}{9}x^{1/2}$ and the domain is $x > 0$; $(fg)(9) = 972$; $\left(\dfrac{f}{g}\right)(9) = \dfrac{4}{3}$

11. $(fg)(x) = -98x^{11/6}$ and the domain is $x \geq 0$; $\left(\dfrac{f}{g}\right)(x) = -\dfrac{1}{2}x^{7/6}$; and the domain is $x > 0$; $(fg)(64) = -200{,}704$; $\left(\dfrac{f}{g}\right)(64) = -64$

13. 2541.04; 2458.96; 102,598.56; 60.92

15. 7.76; −14.60; −38.24; −0.31

17. Because the functions have an even index, the domain is restricted; The domain of $(fg)(x)$ is $x \geq 0$.

19. a. $(F + M)(t) = 0.0001t^3 - 0.016t^2 + 0.21t + 7.4$

b. the total number of employees from the ages of 16 to 19 in the United States

21. yes; When adding or multiplying functions, the order in which they appear does not matter.

23. $(f + g)(3) = -21$; $(f - g)(1) = -1$; $(fg)(2) = 0$; $\left(\dfrac{f}{g}\right)(0) = 2$

25. $r(x) = x^2 - \frac{1}{2}x^2 = \frac{1}{2}x^2$

27. a. $r(x) = \dfrac{20 - x}{6.4}$; $s(x) = \dfrac{\sqrt{x^2 + 144}}{0.9}$

b. $t(x) = \dfrac{20 - x}{6.4} + \dfrac{\sqrt{x^2 + 144}}{0.9}$

c. $x \approx 1.7$; If Elvis runs along the shore until he is about 1.7 meters from point C then swims to point B, the time taken to get there will be a minimum.

5.5 Maintaining Mathematical Proficiency *(p. 274)*

29. $n = \dfrac{5z}{7 + 8z}$ **31.** $n = \dfrac{3}{7b - 4}$

33. no; −1 has two outputs. **35.** no; 2 has two outputs.

5.6 Vocabulary and Core Concept Check *(p. 281)*

1. Inverse functions are functions that undo each other.

3. x; x

5.6 Monitoring Progress and Modeling with Mathematics *(pp. 281–284)*

5. $x = \dfrac{y - 5}{3}$; $-\dfrac{8}{3}$ **7.** $x = 2y + 6$; 0 **9.** $x = \sqrt[3]{\dfrac{y}{3}}$; −1

11. $x = 2 \pm \sqrt{y + 7}$; 0, 4

13. $g(x) = \frac{1}{6}x$;

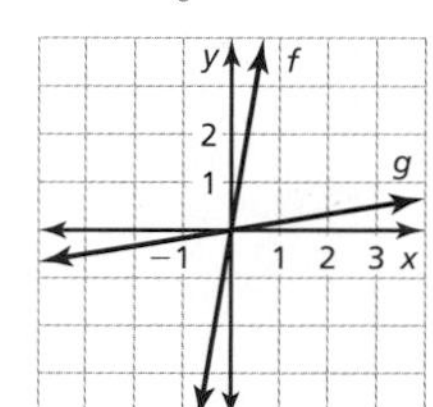

Selected Answers

15. $g(x) = \frac{x - 5}{-2}$;

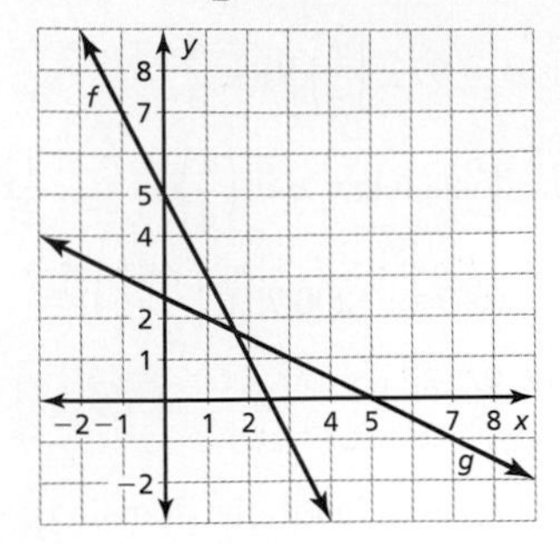

17. $g(x) = -2x + 8$;

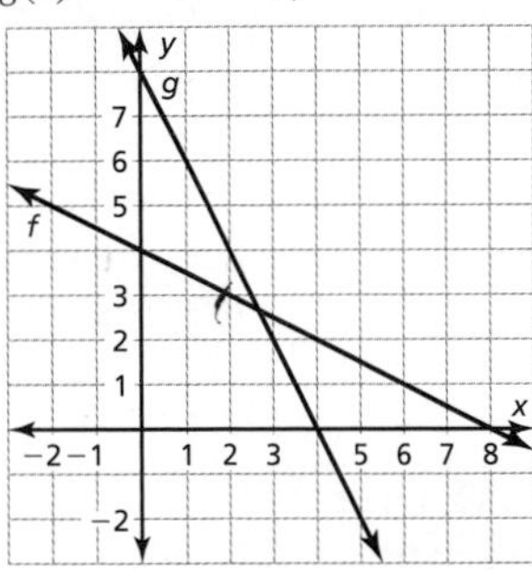

19. $g(x) = \frac{3x + 1}{2}$;

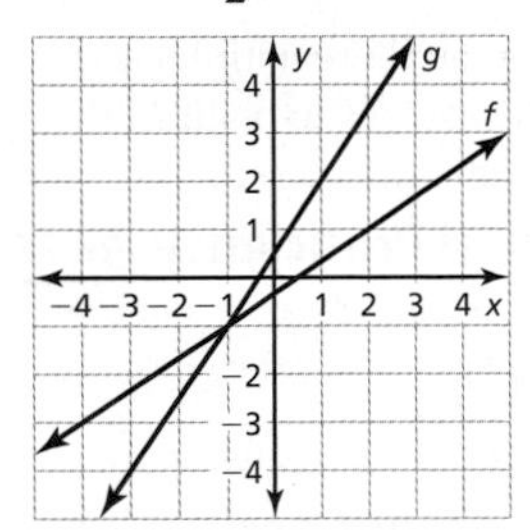

21. $g(x) = \frac{x - 4}{-3}$; *Sample answer:* switching x and y; You can graph the inverse to check your answer.

23. $g(x) = -\frac{\sqrt{x}}{2}$;

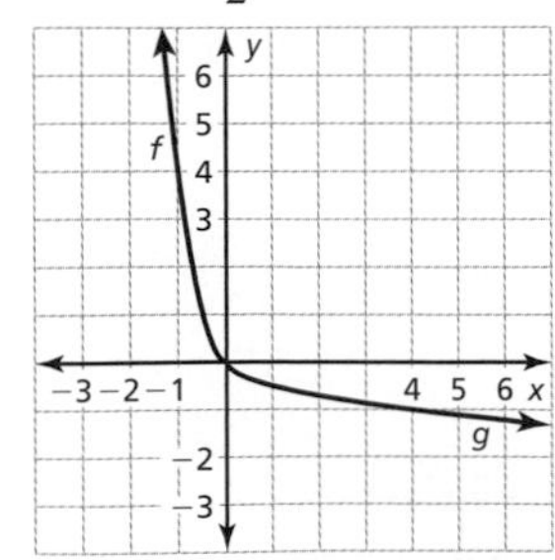

25. $g(x) = \sqrt[3]{x} + 3$

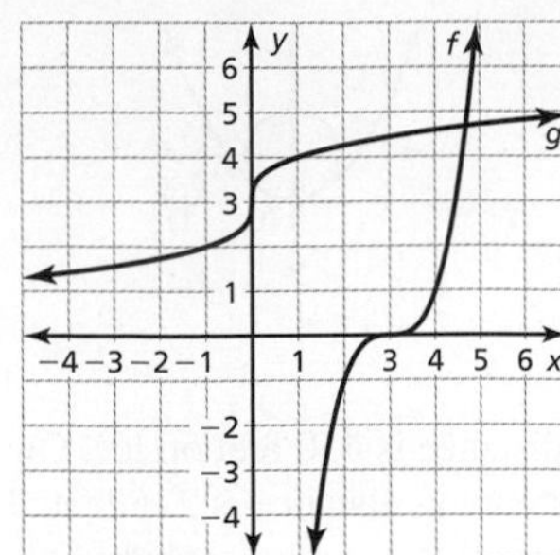

27. $g(x) = \sqrt[4]{\frac{x}{2}}$;

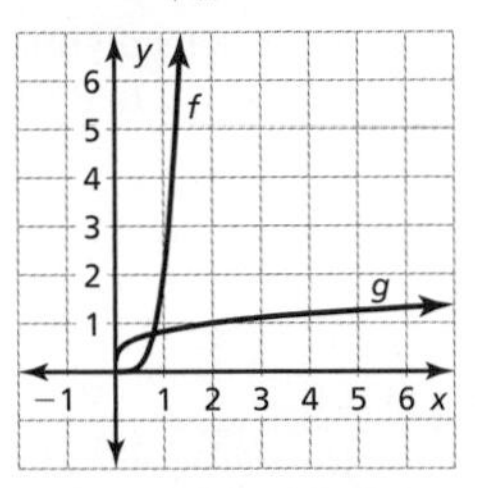

29. When switching x and y, the negative should not be switched with the variables;
$y = -x + 3$
$x = -y + 3$
$-x + 3 = y$

31. no; The function does not pass the horizontal line test.

33. no; The function does not pass the horizontal line test.

35. yes; $g(x) = \sqrt[3]{x + 1}$

37. yes; $g(x) = x^2 - 4$, where $x \geq 0$ **39.** yes; $g(x) = \frac{x^3}{8} + 5$

41. no; $y = \pm\sqrt[4]{x - 2}$ **43.** yes; $g(x) = \frac{x^3}{27} - 1$

45. yes; $g(x) = \sqrt[5]{2x}$ **47.** B

49. The functions are not inverses.

51. The functions are inverses.

53. $\ell = \left(\frac{v}{1.34}\right)^2$; about 31.3 ft **55.** B **57.** A

59. 5; When $x = 5$, $2x^2 + 3 = 53$.

61. **a.** $w = 2\ell - 6$; the weight of an object on a stretched spring of length ℓ
b. 5 lb **c.** $0.5(2\ell - 6) + 3 = \ell$; $2(0.5w + 3) - 6 = w$

63. **a.** $F = \frac{9}{5}C + 32$; The equation converts temperatures in Celsius to Fahrenheit.
b. start: 41° F; end: 14° F **c.** −40°

65. B **67.** A

69. **a.** false; All functions of the form $f(x) = x^n$, where n is an even integer, fail the horizontal line test.
b. true; All functions of the form $f(x) = x^n$, where n is an odd integer, pass the horizontal line test.

71. The inverse $y = \frac{1}{m}x - \frac{b}{m}$ has a slope of $\frac{1}{m}$ and a y-intercept of $-\frac{b}{m}$.

5.6 Maintaining Mathematical Proficiency *(p. 284)*

73. $-\frac{1}{3^3}$ **75.** 4^2

77. The function is increasing when $x > 1$ and decreasing when $x < 1$. The function is positive when $x < 0$ and when $x > 2$, and negative when $0 < x < 2$.

79. The function is increasing when $-2.89 < x < 2.89$ and decreasing when $x < -2.89$ and $x > 2.89$. The function is positive when $x < -5$ and $0 < x < 5$ and negative when $-5 < x < 0$ and $x > 5$.

Chapter 5 Review *(pp. 286–288)*

1. 128 **2.** 243 **3.** $\frac{1}{9}$ **4.** $x \approx 1.78$ **5.** $x = 3$

6. $x = -10$ and $x = -6$ **7.** $\frac{6^{2/5}}{6}$ **8.** 4 **9.** $2 + \sqrt{3}$

10. $7\sqrt[5]{8}$ **11.** $7\sqrt{3}$ **12.** $5^{1/3} \cdot 2^{3/4}$ **13.** $5z^3$

14. $\frac{\sqrt[4]{2z}}{6}$ **15.** $-z^2\sqrt{10z}$

16. The graph of g is a vertical stretch by a factor of 2 followed by a reflection in the x-axis of the graph of f;

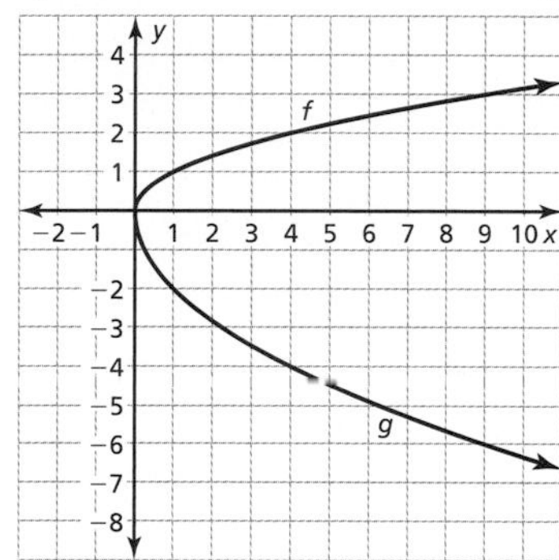

17. The graph of g is a reflection in the y-axis followed by a translation 6 units down of the graph of f.

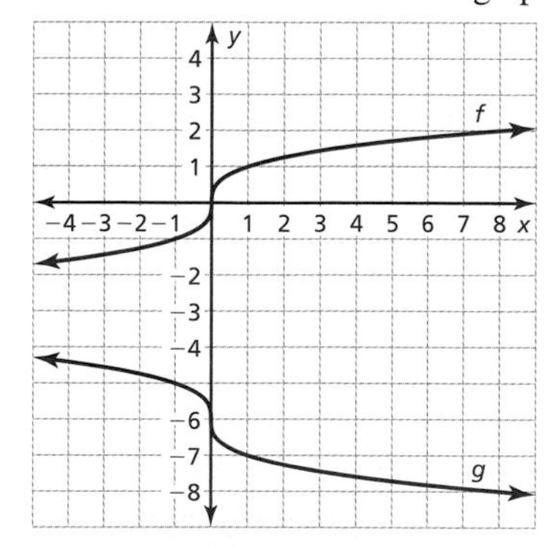

18. $g(x) = \sqrt[3]{-x + 7}$

19.

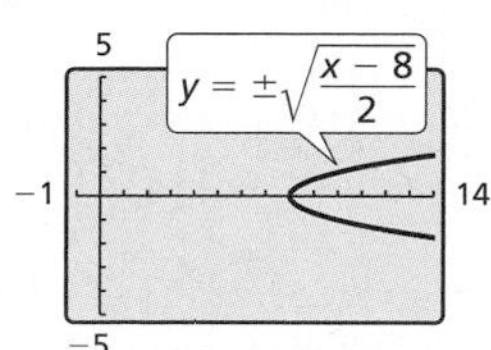

(8, 0); right

20.

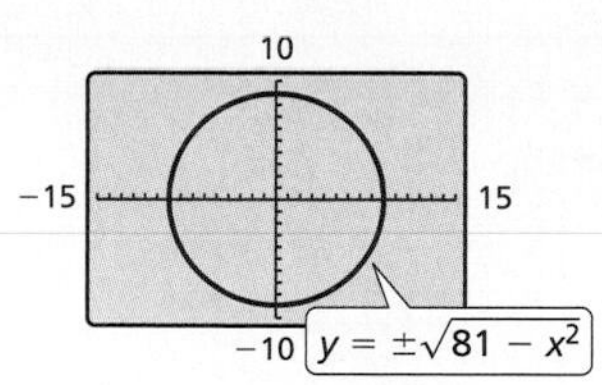

The radius is 9. The x-intercepts are ± 9. The y-intercepts are ± 9.

21. $x = 62$ **22.** $x = 2$ and $x = 10$ **23.** $x = \pm 36$

24. $x > 9$ **25.** $8 \leq x < 152$ **26.** $x \geq 30$

27. about 4082 m

28. $(fg)(x) = 8(3 - x)^{5/6}$ and the domain is $x \leq 3$; $\left(\frac{f}{g}\right)(x) = \frac{1}{2}(3 - x)^{1/6}$ and the domain is $x < 3$; $(fg)(2) = 8$; $\left(\frac{f}{g}\right)(2) = \frac{1}{2}$

29. $(f + g)(x) = 3x^2 + x + 5$ and the domain is all real numbers; $(f - g)(x) = 3x^2 - x - 3$ and the domain is all real numbers; $(f + g)(-5) = 75$; $(f - g)(-5) = 77$

30. $g(x) = -2x + 20$;

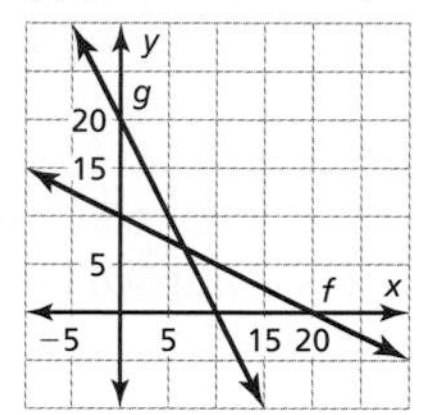

31. $g(x) = \sqrt{x - 8}$;

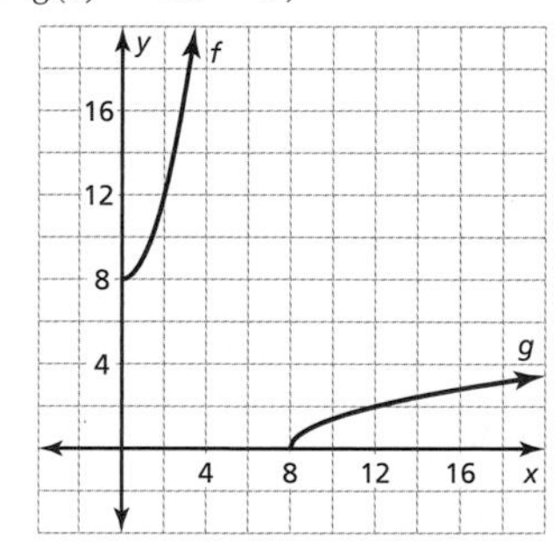

32. $g(x) = \sqrt[3]{-x - 9}$;

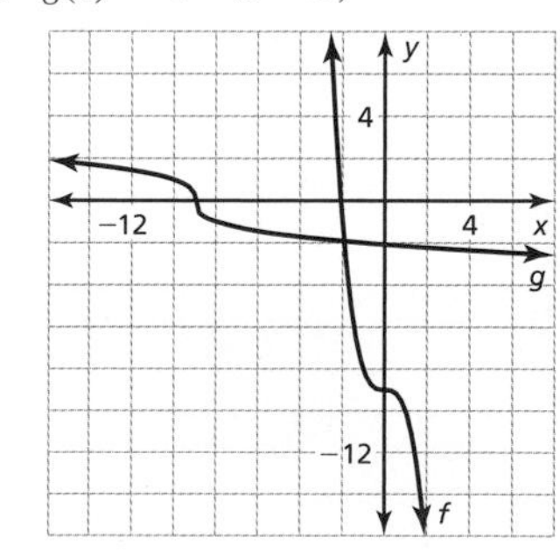

33. $g(x) = \frac{1}{9}(x - 5)^2, x \geq 5$;

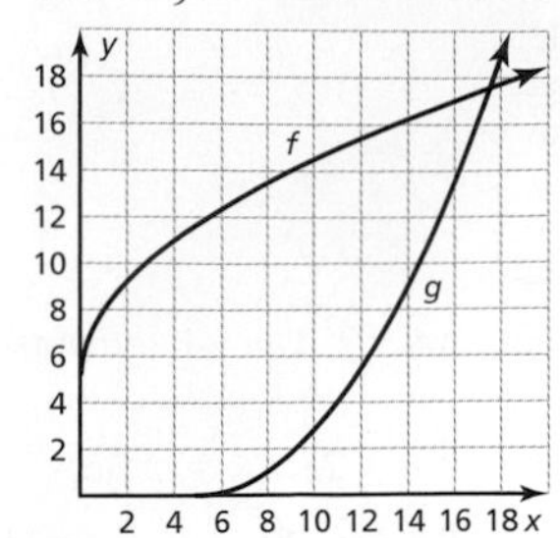

34. no **35.** yes **36.** $p = \dfrac{d}{1.587}$; about 63£

Chapter 6

Chapter 6 Maintaining Mathematical Proficiency *(p. 293)*

1. 48 **2.** -32 **3.** $-\frac{25}{36}$ **4.** $\frac{27}{64}$
5. domain: $-5 \leq x \leq 5$, range: $0 \leq y \leq 5$
6. domain: $\{-2, -1, 0, 1, 2\}$, range: $\{-5, -3, -1, 1, 3\}$
7. domain: all real numbers, range: $y \leq 0$
8. all values, odd values; no values, even values; The exponent of -4^n is evaluated first, then the result is multiplied by -1, so the value will always remain negative. The product of an odd number of negative values is negative. After the exponent of -4^n is evaluated, the result is multiplied by -1, so it will never be positive. The product of an even number of negative values is positive.

6.1 Vocabulary and Core Concept Check *(p. 300)*

1. The initial amount is 2.4, the growth factor is 1.5, and the percent increase is 0.5 or 50%.

6.1 Monitoring Progress and Modeling with Mathematics *(pp. 300–302)*

3. **a.** $\frac{1}{4}$ **b.** 8 **5.** **a.** $\frac{8}{9}$ **b.** 216

7. **a.** $\frac{46}{9}$ **b.** 32

9. exponential growth

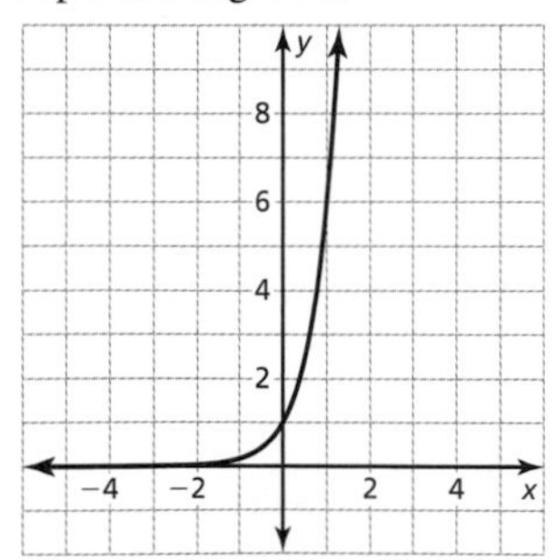

11. exponential decay

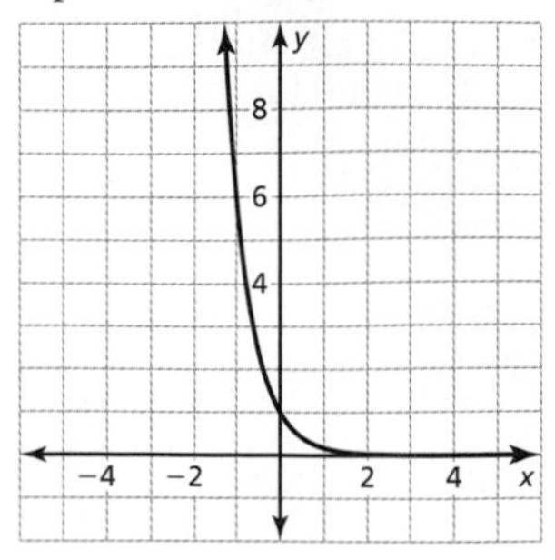

13. exponential growth

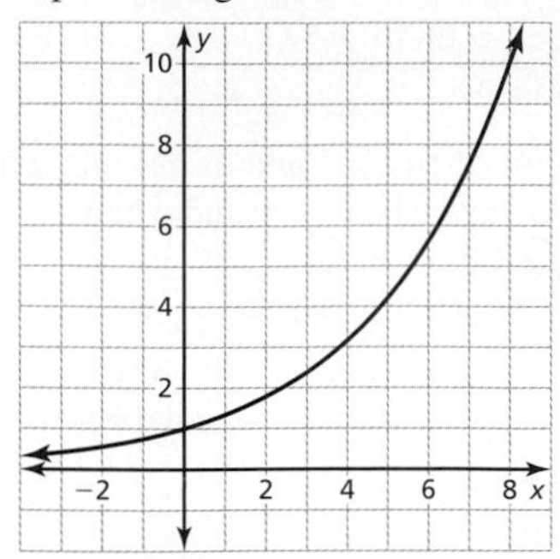

15. exponential growth

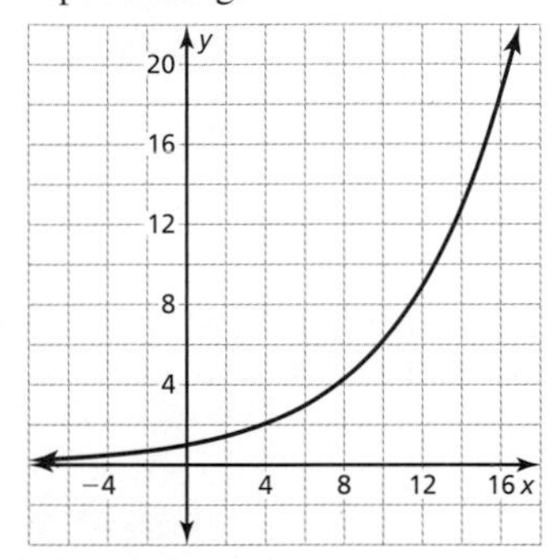

17. exponential decay

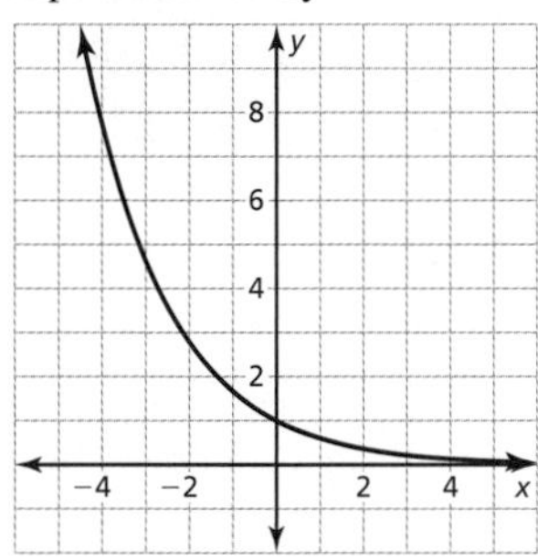

19. $b = 3$

21. **a.** exponential decay **b.** 25% decrease
c. in about 4.8 years

23. **a.** $y = 233(1.06)^t$; about 261.8 million **b.** 2009

25. Power of a Power Property; Evaluate power; Rewrite in form $y = a(1 + r)^t$.

27. about 0.01% **29.** $y = a(1 + 0.26)^t$; 26% growth

31. $y = a(1 - 0.06)^t$; 6% decay

33. $y = a(1 - 0.04)^t$; 4% decay

35. $y = a(1 + 255)^t$; 25,500% growth **37.** \$5593.60

39. The percent decrease needs to be subtracted from 1 to produce the decay factor;
$y = \begin{pmatrix}\text{Initial}\\ \text{amount}\end{pmatrix}\begin{pmatrix}\text{Decay}\\ \text{factor}\end{pmatrix}^t$; $y = 500(1 - 0.02)^t$; $y = 500(0.98)^t$

41. \$3982.92 **43.** \$3906.18

45. a represents the number of referrals it received at the start of the model. b represents the growth factor of the number of referrals each year; 50%; 1.50 can be rewritten as $(1 + 0.50)$, showing the percent increase of 50%.

47. no; $f(x) = 2^x$ eventually increases at a faster rate than $g(x) = x^2$ but not for all $x \geq 0$.

49. 221.5; The curve contains the points (0, 6850) and (6, 8179.26) and $\frac{8179.26 - 6850}{6 - 0} \approx 221.5$.

51. **a.** The decay factor is 0.9978. The percent decrease is 0.22%.

b.

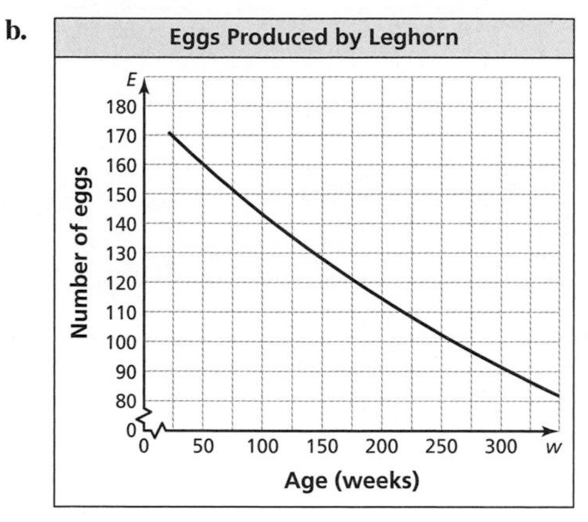

c. about 134 eggs per year

d. Replace $\frac{w}{52}$ with y, where y represents the age of the chicken in years.

6.1 Maintaining Mathematical Proficiency *(p. 302)*

53. x^{11} **55.** $24x^2$ **57.** $2x$ **59.** $3 + 5x$

6.2 Vocabulary and Core Concept Check *(p. 307)*

1. an irrational number that is approximately 2.718281828

6.2 Monitoring Progress and Modeling with Mathematics *(pp. 307–308)*

3. e^8 **5.** $\frac{1}{2e}$ **7.** $625e^{28x}$ **9.** $3e^{3x}$ **11.** $e^{-5x + 8}$

13. The 4 was not squared; $(4e^{3x})^2 = 4^2e^{(3x)(2)} = 16e^{6x}$

15. exponential growth

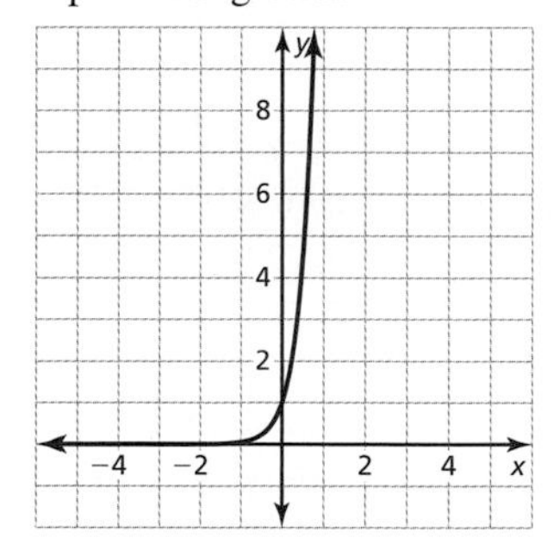

17. exponential decay

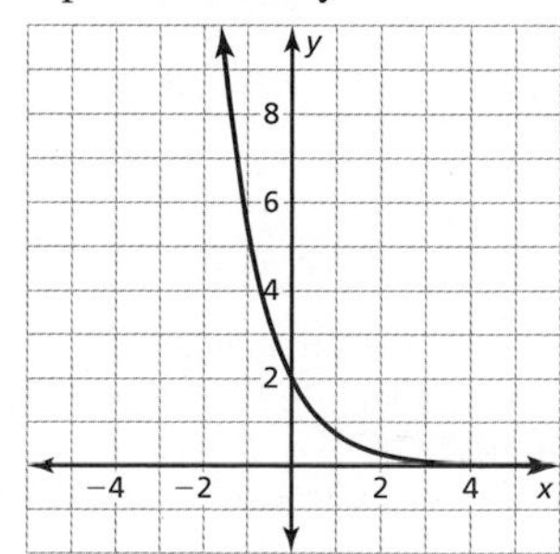

19. exponential growth

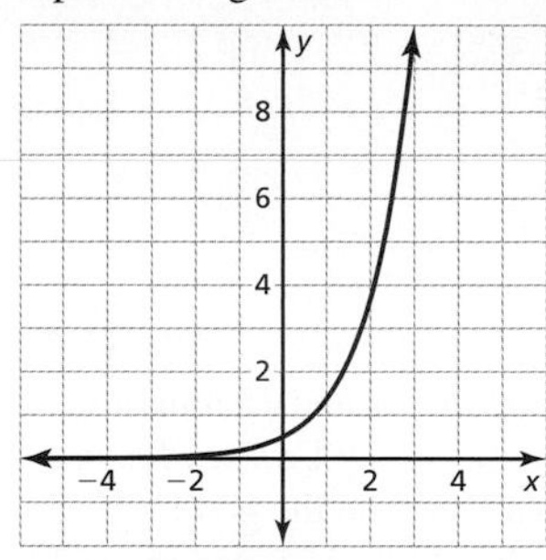

21. exponential decay

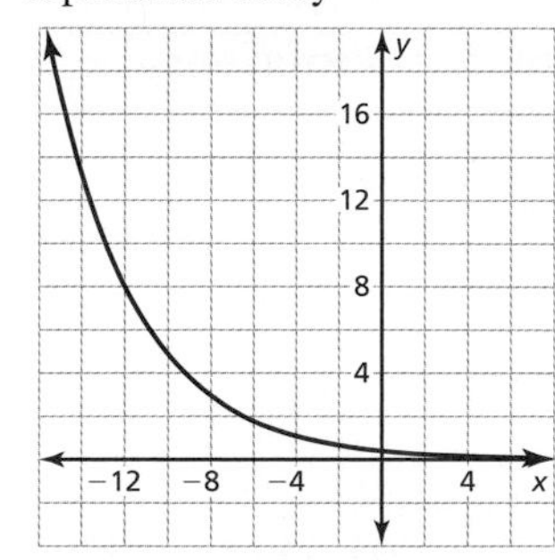

23. D; The graph shows growth and has a y-intercept of 1.

25. B; The graph shows decay and has a y-intercept of 4.

27. $y = (1 - 0.221)^t$; 22.1% decay

29. $y = 2(1 + 0.492)^t$; 49.2% growth

31.

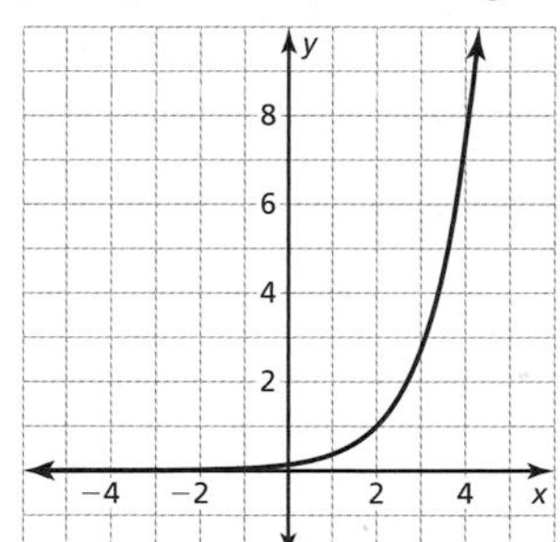

domain: all real numbers, range: $y > 0$

33.

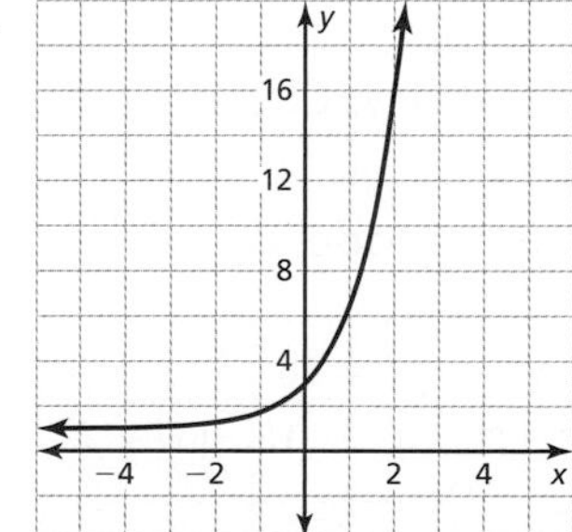

domain: all real numbers, range: $y > 1$

35. the education fund; the education fund

37. *Sample answer:* $a = 6, b = 2, r = -0.2, q = -0.7$

39. no; e is an irrational number. Irrational numbers cannot be expressed as a ratio of two integers.

41. account 1; With account 1, the balance would be $A = 2500\left(1 + \frac{0.06}{4}\right)^{4 \cdot 10} = \$4535.05$. With account 2, the balance would be $A = 2500e^{0.04 \cdot 10} = \$3729.56$.

43. a. $N(t) = 30e^{0.166t}$

b.

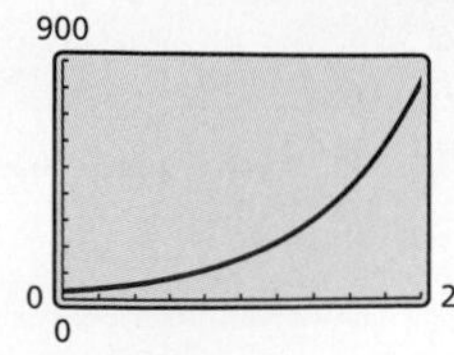

c. At 3:45 P.M., it has been 2 hours and 45 minutes, or 2.75 hours, since 1:00 P.M. Using the *trace* feature of the calculator, type 2.75 to find the point (2.75, 47.356183). At 3:45 P.M., there are about 47 cells.

6.2 Maintaining Mathematical Proficiency *(p. 308)*

45. 5×10^3 **47.** 4.7×10^{-8}

49. $y = -\sqrt{x+1}$

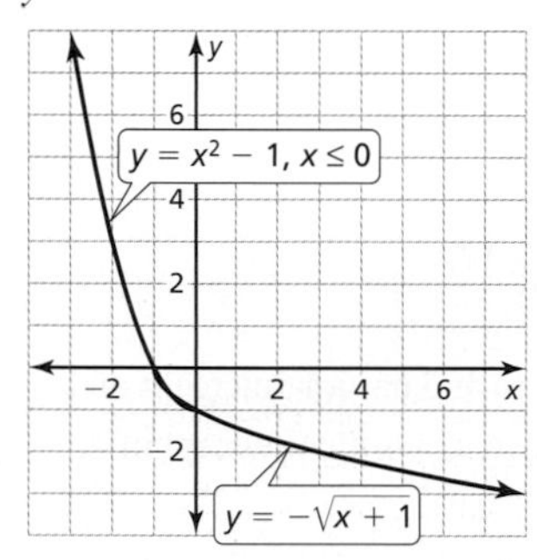

51. $y = \sqrt[3]{x+2}$

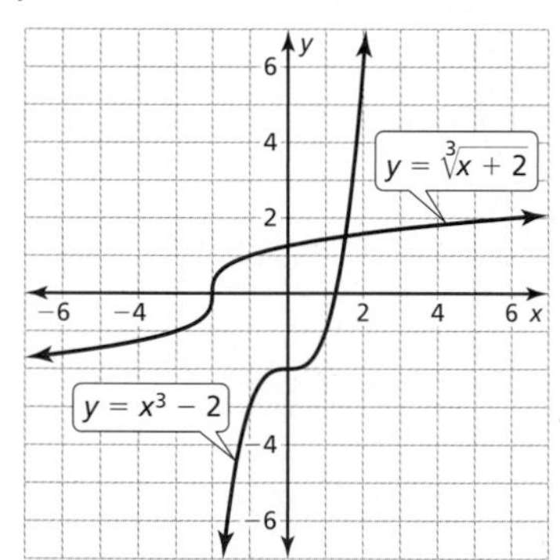

6.3 Vocabulary and Core Concept Check *(p. 314)*

1. common **3.** They are inverse equations.

6.3 Monitoring Progress and Modeling with Mathematics *(pp. 314–316)*

5. $3^2 = 9$ **7.** $6^0 = 1$ **9.** $\left(\frac{1}{2}\right)^{-4} = 16$

11. $\log_6 36 = 2$ **13.** $\log_{16} \frac{1}{16} = -1$ **15.** $\log_{125} 25 = \frac{2}{3}$

17. 4 **19.** 1 **21.** -4 **23.** -1

25. $\log_7 8, \log_5 23, \log_6 38, \log_2 10$ **27.** 0.778

29. -1.099 **31.** -2.079 **33.** 4603 m **35.** x

37. 4 **39.** $2x$

41. -3 and $\frac{1}{64}$ are in the wrong position; $\log_4 \frac{1}{64} = -3$

43. $y = \log_{0.3} x$ **45.** $y = 2^x$ **47.** $y = e^x + 1$

49. $y = \frac{1}{3} \ln x$ **51.** $y = \log_5(x + 9)$

53. a. about 283 mi/h

b. $d = 10^{(s-65)/93}$; The inverse gives the distance a tornado will travel given the wind speed, s.

55.

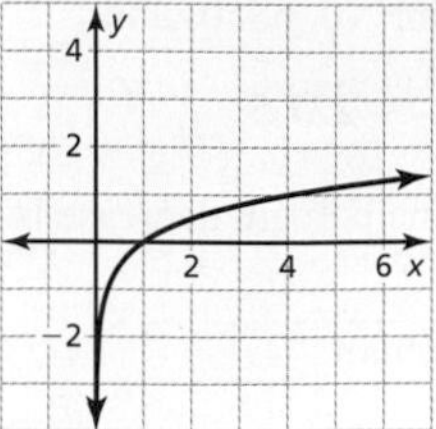

57.

59.

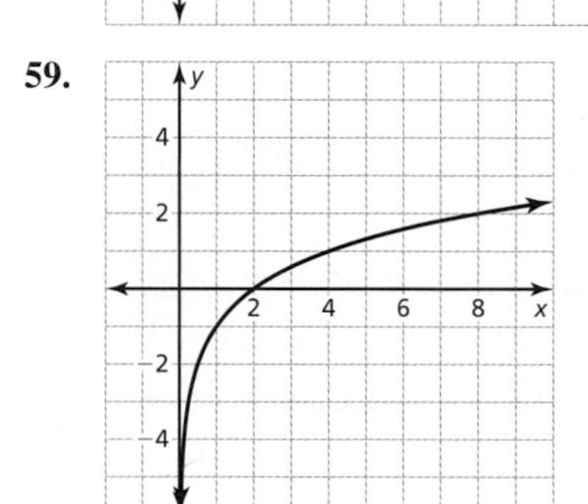

61.

domain: $x > -2$, range: all real numbers, asymptote: $x = -2$

63.

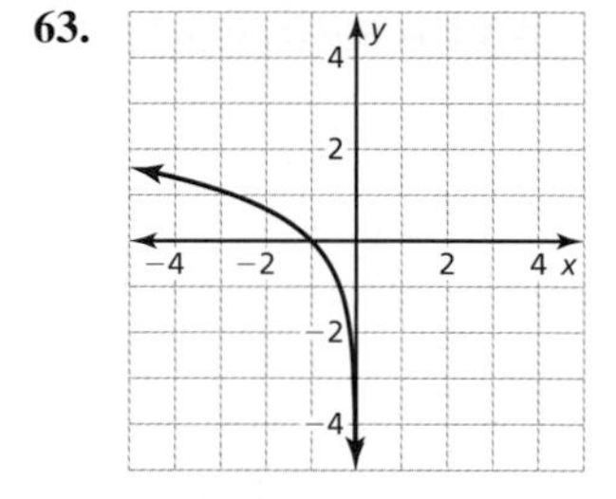

domain: $x < 0$; range: all real numbers, asymptote: $x = 0$

65. no; Any logarithmic function of the form $g(x) = \log_b x$ will pass through (1, 0), but if the function has been translated or reflected in the x-axis, it may not pass through (1, 0).

67. a.

180, 0, 0, 200

b. about 281 lb

c. (3.4, 0); no; The x-intercept shows that an alligator with a weight of 3.4 pounds has no length. If an object has weight, it must have length.

69. a.

20, 200, 0, 35,000

b. 15 species **c.** about 3918 m^2

d. The number of species of fish increases; *Sample answer:* This makes sense because in a smaller pool or lake, one species could dominate another more easily and feed on the weaker species until it became extinct.

71. a. $\frac{2}{3}$ **b.** $\frac{5}{3}$ **c.** $\frac{4}{3}$ **d.** $\frac{7}{2}$

6.3 Maintaining Mathematical Proficiency *(p. 316)*

73. $g(x) = \sqrt[3]{\frac{1}{2}x}$ **75.** $g(x) = \sqrt[3]{x + 2}$

77. quadratic; The graph is a translation 2 units left and 1 unit down of the parent quadratic function.

6.4 Vocabulary and Core Concept Check *(p. 322)*

1. Positive values of a vertically stretch ($a > 1$) or shrink ($a < 1$) the graph of f, h translates the graph of f left ($h < 0$) or right ($h > 0$), and k translates the graph of f up ($k > 0$) or down ($k < 0$). When a is negative, the graph of f is reflected in the x-axis.

6.4 Monitoring Progress and Modeling with Mathematics *(pp. 322–324)*

3. C; The graph of f is a translation 2 units left and 2 units down of the graph of the parent function $y = 2^x$.

5. A; The graph of h is a translation 2 units right and 2 units down of the graph of the parent function $y = 2^x$.

7. The graph of g is a translation 5 units up of the graph of f.

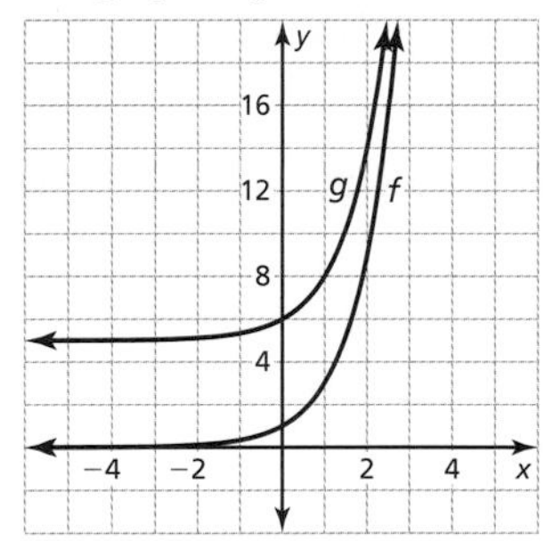

9. The graph of g is a translation 1 unit down of the graph of f.

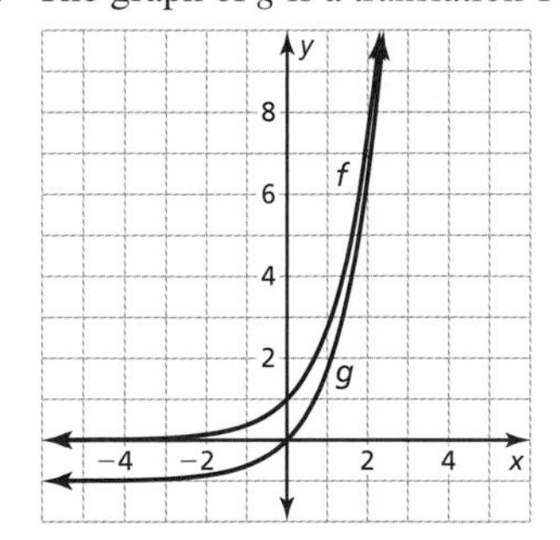

11. The graph of g is a translation 7 units right of the graph of f.

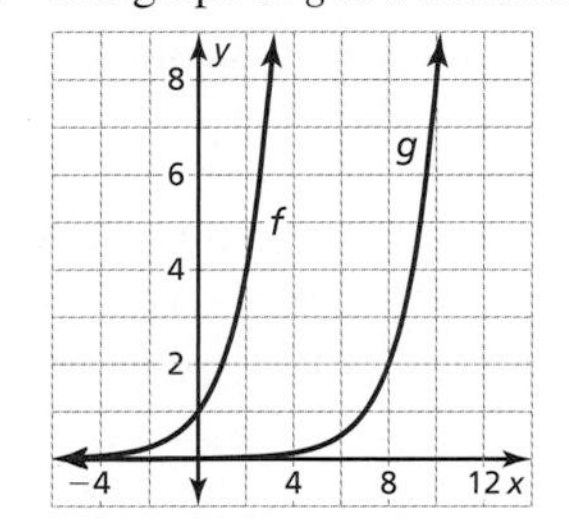

13. The graph of g is a translation 6 units up of the graph of f.

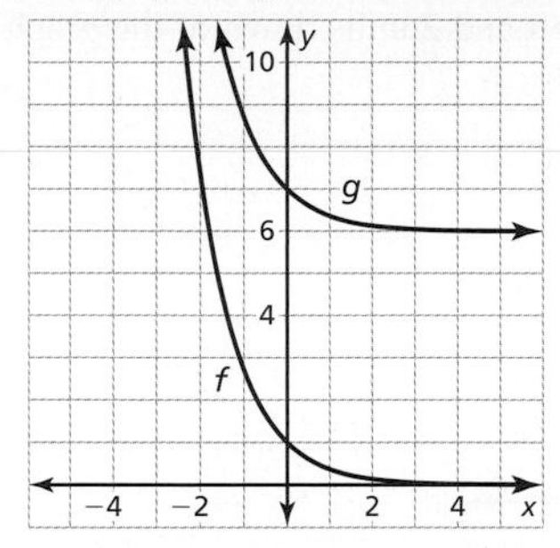

15. The graph of g is a translation 3 units right and 12 units up of the graph of f.

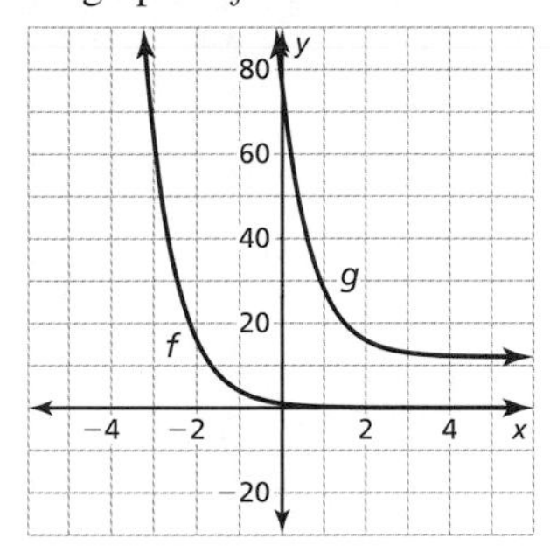

17. The graph of g is a horizontal shrink by a factor of $\frac{1}{2}$ of the graph of f.

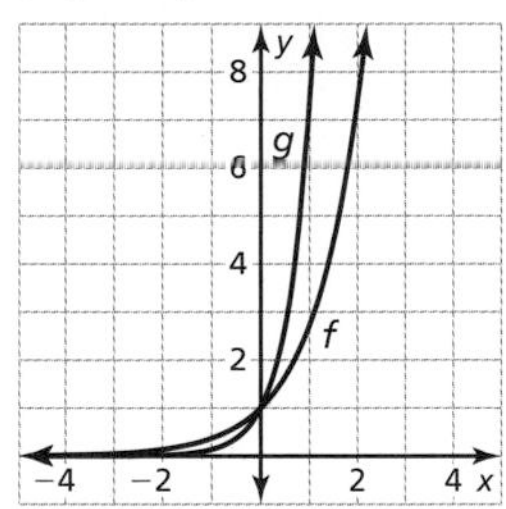

19. The graph of g is reflection in the x-axis followed by a translation 3 units right of the graph of f.

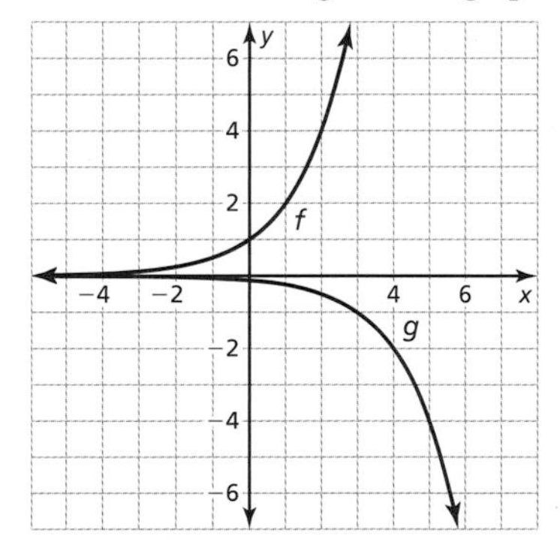

21. The graph of g is a horizontal shrink by a factor of $\frac{1}{6}$ followed by a vertical stretch by a factor of 3 of the graph of f.

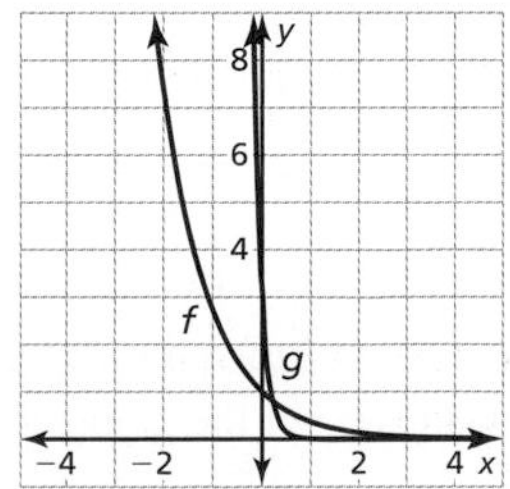

23. The graph of g is a vertical stretch by a factor of 6 followed by a translation 5 units left and 2 units down of the graph of f.

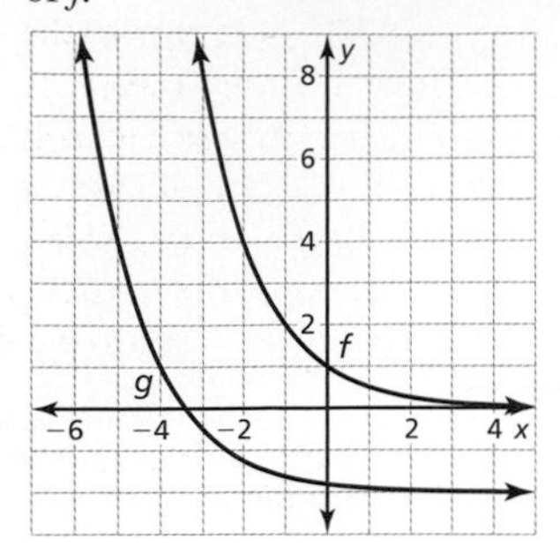

25. The graph of the parent function $f(x) = 2^x$ was translated 3 units left instead of up.

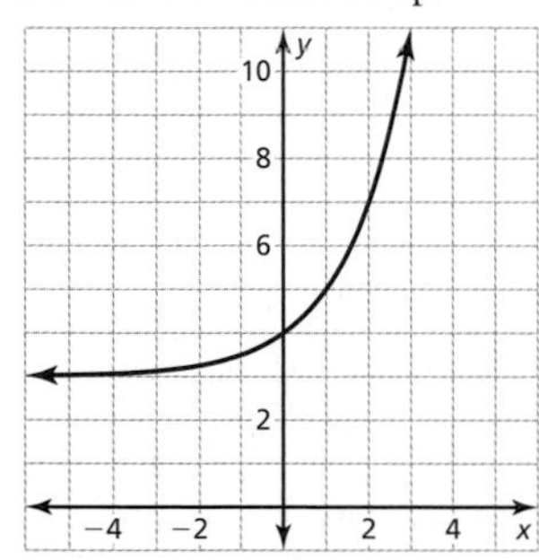

27. The graph of g is a vertical stretch by a factor of 3 followed by a translation 5 units down of the graph of f.

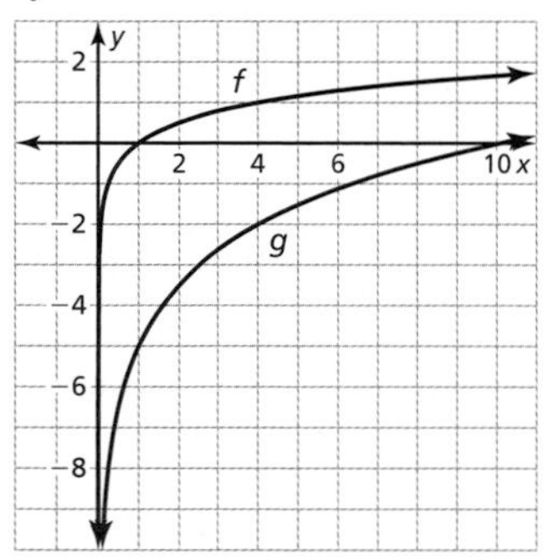

29. The graph of g is a reflection in the x-axis followed by a translation 7 units right of the graph of f.

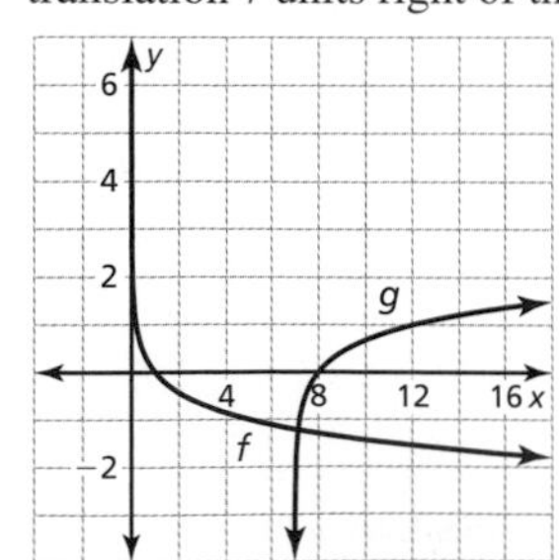

31. A; The graph of f has been translated 2 units right.

33. C; The graph of f has been stretched vertically by a factor of 2.

35. $g(x) = 5^{-x} - 2$ **37.** $g(x) = e^{2x} + 5$

39. $g(x) = 6 \log_6 x - 5$ **41.** $g(x) = \log_{1/2}(-x + 3) + 2$

43. Multiply the output by -1; Substitute $\log_7 x$ for $f(x)$. Subtract 6 from the output; Substitute $-\log_7 x$ for $h(x)$.

45. The graph of g is a translation 4 units up of the graph of f; $y = 4$

47. The graph of g is a translation 6 units left of the graph of f; $x = -6$

49. The graph of S is a vertical shrink by a factor of 0.118 followed by a translation 0.159 unit up of the graph of f; For fine sand, the slope of the beach is about 0.05. For medium sand, the slope of the beach is about 0.09. For coarse sand, the slope of the beach is about 0.12. For very coarse sand, the slope of the beach is about 0.16.

51. yes; *Sample answer:* If the graph is reflected in the y-axis, the graphs will never intersect because there are no values of x where $\log x = \log(-x)$.

53. **a.** never; The asymptote of $f(x) = \log x$ is a vertical line and would not change by shifting the graph vertically.

b. always; The asymptote of $f(x) = e^x$ is a horizontal line and would be changed by shifting the graph vertically.

c. always; The domain of $f(x) = \log x$ is $x > 0$ and would not be changed by a horizontal shrink.

d. sometimes; The graph of the parent exponential function does not intersect the x-axis, but if it is shifted down, the graph would intersect the x-axis.

55. The graph of h is a translation 2 units left of the graph of f; The graph of h is a reflection in the y-axis followed by a translation 2 units left of the graph of g; x has been replaced with $x + 2$. x has been replaced with $-(x + 2)$.

6.4 Maintaining Mathematical Proficiency *(p. 324)*

57. $(fg)(x) = x^6$; $(fg)(3) = 729$

59. $(f + g)(x) = 14x^3$; $(f + g)(2) = 112$

6.5 Vocabulary and Core Concept Check *(p. 331)*

1. Product

6.5 Monitoring Progress and Modeling with Mathematics *(pp. 331–332)*

3. 0.565 **5.** 1.424 **7.** −0.712

9. B; Quotient Property **11.** A; Power Property

13. $\log_3 4 + \log_3 x$ **15.** $1 + 5 \log x$

17. $\ln x - \ln 3 - \ln y$ **19.** $\log_7 5 + \frac{1}{2} \log_7 x$

21. The two expressions should be added, not multiplied; $\log_2 5x = \log_2 5 + \log_2 x$

23. $\log_4 \frac{7}{10}$ **25.** $\ln x^6y^4$ **27.** $\log_5 4\sqrt[3]{x}$

29. $\ln 32x^7y^4$

31. B;

$\log_5 \frac{y^4}{3x} = \log_5 y^4 - \log_5 3x$ Quotient Property

$= 4 \log_5 y - (\log_5 3 + \log_5 x)$ Power and Product Properties

$= 4 \log_5 y - \log_5 3 - \log_5 x$ Distributive Property

33. 1.404 **35.** 1.232 **37.** 1.581 **39.** −0.860

41. yes; Using the change-of-base formula, the equation can be graphed as $y = \frac{\log x}{\log 3}$.

43. 60 decibels

45. a. $2 \ln 2 \approx 1.39$ knots

b.
$$
\begin{aligned}
s(h) &= 2 \ln 100h \\
s(h) &= \ln(100h)^2 \\
e^{s(h)} &= e^{\ln(100h)^2} \\
e^{s(h)} &= (100h)^2 \\
\log e^{s(h)} &= \log(100h)^2 \\
s(h) \log e &= 2 \log(100h) \\
s(h) \log e &= 2(\log 100 + \log h) \\
s(h) \log e &= 2(2 + \log h) \\
s(h) &= \frac{2}{\log e}(\log h + 2)
\end{aligned}
$$

47. Rewrite each logarithm in exponential form to obtain $a = b^x$, $c = b^y$, and $a = c^z$. So,

$$\frac{\log_b a}{\log_b c} = \frac{\log_b c^z}{\log_b c} = \frac{z \log_b c}{\log_b c} = z = \log_c a.$$

6.5 Maintaining Mathematical Proficiency *(p. 332)*

49. $x < -2$ or $x > 2$ **51.** $-7 < x < -6$

53. $x \approx -0.76$ and $x \approx 2.36$ **55.** $x \approx -1.79$ and $x \approx 1.12$

6.6 Vocabulary and Core Concept Check *(p. 338)*

1. exponential

3. The domain of a logarithmic function is positive numbers only, so any quantity that results in taking the log of a non-positive number will be an extraneous solution.

6.6 Monitoring Progress and Modeling with Mathematics *(pp. 338–340)*

5. $x = -1$ **7.** $x = 7$ **9.** $x \approx 1.771$ **11.** $x = -\frac{5}{3}$

13. $x \approx 0.255$ **15.** $x \approx 0.173$ **17.** about 17.6 years old

19. about 50 min **21.** $x = 6$ **23.** $x = 3$ **25.** $x = 6$

27. $x = 10$ **29.** $x = 1$

31. $x = \frac{1 + \sqrt{41}}{2} \approx 3.7$ and $x = \frac{1 - \sqrt{41}}{2} \approx -2.7$

33. $x = 4$ **35.** $x \approx 6.04$ **37.** $x = \pm 1$

39. $x \approx 10.24$

41. 3 should be the base on both sides of the equation;
$$
\begin{aligned}
\log_3(5x - 1) &= 4 \\
3^{\log_3(5x - 1)} &= 3^4 \\
5x - 1 &= 81 \\
5x &= 82 \\
x &= 16.4
\end{aligned}
$$

43. a. 39.52 years **b.** 38.66 years **c.** 38.38 years
d. 38.38 years

45. a. $x \approx 3.57$ **b.** $x = 0.8$

47. $x > 1.815$ **49.** $x \geq 20.086$ **51.** $x < 1.723$

53. $x \geq \frac{1}{5}$

55. $0 < x < 25$; *Sample answer:* algebraically; Converting the equation to exponential form is the easiest method because it isolates the variable.

57. $r > 0.0718$ or $r > 7.18\%$ **59.** $x \approx 1.78$

61. no solution

63. a. $a = -\frac{1}{0.09} \ln\left(\frac{45 - \ell}{25.7}\right)$

b. 36 cm footprint: 11.7 years old; 32 cm footprint: 7.6 years old; 28 cm footprint: 4.6 years old; 24 cm footprint: 2.2 years old

65. *Sample answer:* $2^x = 16$; $\log_3(-x) = 1$ **67.** $x \approx 0.89$

69. $x \approx 10.61$ **71.** $x = 2$ and $x = 3$

73. To solve exponential equations with different bases, take a logarithm of each side. Then use the Power Property to move the exponent to the front of the logarithm, and solve for x. To solve logarithmic equations of different bases, find a common multiple of the bases, and exponentiate each side with this common multiple as the base. Rewrite the base as a power that will cancel out the given logarithm and solve the resulting equation.

6.6 Maintaining Mathematical Proficiency *(p. 340)*

75. $y + 2 = 4(x - 1)$ **77.** $y + 8 = -\frac{1}{3}(x - 3)$

79. 3; $y = 2x^3 - x + 1$

81. 4; $y = -3x^4 + 2x^3 - x^2 + 5x - 6$

6.7 Vocabulary and Core Concept Check *(p. 346)*

1. exponential

6.7 Monitoring Progress and Modeling with Mathematics *(pp. 346–348)*

3. exponential; The data have a common ratio of 4.

5. quadratic; The second differences are constant.

7. $y = 0.75(4)^x$ **9.** $y = \frac{1}{8}(2)^x$ **11.** $y = \frac{2}{5}(5)^x$

13. $y = 5(0.5)^x$ **15.** $y = 0.25(2)^x$

17. Data are linear when the first differences are constant; The outputs have a common ratio of 3, so the data represents an exponential function.

19. *Sample answer:* $y = 7.20(1.39)^x$

21. yes; *Sample answer:* $y = 8.88(1.21)^x$

23. no; *Sample answer:* $y = -0.8x + 66$

25.

Sample answer: $y = 3.25(1.052)^x$

27.

yes; $y = 9.14(1.99)^x$

29.

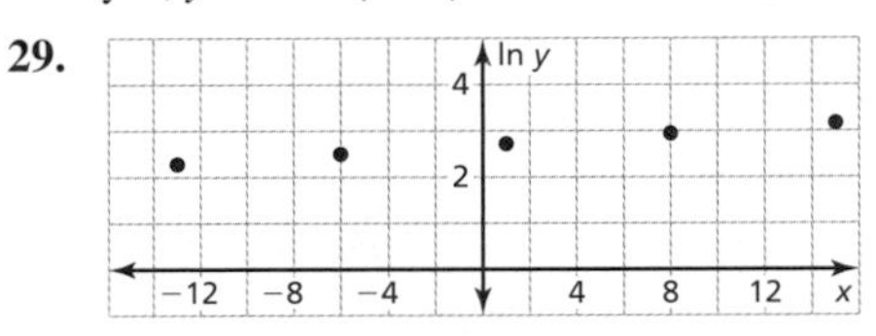

yes; $y = 14.73(1.03)^x$

31. $y = 6.70(1.41)^x$; about 208 scooters

33. $t = 12.59 - 2.55 \ln d$; 2.6 h

Selected Answers

35. a.

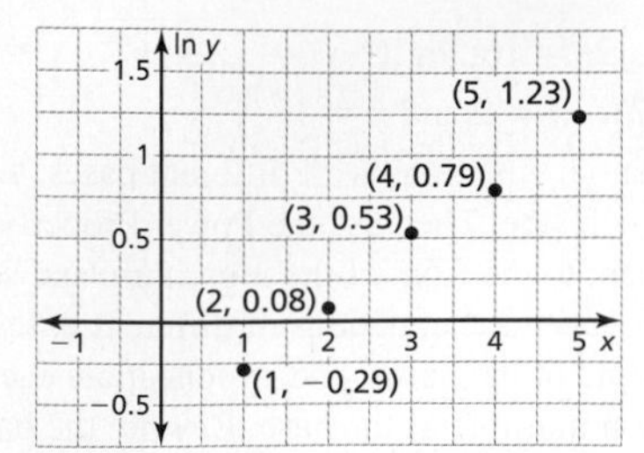

Sample answer: $y = 0.50(1.47)^x$

b. about 47%; The base is 1.47 which means that the function shows 47% growth.

37. no; When d is the independent variable and t is the dependent variable, the data can be modeled with a logarithmic function. When the variables are switched, the data can be modeled with an exponential function.

39. a. 5.9 weeks

b.

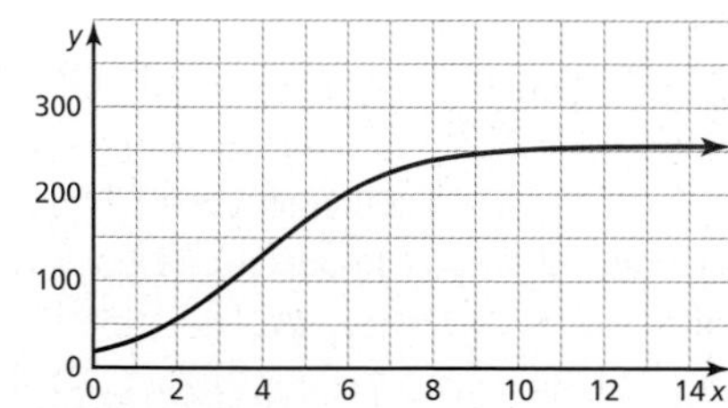

The asymptote is the line $y = 256$ and represents the maximum height of the sunflower.

6.7 Maintaining Mathematical Proficiency *(p. 348)*

41. no; When one variable is increased by a factor, the other variable does not increase by the same factor.

43. yes; When one variable is increased by a factor, the other variable increases by the same factor.

45. The focus is $\left(0, \frac{1}{16}\right)$, the directrix is $y = -\frac{1}{16}$, and the axis of symmetry is $x = 0$.

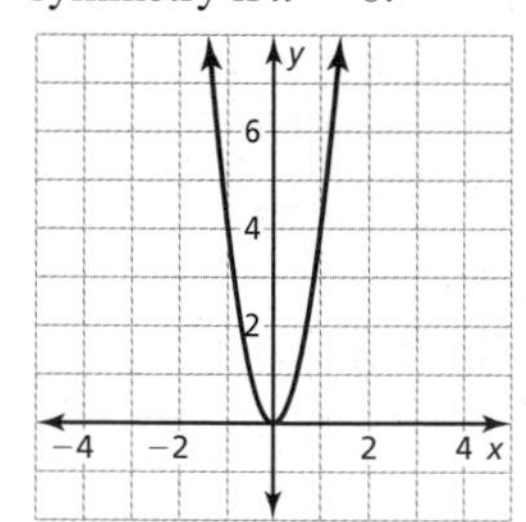

47. The focus is (0.1, 0), the directrix is $x = -0.1$, and the axis of symmetry is $y = 0$.

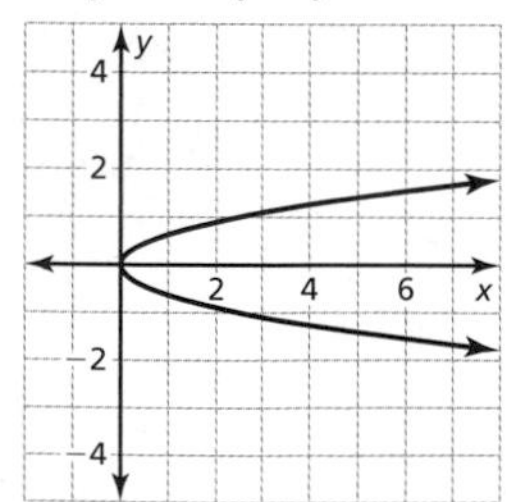

Chapter 6 Review *(pp. 350–352)*

1. exponential decay; 66.67% decrease

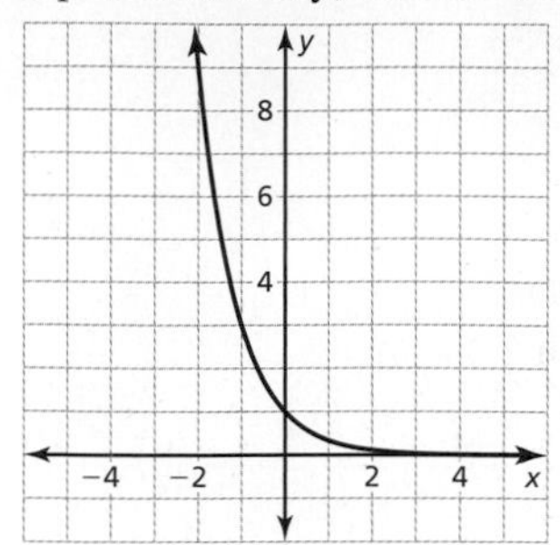

2. exponential growth; 400% increase

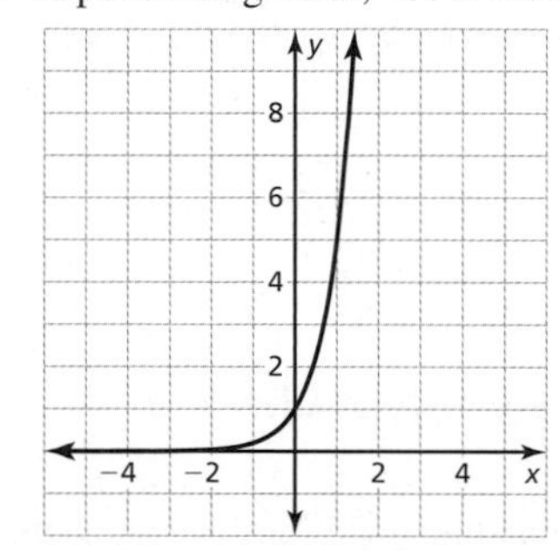

3. exponential decay; 80% decrease

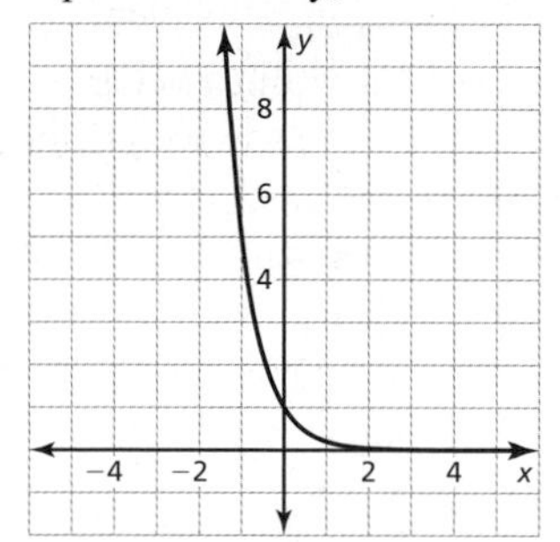

4. \$1725.39 **5.** e^{15} **6.** $\frac{2}{e^3}$ **7.** $\frac{9}{e^{10x}}$

8. exponential growth

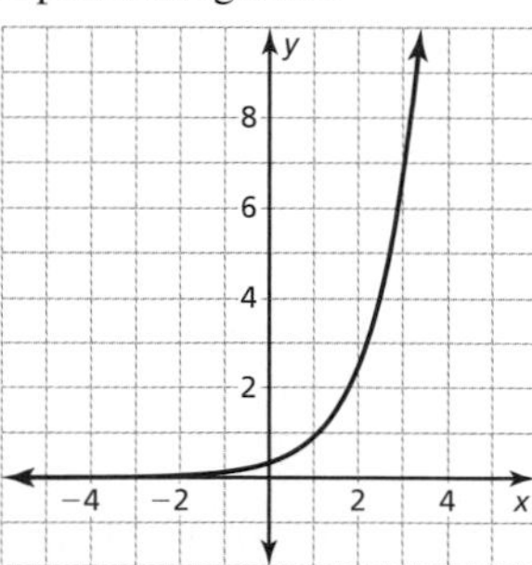

9. exponential decay

10. exponential decay

11. 3 **12.** −2 **13.** 0

14. $g(x) = \log_8 x$ **15.** $y = e^x + 4$ **16.** $y = 10^x - 9$

17.

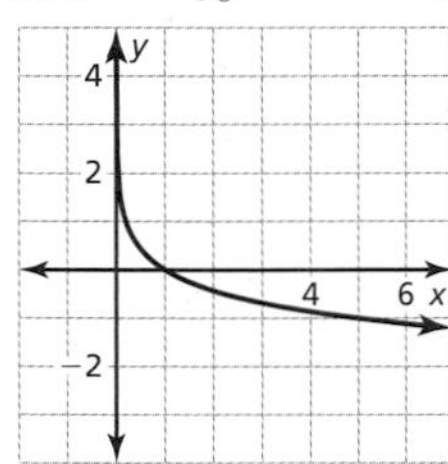

18. The graph of g is a horizontal shrink by a factor of $\frac{1}{5}$ followed by a translation 8 units down of the graph of f.

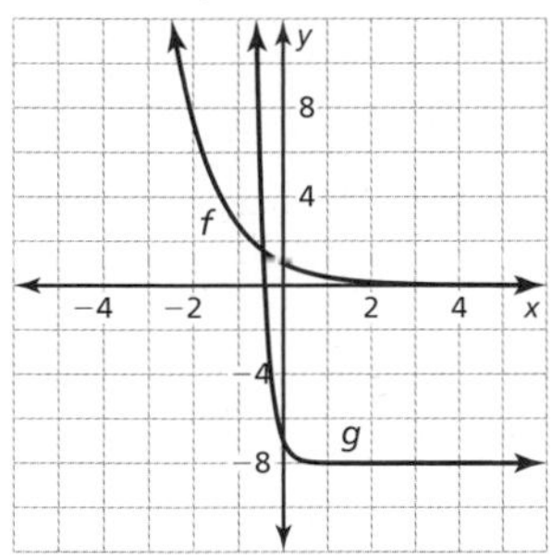

19. The graph of g is a vertical shrink by a factor of $\frac{1}{2}$ followed by a translation 5 units left of the graph of f.

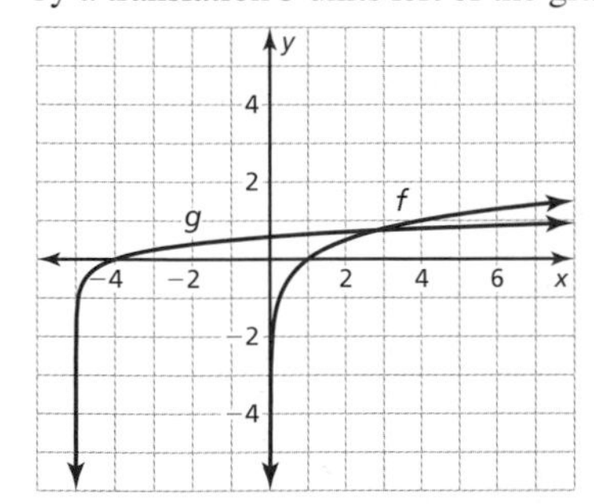

20. $g(x) = 3e^{x+6} + 3$ **21.** $g(x) = \log(-x) - 2$

22. $\log_8 3 + \log_8 x + \log_8 y$ **23.** $1 + 3 \log x + \log y$

24. $\ln 3 + \ln y - 5 \ln x$ **25.** $\log_7 384$ **26.** $\log_2 \frac{12}{x^2}$

27. $\ln 4x^2$ **28.** about 3.32 **29.** about 1.13

30. about 1.19 **31.** $x \approx 1.29$ **32.** $x = 7$

33. $x \approx 3.59$ **34.** $x > 1.39$ **35.** $0 < x \leq 8103.08$

36. $x \geq 1.19$ **37.** $y = 64\left(\frac{1}{2}\right)^x$

38. *Sample answer:* $y = 3.60(1.43)^x$

39. $s = 3.95 + 27.48 \ln t$; 53 pairs

Chapter 7

Chapter 7 Maintaining Mathematical Proficiency *(p. 357)*

1. $\frac{19}{15}$, or $1\frac{4}{15}$ **2.** $-\frac{17}{42}$ **3.** $\frac{1}{3}$ **4.** $\frac{11}{12}$ **5.** $-\frac{3}{7}$

6. $-\frac{1}{20}$ **7.** $\frac{9}{20}$ **8.** $-\frac{7}{20}$ **9.** $\frac{8}{11}$

10. 0; Division by zero is not possible.

7.1 Vocabulary and Core Concept Check *(p. 363)*

1. The ratio of the variables is constant in a direct variation equation, and the product of the variables is constant in an inverse variation equation.

7.1 Monitoring Progress and Modeling with Mathematics *(pp. 363–364)*

3. inverse variation **5.** direct variation **7.** neither

9. direct variation **11.** direct variation

13. inverse variation **15.** $y = -\frac{20}{x}$; $y = -\frac{20}{3}$

17. $y = -\frac{24}{x}$; $y = -8$ **19.** $y = \frac{21}{x}$; $y = 7$

21. $y = \frac{2}{x}$; $y = \frac{2}{3}$

23. The equation for direct variation was used; Because $5 = \frac{a}{8}$, $a = 40$. So, $y = \frac{40}{x}$.

25. a.

Size	2	2.5	3	5
Number of songs	5000	4000	3333	2000

b. The number of songs decreases.

27. $A = \frac{26{,}000}{c}$; about 321 chips per wafer

29. yes; The product of the number of hats and the price per hat is \$50, which is constant.

31. *Sample answer:* As the speed of your car increases, the number of minutes per mile decreases.

33. cat: 4 ft, dog: 2 ft; The inverse equations are $d = \frac{a}{7}$ and $d - 6 = \frac{a}{14}$. Because the constant is the same, solve the equation $7d = 14(6 - d)$ for d.

7.1 Maintaining Mathematical Proficiency *(p. 364)*

35. $x^2 - 6$

37.

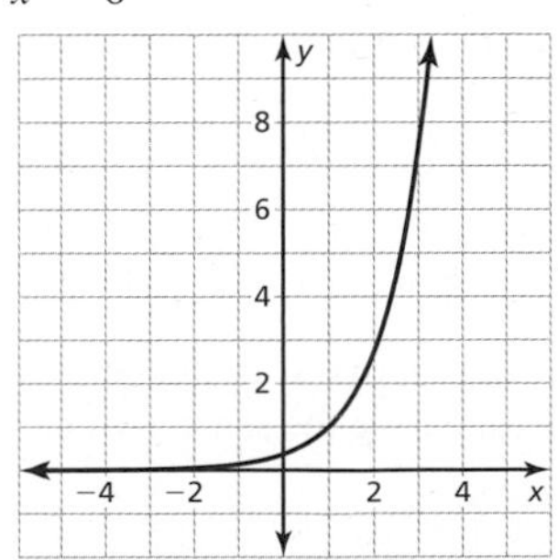

domain: all real numbers, range: $y > 0$

Selected Answers

39.

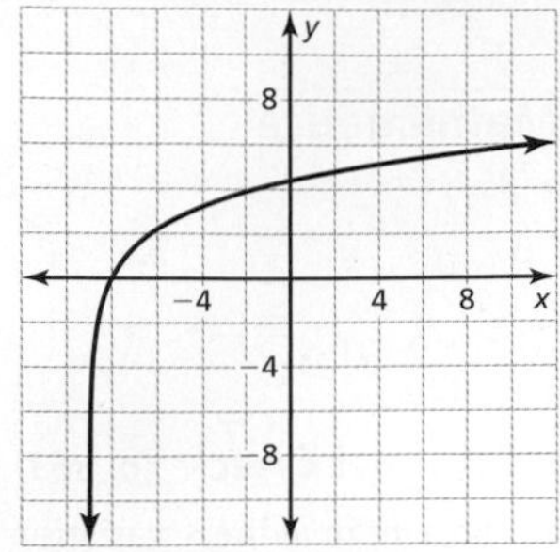

domain: $x > -9$, range: all real numbers

7.2 Vocabulary and Core Concept Check *(p. 370)*

1. range; domain

7.2 Monitoring Progress and Modeling with Mathematics *(pp. 370–372)*

3.

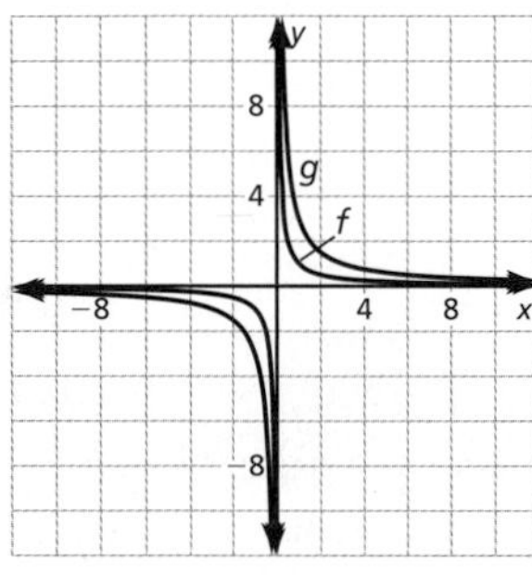

The graph of g lies farther from the axes. Both graphs lie in the first and third quadrants and have the same asymptotes, domain, and range.

5.

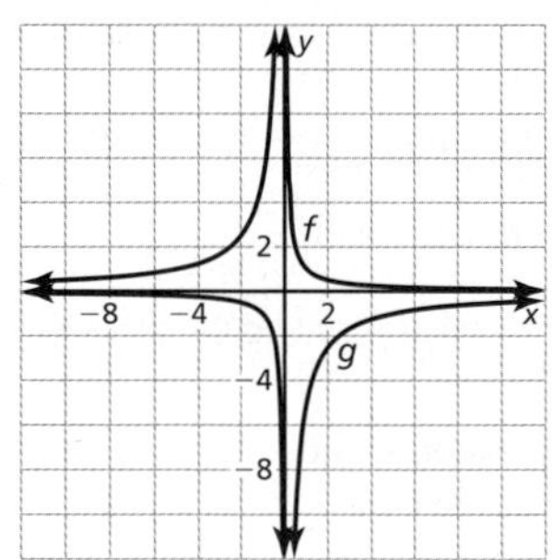

The graph of g lies farther from the axes and is reflected over the x-axis. Both graphs have the same asymptotes, domain, and range.

7.

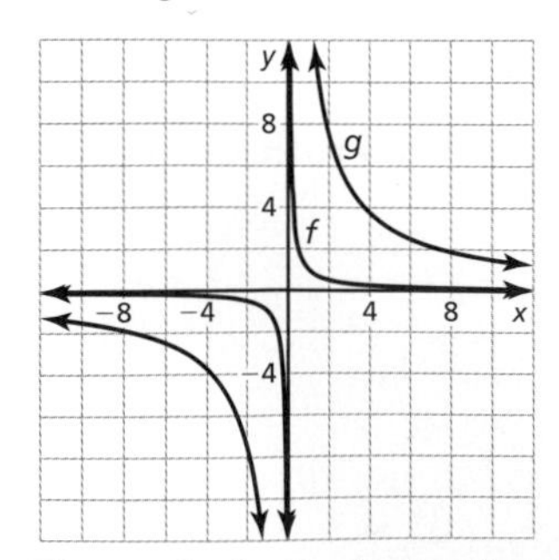

The graph of g lies farther from the axes. Both graphs lie in the first and third quadrants and have the same asymptotes, domain, and range.

9.

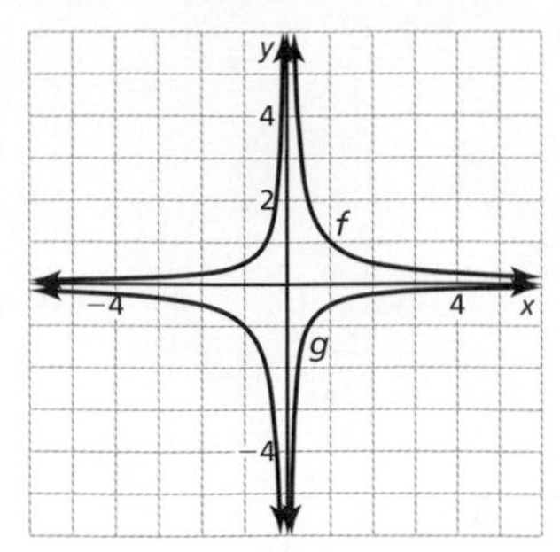

The graph of g lies closer to the axes and is reflected over the x-axis. Both graphs have the same asymptotes, domain, and range.

11.

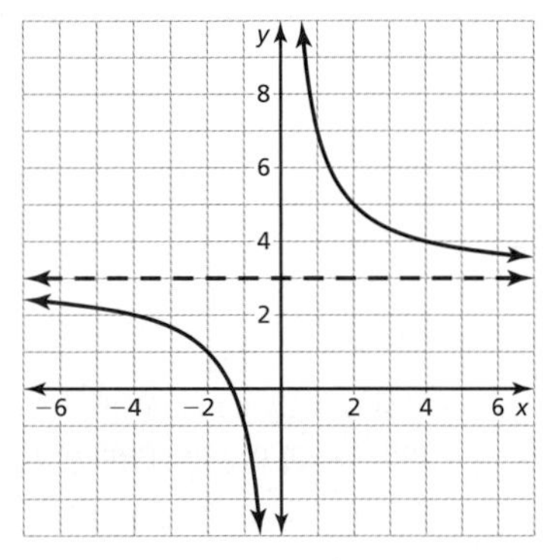

domain: all real numbers except 0; range: all real numbers except 3

13.

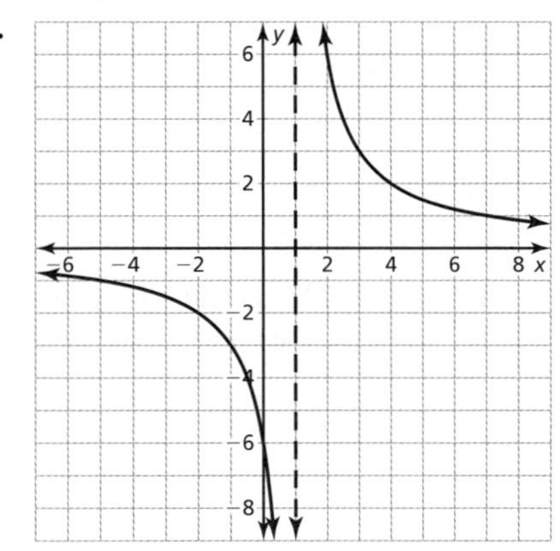

domain: all real numbers except 1; range: all real numbers except 0

15.

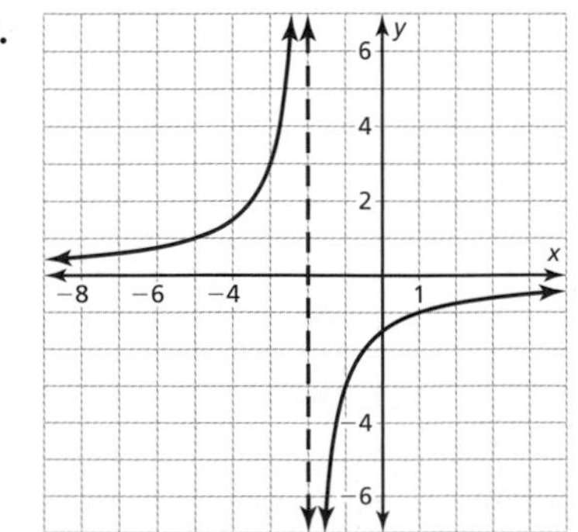

domain: all real numbers except -2; range: all real numbers except 0

17.

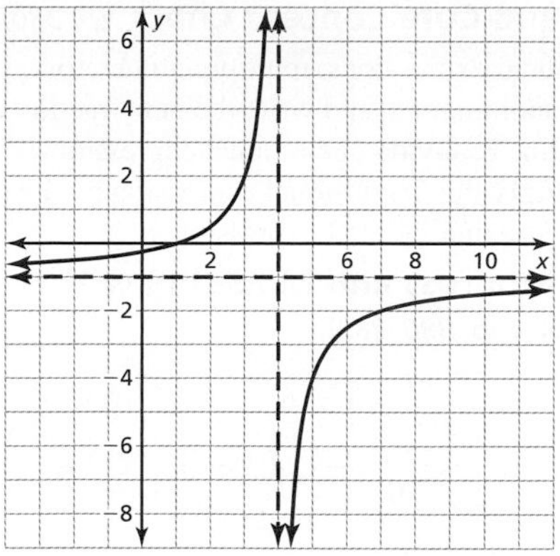

domain: all real numbers except 4; range: all real numbers except -1

19.

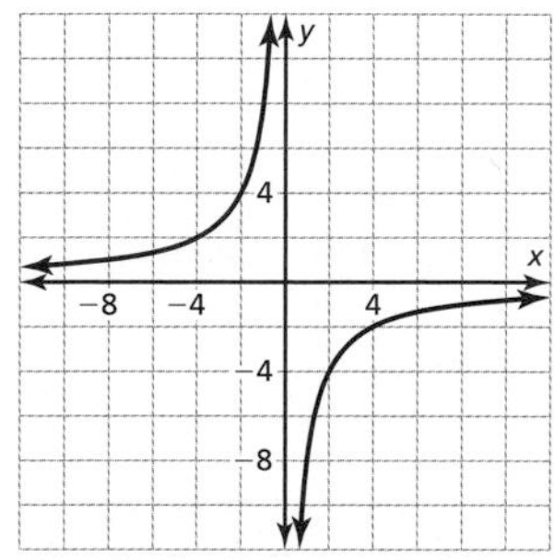

The graph should lie in the second and fourth quadrants instead of the first and third quadrants.

21. A; The asymptotes are $x = 3$ and $y = 1$.

23. B; The asymptotes are $x = 3$ and $y = -1$.

25.

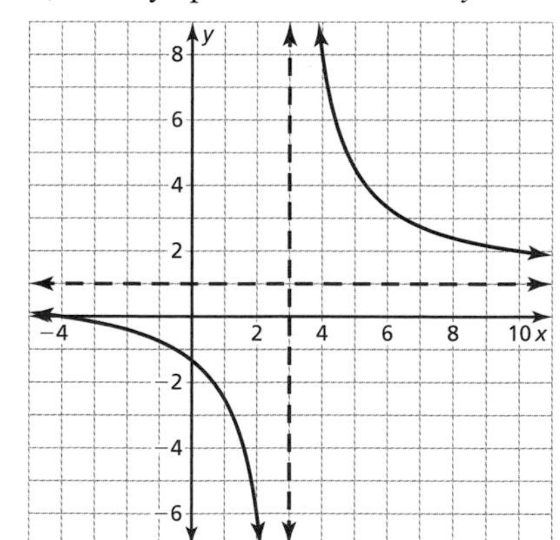

domain: all real numbers except 3; range: all real numbers except 1

27.

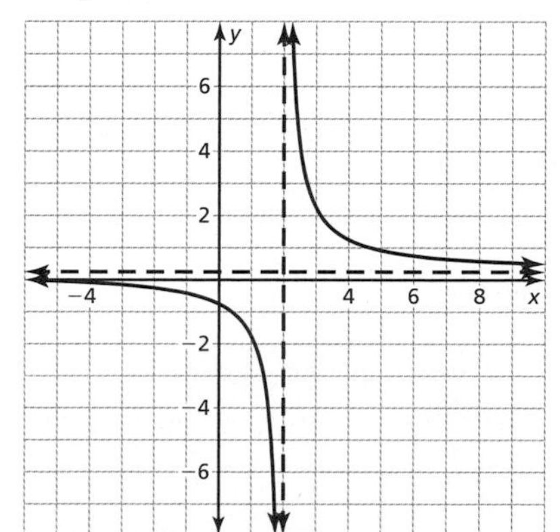

domain: all real numbers except 2; range: all real numbers except $\frac{1}{4}$

29.

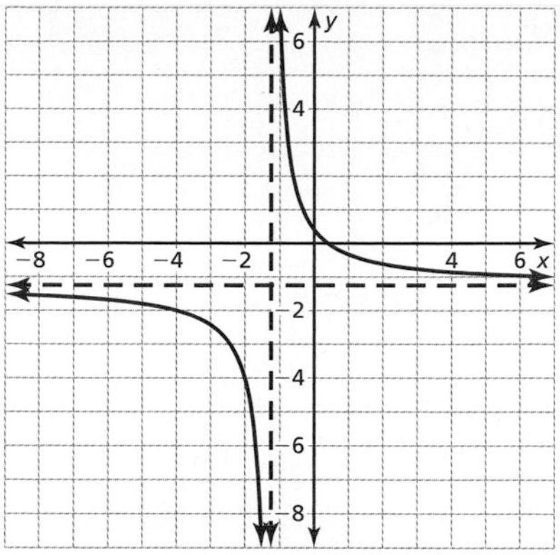

domain: all real numbers except $-\frac{5}{4}$; range: all real numbers except $-\frac{5}{4}$

31.

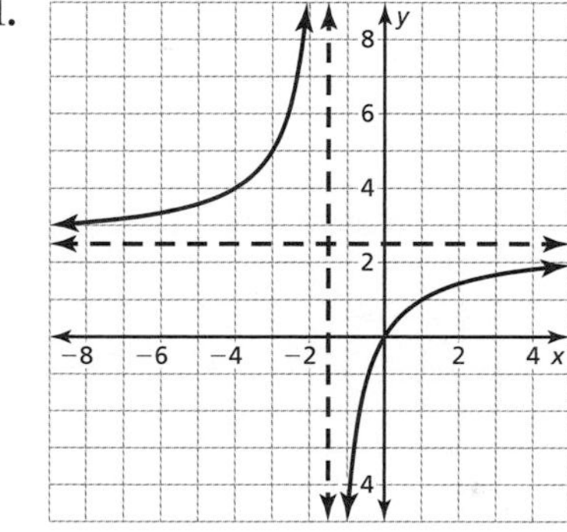

domain: all real numbers except $-\frac{3}{2}$; range: all real numbers except $\frac{5}{2}$

33. $g(x) = \dfrac{1}{x + 1} + 5$

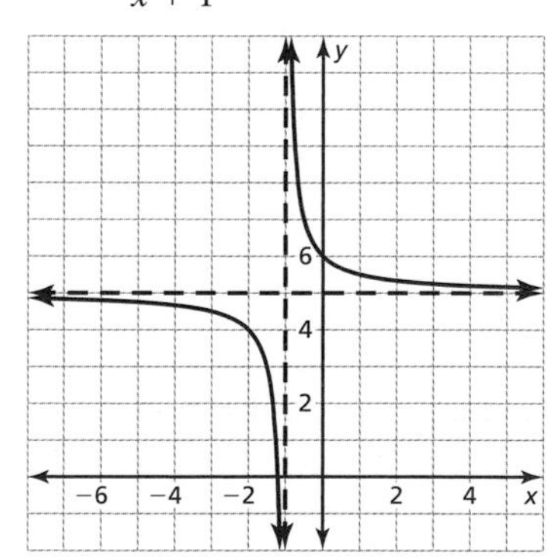

translation 1 unit left and 5 units up

35. $g(x) = \dfrac{6}{x - 5} + 2$

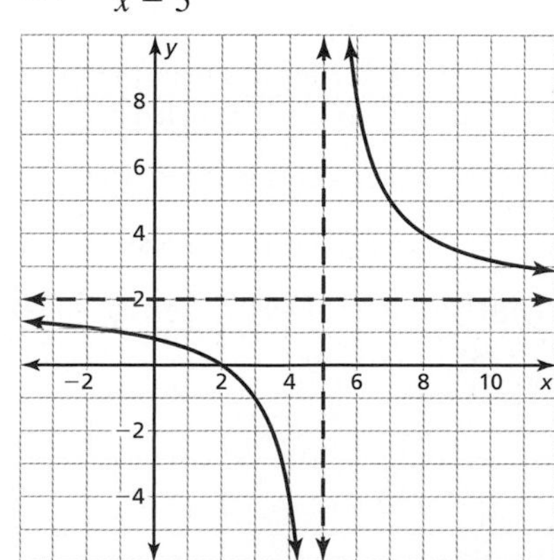

translation 5 units right and 2 units up

37. $g(x) = \dfrac{24}{x-6} + 1$

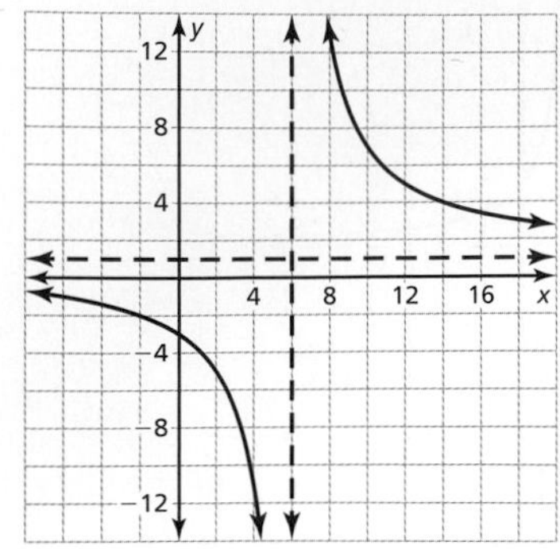

translation 6 units right and 1 unit up

39. $g(x) = \dfrac{-111}{x+13} + 7$

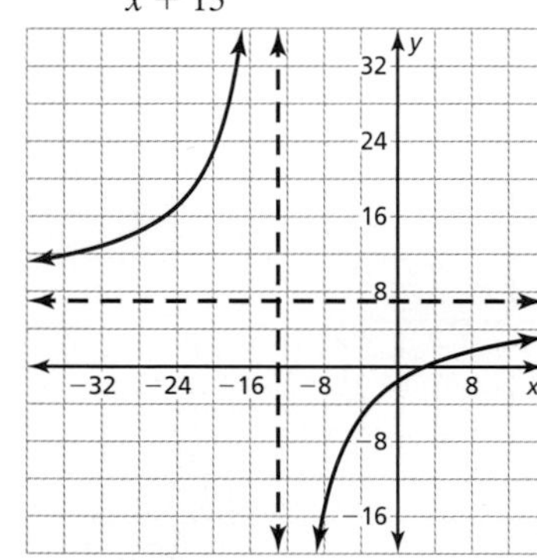

translation 13 units left and 7 units up

41. **a.** 50 students

b. The average cost per student approaches \$20.

43. B **45.** **a.** about 23°C **b.** -0.005 sec/°C

47.

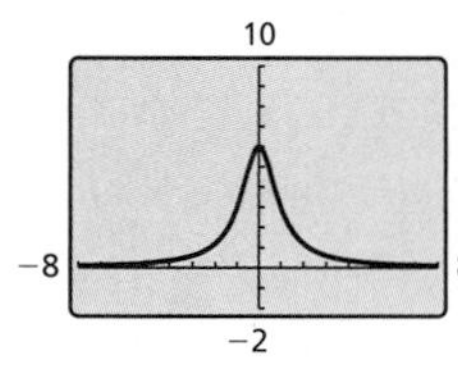

even

49.

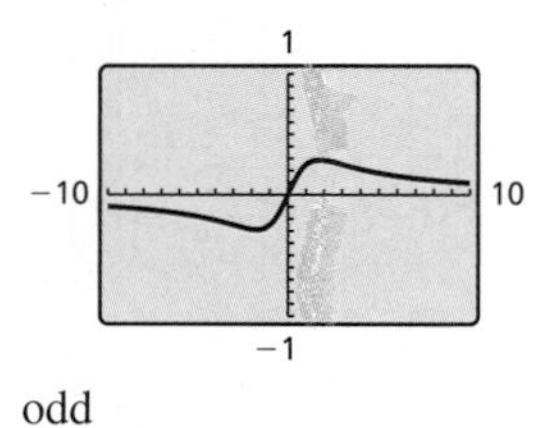

odd

51. yes; A rational function can have more than one vertical asymptote when the denominator is zero for more than one value of x, such as $y = \dfrac{3}{(x+1)(x-1)}$.

53. $y = x$, $y = -x$; The function and its inverse are the same.

55. (4, 3); The point (2, 1) is one unit left and one unit down from (3, 2), so a point on the other branch is one unit right and one unit up from (3, 2).

57. The competitor is a better choice for less than 18 months of service; The cost of Internet service is modeled by $C = \dfrac{50 + 43x}{x}$. The competitor's cost is lesser when $x = 6$ and $x = 12$, and greater when $x = 18$ and $x = 24$.

7.2 Maintaining Mathematical Proficiency *(p. 372)*

59. $4(x - 5)(x + 4)$ **61.** $2(x - 3)(x + 2)$

63. 3^6 **65.** $6^{2/3}$

7.3 Vocabulary and Core Concept Check *(p. 380)*

1. To multiply rational expressions, multiply numerators, then multiply denominators, and write the new fraction in simplified form. To divide one rational expression by another, multiply the first rational expression by the reciprocal of the second rational expression.

7.3 Monitoring Progress and Modeling with Mathematics *(pp. 380–382)*

3. $\dfrac{2x}{3x-4}, x \neq 0$ **5.** $\dfrac{x+3}{x-1}, x \neq 6$

7. $\dfrac{x+9}{x^2-2x+4}, x \neq -2$ **9.** $\dfrac{2(4x^2+5)}{x-3}, x \neq \pm\sqrt{\dfrac{5}{4}}$

11. $\dfrac{y^3}{2x^2}, y \neq 0$ **13.** $\dfrac{(x-4)(x+6)}{x}, x \neq 3$

15. $(x - 3)(x + 3), x \neq 0, x \neq 2$ **17.** $\dfrac{2x(x+4)}{(x+2)(x-3)}, x \neq 1$

19. $\dfrac{(x+9)(x-4)^2}{(x+7)}, x \neq 7$

21. The polynomials need to be factored first, and then the common factors can divide out; $\dfrac{x+12}{x+4}$

23. B

25. The expressions have the same simplified form, but the domain of f is all real numbers except $x \neq \frac{7}{3}$, and the domain of g is all real numbers.

27. $\dfrac{256x^7}{y^{14}}, x \neq 0$ **29.** $2, x \neq -2, x \neq 0, x \neq 3$

31. $\dfrac{(x+2)}{(x+4)(x-3)}$ **33.** $\dfrac{(x+6)(x-2)}{(x+2)(x-6)}, x \neq -4, x \neq -3$

35. **a.** $\dfrac{2(r+h)}{rh}$

b. soup: about 0.784, coffee: about 0.382, paint: about 0.341

From most efficient to least efficient, paint can, coffee can, and soup can

37. $M = \dfrac{171{,}000t + 1{,}361{,}000}{(1 + 0.018t)(2.96t + 278.649)}$; \$8443

39. **a.** The population is increasing by 2,960,000 people each year.

b. The population was 278,649,000 people in 2000.

41.

x	y
−3.5	−0.1333
−3.8	−0.1282
−3.9	−0.1266
−4.1	−0.1235
−4.2	−0.1220

The graph does not have a value for y when $x = -4$ and approaches $y = -0.125$.

43. $\dfrac{4}{7x}$ **45.** $9(x + 3), x \neq -\frac{3}{2}, x \neq \frac{5}{2}, x \neq 7$

47. Galapagos: about 0.371, King: about 0.203; King; The King penguin has a smaller surface area to volume ratio, so it is better equipped to live in a colder environment.

49. $f(x) = \frac{x(x-1)}{x+2}$, $g(x) = \frac{x(x+2)}{x-1}$

7.3 Maintaining Mathematical Proficiency *(p. 382)*

51. $x = -\frac{24}{5}$ 53. $x = \frac{32}{15}$ 55. $7 \cdot 13$ 57. prime

7.4 Vocabulary and Core Concept Check *(p. 388)*

1. complex fraction

7.4 Monitoring Progress and Modeling with Mathematics *(pp. 388–390)*

3. $\frac{5}{x}$ 5. $\frac{9-2x}{x+1}$ 7. $5, x \neq -3$ 9. $3x(x-2)$

11. $2x(x-5)$ 13. $(x+5)(x-5)$

15. $(x-5)(x+8)(x-8)$

17. The LCM of $5x$ and x^2 is $5x^2$, so multiply the first term by $\frac{x}{x}$ and the second term by $\frac{5}{5}$ before adding the numerators; $\frac{2(x+10)}{5x^2}$

19. $\frac{37}{30x}$ 21. $\frac{2(x+7)}{(x+4)(x+6)}$ 23. $\frac{3(x+12)}{(x+8)(x-3)}$

25. $\frac{8x^3 - 9x^2 - 28x + 8}{x(x-4)(3x-1)}$

27. sometimes; When the denominators have no common factors, the product of the denominators is the LCD. When the denominators have common factors, use the LCM to find the LCD.

29. A

31. $g(x) = \frac{-2}{x-1} + 5$

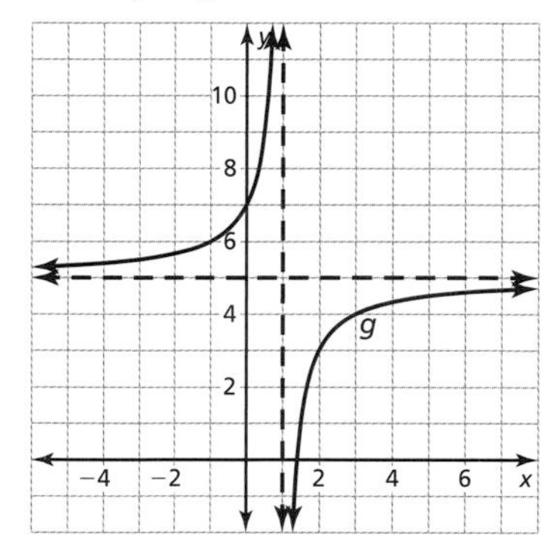

The graph of g is a translation 1 unit to the right and 5 units up of the graph of $f(x) = \frac{-2}{x}$.

33. $g(x) = \frac{60}{x-5} + 12$

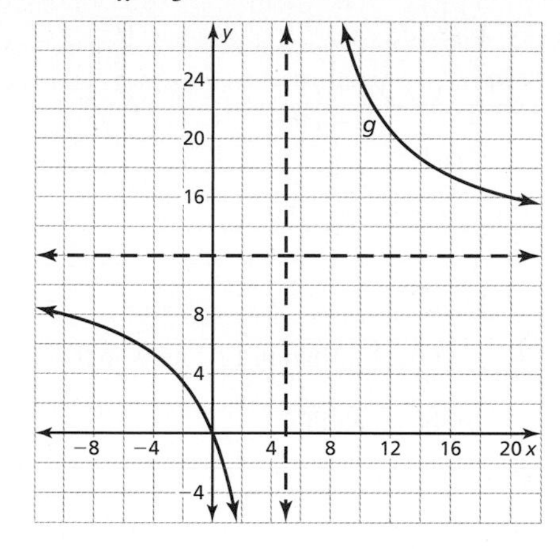

The graph of g is a translation 5 units to the right and 12 units up of the graph of $f(x) = \frac{60}{x}$.

35. $g(x) = \frac{3}{x} + 2$

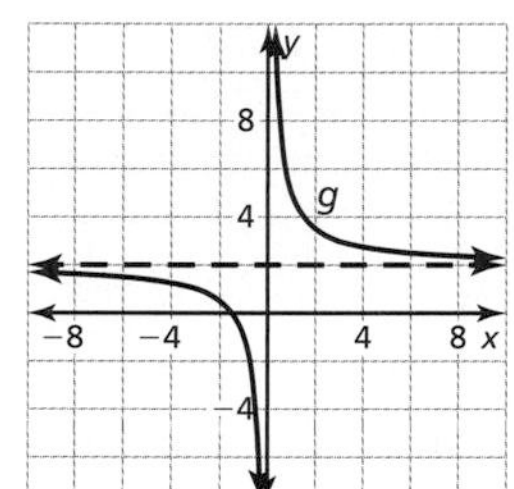

The graph of g is a translation 2 units up of the graph of $f(x) = \frac{3}{x}$.

37. $g(x) = \frac{20}{x-3} + 3$

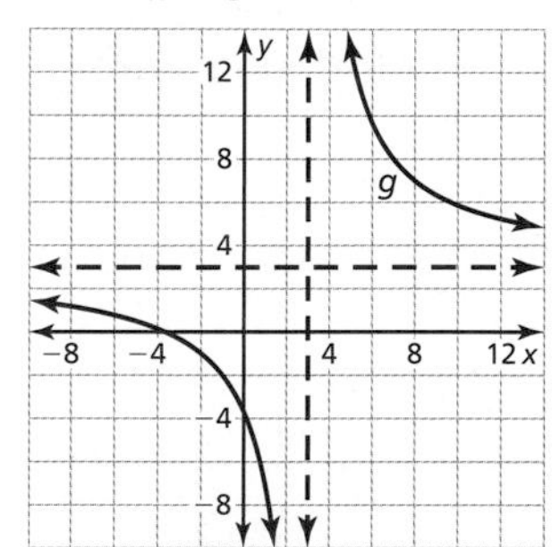

The graph of g is a translation 3 units to the right and 3 units up of the graph of $f(x) = \frac{20}{x}$.

39. $\frac{x(x-18)}{6(5x+2)}, x \neq 0$ 41. $-\frac{3}{4x}, x \neq \frac{5}{2}$

43. $\frac{x-4}{12(x-6)(x-1)}, x \neq -1, x \neq 4$

45. $T = \frac{2ad}{(a+j)(a-j)}$; about 10.2 h 47. $y = \frac{20(7x+60)}{x(x+30)}$

49. no; The LCM of 2 and 4 is 4, which is greater than one number and equal to the other number.

51. a. $M = \dfrac{Pi}{1 - \left(\dfrac{1}{1+i}\right)^{12t}}$

$= \dfrac{Pi}{1 - \dfrac{1}{(1+i)^{12t}}} \cdot \dfrac{(1+i)^{12t}}{(1+i)^{12t}}$

$= \dfrac{Pi(1+i)^{12t}}{(1+i)^{12t} - 1}$

b. \$364.02

53. $g(x) = \dfrac{2.3058}{x + 12.2} + 0.003$; translation 12.2 units to the left and 0.003 unit up of the graph of f

55. a. $R_1 = \dfrac{1}{40}, R_2 = \dfrac{1}{x}, R_3 = \dfrac{1}{x + 10}$

b. $R = \dfrac{x^2 + 90x + 400}{40x(x + 10)}$

c. about 0.0758 car/min; about 4.5 cars/h; Multiply the number of cars washed per minute by the rate 60 min/h to obtain an answer in cars per hour.

57. $1 + \cfrac{1}{2 + \cfrac{1}{2 + \cfrac{1}{2 + \cfrac{1}{2 + \cfrac{1}{2}}}}}, 1 + \cfrac{1}{2 + \cfrac{1}{2 + \cfrac{1}{2 + \cfrac{1}{2 + \cfrac{1}{2 + \cfrac{1}{2}}}}}}$

1.4, about 1.4167, about 1.4138, about 1.4143, about 1.4142; $\sqrt{2}$

7.4 Maintaining Mathematical Proficiency *(p. 390)*

59. $\left(\frac{1}{2}, -1\right)$ and $\left(\frac{9}{4}, \frac{27}{8}\right)$ **61.** no solution

7.5 Vocabulary and Core Concept Check *(p. 396)*

1. when each side of the equation is a single rational expression; *Sample answer:* The equation is a proportion.

7.5 Monitoring Progress and Modeling with Mathematics *(pp. 396–398)*

3. $x = 4$ **5.** $x = 5$ **7.** $x = -5, x = 7$

9. $x = -1, x = 0$ **11.** 26 serves **13.** 20.5 oz

15. $x(x + 3)$ **17.** $2(x + 1)(x + 4)$ **19.** $x = 2$

21. $x = \frac{7}{2}$ **23.** $x = -\frac{3}{2}, x = 2$ **25.** no solution

27. $x = -2, x = 3$ **29.** $x = \dfrac{-3 \pm \sqrt{129}}{4}$

31. Both sides of the equation should be multiplied by the same expression;

$3x^3 \cdot \dfrac{5}{3x} + 3x^3 \cdot \dfrac{2}{x^2} = 3x^3 \cdot 1$

33. a.

	Work rate	Time	Work done
You	$\dfrac{1 \text{ room}}{8 \text{ hours}}$	5 hours	$\dfrac{5}{8}$ room
Friend	$\dfrac{1 \text{ room}}{t \text{ hours}}$	5 hours	$\dfrac{5}{t}$ room

b. The sum is the amount of time it would take for you and your friend to paint the room together; $\dfrac{5}{8} + \dfrac{5}{t} = 1$, $t = 13.\overline{3}$ h = 13 h 20 min

35. *Sample answer:* $\dfrac{x+1}{x+2} = \dfrac{3}{x+4}$, Cross multiplication can be used when each side of the equation is a single rational expression; *Sample answer:* $\dfrac{x+1}{x+2} + \dfrac{3}{x+4} = \dfrac{1}{x+3}$; Multiplying by the LCD can be used when there is more than one rational expression on one side of the equation.

37. yes; $y = \dfrac{2}{x} + 4$ **39.** yes; $y = \dfrac{3}{x+2}$

41. yes; $y = \dfrac{-2}{x} + \dfrac{11}{2}$ **43.** no; $y = \pm\sqrt{\dfrac{1}{x-4}}$

45. a. about 21 mi/gal **b.** about 21 mi/gal

47. $x \approx \pm 0.8165$ **49.** $x \approx 1.3247$

51. $\dfrac{1 + \sqrt{5}}{2}$ **53.** $g(x) = \dfrac{4x + 1}{x - 3}$ **55.** $y = \dfrac{b - xd}{xc - a}$

57. a. always true; When $x = a$, the denominators of the fractions are both zero.

b. sometimes true; The equation will have exactly one solution except when $a = 3$.

c. always true; $x = a$ is an extraneous solution, so the equation has no solution.

7.5 Maintaining Mathematical Proficiency *(p. 398)*

59. discrete; The number of quarters in your pocket is an integer.

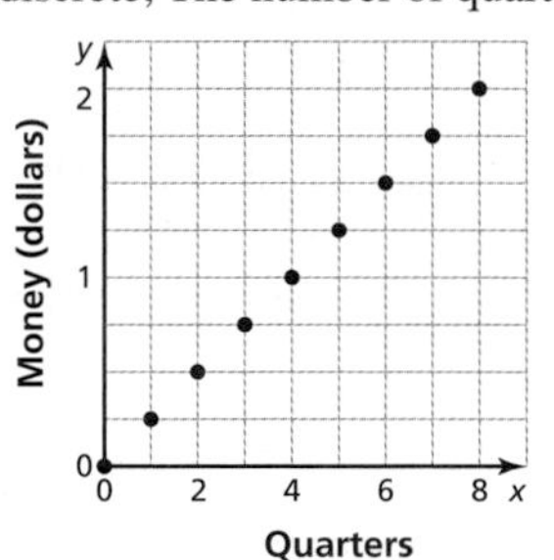

61. 3 **63.** 15

Chapter 7 Review *(pp. 400–402)*

1. inverse variation **2.** direct variation

3. direct variation **4.** neither **5.** direct variation

6. inverse variation **7.** $y = \dfrac{5}{x}$; $y = -\dfrac{5}{3}$

8. $y = \dfrac{24}{x}$; $y = -8$ **9.** $y = \dfrac{45}{x}$; $y = -15$

10. $y = \dfrac{-8}{x}$; $y = \dfrac{8}{3}$

11.

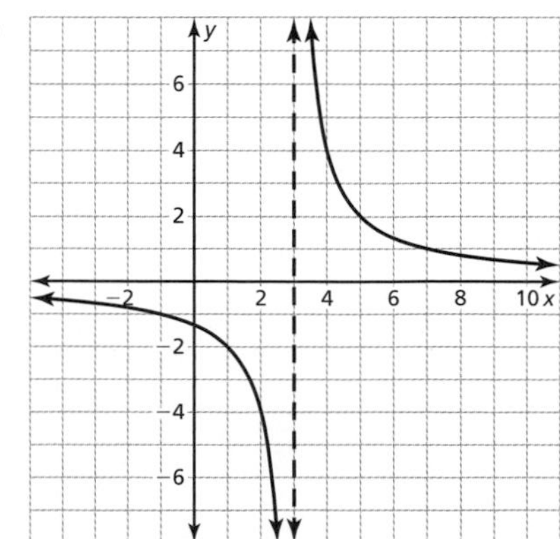

domain: all real numbers except 3; range: all real numbers except 0

12.

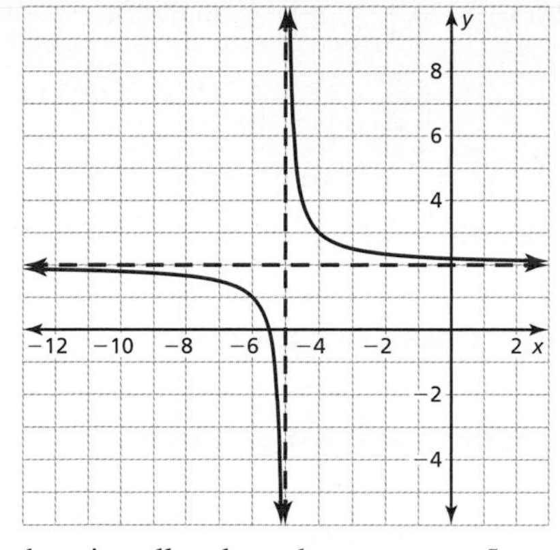

domain: all real numbers except −5; range: all real numbers except 2

13.

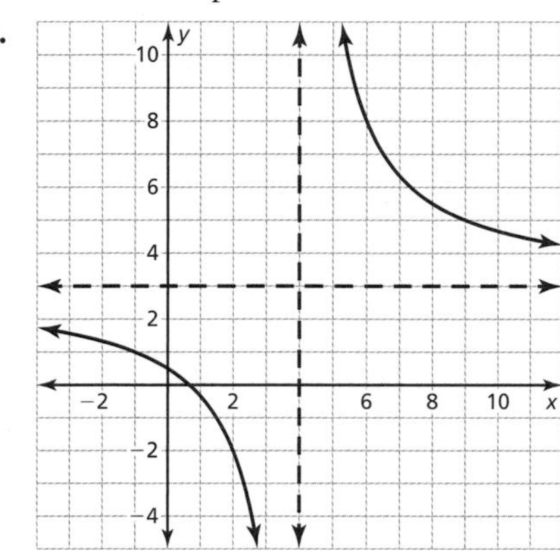

domain: all real numbers except 4; range: all real numbers except 3

14. $\dfrac{16x^3}{y^2}, x \neq 0$ **15.** $\dfrac{3(x+4)}{x+3}, x \neq 3, x \neq 4$

16. $\dfrac{3x(4x-1)}{(x-4)(x-3)}, x \neq 0, x \neq \dfrac{1}{4}$

17. $\dfrac{1}{(x+3)^2}, x \neq 5, x \neq 8$ **18.** $\dfrac{3x^2+26x+36}{6x(x+3)}$

19. $\dfrac{5x^2-11x-9}{(x+8)(x-3)}$ **20.** $\dfrac{-2(2x^2+3x+3)}{(x-3)(x+3)(x+1)}$

21. $g(x) = \dfrac{16}{x-3} + 5$

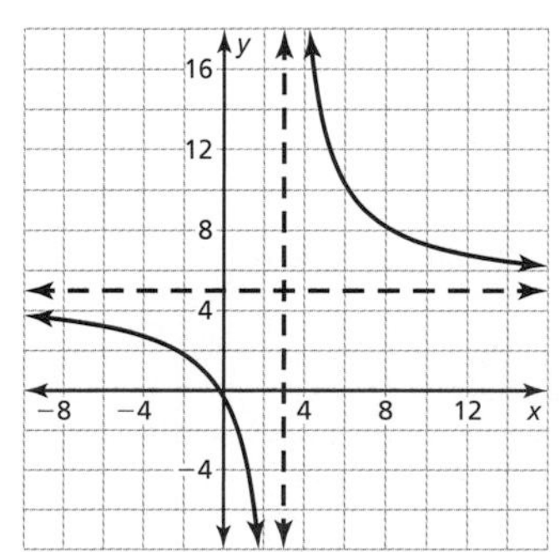

translation 3 units right and 5 units up of the graph of f

22. $g(x) = \dfrac{-26}{x+7} + 4$

translation 7 units left and 4 units up of the graph of f

23. $g(x) = \dfrac{-1}{x-1} + 9$

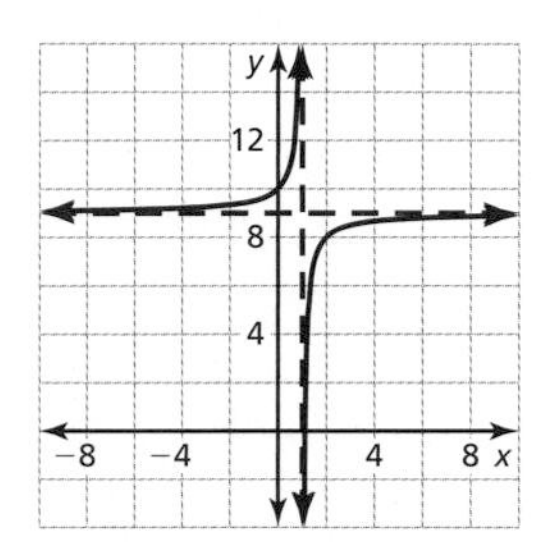

translation 1 unit right and 9 units up of the graph of f

24. $\dfrac{pq}{p+q}, p \neq 0, q \neq 0$ **25.** $x = 5$ **26.** $x = 0$

27. no solution **28.** yes; $g(x) = \dfrac{3}{x} - 6$

29. yes; $g(x) = \dfrac{10}{x} + 7$ **30.** yes; $g(x) = \dfrac{1}{x-8}$

31. **a.** 4 games **b.** 4 games

Chapter 8

Chapter 8 Maintaining Mathematical Proficiency *(p. 407)*

1.

x	y
1	1
2	−1
3	−5

2.

x	y
2	21
3	46
4	81

3.

x	y
5	4
10	−16
15	−36

4. $x = 4$ **5.** $x = 6$ **6.** $x = 66$ **7.** $x = 7$

8. $x = 100$ **9.** $x = 3$

10. *Sample answer:* The points on the scatterplot are increasing and f is decreasing; Both level off as x increases.

8.1 Vocabulary and Core Concept Check *(p. 414)*

1. sigma notation

3. A sequence is an ordered list of numbers and a series is the sum of the terms of a sequence.

8.1 Monitoring Progress and Modeling with Mathematics *(pp. 414–416)*

5. 3, 4, 5, 6, 7, 8 **7.** 1, 4, 9, 16, 25, 36

9. 1, 4, 16, 64, 256, 1024 **11.** −4, −1, 4, 11, 20, 31

13. $\frac{2}{3}, 1, \frac{6}{5}, \frac{4}{3}, \frac{10}{7}, \frac{3}{2}$

15. arithmetic; $a_5 = 5(5) - 4 = 21$; $a_n = 5n - 4$

17. arithmetic; $a_5 = 0.7(5) + 2.4 = 5.9$; $a_n = 0.7n + 2.4$

19. arithmetic; $a_5 = -1.6(6) + 7.4 = -2.2$; $a_n = -1.6n + 7.4$

21. arithmetic; $a_5 = \frac{1}{4}(5) = \frac{5}{4}$; $a_n = \frac{n}{4}$

23. $\frac{2}{3(1)}, \frac{2}{3(2)}, \frac{2}{3(3)}, \frac{2}{3(4)}$; $a_5 = \frac{2}{3(5)} = \frac{2}{15}$; $a_n = \frac{2}{3n}$

25. $(1)^3 + 1, (2)^3 + 1, (3)^3 + 1, (4)^3 + 1$; $a_5 = 5^3 + 1 = 126$; $a_n = n^3 + 1$

27. D; The number of squares in the nth figure is equal to the sum of the first positive n integers which is equal to the equation shown in D.

29. $a_n = 4n + 2$

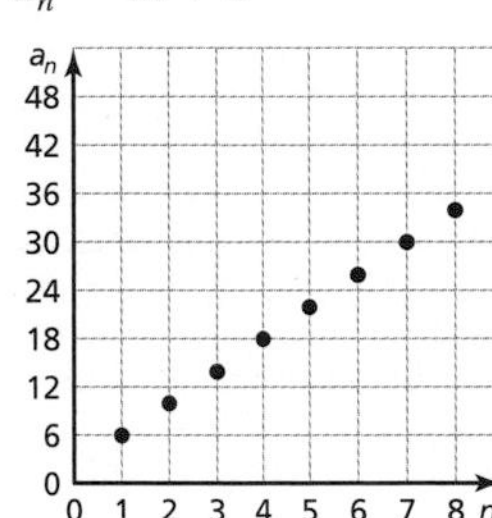

31. $\sum_{i=1}^{5} (3i + 4)$ **33.** $\sum_{i=1}^{\infty} (i^2 + 3)$ **35.** $\sum_{i=1}^{\infty} \frac{1}{3^i}$

37. $\sum_{i=1}^{5} (-1)^i(i + 2)$ **39.** 42 **41.** 100 **43.** 82

45. $\frac{481}{140}$ **47.** 35 **49.** 280

51. There should be ten terms in the series;

$$\sum_{n=1}^{10} (3n - 5) = -2 + 1 + 4 + 7 + 10 + 13 + 16 + 19 + 22 + 25 = 115$$

53. **a.** \$50.50 **b.** 316 days

55. $a_n = \frac{1}{2}(n)(n + 1)$

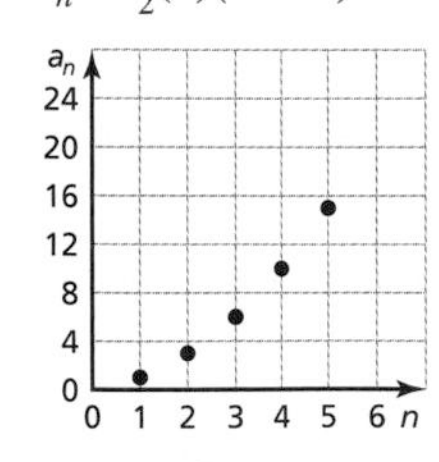

57. yes; Subtract 3 from the sum.

59. **a.** true;

$$\sum_{i=1}^{n} ca_i = ca_1 + ca_2 + ca_3 + \cdots + ca_n = c(a_1 + a_2 + a_3 + \cdots + a_n) = c\sum_{i=1}^{n} a_i$$

b. true;

$$\sum_{i=1}^{n} (a_i + b_i) = (a_1 + b_1) + (a_2 + b_2) + \cdots + (a_n + b_n) = a_1 + a_2 + \cdots + a_n + b_1 + b_2 + \cdots + b_n = \sum_{i=1}^{n} a_i + \sum_{i=1}^{n} b_i$$

c. false; $\sum_{i=1}^{2} (2i)(3i) = 30$, $\left(\sum_{i=1}^{2} 2i\right)\left(\sum_{i=1}^{2} 3i\right) = 54$

d. false; $\sum_{i=1}^{2} (2i)^2 = 20$, $\left(\sum_{i=1}^{2} 2i\right)^2 = 36$

61. **a.** $a_n = 2^n - 1$ **b.** 63; 127; 255

8.1 Maintaining Mathematical Proficiency *(p. 416)*

63. (3, 1, 1)

8.2 Vocabulary and Core Concept Check *(p. 422)*

1. common difference

8.2 Monitoring Progess and Modeling with Mathematics *(pp. 422–424)*

3. arithmetic; The common difference is −2.

5. not arithmetic; The differences are not constant.

7. not arithmetic; The differences are not constant.

9. arithmetic; The common difference is $\frac{1}{4}$.

11. **a.** $a_n = -6n + 3$ **b.** $a_n = 5n + 2$

13. $a_n = 8n + 4$; 164 **15.** $a_n = -3n + 54$; −6

17. $a_n = \frac{2}{3}n - \frac{5}{3}$; $\frac{35}{3}$ **19.** $a_n = -0.8n + 3.1$; −12.9

21. The formula should be $a_n = a_1 + (n - 1)d$; $a_n = 35 - 13n$

23. $a_n = 5n - 12$ **25.** $a_n = -2n + 13$

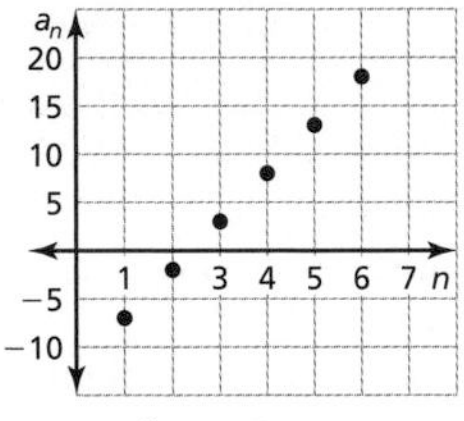

27. $a_n = -\frac{1}{2}n + \frac{7}{2}$

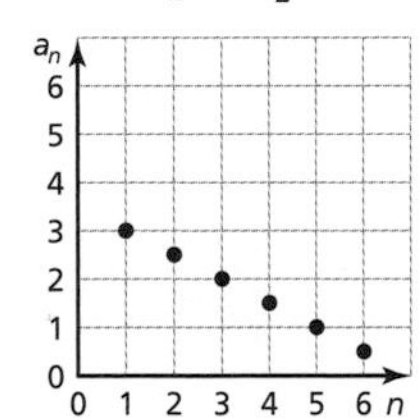

29. C **31.** $a_n = 11n - 14$ **33.** $a_n = -6n + 28$

35. $a_n = -4n + 13$ **37.** $a_n = \frac{5}{4}n + 2$

39. $a_n = -3n + 12$ **41.** $a_n = 3n - 7$

43. $a_n = 4n + 9$

45. The graph of a_n consists of discrete points and the graph of f consists of a continuous line.

47. 360 **49.** −924 **51.** −8.2 **53.** −1026

55. **a.** $a_n = 2n + 1$ **b.** 63 band members

57. $1 + \sum_{i=1}^{4} 8i$; 81

59. no; Doubling the difference does not necessarily double the terms.

61. 22,500; $\sum_{i=1}^{150} (2i - 1) = 150\left(\frac{1 + 299}{2}\right)$

63. $\left(\frac{2y}{n} - x\right)$ seats

65. $\frac{7}{16}, \frac{9}{16}, \frac{11}{16}, \frac{13}{16}, \frac{15}{16}, \frac{17}{16}, \frac{19}{16}, \frac{21}{16}, \frac{23}{16}$, and $\frac{25}{16}$

8.2 Maintaining Mathematical Proficiency *(p. 424)*

67. 3^2 **69.** $5^{3/4}$

71. exponential decay

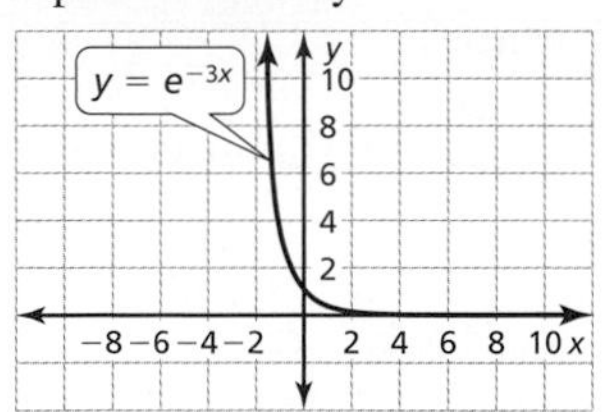

73. exponential growth

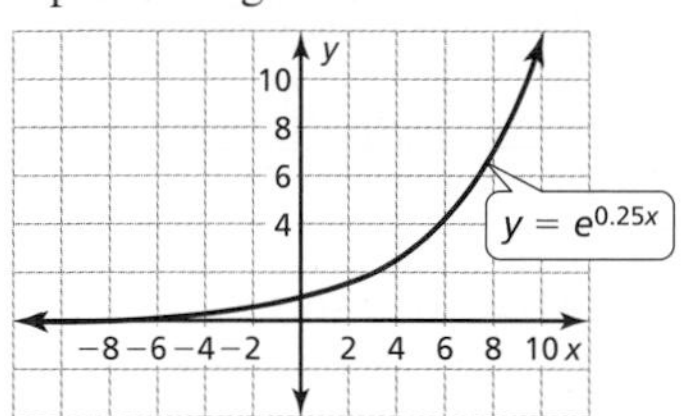

8.3 Vocabulary and Core Concept Check *(p. 430)*

1. common ratio **3.** a_1r^{n-1}

8.3 Monitoring Progress and Modeling with Mathematics *(pp. 430–432)*

5. geometric; The common ratio is $\frac{1}{2}$.

7. not geometric; The ratios are not constant.

9. not geometric; The ratios are not constant.

11. geometric; The common ratio is $\frac{1}{3}$.

13. a. $a_n = -3(5)^{n-1}$ **b.** $a_n = 72\left(\frac{1}{3}\right)^{n-1}$

15. $a_n = 4(5)^{n-1}$; $a_7 = 62{,}500$

17. $a_n = 112\left(\frac{1}{2}\right)^{n-1}$; $a_7 = \frac{7}{4}$ **19.** $a_n = 4\left(\frac{3}{2}\right)^{n-1}$; $a_7 = \frac{729}{16}$

21. $a_n = 1.3(-3)^{n-1}$; $a_7 = 947.7$

23. $a_n = 2^{n-1}$ **25.** $a_n = 60\left(\frac{1}{2}\right)^{n-1}$

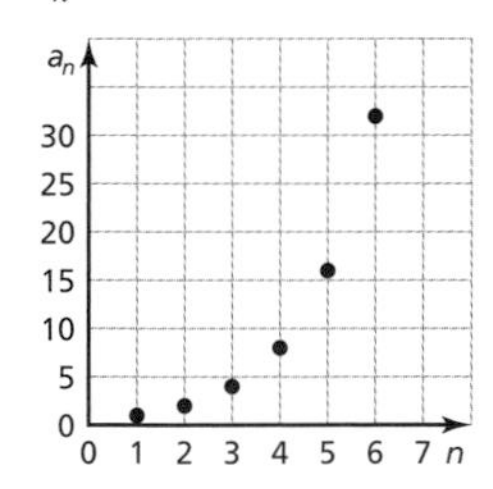

27. $a_n = -3(4)^{n-1}$ **29.** $a_n = 243\left(-\frac{1}{3}\right)^{n-1}$

31. The formula should be $a_n = a_1r^{n-1}$; $a_n = 8(6)^{n-1}$

33. $a_n = 7(4)^{n-1}$ **35.** $a_n = -6(3)^{n-1}$ or $a_n = -6(-3)^{n-1}$

37. $a_n = 512\left(\frac{1}{8}\right)^{n-1}$ or $a_n = -512\left(-\frac{1}{8}\right)^{n-1}$

39. $a_n = -432\left(\frac{1}{6}\right)^{n-1}$ or $a_n = 432\left(-\frac{1}{6}\right)^{n-1}$

41. $a_n = 4(2)^{n-1}$ **43.** $a_n = 5\left(\frac{1}{2}\right)^{n-1}$

45. $a_n = 6(-2)^{n-1}$ **47.** 40,353,606 **49.** $\frac{989{,}527}{65{,}536}$

51. $\frac{32{,}312}{6561}$ **53.** $-262{,}140$

55. The graph of a_n consists of discrete points and the graph of f is continuous.

57. $276.25

59. a. $a_n = 32\left(\frac{1}{2}\right)^{n-1}$; $1 \le n \le 6$; The number of games must be a whole number.

b. 63 games

61. a. $a_n = 8^{n-1}$; 2,396,745 squares

b. $b_n = \left(\frac{8}{9}\right)^n$; about 0.243 square units

63. $132,877.70

65. no; The total amount repaid for loan 1 is about $205,000 and the total amount repaid for loan 2 is about $284,000.

8.3 Maintaining Mathematical Proficiency *(p. 432)*

67. domain: all real numbers except 3; range: all real numbers except 0

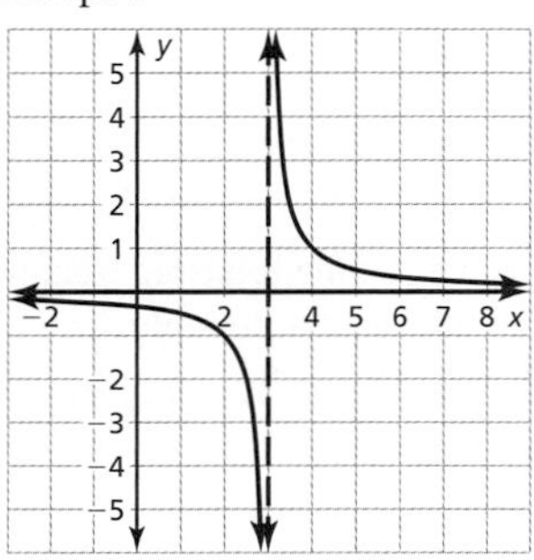

69. domain: all real numbers except 2; range: all real numbers except 1

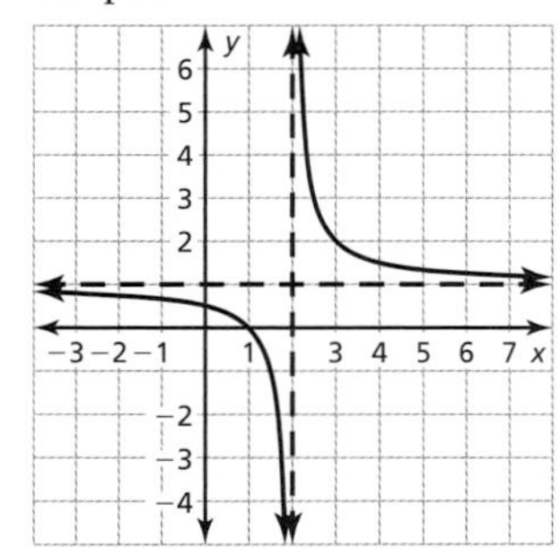

8.4 Vocabulary and Core Concept Check *(p. 439)*

1. partial sum

8.4 Monitoring Progress and Modeling with Mathematics *(pp. 439–440)*

3. $S_1 = 0.5$, $S_2 = 0.67$, $S_3 \approx 0.72$, $S_4 \approx 0.74$, $S_5 \approx 0.75$; S_n appears to approach 0.75.

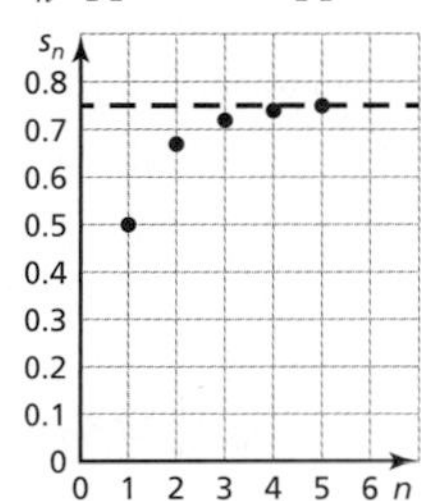

Selected Answers

5. $S_1 = 4, S_2 = 6.4, S_3 = 7.84, S_4 \approx 8.70, S_5 \approx 9.22$; S_n appears to approach 10.

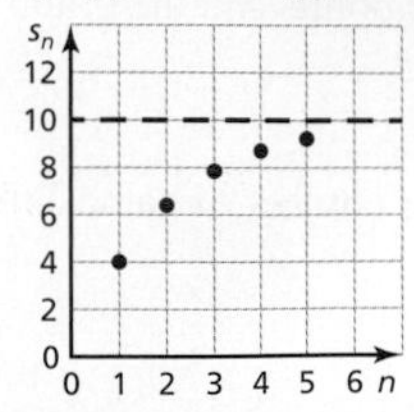

7. 10 9. $\frac{88}{15}$ 11. 8 13. 18

15. Because $\left|\frac{7}{2}\right| > 1$, the sum does not exist. 17. 56 ft

19. $\frac{2}{9}$ 21. $\frac{16}{99}$ 23. $\frac{3200}{99} = 32\frac{32}{99}$

25. *Sample answer:* $\sum_{i=1}^{\infty} 3\left(\frac{1}{2}\right)^{i-1}$; $\sum_{i=1}^{\infty} 2\left(\frac{2}{3}\right)^{i-1}$; $\frac{3}{1-\frac{1}{2}} = 6$ and $\frac{2}{1-\frac{2}{3}} = 6$

27. \$5000

29. yes; At 2 seconds, both distances are 40 feet.

31. a. $a_n = \frac{1}{4}\left(\frac{3}{4}\right)^{n-1}$
 b. 1 ft^2; As n increases, the area of the removed triangles gets closer to the area of the original triangle.

8.4 Maintaining Mathematical Proficiency *(p. 440)*

33. quadratic 35. neither

8.5 Vocabulary and Core Concept Check *(p. 447)*

1. equation

8.5 Monitoring Progress and Modeling with Mathematics *(pp. 447–450)*

3. $a_1 = 1, a_2 = 4, a_3 = 7, a_4 = 10, a_5 = 13, a_6 = 16$

5. $f(0) = 4, f(1) = 8, f(2) = 16, f(3) = 32, f(4) = 64, f(5) = 128$

7. $a_1 = 2, a_2 = 5, a_3 = 26, a_4 = 677, a_5 = 458{,}330, a_6 = 210{,}066{,}388{,}901$

9. $f(0) = 2, f(1) = 4, f(2) = 2, f(3) = -2, f(4) = -4, f(5) = -2$

11. $a_1 = 21, a_n = a_{n-1} - 7$ 13. $a_1 = 3, a_n = 4a_{n-1}$

15. $a_1 = 44, a_n = \frac{a_{n-1}}{4}$

17. $a_1 = 2, a_2 = 5, a_n = a_{n-2} \cdot a_{n-1}$

19. $a_1 = 1, a_2 = 4, a_n = a_{n-2} + a_{n-1}$

21. $a_1 = 6, a_n = n \cdot a_{n-1}$ 23. $f(1) = 1, f(n) = f(n-1) + 1$

25. $f(1) = -2, f(n) = f(n-1) + 3$

27. A recursive rule needs to include the values of the first terms; $a_1 = 5, a_2 = 2, a_n = a_{n-2} - a_{n-1}$

29. $a_1 = 7, a_n = a_{n-1} + 4$ 31. $a_1 = 2, a_n = a_{n-1} - 10$

33. $a_1 = 12, a_n = 11a_{n-1}$ 35. $a_1 = 1.9, a_n = a_{n-1} - 0.6$

37. $a_1 = -\frac{1}{2}, a_n = \frac{1}{4}a_{n-1}$ 39. $a_1 = 112, a_n = a_{n-1} + 30$

41. $a_n = -6n + 9$ 43. $a_n = -2(3)^{n-1}$

45. $a_n = 9.1n - 21.1$ 47. $a_n = -\frac{1}{3}n + \frac{16}{3}$

49. $a_n = -2n + 22$ 51. B; An explicit rule is $a_n = 6n - 2$.

53. a. $a_1 = 50{,}000, a_n = 0.8a_{n-1} + 5000$
 b. 35,240 members
 c. The number stabilizes at about 25,000 people.

55. *Sample answer:* You have saved \$100 for a vacation. Each week, you save \$5 more. $a_1 = 100, a_n = a_{n-1} + 5$

57. a. \$1612.38 b. \$91.39

59. 144 rabbits; When $n = 12$, each formula produces 144.

61. a. $a_1 = 9000, a_n = 0.9a_{n-1} + 800$
 b. The number stabilizes at 8000 trees.

63. a. 1, 2, 4, 8, 16, 32, 64; geometric
 b. $a_n = 2^{n-1}$; $a_1 = 1, a_n = 2a_{n-1}$

65. 15 months; \$213.60; $a_1 = 3000$, $a_n = \left(1 + \frac{0.1}{12}\right)a_{n-1} - 213.59$

67. a. 3, 10, 21, 36, 55 b. quadratic c. $a_n = 2n^2 + n$

69. a. $T_n = \frac{1}{2}n^2 + \frac{1}{2}n$; $S_n = n^2$
 b. $T_1 = 1, T_n = T_{n-1} + n$; $S_1 = 1, S_n = S_{n-1} + 2n - 1$
 c. $S_n = T_{n-1} + T_n$

8.5 Maintaining Mathematical Proficiency *(p. 450)*

71. $x = 25$ 73. $x = 27$

75. $y = \frac{18}{x}$; $y = \frac{9}{2}$ 77. $y = \frac{320}{x}$; $y = 80$

Chapter 8 Review *(pp. 452–454)*

1. $a_n = n^2 + n$ 2. $\sum_{i=1}^{12} (3i + 4)$ 3. $\sum_{i=0}^{\infty} (i^2 + i)$

4. −729 5. 1081 6. 650 7. 15

8. yes; The terms have a common difference of −8.

9. $a_n = 6n - 4$ 10. $a_n = 3n$

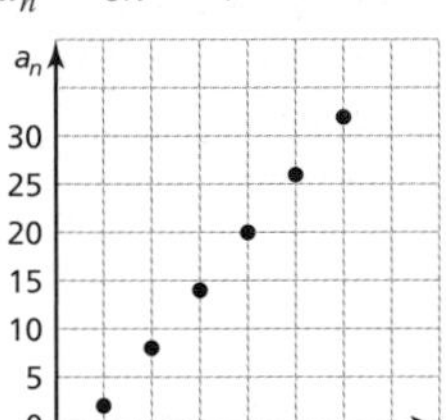

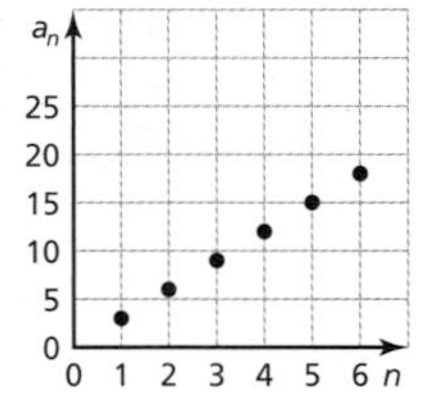

11. $a_n = -4n + 12$

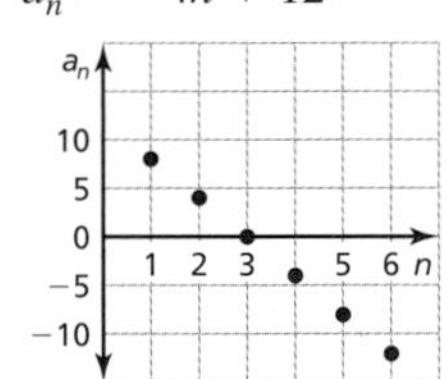

12. 2070 13. $a_n = 1500n + 35{,}500$; \$244,500

14. yes; The terms have a common ratio of 2.

15. $a_n = 25\left(\frac{2}{5}\right)^{n-1}$ 16. $a_n = 2(-3)^{n-1}$

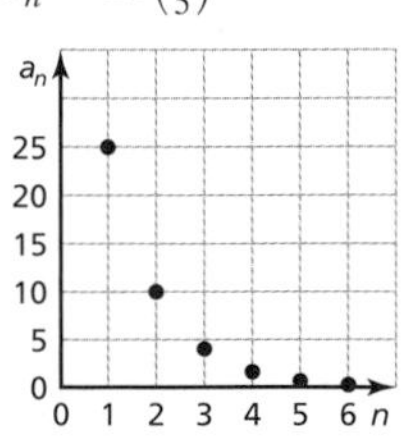

17. $a_n = 4^{n-1}$ or $a_n = (-4)^{n-1}$

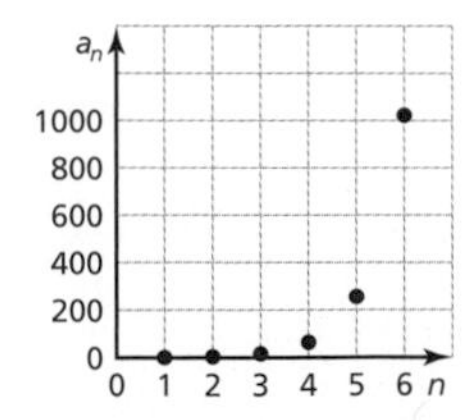

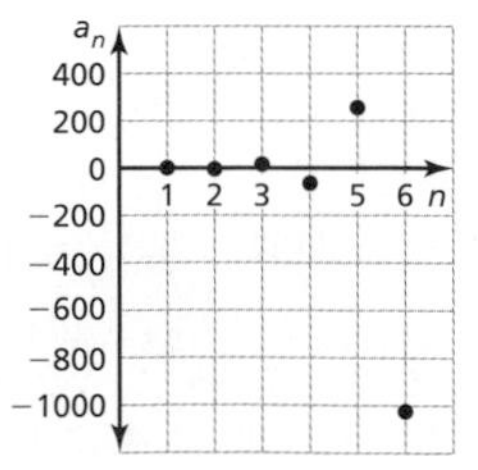

18. 855

19. $S_1 = 1, S_2 = 0.75, S_3 \approx 0.81, S_4 \approx 0.80, S_5 \approx 0.80$; S_n approaches 0.80.

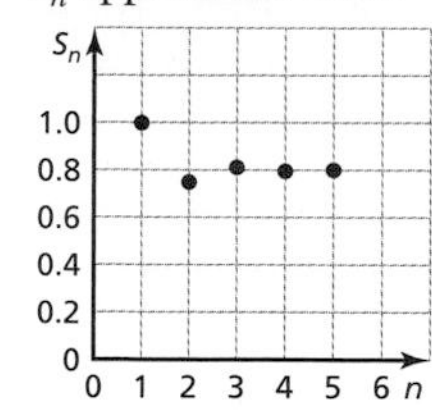

20. -1.6 21. $\frac{4}{33}$

22. $a_1 = 7, a_2 = 18, a_3 = 29, a_4 = 40, a_5 = 51, a_6 = 62$

23. $a_1 = 6, a_2 = 24, a_3 = 96, a_4 = 384, a_5 = 1536, a_6 = 6144$

24. $f(0) = 4, f(1) = 6, f(2) = 10, f(3) = 16, f(4) = 24, f(5) = 34$

25. $a_1 = 9, a_n = \frac{2}{3}a_{n-1}$ 26. $a_1 = 2, a_n = a_{n-1}(n-1)$

27. $a_1 = 7, a_2 = 3, a_n = a_{n-2} - a_{n-1}$

28. $a_1 = 105, a_n = \frac{3}{5}a_{n-1}$ 29. $a_n = 26n - 30$

30. $a_n = 8(-5)^{n-1}$ 31. $a_n = 26\left(\frac{2}{5}\right)^{n-1}$

32. $P_1 = 11{,}120, P_n = 1.04P_{n-1}$

33. $a_1 = 1, a_n = a_{n-1} + 4n - 3$

Chapter 9

Chapter 9 Maintaining Mathematical Proficiency *(p. 459)*

1. $-|7|, |4|, |2-9|, |6+4|$ 2. $|0|, \frac{|-5|}{|2|}, |-4|, |9-3|$

3. $|9-1|, |9| + |-2| - |1|, |-2 \cdot 8|, |-8^3|$

4. $-|4^2|, |5| - |3 \cdot 2|, |-15|, |-4+20|$ 5. 13 m 6. 24 ft

7. 12 mm 8. 28 km 9. $11\frac{2}{3}$ in. 10. 0.4 yd

11. yes; The line passing through the points (x_1, y_1) and (x_2, y_1) is horizontal. The line passing through the points (x_2, y_1) and (x_2, y_2) is vertical. Horizontal and vertical lines are perpendicular, so the triangle formed by the line segments connecting (x_1, y_1), (x_2, y_1), and (x_2, y_2) contains a right angle.

9.1 Vocabulary and Core Concept Check *(p. 466)*

1. cosine and secant

3. To solve a right triangle, the missing angles and side lengths must be found.

9.1 Monitoring Progress and Modeling with Mathematics *(pp. 466–468)*

5. $\sin\theta = \frac{4}{5}, \cos\theta = \frac{3}{5}, \tan\theta = \frac{4}{3}, \csc\theta = \frac{5}{4}, \sec\theta = \frac{5}{3}, \cot\theta = \frac{3}{4}$

7. $\sin\theta = \frac{5}{7}, \cos\theta = \frac{2\sqrt{6}}{7}, \tan\theta = \frac{5\sqrt{6}}{12}, \csc\theta = \frac{7}{5}, \sec\theta = \frac{7\sqrt{6}}{12}, \cot\theta = \frac{2\sqrt{6}}{5}$

9. $\sin\theta = \frac{2\sqrt{14}}{9}, \cos\theta = \frac{5}{9}, \tan\theta = \frac{2\sqrt{14}}{5}, \csc\theta = \frac{9\sqrt{14}}{28}, \sec\theta = \frac{9}{5}, \cot\theta = \frac{5\sqrt{14}}{28}$

11. $\sin\theta = \frac{4\sqrt{97}}{97}, \cos\theta = \frac{9\sqrt{97}}{97}, \csc\theta = \frac{\sqrt{97}}{4}, \cot\theta = \frac{9}{4}$

13. $\cos\theta = \frac{6\sqrt{2}}{11}, \tan\theta = \frac{7\sqrt{2}}{12}, \csc\theta = \frac{11}{7}, \sec\theta = \frac{11\sqrt{2}}{12}, \cot\theta = \frac{6\sqrt{2}}{7}$

15. $\sin\theta = \frac{7\sqrt{85}}{85}, \cos\theta = \frac{6\sqrt{85}}{85}, \csc\theta = \frac{\sqrt{85}}{7}, \sec\theta = \frac{\sqrt{85}}{6}, \cot\theta = \frac{6}{7}$

17. $\sin\theta = \frac{\sqrt{115}}{14}, \cos\theta = \frac{9}{14}, \tan\theta = \frac{\sqrt{115}}{9}, \csc\theta = \frac{14\sqrt{115}}{115}, \cot\theta = \frac{9\sqrt{115}}{115}$

19. The adjacent side was used instead of the opposite; $\sin\theta = \frac{\text{opp}}{\text{hyp}} = \frac{8}{17}$

21. $x = 4.5$ 23. $x = 6$ 25. $x = 8$

27. 0.9703 29. 1.1666 31. 9.5144

33. $A = 54°, b \approx 16.71, c \approx 28.43$

35. $B = 35°, b \approx 11.90, c \approx 20.75$

37. $B = 47°, a \approx 28.91, c \approx 42.39$

39. $A = 18°, a \approx 3.96, b \approx 12.17$ 41. $w \approx 514$ m

43. about 427 m 45. **a.** about 451 ft **b.** about 5731 ft

47. **a.** about 22,818 mi **b.** about 7263 mi

49. **a.** about 59,155 ft **b.** about 53,613 ft
c. about 39,688 ft; Use the tangent function to find the horizontal distance, $x + y$, from the airplane to the second town to be about 93,301 ft. Subtract 53,613 ft to find the distance between the two towns.

51. yes; The triangle must be a 45-45-90 triangle because both acute angles would be the same and have the same cosine value.

53. **a.** $x = 0.5$; 6 units
b. *Sample answer:* Each side is part of two right triangles, with opposing angles $\left(\frac{180°}{n}\right)$. So, each side length is $2\sin\left(\frac{180°}{n}\right)$, and there are n sides.
c. $n \cdot \sin\left(\frac{180°}{n}\right)$; about 3.14

9.1 Maintaining Mathematical Proficiency *(p. 468)*

55. 1.5 gal 57. $C \approx 37.7$ cm, $A \approx 113.1$ cm^2

59. $C \approx 44.0$ ft, $A \approx 153.9$ ft^2

9.2 Vocabulary and Core Concept Check *(p. 474)*

1. origin; initial side

3. *Sample answer:* A radian is a measure of an angle that is approximately equal to 57.3° and there are 2π radians in a circle.

9.2 Monitoring Progress and Modeling with Mathematics *(pp. 474–476)*

5.

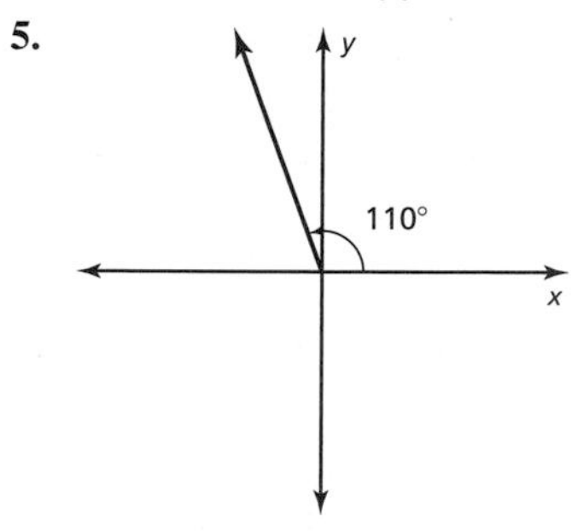

7.

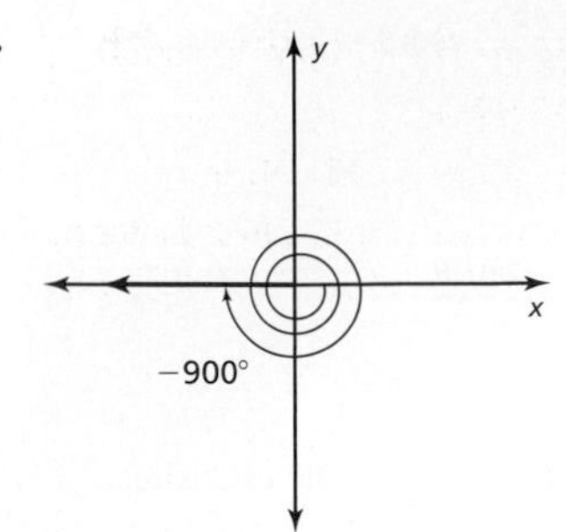

9. 430°; −290° **11.** 235°; −485° **13.** $\frac{2\pi}{9}$

15. $-\frac{13\pi}{9}$ **17.** 20° **19.** about −286.5°

21. A full revolution is 360° or 2π radians. The terminal side rotates one-sixth of a revolution from the positive x-axis, so multiply by $\frac{1}{6}$ to get $\frac{1}{6} \cdot 360° = 60°$ and $\frac{1}{6} \cdot 2\pi = \frac{\pi}{3}$.

23. B **25.** A **27.** about 15.7 yd, about 78.5 yd^2

29. The wrong conversion was used;

$$24° = 24 \text{ degrees}\left(\frac{\pi \text{ radians}}{180 \text{ degrees}}\right)$$

$$= \frac{24\pi}{180} \text{ radians} \approx 0.42 \text{ radians}$$

31. 72,000°, 400π **33.** −0.5 **35.** 3.549

37. −0.138 **39.** 528 in.^{2} **41.** 60°, $\frac{\pi}{3}$

43. about 6.89 in.^{2}, about 0.76 in.^{2}, about 0.46 in.^{2}

45. yes; When the arc length is equal to the radius, the equation $s = r\theta$ shows that $\theta = 1$ and $A = \frac{1}{2}r^2\theta$ is equivalent to $A = \frac{s^2}{2}$ for $r = s$ and $\theta = 1$.

47. a. 70°33′ **b.** 110.76°; $110 + \frac{45}{60} + \frac{30}{3600} \approx 110.76°$

9.2 Maintaining Mathematical Proficiency *(p. 476)*

49. about 27.02 **51.** about 18.03 **53.** about 18.68

9.3 Vocabulary and Core Concept Check *(p. 482)*

1. quadrantal angle

9.3 Monitoring Progress and Modeling with Mathematics *(pp. 482–484)*

3. $\sin\theta = -\frac{3}{5}$, $\cos\theta = \frac{4}{5}$, $\tan\theta = -\frac{3}{4}$, $\csc\theta = -\frac{5}{3}$, $\sec\theta = \frac{5}{4}$, $\cot\theta = -\frac{4}{3}$

5. $\sin\theta = -\frac{4}{5}$, $\cos\theta = -\frac{3}{5}$, $\tan\theta = \frac{4}{3}$, $\csc\theta = -\frac{5}{4}$, $\sec\theta = -\frac{5}{3}$, $\cot\theta = \frac{3}{4}$

7. $\sin\theta = -\frac{3}{5}$, $\cos\theta = -\frac{4}{5}$, $\tan\theta = \frac{3}{4}$, $\csc\theta = -\frac{5}{3}$, $\sec\theta = -\frac{5}{4}$, $\cot\theta = \frac{4}{3}$

9. $\sin\theta = 0$, $\cos\theta = 1$, $\tan\theta = 0$, $\csc\theta =$ undefined, $\sec\theta = 1$, $\cot\theta =$ undefined

11. $\sin\theta = 1$, $\cos\theta = 0$, $\tan\theta =$ undefined, $\csc\theta = 1$, $\sec\theta =$ undefined, $\cot\theta = 0$

13. $\sin\theta = 1$, $\cos\theta = 0$, $\tan\theta =$ undefined, $\csc\theta = 1$, $\sec\theta =$ undefined, $\cot\theta = 0$

15. ; 80°

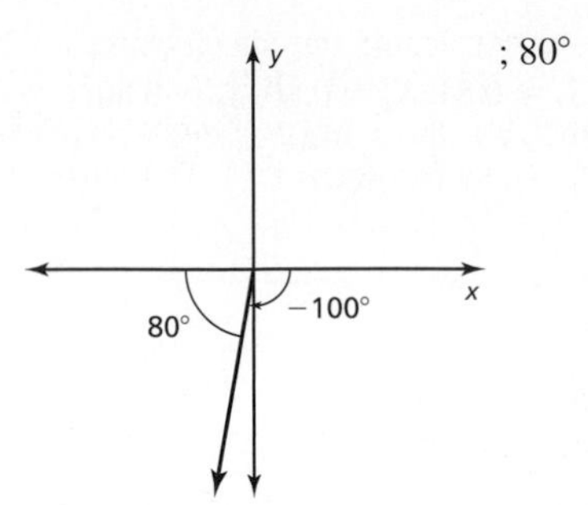

17. ; 40°

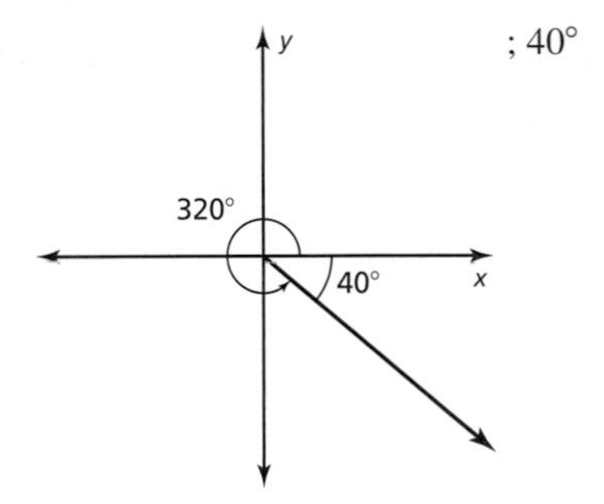

19. ; $\frac{\pi}{4}$

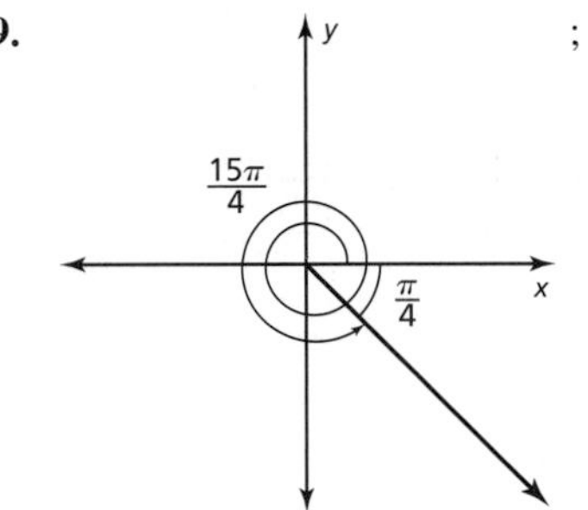

21. ; $\frac{\pi}{6}$

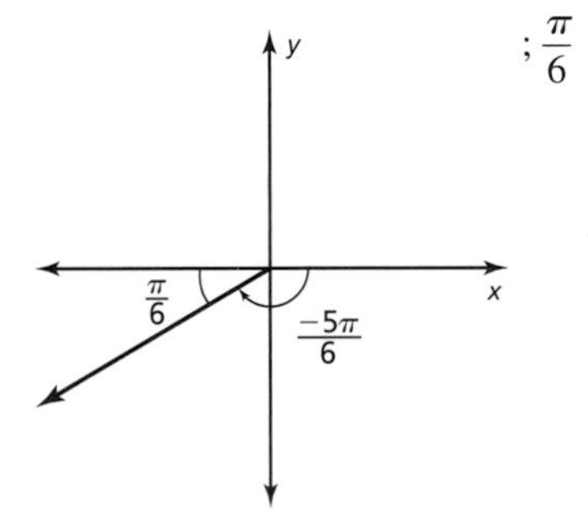

23. The equation for tangent is $\tan\theta = \frac{y}{x}$; $\tan\theta = \frac{y}{x} = -\frac{2}{3}$

25. $-\sqrt{2}$ **27.** $-\frac{1}{2}$ **29.** 1 **31.** $\frac{\sqrt{2}}{2}$ **33.** 65 ft

35. about 16.5 ft/sec **37.** about 10.7 ft

39. a.

Angle of sprinkler, θ	Horizontal distance water travels, d
30°	16.9
35°	18.4
40°	19.2
45°	19.5
50°	19.2
55°	18.4
60°	16.9

b. 45°; Because $\frac{v^2}{32}$ is constant in this situation, the maximum distance traveled will occur when $\sin 2\theta$ is as large as possible. The maximum value of $\sin 2\theta$ occurs when $2\theta = 90°$, that is, when $\theta = 45°$.

c. The distances are the same.

41.

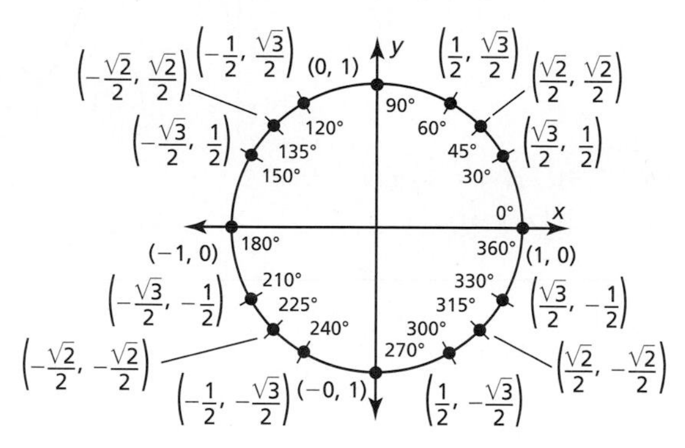

43. $\tan\theta = \frac{\sin\theta}{\cos\theta}$; $\sin 90° = 1$ and $\cos 90° = 0$, so $\tan 90°$ is undefined because you cannot divide by 0, but $\cot 90° = \frac{0}{1} = 0$.

45. $m = \tan\theta$ **47. a.** $(-58.1, 114)$ **b.** about 218 pm

9.3 Maintaining Mathematical Proficiency *(p. 484)*

49. $x = -3$ and $x = 1$

51.

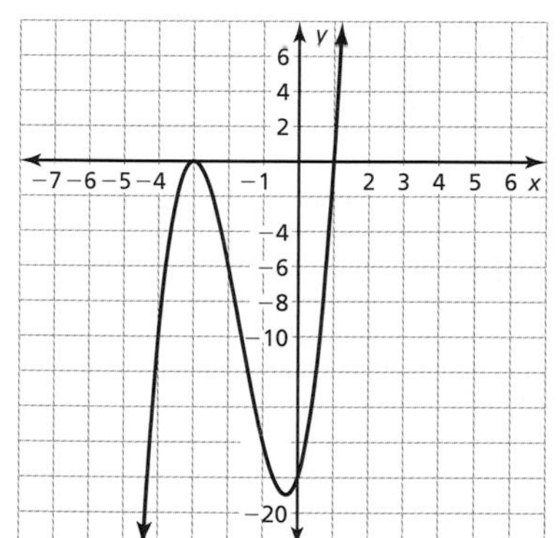

53.

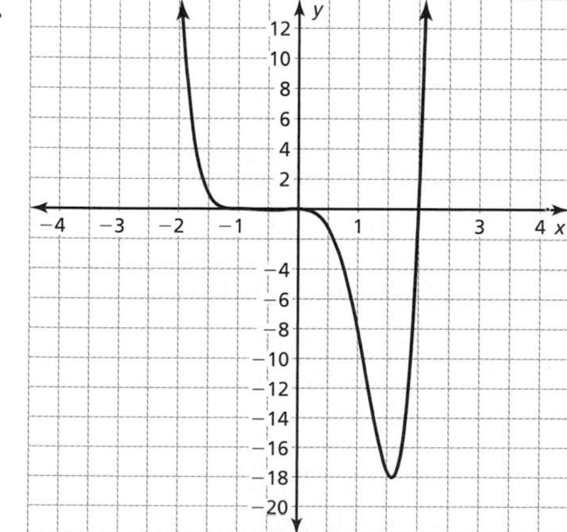

9.4 Vocabulary and Core Concept Check *(p. 491)*

1. cycle

3. A phase shift is a horizontal translation of a periodic function; *Sample answer:* $y = \sin\left(x - \frac{\pi}{2}\right)$

9.4 Monitoring Progress and Modeling with Mathematics *(pp. 491–494)*

5. yes; 2 **7.** no **9.** $1, 6\pi$ **11.** $4, \pi$

13. $3, 2\pi$; The graph of g is a vertical stretch by a factor of 3 of the graph of $f(x) = \sin x$.

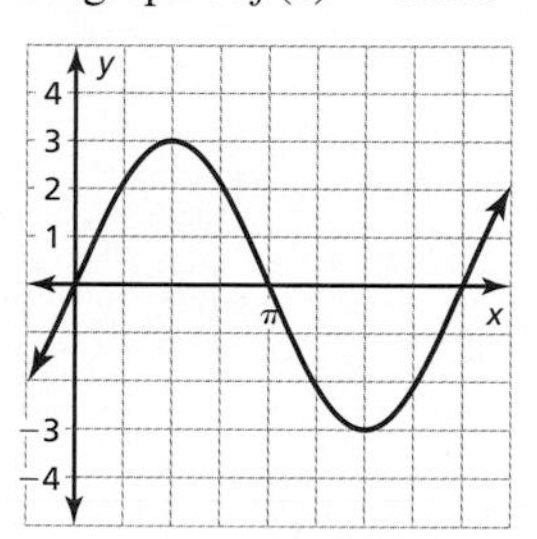

15. $1, \frac{2\pi}{3}$; The graph of g is a horizontal shrink by a factor of $\frac{1}{3}$ of the graph of $f(x) = \cos x$.

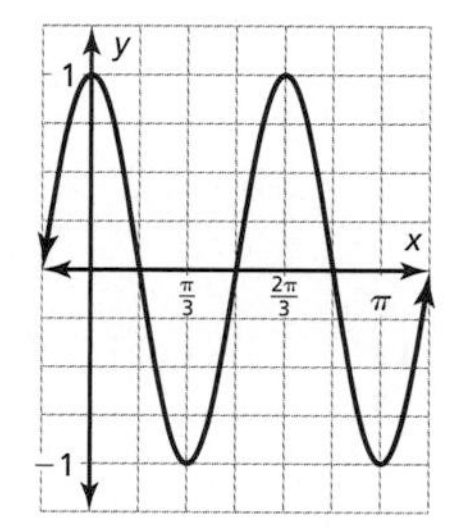

17. 1, 1; The graph of g is a horizontal shrink by a factor of $\frac{1}{2\pi}$ of the graph of $f(x) = \sin x$.

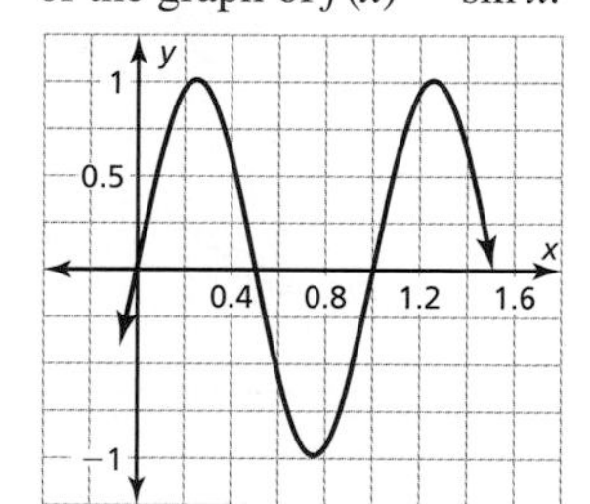

19. $\frac{1}{3}, \frac{\pi}{2}$; The graph of g is a horizontal shrink by a factor of $\frac{1}{4}$ and a vertical shrink by a factor of $\frac{1}{3}$ of the graph of $f(x) = \cos x$.

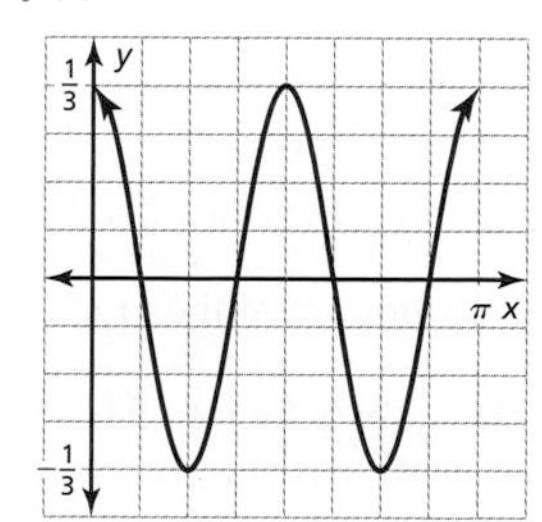

21. B, D

23. The period is $\frac{1}{4}$ and represents the amount of time, in seconds, that it takes for the pendulum to go back and forth and return to the same position. The amplitude is 4 and represents the maximum distance, in inches, the pendulum will be from its resting position.

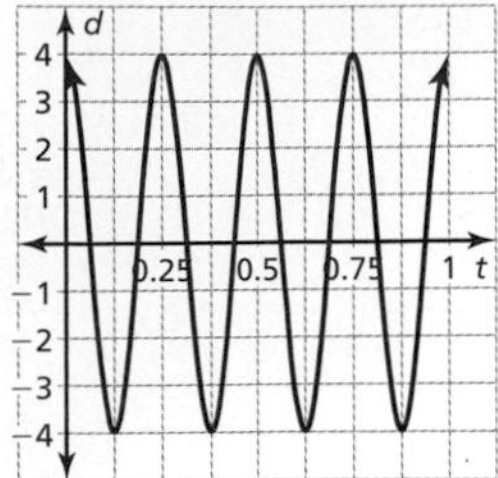

25.

27. **29.**

31. **33.**

35. To find the period, use the expression $\frac{2\pi}{|b|}$;

Period: $\frac{2\pi}{|b|} = \frac{2\pi}{\frac{2}{3}} = 3\pi$

37. The graph of g is a vertical stretch by a factor of 2 followed by a translation $\frac{\pi}{2}$ units right and 1 unit up of the graph of f.

39. The graph of g is a horizontal shrink by a factor of $\frac{1}{3}$ followed by a translation 3π units left and 5 units down of the graph of f.

41.

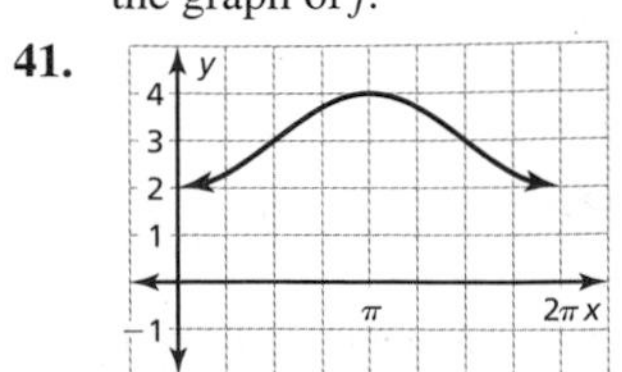

43.

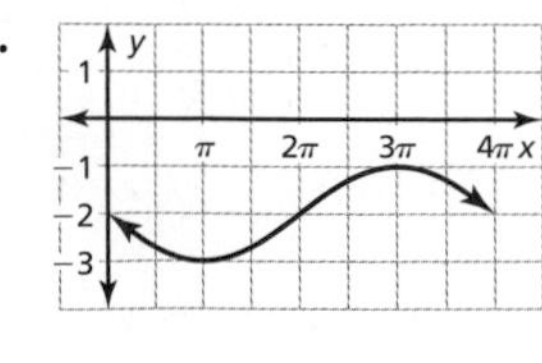

45. **47.**

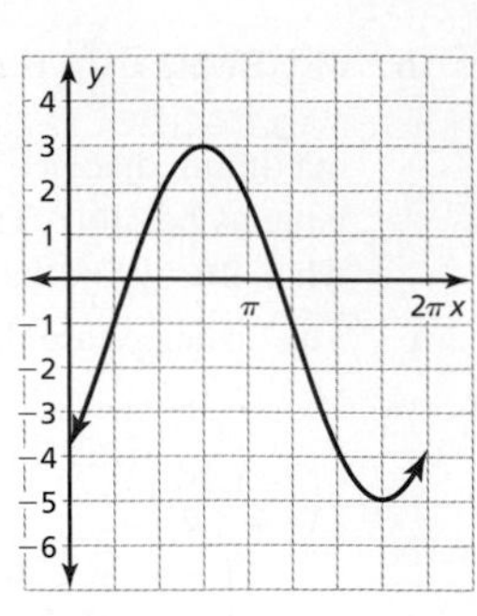

49. A **51.** $g(x) = 3 \sin(x - \pi) + 2$

53. $g(x) = -\frac{1}{3} \cos \pi x - 1$

55. ; 4.3 ft

57. days 205 and 328; When the function is graphed with the line $y = 10$, the two points of intersection are (205.5, 10) and (328.7, 10).

59. **a.** about -1.27 **b.** about 0.64 **c.** about 0.64

61. **a.**

b. 4.5 **c.** 175 ft, 5 ft

63. The x-intercepts occur when $x = \pm\frac{\pi}{4}, \pm\frac{3\pi}{4}, \pm\frac{5\pi}{4}, \ldots$.

Sample answer: The x-intercepts can be represented by the expression $(2n + 1)\frac{\pi}{4}$, where n is an integer.

65. The graph of $g(x) = \cos x$ is a translation $\frac{\pi}{2}$ units to the right of the graph of $f(x) = \sin x$.

67. 80 beats per minute

9.4 Maintaining Mathematical Proficiency *(p. 494)*

69. $x - 2, x \neq -3$ **71.** $\frac{(x - 5)(x + 1)}{(x + 5)(x - 1)}$

73. $2x(x - 5)$ **75.** $(x + 6)(x + 2)$

9.5 Vocabulary and Core Concept Check *(p. 502)*

1. The graphs of the tangent, cotangent, secant and cosecant functions have no amplitude because the ranges do not have minimum or maximum values.

3. 2π; π

9.5 Monitoring Progress and Modeling with Mathematics (pp. 502–504)

5.

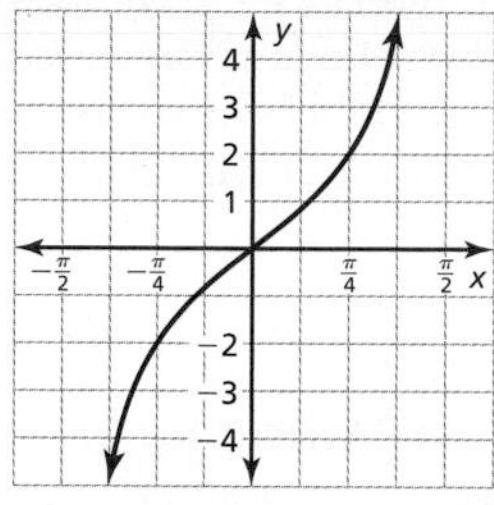

The graph of g is a vertical stretch by a factor of 2 of the graph of $f(x) = \tan x$.

7.

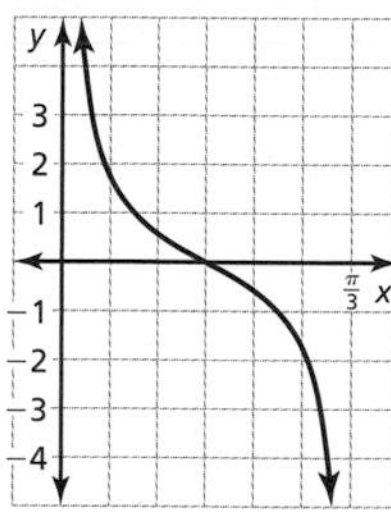

The graph of g is a horizontal shrink by a factor of $\frac{1}{3}$ of the graph of $f(x) = \cot x$.

9.

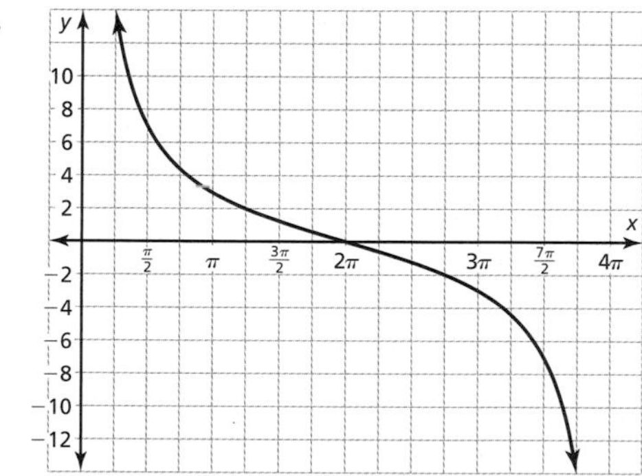

The graph of g is a horizontal stretch by a factor of 4 and a vertical stretch by a factor of 3 of the graph of $f(x) = \cot x$.

11.

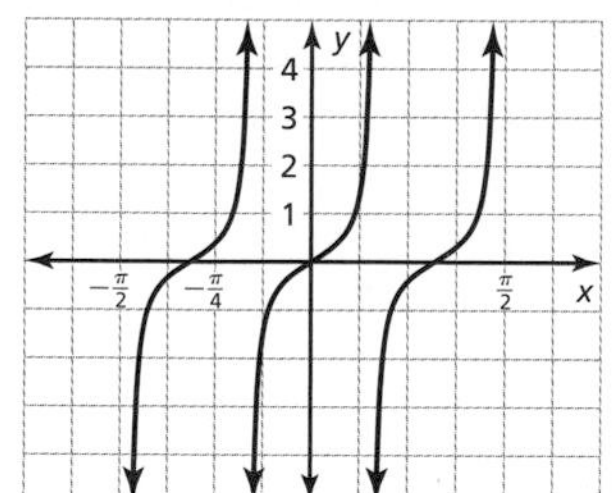

The graph of g is a horizontal shrink by a factor of $\frac{1}{\pi}$ and a vertical shrink by a factor of $\frac{1}{2}$ of the graph of $f(x) = \tan x$.

13. To find the period, use the expression $\frac{\pi}{|b|}$; Period: $\frac{\pi}{|b|} = \frac{\pi}{3}$

15. a.

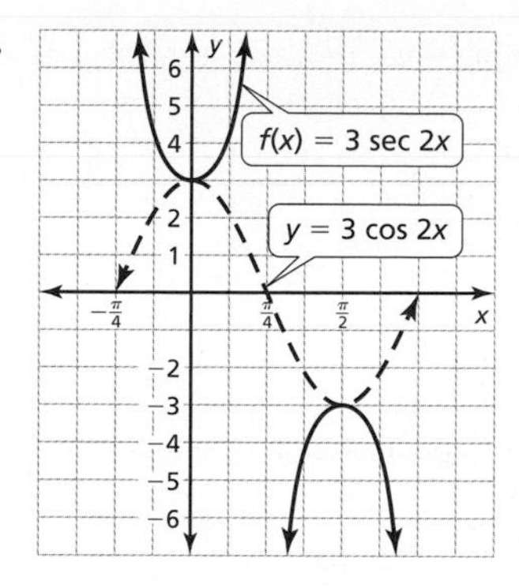

b.

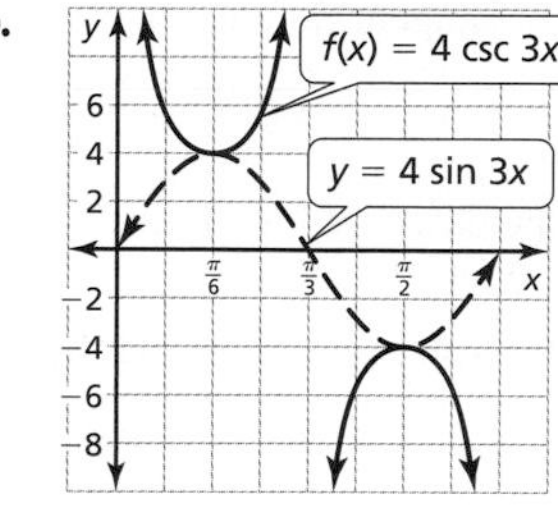

17.

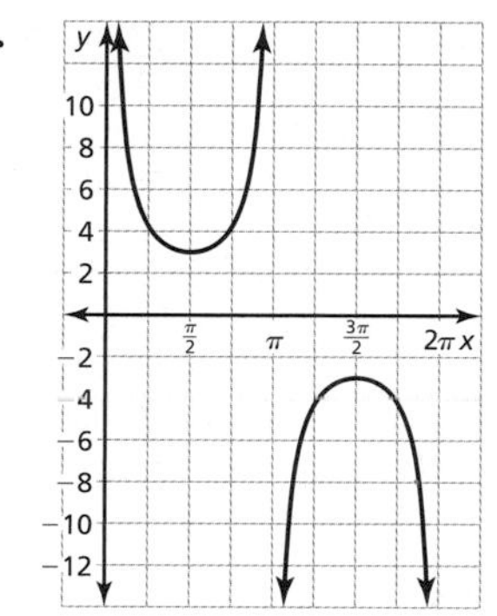

The graph of g is a vertical stretch by a factor of 3 of the graph of $f(x) = \csc x$.

19.

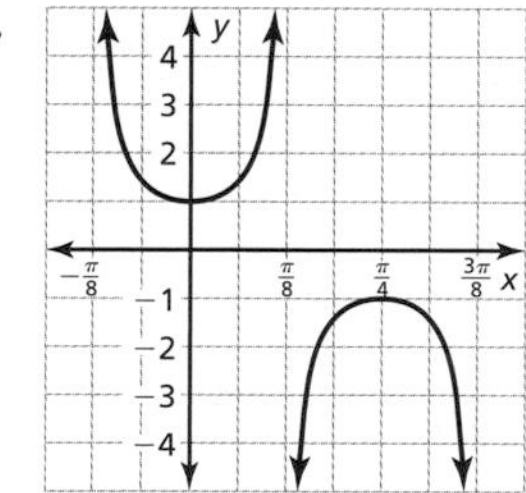

The graph of g is a horizontal shrink by a factor of $\frac{1}{4}$ of the graph of $f(x) = \sec x$.

21.

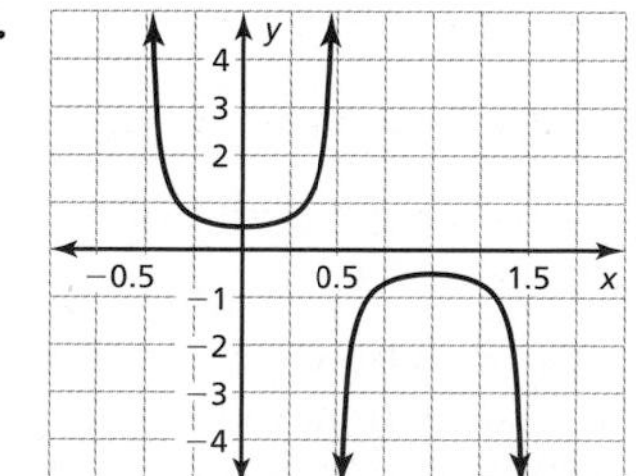

The graph of g is a horizontal shrink by a factor of $\frac{1}{\pi}$ and a vertical shrink by a factor of $\frac{1}{2}$ of the graph of $f(x) = \sec x$.

Selected Answers

23. 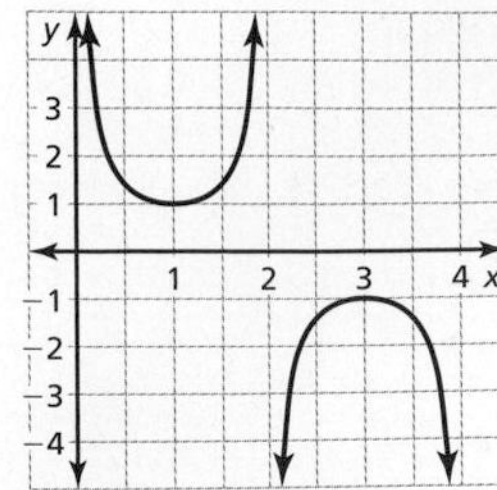

The graph of g is a horizontal stretch by a factor of $\frac{2}{\pi}$ of the graph of $f(x) = \csc x$.

25. $y = 6 \tan x$ **27.** $y = 2 \tan \pi x$

29. B; The parent function is the tangent function and the graph has an asymptote at $x = \frac{\pi}{2}$.

31. D; The parent function is the cosecant function and the graph has an asymptote at $x = 1$.

33. A; The parent function is the secant function and the graph has an asymptote at $x = \frac{\pi}{4}$.

35. The tangent function that passes through the origin and has asymptotes at $x = \pi$ and $x = -\pi$ can be stretched or shrunk vertically to create more tangent functions with the same characteristics.

37. $g(x) = \cot\left(2x + \frac{\pi}{2}\right) + 3$ **39.** $g(x) = -5 \sec(x - \pi) + 2$

41. Function B has a local maximum value of -5 so Function A's local maximum value of $-\frac{1}{4}$ is greater. Function A has a local minimum of $\frac{1}{4}$ so Function B's local minimum value of 5 is greater.

43. 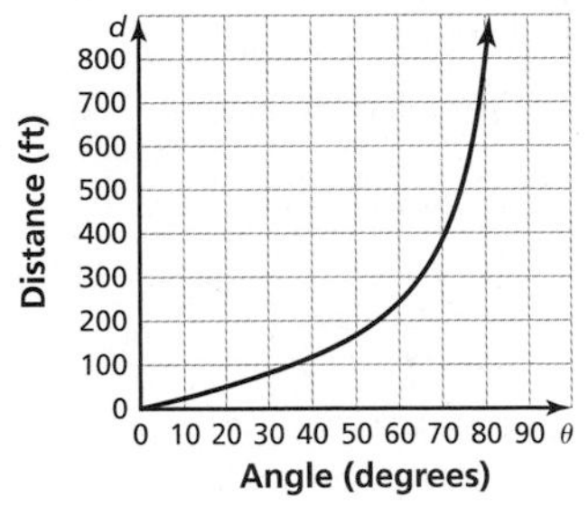

As d increases, θ increases because, as the car gets farther away, the angle required to see the car gets larger.

45. a. $d = 260 - 120 \tan \theta$

b. 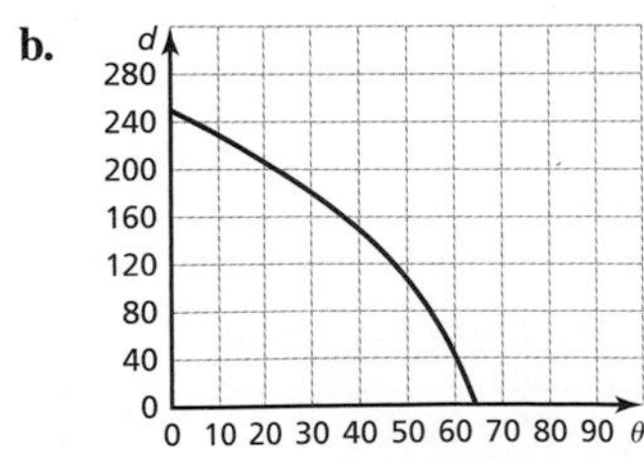

The graph shows a negative correlation meaning that as the angle gets larger, the distance from your friend to the top of the building gets smaller. As the angle gets smaller, the distance from your friend to the top of the building gets larger.

47. no; The graph of cosecant can be translated $\frac{\pi}{2}$ units right to create the same graph as $y = \sec x$.

49. $a \sec bx = \frac{a}{\cos bx}$

Because the cosine function is at most 1, $y = a \cos bx$ will produce a maximum when $\cos bx = 1$ and $y = a \sec bx$ will produce a minimum. When $\cos bx = -1$, $y = a \cos bx$ will produce a minimum and $y = a \sec bx$ will produce a maximum.

51. *Sample answer:* $y = 5 \tan\left(\frac{1}{2}x - \frac{3\pi}{4}\right)$

9.5 Maintaining Mathematical Proficiency *(p. 504)*

53. $y = -x^3 + 2x^2 + 5x - 6$ **55.** $y = \frac{1}{5}x^3 + \frac{1}{5}x^2 - \frac{9}{5}x - \frac{9}{5}$

57. $3, \pi$

9.6 Vocabulary and Core Concept Check *(p. 510)*

1. sinusoids

9.6 Monitoring Progress and Modeling with Mathematics *(pp. 510–512)*

3. $\frac{1}{2\pi}$ **5.** $\frac{2}{\pi}$ **7.** $\frac{3}{2}$ **9.** $\frac{3}{8\pi}$

11. $P = 0.02 \sin 40\pi t$

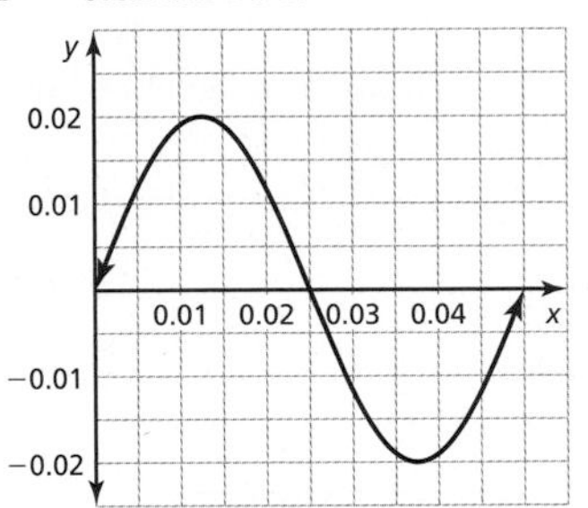

13. $y = 3 \sin 2x$ **15.** $y = -2 \cos \frac{\pi}{2}(x + 4)$

17. To find the amplitude, take half of the difference between the maximum and the minimum; $\frac{10 - (-6)}{2} = 8$

19. $h = -2.5 \cos \pi t + 6.5$

21. $D = 19.81 \sin(0.549t - 2.40) + 79.8$; The period of the graph represents the amount of time it takes for the weather to repeat its cycle, which is about 11.4 months.

23. $V = 100 \sin 4\pi t$

25. a. $N = 3.68 \sin(0.776t - 0.70) + 20.4$

b. about 23,100 employees

27. a. and b. A cosine function because it does not require determining a horizontal shift.

c. A sine function because it does not require determining a horizontal shift.

29. $y = 2.5 \sin 4\left(x - \frac{\pi}{8}\right) + 5.5$, $y = -2.5 \cos 4x + 5.5$

31. a. $d = -6.5 \cos \frac{\pi}{6}t + 10$

b. low tide: 12:00 A.M., 12:00 P.M., high tide: 6:00 A.M., 6:00 P.M.

c. It is a horizontal shift to the left by 3.

9.6 Maintaining Mathematical Proficiency *(p. 512)*

33. $\frac{6 + 3\sqrt{6}}{2}$ **35.** $\frac{13\sqrt{11} - 13\sqrt{3}}{8}$ **37.** $\ln 2 + \ln x$

39. $\ln 4 + 6 \ln x - \ln y$

9.7 Vocabulary and Core Concept Check *(p. 517)*

1. A trigonometric equation is true for some values of a variable but a trigonometric identity is true for all values of the variable for which both sides of the equation are defined.

9.7 Monitoring Progress and Modeling with Mathematics *(pp. 517–518)*

3. $\cos\theta = \frac{2\sqrt{2}}{3}$, $\tan\theta = \frac{\sqrt{2}}{4}$, $\csc\theta = 3$, $\sec\theta = \frac{3\sqrt{2}}{4}$, $\cot\theta = 2\sqrt{2}$

5. $\sin\theta = \frac{3\sqrt{58}}{58}$, $\cos\theta = -\frac{7\sqrt{58}}{58}$, $\csc\theta = \frac{\sqrt{58}}{3}$, $\sec\theta = -\frac{\sqrt{58}}{7}$, $\cot\theta = -\frac{7}{3}$

7. $\sin\theta = -\frac{\sqrt{11}}{6}$, $\tan\theta = \frac{\sqrt{11}}{5}$, $\csc\theta = -\frac{6\sqrt{11}}{11}$, $\sec\theta = -\frac{6}{5}$, $\cot\theta = \frac{5\sqrt{11}}{11}$

9. $\sin\theta = -\frac{\sqrt{10}}{10}$, $\cos\theta = \frac{3\sqrt{10}}{10}$, $\tan\theta = -\frac{1}{3}$, $\csc\theta = -\sqrt{10}$, $\sec\theta = \frac{\sqrt{10}}{3}$

11. $\cos x$ 13. $-\tan\theta$ 15. $\sin^2 x$ 17. $-\sec x$

19. 1

21. $\sin^2\theta = 1 - \cos^2\theta$;
$1 - \sin^2\theta = 1 - (1 - \cos^2\theta) = 1 - 1 + \cos^2\theta = \cos^2\theta$

23. $\sin x \csc x = \sin x \cdot \frac{1}{\sin x} = 1$

25. $\cos\left(\frac{\pi}{2} - x\right)\cot x = \sin x \cdot \frac{\cos x}{\sin x} = \cos x$

27.
$$\frac{\cos\left(\frac{\pi}{2} - \theta\right) + 1}{1 - \sin(-\theta)} = \frac{\sin\theta + 1}{1 - \sin(-\theta)}$$
$$= \frac{\sin\theta + 1}{1 - (-\sin\theta)}$$
$$= \frac{\sin\theta + 1}{1 + \sin\theta}$$
$$= 1$$

29.
$$\frac{1 + \cos x}{\sin x} + \frac{\sin x}{1 + \cos x} = \frac{1 + \cos x}{\sin x} + \frac{\sin x(1 - \cos x)}{(1 + \cos x)(1 - \cos x)}$$
$$= \frac{1 + \cos x}{\sin x} + \frac{\sin x(1 - \cos x)}{1 - \cos^2 x}$$
$$= \frac{1 + \cos x}{\sin x} + \frac{\sin x(1 - \cos x)}{\sin^2 x}$$
$$= \frac{\sin x(1 + \cos x)}{\sin^2 x} + \frac{\sin x(1 - \cos x)}{\sin^2 x}$$
$$= \frac{\sin x(1 + \cos x) + \sin x(1 - \cos x)}{\sin^2 x}$$
$$= \frac{\sin x(1 + \cos x + 1 - \cos x)}{\sin^2 x}$$
$$= \frac{\sin x(2)}{\sin^2 x}$$
$$= \frac{2}{\sin x}$$
$$= 2\csc x$$

31. $\sin x$, $\csc x$, $\tan x$, $\cot x$; $\cos x$, $\sec x$;
$\sin(-\theta) = -\sin\theta$
$\csc(-\theta) = \frac{1}{\sin(-\theta)} = -\frac{1}{\sin\theta} = -\csc\theta$
$\tan(-\theta) = -\tan\theta$
$\cot(-\theta) = \frac{1}{\tan(-\theta)} = -\frac{1}{\tan\theta} = -\cot\theta$

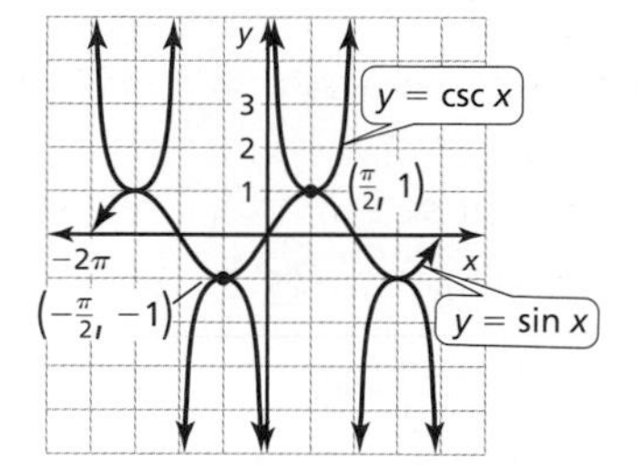

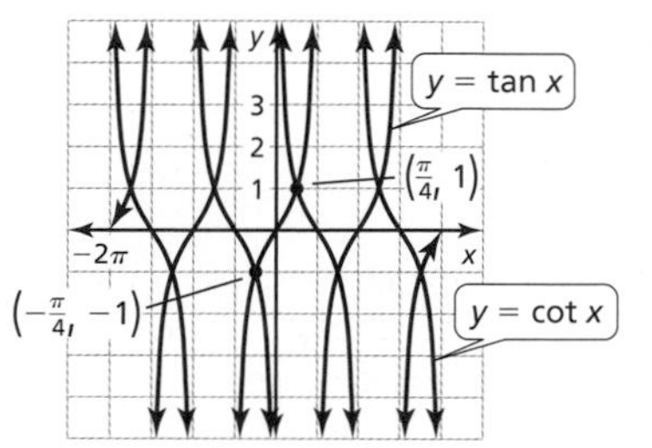

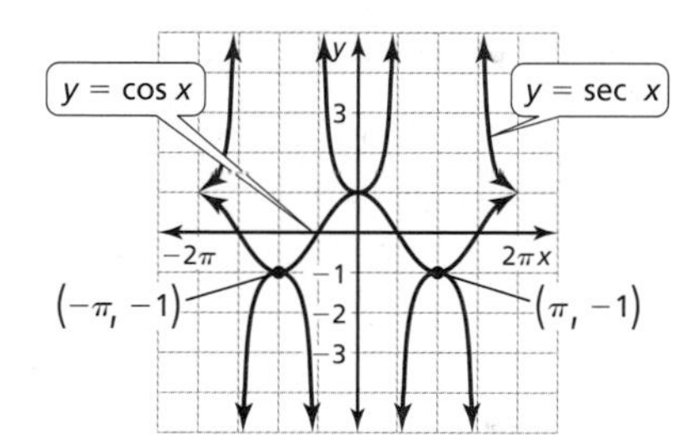

33. yes;
$$\sec x \tan x - \sin x = \frac{1}{\cos x} \cdot \frac{\sin x}{\cos x} - \sin x$$
$$= \frac{\sin x}{\cos^2 x} - \sin x$$
$$= \sec^2 x \sin x - \sin x$$
$$= \sin x(\sec^2 x - 1)$$
$$= \sin x \tan^2 x$$

35. $s = \frac{h\sin(90° - \theta)}{\sin\theta}$
$s = \frac{h\cos\theta}{\sin\theta}$
$s = h\cot\theta$

37. a. $u = \tan\theta$
b. u starts at 0 and increases without bound.

39. You can obtain the graph of $y = \cos x$ by reflecting the graph of $f(x) = \sin x$ in the y-axis and translating it $\frac{\pi}{2}$ units right.

9.7 Maintaining Mathematical Proficiency *(p. 518)*

41. $x = 11$ 43. $x = \frac{7\sqrt{3}}{3}$

9.8 Vocabulary and Core Concept Check *(p. 523)*

1. cos 170°

9.8 Monitoring Progress and Modeling with Mathematics *(pp. 523–524)*

3. $\sqrt{3} - 2$ **5.** $\dfrac{\sqrt{2} - \sqrt{6}}{4}$ **7.** $\dfrac{\sqrt{2} - \sqrt{6}}{4}$ **9.** $\sqrt{3} + 2$

11. $-\frac{36}{85}$ **13.** $-\frac{13}{85}$ **15.** $-\frac{36}{77}$ **17.** $\tan x$

19. $\cos x$ **21.** $\cos x$

23. The sign in the denominator should be negative when using the sum formula;

$$\frac{\tan x + \tan \frac{\pi}{4}}{1 - \tan x \tan \frac{\pi}{4}} = \frac{\tan x + 1}{1 - \tan x}$$

25. B, D **27.** $x = \frac{\pi}{3}, \frac{5\pi}{3}$ **29.** $x = \frac{3\pi}{2}$ **31.** $x = 0, \pi$

33. $\sin\left(\frac{\pi}{2} - \theta\right) = \sin \frac{\pi}{2} \cos \theta - \cos \frac{\pi}{2} \sin \theta$

$= (1) \cos \theta - (0) \sin \theta$

$= \cos \theta$

35. $\dfrac{35 \tan (\theta - 45°) + 35 \tan 45°}{h \tan \theta}$

$= \dfrac{35\left(\dfrac{\tan \theta - \tan 45°}{1 + \tan \theta \tan 45°}\right) + 35 \tan 45°}{h \tan \theta}$

$= \dfrac{35\left(\dfrac{\tan \theta - 1}{1 + \tan \theta}\right) + 35}{h \tan \theta}$

$= \dfrac{35(\tan \theta - 1) + 35(1 + \tan \theta)}{h \tan \theta(1 + \tan \theta)}$

$= \dfrac{35 \tan \theta - 35 + 35 + 35 \tan \theta}{h \tan \theta(1 + \tan \theta)}$

$= \dfrac{70 \tan \theta}{h \tan \theta(1 + \tan \theta)}$

$= \dfrac{70}{h(1 + \tan \theta)}$

37. $y_1 + y_2 = \cos 960\pi t + \cos 1240\pi t$

$= \cos(1100\pi t - 140\pi t) + \cos(1100\pi t + 140\pi t)$

$= \cos 1100\pi t \cos 140\pi t + \sin 1100\pi t \sin 140\pi t$

$+ \cos 1100\pi t \cos 140\pi t - \sin 1100\pi t \sin 140\pi t$

$= \cos 1100\pi t \cos 140\pi t + \cos 1100\pi t \cos 140\pi t$

$= 2 \cos 1100\pi t \cos 140\pi t$

39. a. $\tan(\theta_2 - \theta_1) = \dfrac{m_2 - m_1}{1 + m_2 m_1}$ **b.** 60°

9.8 Maintaining Mathematical Proficiency *(p. 524)*

41. $x = 4$ **43.** $x = -\frac{2}{3}$

Chapter 9 Review *(pp. 526–530)*

1. $\sin \theta = \dfrac{\sqrt{85}}{11}$, $\tan \theta = \dfrac{\sqrt{85}}{6}$, $\csc \theta = \dfrac{11\sqrt{85}}{85}$, $\sec \theta = \dfrac{11}{6}$,

$\cot \theta = \dfrac{6\sqrt{85}}{85}$

2. about 15 ft **3.** 22°; −338° **4.** $\dfrac{\pi}{6}$ **5.** $\dfrac{5\pi}{4}$

6. 135° **7.** 300°

8.

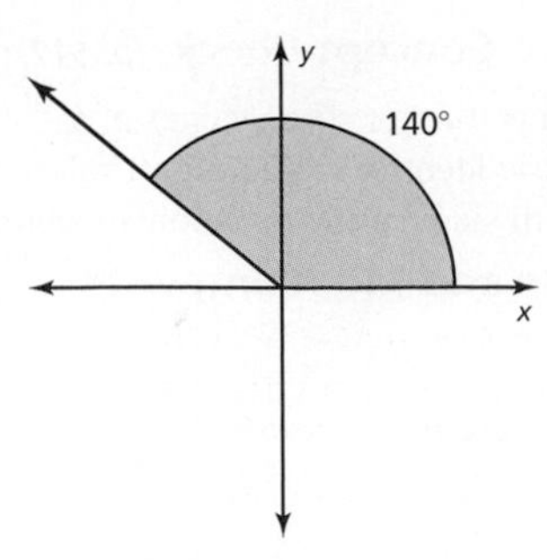

about 1497 m^2

9. $\sin \theta = 1$, $\cos \theta = 0$, $\tan \theta =$ undefined, $\csc \theta = 1$, $\sec \theta =$ undefined, $\cot \theta = 0$

10. $\sin \theta = -\frac{7}{25}$, $\cos \theta = \frac{24}{25}$, $\tan \theta = -\frac{7}{24}$, $\csc \theta = -\frac{25}{7}$, $\sec \theta = \frac{25}{24}$, $\cot \theta = -\frac{24}{7}$

11. $\sin \theta = \dfrac{3\sqrt{13}}{13}$, $\cos \theta = -\dfrac{2\sqrt{13}}{13}$, $\tan \theta = -\dfrac{3}{2}$, $\csc \theta = \dfrac{\sqrt{13}}{3}$,

$\sec \theta = -\dfrac{\sqrt{13}}{2}$, $\cot \theta = -\dfrac{2}{3}$

12. $-\dfrac{\sqrt{3}}{3}$ **13.** $\sqrt{2}$ **14.** $\dfrac{1}{2}$ **15.** 2

16. 8, 2π; The graph of g is a vertical stretch by a factor of 8 of the graph of $f(x) = \cos x$;

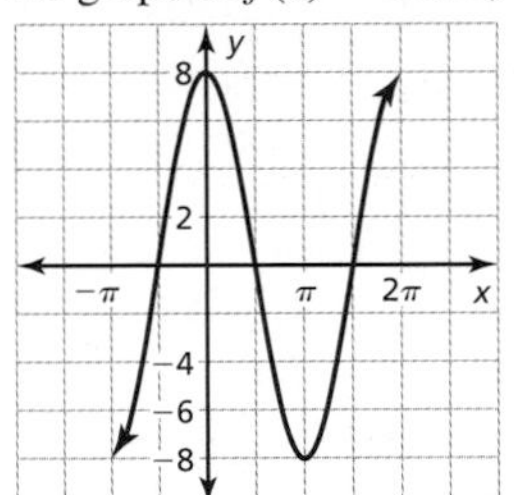

17. 6, 2; The graph of g is a horizontal shrink by a factor of $\dfrac{1}{\pi}$ and a vertical stretch by a factor of 6 of the graph of $f(x) = \sin x$;

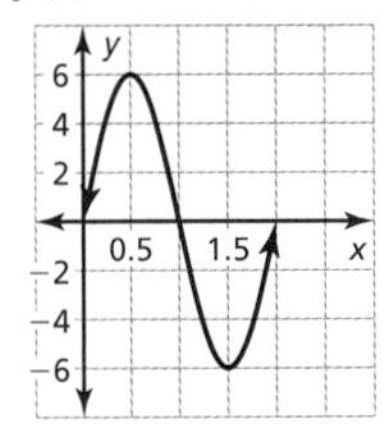

18. $\dfrac{1}{4}, \dfrac{\pi}{2}$; The graph of g is a horizontal shrink by a factor of $\dfrac{1}{4}$ and a vertical shrink by a factor of $\dfrac{1}{4}$ of the graph of $f(x) = \cos x$;

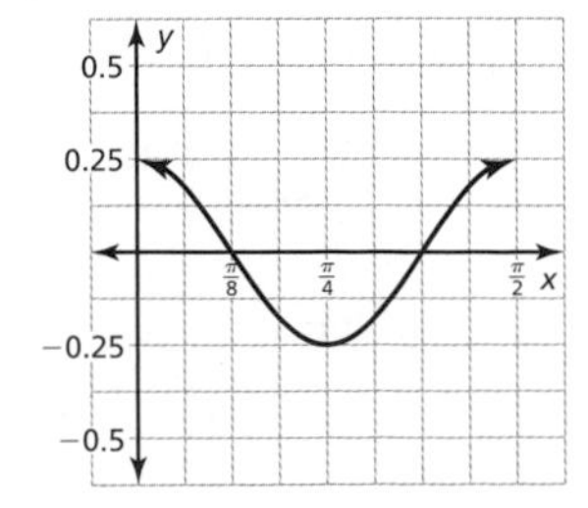

19.

20. 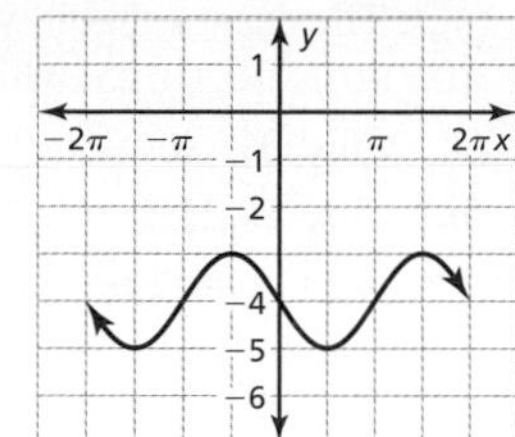

27. 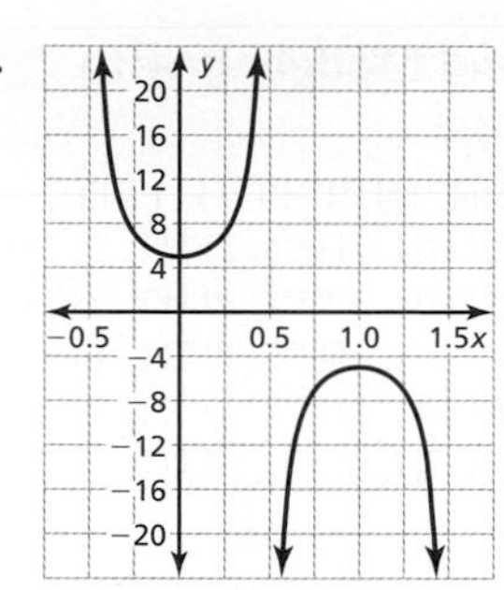

28. 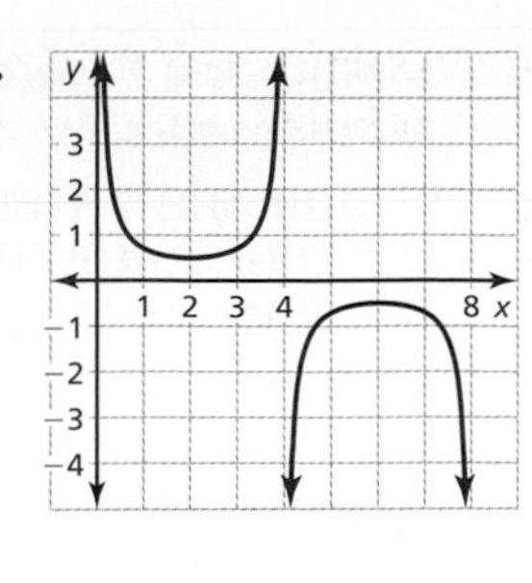

21. 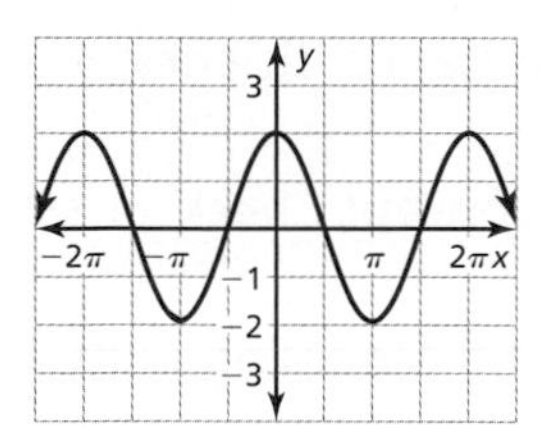

22. 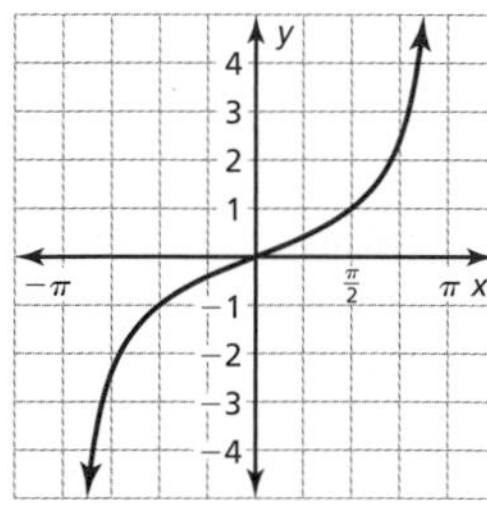

The graph of g is a horizontal stretch by a factor of 2 of the graph of $f(x) = \tan x$.

23. 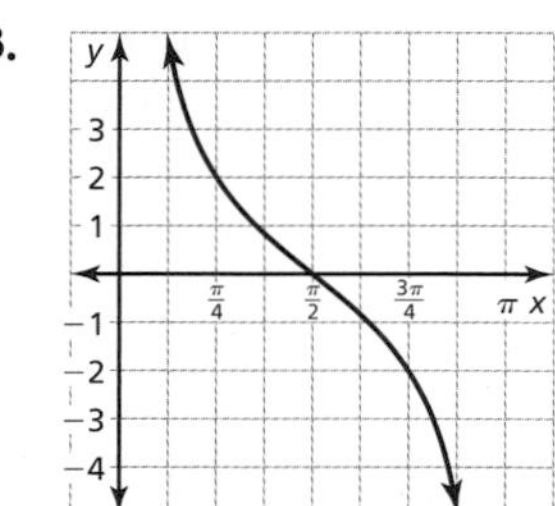

The graph of g is a vertical stretch by a factor of 2 of the graph of $f(x) = \cot x$.

24. 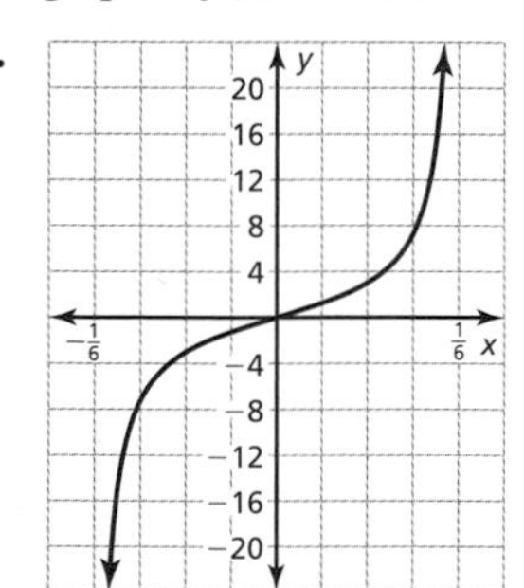

The graph of g is a horizontal shrink by a factor of $\frac{1}{3\pi}$ and a vertical stretch by a factor of 4 of the graph of $f(x) = \tan x$.

25. 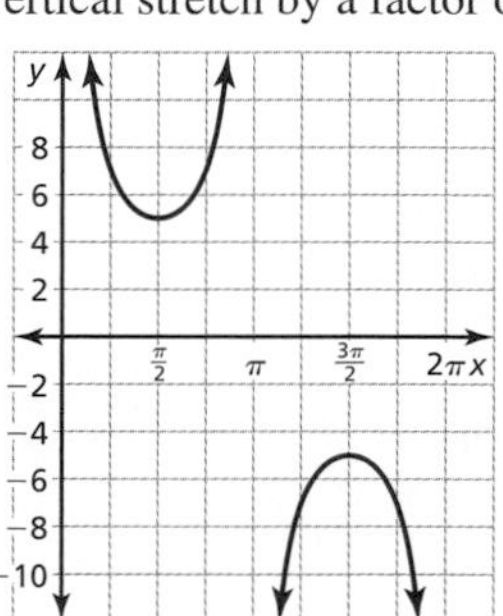

26. 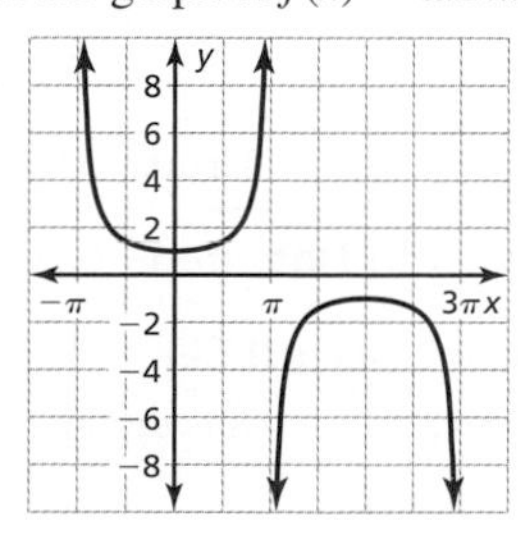

29. *Sample answer:* $y = -\sin \frac{1}{2}x$

30. *Sample answer:* $y = \cos \pi x - 2$

31. $h = -11.5 \cos 2\pi t + 13.5$

32. $P = 1.08 \sin(0.585t - 2.33) + 1.5$; The period represents the amount of time it takes for the precipitation level to complete one cycle, which is about 10.7 months.

33. $\cos^2 x$ **34.** $\tan x$ **35.** $\sin x$

36.
$$\frac{\cos x \sec x}{1 + \tan^2 x} = \frac{\cos x \sec x}{\sec^2 x} = \frac{\cos x}{\sec x} = \cos x \cos x = \cos^2 x$$

37.
$$\tan\left(\frac{\pi}{2} - x\right) \cot x = \cot x \cot x = \cot^2 x = \csc^2 x - 1$$

38. $\frac{\sqrt{2} + \sqrt{6}}{4}$ **39.** $\sqrt{3} - 2$ **40.** $\frac{\sqrt{6} + \sqrt{2}}{4}$

41. $\frac{19}{25}$ **42.** $x = \frac{3\pi}{4}, \frac{5\pi}{4}$ **43.** $x = 0, \pi$

Chapter 10

Chapter 10 Maintaining Mathematical Proficiency *(p. 535)*

1. $\frac{6}{30} = \frac{p}{100}$, 20% **2.** $\frac{a}{25} = \frac{68}{100}$, 17

3. $\frac{34.4}{86} = \frac{p}{100}$, 40%

4. 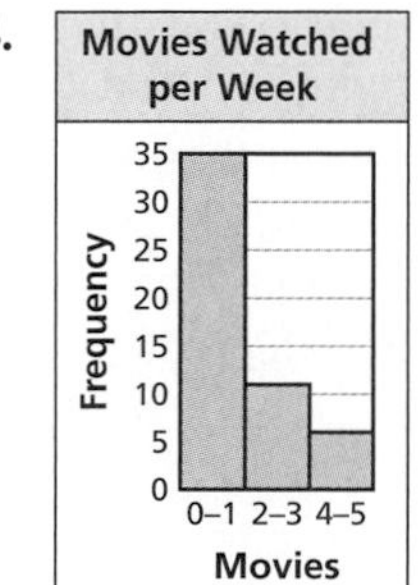

5. no; The sofa will cost 80% of the retail price and the arm chair will cost 81% of the retail price.

10.1 Vocabulary and Core Concept Check *(p. 542)*

1. probability

10.1 Monitoring Progress and Modeling with Mathematics (pp. 542–544)

3. 48; 1HHH, 1HHT, 1HTH, 1THH, 1HTT, 1THT, 1TTH, 1TTT, 2HHH, 2HHT, 2HTH, 2THH, 2HTT, 2THT, 2TTH, 2TTT, 3HHH, 3HHT, 3HTH, 3THH, 3HTT, 3THT, 3TTH, 3TTT, 4HHH, 4HHT, 4HTH, 4THH, 4HTT, 4THT, 4TTH, 4TTT, 5HHH, 5HHT, 5HTH, 5THH, 5HTT, 5THT, 5TTH, 5TTT, 6HHH, 6HHT, 6HTH, 6THH, 6HTT, 6THT, 6TTH, 6TTT

5. 12; R1, R2, R3, R4, W1, W2, W3, W4, B1, B2, B3, B4

7. $\frac{5}{16}$, or about 31.25%

9. **a.** $\frac{11}{12}$, or about 92% **b.** $\frac{13}{18}$, or about 72%

11. There are 4 outcomes, not 3; The probability is $\frac{1}{4}$.

13. about 0.56, or about 56% **15.** 4

17. **a.** $\frac{9}{10}$, or 90% **b.** $\frac{2}{3}$, or about 67%

c. The probability in part (b) is based on trials, not possible outcomes.

19. about 0.08, or about 8% **21.** C, A, D, B

23. **a.** 2, 3, 4, 5, 6, 7, 8, 9, 10, 11, 12

b. 2: $\frac{1}{36}$, 3: $\frac{1}{18}$, 4: $\frac{1}{12}$, 5: $\frac{1}{9}$, 6: $\frac{5}{36}$, 7: $\frac{1}{6}$, 8: $\frac{5}{36}$, 9: $\frac{1}{9}$, 10: $\frac{1}{12}$, 11: $\frac{1}{18}$, 12: $\frac{1}{36}$

c. *Sample answer:* The probabilities are similar.

25. $\frac{\pi}{6}$, or about 52%

27. $\frac{3}{400}$, or 0.75%; about 113; (0.0075)15,000 = 112.5

10.1 Maintaining Mathematical Proficiency (p. 544)

29. $\frac{6x^4}{y^3}$ **31.** $\frac{x^3 - 4x^2 - 15x + 18}{x^4 - 2}$

33. $\frac{15x^2}{12x^2 + x - 11}, x \neq 0$

10.2 Vocabulary and Core Concept Check (p. 550)

1. When two events are dependent, the occurrence of one event affects the other. When two events are independent, the occurrence of one event does not affect the other. *Sample answer:* choosing two marbles from a bag without replacement; rolling two dice

10.2 Monitoring Progress and Modeling with Mathematics (pp. 550–552)

3. dependent; The occurrence of event A affects the occurrence of event B.

5. dependent; The occurrence of event A affects the occurrence of event B.

7. yes **9.** yes **11.** about 2.8% **13.** about 34.7%

15. The probabilities were added instead of multiplied; $P(A \text{ and } B) = (0.6)(0.2) = 0.12$

17. 0.325

19. **a.** about 1.2% **b.** about 1.0%

You are about 1.2 times more likely to select 3 face cards when you replace each card before you select the next card.

21. **a.** about 17.1% **b.** about 81.4% **23.** about 53.5%

25. **a.** *Sample answer:* Put 20 pieces of paper with each of the 20 students' names in a hat and pick one; 5%

b. *Sample answer:* Put 45 pieces of paper in a hat with each student's name appearing once for each hour the student worked. Pick one piece; about 8.9%

27. yes; The chance that it will be rescheduled is (0.7)(0.75) = 0.525, which is a greater than a 50% chance.

29. **a.** wins: 0%; loses: 1.99%; ties: 98.01%

b. wins: 20.25%; loses: 30.25%; ties: 49.5%

c. yes; Go for 2 points after the first touchdown, and then go for 1 point if they were successful the first time or 2 points if they were unsuccessful the first time; winning: 44.55%; losing: 30.25%

10.2 Maintaining Mathematical Proficiency (p. 552)

31. $x = 0.2$ **33.** $x = 0.15$

10.3 Vocabulary and Core Concept Check (p. 558)

1. two-way table

10.3 Monitoring Progress and Modeling with Mathematics (p. 558–560)

3. 34; 40; 4; 6; 12

5.

		Gender		
		Male	Female	Total
Response	Yes	132	151	283
	No	39	29	68
	Total	171	180	351

351 people were surveyed, 171 males were surveyed, 180 females were surveyed, 283 people said yes, 68 people said no.

7.

		Dominant Hand		
		Left	Right	Total
Gender	Female	0.048	0.450	0.498
	Male	0.104	0.398	0.502
	Total	0.152	0.848	1

9.

		Gender		
		Male	Female	Total
Response	Yes	0.376	0.430	0.806
	No	0.111	0.083	0.194
	Total	0.487	0.513	1

11.

		Breakfast	
		Ate	Did Not Eat
Feeling	Tired	0.091	0.333
	Not Tired	0.909	0.667

13. **a.** about 0.789 **b.** 0.168

c. The events are independent.

15. The value for P(yes) was used in the denominator instead of the value for P(Tokyo);

$$\frac{0.049}{0.39} \approx 0.126$$

17. Route B; It has the best probability of getting to school on time.

19. *Sample answer:*

		Transportation to School			
		Rides Bus	Walks	Car	Total
Gender	Male	6	9	4	19
	Female	5	2	4	11
	Total	11	11	8	30

		Transportation to School			
		Rides Bus	Walks	Car	Total
Gender	Male	0.2	0.3	0.133	0.633
	Female	0.167	0.067	0.133	0.367
	Total	0.367	0.367	0.266	1

21. Routine B is the best option, but your friend's reasoning of why is incorrect; Routine B is the best choice because there is a 66.7% chance of reaching the goal, which is higher than the chances of Routine A (62.5%) and Routine C (63.6%).

23. a. about 0.438 **b.** about 0.387

25. a. More of the current consumers prefer the leader, so they should improve the new snack before marketing it.

b. More of the new consumers prefer the new snack than the leading snack, so there is no need to improve the snack.

10.3 Maintaining Mathematical Proficiency *(p. 560)*

27.

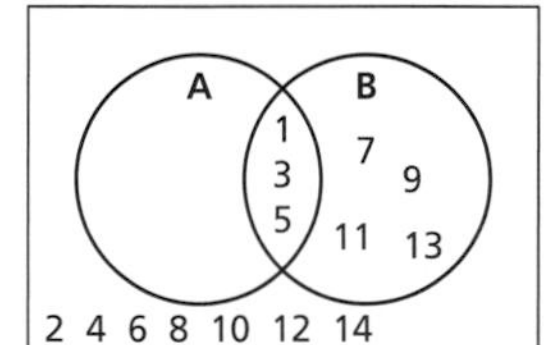

29.

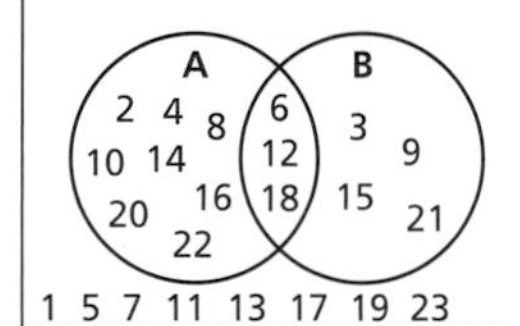

10.4 Vocabulary and Core Concept Check *(p. 567)*

1. yes; $\overline{A}$ is everything not in A; *Sample answer:* event A: you win the game, event $\overline{A}$: you do not win the game

10.4 Monitoring Progress and Modeling with Mathematics *(p. 567–568)*

3. 0.4 **5.** $\frac{7}{12}$, or about 0.58 **7.** $\frac{9}{20}$, or 0.45

9. $\frac{7}{10}$, or 0.7

11. forgot to subtract P(heart and face card); $P(\text{heart}) + P(\text{face card}) - P(\text{heart and face card}) = \frac{11}{26}$

13. $\frac{2}{3}$ **15.** 10% **17.** 0.4742, or 47.42% **19.** $\frac{13}{18}$

21. $\frac{3}{20}$

23. no; Until all cards, numbers, and colors are known, the conclusion cannot be made.

10.4 Maintaining Mathematical Proficiency *(p. 568)*

25. $a_1 = 1, a_2 = 2, a_3 = 3, a_4 = 4, a_5 = 5, a_6 = 6$

10.5 Vocabulary and Core Concept Check *(p. 575)*

1. permutation

10.5 Monitoring Progress and Modeling with Mathematics *(p. 575–578)*

3. a. 2 **b.** 2 **5. a.** 24 **b.** 12

7. a. 720 **b.** 30 **9.** 20 **11.** 9 **13.** 20,160

15. 870 **17.** 990 **19.** $\frac{1}{56}$ **21.** 4 **23.** 20

25. 5 **27.** 1 **29.** 220 **31.** 6435 **33.** 635,376

35. The factorial in the denominator was left out;

$$_{11}P_7 = \frac{11!}{(11-7)!} = 1{,}663{,}200$$

37. combinations; The order is not important; 45

39. permutations; The order is important; 132,600

41. $_{50}C_9 = {}_{50}C_{41}$; For each combination of 9 objects, there is a corresponding combination of the 41 remaining objects.

43. a. neither, they are the same; $_4P_4 = {}_4P_3 = 24$

b. 3; $_4C_4 = 1, {}_4C_3 = 4$

c. $_nP_n = {}_nP_{n-1}$, but $_nC_n < {}_nC_{n-1}$ when $n > 1$, and $_nC_n = {}_nC_{n-1}$ when $n = 1$.

45.

	$r = 0$	$r = 1$	$r = 2$	$r = 3$
$_3P_r$	1	3	6	6
$_3C_r$	1	3	3	1

$_nP_r \geq {}_nC_r$; Because $_nP_r = \frac{n!}{(n-r)!}$ and $_nC_r = \frac{n!}{(n-r)! \cdot r!}$, $_nP_r > {}_nC_r$ when $r > 1$ and $_nP_r = {}_nC_r$ when $r = 0$ or $r = 1$.

47. $\frac{1}{44{,}850}$ **49.** $\frac{1}{15{,}890{,}700}$ **51.** $x^3 + 6x^2 + 12x + 8$

53. $a^4 + 12a^3b + 54a^2b^2 + 108ab^3 + 81b^4$

55. $w^{12} - 12w^9 + 54w^6 - 108w^3 + 81$

57. $729u^6 + 1458u^5v^2 + 1215u^4v^4 + 540u^3v^6 + 135u^2v^8 + 18uv^{10} + v^{12}$

59. -8064 **61.** $-13{,}608$ **63.** 316,800,000

65. $-337{,}920$

67. $_8C_0, {}_8C_1, {}_8C_2, {}_8C_3, {}_8C_4, {}_8C_5, {}_8C_6, {}_8C_7, {}_8C_8$; 1, 8, 28, 56, 70, 56, 28, 8, 1

69. a. $_nC_{n-2} - n$ **b.** $\frac{n(n-3)}{2}$ **71.** 30 **73.** $\frac{1061}{1250}$

75. a. $\frac{1}{90}$ **b.** $\frac{9}{10}$ **77. a.** 2,598,960 **b.** 5148

79. a. about 0.04; about 0.12 **b.** $1 - \frac{_{365}P_x}{365^x}$ **c.** 23 people

10.5 Maintaining Mathematical Proficiency *(p. 578)*

81. TH

10.6 Vocabulary and Core Concept Check *(p. 583)*

1. a variable whose value is determined by the outcomes of a probability experiment

10.6 Monitoring Progress and Modeling with Mathematics *(pp. 583–584)*

3.

x (value)	1	2	3
Outcomes	5	3	2
P(x)	$\frac{1}{2}$	$\frac{3}{10}$	$\frac{1}{5}$

Drawing a Ball

5.

w (value)	1	2
Outcomes	5	21
P(w)	$\frac{5}{26}$	$\frac{21}{26}$

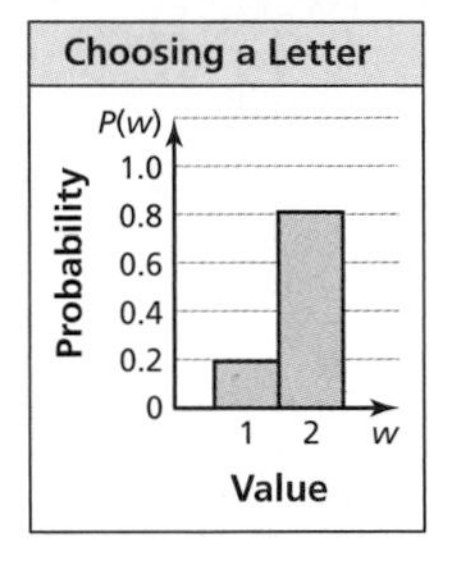

7. a. 2 b. $\frac{5}{8}$ 9. about 0.00002 11. about 0.00018

13. a.

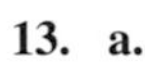

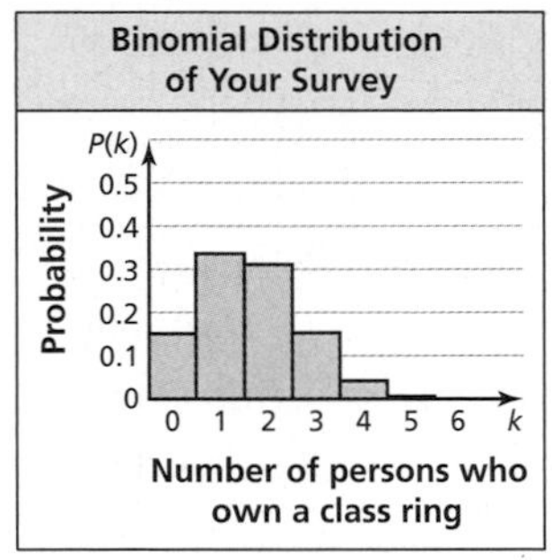

b. The most likely outcome is that 1 of the 6 students owns a ring.

c. about 0.798

15. The exponents are switched;
$P(k = 3) = {}_5C_3\left(\frac{1}{6}\right)^3\left(\frac{5}{6}\right)^{5-3} \approx 0.032$

17. a. $P(0) \approx 0.099$, $P(1) \approx 0.271$, $P(2) \approx 0.319$, $P(3) \approx 0.208$, $P(4) \approx 0.081$, $P(5) \approx 0.019$, $P(6) \approx 0.0025$, $P(7) \approx 0.00014$

b.

x	0	1	2	3	4
P(x)	0.099	0.271	0.319	0.208	0.081

x	5	6	7
P(x)	0.019	0.0025	0.00014

c.

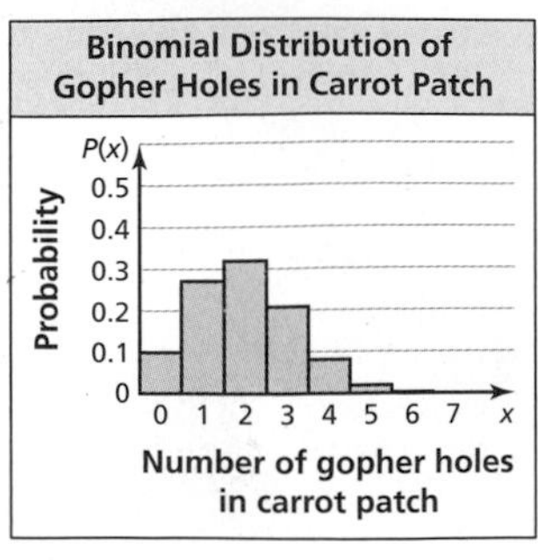

19. no; The data is skewed right, so the probability of failure is greater.

21. a. The statement is not valid, because having a male and having a female are independent events.

b. 0.03125

c.

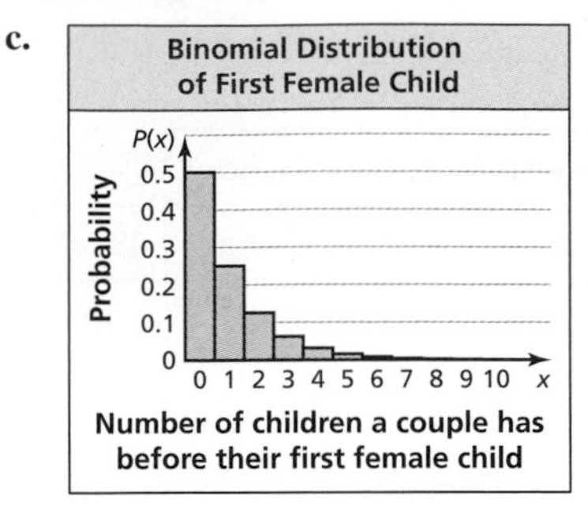

skewed right

10.6 Maintaining Mathematical Proficiency *(p. 584)*

23. FFF, FFM FMF, FMM, MMM, MMF, MFM, MFF

Chapter 10 Review *(pp. 586–588)*

1. $\frac{2}{9}, \frac{7}{9}$ 2. 20 points

3. a. 0.15625 b. about 0.1667
You are about 1.07 times more likely to pick a red then a green if you do not replace the first marble.

4. a. about 0.0586 b. 0.0625
You are about 1.07 times more likely to pick a blue then a red if you do not replace the first marble.

5. a. 0.25 b. about 0.2333
You are about 1.07 times more likely to pick a green and then another green if you replace the first marble.

6. about 0.529

7.

		Gender		
		Men	Women	Total
Response	Yes	200	230	430
	No	20	40	60
	Total	220	270	490

About 44.9% of responders were men, about 55.1% of responders were women, about 87.8% of responders thought it was impactful, about 12.2% of responders thought it was not impactful.

8. 0.68 9. 0.02 10. 5040 11. 1,037,836,800

12. 15 13. 70 14. $16x^4 + 32x^3y^2 + 24x^2y^4 + 8xy^6 + y^8$

15. $\frac{1}{84}$ 16. about 0.12

17.

Binomial Distribution for Made Free Throws

The most likely outcome is that 4 of the 5 free throw shots will be made.

Chapter 11

Chapter 11 Maintaining Mathematical Proficiency *(p. 593)*

1. about 77.2, 82.5, 82; median or mode; The mean is less than most of the data.
2. about 73.7, 70.5, 70; median or mode; The mean is greater than most of the data.
3. about 19.8, 16, 44; median; The mean and mode are both greater than most of the data.
4. about 3.85; The typical data value differs from the mean by about 3.85 units.
5. about 7.09; The typical data value differs from the mean by about 7.09 units.
6. 6.5; The typical data value differs from the mean by 6.5 units.
7. All the data values are the same; no; The formula for standard deviation includes taking only the positive square root.

11.1 Vocabulary and Core Concept Check *(p. 600)*

1. Find the value where row 1 and column 4 intersect.

11.1 Monitoring Progress and Modeling with Mathematics *(pp. 600–602)*

3. 50% **5.** 2.5% **7.** 0.16 **9.** 0.025

11. 0.68 **13.** 0.68 **15.** 0.975 **17.** 0.84

19. **a.** 81.5% **b.** 0.15%

21. The values on the horizontal axis show a standard deviation of 1 instead of 2.

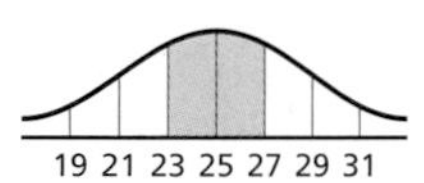

The probability that x is between 23 and 27 is 0.68.

23. 0.0668 **25.** no

27.

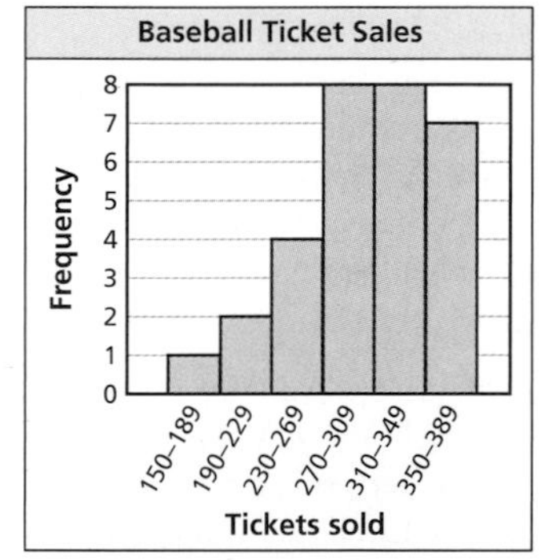

no; The histogram is skewed left, not bell-shaped.

29. **a.** about 4.52×10^{-9}

b. yes; The probability that a box contains an amount of cereal significantly less than the mean is very small.

31. one standard deviation above the mean

33. **a.** 88th percentile **b.** 93rd percentile

c. ACT; Your percentile on the ACT was higher than your percentile on the SAT.

35. no; When the mean is greater than the median, the distribution is skewed right.

11.1 Maintaining Mathematical Proficiency *(p. 602)*

37.

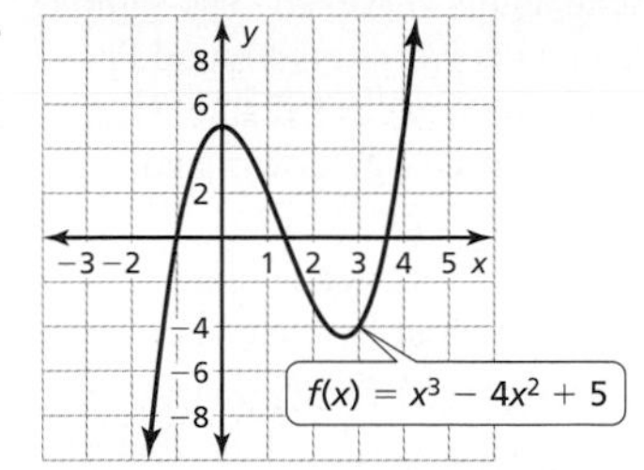

x-intercepts: -1, about 1.4, and about 3.6; local maximum: $(0, 5)$; local minimum: $(2.67, -4.48)$; increasing when $x < 0$ and $x > 2.67$; decreasing when $0 < x < 2.67$

39.

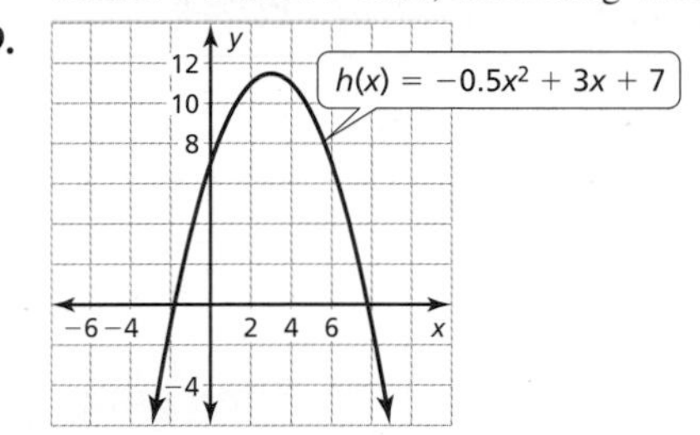

x-intercepts: about -1.8 and about 7.8; maximum: $(3, 11.5)$; no local minimum; increasing when $x < 3$; decreasing when $x > 3$

11.2 Vocabulary and Core Concept Check *(p. 607)*

1. sample
3. a claim about a characteristic of a population

11.2 Monitoring Progress and Modeling with Mathematics *(pp. 607–608)*

5. population; Every high school student is counted.

7. sample; The survey is given to a subset of the population of spectators.

9. population: every adult age 18 and over in the United States, sample: the 1152 adults age 18 and over who were surveyed; The sample consists of 403 adults who pretend to use their smartphone to avoid talking to someone, and 749 adults who do not.

11. population: every high school student in the district, sample: the 1300 high school students in the district who were surveyed; The sample consists of 1001 high school students who like the new healthy cafeteria food choices, and 299 high school students who do not.

13. statistic; The average annual salary of a subset of the population was calculated.

15. parameter; The percentage of every student in the school was calculated.

17. The sample number in the statement is not the size of the entire sample; The population consists of all the students in the high school. The sample consists of the 1270 students that were surveyed.

19. **a.** The maker's claim is most likely true.

b. The maker's claim is most likely false.

21. possibly, but extremely unlikely; The result is unlikely to occur by chance. The sample size of the population is too small to make such a conclusion.

23. *Sample answer:* population: all American adults, sample: the 801 American adults surveyed; The sample consists of 606 American adults who say the world's temperature will go up over the next 100 years, 174 American adults who say it will go down, and 21 American adults who have no opinion.

25. simulation 2; Simulation 2 gives a better indication of outcomes that are not likely to occur by chance.

11.2 Maintaining Mathematical Proficiency *(p. 608)*

27. $x = 5 \pm \sqrt{29}$ or $x \approx 10.39, x \approx -0.39$

29. $s = -5 \pm \sqrt{17}$ or $s \approx -0.88, s \approx -9.12$

31. $z = \frac{1}{2}, z = -\frac{15}{2}$

11.3 Vocabulary and Core Concept Check *(p. 614)*

1. In a stratified sample, after the groups are formed, a random sample is selected from each group. In a cluster sample, after the groups are formed, all the members of one or more groups are randomly selected.

3. *Sample answer:* to determine how quickly an oil spill would spread through a lake

11.3 Monitoring Progress and Modeling with Mathematics *(pp. 614–616)*

5. convenience sample 7. systematic sample

9. convenience sample; Dog owners probably have a strong opinion about an off-leash area for dogs.

11. cluster sample; Booth holders in section 5 are likely to have a different opinion than booth holders in other sections about the location of their booth.

13. Not every survey that was mailed out will be returned, so it is not a systematic sample; Because households in the neighborhood can choose whether or not to return the survey, the sample is a self-selected sample.

15. no; The sample represents the population.

17. yes; Only customers with a strong opinion about their experience are likely to complete the survey.

19. *Sample answer:* Assign each student in the school a different integer from 1 to 1225. Generate 250 unique random integers from 1 to 1225 using the random number function in a spreadsheet program. Choose the 250 students who correspond to the 250 integers generated.

21. simulation 23. observational study

25. encourages a yes response; *Sample answer:* Reword the question, for example: Should the budget of our city be cut?

27. implies that the arsenic level is a health risk; *Sample answer:* Reword the question, for example: Do you think the government should address the issue of arsenic in tap water?

29. no; Responses to the question will accurately reflect the opinions of those being surveyed.

31. yes; Visitors are unlikely to admit to a police officer that they do not wear their seatbelt.

33. **a.** *Sample answer:* The researcher did not take into account previous heart conditions.
    b. *Sample answer:* Divide the population into groups based on past heart conditions and whether or not they take fiber supplements. Select a random sample from each group.

35. self-selected sample and convenience sample; In a self-selected sample, only people with strong opinions are likely to respond. In a convenience sample, parts of the population have no chance of being selected for the survey.

37. **a.** to determine the employment rate of graduates in their field of study
    b. all graduating seniors of the college
    c. *Sample answer:* Are you employed? If yes, is your job in your field of study?

39. no; *Sample answer:* Some groups in the population, like the homeless, are difficult to contact.

41. **a.** self-selected sample
    b. people who spend a lot of time on the Internet and visit that particular site; The survey is probably biased.

11.3 Maintaining Mathematical Proficiency *(p. 616)*

43. 9 45. $\frac{1}{4}$ 47. $\frac{\sqrt[3]{18}}{18}$ 49. 3

11.4 Vocabulary and Core Concept Check *(p. 623)*

1. replication

11.4 Monitoring Progress and Modeling with Mathematics *(pp. 623–624)*

3. The study is a randomized comparative experiment; The treatment is the drug for insomnia. The treatment group is the individuals who received the drug. The control group is the individuals who received the placebo.

5. The individuals who do not use either of the conditioners were not monitored; The control group is the individuals who use the regular conditioner.

7. observational study; *Sample answer:* Randomly choose one group of individuals who smoke. Then, randomly choose one group of individuals who do not smoke. Find the body mass index of the individuals in each group.

9. experiment; *Sample answer:* Randomly select the same number of strawberry plants to be put in each of two groups. Use the new fertilizer on the plants in one group, and use the regular fertilizer on plants in the other group. Keep all other variables constant and record the weight of the fruit produced by each plant.

11. **a.** *Sample answer:* Because the heart rates are monitored for two different types of exercise, the groups cannot be compared. Running on a treadmill may have a different effect on heart rate than lifting weights; Check the heart rates of all the athletes after the same type of exercise.
    b. no potential problems

13. *Sample answer:* The sample size is not large enough to provide valid results; Increase the sample size.

15. no; Your friend would have to perform an observational study, and an observational study can show correlation, but not causality.

17. *Sample answer:* The placebo effect is response to a dummy treatment that may result from the trust in the researcher or the expectation of a cure; It can be minimized by comparing two groups so the placebo effect has the same effect on both groups.

19. yes; Repetition reduces the effect of unusual results that may occur by chance.

11.4 Maintaining Mathematical Proficiency *(p. 624)*

21.

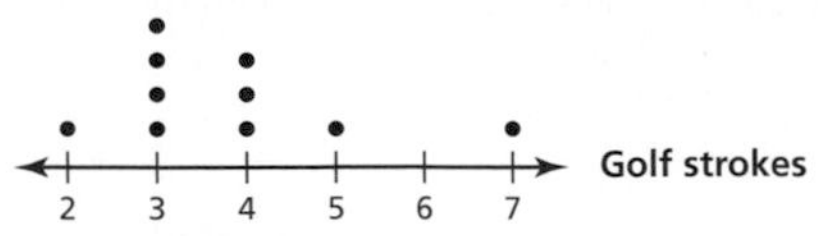

Golf strokes

2 3 4 5 6 7

skewed right

23. exponential decay

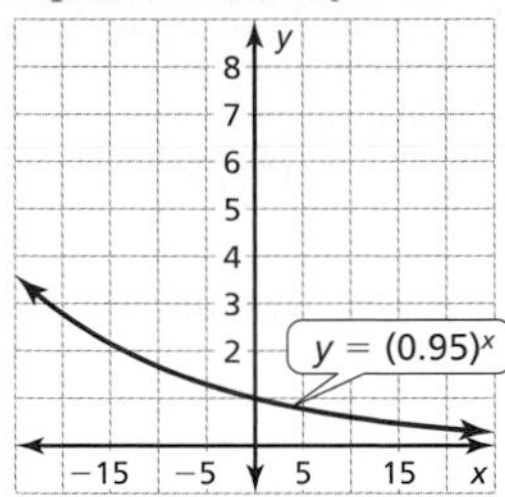

25. exponential growth

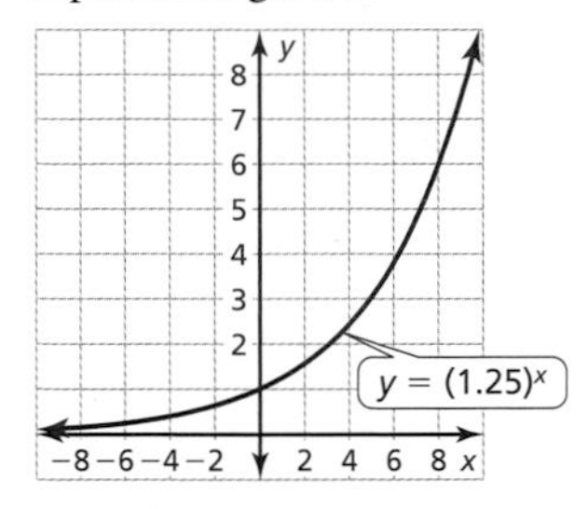

11.5 Vocabulary and Core Concept Check *(p. 630)*

1. margin of error

11.5 Monitoring Progress and Modeling with Mathematics *(pp. 630–632)*

3. 60.4 **5. a.** about 0.267 **b.** about 0.267

7. a. yes; The first 2 surveys show more than the 66.7% of votes needed to override the veto.

b. no; As the sample size increases, the percent of votes approaches 55.1%, which is not enough to override the veto.

9. a. The company's claim is probably accurate.

b. The company's claim is probably not accurate.

c. *Sample answer:* 0.42 to 0.68

11. about ±6.2% **13.** about ±2.2% **15.** about ±1.7%

17. a. about ±3.1% **b.** between 37.9% and 44.1%

19. The wrong percentage was substituted in the formula;

$\pm 0.04 = \pm \frac{1}{\sqrt{n}}$; $0.0016 = \frac{1}{n}$; $n = 625$

21. no; A sample size of 1 would have a margin of error of 100%.

23. about 453 residents

25. a. 500 voters **b.** about ±4.5%

c. candidate A: between 42.5% and 51.5%, candidate B: between 48.5% and 57.5%

d. no; 273 voters

27. more than 2500; To be confident that sports drink X is preferred, the margin of error would need to be less than 2%.

11.5 Maintaining Mathematical Proficiency *(p. 632)*

29. $y = \log_2(x + 5)$ **31.** $y = 6^{x+1}$

33. geometric; $a_n = 3(2)^{n-1}$

11.6 Vocabulary and Core Concept Check *(p. 637)*

1. resampling

11.6 Monitoring Progress and Modeling with Mathematics *(pp. 637–638)*

3. a. 46 **b.** 40.125 **c.** −5.875

d.

Control group

Treatment group

36 38 40 42 44 46 48 50

Depression Score

e. The music therapy may be effective in reducing depression scores of college students.

5. The order of the subtraction is reversed; $\bar{x}_{\text{treatment}} - \bar{x}_{\text{control}} = 11 - 16 = -5$; So, you can conclude the treatment decreases the score.

7. *Sample answer:* −1.75

9. The hypothesis is most likely false; Music therapy decreases depression scores.

11. The histogram in Exercise 9 has a roughly normal distribution and shows the mean differences from 200 resamplings. The histogram in Exercise 11 is random and shows the mean differences from 20 resamplings; the histogram in Exercise 9 because it uses a large number of resamplings and the roughly normal distribution suggests music therapy decreases depression scores

13. yes; As the number of samplings increase, the individual values should end up in each group approximately the same number of times, so the positive and negative differences in the means should balance out to 0.

15. 12,870; The number of combinations of 16 items in groups of 8 amounts to 12,870.

11.6 Maintaining Mathematical Proficiency *(p. 638)*

17. $(y - 2)(y^2 + 2y + 4)$ **19.** $(9w^2 + 4)(3w + 2)(3w - 2)$

21. yes; $g(x) = \frac{1}{2x} + \frac{1}{2}$ **23.** no; $y = \pm\sqrt{\frac{3}{x - 1}}$

Chapter 11 Review *(pp. 640–642)*

1. 0.0015 **2.** 0.0082

3. population: all U.S. motorists, sample: the 1000 drivers surveyed

4. statistic; The mean was calculated from a sample.

5. The host's claim is most likely false.

6. stratified sample; not biased **7.** observational study

8. It encourages a yes response; *Sample answer:* Reword the question, for example: Should the city replace the police cars it is currently using?

9. experiment; *Sample answer:* Randomly select the same number of customers to give each type of bread to. Record how many customers from each group return.

10. *Sample answer:* The volunteers may not be representative of the population; Randomly select from members of the population for the study.

11. The study is a randomized comparative experiment; The treatment is using the new design of the car wash. The treatment group is the individuals who use the new design of the car wash. The control group is the individuals who use the old design of the car wash.

12. between 58.9% and 65.1%

13. no; As the sample size increases, the percent of votes approaches 46.8%, which is not enough to win.

14. *Sample answer:* Combine the measurements from both groups and assign a number to each value. Let the numbers 1 through 10 represent the data in the original control group, and let the numbers 11 through 20 represent the data in the original treatment group. Use a random number generator. Randomly generate 20 numbers from 1 through 20 without repeating a number. Use the first 10 numbers to make the new control group, and the next 10 to make the new treatment group; Repeatedly make new control and treatment groups and see how often you get differences between the new groups that are at least as large as the one you measured.

English-Spanish Glossary

English / Spanish

A

amplitude *(p. 486)* One-half the difference of the maximum value and the minimum value of the graph of a trigonometric function

amplitud *(p. 486)* La mitad de la diferencia del valor máximo y el valor mínimo del gráfico de una función trigonométrica

arithmetic sequence *(p. 418)* A sequence in which the difference of consecutive terms is constant

secuencia aritmética *(p. 418)* Una secuencia en la que la diferencia de términos consecutivos es constante

arithmetic series *(p. 420)* An expression formed by adding the terms of an arithmetic sequence

serie aritmética *(p. 420)* Una expresión formada al sumar los términos de una secuencia aritmética

asymptote *(p. 296)* A line that a graph approaches more and more closely

asíntota *(p. 296)* Una recta a la que una gráfica se acerca cada vez más

axis of symmetry *(p. 56)* A line that divides a parabola into mirror images and passes through the vertex

eje de simetría *(p. 56)* Una recta que divide una parábola en imágenes reflejo y que pasa a través del vértice

B

bias *(p. 611)* An error that results in a misrepresentation of a population

sesgo *(p. 611)* Un error que da como resultado una representación errónea de una población

biased question *(p. 613)* A question that is flawed in a way that leads to inaccurate results

pregunta sesgada *(p. 613)* Una pregunta imperfecta que lleva a obtener resultados inexactos

biased sample *(p. 611)* A sample that overrepresents or underrepresents part of the population

muestra sesgada *(p. 611)* Una muestra que representa excesiva o insuficientemente parte de la población

binomial distribution *(p. 581)* A type of probability distribution that shows the probabilities of the outcomes of a binomial experiment

distribución del binomio *(p. 581)* Un tipo de distribución de probabilidades que muestra las probabilidades de los resultados posibles de un experimento del binomio

binomial experiment *(p. 581)* An experiment in which there are a fixed number of independent trials, exactly two possible outcomes for each trial, and the probability of success is the same for each trial

experimento del binomio *(p. 581)* Un experimento en el que hay un número fijo de pruebas independientes, exactamente dos resultados posibles para cada prueba, y la probabilidad de éxito es la misma para cada prueba

Binomial Theorem *(p. 574)* For any positive integer n, the binomial expansion of $(a + b)^n$ is

$$(a + b)^n = {}_nC_0a^nb^0 + {}_nC_1a^{n-1}b^1 + {}_nC_2a^{n-2}b^2 + \cdots + {}_nC_na^0b^n.$$

teorema del binomio *(p. 574)* Por cada número entero positivo n, la expansión del binomio de $(a + b)^n$ es

$$(a + b)^n = {}_nC_0a^nb^0 + {}_nC_1a^{n-1}b^1 + {}_nC_2a^{n-2}b^2 + \cdots + {}_nC_na^0b^n.$$

C

central angle *(p. 472)* The angle measure of a sector of a circle formed by two radii

ángulo central *(p. 472)* La medida del ángulo de un sector de un círculo formado por dos radios

English-Spanish Glossary

cluster sample *(p. 610)* A sample in which a population is divided into groups, called clusters, and all of the members in one or more of the clusters are randomly selected

muestra de cluster *(p. 610)* Una muestra en la que una población se divide en grupos, llamados cluster en inglés, y todos los miembros de uno o más de los cluster son seleccionados en forma aleatoria

combination *(p. 572)* A selection of objects in which order is not important

combinación *(p. 572)* Una selección de objetos en la que el orden no es importante

common difference *(p. 418)* The constant difference d between consecutive terms of an arithmetic sequence

diferencia común *(p. 418)* La diferencia constante d entre términos consecutivos de una secuencia aritmética

common logarithm *(p. 311)* A logarithm with base 10, denoted as $\log_{10}$ or simply by log

logaritmo común *(p. 311)* Un logaritmo de base 10, denotado como $\log_{10}$ o simplemente como log

common ratio *(p. 426)* The constant ratio r between consecutive terms of a geometric sequence

razón común *(p. 426)* La razón constante r entre términos consecutivos de una secuencia geométrica

completing the square *(p. 112)* To add a term c to an expression of the form $x^2 + bx$ such that $x^2 + bx + c$ is a perfect square trinomial

completando el cuadrado *(p. 112)* Agregar un término c a una expresión de la forma $x^2 + bx$ para que $x^2 + bx + c$ sea un trinomio de cuadrado perfecto

complex conjugates *(p. 199)* Pairs of complex numbers of the forms $a + bi$ and $a - bi$, where $b \neq 0$

conjugados complejos *(p. 199)* Pares de números complejos de las formas $a + bi$ y $a - bi$, donde $b \neq 0$

complex fraction *(p. 387)* A fraction that contains a fraction in its numerator or denominator

fracción compleja *(p. 387)* Una fracción que contiene una fracción en su numerador o denominador

complex number *(p. 104)* A number written in the form $a + bi$, where a and b are real numbers

número complejo *(p. 104)* Un número escrito en la forma $a + bi$, donde a y b son números reales

compound event *(p. 564)* The union or intersection of two events

evento compuesto *(p. 564)* La unión o intersección de dos eventos

conditional probability *(p. 547)* The probability that event B occurs given that event A has occurred, written as $P(B|A)$

probabilidad condicional *(p. 547)* La probabilidad de que el evento B ocurra dado que el evento A ha ocurrido, escrito como $P(B|A)$

conditional relative frequency *(p. 555)* The ratio of a joint relative frequency to the marginal relative frequency in a two-way table

frecuencia relativa condicional *(p. 555)* La razón de una frecuencia relativa conjunta a la frecuencia relativa marginal en una tabla de doble entrada

conjugate *(p. 246)* Binomials of the form $a\sqrt{b} + c\sqrt{d}$ and $a\sqrt{b} - c\sqrt{d}$, where a, b, c, and d are rational numbers

conjugado *(p. 246)* Binomios de la forma $a\sqrt{b} + c\sqrt{d}$ y $a\sqrt{b} - c\sqrt{d}$, donde a, b, c y d son números racionales

constant of variation *(p. 360)* The constant a in the inverse variation equation $y = \frac{a}{x}$, where $a \neq 0$

constante de variación *(p. 360)* La constante a en la ecuación de variación inversa $y = \frac{a}{x}$, dondea $\neq 0$

control group *(p. 620)* The group under ordinary conditions that is subjected to no treatment during an experiment

grupo de control *(p. 620)* El grupo bajo condiciones ordinarias, que no se ve sometido a tratamiento durante un experimento

controlled experiment *(p. 620)* An experiment in which two groups are studied under identical conditions with the exception of one variable

experimento controlado *(p. 620)* Un experimento en el que dos grupos son estudiados bajo condiciones idénticas, con la excepción de una variable

convenience sample *(p. 610)* A sample in which only members of a population who are easy to reach are selected

muestra de conveniencia *(p. 610)* Una muestra en la que únicamente se seleccionan los miembros de una población a los que es fácil de llegar

correlation coefficient *(p. 25)* A number r from -1 to 1 that measures how well a line fits a set of data pairs (x, y)

coeficiente de correlación *(p. 25)* Un número r de -1 a 1 que mide cuán bien ajusta una recta a un conjunto de pares de datos (x, y)

cosecant *(p. 462)* A trigonometric function for an acute angle θ of a right triangle, denoted by $\csc \theta = \frac{\text{hypotenuse}}{\text{opposite}}$

cosecante *(p. 462)* Una ecuación trigonométrica de un ángulo agudo θ de un triángulo recto, denotado por $\csc \theta = \frac{\text{hipotenusa}}{\text{opuesto}}$

cosine *(p. 462)* A trigonometric function for an acute angle θ of a right triangle, denoted by $\cos \theta = \frac{\text{adjacent}}{\text{hypotenuse}}$

coseno *(p. 462)* Una ecuación trigonométrica de un ángulo agudo θ de un triángulo recto, denotado por $\cos \theta = \frac{\text{adyacente}}{\text{hipotenusa}}$

cotangent *(p. 462)* A trigonometric function for an acute angle θ of a right triangle, denoted by $\cot \theta = \frac{\text{adjacent}}{\text{opposite}}$

cotangente *(p. 462)* Una ecuación trigonométrica de un ángulo agudo θ de un triángulo recto, denotado por $\cot \theta = \frac{\text{adyacente}}{\text{opuesto}}$

coterminal *(p. 471)* Two angles whose terminal sides coincide

coterminal *(p. 471)* Dos ángulos cuyos lados terminales coinciden

cross multiplying *(p. 392)* A method used to solve a rational equation when each side of the equation is a single rational expression

multiplicación cruzada *(p. 392)* Un método utilizado para resolver una ecuación racional cuando cada lado de la ecuación es una sola expresión racional

cycle *(p. 486)* The shortest repeating portion of the graph of a periodic function

ciclo *(p. 486)* La porción más corta que se repite en el gráfico de una función periódica

D

decay factor *(p. 296)* The value of b in an exponential decay function of the form $y = ab^x$, where $a > 0$ and $0 < b < 1$

factor de decaimiento *(p. 296)* El valor de b en una función de decaimiento exponencial de la forma $y = ab^x$, donde $a > 0$ y $0 < b < 1$

dependent events *(p. 547)* Two events in which the occurrence of one event does affect the occurrence of the other event

eventos dependientes *(p. 547)* Dos eventos en los que la ocurrencia de un evento afecta la ocurrencia del otro evento

descriptive statistics *(p. 626)* The branch of statistics that involves the organization, summarization, and display of data

estadística descriptiva *(p. 626)* La rama de la estadística que implica la organización, resumen y presentación de datos

directrix *(p. 68)* A fixed line perpendicular to the axis of symmetry, such that the set of all points (x, y) of the parabola are equidistant from the focus and the directrix

directriz *(p. 68)* Una recta fija perpendicular al eje de simetría de modo tal, que el conjunto de todos los puntos (x, y) de la parábola sean equidistantes del foco y la directriz

discriminant *(p. 124)* The expression $b^2 - 4ac$ in the Quadratic Formula

discriminante *(p. 124)* La expresión $b^2 - 4ac$ en la Fórmula Cuadrática

disjoint events *(p. 564)* Two events that have no outcomes in common

eventos disjunto *(p. 564)* Dos eventos que no tienen resultados en común

E

end behavior *(p. 159)* The behavior of the graph of a function as x approaches positive infinity or negative infinity

comportamiento final *(p. 159)* El comportamiento del gráfico de una función a medida que x se aproxima al infinito positivo o negativo

even function *(p. 215)* For a function f, $f(-x) = f(x)$ for all x in its domain

función par *(p. 215)* Para una función f, $f(-x) = f(x)$ para toda x en su dominio

event *(p. 538)* A collection of one or more outcomes in a probability experiment

evento *(p. 538)* Una colección de uno o más resultados en un experimento de probabilidades

experiment *(p. 612)* A method that imposes a treatment on individuals in order to collect data on their response to the treatment

experimento *(p. 612)* Un método que impone un tratamiento a individuos para recoger datos con respecto a su respuesta al tratamiento

experimental probability *(p. 541)* The ratio of the number of successes, or favorable outcomes, to the number of trials in a probability experiment

probabilidad experimental *(p. 541)* La razón del número de éxitos, o resultados favorables, con respecto al número de pruebas en un experimento de probabilidades

explicit rule *(p. 442)* A rule that gives a_n as a function of the term's position number n in the sequence

regla explícita *(p. 442)* Una regla que da a_n como una función del número de posición n del término en la secuencia

exponential decay function *(p. 296)* A function of the form $y = ab^x$, where $a > 0$ and $0 < b < 1$

función de decaimiento exponencial *(p. 296)* Una función de la forma $y = ab^x$, donde $a > 0$ y $0 < b < 1$

exponential equations *(p. 334)* Equations in which variable expressions occur as exponents

ecuaciones exponenciales *(p. 334)* Ecuaciones en donde las expresiones de una variable ocurren como exponentes

exponential function *(p. 296)* A function of the form $y = ab^x$, where $a \neq 0$ and the base b is a positive real number other than 1

función exponencial *(p. 296)* Una función de la forma $y = ab^x$, donde $a \neq 0$ y la base b es un número real positivo distinto de 1

exponential growth function *(p. 296)* A function of the form $y = ab^x$, where $a > 0$ and $b > 1$

función de crecimiento exponencial *(p. 296)* Una función de la forma $y = ab^x$, dónde $a > 0$ y $b > 1$

extraneous solutions *(p. 263)* Solutions that are not solutions of the original equation

soluciones externas *(p. 263)* Soluciones que no son soluciones de la ecuación original

F

factor by grouping *(p. 181)* A method of factoring a polynomial by grouping pairs of terms that have a common monomial factor

factorización por agrupación *(p. 181)* Un método de factorización de un polinomio al agrupar pares de términos que tienen un factor monomio común

factored completely *(p. 180)* A polynomial written as a product of unfactorable polynomials with integer coefficients

factorizado completamente *(p. 180)* Un polinomio escrito como un producto de polinomios no factorizables con coeficientes de números enteros

finite differences *(p. 220)* The differences of consecutive y-values in a data set when the x-values are equally spaced

diferencias finitas *(p. 220)* Las diferencias de valores consecutivos y en un conjunto de datos cuando los valores x están igualmente espaciados

focus *(p. 68)* A fixed point in the interior of a parabola, such that the set of all points (x, y) of the parabola are equidistant from the focus and the directrix

foco *(p. 68)* Un punto fijo en el interior de una parábola, de tal forma que el conjunto de todos los puntos (x, y) de la parábola sean equidistantes del foco y la directriz

frequency *(p. 506)* The number of cycles per unit of time, which is the reciprocal of the period

frecuencia *(p. 506)* El número de ciclos por unidad de tiempo, que es el recíproco del período

G

geometric probability *(p. 540)* A probability found by calculating a ratio of two lengths, areas, or volumes

probabilidad geométrica *(p. 540)* Una probabilidad hallada al calcular la razón de dos longitudes, áreas o volúmenes

geometric sequence *(p. 426)* A sequence in which the ratio of any term to the previous term is constant

secuencia geométrica *(p. 426)* Una secuencia en donde la razón de cualquier término con respecto al término anterior es constante

geometric series *(p. 428)* The expression formed by adding the terms of a geometric sequence

serie geométrica *(p. 428)* La expresión formada al sumar los términos de una secuencia geométrica

growth factor *(p. 296)* The value of b in an exponential growth function of the form $y = ab^x$, where $a > 0$ and $b > 1$

factor de crecimiento *(p. 296)* El valor de b en una función de crecimiento exponencial de la forma $y = ab^x$, donde $a > 0$ y $b > 1$

H

horizontal shrink *(p. 14)* A transformation that causes the graph of a function to shrink toward the y-axis when all the x-coordinates are multiplied by a factor a, where $a > 1$

reducción horizontal *(p. 14)* Una transformación que hace que el gráfico de una función se reduzca hacia el eje y cuando todas las coordenadas x se multiplican por un factor a, donde $a > 1$

horizontal stretch *(p. 14)* A transformation that causes the graph of a function to stretch away from the y-axis when all the x-coordinates are multiplied by a factor a, where $0 < a < 1$

ampliación horizontal *(p. 14)* Una transformación que hace que el gráfico de una función se amplíe desde el eje y cuando todas las coordenadas x se multiplican por un factor a, donde $0 < a < 1$

hypothesis *(p. 605)* A claim about a characteristic of a population

hipótesis *(p. 605)* Una declaración acerca de una característica de una población

I

imaginary number *(p. 104)* A number written in the form $a + bi$, where a and b are real numbers and $b \neq 0$

número imaginario *(p. 104)* Un número escrito de la forma $a + bi$, donde a y b son números reales y $b \neq 0$

imaginary unit *i* *(p. 104)* The square root of -1, denoted $i = \sqrt{-1}$

unidad imaginaria *i* *(p. 104)* La raíz cuadrada de -1, denotado $i = \sqrt{-1}$

independent events *(p. 546)* Two events in which the occurrence of one event does not affect the occurrence of another event

eventos independientes *(p. 546)* Dos eventos en los que la ocurrencia de un evento no afecta la ocurrencia de otro evento

index of a radical *(p. 238)* The value of n in the radical $\sqrt[n]{a}$

índice de un radical *(p. 238)* El valor de n en el radical $\sqrt[n]{a}$

inferential statistics *(p. 626)* The branch of statistics that involves using a sample to draw conclusions about a population

estadística inferencial *(p. 626)* La rama de la estadística que implica el uso de una muestra para sacar conclusiones acerca de una población

information design *(p. 594)* The designing of data and information so it can be understood and used

diseño de información *(p. 594)* El diseño de datos e información, de manera que puedan ser comprendidos y utilizados

initial side *(p. 470)* The fixed ray of an angle in standard position in a coordinate plane

lado inicial *(p. 470)* El rayo fijo de un ángulo en posición normal en un plano coordenado

intercept form *(p. 59)* A quadratic function written in the form $f(x) = a(x - p)(x - q)$, where $a \neq 0$

forma de intersección *(p. 59)* Una ecuación cuadrática escrita en la forma $f(x) = a(x - p)(x - q)$, donde $a \neq 0$

inverse functions *(p. 277)* Functions that undo each other

funciones inversas *(p. 277)* Funciones que se anulan entre sí

inverse variation *(p. 360)* Two variables x and y show inverse variation when $y = \frac{a}{x}$, where $a \neq 0$.

variación inversa *(p. 360)* Dos variables x e y muestran variación inversa cuando $y = \frac{a}{x}$, donde $a \neq 0$.

J

joint frequency *(p. 554)* Each entry in a two-way table

frecuencia conjunta *(p. 554)* Cada valor en una tabla de doble entrada

joint relative frequency *(p. 555)* The ratio of a frequency that is not in the total row or the total column to the total number of values or observations in a two-way table

frecuencia relativa conjunta *(p. 555)* La razón de una frecuencia que no está en la hilera total o columna total del número total de valores u observaciones en una tabla de doble entrada

L

like radicals *(p. 246)* Radical expressions with the same index and radicand

radicales semejantes *(p. 246)* Expresiones radicales con el mismo índice y radicando

line of best fit *(p. 25)* A line that lies as close as possible to all of the data points in a scatter plot

recta de mejor ajuste *(p. 25)* Una recta que se acerca lo más posible a todos los puntos de datos en un diagrama de dispersión

line of fit *(p. 24)* A line that models data in a scatter plot

recta de ajuste *(p. 24)* Una recta que modela datos en un diagrama de dispersión

linear equation in three variables *(p. 30)* An equation of the form $ax + by + cz = d$, where x, y, and z are variables and a, b, and c are not all zero

ecuación lineal en tres variables *(p. 30)* Una ecuación de la forma $ax + by + cz = d$, donde x, y, y z son variables y a, b, y c no son todas cero

local maximum *(p. 214)* The y-coordinate of a turning point of a function when the point is higher than all nearby points

máximo local *(p. 214)* La coordenada y de un punto de inflexión de una función cuando el punto es mayor que todos los puntos cercanos

local minimum *(p. 214)* The y-coordinate of a turning point of a function when the point is lower than all nearby points

mínimo local *(p. 214)* La coordenada y de un punto de inflexión de una función cuando el punto es menor que todos los puntos cercanos

logarithm of *y* with base *b* *(p. 310)* The function $\log_b y = x$ if and only if $b^x = y$, where $b > 0$, $y > 0$, and $b \neq 1$

logaritmo de *y* con base *b* *(p. 310)* La función $\log_b y = x$ si y solo si $b^x = y$, donde $b > 0$, $y > 0$, y $b \neq 1$

logarithmic equations *(p. 335)* Equations that involve logarithms of variable expressions

ecuaciones logarítmicas *(p. 335)* Ecuaciones que implican logaritmos de expresiones variables

M

margin of error *(p. 629)* The limit on how much the responses of the sample would differ from the responses of the population

margen de error *(p. 629)* El límite de cuánto habrían de diferir las respuestas de la muestra de las respuestas de la población

marginal frequency *(p. 554)* The sums of the rows and columns in a two-way table

frecuencia marginal *(p. 554)* Las sumas de las hileras y columnas en una tabla de doble entrada

marginal relative frequency *(p. 555)* The sum of the joint relative frequencies in a row or a column in a two-way table

frecuencia relativa marginal *(p. 555)* La suma de las frecuencias relativas conjuntas en una hilera o columna en una tabla de doble entrada

maximum value *(p. 58)* The y-coordinate of the vertex of the quadratic function $f(x) = ax^2 + bx + c$, when $a < 0$

valor máximo *(p. 58)* La coordenada y del vértice de la función cuadrática $f(x) = ax^2 + bx + c$, cuando $a < 0$

midline *(p. 488)* The horizontal line $y = k$ in which the graph of a periodic function oscillates

línea media *(p. 488)* La línea horizontal $y = k$ en la que oscila el gráfico de una función periódica

minimum value *(p. 58)* The y-coordinate of the vertex of the quadratic function $f(x) = ax^2 + bx + c$, when $a > 0$

valor mínimo *(p. 58)* La coordenada y del vértice de la función cuadrática $f(x) = ax^2 + bx + c$, cuando $a > 0$

mutually exclusive events *(p. 564)* Two events that have no outcomes in common

eventos mutuamente exclusivos *(p. 564)* Dos eventos que no tienen resultados en común

N

***n* factorial** *(p. 570)* The product of the integers from 1 to n, for any positive integer n

factorial de *n* *(p. 570)* El producto de los números enteros de 1 a n, para cualquier número entero positivo n

natural base *e* *(p. 304)* An irrational number approximately equal to 2.71828…

base natural *e* *(p. 304)* Un número irracional aproximadamente equivalente a 2.71828…

natural logarithm *(p. 311)* A logarithm with base e, denoted by $\log_e$ or ln

logaritmo natural *(p. 311)* Un logaritmo con base e, denotado como $\log_e$ o ln

normal curve *(p. 596)* The graph of a normal distribution that is bell-shaped and is symmetric about the mean

curva normal *(p. 596)* El gráfico de una distribución normal con forma acampanada y es simétrica con respecto a la media

normal distribution *(p. 596)* A type of probability distribution in which the graph is a bell-shaped curve that is symmetric about the mean

distribución normal *(p. 596)* Un tipo de distribución de probabilidades en la que el gráfico es una curva acampanada que es simétrica con respecto a la media

n*th root of *a *(p. 238)* For an integer n greater than 1, if $b^n = a$, then b is an nth root of a.

raíz de orden *n* de *a* *(p. 238)* Para un número entero n mayor que 1, si $b^n = a$, entonces b es una raíz de orden n de a.

O

observational study *(p. 612)* Individuals are observed and variables are measured without controlling the individuals or their environment.

estudio de observación *(p. 612)* Se observan individuos y se miden variables sin controlar a los individuos o a su entorno.

odd function *(p. 215)* For a function f, $f(-x) = -f(x)$ for all x in its domain

función impar *(p. 215)* Para una función f, $f(-x) = -f(x)$ para toda x en su dominio

ordered triple *(p. 30)* A solution of a system of three linear equations represented by (x, y, z)

triple ordenado *(p. 30)* Un solución de un sistema de tres ecuaciones lineales representadas por (x, y, z)

outcome *(p. 538)* The possible result of a probability experiment

resultado *(p. 538)* El resultado posible de un experimento de probabilidad

overlapping events *(p. 564)* Two events that have one or more outcomes in common

eventos superpuestos *(p. 564)* Dos eventos que tienen uno o más resultados en común

P

parabola *(p. 48)* The U-shaped graph of a quadratic function

parábola *(p. 48)* El gráfico con forma de "U" de una función cuadrática

parameter *(p. 605)* A numerical description of a population characteristic

parámetro *(p. 605)* Una descripción numérica de una característica de la población

parent function *(p. 4)* The most basic function in a family of functions

función principal *(p. 4)* La función más básica en una familia de funciones

partial sum *(p. 436)* The sum S_n of the first n terms of an infinite series

sumatoria parcial *(p. 436)* La sumatoria parcial S_n de los primeros términos n de una serie infinita

Pascal's Triangle *(p. 169)* A triangular array of numbers such that the numbers in the nth row are the coefficients of the terms in the expansion of $(a + b)^n$ for whole number values of n

triángulo de Pascal *(p. 169)* Una disposición triangular de números, de tal manera que los números en la fila n son los coeficientes de los términos en la expansión de $(a + b)^n$ para los valores de números enteros de n

period *(p. 486)* The horizontal length of each cycle of a periodic function

período *(p. 486)* La longitud horizontal de cada ciclo de una función periódica

periodic function *(p. 486)* A function whose graph has a repeating pattern

función periódica *(p. 486)* Una función cuyo gráfico tiene un patrón de repetición

permutation *(p. 570)* An arrangement of objects in which order is important

permutación *(p. 570)* Una disposición de objetos en la que el orden es importante

phase shift *(p. 488)* A horizontal translation of a periodic function

desplazamiento de fase *(p. 488)* Una traslación horizontal de una función periódica

placebo *(p. 620)* A harmless, unmedicated treatment that resembles the actual treatment

placebo *(p. 620)* Un tratamiento no medicado e inofensivo que se asemeja al tratamiento real

polynomial *(p. 158)* A monomial or a sum of monomials

polinomio *(p. 158)* Un monomio o una suma de monomios

polynomial function *(p. 158)* A function of the form $f(x) = a_nx^n + a_{n-1}x^{n-1} + \cdots + a_1x + a_0$, where $a_n \neq 0$, the exponents are all whole numbers, and the coefficients are all real numbers

función polinómica *(p. 158)* Una función de la forma $f(x) = a_nx^n + a_{n-1}x^{n-1} + \cdots + a_1x + a_0$, donde $a_n \neq 0$, todos los exponentes son números enteros y todos los coeficientes son números reales

polynomial long division *(p. 174)* A method to divide a polynomial $f(x)$ by a nonzero divisor $d(x)$ to yield a quotient polynomial $q(x)$ and a remainder polynomial $r(x)$

división larga de polinomios *(p. 174)* Un método para dividir un polinomio $f(x)$ por un divisor distinto de cero $d(x)$ para obtener un polinomio de cociente $q(x)$ y un polinomio de resto $r(x)$

population *(p. 604)* The collection of all data, such as responses, measurements, or counts, that you want information about

población *(p. 604)* La recolección de datos, tales como respuestas, medidas o conteos, sobre los que se quiere información

probability distribution *(p. 580)* A function that gives the probability of each possible value of a random variable

distribución de probabilidad *(p. 580)* Una función que da la probabilidad de cada valor posible de una variable aleatoria

probability of an event *(p. 538)* A measure of the likelihood, or chance, that an event will occur

probabilidad de un evento *(p. 538)* Una medida de la probabilidad o posibilidad de que ocurrirá un evento

probability experiment *(p. 538)* An action, or trial, that has varying results

experimento de probabilidad *(p. 538)* Una acción o prueba que tiene resultados variables

pure imaginary number *(p. 104)* A number written in the form $a + bi$, where $a = 0$ and $b \neq 0$

número imaginario puro *(p. 104)* Un número escrito en la forma $a + bi$, donde $a = 0$ y $b \neq 0$

Q

quadrantal angle *(p. 479)* An angle in standard position whose terminal side lies on an axis

ángulo cuadrantal *(p. 479)* Un ángulo en posición estándar cuyo lado terminal descansa en un eje

quadratic equation in one variable *(p. 94)* An equation that can be written in the standard form $ax^2 + bx + c = 0$, where a, b, and c are real numbers and $a \neq 0$

ecuación cuadrática en una variable *(p. 94)* Una ecuación que puede escribirse en la forma estándar $ax^2 + bx + c = 0$, donde a, b, y c son números reales y $a \neq 0$

quadratic form *(p. 181)* An expression of the form $au^2 + bu + c$, where u is an algebraic expression

forma cuadrática *(p. 181)* Una expresión de la forma $au^2 + bu + c$, donde u es una expresión algebraica

Quadratic Formula *(p. 122)* The solutions of the quadratic equation $ax^2 + bx + c = 0$ are $x = \dfrac{-b \pm \sqrt{b^2 - 4ac}}{2a}$, where a, b, and c are real numbers and $a \neq 0$.

Formula Cuadrática *(p. 122)* Las soluciones de la expresión cuadrática $ax^2 + bx + c = 0$ son $x = \dfrac{-b \pm \sqrt{b^2 - 4ac}}{2a}$, donde a, b, y c son números reales y $a \neq 0$.

quadratic function *(p. 48)* A function that can be written in the form $f(x) = a(x - h)^2 + k$, where $a \neq 0$

función cuadrática *(p. 48)* Una función que puede escribirse en la forma $f(x) = a(x - h)^2 + k$, donde $a \neq 0$

quadratic inequality in one variable *(p. 142)* An inequality of the form $ax^2 + bx + c < 0$, $ax^2 + bx + c > 0$, $ax^2 + bx + c \leq 0$, or $ax^2 + bx + c \geq 0$, where a, b, and c are real numbers and $a \neq 0$

desigualdad cuadrática en una variable *(p. 142)* Una desigualdad de la forma $ax^2 + bx + c < 0$, $ax^2 + bx + c > 0$, $ax^2 + bx + c \leq 0$, o $ax^2 + bx + c \geq 0$, donde a, b, y c son números reales y $a \neq 0$

quadratic inequality in two variables *(p. 140)* An inequality of the form $y < ax^2 + bx + c$, $y > ax^2 + bx + c$, $y \leq ax^2 + bx + c$, or $y \geq ax^2 + bx + c$, where a, b, and c are real numbers and $a \neq 0$

desigualdad cuadrática en dos variables *(p. 140)* Una desigualdad de la forma $y < ax^2 + bx + c$, $y > ax^2 + bx + c$, $y \leq ax^2 + bx + c$, o $y \geq ax^2 + bx + c$, donde a, b, y c son números reales y $a \neq 0$

R

radian *(p. 471)* For a circle with radius r, the measure of an angle in standard position whose terminal side intercepts an arc of length r is one radian.

radián *(p. 471)* Para un círculo con radio r, la medida de un ángulo en posición estándar cuyo lado terminal intercepta un arco de longitud r es un radián.

radical equation *(p. 262)* An equation with a radical that has a variable in the radicand

ecuación radical *(p. 262)* Una ecuación con un radical que tiene una variable en el radicando

radical function *(p. 252)* A function that contains a radical expression with the independent variable in the radicand

función radical *(p. 252)* Una función que contiene una expresión radical con la variable independiente en el radicando

random sample *(p. 610)* A sample in which each member of a population has an equal chance of being selected

muestra aleatoria *(p. 610)* Una muestra en la que cada miembro de una población tiene igual posibilidad de ser seleccionado

random variable *(p. 580)* A variable whose value is determined by the outcomes of a probability experiment

variable aleatoria *(p. 580)* Una variable cuyo valor está determinado por los resultados de un experimento de probabilidad

randomization *(p. 620)* A process of randomly assigning subjects to different treatment groups

aleatorización *(p. 620)* Un proceso de asignación aleatoria de sujetos a distintos grupos de tratamiento

randomized comparative experiment *(p. 620)* An experiment in which subjects are randomly assigned to the control group or the treatment group

experimento comparativo aleatorizado *(p. 620)* Un experimento en el que los sujetos son asignados aleatoriamente al grupo de control o al grupo de tratamiento

rational expression *(p. 376)* A fraction whose numerator and denominator are nonzero polynomials

expresión racional *(p. 376)* Una fracción cuyo numerador y denominador son polinomios distintos a cero

rational function *(p. 366)* A function that has the form $f(x) = \frac{p(x)}{q(x)}$, where $p(x)$ and $q(x)$ are polynomials and $q(x) \neq 0$

función racional *(p. 366)* Una función que tiene la forma $f(x) = \frac{p(x)}{q(x)}$, donde $p(x)$ y $q(x)$ son polinomios y $q(x) \neq 0$

recursive rule *(p. 442)* A rule that gives the beginning term(s) of a sequence and a recursive equation that tells how a_n is related to one or more preceding terms

regla recursiva *(p. 442)* Una regla para definir el(los) primer(os) término(s) de una secuencia y una ecuación recursiva que indica cómo se relaciona a_n a uno o más términos precedentes

reference angle *(p. 480)* The acute angle formed by the terminal side of an angle in standard position and the x-axis

ángulo de referencia *(p. 480)* El ángulo agudo formado por el lado terminal de un ángulo en posición normal y el eje x

reflection *(p. 5)* A transformation that flips a graph over the line of reflection

reflexión *(p. 5)* Una transformación que voltea un gráfico sobre una recta de reflexión

repeated solution *(p. 190)* A solution of an equation that appears more than once

solución repetida *(p. 190)* Una solución de una ecuación que aparece más de una vez

replication *(p. 622)* The repetition of an experiment under the same or similar conditions to improve the validity of the experiment

réplica *(p. 622)* La repetición de un experimento bajo las mismas o similares condiciones para mejorar la validez del experimento

root of an equation *(p. 94)* A solution of an equation

raíz de una ecuación *(p. 94)* Una solución de una ecuación

S

sample *(p. 604)* A subset of a population

muestra *(p. 604)* Un subconjunto de una población

sample space *(p. 538)* The set of all possible outcomes for an experiment

espacio de muestra *(p. 538)* El conjunto de todos los resultados posibles de un experimento

secant *(p. 462)* A trigonometric function for an acute angle θ of a right triangle, denoted by $\sec \theta = \frac{\text{hypotenuse}}{\text{adjacent}}$

secante *(p. 462)* Una ecuación trigonométrica de un ángulo agudo θ de un triángulo recto, denatado por $\sec \theta = \frac{\text{hipotenusa}}{\text{adyacente}}$

sector *(p. 472)* A region of a circle that is bounded by two radii and an arc of the circle

sector *(p. 472)* Una región de un círculo conformada por dos radios y un arco del círculo

self-selected sample *(p. 610)* A sample in which members of a population can volunteer to be in the sample

muestra autoseleccionada *(p. 610)* Una muestra en la que los miembros de una población pueden ofrecerse voluntariamente para formar parte de la misma

sequence *(p. 410)* An ordered list of numbers

secuencia *(p. 410)* Una lista ordenada de números

series *(p. 412)* The sum of the terms of a sequence

serie *(p. 412)* La suma de los términos de una secuencia

sigma notation *(p. 412)* For any sequence $a_1, a_2, a_3, \ldots,$ the sum of the first k terms may be written as $\sum_{n=1}^{k} a_n = a_1 + a_2 + a_3 + \cdots + a_k$, where k is an integer.

notación sigma *(p. 412)* Para cualquier secuencia $a_1, a_2, a_3, \ldots,$ la suma de los primeros términos k puede escribirse como $\sum_{n=1}^{k} a_n = a_1 + a_2 + a_3 + \cdots + a_k$, donde k es un número entero.

simplest form of a radical *(p. 245)* An expression involving a radical with index n that has no radicands with perfect nth powers as factors other than 1, no radicands that contain fractions, and no radicals that appear in the denominator of a fraction

mínima expresión de un radical *(p. 245)* Una expresión que conlleva un radical con índice n que no tiene radicandos con potencias perfectas de orden n como factores distintos a 1, que no tiene radicandos que contengan fracciones y que no tiene radicales que aparezcan en el denominador de una fracción

simplified form of a rational expression *(p. 376)* A rational expression whose numerator and denominator have no common factors (other than ± 1)

forma simplificada de una expresión racional *(p. 376)* Una expresión racional cuyo numerador y denominador no tienen factores comunes (distintos a ± 1)

simulation *(p. 612)* The use of a model to reproduce the conditions of a situation or process so that the simulated outcomes closely match the real-world outcomes

simulación *(p. 612)* El uso de un modelo para reproducir las condiciones de una situación o proceso, de tal manera que los resultados posibles simulados coincidan en gran medida con los resultados del mundo real

sine *(p. 462)* A trigonometric function for an acute angle θ of a right triangle, denoted by $\sin \theta = \frac{\text{opposite}}{\text{hypotenuse}}$

seno *(p. 462)* Una ecuación trigonométrica de un ángulo agudo θ de un triángulo recto, denotado por $\sin \theta = \frac{\text{opuesto}}{\text{hipotenusa}}$

sinusoid *(p. 507)* The graph of a sine or cosine function

sinusoide *(p. 507)* El gráfico de una función seno o coseno

solution of a system of three linear equations *(p. 30)* An ordered triple (x, y, z) whose coordinates make each equation true

solución de un sistema de tres ecuaciones lineales *(p. 30)* Un triple ordenado (x, y, z) cuyas coordenadas hacen verdadera cada ecuación

standard form *(p. 56)* A quadratic function written in the form $f(x) = ax^2 + bx + c$, where $a \neq 0$

forma estándar *(p. 56)* Una función cuadrática escrita en la forma $f(x) = ax^2 + bx + c$, donde $a \neq 0$

standard normal distribution *(p. 597)* The normal distribution with mean 0 and standard deviation 1

distribución normal estándar *(p. 597)* La distribución normal con una media de 0 y desviación estándar 1

standard position *(p. 470)* An angle in a coordinate plane such that its vertex is at the origin and its initial side lies on the positive x-axis

posición estándar *(p. 470)* Un ángulo en un plano coordenado de tal manera que su vértice esté en el origen y que su lado inicial descanse en el eje x positivo

statistic *(p. 605)* A numerical description of a sample characteristic

estadística *(p. 605)* Una descripción numérica de una característica de la muestra

stratified sample *(p. 610)* A sample in which a population is divided into smaller groups that share a similar characteristic and a sample is then randomly selected from each group

muestra estratificada *(p. 610)* Una muestra en la que una población se divide en grupos más pequeños que comparten una característica similar, y una muestra se selecciona en forma aleatoria de cada grupo

summation notation *(p. 412)* For any sequence $a_1, a_2, a_3, \ldots$, the sum of the first k terms may be written as $\sum_{n=1}^{k} a_n = a_1 + a_2 + a_3 + \cdots + a_k$, where k is an integer.

notación de sumatoria *(p. 412)* Para cualquier secuencia $a_1, a_2, a_3, \ldots$, la sumatoria de los primeros términos k puede escribirse como $\sum_{n=1}^{k} a_n = a_1 + a_2 + a_3 + \cdots + a_k$, donde k es un número entero.

survey *(p. 612)* An investigation of one or more characteristics of a population

encuesta *(p. 612)* Una investigación de una o más características de una población

synthetic division *(p. 175)* A shortcut method to divide a polynomial by a binomial of the form $x - k$

división sintética *(p. 175)* Un método abreviado para dividir un polinomio por un binomio de la forma $x - k$

system of nonlinear equations *(p. 132)* A system of equations where at least one of the equations is nonlinear

sistema de ecuaciones no lineales *(p. 132)* Un sistema de ecuaciones en donde al menos una de las ecuaciones no es lineal

system of three linear equations *(p. 30)* A set of three equations of the form $ax + by + cz = d$, where x, y, and z are variables and a, b, and c are not all zero

sistema de tres ecuaciones lineales *(p. 30)* Un conjunto de tres ecuaciones de la forma $ax + by + cz = d$, donde x, y, y z son variables y a, b, y c no son todos cero

systematic sample *(p. 610)* A sample in which a rule is used to select members of a population

muestra sistemática *(p. 610)* Una muestra en la que se usa una regla para seleccionar miembros de una población

T

tangent *(p. 462)* A trigonometric function for an acute angle θ of a right triangle, denoted by $\tan \theta = \frac{\text{opposite}}{\text{adjacent}}$

tangente *(p. 462)* Una ecuación trigonométrica de un ángulo agudo θ de un triángulo recto, denotado por $\tan \theta = \frac{\text{opuesto}}{\text{adyacente}}$

terminal side *(p. 470)* A ray of an angle in standard position that has been rotated about the vertex in a coordinate plane

lado terminal *(p. 470)* Un rayo de un ángulo en posición normal que ha sido rotado con respecto al vértice en un plano coordenado

terms of a sequence *(p. 410)* The values in the range of a sequence

término de una secuencia *(p. 410)* Los valores en el rango de una secuencia

theoretical probability *(p. 539)* The ratio of the number of favorable outcomes to the total number of outcomes when all outcomes are equally likely

probabilidad teórica *(p. 539)* La razón del número de resultados favorables con respecto al número total de resultados cuando todos los resultados son igualmente probables

transformation *(p. 5)* A change in the size, shape, position, or orientation of a graph

transformación *(p. 5)* Un cambio en el tamaño, forma, posición u orientación de un gráfico

translation *(p. 5)* A transformation that shifts a graph horizontally and/or vertically but does not change its size, shape, or orientation

traslación *(p. 5)* Una transformación que desplaza un gráfico horizontal y/o verticalmente, pero no cambia su tamaño, forma u orientación

treatment group *(p. 620)* The group that is subjected to the treatment in an experiment

grupo de tratamiento *(p. 620)* El grupo que está sometido al tratamiento en un experimento

trigonometric identity *(p. 514)* A trigonometric equation that is true for all values of the variable for which both sides of the equation are defined

identidad trigonométrica *(p. 514)* Una ecuación trigonométrica verdadera para todos los valores de la variable por la cual se definen ambos lados de la ecuación

two-way table *(p. 554)* A frequency table that displays data collected from one source that belong to two different categories

tabla de doble entrada *(p. 554)* Una tabla de frecuencia que muestra los datos recogidos de una fuente que pertenece a dos categorías distintas

unbiased sample *(p. 611)* A sample that is representative of the population that you want information about

muestra no sesgada *(p. 611)* Una muestra que es representativa de la población de la que se quiere información

unit circle *(p. 479)* The circle $x^2 + y^2 = 1$, which has center (0, 0) and radius 1

círculo unitario *(p. 479)* El círculo $x^2 + y^2 = 1$, que tiene como centro (0, 0) y radio 1

vertex form *(p. 50)* A quadratic function written in the form $f(x) = a(x - h)^2 + k$, where $a \neq 0$

fórmula de vértice *(p. 50)* Una función cuadrática escrita en la forma $f(x) = a(x - h)^2 + k$, donde $a \neq 0$

vertex of a parabola *(p. 50)* The lowest point on a parabola that opens up or the highest point on a parabola that opens down

vértice de una parábola *(p. 50)* El punto más bajo de una parábola que se abre hacia arriba o el punto más alto de una parábola que se abre hacia abajo

vertical shrink *(p. 6)* A transformation that causes the graph of a function to shrink toward the x-axis when all the y-coordinates are multiplied by a factor a, where $0 < a < 1$

reducción vertical *(p. 6)* Una transformación que hace que el gráfico de una función se reduzca hacia el eje x cuando todas las coordenadas y se multiplican por un factor a, donde $0 < a < 1$

vertical stretch *(p. 6)* A transformation that causes the graph of a function to stretch away from the x-axis when all the y-coordinates are multiplied by a factor a, where $a > 1$

ampliación vertical *(p. 6)* Una transformación que hace que el gráfico de una función se amplíe desde el eje x cuando todas las coordenadas y se multiplican por un factor a, donde $a > 1$

Z

***z*-score** *(p. 597)* The z-value for a particular x-value which is the number of standard deviations the x-value lies above or below the mean

puntaje *z* *(p. 597)* El valor z para un valor particular x que es el número de desviaciones estándar que el valor x tiene por encima o por debajo de la media

zero of a function *(p. 96)* An x-value of a function f for which $f(x) = 0$

cero de una función *(p. 96)* Un valor x de una funcíon f para el cual $f(x) = 0$

Index

B

D

Index

Index

Q

R

S

T

U

Reference

Properties

Properties of Exponents

Let a and b be real numbers and let m and n be rational numbers.

Zero Exponent
$a^0 = 1$, where $a \neq 0$

Negative Exponent
$a^{-n} = \frac{1}{a^n}$, where $a \neq 0$

Product of Powers Property
$a^m \cdot a^n = a^{m+n}$

Quotient of Powers Property
$\frac{a^m}{a^n} = a^{m-n}$, where $a \neq 0$

Power of a Power Property
$(a^m)^n = a^{mn}$

Power of a Product Property
$(ab)^m = a^m b^m$

Power of a Quotient Property
$\left(\frac{a}{b}\right)^m = \frac{a^m}{b^m}$, where $b \neq 0$

Rational Exponents
$a^{m/n} = (a^{1/n})^m = (\sqrt[n]{a})^m$

Rational Exponents
$a^{-m/n} = \frac{1}{a^{m/n}} = \frac{1}{(a^{1/n})^m} = \frac{1}{(\sqrt[n]{a})^m}$,
where $a \neq 0$

Properties of Radicals

Let a and b be real numbers and let n be an integer greater than 1.

Product Property of Radicals
$\sqrt[n]{ab} = \sqrt[n]{a} \cdot \sqrt[n]{b}$

Quotient Property of Radicals
$\sqrt[n]{\frac{a}{b}} = \frac{\sqrt[n]{a}}{\sqrt[n]{b}}$, where $b \neq 0$

Square Root of a Negative Number
1. If r is a positive real number, then $\sqrt{-r} = i\sqrt{r}$.
2. By the first property, it follows that $(i\sqrt{r})^2 = -r$.

Properties of Logarithms

Let b, m, and n be positive real numbers with $b \neq 1$.

Product Property
$\log_b mn = \log_b m + \log_b n$

Quotient Property
$\log_b \frac{m}{n} = \log_b m - \log_b n$

Power Property
$\log_b m^n = n \log_b m$

Other Properties

Zero-Product Property
If A and B are expressions and $AB = 0$, then $A = 0$ or $B = 0$.

Property of Equality for Exponential Equations
If $b > 0$ and $b \neq 1$, then $b^x = b^y$ if and only if $x = y$.

Property of Equality for Logarithmic Equations
If b, x, and y are positive real numbers with $b \neq 1$, then $\log_b x = \log_b y$ if and only if $x = y$.

Patterns

Square of a Binomial Pattern
$(a + b)^2 = a^2 + 2ab + b^2$
$(a - b)^2 = a^2 - 2ab + b^2$

Cube of a Binomial
$(a + b)^3 = a^3 + 3a^2b + 3ab^2 + b^3$
$(a - b)^3 = a^3 - 3a^2b + 3ab^2 - b^3$

Difference of Two Squares Pattern
$a^2 - b^2 = (a + b)(a - b)$

Sum of Two Cubes
$a^3 + b^3 = (a + b)(a^2 - ab + b^2)$

Sum and Difference Pattern
$(a + b)(a - b) = a^2 - b^2$

Completing the Square
$x^2 + bx + \left(\frac{b}{2}\right)^2 = \left(x + \frac{b}{2}\right)^2$

Perfect Square Trinomial Pattern
$a^2 + 2ab + b^2 = (a + b)^2$
$a^2 - 2ab + b^2 = (a - b)^2$

Difference of Two Cubes
$a^3 - b^3 = (a - b)(a^2 + ab + b^2)$

Theorems

The Remainder Theorem
If a polynomial $f(x)$ is divided by $x - k$, then the remainder is $r = f(k)$.

The Factor Theorem
A polynomial $f(x)$ has a factor $x - k$ if and only if $f(k) = 0$.

The Rational Root Theorem
If $f(x) = a_nx^n + \cdots + a_1x + a_0$ has *integer* coefficients, then every rational solution of $f(x) = 0$ has the form
$\frac{p}{q} = \frac{\text{factor of constant term } a_0}{\text{factor of leading coefficient } a_n}$.

The Irrational Conjugates Theorem
Let f be a polynomial function with rational coefficients, and let a and b be rational numbers such that $\sqrt{b}$ is irrational.
If $a + \sqrt{b}$ is a zero of f, then $a - \sqrt{b}$ is also a zero of f.

The Fundamental Theorem of Algebra

Theorem If $f(x)$ is a polynomial of degree n where $n > 0$, then the equation $f(x) = 0$ has at least one solution in the set of complex numbers.

Corollary If $f(x)$ is a polynomial of degree n where $n > 0$, then the equation $f(x) = 0$ has exactly n solutions provided each solution repeated twice is counted as 2 solutions, each solution repeated three times is counted as 3 solutions, and so on.

The Complex Conjugates Theorem
If f is a polynomial function with real coefficients, and $a + bi$ is an imaginary zero of f, then $a - bi$ is also a zero of f.

Descartes's Rule of Signs
Let $f(x) = a_nx^n + a_{n-1}x^{n-1} + \cdots + a_2x^2 + a_1x + a_0$ be a polynomial function with real coefficients.

- The number of positive real zeros of f is equal to the number of changes in sign of the coefficients of $f(x)$ or is less than this by an even number.
- The number of negative real zeros of f is equal to the number of changes in the sign of the coefficients of $f(-x)$ or is less than this by an even number.

Formulas

Algebra

Slope

$m = \frac{y_2 - y_1}{x_2 - x_1}$

Slope-intercept form

$y = mx + b$

Point-slope form

$y - y_1 = m(x - x_1)$

Standard form of a quadratic function

$f(x) = ax^2 + bx + c$, where $a \neq 0$

Vertex form of a quadratic function

$f(x) = a(x - h)^2 + k$, where $a \neq 0$

Intercept form of a quadratic function

$f(x) = a(x - p)(x - q)$, where $a \neq 0$

Quadratic Formula

$x = \frac{-b \pm \sqrt{b^2 - 4ac}}{2a}$, where $a \neq 0$

Standard equation of a circle

$x^2 + y^2 = r^2$

Standard form of a polynomial function

$f(x) = a_n x^n + a_{n-1} x^{n-1} + \cdots + a_1 x + a_0$

Exponential growth function

$y = ab^x$, where $a > 0$ and $b > 1$

Exponential decay function

$y = ab^x$, where $a > 0$ and $0 < b < 1$

Logarithm of *y* with base *b*

$\log_b y = x$ if and only if $b^x = y$

Change-of-base formula

$\log_c a = \frac{\log_b a}{\log_b c}$, where a, b, and c are positive real numbers with $b \neq 1$ and $c \neq 1$.

Sum of *n* terms of 1

$\sum_{i=1}^{n} 1 = n$

Sum of first *n* positive integers

$\sum_{i=1}^{n} i = \frac{n(n+1)}{2}$

Sum of squares of first *n* positive integers

$\sum_{i=1}^{n} i^2 = \frac{n(n+1)(2n+1)}{6}$

Explicit rule for an arithmetic sequence

$a_n = a_1 + (n - 1)d$

Sum of first *n* terms of an arithmetic series

$S_n = n\left(\frac{a_1 + a_n}{2}\right)$

Explicit rule for a geometric sequence

$a_n = a_1 r^{n-1}$

Sum of first *n* terms of a geometric series

$S_n = a_1\left(\frac{1 - r^n}{1 - r}\right)$, where $r \neq 1$

Sum of an infinite geometric series

$S = \frac{a_1}{1 - r}$ provided $|r| < 1$

Recursive equation for an arithmetic sequence

$a_n = a_{n-1} + d$

Recursive equation for a geometric sequence

$a_n = r \cdot a_{n-1}$

Statistics

Sample mean

$\overline{x} = \frac{\Sigma x}{n}$

Standard deviation

$\sigma = \sqrt{\frac{(x_1 - \mu)^2 + (x_2 - \mu)^2 + \cdots + (x_n - \mu)^2}{n}}$

***z*-Score**

$z = \frac{x - \mu}{\sigma}$

Margin of error for sample proportions

$\pm\frac{1}{\sqrt{n}}$

Reference

Trigonometry

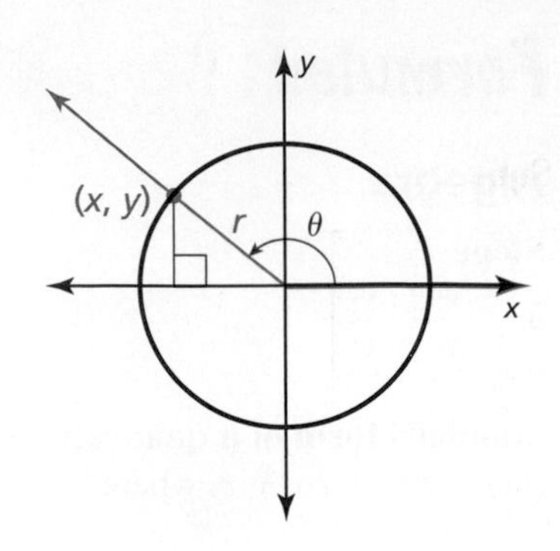

General definitions of trigonometric functions

Let θ be an angle in standard position, and let (x, y) be the point where the terminal side of θ intersects the circle $x^2 + y^2 = r^2$. The six trigonometric functions of θ are defined as shown.

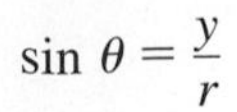

$\sin \theta = \frac{y}{r}$ $\quad \cos \theta = \frac{x}{r}$ $\quad \tan \theta = \frac{y}{x}, x \neq 0$

$\csc \theta = \frac{r}{y}, y \neq 0$ $\quad \sec \theta = \frac{r}{x}, x \neq 0$ $\quad \cot \theta = \frac{x}{y}, y \neq 0$

Conversion between degrees and radians

$180° = \pi$ radians

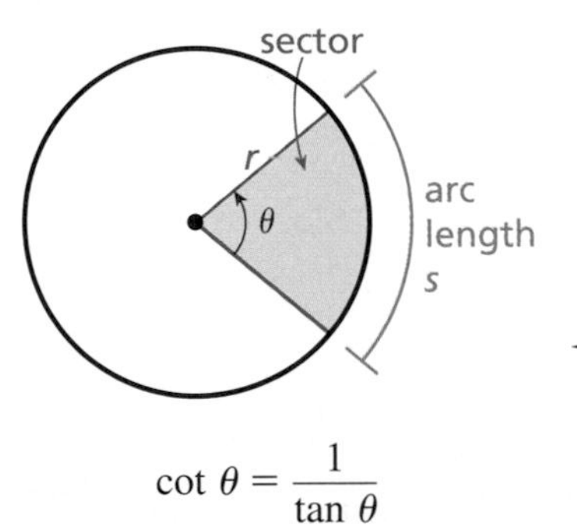

Arc length of a sector

$s = r\theta$

Area of a sector

$A = \frac{1}{2}r^2\theta$

Reciprocal Identities

$\csc \theta = \frac{1}{\sin \theta}$ $\quad \sec \theta = \frac{1}{\cos \theta}$ $\quad \cot \theta = \frac{1}{\tan \theta}$

Tangent and Cotangent Identities

$\tan \theta = \frac{\sin \theta}{\cos \theta}$ $\quad \cot \theta = \frac{\cos \theta}{\sin \theta}$

Pythagorean Identities

$\sin^2 \theta + \cos^2 \theta = 1$

$1 + \tan^2 \theta = \sec^2 \theta$

$1 + \cot^2 \theta = \csc^2 \theta$

Negative Angle Identities

$\sin(-\theta) = -\sin \theta$

$\cos(-\theta) = \cos \theta$

$\tan(-\theta) = -\tan \theta$

Cofunction Identites

$\sin\left(\frac{\pi}{2} - \theta\right) = \cos \theta$

$\cos\left(\frac{\pi}{2} - \theta\right) = \sin \theta$

$\tan\left(\frac{\pi}{2} - \theta\right) = \cot \theta$

Sum Formulas

$\sin(a + b) = \sin a \cos b + \cos a \sin b$

$\cos(a + b) = \cos a \cos b - \sin a \sin b$

$\tan(a + b) = \frac{\tan a + \tan b}{1 - \tan a \tan b}$

Difference Formulas

$\sin(a - b) = \sin a \cos b - \cos a \sin b$

$\cos(a - b) = \cos a \cos b + \sin a \sin b$

$\tan(a - b) = \frac{\tan a - \tan b}{1 + \tan a \tan b}$

Probability and Combinatorics

Theoretical Probability $= \frac{\text{Number of favorable outcomes}}{\text{Total number of outcomes}}$

Experimental Probability $= \frac{\text{Number of successes}}{\text{Number of trials}}$

Probability of the complement of an event

$P(\overline{A}) = 1 - P(A)$

Probability of independent events

$P(A \text{ and } B) = P(A) \cdot P(B)$

Probability of dependent events

$P(A \text{ and } B) = P(A) \cdot P(B \mid A)$

Probability of compound events

$P(A \text{ or } B) = P(A) + P(B) - P(A \text{ and } B)$

Permutations

${}_nP_r = \frac{n!}{(n - r)!}$

Combinations

${}_nC_r = \frac{n!}{(n - r)! \cdot r!}$

Binomial experiments

$P(k \text{ successes}) = {}_nC_k p^k (1 - p)^{n - k}$

The Binomial Theorem

$(a + b)^n = {}_nC_0 a^n b^0 + {}_nC_1 a^{n-1} b^1 + {}_nC_2 a^{n-2} b^2 + \cdots + {}_nC_n a^0 b^n$, where n is a positive integer.

Perimeter, Area, and Volume Formulas

Square

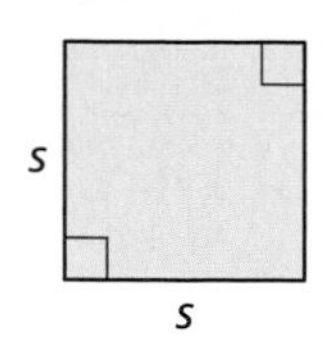

$P = 4s$
$A = s^2$

Rectangle

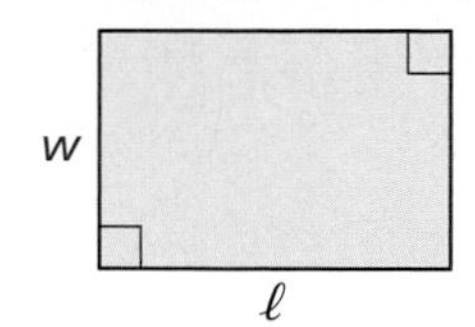

$P = 2\ell + 2w$
$A = \ell w$

Triangle

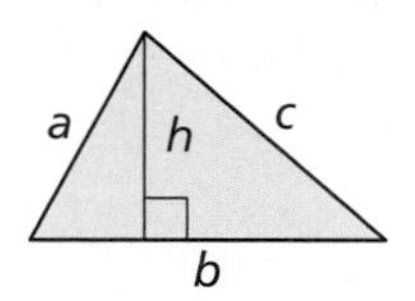

$P = a + b + c$
$A = \frac{1}{2}bh$

Circle

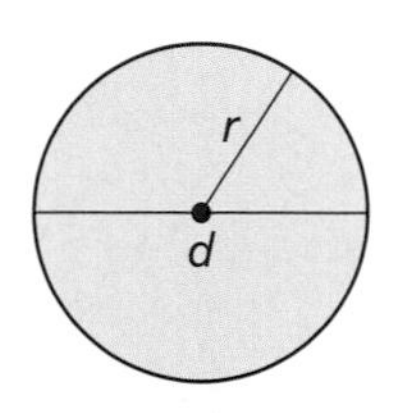

$C = \pi d$ or $C = 2\pi r$
$A = \pi r^2$

Parallelogram

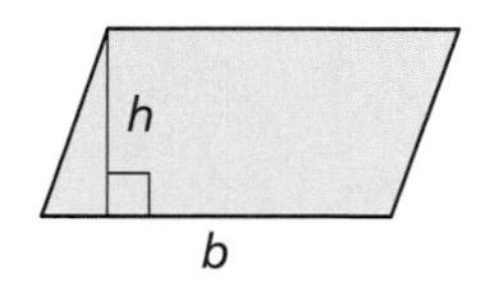

$A = bh$

Trapezoid

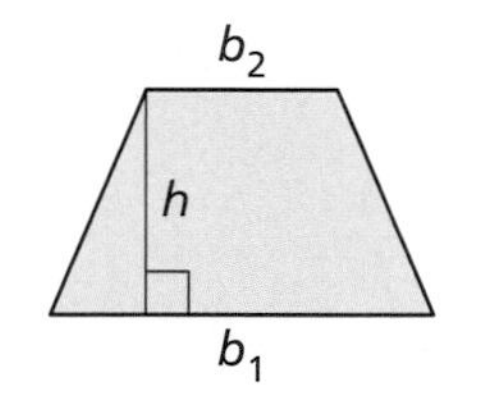

$A = \frac{1}{2}h(b_1 + b_2)$

Rhombus/Kite

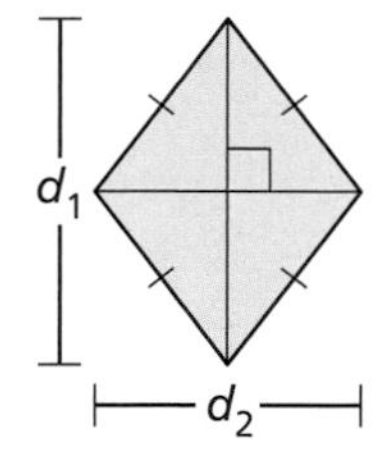

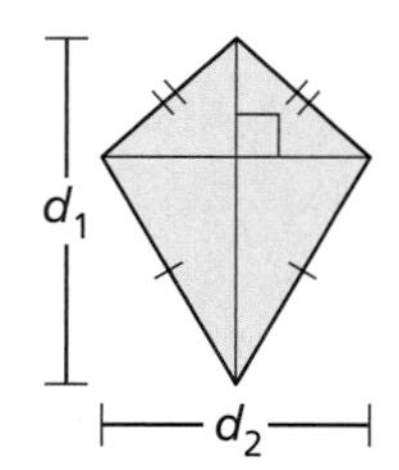

$A = \frac{1}{2}d_1 d_2$

Regular *n*-gon

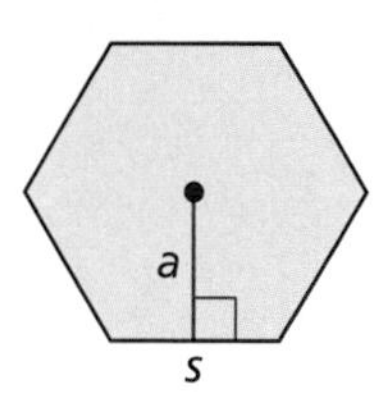

$A = \frac{1}{2}aP$ or $A = \frac{1}{2}a \cdot ns$

Prism

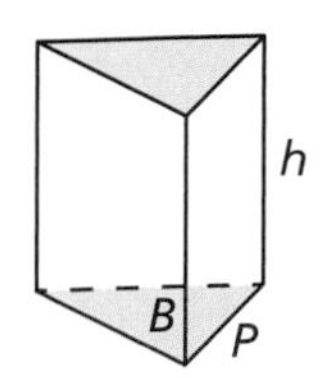

$L = Ph$
$S = 2B + Ph$
$V = Bh$

Cylinder

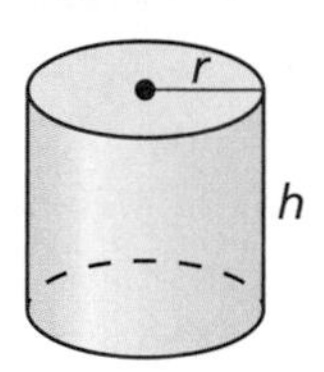

$L = 2\pi rh$
$S = 2\pi r^2 + 2\pi rh$
$V = \pi r^2 h$

Pyramid

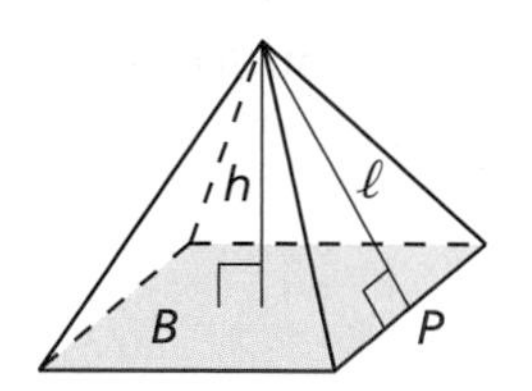

$L = \frac{1}{2}P\ell$
$S = B + \frac{1}{2}P\ell$
$V = \frac{1}{3}Bh$

Cone

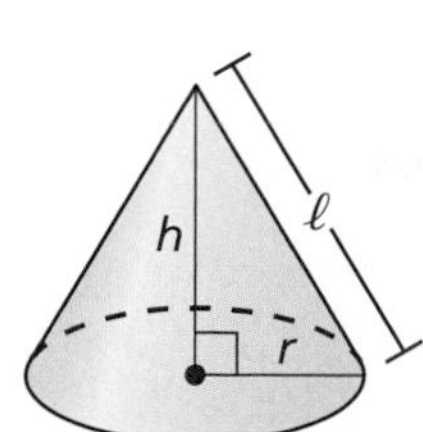

$L = \pi r\ell$

$S = \pi r^2 + \pi r\ell$

$V = \frac{1}{3}\pi r^2 h$

Sphere

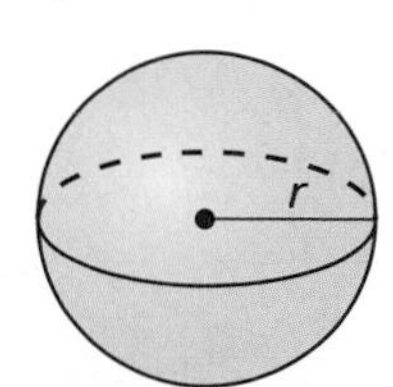

$S = 4\pi r^2$

$V = \frac{4}{3}\pi r^3$

Other Formulas

Pythagorean Theorem
$a^2 + b^2 = c^2$

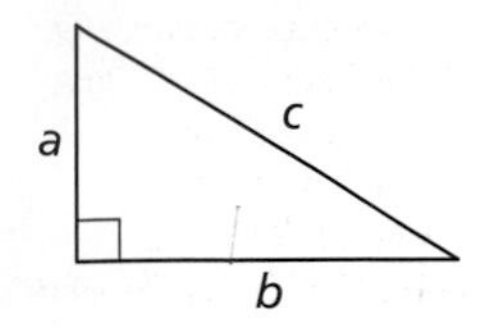

Simple Interest
$I = Prt$

Compound Interest
$A = P\left(1 + \frac{r}{n}\right)^{nt}$

Continuously Compounded Interest
$A = Pe^{rt}$

Distance
$d = rt$

Conversions

U.S. Customary
1 foot = 12 inches
1 yard = 3 feet
1 mile = 5280 feet
1 mile = 1760 yards
1 acre = 43,560 square feet
1 cup = 8 fluid ounces
1 pint = 2 cups
1 quart = 2 pints
1 gallon = 4 quarts
1 gallon = 231 cubic inches
1 pound = 16 ounces
1 ton = 2000 pounds

U.S. Customary to Metric
1 inch = 2.54 centimeters
1 foot ≈ 0.3 meter
1 mile ≈ 1.61 kilometers
1 quart ≈ 0.95 liter
1 gallon ≈ 3.79 liters
1 cup ≈ 237 milliliters
1 pound ≈ 0.45 kilogram
1 ounce ≈ 28.3 grams
1 gallon ≈ 3785 cubic centimeters

Time
1 minute = 60 seconds
1 hour = 60 minutes
1 hour = 3600 seconds
1 year = 52 weeks

Temperature
$C = \frac{5}{9}(F - 32)$
$F = \frac{9}{5}C + 32$

Metric
1 centimeter = 10 millimeters
1 meter = 100 centimeters
1 kilometer = 1000 meters
1 liter = 1000 milliliters
1 kiloliter = 1000 liters
1 milliliter = 1 cubic centimeter
1 liter = 1000 cubic centimeters
1 cubic millimeter = 0.001 milliliter
1 gram = 1000 milligrams
1 kilogram = 1000 grams

Metric to U.S. Customary
1 centimeter ≈ 0.39 inch
1 meter ≈ 3.28 feet
1 meter ≈ 39.37 inches
1 kilometer ≈ 0.62 mile
1 liter ≈ 1.06 quarts
1 liter ≈ 0.26 gallon
1 kilogram ≈ 2.2 pounds
1 gram ≈ 0.035 ounce
1 cubic meter ≈ 264 gallons

Credits

Front Matter

viii Rodrigo Garrido/Shutterstock.com; **ix** J.D.S/Shutterstock.com; **x** rickyd/Shutterstock.com; **xi** Doug Matthews/Shutterstock.com; **xii** Reinhold Leitner/Shutterstock.com; **xiii** Ariwasabi/Shutterstock.com; **xiv** Miguel Navarro; **xv** Charles Knowles/Shutterstock.com; **xvi** Guy Shapira/Shutterstock.com, **xvii** Nadiia Gerbish/Shutterstock.com; **xviii** © Pniesen | Dreamstime.com; **xix** Monkey Business Images/Shutterstock.com

Chapter 1

0 *top left* Daniel Korzeniewski/Shutterstock.com; *top right* bikeriderlondon/Shutterstock.com; *center left* © Farang | Dreamstime.com; *bottom right* Rodrigo Garrido/Shutterstock.com; *bottom left* bikeriderlondon/Shutterstock.com; **7** Rodrigo Garrido/Shutterstock.com; **10** bikeriderlondon/Shutterstock.com; **16** © Farang | Dreamstime.com; **19** Aleksander Erin/Shutterstock.com; **23** bikeriderlondon/Shutterstock.com; **25** Potapov Alexander/Shutterstock.com; **28** iPortret/Shutterstock.com; **34** Daniel Korzeniewski/Shutterstock.com; **35** Gemenacom/Shutterstock.com; **37** zimmytws/Shutterstock.com

Chapter 2

44 *top left* Lisa F. Young/Shutterstock.com; *top right* J.D.S/Shutterstock.com; *center left* © Redeyed | Dreamstime.com; *bottom right* Fotokostic/Shutterstock.com; *bottom left* ©iStockphoto.com/Dirk Freder; **51** Mike Brake/Shutterstock.com; **53** ©iStockphoto.com/Dirk Freder; **54** feathercollector/Shutterstock.com; **60** hugolaeasse/Shutterstock.com, passpunShutterstock.com; **62** Mitch Gunn/Shutterstock.com; **63** Fotokostic/Shutterstock.com; **64** *bottom left* © Redeyed | Dreamstime.com; *center right* Garsya/Shutterstock.com; **65** Aleksander Erin/Shutterstock.com; **66** Steve Mann/Shutterstock.com; **73** inxti/Shutterstock.com; **77** Lisa F. Young/Shutterstock.com; **83** Timothy Large/Shutterstock.com; **86** KID_A/Shutterstock.com, Adam Fahey Designs/Shutterstock.com

Chapter 3

90 *top left* Pavel L Photo and Video/Shutterstock.com; *top right* Jakrit Jiraratwaro/Shutterstock.com; *center left* paul cowell/Shutterstock.com; *bottom right* rickyd/Shutterstock.com; *bottom left* Aspen Photo/Shutterstock.com; **97** *top right* Kovalchuk Oleksandr, Subbotina Anna/Shutterstock.com; *center left* wavebreakmedia/Shutterstock.com; **100** L.Watcharapol/Shutterstock.com; **101** *top left* Luciano Mortula/Shutterstock.com; *center left* David W. Leindecker / Shutterstock.com; *bottom left* f9photos/Shutterstock.com; *center right* Antony McAulay/Shutterstock.com; **106** rickyd/Shutterstock.com; **115** Aspen Photo/Shutterstock.com; **117** Wesley Guijt/Shutterstock.com; **119** Aleksander Erin/Shutterstock.com; **126** *top left* © Matthew Apps | Dreamstime.com; *top center* muzsy/Shutterstock.com; *top right* © Vlue | Dreamstime.com; **128** Haslam Photography/Shutterstock.com; **129** *top left* ©iStockphoto.com/AtomA; *bottom left* Oleksiy Mark/Shutterstock.com; *top right* paul cowell/Shutterstock.com; **130** Jorg Hackemann/Shutterstock.com; **137** Jakrit Jiraratwaro/Shutterstock.com; **145** *bottom left* © Tammygaffney | Dreamstime.com; *top right* Dan Breckwoldt/Shutterstock.com; *center right* Pavel L Photo and Video/Shutterstock.com; **146** *center left* BlueRingMedia/Shutterstock.com; *center right* © Tatjana Keisa | Dreamstime.com; **147** Seanika/Shutterstock.com

Chapter 4

154 *top left* Doug Matthews/Shutterstock.com; *top right* © Oseland | Dreamstime.com; *center left* Anton_Ivanov/Shutterstock.com; *bottom right* Aspen Photo/Shutterstock.com; *bottom left* clean_fotos/Shutterstock.com; **161** clean_fotos/Shutterstock.com; **163** *center right* © Lightvision | Dreamstime.com; *bottom right* tawan/Shutterstock.com; **164** JRB67/Shutterstock.com; **170** Flashon Studio/Shutterstock.com; **171** balein/Shutterstock.com; **178** Aspen Photo/Shutterstock.com; **183** Xavier Pironet/Shutterstock.com; **185** GoodMood Photo/Shutterstock.com; **187** Aleksander Erin/Shutterstock.com; **195** Anton_Ivanov/Shutterstock.com; **203** *top right* © Oseland | Dreamstime.com; *center right* Tyler Olson/Shutterstock.com; **210** Steve Byland/Shutterstock.com; **217** bikeriderlondon/Shutterstock.com; **218** Doug Matthews/Shutterstock.com; **222** nostal6ie/Shutterstock.com; **225** Mihai Dancaescu/Shutterstock.com; **231** Sergio Bertino/Shutterstock.com

Chapter 5

234 *top left* © Irur | Dreamstime.com; *top right* Reinhold Leitner/Shutterstock.com; *center left* katatonia82/Shutterstock.com; *bottom right* NASA/JPL-Caltech/MSSS; *bottom left* Sarun T/Shutterstock.com; **240** Nerthuz/Shutterstock.com; **250** Sarun T/Shutterstock.com; **254** NASA/JPL-Caltech/MSSS; **258** Krzysztof Gorski/Shutterstock.com; **259** Aleksander Erin/Shutterstock.com; **263** Glynnis Jones/Shutterstock.com; **266** Xiebiyun/Shutterstock.com; **267** *Exercise 46 left* Christopher Meder/Shutterstock.com; *Exercise 46 right* Samot/Shutterstock.com; **268** katatonia82/Shutterstock.com; **272** Reinhold Leitner/Shutterstock.com; **282** © Irur | Dreamstime.com; **283** *top left* Piotr Marcinski/Shutterstock.com; *bottom right* Dmitry Kalinovsky/Shutterstock.com; **285** Andresr/Shutterstock.com; **289** Aspen Photo/Shutterstock.com

Chapter 6

292 *top left* NASA; *top right* Ariwasabi/Shutterstock.com; *center left* UnaPhoto/Shutterstock.com; *bottom right* Olga Kovalenko/Shutterstock.com; *bottom left* Minerva Studio/Shutterstock.com; **301** Olga Kovalenko/Shutterstock.com; **302** Egon Zitter/Shutterstock.com; **314** Joggie Botma/Shutterstock.com, Germanskydiver/Shutterstock.com; **315** *bottom left* Minerva Studio/Shutterstock.com; *top right* Lightspring/Shutterstock.com; **316** *top left* Richard A McMillin/Shutterstock.com; *top right* Arkorn/Shutterstock.com; **325** Aleksander Erin/Shutterstock.com; **330** UnaPhoto/Shutterstock.com; **332** *bottom left* Johan_R/Shutterstock.com; *top right* vasabii/Shutterstock.com, Suat Gursozlu/Shutterstock.com, ©iStockphoto.com/claudiaveja; **335** Ariwasabi/Shutterstock.com; **338** *bottom left* Ian Scott/Shutterstock.com; *center right* ©iStockphoto.com/joebelanger; **345** Edward Haylan/Shutterstock.com; **346** bikeriderlondon/Shutterstock.com; **347** *bottom left* Anton Zabielskyi/Shutterstock.com, hotobank.ch/Shutterstock.com, Markus Gann/Shutterstock.com, Gladskikh Tatiana/Shutterstock.com, VladGavriloff/Shutterstock.com, Vasilyev Alexandr/Shutterstock.com; *bottom right* NASA; **348** rawcaptured/Shutterstock.com; **349** Baloncici/Shutterstock.com; **353** escova/Shutterstock.com, Inna Petyakina/Shutterstock.com

Chapter 7

356 *top left* Luna Vandoorne/Shutterstock.com; *top right* javarman/Shutterstock.com; *center left* Minerva Studio/Shutterstock.com; *bottom right* Miguel Navarro; *bottom left* pan_kung/Shutterstock.com; **362** pan_kung/Shutterstock.com; **364** *Exercise 26 left* © Lois Mcartney | Dreamstime.com; *Exercise 26 right* Igor Sokolov (breeze)/Shutterstock.com; **369** Miguel Navarro; **371** Minerva Studio/Shutterstock.com; **372** Tilil/Shutterstock.com, Genestro/Shutterstock.com; **373** Aleksander Erin/Shutterstock.com; **374** Richard Paul Kane/Shutterstock.com; **379** ndoeljindoel/Shutterstock.com; **381** Lucky Business/Shutterstock.com; **382** *Exercise 47 left* Andreas Meyer/Shutterstock.com; *Exercise 47 right* Anna Kucherova/Shutterstock.com; **389** AridOcean/Shutterstock.com; **390** wanpatsorn/Shutterstock.com; **396** ©iStockphoto.com/Nikada; **397** Luna Vandoorne/Shutterstock.com; **399** José Carlos González Sánchez / Wikimedia Commons / Public Domain

Chapter 8

406 *top left* Photographee.eu/Shutterstock.com; *top right* Charles Knowles/Shutterstock.com; *center left* Rainbow Skydive; *bottom right* Debby Wong/Shutterstock.com; *bottom left* © Inigocia | Dreamstime.com; **411** Halina Yakushevich/Shutterstock.com; **415** *top left* ©iStockphoto.com/colematt; *center right* Tomasz Trojanowski/Shutterstock.com; *bottom right* irin-k/Shutterstock.com; **416** *bottom left* © Inigocia | Dreamstime.com; *bottom right* © Dmitry Elagin | Dreamstime.com; **421** ©iStockphoto.com/ZernLiew; **423** Jeff Kinsey/Shutterstock.com; **431** *center right* Andy Dean Photography/Shutterstock.com; *bottom right* Rainbow Skydive; **433** Aleksander Erin/Shutterstock.com; **439** Rashevskyi Viacheslav/Shutterstock.com; **440** Angela Hawkey/Shutterstock.com, xpixel/Shutterstock.com, Pavel K/Shutterstock.com; **445** Charles Knowles/Shutterstock.com; **448** Horiyan/Shutterstock.com; **449** Photographee.eu/Shutterstock.com; **451** wandee007/Shutterstock.com

Chapter 9

458 *top left* Piotr Wawrzyniuk/Shutterstock.com; *top right* Charles Knowles/Shutterstock.com; *center left* artincamera/Shutterstock.com; *bottom right* ©iStockphoto.com/alexsl; *bottom left* Guy Shapira/Shutterstock.com; **465** *top right* Glenn Young/Shutterstock.com; *bottom left* Pavel L Photo and Video/Shutterstock.com; **467** ©iStockphoto.com/Kubrak78; **468** Skalapendra/Shutterstock.com; **475** *center left* Evangelos Thomaidis; *bottom right* NASA; **476** ©iStockphoto.com/alexsl; **481** Iaremenko Sergii/Shutterstock.com; **483** *top left* Courtesy NASA/JPL-Caltech; *center left* richsouthwales/Shutterstock.com, Brian Hook, *http://www.flickr.com/photos/brianhook*; **484** leonello calvetti/Shutterstock.com; **492** © Joseph Gough | Dreamstime.com; **493** *center left* Matthew Cole/Shutterstock.com; *Exercise 58 left* ©Albertoloyo | Dreamstime.com; *Exercise 58 right* © Robseguin | Dreamstime.com; **494** *top left* artincamera/Shutterstock.com; *center right* Levent Konvk/Shutterstock.com; **495** Aleksander Erin/Shutterstock.com; **496** fuyu liu/Shutterstock.com, ©iStockphoto.com/4x6, ©iStockphoto.com/knape; **504** *top left* IR Stone/Shutterstock.com; *center left* bikeriderlondon/Shutterstock.com; **506** Eliks/Shutterstock.com; **508** ©iStockphoto.com/wdstock, ©iStockphoto.com/McIninch; **509** EdgeOfReason/Shutterstock.com; **510** Charles Knowles/Shutterstock.com; **511** © Graham Taylor | Dreamstime.com; **518** Piotr Wawrzyniuk/Shutterstock.com; **525** Peter Elvidge/Shutterstock.com; **531** Petr Student/Shutterstock.com

Chapter 10

534 *top left* Margaret M Stewart/Shutterstock.com; *top right* Mikhail Pogosov/Shutterstock.com; *center left* Nadiia Gerbish/Shutterstock.com; *bottom right* Tyler Olson/Shutterstock.com; *bottom left* bikeriderlondon/Shutterstock.com; **537** *top* Sascha Burkard/Shutterstock.com; *center* Ruslan Gi/Shutterstock.com; *bottom right* cobalt88/Shutterstock.com, Sanchai Khudpin/Shutterstock.com; **544** ©iStockphoto.com/hugolacasse; **545** *top right* cTermit/Shutterstock.com; *top center* Picsfive/Shutterstock.com; **548** ©iStockphoto.com/andriikoval; **550** *center left* Marques/Shutterstock.com, Laurin Rinder/Shutterstock.com; *bottom left* oliveromg/Shutterstock.com; **552** *center left* cobalt88/Shutterstock.com; *center right* bikeriderlondon/Shutterstock.com; **557** Tyler Olson/Shutterstock.com; **558** *bottom left* Andrey Bayda/Shutterstock.com; *bottom right* Image Point Fr/Shutterstock.com; **561** Aleksander Erin/Shutterstock.com; **563** CrackerClips Stock Media/Shutterstock.com; **567** ©iStockphoto.com/Andy Cook; **568** *top left* Nadiia Gerbish/Shutterstock.com; *center right* bikeriderlondon/Shutterstock.com; **571** Mikhail Pogosov/Shutterstock.com; **573** *center top* Alan Bailey/Shutterstock.com; *center bottom* ©iStockphoto.com/zorani; **575** ©iStockphoto.com/Rich Legg; **576** *top left* William Stall/Shutterstock.com; *bottom right* CBS via Getty Images; **577** Tony Oshlick/Shutterstock.com; **578** *top left* Telnov Oleksii/Shutterstock.com; *top right* Suat Gursozlu/Shutterstock.com; **579** Sascha Burkard/Shutterstock.com; **581** claudiofichera/Shutterstock.com; **583** Margaret M Stewart/Shutterstock.com; **584** ecco/Shutterstock.com; **585** Dmitry Melnikov/Shutterstock.com; **589** Christos Georghiou/Shutterstock.com

Chapter 11

592 *top left* Penka Todorova Vitkova/Shutterstock.com; *top right* Andresr/Shutterstock.com; *center left* © Pniesen | Dreamstime.com; *bottom right* Monkey Business Images/Shutterstock.com; *bottom left* Kalinina Alisa/Shutterstock.com; **598** Kalinina Alisa/Shutterstock.com; **600** irin-k/Shutterstock.com; **601** Erwin Niemand/Shutterstock.com; **602** © Biswarup Ganguly / Wikimedia Commons / GFDL; **603** Skylines/Shutterstock.com; **605** Monkey Business Images/Shutterstock.com; **607** *center left* ©iStockphoto.com/linzyslusher; *bottom left* wavebreakmedia/Shutterstock.com; *top right* CandyBox Images/Shutterstock.com; **610** Denis Cristo/Shutterstock.com; **611** Goodluz/Shutterstock.com; **614** Lenka_N/Shutterstock.com; **615** *center left* ©iStockphoto.com/Pixsooz; *bottom left* © Pniesen | Dreamstime.com; *bottom right* Sergey Nivens/Shutterstock.com; **616** ©iStockphoto.com/gchutka; **617** Aleksander Erin/Shutterstock.com; **619** sauletas/Shutterstock.com; **621** Pressmaster/Shutterstock.com; **622** sagir/Shutterstock.com; **624** Andresr/Shutterstock.com; **630** Monkey Business Images/Shutterstock.com; **631** Penka Todorova Vitkova/Shutterstock.com; **632** Danny E Hooks/Shutterstock.com; **637** Edw/Shutterstock.com; **639** Constantine Pankin/Shutterstock.com